International Congress of Mathematicians (ICM 2022)

Plenary Lectures
Volume 2 of 7

Online
6-14 July 2022

Editors:

Dmitry Beliaev
Stanislav Smirnov

ISBN: 978-1-7138-9869-6

ICM INTERNATIONAL CONGRESS OF MATHEMATICIANS 2022 JULY 6–14

PLENARY LECTURES

EDITED BY D. BELIAEV AND S. SMIRNOV

INTERNATIONAL MATHEMATICAL UNION IMU

EMS PRESS

Editors

Dmitry Beliaev
Mathematical Institute
University of Oxford
Andrew Wiles Building
Radcliffe Observatory Quarter
Woodstock Road
Oxford OX2 6GG, UK

Email: belyaev@maths.ox.ac.uk

Stanislav Smirnov
Section de mathématiques
Université de Genève
rue du Conseil-Général 7–9
1205 Genève, Switzerland

Email: stanislav.smirnov@unige.ch

2020 Mathematics Subject Classification: 00B25

ISBN 978-3-98547-058-7, eISBN 978-3-98547-558-2, DOI 10.4171/ICM2022

Volume 1. Prize Lectures
ISBN 978-3-98547-059-4, eISBN 978-3-98547-559-9, DOI 10.4171/ICM2022-1
⟶ Volume 2. Plenary Lectures
ISBN 978-3-98547-060-0, eISBN 978-3-98547-560-5, DOI 10.4171/ICM2022-2
Volume 3. Sections 1–4
ISBN 978-3-98547-061-7, eISBN 978-3-98547-561-2, DOI 10.4171/ICM2022-3
Volume 4. Sections 5–8
ISBN 978-3-98547-062-4, eISBN 978-3-98547-562-9, DOI 10.4171/ICM2022-4
Volume 5. Sections 9–11
ISBN 978-3-98547-063-1, eISBN 978-3-98547-563-6, DOI 10.4171/ICM2022-5
Volume 6. Sections 12–14
ISBN 978-3-98547-064-8, eISBN 978-3-98547-564-3, DOI 10.4171/ICM2022-6
Volume 7. Sections 15–20
ISBN 978-3-98547-065-5, eISBN 978-3-98547-565-0, DOI 10.4171/ICM2022-7

This edition printed by Curran Associates, Inc.
ISBN 978-1-7138-9869-6

e-Book available: https://ems.press/books/standalone/274

No changes have been made to the content of these proceedings. There may be changes to pagination and minor adjustments for aesthetics.

Bibliographic information published by the Deutsche Nationalbibliothek
The Deutsche Nationalbibliothek lists this publication in the Deutsche Nationalbibliografie; detailed bibliographic data are available on the Internet at http://dnb.dnb.de.

Published by EMS Press, an imprint of the

European Mathematical Society – EMS – Publishing House GmbH
Institut für Mathematik
Technische Universität Berlin
Straße des 17. Juni 136
10623 Berlin, Germany

https://ems.press

CONTENTS

VOLUME 1

THE WORK OF THE FIELDS MEDALISTS AND THE IMU PRIZE WINNERS

PRIZE LECTURES

POPULAR SCIENTIFIC EXPOSITIONS

SUMMARIES OF PRIZE WINNERS' WORK

VOLUME 2

SPECIAL PLENARY LECTURES

PLENARY LECTURES

VOLUME 3

1. LOGIC

2. ALGEBRA

3. NUMBER THEORY – SPECIAL LECTURE

3. NUMBER THEORY

4. ALGEBRAIC AND COMPLEX GEOMETRY – SPECIAL LECTURE

4. ALGEBRAIC AND COMPLEX GEOMETRY

VOLUME 4

5. GEOMETRY – SPECIAL LECTURES

5. GEOMETRY

6. TOPOLOGY

7. LIE THEORY AND GENERALIZATIONS

8. ANALYSIS – SPECIAL LECTURE

8. ANALYSIS

VOLUME 5

9. DYNAMICS

10. PARTIAL DIFFERENTIAL EQUATIONS

11. MATHEMATICAL PHYSICS – SPECIAL LECTURE

11. MATHEMATICAL PHYSICS

VOLUME 6

12. PROBABILITY – SPECIAL LECTURE

12. PROBABILITY

13. COMBINATORICS – SPECIAL LECTURE

13. COMBINATORICS

14. MATHEMATICS OF COMPUTER SCIENCE – SPECIAL LECTURES

14. MATHEMATICS OF COMPUTER SCIENCE

VOLUME 7

15. NUMERICAL ANALYSIS AND SCIENTIFIC COMPUTING

18. STOCHASTIC AND DIFFERENTIAL MODELLING

19. MATHEMATICAL EDUCATION AND POPULARIZATION OF MATHEMATICS

20. HISTORY OF MATHEMATICS

SPECIAL PLENARY LECTURES

WHAT IS THE POINT OF COMPUTERS? A QUESTION FOR PURE MATHEMATICIANS

KEVIN BUZZARD

ABSTRACT

We discuss the idea that computers might soon help mathematicians prove theorems in areas where they have not previously been useful. Furthermore, we argue that these same computer tools will also help us in the communication and teaching of mathematics.

MATHEMATICS SUBJECT CLASSIFICATION 2020

68V20

KEYWORDS

Theorem prover, ITP, proof assistant.

INTERNATIONAL CONGRESS OF MATHEMATICIANS 2022 JULY 6–14

Proc. Int. Cong. Math. 2022, Vol. 2, pp. 578–608
DOI 10.4171/ICM2022/166

Published by EMS Press

1. INTRODUCTION

Computers in 2021 are phenomenal. They can do billions of calculations in a second. They are extremely good at obeying precise instructions accurately. Mathematics is a game with precise rules. One can thus ask in what ways computers can be used to help us[1] mathematicians to do our job.

Of course, computers have been used to help some mathematicians to do their job ever since computers have existed. Birch and Swinnerton-Dyer used an early computer (which was the size of a large room and had 20 kilobytes of memory) to compute many examples of solutions to cubic equations in two variables modulo prime numbers [5]. Graphing the output data in the right way led to new insights in the theory of elliptic curves which ultimately became the Birch and Swinnerton-Dyer conjecture, one of the Clay Millennium problems. At the time of writing, this conjecture is still open, although regular breakthroughs (most recently in noncommutative Iwasawa theory) provide us with incremental progress.

This article is not about using computers in that way. This article is an attempt to explain to *all* researchers in mathematics that, thanks to breakthroughs in computer science, computers can now be used to help us not just with computations, but with *reasoning*. In other words, it is about the possibility that computers might soon be helping us *prove theorems*, whether they be about "computable" objects such as elliptic curves, or about more intractable objects such as Banach spaces, schemes, abelian categories or perfectoid spaces, things which cannot be listed or classified, or in general stored in a traditional computer algebra package in any meaningful way. In particular, it is about the possibility that computer proof assistants can help mathematicians who up until this point have had no need for computation in their research and might hence incorrectly deduce that computers have nothing at all to offer them. I should also stress that the applications are not limited to people interested in foundational subjects such as set theory or type theory; I am thinking about applications in geometry, topology, combinatorics, number theory, algebra, analysis,…

I end this introduction with a summary of what to expect, and what not to expect, from this fast-growing area within the next decade. The first thing to stress is that computers will not be putting us out of a job. Computer proof assistants can now understand the *statement* of the Riemann hypothesis, but I will eat my hat if a computer, all by itself, comes up with a *proof* of the Riemann hypothesis (or indeed a proof of any open problem of interest to mainstream pure mathematicians) within the next 10 years.[2]

What I do believe is going to happen within the next 10 years: tools will be created which will *help* mathematicians prove theorems. Digitized and semantically searchable databases of mathematics are appearing. Computers are going to start doing diagram chases for us, filling in the proofs of lemmas, pointing out counterexamples to our ideas, and suggesting results which might be helpful to us. The technology to make such tools is already

1 Throughout this article, by "us" and "we" I am referring to the community of people who, like myself, identify themselves as pure mathematicians.

2 Conjectures which stretch beyond a 10-year period are, I think, very unwise; like mathematics, sometimes computer science moves very quickly.

coming; it is viable. Furthermore, the databases of theorem statements and proofs which are appearing will not only have applications in research; we will be able to use them for teaching and for communicating mathematics in new ways. Undergraduates will be able to get instant feedback on their work. PhD students will be able to search for theorems and counterexamples in databases. Researchers will be able to write next-generation error-free papers where details can be folded and unfolded by the user. Patrick Massot has written a thoughtful piece **[44]** explaining these and other ideas in more detail. Computers are going to be able to understand *your area* of mathematics, and even keep up with it as it develops. But there is a catch. Who is going to make the database of important results in noncommutative Iwasawa theory, or whatever area you are interested in, which will power these tools? It is not going to be the computer scientists, because most of them know nothing about noncommutative Iwasawa theory. *It has to be us.*

If you want to see progress within this domain in your own area of mathematics, I would *urge* you to take some time working through some tutorials and learning one of these computer proof assistant languages. It is not difficult to do so—I teach a popular course to final year mathematics students where we learn how to do undergraduate level mathematics (topology, analysis, group theory, and so on) using the Lean theorem prover.[3] Engaging with harder mathematics is not at all difficult *once you know the language*. If you want to learn Lean's language, a good place to start is the Lean prover community's website **[58]**. Coq and Isabelle/HOL are two other well-established theorem provers with big mathematics libraries, and there are plenty of others. If you can get to the point where you are able to explain the *statements* of your own theorems to a computer proof assistant, then these statements can be added to databases, and, furthermore, you have learnt a new skill. If, however, you can get to the point where you can explain the *proofs*, then the AI people will be extremely interested, as will the people building huge formalized mathematical libraries which represent a 21st century Bourbaki. Furthermore, you will be having fun: formalization of proofs is mathematics reinterpreted as an interesting computer puzzle game. If you do not have the time, then find a student who does. Instead of the traditional "do a project consisting of reading a paper and then writing a paper showing that you understood the paper," why not get a student to write some code which proves that they understood the paper? They can learn the language of the prover themselves, and then teach it to you as you teach them the mathematics.

The files which computer proof assistants can read and write represent a way of digitizing mathematical ideas. Digitizing something *completely* changes (in fact, it vastly augments) the ways in which it can be used. Consider, for example, the digitization of music, with the CD and the mp3 file. This has revolutionized how music is consumed and delivered. My collection of music consists of hundreds of vinyl records, tapes, and CDs in my office and loft. My children's collection is in the cloud, has essentially zero mass and volume, and is accessible anywhere. Not only that, but cloud-based music platforms have also fundamentally changed the way the modern musicians communicate with their fans, bypassing the traditional process completely. The music industry was turned upside-down by digitization.

---

3 If you have Lean installed then you can take the course yourself; the materials are here **[8]**.

Mathematics has been done in the same "pencil and paper" way for millennia, but now there is a true opportunity to rethink and enhance this approach. I do not dare to dream what the ultimate consequences of digitizing mathematics will be, but I firmly believe that it will make mathematics more accessible—and easier for us to do, to communicate, and to play with. The ball is in our court.

2. OVERVIEW OF THE PAPER

This paper describes a "new" way in which computers can be used by mathematicians. As mathematicians, our typical experience with computers is that we can use traditional programming languages like Python or traditional computer algebra packages like `sage` to do things like compute the sum of the first 100 prime numbers. We know equally well that these traditional tools, even though they can compute as many prime numbers as you like (within reason), are not capable of *proving* that there are infinitely many primes; the infinite is our domain, not the domain of the computer.

However, this is no longer the case. Computer proof assistants are programs which know the axioms of mathematics. A consequence of this is that they can do both computing in the traditional sense, and also *reasoning*. In practice this means that one can write some computer code in a proof assistant which corresponds to the proof that there are infinitely many primes (https://leanprover-community.github.io/lean-web-editor/#url=https%3A%2F%2Fraw.githubusercontent.com%2Fkbuzzard%2Fxena%2Fmaster%2Fsrc%2FICM%2Finfinitude_primes.lean), or even to a proof **[22]** of the main result in a recent Annals paper **[26]**.

I wrote "new" in quotes above because it is not new at all; computer scientists have been creating tools like this for decades now. Indeed, the first computer proof assistants appeared in the 1960s. However, more recently three things have happened. First, the technology has now reached the point where research level results across all of the traditional mainstream areas of pure mathematics are now simultaneously accessible to these systems, at least in theory, and, increasingly, in practice. Secondly, the systems are far more autonomous than they used to be. Tactics are commands which can be designed by users and which are capable of putting together hundreds if not thousands of tedious axiomatic steps, enabling mathematicians to communicate with these machines in a high-level way, similar to the way which they communicate with each other. Finally, and crucially, research level mathematicians are finally beginning to get involved; we are seeing material at MSc level and beyond being formalized, by mathematicians, across many areas of mathematics now. These developments mean that teaching research level material to a computer proof system in all areas of mathematics is now becoming a feasible possibility—indeed, it is already happening right now, and shows no signs of stopping.

The main body of this paper consists of 4 sections (numbered sections 3 to 6), which are independent of one another, and can be read in any order.

The first is historical; it consists of descriptions of the systems which are being, or have been used, to formalize mathematics, and discussions of results which have been taught by humans to computers over the last 20 years. It also notes various historical technical advances.

The second is an overview of one of the largest currently available monolithic mathematical libraries in existence, namely Lean's mathematics library `mathlib`. Lean [23] is a free and open source computer proof assistant written primarily by Leonardo de Moura at Microsoft Research. Lean's maths library `mathlib` [60] is a free and open source library for Lean, developed by a community of users across the world, ranging from undergraduates to professional mathematicians. `mathlib` is the library which has powered several of the most recent significant results in the area.

The third section consists of an introduction to type theory as a foundation for mathematics; it explains how mathematical structures, theorems, and proofs can be encoded within these foundations. Note that many of the modern computer proof systems where nonfoundational mathematics is happening (Lean, Coq, Isabelle/HOL) use type theory rather than set theory; however, type theory proves the same theorems as set theory. Furthermore, mathematicians who can prove theorems but who do not know the axioms of ZFC set theory can happily write code in a type theory proof system corresponding to these theorems without knowing the axioms of type theory either.

Finally, a speculative final section describes in more detail some personal ideas of the author and others about the kinds of things which software such as this can be used for, and how it might help us to do our jobs.

I thank the Lean prover community for welcoming me, a mathematician with very little programming experience, into their community back in 2017, and also for reading and giving extensive comments on a preliminary version of this article. Patrick Massot in particular sent many helpful comments on a first draft. I thank Assia Mahboubi and Manuel Eberl for giving advice on the Coq and Isabelle/HOL code in this paper, and to both them and Jeremy Avigad for helpful historical comments. Finally, I would like to effusively thank Leonardo de Moura for writing my favorite computer game, and Mario Carneiro for teaching me how to play it.

3. A BRIEF HISTORY OF FORMALLY VERIFIED THEOREMS

In this section I will talk about the previous successes of computer proof assistants—computer programs which check human proofs—in mathematics. There are far more projects here which I could have mentioned, and I apologize to those who have undertaken major mathematical formalization projects which I have not cited. Examples of computer proof assistants in which a substantial amount of mathematics has been formalized include Lean [23], Coq [19], Isabelle/HOL [48], HOL Light [37], Metamath [45], and Mizar [47].

For a computer to formally verify a theorem, it ultimately needs to be able to deduce the theorem from the axioms of the foundational system (typically, set theory or type theory) which the proof assistant has been designed to use. I will use the below discussion of historical results to introduce some conceptual breakthroughs which have over the years enabled the formalization of mathematics to become feasible.

This section cannot do justice to all of the work which has occurred in the area; I thoroughly recommend Hales' paper "Mathematics in the age of the Turing machine" [33] for more background and examples, although much has happened since that paper was written in 2014.

3.1. The 20th century

Consider the problem of proving from first principles that if x and y are real numbers, then $(x+y)(x+2y)(x+3y) = x^3 + 6x^2y + 11xy^2 + 6y^3$. We all know that the real numbers are a commutative ring, so let us assume that fact. The question now becomes how to use the axioms of a commutative ring to prove the equality that we want. How many lines would a proof from first principles be? Surely not too many! We apply distributivity a few times to expand out the brackets on the left-hand side, and then, of course, it just becomes a matter of tidying up and equating terms. As humans we do not think too much about the tidying-up process; however, if you try proving this in a theorem prover then you will discover that actually it is a combinatorial nightmare. For example, there is a step in the proof where where we need to prove something of the form

$$\begin{aligned}&\big((A+B)+(C+E)\big)+\big((D+F)+(G+H)\big)\\&=\big(\big(\big(\big(\big((A+B)+C\big)+D\big)+E\big)+F\big)+G\big)+H\end{aligned}$$

using only the laws of commutativity and associativity of addition. Humans apply a *principle* to justify this step, not an axiom, and indeed proving such a triviality using only the axioms of a ring is surprisingly fiddly. There is also the issue of turning things like $x((2y)x)$ into $(2(x^2))y$ and so on.

The very early theorem provers had very limited ability to apply principles, meaning that proving results such $(x+y)(x+2y)(x+3y) = x^3 + 6x^2y + 11xy^2 + 6y^3$ would need to be done manually, meaning something like a 30 line proof. If such a triviality hides 30 lines of axiomatic mathematics, imagine what is hidden behind claims of the form "The function f is clearly $O(x^{-2})$ for x large"? It is one thing writing a computer proof assistant—it is quite another one to write one which scales to do the kind of things which we humans do intuitively. For this and other reasons, many of the earlier formalization achievements of the 20th century were mathematically trivial. In particular, there were many proofs of the irrationality of $\sqrt{2}$ and of the infinitude of primes, but these were being used as benchmarks for the systems.

In the final two decades of the 20th century, computer provers began to appear which had new functionality. In these later systems, users could write "tactics." Tactics are computer code which assembles axiom applications together into principles. For example, in a modern prover like Lean, $(x+y)(x+2y)(x+3y) = x^3 + 6x^2y + 11xy^2 + 6y^3$ can now be proved in one line by invoking the `ring` tactic.[4] Tactics allow formalized mathematics to more closely resemble ordinary mathematical practice by making "obvious" things automatic.

---

4 See [31] for a description of the sort of issues which arise when writing such a tactic.

3.2. The prime number theorem

In 2004, a team comprising Jeremy Avigad, Kevin Donnelly, David Gray, and Paul Raff formally verified the prime number theorem, in the Isabelle/HOL system. The proof they formalized was the Erdős–Selberg "elementary proof." The work used inputs from both arithmetic and basic real analysis. Of course, calculations involving growth of functions which look easy on paper still took time and effort to formalize. Manipulation of inequalities which to humans look easy need to be done either by hand or via a Fourier–Motzkin elimination tactic in a theorem prover. The reason that the Erdős–Selberg proof was preferred to the traditional complex analysis proof was that at that time Isabelle/HOL had no complex analysis library at all.

What we conclude is that by 2004, more serious undergraduate and MSc level material was now in theory accessible to these systems, at least in some areas of mathematics. We also see that we are at a stage where libraries of proofs in distinct areas of mathematics are able to interact with one another.

In 2009 John Harrison formalized the complex-analytic proof of the prime number theorem in the HOL Light theorem prover [36], motivated in part by the fact that HOL Light already had a theory of complex analysis including Cauchy's integral formula. In 2016 Mario Carneiro formalized the Erdős–Selberg proof in Metamath, a set theory based prover which has essentially no tactics; as you can imagine this was a heroic effort.

Thus the Prime Number Theorem became some kind of a poster child for formalization. One can understand why—it was a celebrated theorem in mathematics, the proof is not at all trivial, and any formalization in a theorem prover demonstrates that the prover is capable of reasoning about both the discrete and the continuous simultaneously.

As may be becoming apparent to the reader, however, one reason that the result was being independently formalized in several theorem provers was that it is extremely difficult to translate a proof written in one of the systems to a proof in another system. One issue is that different systems might have different foundations; for example, HOL Light and Isabelle/HOL are type theory systems, and Metamath is a set theory system. Another issue is that even if two proof systems have very similar foundations, they might have different *idioms*; different libraries in different systems could be set up to do the same thing in very different ways. Without getting too technical, in order for these computer proof systems to work, one has to have some kind of a method for moving between structures "behind the scenes"—for example, the reals are a field, and hence they are an additive group (and a multiplicative monoid), and in particular one wants all theorems about additive groups such as $0 + a = a$ to apply instantly to fields such as the reals without any fuss. Humans, of course, have no problems with this, but in a computer proof system one needs some kind of infrastructure which is making this happen automatically, and if different systems are doing this in different ways then, of course, this makes automatic proof translation much harder.

Thus it came as a shock to me when in 2020 Mario Carneiro announced that he had used his Metamath Zero project [14] to port the Metamath proof of the prime number theorem to Lean. The two systems are about as far apart as it is possible to be—Metamath uses set theory as a foundation and Lean uses type theory, for example. Metamath proofs are typically far more low-level, with limited automation available, whereas typical Lean proofs are very tactic-heavy. However, the system worked, and produced code which compiled; it was, of course, also unreadable. It was tens of thousands of lines of completely unmotivated primitive code defining variables and applying basic principles of logic, with no comments. In fact, it was a wonderful example of something which satisfied a formal definition of "being a proof," whilst in some sense imparting no information whatsoever to the human reader other than the fact that the theorem was true.

Of course, if computers begin to write proofs by themselves, they might all look like this, at least at first.

3.3. The four color theorem

The four color theorem (formerly the four color conjecture) was a notorious problem in graph theory raised in the 1850s and which remained unresolved for over 100 years. One formulation of it is the assertion that the vertices of every planar graph can be colored with four colors in such a way that no two adjacent vertices share a color. The statement is an elegant combinatorial problem, and it came as a shock to some in the mathematical community that the proof, announced by Appel and Haken in 1976, used a computer in an essential way. Appel and Haken constructed a collection of 1834 graphs with the property that a minimal counterexample must contain one of these graphs as a subgraph, but that conversely no graph containing one of these 1834 graphs as a subgraph can be a minimal counterexample. The verification of these claims was done using a bespoke computer program which, in those days, took over a month to finish running. The Appel–Haken proof was an outlier because whilst the principle of the proof was possible to understand, the details were too difficult for a human to follow in practice; one billion case splits (this is what the computer part of the proof looks like) is not something which humans can do manually and accurately within a reasonable time frame. The proof relies, essentially, on a computer calculation and hence it relies on the correctness of the computer code. Small bugs in computer code are, of course, commonplace, although it could be added that small bugs in human-written proofs are also commonplace. However, the mathematical community is well-equipped to discover and fix small bugs in human-written proofs, and was perhaps rather less well equipped to verify the correctness of computer code, especially in 1976.

In 2004 Georges Gonthier finished a formal verification of the Appel–Haken result—more precisely he formalized the 1997 Robertson–Sanders–Seymour–Thomas variant of the argument [27]. The work comprised 60,000 lines of code written in the Coq proof assistant. In particular, it completely dwarfs the prime number project discussed in the previous subsection. It contains a complete formalization of the theoretical part of the work—formal proofs of results in topology (to reduce the statement about arbitrary planar graphs to one of a discrete combinatorial nature) and graph theory—whilst also formally verifying the computer calculation necessary to finish the proof. Note in particular that (as in the proof of the prime number theorem) much of the work comprised writing foundational material rather than formalizing the proof itself.

It is interesting to note that the process of formalization led to simplifications in the argument. For example, Gonthier developed a theory of what he called combinatorial hypermaps, which greatly reduced the amount of topology needed in the proof, and in particular removed the dependency of the argument on the Jordan Curve theorem. Gonthier developed some original mathematics as part of the work—for example, he isolated a combinatorial criterion for his hypermaps which was equivalent to planarity.

Naively, it looks like in this case we are replacing one "proof by computer" with another one; however, this is missing the point. Firstly, the Coq formalization covers not just the Appel–Haken computer code, but also all of the rest of the Appel–Haken argument. Secondly, one can view the formal verification as an independent check of the proof. Finally, instead of having to trust the code written by Appel and Haken and which few people have read, we are instead having to trust code written by the authors of Coq. Coq has been around for a long time (the first version was written in 1984), has a small kernel, and the system has many users. A bug which meant that Coq could incorrectly claim that an unproved theorem was true would be unlikely to manifest itself in just one project and is far more likely to ultimately be discovered. In contrast, the Appel–Haken code is a bespoke piece of code with few users so arguably bugs are more likely.

Gonthier wrote a very informative piece [28] about his work for the Notices of the American Mathematical Society (including an exposition of the theory of hypermaps), as part of the November 2008 issue; this issue was devoted to formal verification of mathematics in a computer proof system and provides an excellent survey of the field as it then stood.

3.4. The odd order theorem

The odd order theorem is the theorem that any finite group of odd order is solvable. In 2013 a team of 15 people led by Gonthier formally verified a proof of this theorem in Coq [29]. This piece of work is notable for several reasons. Firstly, the proof is very long; a complete argument (modulo the basics in group and representation theory) is presented in the two volumes [2] and [50]. Secondly, we have moved way beyond MSc level mathematics here—this work was one of the reasons that Thompson was awarded the Fields Medal in 1970. The proof is a very delicate argument in finite group theory, much of which involves analyzing the structure of a minimal counterexample and ultimately showing that it cannot exist. The Coq proof involved formalizing both of the books mentioned above, plus, of course, all the background material in group theory, representation theory, Galois theory, and number theory; indeed, formalization of the background material took up much of the six years which the authors spent on the proof. Figuring out how to handle such a large-scale formalization project was also a nontrivial task.

It is perhaps worth stepping back and asking how work like this contributes to human understanding. The naive answer to this is "it guarantees that the human proof is correct." However, in my opinion, this is not the main contribution. Humans were well aware even back in the 1960s that the proof was correct—had there been any doubt, Thompson would not have got the Fields Medal. What the formalization work shows us that theorem provers have now become able to operate at this kind of scale. Entire books of mathematics can now be formalized in one system without the system running out of memory or grinding to a halt. On average, one line of mathematics in [2] or [50] corresponded to five lines of computer code, so we learn that by 2013 the so-called "De Bruijn factor" for this kind of mathematics is around 5. However, this ratio should not be taken too seriously: in parts of the argument, the ratio is essentially one, and in other parts it is much larger. Note also that this factor may vary considerably between theorem provers.

We also learn that large formalization projects such as this are a very effective way to motivate development of foundational mathematics libraries. One consequence of this formalization project was that Coq developed a very solid library of undergraduate-level algebra which can, of course, be used (and is used) for other projects.

The write-up [29] of the odd order work is an interesting read. Some sections concentrate on the mathematics or the history, but there is also a discussion about constructive mathematics, something which I felt should have nothing to do with the work, and also about implementation issues, something else which mathematicians typically do not ever have to think about. For example, one observation made in section 3 was that many theorems involving two or more finite groups would usually be formalised assuming that these groups were both subgroups of some larger ambient finite group X. This can be done without any loss of generality of course, because given two groups G and H, they are both subgroups of $G \times H$. Why is this observation important? This is an *implementation issue* – the domain of the computer scientist. Working with subgroups rather than groups might be easier, or nicer, when it comes to actually implement certain theorems in the theorem prover. It is worth noting however that such a trick does not work more generally: for example in algebraic geometry one uses the category of commutative rings with 1, and morphisms by definition send 1 to 1. If R and S are general commutative rings with 1 then there is in general no morphism of rings from R to $R \times S$ sending 1 to 1, so one is forced to implement commutative ring theory in a more "traditional" manner. See [63] for how this was done in Lean.

Regarding constructivism – the authors of the work put in a lot of effort to keep their proof "constructive", for example the avoidance of all uses of the complex numbers when setting up the basics of representation theory. The complex numbers do not have decidable equality, meaning that there is in general no algorithm for proving that two constructively defined complex numbers are equal (for example, one can evaluate a definite integral numerically and observe that it seems to be 0 to 1000 decimal places, but there is not some generic algorithm which we can apply to an arbitrary integral in order to decide whether or not it equals 0). This means that in constructive mathematics, where the law of the excluded middle cannot be assumed, one cannot do a case split on whether $z = 0$ or not, if z is a complex number, and more generally plenty of constructions become noncomputable and hence much harder to reason with constructively. These design choices thus increase the amount of work needed to get representation theory working. I had thought that constructivists had died out in the early part of the 20th century. It turns out that they are alive and well, and typically working nowadays in computer science departments. One reason for this is that constructivism plays an important role in the theory of programming languages. Reluctance to use the law of the excluded middle is to a certain extent a cultural decision. However there are also situations where working constructively enables a computer proof system to prove certain results "automatically" (for example by an explicit computation). Whilst working constructively may have been feasible for a project about finite groups, the law of the excluded middle is used throughout most modern research level mathematics and it is not really feasible to work constructively when doing the kind of mathematics which is happening nowadays in mathematics departments. However it is also worth stressing that most modern proof assistants have no problems with the law of the excluded middle, the axiom of choice, and other non-constructive axioms – they are available, if you want them. Certain proof strategies are ruled out if one chooses to work nonconstructively, but one can counter this by writing new tactics specifically designed to do computations in fields such as the real and complex numbers. Nonconstructive axioms are used extensively in Lean's mathematics library `mathlib`, for example.

3.5. The Kepler conjecture

The Kepler conjecture states that the face-centered cubic packing is the densest way to pack congruent spheres in 3-space. Hales and Ferguson proved the conjecture in 1998; it had at that point been open for over 350 years (it was raised by Kepler before Fermat proposed his Last Theorem). Part of the Hales–Ferguson proof involved the checking of over 23000 nonlinear inequalities on a computer; another part involved a computer classification of all tame graphs. Other computer calculations were also involved. In this respect the proof is similar to the Appel–Haken proof of the four color theorem; computations need to be carried out which are simply far too great for humans to do in a reasonable time frame.

Because the result was regarded as important, the referees felt duty-bound to attempt to check the computer part of the proof in some way; however, ultimately they gave up, and in [34] Hales states that the paper was published (in the Annals) without complete certification from the referees. In 2003 Hales announced a project to formally verify the proof using computer proof systems. Hales used a combination of HOL Light and Isabelle/HOL, and the project turned into an international collaboration, with 22 authors listed on the final paper. The formalization project took around 12 years to complete, and comprised over half a million lines of code. Just as for the other projects in this section, one of the main benefits of the work to the formal proof community is that HOL Light's standard library grew to include theorems such as the Brouwer fixed-point theorem, the Krein–Milman theorem, and the Stone–Weierstrass theorem.

In 2017 Hales gave a talk [32] at the Newton Institute where he told the story of the Kepler proof, and explained a vision for the future of formalized mathematics. This talk was, for me, the turning point, and was one of the main motivations behind the work described in the following subsection.

3.6. Perfectoid spaces

The previous formalized results all have something in common. Whilst some of them represent truly deep mathematics, all of the formalized proofs involve reasoning about objects which are in some sense *elementary* (planar graphs, prime numbers, finite groups, spheres). Furthermore, most (but not all) of the formalizing done prior to 2017 was being done by computer scientists. In Hales' talk linked to above, he coherently argued that for further progress in this area, this state of affairs had to change. At that time I had only just begun to dabble with computer proof assistants and my initial plans were to attempt to integrate them into my undergraduate teaching. However, Hales' arguments resonated with me, and within a few months I found myself working with undergraduates at Imperial College, formalizing the definition and basic properties of schemes in the Lean theorem prover. This project involved developing basic theories of localization of rings and of sheaves on topological spaces; however, it was relatively straightforward (modulo poor design decisions; the reader interested in more details can see them in **[12]**). I was thus shocked to discover afterwards that schemes—such a basic notion in algebraic geometry—had not been previously formalized in *any* other computer proof system! Furthermore, the project made it quite clear to me that formalizing far more heavyweight mathematical objects should easily be possible.

In late 2017 Patrick Massot (a topologist) and myself independently came up with the idea of formalizing perfectoid spaces; the topic was in the air because it was at that time an open secret that Scholze was going to be awarded a Fields Medal for his invention/discovery of the concept and its applications to arithmetic geometry. I knew the mathematical definition, having dabbled in the area myself, and when Johan Commelin, another arithmetic geometer, appeared in the Lean Zulip chatroom in 2018 the three of us decided to go for it. Around 16000 lines of code and eight months later, we had a formalized definition; one could summarize the work as a computer formalization of the single line of mathematics "let X be a perfectoid space".[5]

The work was, of course, partly intended as a public relations stunt; computer scientists were well aware of the existence of computer theorem provers, however, mathematicians seemed not to be, and this was an attempt to make them notice. The plan was a success—the project did seem to raise the profile of computer theorem provers within the mathematics community. Note, however, that we did not construct any examples of perfectoid spaces other than the empty perfectoid space,[6] and all three of us were well aware of the problems preventing us from formalizing any of Scholze's serious theorems about perfectoid spaces at that time; we were missing so many of the prerequisites. As with previous projects, one tangible gain from the work was the growth of the mathematics library of the system in question. Most of the results in Bourbaki's General Topology ended up as part of Lean's mathematics library `mathlib` as a result of this project, as well as plenty of results in topological algebra, and it also motivated the beginnings of a theory of valuations and discrete valuation fields.

One can consider the perfectoid space work as in some sense being orthogonal to what was usually being attempted in a theorem prover. Many of the prior results highlighted

---

5 In the odd order formalization, the de Bruijn factor (ratio of lines of computer code to lines of human text) was around 5. Here one could argue that it is 16000. However, one could, of course, also argue that it might well take several thousand lines of human text to define a perfectoid space in full.

6 To prove that the empty set can be given the structure of a perfectoid space, one needs to check that an arbitrary product of trivial topological rings is the trivial topological ring.

in this section are proofs of long and complex theorems about relatively simple objects. The proof that the empty set can be given the structure of a perfectoid space is a very simple theorem about a much more complex concept. Of course, the natural next question is whether computer proof systems can prove complex theorems about complex objects. One year after the perfectoid space project, we began to find out.

3.7. Condensed mathematics

Clausen and Scholze have been developing a theory of condensed mathematics. A condensed set is a variant of a topological space. The main insight is that condensed objects may have better homological properties than topological objects (for example, the category of condensed abelian groups is an abelian category, whereas the category of topological abelian groups is not). They hope that these ideas will enable techniques in homological algebra to apply to new areas of analytic geometry. At the end of 2020, Scholze approached me and asked if we had had a study group on the work at Imperial; I answered that we had. Scholze then asked whether we had looked through all the details of the proof of Theorem 9.1 of [51]; I answered that we had not. Scholze then remarked that he had had the same response from other mathematicians, and raised the possibility that perhaps nobody other than himself and Clausen had ever read the proof carefully. Furthermore, he suggested that perhaps this might remain true even after the refereeing process. The reason he was concerned about this was that, for Scholze, this was the theorem that the entire theory stood upon. The proof was very technical; it built upon a more "elementary" but rather unwieldy intermediate result, Theorem 9.4 of [51]. Scholze agreed to challenge the formalization community to prove his Theorem 9.1 in a blog post [55], later published as [53]. Although the challenge was to the formalization community in general, it seems that only the Lean community responded; this is perhaps unsurprising, as (for perhaps only for sociological reasons) it has come to be the case that mathematicians interested in "the kind of mathematics which wins Fields Medals" and also interested in theorem provers tend to gravitate towards Lean.

Johan Commelin became the de facto leader of the formalization process, with Patrick Massot supporting him in making a blueprint [18] of the strategy (that is, a carefully-written roadmap) and a team of algebraic number theorists, arithmetic geometers and other mathematicians (Riccardo Brasca, Damiano Testa, Filippo Nuccio, Adam Topaz, myself, Patrick Massot, Bhavik Mehta,...) then began working on the project, with the occasional help from people with a computer science background such as Mario Carneiro. Within six months the team had grown to over ten people and we had formalized a complete proof of Theorem 9.4 (see [15]). At the time of writing, we have not deduced Theorem 9.1, but it is only a matter of time. A second blog post [54] by Scholze indicates his thoughts on the matter; in particular, we see that he is now far less concerned about the situation regarding the correctness of the results. Furthermore, Scholze has indicated (personal communication) that the process has enabled him to better understand what powers the proof, and Commelin not only learnt the mathematics as part of the process, but also simplified the argument in several places, most notably in the removal of the dependency of the argument on prior work of Breen and Deligne.

For me, this represents substantial evidence that now *any pure mathematics* can be formalized in theorem provers—both in theory, and in practice. It takes time, but it is possible. The formalization of the work led both to better understanding of it, and to simplifications of the argument. Also worth mentioning is that, as in many other formalization projects, a substantial amount of time was spent formalizing background material (for example, the theory of normed groups and the theory of profinite spaces). As the libraries of the provers get better and start to contain the kind of material which working mathematicians take for granted, there will be fewer of these "startup costs."

3.8. Other results

There are plenty more examples of serious formalization efforts which we do not have the space to cover. We list some examples here. Gouëzel formalized the basic definitions of C^k and C^∞ manifolds in Lean, extending earlier work done in Isabelle/HOL. Mahboubi and Sibut-Pinote proved irrationality of $\zeta(3)$ in Coq [16] and Eberl proved it in Isabelle/HOL [24]. Mahboubi has also done extensive work on rigorous numerical values of integrals in Coq, and also in Coq Bertot, Rideau, and Théry formally verified the first one million decimal digits of π [4]. Eberl has formalized much of Apostol's textbook on analytic number theory in Isabelle/HOL [25]. Han and van Doorn proved independence of the continuum hypothesis in Lean [35]. Immler formally verified Tucker's calculations used to verify the existence of the strange attractor [38]. Mehta and Dillies formally verified Szemerédi's regularity lemma and Roth's theorem on arithmetic progressions in Lean, and Edmonds, Koutsoukou-Argyraki and Paulson verified them in Isabelle/HOL. The Poincaré–Bendixson theorem was formalized by Immler and Tan [39] in Isabelle/HOL; note that the usual proof as understood by mathematicians relies on drawings, and formalizing drawings can be hard work. The Ellenberg–Gijswijt resolution of the cap set conjecture was verified in Lean by Dahmen, Hölzl, and Lewis [22]. Commelin and Lewis constructed Witt vectors and showed that $W(\mathbb{F}_p) = \mathbb{Z}_p$ in [17]; this work is interesting because not only did they formalize the delicate mathematics involved, they also wrote tactics which would enable them to reduce various questions to the universal case in a painless manner. Finiteness of the class group of a global field was proved in Lean by Baanen, Dahmen, Narayanan, and Nuccio in [1] (it still astonishes me that this result, special cases of which were known to Gauss and which is a standard theorem in an undergraduate mathematics degree, was formalized in a proof assistant for the first time in 2021).

There is also work in progress (at the time of writing). Teams of people who collaborate on the Lean Zulip chat [65] are currently working on a proof of Fermat's Last Theorem for regular primes, and on Smale's theorem that it is possible to evert a sphere. A general project to formalize many basic results in the theory of schemes is also underway.

4. `mathlib`

In this section I will give an overview of Lean's mathematics library, one of the largest monolithic collections of formalized mathematics in existence and, more importantly, one which is currently experiencing rapid growth. To a certain extent it is a personal perspective; a different point of view, which talks more about the computer science powering the library, is presented in [60].

The principal developer of the Lean Theorem Prover is Leonardo de Moura, who started the project in 2013. At the 2017 Big Proof conference in Cambridge, it was decided to split off most of the "mathematical" part of the prover from the "core" part, and move the mathematics into a library of its own. Thus `mathlib` was born. At the time `mathlib` contained definitions of groups, rings, and topological spaces, filters, a construction of the rational numbers (the naturals and integers remained in core Lean), and little else. Johannes Hölzl and Mario Carneiro became the maintainers of the library, and between them they began to slowly build more mathematics, for example, the real numbers. Hölzl had written a lot of the topology part of the repository, following the Isabelle/HOL approach which relied heavily on the concept of a filter. Carneiro wrote a robust theory of finiteness, and slowly the library began to become relevant to the "working mathematician."

The library is a free and open source project. It is monolithic in the sense that there is one definition of a group, one definition of a ring, one definition of the real numbers, and so on, and all of these definitions can be imported simultaneously and interact with each other. Initially it was not clear what its goals were, other than being a place where people could experiment with doing mathematics in Lean. Mathematicians such as Scott Morrison, myself, and Patrick Massot got involved at a very early stage, and because our background was in mathematics which relied on classical logic (i. e., the law of the excluded middle) and other nonconstructive axioms such as the axiom of choice, the library developed with these classical assumptions at its core. Each successful mathematics project written in Lean and powered by `mathlib` seemed to attract more mathematicians to its chatroom, which in turn led to more projects. Within a couple of years Lewis had formalized the p-adic numbers [41], myself and a team of undergraduates (Lau, Hughes, Livingston, and Fernández Mir) formalized schemes [12], Dahmen, Hölzl, and Lewis formalized the 2017 Ellenberg–Gijswijt Annals proof of the cap set conjecture [22], and Massot, Commelin, and myself formalized the definition of a perfectoid space [10]. Each of these projects could not have happened without `mathlib`; conversely, each of these projects contributed to the growth of `mathlib`.

Plenty of developments were also taking place which were not written up as papers, and whose main purpose was simply to grow `mathlib`. I supervised student projects where undergraduates could formalize material they were learning in class and add it to the library; for example, Sylow's theorems (Chris Hughes), nilpotent groups (Ines Wright), conformal maps (Yourong Zang), and the Radon–Nikodym theorem (Kexing Ying) were added this way. Amelia Livingston developed a theory of localization of monoids and rings which we needed for algebraic geometry. I pushed undergraduates (Hughes, Lau, Lee) to formalize a standard Galois theory course in Lean; they developed a theory of field extensions, and the project was then taken up by a group of mathematics PhD students in Berkeley (Miller, Browning, Lutz) who finished the job, proving the fundamental theorem of Galois theory and the insolvability of the quintic [7] (note that this was coincidentally formalized in Coq just a couple of months beforehand [3]). Baanen, Dahmen, Narayanan, and Nuccio formalized a proof of the finiteness of the class group of a global field [1]. I was pushing algebra, but others were pushing geometry and analysis. Gouëzel and Macbeth developed a theory of manifolds, and Gouëzel and Kudryashov developed an extensive theory of single and multivariable calculus, including the implicit function theorem and the Picard–Lindelöf theorem. Gouëzel also formalized the Gromov–Hausdorff space: a metric space parametrizing nonempty compact Hausdorff metric spaces up to isometry.

Morrison has developed a huge amount of category theory, and he and Topaz have now formalized the definitions of abelian categories and the beginning of the development of derived functors and homological algebra. Massot has developed valuation theory and a theory of completions of uniform spaces and of topological groups and rings. Tuma developed the theory of Jacobson rings, and I developed some of the basics of other standard ideas in commutative algebra (projective and flat modules, discrete valuation rings), and Springer, Kuelshammer, and many others have also contributed to algebra. Hölzl developed the theory of Lebesgue measure, and van Doorn formalized Haar measure. There are many more people who have made contributions (`mathlib` now has over 200 contributors) and new contributions are always welcome. Contributions are reviewed by the maintainers. One of the principles of the library is to do things "in the correct generality." This meant, for example, that multivariable calculus and some exotic integrals taking values in Banach vector spaces was developed first, and single variable calculus was deduced as a corollary. The library is not optimized for pedagogy or readability; the idea is to continue to make a solid foundation for the kind of mathematics which is happening in a contemporary mathematics department.

It is interesting to note that Lean seems to be learning mathematics at around the same speed as an undergraduate. In the four years which the library has been growing, it has gone from essentially zero to a solid MSc level coverage in number theory and commutative algebra, and BSc level real analysis. In complex analysis, differential geometry, and representation theory it is perhaps not quite yet at final year BSc level, but things move fast and this sentence, written in 2021, will quickly date. For an up to date idea of the current status of `mathlib`, the best idea is to take a look at the Lean community's full overview of `mathlib` [57], or its summary of the undergraduate level mathematics it contains [59].

5. A BRIEF GUIDE TO TYPE THEORY

In this section we explain the basics of type theory and how it can be used as a foundation of mathematics. Many modern theorem provers use some version of type theory as their foundations. For example, Isabelle/HOL and the other HOL systems use simple type theory, Lean and Coq use dependent type theory, and the various HoTT systems developed by Voevodsky and others use homotopy type theory. There are a few computer proof assistants which use set theory—Metamath and Mizar are the two most prominent—however, it is not unfair to say that what nowadays most mathematics have done in theorem provers is done in a type theory system, so a mathematician interested in dabbling in formal proofs should at least know something about the basics, which is what this section attempts to describe.

5.1. What is a type?

Mathematicians nowadays are used to seeing the word "set" floating around when it comes to basic definitions. For example, we are told that a group is a set equipped with a multiplication such that some axioms hold. We are not told what a set is though; a course on ZFC set theory tells us a list of properties which sets *have*, but they do not tell us what a set *is*. Indeed, in this context the word "set" has no formal definition; it is simply the generic term for an object in our model of the axioms of mathematics, and we build other mathematical objects on top of this basic object.

In definitions such as the definition of a group, the word "set" is being used to mean no more than "collection of elements." In type theory, the role of a "collection of elements" is played by the *type*. A type is a collection of terms. The definition of a group in type theory: a group is a type equipped with a multiplication such that some axioms hold. The only difference is the notation: the set-theoretic $a \in X$ is replaced by the type-theoretic `a : X`.

As mentioned above, those of us who have been to a set theory class will know that, when using set theory as a foundation of mathematics, *everything* is a set. For example, the elements of a group are, strictly speaking, also sets, so one could in theory talk about their elements too, although within the context of group theory such questions would not be mathematically meaningful, as they are not isomorphism-invariant. In type theory this is not possible; the elements of a type are called *terms*, and in general terms are not types. In type theory, everything is a term, and every term *has* a type, but not every term *is* a type. For example, in type theory $37\pi^2$ is a term, whose type is $\mathbb{R}$, the type of real numbers. We write $37\pi^2 : \mathbb{R}$. However, $x : 37\pi^2$ *does not make sense*, because $37\pi^2$ is not a type. In a set-theory based theorem prover, questions such as asking if the trivial group is an element of the Riemann zeta function would make sense but its meaning would be unmathematical—it would depend on implementation decisions. Type theory thus provides a basic check that what you are writing has mathematical meaning.

In a type theory system, the type $\mathbb{R}$ is still built from $\mathbb{Q}$ as equivalence classes of Cauchy sequences, or via Dedekind cuts, or as another of the standard constructions; the mathematical part of the story is identical to the set theory setup, it is just that the language used is slightly different (types and terms, rather than sets and elements).

One difference between types and sets, however, is that *types do not mix*: distinct types are disjoint. This has practical advantages when formalizing mathematics because it provides a strong check that the mathematics you are typing *makes sense*: in type theory, if g is an element of the group G, then the only type that g can ever be a term of is G.

This approach does, however, have consequences which can initially come as a shock to a mathematician. For example, one could make a type representing the positive reals $\mathbb{R}_{>0}$ and a type representing the reals $\mathbb{R}$, but if a term x had type $\mathbb{R}_{>0}$ then x itself would *not*, strictly speaking, have type $\mathbb{R}$; I stress again that every term has a *unique* type. To make a term of type $\mathbb{R}_{>0}$, one has to give *two* pieces of data: a real number, and a proof that it is positive. A term of type $\mathbb{R}_{>0}$ is an object corresponding to this pair, so, strictly speaking, it is not a real number, and a type theory based system will hold you to this. However, of course, there is a canonical map from $\mathbb{R}_{>0}$ to $\mathbb{R}$—you just throw away the proof. More generally, a type theory system could well have a *coercion system*, consisting of a collection of "invisible functions" mapping types to other types in the way which mathematicians would expect. For example, given a term of type $\mathbb{R}_{>0}$, it might well be possible to feed it into a function which is expecting a term of type $\mathbb{R}$; the system will just throw away the proof of positivity and use the underlying real number anyway. Mathematicians use these invisible functions everywhere, often without noticing. We have already mentioned above that in a foundational system the real numbers need to be built using one of the standard constructions, for example, via Cauchy sequences. In particular, a rational number is not literally a real number. However, taking Lean as an example, if one has a term `x : ℚ` then one can simply write `x : ℝ` to get the corresponding real number, although a careful inspection of the corresponding term will unearth the fact that the real number is actually called `↑x`, indicating that a coercion has been applied. The coercion is a ring homomorphism, and Lean has a "normalize casts" tactic [42] which knows this and will apply theorems such as `↑(x+y)=↑x+↑y` and `↑(x*y)=↑x*↑y` automatically (before this tactic had been written, doing mathematics which involved switching between the naturals, integers, and rationals could be quite frustrating because of these invisible maps). In summary then, type theory forces you to think more precisely about the actual objects you are working with, however, tactics can be used to manipulate these objects the way we usually manipulate them. Learning how to "steer" mathematics in a theorem prover this way simply comes from practice.

5.2. Inductive types

I have already mentioned that in a type theory system the definition of the real numbers is the same as in a set theory system—it is Cauchy sequences, or Dedekind cuts, or whatever your favorite construction of the reals is. Similarly, the usual definitions of the rationals and integers as quotients work just as well in type theory as they do in set theory. But one place where the type-theoretic and set-theoretic foundations of mathematics differ is in the definition of the natural numbers. The natural numbers are a foundational object in mathematics—they are typically the first example of an infinite object to be born—so it is perhaps unsurprising that different foundational systems will treat them in different ways.

In ZFC set theory, the existence of the set of natural numbers is postulated as an axiom, namely the axiom of infinity. Type theories such as Lean's instead allow the user to define custom *inductive types*. Such types include the naturals and other recursively-defined constructions. Implementation details of this so-called calculus of inductive constructions [20] differ between systems; the rest of this section explains details which are specific to Lean's type theory, but much of what I say applies to Coq and Agda, other popular type theory provers.

In Lean, the definition of the naturals looks like this:

```
inductive nat
| zero : nat
| succ (n : nat) : nat
```

This definition says "zero is a natural number, the successor of a natural number is a natural number, and that is it." As one might guess, this inductive construction can be used to construct far more exotic types, but one can show that any type which can be defined using the rules of the calculus of inductive constructions corresponds to a set which can be built using the usual axioms of set theory.

Let us see what goes on under the hood when the naturals are defined as an inductive type. When such a definition is made, a new type `nat` appears in the system, as does the term `nat.zero` and the function `nat.succ : nat → nat`. The latter terms are called *constructors*: they are ways to make natural numbers. However, one more thing also appears, namely the *eliminator* for the type—the object which enables the user to construct functions whose domain is the naturals and whose codomain is something else. It represents the idea that the only way that one can construct naturals is via `nat.zero` and `nat.succ`, and it states that to define a function out of the naturals, it suffices to (1) say where `nat.zero` goes, and (2) to say where `nat.succ n` goes, given where `n` went. In other words, it is the principle of recursion.

So this is how new inductive types are born in Lean; after their definition they, together with their constructors and eliminator, are automatically added by the proof assistant to the system as new constants, or axioms, or however you would like to think of them. There are, of course, precise rules telling us the exact form of the eliminator for a given inductive type; we do not go into these here. From a foundational point of view, this approach, where new axioms appear "by magic" as types are constructed, is very different to the set-theoretic viewpoint; however, in **[62]** it is shown that type theory with these constructions is equiconsistent with set theory. The strategy of the proof is to make a model of set theory within type theory, and to make a model of type theory within set theory. For a more precise statement, one has to be more precise about exactly what kind of type theory one is working with. For example, Mario Carneiro's MSc thesis **[13]** shows that Lean's type theory is equiconsistent with ZFC plus countably many inaccessible cardinals.

It is worth noting, and quite amusing, that equality itself is defined as an inductive type in many type theory systems. This is in contrast to set theory, where equality is typically considered as part of the logic. Indeed, equality in type theory is generally more subtle than in set theory. Here is Lean's definition of equality:

```
inductive eq {X : Type} : X → X → Prop
| refl (a : X) : eq a a
```

The slightly unnerving `X → X → Prop`, bracketed as `X → (X → Prop)`, means that equality is a function which takes in an element of `X` and outputs a function which takes in an element of `X` and outputs a Proposition, that is, a true–false statement. In other words, if `a` and `b` are terms of type `X` then `eq a b` is a true–false statement. Using the usual notation `a = b` for `eq a b`, we see that equality of terms of a type `X` is an inductive type with one constructor, namely `eq.refl a`, a proof that `a = a`. It turns out that from this definition we can *prove* all the usual properties of equality! The eliminator for the equality type is the *substitution property*, that if $a = b$ then given a term of type $P(a)$ we can get a term of type $P(b)$. It is a rather pleasant game to go on from this to deduce that equality is both symmetric and transitive (for more details on this, see, for example, [9]). Of course, whilst it is of interest to some to see how basic properties of equality can be proved within a type theory system, it is also worth stressing that to use a computer theorem prover one does not have to know anything about them.

5.3. Dependent types

Lean and Coq both use a version of type theory called dependent type theory, so it is perhaps worth taking some time to explain what a dependent type is. Imagine X is a geometric object, for example, a real manifold. Say that we have a vector bundle on X, that is, for each point x of X a vector space V_x (which varies smoothly with x in some appropriate sense). A section of this bundle is a function which takes as input a point x in X and outputs an element of V_x. From a foundational point of view, there are two ways to think about such a section. One could regard this section as a function from X to the disjoint union of the V_x, sending $x \in X$ to an element of V_x. Alternatively, one could regard it as a slightly stranger kind of "function" which has domain X but whose codomain varies according to the input. There are times in mathematics when taking the disjoint union of the codomains is a natural thing to do—for example, in the example above, the disjoint union of the V_x is naturally a space V sitting above X. However, there are also times when taking the disjoint union is quite unnatural. For example, in algebraic geometry one way of defining the sections of the structure sheaf on an affine scheme $\mathrm{Spec}(R)$ is functions which send a prime ideal P of R to an element of the localization R_P of R at P, and the disjoint union of the R_P as P varies over the prime ideals of R has no natural algebraic structure. The set or type consisting of the disjoint union of these local rings is typically not part of the mental model which an algebraic geometer has when describing these sections.

These kinds of "functions" which have a well-defined domain, but a codomain which can vary according to the input, are called dependent functions. Not all proof assistants have such functions; for example, Isabelle/HOL (a powerful proof assistant which contains a lot of analysis and analytic number theory) and various other HOL systems do not have them, which means that certain constructions in geometry are more convoluted than in Coq or Lean. See, for example, [6], which defines schemes in Isabelle/HOL but which has to build a new implementation of ring theory from scratch in order to do so.

5.4. Examples

Let us take a look at some examples of what mathematics looks like in a theorem prover based on type theory. I give these examples mainly to convince the reader who has been brought up using the language of set theory that there really is very little difference.

Here is what the claim that $\sqrt{2}$ is not rational looks like in Isabelle/HOL:

```
theorem sqrt2_not_rational:
  "sqrt 2 ∉ ℚ"
```

You can see the proof on Isabelle's Wikipedia article [64]. The fact that 2 is a term of a type and not a set, or an element of a set, is invisible.

Here is some more advanced mathematics, written in Coq:

```
Lemma prod_Cyclotomic n :
(n > 0)%N -> \prod_(d <- divisors n) 'Phi_d = 'X^n - 1.
```

This is the statement that the product of the dth cyclotomic polynomials over $d \mid n$ is $X^n - 1$. Note the hypothesis $n > 0$, an assumption which a human would typically omit; computers are very picky with such "edge cases."

Here is the definition of a perfectoid ring in Lean, taken from the Lean perfectoid spaces website [11] which accompanies the article [10].

```
/-- A perfectoid ring is a Huber ring that is complete, uniform,
that has a pseudo-uniformizer whose p-th power divides p in the
    power bounded subring,
and such that Frobenius is a surjection on the reduction
    modulo p.-/
structure perfectoid_ring (R : Type) [Huber_ring R] extends
    Tate_ring R : Prop :=
(complete  : is_complete_hausdorff R)
(uniform   : is_uniform R)
(ramified  : ∃ ϖ : pseudo_uniformizer R, ϖ^p ∣ p in R°)
(Frobenius : surjective (Frob R°/p))
```

The comment at the top of the code is the "docstring" for the code—this is the human-readable explanation of what the Lean definition `perfectoid_ring` represents, and this docstring is visible when you hover your cursor on the word `perfectoid_ring` in some Lean code; if you are running the code in an IDE such as Microsoft VS Code then right-clicking on this word will jump you to the definition.

The Lean definition pretty much coincides with the human definition. If R is a Huber ring which is a Tate ring (these are technical properties of topological rings), then we say R is a perfectoid ring if it is complete, uniform, and satisfies a couple of technical properties. The point to observe is that the computer code is no more or less difficult than the human definition.

5.5. Foundations

In my experience, mathematicians often have very little interest in the technicalities of the logical foundations of their subject—they cannot list the axioms of set theory, but they know from experience what is "legal mathematics." The controversies of the early 20th century about whether nonconstructive methods are allowed in mathematical proofs have long ago died down; working mathematicians use the law of the excluded middle all over the place, and many use the axiom of choice in some form or another (indeed, countable dependent choice can be invoked almost without one noticing). A typical research mathematician will have gone to at most one course on the foundations of mathematics; in such a course one typically learns that Zermelo–Frankel set theory with the axiom of choice, or ZFC, can be used as a foundation for much of mathematics. Indeed, it can be used for essentially all of mathematics up until the 1960s; however, Grothendieck's supergeneral cohomology theories developed in SGA4 introduced a new "axiom of universes" (the assertion that every set is an element of a set which is a model of ZFC). This axiom cannot be proved from the axioms of ZFC, by Gödel's theorem. The original proofs of the Weil conjectures in theory used this axiom in the weak sense that at the time the only reference for étale cohomology was SGA4. However, Deligne and others point out in SGA4$\frac{1}{2}$ that the theory of étale cohomology, and hence the proof of the Weil conjectures, can be set up within ZFC alone. Readers interested in the contortions that one has to go through in order to do this can look at the Set Theory section of the Stacks project, for example, here [61, HTTPS://STACKS.MATH.COLUMBIA.EDU/TAG/000H]. For a more extreme example, see Section 4 of [52], where we see a Fields Medallist forcing a more elaborate theory into ZFC.

My personal opinion is that whilst ZFC was a wonderful foundation for much of early 20th century mathematics, the lack of a universe axiom now means that it is becoming more and more of an effort to get parts of modern mathematics to fit into it. In books and papers dealing with infinity categories or condensed mathematics, it is not at all uncommon to see universes showing up, and I do wonder whether now it is time for mathematicians to begin embracing universes, as Grothendieck was encouraging us to do since the 1960s. Coq's type theory and Lean's type theory both contain universes as part of the foundations; however, mathematicians can choose not to use them if they so desire.

6. THE FUTURE

In this section I describe some of the plausible consequences of formalizing mathematics in a computer theorem prover. I also highlight some things which I believe will remain out of reach for some time yet. Patrick Massot's more extensive observations [44] are also well worth a read (indeed, several of my ideas here were formed after conversations with Massot).

6.1. A new kind of mathematical document

Right now, an author of a textbook or research paper has to decide how much background material to assume, and which techniques they will regard as standard in the arguments they present. In other words, they have to decide where to start, and how fast to go. If a potential reader (for example, a new PhD student, or an undergraduate interested in the area) does not have the necessary prerequisites then it will be far more difficult for them to get anything out of the paper.

Computer formalization offers the possibility of a new kind of mathematical document, where the *reader* can make the decisions about how much detail is visible. Patrick

Massot has been experimenting with such documents. A preliminary version of his vision can be seen at his Sphere Eversion Project web pages [43]. This is a project whose main goal is to formalize in Lean a proof of Smale's theorem saying that a sphere can be turned inside out (or more formally, that there is a homotopy of immersions between the identity immersion of S^2 in $\mathbb{R}^3$ and the antipodal immersion). At the time of writing, the proof is not yet fully formalized, but it is only a matter of time. The blueprint is written in LaTeX, but using plasTeX it has been converted into a web page with live Lean links. Right now these links take you to static web pages containing Lean code, but tools are currently being developed which will change this. Alectryon is a program available for Coq and Lean which can turn compiled code into web pages. Tools like Alectryon will enable us to make documents which will allow links to dynamic web pages displaying anything from mathematical details to interactive pictures, in a human-readable form, and which will allow one to keep digging right down to the axioms, although, of course, it is unlikely that anyone would like to go down this far.

There are already variants of this idea in existence, Lamport's idea of a "structured proof" came from a desire to encourage mathematicians to write far more details down in their papers, but one can see why such a proposal would not go down well. Here we can let automation do part of the work for us. The Metamath proof assistant also offers similar functionality already, because Metamath has very little automation and hence drilling down to the axioms is essentially the same as inspecting the proof.

One could also imagine error-free undergraduate textbooks also written in this way, where statements which a student cannot understand (perhaps because they are ambiguous) can be inspected in more details until difficulties are resolved.

6.2. Semantic search in a mathematical database

One thing that is not going to happen any time soon is some kind of revolution where all mathematicians start writing all their papers in a formal proof assistant. Whilst one might expect a future where *some* papers are partially, or even completely formalized in a theorem prover (see, for example, [30, 40, 56]), this kind of approach will not become the norm any time soon. Faced with this reality, how will formalized mathematics be able to keep up with the frontiers of mathematics?

I have already mentioned Tom Hales' 2017 "Big Conjectures" talk at the Newton Institute in Cambridge. In the talk [32], Hales argued for a formalized version of Math Reviews/Zentralblatt. That is, a website whose role is to formally *state* the results being announced in the main mathematical journals. Note that such a project is nowhere near as far-fetched as the idea of formalizing mathematical *proofs* in real time; theorem *statements* are far easier to formalize.

The issue with Hales' plan, as he points out in the talk, is that to be able to formalize statements of theorems in even a part of modern mathematics such as the Langlands philosophy, one would have to define all of the basic objects which mathematicians in this area use. In the Langlands philosophy this would include, but be by no means limited to, definitions of automorphic forms and automorphic representations, Galois representations, abelian varieties, the rings defined by Fontaine and used to do p-adic Hodge theory, schemes, all the cohomology theories used in the area, perfectoid spaces, adeles and ideles,.... The Lean community has over the last few years pushed hard to get some of the main definitions of modern research mathematics into `mathlib`. At Imperial College alone we currently have Oliver Nash developing the basics of the theory of Lie Algebras so we can talk about centers of universal enveloping algebras, María Inés de Frutos-Fernández developing the theory of adeles and ideles of global fields with an eye on the statements of class field theory, Amelia Livingston developing group and Galois cohomology, Jujian Zhang developing sheaf cohomology with an eye on GAGA, and Ashvni Narayanan developing the basics of Iwasawa theory in her PhD thesis. I have already mentioned the work of myself, Massot and Commelin defining perfectoid spaces. The work of Scott Morrison, Bhavik Mehta, Justus Springer, and Adam Topaz has recently enabled us to start developing the theory of sheaves on sites and homological algebra, so cohomology theories are now not too far away. Of course, much remains to be done, but we are hoping that the idea of being able to formally *state* the theorems of Annals and Inventiones algebraic number theory papers in Lean will soon become a reality.

A related project is formalizing tags in the Stacks Project **[61]**, which is a gigantic online database of algebraic geometry, freely accessible online. When printed out, it fills over 7000 pdf pages. Formalizing all the proofs in the database would be an extremely arduous task involving many person-decades of work with current technology. In theory it is possible, however, one would need a team who were experts in both algebraic geometry and in formalization. Furthermore, for it to actually happen, the incentive structure in academic mathematics would have to change drastically. Publishing papers in prestigious computer science conference proceedings explaining how you developed the basic theory of Cohen–Macauley rings and modules in a theorem prover (and, of course, such work would be publishable in a prestigious computer science conference proceedings—nobody has ever done it before) is perhaps not something which is recognized by promotions committees.

However, there is a solution available to us right now. Formalizing just the *definitions* and theorem *statements* in the Stacks Project is a *much* simpler task. Anybody interested in algebraic geometry would be more than welcome to learn Lean by attempting to formalize statements in Stacks Project tags. Point your web browser to the Lean Zulip instance **[65]** and ask where to get started in the `#new members` stream.

The reason that building such databases is important is that they will enable the community to build tools the likes of which mathematicians have never seen before. Let us imagine that all the definitions and theorem statements in the Stacks Project have been formalized in Lean or some other theorem prover. A "hammer" is a tool which runs inside a theorem prover and which can attempt to construct mathematical arguments by piecing together results in a database. The original hammer was Isabelle/HOL's *Sledgehammer* **[49]**. The cleverness behind such tools is the ability to isolate which of the many results in the database look the most useful, and to concentrate on these when attempting to prove the required result. Now consider a PhD student who is beginning to learn algebraic geometry.

Such a student would then be able to ask the theorem prover a question, and the prover could attempt to use the database to answer the question positively (by piecing together a proof) or negatively (by producing a counterexample, like the website π-base [21] is doing for counterexamples in topology). The resulting output of the computer would be able to explicitly point to references in the literature, or direct proofs of the claims it is making in its argument. This sort of tool—computer assisted learning—has the potential to beat the techniques currently used by PhD students ("google hopefully," "page through a textbook/paper hopefully," "ask on a maths website and then wait," "ask another human") hands down. But as I have stressed before, the main thing which is missing is the database of theorems, and it is up to us to construct it. The sooner it is there, the sooner the tools will appear. And the bigger the database gets, the more powerful the tools will become.

6.3. Checking proofs

Some computer scientists have argued that mathematicians are sloppy, and our literature has errors in, and that this problem can be solved with computer proof assistants. Such an argument might initially look plausible, and I myself was a proponent of it a few years ago, but it does not stand up to much scrutiny. Firstly, the experts in our community know which results can be relied upon. Secondly, many errors are not serious and can be fixed. Thirdly, the more serious instances of this problem cannot be solved with computer proof assistants right now anyway. A great example is Mochizuki's claimed proof of the ABC conjecture [46]. This proof has now been published in a serious research journal; however, it is clear that it is not accepted by the mathematical community in general. One could challenge Mochizuki, or indeed anyone, to formalize the proof in a computer theorem prover. However, this would be a completely unreasonable thing to do. A computer formalization is not expected of other proofs appearing in our literature. Furthermore, the key sticking point right now is that the unbelievers argue that more details are needed in the proof of Corollary 3.12 in the main paper, and the state-of-the-art right now is simply that one cannot begin to formalize this corollary without access to these details in some form (for example, a paper proof containing far more information about the argument).

What *would*, however, be feasible is for mathematicians to formalize *parts* of technical work, or to get others to do so. There might be several reasons to do such a thing—Commelin and his team have already shown that theorem provers can be used to check parts of complex proofs which humans might find it difficult to plough through, whilst learning about the mathematics in the process.

6.4. Teaching

I have heard students say "I think my proof is OK" when talking about their homework. Computer proof assistants are able to tell them immediately if this is so—as long as the student has taken the trouble to learn the language of the proof assistant. Should we be teaching undergraduate mathematicians how to use computer proof assistants? I certainly think so. Patrick Massot in Orsay and myself at Imperial College London are both teaching undergraduate-level courses which do precisely this.

Students want feedback on their work as soon as possible. A computer proof assistant can supply it immediately.

Beginner students can be confused about the basics. What is the difference between $\forall\epsilon > 0, \exists\delta > 0, \ldots$ and $\exists\delta > 0, \forall\epsilon > 0, \ldots$? Once these systems become easier for mathematicians to use, students can experiment for themselves with well-chosen examples supplied by a lecturer and begin to understand what is going on. I was once told by a student "I did not understand equivalence relations, so I formalized them in Lean, and then I understood them." Forcing students to think pedantically and logically can be good for them.

It is, however, worth stressing that asking a weak student to both keep up with your course and to simultaneously learn how to use a computer theorem prover is clearly asking too much from that student. The provers need to become easier to use, perhaps with graphical interfaces and documentation more appropriate for mathematicians. Asking people to change the way they teach is, of course, asking a lot. However, mathematics education experts will be only too happy to tell us that our preferred medium—"write for an hour on a board"—is becoming less and less appropriate for our students, who like to learn things by watching 5 minute videos or playing with interactive toys. Can we make abstract mathematics more interactive? I suspect that we can. The more people who understand how to use these machines, the sooner the new ideas will come.

6.5. Other ideas

I do not claim to have exhausted the possibilities here. The people who designed the CD in the 1980s surely could not envisage music services like YouTube and Spotify, or the audiobook. The people who started to think about how to make typesetting of books look good on a computer screen surely did not envisage devices like the Kindle. It is time to look beyond how we usually teach and learn mathematics, and try to understand how we as a community of mathematicians can use the inevitable digitization of mathematical material as a tool to make our lives, and the lives of our students, better. As Carneiro once said, you cannot stop progress.

ACKNOWLEDGMENTS

I thank the Lean prover community for welcoming me, a mathematician with very little programming experience, into their community back in 2017, and also for reading and giving extensive comments on a preliminary version of this article. Patrick Massot in particular sent many helpful comments on a first draft. I thank Assia Mahboubi and Manuel Eberl for giving advice on the Coq and Isabelle/HOL code in this paper, and to both them and Jeremy Avigad for helpful historical comments. Finally, I would like to effusively thank Leonardo de Moura for writing my favorite computer game, and Mario Carneiro for teaching me how to play it.

REFERENCES

[1] A. Baanen, S. R. Dahmen, A. Narayanan, and F. A. E. Nuccio, Mortarino Majno di Capriglio, A formalization of Dedekind domains and class groups of global fields. In *12th International Conference on Interactive Theorem Proving, ITP 2021, June 29 to July 1, 2021, Rome, Italy (virtual conference)*, edited by L. Cohen and C. Kaliszyk, pp. 5:1–5:19, LIPIcs 193, Schloss Dagstuhl – Leibniz-Zentrum für Informatik, 2021.

[2] H. Bender and G. Glauberman, *Local analysis for the odd order theorem. With the assistance of Walter Carlip*. London Math. Soc. Lecture Note Ser. 188, Cambridge University Press, Cambridge, 1994.

[3] S. Bernard, C. Cohen, A. Mahboubi, and P.-Y. Strub, Unsolvability of the quintic formalized in dependent type theory. In *12th International Conference on Interactive Theorem Proving, ITP 2021, June 29 to July 1, 2021, Rome, Italy (virtual conference)*, edited by L. Cohen and C. Kaliszyk, pp. 8:1–8:18, LIPIcs 193, Schloss Dagstuhl – Leibniz-Zentrum für Informatik, 2021.

[4] Y. Bertot, L. Rideau, and L. Théry, Distant decimals of π: formal proofs of some algorithms computing them and guarantees of exact computation. *J. Automat. Reason.* **61** (2018), no. 1–4, 33–71.

[5] B. J. Birch and H. P. F. Swinnerton-Dyer, Notes on elliptic curves. II. *J. Reine Angew. Math.* **218** (1965), 79–108.

[6] A. Bordg, L. Paulson, and W. Li, Grothendieck's schemes in algebraic geometry. 2021, https://isa-afp.org/entries/Grothendieck_Schemes.html, Formal proof development.

[7] T. Browning and P. Lutz, Formalizing Galois theory, 2021. Exp. Math. published online, 2021.

[8] K. Buzzard, Formalising mathematics: a first course for mathematicians. https://github.com/ImperialCollegeLondon/formalising-mathematics, accessed: 30-11-2021.

[9] K. Buzzard, Induction on equality. https://xenaproject.wordpress.com/2021/04/18/induction-on-equality/, accessed: 30-11-2021.

[10] K. Buzzard, J. Commelin, and P. Massot, Formalising perfectoid spaces. In *Proceedings of the 9th ACM SIGPLAN International Conference on Certified Programs and Proofs, CPP 2020, New Orleans, LA, USA, January 20–21, 2020*, edited by J. Blanchette and C. Hritcu, pp. 299–312, ACM, 2020.

[11] K. Buzzard, J. Commelin, and P. Massot, Lean perfectoid spaces. https://leanprover-community.github.io/lean-perfectoid-spaces/, accessed: 30-11-2021.

[12] K. Buzzard, C. Hughes, K. Lau, A. Livingston, R. Fernández Mir, and S. Morrison, Schemes in Lean, 2021. Exp. Math. published online, 2021.

[13] M. Carneiro, The type theory of Lean. https://github.com/digama0/lean-type-theory/releases/download/v1.0/main.pdf, accessed: 30-11-2021.

[14] M. Carneiro, Metamath Zero. 2021, https://github.com/digama0/mm0.

[15] D. Castelvecchi, Mathematicians welcome computer-assisted proof in 'grand unification'. https://www.nature.com/articles/d41586-021-01627-2, accessed: 30-11-2021.

[16] F. Chyzak, A. Mahboubi, T. Sibut-Pinote, and E. Tassi, A computer-algebra-based formal proof of the irrationality of ζ (3). In *International Conference on Interactive Theorem Proving*, pp. 160–176, Springer, 2014.

[17] J. Commelin and R. Y. Lewis, Formalizing the ring of Witt vectors. In *CPP'21: 10th ACM SIGPLAN International Conference on Certified Programs and Proofs, virtual event, Denmark, January 17–19, 2021*, edited by C. Hritcu and A. Popescu, pp. 264–277, ACM, 2021.

[18] J. Commelin and P. Massot, Blueprint for the Liquid Tensor Experiment. https://leanprover-community.github.io/liquid/, accessed: 30-11-2021.

[19] Coq Development Team, The coq proof assistant. http://coq.inria.fr, accessed: 11-12-2021.

[20] T. Coquand and C. Paulin, Inductively defined types. In *COLOG-88, International Conference on Computer Logic, Tallinn, USSR, December 1988, Proceedings*, edited by G. Mints, pp. 50–66, Lecture Notes in Comput. Sci. 417, Springer, 1988.

[21] J. Dabbs and S. Clontz, π-base. https://topology.pi-base.org/, accessed: 30-11-2021.

[22] S. R. Dahmen, J. Hölzl, and R. Y. Lewis, Formalizing the solution to the cap set problem. In *10th International Conference on Interactive Theorem Proving*, pp. 15–19, LIPIcs. Leibniz Int. Proc. Inform. 141, Schloss Dagstuhl – Leibniz-Zent. Inform., Wadern 2019.

[23] L. de Moura, S. Kong, J. Avigad, F. van Doorn, and J. von Raumer, The Lean theorem prover (system description). In *Automated Deduction – CADE-25 (Cham)*, edited by A. P. Felty and A. Middeldorp, pp. 378–388, Springer, 2015.

[24] M. Eberl, The irrationality of $\zeta(3)$. 2019, https://www.isa-afp.org/entries/Zeta_3_Irrational.html, accessed: 30-11-2021.

[25] M. Eberl, Nine chapters of analytic number theory in Isabelle/HOL. In *10th International Conference on Interactive Theorem Proving, ITP 2019, September 9–12, 2019, Portland, OR, USA*, edited by J. Harrison, J. O'Leary, and A. Tolmach, pp. 16:1–16:19, LIPIcs 141, Schloss Dagstuhl – Leibniz-Zentrum für Informatik, 2019.

[26] J. S. Ellenberg and D. Gijswijt, On large subsets of $\mathbb{F}_q^n$ with no three-term arithmetic progression. *Ann. of Math. (2)* **185** (2017), no. 1, 339–343.

[27] G. Gonthier, The four colour theorem: Engineering of a formal proof. In *Computer Mathematics, 8th Asian Symposium, ASCM 2007, Singapore, December 15–17, 2007. Revised and invited papers*, edited by D. Kapur, p. 333, Lecture Notes in Comput. Sci. 5081, Springer, 2007.

[28] G. Gonthier, Formal proof—the four-color theorem. *Notices Amer. Math. Soc.* **55** (2008), no. 11, 1382–1393.

[29] G. Gonthier, A. Asperti, J. Avigad, Y. Bertot, C. Cohen, F. Garillot, S. Le Roux, A. Mahboubi, R. O'Connor, S. Ould Biha, I. Pasca, L. Rideau, A. Solovyev, E. Tassi, and L. Théry, A machine-checked proof of the odd order theorem. In *Interactive Theorem Proving – 4th International Conference, ITP 2013, Rennes, France, July 22–26, 2013. Proceedings*, edited by S. Blazy, C. Paulin-Mohring, and D. Pichardie, pp. 163–179, Lecture Notes in Comput. Sci. 7998, Springer, 2013.

[30] S. Gouëzel and V. Shchur, Corrigendum: A corrected quantitative version of the Morse lemma [MR 3003738]. *J. Funct. Anal.* **277** (2019), no. 4, 1258–1268.

[31] B. Grégoire and A. Mahboubi, Proving equalities in a commutative ring done right in Coq. In *Theorem Proving in Higher Order Logics, 18th International Conference, TPHOLs 2005, Oxford, UK, August 22–25, 2005, Proceedings*, edited by J. Hurd and T. F. Melham, pp. 22–25, Lecture Notes in Comput. Sci. 3603, Springer, Oxford, UK, 2005.

[32] T. Hales, Big conjectures. https://www.newton.ac.uk/seminar/21474/, accessed: 30-11-2021.

[33] T. C. Hales, Mathematics in the age of the Turing machine. In *Turing's legacy: developments from Turing's ideas in logic*, pp. 253–298, Lect. Notes Log. 42, Assoc. Symbol. Logic, La Jolla, CA, 2014.

[34] T. Hales, M. Adams, G. Bauer, T. D. Dang, J. Harrison, L. T. Hoang, C. Kaliszyk, V. Magron, S. McLaughlin, T. T. Nguyen, Q. T. Nguyen, T. Nipkow, S. Obua, J. Pleso, J. Rute, A. Solovyev, T. Hoai An Ta, N. T. Tran, T. D. Trieu, J. Urban, K. Vu, and R. Zumkeller, A formal proof of the Kepler conjecture. *Forum Math. Pi* **5** (2017), e2, 29.

[35] J. M. Han and F. van Doorn, A formal proof of the independence of the continuum hypothesis. In *Proceedings of the 9th ACM SIGPLAN International Conference on Certified Programs and Proofs, CPP 2020, New Orleans, LA, USA, January 20–21, 2020*, edited by J. Blanchette and C. Hritcu, pp. 353–366, ACM, 2020.

[36] J. Harrison, Formalizing an analytic proof of the prime number theorem. *J. Automat. Reason.* **43** (2009), no. 3, 243–261.

[37] J. Harrison, HOL light: An overview. In *Theorem Proving in Higher Order Logics, 22nd International Conference, TPHOLs 2009, Munich, Germany, August 17–20, 2009. Proceedings*, edited by S. Berghofer, T. Nipkow, C. Urban, and M. Wenzel, pp. 60–66, Lecture Notes in Comput. Sci. 5674, Springer, 2009.

[38] F. Immler, A verified ODE solver and the Lorenz attractor. *J. Automat. Reason.* **61** (2018), no. 1, 73–111.

[39] F. Immler and Y. K. Tan, The Poincaré–Bendixson theorem in Isabelle/HOL. In *Proceedings of the 9th ACM SIGPLAN International Conference on Certified Programs and Proofs (New York, NY, USA), CPP 2020*, pp. 338–352, Association for Computing Machinery, 2020.

[40] B. Kjos-Hanssen, S. Niraula, and S. Yoon, A parametrized family of Tversky metrics connecting the Jaccard distance to an analogue of the normalized information distance. 2021, arXiv:2111.02498.

[41] R. Y. Lewis, A formal proof of Hensel's lemma over the p-adic integers. In *Proceedings of the 8th ACM SIGPLAN International Conference on Certified Programs and Proofs (New York, NY, USA), CPP 2019*, pp. 15–26, Association for Computing Machinery, 2019.

[42] R. Y. Lewis and P-N. Madelaine, Simplifying casts and coercions (extended abstract). In *Joint Proceedings of the 7th Workshop on Practical Aspects of Automated Reasoning (PAAR) and the 5th Satisfiability Checking and Symbolic Computation Workshop (SC-Square) Workshop, 2020 co-located with the 10th International Joint Conference on Automated Reasoning (IJCAR 2020), Paris, France, June–July, 2020 (virtual)*, edited by K. Korovin, I. S. Kotsireas, P. Rümmer, and S. Tourret, pp. 53–62, CEUR Workshop Proceedings 2752, CEUR-WS.org, 2020.

[43] P. Massot, The sphere eversion project. https://leanprover-community.github.io/sphere-eversion/blueprint/index.html, accessed: 30-11-2021.

[44] P. Massot, Why formalize mathematics? https://www.imo.universite-paris-saclay.fr/~pmassot/files/exposition/why_formalize.pdf, accessed: 11-12-2021.

[45] N. D. Megill and D. A. Wheeler, *Metamath: A computer language for mathematical proofs*. Lulu Press, Morrisville, North Carolina, 2019. http://us.metamath.org/downloads/metamath.pdf.

[46] S. Mochizuki, Inter-universal Teichmüller theory III: Canonical splittings of the log-theta-lattice. *Publ. Res. Inst. Math. Sci.* **57** (2021), no. 1, 403–626.

[47] A. Naumowicz and A. Korniłowicz, A brief overview of mizar. In *Theorem Proving in Higher Order Logics (Berlin, Heidelberg)*, edited by S. Berghofer, T. Nipkow, C. Urban, and M. Wenzel, pp. 67–72, Springer, Berlin, Heidelberg, 2009.

[48] T. Nipkow, L. C. Paulson, and M. Wenzel, *Isabelle/HOL – A proof assistant for higher-order logic*. Lecture Notes in Comput. Sci. 2283, Springer, 2002.

[49] L. C. Paulson and J. C. Blanchette, Three years of experience with Sledgehammer, a practical link between automatic and interactive theorem provers. In *The 8th International Workshop on the Implementation of Logics, IWIL 2010, Yogyakarta, Indonesia, October 9, 2011*, edited by G. Sutcliffe, S. Schulz, and E. Ternovska, pp. 1–11, EPiC Series in Computing 2, EasyChair, 2011.

[50] T. Peterfalvi, *Character theory for the odd order theorem*. London Math. Soc. Lecture Note Ser. 272, Cambridge University Press, Cambridge, 2000. Translated from the 1986 French original by Robert Sandling and revised by the author.

[51] P. Scholze, Lectures on analytic geometry. https://www.math.uni-bonn.de/people/scholze/Analytic.pdf, accessed: 30-11-2021.
[52] P. Scholze, Etale cohomology of diamonds. 2017, arXiv:1709.07343.
[53] P. Scholze, Liquid tensor experiment, 2021. Exp. Math. published online, 2021.
[54] P. Scholze, Half a year of the Liquid Tensor Experiment: Amazing developments. https://xenaproject.wordpress.com/2021/06/05/half-a-year-of-the-liquid-tensor-experiment-amazing-developments/, accessed: 30-11-2021.
[55] P. Scholze, Liquid tensor experiment. https://xenaproject.wordpress.com/2020/12/05/liquid-tensor-experiment/, accessed: 30-11-2021.
[56] N. Strickland and N. Bellumat, Iterated chromatic localisation. 2019, arXiv:1907.07801.
[57] The Lean prover community, A mathlib overview. https://leanprover-community.github.io/mathlib-overview.html, accessed: 30-11-2021.
[58] The Lean prover community, The Lean community website. https://leanprover-community.github.io/index.html, accessed: 30-11-2021.
[59] The Lean prover community, Undergraduate mathematics in mathlib. https://leanprover-community.github.io/undergrad.html, accessed: 30-11-2021.
[60] The mathlib community, The Lean mathematical library. In *Proceedings of the 9th ACM SIGPLAN International Conference on Certified Programs and Proofs (New York, NY, USA), CPP 2020*, pp. 367–381, Association for Computing Machinery, 2020.
[61] The Stacks Project Authors, Stacks Project. 2018, https://stacks.math.columbia.edu.
[62] B. Werner, Sets in types, types in sets. In *Theoretical aspects of computer software (Sendai, 1997)*, pp. 530–546, Lecture Notes in Comput. Sci. 1281, Springer, Berlin, 1997.
[63] E. Wieser, Scalar actions in Lean's mathlib. 2021.
[64] Wikipedia, Isabelle (proof assistant). https://en.wikipedia.org/wiki/Isabelle_(proof_assistant), accessed: 30-11-2021.
[65] Zulip, The Lean community Zulip chatroom. https://leanprover.zulipchat.com, accessed: every day since 2018.

KEVIN BUZZARD

Department of Mathematics, Imperial College London, London, UK,
k.buzzard@imperial.ac.uk

This Page Intentionally Left Blank

RECIPROCITY IN THE LANGLANDS PROGRAM SINCE FERMAT'S LAST THEOREM

FRANK CALEGARI

ABSTRACT

The *reciprocity conjecture* in the Langlands program links motives to automorphic forms. The proof of Fermat's Last Theorem by Wiles [171, 181] introduced new tools to study reciprocity. This survey reports on developments using these ideas (and their generalizations) in the last three decades.

MATHEMATICS SUBJECT CLASSIFICATION 2020

Primary 11F80; Secondary 11F33, 11F75, 11S37

KEYWORDS

Galois representations, Langlands program, modularity lifting

INTERNATIONAL CONGRESS OF MATHEMATICIANS 2022 JULY 6–14

Proc. Int. Cong. Math. 2022, Vol. 2, pp. 610–651
DOI 10.4171/ICM2022/101

1. INTRODUCTION

The *reciprocity conjecture* in the Langlands program predicts a relationship between pure motives[1] and automorphic representations. The simplest version (as formulated by Clozel [48, CONJ. 2.1]) states that there should be a bijection between irreducible motives M over a number field F with coefficients in $\overline{\mathbf{Q}}$ and cuspidal algebraic representations π of $\mathrm{GL}_n(\mathbf{A}_F)$ satisfying a number of explicit additional compatibilities, including the equality of algebraic and analytic L-functions $L(M, s) = L(\pi, s)$. In light of multiplicity-one theorems [105], this pins down the correspondence uniquely. There is also a version of this conjecture for more general reductive groups, although its formulation requires some care (as was done by Buzzard and Gee [32]). Beyond the spectacular application by Wiles to Fermat's Last Theorem [181, THEOREM 0.5], the Taylor–Wiles method [171, 181] gave a completely new technique—and to this date the most successful one—for studying the problem of reciprocity. The ideas in these two papers have sustained progress in the field for almost[2] 30 years. In this survey, we explain how the Taylor–Wiles method has evolved over this period and where it stands today. *One warning:* the intended audience for this document is entirely complementary to the audience for my talk—I shall assume more than a passing familiarity with the arguments of [171, 181]. Moreover, this survey is as much a personal and historical[3] discussion as a mathematical one—giving anything more than hints on even a fraction of what is discussed here would be close to impossible given the space constraints and the competence of the author. Even with the absence of any real mathematical details in this paper, the sheer amount of activity in this field has led me to discard any discussion of advances not directly related to $R = \mathbf{T}$ theorems, which necessitates the omission of a lot of closely related beautiful mathematics.

1.1. The Fontaine–Mazur conjecture

Let F be a number field. The Fontaine–Mazur conjecture[4] [83] predicts that any continuous irreducible p-adic Galois representation

$$\rho : G_F \to \mathrm{GL}_n(\overline{\mathbf{Q}}_p)$$

1 Here (in light of the standard conjectures [124]) one may take pure motives up to numerical or homological equivalence. Conjecturally, one can also substitute (for irreducible motive) the notion of an irreducible weakly compatible system of Galois representations [167] or an irreducible geometric Galois representation in the sense of Fontaine–Mazur [83].

2 Wiles in [181] dates the completion of the proof to September 19, 1994.

3 A whiggish history, naturally. Even with this caveat, it should be clear that the narrative arc of progress presented here at best represents my own interpretation of events. I have added a few quotes from first hand sources when I felt they conveyed a sense of what the experts were thinking in a manner not easily obtainable from other sources. For other survey articles on similar topics, see [24, 30].

4 Fontaine told me (over a *salad de gésiers* in Roscoff in 2009) that he and Mazur formulated their conjecture in the mid-1980s. (Colmez pointed me towards these notes [81] from a talk given by Fontaine at the 1988 Mathematische Arbeitstagung in Bonn.) He noted that Serre had originally been skeptical, particularly of the claim that any everywhere unramified representation inside $\mathrm{GL}_n(\overline{\mathbf{Q}}_p)$ must have finite image, and set off to find a counterexample (using the construction of Golod–Shavarevich [91]). He (Serre) did not succeed!

which is both unramified outside finitely many primes and potentially semistable (equivalently, de Rham [62]) at all places $v|p$ should be associated to a motive M/F with coefficients in $\overline{\mathbf{Q}}$. Any such ρ is automatically conjugate to a representation in $\mathrm{GL}_n(E)$ for some finite field $E/\mathbf{Q}_p$ and further stabilizes an $\mathcal{O}_E$-lattice. The corresponding residual representation $\overline{\rho} : G_F \to \mathrm{GL}_n(k)$ where $k = \mathcal{O}_E/\pi_E$ is the residue field of E is unique up to semisimplification. Let us assume here for expositional convenience that $\overline{\rho}$ is absolutely irreducible. Following Mazur [129], one may define a universal deformation ring which parameterizes all deformations of $\overline{\rho}$ unramified outside a finite set S. One can then further impose local conditions to define deformation rings R whose $\overline{\mathbf{Q}}_p$-valued points are associated to Galois representations which are de Rham at $v|p$ with fixed Hodge–Tate weights. Assuming the Fontaine–Mazur conjecture, these $\overline{\mathbf{Q}}_p$-valued points correspond to all pure motives M unramified outside S whose p-adic realizations are Galois representations with the same local conditions at p and the same fixed residual representation $\overline{\rho}$. Assuming the reciprocity conjecture, these motives should then be associated to a finite dimensional space of automorphic forms. This leads to the extremely nontrivial prediction that R has finitely many $\overline{\mathbf{Q}}_p$-valued points. The problem of reciprocity is now to link these $\overline{\mathbf{Q}}_p$-valued points of R to automorphic forms.

1.2. $R = \mathbf{T}$ theorems

Associated to the (conjectural) space of automorphic forms corresponding to $\overline{\mathbf{Q}}_p$-valued points of R is a ring of endomorphisms generated by Hecke operators. The naïve version of $\mathbf{T}$ is defined to be the completion of this ring with respect to a maximal ideal $\mathfrak{m}$ defined in terms of $\overline{\rho}$. The mere existence of $\mathfrak{m}$ is itself conjectural, and amounts—in the special case of odd absolutely irreducible 2-dimensional representations $\overline{\rho}$ of $G_{\mathbf{Q}}$—to Serre's conjecture [156]. Hence, in the Taylor–Wiles method, one usually assumes the existence of a suitable $\mathfrak{m}$ as a hypothesis. The usual shorthand way of describing what comes out of the Taylor–Wiles method is then an "$R = \mathbf{T}$ theorem." Proving an $R = \mathbf{T}$ theorem can more or less be divided into three different problems:

(1) Understanding $\mathbf{T}$. Why does there exist[5] a map $R \to \mathbf{T}$? This is the problem of the "existence of Galois representations." Implicit here is the problem of showing that those Galois representations not only exist but have the "right local properties" at the ramified primes, particularly those dividing p.

(2) Understanding R. Wiles introduced a mechanism for controlling R via its tangent space using Galois cohomology (in particular Poitou–Tate duality [131]), and this idea has proved remarkably versatile. What has changed, however, is our understanding of local Galois representations and how this information can be leveraged to understand the structure of R.

(3) Understanding why the map $R \to \mathbf{T}$ is an isomorphism.

5 At the time of Wiles' result, this was seen as the easier direction (if not easy), although, in light of the success of the Taylor–Wiles method, it may well be the harder direction in general.

We begin by summarizing the original $R = \mathbf{T}$ theorem from this viewpoint (or more precisely, the modification by Faltings which appears as an appendix to **[171]**). We only discuss for now the so-called "minimal case[6]" since this is most relevant for subsequent generalizations (see Section 6.2). Our summary is cursory, but see **[67,68]** for excellent expositional sources on early versions of the Taylor–Wiles method. We start with a representation $\overline{\rho}: G_{\mathbf{Q}} \to \mathrm{GL}_2(\overline{\mathbf{F}}_p)$ for $p > 2$ which (say) comes from a semistable elliptic curve E and which we assume to be modular. By a theorem of Ribet **[149]**, we may assume it is modular of level either $N = N(\overline{\rho})$ or $N = N(\overline{\rho})p$ where $N(\overline{\rho})$ is the Serre weight **[156]** of $\overline{\rho}$.

(1) Understanding $\mathbf{T}$: The construction of Galois representations associated to modular forms has its own interesting history (omitted here), but (in the form originally needed by Wiles) was more or less complete for modular forms (and even Hilbert modular forms) by 1990. The required local properties at primes different from p followed from work of Carayol **[47]**, and the local properties at p were well understood either by Fontaine–Laffaille theory **[82]**, or, in the ordinary case, by Mazur–Wiles **[130]** (see also work of Hida **[101,102]**).

(2) Understanding R: Here R is a deformation ring of $\overline{\rho}$ subject to precise local deformation conditions at p and the primes dividing $N(\overline{\rho})$. For the prime p, the local conditions amount either to an "ordinary" or "finite-flat" restriction. One then interprets the dual of the reduced tangent space $\mathfrak{m}_R/(\mathfrak{m}_R^2, p)$ of R in terms of Galois cohomology, in particular as a subgroup (Selmer group) of classes in $H^1(\mathbf{Q}, \mathrm{ad}^0(\overline{\rho}))$ satisfying local conditions. This can be thought of as analogous to a class group, and one does not have any *a priori* understanding of how large it can be although it has some finite dimension d. Using the Greenberg–Wiles formula, the obstructions in $H^2(\mathbf{Q}, \mathrm{ad}^0(\overline{\rho}))$ can be related to the reduced tangent space, and allow one to realize R as a quotient of $W(k)[[x_1, \ldots, x_d]]$ by d relations. In particular, if R was finite and free as a $W(k)$-module (as would be the case if $R = \mathbf{T}$) then R would be a complete intersection.

(3) Understanding why the map $R \to \mathbf{T}$ is an isomorphism. Here lies the heart of the Taylor–Wiles method. The ring $\mathbf{T}$ acts on a natural module M of modular forms. One shows—under a mild hypothesis on $\overline{\rho}$—the existence of (infinitely many) sets $Q = Q_N$ for any natural number N of cardinality $|Q| = d$—so-called Taylor–Wiles primes—with a number of pleasant properties:

    (i) The primes $q \in Q$ are congruent to 1 mod p^N.

    (ii) Let R_Q be the deformation ring capturing the same local properties as R but modified so that the representations at primes in Q may now be ramified of degree p^N. There is naturally a surjection $R_Q \to R$, but for

6 The case when the Galois representations attached to R and $\mathbf{T}$ have minimal level N as determined by the residual representation.

Taylor–Wiles primes, this modification does not increase the size of the tangent space. In particular, for a fixed ring $R_\infty = W(k)[[x_1, \dots, x_d]]$ there are surjections $R_\infty \to R_Q \to R$ for every Q.

(iii) The corresponding rings $\mathbf{T}$ and $\mathbf{T}_Q$ act naturally on spaces of modular forms M and M_Q, respectively. Using multiplicity one theorems, Wiles proves (see [181, THEOREM 2.1]) that M and M_Q are free of rank one over $\mathbf{T}$ and $\mathbf{T}_Q$, respectively. The space M can be interpreted as a space of modular forms for a particular modular curve X. The second key property of Taylor–Wiles primes is that there are no new modular forms associated to $\overline{\rho}$ at level $X_0(Q)$, and hence M can *also* be interpreted as a space of modular forms for $X_0(Q)$. There is a Galois cover $X_1(Q) \to X_0(Q)$ with Galois group $(\mathbf{Z}/Q\mathbf{Z})^\times$, and hence an intermediate cover $X_H(Q) \to X_0(Q)$ with Galois group $\Delta_N = (\mathbf{Z}/p^N\mathbf{Z})^d$ acting via diamond operators. The space M_Q is essentially a localization of a certain space of modular forms for $X_H(Q)$ (with some care taken at the Hecke operators for primes dividing Q). Since the cohomology of modular curves (localized at the maximal ideal corresponding to $\mathfrak{m}$) is concentrated in degree one, the module M_Q turns out to be free over an auxiliary ring $S_N = W(k)[\Delta_N]$ of diamond operators, and the quotient $M_Q/\mathfrak{a}_Q$ for the augmentation ideal $\mathfrak{a}_Q$ of S_N is isomorphic to M. It follows that $\mathbf{T}_Q/\mathfrak{a}_Q = \mathbf{T}$.

(iv) The diamond operators have an interpretation on the Galois deformation side, and there is a identification $R_Q/\mathfrak{a}_Q = R$ where R_Q and $\mathbf{T}_Q$ can be viewed compatibly as S_N-modules.

(4) Finally, one "patches" these constructions together for larger and larger Q. This is somewhat counterintuitive, since for different Q the Galois representations involved are not compatible. However, one forgets the Galois representations and only remembers the structures relative to both the diamond operators S_N and R_∞, giving the data of a surjection

$$R_\infty \to \mathbf{T}_\infty$$

with a compatible action of $S_\infty = \operatorname{proj\,lim} S_N \simeq W(k)[[t_1, \dots, t_d]]$. Using the fact that $\mathbf{T}_\infty$ is free of finite rank over S_∞, and that R_∞ and S_∞ are formally smooth of the same dimension, one deduces that $R_\infty = \mathbf{T}_\infty$ and then $R = \mathbf{T}$ after quotienting out by the augmentation ideal of S_∞.

2. THE EARLY YEARS

2.1. The work of Diamond and Fujiwara

Wiles made essential uses of multiplicity one theorems in order to deduce that M_Q was free over $\mathbf{T}_Q$. Diamond [72] and Fujiwara [85] (independently) had the key insight that one could instead patch the modules M_Q directly—and then argue directly with the resulting object M_∞ as a module over R_∞ which was also free over S_∞. Using the fact that R_∞ is formally smooth, this allowed one to deduce *a posteriori* that M_∞ was free over R_∞ using the Auslander–Buchsbaum formula **[9]**. This not only removed the necessity of proving difficult multiplicity-one results but gave new proofs of these results[7] which could then be generalized to situations where the known methods (often using the q-expansion principle) were unavailable.[8] Diamond had the following to say about how he came up with the idea to patch modules rather than use multiplicity-one theorems:

> *My vague memory is that I was writing down examples of ring homomorphisms and modules, subject to some constraints imposed by a Taylor–Wiles setup, and I could not break "M free over the group ring implies M free over R." (I still have the notebook with the calculations somewhere, mostly done during a short trip with some friends to Portugal.) I did not know* what *commutative algebra statement I needed, but I knew I needed to learn more commutative algebra and found my way to Bruns and Herzog's "Cohen–Macaulay Rings"* **[28]** *(back in the library in Cambridge, UK by then). When I saw the statement of Auslander–Buchsbaum, it just clicked.*

Diamond made a second improvement **[70, 71]** dealing with primes away from p in situations where the corresponding minimal local deformation problem was not controlled by the Serre level $N(\overline{\rho})$ alone.

2.2. Integral p-adic Hodge Theory, part I: Conrad–Diamond–Taylor

One early goal after Fermat was the resolution of the full Taniyama–Shimura conjecture, namely, the modularity of all elliptic curves over $\mathbf{Q}$. After the improvements of Diamond, the key remaining problem was understanding deformation rings associated to local Galois representations at p coming from elliptic curves with bad reduction at p. Since Wiles' method (via Langlands–Tunnell **[127, 178]**) was ultimately reliant on working with the prime $p = 3$, this meant understanding deformations at p of level p^2 and level p^3, since any elliptic curve over $\mathbf{Q}$ has a twist such that the largest power of 3 dividing the conductor is at most 27. Ramakrishna in his thesis **[145]** had studied the local deformation problem for finite flat representations (the case when $(N, p) = 1$) and proved that the corresponding local deformation rings were formally smooth. The case when p exactly divides N was subsumed into the ordinary case, also treated by Wiles. In level p^2, one can show that the Galois represen-

7 There is an intriguing result of Brochard **[27]** which weakens the hypotheses of Diamond's freeness criterion even further, although this idea has not yet been fully exploited.

8 The history of the subject involves difficult theorems in the arithmetic geometry of Shimura varieties being replaced by insights from commutative algebra, paving the way to generalizations where further insights from the arithmetic geometry of Shimura varieties are required.

tations associated to the relevant modular forms[9] of level p^2 become finite flat after passing to a finite extension $L/\mathbf{Q}_p$ with ramification degree $e \leq p-1$. In this range, Conrad [64,65] was able to adapt ideas of Fontaine [80] to give an equivalence between the local Galois deformations (assuming $\overline{\rho}|G_{\mathbf{Q}_p}$ was irreducible) and linear algebra data. In particular, as in the work of Ramakrishna, one can show that the relevant local deformation rings are formally smooth, and so Conrad, Diamond, and Taylor were able to adapt the Taylor–Wiles method to this setting [66].

2.3. Integral p-adic Hodge Theory, part II: Breuil–Conrad–Diamond–Taylor

A central technical ingredient in all of the arguments so far has been some use of integral p-adic Hodge Theory, and in particular the theory of finite flat group schemes and Barsotti–Tate groups as developed by Fontaine and others. All integral versions of this theory required a hypothesis on either the weight or the ramification index e relative to the bound $p-1$. However, around this time, Christophe Breuil made a breakthrough[10] by finding a new way to understand the integral theory of finite flat group schemes over arbitrarily ramified bases [19]. This was just the technical tool required to push the methods of [66] to level p^3. Using these results, Breuil, Conrad, Diamond, and Taylor [25] were able to show that enough suitably chosen local deformation rings were formally smooth to prove the modularity of all elliptic curves.

2.4. Higher weights, totally real fields, and base change

Many of the methods which worked for modular forms were directly adaptable to the case both of general rank 2 motives over $\mathbf{Q}$ with distinct Hodge–Tate weights (corresponding to modular forms of weight $k \geq 2$ rather than $k=2$) and also to such motives over totally real fields (which are related to Hilbert modular forms), see in particular the work of Fujiwara [85] (and more recently Frietas–Le Hung–Siksek [84]). Another very useful innovation was a base change idea of Skinner–Wiles [161] which circumvented the need to rely on Ribet's level lowering theorem. The use of cyclic base change ([127] in this case and [5] in general) subsequently became a standard tool in the subject. For example, it meant that one could always reduce to a situation where the ramification at all primes $v \nmid p$ was unipotent. The paper [161] was related to a more ambitious plan by Wiles to prove modularity for all totally real fields:

> *After Fermat I started to work with Taylor and then Diamond on the general case but decided very soon that I would rather try to do the totally real case for* GL(2).

---

9 This is not true for all modular forms of level p^2 and weight 2, but only for those whose conductor at p remains divisible by p^2 after any quadratic twist.

10 Much of the development of integral p-adic Hodge theory over the last 20 years since [25] has been inspired by its use in the Taylor–Wiles method. However, the timing of Breuil's work was more of a happy coincidence, although Breuil was certainly aware of the fact that a computable theory of finite flat group schemes over highly ramified bases could well have implications in the Langlands program.

I think this was while I was getting back into other kinds of problems but I thought I should still earn my bread and butter. One lunch time at the IAS in 1996 Florian Pop spoke to me and explained to me about finding points over fields totally split at some primes (e.g., real places) as he had written a paper **[92]** *about this with some others. Was this any use for the Tate–Shafarevich group? I immediately saw that whether or not it was any use for TS (I doubted it) it should certainly give potential modularity. This gave some kinds of lifting so I worked on the other half (i.e., descent) thinking that just needed a similar insight. At some point I suggested to Chris that we try to do Ribet's theorem using cyclic base change as that would be useful in proving modularity and was buying time while I waited to get the right idea. Unfortunately, I completely misjudged the difficulty of descent and the problem is still there. I think it is both much harder than I thought and also more important. I hope still to prove it! Of course, Taylor found potential modularity and then, what I had assumed was much harder, a way to think about* $\mathrm{GL}(n)$.

3. REDUCIBLE REPRESENTATIONS: SKINNER–WILES

One of the key hypotheses in the Taylor–Wiles method concerns restrictions on the representation $\overline{\rho}$, in particular the hypothesis that $\overline{\rho}|_{G_{\mathbf{Q}(\zeta_p)}}$ is absolutely irreducible. In **[159, 160, 162]**, Skinner and Wiles introduced a new argument in which this hypothesis was relaxed, at least assuming the representations were ordinary at p. In the ordinary setting, one can replace the rings R and $\mathbf{T}$ (which in the original setting are finite over $W(k)$) by rings which are finite (and typically flat) over Iwasawa algebras $\Lambda = W(k)[\![(\mathbf{Z}_p)^d]\!]$ for some d which arise as weight spaces, the point being that the ordinary deformations of varying weight admit a good integral theory. The first innovation (in part) involves making a base change so that the reducible locus is (relatively) "small," (measured in terms of the codimension over Λ). The second idea is then to apply a variant of the Taylor–Wiles method to representations $\varrho : G_F \to \mathrm{GL}_2(\mathbf{T}/\mathfrak{p})$ for nonmaximal prime ideals $\mathfrak{p}$ of F.[11] Wiles again:

We had worked out a few cases we could do without big Hecke rings in some other papers and I would say it was more a feat of stamina and technique to work through it. Of course, the use of these primes was much more general and systematic than anything that went before. There is also an amusing point in this paper where we use a result from commutative algebra. It seemed crucial then though I don't know if it still is. This is Proposition A.1 of Raynaud **[148]**. *I had thought at some point during the work on Fermat that this result might be needed and had asked Michel Raynaud about it. He said he would think about it. A week later he came back to me, somewhat embarrassed that he had not known right away, to say*

---

11 Representations ϱ to infinite quotients $\mathbf{T}/\mathfrak{p}$ had also arisen in Wiles' paper on Galois representations associated to ordinary modular forms **[180]** where the concept of pseudodeformation was also first introduced.

that it was a result in his wife's thesis. So the reference to M. Raynaud is actually to his wife, Michèle Raynaud, though he gave the reference.

Allen **[1]** was later able to adapt these arguments to the $p = 2$ dihedral case, which (in a certain sense) realized the original desire[12] of Wiles to work at the prime $p = 2$.

4. THE ARTIN CONJECTURE

While the approach of **[171, 181]** applied (in principle) to all Galois representations associated to modular forms of weight $k \geq 2$, the case of modular forms of weight $k = 1$ is qualitatively quite different (see also Section 10.1). It was therefore quite surprising when Buzzard–Taylor **[33]** proved weight one modularity lifting theorems for odd continuous representations $\rho : G_{\mathbf{Q}} \to \mathrm{GL}_2(\overline{\mathbf{Q}}_p)$ which were unramified at p. Using this, Buzzard–Dickinson–Shepherd-Barron–Taylor **[31]** proved the Artin conjecture for a positive proportion of all odd A_5 representations, which had previously only been known in a finite number of cases[13] up to twist. Standard ordinary modularity theorems showed the existence of ordinary modular forms associated to such representations ρ—however, the classicality theorems of Hida **[101]** do not apply (and are not true!) in weight one. The main idea of **[33]** was to exploit the fact that ρ is unramified to construct *two* ordinary modular forms each corresponding to a choice of eigenvalue of $\rho(\mathrm{Frob}_p)$ assuming these eigenvalues are distinct.[14] One then has to argue **[33]** that these two ordinary forms are the oldforms associated to a classical eigenform of weight one, which one can do by exploiting both the rigid geometry of modular curves and the q-expansion principle.

Although the original version of this argument required a number of improvements to the usual Taylor–Wiles method (Dickinson overcame some technical issues when $p = 2$ **[73]** and Shepherd-Barron–Taylor proved some new cases of Serre's conjecture for $\mathrm{SL}_2(\mathbf{F}_4)$ and $\mathrm{SL}_2(\mathbf{F}_5)$-representations in **[157]**), it was ripe for generalization to totally real fields.[15] After a key early improvement by Kassaei **[106]**, the $n = 2$ Artin conjecture for totally real fields is now completely resolved under the additional assumption that the representation is odd by a number of authors, including Kassaei–Sasaki–Tian and Pilloni–Stroh **[107–109, 141,**

---

12 As far as primary historical sources go, the introduction of Wiles' paper **[181]** is certainly worth reading.

13 In a computational *tour de force* for the time, Buhler **[29]** in his thesis had previously established the modularity of an explicit odd projective A_5 representation of conductor 800.

14 This argument can be modified to deal with the case when the eigenvalues of $\overline{\rho}(\mathrm{Frob}_p)$ coincide by modifying R and $\mathbf{T}$ to include operators corresponding (on the Hecke side) to U_p. Geraghty and I discovered an integral version of this idea ourselves ("doubling," following Wiese's paper **[179]**) during the process of writing **[38]**, although it turned out that, at least in characteristic zero, Taylor already had the idea in his back pocket in the early 2000s.

15 The proof all that finite odd 2-dimensional representations over $\mathbf{Q}$ are modular was completed by Khare and Wintenberger as a consequence of their proof of Serre's conjecture, see Section 8.

143, 152]. On the other hand, the reliance on q-expansions in this argument has proved an obstruction to extending this to other groups. (See also Section 11.2.)

5. POTENTIAL MODULARITY

One new idea which emerged in Taylor's paper [166] was the concept of *potential modularity*. Starting with a representation $\rho : G_F \to \mathrm{GL}_2(\overline{\mathbf{Q}}_p)$ for a totally real field F, one could sidestep the (difficult) problem of proving the modularity of $\overline{\rho}$ by proving it was modular over some finite totally real extension F'/F. In the original paper [181], Wiles employed a 3–5 switch to deduce the modularity of certain mod 5 representations from the modularity of mod 3 representations. More generally, one can prove the modularity of a mod p representation $\overline{\rho}_p$ from the modularity of a mod q representation $\overline{\rho}_q$ if one can find both of them occurring as the residual representation of a compatible family where the Taylor–Wiles hypotheses apply to $\overline{\rho}_q$. For example, if $\overline{\rho}_p$ and $\overline{\rho}_q$ are representations valued in $\mathrm{GL}_2(\mathbf{F}_p)$ and $\mathrm{GL}_2(\mathbf{F}_q)$ respectively, one can try to find the compatible family by finding an elliptic curve with a given mod p and mod q representation. The obstruction to doing such a p–q switch over F is that the corresponding moduli spaces (which in this case are twists of the modular curve $X(pq)$) are not in general rational, and hence have no reason to admit rational points. However, exploiting an idea due to Moret-Bailly [132], Taylor showed that these moduli spaces at least had many points over totally real fields where one could additionally ensure that the Taylor–Wiles hypothesis applies at the prime q. At the cost of proving a weaker result, this gives a huge amount of extra flexibility that has proved remarkably useful. Taylor's first application of this idea was to prove the Fontaine–Mazur conjecture for many 2-dimensional representations, since the potential modularity of these representations was enough to prove (for example) that they come from compatible families of Galois representations (even over the original field F!), and that they satisfy purity (which is known for Hilbert modular forms of regular weight). The concept of potential modularity, however, has proved crucial for other applications, not least of which is the proof of the Sato–Tate conjecture (see Section 9.2).

6. THE WORK OF KISIN

A key ingredient in the work of Breuil–Conrad–Diamond–Taylor (Sections 2.2 and 2.3) (and subsequent work of Savitt [153, 154]) was the fact that a certain local deformation ring R^{fl} defined in terms of integral p-adic Hodge theory was formally smooth. The calculations of [25, 154], however, applied only to some (very) carefully chosen situations sufficient for elliptic curves but certainly not for all 2-dimensional representations. In the 2000s, Kisin made a number of significant contributions, both to the understanding of local deformation rings but also to the structure of the Taylor–Wiles argument itself [117–119, 121–123].

6.1. Local deformation rings at $v = p$

One difficulty with understanding local deformation rings R^{fl} associated to finite flat group schemes over highly ramified bases is that the group schemes themselves are not uniquely defined by their generic fibers. Kisin [122] had the idea that one could also define the moduli space of the group schemes themselves, giving a projective resolution $\mathscr{G}\mathscr{R} \to \mathrm{Spec}(R^{\mathrm{fl}})$ (this map is an isomorphism after inverting p). Kisin further realized that the geometry of $\mathscr{G}\mathscr{R}$ was related to local models of Shimura varieties, for which one had other available techniques to analyze their structure and singularities. Later, Kisin was also able [119] to construct local deformation rings R capturing deformations of a fixed local representation $\overline{\rho}$ which become semistable over a fixed extension $L/\mathbf{Q}_p$ and had Hodge–Tate weights in any fixed finite range $[a, b]$, absent a complete integral theory of such representations. (There are also are constructions where one fixes the inertial type of the corresponding representation.) Kisin further proved that the generic fibers of these rings were indeed of the expected dimension and often formally smooth.

6.2. Kisin's modification of Taylor–Wiles

Beyond analyzing the local deformation rings themselves, Kisin crucially found a way [122] to modify the Taylor–Wiles method to avoid the requirement that these rings are formally smooth, thus greatly expanding the scope of the method. First of all, Kisin reimagined the global deformation ring R as an algebra over a (completed tensor product)

$$R^{\mathrm{loc}} = \widehat{\bigotimes}_{v \in S} R_v$$

of local deformation rings R_v for sets of places $v \in S$, in particular including the prime p.[16] Now, after a Taylor–Wiles patching argument, one constructs a big module M_∞ over R_∞ (and free over the auxiliary ring of diamond operators S_∞) but where R_∞ is no longer a power series ring over $W(k)$ but a power series ring over R^{loc}. If the algebras R_v for $v \in S$ are themselves power series rings, one is reduced precisely to the original Taylor–Wiles setting as modified by Diamond. On the other hand, if the R_v are (for example) not power series rings but are integral domains over $W(k)$ of the expected dimension, then Kisin explained how one could still deduce that $M[1/p]$ was a faithful $R[1/p]$-module, which proves that $R[1/p] = \mathbf{T}[1/p]$ and suffices for applications to modularity. More generally, assuming only that the R_v are flat over $W(k)$ and that the generic fiber $R_v[1/p]$ is equidimensional of the expected dimension, the modularity of any point of R reduces to showing that there is at least one modular point which lies on the same component of $R_v[1/p]$.[17]

16 Since the local residual representations are typically reducible, Kisin also introduced the notion of *framed* deformation rings which are always well defined, and which (properly taking into account the extra variables) are compatible with the Taylor–Wiles argument.

17 There are some subtleties to understanding $R[1/p]$ for complete local Noetherian $W(k)$-algebras that are not obvious on first consideration. The first and most obvious blunder to avoid is to recognize that $R[1/p]$ is usually far from being a local ring. Similarly, the ring $R[1/p]$ can be regular and still have multiple components, as can be seen in an example as simple as $R = \mathbf{Z}_p[\![X]\!]/X(X-p)$. Perhaps more importantly, however, the ring $R[1/p]$ "behaves" in some important ways like a finitely generated algebra over a field.

In the original modularity lifting arguments, one treated the minimal case first and then deduced the nonminimal cases using a subtle commutative algebra criterion which detected isomorphisms between complete intersections. From the perspective of Kisin's modification, all that is required is to show that there exists a single modular point with the right nonminimal local properties. In either case, both Wiles and Kisin used Ihara's Lemma to establish the existence of congruences between old and new forms, but Kisin's argument is much softer and thus more generalizable to other situations.[18] Kisin had the following to say about his thought process:

> *The idea of thinking of R as an R^{loc} algebra just popped into my head, after I'd been thinking about the Wiles–Poitou–Tate formula, and how it fit into the Taylor–Wiles patching argument. This was in Germany, I think in 2002. I had the idea about moduli of finite flat group schemes in the Fall of 2003, after I arrived in Chicago. It was entirely motivated by modularity. I had been trying to compute these deformation rings, by looking at deformations of finite flat group schemes. For $e < p - 1$, the finite flat model is unique, so I knew this gave the deformation ring in this case; this already gave some new cases. However, I was stuck about the meaning of these calculations in general for quite some time. At some point I thought I'd better write up what I had, but as soon as I started thinking about that—within a day—I realized what the correct picture was with the families of finite flat group schemes resolving the deformation ring. I already knew about Breuil's unpublished note* [18]*, and quite quickly was able to prove the picture was correct. It was remarkable that prior to coming to Chicago, I didn't even know the definition of the affine Grassmannian, but within a few months of arriving, it actually showed up in my own work.*
>
> *To me the whole project was incredibly instructive. If I had known more about what was (thought to be) essential in the Taylor–Wiles method, I never would have started the project. Not having fixed ideas gave me time to build up intuition. I also should have gotten the idea about moduli of finite flat group schemes much sooner if I'd been more attentive to what the geometry was trying to tell me.*

7. p-ADIC LOCAL LANGLANDS

7.1. The Breuil–Mézard conjecture

Prior to Kisin's work, Breuil and Mézard [26] undertook a study of certain low weight potentially semistable deformation rings, motivated by [25]. They discovered (in part conjecturally) a crucial link between the geometry of these Galois deformation rings (in

---

18 In particular, Wiles' numerical criterion [68, THM. 5.3] relies on certain rings being complete intersections, and Kisin's local deformation rings are not complete intersections (or even Gorenstein) in general—see [163].

particular, the Hilbert–Samuel multiplicities of their special fibers) with the mod-p reductions (and corresponding irreducible constituents) of lattices inside locally algebraic p-adic representations of $\mathrm{GL}_2(\mathbf{Z}_p)$. In the subsequent papers [20, 21], Breuil raised the hope that there could exist a p-adic Langlands correspondence relating certain mod-p (or p-adic Banach space) representations of $\mathrm{GL}_2(\mathbf{Q}_p)$ to geometric 2-dimensional p-adic representations of $G_{\mathbf{Q}_p}$.[19] Breuil recounts the origins of these conjectures as follows:

> *The precise moment I became 100% sure that there would be a non-trivial p-adic correspondence for* $\mathrm{GL}_2(\mathbf{Q}_p)$ *was in the computations of* [21]. *In these computations, I reduced mod p certain* $\overline{\mathbf{Z}}_p$*–lattices in certain locally algebraic representations of* $\mathrm{GL}_2(\mathbf{Q}_p)$*, and at some point, I found out that this reduction mod p had a really nice behaviour, so nice that clearly, it was predicting (via the mod-p correspondence) what the reduction mod-p would be on the* $\mathrm{Gal}(\overline{\mathbf{Q}}_p/\mathbf{Q}_p)$*-side.*

These ideas were further developed by Colmez in [57–59] amongst other papers[20]: Colmez studied various Banach space completions defined by Breuil and proved they were nonzero using the theory of (φ, Γ)-modules. Since the theory of (φ, Γ)-modules applies to all Galois representations and not just potentially semistable ones, this led Colmez to propose a p-adic local Langlands correspondence for *arbitrary* 2-dimensional representations $G_{\mathbf{Q}_p} \to \mathrm{GL}_2(E)$, and he was ultimately able to construct a functor from suitable $\mathrm{GL}_2(\mathbf{Q}_p)$-representations to Galois representations of $G_{\mathbf{Q}_p}$. Colmez gave a talk on his construction at a conference in Montreal in September 2005. At the same conference, Kisin gave a talk presenting a proof of the Breuil–Mézard conjecture by relating it directly to $R = \mathbf{T}$ theorems and the Fontaine–Mazur conjecture for odd 2-dimensional representations of $G_{\mathbf{Q}}$ with distinct Hodge–Tate weights. While Kisin's argument exploited results of Berger–Breuil [14] and Colmez, it was realized by the key participants (perhaps in real time) that Colmez' p-adic local Langlands correspondence should be viewed as taking place over

19 The starting observation [22] is as follows: if $\pi = \otimes' \pi_v$ is the automorphic representation associated to a modular form f, then π_v determines (and is determined by) $\rho_f|_{G_{\mathbf{Q}_v}}$ for all $v \neq p$ (at least up to Frobenius semisimplification). On the other hand, π_p does not determine the p-adic representation $\rho_f|_{G_{\mathbf{Q}_p}}$ (except in the exceptional setting where π_p is spherical and a_p is not a p-adic unit), raising the question of what extra $\mathrm{GL}_2(\mathbf{Q}_p)$ structure associated to f should determine (and be determined by) $\rho_f|_{G_{\mathbf{Q}_p}}$.

20 In [116], Kisin had shown that the p-adic representations V associated to nonclassical finite slope overconvergent modular forms with U_p-eigenvalue a_p satisfied $\dim D_{\mathrm{cris}}(V) = 1$, and moreover that crystalline Frobenius acted on this space by a_p. (This paper was itself apparently motived by the goal of *disproving* the Fontaine–Mazur conjecture!) On the way to the 2004 Durham symposia on L-functions and Galois representations, Fontaine raised the question to Colmez to what extent this determined the corresponding Galois representation. Colmez worked out the answer the evening before his talk and incorporated it into his lecture the following day, ultimately leading to the notion of trianguline representations [57].

the entire local deformation ring. Subsequently Colmez was able to construct the inverse functor.[21] Colmez writes:

> *I received a paper of Breuil (a former version of* [23]*) during my stay at the Tata Institute in December 2003–January 2004. In December, I was spending Christmas under Goa's palm trees with my daughter when Breuil's paper arrived in my email. That paper contained a conjecture (in the semi-stable case) that I was sure I could prove using* (φ, Γ)*-modules (if it was true…). I spent January 2004 working on it and after 15 days of computations in the dark, I finally found a meaning to some part of a painful formula (you can find some shadow of all of this in (iii) of Remark 0.5 of my unpublished* [56]*). By the end of the month, I was confident that the conjecture was proved and I told so to Breuil who adapted the computations to the crystalline case, and wrote them down with the help of Berger (which developed into* [14]*). (One thing that makes computations easier and more conceptual in the crystalline case is that you end up with the universal completion of the locally algebraic representation you start with; something that is crucial in Matthew [Emerton]'s proof of the FM conjecture.) Durham was in August of that year and Berger–Breuil had notes from a course they had given in China* [13]*. Those notes were instrumental in my dealing with trianguline representations at Durham (actually, I did some small computation and the theory just developed by itself during the night before my talk which was supposed to be on something else…I think I came up with the concept of trianguline representations later, to justify the computations, I don't remember what language I used in my talk which had some part on Banach–Colmez spaces as far as I can remember.*

7.2. Local–global compatibility for completed cohomology

From a different perspective, Emerton had introduced the completed cohomology groups [77] as an alternative means for constructing the Coleman–Mazur eigencurve [55]. Inspired by Breuil's work, Emerton formulated [76] a local–global compatibility conjecture for completed cohomology in the language of the then nascent p-adic Langlands correspondence. After the construction of the correspondence for $\mathrm{GL}_2(\mathbf{Q}_p)$ by Colmez and Kisin,

---

21 To add some further confusion to the historical chain of events, the published version of [120] incorporates some of these subsequent developments. Note also that the current state of affairs is that the proof of the full p-adic local Langlands correspondence for $\mathrm{GL}_2(\mathbf{Q}_p)$ (for example as proved in [63] but see also [59, REMARQUE VI.6.51]) still relies on the global methods of [78], which in turn relies on [59]. These mutual dependencies, however, are not circular! The difficulty arises in the supercuspidal case. One philosophical reason that global methods are useful here is that all global representations are yoked together by an object (the completed cohomology group $\tilde{H}^1(\mathbf{Z}_p)$) with good finiteness properties. One can then exploit the fact that crystabeline representations (for which the p-adic local Langlands correspondence is known by [59]) are Zariski dense inside unrestricted global deformation rings ([78, THEOREM 1.2.3], using arguments going back to Böckle [15]).

Emerton was able to prove most of his conjecture, leading to a new proof of (many cases of) the Fontaine–Mazur conjecture. The results of Kisin [120] and Emerton fell short of proving the full version of this conjecture for two reasons. The first was related to some technical issues with the p-adic local Langlands correspondence, both at the primes $p = 2$ and 3 but also when the residual representation locally had the shape $1 \oplus \overline{\varepsilon}$ for the cyclotomic character ε. (The local issues have now more or less all been resolved [63]. The most general global results for $p = 2$ are currently due to Tung [177].) A second restriction was the Taylor–Wiles hypothesis that $\overline{\rho}$ was irreducible. Over the intervening years, a number of key improvements to the local story have been found, in particular by Colmez, Dospinescu, Hu, and Paškūnas [63, 104, 139]. Very recently, Lue Pan [137] found a way to marry techniques from Skinner–Wiles in the reducible case (Section 3) to techniques from p-adic local Langlands to completely prove the modularity (up to twist) of any geometric representation $\rho : G_{\mathbf{Q}} \to \mathrm{GL}_2(\overline{\mathbf{Q}}_p)$ for $p \geq 5$ only assuming the hypotheses that ρ has distinct Hodge–Tate weights and that ρ is odd.[22]

8. SERRE'S CONJECTURE

In Wiles' original lectures in Cambridge in 1993, he introduced his method with the statement that it was *orthogonal* to Serre's conjecture [158]. In some senses, this viewpoint turned out to be the opposite of prophetic, in that the ultimate resolution of Serre's conjecture used the Taylor–Wiles method as its central core. The proof of Serre's conjecture by Khare and Wintenberger [111, 113–115] introduced a new technique for lifting residual Galois representations to characteristic zero (see §8.2) which has proved very useful for subsequent modularity lifting theorems.

8.1. Ramakrishna lifting

Ramakrishna, in a series of papers in the late 1990s [146, 147], studied the question of lifting an odd Galois representation

$$\overline{\rho} : G_{\mathbf{Q}} \to \mathrm{GL}_2(\overline{\mathbf{F}}_p)$$

to a global potentially semistable representation in characteristic zero unramified outside finitely many primes. This is a trivial consequence of Serre's conjecture[23] but is highly nonobvious without such an assumption. Ramakrishna succeeded in proving the existence of lifts by an ingenious argument involving adding auxiliary primes and modifying the local deformation problem to a setting where there all global obstructions vanished. The resulting lifts had the added property that they were valued in $\mathrm{GL}_2(W(k))$ whenever $\overline{\rho}$ was valued in $\mathrm{GL}_2(k)$. Adaptations of Ramakrishna's method had a number of important applications

---

22 The assumption on the Hodge–Tate weights is almost certainly removable using recent progress on the ideas discussed in Section 4 (Sasaki has announced such a result). Moreover, Pan has found a different approach to this case as well, see [138, THEOREM 1.0.5] and the subsequent comments. The hypothesis that ρ is odd is more troublesome—see Section 9.7.

23 Trivial only assuming the results of Tsuji [176] and Saito [151], of course.

even under the assumption of residual modularity, including in [58] where it was used to produce characteristic zero lifts with Steinberg conditions at some auxiliary primes. There is also recent work of Fakhruddin, Khare, and Patrikis [79] which considerably extends these results in a number of directions.

8.2. The Khare–Wintenberger method

One disadvantage of Ramakrishna's method was that it required allowing auxiliary ramification which (assuming Serre's conjecture) should not be necessary.[24] Khare and Wintenberger found a new powerful method for avoiding this. The starting point is the idea that, given an odd representation $\overline{\rho} : G_F \to \mathrm{GL}_2(\overline{\mathbf{F}}_p)$ for a totally real field F satisfying the Taylor–Wiles hypotheses, one could find a finite extension H/F where $\overline{\rho}$ is modular (exactly as in Section 5). Then, using an $R = \mathbf{T}$ theorem over H, one proves that the corresponding deformation ring R_H of $\overline{\rho}|G_H$ is finite over $W(k)$. However, for formal reasons, there is a map $R_H \to R_F$ (where R is the deformation ring corresponding to the original representation $\overline{\rho}$) which is a finite morphism, and hence the ring R_F/p is Artinian. Then, by Galois cohomological arguments, one proves the ring R_F has dimension at least one, from which one deduces that R_F has $\overline{\mathbf{Q}}_p$-valued points. Even more can be extracted from this argument, however,—the $\overline{\mathbf{Q}}_p$-valued point of R_F certainly comes from a $\overline{\mathbf{Q}}_p$-valued point of R_H, and hence comes from a compatible family of Galois representations over H. Using the fact that one member of the family extends to G_F, it can be argued that the entire family descends to a compatible family over F. This one can then hope to prove is modular by working at a different (possibly smaller) prime, where (hopefully) one can prove the associated residual representation is modular. In this way, one can inductively reduce Serre's conjecture [156] to the case $p = 2$ and $N(\overline{\rho}) = 1$, where Tate had previously proved in a letter to Serre [61, JULY 2, 1973] (also [164]) that all such absolutely irreducible representations are modular by showing that no such representations exist. The entire idea is very clean, although in practice the difficulty reduces to the step of proving modularity lifting theorems knowing either that $\overline{\rho}$ is either modular and absolutely irreducible or is reducible. Khare and Wintenberger's timing was such that the automorphy lifting technology was just good enough for the proof to work, although this required some extra effort at the prime $p = 2$ (both in their own work and in a key assist by Kisin [121]). As with Ramakrishna's method, the Khare–Wintenberger lifting method has also been systematically exploited for modularity lifting applications (for example, in [11] (see Section 9.6) building on ideas of Gee [87]).

9. HIGHER DIMENSIONS

Parallel to the developments of p-adic Langlands for $n = 2$, the first steps were made to generalize the theory to higher dimensional representations. Unlike in the case of

24 If one insists on finding a lift valued in $\mathrm{GL}_2(W(k))$ rather than $\mathrm{GL}_2(\mathcal{O}_E)$ for some ramified extension $E/W(k)[1/p]$, then some auxiliary ramification is necessary in general, at least in fixed weight.

modular forms, substantially less was known about the existence of Galois representations until the 1990s.

9.1. Construction of Galois representations, part I: Clozel–Kottwitz

The first general construction[25] of Galois representations in dimension $n > 2$ was made by Clozel **[48]** (see also the work of Kottwitz **[125]**). Clozel's theorem applies to certain automorphic forms for $\mathrm{GL}_n(\mathbf{A}_L)$ for CM fields L/L^+. The construction requires three important hypotheses on π: The first is that π is conjugate self-dual, that is, $\pi^\vee \simeq \pi^c$. If π is a base change from an algebraic representation of L^+ and $n = 2$ then this condition is automatic,[26] but it is far from automatic when $n > 2$. The second condition is an assumption on the infinitesimal character which (in the case of modular forms) is equivalent to the condition that the weight k is ≥ 2. Finally, there is a technical condition that for some finite place x the representation π_x is square integrable. A number of improvements (particularly at the bad primes) were made by Harris–Taylor in **[100, THEOREM C]**, and later by Taylor–Yoshida and Caraiani **[42, 43, 172]**, bringing the theory roughly in line with that of modular forms at the time of Wiles, and in particular primed for possible generalizations of the Taylor–Wiles method to higher dimensions.

9.2. The Sato–Tate conjecture, part I

Harris and Taylor (as early as 1996) started the work of generalizing the Taylor–Wiles machinery to the setting of n-dimensional representations. They quickly understood that the natural generalization of these ideas in n-dimensions required the hypothesis that the Galois representations were self-dual up to a twist. This meant that one should not consider general automorphic forms on the group $\mathrm{GL}_n(\mathbf{A}_\mathbf{Q})$ but rather groups of symplectic or orthogonal type depending on the parity of n. If one replaced Galois representations over totally real fields by Galois representations over imaginary CM fields and then further imposed the condition that the Galois representations are conjugate self-dual, the relevant automorphic forms should then come from unitary groups. There were two benefits of working with these hypotheses. First of all, the relevant automorphic representations for unitary groups were, as with modular forms, associated to cohomology classes on Shimura varieties. In particular, under the assumption that there existed an auxiliary prime x such that π_x was square integrable, they could be seen inside the "simple" Shimura varieties of type $U(n-1,1)$ considered by Kottwitz **[125]**. On the other hand, the same Hecke eigenclasses (if not Galois representations) also came from a compact form of the group and thus inside the cohomology of zero-dimensional varieties.[27] The advantage of working in this setting is that

---

25 Clozel's paper is from 1991 and thus not strictly "post-Fermat" as is the remit of this survey. However, it can be considered a natural starting point for the "modern" arithmetic theory of automorphic forms for $\mathrm{GL}(n)$ and so it seems reasonable to mention it here.

26 At least after a twist which is always possible to achieve in practice, see **[50, LEMMA 4.1.4]**. More generally, one can work with unitary similitude groups and consider π with $\pi^\vee \simeq \pi^c \otimes \chi$ for suitable characters χ.

27 Inside H^0, of course.

the freeness of M_Q over the ring of diamond operators is immediate.[28] In the fundamental paper **[58]**, Clozel, Harris, and Taylor succeeded in overcoming many of the technical difficulties generalizing the arguments of **[171, 181]** to these representations. Although the argument in spirit was very much the same, there are a number of points for GL_2 where things are much easier. One representative example of this phenomenon is understanding Taylor–Wiles primes. While the Galois side generalizes readily, the automorphic side requires many new ideas and some quite subtle arguments concerning the mod p structure of certain $GL_n(\mathbf{Q}_q)$-representations of conductor 1 and conductor q. In order to prove the Sato–Tate conjecture for a modular form f, it was already observed by Langlands that it sufficed to prove the modularity of all the symmetric powers of f. However, it turns out that the weaker assumption that each of these symmetric powers is *potentially modular* suffices, and by some subterfuge only the even powers are required **[95]**. In order to prove potential modularity theorems, one needs to be able to carry out some version of the p–q switch (Section 5). In order to do *this*, one needs a source of motives which both generate Galois representations of the right shape (conjugate self dual and with distinct Hodge–Tate weights) and yet also come in positive dimensional families. It turned out that there already existed such motives in the literature, namely, the so-called Dwork family. However, given the strength of the automorphy lifting theorems in **[58]**, considerable effort had to be made in studying the geometry of the Dwork family to ensure that the p–q switch would produce geometric Galois representations with the right local properties. These issues were precisely addressed in the companion paper by Harris, Shepherd-Barron, and Taylor **[97]**. Taken together, these papers contained all the ingredients to prove the potential modularity of higher symmetric powers of modular forms (satisfying a technical square integrable condition at some auxiliary prime) with one exception. As mentioned earlier, the work of Kisin had simplified the passage from the minimal case to the nonminimal case—"all" that was required was to produce congruences between the original form and forms of higher level rather than to compute a precise congruence number as in **[181]**. However, even applying Kisin's approach seemed to require Ihara's Lemma, and despite several years of effort, the authors of **[58]** were not able to overcome this obstacle.[29] Here is Michael Harris' recollection of the process:

> *In the spring of 1995, I was at Brandeis, Richard was at MIT, and I wanted to understand the brand new proof of Fermat's Last Theorem. So I asked Richard if he would help me learn by collaborating on modularity for higher-dimensional groups. The collaboration took off a year later, when Richard wrote to tell me about the Diamond–Fujiwara argument and suggested that we work out the Taylor–Wiles method for unitary groups. This developed over the next 18 months or so into the early version of what eventually became the IHES paper with Clozel. But it had no punch line. I was hoping to work out some non-trivial exam-*

28 In more general contexts, the freeness of M_Q is closely related to the vanishing of cohomology localized at $\mathfrak{m}$ in all but one degree.

29 The issue remains unresolved to this day.

ples of tensor product functoriality for $\mathrm{GL}(n) \times \mathrm{GL}(m)$*, where one of the two representations was congruent mod l to one induced from a CM Hecke character. This would have required some numerical verification. In the meantime we got sidetracked into proving the local Langlands conjecture* [100].

The manuscript on automorphy lifting went through several drafts and was circulated; you can still read it on my home page [99]. *Genestier and Tilouine* [88] *quoted it when they proved modularity lifting for Siegel modular forms. When Clozel saw the draft he told me we should try to prove the Sato–Tate Conjecture. Although this was in line with my hope for examples of tensor product functoriality, it seemed completely out of reach, because I saw no way to prove residual modularity of symmetric powers.*

When I heard about the Skinner–Wiles paper I came up with a quixotic plan to prove symmetric power functoriality for Eisenstein representations, using the main conjecture of Iwasawa theory to control the growth of the deformation rings. This was in the spring of 2000, at the IHP special semester on the Langlands program, where I first met Chris Skinner.

One day Chris told me that Richard had invented potential modularity. This led me to a slightly less hopeless plan to prove potential symmetric power functoriality by proving it for 2*-dimensional representations congruent to potentially abelian representations, as in the potential modularity argument. I told Richard about this idea, probably the day he arrived in Paris. He asked: why apply potential modularity to the* 2*-dimensional representation; why not instead apply it to the symmetric power representations directly? I then replied: that would require a variation of Hodge structures with a short list of properties: mainly, the correct* $h^{p,q}$*'s and large monodromy groups. We checked that potential modularity was sufficient for Sato–Tate. We then resolved to ask our contacts if they knew of VHS with the required properties. The whole conversation lasted about 20 minutes.*

I asked a well-known algebraic geometer, who said he did not know of any such VHS. Richard asked Shepherd-Barron, who immediately told him about the Calabi–Yau hypersurfaces that had played such an important role in the mirror symmetry program. (And if my algebraic geometer hadn't wanted to be dismissive, for whatever reason, he would have realized this as well.) The $h^{p,q}$*'s were fine but we didn't know about the monodromy. However, Richard was staying at the IHES, and by a happy accident so was Katz, and when Richard asked Katz about the monodromy for this family of hypersurfaces Katz told him they were called the Dwork family and gave him the page numbers in one of his books.*

So within a week or two of our first conversation, we found ourselves needing only one more result to complete the proof of Sato–Tate. This was Ihara's lemma, which occupied our attention over the next five years. In the meantime, Clozel had written a manuscript on symmetric powers, based on the reducibility mod ell of symmetric powers. The argument was incomplete but he had several ideas that led to his joining the project, and he also hoped to use ergodic theory to prove Ihara's lemma. In the summer of 2003 Clozel and I joined Richard in "old" Cambridge to try to work this out. The rest you know. We finally released a proof conditional on Ihara's lemma in the fall of 2005. A few months later Richard found his local deformation argument, and the proof was complete.

9.3. Taylor's trick: Ihara avoidance

Shortly after the preprints **[50, 157]** appeared, Taylor **[168]** found a way to overcome the problem of Ihara's lemma. Inspired by Kisin's formulation of the Taylor–Wiles method (Section 6.2), Taylor had the idea of comparing two global deformation rings R^1 and R^2. Here (for simplicity) the local deformation problems associated to R^1 and R^2 are formally smooth at all but a single prime q. At the prime q, however, the local deformation problem associated to R^1 consists of tamely ramified representations where a generator σ of tame inertia has characteristic polynomial $(X-1)^n$, and for R^2 the characteristic polynomial has the shape $(X-\zeta_1)\cdots(X-\zeta_n)$ for some fixed distinct roots of unity $\zeta_i \equiv 1 \bmod p$. On the automorphic side, there are two patched modules H_1 and H_2, and there is an equality $H_1/p = H_2/p$. The local deformation ring R^1_q associated to R^1 at q is reducible and has multiple components in the generic fiber, although the components in characteristic zero are in bijection to the components in the special fiber. On the other hand, the local deformation ring associated to R^2 at q consists of a single component, and so using Kisin's argument one deduces that H_2 has full support. Now a commutative algebra argument using the identity $H_1/p = H_2/p$ and the structure of R^1_q implies that H_1 has sufficiently large support over R_1, enabling one to deduce the modularity of every $\overline{\mathbf{Q}}_p$-valued point of R_1.[30]

9.4. The Sato–Tate conjecture, part II

After Taylor's trick, one was *almost* in a position to complete the proof of Sato–Tate for all classical modular forms. A few more arguments were required. One was the tensor product trick due to Harris which enabled one to pass from conjugate self-dual motives with weights in an arithmetic progression to conjugate self-dual motives with consecutive Hodge–Tate weights by a judicious twisting argument using CM characters. A second ingredient was the analysis of the ordinary deformation ring by Geraghty **[89]**. One of the requirements of the p–q trick was the condition that certain moduli spaces (the Dwork family in this case) had points over various local extensions E of $\mathbf{Q}_p$, in order to construct a motive M over a number field F with $F_v = E$ for $v|p$. For the purposes of modularity lifting, one wants strong control over the local deformation ring at p, and the choice of local deformation ring is more or less forced by the geometric properties of the p-adic representations associated to M. One way to achieve this would be to work in the Fontaine–Laffaille range where the

30 Taylor's argument proves theorems of the form $R[1/p]^{\mathrm{red}} = \mathbf{T}[1/p]$ rather than $R = \mathbf{T}$. This is still perfectly sufficient for proving modularity lifting results, but not always other interesting corollaries associated to $R = \mathbf{T}$ theorems like finiteness of the corresponding adjoint Selmer groups (though see **[2, 133]**).

local deformation rings were smooth. But this requires both that M is smooth at p and that the ramification degree e of $E/\mathbf{Q}_p$ is one. It is not so clear, however, that the Dwork family contains suitable points (for a fixed residual representation $\overline{\rho}$) which lie in any unramified extension of $\mathbf{Q}_p$. What Geraghty showed, however, was that certain ordinary deformation rings[31] were connected over arbitrarily ramified bases. The final piece, however, was the construction of Galois representations for all conjugate self-dual regular algebraic cuspidal π without the extra condition that π_q was square integrable for some q. This story merits it own separate discussion; suffice to say that it required the combined efforts of many people and the resolution of many difficult problems, not least of which was the fundamental lemma by Laumon and Ngô [128, 136] (see also the Paris book project [49, 93], the work of Shin [158], and many more references which if I attempted to make complete would weigh down the bibliography and still contain grievous omissions).

9.5. Big image conditions

The original arguments in [171, 181] required a "big image" hypothesis, namely that $\overline{\rho}$ was absolutely irreducible after restriction to the Galois group of $\mathbf{Q}(\zeta_p)$. Wiles' argument also required the vanishing of certain cohomology groups associated to the adjoint representation of the image of $\overline{\rho}$. These assumptions had natural analogues in [50] (so-called "big image" hypotheses) although they were quite restrictive, and it wasn't clear that they would even apply to most residual representations coming from some irreducible compatible family. In the setting of 2-dimensional representations, the Taylor–Wiles hypothesis guarantees the existence of many primes q such that $q \equiv 1 \bmod p$ and such that $\overline{\rho}(\mathrm{Frob}_q)$ has distinct eigenvalues. This ensures, for example, that there cannot be any Steinberg deformations at q because the ratio of the eigenvalues of any Steinberg deformation must be q. In dimension n, one natural way to generalize this might be to say that $\overline{\rho}(\mathrm{Frob}_q)$ has distinct eigenvalues, although this is not always possible to achieve for many irreducible representations $\overline{\rho}$. A weaker condition is that $\overline{\rho}(\mathrm{Frob}_q)$ has an eigenvalue α with multiplicity one. For such q, there will be no deformations which are unipotent on inertia at q for which the generalized α eigenspace is not associated to a 1-dimensional block. The translation of this into an automorphic condition on U_q-eigenvalues is precisely what is done in [50] (there are additional technical conditions on Frob_q with respect to the adjoint representation $\mathrm{ad}(\overline{\rho})$ which we omit here). In [173], however, Thorne finds a way to allow $\overline{\rho}(\mathrm{Frob}_q)$ to have an eigenvalue α with higher multiplicity, and yet still cut out (integrally) the space of automorphic forms whose Galois representations decompose at q as an unramified representation plus a one dimensional representation which is tamely ramified of p-power order. This technical improvement is very important because (as proved in the appendix by Guralnick, Herzig, Taylor, and Thorne [173]) it imposes no restrictions on $\overline{\rho}$ when $p \geq 2n + 1$ beyond the con-

31 In Geraghty's setting, the residual representations $\overline{\rho}$ were locally trivial. Hence the definition of "ordinary" was not something that could be defined on the level of Artinian rings, and the construction (as with Kisin's construction of local deformation rings associated to certain types) is therefore indirect.

dition that $\overline{\rho}$ is absolutely irreducible after restriction to $G_{\mathbf{Q}(\zeta_p)}$. This improvement is very useful for applications.

9.6. Potentially diagonalizable representations

After the proof of Sato–Tate for modular forms, Barnet-Lamb, Gee, and Geraghty turned their attention to proving the analogous theorem for Hilbert modular forms of regular weight. The methods developed so far were well suited both to representations ρ which were either ordinary or when ρ was not ordinary but still Fontaine–Laffaille. (The latter implies that $\overline{\rho}|G_{\mathbf{Q}_p}$ is absolutely irreducible of some particular shape.) For a modular form over $\mathbf{Q}$, one easily sees that ρ takes one of one of these forms for any sufficiently large p. For Hilbert modular forms, one certainly expects that the ordinary hypothesis should hold for all $v|p$ and infinitely many p, but this remains open. The difficulty arises when, for some prime p (that splits completely, say) the p-adic representation is ordinary at some $v|p$ but nonordinary other $v|p$. The reason that this causes issues is that, when applying the Moret-Bailly argument in the p–q switch, one wants to avoid any ramification at p for the nonordinary case, and yet have large ramification at the ordinary case to make $\overline{\rho}$ locally trivial, and these desires are not compatible. The resolution in **[10]** involved a clever refinement of the Harris tensor product trick. These ideas were further refined in **[11]** and led to the concept of a *potentially diagonalizable* representation $\rho : G_E \to \mathrm{GL}_n(\overline{\mathbf{Q}}_p)$ for some finite extension of $E/\mathbf{Q}_p$. Recall from Section 6.2 that, in the modified form of the Taylor–Wiles method, proving modularity of some lift of $\overline{\rho}$ often comes down to showing the existence of a modular lift lying on a smooth point of the corresponding component of the generic fiber of R^{loc}. In light of Taylor's Ihara avoidance trick (Section 9.3), the difficulty in this problem is mostly at the prime p, and in particular the fact that one knows very little about the components of general Kisin potentially crystalline deformation rings. A potentially diagonalizable representation is one for which, after some finite (necessarily solvable!) extension E'/E, the representation $\rho|G_{E'}$ is crystalline and lies on the same generic irreducible component as a diagonal representation. This notion has a number of felicitous properties. First, it includes Fontaine–Laffaille representations and ordinary potentially crystalline representations. Second, it is clearly invariant under base change. Third, it is compatible with the tensor product trick of Harris. These features make it supremely well adapted to the current forms of the Taylor–Wiles method. By combining this notion with methods of **[10,12]**, as well as extensive use of Khare–Wintenberger lifting (Section 8.2), Barnet–Lamb, Gee, Geraghty, and Taylor in **[11]** proved the potential automorphy of *all* conjugate self-dual irreducible[32] odd[33] compatible systems of Galois representations over a totally real field.

---

32 One variant proved shortly thereafter by Patrikis–Taylor **[140]** replaced the irreducibility condition by a purity condition (which is automatically satisfied by representations coming from pure motives).

33 Although there is no longer a nontrivial complex conjugation in the Galois group of a CM field, there is still an oddness condition related to the conjugate self-duality of the representation and the fact that there are two ways for an irreducible representation to be self-dual (orthogonal and symplectic).

9.7. Even Galois representations

The Fontaine–Mazur conjecture for geometric Galois representations $\rho : G_{\mathbf{Q}} \rightarrow \mathrm{GL}_2(\overline{\mathbf{Q}}_p)$ predicts that, up to twist, either ρ is modular *or* ρ is even with finite image. The methods of [171,181] required the assumption that $\overline{\rho}$ was modular and so *a priori* the assumption that ρ was odd (at least when $p > 2$). Nothing at all was known about the even case before the papers [34,35] in which a very simple trick made the problem accessible to modularity lifting machinery under the assumption that the Hodge–Tate weights are distinct. The punch line is that, for any CM field F/F^{+}, the restriction $\mathrm{Sym}^2(\rho) : G_F \rightarrow \mathrm{GL}_3(\overline{\mathbf{Q}}_p)$ is conjugate self-dual and no longer sees the "evenness" of ρ.[34] Hence one can hope to prove it is potentially modular for some CM extension L/L^{+}, and then by cyclic base change [5] potentially modular for the totally real field L^{+}. But Galois representations coming from regular algebraic automorphic forms for totally real fields will not be even,[35] and thus one obtains a contradiction. These ideas are already enough to deduce the main result of [34] directly from [11], although in contrast [35] uses (indirectly) the full strength of the p-adic local Langlands correspondence via theorems of Kisin [120]. The papers [34,35] still fall short of completely resolving the Fontaine–Mazur in this case even for $p > 7$, since there remain big image hypotheses on $\overline{\rho}$. On the other hand, this trick has nothing to say about the case when the Hodge–Tate weights are equal (see Section 12).

9.8. Modularity of higher symmetric powers

Another parallel development in higher dimensions was the extension of Skinner–Wiles (Section 3) to higher dimensions. Many of the arguments of Skinner–Wiles relied heavily on the fact that any proper submodule of a 2-dimensional representation must have dimension 1, and one-dimensional representations are very well understood by class field theory. Nonetheless, in [174], Thorne proved a residually reducible modularity theorem for higher dimensional representations. In order to overcome the difficulty of controlling reducible deformations, he imposed a Steinberg condition at some auxiliary prime. Although this is a definite restriction, it does apply (for example) to the Galois representation coming from the symmetric power of a modular form which also satisfies this condition. In a sequence of papers [51–53], Clozel and Thorne applied this modularity lifting theorem to prove new cases of symmetric power functoriality (see also the paper of Dieulefait [74]). A key difficulty here is again the absence of Ihara's lemma in order to find automorphic forms with the correct local properties. Very recently (using a number of new ideas), Newton and Thorne [134,135] were able to (spectacularly!) complete this program and prove the full modularity of all symmetric powers of all modular forms.

---

34 The representation ρ itself restricted to G_F will not be odd in the required sense—one exploits the fact here that 3 is odd whereas symplectic representations are always even dimensional.

35 I managed to twist Taylor's arm into writing the paper [169] which proved this for odd n, which sufficed for my purposes where n was either 3 or 9. This is now also known for general n, see Caraiani–Le Hung [44].

10. BEYOND SELF-DUALITY AND SHIMURA VARIETIES

All the results discussed so far—with the exception of those discussed in Section 4—apply only to Galois representations which are both regular and satisfy some form of self-duality. Moreover, they all correspond to automorphic forms which can be detected by the (étale) cohomology of Shimura varieties. Once one goes beyond these representations, many of the established methods begin to break immediately.[36]

An instructive case to consider is the case of 2-dimensional geometric Galois representations of an imaginary quadratic field F with distinct Hodge–Tate weights. The corresponding automorphic forms for $\mathrm{GL}_2(\mathbf{A}_F)$ contribute to the cohomology of locally symmetric spaces X which are arithmetic hyperbolic 3-manifolds.[37] These spaces are certainly not algebraic varieties and their cohomology is hard to access via algebraic methods. One of the first new questions to arise in this context is the relationship between torsion classes and Galois representations. Some speculations about this matter were made by Elstrodt, Grunewald, and Mennicke at least as far back as 1981 **[75]**, but the most influential conjecture was due to Ash **[6]**, who conjectured that eigenclasses in the cohomology of congruence subgroups of $\mathrm{GL}_n(\mathbf{Z})$ over $\overline{\mathbf{F}}_p$ (which need not lift to characteristic zero) should give rise to n-dimensional Galois representations over finite fields. Later, conjectures were made **[7, 8]** in the converse conjecture in the spirit of Serre **[156]** linking Galois representations to classes in cohomology modulo p. Certainly around 2004, however, it was not at all clear what exactly one should expect the landscape to be,[38] and so it was around this time I decided to start thinking about this question[39] in earnest. I became convinced very soon (for aesthetic reasons if not anything else) that if one modified $\mathbf{T}$ to be the ring of endomorphisms acting on integral cohomology (so that it would see not only the relevant automorphic forms but also the torsion classes) then there should still be an isomorphism $R = \mathbf{T}$. Moreover, this equality would not only be a form of reciprocity which moved beyond the conjecture linking motives to automorphic forms, but it suggested that the integral cohomology of arithmetic groups (including the torsion classes) were themselves the fundamental object of interest. Various

---

36 I should warn the reader that this section and the next (even more than the rest of this paper) is filtered through the lens of my own personal research journey—*caveat lector*!

37 Already by 1970, Serre (following ideas of Langlands) was trying to link Mennicke's computation that $\mathrm{GL}_2(\mathbf{Z}[\sqrt{-109}])^{\mathrm{ab}}$ is infinite to the possible existence of elliptic curves over $\mathbf{Q}(\sqrt{-109})$ with good reduction everywhere **[60, JAN 14, 1970]**.

38 I recall conversations with a number of experts at the 2004 Durham conference, where nobody seemed quite sure even what the dimension of the ordinary deformation ring R of a 3-dimensional representation $\rho : G_{\mathbf{Q}} \to \mathrm{GL}_3(\overline{\mathbf{F}}_p)$ should be. Ash, Pollack, and Stevens had computed numerical examples where a regular algebraic ordinary cuspidal form for $\mathrm{GL}_3(\mathbf{A}_{\mathbf{Q}})$ not twist-equivalent to a symmetric square did not appear to admit classical deformations. (I learnt about this example from Stevens at a talk at Banff in December 2003.) This would be easily explained if R had (relative) dimension 0 over $\mathbf{Z}_p$ but be more mysterious otherwise.

39 One great benefit to me at the time of thinking about Galois representations over imaginary quadratic fields was that it did not require me to understand the geometry of Shimura varieties which I have always found too complicated to understand. The irony, of course, is that the results of **[4, 17]** ultimately rely on extremely intricate properties of Shimura varieties.

developments served only to confirm this point of view. In my paper with Mazur [40], we gave some theoretical evidence for why ordinary families of Galois representations of imaginary quadratic fields might on the one hand be positive dimensional and explained completely by torsion classes and yet not contain any classical automorphic points at all. During the process of writing [36], Dunfield (numerically) compared the torsion classes in the cohomology of inner forms of GL_2 and the data was in perfect agreement with a conjectural Jacquet–Langlands correspondence for torsion (later taken up in joint work with Venkatesh [41]). Emerton and I had the idea of working with completed cohomology groups both to construct Galois representations and even possibly to approach questions of modularity. The first idea was to exploit the well-known relationship between the cohomology of these manifolds and the cohomology of the boundary of certain Shimura varieties. We realized that if we could control the codimension of the completed cohomology groups over the noncommutative Iwasawa algebra, the Hecke eigenclasses would be forced to be seen by eigenclasses coming from the middle degree of these Shimura varieties where one had access to Galois representations.[40] On the automorphy lifting side, we had even vaguer ideas [37, §1.8][41] on how to proceed. A different (and similarly unsuccessful) approach[42] was to work with ordinary deformations over a partial weight space for a split prime $p = vw$ in an imaginary quadratic field F. That is, deformations of $\overline{\rho}$ which had an unramified quotient at v and w but with varying weight at v and fixed weight at w. Here the yoga of Galois deformations suggested that R should be finite flat over $W(k)$ in this case (and even a complete intersection). Moreover, one had access to $\mathbf{T}$ using an overlooked[43] result of Hida [103], and in particular one could deduce that $\mathbf{T}$ has dimension at least one. If one could show that $\mathbf{T}$ was flat over $W(k)$, then one could plausibly apply (assuming the existence of Galois representations) the original argument of [171, 181]. The flatness of $\mathbf{T}$, however, remains an open problem.[44]

10.1. The Taylor–Wiles method when $l_0 > 0$, part I: Calegari–Geraghty

Shortly before (and then during) the special year on Galois representations at the IAS in 2010–2011, I started to work with Geraghty in earnest on the problem of proving $R = \mathbf{T}$ in the case of imaginary quadratic fields, assuming the existence of a surjection $R \to \mathbf{T}$. A computation in Galois cohomology shows that the expected "virtual" dimension of R over $W(k)$ should be -1, and hence the patched module M_∞ should have codimension 1

---

40 Unfortunately, these conjectures [37, CONJ. 1.5] remain all open in more or less all cases except for Scholze's results in the case of certain Shimura varieties [155, COR. 4.2.3].

41 Pan's remarkable paper [137] turned some of these pipe dreams into reality.

42 This is taken from my 2006 NSF proposal, and I believe influenced by my conversations with Taylor at Harvard around that time.

43 One should never overlook results of Hida. I only learnt about this paper when Hida pointed it out to me (with a characteristic smile on his face) after my talk in Montreal in 2005. I was pleased at least that the idea that these families were genuinely nonclassical was not anticipated either in [103] or in Section 4 of Taylor's thesis [170].

44 One might even argue that there is no compelling argument to believe it is true—the problem is analogous to the vanishing of the μ-invariant in Iwasawa-theoretic settings.

over the ring of diamond operators S_∞. We realized this was a consequence of the fact that, after localizing the cohomology at a non-Eisenstein maximal ideal, the cohomology should be nonzero in exactly two degrees. More precisely, patching the presentations of these S_N-modules would result in a balanced presentation of M_∞ as an S_∞-module with the same (finite) number of generators as relations. We then realized that the same principle held more generally for n-dimensional representations over any number field. In characteristic zero, the localized cohomology groups were nonzero exactly in a range $[q_0, q_0 + l_0]$ (with q_0 and l_0 as defined in **[26]**) where $-l_0$ coincided with the expected virtual dimension of R over $W(k)$ coming from Galois cohomology. We could thus show—assuming the localized torsion cohomology also vanished in this range—that by patching *complexes* P_Q (rather than modules M_Q), one arrives at a complex P_∞ of free S_∞ modules in degrees $[q_0, q_0 + l_0]$. Because the ring R_∞ of dimension $\dim R_\infty = (\dim S_\infty) - l_0$ acts by patching on $H^*(P_\infty)$, a simple commutative algebra lemma then shows that $M_\infty = H^*(P_\infty)$ has codimension l_0 over S_∞ and must be concentrated in the final degree. In particular, the Taylor–Wiles method (as modified by Diamond) could be happily adapted to this general setting.[45] Moreover, the arguments were compatible with all the other improvements, including Taylor's Ihara avoidance argument amongst other things.[46] We also realized that the same idea applied to Galois representations coming from the coherent cohomology of Shimura varieties even when the corresponding automorphic forms were not discrete series. While our general formulation involved a number of conjectures we considered hopeless, the coherent case had at least one setting in which many more results were available, namely the case of modular forms of weight one, where the required vanishing conjecture was obvious, and where we were able to establish the existence of the required map $R \to \mathbf{T}$ with all the required local properties by direct arguments. Although the state of knowledge concerning Galois representations increased tremendously between the original conception of **[38]** and its final publication, by early 2016 it still seemed out of reach to make any of the results in **[38]** unconditional.

10.2. Construction of Galois representations, part II

Before one can hope to prove $R = \mathbf{T}$ theorems, one needs to be able to associate Galois representations to the corresponding automorphic forms. There are two contexts in which one might hope to make progress. The first is in situations where the automorphic forms contribute to the Betti cohomology of some locally symmetric space—for example, tempered algebraic cuspidal automorphic representations for $\mathrm{GL}_n(\mathbf{A}_F)$ and any F. The second is in situations where the automorphic forms contribute to the coherent cohomology

45 David Hansen came up with a number of these ideas independently **[94]**.

46 These methods only prove $R[1/p]^{\mathrm{red}} = \mathbf{T}[1/p]^{\mathrm{red}}$, of course. In situations where $\mathbf{T} \otimes \mathbf{Q} = 0$, the methods of **[38]** in the minimal case prove not only that $R = \mathbf{T}$ but also that (both) rings are complete intersections. Moreover, one also has access to level raising (on the level of complexes) and Ihara's lemma **[41, §4]**, and I tried for some time (unsuccessfully) to adapt the original minimal $\Rightarrow$ nonminimal arguments of **[181]** to this setting. There certainly seems to be some rich ideas in commutative algebra in these situations to explore, see, for example, recent work of Tilouine–Urban **[175]**.

of some Shimura variety. Here the first and easiest case corresponds to weight one modular forms, where the Galois representations were first constructed by Deligne and Serre **[69]**.

In work of Harris–Soudry–Taylor **[98, 165]**, Galois representations were constructed for regular algebraic forms for $\mathrm{GL}_2(\mathbf{A}_F)$ for an imaginary quadratic field F and satisfying a further restriction on the central character. Harris, Soudry, and Taylor exploited (more or less) the fact that the automorphic induction of such forms are self-dual (although not regular) and still contribute to coherent cohomology, so one can construct Galois representations using a congruence argument as in the paper of Deligne and Serre **[69]**. On the other hand, this does not prove the expected local properties of the Galois representation at $v|p$.

It was well-known for many years that the Hecke eigenclasses associated to regular algebraic cuspidal automorphic forms for $\mathrm{GL}_n(\mathbf{A}_F)$ for a CM field F could be realized as eigenclasses coming from the boundary of certain unitary Shimura varieties of type $U(n, n)$. It was, however, also well known that the corresponding étale cohomology classes did not realize the desired Galois representations.[47] Remarkably, this problem was completely and unexpectedly resolved in 2011 in **[96]** by Harris–Lan–Taylor–Thorne. Richard Taylor writes:

> *For* **[96]** *I knew that the Hecke eigenvalues we were interested in contributed to Betti cohomology of $U(n, n)$. The problem was to show that they contributed to overconvergent p-adic cusp forms. I was convinced on the basis of Coleman's paper "classical and overconvergent modular forms"* **[54]** *that this must be so. I can't now reconstruct exactly why Coleman's paper convinced me of this, and it is possible, even probable, that my reasoning didn't really make any sense. However, it was definitely this that kept me working at the problem, when we weren't really getting anywhere.*

Amazingly, this breakthrough immediately inspired the next development:

10.3. Construction of Galois representations, part III: Scholze

In **[155]**, Scholze succeeded in constructing Galois representations associated to torsion classes in the setting of $\mathrm{GL}_n(\mathbf{A}_F)$ for a CM field F. Scholze had the idea after seeing some lectures on **[96]**:

> *During a HIM trimester at Bonn, Harris and Lan gave some talks about their construction of Galois representations. At the time, I had some ideas in my head that I didn't have any use for: That Shimura varieties became perfectoid at infinite level, and that there is a Hodge–Tate period map defined on them. The only consequence I could draw from this were certain vanishing results for completed cohomology as conjectured in your work with Emerton; so at least I knew that the*

47 For a more basic example of what can go wrong, note that the Hecke eigenvalues of T_l on $H^0(X, \mathbf{Q}_p)$ of a modular curve are $1 + l$, which corresponds to the Galois representation $\mathbf{Q}_p \oplus \mathbf{Q}_p(1)$. However, only the piece $\mathbf{Q}_p$ occurs inside H^0.

methods were able to say something nontrivial about torsion classes in the cohomology. After hearing Harris' and Lan's talks, I was trying to see whether these ideas could help in extending their results to torsion classes. After a little bit of trying, I found the fake-Hasse invariants, and then it was clear how the argument would go.

Even after this breakthrough, Scholze's construction still fell short of the conjectures in [38] in two ways. The first was that the Galois representation (ignoring here issues of pseudorepresentations) was valued not in $\mathbf{T}$ but in $\mathbf{T}/I$ for some ideal I of fixed nilpotence. This is not a crucial obstruction to the methods of [38]. The second issue, however, was that the Galois representations were constructed (in the end) via p-adic congruences, and thus one did not have control over their local properties at p which *are* crucial for modularity applications.

10.4. The Taylor–Wiles method when $l_0 > 0$, part II: DAG

Although not directly related to new $R = \mathbf{T}$ theorems, one new recent idea in the subject has been the work of Galatius–Venkatesh [86] on derived deformation rings in the context of Venkatesh's conjectures over $\mathbf{Z}$. This work (in part) reinterprets the arguments of [38] in terms of a derived Hecke action. The authors define a derived version $\mathcal{R}$ of R with $\pi_0(\mathcal{R}) = R$. Under similar hypotheses to [38], the higher homotopy groups of $\mathcal{R}$ are shown to exist precisely in degrees 0 to l_0. One viewpoint of the minimal case of [38] is that one constructs a (highly noncanonical) formally smooth ring R_∞ of dimension $n - l_0$ with an action of a formally smooth ring S_∞ of dimension n such that the minimal deformation ring R is $R_\infty \otimes_{S_\infty} S_\infty/\mathfrak{a}$ for the augmentation ideal $\mathfrak{a}$. Moreover, the ring R is identified both with the action of $\mathbf{T}$ on the entire cohomology and simultaneously on the cohomology in degree $q_0 + l_0$. On the other hand, when $l_0 > 1$, the intersection of R_∞ and $S_\infty/\mathfrak{a}$ over S_∞ is never transverse,[48] and homotopy groups of the derived intersection recover the cohomology in all degrees (under the running assumption, one also knows that the patched cohomology is free). On the other hand, there is a more canonical way to define $\mathcal{R}$, namely to take the unrestricted global deformation ring R^{glob} (which has no derived structure) and intersect it with a suitable local crystalline deformation ring as algebras over the unrestricted local deformation ring. The expected dimension of this intersection is also $-l_0$ over $W(k)$, although this is not so clear from this construction. Hence [86] can be viewed as giving an intrinsic definition of $\mathcal{R}$ independent of any choices of Taylor–Wiles primes and showing that its homotopy groups are related (as with $R_\infty \otimes^{\mathbf{L}}_{S_\infty} S_\infty/\mathfrak{a}$) to the cohomology.[49] These

48 When $l_0 = 1$, the intersection *can* be transverse when R is a finite ring. In this setting, the relevant cohomology is also nonzero and finite in exactly one degree. On the other hand, as soon as $\mathrm{Hom}(R, \overline{\mathbf{Q}}_p)$ is nonzero (for example, when there exists an associated motive) and $l_0 > 0$, the intersection will always be nontransverse.

49 There are some subtleties as to what the precise statement should be in the presence of global congruences, but already this author gets confused at the best of times between homology and cohomology, so I will not try to unentangle these issues here.

ideas have hinted at a closer connection between the Langlands program in the arithmetic case and the function field case than was previously anticipated,[50] see, for example, the work of Zhu [182].

11. RECENT PROGRESS

11.1. Avoiding conjectures involving torsion I: the 10-author paper

As mentioned in Section 10.3, even after the results of [155] there remained a significant gap to make the results of [38] unconditional, namely, the conjecture that these Galois representations had the right local properties at p and a second conjecture predicting the vanishing of (integral) cohomology localized at a nonmaximal ideal $\mathfrak{m}$ outside a certain precise range (corresponding to known results in characteristic zero). It should be noted that the second conjecture was still open (in all but the easiest cases) in the simpler setting of Shimura varieties. The first hints that one could possibly make progress on this second conjecture (at least for Shimura varieties) was given in an informal talk by Scholze in Bellairs[51] in 2014. This very quickly led to a long term collaboration between Scholze and Caraiani [45, 46], which Caraiani describes as follows:

> *At the Barbados conference in May 2014, Peter gave a lecture on how one might compute the cohomology of compact unitary Shimura varieties with torsion coefficients. The key was to have some control for $R\pi_{HT*}\mathbb{F}_\ell$ restricted to any given Newton stratum. He was expressing this in terms of a conjecture that had grown out of his work on local Langlands using the Langlands–Kottwitz method. After his talk, I went to ask him some questions about this conjecture and it sounded like there were some things that still needed to be made precise. He asked if I wanted to help him make his strategy work. After some hesitation (because I didn't think I knew enough or was strong enough to work with him), I accepted. Later that evening, I suggested switching from the Langlands–Kottwitz approach to understanding $R\pi_{HT*}\mathbb{F}_\ell$ to an approach more in the style of Harris–Taylor. This relies on the beautiful Mantovan product formula that describes Newton strata in terms of Rapoport–Zink spaces and Igusa varieties. Maybe something like this could help illuminate the geometry of the Hodge–Tate period morphism? Peter*

50 Not anticipated by many people, at least; Michael Harris has been proselytizing the existence of a connection for quite some time.

51 I was invited to give the lecture series in Bellairs after Matthew Emerton did not respond to his emails. Through some combination of the appeal of my own work and the fact that the lectures were given on a beach in Barbados, I managed to persuade Patrick Allen, George Boxer, Ana Caraiani, Toby Gee, Vincent Pilloni, Peter Scholze, and Jack Thorne to come, all of whom are now my coauthors, and all of whom (if they were not already at the time) are now more of an expert in this subject than I am. The thought that I managed to teach any of them something about the subject is pleasing indeed.

immediately saw that this should work and we made plans for me to visit Bonn that summer to continue the collaboration.

As Peter and I were finishing writing up the compact case, it became clear to us that the vanishing theorem would give a way to construct Galois representations associated to generic mod p classes that preserves the desired information at p. Peter started thinking about the non-compact case and how that might apply to the local-global compatibility needed for Calegari–Geraghty. I remember discussing this with him at the Clay Research conference in Oxford in September 2015. By spring 2016, Richard started floating the idea of a working group on Calegari–Geraghty and found out that Peter and I had an approach to local-global compatibility. Around June 2016, Richard suggested to me to organize the working group with him. Peter was very excited about the idea, but wasn't sure he would be able to attend for family reasons. In the end, we found a date in late October 2016 that worked for everyone.

The working group met under the auspices of the first "emerging topics" workshop[52] at the IAS to determine the extent to which the expected consequences could be applied to modularity lifting: A clear stumbling point was the vanishing of integral cohomology after localization outside the range of degrees $[q_0, q_0 + l_0]$. On the other hand, Khare and Thorne had already observed in **[112]** by a localization argument that this could sometimes be avoided in certain minimal cases. It was this argument we were able to modify for the general case, thus avoiding the need to prove the (still open) vanishing conjectures for torsion classes.[53] The result of the workshop was a success beyond what we could have reasonably anticipated—we ended up with more or less[54] the outline of a plan to prove all the main modularity lifting theorems which finally appeared in **[4]**, namely the Ramanujan conjecture for regular algebraic automorphic forms for $\mathrm{GL}_2(\mathbf{A}_F)$ of weight zero for any CM field F, and potential modularity (and the Sato–Tate conjecture) for elliptic curves over CM fields.

There have already been a number of advancements beyond **[4]** including in particular by Allen, Khare, and Thorne **[3]** proving the modularity of many elliptic curves over CM

---

52 Although later described as a "secret" workshop, it was an "invitation-only working group."

53 I regard my main contribution to **[4]** as explaining how the arguments in **[38]** using Taylor's Ihara avoidance (Section 9.3) were incompatible with any characteristic zero localization argument in the absence of (unknown) integral vanishing results in cohomology. The objection (even in the case $l_0 = 0$) was that it was easy to construct complexes P^1 and P^2 of free S_∞ modules so that the support of $H^*(P^1/p)$ and $H^*(P^2/p)$ coincided (as they must) but that (for example) $H^*(P^1)[1/p]$ was zero even though $H^*(P^2)[1/p]$ was not. The objection to this objection, however, which was resolved during the workshop (and which to be clear I played no part in resolving!) is to not merely to compare the support of the complexes P^i/p but to consider the entire complex in the derived category. In particular, even (say) for a finite $\mathbf{Z}_p$-module M, the module $M[1/p]$ is nonzero exactly when $M \otimes^{\mathbf{L}} \mathbf{F}_p$ has nonzero Euler characteristic.

54 It is worth emphasizing that an incredible amount of work was required to turn these ideas into reality, and that this intellectual effort was by and large carried out by the younger members of the collaboration.

fields and a potential automorphy theorem for ordinary representations by Qian [144]. It does not seem completely implausible that results of the strength of [11] for n-dimensional regular Galois representations of $G_{\mathbf{Q}}$ are within reach.

11.2. Avoiding conjectures involving torsion II: abelian surfaces

A second example that Geraghty and I had considered during the 2010–2011 IAS special year was the case of abelian surfaces, corresponding to low (irregular) weight Siegel modular forms of genus $g = 2$. It was clear that a key difficulty was proving the vanishing of $H^2(X, \omega^2)_{\mathfrak{m}}$ where X was a (compactified) Siegel 3-fold with good reduction at p, where $\mathfrak{m}$ is maximal ideal of the Hecke algebra corresponding to an absolutely irreducible representation, and where $\omega|_Y = \det \pi_* \Omega^1_{\mathcal{A}/Y}$ on the open moduli space $Y \subset X$ admitting a corresponding universal abelian surface $\mathcal{A}/Y$. In other irregular weights (corresponding to motives with Hodge–Tate weights $[0, 0, k-1, k-1]$ for $k \geq 4$) the vanishing of the corresponding cohomology groups was known by Lan and Suh [126]. The vanishing of $H^2(X, \omega^2)_{\mathfrak{m}}$ was more subtle, however, because the corresponding group does *not* vanish in general before localization in contrast to the previous cases. In [39], we proved a minimal modularity theorem for these higher weight representations and a minimal modularity theorem in the abelian case contingent on the vanishing conjecture above which we did not manage to resolve (and which remains unresolved). I finished and then submitted the paper after I had moved to Chicago and Geraghty had moved to Facebook in 2015. I then started working with Boxer and Gee[55] on this vanishing question under certain supplementary local hypotheses. (By this point, Galois representations associated to torsion classes in coherent cohomology had been constructed by Boxer [16] and Goldring–Koskivirta [90].) But then in November of 2016 (one week after the IAS workshop!), Pilloni's paper on higher Hida theory [142] was first posted. It was apparent to us that Pilloni's ideas would be extremely useful, and the four of us began a collaboration almost immediately. Just as in [4], we were ultimately able to avoid proving any vanishing conjectures. However, unlike [4], the way around this problem was not purely by commutative algebra, but instead by working with ideas from [142]. Namely, instead of working with the cohomology of the full Siegel modular variety X, one could work with the coherent cohomology of a certain open variety of X with cohomological dimension one whose (infinite dimensional) cohomology could still be tamed using the methods of higher Hida theory [142] in a way analogous to how Hida theory controls the (infinite dimensional) cohomology of the affine variety (with cohomological dimension zero) corresponding to the ordinary locus. Generalizing this to a totally real field, one could then *combine* these ideas with the Taylor–Wiles method as modified in [38] to

55 George Boxer had also arrived at Chicago in 2015, and was collaborating with Gee on companion form results for Siegel modular forms, with the hope (in part) of deducing the modularity of abelian surfaces from Serre's conjecture for GSp_4 in a manner analogous to the deduction by of the Artin conjecture from Serre's conjecture for GL_2 in [110, 114]. They usually worked together at Plein Air cafe. Since I had thought about similar questions with Geraghty and frequently went to Plein Air for 6 oz cappuccinos, it was not entirely surprising for us to start working together.

prove the potential modularity of abelian surfaces over totally real fields [17]. This coincidentally gives a second proof of the potential modularity of elliptic curves over CM fields proven in [4]. (The papers [4] and [17] both were conceived of and completed within a week or so of each other.)

12. THE DEPTHS OF OUR IGNORANCE

Despite what can reasonably be considered significant progress in proving many cases of modularity since 1993, it remains the case that many problems appear just as hopeless as they did then.[56] Perhaps most embarrassing is the case of even Galois representations $G_{\mathbf{Q}} \to \mathrm{GL}_2(\mathbf{C})$ with nonsolvable image (equivalently, projective image A_5). For example, we cannot establish the Artin conjecture for a *single* Galois representation whose image is the binary icosahedral group $\mathrm{SL}_2(\mathbf{F}_5)$ of order 120. The key problem is that the automorphic forms (Maass forms with eigenvalue $\lambda = 1/4$ in this case) are very hard to access—given an even (projective) A_5 Galois representation, we do not even know how to prove that there exists a corresponding Maass form with the right Laplacian eigenvalue, let alone one whose Hecke eigenvalues correspond to the Galois representation.[57] In many ways, we have made no real progress on this question. The case of curves of genus $g > 2$ whose Jacobians have no extra endomorphisms seems equally hopeless. One can only take solace in the fact that the Shimura–Taniyama conjecture seemed equally out of reach before Wiles' announcement in Cambridge in 1993.

ACKNOWLEDGMENTS

The arithmetic Langlands program has been a bustling field of activity in the past 30 years with many deep results by a number of extraordinary mathematicians. To be part of this field (mostly as an observer but occasionally as a contributor) has been an amazing experience. I would like to thank my collaborators all of whom have taught me so much, with special thanks to Matthew Emerton and Toby Gee who have in addition been my close friends. Thanks to Toby Gee and Ravi Ramakrishna for a number of useful suggestions and corrections on the first version of this survey. Finally, special thanks to those mathematicians who allowed me to share some of their personal recollections in this paper,

---

56 Or perhaps harder, since there has been almost 30 years without any progress whatsoever.

57 Motives can be divided according to a tetrachotomy. The first form are the Tate (and potentially Tate) motives, whose automorphy was known to Riemann and Hecke. The second form are the motives (conjecturally) associated to automorphic representations which are discrete series at infinity and thus amenable to the Taylor–Wiles method. The third form are the motives (conjecturally) associated to automorphic representations which are at least seen by some flavor of cohomology, either by the Betti cohomology of locally symmetric spaces or the coherent cohomology of Shimura varieties (possibly in degrees greater than zero) which are amenable in principle to the modified Taylor–Wiles method. The fourth form consist of the rest, which (besides a few that can be accessed by cyclic base change) are a complete mystery.

in particular Christophe Breuil, Ana Caraiani, Pierre Colmez, Fred Diamond, Michael Harris, Mark Kisin, Peter Scholze, Richard Taylor, and Andrew Wiles.

FUNDING

The author was supported in part by NSF Grant DMS-2001097

REFERENCES

[1] P. B. Allen, Modularity of nearly ordinary 2-adic residually dihedral Galois representations. *Compos. Math.* **150** (2014), no. 8, 1235–1346.

[2] P. B. Allen, Deformations of polarized automorphic Galois representations and adjoint Selmer groups. *Duke Math. J.* **165** (2016), no. 13, 2407–2460.

[3] P. B. Allen, C. Khare, and J. A. Thorne, Modularity of $\mathrm{GL}_2(\mathbb{F}_p)$-representations over CM fields. 2019, arXiv:1910.12986.

[4] P. Allen, F. Calegari, A. Caraiani, T. Gee, D. Helm, B. Le Hung, J. Newton, P. Scholze, R. Taylor, and J. Thorne, Potential automorphy over CM fields. 2018, arXiv:1812.09999.

[5] J. Arthur and L. Clozel, *Simple algebras, base change, and the advanced theory of the trace formula*. Ann. of Math. Stud. 120, Princeton University Press, Princeton, NJ, 1989.

[6] A. Ash, Galois representations attached to mod p cohomology of $\mathrm{GL}(n, \mathbf{Z})$. *Duke Math. J.* **65** (1992), no. 2, 235–255.

[7] A. Ash, D. Doud, and D. Pollack, Galois representations with conjectural connections to arithmetic cohomology. *Duke Math. J.* **112** (2002), no. 3, 521–579.

[8] A. Ash and W. Sinnott, An analogue of Serre's conjecture for Galois representations and Hecke eigenclasses in the mod p cohomology of $\mathrm{GL}(n, \mathbf{Z})$. *Duke Math. J.* **105** (2000), no. 1, 1–24.

[9] M. Auslander and D. A. Buchsbaum, Homological dimension in local rings. *Trans. Amer. Math. Soc.* **85** (1957), 390–405.

[10] T. Barnet-Lamb, T. Gee, and D. Geraghty, The Sato–Tate conjecture for Hilbert modular forms. *J. Amer. Math. Soc.* **24** (2011), no. 2, 411–469.

[11] T. Barnet-Lamb, T. Gee, D. Geraghty, and R. Taylor, Potential automorphy and change of weight. *Ann. of Math. (2)* **179** (2014), no. 2, 501–609.

[12] T. Barnet-Lamb, D. Geraghty, M. Harris, and R. Taylor, A family of Calabi–Yau varieties and potential automorphy II. *Publ. Res. Inst. Math. Sci.* **47** (2011), no. 1, 29–98.

[13] L. Berger and C. Breuil, Towards a p-adic Langlands programme, unpublished. 2004, http://perso.ens-lyon.fr/laurent.berger/autrestextes/hangzhou.pdf.

[14] L. Berger and C. Breuil, Sur quelques représentations potentiellement cristallines de $\mathrm{GL}_2(\mathbf{Q}_p)$. *Astérisque* **330** (2010), 155–211.

[15] G. Böckle, On the density of modular points in universal deformation spaces. *Amer. J. Math.* **123** (2001), no. 5, 985–1007.

[16] G. Boxer, Torsion in the coherent cohomology of Shimura varieties and Galois representations. 2015, arXiv:1507.05922.

[17] G. Boxer, F. Calegari, T. Gee, and V. Pilloni, Abelian surfaces over totally real fields are potentially modular. *Publ. Math. Inst. Hautes Études Sci.* **134** (2021), 153–501.

[18] C. Breuil, Schémas en groupe et corps des normes. 1998, https://www.imo.universite-paris-saclay.fr/~breuil/PUBLICATIONS/groupesnormes.pdf.

[19] C. Breuil, Groupes p-divisibles, groupes finis et modules filtrés. *Ann. of Math. (2)* **152** (2000), no. 2, 489–549.

[20] C. Breuil, Sur quelques représentations modulaires et p-adiques de $\mathrm{GL}_2(\mathbf{Q}_p)$. I. *Compos. Math.* **138** (2003), no. 2, 165–188.

[21] C. Breuil, Sur quelques représentations modulaires et p-adiques de $\mathrm{GL}_2(\mathbf{Q}_p)$. II. *J. Inst. Math. Jussieu* **2** (2003), no. 1, 23–58.

[22] C. Breuil, Introduction générale. *Astérisque* **319** (2008), 1–12.

[23] C. Breuil, Série spéciale p-adique et cohomologie étale complétée. *Astérisque* **331** (2010), 65–115.

[24] C. Breuil, Correspondance de Langlands p-adique, compatibilité local-global et applications [d'après Colmez, Emerton, Kisin, ...]. *Astérisque* **348, Exp. No. 1031** (2012), viii, 119–147.

[25] C. Breuil, B. Conrad, F. Diamond, and R. Taylor, On the modularity of elliptic curves over **Q**: wild 3-adic exercises. *J. Amer. Math. Soc.* **14** (2001), no. 4, 843–939.

[26] C. Breuil and A. Mézard, Multiplicités modulaires et représentations de $\mathrm{GL}_2(\mathbf{Z}_p)$ et de $\mathrm{Gal}(\overline{\mathbf{Q}}_p/\mathbf{Q}_p)$ en $l = p$. *Duke Math. J.* **115** (2002), no. 2, 205–310.

[27] S. Brochard, Proof of de Smit's conjecture: a freeness criterion. *Compos. Math.* **153** (2017), no. 11, 2310–2317.

[28] W. Bruns and J. Herzog, *Cohen–Macaulay rings*. Cambridge Stud. Adv. Math. 39, Cambridge University Press, Cambridge, 1993.

[29] J. P. Buhler, *Icosahedral Galois representations*. Lecture Notes in Math. 654, Springer, Berlin–New York, 1978.

[30] K. Buzzard, Potential modularity—a survey. In *Non-abelian fundamental groups and Iwasawa theory*, pp. 188–211, London Math. Soc. Lecture Note Ser. 393, Cambridge Univ. Press, Cambridge, 2012.

[31] K. Buzzard, M. Dickinson, N. Shepherd-Barron, and R. Taylor, On icosahedral Artin representations. *Duke Math. J.* **109** (2001), no. 2, 283–318.

[32] K. Buzzard and T. Gee, The conjectural connections between automorphic representations and Galois representations. In *Automorphic forms and Galois representations. Vol. 1*, pp. 135–187, London Math. Soc. Lecture Note Ser. 414, Cambridge Univ. Press, Cambridge, 2014.

[33] K. Buzzard and R. Taylor, Companion forms and weight one forms. *Ann. of Math. (2)* **149** (1999), no. 3, 905–919.

[34] F. Calegari, Even Galois representations and the Fontaine–Mazur conjecture. *Invent. Math.* **185** (2011), no. 1, 1–16.

[35] F. Calegari, Even Galois Representations and the Fontaine–Mazur conjecture II. *J. Amer. Math. Soc.* **25** (2012), no. 2, 533–554.

[36] F. Calegari and N. M. Dunfield, Automorphic forms and rational homology 3-spheres. *Geom. Topol.* **10** (2006), 295–329.

[37] F. Calegari and M. Emerton, Completed cohomology—a survey. In *Non-abelian fundamental groups and Iwasawa theory*, pp. 239–257, London Math. Soc. Lecture Note Ser. 393, Cambridge Univ. Press, Cambridge, 2012.

[38] F. Calegari and D. Geraghty, Modularity lifting beyond the Taylor–Wiles method. *Invent. Math.* **211** (2018), no. 1, 297–433.

[39] F. Calegari and D. Geraghty, Minimal modularity lifting for nonregular symplectic representations. *Duke Math. J.* **169** (2020), no. 5, 801–896.

[40] F. Calegari and B. Mazur, Nearly ordinary Galois deformations over arbitrary number fields. *J. Inst. Math. Jussieu* **8** (2009), no. 1, 99–177.

[41] F. Calegari and A. Venkatesh, A torsion Jacquet–Langlands correspondence. *Astérisque* (2019), no. 409, x+226.

[42] A. Caraiani, Local-global compatibility and the action of monodromy on nearby cycles. *Duke Math. J.* **161** (2012), no. 12, 2311–2413.

[43] A. Caraiani, Monodromy and local-global compatibility for $l = p$. *Algebra Number Theory* **8** (2014), no. 7, 1597–1646.

[44] A. Caraiani and B. V. Le Hung, On the image of complex conjugation in certain Galois representations. *Compos. Math.* **152** (2016), no. 7, 1476–1488.

[45] A. Caraiani and P. Scholze, On the generic part of the cohomology of compact unitary Shimura varieties. *Ann. of Math. (2)* **186** (2017), no. 3, 649–766.

[46] A. Caraiani and P. Scholze, On the generic part of the cohomology of non-compact unitary Shimura varieties. 2019, arXiv:1909.01898.

[47] H. Carayol, Sur les représentations l-adiques associées aux formes modulaires de Hilbert. *Ann. Sci. Éc. Norm. Supér. (4)* **19** (1986), no. 3, 409–468.

[48] L. Clozel, Représentations galoisiennes associées aux représentations automorphes autoduales de $\mathrm{GL}(n)$. *Publ. Math. Inst. Hautes Études Sci.* **73** (1991), 97–145.

[49] L. Clozel, M. Harris, J.-P. Labesse, and B.-C. Ngô (eds.), *On the stabilization of the trace formula, Vol. I*. Stab. Trace Formula Shimura Var. Arith. Appl., International Press, Somerville, MA, 2011.

[50] L. Clozel, M. Harris, and R. Taylor, Automorphy for some l-adic lifts of automorphic mod l Galois representations. *Publ. Math. Inst. Hautes Études Sci.* **108** (2008), 1–181.

[51] L. Clozel and J. A. Thorne, Level-raising and symmetric power functoriality. I. *Compos. Math.* **150** (2014), no. 5, 729–748.

[52] L. Clozel and J. A. Thorne, Level raising and symmetric power functoriality, II. *Ann. of Math. (2)* **181** (2015), no. 1, 303–359.

[53] L. Clozel and J. A. Thorne, Level-raising and symmetric power functoriality, III. *Duke Math. J.* **166** (2017), no. 2, 325–402.

[54] R. F. Coleman, Classical and overconvergent modular forms. *Invent. Math.* **124** (1996), no. 1–3, 215–241.

[55] R. Coleman and B. Mazur, The eigencurve. In *Galois representations in arithmetic algebraic geometry (Durham, 1996)*, pp. 1–113, London Math. Soc. Lecture Note Ser. 254, Cambridge Univ. Press, Cambridge, 1998.

[56] P. Colmez, Une correspondance de langlands locale p-adique pour les représentations semi-stables de dimension 2, unpublished. 2004, https://webusers.imj-prg.fr/~pierre.colmez/sst.pdf.

[57] P. Colmez, Représentations triangulines de dimension 2. *Astérisque* **319** (2008), 213–258.

[58] P. Colmez, La série principale unitaire de $\mathrm{GL}_2(\mathbf{Q}_p)$. *Astérisque* **330** (2010), 213–262.

[59] P. Colmez, Représentations de $\mathrm{GL}_2(\mathbf{Q}_p)$ et (ϕ, Γ)-modules. *Astérisque* **330** (2010), 281–509.

[60] P. Colmez (ed.), *Correspondance Serre-Tate. Vol. I*, Doc. Math. (Paris) [Math. Doc. (Paris)] 13, Société Mathématique de France, Paris, 2015.

[61] P. Colmez (ed.), *Correspondance Serre-Tate. Vol. II*, Doc. Math. (Paris) [Math. Doc. (Paris)] 14, Société Mathématique de France, Paris, 2015.

[62] P. Colmez, Les conjectures de monodromie p-adiques, séminaire Bourbaki. Vol. 2001/2002, 2003.

[63] P. Colmez, G. Dospinescu, and V. Paškūnas, The p-adic local Langlands correspondence for $GL_2(\mathbb{Q}_p)$. *Cambridge J. Math.* **2** (2014), no. 1, 1–47.

[64] B. Conrad, Finite group schemes over bases with low ramification. *Compos. Math.* **119** (1999), no. 3, 239–320.

[65] B. Conrad, Ramified deformation problems. *Duke Math. J.* **97** (1999), no. 3, 439–513.

[66] B. Conrad, F. Diamond, and R. Taylor, Modularity of certain potentially Barsotti–Tate Galois representations. *J. Amer. Math. Soc.* **12** (1999), no. 2, 521–567.

[67] G. Cornell, J. H. Silverman, and G. Stevens (eds.), *Modular forms and Fermat's last theorem*, Springer, New York, 1997.

[68] H. Darmon, F. Diamond, and R. Taylor, Fermat's last theorem. In *Elliptic curves, modular forms & Fermat's last theorem (Hong Kong, 1993)*, pp. 2–140, Int. Press, Cambridge, MA, 1997.

[69] P. Deligne and J.-P. Serre, Formes modulaires de poids 1. *Ann. Sci. Éc. Norm. Supér. (4)* **7** (1974), 507–530.

[70] F. Diamond, On deformation rings and Hecke rings. *Ann. of Math. (2)* **144** (1996), no. 1, 137–166.

[71] F. Diamond, An extension of Wiles' results. In *Modular forms and Fermat's last theorem (Boston, MA, 1995)*, pp. 475–489, Springer, New York, 1997.

[72] F. Diamond, The Taylor–Wiles construction and multiplicity one. *Invent. Math.* **128** (1997), no. 2, 379–391.

[73] M. Dickinson, On the modularity of certain 2-adic Galois representations. *Duke Math. J.* **109** (2001), no. 2, 319–382.

[74] L. Dieulefait, Automorphy of $\mathrm{Symm}^5(\mathrm{GL}(2))$ and base change. *J. Math. Pures Appl. (9)* **104** (2015), no. 4, 619–656.

[75] J. Elstrodt, F. Grunewald, and J. Mennicke, PSL(2) over imaginary quadratic integers. In *Arithmetic Conference (Metz, 1981)*, pp. 43–60, Astérisque 94, Soc. Math. France, Paris, 1982.

[76] M. Emerton, A local-global compatibility conjecture in the p-adic Langlands programme for $\mathrm{GL}_{2/\mathbb{Q}}$. *Pure Appl. Math. Q.* **2** (2006), no. 2. Special Issue: In honor of John H. Coates. Part 2.

[77] M. Emerton, On the interpolation of systems of eigenvalues attached to automorphic Hecke eigenforms. *Invent. Math.* **164** (2006), no. 1, 1–84.

[78] M. Emerton, Local-global compatibility in the p-adic Langlands programme for $\mathrm{GL}_2/\mathbb{Q}$. Preprint, 2010.

[79] N. Fakhruddin, C. Khare, and S. Patrikis, Lifting and automorphy of reducible mod p galois representations over global fields. 2020, arXiv:2008.12593.

[80] J.-M. Fontaine, Groupes finis commutatifs sur les vecteurs de Witt. *C. R. Acad. Sci. Paris Sér. A–B* **280** (1975), Ai, A1423–A1425.

[81] J.-M. Fontaine, Semi-stable Galois representations. 1988, https://webusers.imj-prg.fr/~pierre.colmez/arbeitstagung-1988.pdf, notes from the 1988 Mathematische Arbeitstagung.

[82] J.-M. Fontaine and G. Laffaille, Construction de représentations p-adiques. *Ann. Sci. Éc. Norm. Supér. (4)* **15** (1982), no. 4, 547–608.

[83] J.-M. Fontaine and B. Mazur, Geometric Galois representations. In *Elliptic curves, modular forms, & Fermat's last theorem (Hong Kong, 1993)*, pp. 41–78, Ser. Number Theory I, Int. Press, Cambridge, MA, 1995.

[84] N. Freitas, B. V. Le Hung, and S. Siksek, Elliptic curves over real quadratic fields are modular. *Invent. Math.* **201** (2015), no. 1, 159–206.

[85] K. Fujiwara, Deformation rings and Hecke algebras in the totally real case. 2006, arXiv:math/0602606.

[86] S. Galatius and A. Venkatesh, Derived Galois deformation rings. *Adv. Math.* **327** (2018), 470–623.

[87] T. Gee, Automorphic lifts of prescribed types. *Math. Ann.* **350** (2011), no. 1, 107–144.

[88] A. Genestier and J. Tilouine, Systèmes de Taylor–Wiles pour GSp(4). *Astérisque* **302** (2005), 177–290.

[89] D. Geraghty, Modularity lifting theorems for ordinary Galois representations. *Math. Ann.* **373** (2019), no. 3–4, 1341–1427.

[90] W. Goldring and J.-S. Koskivirta, Strata Hasse invariants, Hecke algebras and Galois representations. *Invent. Math.* **217** (2019), no. 3, 887–984.

[91] E. S. Golod and I. R. Šafarevič, On the class field tower. *Izv. Akad. Nauk SSSR Ser. Mat.* **28** (1964), 261–272.

[92] B. Green, F. Pop, and P. Roquette, On Rumely's local–global principle. *Jahresber. Dtsch. Math.-Ver.* **97** (1995), no. 2, 43–74.

[93] T. Haines and M. Harris (eds.), *Stabilization of the trace formula, Shimura varieties, and arithmetic applications, volume II: Shimura varieties*, London Math. Soc. Lecture Note Ser., Cambridge University Press, 2020.

[94] D. Hansen, Minimal modularity lifting for GL_2 over an arbitrary number field. 2012, arXiv:1209.5309.

[95] M. Harris, Potential automorphy of odd-dimensional symmetric powers of elliptic curves and applications. In *Algebra, arithmetic, and geometry: in honor of Yu. I. Manin. Vol. II*, pp. 1–21, Progr. Math. 270, Birkhäuser Boston, Boston, MA, 2009.

[96] M. Harris, K.-W. Lan, R. Taylor, and J. Thorne, On the rigid cohomology of certain Shimura varieties. *Res. Math. Sci.* **3** (2016), no. 3, 37.

[97] M. Harris, N. Shepherd-Barron, and R. Taylor, A family of Calabi–Yau varieties and potential automorphy. *Ann. of Math. (2)* **171** (2010), no. 2, 779–813.

[98] M. Harris, D. Soudry, and R. Taylor, l-adic representations associated to modular forms over imaginary quadratic fields. I. Lifting to $\mathrm{GSp}_4(\mathbf{Q})$. *Invent. Math.* **112** (1993), no. 2, 377–411.

[99] M. Harris and R. Taylor, Deformations of automorphic Galois representations. 1998–2003, http://www.math.columbia.edu/~harris/website/publications.

[100] M. Harris and R. Taylor, *The geometry and cohomology of some simple Shimura varieties*. Ann. of Math. Stud. 151, Princeton University Press, Princeton, NJ, 2001.

[101] H. Hida, Galois representations into $\mathrm{GL}_2(\mathbf{Z}_p[[X]])$ attached to ordinary cusp forms. *Invent. Math.* **85** (1986), no. 3, 545–613.

[102] H. Hida, Nearly ordinary Hecke algebras and Galois representations of several variables. In *Algebraic analysis, geometry, and number theory (Baltimore, MD, 1988)*, pp. 115–134, Johns Hopkins Univ. Press, Baltimore, MD, 1989.

[103] H. Hida, p-adic ordinary Hecke algebras for $\mathrm{GL}(2)$. *Ann. Inst. Fourier (Grenoble)* **44** (1994), no. 5, 1289–1322.

[104] Y. Hu and V. Paškūnas, On crystabelline deformation rings of $\mathrm{Gal}(\overline{\mathbb{Q}}_p/\mathbb{Q}_p)$. *Math. Ann.* **373** (2019), no. 1–2, 421–487.

[105] H. Jacquet and J. A. Shalika, On Euler products and the classification of automorphic forms. II. *Amer. J. Math.* **103** (1981), no. 4, 777–815.

[106] P. L. Kassaei, A gluing lemma and overconvergent modular forms. *Duke Math. J.* **132** (2006), no. 3, 509–529.

[107] P. L. Kassaei, Modularity lifting in parallel weight one. *J. Amer. Math. Soc.* **26** (2013), no. 1, 199–225.

[108] P. L. Kassaei, Analytic continuation of overconvergent Hilbert modular forms. *Astérisque* **382** (2016), 1–48.

[109] P. L. Kassaei, S. Sasaki, and Y. Tian, Modularity lifting results in parallel weight one and applications to the Artin conjecture: the tamely ramified case. *Forum Math. Sigma* **2** (2014), e18, 58 pp.

[110] C. Khare, Remarks on mod p forms of weight one. *Int. Math. Res. Not.* **3** (1997), 127–133.

[111] C. Khare, Serre's modularity conjecture: the level one case. *Duke Math. J.* **134** (2006), no. 3, 557–589.

[112] C. B. Khare and J. A. Thorne, Potential automorphy and the Leopoldt conjecture. *Amer. J. Math.* **139** (2017), no. 5, 1205–1273.

[113] C. Khare and J.-P. Wintenberger, On Serre's conjecture for 2-dimensional mod p representations of $\mathrm{Gal}(\overline{\mathbb{Q}}/\mathbb{Q})$. *Ann. of Math. (2)* **169** (2009), no. 1, 229–253.

[114] C. Khare and J.-P. Wintenberger, Serre's modularity conjecture. I. *Invent. Math.* **178** (2009), no. 3, 485–504.

[115] C. Khare and J.-P. Wintenberger, Serre's modularity conjecture. II. *Invent. Math.* **178** (2009), no. 3, 505–586.

[116] M. Kisin, Overconvergent modular forms and the Fontaine–Mazur conjecture. *Invent. Math.* **153** (2003), no. 2, 373–454.

[117] M. Kisin, Modularity of 2-dimensional Galois representations. In *Current developments in mathematics, 2005*, pp. 191–230, Int. Press, Somerville, MA, 2007.

[118] M. Kisin, Modularity for some geometric Galois representations. In *L-functions and Galois representations*, pp. 438–470, London Math. Soc. Lecture Note Ser. 320, Cambridge Univ. Press, Cambridge, 2007.

[119] M. Kisin, Potentially semi-stable deformation rings. *J. Amer. Math. Soc.* **21** (2008), no. 2, 513–546.

[120] M. Kisin, The Fontaine–Mazur conjecture for GL_2. *J. Amer. Math. Soc.* **22** (2009), no. 3, 641–690.

[121] M. Kisin, Modularity of 2-adic Barsotti–Tate representations. *Invent. Math.* **178** (2009), no. 3, 587–634.

[122] M. Kisin, Moduli of finite flat group schemes, and modularity. *Ann. of Math. (2)* **170** (2009), no. 3, 1085–1180.

[123] M. Kisin, The structure of potentially semi-stable deformation rings. In *Proceedings of the International Congress of Mathematicians. Volume II*, pp. 294–311, Hindustan Book Agency, New Delhi, 2010.

[124] S. L. Kleiman, The standard conjectures. In *Motives (Seattle, WA, 1991)*, pp. 3–20, Proc. Sympos. Pure Math. 55, Amer. Math. Soc., Providence, RI, 1994.

[125] R. E. Kottwitz, On the λ-adic representations associated to some simple Shimura varieties. *Invent. Math.* **108** (1992), no. 3, 653–665.

[126] K.-W. Lan and J. Suh, Vanishing theorems for torsion automorphic sheaves on general PEL-type Shimura varieties. *Adv. Math.* **242** (2013), 228–286.

[127] R. P. Langlands, *Base change for* GL(2). Ann. of Math. Stud. 96, Princeton University Press, Princeton, NJ; University of Tokyo Press, Tokyo, 1980.

[128] G. Laumon and B. C. Ngô, Le lemme fondamental pour les groupes unitaires. *Ann. of Math. (2)* **168** (2008), no. 2, 477–573.

[129] B. Mazur, Deforming Galois representations. In *Galois groups over* **Q** *(Berkeley, CA, 1987)*, pp. 385–437, Math. Sci. Res. Inst. Publ. 16, Springer, New York, 1989.

[130] B. Mazur and A. Wiles, On p-adic analytic families of Galois representations. *Compos. Math.* **59** (1986), no. 2, 231–264.

[131] J. S. Milne, *Arithmetic duality theorems*. Perspekt. Math. 1, Academic Press, Inc., Boston, MA, 1986.

[132] L. Moret-Bailly, Groupes de Picard et problèmes de Skolem II. *Ann. Sci. Éc. Norm. Supér.* **22** (1989), 181–194.

[133] J. Newton and J. A. Thorne, Adjoint selmer groups of automorphic Galois representations of unitary type. 2019, arXiv:1912.11265.

[134] J. Newton and J. A. Thorne, Symmetric power functoriality for holomorphic modular forms. *Publ. Math. Inst. Hautes Études Sci.* **134** (2021), 1–116.

[135] J. Newton and J. A. Thorne, Symmetric power functoriality for holomorphic modular forms, II. *Publ. Math. Inst. Hautes Études Sci.* **134** (2021), 117-152.

[136] B. C. Ngô, Le lemme fondamental pour les algèbres de Lie. *Publ. Math. Inst. Hautes Études Sci.* **111** (2010), 1–169.

[137] L. Pan, The Fontaine–Mazur conjecture in the residually reducible case. 2019, arXiv:1901.07166.

[138] L. Pan, On locally analytic vectors of the completed cohomology of modular curves. 2020, arXiv:2008.07099.

[139] V. Paškūnas, The image of Colmez's Montreal functor. *Publ. Math. Inst. Hautes Études Sci.* **118** (2013), 1–191.

[140] S. Patrikis and R. Taylor, Automorphy and irreducibility of some l-adic representations. *Compos. Math.* **151** (2015), no. 2, 207–229.

[141] V. Pilloni, Formes modulaires p-adiques de Hilbert de poids 1. *Invent. Math.* **208** (2017), no. 2, 633–676.

[142] V. Pilloni, Higher coherent cohomology and p-adic modular forms of singular weights. *Duke Math. J.* **169** (2020), no. 9, 1647–1807.

[143] V. Pilloni and B. Stroh, Surconvergence, ramification et modularité. *Astérisque* **382** (2016), 195–266.

[144] L. Qian, Potential automorphy for GL_n. 2021, arXiv:2104.09761.

[145] R. Ramakrishna, On a variation of Mazur's deformation functor. *Compos. Math.* **87** (1993), no. 3, 269–286.

[146] R. Ramakrishna, Lifting Galois representations. *Invent. Math.* **138** (1999), no. 3, 537–562.

[147] R. Ramakrishna, Deforming Galois representations and the conjectures of Serre and Fontaine–Mazur. *Ann. of Math. (2)* **156** (2002), no. 1, 115–154.

[148] M. Raynaud, *Théorèmes de Lefschetz en cohomologie cohérente et en cohomologie étale*. Suppl. Bull. Soc. Math. France 103, Société Mathématique de France, Paris, 1975.

[149] K. A. Ribet, On modular representations of Gal($\overline{\mathbf{Q}}/\mathbf{Q}$) arising from modular forms. *Invent. Math.* **100** (1990), no. 2, 431–476.

[150] K. A. Ribet, Galois representations and modular forms. *Bull. Amer. Math. Soc. (N.S.)* **32** (1995), no. 4, 375–402.

[151] T. Saito, Modular forms and p-adic Hodge theory. *Invent. Math.* **129** (1997), no. 3, 607–620.

[152] S. Sasaki, Integral models of Hilbert modular varieties in the ramified case, deformations of modular Galois representations, and weight one forms. *Invent. Math.* **215** (2019), no. 1, 171–264.

[153] D. Savitt, Modularity of some potentially Barsotti–Tate Galois representations. *Compos. Math.* **140** (2004), no. 1, 31–63.

[154] D. Savitt, On a conjecture of Conrad, Diamond, and Taylor. *Duke Math. J.* **128** (2005), no. 1, 141–197.

[155] P. Scholze, On torsion in the cohomology of locally symmetric varieties. *Ann. of Math. (2)* **182** (2015), no. 3, 945–1066.

[156] J.-P. Serre, Sur les représentations modulaires de degré 2 de Gal($\overline{\mathbf{Q}}/\mathbf{Q}$). *Duke Math. J.* **54** (1987), no. 1, 179–230.

[157] N. Shepherd-Barron and R. Taylor, On degree 2 Galois representations over $\mathbf{F}_4$. *Proc. Natl. Acad. Sci. USA* **94** (1997), no. 21, 11147–11148.

[158] S. W. Shin, Galois representations arising from some compact Shimura varieties. *Ann. of Math. (2)* **173** (2011), no. 3, 1645–1741.

[159] C. M. Skinner and A. J. Wiles, Ordinary representations and modular forms. *Proc. Natl. Acad. Sci. USA* **94** (1997), no. 20, 10520–10527.

[160] C. M. Skinner and A. J. Wiles, Residually reducible representations and modular forms. *Publ. Math. Inst. Hautes Études Sci.* **89** (1999), 5–126.

[161] C. M. Skinner and A. J. Wiles, Base change and a problem of Serre. *Duke Math. J.* **107** (2001), no. 1, 15–25.

[162] C. M. Skinner and A. J. Wiles, Nearly ordinary deformations of irreducible residual representations. *Ann. Fac. Sci. Toulouse Math. (6)* **10** (2001), no. 1, 185–215.

[163] A. Snowden, Singularities of ordinary deformation rings. *Math. Z.* **288** (2018), no. 3–4, 759–781.

[164] J. Tate, The non-existence of certain Galois extensions of $\mathbf{Q}$ unramified outside 2. In *Arithmetic geometry (Tempe, AZ, 1993)*, pp. 153–156, Contemp. Math. 174, Amer. Math. Soc., Providence, RI, 1994.

[165] R. Taylor, l-adic representations associated to modular forms over imaginary quadratic fields. II. *Invent. Math.* **116** (1994), no. 1–3, 619–643.

[166] R. Taylor, Remarks on a conjecture of Fontaine and Mazur. *J. Inst. Math. Jussieu* **1** (2002), no. 1, 125–143.

[167] R. Taylor, Galois representations. *Ann. Fac. Sci. Toulouse Math. (6)* **13** (2004), no. 1, 73–119.

[168] R. Taylor, Automorphy for some l-adic lifts of automorphic mod l Galois representations. II. *Publ. Math. Inst. Hautes Études Sci.* **108** (2008), 183–239.

[169] R. Taylor, The image of complex conjugation in l-adic representations associated to automorphic forms. *Algebra Number Theory* **6** (2012), no. 3, 405–435.

[170] R. L. Taylor, *On congruences between modular forms*. ProQuest LLC, Ann Arbor, MI, 1988.

[171] R. Taylor and A. Wiles, Ring-theoretic properties of certain Hecke algebras. *Ann. of Math. (2)* **141** (1995), no. 3, 553–572.

[172] R. Taylor and T. Yoshida, Compatibility of local and global Langlands correspondences. *J. Amer. Math. Soc.* **20** (2007), no. 2, 467–493.

[173] J. A. Thorne, On the automorphy of l-adic Galois representations with small residual image. *J. Inst. Math. Jussieu* **11** (2012), no. 4, 855–920.

[174] J. A. Thorne, Automorphy lifting for residually reducible l-adic Galois representations. *J. Amer. Math. Soc.* **28** (2015), no. 3, 785–870.

[175] J. Tilouine and E. Urban, Integral period relations and congruences. 2018, arXiv:1811.11166.

[176] T. Tsuji, p-adic étale cohomology and crystalline cohomology in the semi-stable reduction case. *Invent. Math.* **137** (1999), no. 2, 233–411.

[177] S.-N. Tung, On the modularity of 2-adic potentially semi-stable deformation rings. *Math. Z.* **298** (2021), no. 1–2, 107–159.

[178] J. Tunnell, Artin's conjecture for representations of octahedral type. *Bull. Amer. Math. Soc. (N.S.)* **5** (1981), no. 2, 173–175.

[179] G. Wiese, On Galois representations of weight one. *Doc. Math.* **19** (2014), 689–707.

[180] A. Wiles, On ordinary λ-adic representations associated to modular forms. *Invent. Math.* **94** (1988), no. 3, 529–573.

[181] A. Wiles, Modular elliptic curves and Fermat's last theorem. *Ann. of Math. (2)* **141** (1995), no. 3, 443–551.

[182] X. Zhu, Coherent sheaves on the stack of Langlands parameters. 2021, arXiv:2008.02998.

FRANK CALEGARI

The University of Chicago, 5734 S University Ave, Chicago, IL 60637, USA, fcale@uchicago.edu

A SURVEY OF GRAVITATIONAL WAVES

FRANS PRETORIUS

ABSTRACT

We review the state of the field of gravitational wave astrophysics, framing the challenges, current observations, and future prospects within the context of the predictions of Einstein's theory of general relativity.

MATHEMATICS SUBJECT CLASSIFICATION 2020

Primary 83C35; Secondary 83B05, 83C57, 83C10, 83F05

KEYWORDS

General relativity, gravitational waves, compact object mergers

ICM INTERNATIONAL CONGRESS OF MATHEMATICIANS 2022 JULY 6–14

Proc. Int. Cong. Math. 2022, Vol. 2, pp. 652–676
DOI 10.4171/ICM2022/168

Published by EMS Press

1. INTRODUCTION

This article is meant to serve as an overview of the current state of the field of gravitational wave astrophysics. It is not meant to be comprehensive, or a reference for experts, but rather an introduction to this nascent field of observational science, targeted toward mathematicians and scientists. The three primary goals are (a) to give a sufficient introduction to the physics of general relativity to appreciate the challenges of gravitational wave detection, as well as the remarkable nature of sources of gravitational waves in the dynamical, strong field regime of the theory, (b) to review what has been learnt about the Universe from the gravitational wave signals detected to date by the LIGO (Laser Interferometer Gravitational-Wave Observatory)/Virgo detectors, and (c) to briefly speculate about future discoveries that will unfold over the coming decades as a variety of observational campaigns are undertaken. To set the stage then, in Section 2 we review the underlying theoretical framework, Einstein's theory of general relativity, focusing on the nature of gravitational waves and how they are produced. In Section 3 we briefly survey the current detectors and observational campaigns, either in operation today or planned for the coming decade or two: ground-based detectors (as LIGO/Virgo), the space-based mission LISA (Laser Interferometer Space Antenna), pulsar timing arrays, and the search for B-mode polarization of the cosmic microwave background (CMB).

LIGO measured the first gravitational wave signal, GW150914, in 2015, which is interpreted as originating from the merger of two black holes [1]. Since then, LIGO/Virgo has observed almost 100 additional signals, most also from binary black hole mergers, though a small handful likely coming from black hole/neutron star or binary neutron star mergers. However, the loudest event to date, GW170817, was a binary neutron star merger, as confirmed by a spectacular suite of electromagnetic observations of its aftermath. In Section 4 we review these observations, and what they have so far taught us about the Universe. Highlights are the first quantitative evidence that black holes as described by Einstein's theory do in fact exist, that the speed of gravitational waves is equal to that of the speed of light to within ~ 1 part in 10^{15}, and that neutron star mergers are responsible for at least a class of the mysterious so-called short gamma ray bursts (observed at a rate of about one every 3 days by special purpose satellites designed for this).

We conclude in Section 5 with speculations on the coming two decades of gravitational wave astronomy.

2. EINSTEIN GRAVITY

The working hypothesis upon which the science of gravitational wave astrophysics is built is that "gravity" is described by Einstein's classical theory of general relativity. This begins by positing that space and time taken together, or spacetime for short, has the structure of a 4-dimensional Lorentzian geometry. A convenient way to describe this geometry is via the metric tensor g_{ab}, defined in a coordinate basis through the line element

$$ds^2 = g_{ab} dx^a dx^b, \tag{1}$$

which gives the local, infinitesimal proper distance-squared ds^2 as a quadratic form of an arbitrary infinitesimal coordinate displacement dx^a (we use the Einstein summation convention where repeated indices in a tensor expression imply summation). The phrase *proper distance* means the coordinate invariant, physically measurable length or time interval, in contrast to a coordinate distance in some (arbitrary) coordinate system. The Lorentzian (indefinite) character of the metric is crucial, as it allows one to define causality through geometry: two different events are causally related if and only if there exists at least one curve connecting them where the proper distance along the curve is everywhere timelike, $ds^2 < 0$, and/or null, $ds^2 = 0$ (the sign convention for timelike $ds^2 < 0$ versus spacelike $ds^2 > 0$ is arbitrary).

The second key postulate of general relativity is that the geometry of spacetime relevant to the Universe is not a fixed structure given *a priori*, but instead is a dynamical entity governed by the Einstein field equations:

$$G_{ab} \equiv R_{ab} - \frac{1}{2} R g_{ab} = \frac{8\pi G}{c^4} T_{ab}, \tag{2}$$

where the Einstein tensor G_{ab} is defined as above in terms of the Ricci tensor R_{ab} and Ricci scalar R, T_{ab} is the stress–energy–momentum tensor of the matter content of the Universe, G is Newton's constant, and c is the speed of light. General relativity ignores torsion, which is thought would only be needed to describe matter with intrinsic spin, and is expected to be irrelevant for macroscopic distributions of matter in the classical limit. Thus all tensors appearing in (2) are symmetric. In terms of practically solving this equation, one views the Einstein tensor as a second order, quasilinear partial differential operator acting on the metric tensor g_{ab}. In 4 spacetime dimensions, this gives a set of 10 coupled equations for the independent components of g_{ab}, and must be solved simultaneously with the additional equations governing the matter fields in T_{ab}. It is obvious from (2) then that matter (T_{ab}) will influence the dynamics and curvature of spacetime. Less obvious is even in the absence of matter ($T_{ab} = 0$) nontrivial, dynamical solutions exist: most interesting among these are those describing black holes and gravitational waves.

It is often stated that a third key postulate of general relativity is the *geodesic hypothesis*: a test body not subject to any force follows a geodesic of the spacetime (a test body is one with insufficient energy to cause any noticeable perturbation on the surrounding geometry). However, perhaps more fundamentally, geodesic motion in the test body limit can be viewed as coming from energy/momentum conservation, which is already built into the Einstein equations and does not need to be imposed as a separate hypotheses. This follows from the contracted Bianchi identities, showing that the Einstein tensor necessarily has vanishing divergence $\nabla_a G^{ab} = 0$. Thus, any matter that can self-consistently be coupled to spacetime through the Einstein equations (2) must have a divergenceless stress tensor $\nabla_a T^{ab} = 0$, the latter equation being the covariant statement of the conservation of energy/momentum of the matter. Likewise, pure spacetime energy, whether in the form of gravitational waves, or confined to black holes, will exhibit similar dynamics in an equivalent test body limit. For example, in vacuum an infinitesimal mass black hole will orbit a large (finite mass) black hole following a geodesic of the latter's spacetime by virtue of the vacuum Einstein equations alone, and not any additional hypothesis one needs to supply.

If the nature of spacetime is as described by general relativity, the most immediate consequences of this are well described by Newtonian's theory of gravity in the weak field limit (for example, our environment here on Earth and in the solar system). This is why Einstein's theory is also called a theory of gravity despite there being no gravitational force in general relativity.

2.1. Gravitational waves

It is not possible to precisely define what a gravitational wave is in all scenarios. For our purposes, it suffices to think of gravitational waves as small, local disturbances in spacetime that propagate at the speed of light. In an asymptotically flat space time (the metric at large distances from any source of curvature approaches that of special relativity—Minkowski spacetime) the properties of gravitational waves can be defined more precisely. Our Universe is *not* asymptotically flat, though with appropriate accommodation for the overall cosmic expansion with time, to good approximation we can consider ourselves to be in an asymptotically flat region relative to any source we expect to observe.

Regarding sources of gravitational waves, there are two broad classes. First is what one traditionally thinks of as a source: at some place a localized event occurs that produces gravitational waves over a period of time, and these waves then stream outward away from the source. Second are "primordial" gravitational waves, namely an overall background of gravitational waves filling all of space, having been produced in an earlier epoch of the evolution of the Universe. In some cases the distinction between these classes is blurred; for example, a sufficiently high density of localized sources emitting over a long period of time will eventually also fill the observable Universe with a background of gravitational waves. In these settings then, we next review some of the basic properties of gravitational waves, and how they are produced.

2.1.1. Basic properties of gravitational waves in the weak field limit

Consider a metric perturbation h_{ab} about a background Minkowski spacetime η_{ab}, i.e., $g_{ab} = \eta_{ab} + h_{ab}$ with $\eta_{ab}dx^a dx^b = -c^2dt^2 + dx^2 + dy^2 + dz^2$ in Cartesian coordinates. Then the linearized Einstein equations show that general relativity allows two linearly independent gravitational wave solutions for h_{ab}, or so-called *polarizations,*[1] propagating in any given direction. Even restricting the background metric to be in Cartesian form, there is still much coordinate (or "gauge") freedom to choose the representation of the solution. A gauge commonly used is the so-called *transverse traceless* gauge, and in these coordinates a wave propagating in the $+z$ direction (for example) takes the form (e.g., [18, 29])

$$h_{ab}dx^a dx^b = h_+(t - z/c)\left[dx^2 - dy^2\right] + h_\times(t - z/c)[2dxdy]. \qquad (3)$$

---

1 In principle, a general metric theory of gravity can allow up to 6 linearly independent polarizations; see, e.g., [53].

Here h_+ and $h_\times$ are arbitrary (but small amplitude) functions of their arguments, and describe the so-called *plus* and *cross* polarized waves, respectively. From (3) one can see that gravitational waves in general relativity are transverse, namely they only perturb the background metric along a plane (the (x, y) plane in this example) orthogonal to the direction of propagation (z here). Equation (3) also shows that as a plus polarized wave passes a given point, when $h_+ > 0$, it will stretch proper distances in x by $\sqrt{1 + h_+}$ while simultaneously squeezing distances in y by $\sqrt{1 - h_+}$, and the opposite when $h_+ < 0$. The effect of the cross-polarized wave on the transverse geometry is qualitatively the same, except the directions of stretching/squeezing are rotated by 45° about the z axis relative to that of the plus polarized wave.

The energy flux density carried by these waves is

$$\frac{dE}{dAdt} = \frac{c^3}{16\pi G}\left\langle \left(\frac{dh_+}{dt}\right)^2 + \left(\frac{dh_\times}{dt}\right)^2 \right\rangle, \tag{4}$$

where dA is the transverse area element, and the angle brackets denote a time average over a characteristic period of the wave (the reason for the averaging is that gravitational wave energy cannot be localized—see, e.g., [41]). A truly infinite plane wave such as (3) will have infinite total energy, which is not consistent with an asymptotically flat space time when backreaction is taken into account. However, sufficiently far from a local source (as discussed below) the outgoing spherical wavefronts are locally well approximated by these plane wave solutions. Similarly, an on-average homogeneous, primordial stochastic background that fills all of spacetime cannot be asymptotically flat when backreaction is considered,[2] but still the above (generalizing to superpositions of plane waves traveling in all directions) can give a good description of the geometry in any local patch of the spacetime.

Notice the way G and c appear in (4), and hence the dimension-full constant relating energy flux on the left-hand side to the time derivative of metric strain on the right-hand side: in SI units $c^3/G \sim 10^{36}\mathrm{J}\cdot\mathrm{s}/\mathrm{m}^2$. This implies, at least from the perspective of our everyday intuition of energy and length scales, that it requires an enormous amount of energy to perturb spacetime by a comparatively miniscule amount. This is the reason why it is completely impractical to study gravitational waves by building transmitters/receivers on Earth in analogy with electromagnetic waves. Instead, we must look to cataclysmic gravitational wave "explosions" in the cosmos, such as those produced by black hole mergers, and even then, despite the astonishing sensitivity of the LIGO/Virgo detectors, we are now just barely able to observe them.

Regarding localized sources of gravitational waves, good insight can again be obtained from linearized theory, resulting in the so-called quadrupole formula. Here, one assumes a weak field, slowly varying distribution of energy density $\rho(t, x, y, z)$. This will emit gravitational waves propagating outward that at a large distance r from the source takes

---

2 Instead, then one obtains the Friedmann–Robertson–Lemaître–Walker (FRLW) asymptotics that observations indicate describe our Universe on very large scales.

the following form in terms of the spatial components of the metric perturbation h_{ij}:

$$h_{ij}(t,r) = \frac{1}{r}\frac{2G}{c^4}\frac{d^2 \mathcal{I}_{kl}(t-r/c)}{dt^2}\left[P_i{}^k P_j{}^l - \frac{1}{2}P^{kl}P_{ij}\right], \tag{5}$$

with all indices here only running over the spatial coordinates $x_i \in (x, y, z)$ (in transverse traceless gauge there are no time–time or space–time propagating components of h_{ab}), $\mathcal{I}_{ij}$ is the reduced quadrupole moment tensor of the source, and the projection tensor $P_{ij} \equiv \delta_{ij} - n_i n_j$, where n^i is a unit spatial vector pointing from the origin $r = 0$ to the observer location $r = \sqrt{x^2 + y^2 + z^2}$; $\mathcal{I}_{ij}$ is defined in terms of the quadrupole moment tensor I_{ij} as

$$\mathcal{I}_{ij} \equiv I_{ij} - \frac{1}{3}I^k{}_k\delta_{ij}, \quad I_{ij}(t) \equiv \int x_i x_j \rho(t, x, y, z) dV, \tag{6}$$

where the integration is over all of space at some instant of time t, but note that in deriving (5) the source is assumed to be localized in space around $r = 0$, and the observer location $r \gg 0$ is assumed to be in vacuum.

Several properties of gravitational wave emission are evident from (5). First, unsurprisingly, the outgoing wave propagates at the speed of light, and its amplitude decays with distance like $1/r$ from the source. Second, similar to that implied by the energy expression in (4), the factor of $G/c^4 \sim 10^{-44}\mathrm{s}^2/\mathrm{kg/m}$ illustrates what extreme dynamics, in the form of rapid accelerations of large energy densities, need to be present in the source to produce nonnegligible metric perturbations. Third, it is only the acceleration of *asymmetric* concentrations of energy that produce gravitational waves in general relativity; for example, a spherically symmetric pulsating star cannot produce any gravitational waves.

2.1.2. Weak field emission from a compact object binary

Though it is not obvious from the discussion above, it turns out that the quadrupole formula (5) gives a good approximation to the gravitational wave emission even for certain strong field sources, and even if the energy density is purely gravitational, such as with black hole binaries. Another property of binary systems in general relativity that we will simply mention without giving further details is that backreaction from the loss of energy to gravitational wave emission not only causes the semimajor axis of the binary to decrease (as anticipated by Newtonian energy balance), but it also reduces the eccentricity of the binary with time. LIGO/Virgo is only sensitive to the very last stages of binary inspiral, and the majority of observable systems are thus expected to have close to zero eccentricity. In all then, to get a good understanding of the structure of gravitational waves emitted by such a so-called quasicircular inspiral, we can evaluate the quadrupole formula (5) for two point masses m_1 and m_2 orbiting each other on a circle separated by a distance D, with orbital frequency ω, which for large separations is well approximated by the Keplerian result $\omega = \sqrt{GM/D^3}$, with $M = m_1 + m_2$. For a binary orbiting in the $z = 0$ plane about $r = 0$, using spherical polar coordinates to label the observer location $(x, y, z) = (r\cos\phi\sin\theta, r\sin\phi\sin\theta, r\cos\theta)$, and expressing the answer in terms of the two polarization amplitudes in the plane orthogonal to the propagation vector n^i, gives

$$h_+(t,r,\theta,\phi) = \frac{1}{r}\frac{4G}{c^4}\mu D^2\omega^2 \cos\big(2\omega(t-r/c)-2\phi\big)\left[\frac{1+\cos^2\theta}{2}\right], \tag{7}$$

$$h_\times(t,r,\theta,\phi) = \frac{1}{r}\frac{4G}{c^4}\mu D^2\omega^2 \sin\big(2\omega(t-r/c)-2\phi\big)\cos\theta, \tag{8}$$

where $\mu = m_1 m_2/(m_1+m_2)$ is the reduced mass of the binary, and an arbitrary initial phase was set to zero. The corresponding orbit-averaged energy fluxes, from (4), are

$$\frac{dE_+}{dAdt} = \frac{2}{\pi r^2}\frac{G}{c^5}\mu^2 D^4\omega^6\left[\frac{1+\cos^2\theta}{2}\right]^2, \tag{9}$$

$$\frac{dE_\times}{dAdt} = \frac{2}{\pi r^2}\frac{G}{c^5}\mu^2 D^4\omega^6\cos^2\theta. \tag{10}$$

Integrating these over the sphere gives the net radiated power in the two modes

$$\frac{dE_+}{dt} = \frac{56}{15}\frac{G}{c^5}\mu^2 D^4\omega^6, \tag{11}$$

$$\frac{dE_\times}{dt} = \frac{8}{3}\frac{G}{c^5}\mu^2 D^4\omega^6. \tag{12}$$

Several interesting properties are apparent from expressions (7)–(12): the observed gravitational wave frequency is twice the orbital frequency, the orbit averaged amplitudes (hence energy fluxes) are not isotropic in latitude, nor is emission equally balanced between the plus and cross polarizations. Also, as expected, the emission vanishes in the test body limit $\mu \to 0$. The total energy flux $dE/dt = 32G\mu^2 D^4\omega^6/5c^5$, or using the Kepler relation for $\omega(D)$, is

$$\frac{dE}{dt} = \frac{32G^4M^3\mu^2}{5c^5D^5} = \frac{32G^{7/3}M^{4/3}\omega^{10/3}\mu^2}{5c^5}. \tag{13}$$

This illustrates how sensitive the luminosity is to orbital separation D or frequency ω.

Note again that equations (7)–(13) do *not* include back reaction; we have simply evaluated the quadrupole formula for two point masses moving in a circular orbit. To obtain the so-called Newtonian quasicircular approximation to estimate the radiation reaction on the orbit, one elevates the frequency (or equivalently separation) to a function of time $\omega(t)$, assumes the Newtonian expression for the energy of the orbit, and uses the latter together with the total luminosity of the binary to derive an equation for the evolution of $\omega(t)$ consistent with total energy conservation. The result is

$$\omega^{-11/3}\frac{d\omega}{dt} = \frac{96}{5}\nu\left(\frac{GM}{c^3}\right)^{5/3}, \tag{14}$$

where $\nu = \mu/M$ is the symmetric mass ratio of the binary. It is essentially a measurement of (14) from the famous Hulse–Taylor binary pulsar that gave the first (indirect) evidence for the existence of gravitational waves, and that the weak-field description of the emission process is consistent with general relativity.

2.1.3. Strong field gravitational wave emission

In contrast to the other fundamental laws of physics, the strongly interacting, or strong field, regime of *classical* general relativity is not associated with any particular scale within the theory. Or said another way, general relativity is a geometric theory, but there is no fundamental constant of dimension length in the field equations that would describe a radius of curvature to demark a scale where a qualitative change in the character of solutions might occur. Despite that, general relativity *does* have a strong field regime, essentially because the field equations are nonlinear. In contrast, Newtonian gravity, a scale-free linear theory, does not have a strong field regime: the Newtonian gravitational force can certainly be "strong," but it is not qualitatively different from a "weak" Newtonian gravitational force—they only differ in magnitude.

In general relativity there is no universal criteria for when nonlinear effects become significant enough to qualitatively change solutions, though for spherical-like compact objects in asymptotically flat spacetime there is a good heuristic understanding: if an amount of energy Mc^2 is confined to a region within a radius (roughly) smaller than its so-called Schwarzschild radius $R_s = 2GM/c^2$, the geometry of spacetime qualitatively changes character compared to a less compact distribution of energy. In particular, spacetime necessarily becomes dynamical, undergoing what is called *gravitational collapse*, and some kind of spacetime singularity forms in the interior. A version of Penrose's cosmic censorship conjecture argues that generically one expects an event horizon to form about the collapsing region of spacetime [43], i.e., from an exterior observer's perspective a black hole forms. If the collapsing region is much more elongated (more cylindrical rather than spherical), Thorne's hoop conjecture argues a naked singularity would form instead [52], though there are comparatively few studies of such asymmetric collapse, nor indications that such scenarios arise in astrophysical settings.

Regarding sources of gravitational waves, again it is not easy to define when we are in the strong versus weak field emission regime, though for binary inspiral we can heuristically characterize the differences. In the weak field the linearized results described in the previous section are quite accurate. Somewhat surprisingly, as mentioned, the weak field description can still be good even if the individual members of the binary by themselves require strong field gravity to describe their local geometries (case in point the Hulse–Taylor binary pulsar, as a neutron star's radius is only a factor of 3 or so larger than its Schwarzschild radius). A strong field description for a compact binary interaction is necessary either when the two objects come close enough that the local geometry of one object is significantly perturbed by the other (i.e., the point mass approximation breaks down), or the metric perturbation h_{ab} of the observed radiation, when "scaled back" by r to the location of the source, becomes of order unity.

Interestingly, the radiative perturbation reaching of order unity coincides with the gravitational wave luminosity approaching the Planck luminosity, $L_p = c^5/G$. Planck units are a set of units based on the dimensionful constants one can obtain from the simplest products of powers of the fundamental constants of nature, in particular G, c, Planck's constant h, and the Boltzmann constant k. It is theorized that "quantum gravity" effects become important when any relevant physical scale in a process becomes of order unity when measured in Planck units. The Planck luminosity does not involve Planck's constant, the hallmark of quantum processes, yet still, exceeding L_p in a local interaction does seem to anticipate evolution to a regime where quantum gravity would be necessary. The reason is based on dimensional analysis, together with the above heuristic for when one expects gravity to be so strong that a black hole would be present, as follows.[3] Consider a causal process confined to a volume of characteristic size $2R$, emitting gravitational waves with total energy E. For the gravitational waves by themselves to not have enough energy to form a black hole requires R to be larger than the effective Schwarzschild radius $2GE/c^4$ of the gravitational wave energy, or $E < c^4 R/2G$. If not confined to a black hole, these gravitational waves will leak out on a light crossing time of the system $T = 2R/c$, implying a luminosity limit of $L = E/T < L_p/4$. Or conversely, a process emitting at super-Planck luminosity is necessarily confined to a black hole, hence censored from exterior observation, and whose interior would require some form of quantum gravity for a complete description.

2.1.4. Strong field emission from a compact object binary

For a quasicircular binary black hole merger, the weak field description breaks down primarily because of finite size effects (the two horizons fuse together), and less so because of high gravitational wave luminosity, which "only" reaches up to around $10^{-3} L_p$ for equal mass mergers ($\mu = M/4$) as computed via full numerical solutions [45].[4] To put this number in context, the sun's luminosity in light is $\sim 10^{-26} L_p$; thus, for the brief moment about the time of merger, a binary black hole radiates as much power in gravitational waves as 10^{23} suns do in light—that is comparable to the current estimated luminosity of *all* stars in the visible Universe combined. That gravitational wave energy liberated from black hole mergers does not dominate the energy content of the Universe is in part because they are so rare, and in part because this incredible luminosity only lasts for a short time. For example, with GW150914, the merger of two black holes each roughly 30 times the mass of the sun $M_\odot$, the luminosity integrated over the entire inspiral and merger came to about $3M_\odot c^2$; the majority of this was emitted within a few tens of milliseconds [1].

The quadrupole formula based calculation (13) does a decent job of anticipating these properties, both the rapid increase in luminosity approaching merger, and the ballpark

3 To our knowledge, arguments like this were first proposed by Dyson in thought experiments on whether a single "graviton," the hypothetical quantum particle of geometry, could be detected [25].

4 A binary neutron star merger has a peak luminosity a couple of orders of magnitude lower than that of a binary black hole merger. Finite size effects are more pronounced for neutron stars at late stages of the inspiral due to their higher tidal deformability, and, of course, when they finally collide the point mass approximation used in (7)–(13) completely breaks down. If the neutron star does not promptly collapse to a black hole, the gravitational wave emission of the remnant can still qualitatively be understood using weak field/quadrupole-formula type analysis, though the complicated dynamics of the matter in the remnant would not be easy to compute without a numerical solution.

maximum, if for the latter we take some liberty in interpreting when the inspiral should terminate. Rewriting the distance D between the two point masses in (13) as a fraction D_s of the Schwarzschild radius $2GM/c^2$ of the combined mass M of the system, i.e., $D = D_s \cdot 2GM/c^2$, for the equal mass $\mu = M/4$ case gives

$$\frac{dE}{dt} = \frac{L_p}{80 D_s^5}. \tag{15}$$

Clearly, the maximum inspiral luminosity depends quite sensitively on D_s. For an upper limit, one would not expect this to be remotely accurately if $D_s < 1$, as then the two horizons of the individual black holes would already be overlapping. With $D_s \sim 1, dE/dt \sim 10^{-2} L_p$. For a lower limit estimate, one can appeal to a result from circular geodesics, where the inner most stable orbit is at $R = 3R_s$, and then a small loss of angular momentum will cause the geodesic to plunge into the black hole. Setting $D_s \sim 3$ for the maximum in (15) thus amounts to assuming that for comparable mass mergers a similar instability sets in that accelerates the merger beyond what radiation reaction does by itself; this gives $dE/dt \sim 10^{-4} L_p$.

Of course, for these back-of-the-envelope estimates to have any relevance to the maximum merger luminosity requires that the actual collision of two black holes is not much more violent than the last stages of inspiral. In fact, before numerical solutions become available, it was unknown whether black hole collisions would generically even adhere to cosmic censorship, let alone how bright they ultimately were. If a merger does satisfy cosmic censorship, the no-bifurcation theorem of Hawking would apply, telling us two black holes must fuse into a larger one [35]; then also, by Hawking's area theorem [34], one can place limits on the maximum amount of energy that could be liberated in this most nonlinear phase of the merger. If a naked singularity is produced, classical general relativity will cease to predict the spacetime to the causal future of this event, and we would have no idea what the remnant of such a black hole collision is. Fortunately for our ability to predict waveforms to interpret LIGO events, but unfortunately for our ability to use black hole mergers to give an observational glimpse into the mystery of quantum gravity, there is no example yet from a merger simulation that shows any violation of cosmic censorship, or anomalously large curvatures forming exterior to the existing horizons.[5]

Though likely not relevant to the kind of black hole mergers that occur in the Universe, there is a regime of the two body problem where it *is* large gravitational energy that pushes the interaction to the nonlinear regime, and not any finite size effects: the ultrarelativistic scattering problem. Here, one imagines shooting two black holes toward each other at very high velocities, so that in the center of mass frame of the interaction the kinetic energy of either black hole is much greater than its rest mass energy: $(\gamma_i - 1)m_i c^2 \gg m_i c^2$, with

5 In spacetime dimensions above four, there are examples of (apparent) naked singularity formation from fragmentation of unstable horizons [40], and hints that certain collisions may also lead to naked singularities [42]. Though the kind of microscopic extra dimensions that could exist while still evading experimental detection will not cause instabilities in astrophysically sized black holes, and then the effective four dimensional simulations used to study the latter should be quite accurate.

$\gamma_i = 1/\sqrt{1 - v_i^2/c^2}$. Though few detailed results are available for the case with generic impact parameter b, it is expected that when b is of order a few times or less than that of the Schwarzschild radius $R_s = 2GE/c^4 = 2G(\gamma_1 m_1 + \gamma_2 m_2)/c^2$ of the system (and note that this scale is much larger than the Schwarzschild radii of either black hole when $\gamma_i \gg 1$), a sizable fraction of the kinetic energy can be converted to gravitational wave energy on a time scale R_s/c. Moreover, for $b \lesssim R_s$, an encompassing black hole forms, trapping most of the kinetic/gravitational wave energy. Again, exactly how much is not known for generic b, though for $b = 0$ numerical simulations show $\sim 14\%$ of E is liberated as gravitational wave energy, with the remainder trapped [51]. It has been conjectured that the highest luminosity will be reached at the critical impact parameter b_{crit} marking the threshold of formation of a central black hole (for larger impact parameters the two black holes will fly apart again) [47]. Then, essentially all of the kinetic energy ($\approx E$) is expected to be converted to gravitational wave energy, though due to how strongly this seems to be focused inward when produced, only about half of this energy may likely escape as gravitational waves [33,50]. The other half will then be trapped in the central black hole for $b < b_{crit}$, or the two individual black holes for $b \gtrsim b_{crit}$ (whose local Schwarzschild radii would consequently grow by an enormous amount).

A fascinating conjectured aspect of the ultrarelativistic scattering problem is that it actually does not matter what the source of the kinetic energy is, be it black holes, or some compact distribution of matter, such as a neutron star, or even a fundamental particle. It is this conjecture behind the arguments that the Large Hadron Collider(LHC) [24,32], or cosmic ray collisions with the earth's atmosphere [27], could produce black holes in certain extra dimension scenarios which give a much lower Planck luminosity than our (then erroneous) 4-dimensional analysis predicts. To date, numerical evidence in favor of this "matter does not matter" conjecture has only been obtained for a few select matter models in the head-on collision limit [23,26,46].

2.1.5. The ringdown

Due to the uniqueness, or "no hair" theorems of general relativity [21,35,37,48], the two-parameter (mass and angular momentum) Kerr family of metrics are the only vacuum, stationary, asymptotically flat black hole solutions without any exterior (naked) singularities allowed by general relativity in four spacetime dimensions. Taken by itself, this would suggest that either black holes are sets of measure zero and not relevant to realistic gravitational collapse (the Kerr solutions being axisymmetric and stationary), or Kerr black holes are in a sense dynamical attractors where once an asymmetric, dynamical horizon forms, evolution causes the exterior spacetime to "loose its hair" and settle down to a Kerr solution. The latter is a special case of Penrose's *final state conjecture* [44]: the generic endstate of evolution governed by general relativity, beginning with naked-singularity free vacuum initial data on a Cauchy slice of an asymptotically flat spacetime, is a set black holes flying apart, the local geometry of each approaching that of a given member of the Kerr family, together with gravitational waves streaming outward to null infinity. Indeed, this is what seems to generically happen in gravitational collapse studies and merger simulations to date. In particular, for both quasicircular inspirals and ultrarelativistic scattering with $b < b_{crit}$, once a single common horizon forms, the spacetime rapidly settles down to a Kerr black hole. This is accompanied by the emission of gravitational waves, whose characteristics are largely determined by the quasinormal mode oscillation spectrum of the remnant black hole. In analogy with a bell emitting decaying sound waves after it is hit, this is called the *ringdown* of the black hole. The least damped mode of a Kerr black hole is the $\ell = m = 2$ spherical harmonic mode. The damping rate decreases with the spin of the black hole, approaching zero for the maximally spinning (extremal) black holes allowed in general relativity. However, the spins of remnants produced in comparable mass mergers, as observed by LIGO/Virgo, are sufficiently far from extremal that their ringdown phases are very short, damping exponentially with a characteristic e-fold time on order-of-magnitude the light-crossing time R_s/c of the remnant.

3. GRAVITATIONAL WAVE OBSERVATIONAL LANDSCAPE

In this section we outline what the current and planned near future observational campaigns to witness the Universe in gravitational waves are. Gravitational wave "observatories" fall into two categories: those that people have built specifically for this purpose, and those that the Universe has fortuitously provided us. The former include earlier resonant bar detectors pioneered by Joseph Weber, the LIGO/Virgo and Kagra ground-based detectors, and various planned future ground- and space-based detectors. The latter include a network of millisecond pulsars in our galaxy, and the cosmic microwave background (CMB). We will not cover the history of any of these endeavors, instead we will comment on properties/challenges common to any of them that can be appreciated with knowledge of the properties of gravitational waves outlined in the previous section.

Given that general relativity is a theory about the geometric nature of space and time, and that gravitational waves are propagating distortions in the geometry, it should not be surprising that essentially all gravitational wave detectors are composed of elements that are sensitive to changing distances or times. Moreover, the most sensitive measurements are those adapted to the plus and cross-polarized transverse disturbances allowed in general relativity. This informs the "L" shape of the current ground-based detectors, that measure relative changes in distances along the two arms of the detector through laser interferometry. Pulsar timing relies on the remarkably stable rotational periods of certain pulsars, where models can be built to predict the arrival time of radio pulses from them to within tens of nanoseconds over a year of observation. Long wavelength gravitational waves between the earth and the pulsar will change the arrival times, and the most subtle signals can be extracted from correlations between changes in arrival times between pairs of pulsars. Regarding the CMB, this is an image of the "surface of last scattering," where photons were last able to Thompson-scatter off free electrons (afterward the temperature of the Universe dropped below a threshold allowing the electrons to recombined with free protons to form neutral hydrogen). The photons can pick up a net polarization after Thompson scattering if the background radiation field is anisotropic. The ability to use polarization measurements of the CMB to detect gravitational waves present then is due to the fact that of the known sources of anisotropy in the early Universe, only gravitational waves are able to produce anisotropy that creates a so-called "B-mode" polarization pattern over the CMB (as opposed to an "E-mode" pattern, that both matter anisotropies and gravitational waves can create).

Most sources produce gravitational waves at some characteristic length or frequency scale. Gravitational wave detectors tend to be most sensitive to a frequency/length associated with some scale of the detector. Therefore, since the different detectors operate at very different scales, they are sensitive to a correspondingly broad spectrum of potential sources. The ground-based detectors are km-scale instruments, and are most sensitive to physical processes associated with km-scale sources: stellar mass black holes, neutron stars, and the inner core of a star undergoing a supernova explosion. The space-based LISA instrument is planned to be a triangular configuration of satellites with 2.5 million km length arms; this is the scale of the smaller of the so-called supermassive black holes thought to exist in the centers of most galaxies, as well as the orbits of many close binaries containing white dwarfs, neutron stars, and black holes. Pulsar timing is most sensitive to gravitational waves with periods close to the years-to-decades long observation time of the pulsars. This translates to physical scales on the order of a few light years, and one of the most promising sources on this scale is an effective stochastic background from the population of supermassive black hole binaries in their last stages of inspiral. Gravitational waves from the early Universe would likely leave a most pronounced effect on the CMB on scales of order the Hubble radius at the surface of last scattering, which is roughly $1/1000$ that of the present day Hubble radius $R_{H_0} \sim 10^{26}$ m.

A common problem for all detectors is how weak the gravitational waves are expected to be when they reach the detectors. This is true even for the strongest known source—a binary black hole merger—when factoring in how far away the event is expected to occur from the earth. For stellar mass black hole binaries, the observed merger rate is ~ 10 per cubic gigaparsec (Gpc) per year [12]. In fact, the first event ever detected, GW150914, is still one of the closest black hole mergers seen to date, at an estimated distance of 0.4 Gpc $\sim 10^{25}$ m. Since gravitational waves decay like 1/distance from the source, what were metric perturbations of magnitude $h \sim 1/10$ on the $\sim 10^5$ m scale of GW150914's last orbit, caused a metric perturbation $h \sim 10^{-21}$ as it passed the earth, resulting in a maximum change in distance along LIGO's 4 km long arms of $\sim 10^{-17}$ m, or about $1/100$th the diameter of a proton![6] It is not surprising then that one of the most significant challenges facing all

6 Though the 1/distance decay seems like a curse, and it is for being able to detect rare events like black hole mergers relatively frequently on a human timescale, once the tremendous experimental effort needed to cross that threshold has been met, the 1/distance decay also means it does not take that much more effort to open up a significantly larger volume of spacetime to observation. For example, the next (third) generation of ground-based detectors are planned to be about 10 times more sensitive than Advanced LIGO's design sensitivity. Being able to see 10 times further is enough that GW150914-like black hole mergers could be seen throughout the visible Universe!

detectors is a thorough understanding and mitigation of sources of noise that could otherwise swamp or masquerade as gravitational waves. This is one of the primary reasons why LIGO consists of *two* detectors with nearly the same orientation relative to the sky, but separated by a few thousand kilometers: a true gravitational wave must produce signals with similar characteristics in both detectors, separated in time by at most the few ms of light travel time between them; conversely, the probability that noise could mimic such a correlated signal is much less than noise being able to mimic a gravitational wave in a single detector alone.

A second issue with most gravitational wave detectors is how to interpret an observed signal once it is confirmed to be of likely astrophysical origin. Except for the CMB, the difficulty here is that the signal is a one-dimensional time series, and so these detectors are more akin to seismometers than telescopes (with the CMB a two-dimensional polarization map over the sky can be obtained). Without theoretical *templates* of waveforms from expected sources to compare against, there is very little other than broad temporal/spectral characteristics that could be inferred from a novel, or unmodeled, source. Thus a crucial part of the gravitational wave astronomy endeavor is to have banks of template waveforms from expected sources. For compact object mergers, the issue of source interpretation is also closely tied to detection: current instruments are still not sensitive enough for the vast majority of mergers to be clearly evident above the detector noise, and matched filtering is essential to extract such weak signals from the noise.[7] This is why solving the two-body problem in general relativity became such a focused effort within the theoretical general relativity community beginning in the early 1990s. Due to the complexities of the Einstein field equations, no analytical solution seems possible, and currently a full solution (for a given set of orbital parameters) needs to be computed numerically, which introduces some numerical truncation error. Moreover, since numerical solutions are currently too computationally expensive to use to produce template banks that densely sample parameter space, template banks of practical use are constructed using various approximation methods; these include the effective one body (EOB) approach **[19]**, modern versions of which use select numerical results to calibrate the stitching together of perturbative post-Newtonian inspiral calculations with linear quasinormal mode ringdown calculations, and reduced basis models constructed from a set of numerical waveforms **[28]** (see **[16]** for a review of these and other approaches). In the future, as more sensitive detectors come online, templates will not be needed as much for detection, though will still be crucial for source identification and parameter estimation, which would be hampered if systematic modeling errors are present in the template libraries. Thus, even though the first numerical solution to a general relativistic two body problem describing inspiral, merger, and ringdown was obtained almost two decades ago **[45]**, it is still an active area of research to calculate ever more accurate binary merger waveforms.

---

7 Matched filtering refers to convolving the detector signal with a template waveform. If a nearly periodic signal with many cycles is present, such as the inspiral phase of a merger, and an accurate template is phase aligned with the signal, then with time the convolution will increase the signal-to-noise ratio, as the signal will add coherently while typical noise will not.

4. SURVEY OF WHAT HAS BEEN OBSERVED TO DATE

In this section we give an overview of the three most important (in our opinion) scientific advances to date coming from gravitational wave observation of the Universe: testing dynamical strong field gravity, multimessenger observation of neutron star mergers, and obtaining the first glimpses of the demographics of black holes in our Universe. Amongst the observatories mentioned in the previous section, only LIGO/Virgo have made actual detections, and we will only comment on these.[8]

First quantifiable evidence for the existence of black holes as governed by the theory of general relativity. Though the evidence for the existence of black holes has steadily grown since the first candidates where identified beginning in the 1960s—the first stellar mass black hole candidate Cygnus X-1, the suggested connection between quasars and supermassive black holes, our own Milky Way supermassive black hole Sagittarius A*—before GW150914 the evidence was all circumstantial. In other words, the only scientifically sound statement one could have made is that the Universe definitely harbors a few ultracompact objects and has some unusual sources of electromagnetic emission, and none of these observations can readily be explained using conventional physics if Kerr black holes are not involved.

The gravitational wave data from black hole mergers is fundamentally different in this regard, as it is coming from the strong field dynamics of spacetime itself, and there is already enough signal in some of the loudest events, such as GW150914, that quantifiable self-consistency tests can be performed. Most notable in this regard is the consistency between the inspiral and ringdown portions of the waveforms. From the inspiral signal alone, an estimate of the progenitor black holes in the binary can be made, and from this, together with predicted dynamics of the merger using numerical solutions of the field equations, the mass and spin of the remnant can be computed. From the observed decay and frequency of the ringdown signal alone, and using black hole perturbation theory calculations, the mass and spin of the remnant black hole can also be determined. These two independent measures of properties of the final black hole must agree if the signal comes from two Kerr black holes colliding and forming a remnant Kerr black hole, as described by general relativity. So far all the LIGO/Virgo data is consistent in this regard [2, 13, 14], albeit the error bars are quite large, as the signal-to-noise ratios (SNRs) of current events are still quite small for making precise tests of this kind. As an illustrative example to put this data and its veracity in context compared to that obtained using the Event Horizon Telescope images of M87, or the Nobel prize winning data of stellar orbits around Sagittarius A* used to measure its gravitational mass and confirm its ultracompact nature: we still cannot rule out that M87 or Sagittarius

---

8 Of course, that is not to say that the absence of a signal does not provide useful information, e.g., the negative results from the CMB and pulsar timing place constraints on the magnitude of stochastic backgrounds, and the absence of long-lived periodic signals in LIGO/Virgo data from known pulsars place limits on the size of quadrupolar deformations ("mountains") of those neutron stars.

A* are ultracompact boson stars[9]; nor can we exclude that the *progenitors* in GW150914 were ultracompact boson stars. However, for the latter, if they were boson stars, the ring-down part of the signal shows they promptly collapsed and *formed* a Kerr black hole, with mass and angular momentum consistent with that of the binary just prior to merger. In other words, even in this hypothetical scenario GW150194 still gives evidence for the existence of Kerr black holes—exotic compact objects more "bizarre" than boson stars would need to be invoked to avoid that conclusion **[54]**.

Because of the uniqueness properties of black holes in general relativity, and if general relativity does accurately describe strong field gravity on astrophysical scales, then unfortunately we cannot learn anything more about the *physics* of black holes from more precise merger observations (the *astrophysics* of black holes is a different issue, discussed below). In other words, all black holes in the Universe are then Kerr black holes to within environmental perturbations, and perhaps future ultraprecise measurements of mergers could show imprints of a circumbinary environment, but there are no novel classes, shapes, or topologies of black holes to discover. Then, the utility of black hole merger observations for fundamental physics is essentially entirely to provide detailed tests of nonlinear general relativity as outlined in the previous paragraph. Of course, as the scientific method requires such tests for the health of its theories this is a useful endeavor, and we do not need a motivation other than that. However, there is at least one observationally driven motivation for why one might be skeptical about the precise nature of strong field gravity as described by general relativity, namely dark energy.

On large scales, the Universe is observed to be in an epoch of accelerated expansion; *interpreting* this as being due to dark energy comes from assuming that Einstein gravity accurately describes the geometry of the Universe on such scales. Specifically, on large scales it is assumed that, with an appropriate time slicing, the spatial metric of the Universe is nearly homogeneous and isotropic, and its time evolution is driven by a stress energy tensor characterizing the average energy densities and pressures of all the matter/energy in the Universe. It is sometimes stated that today (i.e., away from any "big bang" singularities) gravity on average in the Universe is weak, and certainly on small scales like our solar system, galaxy, or even that of galaxy clusters it is weak (except near the rare black hole or neutron star). However, as described in Section 2.1.3, the strong field regime of general relativity is not associated with any physical length scale per se, but rather manifests when some physical scale in the problem becomes commensurate with the radius of curvature of spacetime. And by that measure, our Universe is *always* in the strong field regime on scales of the Hubble radius R_H, i.e., R_H is of the same order of magnitude as the Schwarzschild radius $R_s \sim \sqrt{3c^2/8\pi G\rho}$ of a spherical distribution of matter with the same average energy density ρ as the matter in the Universe. For example, today ρ_0 is roughly that of six hydrogen atoms

---

9 Boson stars are hypothetical star-like objects formed from exotic (i.e., not part of the standard model of particle physics) self-interacting bosonic matter, in contrast to neutron stars which are largely composed of fermionic matter. A boson star's gravitational dynamics is still governed by general relativity, so it is not an "alternative" to a black hole, but could be a novel class of compact object.

per cubic meter, giving $R_{s0} \sim 10^{26}$ m $\sim R_{H_0}$. One might complain that the Schwarzschild radius argument does not apply to our Universe because the latter is not asymptotically flat. Perhaps, though the point here is not to argue whether or not we are inside a Schwarzschild black hole, but instead that on scales of the Hubble radius the Universe must be in the nonlinear regime of general relativity for an entirely different class of solution (the FRLW metrics) to be possible. Bringing the discussion back to testing gravity on stellar mass black holes scales, if dark energy is telling us general relativity gets things wrong on the scale of the Hubble radius, we should be cautious about immediately accepting its predictions for black holes, as the scale-free nature of general relativity implies cosmological horizons and event horizons reside in a related regime of the theory.[10] The current LIGO/Virgo observations are therefore an important step toward quantitative verification of the physics of horizons.

The wealth of knowledge gained from GW170817, the first binary neutron star merger detected [4]. That so much information was garnered from this event is because a host of electromagnetic counterpart emission was also seen—the first, and to date only gravitational wave–electromagnetic "multimessenger" event [5]. Here we briefly comment on the highlights. The first is that a short Gamma Ray Burst (sGRB) was detected ~ 1.7 s after the observed gravitational wave inspiral, the latter which ended a few ms before the presumed collision of the two neutron stars (this and any postcollision gravitational waves were not seen by LIGO/Virgo, which is as expected as they occur at frequencies several times higher than what LIGO/Virgo is sensitive to). The origin of sGRBs has long been a mystery, though one of the leading hypothesis for their formation is that they are produced in polar jets powered by accretion onto the remnant of a binary neutron star merger (whether it be a hypermassive neutron star or a black hole that formed, though the latter seems to be a more favorable environment for jet formation). The coincidence of the gravitational wave emission and sGRB, both in terms of time and region of the sky where both fluxes appeared to come from, gives the first solid evidence that at least a class of sGRBs are produced following a binary neutron star merger. Assuming this connection, together with the estimated distance to the event of ~ 40 Mpc, then also gives a direct measurement of the speed of gravitational waves relative to the speed of light, and a remarkably tight constraint for a first measurement: the two speeds are the same to within approximately 1 part in 10^{15} [3].

Almost immediately after GW170817 was detected, a worldwide effort was undertaken by astronomers to search for other electromagnetic counterparts, and within 11 hours a bright, but fading, optical transient was identified in the galaxy NGC 4993. Follow-up observation over the subsequent weeks saw the event in radio, X-ray, infrared, and the ultraviolet. The observed properties of the emission are consistent with the neutron star merger having produce a so-called kilonova (or macronova) [39]. During merger, a small fraction (~ 0.1–1%) of the neutron star's material is tidally ejected from the system at mildly relativis-

10 The majority of proposals to explain dark energy using modified gravity specifically introduce a new physical length scale into the problem, and if that scale is tuned to the Hubble radius it would avoid the conclusion that altering gravity on present day cosmological horizon scales could have consequences for stellar mass or supermassive black holes.

tic speeds ($\sim c/3$), and over the subsequent few seconds following merger a similar amount of material can be blown away from a hot accretion disk formed around the remnant, at similar but slightly lower velocities. This initially high density material is very neutron rich, and as it expands heavy elements (with atomic number in the range $Z \in 28..90$) are formed through the r-process. Many of these elements are radioactive with relatively short lifetimes, and it is their decay over the subsequent days that produces the light of the kilonova. This also confirms that neutron star mergers are one of the sites where a significant fraction of the Universe's heavy elements are produced—it is quite likely that the gold and platinum we humans so love to adorn ourselves with are the ashes of ancient galactic neutron star mergers.

The late stages of the gravitational wave emission in GW170817 also showed mild deviation from the predictions of a black hole inspiral, indicative of tidal deformations occurring in both neutron stars. The strength of the tidal deformation is governed by the equation of state of matter at nuclear density, which is not theoretically well understood today, or accessible to experiments on the earth to investigate. Thus neutron star mergers offer an avenue to explore this extreme state of matter, and though this first event did not provide strong constraints on competing models, this is one of the subjects future observations are expected to bring increasing clarity to.

Another subject that GW170817 allowed gravitational wave astronomy to take a first step in, but will also require more future observations to make a useful contribution toward, is measuring the local expansion rate of the Universe. This is typically done by measuring both the distance d and redshift z to a set of sources in galaxies, and the expansion history can be inferred from the relationship $z(d)$ (for small redshifts, so nearby galaxies, $z \approx H_0 d/c$, where H_0 is the Hubble constant). Measuring the distance to a source is quite challenging. One method relies on a so-called standard candle, where the intrinsic luminosity L of a source is assumed known, and hence the observed flux is simply $L/4\pi d^2$. Type Ia supernovae are the most well-known standard candles, though inferring their intrinsic luminosity relies on several calibration steps, including the cosmic distance ladder. With a binary neutron star merger where a counterpart is seen (and hence the host galaxy identified for a redshift measurement), a luminosity distance–redshift measurement can be obtained that bypasses all of these calibration steps, since the intrinsic luminosity of the merger is known from the general relativity waveform calculation. This makes a binary neutron star merger a standard "siren" (siren is used here instead of candle as the last stages of inspiral emit waves in the audio frequency range).[11] GW170817 has already by itself allowed a measurement of H_0 to within about 10%; though this is not an improvement over other existing measurements, the more multimessenger binary neutron star events that are observed, the tighter the standard siren based value will become. Eventually, this might prove to be instrumental to help resolve the present "Hubble tension": measurements of H_0 inferred by the Planck satel-

11 Binary black holes are also standard sirens, and better ones in fact, as some uncertainty will be present in the neutron star measurements until the nuclear equation of state is known. However, binary black holes are not typically expected to be in an environment where a strong electromagnetic counterpart will be produced, and none have been observed to date.

lite's observation of the CMB show a small, but statistically significant, mismatch with H_0 obtained using supernovae data (see, e.g., [30]).

Tentative hints pointing to an "unusual" stellar mass black hole population. Of the almost 100 signals LIGO/Virgo have so far detected that are of likely astrophysical origin, the vast majority are consistent with binary black hole merger templates [7,10,11].[12] As discussed above, if general relativity is correct, then we know these are all merging Kerr black holes, forming remnant Kerr black holes. The utility then in having this large number of events, and anticipating even more in the years to come, is to learn what the distribution of masses and spins of this subpopulation of black holes in the Universe is as a function of time (redshift). This will provide information on the fates of the most massive stars that are expected to form black holes at the ends of their lives, as well as binary formation channels. Regarding the latter, the two thought to be predominant are from stellar binaries where both stars are massive enough to form black holes, and dynamical assembly in dense cluster environments (following chance encounters between either two isolated black holes, a binary containing a black hole and a single black hole, or a binary–binary interaction where each contains a black hole). Though even 100 events is not yet enough to give definitive answers to some of these population questions, there are already some interesting trends, and a few outliers that are somewhat puzzling or surprising (at least without hindsight to select amongst the many reasonable arguments present in the prior body of literature speculating about the unknown).

The first surprise came with GW150914, in that both progenitor black holes had masses ($\sim 29M_{\odot}$ and $36M_{\odot}$) at least twice that of any known stellar mass black hole candidate in the Milky Way (see, e.g., [22]). Subsequent detections showed that GW150914 is not an outlier in this regard, and most (though not all) LIGO/Virgo black hole progenitors are more massive than known galactic black holes. This could partly be a selection effect, as LIGO/Virgo is more sensitive to higher mass mergers, and also that the X-ray binary systems that have been used to identify galactic black holes might be a distinct population of binaries from those that lead to black holes that merge within a Hubble time.

A second puzzle is that the vast majority of progenitor black holes seem to have very low spin (the remnants acquire higher spin, around 60–80% that of the maximum allowed for Kerr black holes). Or to be more technically precise, given the detector's current sensitivities, with most inspirals a confident measurement can only be made of the net spin angular momentum aligned with the orbital angular momentum—for most mergers detected to date this result is consistent with zero (to within error bars). There are three primary configurations that can achieve this: (1) the individual black holes actually do have close to zero spins, (2) the individual black holes have roughly equal but opposite spin angular momenta,

---

12 The remainder also match binary black hole templates, but when one or both companions have masses less than $\sim 2.5M_{\odot}$, the event is classified as a black hole–neutron star or binary neutron star merger, respectively. To be able to distinguish between black holes and neutron stars from the gravitational waves alone would require observation of the higher frequency late stages of inspiral/merger, or a high enough SNR event that the effect of tidal deformation is already evident in the earlier lower frequency inspiral that can be observed with present detectors.

one aligned, the other antialigned with the orbital angular momentum, (3) the black holes have arbitrary spins (less than extremal) but the spin vectors are mostly *within* the orbital plane. Both options (2) and (3) are difficult to explain with a binary formed from a stellar binary, where one would typically expect the spin vectors to be almost aligned with the orbital angular momentum vector. Options (2) and (3) are consistent with the occasional dynamically assembled binary, as there is no preferential orientation for an essentially random close encounter, but one would not expect this for the majority of events as currently observed. Thus (1) seems the most plausible explanation at the moment. Given how challenging it is to simulate stellar collapse at present, hence have robust predictions for what the initial spin distributions of black holes should be, the observations will serve as useful guide posts for ongoing theoretical studies of collapse.

There are more speculative suggestions for why the progenitors have low spin. One is that many of these low-spinning black holes are primordial in nature, meaning the black holes might have formed at a very early epoch in the Universe (well before structure formation) from rare superhigh density fluctuations in the background radiation field. The concrete mechanisms people have proposed for this typically produce very low spin black holes (see, e.g., [20] for a review). Another possibility is that there are as of yet undiscovered "ultralight" particles, with Compton wavelengths on the order of the tens of kilometer scale of the Schwarzschild radii of stellar mass black holes. Such particles can form bound states around the black holes, and if the black hole is spinning, these bound states can grow by a so-called superradiant interaction with the surrounding spacetime [17]. In reaction, the black hole spins down, possibly quite rapidly on astrophysical timescales (much less than the relevant gigayear timescale, which is order of magnitude the maximum time between a black hole's birth and when it should suffer a collision with another to be visible to LIGO/Virgo). Of course, even if such particles exists, they might not be the reason for the low spin black hole population—that could still just be due to properties of stellar collapse and black hole formation.

The third surprise relates to several outlier events, the two most prominent being GW190521 and GW190814, that seem to have progenitor compact objects in the so-called "mass gaps." GW190814 is the merger of a $\sim 23 \pm 1 M_\odot$ (presumed) black hole with a $\sim 2.6 \pm 0.1 M_\odot$ compact object [9]. GW190521 is the merger of a $\sim 85 \pm 20 M_\odot$ black hole with a $\sim 66 \pm 18 M_\odot$ black hole [8]. Regarding GW190814, arguments from stellar collapse studies, as well as a dearth of candidates from our known galactic compact object population, suggest objects with masses in the range $\sim 2.5 M_\odot$–$5 M_\odot$ do not typically form in stellar collapse. Moreover, it is currently unknown if the maximum allowed mass for a neutron star can reach $2.5 M_\odot$; if it turns out to be less than $2.5 M_\odot$, the lower mass companion of GW190814 would be challenging to explain (or be an exceedingly rare object, for example, a low mass black hole formed via a prior binary neutron star merger). Regarding GW190521, stellar structure theory suggests stars with cores in the mass range $\sim 65 M_\odot$–$135 M_\odot$ are subject to the so-called pulsational pair-instability supernova processes, which blows the cores apart leaving behind no remnant. However, similar to the issue of the spin of a black hole at birth, there is a fair amount of uncertainty to the exact range of this mass gap, and given the error bars in the mass measurements, there is only mild tension between GW190814 and conventional theories.

5. THE FUTURE OF GRAVITATIONAL WAVE ASTRONOMY

Einstein's theory of general relativity is over 100 years old, and the quest to observe the Universe in gravitational waves is over 50 years old, beginning with Joseph Weber's pioneering attempts in the 1960s. Despite these long histories, the field of gravitational wave astronomy is in its infancy, with the first detection only 6 years ago. Though many signals observed to date are solidly above the threshold for confident assertion that they are gravitational waves coming from astrophysical sources, they are still not loud enough for high precision tests of strong field gravity, or for high accuracy estimation of all source parameters. Moreover, most of these detections have relied on theoretical templates of expected sources, which improves the effective sensitivity of the detectors. Thus any truly novel source will likely only be discovered once the detector sensitivities are well above the threshold the new source could otherwise have been seen using templates. The one exception here is a source that emits a short burst well approximated by a sine-Gaussian, as LIGO/Virgo do employ searches using such templates (this can be thought of as an "unmodeled" search in the sense that there is no particular astrophysical source from which the template is derived).

To realize a future where a detailed picture of the Universe in gravitational waves is attained will thus require more sensitive detectors that cover a broader range of frequencies than at present. These are being planned, and within the next decade or two we can expect an order of magnitude improvement over essentially the entire slate of observational campaigns. LIGO is within a factor of two of the original "Advanced LIGO" design sensitivity, which should be reached during the next observing campaign (beginning late 2022–early 2023), when the KAGRA detector in Japan will also join the LIGO/Virgo network [6]. Following that, the plan is for an "A+" upgrade that will improve sensitivity by another factor of two, and LIGO India will join the network (anticipated to start in 2025). To improve sensitivities significantly beyond this will require new facilities, and several third generation designs are being planned for the 2030s, including Cosmic Explorer and the Einstein Telescope [38]. These could further increase sensitivity by a factor of 10, as well as offer improved frequency bandwidth over both lower (earlier in the inspiral for binary compact objects) and higher frequencies (merger regime for binary neutron stars). New technologies are also being considered, most promising among these are atom interferometers [31], though it is less clear what the timeline for their deployment is. The space-based LISA mission is expected to launch in the late 2030s. Both LISA and third generation ground-based detectors could see black hole mergers with SNR close to a thousand (the current SNR record holder is GW170817, at $\sim$ 32). CMB measurements of B-mode polarization over the next decade (e.g., with the Simons Observatory [15] currently under construction, and the LiteBIRD satellite planned to be launched by the end of the decade [49]), should lower the threshold above which cosmic gravitational waves would be observed by about an order of magnitude. The sensitivity of the pulsar timing network increases roughly with the square-root of the observation time, and could be accelerated with the discovery of more highly stable pulsars clocks to add to the network (see, e.g., [36]).

We conclude with a brief discussion of what we can hope/expect to learn from these observatories if everything goes according to plan. At the very least we can expect an ever clearer picture of the demographics of compact objects in our Universe unfolding, improved tests of the dynamical strong field regime of general relativity, tighter constraints on the Hubble constant H_0 from gravitational wave standard sirens, first detection of a stochastic background of gravitational waves from unresolved supermassive black hole binaries, and either a first measurement of a primordial gravitational wave background from an inflationary epoch in the early Universe, or a bound on the latter that would severely challenge the inflationary paradigm. If we are fortunate, a binary neutron star merger as close or closer than GW170817 will occur during the era of the third generation of ground-based detectors, which would provide unprecedented insight into the nature of matter at the extreme nuclear densities present in the interior of neutron stars. If we are very fortunate, a star will go supernova (while the detectors are on!) in our neighborhood of the Milky Way, which should be close enough for us to be able to hear it in gravitational waves.

A wish opening up our view of the Universe to the medium of gravitational waves has always been that new, unexpected, and surprising sources will be discovered. Though, of course, we cannot make a list of the truly unexpected, there are sources that people have speculated about that would be surprising, and some quite revolutionary, if discovered. These include cosmic strings, ultralight particles driving black hole superradiance, new kinds of compact objects such as boson stars, and various "exotic" horizonless compact object alternatives to black holes. The latter include fuzzballs, gravastars, and AdS (anti-de Sitter) black bubbles, all inspired by ideas on how "quantum gravity" could resolve the singularities of general relativity and apparent information loss paradox associated with black holes that evaporate via the Hawking process. But perhaps the biggest surprise of all would be if, once all is said and done, there are no surprises beyond a few black holes having been born with their two strands of Kerr hair standing mildly out of place.

FUNDING

The author acknowledges support from NSF Grant No. PHY-1912171, the Simons Foundation, and the Canadian Institute For Advanced Research (CIFAR).

REFERENCES

[1] B. P. Abbott, et al., Observation of gravitational waves from a binary black hole merger. *Phys. Rev. Lett.* **116** (2016), no. 6, 061102.

[2] B. P. Abbott, et al., Tests of general relativity with GW150914. *Phys. Rev. Lett.* **116** (2016), no. 22, 221101. [Erratum: *Phys. Rev. Lett.* **121** (2018), 129902].

[3] B. P. Abbott, et al., Gravitational waves and Gamma-rays from a binary neutron star merger: GW170817 and GRB 170817A. *Astrophys. J. Lett.* **848** (2017), no. 2, L13.

[4] B. P. Abbott, et al., GW170817: observation of gravitational waves from a binary neutron star inspiral. *Phys. Rev. Lett.* **119** (2017), no. 16, 161101.

[5] B. P. Abbott, et al., Multi-messenger observations of a binary neutron star merger. *Astrophys. J. Lett.* **848** (2017), no. 2, L12.

[6] B. P. Abbott, et al., Prospects for observing and localizing gravitational-wave transients with Advanced LIGO, Advanced Virgo and KAGRA. *Living Rev. Relativ.* **21** (2018), no. 1, 3.

[7] B. P. Abbott, et al., GWTC-1: a gravitational-wave transient catalog of compact binary mergers observed by LIGO and Virgo during the first and second observing runs. *Phys. Rev. X* **9** (2019), no. 3, 031040.

[8] R. Abbott, et al., GW190521: a binary black hole merger with a total mass of $150M_{\odot}$. *Phys. Rev. Lett.* **125** (2020), no. 10, 101102.

[9] R. Abbott, et al., GW190814: gravitational waves from the coalescence of a 23 solar mass black hole with a 2.6 solar mass compact object. *Astrophys. J. Lett.* **896** (2020), no. 2, L44.

[10] R. Abbott, et al., GWTC-2.1: deep extended catalog of compact binary coalescences observed by LIGO and Virgo during the first half of the third observing run. 2021.

[11] R. Abbott, et al., GWTC-3: compact binary coalescences observed by LIGO and Virgo during the second part of the third observing run. 2021.

[12] R. Abbott, et al., The population of merging compact binaries inferred using gravitational waves through GWTC-3. 2021.

[13] R. Abbott, et al., Tests of general relativity with binary black holes from the second LIGO-Virgo gravitational-wave transient catalog. *Phys. Rev. D* **103** (2021), no. 12, 122002.

[14] R. Abbott, et al., Tests of general relativity with GWTC-3. 2021.

[15] P. Ade, et al., The Simons Observatory: science goals and forecasts. *J. Cosmol. Astropart. Phys.* **02** (2019), 056.

[16] L. Barack, et al., Black holes, gravitational waves and fundamental physics: a roadmap. *Classical Quantum Gravity* **36** (2019), no. 14, 143001.

[17] R. Brito, V. Cardoso, and P. P. Superradiance, New frontiers in black hole physics. *Lecture Notes in Phys.* **906** (2015), 1–237.

[18] A. Buonanno, Gravitational waves. In *Les Houches Summer School – Session 86: particle physics and cosmology: the fabric of spacetime*, Les Houches, France, 2007.

[19] A. Buonanno and T. Damour, Effective one-body approach to general relativistic two-body dynamics. *Phys. Rev. D* **59** (1999), 084006.

[20] B. Carr and F. Kuhnel, Primordial black holes as dark matter candidates. In *Les Houches summer school on Dark Matter*, Les Houches, France, 2021.

[21] B. Carter, Axisymmetric black hole has only two degrees of freedom. *Phys. Rev. Lett.* **26** (1971), 331–333.

[22] J. Casares, P. G. Jonker, and G. Israelian, X-ray binaries. 2017.

[23] M. W. Choptuik and F. Pretorius, Ultra relativistic particle collisions. *Phys. Rev. Lett.* **104** (2010), 111101.

[24] S. Dimopoulos and G. L. Landsberg, Black holes at the LHC. *Phys. Rev. Lett.* **87** (2001), 161602.

[25] F. Dyson, Is a graviton detectable? *Internat. J. Modern Phys. A* **28** (2013), 1330041.

[26] W. E. East and F. Pretorius, Ultrarelativistic black hole formation. *Phys. Rev. Lett.* **110** (2013), no. 10, 101101.

[27] J. L. Feng and A. D. Shapere, Black hole production by cosmic rays. *Phys. Rev. Lett.* **88** (2002), 021303.

[28] S. E. Field, C. R. Galley, F. Herrmann, J. S. Hesthaven, E. Ochsner, and M. Tiglio, Reduced basis catalogs for gravitational wave templates. *Phys. Rev. Lett.* **106** (2011), 221102.

[29] E. E. Flanagan and S. A. Hughes, The Basics of gravitational wave theory. *New J. Phys.* **7** (2005), 204.

[30] W. L. Freedman, Measurements of the Hubble constant: tensions in perspective. *Astrophys. J.* **919** (2021), no. 1, 16.

[31] R. Geiger, In *Future gravitational wave detectors based on atom interferometry*, edited by G. Augar and E. Plagnol, pp. 285–313, World Scientific Publishing Co. Pte. Ltd., 2017.

[32] S. B. Giddings and S. D. Thomas, High-energy colliders as black hole factories: The End of short distance physics. *Phys. Rev. D* **65** (2002), 056010.

[33] C. Gundlach, S. Akcay, L. Barack, and A. Nagar, Critical phenomena at the threshold of immediate merger in binary black hole systems: the extreme mass ratio case. *Phys. Rev. D* **86** (2012), 084022.

[34] S. W. Hawking, Gravitational radiation from colliding black holes. *Phys. Rev. Lett.* **26** (1971), 1344–1346.

[35] S. W. Hawking, Black holes in general relativity. *Comm. Math. Phys.* **25** (1972), 152–166.

[36] G. Hobbs and S. Dai, Gravitational wave research using pulsar timing arrays. *Nat. Sci. Rev.* **4** (2017), no. 5, 707–717.

[37] W. Israel, Event horizons in static vacuum space-times. *Phys. Rev.* **164** (1967), 1776–1779.

[38] V. Kalogera, et al., The next generation global gravitational wave observatory: the science book. 2021, arXiv:2111.06990.

[39] D. Kasen, B. Metzger, J. Barnes, E. Quataert, and E. Ramirez-Ruiz, Origin of the heavy elements in binary neutron-star mergers from a gravitational wave event. *Nature* **551** (2017), 80.

[40] L. Lehner and F. Pretorius, Black strings, low viscosity fluids, and violation of cosmic censorship. *Phys. Rev. Lett.* **105** (2010), 101102.

[41] C. W. Misner, K. S. Thorne, and J. A. Wheeler, *Gravitation*. W. H. Freeman, San Francisco, 1973.

[42] H. Okawa, K-i. Nakao, and M. Shibata, Is super-Planckian physics visible? – Scattering of black holes in 5 dimensions. *Phys. Rev. D* **83** (2011), 121501.

[43] R. Penrose, Gravitational collapse: the role of general relativity. *Nuovo Cimento Riv.* **1** (1969).

[44] R. Penrose, In *Seminar on differential geometry*, edited by S. T. Yau, Princeton University Press, Princeton, New Jersey, 1982.

[45] F. Pretorius, Evolution of binary black hole spacetimes. *Phys. Rev. Lett.* **95** (2005), 121101.

[46] F. Pretorius and W. E. East, Black hole formation from the collision of plane-fronted gravitational waves. *Phys. Rev. D* **98** (2018), no. 8, 084053.

[47] F. Pretorius and D. Khurana, Black hole mergers and unstable circular orbits. *Classical Quantum Gravity* **24** (2007), S83–S108.

[48] D. C. Robinson, Uniqueness of the Kerr black hole. *Phys. Rev. Lett.* **34** (1975), 905–906.

[49] Y. Sekimoto, et al., Concept design of low frequency telescope for CMB B-mode polarization satellite LiteBIRD. *Proc. SPIE Int. Soc. Opt. Eng.* **11453** (2020), 1145310.

[50] U. Sperhake, E. Berti, V. Cardoso, and F. Pretorius, Universality, maximum radiation and absorption in high-energy collisions of black holes with spin. *Phys. Rev. Lett.* **111** (2013), no. 4, 041101.

[51] U. Sperhake, V. Cardoso, F. Pretorius, E. Berti, and J. A. Gonzalez, The High-energy collision of two black holes. *Phys. Rev. Lett.* **101** (2008), 161101.

[52] K. Thorne, In *Magic without magic*, edited by J. Klauder, W. H. Freeman, San Francisco, 1972.

[53] C. M. Will, The confrontation between general relativity and experiment. *Living Rev. Relativ.* **17** (2014), no. 4.

[54] N. Yunes, K. Yagi, and F. Pretorius, Theoretical physics implications of the binary black hole mergers GW150914 and GW151226. *Phys. Rev. D* **94** (2016), no. 8, 084002.

FRANS PRETORIUS

Department of Physics, Princeton University, Princeton, NJ 08544, USA,
fpretori@princeton.edu

PLENARY LECTURES

GROUPS ACTING ON HYPERBOLIC SPACES—A SURVEY

MLADEN BESTVINA

ABSTRACT

This is a (very subjective) survey paper for nonspecialists, covering group actions on Gromov hyperbolic spaces. The first section is about hyperbolic groups themselves, while the rest of the paper focuses on mapping class groups and $\mathrm{Out}(F_n)$, and the way to understand their large scale geometry using their actions on various hyperbolic spaces constructed using projection complexes. This understanding for $\mathrm{Out}(F_n)$ significantly lags behind that of mapping class groups, and the paper ends with a few open questions.

MATHEMATICS SUBJECT CLASSIFICATION 2020

Primary 20F65; Secondary 20F67, 20F69, 57K20

INTERNATIONAL CONGRESS OF MATHEMATICIANS 2022 JULY 6–14

Proc. Int. Cong. Math. 2022, Vol. 2, pp. 678–711
DOI 10.4171/ICM2022/110

1. INTRODUCTION

The goal of this paper is to give a flavor of the developments in geometric group theory in the last 35 years, focusing on groups acting on Gromov hyperbolic spaces. The field of geometric group theory is relatively young and its birth can be attributed to Gromov's paper [71] in 1987, when the subject exploded and attracted many mathematicians. The term itself was coined by Niblo and Roller, who organized and named a very influential conference in 1991 [107,108] (though it was possibly used informally before). Loosely speaking, geometric group theory studies groups by looking at their actions on metric spaces and the geometry and topology of these spaces. Increasingly, methods of other branches of mathematics, such as dynamics and analysis, are also brought to bear.

There were, of course, significant developments that can be comfortably placed within this subject even long before Gromov's paper. Works of Klein, Dehn, Nielsen, Stallings, and others in some sense form the backbone of the subject. The theory of groups acting on trees, i.e., Bass–Serre theory [6,131] and its language, will be used freely in these notes. Gromov's celebrated theorem that groups of polynomial growth are virtually nilpotent [69] appeared in 1981, and Gromov's basic philosophy of viewing groups as metric spaces was eloquently explained in [70]. Of course, the influence on this subject of the work of Thurston cannot be overstated. Perhaps the development of combinatorial group theory, focusing on the combinatorics of the words in a finitely presented group, distracted from a more geometric approach to group theory.

This paper will focus on the part of geometric group theory that studies groups acting on (Gromov) hyperbolic spaces. In the early days, right after Gromov's paper, this meant studying (Gromov) hyperbolic groups. Around 2000, the work of Masur and Minsky [95,96] shifted the focus to groups that are not hyperbolic but admit interesting actions on hyperbolic spaces. The main examples of such groups are mapping class groups of compact surfaces (the subject of the papers by Masur and Minsky) and $\mathrm{Out}(F_n)$, the outer automorphism group of a finite rank free group. This survey will concentrate on these two classes of groups.

The definition of Gromov hyperbolic spaces is modeled on the standard hyperbolic spaces by "coarsification" and captures the fact that geodesic triangles in the hyperbolic plane are "thin." For a wonderful survey of the history of hyperbolic geometry from Lobachevsky to 1980, see Milnor's paper [99]. For much more about this subject, see Bridson–Haefliger [44], Ghys–de la Harpe [68], or Druţu–Kapovich [58]. There are many important topics this survey will not cover, e.g., relative hyperbolicity [61], hyperbolic Dehn filling [74,114], small cancelation [1,112], uniform embeddings in Hilbert spaces [120], the celebrated work of Agol and Wise, see, e.g., [16], random walk [93], Cannon–Thurston maps [100], and many others.

2. HYPERBOLIC GROUPS

Every finitely generated group G can be viewed as a metric space. Fix a finite generating set S which is symmetric, i.e., $S^{-1} = S$. The *word norm* $|g|_S$ of $g \in G$ is the smallest n such that g can be written as $g = s_1 s_2 \cdots s_n$ for $s_i \in S$. Then $d_S(g,h) = |g^{-1}h|_S$ is the *word metric* on G, and left translations $L_x : g \mapsto xg$ are isometries. More geometrically, this is the distance function on the vertices of the Cayley graph Γ_S, with vertex set G, and edges of length 1 between g and gs for $g \in G$ and $s \in S$. If S' is a different finite symmetric generating set for G, the identity map $G \to G$ is bilipschitz with respect to the two word metrics, and are considered equivalent. There is a more general equivalence relation between metric spaces that is very convenient in the subject. Let (X, d_X) and (Y, d_Y) be metric spaces. A (not necessarily continuous) function $f : X \to Y$ is a *quasiisometry* if there is a number $A > 0$ such that

$$\frac{1}{A} d_X(a,b) - A \leq d_Y\big(f(a), f(b)\big) \leq A d_X(a,b) + A$$

for all $a, b \in X$, and every metric ball of radius A in Y intersects the image of f. Without the second condition, f is a *quasiisometric embedding* (when we want to refer to the constant A, we say A-quasiisometric embedding). Two metric spaces are *quasiisometric* if there is a quasiisometry between them, and this is an equivalence relation. For example, inclusion $\mathbb{Z} \hookrightarrow \mathbb{R}$ is a quasiisometry, as is any bilipschitz homeomorphism or a finite index inclusion between finitely generated groups equipped with word metrics. More generally, the following is considered to be the Fundamental Theorem of Geometric Group Theory.

Theorem 2.1 (Milnor [98], Švarc [135]). *Suppose a group G acts properly and cocompactly by isometries on a proper geodesic metric space X. Then G is finitely generated and any orbit map $G \to X$ is a quasiisometry.*

A metric space is proper if closed metric balls are compact, and it is *geodesic* if any two distinct points a, b are joined by a subset isometric to the closed interval $[0, d(a,b)]$. For example, cocompact lattices in a simple Lie group are quasiisometric to each other. The "Gromov program" is to classify groups, at least in a given class, up to quasiisometry.

According to Gromov, the following definition was given by Rips. There are several other definitions, all of which are equivalent up to changing the value of δ, see [44, 58].

Definition 2.2. Let $\delta \geq 0$. A geodesic metric space X is *δ-hyperbolic* if in any geodesic triangle each side is contained in the δ-neighborhood of the other two sides. We say X is *hyperbolic* if it is δ-hyperbolic for some $\delta \geq 0$. See Figure 1.

For example, trees are 0-hyperbolic and so are complete simply-connected Riemannian manifolds of sectional curvature $\leq -\varepsilon < 0$. A fundamental property of hyperbolic spaces is the Morse Lemma, proved by Morse [105], Busemann [48], and Gromov [71] in increasing generality.

Lemma 2.3 (Morse Lemma). *There is a number $D = D(\delta, A)$ such that for any δ-hyperbolic space X and any A-quasiisometric embedding $f : [a,b] \to X$ the image of f is contained in the D-neighborhood of any geodesic from $f(a)$ to $f(b)$.*

It then quickly follows that if two geodesic spaces are quasiisometric and one is hyperbolic, so is the other. In particular, groups that act properly and cocompactly by isometries on proper hyperbolic spaces are hyperbolic.

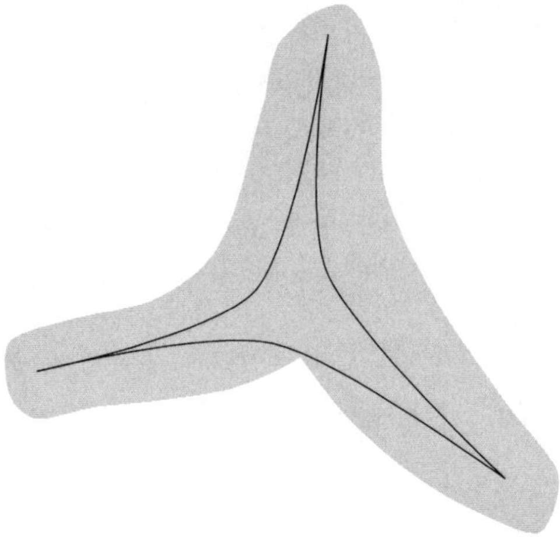

FIGURE 1
The union of the δ-neighborhoods of two sides contains the third.

Hyperbolic groups are well behaved, both topologically and geometrically, and they are generic, so they form a model class of groups in geometric group theory. We now elaborate.

2.1. Classification of elements

Let G be a hyperbolic group. If $g \in G$ has finite order, then there is a coset $\langle g \rangle x$ that has diameter $\leq 4\delta + 2$, so in particular there is an a priori bound on the order in terms of δ and the number of generators. This is proved by a coarse version of the standard argument that a bounded set in $\mathbb{R}^n$ (or any Hadamard manifold) is contained in a unique closed ball of smallest radius. If g has infinite order, then $k \mapsto g^k x$ is a quasiisometric embedding for every $x \in G$, and g is *loxodromic*.

2.2. The Rips complex

The classical Cartan–Hadamard theorem states that closed manifolds of nonpositive sectional curvature have contractible universal cover. In a similar way, every hyperbolic group G acts properly and cocompactly on a contractible simplicial complex, called the *Rips complex*. It is constructed as follows. Fix a number $d > 0$ and form the complex $P_d(G)$: the set of vertices is G, and a set $\{v_0, v_1, \ldots, v_n\}$ of distinct vertices forms a simplex if $d(v_i, v_j) \leq d$ for all i, j. This is a version of the Vietoris approximation of a metric space by a simplicial complex, except here we think of d as being large.

Theorem 2.4. *For $d > 4\delta + 6$, $P_d(G)$ is contractible.*

So, for example, if G is torsion-free, the quotient $P_d(G)/G$ is a finite classifying space for G, and in any case G is finitely presented, and has a classifying space with finitely many cells in each dimension. Every finite subgroup of G fixes a point of $P_d(G)$ (for d large), so it follows that G has finitely many conjugacy classes of finite subgroups. Interestingly, it is not known whether every infinite hyperbolic group is virtually torsion-free, or even if it always has a proper subgroup of finite index.

2.3. Subgroups

If $g \in G$ has infinite order, there is a unique maximal virtually cyclic subgroup $E(g)$ of G that contains g, and $E(g)$ also contains the normalizer of g. It follows that G cannot contain $\mathbb{Z}^2$ as a subgroup. Translation length considerations show that G cannot contain any Baumslag–Solitar groups $B(m,n) = \{a,t \mid ta^m t^{-1} = a^n\}, m,n \neq 0$, as subgroups. The long standing open question whether every group with finite classifying space and not containing any $B(m,n)$ is necessarily hyperbolic was recently answered in the negative [86].

2.4. Boundary

Inspired by the visual boundary of Hadamard manifolds, Gromov defined a boundary ∂G of a hyperbolic group (or a proper geodesic metric space which is hyperbolic). It is a compact metrizable space and a point is represented by a quasigeodesic ray $\mathbb{Z}_+ \to G$, with two rays representing the same boundary point if their images stay a bounded distance apart. The topology is based on the principle that rays issuing from a basepoint and with fixed quasigeodesic constants will stay longer together if they represent points that are closer together. If G is infinite and virtually cyclic then ∂G consists of two points, and if G is not virtually cyclic (termed "nonelementary") ∂G has no isolated points.

There is also a natural topology on the union

$$\overline{X} = P_d(G) \sqcup \partial G$$

of the Rips complex and the Gromov boundary that makes it into a compact metrizable space, and G acts naturally by homeomorphisms. Loxodromic elements act by north–south dynamics on $\overline{X}$. The most important property of the boundary, used, for example, in the proof of Mostow rigidity [106], is the following:

Theorem 2.5. *Let $f : X \to Y$ be a quasiisometry between two hyperbolic proper geodesic mmetric spaces. Then f extends to a homeomorphism $\partial X \to \partial Y$.*

Theorem 2.6 ([38]). *$\overline{X}$ is a Euclidean retract, i.e., it is contractible, locally contractible, and finite-dimensional. The covering dimension of ∂G can be computed from the cohomology of G and, in particular, if G is torsion-free,* $\dim \partial G$ *equals the cohomological dimension of G minus 1, and in any case the rational cohomological dimension of ∂G equals the rational cohomological dimension of G minus 1.*

2.5. Asymptotic dimension

In [72] Gromov introduced many quasiisometric invariants of groups and spaces. Here we focus on *asymptotic dimension*. Let X be any metric space. For an integer $n \geq 0$, we write $\operatorname{asdim}(X) \leq n$ provided that for every $R > 0$ there exists a cover of X by uniformly bounded sets such that every ball of radius R in X intersects at most $n + 1$ elements of the cover. This is the "large scale" analog of the usual covering dimension. For example, asdim($\mathbb{R}^n$) $= n$ and asdim(T) ≤ 1 for a tree T with the geodesic metric. This is a quasiisometric invariant, so it is well defined for finitely generated groups as well. See [12] for the basic properties of asdim. There are many groups that contain $\mathbb{Z}^n$ for every n, and they will have infinite asymptotic dimension. However, Gromov proved:

Theorem 2.7 ([72]). *Every hyperbolic group has finite asymptotic dimension.*

One can hardly make a claim that one understands the large-scale geometry of a group if its asymptotic dimension is not known to be finite or infinite. However, the significance of the theorem became particularly clear with the work of Guoliang Yu [143] (see also [57]), who proved that groups with finite asdim and finite classifying space satisfy the Novikov conjecture (this predicts the possible placement of Pontrjagin classes in the cohomology ring of a closed oriented manifold with the given fundamental group).

An even stronger conjecture in manifold topology is the Farrell–Jones conjecture. If it holds for a (torsion-free) group G then one can in principle compute the set of closed manifolds homotopy equivalent to a given closed manifold of dimension ≥ 5 and fundamental group G. Following the work of Farrell and Jones, there has been a great progress in proving the Farrell–Jones conjecture for many groups. For hyperbolic groups, this was done by Bartel, Lück, and Reich [5], see also [3] for a proof using coarse methods that generalize to other groups.

2.6. JSJ decomposition

For simplicity, we now assume that G is a torsion-free hyperbolic group. By Grushko's theorem [75,132], G can be decomposed as a free product $G = G_1 * G_2 * \cdots * G_k * F_r$ where each G_i is noncyclic and freely indecomposable and F_r is a free group. Each G_i is a 1-ended group by the celebrated theorem of Stallings [133], meaning that the Cayley graph of G_i has one end (every finite subgraph has only one unbounded complementary component). Quite unexpectedly, Rips–Sela [119] discovered a further structure theorem for 1-ended hyperbolic groups (the theorem applies to many groups that are not hyperbolic as well). The theorem is motivated by the Jaco–Shalen–Johanssen torus decomposition theorem for 3-manifolds, which provides a canonical decomposition of an aspherical closed orientable 3-manifold by cutting along pairwise disjoint tori so that each piece either has many tori (it is Seifert fibered), or it is not an I-bundle and has no essential tori (except on the boundary, and then by Thurston's hyperbolization theorem it is hyperbolic), or it is an I-bundle. The Rips–Sela theorem can be stated as follows:

Theorem 2.8. *Let G be a 1-ended torsion-free hyperbolic group. Then G is a finite graph of groups with all edge group infinite cyclic, and with vertex groups V coming in three types:*

(QH) *V is the fundamental group of a compact surface (with a pair of intersecting 2-sided simple closed curves) and the incident edge groups correspond exactly to the boundary components,* (rigid) *V is not cyclic and does not admit a nontrivial splitting over a cyclic group such that all incident edge groups are elliptic, and*

(cyclic) *V is cyclic.*

See also [59,64,76] for different proofs and generalizations, and [40] for how to read off the JSJ decomposition purely from the boundary of G. For example, a splitting over $\mathbb{Z}$ gives a pair of points in ∂G that together separate ∂G, and Bowditch shows how to go in the other direction. Thus the QH vertices give rise to many splittings of G over cyclic groups (one for every simple closed curve), while rigid vertices give rise to none.

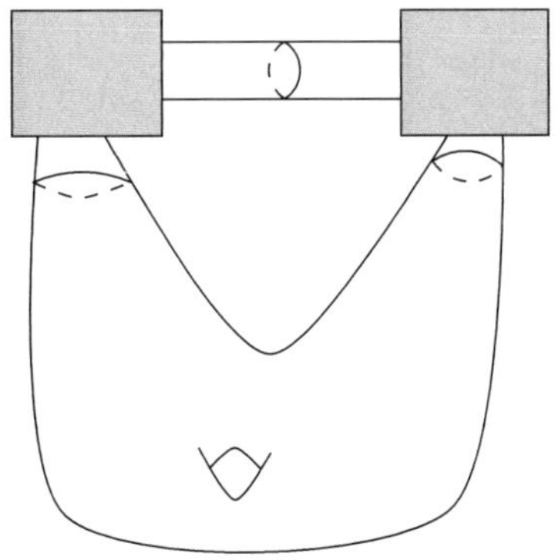

FIGURE 2
A possible JSJ decomposition of a group G, with two rigid vertices and one QH vertex.

We can picture G as the fundamental group of the space obtained from a disjoint union of compact surfaces, "black boxes" and circles by attaching cylinders according to the graph of groups. See Figure 2. The JSJ decomposition is not quite unique, but there are standard moves that transform one such decomposition to another. For example, sometimes one can slide one cylinder over another if they meet at a common circle. The main feature of a JSJ decomposition is that splittings over cyclic groups can be "read off," at least up the standard moves, just like all essential tori in a 3-manifold can be read off from its JSJ decomposition.

2.7. The combination theorem

This is also motivated by 3-manifold theory. The classical Klein–Maskit combination theorem gives conditions under which two discrete groups A, B of isometries of hyperbolic space $\mathbb{H}^3$ with intersection $C = A \cap B$ generate the amalgam $A *_C B$. Thurston's Hyperbolization Theorem [101,138] is proved by cutting the 3-manifold into pieces, and then inductively constructing a hyperbolic structure when gluing the pieces together. There are two opposite extremes in the kinds of gluings, when the intersection of the pieces is quasi-isometrically embedded on both sides, and when it is exponentially distorted. The latter arises when the 3-manifold fibers over the circle and the monodromy is pseudo-Anosov. The following is the hyperbolic group analog.

Theorem 2.9 ([25,26]). *Let G be the fundamental group of a finite graph of hyperbolic groups so that each edge group is quasiisometrically embedded in both vertex groups (but not necessarily in G). Assume the "annuli flare" condition. Then G is a hyperbolic group.*

The precise definition of the annuli flare condition is a bit technical, but let us mention two special cases. The first is when the graph of groups is *acylindrical*, that is, for some $M > 0$ the stabilizer of every segment of length M in the associated Bass–Serre tree is finite. In this case there are no (long) annuli at all. The other case is that of a hyperbolic automorphism $\phi : H \to H$ of a hyperbolic group H. This means that there is $M > 0$ such that for every element $h \in H$ of sufficiently large word length $|h|$ we have

$$\max\{|\phi^M(h)|, |\phi^{-M}(h)|\} \geq 2|h|,$$

so in this case the induced infinite annulus defined on $S^1 \times \mathbb{R}$ sending $S^1 \times \{K\}$ to the loop determined by $\phi^K(h)$ flares exponentially. Aside from automorphisms of closed surface groups induced by pseudo-Anosov homeomorphisms, there are many examples (in fact, they are generic in the sense of random walk [87]) of hyperbolic automorphisms of free groups coming from train track theory [34]. The combination theorem then implies that the mapping torus $H \rtimes_\phi \mathbb{Z}$ is hyperbolic.

The combination theorem has also been used to study hyperbolicity of extensions of free or surface groups in terms of the monodromy homomorphism from the quotient group to the mapping class group or $\mathrm{Out}(F_n)$, giving rise to *convex cocompact subgroups* of these groups [56,63,78,89].

2.8. Random groups are hyperbolic

The most straightforward way to talk about "random groups" is the following model. Fix integers $k \geq 2$ and $m \geq 1$, and for integers $n_1, \ldots, n_m$ consider the finite set

$$N(k, m; n_1, \ldots, n_m)$$

of all group presentations with k generators and m relators of lengths $n_1, \ldots, n_m$. We say that a *random group has property P* if the fraction of groups in $N(k, m; n_1, \ldots, n_m)$ that have P goes to 1 as $\min\{n_1, \ldots, n_m\} \to \infty$.

Theorem 2.10 ([50,110]). *A random group is hyperbolic and its boundary is the Menger curve.*

Thus a random group has rational cohomological dimension 2 and does not split over a finite or a 2-ended group.

Gromov [73] introduced a more sophisticated random model for groups, called the *density model*, that depends on a parameter $d \in (0, 1)$ and properties of random group depend on the chosen range of d. For more information, see [67,111].

2.9. $\mathbb{R}$-trees and applications

$\mathbb{R}$-trees are metric spaces such that any two distinct points x, y are contained in a unique subspace homeomorphic to a closed interval in $\mathbb{R}$ with x, y corresponding to the endpoints, and this subspace is isometric to a closed interval. Simplicial trees with the length metric induced by identifying edges with closed intervals are examples of $\mathbb{R}$-trees. More generally, $\mathbb{R}$-trees can have a dense set of "vertices" (points whose complement has more than two components). For example, let $T = \mathbb{R}^2$ as the underlying set, and define the metric d as follows: $d(x, y) = |x - y|$ is the Euclidean distance if x, y are on the same vertical line, and otherwise if $x = (x_1, x_2)$, $y = (y_1, y_2)$, then $d(x, y) = |y_1| + |y_2| + |x_1 - x_2|$. Thus one imagines train lines running along all vertical lines and along the x-axis, with the distance function being the shortest train trip.

$\mathbb{R}$-trees were put to good use by Morgan and Shalen **[102–104]** in their work on hyperbolization of 3-manifolds following Thurston's work.

The importance of $\mathbb{R}$-trees in geometric group theory comes from two principles that we briefly review. Let X be a proper hyperbolic space with the isometry group of X acting with coarsely dense orbits.

(1) A sequence of actions of a finitely generated group G on X either, after taking a subsequence, converges (after conjugations) to an isometric action on X, or else it converges to an isometric action on an $\mathbb{R}$-tree.

(2) There is a theory analogous to the Bass–Serre theory, called the "Rips machine," that explains the structure of a group acting isometrically on an $\mathbb{R}$-tree from the stabilizers of the action (under some technical conditions).

2.10. Hyperbolic spaces degenerate to $\mathbb{R}$-trees

This construction is due to F. Paulin **[116]** and the author **[13]**. See also **[14]**. We fix a group G and a finite generating set $a_1, \ldots, a_n$. Suppose we are given an isometric action $\rho : g \mapsto \rho(g) : X \to X$ of G on a proper δ-hyperbolic space X, defined up to conjugation by an isometry of X. We impose the mild assumption that the action is *nonelementary*, i.e., the function

$$x \mapsto \max_j \{d_X(x, a_j(x))\}$$

is a proper function $X \to [0, \infty)$. We then choose a basepoint $x_\rho \in X$ where the minimum is attained. Identifying G with the orbit of x_ρ, this induces a left-invariant (pseudo)metric on G via

$$d_\rho(g, h) = d_X(g(x_\rho), h(x_\rho)).$$

This metric is "hyperbolic," although G as a discrete set is not a geodesic metric space. To make this precise, it is convenient to give Gromov's reformulation of δ-hyperbolicity, in terms of the "4-point condition." For $a, b \in X$, define the "Gromov product"

$$(a \cdot b) = \frac{1}{2}\big(d_X(x_\rho, a) + d_X(x_\rho, b) - d_X(a, b)\big).$$

Thus, when X is a tree, $(a \cdot b)$ is the distance between x_ρ and $[a, b]$, and in general it is within 2δ of it. If $a, b, c \in X$ then consider the 3 numbers $(a \cdot b)$, $(b \cdot c)$, and $(c \cdot a)$. When X is a tree, the two smaller numbers are equal. Gromov's 4-point condition is that the two smaller numbers are within δ of each other. Up to changing the value of δ, a geodesic metric space is hyperbolic if and only if it satisfies the 4-point condition. Moreover, if the 4-point condition holds with $\delta = 0$, then the space can be isometrically embedded in an $\mathbb{R}$-tree.

Returning to our setup, assume now that ρ_i is a sequence of isometric actions of G on X, x_{ρ_i} are the corresponding basepoints, and d_{ρ_i} the induced metrics on G. They all satisfy the 4-point condition with a fixed δ. There are now two cases, up to passing to a subsequence. Define $D_i = \max_j \{x_{\rho_i}, a_j(x_{\rho_i})\}$.

Case 1. $D_i \to \infty$. Then rescale the metrics d_{ρ_i} by D_i, i.e., consider d_{ρ_i}/D_i. After a subsequence, this will converge to a (pseudo)metric on G which will now satisfy the 4-point condition with $\delta = 0$. Thus (G, d) can be isometrically embedded into a (unique) $\mathbb{R}$-tree T and there will be an induced isometric action of G on T. Thanks to the careful choice of basepoints, this action will not have a global fixed point.

Case 2. D_i stays bounded. Under the mild condition that the isometry group of X acts with coarsely dense orbits, we can conjugate the given actions so that all x_{ρ_i} belong to a fixed bounded set. Since X is proper, there is a further subsequence so that ρ_i converge to an isometric action ρ of G on X.

2.11. The Rips machine

If a group acts freely on a simplicial tree, it is necessarily free. This simple instance of Bass–Serre theory follows quickly from covering space theory. However, this is not true for $\mathbb{R}$-trees. For example, $\mathbb{Z}^n$ acts freely on $\mathbb{R}$ by letting basis elements act by n rationally independent translations. More interestingly, closed surfaces of Euler characteristic < -1 admit measured foliations with simple singularities and with all leaves being trees (and all but finitely many are lines), see [140]. Lifting to the universal cover, the transverse measure turns the leaf space to an $\mathbb{R}$-tree and the deck group induces a free action of the fundamental group of the surface on this $\mathbb{R}$-tree.

Suppose now we are given an isometric action of a finitely presented group G on an $\mathbb{R}$-tree T. We make a technical condition that the action is *stable* meaning that for every arc $I \subset T$ there is a subarc $J \subset I$ such that the stabilizer of J is equal to the stabilizer of any further subarc of J. This property is frequently satisfied for actions on $\mathbb{R}$-trees obtained by degenerating δ-hyperbolic spaces described above. We then fix a finite simplicial 2-complex K with $G = \pi_1(K)$ and construct a G-equivariant map $\tilde{K} \to T$, called a *resolution* of T. Point inverses form a foliation of $\tilde{K}$ (with certain standard singularities) which descends to K. The Rips machine transforms K with this foliation, changing neither the fundamental group nor the fact that the universal cover resolves T, and puts it in a certain "normal form." The pieces of this normal form are foliated subcomplexes that occur, very surprisingly, in only the following four types:

(simplicial) leaves are compact and the piece resolves a simplicial tree,

(surface) the piece is a surface (perhaps with boundary) and the nonboundary leaves are trees as above, (axial) the piece resolves the tree which is a line, and

(Levitt) the piece is of *Levitt type*.

Levitt-type foliations were first constructed by G. Levitt [91]. Generic leaves are 1-ended graphs, and in fact they are quasiisometric to 1-ended trees with finite graphs attached. In addition to proving this classification, the Rips machine also provides the structure of the group corresponding to these cases, and particularly in the Levitt case. It turns out that if there is a Levitt piece then G always splits along a subgroup which fixes an arc in T. The other three cases are classical, with the simplicial case amounting to Bass–Serre theory. As an example, Rips proved the conjecture of Morgan and Shalen that any finitely generated group acting freely on an $\mathbb{R}$-tree is isomorphic to the free product of surface groups and free abelian groups. For more details, see [28, 66].

2.12. Applications

We mention some of the applications of $\mathbb{R}$-trees; for more see [14]. They are a basic tool in the theory of $\mathrm{Out}(F_n)$. Zlil Sela used them extensively in his seminal work on the Tarski problems [124–130].

2.12.1. Automorphisms of hyperbolic groups

Let G be a 1-ended hyperbolic group, and for simplicity assume it is torsion-free. Combining Paulin's construction [117] with the Rips machine, we get

Theorem 2.11. *If G does not split over $\mathbb{Z}$ then* $\mathrm{Out}(G)$ *is finite.*

This is analogous to a consequence of Mostow Rigidity that $\mathrm{Out}(G)$ is finite when G is the fundamental group of a closed hyperbolic n-manifold with $n \geq 3$.

The proof goes like this. Assuming $\mathrm{Out}(G)$ is infinite, choose a sequence f_i of automorphisms in distinct classes and consider isometric actions ρ_i of G on itself given by left translations twisted by f_i, i.e., $g \mapsto (h \mapsto f_i(g)h)$. Since f_i are distinct in $\mathrm{Out}(G)$, we see that we are in Case 2 of the construction outlined above and we obtain an isometric action of G on an $\mathbb{R}$-tree and with arc stabilizers cyclic (or trivial). The Rips machine now yields a splitting of G over a cyclic group.

A proper generalization of this theorem was given by Z. Sela. Fix a JSJ decomposition of G. There are now "visible" automorphisms of G realized as compositions of powers of Dehn twists in the cylinders and homeomorphisms of the QH vertices, which are surfaces.

Theorem 2.12 ([118]). *The subgroup of visible automorphisms has finite index in* $\mathrm{Out}(G)$.

The proof is quite a bit harder. The idea is that if the index is infinite, one can choose a sequence of automorphisms f_i in distinct cosets of the visible subgroup. In addition, one chooses the f_i's to be the "shortest" in their cosets. Then one argues that the action in the limit produces a "new" splitting of G, one not explained by the JSJ, or else the f_i could be shortened for large i.

Recall that a group G is *Hopfian* if every surjective endomorphism of G is an automorphism and it is co-Hopfian if every injective endomorphism is an automorphism. For example, nontrivial free groups are not co-Hopfian. By adapting the above methods to endomorphisms, Sela proved:

Theorem 2.13 ([122, 123]). *Let G be torsion-free hyperbolic. Then G is Hopfian. If G is 1-ended it is also co-Hopfian.*

In 1911 Max Dehn proposed three algorithmic problems about groups: the word problem (decide if a word in the generators represents the trivial element), the conjugacy problem (decide if two words in the generators represent conjugate elements), and the isomorphism problem (decide if two groups given by presentations are isomorphic). Dehn solved the word problem for surface groups and his solution generalizes to hyperbolic groups. There is also a similar solution of the conjugacy problem for hyperbolic groups, see [71]. The isomorphism problem takes more work and uses $\mathbb{R}$-trees. For torsion-free hyperbolic groups that do not split over cyclic subgroups, the isomorphism problem was solved by Sela [121], and for general hyperbolic groups by Dahmani–Guirardel [54].

Even though hyperbolic groups are generally very well behaved, they also contain a certain amount of pathologies, see, e.g., [46].

2.12.2. Local connectivity of ∂G

The use of $\mathbb{R}$-trees completed the proof of the following theorem.

Theorem 2.14. *If G is a 1-ended hyperbolic group, then ∂G is locally connected (as well as connected).*

There are several ingredients in the proof. First, [38] shows that if ∂G is not locally connected then it has (many) cut points. Bowditch [40] then shows that G acts on an $\mathbb{R}$-tree constructed as a kind of a "dual" tree, which does not come with a metric but can be endowed with one using [92]. The Rips machine then yields a splitting of G over a 2-ended group, finishing the proof if such splittings do not exist. Swarup [136] finished the proof in the general case by showing how to continue refining these splittings (in the presence of cut points in ∂G) until the full JSJ decomposition is obtained, at which point a contradiction arises with any further splitting.

2.12.3. Thurston's compactness theorem

With the machinery of $\mathbb{R}$-trees one can give a quick proof of the following theorem.

Theorem 2.15 ([139]). *Let M be a compact aspherical 3-manifold whose fundamental group does not split over a cyclic group. Then the space of hyperbolic structures $H(M)$ on M is compact.*

The space $H(M)$ is the space of discrete and faithful representations of $G = \pi_1(M)$ into the orientation isometry group $\mathrm{PSL}_2(\mathcal{C})$ of hyperbolic 3-space $\mathbb{H}^3$, up to conjugacy (it takes some work to see that the quotient of $\mathbb{H}^3$ by such a group is homeomorphic to the interior of M). Indeed, to rule out Case 2 above, one shows that the limiting action on an $\mathbb{R}$-tree is stable and has abelian arc stabilizers (which follows from discreteness and faithfulness).

3. MAPPING CLASS GROUPS

A fundamental shift in the subject occurred after the work of Masur and Minsky **[95, 96]** on mapping class groups, the work that set the foundations for an eventual understanding of the large scale geometry of these groups. Mapping class groups are not hyperbolic (except for some sporadic surfaces) but naturally act on hyperbolic spaces.

We start by recalling some definitions. Let S be an orientable surface of finite type, i.e., one obtained from a closed orientable surface by removing finitely many points (called punctures). The group $\mathrm{Homeo}_+(S)$ of orientation preserving homeomorphisms of S has the natural compact-open topology which makes it locally path-connected, and the mapping class group (or the Teichmüller modular group) $\mathrm{Mod}(S)$ is the discrete group of (path) components of $\mathrm{Homeo}_+(S)$. Classically, this group has been studied since the early 20th century. A very nice introduction to the subject is the book **[62]**, and we will freely use the standard concepts. For example, the subgroup $\mathrm{PMod}(S)$ of "pure" mapping classes (those that fix the punctures) is generated by finitely many Dehn twists and the group will not be hyperbolic if S is big enough to contain two essential (not bounding a disk or a punctured disk) nonparallel (not cobounding an annulus) disjoint simple closed curves.

To the surface S Harvey **[83]** associates a simplicial complex $\mathcal{C} = \mathcal{C}(S)$, called the *curve complex* of S. A vertex is an isotopy class of essential simple closed curves. A collection of distinct vertices spans a simplex if each pair can be represented by curves that intersect minimally (most of the time this means "disjointly," but in a torus punctured at most once it means "once" and in a four times punctured sphere it means "twice"). For the purposes of this discussion, we restrict to the 1-skeleton (called the *curve graph*), which we equip with the length metric with all edges of length 1. The group $\mathrm{Mod}(S)$ acts naturally on $\mathcal{C}(S)$. For some very small surfaces, like a 3 times punctured sphere, the curve complex is empty, but otherwise it is infinite, and even locally infinite, a big contrast with Cayley graphs of hyperbolic groups. In a similar way, one can define the arc complex of a surface with punctures.

Theorem 3.1 (**[95]**). *$\mathcal{C}(S)$ is hyperbolic. An element of* $\mathrm{Mod}(S)$ *acts loxodromically if and only if it is pseudo-Anosov.*

Here are some ideas in the original proof, which uses Teichmüller theory. Let $\mathcal{T} = \mathcal{T}(S)$ be the Teichmüller space of S, i.e., the space of all (marked) hyperbolic structures on S. There is a natural coarse map $\pi : \mathcal{T} \to \mathcal{C}$ that to a hyperbolic metric on S assigns (the isotopy class of) a shortest simple closed geodesic. Any two points in $\mathcal{T}$ are joined by a unique Teichmüller geodesic, and their images under π form a family of coarse paths in $\mathcal{C}$ satisfying (and this needs proof):

- any two points in $\mathcal{C}$ are connected by some such path,
- the family is closed under taking subpaths,
- any two paths in the collection starting at nearby points are contained in each other's uniform Hausdorff neighborhood (i.e., they *fellow travel*), and
- triangles formed by these paths are uniformly thin.

Thus the collection behaves like the collection of geodesics in a hyperbolic space. Remarkably, the existence of such a collection of paths implies that the space is hyperbolic and the paths are (reparametrized) quasigeodesics with uniform constants. See [97], which proves that arc complexes are hyperbolic, and [42].

Since the original proof of hyperbolicity of $\mathcal{C}(S)$, there have been others, the simplest being [84], not using Teichmüller theory at all but constructing a family of paths in $\mathcal{C}(S)$ directly using surgeries on curves. Perhaps surprisingly, the more recent proofs also show that curve graphs are *uniformly* hyperbolic, i.e., δ can be taken independently of the surface.

3.1. The boundary of the curve complex

If X is a hyperbolic space which is not proper, its boundary ∂X may not be compact. For example, the boundary of the wedge of countably many rays joined at the initial point is a discrete countable set, and the boundary of a tree all of whose vertices have countable valence is homeomorphic to the irrationals.

In [90] E. Klarreich identified the boundary $\partial\mathcal{C}$ of the curve complex as a proper quotient of a subspace of Thurston's boundary of Teichmüller space $\mathcal{T}$. This description serves as a model for boundaries of other hyperbolic complexes.

3.2. WPD, acylindrically hyperbolic groups, quasimorphisms

In the absence of properness of the action, one needs some kind of a substitute. The property WPD (for "weak proper discontinuity") was introduced in [32].

Definition 3.2. Suppose a group G acts by isometries on a hyperbolic space X. A loxodromic element $g \in G$ is WPD if for every $x \in X$ and $C > 0$ there is $N > 0$ such that the set

$$\{h \in G \mid d(x, h(x)) \le C, d(g^N(x), hg^N(x)) \le C\}$$

is finite. The action of G on X is WPD if G is not virtually cyclic and every loxodromic element is WPD.

The WPD condition says that the collection of translates of an axis (or an orbit) of a loxodromic element is discrete: any two translates are either parallel or else they are in a bounded Hausdorff neighborhood of each other only along a bounded length interval.

Theorem 3.3 ([32]). *The action of* $\mathrm{Mod}(S)$ *on* $\mathcal{C}(S)$ *is WPD. If a nonvirtually cyclic group acts isometrically on a hyperbolic space with a WPD element then the space* $\widetilde{\mathrm{QH}}(G)$ *of (reduced) quasimorphisms on* G *is infinite-dimensional.*

A *quasimorphism* is a function $f : G \to \mathbb{R}$ such that

$$\sup_{a,b\in G} \left| f(ab) - f(a) - f(b) \right| < \infty.$$

Denote by $\mathrm{QH}(G)$ the vector space of all quasimorphisms on G and note the vector subspaces $\mathrm{Hom}(G, \mathbb{R})$ of homomorphisms $G \to \mathbb{R}$ and $B(G)$ of bounded functions on G. Then the space $\widetilde{\mathrm{QH}}(G)$ is defined as the quotient

$$\widetilde{\mathrm{QH}}(G) = \mathrm{QH}(G) / \big(\mathrm{Hom}(G, \mathbb{R}) + B(G)\big)$$

and it can also be identified with the kernel of the natural homomorphism $H_b^2(G; \mathbb{R}) \to H^2(G; \mathbb{R})$ from bounded cohomology of G. For more on bounded cohomology, see [49].

The basic method for showing $\widetilde{\mathrm{QH}}(G)$ is infinite-dimensional is due to Brooks [47] in the case of free groups. Fix a free group F with a basis $a_1, a_2, \ldots$ Let w be any cyclically reduced word in the basis. Define $f_w : F \to \mathbb{Z} \subset \mathbb{R}$ as $f_w(x) = C_w(x) - C_{w^{-1}}(x)$, where $C_w(x)$ is the number of occurrences of w as a subword of x, written as a reduced word. That f_w is a quasimorphism can be seen by considering the tripod in the Cayley tree of F spanned by $1, a$, and ab, and marking all occurrences of $w^{\pm 1}$ along it. All such occurrences that do not contain the central vertex will be counted twice, with opposite signs, in the expression $f(ab) - f(a) - f(b)$, and, of course, the number occurrences that do contain the central vertex is uniformly bounded. With a bit more work, one can show that for a suitable choice of w_i's the quasimorphisms f_{w_i} will yield linearly independent elements of $\widetilde{\mathrm{QH}}(F)$. The proof of the second half of Theorem 3.3 is a coarse version of this method, where w is replaced by a long segment along an axis of a WPD element, and the discreteness of the set of translates guarantees that the counting function is finite.

A quick application is the following statement, suggesting that pseudo-Anosov elements of $\mathrm{Mod}(S)$ are "generic."

Corollary 3.4. *Fix a finite generating set and the corresponding word metric on* $\mathrm{Mod}(S)$. *For any* $R > 0$, *there exists* $M > 0$ *such that every ball of radius* M *contains a ball of radius* R *that consist entirely of pseudo-Anosov mapping classes.*

This follows quickly from the feature of the quasimorphisms on $\mathrm{Mod}(S)$ constructed above that they are uniformly bounded on all elements of $\mathrm{Mod}(S)$ which are not pseudo-Anosov.

Bowditch noticed that the action of $\mathrm{Mod}(S)$ on $\mathcal{C}(S)$ satisfies a property stronger than WPD.

Definition 3.5. An isometric action of G on a hyperbolic space X is *acylindrical* if for all $r > 0$ there exist $R, N > 0$ so that whenever $a, b \in X$ with $d(a, b) \geq R$, then there are at most N elements h of G such that $d(a, h(a)) \leq r$ and $d(b, h(b)) \leq r$.

Thus acylindricity gives control in all directions, not only along axes of loxodromic elements.

Theorem 3.6 ([41]). *The action of* $\mathrm{Mod}(S)$ *on* $\mathcal{C}(S)$ *is acylindrical.*

These results motivated Denis Osin to propose acylindrically hyperbolic groups as a generalization of hyperbolic groups. A group is *acylindrically hyperbolic* if it is not virtually cyclic and admits an acylindrical action on a hyperbolic space with unbounded orbits. This class contains many groups of interest (e.g., mapping class groups and $\mathrm{Out}(F_n)$) and many constructions on hyperbolic groups carry over to this larger class, e.g., small cancelation theory, or quasimorphisms indicated above; see [113,115].

3.3. Subsurface projections

The main drawback of acylindrically hyperbolic groups is that in general one does not have access to elements that do not act loxodromically. In the case of mapping class groups, this problem is resolved through *subsurface projections* of Masur and Minsky [95,96].

Let S be a surface as before and $X \subset S$ a connected π_1-injective subsurface which is closed as a subset. Let α be a simple closed curve in S which cannot be homotoped in the complement of X and which is in minimal position with respect to ∂X. Then the intersection $\alpha \cap X$ consists of finitely many disjoint arcs (or just α if $\alpha \subset X$). For each such arc J, consider one or two curves obtained as follows. If the endpoints of J are contained in the same boundary component b of X, there are two ways of closing up J to a closed curve by adding an arc in b; take both of these curves. If the endpoints of J are on distinct boundary components b, b' then form a curve by taking two parallel copies of J and connect them by adding "long" arcs in b and b'. It is not hard to see that taking the collection of all these curves for all arcs J produces a uniformly bounded set $\pi_X(\alpha) \subset \mathcal{C}(X)$ (we collapse all boundary components of X to punctures). This construction makes sense whenever $\mathcal{C}(X)$ is defined (so notably a pair of pants is excluded). It also makes sense when X is an annulus, in which case the curve complex is formed by arcs joining the boundary components, but we will not describe this case in detail. If α is disjoint from X then $\pi_X(\alpha)$ is not defined and we set it to be empty.

Now fix a finite collection of curves $\vec{\alpha} = \{\alpha_1, \dots, \alpha_n\}$ in S that "fill" the surface, i.e., every (essential) curve intersects at least one of them. By the classical fact that the distance in the curve complex is bounded by a function of the intersection number, if $\pi_X(\alpha_i)$ and $\pi_X(\alpha_j)$ are both defined then their union has uniformly bounded diameter (with the bound depending on the intersection number between α_i and α_j). We then define

$$\pi_X(\vec{\alpha}) = \bigcup_i \pi_X(\alpha_i).$$

This is always a nonempty, uniformly bounded subset of $\mathcal{C}(X)$.

The following is the fundamental result of Masur and Minsky, expressing (coarsely) the word metric in $\mathrm{Mod}(S)$ in terms of subsurface projections. For $K > 0$ and $x \geq 0$, define $\{\{x\}\}_K$ as 0 if $x < K$ and as x if $x \geq K$.

Theorem 3.7 (The distance formula, [96]). *For all sufficiently large K (depending on $\vec{\alpha}$) and for all $g, h \in \mathrm{Mod}(S)$, we have*

$$d(g,h) \asymp \sum_X \{\{d_X(g(\vec{\alpha}), h(\vec{\alpha}))\}\}_K.$$

The left-hand side is the distance in the word metric. The summation is over all (isotopy classes of) connected, π_1-injective subsurfaces X with $\mathcal{C}(X) \neq \emptyset$, and the displayed summand is the diameter of the set $\pi_X(g(\vec{\alpha})) \cup \pi_X(h(\vec{\alpha}))$. The symbol $\asymp$ means that there is a linear function (depending on K and the finite generating set of $\mathrm{Mod}(S)$) $f(x) = Ax + B$ such that the left-hand side is bounded by the f-value of the right-hand side, and vice versa. In particular, only finitely many terms are $\geq K$.

The distance formula is a powerful tool in the study of large-scale geometry of mapping class groups. It is used in an essential way in the following remarkable theorem, establishing quasiisometric rigidity of mapping class groups. To state the theorem, let $\mathrm{Mod}^{\pm}(S)$ denote the *extended* mapping class group, i.e., allowing orientation-reversing homeomorphisms (this is an index 2 extension of $\mathrm{Mod}(S)$). If G is a finitely generated group with a word metric, denote by $\mathrm{QI}(G)$ the group of quasiisometries $G \to G$ with the equivalence relation $f_1 \sim f_2$ if $\sup_g d(f_1(g), f_2(g)) < \infty$. There is a natural homomorphism $G \to \mathrm{QI}(G)$ sending g to the left translation by g.

Theorem 3.8 ([10, 79]). *Let S be a surface of finite type. Except for a small number of sporadic surfaces, the natural homomorphism $\mathrm{Mod}^{\pm}(S) \to \mathrm{QI}(\mathrm{Mod}^{\pm}(S))$ is an isomorphism. In particular, if G is any group quasiisometric to $\mathrm{Mod}(S)$, then there is a homomorphism $G \to \mathrm{Mod}^{\pm}(S)$ with finite kernel and finite index image.*

4. PROJECTION COMPLEXES

It is tempting to view the distance formula as saying that the coarse map

$$\mathrm{Mod}(S) \to \prod_X \mathcal{C}(X)$$

defined by $g \mapsto \pi_X(g(\vec{\alpha}))$ is a quasiisometric embedding, where we equip the right-hand side with the ℓ_1-metric. The trouble is that this is not really a metric, and "cutting off" at K in each coordinate would not satisfy the triangle inequality. Up to modifying each coordinate a bounded amount, the image of this map was identified in [7, 10]. The main restriction on the image is the following inequality.

Theorem 4.1 (Behrstock inequality, [7]). *There is a $\theta \geq 0$ such that the following holds. Suppose $X, Y \subset S$ are two subsurfaces such that the boundary of each intersects the other. Then at least one of $d_X(\partial Y, \vec{\alpha})$ and $d_Y(\partial X, \vec{\alpha})$ is $\leq \theta$.*

There is a simple proof of the Behrstock inequality, due to Chris Leininger, see [94]. If we focus on the two coordinates $\mathcal{C}(X) \times \mathcal{C}(Y)$, the inequality says that the image is contained in a Hausdorff neighborhood of the "wedge" of $\mathcal{C}(X) \times \{y\} \cup \{x\} \times \mathcal{C}(Y)$ where

$x = \pi_X(\partial Y)$ and $y = \pi_Y(\partial X)$. This suggests taking wedges instead of products for the right-hand side in order to fix the metrizability problem, and leads to the following construction that can be axiomatized.

Let $\mathcal{Y}$ be a collection of metric spaces (technically we allow the distance to be infinite, for example, we might have disconnected graphs with the path metric). Suppose that for distinct $X, Y \in \mathcal{Y}$ we are given a subset $\pi_X(Y) \subset X$. If $Z \in \mathcal{Y}$, $Z \neq X$, we define

$$d_X(Y, Z) = \operatorname{diam}\big(\pi_X(Y) \cup \pi_X(Z)\big).$$

We will assume that the following axioms hold for some fixed $\theta \geq 0$:

(P1) $d_X(Y, Y) \leq \theta$,

(P2) if $d_X(Y, Z) > \theta$ then $d_Y(X, Z) \leq \theta$, and

(P3) for $X \neq Z$, the set

$$\big\{Y \in \mathcal{Y} \mid d_Y(X, Z) > \theta\big\}$$

is finite.

There are many natural situations where these axioms hold.

Examples 4.2. (1) Let T be a simplicial tree and $\mathcal{Y}$ a collection of pairwise disjoint simplicial subtrees. The projection $\pi_X(Y)$ is the point of X nearest to Y. The axioms hold with $\theta = 0$. See Figure 3.

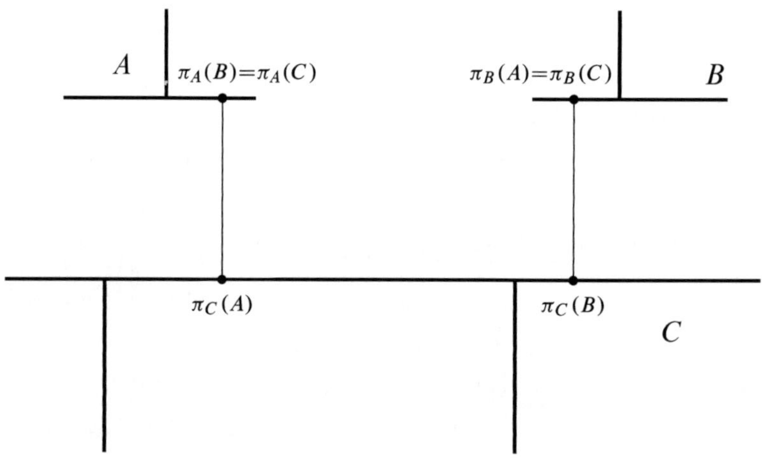

FIGURE 3
The situation of Example 4.2(1), $d_C(A, B) > 0$ while $d_A(B, C) = d_B(A, C) = 0$.

(2) Let S be a closed hyperbolic surface and γ an immersed closed geodesic which is not a multiple. In the universal cover $\tilde{S} = \mathbb{H}^2$ consider the set $\mathcal{Y}$ of all lifts of γ, and define projections as nearest point projections. A similar construction

can be performed with a group acting on a hyperbolic space and a maximal virtually cyclic subgroup that contains a WPD element.

(3) Let S be a complete hyperbolic surface of finite area and a cusp. In the universal cover $\tilde{S} = \mathbb{H}^2$, consider the set $\mathcal{Y}$ of all lifts of a fixed horocyclic curve in the cusp (with either the intrinsic or the induced metric). Again the projection is the nearest point projection. A similar construction can be performed with relatively hyperbolic groups.

(4) Let G be a group acting on a simplicial hyperbolic graph X and let H be the stabilizer of a vertex $v \in X$. Assume that H acts simply transitively on the edges incident to v, and that the metric on the link $\mathrm{Lk}(v, X)$ (which can be identified with H) induced by the path metric on $X \smallsetminus \{v\}$ is proper (finite radius balls contain finitely many points; here we allow distances to be infinite). Let $\mathcal{Y}$ be the collection of links of vertices in the orbit of v with this proper metric on each. If u, w are two distinct vertices in the orbit of v the projection of $\mathrm{Lk}(u, X)$ to $\mathrm{Lk}(w, X)$ is the set of points in $\mathrm{Lk}(w, X)$ that belong to a geodesic between u and w. If (G, H) admit such an action, H is said to be *hyperbolically embedded* in G; see [55]. For example, parabolic subgroups of hyperbolic groups, or maximal virtually cyclic subgroups containing a WPD element as in (2) are hyperbolically embedded, as can be seen by building the projection complex below.

(5) Let S be an orientable surface of finite type and let $\mathcal{Y}$ be a collection of isotopy classes of π_1-injective subsurfaces where subsurface projections are defined, and assume that if $X, Y \in \mathcal{Y}$ and $X \neq Y$ then ∂X is not disjoint from Y (up to isotopy). Define $\pi_Y(X) = \pi_Y(\partial X)$.

The construction of a projection complex $\mathcal{P}(\mathcal{Y})$ (and the blow-up version $\mathcal{C}(\mathcal{Y})$) is kind of a converse to Example 4.2(2) above, where one tries to "reconstruct" the ambient space from the projection data (though usually one gets a different ambient space).

Theorem 4.3 ([19], for a simpler construction see [22]). *Suppose the projection data*

$$\big(\mathcal{Y}, \pi_X(Y), \theta\big)$$

satisfy (P1)–(P3). There is a metric space $\mathcal{C}(\mathcal{Y})$ containing metric spaces in $\mathcal{Y}$ as pairwise disjoint isometrically embedded subspaces and so that $\pi_X(Y)$ agrees, up to a bounded error, with the nearest point projection of Y to X within $\mathcal{C}(\mathcal{Y})$. Moreover,

- *If each $Y \in \mathcal{Y}$ is δ-hyperbolic for some $\delta \geq 0$ then $\mathcal{C}(\mathcal{Y})$ is hyperbolic.*
- *If each $Y \in \mathcal{Y}$ is quasiisometric to a tree (a "quasitree") with fixed QI constants, then $\mathcal{C}(\mathcal{Y})$ is also a quasitree.*
- *If the collection $\mathcal{Y}$ consists of finitely many isometry types of metric spaces and they all have asymptotic dimension $\leq n$ then* $\operatorname{asdim} \mathcal{C}(\mathcal{Y}) \leq n + 1$.

- *The space $\mathcal{P}(\mathcal{Y})$ obtained from $\mathcal{C}(\mathcal{Y})$ by collapsing all embedded copies of spaces in $\mathcal{Y}$ is a quasitree.*
- *If a group G acts by isometries on $\bigsqcup_{Y\in\mathcal{Y}} Y$ preserving the projections (i.e., $g(\pi_X(Z) = \pi_{g(X)}(g(Z))$ for all $g \in G$) then G acts by isometries on $\mathcal{C}(\mathcal{Y})$ extending the action on $\bigsqcup_{Y\in\mathcal{Y}} Y$, and it also acts isometrically on $\mathcal{P}(\mathcal{Y})$.*

We briefly outline the construction. As indicated above, the idea is to start with the disjoint union of all $Y \in \mathcal{Y}$ and then for certain pairs (X, Z) add edges joining points in $\pi_X(Z)$ to points in $\pi_Z(X)$.

Step 1 is to promote (P2) to a stronger property (P2++):

(P2++) If $d_Y(X, Z) > \theta$ then $\pi_Y(X) = \pi_Y(Z)$.

This can be done by modifying the projection $\pi_X(Y)$ by a bounded amount and replacing θ by a larger constant. This modification preserves group equivariance.

In step 2, assuming (P1), (P2++), and (P3), one chooses a constant $K \geq 2\theta$ and posits that X and Z are connected by edges as above provided $d_Y(X, Z) \leq K$ for all $Y \neq X, Z$. The key property that makes the proof of Theorem 4.3 possible is that the set

$$\{X\} \cup \{Y \mid d_Y(X, Z) > K\} \cup \{Z\}$$

is finite (by (P3)) and is naturally linearly ordered giving a path from X to Z, called a *standard path*, in $\mathcal{P}_K(\mathcal{Y})$. These standard paths are quasigeodesics and behave very nicely. The construction depends on the choice of the constant K: when K is enlarged, there will be more edges attached.

We mention a few applications of this construction to mapping class groups.

Theorem 4.4 ([19]). asdim(Mod(S)) $< \infty$.

The basic idea is to replace the infinite product of curve complexes by a smaller space. The collection of all subsurfaces $\mathcal{Y}$ does not satisfy the assumptions of Example 4.2(5) above since subsurfaces can be disjoint or nested. However, one shows that there is a way to write $\mathcal{Y}$ equivariantly as a finite disjoint union $\sqcup \mathcal{Y}_i$ so that each collection $\mathcal{Y}_i$ satisfies Example 4.2(5). Thus one gets the spaces $\mathcal{C}(\mathcal{Y}_i)$. These are all hyperbolic, and crucially, have finite asymptotic dimension by Theorem 4.3 and the theorem of Bell–Fujiwara [11] that curve complexes have finite asymptotic dimension. Then we have a quasiisometric embedding

$$\mathrm{Mod}(S) \to \prod_i \mathcal{C}(\mathcal{Y}_i)$$

which finishes the proof since passing to finite products and subspaces preserves finiteness of asymptotic dimension.

There is quite a bit of inefficiency when we take the product of the blown-up projection complexes over the families $\mathcal{Y}_i$. There is a more involved system of axioms that keeps track of pairs of surfaces that are disjoint or nested leading to the notion of a *hierarchically hyperbolic group*, due to J. Behrstock, M. Hagen, and A. Sisto. For example, in [8] they derive a bound on asdim(Mod(S)), using [18], which is quadratic in the complexity of the surface. There are other applications of this theory, for example, in [9] they show how to understand quasiflats in mapping class groups and how to approximate a "hull" of a finite set by a CAT(0) cube complex.

Theorem 4.5 ([21]). *There is a classification, in terms of the Nielsen–Thurston normal form, of those elements g of* $\mathrm{Mod}(S)$ *that have stable commutator length* $\mathrm{scl}(g) = 0$.

Recall that for $g \in [G, G]$ $\mathrm{cl}(g)$ is the smallest k such that g can be written as a product of k commutators, and $\mathrm{scl}(g) = \lim_n \frac{\mathrm{cl}(g^n)}{n}$. By Bavard duality (see [49]), $\mathrm{scl}(g) > 0$ is equivalent to having a quasimorphism $G \to \mathbb{R}$ which is unbounded on the powers of g. Projection complexes are used to construct actions of finite index subgroups of Mod(S) on hyperbolic spaces with a power of a given element acting loxodromically, and then the Brooks method can be used to construct such quasimorphisms. It is worth stating this fact:

Theorem 4.6 ([21]). *Let S be a finite type surface. There is a torsion-free finite index subgroup $G <$ Mod(S) such that for every element $g \in G$ of infinite order there an action of G on a hyperbolic space such that g is loxodromic.*

For example, this applies to (powers of) Dehn twists. By contrast, a theorem of Bridson [43] says that whenever Mod(S) (with S of genus ≥ 3) acts on a CAT(0) space, Dehn twists have translation length 0.

Projection complexes are useful more generally for constructing quasicocycles on groups G with coefficients in orthogonal representations on strictly convex Banach spaces (such as $l^p(G)$ for $1 < p < \infty$); see [20].

Theorem 4.7 (Balasubramanya [2]). *If a group G acts on a hyperbolic space with a WPD element, then it admits a cobounded acylindrical action on a quasitree.*

Another proof of Balasubramanya's theorem is given in [22]. The quasitree is the projection complex applied to Example 4.2(2) and acylindricity is proved using the geometry of standard paths.

F. Dahmani, V. Guirardel, and D. Osin solved a long standing open problem when they proved the following.

Theorem 4.8 ([55]). *Let $\phi \in$ Mod(S) be a pseudo-Anosov mapping class. Then for a suitable power ϕ^n with $n > 0$ the subgroup normally generated by ϕ^n is free.*

They derive this theorem using the method of *rotating families*.

Theorem 4.9 ([55]). *For every $\delta \geq 0$ there is $R > 0$ such that the following holds. Let X be a δ-hyperbolic space and G a group of isometries of X. Let $C \subset X$ be a G-invariant set which is R-separated (meaning that $d(c, c') > R$ if $c, c' \in C$ are distinct). Suppose for every $c \in C$ we are given a subgroup G_c of the stabilizer* $\mathrm{Stab}_G(c)$ *such that*

(i) $G_{g(c)} = gG_c g^{-1}$ *for* $c \in C$ *and* $g \in G$, *and*

(ii) *if* $g \in G_c \smallsetminus \{1\}$, $c' \in C$ *and* $c' \neq c$ *then every geodesic from* c' *to* $g(c')$ *passes through* c.

Then the subgroup of G *generated by* $\cup_{c \in C} G_c$ *is the free product of a subcollection of the family* $\{G_c\}_{c \in C}$.

To prove Theorem 4.8, they apply this theorem to the space obtained from the curve complex $\mathcal{C}(S)$ by equivariantly coning off an orbit of the elementary closure $\mathrm{EC}(\phi)$. Pretending that this orbit is in an isometrically embedded line, one would attach the universal cover of a disk of large radius in $\mathbb{H}^2$ punctured at the center, and then completed to add the cone point back in. The set of these cone points is the set C from the theorem, and G_c is the cyclic group generated by (a conjugate of) a suitable power ϕ^n.

More recently, M. Clay, J. Mangahas, and D. Margalit proved a version of Theorem 4.9 that applies to projection complexes. Rotating families are replaced by *spinning families*.

Theorem 4.10 ([51]). *For every* θ *and* K, *there is* L *so that the following holds. Suppose a group* G *acts on the projection data and on the associated projection complex* $\mathcal{P} = \mathcal{P}_K(\mathcal{Y})$. *Suppose for every vertex* $v \in \mathcal{P}$ *we are given a subgroup* G_v *of the stabilizer* $Stab_G(v)$ *such that*

(i) $G_{g(v)} = gG_v g^{-1}$ *for any vertex* v *and* $g \in G$, *and*

(ii) *if* v, v' *are distinct vertices and* $g \in G_v \smallsetminus \{1\}$ *then* $d_v(v', g(v')) > L$.

Then the subgroup of G *generated by* $\bigcup_v G_v$ *is the free product of a subcollection of the family* $\{G_v\}_{v \in \mathcal{P}^{(0)}}$.

They derive Theorem 4.8 directly from Theorem 4.10 using the projection complex as in Example 4.2(2). They also prove several statements about normal closures of powers of other kinds of elements, or collections of elements, in $\mathrm{Mod}(S)$. One extreme behavior is that the normal closure is free, another that it is the whole $\mathrm{Mod}(S)$, but surprisingly there are examples when the normal closure turns out to be a certain kind of (infinitely generated) right angled Artin groups.

In [24] the two theorems above are revisited, and in particular the paper shows how to derive Theorem 4.9 from Theorem 4.10.

Here are two more applications of projection complexes to mapping class groups, though we will not comment on the proofs.

Theorem 4.11 ([4]). *Mapping class groups satisfy the Farrell–Jones conjecture.*

Theorem 4.12 ([60]). *Mapping class groups are semihyperbolic.*

This means that one can equivariantly choose uniform quasigeodesics connecting any pair of points in $\text{Mod}(S)$ so that they fellow-travel, i.e., if the endpoints are at distance ≤ 1 then each is in the other's uniform Hausdorff neighborhood.

5. GROUP $\text{Out}(F_n)$

Let F_n be the free group of rank $n \geq 2$, $\text{Aut}(F_n)$ its automorphism group, and $\text{Out}(F_n) = \text{Aut}(F_n)/F_n$ the *outer automorphism group* of F_n, obtained by quotienting out the inner automorphisms. This group has been studied for over a century, see Nielsen's paper [109] where he proves that $\text{Out}(F_n)$ is generated by $n+1$ involutions. A big impediment in the study of $\text{Out}(F_n)$, and free groups in general, was the tendency to think of elements of free groups as words in a basis. A much more flexible approach is to think of a free group as the fundamental group of a graph, which is not necessarily a rose R_n (a wedge of n circles). For example, the proof that subgroups of free groups are free is essentially trivial using covering spaces and general graphs, while the more algebraic proof is much less transparent. In [134] J. Stallings introduced the operation of *folding* graphs and used it to show that many standard algorithmic problems about free groups have easy solutions.

5.1. Outer space

Given this philosophy, the definition of Culler–Vogtmann's Outer space CV_n should seem very natural. Fix the rose R_n. A point in CV_n is represented by a homotopy equivalence $h : R_n \to \Gamma$, called *marking*, where Γ is a finite graph with all vertices of valence > 2 equipped with a *metric* of volume 1, i.e., an assignment of positive numbers to its edges that add to 1. Two such markings $h : R_n \to \Gamma$ and $h' : R_n \to \Gamma'$ represent the same point in CV_n if there is an isometry $\phi : \Gamma \to \Gamma'$ such that ϕh is homotopic to h'. Formally, the definition is analogous to the definition of Teichmüller space, where metric graphs are replaced by hyperbolic surfaces. There are many useful analogies between mapping class groups and $\text{Out}(F_n)$, perhaps stemming from the classical theorem of Dehn–Nielsen–Baer (see [62]) that when G is the fundamental group of a closed orientable surface S then $\text{Out}(G) \cong \text{Mod}^{\pm}(S)$. While Teichmüller space is diffeomorphic to Euclidean space, Outer space is a contractible polyhedron and the study of $\text{Out}(F_n)$ is decidedly more combinatorial compared to the study of mapping class groups. The group $\text{Out}(F_n)$ acts naturally on CV_n by changing the marking. The action is proper. For more on Outer space and the consequences to the structure of $\text{Out}(F_n)$, see the original paper [53], as well as the excellent survey [141], and also [15].

5.2. The boundary of Outer space

By taking universal covers, another way to think about a point $h : R_n \to \Gamma$ in CV_n is as a free action of F_n on a simplicial metric tree. The construction in Section 2.10 then yields a compactification of CV_n with the points in the ideal boundary $\partial\,\text{CV}_n$ represented by actions of F_n on $\mathbb{R}$-trees (which are either nonsimplicial or non-free). This construction was carried out in [52]. Exactly which trees arise in $\partial\,\text{CV}_n$ was identified in [27, 85].

5.3. Lipschitz metric and train-track maps

There is a natural notion of a *Lipschitz distance* between two points $h_i : R_n \to \Gamma_i$, $i = 1, 2$. It is defined by

$$d(\Gamma_1, \Gamma_2) = \log \lambda$$

where $\lambda \geq 1$ is the smallest possible Lipschitz constant of all maps $f : \Gamma_1 \to \Gamma_2$ that commute with markings, i.e., $h_2 f$ is homotopic to h_1 (and Γ_i are viewed as geodesic metric spaces). This "metric" is not symmetric, but satisfies the triangle inequality $d(\Gamma_1, \Gamma_3) \leq d(\Gamma_1, \Gamma_2) + d(\Gamma_2, \Gamma_3)$, and $d(\Gamma, \Gamma') \geq 0$ with equality only for $\Gamma = \Gamma'$. This metric has interesting properties and displays a mixture of behaviors of the well-studied metrics on Teichmüller space (Teichmüller, Weil–Petersson, and Thurston metrics). It can be used in the $\mathrm{Out}(F_n)$ setting in a way similar to the Bers' proof of the Nielsen–Thurston classification of mapping classes (see [62]) to give a proof of the following train-track theorem; see [17].

Theorem 5.1 ([35]). *Every irreducible automorphism $\phi \in \mathrm{Out}(F_n)$ can be represented by a train-track map $f : \Gamma \to \Gamma$ for some $\Gamma \in \mathrm{CV}_n$.*

A marking gives an identification between $\pi_1(\Gamma)$ and F_n and $f : \Gamma \to \Gamma$ "represents" ϕ if the induced endomorphism on $\pi_1(\Gamma)$ is ϕ. We say that ϕ is *irreducible* if it cannot be represented by some $f : \Gamma \to \Gamma$ that leaves a proper subgraph with nontrivial π_1 invariant. The map f is a *train-track map* if all positive powers of f are locally injective on all edges of Γ. It is easy to control the growth of lengths of loops under iteration by train-track maps, which makes them important in the study of the dynamics of an automorphism. More generally, when ϕ is not irreducible, there are *relative* train-track representatives.

The Lipschitz metric admits geodesic paths, called *folding paths*, which are induced, in the spirit of Stallings, by identifying segments of the same length and issuing from the same vertex. For more on this, see [17].

5.4. Hyperbolic complexes

By analogy with the arc and curve complexes, there are several complexes where $\mathrm{Out}(F_n)$ acts.

5.4.1. The free splitting complex FS_n

This one is analogous to the arc complex. A k-simplex is a $(k+1)$-edge free splitting of F_n, i.e., it is a minimal action of F_n on a simplicial tree with vertices of valence > 2, with trivial edge stabilizers and with $(k+1)$ orbits of edges. Passing to a face is induced by equivariantly collapsing an orbit of edges. Outer space CV_n is naturally a subset of FS_n, which can be viewed as a "simplicial completion" of CV_n.

5.4.2. The cyclic splitting complex FZ_n

This is defined the same way, except that the edge stabilizers can be cyclic subgroups. It is analogous to the curve complex.

5.4.3. The free factor complex FF_n

This one is different from FZ_n but can also be viewed as an analog of the curve complex. A vertex of FF_n is a *proper free factor* $A < F_n$, i.e., a subgroup such that $F_n = A * B$ for some $A \neq 1 \neq B$, defined up to conjugation. A k-simplex is a k-tuple of distinct conjugacy classes of proper free factors that are nested after suitable conjugation.

There are natural coarse equivariant maps

$$\mathrm{CV}_n \to \mathrm{FS}_n \to \mathrm{FZ}_n \to \mathrm{FF}_n .$$

For example, $\mathrm{FS}_n \to \mathrm{FF}_n$ sends a free splitting to a nontrivial vertex group (or if they are all trivial, to a free factor represented by a subgraph of the quotient graph).

Now, it turns out that all three of these complexes are hyperbolic, and there are several others that this survey is not mentioning. The first hyperbolic $\mathrm{Out}(F_n)$-complex was constructed in [29], though it is not canonical. The hyperbolicity of FF_n was established in [30] along the lines of the Masur–Minsky's argument for the curve complex, by projecting folding paths from CV_n to FF_n. A novel argument by Handel–Mosher [80] established hyperbolicity of FS_n, by considering folding paths directly in FS_n. Kapovich–Rafi [88] found a general criterion that a Lipschitz map $X \to Y$ has to satisfy in order for the hyperbolicity of X to imply the hyperbolicity of Y. Essentially, Lipschitz images of thin triangles are thin triangles. The maps $\mathrm{FS}_n \to \mathrm{FZ}_n \to \mathrm{FF}_n$ satisfy the Kapovich–Rafi criterion, so the hyperbolicity of FS_n implies the hyperbolicity of the other two. Loxodromic elements in FF_n are precisely the *fully irreducible automorphisms* (those whose positive powers are irreducible) and they are all WPD (in FS_n there are more loxodromic elements and they are not all WPD). Thus the space of quasimorphisms on $\mathrm{Out}(F_n)$ is infinite-dimensional and $\mathrm{Out}(F_n)$ is acylindrically hyperbolic. Handel and Mosher [81, 82] extended this and proved the H_b^2-alternative: any subgroup of $\mathrm{Out}(F_n)$ which is not virtually abelian has an infinite-dimensional space of quasimorphisms. This recovers the theorem of Bridson and Wade [45] that no higher rank lattice embeds as a subgroup of $\mathrm{Out}(F_n)$. The proof is much more involved than the H_b^2-alternative for mapping class groups [32].

The boundary of FF_n was identified with a proper quotient of a subspace of $\partial\,\mathrm{CV}_n$ in [39] and in [77].

5.5. Subfactor projections

By analogy with the Masur–Minsky subsurface projections, there are *subfactor projections*, see [31, 137]. Let A, B be two proper free factors in F_n. Our goal is to define $\pi_A(B) \in \mathrm{FS}(A)$, the projection of B to the free splitting complex of A. Choose $\Gamma \in \mathrm{CV}_n$ so that B is represented by a subgraph Γ_B of Γ. Then represent A by an immersion $\Gamma_A \to \Gamma$. Thus Γ_A determines a simplex in Outer space for A, and can be projected to $\mathrm{FS}(A)$ (or $\mathrm{FF}(A)$). It takes some work to show that coarsely this projection does not depend on the choice of Γ, at least when A and B are sufficiently far apart in FF_n. Moreover, the set $\mathcal{Y}$ of all free factors can be equivariantly and finitely partitioned into $\sqcup \mathcal{Y}_i$ so that projection is defined within each $\mathcal{Y}_i$, and this projection satisfies the projection axioms. One then gets a map

$$\mathrm{Out}(F_n) \to \prod_i \mathcal{C}(\mathcal{Y}_i)$$

in the same way as for mapping class groups (see the discussion after Theorem 4.4). However, here the map is *not* a quasiisometric embedding. The main issue is that there is no analog of annulus projections: when A has rank 1, the corresponding complex $\mathrm{FS}(A)$ is a single point. For example, the orbits on the right-hand side under the powers of any polynomially growing automorphism are bounded. For more on this, see [142].

5.6. Questions

The following is the key question, if one hopes to understand $\mathrm{Out}(F_n)$ using hyperbolic methods. The other questions reiterate the state of affairs that the large scale geometry of $\mathrm{Out}(F_n)$ is lagging behind the one of mapping class groups.

(1) Given $\phi \in \mathrm{Out}(F_n)$ of infinite order, is there a finite index subgroup $G < \mathrm{Out}(F_n)$ and an isometric action of G on a hyperbolic space so that a positive power of ϕ that belongs to G acts loxodromically?

This is true for mapping class groups (see Theorem 4.6), and it is also true for automorphisms ϕ that grow exponentially.

(2) Do any of hyperbolic $\mathrm{Out}(F_n)$-complexes admit *tight (quasi-)geodesics*?

These were defined for curve complexes by Masur and Minsky, and a very strong finiteness property was established by Bowditch [41]. Thus the question is asking for an equivariant collection of uniform quasigeodesics so that any two are connected by at least one, but only finitely many of these.

Bowditch used his strong finiteness of tight geodesics to show that translation lengths in the curve complex are rational, and Bell–Fujiwara [11] used it to show that curve complexes have finite asymptotic dimension.

(3) Do the hyperbolic $\mathrm{Out}(F_n)$-complexes $\mathrm{FS}_n, \mathrm{FZ}_n, \mathrm{FF}_n$ have finite asymptotic dimension? Are the translation lengths always rational? Does $\mathrm{Out}(F_n)$ have finite asymptotic dimension?

We remark that the Novikov conjecture is known for $\mathrm{Out}(F_n)$ [33].

The following seems out of reach with the present methods, although [36] is a promising start:

(4) Does $\mathrm{Out}(F_n)$ satisfy the Farrell–Jones conjecture?

(5) Does the local and global connectivity of $\partial\,\mathrm{FF}_n$ go to infinity as $n \to \infty$?

By the work of Gabai [65], the answer is yes for the boundary of the curve complex. Each $\partial\,\mathrm{FF}_n$ is finite0dimensional [37], and [23] is a start. Of course, the same question can be asked about the boundaries of FZ_n and FS_n.

ACKNOWLEDGMENTS

I would like to thank all my collaborators over the years, and particularly Ken Bromberg, Mark Feighn, and Koji Fujiwara with whom I wrote many papers. I had a lot of fun and learned many things from you. Here is to the future papers!

FUNDING

This work was partially supported by the National Science Foundation, DMS-1905720.

REFERENCES

[1] G. N. Arzhantseva, C. H. Cashen, D. Gruber, and D. Hume, Negative curvature in graphical small cancellation groups. *Groups Geom. Dyn.* **13** (2019), no. 2, 579–632.

[2] S. H. Balasubramanya, Acylindrical group actions on quasi-trees. *Algebr. Geom. Topol.* **17** (2017), no. 4, 2145–2176.

[3] A. Bartels, Coarse flow spaces for relatively hyperbolic groups. *Compos. Math.* **153** (2017), no. 4, 745–779.

[4] A. Bartels and M. Bestvina, The Farrell–Jones conjecture for mapping class groups. *Invent. Math.* **215** (2019), no. 2, 651–712.

[5] A. Bartels, W. Lück, and H. Reich, The K-theoretic Farrell–Jones conjecture for hyperbolic groups. *Invent. Math.* **172** (2008), no. 1, 29–70.

[6] H. Bass, Covering theory for graphs of groups. *J. Pure Appl. Algebra* **89** (1993), no. 1–2, 3–47.

[7] J. A. Behrstock, Asymptotic geometry of the mapping class group and Teichmüller space. *Geom. Topol.* **10** (2006), 1523–1578.

[8] J. Behrstock, M. F. Hagen, and A. Sisto, Asymptotic dimension and small-cancellation for hierarchically hyperbolic spaces and groups. *Proc. Lond. Math. Soc. (3)* **114** (2017), no. 5, 890–926.

[9] J. Behrstock, M. F. Hagen, and A. Sisto, Quasiflats in hierarchically hyperbolic spaces. *Duke Math. J.* **170** (2021), no. 5, 909–996.

[10] J. Behrstock, B. Kleiner, Y. Minsky, and L. Mosher, Geometry and rigidity of mapping class groups. *Geom. Topol.* **16** (2012), no. 2, 781–888.

[11] G. C. Bell and K. Fujiwara, The asymptotic dimension of a curve graph is finite. *J. Lond. Math. Soc. (2)* **77** (2008), no. 1, 33–50.

[12] G. Bell and A. Dranishnikov, Asymptotic dimension. *Topology Appl.* **155** (2008), no. 12, 1265–1296.

[13] M. Bestvina, Degenerations of the hyperbolic space. *Duke Math. J.* **56** (1988), no. 1, 143–161.

[14] M. Bestvina, $\mathbb{R}$-trees in topology, geometry, and group theory. In *Handbook of geometric topology*, pp. 55–91, North-Holland, Amsterdam, 2002.

[15] M. Bestvina, The topology of Out(F_n). In *Proceedings of the International Congress of Mathematicians, Vol. II (Beijing, 2002)*, pp. 373–384, Higher Ed. Press, Beijing, 2002.

[16] M. Bestvina, Geometric group theory and 3-manifolds hand in hand: the fulfillment of Thurston's vision. *Bull. Amer. Math. Soc. (N.S.)* **51** (2014), no. 1, 53–70.

[17] M. Bestvina, Geometry of outer space. In *Geometric group theory*, pp. 173–206, IAS/Park City Math. Ser. 21, Amer. Math. Soc., Providence, RI, 2014.

[18] M. Bestvina and K. Bromberg, On the asymptotic dimension of the curve complex. *Geom. Topol.* **23** (2019), no. 5, 2227–2276.

[19] M. Bestvina, K. Bromberg, and K. Fujiwara, Constructing group actions on quasitrees and applications to mapping class groups. *Publ. Math. Inst. Hautes Études Sci.* **122** (2015), 1–64.

[20] M. Bestvina, K. Bromberg, and K. Fujiwara, Bounded cohomology with coefficients in uniformly convex Banach spaces. *Comment. Math. Helv.* **91** (2016), no. 2, 203–218.

[21] M. Bestvina, K. Bromberg, and K. Fujiwara, Stable commutator length on mapping class groups. *Ann. Inst. Fourier (Grenoble)* **66** (2016), no. 3, 871–898.

[22] M. Bestvina, K. Bromberg, K. Fujiwara, and A. Sisto, Acylindrical actions on projection complexes. *Enseign. Math.* **65** (2019), no. 1–2, 1–32.

[23] M. Bestvina, J. Chaika, and S. Hensel, Connectivity of the Gromov boundary of the free factor complex. 2021, arXiv:2105.01537v2.

[24] M. Bestvina, R. Dickmann, G. Domat, S. Kwak, P. Patel, and E. Stark, Free products from spinning and rotating families. 2020, arXiv:2010.10735v2.

[25] M. Bestvina and M. Feighn, Addendum and correction to: "A combination theorem for negatively curved groups" [*J. Differential Geom.* **35** (1992), no. 1, 85–101]. *J. Differential Geom.* **43** (1996), no. 4, 783–788.

[26] M. Bestvina and M. Feighn, A combination theorem for negatively curved groups. *J. Differential Geom.* **35** (1992), no. 1, 85–101.

[27] M. Bestvina and M. Feighn, Outer limits, 1992, revised 1994, preprint, http://www.math.utah.edu/~bestvina/research.html.

[28] M. Bestvina and M. Feighn, Stable actions of groups on real trees. *Invent. Math.* **121** (1995), no. 2, 287–321.

[29] M. Bestvina and M. Feighn, A hyperbolic Out(F_n)-complex. *Groups Geom. Dyn.* **4** (2010), no. 1, 31–58.

[30] M. Bestvina and M. Feighn, Hyperbolicity of the complex of free factors. *Adv. Math.* **256** (2014), 104–155.

[31] M. Bestvina and M. Feighn, Subfactor projections. *J. Topol.* **7** (2014), no. 3, 771–804.

[32] M. Bestvina and K. Fujiwara, Bounded cohomology of subgroups of mapping class groups. *Geom. Topol.* **6** (2002), 69–89.

[33] M. Bestvina, V. Guirardel, and C. Horbez, Boundary amenability of Out(F_N). 2021, arXiv:1705.07017v2.

[34] M. Bestvina and M. Handel, Train tracks and automorphisms of free groups. *Ann. of Math. (2)* **135** (1992), no. 1, 1–51.

[35] M. Bestvina and M. Handel, Train tracks and automorphisms of free groups. *Ann. of Math. (2)* **135** (1992), no. 1, 1–51.

[36] M. Bestvina and C. Horbez, A compactification of outer space which is an absolute retract. *Ann. Inst. Fourier (Grenoble)* **69** (2019), no. 6, 2395–2437.

[37] M. Bestvina, C. Horbez, and R. D. Wade, On the topological dimension of the Gromov boundaries of some hyperbolic Out(F_N)-graphs. *Pacific J. Math.* **308** (2020), no. 1, 1–40.

[38] M. Bestvina and G. Mess, The boundary of negatively curved groups. *J. Amer. Math. Soc.* **4** (1991), no. 3, 469–481.

[39] M. Bestvina and P. Reynolds, The boundary of the complex of free factors. *Duke Math. J.* **164** (2015), no. 11, 2213–2251.

[40] B. H. Bowditch, Cut points and canonical splittings of hyperbolic groups. *Acta Math.* **180** (1998), no. 2, 145–186.

[41] B. H. Bowditch, Tight geodesics in the curve complex. *Invent. Math.* **171** (2008), no. 2, 281–300.

[42] B. H. Bowditch, Uniform hyperbolicity of the curve graphs. *Pacific J. Math.* **269** (2014), no. 2, 269–280.

[43] M. R. Bridson, Semisimple actions of mapping class groups on CAT(0) spaces. In *Geometry of Riemann surfaces*, pp. 1–14, London Math. Soc. Lecture Note Ser. 368, Cambridge Univ. Press, Cambridge, 2010.

[44] M. R. Bridson and A. Haefliger, *Metric spaces of non-positive curvature.* Grundlehren Math. Wiss. [Fundam. Principles Math. Sci.] 319, Springer, Berlin, 1999.

[45] M. R. Bridson and R. D. Wade, Actions of higher-rank lattices on free groups. *Compos. Math.* **147** (2011), no. 5, 1573–1580.

[46] M. R. Bridson and D. T. Wise, Malnormality is undecidable in hyperbolic groups. *Israel J. Math.* **124** (2001), 313–316.

[47] R. Brooks, Some remarks on bounded cohomology. In *Riemann surfaces and related topics: proceedings of the 1978 Stony Brook Conference (State Univ. New York, Stony Brook, NY, 1978)*, pp. 53–63, Ann. of Math. Stud. 97, Princeton Univ. Press, Princeton, NJ, 1981.

[48] H. Busemann, Extremals on closed hyperbolic space forms. *Tensor (N.S.)* **16** (1965), 313–318.

[49] D. Calegari, *scl.* MSJ Mem. 20, Mathematical Society of Japan, Tokyo, 2009.

[50] C. Champetier, Propriétés statistiques des groupes de présentation finie. *Adv. Math.* **116** (1995), no. 2, 197–262.

[51] M. Clay, J. Mangahas, and D. Margalit, Right-angled Artin groups as normal subgroups of mapping class groups. *Compos. Math.* **157** (2021), no. 8, 1807–1852.

[52] M. Culler and J. W. Morgan, Group actions on $\mathbb{R}$-trees. *Proc. Lond. Math. Soc. (3)* **55** (1987), no. 3, 571–604.

[53] M. Culler and K. Vogtmann, Moduli of graphs and automorphisms of free groups. *Invent. Math.* **84** (1986), no. 1, 91–119.

[54] F. Dahmani and V. Guirardel, The isomorphism problem for all hyperbolic groups. *Geom. Funct. Anal.* **21** (2011), no. 2, 223–300.

[55] F. Dahmani, V. Guirardel, and D. Osin, Hyperbolically embedded subgroups and rotating families in groups acting on hyperbolic spaces. *Mem. Amer. Math. Soc.* **245** (2017), no. 1156, v+152 pp.

[56] S. Dowdall and S. J. Taylor, Hyperbolic extensions of free groups. *Geom. Topol.* **22** (2018), no. 1, 517–570.

[57] A. N. Dranishnikov, S. C. Ferry, and S. A. Weinberger, An etale approach to the Novikov conjecture. *Comm. Pure Appl. Math.* **61** (2008), no. 2, 139–155.

[58] C. Druţu and M. Kapovich, *Geometric group theory*. Amer. Math. Soc. Colloq. Publ. 63, American Mathematical Society, Providence, RI, 2018.

[59] M. J. Dunwoody and M. E. Sageev, JSJ-splittings for finitely presented groups over slender groups. *Invent. Math.* **135** (1999), no. 1, 25–44.

[60] M. G. Durham, Y. N. Minsky, and A. Sisto, Stable cubulations, bicombings, and barycenters. 2020, arXiv:2009.13647.

[61] B. Farb, Relatively hyperbolic groups. *Geom. Funct. Anal.* **8** (1998), no. 5, 810–840.

[62] B. Farb and D. Margalit, *A primer on mapping class groups*. Princeton Math. Ser. 49, Princeton University Press, Princeton, NJ, 2012.

[63] B. Farb and L. Mosher, Convex cocompact subgroups of mapping class groups. *Geom. Topol.* **6** (2002), 91–152.

[64] K. Fujiwara and P. Papasoglu, JSJ-decompositions of finitely presented groups and complexes of groups. *Geom. Funct. Anal.* **16** (2006), no. 1, 70–125.

[65] D. Gabai, On the topology of ending lamination space. *Geom. Topol.* **18** (2014), no. 5, 2683–2745.

[66] D. Gaboriau, G. Levitt, and F. Paulin, Pseudogroups of isometries of R and Rips' theorem on free actions on R-trees. *Israel J. Math.* **87** (1994), no. 1–3, 403–428.

[67] E. Ghys, Groupes aléatoires (d'après Misha Gromov, . . .). *Astérisque* (2004), no. 294, viii, 173–204.

[68] E. Ghys and P. de la Harpe (eds.), *Sur les groupes hyperboliques d'après Mikhael Gromov*, Progr. Math. 83, Birkhäuser Boston, Inc., Boston, MA, 1990.

[69] M. Gromov, Groups of polynomial growth and expanding maps. *Publ. Math. Inst. Hautes Études Sci.* **53** (1981), 53–73.

[70] M. Gromov, Infinite groups as geometric objects. In *Proceedings of the International Congress of Mathematicians, Vol. 1, 2 (Warsaw, 1983)*, pp. 385–392, PWN, Warsaw, 1984.

[71] M. Gromov, Hyperbolic groups. In *Essays in group theory*, pp. 75–263, Math. Sci. Res. Inst. Publ. 8, Springer, New York, 1987.

[72] M. Gromov, Asymptotic invariants of infinite groups. In *Geometric group theory, Vol. 2 (Sussex, 1991)*, pp. 1–295, London Math. Soc. Lecture Note Ser. 182, Cambridge Univ. Press, Cambridge, 1993.

[73] M. Gromov, Random walk in random groups. *Geom. Funct. Anal.* **13** (2003), no. 1, 73–146.

[74] D. Groves and J. F. Manning, Dehn filling in relatively hyperbolic groups. *Israel J. Math.* **168** (2008), 317–429.

[75] I. Gruschko, Über die Basen eines freien Produktes von Gruppen. *Rec. Math. [Mat. Sb.] N.S.* **8** (1940), no. 50, 169–182.

[76] V. Guirardel and G. Levitt, JSJ decompositions of groups. *Astérisque* (2017), no. 395, vii+165 pp.

[77] U. Hamenstädt, The boundary of the free splitting graph and the free factor graph. 2014, arXiv:1211.1630v5.

[78] U. Hamenstaedt, Word hyperbolic extensions of surface groups. 2005, arXiv:math/0505244v2.

[79] U. Hamenstaedt, Geometry of the mapping class groups III: quasi-isometric rigidity. 2007, arXiv:math/0512429v2.

[80] M. Handel and L. Mosher, The free splitting complex of a free group, I: hyperbolicity. *Geom. Topol.* **17** (2013), no. 3, 1581–1672.

[81] M. Handel and L. Mosher, Hyperbolic actions and 2nd bounded cohomology of subgroups of $\mathrm{Out}(F_n)$. Part I: infinite lamination subgroups. 2019, arXiv:1511.06913v6.

[82] M. Handel and L. Mosher, Hyperbolic actions and 2nd bounded cohomology of subgroups of $\mathrm{Out}(F_n)$. Part II: finite lamination subgroups. 2019, arXiv:1702.08050v5.

[83] W. J. Harvey, Boundary structure of the modular group. In *Riemann surfaces and related topics: proceedings of the 1978 Stony Brook Conference (State Univ. New York, Stony Brook, NY, 1978)*, pp. 245–251, Ann. of Math. Stud. 97, State Univ., New York, Stony Brook, NY, 1981.

[84] S. Hensel, P. Przytycki, and R. C. H. Webb, 1-slim triangles and uniform hyperbolicity for arc graphs and curve graphs. *J. Eur. Math. Soc. (JEMS)* **17** (2015), no. 4, 755–762.

[85] C. Horbez, The boundary of the outer space of a free product. *Israel J. Math.* **221** (2017), no. 1, 179–234.

[86] G. Italiano, B. Martelli, and M. Migliorini, Hyperbolic 5-manifolds that fiber over S^1. 2021, arXiv:2105.14795v3.

[87] I. Kapovich, J. Maher, C. Pfaff, and S. J. Taylor, Random trees in the boundary of Outer space. 2021, arXiv:1904.10026v3.

[88] I. Kapovich and K. Rafi, On hyperbolicity of free splitting and free factor complexes. *Groups Geom. Dyn.* **8** (2014), no. 2, 391–414.

[89] A. E. Kent and C. J. Leininger, Shadows of mapping class groups: capturing convex cocompactness. *Geom. Funct. Anal.* **18** (2008), no. 4, 1270–1325.

[90] E. Klarreich, The boundary at infinity of the curve complex and the relative Teichmüller space. 1999, arXiv:1803.10339. To appear in *Groups Geom. Dyn.*

[91] G. Levitt, Constructing free actions on $\mathbb{R}$-trees. *Duke Math. J.* **69** (1993), no. 3, 615–633.

[92] G. Levitt, Non-nesting actions on real trees. *Bull. Lond. Math. Soc.* **30** (1998), no. 1, 46–54.

[93] J. Maher and G. Tiozzo, Random walks on weakly hyperbolic groups. *J. Reine Angew. Math.* **742** (2018), 187–239.

[94] J. Mangahas, A recipe for short-word pseudo-Anosovs. *Amer. J. Math.* **135** (2013), no. 4, 1087–1116.

[95] H. A. Masur and Y. N. Minsky, Geometry of the complex of curves. I. Hyperbolicity. *Invent. Math.* **138** (1999), no. 1, 103–149.

[96] H. A. Masur and Y. N. Minsky, Geometry of the complex of curves. II. Hierarchical structure. *Geom. Funct. Anal.* **10** (2000), no. 4, 902–974.

[97] H. Masur and S. Schleimer, The geometry of the disk complex. *J. Amer. Math. Soc.* **26** (2013), no. 1, 1–62.

[98] J. Milnor, A note on curvature and fundamental group. *J. Differential Geom.* **2** (1968), 1–7.

[99] J. Milnor, Hyperbolic geometry: the first 150 years. *Bull. Amer. Math. Soc. (N.S.)* **6** (1982), no. 1, 9–24.

[100] M. Mj, Cannon–Thurston maps. In *Proceedings of the International Congress of Mathematicians—Rio de Janeiro 2018. Vol. II. Invited lectures*, pp. 885–917, World Sci. Publ., Hackensack, NJ, 2018.

[101] J. W. Morgan, On Thurston's uniformization theorem for three-dimensional manifolds. In *The Smith conjecture (New York, 1979)*, pp. 37–125, Pure Appl. Math. 112, Academic Press, New York, 1979.

[102] J. W. Morgan and P. B. Shalen, Valuations, trees, and degenerations of hyperbolic structures. I. *Ann. of Math. (2)* **120** (1984), no. 3, 401–476.

[103] J. W. Morgan and P. B. Shalen, Degenerations of hyperbolic structures. II. Measured laminations in 3-manifolds. *Ann. of Math. (2)* **127** (1988), no. 2, 403–456.

[104] J. W. Morgan and P. B. Shalen, Degenerations of hyperbolic structures. III. Actions of 3-manifold groups on trees and Thurston's compactness theorem. *Ann. of Math. (2)* **127** (1988), no. 3, 457–519.

[105] H. M. Morse, A fundamental class of geodesics on any closed surface of genus greater than one. *Trans. Amer. Math. Soc.* **26** (1924), no. 1, 25–60.

[106] G. D. Mostow, Quasi-conformal mappings in n-space and the rigidity of hyperbolic space forms. *Publ. Math. Inst. Hautes Études Sci.* **34** (1968), 53–104.

[107] G. A. Niblo and M. A. Roller (eds.), *Geometric group theory, Vol. 1*. London Math. Soc. Lecture Note Ser. 181, Cambridge University Press, Cambridge, 1993.

[108] G. A. Niblo and M. A. Roller (eds.), *Geometric group theory, Vol. 2*. London Math. Soc. Lecture Note Ser. 182, Cambridge University Press, Cambridge, 1993.

[109] J. Nielsen, Über die Isomorphismen unendlicher Gruppen ohne Relation. *Math. Ann.* **79** (1918), no. 3, 269–272.

[110] A. Y. Ol'shanskiĭ, Almost every group is hyperbolic. *Internat. J. Algebra Comput.* **2** (1992), no. 1, 1–17.

[111] Y. Ollivier, *A January 2005 invitation to random groups*. Ensaios Mat. [Math. Surveys] 10, Sociedade Brasileira de Matemática, Rio de Janeiro, 2005.

[112] D. Osajda, Small cancellation labellings of some infinite graphs and applications. *Acta Math.* **225** (2020), no. 1, 159–191.

[113] D. Osin, Acylindrically hyperbolic groups. *Trans. Amer. Math. Soc.* **368** (2016), no. 2, 851–888.

[114] D. V. Osin, Peripheral fillings of relatively hyperbolic groups. *Invent. Math.* **167** (2007), no. 2, 295–326.

[115] D. V. Osin, Groups acting acylindrically on hyperbolic spaces. In *Proceedings of the International Congress of Mathematicians—Rio de Janeiro 2018. Vol. II. Invited lectures*, pp. 919–939, World Sci. Publ., Hackensack, NJ, 2018.

[116] F. Paulin, Topologie de Gromov équivariante, structures hyperboliques et arbres réels. *Invent. Math.* **94** (1988), no. 1, 53–80.

[117] F. Paulin, Outer automorphisms of hyperbolic groups and small actions on **R**-trees. In *Arboreal group theory (Berkeley, CA, 1988)*, pp. 331–343, Math. Sci. Res. Inst. Publ. 19, Springer, New York, 1991.

[118] E. Rips and Z. Sela, Structure and rigidity in hyperbolic groups. I. *Geom. Funct. Anal.* **4** (1994), no. 3, 337–371.

[119] E. Rips and Z. Sela, Cyclic splittings of finitely presented groups and the canonical JSJ decomposition. *Ann. of Math. (2)* **146** (1997), no. 1, 53–109.

[120] Z. Sela, Uniform embeddings of hyperbolic groups in Hilbert spaces. *Israel J. Math.* **80** (1992), no. 1–2, 171–181.

[121] Z. Sela, The isomorphism problem for hyperbolic groups. I. *Ann. of Math. (2)* **141** (1995), no. 2, 217–283.

[122] Z. Sela, Structure and rigidity in (Gromov) hyperbolic groups and discrete groups in rank 1 Lie groups. II. *Geom. Funct. Anal.* **7** (1997), no. 3, 561–593.

[123] Z. Sela, Endomorphisms of hyperbolic groups. I. The Hopf property. *Topology* **38** (1999), no. 2, 301–321.

[124] Z. Sela, Diophantine geometry over groups. I. Makanin–Razborov diagrams. *Publ. Math. Inst. Hautes Études Sci.* **93** (2001), 31–105.

[125] Z. Sela, Diophantine geometry over groups. II. Completions, closures and formal solutions. *Israel J. Math.* **134** (2003), 173–254.

[126] Z. Sela, Diophantine geometry over groups. IV. An iterative procedure for validation of a sentence. *Israel J. Math.* **143** (2004), 1–130.

[127] Z. Sela, Diophantine geometry over groups. V_1. Quantifier elimination. I. *Israel J. Math.* **150** (2005), 1–197.

[128] Z. Sela, Diophantine geometry over groups. V_2. Quantifier elimination. II. *Geom. Funct. Anal.* **16** (2006), no. 3, 537–706.

[129] Z. Sela, Diophantine geometry over groups. VII. The elementary theory of a hyperbolic group. *Proc. Lond. Math. Soc. (3)* **99** (2009), no. 1, 217–273.

[130] Z. Sela, Diophantine geometry over groups VIII: stability. *Ann. of Math. (2)* **177** (2013), no. 3, 787–868.

[131] J.-P. Serre, *Trees*. Springer, Berlin–New York, 1980.

[132] J. R. Stallings, A topological proof of Grushko's theorem on free products. *Math. Z.* **90** (1965), 1–8.

[133] J. R. Stallings, On torsion-free groups with infinitely many ends. *Ann. of Math. (2)* **88** (1968), 312–334.

[134] J. R. Stallings, Topology of finite graphs. *Invent. Math.* **71** (1983), no. 3, 551–565.

[135] A. S. Švarc, A volume invariant of coverings. *Dokl. Akad. Nauk SSSR* **105** (1955), 32–34.

[136] G. A. Swarup, On the cut point conjecture. *Electron. Res. Announc. Am. Math. Soc.* **2** (1996), no. 2, 98–100.

[137] S. J. Taylor, A note on subfactor projections. *Algebr. Geom. Topol.* **14** (2014), no. 2, 805–821.

[138] W. P. Thurston, The geometry and topology of 3-manifolds, 1980, electronic version http://library.msri.org/books/gt3m/.

[139] W. P. Thurston, Hyperbolic structures on 3-manifolds. I. Deformation of acylindrical manifolds. *Ann. of Math. (2)* **124** (1986), no. 2, 203–246.

[140] W. P. Thurston, On the geometry and dynamics of diffeomorphisms of surfaces. *Bull. Amer. Math. Soc. (N.S.)* **19** (1988), no. 2, 417–431.

[141] K. Vogtmann, Automorphisms of free groups and outer space. In *Proceedings of the Conference on Geometric and Combinatorial Group Theory, Part I (Haifa, 2000)*, Geometriae Dedicata, pp. 1–31, 2002.

[142] K. Vogtmann, On the geometry of outer space. *Bull. Amer. Math. Soc. (N.S.)* **52** (2015), no. 1, 27–46.

[143] G. Yu, The Novikov conjecture for groups with finite asymptotic dimension. *Ann. of Math. (2)* **147** (1998), no. 2, 325–355.

MLADEN BESTVINA

Department of Mathematics, University of Utah, 155 S 1400 E, RM 233, Salt Lake City, UT 84112, USA, bestvina@math.utah.edu

ALGEBRAIC GEOMETRY IN MIXED CHARACTERISTIC

BHARGAV BHATT

ABSTRACT

Fix a prime number p. We report on some recent developments in algebraic geometry (broadly construed) over p-adically complete commutative rings. These developments include foundational advances within the subject, as well as external applications.

MATHEMATICS SUBJECT CLASSIFICATION 2020

Primary 14F30; Secondary 14F17, 11G25, 11S70, 19F99, 13D99

KEYWORDS

p-adic Hodge theory, prismatic cohomology, topological Hochschild homology, algebraic K-theory, vanishing theorems, Riemann–Hilbert correspondence, minimal model program

ICM INTERNATIONAL CONGRESS OF MATHEMATICIANS 2022 JULY 6–14

Proc. Int. Cong. Math. 2022, Vol. 2, pp. 712–748
DOI 10.4171/ICM2022/113

1. INTRODUCTION

Fix a prime number p. The last decade has witnessed multiple conceptual advances in algebraic geometry over mixed characteristic rings (which, in this article, we take to mean commutative rings that are p-adically complete). These advances have led to the resolution of longstanding questions in different areas of mathematics where p-adic completions appear. Moreover, entirely new and fascinating domains of inquiry have been uncovered. The goal of this survey is to discuss some of these developments, especially in topics close to the author's expertise.

A highlight of the last decade of activity in the area has been its seat as an exchange of ideas across different fields of mathematics. For instance, a central topic of this survey is prismatic cohomology, which is a new cohomology theory for mixed characteristic rings (Sections 2 and 3); its discovery was inspired in part by calculations in homotopy theory and in part by developments in Galois representation theory (Remark 4.5). Prismatic cohomology in turn played a prominent role in the proof of a mixed characteristic analog of the Kodaira vanishing theorem (Theorem 5.7), which then helped develop the minimal model program in the birational geometry of arithmetic threefolds (Theorem 5.10). In the reverse direction, an important flatness lemma discovered in the solution [3] of a longstanding question in commutative algebra facilitated, via prismatic cohomology again, the proof of an analog of Bott's vanishing theorem for algebraic K-theory (Theorem 4.8). The author hopes this survey can convey some of the excitement surrounding this interplay of ideas.

We emphasize right away that the topics covered are chosen somewhat idiosyncratically, and we have not attempted to be comprehensive even in the topics we do cover; to make partial amends, a number of references have been included to help the reader navigate the subject. Moreover, the level of the exposition is uneven across sections; for instance, we have taken a macroscopic view of topics that are reasonably well documented elsewhere, but have gone into more detail and depth while covering very recent ideas that seem promising.

This survey is organized as follows. In Section 2, we discuss relative prismatic cohomology and related developments. The absolute version of this story, which is comparatively new, is the subject of Section 3. We then present applications, covering algebraic K-theory in Section 4 and commutative algebra and birational geometry in Section 5. We end in Section 6 with some relatively recent work on p-adic Riemann–Hilbert problems and their algebro-geometric implications.

All rings that appear are assumed commutative unless otherwise specified.

2. PRISMS AND RELATIVE PRISMATIC COHOMOLOGY

In the last century, especially following the work of Grothendieck, cohomology theories have emerged as extremely important tools in algebraic geometry and number theory: they lie at the heart of some of the deepest theorems and conjectures in both subjects. For example, classical Hodge theory, which studies the singular cohomology with real/complex coefficients for complex varieties, is a central topic in modern algebraic geometry, with applications throughout the subject and beyond. Likewise, p-adic Hodge theory, which studies the p-adic cohomology of p-adic varieties, is an equally fundamental notion in arithmetic geometry: it provides one of the best known tools for understanding Galois representations of the absolute Galois group of $\mathbf{Q}$. Moreover, unlike in the complex setting, there is a large number of cohomology theories in the p-adic world: étale, de Rham, Hodge, crystalline, de Rham–Witt, etc. In this section, we will report on work from the last few years dedicated to finding an organizational framework to better understand p-adic cohomology theories in p-adic arithmetic geometry, especially their relationships with each other.

Remark 2.1 (Why do derived objects appear repeatedly?). Before embarking on our journey, let us explain one reason derived notions (i.e., those with a homological/homotopical flavor) often appear in recent work in the area and consequently also in our exposition.

In classical algebraic geometry, the fundamental objects are smooth algebraic varieties over an algebraically closed field. Similarly, in mixed characteristic algebraic geometry, the basic geometric objects are (p-adic formal[1]) smooth schemes over the ring of integers $\mathcal{O}_C$ of a complete algebraically closed nonarchimedean field $C/\mathbf{Q}_p$. In particular, unlike the classical setting, the rings of functions that appear in mixed characteristic algebraic geometry are often not noetherian: indeed, the ring $\mathcal{O}_C$ is a nonnoetherian valuation ring as its value group is divisible. Replacing $\mathcal{O}_C$ with a discrete valuation ring like $\mathbf{Z}_p$, while quite tempting and important for applications, leads to arithmetic subtleties that one would like to avoid, at least at first pass, in a purely geometric study. Even more exotic nonnoetherian rings are critical to several recent innovations in the area, such as perfectoid geometry [143, 202], descent techniques for extremely fine Grothendieck topologies such as the pro-étale, quasisyntomic, v and arc topologies [36, 38–40, 205], the theory of δ-rings [138], etc.

In the nonclassical situations described above, derived notions often have better stability properties than their classical counterparts. For instance, given a commutative ring R with a finitely generated ideal I, the category of derived I-complete R-modules forms an abelian subcategory (e.g., it is closed under kernels and cokernels) of the category of all R-modules, unlike the subcategory of classically I-adically complete R-modules; moreover, the assignment carrying R to the ∞-category $\mathcal{D}_{I-\mathrm{comp}}(R)$ of derived I-complete R-complexes forms a stack for the flat topology (or even a suitably defined I-completely flat topology), unlike the corresponding assignment at the triangulated category level. For such reasons, the language of higher category theory and derived algebraic geometry [169–171, 223, 224] has played an important role in the developments discussed in this paper.

The work described in this section began with the goal to enhance Fontaine's perspective [93] on p-adic Hodge theory to work well with integral coefficients. A concrete goal was to understand how the torsion in the $\mathbf{Z}_p$-étale cohomology of the geometric generic fiber of a smooth projective scheme over a mixed characteristic discrete valuation ring relates to the torsion in the crystalline cohomology of its special fiber; this question was already

1 A p-adic formal scheme is a formal scheme whose affine opens are given by formal spectra of p-adically complete rings equipped with the p-adic topology.

stressed by Grothendieck in his Algerian letter to Deligne in 1965 (see [60] for a survey on the status of this question 20 years ago, and Section 2 in [60] for history). After initial attempts [37,38] that worked well in important examples, a satisfactory theory was found via the notion of a prism, recalled next. The definition relies on the notion of a δ-ring, which is roughly a commutative ring A equipped with a map $\varphi : A \to A$ lifting the Frobenius endomorphism $f \mapsto f^p$ of A/p (interpreted in a derived sense when A has p-torsion); see [61,138]. The importance of this notion in arithmetic geometry has long been stressed by Borger, see [57].

Definition 2.2 (Prisms, [41]). A *prism* is a pair (A, I), where A is a δ-ring and $I \subset A$ is an invertible ideal such that A is derived (p, I)-complete and $p \in (I, \varphi(I))$. Write $\overline{A} := A/I$.

In practice, we restrict to *bounded* prisms, i.e., those prisms (A, I) where the p-power torsion in $\overline{A}$ is annihilated by p^n for some $n \geq 0$; this restriction allows us to avoid certain derived technicalities without sacrificing the key examples. Two important examples are discussed next; see Remark 2.9 for another key example.

Example 2.3 (Crystalline prisms). If A is any p-complete p-torsion-free δ-ring, then $(A, (p))$ is a bounded prism. For instance, given a reduced $\mathbf{F}_p$-algebra R, we could take $A = W(R)$ to be the ring of p-typical Witt vectors of ring with its natural Frobenius lift.

Example 2.4 (Perfect prisms). A prism (A, I) is called *perfect* if the Frobenius map $\varphi : A \to A$ is an isomorphism; any such prism is bounded and the ring $\overline{A}$ is perfectoid as in [37,101]. In fact, the construction $(A, I) \mapsto \overline{A}$ yields an equivalence of categories between perfect prisms and perfectoid rings; thus, the notion of a prism may be viewed as a "deperfection" of the notion of a perfectoid ring. An important example is the perfect prism (A, I) corresponding to the perfectoid ring $\overline{A} = \mathcal{O}_C$, where $C/\mathbf{Q}_p$ is a complete and algebraically closed extension; we call this *a Fontaine prism*, in homage to its discovery [92, §5].

Given a bounded prism (A, I) as well as an $\overline{A}$-scheme X, the following key definition allows us to extract an A-linear cohomology theory for X.

Definition 2.5 (The relative prismatic site). Fix a bounded prism (A, I) and a p-adic formal $\overline{A}$-scheme X. The relative prismatic site $(X/A)_{\Delta}$ is the category of all bounded prisms (B, J) over (A, I) equipped with an $\overline{A}$-map $\mathrm{Spf}(B/J) \to X$, topologized via the flat topology; write $\mathcal{O}_{\Delta}$, I_{Δ}, and $\overline{\mathcal{O}}_{\Delta}$ for the sheaves obtained by remembering B, J, and B/J, respectively, so there is a natural φ-action on $\mathcal{O}_{\Delta}$ and $\overline{\mathcal{O}}_{\Delta} = \mathcal{O}_{\Delta}/I_{\Delta}$. Write $R\Gamma_{\Delta}(X/A) := R\Gamma((X/A)_{\Delta}, \mathcal{O}_{\Delta}) \circlearrowleft \varphi$ for the resulting cohomology theory.

The main comparison theorems for $R\Gamma_{\Delta}(X/A)$ are informally summarized next:

Theorem 2.6 (Relative prismatic cohomology, [37,41]). *Fix a bounded prism (A, I) and let X be a smooth p-adic formal $\overline{A}$-scheme. The relative prismatic cohomology theory $R\Gamma_{\Delta}(X/A) \circlearrowleft \varphi$ recovers the standard integral p-adic cohomology theories for X/A with their extra structures (e.g., étale, de Rham, Hodge, crystalline, de Rham–Witt) via a specialization procedure, thereby giving new relationships between them.*

For instance, if (A, I) is crystalline, Theorem 2.6 leads to a canonical Frobenius descent of crystalline cohomology [17]; this descent was previously observed on cohomology groups in [189, 210]. On the other hand, the A_{inf}-cohomology of [37] is recovered by specializing to a Fontaine prism; we refer to the surveys [24, 183] for more precise assertions (with pictures!) about the comparisons in this case. An early concrete application of the latter was the following result relating étale and de Rham cohomology integrally; via classical comparisons, this gives a new technique to bound the p-torsion in singular cohomology of complex algebraic varieties via the geometry of their mod p reductions.

Corollary 2.7 (Torsion inequality, [37]). *Let $C/\mathbf{Q}_p$ be a complete and algebraically closed field with residue field k (e.g., we may take $C = \mathbf{C}_p = \widehat{\overline{\mathbf{Q}_p}}$, so $k = \overline{\mathbf{F}_p}$). Let $X/\mathcal{O}_C$ be a proper smooth p-adic formal scheme. Then*

$$\dim_{\mathbf{F}_p} H^i_{et}(X_C, \mathbf{F}_p) \leq \dim_k H^i_{\mathrm{dR}}(X_k) \quad \textit{for all } i \geq 0.$$

More generally, a similar inequality bounds the length of the torsion subgroup of $H^i_{et}(X_C, \mathbf{Z}_p)$ in terms of that of $H^i_{crys}(X_k)$. In particular, if the latter is torsion-free, so is the former.

Since its discovery, the prismatic theory in [37, 41] has found several applications, some of which are discussed below and elsewhere in this paper. Other results featuring this theory include: Hodge theory of classifying spaces of reductive groups [28, 154], vanishing theorems for the cohomology of the moduli space of curves with level structures [196], essential dimension calculations [89], Poincaré duality for $\mathbf{Z}/p^n$-coefficients in rigid geometry [232], calculation of the $\mathbf{Z}_p$-cohomology of Drinfeld's p-adic symmetric spaces [67], a fairly optimal form of Dieudonné theory in mixed characteristic [8], better understanding of the moduli stacks of Breuil–Kisin–Fargues modules [81], and several improvements to known comparisons in integral p-adic Hodge theory [160, 161, 182].

Remark 2.8 (Rational comparison theorems). Specializing part of Theorem 2.6 to a Fontaine prism (A, I) gives a generalization of Fontaine's crystalline comparison conjecture C_{crys} to proper smooth formal schemes $X/\overline{A}$; variants of both this result and its proof have a long history in p-adic Hodge theory, including [15, 50, 68, 83, 85, 95, 187, 226].

Remark 2.9 (q-de Rham cohomology, [41]). Given a smooth $\mathbf{Z}$-algebra R equipped with a choice of étale coordinates (which we call a *framing* and indicate by $\square$), one can define a complex $q\Omega^\bullet_{(R,\square)}$ of $\mathbf{Z}[\![q-1]\!]$-modules by q-deforming the differential of the de Rham complex $\Omega^*_{R/\mathbf{Z}}$ (see [12, 204]); this complex strongly depends on the framing $\square$. Nevertheless, motivated by some local calculations from [37], Scholze had conjectured in [204] that $q\Omega^\bullet_{(R,\square)}$ is independent of the framing $\square$ up to canonical quasiisomorphism. This conjecture was deduced from the existence of prismatic cohomology in [41], as explained next; prior partial progress was made by Pridham [192], also using δ-rings.

By a patching procedure, Scholze's coordinate independence conjecture reduces to its analog when all objects are p-completed. The latter follows from the existence of prismatic cohomology relative to the q-de Rham prism $(A, I) := (\mathbf{Z}_p[\![q-1]\!], (\frac{q^p-1}{q-1}))$ where

$\varphi(q) = q^p$: given a formally smooth p-complete $\mathbf{Z}_p$-algebra R equipped with a framing $\square$ as before, the relative prismatic complex $R\Gamma_{\Delta}(\mathrm{Spf}(R \otimes_{\mathbf{Z}_p} \overline{A})/A)$ (which is visibly independent of the framing $\square$) is naturally quasiisomorphic to the q-dR complex $q\Omega^\bullet_{(R,\square)}$.

The preceding perspective on q-de Rham cohomology also yields a formalism for more systematically discussing related notions such as, e.g., Gauss–Manin q-connections; we refer to **[64, 107, 108, 184]** for more on these and related developments.

Remark 2.10 (Perfections in mixed characteristic, **[41]**). The theory of perfectoid rings can be regarded as a mixed characteristic analog of the theory of perfect $\mathbf{F}_p$-algebras, i.e., $\mathbf{F}_p$-algebras where the Frobenius map is bijective. The utility of this analogy is enhanced by Theorem 2.6: by attaching objects with Frobenius actions to rings in mixed characteristic, this result yields a notion of "perfectoidization" for a large class of mixed characteristic rings. Indeed, given any perfect prism (A, I) and an $\overline{A}$-algebra R, one can naturally construct a "(derived) perfectoidization" $R \to R_{\mathrm{perfd}}$ with excellent formal properties. For instance, if R is integral over $\overline{A}$, then $R \to R_{\mathrm{perfd}}$ is in fact the universal map to a perfectoid ring. This construction has several applications. For instance, **[41]** uses these to prove an optimal generalization of the Faltings' almost purity theorem (extending versions from **[4, 83, 86, 143, 202]**), as well as the result that "Zariski closed = strongly Zariski closed" for affinoid perfectoid spaces; the latter plays an important role in aspects of **[90]**. The perfectoidization functor also powers the construction of the p-adic Riemann–Hilbert functor in Section 6.

Remark 2.11 (A new perspective on de Rham–Witt complexes, **[34]**). The de Rham–Witt complex of Bloch–Deligne–Illusie **[47, 134]** is a fundamental object in characteristic p algebraic geometry with applications transcending algebra (e.g., **[119]**). Its construction traditionally relied on somewhat laborious calculations. The paper **[34]**, inspired by structures on relative prismatic cohomology, offered a new homological perspective on this object.

To explain this, we first recall the isogeny theorem for prismatic cohomology. In the setup of Theorem 2.6, when $X = \mathrm{Spf}(R)$ is affine, one often writes $\Delta_{R/A} = R\Gamma_{\Delta}(X/A)$, regarded as an object of the derived category of A. There is then a natural quasiisomorphism

$$\widetilde{\varphi}_{R/A} : \varphi^* \Delta_{R/A} \simeq L\eta_I \Delta_{R/A} \tag{Isog}$$

induced by the relative Frobenius map, where $L\eta_I$ is a variant of the Berthelot–Ogus–Deligne décalage functor (see **[37, §6]**); the isomorphism $\widetilde{\varphi}_{R/A}$, which is a prismatic avatar of the Berthelot–Ogus isogeny theorem **[18]**, plays a critical organizational role in capturing the additional structures on $\Delta_{R/A}$ (such as the Nygaard filtration).

The paper **[34]** shows that when (A, I) is a perfect crystalline prism (e.g., $(\mathbf{Z}_p, (p))$), one can reconstruct the de Rham–Witt complex $W\Omega^\bullet_R$ from the pair $(\Delta_{R/A}, \widetilde{\varphi}_{R/A})$ by a pure homological algebra construction dubbed "saturation." Moreover, this construction has the potential to offer a better behaved alternative to the de Rham complex for singular varieties in characteristic p, analogous to the du Bois complex in characteristic 0; we refer to **[135, 188]** for more on these developments.

Remark 2.12 (Logarithmic analogs). The smoothness assumption on X in Theorem 2.6 is a "good reduction" hypothesis. While adequate for several purposes, this is often too restrictive for studying the generic fiber: not every proper smooth scheme X_η/C admits such a smooth model $X/\mathcal{O}_C$. A more natural assumption—one that is conjecturally always satisfied, up to replacing models—would be a form of logarithmic smoothness of $X/\mathcal{O}_C$ (e.g., semistability) in the sense of log geometry [139]. Thus, one wants a version of [37,41] in the logarithmic setting. This has been accomplished in [62,150,151]; it is also possible to approach this problem by reduction to the smooth case using the language of infinite root stacks, following ideas of Olsson [190] (work in progress with Mathew).

Remark 2.13 (Nonabelian p-adic Hodge theory). Fix a bounded prism (A, I) and a smooth p-adic formal $\overline{A}$-scheme X. Motivated by the precise form of Theorem 2.6, define a *prismatic F-crystal* on (X/A) to be a vector bundle $\mathcal{E}$ on $((X/A)_{\Delta}, \mathcal{O}_{\Delta})$ equipped with a Frobenius structure $\varphi_{\mathcal{E}} : \varphi^*\mathcal{E}[\frac{1}{I_\Delta}] \simeq \mathcal{E}[\frac{1}{I_\Delta}]$; see Definition 3.2 for a more explicit description in a variant context. Prismatic F-crystals provide a viable notion of "coefficients" in the theory, somewhat analogous to the role played by harmonic bundles in complex nonabelian Hodge theory [212,213]. In particular, given such an F-crystal $(\mathcal{E}, \varphi_{\mathcal{E}})$, the specialization functors used in Theorem 2.6 yield a $\mathbf{Z}_p$-local system $T(\mathcal{E})$ on the rigid generic fiber X_η when A is perfect, a vector bundle $\mathcal{E}_{\mathrm{dR}}$ with flat connection on $X/\overline{A}$, an F-crystal $\mathcal{E}_{\mathrm{crys}}$ on $X \otimes_{\overline{A}} (\overline{A}/p)_{\mathrm{perf}}$, and (under certain auxiliary lifting data) a Higgs bundle $\mathcal{E}_{\mathrm{Higgs}}$ on $X/\overline{A}$. The relationship realized by these functors is rather close and has been investigated by various authors (such as [32,184,222]). When (A, I) is a Fontaine prism, this relationship is part of the p-adic Simpson correspondence pioneered by Faltings [1,87,88]. On the other hand, if (A, I) is a perfect crystalline prism, this relationship yields an alternative perspective on (at least the local aspects of) the nonabelian Hodge theory of [189].

Remark 2.14 (Extension to the singular case via animation). For several applications including most results discussed in this paper, it is important to extend the prismatic cohomology construction $R \mapsto \Delta_{R/A}$ (see Remark 2.11) to possibly singular $\overline{A}$-algebras R. Directly imitating Definition 2.5 does not produce a computable or useable result. Instead, inspired by the construction of the cotangent complex and derived de Rham cohomology [132,133] as well as their utility in a wide variety of problems [14,19,20,26,111,136], one extends the functor $\Delta_{-/A}$ to arbitrary p-complete $\overline{A}$-algebras by Quillen's nonabelian derived functor machinery [194] (dubbed *animation* by Clausen [63]) as reformulated in [169]. The resulting complex $\Delta_{R/A}$ can be fairly efficiently controlled using the cotangent complex $L_{R/\overline{A}}$ thanks to the animated Hodge–Tate comparison, which makes this extension quite useable.

3. ABSOLUTE PRISMATIC COHOMOLOGY

In Section 2, we fixed a base prism (A, I) and discussed results about the relative prismatic cohomology of a smooth p-adic formal $\overline{A}$-scheme X. In this section, we describe the picture that arises if one does not fix a base prism (A, I). This distinction is analogous to that between geometric and absolute étale cohomology in arithmetic, or that between singular cohomology and Deligne–Beilinson cohomology in Hodge theory. The objects considered here are newer than those in Section 2; consequently, some results are surely not optimal, and we have tried to indicate some natural further directions in the exposition.

3.1. Definition and key examples

To begin, let us recall the definition of the absolute prismatic site (obtained roughly from Definition 2.5 by discarding (A, I)).

Definition 3.1 (The absolute prismatic site). Given a p-adic formal scheme X, its absolute prismatic site X_{Δ} is the category of all bounded prisms (B, J) equipped with a map $\mathrm{Spf}(B/J) \to X$, topologized using the flat topology; write $\mathcal{O}_{\Delta}$, I_{Δ}, and $\overline{\mathcal{O}}_{\Delta}$ for the sheaves obtained by remembering B, J, and B/J, respectively. Write $R\Gamma_{\Delta}(X) := R\Gamma(X_{\Delta}, \mathcal{O}_{\Delta}) \circlearrowleft \varphi$ and $R\Gamma_{\overline{\Delta}}(X) := R\Gamma(X_{\Delta}, \overline{\mathcal{O}}_{\Delta})$ for the resulting cohomology theories.

If there exists a perfect prism (A, I) and a map $X \to \mathrm{Spf}(\overline{A})$, the natural map $(X/A)_{\Delta} \to X_{\Delta}$ is an equivalence, so Theorem 2.6 describes $R\Gamma_{\Delta}(X)$ in this case, e.g., $R\Gamma_{\Delta}(\mathrm{Spf}(\overline{A})) \simeq A$. At the other end, $\mathrm{Spf}(\mathbf{Z}_p)_{\Delta}$ is the opposite of the category of all bounded prisms. As this category has no final object, the cohomology theory $R\Gamma_{\Delta}(\mathrm{Spf}(\mathbf{Z}_p))$ is potentially interesting; in fact, we shall see in Section 4 that $R\Gamma_{\Delta}(\mathrm{Spf}(\mathbf{Z}_p))$ is closely related to the p-completed algebraic K-theory of $\mathbf{Z}_p$.

In this section, we shall be interested in the following objects on X_{Δ}:

Definition 3.2 (Crystals). Fix a p-adic formal scheme X. A *prismatic crystal* (resp. *Hodge–Tate crystal*) $\mathcal{E}$ of vector bundles on X is given by an assignment

$$(B, J) \in X_{\Delta} \mapsto \mathcal{E}(B) \in \mathrm{Vect}_B := \{\text{finite projective } B\text{-modules}\}$$
$$\big(\text{resp. } (B, J) \in X_{\Delta} \mapsto \mathcal{E}(B) \in \mathrm{Vect}_{B/J}\big)$$

that is compatible with base change in $(B, J) \in X_{\Delta}$. A *prismatic F-crystal* of vector bundles on X is given by a prismatic crystal $\mathcal{E}$ with an isomorphism $\varphi_{\mathcal{E}} : \varphi^*\mathcal{E}[\frac{1}{I_{\Delta}}] \simeq \mathcal{E}[\frac{1}{I_{\Delta}}]$. Similarly, one has analogous notions of crystals of perfect (or just (p, I_{Δ})-complete) complexes.

As in the relative case (Remark 2.13), there are realization functors carrying a prismatic F-crystal $\mathcal{E}$ on X to a $\mathbf{Z}_p$-local system $T(\mathcal{E})$ on the rigid generic fiber X_η, a vector bundle $\mathcal{E}_{\mathrm{dR}}$ with flat connection on X, and an F-crystal $\mathcal{E}_{\mathrm{crys}}$ on $X \otimes_{\mathbf{Z}_p} \mathbf{F}_p$. The simplest examples of such crystals are as follows:

Example 3.3 (Breuil–Kisin twists). For any prism (B, J), one has a naturally defined invertible B-module $B\{1\}$ given heuristically by

$$B\{1\} := J \otimes \varphi^* J \otimes \varphi^{2,*} J \otimes \cdots .$$

This B-module comes equipped with a natural isomorphism $\varphi_{B\{1\}} : \varphi^* B\{1\} \simeq J^{-1} \otimes B\{1\}$, so the assignment $(B, J) \mapsto (B\{1\}, \varphi_{B\{1\}})$ gives a prismatic F-crystal $(\mathcal{O}_{\Delta}\{1\}, \varphi_{\mathcal{O}_{\Delta}}\{1\})$ on $\mathrm{Spf}(\mathbf{Z}_p)_{\Delta}$ (and thus on X_{Δ} for any X); we refer to this F-crystal as the (first) Breuil–Kisin twist. The étale realization of $\mathcal{O}_{\Delta}\{1\}$ is identified with the usual Tate twist $\mathbf{Z}_p(1)$.

Example 3.4 (Gauss–Manin F-crystals). Fix a proper smooth map $f : Y \to X$ of p-adic formal schemes. The formalism of relative prismatic cohomology yields an F-crystal $Rf_* \mathcal{O}_{\Delta}$ of perfect complexes on X_{Δ}: its value on a prism $(B, J) \in X_{\Delta}$ identifies with the relative prismatic complex $R\Gamma_{\Delta}((Y \times_X \mathrm{Spf}(B/J))/B)$. Similarly, one obtains a Hodge–Tate crystal $Rf_* \overline{\mathcal{O}}_{\Delta}$ of perfect complexes on X_{Δ}. The formation of $Rf_* \mathcal{O}_{\Delta}$ (resp. $Rf_* \overline{\mathcal{O}}_{\Delta}$) is compatible with the aforementioned realization functors. Moreover, if $Y = \mathbf{P}^1 \times X$, then the prismatic logarithm [32] yields a natural isomorphism $\mathcal{H}^2(Rf_* \mathcal{O}_{\Delta}) \simeq \mathcal{O}_{\Delta}\{-1\}$ of F-crystals, giving a geometric description of the Breuil–Kisin twist.

3.2. Hodge–Tate crystals

In this subsection, we fix a perfect field k of characteristic p, and write $W(k)_{\Delta} = \mathrm{Spf}(W(k))_{\Delta}$ for the absolute prismatic site of $W(k)$. Our goal is to explicitly describe the structure of Hodge–Tate crystals on $W(k)_{\Delta}$; we then specialize this description to the Gauss–Manin case to obtain geometric consequences. For the former, we have:

Proposition 3.5 (Sen theory, [32,76,77]). *The ∞-category $\hat{\mathcal{D}}_{\mathrm{crys}}(W(k)_{\Delta}, \overline{\mathcal{O}}_{\Delta})$ of Hodge–Tate crystals $\mathcal{E}$ of p-complete complexes on $W(k)_{\Delta}$ can be identified as the ∞-category of pairs (E, Θ) consisting of a p-complete object $E \in \mathcal{D}(W(k))$ and an endomorphism $\Theta : E \to E$ such that $\Theta^p - \Theta$ is locally nilpotent on $H^*(E/p)$; we refer to such pairs (E, Θ) as* Sen complexes *and Θ as the* Sen operator.

The implicit functor carrying the crystal $\mathcal{E}$ to $E \in \mathcal{D}(W(k))$ in Proposition 3.5 is given by evaluating at the object of $W(k)_{\Delta}$ obtained by base changing to $W(k)$ the $\mathbf{F}_p^*$-fixed points of the q-de Rham prism (Remark 2.9).

Remark 3.6 (The stacky approach to prismatic crystals, [32,76]). Proposition 3.5 is proven via a stacky approach to prismatic cohomology, developed independently in [76] (with a precursor in [75]) and [32]. Using a tiny amount of derived algebraic geometry [171], these works attach a stack WCart_X—the *Cartier–Witt stack of X* (called the *prismatization* X^{Δ} in [76])—on p-nilpotent test rings to any p-adic formal scheme X. This stack comes equipped with an effective Cartier divisor $\mathrm{WCart}_X^{\mathrm{HT}} \subset \mathrm{WCart}_X$ called the *Hodge–Tate locus*. These stacks are devised to geometrize the study of crystals on the prismatic site: for a quasisyntomic X, there is a natural $\otimes$-identification of the ∞-category $\hat{\mathcal{D}}_{\mathrm{crys}}(X_{\Delta}, \mathcal{O}_{\Delta})$ of crystals of (p, I_{Δ})-complete complexes on $(X_{\Delta}, \mathcal{O}_{\Delta})$ with the quasicoherent derived ∞-category $\mathcal{D}_{qc}(\mathrm{WCart}_X)$; similarly the ∞-category $\hat{\mathcal{D}}_{\mathrm{crys}}(X_{\Delta}, \overline{\mathcal{O}}_{\Delta})$ of crystals of p-complete complexes on $(X_{\Delta}, \overline{\mathcal{O}}_{\Delta})$ identifies with the quasicoherent derived ∞-category $\mathcal{D}_{qc}(\mathrm{WCart}_X^{\mathrm{HT}})$. Proposition 3.5 is then deduced from an explicit description of $\mathrm{WCart}_{W(k)}^{\mathrm{HT}}$ as BG for a group scheme $G/W(k)$ whose representations are identified with Sen complexes.

Notation 3.7 (Diffracted Hodge cohomology). Let $f : X \to \mathrm{Spf}(W(k))$ be a smooth map of p-adic formal schemes. Write $(R\Gamma(X, \Omega_X^{\not{D}}), \Theta)$ for the Sen complex corresponding to the Hodge–Tate crystal $Rf_* \overline{\mathcal{O}}_{\Delta} \in \hat{\mathcal{D}}_{\mathrm{crys}}(W(k)_{\Delta}, \overline{\mathcal{O}}_{\Delta})$ via Proposition 3.5; we call $R\Gamma(X, \Omega_X^{\not{D}})$ the *diffracted Hodge complex* of X.

The next result says that $R\Gamma(X, \Omega_X^{D\!\!\!\!D})$ is a slightly twisted form of the Hodge cohomology complex $\bigoplus_i R\Gamma(X, \Omega_X^i[-i])$, justifying the name "diffracted Hodge cohomology."

Theorem 3.8 (The Sen structure of $\Omega_X^{D\!\!\!\!D}$, [32]). *Let $X/W(k)$ be a smooth p-adic formal scheme. Then the Sen complex $(R\Gamma(X, \Omega_X^{D\!\!\!\!D}), \Theta)$ has a natural multiplicative increasing* conjugate *filtration* $\mathrm{Fil}^{\bullet}_{\mathrm{conj}}$ *equipped with natural isomorphisms*

$$\mathrm{gr}^i_{\mathrm{conj}}\big(R\Gamma(X, \Omega_X^{D\!\!\!\!D}), \Theta\big) \simeq \big(R\Gamma(X, \Omega^i_{X/W(k)})[-i], \Theta = -i\big)$$

for all i.

Theorem 3.8 also shows that the assignment $U \mapsto R\Gamma(U, \Omega_U^{D\!\!\!\!D})$ patches to a perfect complex $\Omega_X^{D\!\!\!\!D}$ on X, justifying the notation $R\Gamma(X, \Omega_X^{D\!\!\!\!D})$.

Remark 3.9 (Relation to classical Sen theory, [32]). Fix a proper smooth map $f : X \to \mathrm{Spf}(W(k))$ of p-adic formal schemes; write $K = W(k)[1/p]$, fix a completed algebraic closure C/K, and write G_K for the absolute Galois group of K. Classical results in p-adic Hodge theory [83, 209] show that for each $n \geq 0$, the C-semilinear G_K-representation $H^n(X_C, \mathbf{Q}_p) \otimes_{\mathbf{Q}_p} C$ comes equipped with a canonical semisimple endomorphism θ_n whose eigenvalue decomposition yields the Hodge–Tate decomposition: we have

$$H^n(X_C, \mathbf{Q}_p) \otimes_{\mathbf{Q}_p} C \simeq \bigoplus_{i=0}^{n} H^{n-i}(X, \Omega^i_{X/W(k)}) \otimes_{W(k)} C(-i),$$

with θ_n acting by $-i$ on the ith summand on the right. Using the comparison isomorphisms in Theorem 2.6, one can roughly regard Theorem 3.8 as an integral lift of this assertion: the value of $Rf_*\overline{\mathcal{O}}_{\mathbb{\Delta}}$ on the Fontaine prism for $\mathcal{O}_C$ recovers $R\Gamma(X_C, \mathbf{Q}_p) \otimes_{\mathbf{Q}_p} C$ on inverting p, the Sen operator Θ from Theorem 3.8 induces the Sen operator θ_n on each H^n with the conjugate filtration from Theorem 3.8 yielding the Hodge–Tate decomposition. It was a pleasant surprise to the author that the Sen operator admits a nice integral form.

Remark 3.10 (Drinfeld's refinement of Deligne–Illusie, [32, 76]). In the setup of Theorem 3.8, there is a natural identification $R\Gamma(X, \Omega_X^{D\!\!\!\!D})/p \simeq R\Gamma_{\mathrm{dR}}(X_k)$ compatible with the conjugate filtration via Theorem 2.6. Drinfeld observed that the Sen operator then yields interesting consequences for $R\Gamma_{\mathrm{dR}}(X_k)$. More precisely, there is a $\mathbf{Z}/p$-grading on $R\Gamma_{\mathrm{dR}}(X_k)$ corresponding to the generalized eigenspace decomposition for the Sen operator Θ, and the ith conjugate graded piece $\mathrm{gr}_i^{\mathrm{conj}} \simeq R\Gamma(X_k, \Omega^i_{X_k})[-i]$ contributes only to the generalized eigenspace for $\Theta = -i$ by Theorem 3.8. In particular, if $\dim(X_k) < p$, the conjugate filtration on $R\Gamma_{\mathrm{dR}}(X_k)$ splits canonically. This gives a conceptual new proof—in fact, a refinement—of the seminal Deligne–Illusie result [73] on Hodge-to-de Rham degeneration (itself inspired by [83, 95, 141]). As in [73], one only needs a $W_2(k)$-lift of X_k to obtain the Sen operator—and thus the $\mathbf{Z}/p$-grading—on $R\Gamma_{\mathrm{dR}}(X_k)$; this follows from an analysis of $\mathrm{WCart}^{\mathrm{HT}}_{W_2(k)}$ similar to Proposition 3.5. The results discussed in this paragraph refine those in [2] by one cohomological degree; another stacky proof was recently found in [162].

For $X/W(k)$, a smooth formal scheme without any constraints on $\dim(X_k)$, one now obtains a residual nilpotent operator $\Theta + i$ on the generalized Θ-eigenspace

$R\Gamma_{\mathrm{dR}}(X_k)_i \subset R\Gamma_{\mathrm{dR}}(X_k)$ corresponding to the eigenvalue $-i$; this operator seems to be a new piece of structure that awaits further investigation.

3.3. The Nygaard filtration

The absolute prismatic cohomology[2] $R\Gamma_{\Delta}(X)$ of a p-adic formal scheme X carries an important filtration $\mathrm{Fil}^{\bullet}_N$, called *the Nygaard filtration*. This filtration plays roughly the same role for prismatic cohomology as the Hodge filtration does for de Rham cohomology. Moreover, for applications to algebraic topology (such as Theorem 4.8 below), it is critical to understand this filtration. Its defining feature is that the Frobenius map φ on $R\Gamma_{\Delta}(X)$ carries $\mathrm{Fil}^i_N R\Gamma_{\Delta}(X)$ to $R\Gamma(X_{\Delta}, I^i_{\Delta})$ for all i. The relative version of this filtration is well understood, at least on graded pieces, thanks to the isomorphism (Isog) and the Beilinson t-structure (see **[38, PROPOSITION 5.8]**). For the absolute version, one has a similar description:

Theorem 3.11 (The Nygaard fiber sequence, **[32]**). *For any p-adic formal scheme X and any integer $i \geq 0$, there are natural fiber sequences*

$$\mathrm{gr}^i_N R\Gamma_{\Delta}(X) \to \mathrm{Fil}^{\mathrm{conj}}_i R\Gamma(X, \Omega^{D\!\!\!/}_X) \xrightarrow{\Theta+i} \mathrm{Fil}^{\mathrm{conj}}_{i-1} R\Gamma(X, \Omega^{D\!\!\!/}_X) \tag{Nyg}$$

and

$$R\Gamma_{\overline{\Delta}}(X)\{i\} := R\Gamma\big(X_{\Delta}, \overline{\mathcal{O}}_{\Delta}\{i\}\big) \to R\Gamma(X, \Omega^{D\!\!\!/}_X) \xrightarrow{\Theta+i} R\Gamma(X, \Omega^{D\!\!\!/}_X) \tag{HT}$$

with the convention that $\mathrm{Fil}^{\mathrm{conj}}_{<0} = 0$.

Remark 3.12 (Calculations via the Nygaard fiber sequence). The sequence (Nyg) is quite useful for calculations of the Nygaard filtration. For instance, in conjunction with the THH$(-)$ variant of Theorem 4.2 below (see Remark 4.3 as well), one may use (Nyg) to calculate $\pi_*\mathrm{THH}(R; \mathbf{Z}_p)$ for a p-completely smooth $\mathcal{O}_K$-algebra R, where K is a discretely valued extension of $\mathbf{Q}_p$ with perfect residue field; this recovers calculations of **[54, 163]**. Comparing (Nyg) and (HT) also quantifies the failure of the Frobenius map $\varphi : \mathrm{gr}^i_N R\Gamma_{\Delta}(X) \to R\Gamma_{\overline{\Delta}}(X)\{i\}$ to be an isomorphism in terms of coherent cohomology, thus giving a new mechanism to study the so-called Segal conjecture for THH.

Remark 3.13 (View Spec$(\mathbf{Z})$ as a curve). Several results in mathematics have been inspired by the seemingly nonsensical idea that Spec$(\mathbf{Z})$ is a curve over some nonexistent base $\mathbf{F}$. In p-adic arithmetic geometry, this idea can sometimes lead to useful (and testable!) predictions in conjunction with the following related heuristics:

- Perfectoid rings (e.g., finite fields) are formally étale over $\mathbf{F}$.
- Topologically finite type regular p-complete $\mathbf{Z}_p$-algebras R are smooth over $\mathbf{F}$ of relative dimension $\dim(R)$ (the Krull dimension).

---

2 The complex $R\Gamma_{\Delta}(X)$ as defined in Definition 3.1 works well under mild assumptions on the singularities of X (e.g., for lci X). In general, one modifies the definition of $R\Gamma_{\Delta}(X)$ by a categorical procedure involving quasisyntomic descent and animation; we do not elaborate on this further in this survey and refer to **[11, 32]** for more.

We briefly discuss some examples of such predictions. First, if R is a perfectoid ring, then the p-completion of $L_{R/\mathbf{Z}_p}$ identifies with $R[1]$, which is consistent with the prediction of a transitivity triangle of cotangent complexes for $\mathbf{F} \to \mathbf{Z} \to R$ and the heuristics above; this was already essentially observed in [84].

Next, Theorem 3.11 was partially conceived based on these heuristics: the fiber sequence (Nyg) is obtained as the associated graded of a fiber sequence of filtered complexes allowing one to compute the absolute Nygaard filtration in terms of the relative one; the underlying fiber sequence of complexes for the latter was guessed based on the analogy between $\mathbf{Z}$ and a smooth curve over a perfectoid ring.

Next, if one views the Hodge–Tate locus $\mathrm{WCart}_X^{\mathrm{HT}} \subset \mathrm{WCart}_X$ of the Cartier–Witt stack (Remark 3.6) as a version of the Hodge stack (i.e., the classifying stack of the tangent bundle) over $\mathbf{F}$, then the second heuristic above predicts that $\mathrm{WCart}_X^{\mathrm{HT}}$ is well behaved if X is regular, e.g., the map $\pi : \mathrm{WCart}_X^{\mathrm{HT}} \to X$ should be a gerbe, and $R\pi_*$ must have coherent cohomological dimension at most $\dim(X)$; the first of these predictions is true, while the second is true at least in dimension 1 ([32]). Relatedly, there are some recently defined candidate notions of differential forms relative to $\mathbf{F}$ [80, 128, 198]; it would be interesting to find a direct connection between the stack $\mathrm{WCart}_X^{\mathrm{HT}}$ and these objects.

Finally, let us remark that the philosophy discussed in this remark also featured in Scholze's report for the previous ICM [206], and has paid amazing dividends in geometrizing the local Langlands correspondence in recent years [90, 205, 207].

Remark 3.14 (p-adic Tate twists, [38,41]). An early observable extracted from absolute prismatic cohomology was a good notion of p-adic Tate twists $\mathbf{Z}_p(i)(-)$ in mixed characteristic: these are functors on p-adic formal schemes X defined by a fiber sequence

$$\mathbf{Z}_p(i)(X) \to \mathrm{Fil}_N^i R\Gamma_{\Delta}(X)\{i\} \xrightarrow{\varphi-1} R\Gamma_{\Delta}(X)\{i\} \tag{Syn}$$

for all $i \geq 0$. These functors are often called *syntomic complexes* for mixed characteristic rings as they extend those in characteristic p considered in [140, 181]. One can also regard $\mathbf{Z}_p(i)(-)$ as a form of étale motivic cohomology in weight i (see the forthcoming Remark 4.6). In fact, for formally smooth $\mathcal{O}_K$-schemes with $K/\mathbf{Q}_p$ finite, the syntomic complexes $\mathbf{Z}_p(i)(-)$ agree with the p-adic étale Tate twists of Geisser–Sato–Schneider [103, 199, 200] defined by glueing motivic complexes on the generic and special fibers [35]. We refer the reader to [11, 32, 38, 41] for more on these syntomic complexes.

Remark 3.15 (p-adic Picard and Brauer groups via coherent cohomology). The syntomic complex from Remark 3.14 in weight 1 has the following relationship with $\mathbf{G}_m$ ([38, PROPOSITION 7.17]), as motivic intuition predicts: for any p-adic formal scheme X, we have

$$\mathbf{Z}_p(1)(X) \simeq R\Gamma(X_{et}, \mathbf{G}_m)^\wedge[-1],$$

where the completion is p-adic. Plugging this into the sequence (Syn) gives a fiber sequence

$$R\Gamma(X_{et}, \mathbf{G}_m)^\wedge[-1] \to \mathrm{Fil}_N^1 R\Gamma_{\Delta}(X)\{1\} \xrightarrow{\varphi-1} R\Gamma_{\Delta}(X)\{1\} \tag{Lef}$$

that can be regarded as a weak p-adic analog of the Lefschetz $(1,1)$-theorem, e.g., it enables one to compute the p-completion of $\mathrm{Pic}(X)$ or $\mathrm{Br}(X)$ in terms of absolute prismatic cohomology, and thus ultimately via coherent cohomology.

The idea described in the previous paragraph inspired the eventual proof of (a generalization of) Gabber's purity conjectures for Picard and Brauer groups in [63]. In a global direction, Cotner and Zavyalov have recently used (Lef) to prove the vanishing of $\mathrm{Pic}^{\tau}(X)$ for complete intersection surfaces $X \subset \mathbf{P}^N$ in characteristic p (in progress), settling a question left open since [189]. In a different direction, the sequence (Lef) can be used to prove that $R \mapsto R\Gamma(\mathrm{Spf}(R)_{et}, \mathbf{G}_m)^\wedge$ commutes with sifted colimits in R (in the p-complete world); this allows one to reduce general questions about $R\Gamma(\mathrm{Spf}(R)_{et}, \mathbf{G}_m)^\wedge$ to particularly nice rings, and played an important role in Bragg and Olsson's work [58] on finiteness results for higher direct images of finite flat group schemes along projective morphisms in characteristic p.

3.4. Galois representations

In this subsection, fix a discretely valued field $K/\mathbf{Q}_p$ with perfect residue field k. We discuss the relationship of prismatic F-crystals over $X = \mathrm{Spf}(\mathcal{O}_K)$ and Galois representations of $G_K = \mathrm{Gal}(\overline{K}/K)$.

For a prime $\ell \neq p$, the notion of *unramifiedness* for $\mathbf{Z}_\ell$- or $\mathbf{Q}_\ell$-representations of G_K is a Galois-theoretic analog of the property of having "good reduction" for varieties over K: viewed as an ℓ-adic local system on $\mathrm{Spec}(K)$, an unramified G_K-representation is exactly one that extends to a local system over $\mathrm{Spf}(\mathcal{O}_K)$. In contrast, for $\mathbf{Z}_p$- or $\mathbf{Q}_p$-representations, unramifiedness is too restrictive: even the cyclotomic character—or any nonzero $H^i(Y_{\overline{K}}, \mathbf{Q}_p)$ with $Y/\mathcal{O}_K$ smooth projective and $i > 0$—is not unramified. To remedy this, Fontaine invented [93] the notion of *crystalline* G_K-representations; it has been stunningly successful at capturing the desired "good reduction" intuition. On the other hand, any prismatic F-crystal $\mathcal{E}$ on $\mathrm{Spf}(\mathcal{O}_K)$ gives rise to a G_K-representation $T(\mathcal{E})$, as well an F-crystal $\mathcal{E}_{\mathrm{crys}}$ on k (see Definition 3.2 and following discussion); these have the same rank, so one may view $\mathcal{E}_{\mathrm{crys}}$ as "a special fiber" of $T(\mathcal{E})$, suggesting that the prismatic F-crystal $\mathcal{E}$ itself should be viewed as a witness for a "good reduction" of $T(\mathcal{E})$. The following theorem shows that these two perspectives on good reduction for p-adic representations coincide:

Theorem 3.16 (Prismatic F-crystals and crystalline G_K-representations, [42]). *The étale realization functor $\mathcal{E} \mapsto T(\mathcal{E})$ gives an equivalence of the category of prismatic F-crystals on $\mathrm{Spf}(\mathcal{O}_K)$ with the category of $\mathbf{Z}_p$-lattices in crystalline $\mathbf{Q}_p$-representations of G_K.*

Thus, prismatic F-crystals on $\mathrm{Spf}(\mathcal{O}_K)$ provide a reasonable notion for "local systems on $\mathrm{Spf}(\mathcal{O}_K)$ with $\mathbf{Z}_p$-coefficients".

Remark 3.17. Theorem 3.16 can be viewed as a refinement of Kisin's classification of crystalline G_K-representations [146]; in particular, this refinement attaches prismatic meaning to the integrality properties of a somewhat mysterious connection in [146]. An alternative proof of Theorem 3.16 was since given in [78], relying on the theory in [166]; see also [230].

Various results in the deformation theory of G_K-representations (e.g., [81,147]) indicate it would be fruitful to extend the notion of crystalline G_K-representations to torsion coefficients or even to the derived category. However, as the property of being crystalline is essentially a rational concept, it is not clear how to proceed. Theorem 3.16 points to a way forward, e.g., perhaps prismatic F-crystals with $\mathcal{O}_{\Delta}/p^n$-coefficients are a reasonable candidate for crystalline $\mathbf{Z}/p^n$-representations? While satisfactory for describing $\mathbf{Z}_p$-local systems, this approach does not quite lead to a reasonable derived theory as the definition of a prismatic F-crystal $(\mathcal{E}, \varphi_{\mathcal{E}})$ is not quantitative enough: the isomorphism $\varphi_{\mathcal{E}}$ does not come equipped with bounds on its poles/zeroes, leading to certain poorly behaved Ext-groups. Instead, the correct objects seem to be perfect complexes on an enlargement of the Cartier–Witt stack $\mathrm{WCart}_{\mathcal{O}_K}$ (Remark 3.6) constructed by Drinfeld [76]; we describe one piece of evidence for this correctness assertion in the rest of the subsection.

Write $\mathcal{D}^{\varphi}_{\mathrm{perf}}(\mathrm{WCart}^{+}_{\mathcal{O}_K})$ for the ∞-category of perfect complexes on the stack $\mathrm{Spf}(\mathcal{O}_K)^{\Delta''}$ from [76, §1.8]; let us call such objects *prismatic F-gauges on $\mathcal{O}_K$*[3]. Given such an F-gauge $\mathcal{E}$, write $R\Gamma^{\varphi}(\mathrm{WCart}^{+}_{\mathcal{O}_K}, \mathcal{E})$ for its global sections. To a first approximation, a prismatic F-gauge $\mathcal{E}$ consists of a prismatic F-crystal E of perfect complexes on $\mathrm{WCart}_{\mathcal{O}_K}$ equipped with the additional datum of a Nygaard-style filtration on $R\Gamma(\mathrm{WCart}_{\mathcal{O}_K}, E)$; in fact, this can be made precise if $\mathcal{O}_K$ is replaced by a qrsp ring (work in progress with Lurie). The prismatic F-crystals from Examples 3.3 and 3.4 have natural lifts to prismatic F-gauges. The promised piece of evidence is the following result:

Theorem 3.18 (A Lagrangian property, [33]). *Assume K is unramified. Let*

$$\mathcal{E} \in \mathcal{D}^{\varphi}_{\mathrm{perf}}(\mathrm{WCart}^{+}_{\mathcal{O}_K})$$

be a prismatic F-gauge with $\mathcal{O}$-linear dual $\mathbf{D}(\mathcal{E})$ and étale realization $T(\mathcal{E})$ in an appropriate derived category of continuous $\mathbf{Z}_p$-representations of G_K. Then the natural map

$$R\Gamma^{\varphi}(\mathrm{WCart}^{+}_{\mathcal{O}_K}, \mathcal{E}) \to R\Gamma\big(G_K, T(\mathcal{E})\big)$$

is the exact annihilator of the corresponding map of local Tate duals, i.e., there is a natural fiber sequence

$$R\Gamma^{\varphi}(\mathrm{WCart}^{+}_{\mathcal{O}_K}, \mathcal{E}) \to R\Gamma\big(G_K, T(\mathcal{E})\big) \to \big(R\Gamma^{\varphi}(\mathrm{WCart}^{+}_{\mathcal{O}_K}, \mathbf{D}(\mathcal{E})\{1\}[2])\big)^{\vee} \qquad \text{(Lag)}$$

where $(-)^{\vee} = \mathrm{RHom}_{\mathbf{Z}_p}(-, \mathbf{Z}_p)$ on the rightmost term.

Theorem 3.18 is work in progress with Lurie [33]; the statement is likely not quite optimal yet (e.g., we hope to show it for ramified K as well).

Remark 3.19 (The crystalline part of Galois cohomology). Given a $\mathbf{Q}_p$-representation V of G_K, Bloch–Kato constructed [51, §3] the "crystalline part" $H^1_f(G_K, V) \subset H^1(G_K, V)$ of the Galois cohomology of V, and proved that the crystalline parts for V and $V^{\vee}(1)$ are

3 The definition of $\mathrm{WCart}^{+}_{\mathcal{O}_K}$ in [76] (and denoted $\Sigma'_{\mathcal{O}_K}$ there) is inspired by the Fontaine–Jansen theory [94] of F-gauges in crystalline cohomology.

orthogonal complements under the Tate duality pairing. When V is crystalline, Theorem 3.18 may be viewed as an integral and cochain-level variant of this statement. Such integral refinements have been formulated previously in special cases (e.g., [102]); it would be interesting to compare them to Theorem 3.18.

Remark 3.20 (The analogy with 3-manifolds). In the Mazur(–Mumford) analogy between number rings and 3-manifolds [158, 180], the scheme $\mathrm{Spec}(K)$ corresponds to a Riemann surface Σ while $\mathrm{Spf}(\mathcal{O}_K)$ corresponds to a 3-manifold M with boundary $\partial M = \Sigma$. A standard topological result states that the space $\mathrm{Loc}(\Sigma)$ of local systems on Σ has a symplectic structure induced by Poincaré duality on Σ, and the restriction map $\mathrm{Loc}(M) \to \mathrm{Loc}(\Sigma)$ is Lagrangian (see [96, PROPOSITION 3.27]). The sequence (Lag) may be viewed as an arithmetic analogue of the infinitesimal form of this result, with the role of a local system on $\mathrm{Spf}(\mathcal{O}_K)$ played by prismatic F-gauges; in fact, this picture motivated the discovery of Theorem 3.18.

Remark 3.21 (Lichtenbaum–Quillen for $\mathcal{O}_K$). The Breuil–Kisin prismatic F-gauges $\mathcal{O}\{i\}$ compute the p-adic Tate twists from Remark 3.14, i.e., we have natural identifications $R\Gamma^{\varphi}(\mathrm{WCart}^{+}_{\mathcal{O}_K}, \mathcal{O}\{i\}) \simeq \mathbf{Z}_p(i)(\mathcal{O}_K)$. Using the vanishing of $\mathbf{Z}_p(i)(-)$ for $i < 0$, the sequence (Lag) then implies that the natural map

$$\mathbf{Z}_p(i)(\mathcal{O}_K) \to R\Gamma\big(G_K, \mathbf{Z}_p(i)\big)$$

is an equivalence for $i \geq 2$. Under the relationship of either side to the étale K-theory of $\mathcal{O}_K$ and K, as well as the localization sequence in K-theory, this result was essentially known ([119]); nevertheless, Theorem 3.18 provides a different conceptual explanation.

4. ALGEBRAIC K-THEORY

Quillen's algebraic K-theory [193] functor attaches a space (in fact, a spectrum) $K(X)$ to a scheme X, generalizing the construction of the Grothendieck group $K_0(X)$; the study of these invariants and their generalizations is an important pursuit in modern algebraic topology. In fact, its impact extends far beyond algebraic topology: the higher K-groups $K_i(X)$ feature prominently in some of the deepest conjectures in arithmetic geometry. In this section, we report on some recent progress in understanding the structural features of the p-completed algebraic K-theory spectrum $K(R; \mathbf{Z}_p)$ of a p-complete ring R, with an emphasis on connections to prismatic cohomology; the case of ℓ-adic completions for $\ell \neq p$ is classical, going back to work of Thomason [221], Suslin [217], and Gabber [98]. More complete recent surveys of material covered in this section include [120, 121, 179].

In classical algebraic topology, combining the Atiyah–Hirzeburch spectral sequence with Bott periodicity gives the following structure on the complex K-theory $K^{\mathrm{top}}(X)$ of a (reasonable) topological space X:

The K-theory spectrum $K^{\mathrm{top}}(X)$ admits a natural filtration with gr^i identified with the shifted singular cohomology complex $R\Gamma(X, \mathbf{Z})[2i]$. $\quad (\mathrm{Filt}_K)$

In recent work, motivated in part by conjectures of Beilinson [13] and Hesselholt, a p-adic analog of (Filt_K) for (étale sheafified) algebraic K-theory of p-complete rings has been established, with the role of singular cohomology now played by prismatic cohomology. To explain this better, recall that algebraic topologists often study algebraic K-theory through a "cyclotomic trace" map

$$\mathrm{Tr} : K(-) \to \mathrm{TC}(-),$$

where TC(−) is the *topological cyclic homology* functor; this invariant of rings was invented by contemplating Hochschild homology over the sphere spectrum, goes back to [55], and was recently given a simple ∞-categorical definition in [186]. The trace map is a powerful calculational tool (see [104, 117–119] for some successes), and there are two main reasons for this. First, TC(−) is built, via a rather elaborate homotopical procedure, from objects of coherent cohomology (namely, differential forms) and is thus potentially more computable than K-theory. Secondly, the trace map turns out to yield a very good approximation of K-theory in various situations; in our p-adic context, the state of the art is the following:

Theorem 4.1 (p-adic étale K-theory is TC, [65, 66]). *For p-complete rings R, the trace map $K(R; \mathbf{Z}_p) \to \mathrm{TC}(R; \mathbf{Z}_p)$ identifies the target with the p-completed étale sheafified K-theory of R. Moreover, the étale sheafification is not necessary in sufficiently large degrees if R satisfies mild finiteness conditions.*

Via Theorem 4.1, the following result can be viewed as a p-adic analog of the Atiyah–Hirzebuch part of (Filt_K):

Theorem 4.2 (The motivic filtration on étale K-theory, [11, 38]). *As a functor on p-complete rings, there is a natural "motivic" filtration on* $\mathrm{TC}(-; \mathbf{Z}_p)$ *with* $\mathrm{gr}^i_{\mathrm{mot}}\mathrm{TC}(-; \mathbf{Z}_p)$ *naturally identified with the shifted syntomic complex* $\mathbf{Z}_p(i)(-)[2i]$ *from Remark* 3.14.

Remark 4.3 (Variants for THH and cousins). Let us briefly recall the [186] approach to calculating TC. For a commutative ring R, the p-completed topological Hochschild homology spectrum $\mathrm{THH}(R; \mathbf{Z}_p)$ comes equipped with a natural action of the circle S^1 and a certain Frobenius map. One can then define auxiliary invariants $\mathrm{TC}^-(R; \mathbf{Z}_p) := \mathrm{THH}(R; \mathbf{Z}_p)^{hS^1}$ and $\mathrm{TP}(R; \mathbf{Z}_p) := \mathrm{THH}(R; \mathbf{Z}_p)^{tS^1}$ together with two natural maps can, $\varphi : \mathrm{TC}^-(R; \mathbf{Z}_p) \to \mathrm{TP}(R; \mathbf{Z}_p)$. The paper [186] then proves there is a natural fiber sequence

$$\mathrm{TC}(R; \mathbf{Z}_p) \to \mathrm{TC}^-(R; \mathbf{Z}_p) \xrightarrow{\varphi - \mathrm{can}} \mathrm{TP}(R; \mathbf{Z}_p), \qquad \text{(TC)}$$

thereby yielding a clean modern construction of $\mathrm{TC}(R; \mathbf{Z}_p)$.

The construction of the motivic filtration on $\mathrm{TC}(-; \mathbf{Z}_p)$ in Theorem 4.2 is sufficiently flexible to ensure that similar ideas also yield compatible "motivic filtrations" on $\mathrm{TC}^-(R; \mathbf{Z}_p)$ and $\mathrm{TP}(R; \mathbf{Z}_p)$. In fact, [38] lifts the sequence (TC) to a filtered fiber sequence that recovers the sequence (Syn) on associated graded pieces, up to Nygaard completions. In particular, one recovers (Nygaard completed) absolute prismatic cohomology as the associated graded of a filtration on TP.

Note that while THH(−) and cousins are noncommutative invariants (i.e., can be defined for arbitrary stable ∞-categories), the construction of the motivic filtration crucially uses algebraic geometry; it is unclear if analogous filtrations exist even in slightly more general settings, e.g., for $\mathrm{TC}(\mathcal{C})$ for a symmetric monoidal stable ∞-category $\mathcal{C}$.

Remark 4.4 (Comparison with the Hodge filtration on classical Hochschild homology). The topological Hochschild homology of a commutative ring R is defined as $\mathrm{THH}(R) := \mathrm{HH}(R/\mathbf{S})$, i.e., it is the Hochschild homology relative to the sphere spectrum $\mathbf{S}$. From this optic, Theorem 4.2 and the variants in Remark 4.3 are analogs of known constructions in the Hochschild homology of ordinary rings. For instance, given a smooth algebra R over a commutative ring k, there is a natural filtration on $\mathrm{HC}^-(R/k) := \mathrm{HH}(R/k)^{hS^1}$ with gr^i identified with the Hodge filtration level $\mathrm{Fil}^i_{\mathrm{Hodge}} \Omega^\bullet_{R/k}[2i]$ (see [167, §5.1.2] and [228] for characteristic 0, and [38, §5.2] and [9] in general). Specializing to $k = \overline{A}$ for a Fontaine prism (A, I), this allows one to recover crystalline and de Rham cohomology—but not étale cohomology of the generic fiber—as graded pieces of a natural filtration on classical Hochschild homology and its variants. Theorem 4.2 and the variants in Remark 4.3 thus contain the surprise that working relative to the sphere spectrum permits one to see the étale cohomology of the generic fiber as well: one can in fact recover prismatic cohomology.

Remark 4.5 (Origin story). Let (A, I) be a Fontaine prism (Example 2.4). Write A_{crys} for the ring obtained by formally adjoining divided powers of I to A in the p-complete setting. Given a smooth proper scheme X over $\mathcal{O}_C = \overline{A}$, its absolute crystalline cohomology $R\Gamma_{\mathrm{crys}}(X)$ is a perfect complex of A_{crys}-modules with a Frobenius structure. The relation between the theory of Breuil modules [59] and Breuil–Kisin modules [146] in Galois representation theory strongly suggested that $R\Gamma_{\mathrm{crys}}(X)$ ought to descend naturally along $A \to A_{\mathrm{crys}}$. Separately, Hesselholt had calculated [116] that $\pi_0 \mathrm{TP}(\mathcal{O}_C; \mathbf{Z}_p)$ equals A. Comparing this to the known fact that $\pi_0 \mathrm{HP}(\mathcal{O}_C; \mathbf{Z}_p)$ equals A_{crys} (up to a completion), it was natural to speculate that for any $X/\mathcal{O}_C$ as above, one could find a filtration on $\mathrm{TP}(X; \mathbf{Z}_p)$ whose graded pieces realize the desired descent of $R\Gamma_{\mathrm{crys}}(X)$ along $A \to A_{\mathrm{crys}}$; this eventually led to the $\mathrm{TP}(-)$-variant of Theorem 4.2 and gave a construction of prismatic cohomology over the Fontaine and Breuil–Kisin prisms [38, §11]. In fact, as $\mathrm{TP}(-; \mathbf{Z}_p)$ is independent of the base $\mathcal{O}_C$, this also gave the first construction of absolute prismatic cohomology [38, §7.3].

Remark 4.6 (Étale motivic cohomology). We briefly explain why Theorem 4.2 can be viewed as a p-completed and étale sheafified analog of the filtration of algebraic K-theory by motivic cohomology (defined via Bloch's higher Chow groups [48]). Recall that the latter geometric motivic filtration was conjectured to exist in [13], and established in many cases, including most smooth cases, in [52, 97, 159, 227]; see [129] for a clean construction. It is thus natural to conjecture that the p-completed étale sheafification of this geometric motivic filtration identifies with the one in Theorem 4.2. For smooth varieties over a perfect field k of characteristic p, this is indeed the case by combining [106] and [38]. In mixed characteristic, while we do not know the full story, a positive answer at the associated graded level follows from the comparison result from [35] mentioned in Remark 3.14. Let us also remark that [168] has established the expected relationship of Milnor K-theory (extended as in [145]) and the (i, i)-part of the syntomic complexes, giving a p-adic analog of [185, 225].

The above discussion raises a natural question: as Theorem 4.2 applies to any p-complete ring, can the domain of definition of the geometric motivic filtration of the previous paragraph also be extended to all p-complete rings? In particular, is there a meaningful geometric motivic filtration on $K(R)$ for a non-reduced p-complete ring R? Thanks to a forthcoming result of Mathew on glueing the filtration from Theorem 4.2 with the étale sheafified Postnikov filtration on $K(1)$-local K-theory, there is a variant of Theorem 4.2 for any ring, so one may even reasonably ask these questions for all rings.

Remark 4.7 (Constructing the motivic filtration via quasisyntomic descent). The construction of the motivic filtration in Theorem 4.2 is quite different from that of the geometric motivic filtration mentioned in Remark 4.6. Indeed, the general case of Theorem 4.2 is proven in [11] (see also [32]) by reducing (via animation as in Remark 2.14) to the quasisyntomic case treated in [38]. The latter has two essential ingredients. The topological ingredient is Bökstedt's fundamental periodicity result [53] that $\pi_* \mathrm{THH}(\mathbf{F}_p) = \mathbf{F}_p[u]$ for a degree 2 class u; see [153] for a quick modern proof based on properties of the dual Steenrod algebra, [91] for an overview of other approaches, and [120] for a deduction of Bott periodicity from Bökstedt periodicity. The new algebraic input is the flat descent property ([20, 38]) of the cotangent complex, used in conjunction with the very perfectoid idea (going back in spirit to [95]) that working with certain infinitely ramified covers can "discretize" constructions involving differential forms in the p-adic world.

As Remark 4.5 explains, the first construction of absolute prismatic cohomology was through Theorem 4.2 and variants. However, thanks to the alternative and more direct construction via the prismatic site, one can now use Theorem 4.2 as a tool to study K-theory via prismatic cohomology. For instance, this approach gives the following result:

Theorem 4.8 (The odd vanishing theorem, [41]). *For odd i, the functor $\pi_i K(-;\mathbf{Z}_p)$ is quasi-syntomic locally* 0 *on the category of quasi-syntomic rings.*

Theorem 4.8 can be regarded as a variant of the Bott periodicity part of (Filt_K) in the algebraic setting: while periodicity is known to be false due to geometric phenomena, we still have vanishing in odd degrees. The proof in [41] relies on André's flatness lemma [3], and it would be interesting to find a more explicit description of the necessary covers.

Remark 4.9 (Further relations to p-adic arithmetic geometry). Another application of prismatic cohomology to K-theory was proving that $L_{K(1)}K(R) \simeq L_{K(1)}K(R[1/p])$ for any associative ring R ([27]). This equality is a K-theoretic avatar of the étale comparison from [41] and was proved in [27] via explicit calculations in prismatic cohomology; it has since been reproved and significantly extended using purely homotopy-theoretic methods in [156].

In the reverse direction (and preceding most of the developments reported in this paper), [49] used results from topological cyclic homology [105] to prove the infinitesimal portion of the p-adic variational Hodge conjecture in the unramified case. The extension to the ramified case was recently obtained in [11] as a consequence of a purely K-theoretic assertion called the Beilinson fiber square. Using this square and Theorem 4.2, [11] also gave a simple description of the rationalized syntomic complexes $\mathbf{Z}_p(i)(-)[1/p]$ via derived de Rham cohomology. This description is quite useful as derived de Rham cohomology is more computable in practice than prismatic cohomology; in fact, this description formed an essential ingredient in the classification of crystalline representations given in Theorem 3.16.

The connections discussed above have mostly concerned relative prismatic cohomology. It seems likely that a better understanding of absolute prismatic cohomology (as in Section 3) will lead to more refined applications. For instance, [164] recovers rather conceptually the highly non-trivial calculation [119] of the K-theory of local fields $K/\mathbf{Q}_p$ by exploiting certain covers of the final object in the absolute prismatic topos of $\mathcal{O}_K$ coming from Breuil–Kisin prisms. Other related observations are discussed in Remark 3.12 and Remark 3.21.

5. COMMUTATIVE ALGEBRA AND BIRATIONAL GEOMETRY

The Kodaira vanishing theorem (as well as the generalization by Kawamata–Viehweg) is one of the most important foundational results in complex algebraic geometry; it is especially useful in birational geometry. Its (original) proof relies crucially on Hodge theory, and thus no longer applies in positive/mixed characteristic. In fact, the result is known to be false in those settings [195]; alongside the nonavailability of resolution of singularities in dimensions ≥ 4, this is a major obstacle to progress in birational geometry in positive/mixed characteristic. About a decade ago, Schwede observed [208] that methods from F-singularity theory in positive characteristic commutative algebra can sometimes be used as a substitute for the use of vanishing theorems in positive characteristic algebraic geometry; this eventually led to significant progress in birational geometry in positive characteristic in dimension ≤ 3, such as [113]. In recent years, input from p-adic Hodge theory has made it possible to prove similar vanishing theorems in mixed characteristic algebraic geometry; this has led to solutions of longstanding questions in commutative algebra and also to progress in the minimal model program in mixed characteristic.

5.1. Vanishing theorems in commutative algebra

F-singularity theory is the study of singularities in positive characteristic via the behavior of the Frobenius endomorphism. It was born with a classical theorem of Kunz [155] proving that a noetherian $\mathbf{F}_p$-algebra is regular exactly when its Frobenius endomorphism is flat. This subject was systematically developed by Hochster–Huneke and several others over many decades; see [125] as well as the survey [218]. An important landmark in the subject was a Cohen–Macaulayness result of Hochster–Huneke [126]; see [130] for a fairly recent survey. The following recent result extends this to mixed characteristic:

Theorem 5.1 (Cohen–Macaulayness of R^+, [25, 43]). *Let R be an excellent noetherian domain. Let R^+ be the integral closure of R in an algebraic closure of its fraction field. Then the p-adic completion $\widehat{R^+}$ is Cohen–Macaulay over R.*

Remark 5.2 (A concrete formulation). Despite involving the large ring R^+, Theorem 5.1 is a finitistic statement whose essential content is the following: if R is local and $\underline{x} := \{p, x_1, \ldots, x_d\}$ is a system of parameters, then any relation on $\underline{x}$ becomes a linear combination of the trivial Koszul relations in a finite extension S of R. This formulation explains why Theorem 5.1 can be viewed as a "vanishing theorem up to finite covers": it says that the local cohomology classes on R coming from the potentially nontrivial relations can be annihilated by passing to finite extensions $R \to S$. Moreover, it also highlights the essential difficulty: one must *construct* finite extensions of R from the unwanted relations.

Remark 5.3 (Weakly functorial Cohen–Macaulay algebras). André's recent resolution [3,4] of Hochster's direct summand conjecture led to a lot of activity in mixed characteristic commutative algebra, including [23,115,175,176,178,211]; see [177] for a recent survey. In particular, André [6] and Gabber [99] proved the existence of "weakly functorial Cohen–Macaulay algebras" in the key remaining mixed characteristic case (via rather indirect constructions). This existence result implies many of the "homological conjectures" in commutative algebra (a notable exception being Serre's intersection multiplicity conjecture); see [5,123,124]. Prior to André and Gabber's work, this existence was known [114,122] only in dimension ≤ 3. Theorem 5.1 now yields an alternative and extremely simple construction of such weakly functorial Cohen–Macaulay algebras in mixed characteristic: we may simply use $\widehat{R^+}$.

Remark 5.4 (What was known?). Theorem 5.1 is straightforward in dimension ≤ 2, and is the main result of [126] in the positive characteristic case. In mixed characteristic, Theorem 5.1 is new even in dimension 3: it was previously known [114] in dimension ≤ 3 only in the almost category (in the sense of Faltings' almost mathematics [83,100]); see [25, REMARK 1.9] for an explanation of prior expectations.

Remark 5.5 (Splinters). A noetherian commutative ring R is called a *splinter* if it satisfies the conclusion of the direct summand conjecture, i.e., it splits off as a module from every finite extension. This class of singularities, formally introduced in [172], has recently received renewed attention (e.g., [7,10,70,173]). An external reason to care about this notion is a major conjecture in F-singularity theory (see [126, PAGE 85], [127, PAGE 640]): splinters in characteristic p are expected to be the same as strongly F-regular rings (see [174, END OF §3] for a discussion). This conjecture is known for $\mathbf{Q}$-Gorenstein rings [214]. One consequence of this conjecture is that characteristic p splinters are derived splinters, i.e., they satisfy a derived version of the splinter condition for any proper surjective map and are thus analogous to rational singularities. This consequence was proven unconditionally in [21]. Methods from [29] used in proving Theorem 5.1 give the same result in mixed characteristic. In conjunction with Theorem 5.1 itself, one learns that any mixed characteristic splinter is Cohen–Macaulay and has rational singularities in the sense of [152]; it would be interesting to prove the latter (even just after inverting p) without using p-adic Hodge theory.

Remark 5.6 (Ingredients in the proof of Theorem 5.1). Using essentially elementary methods, [43] reduces Theorem 5.1 to the statement that R^+/p is Cohen–Macaulay over R/p, which is proven in [25]. Despite the simple reformulation highlighted in Remark 5.2, the proof relies on two major theoretical inputs. The first is prismatic cohomology (Theorem 2.6), which gives a substitute for the Frobenius operator in mixed characteristic; this allows one to begin mimicking the cohomological proof of [126] given in [131] in mixed characteristic at the cost of replacing rings with derived rings. The second is the p-adic Riemann–Hilbert functor from Theorem 6.1 below, applied to certain perverse $\mathbf{F}_p$-sheaves on the generic fiber $\mathrm{Spec}(R[1/p])$ arising from finite covers, to facilitate the induction on dimension strategy of [131]. This proof is not effective, and it might be interesting to explicitly constructed the relevant covers in low dimensional examples, such as cones over smooth projective curves and surfaces over a p-adic discrete valuation ring.

5.2. Birational geometry

There is a well-known analogy between projective geometry and local algebra, e.g., the global cohomological properties of a projective variety $X \subset \mathbf{P}^n$ are faithfully reflected in the local cohomological properties of its affine cone $Y \subset \mathbf{A}^{n+1}$ over X near the vertex $0 \in Y$. This analogy suggests that Theorem 5.1 ought to have a global variant; this is indeed the case, and the result can be summarized as follows:

Theorem 5.7 (Kodaira vanishing up to finite covers, [25]). *Let V be a p-adic discrete valuation ring (e.g., $V = \mathbf{Z}_p$). Let X/V be a flat proper scheme equipped with a semiample and big line bundle L. Then any p-power torsion class in $H^*(X, L^{-1})$, $H^*(X, \mathcal{O}_X)$, or $H^*(X, L)$ can be annihilated by pullback to a finite cover of X.*

Analogous results hold true in the relative setting [25], and were previously known in characteristic p ([126] for L ample, and [21] in general).

Remark 5.8 (Relation to Kodaira vanishing). The classical Kodaira vanishing theorem says that $H^{<\dim(Y)}(Y, M^{-1}) = 0$ for a smooth projective variety $Y/\mathbf{C}$ with ample line bundle M. This assertion is false in characteristic p ([195]) and mixed characteristic (by Totaro, see [43, FOOTNOTE 1]). The L^{-1} case of Theorem 5.7 can be viewed as an "up to finite covers" variant of the Kodaira vanishing theorem that *is* true in mixed characteristic: spurious cohomology classes—those that should not be there if Kodaira vanishing were true for (X, L)—can be annihilated by passing to finite covers. This "up to finite covers" perspective was pioneered in characteristic p by [215] in the wake of [126].

For completeness, we remark that an "up to finite covers" version of the more general Kodaira–Akizuki–Nakano vanishing theorem also holds true in the setting of Theorem 5.7: in fact, the cases not covered by Theorem 5.7 are much easier as sheaves of differential forms themselves become p-divisible on passage to finite covers.

Remark 5.9 (Relation to the p-adic Poincaré lemma). The assertion in Theorem 5.7 for $H^*(X, \mathcal{O}_X)$, with finite covers weakened to alterations, was previously known by [14, 22]; in fact, it formed the key geometric ingredient in the proof of the p-adic Poincaré lemma in [14]. Curiously, while the p-adic Poincaré lemma was used in [14] to give a new proof of the fundamental de Rham comparison conjecture in p-adic Hodge theory, the proof of Theorem 5.7 uses the full strength of modern advances in p-adic Hodge theory (such as the primitive comparison theorem of [203] for arbitrarily singular varieties).

We end this section with an application of Theorem 5.7 to birational geometry in mixed characteristic. Briefly, it is possible to use this variant of Kodaira vanishing in a critical lifting argument in an inductive proof of the existence of flips in dimension 3, following [113] (which goes back to ideas of Shokurov). Combining this with Witaszek's recent mixed characteristic analog [229] of Keel's semiampleness theorems [144], it became possible to emulate the ideas of [44,45,69,113] (amongst others) to show the following:

Theorem 5.10 (Minimal model program in mixed characteristic, [43,219]). *One can run the minimal model program for arithmetic threefolds whose residue characteristics are* > 5.

Theorem 5.10 uses ideas from [46,178] and extends [142,228]. Global geometric applications of (the ideas going into) Theorem 5.10 can be found in [43,112,216,219,231].

Remark 5.11 (The $+$-stable sections). We informally discuss a new notion introduced in the proof of Theorem 5.10 in a simple case, and state a question; see [43, §4] or [219, §3.2] for the general notion. For reasonable mixed characteristic rings R, one can define a submodule $B^0(R, \omega_R) \subset \omega_R$ of the dualizing module ω_R: it is the submodule of elements that lift to all alterations of $\mathrm{Spec}(R)$ under the trace maps. If R is regular, then $B^0(R, \omega_R) = \omega_R$, so in general $B^0(R, \omega_R)$ is an invariant measuring the singularities of R. Analogous invariants exist in characteristic 0 (given by the Grauert–Riemenschneider sheaf [157, EXAMPLE 4.3.12]) and characteristic p (given by the parameter test submodule [46, §2.5 & COROLLARY 3.4]). Basic properties of $B^0(R, \omega_R)$, such as its behavior under alterations or restriction to divisors, play a key role in the proof of Theorem 5.10. However, a fundamental question about these invariants remains open: does their formation commute with localization? Due to the infinite intersection implicit in the definition of $B^0(R, \omega_R)$, this question is delicate. Nevertheless, a positive answer (which we expect) would have several geometric applications. As evidence for a positive answer, using Theorem 6.4, one can show the claim for inverting p: the localization $B^0(R, \omega_R)[1/p]$ agrees with the Grauert–Riemenschneider sheaf of $\mathrm{Spec}(R[1/p])$ (work in progress as a sequel to [43]). We refer to [112, §8] for more discussion of this question.

6. p-ADIC RIEMANN–HILBERT

The Riemann–Hilbert problem has a rich history, going back at least to Hilbert's 21st problem. In modern terms, it asked if any $\mathbf{C}$-local system on a smooth complex algebraic curve X could be realized as the solution system of a flat vector bundle on X with regular singularities at ∞; this variant was (precisely formulated and) solved by Deligne [71]. Soon after, this picture was generalized to higher dimensions by Kashiwara and Mebkhout: there is an equivalence of categories between topological objects ($\mathbf{C}$-linear perverse sheaves) and differential objects (regular holonomic $\mathcal{D}$-modules) on any smooth complex variety, see [56].

In this section, we discuss joint work with Lurie towards a p-adic analog of the preceding story; our aim was to extend existing results attaching flat connections to p-adic local systems on p-adic varieties (such as [1,74,87,88,165,203]) to p-adic constructible complexes and in particular, to p-adic perverse sheaves. Unlike the complex picture, there are several meanings one can attach to "p-adic sheaves": one can work with $\mathbf{Z}/p^n$, $\mathbf{Z}_p$ or $\mathbf{Q}_p$-coefficients. Our theorem for $\mathbf{F}_p$-coefficients is the following (the $\mathbf{Z}/p^n$ case is analogous):

Theorem 6.1 (Riemann–Hilbert for torsion coefficients, [29]). *Let $C/\mathbf{Q}_p$ be a complete and algebraically closed extension. Let $X/\mathcal{O}_C$ be a finite type scheme. Then there is a natural exact functor*

$$\mathrm{RH} : D^b_{\mathrm{cons}}(X_C, \mathbf{F}_p) \to D^b_{\mathrm{qc}}(X \otimes_{\mathcal{O}_C} \mathcal{O}_C/p).$$

This functor commutes with proper pushforward, intertwines Verdier and Grothendieck duality in the almost category [83,100], and interacts well with the perverse t-structure.

The functor RH above also almost commutes with tensor products and pullbacks provided the target is refined to $\mathrm{RH}(\mathbf{F}_p)$-modules. In fact, it is possible to refine the target further to Frobenius modules over the tilt $\mathrm{RH}(\mathbf{F}_p)^\flat$; the resulting functor is fully faithful, and agrees with the construction in [31] (which was a dual form of [82] that works for all characteristic p schemes) when X has characteristic p.

Remark 6.2 (Relation to existing work in p-adic geometry). Theorem 6.1 appears to be the first general construction attaching coherent objects to constructible $\mathbf{F}_p$-sheaves on algebraic varieties in characteristic 0. On the other hand, several ingredients that go into the proof have appeared before in p-adic arithmetic geometry. Indeed, the functor RH can be regarded as a generalization of a perfectoidization functor from Remark 2.10 to nonconstant coefficients: one can almost identify $\mathrm{RH}(\mathbf{F}_p)$ with $\mathcal{O}_{X,\mathrm{perfd}}/p$. Moreover, the compatibility with duality with constant coefficients is closely related to the Gabber–Zavyalov approach [232] to Poincaré duality for the $\mathbf{F}_p$-cohomology of rigid spaces. Nevertheless, the flexibility of applying $\mathrm{RH}(-)$ to nonconstant perverse coefficients is immensely useful in applications including Theorems 5.1 and 5.7 or the localization result mentioned in Remark 5.11. Relatedly, let us mention that Theorem 6.1 itself suffices to prove Theorem 5.1 in the almost category, extending Heitmann's almost vanishing theorem [114] to arbitrary dimensions.

Prima facie, Theorem 6.1 looks quite different from the complex Riemann–Hilbert correspondence: the output is a quasicoherent (and in fact almost coherent) complex rather than a $\mathcal{D}$-module. In fact, the functor in Theorem 6.1 is better understood as a p-adic analog of a construction from Saito's fundamental work [197] on mixed Hodge modules. Recall that this theory gives a *filtered* refinement of the classical Riemann–Hilbert functor for many constructible sheaves, including those that are "of geometric origin." More precisely, given a smooth proper complex variety X, any mixed Hodge module on X has an underlying $\mathcal{D}_X$-module equipped with a Hodge filtration and an underlying perverse sheaf; the picture relating them can be summarized in the following commutative diagram:

$$\begin{array}{ccccccc} & & \mathrm{MHM}(X) & & & & \\ & \swarrow^{\text{forget}} & & \searrow^{\text{forget}} & & & \\ D^b_{\mathrm{cons}}(X, \mathbf{C}) & & & & DF_{\mathrm{coh}}(\mathcal{D}_X) \xrightarrow{\mathrm{gr}^*(-)} & D^b_{\mathrm{coh,gr}}(T^*X) \xrightarrow{\Omega^*(-)} & D^b_{\mathrm{coh,gr}}(X) \\ & \searrow_{\mathrm{RH}^{\mathrm{cl}}} & & \swarrow_{\text{forget}} & & & \\ & & D^b(\mathcal{D}_X), & & & & \end{array}$$

where $\mathrm{MHM}(X)$ is Saito's category of mixed Hodge modules, $DF_{\mathrm{coh}}(\mathcal{D}_X)$ is a suitable derived category of $\mathcal{D}_X$-modules equipped with a "good" filtration, the functor $\mathrm{RH}^{\mathrm{cl}}$ is the classical Riemann–Hilbert functor, the functor $\mathrm{gr}^*(-)$ is the associated graded construction carrying a filtered $\mathcal{D}$-module to a graded $\mathcal{O}_X$-module with an action of $\mathrm{gr}^* \mathcal{D}_X = \mathrm{Sym}^*(T_X)$ (i.e., a Higgs module), and the functor $\Omega^*(-)$ is the graded Higgs complex construction. Heuristically, the functor in Theorem 6.1 is an analog of the composite correspondence

$$D^b_{\mathrm{cons}}(X, \mathbf{C}) \xleftarrow{\text{forget}} \mathrm{MHM}(X) \xrightarrow{\Omega^*(-)\circ \mathrm{gr}^*(-)\circ \text{forget}} D^b_{\mathrm{coh,gr}}(X) \qquad (\widetilde{\mathrm{RH}})$$

for $\mathbf{F}_p$-coefficients. Slightly surprisingly, unlike in the complex story, we get an honest functor instead of a correspondence in the p-adic setting. (On the other hand, objects of $\mathrm{MHM}(X)$ also have a weight filtration, which we ignore in our discussion.)

Remark 6.3 (Why is there no grading?). In comparison with the correspondence $(\widetilde{\mathrm{RH}})$, there is no grading in the target of Theorem 6.1. But this is to be expected: the grading on the target of $(\widetilde{\mathrm{RH}})$ reflects the fact that objects in $\mathrm{MHM}(X)$ are fairly motivic in nature, e.g., they give variations of Hodge structures on an open subset of X. In contrast, in Theorem 6.1 we are working with *all* constructible sheaves over the algebraically closed field C, so there is no motivicity or even a Galois action.

The previous discussion suggests it might be useful to lift Theorem 6.1 to $\mathbf{Q}_p$-coefficients and restrict to sheaves defined over a discretely valued field (so there is a Galois action) in order to obtain a p-adic variant of $(\widetilde{\mathrm{RH}})$. This can indeed be done, and the resulting structure seems slightly cleaner than $(\widetilde{\mathrm{RH}})$:

Theorem 6.4 (Riemann–Hilbert for $\mathbf{Q}_p$-coefficients, [30]). *Let $K/\mathbf{Q}_p$ be a finite extension. Let X/K be a smooth proper variety. Then there is a natural exact functor*

$$\mathrm{RH}_{\mathcal{D}} : D^b_{\mathrm{wHT}}(X, \mathbf{Q}_p) \to DF_{\mathrm{coh}}(\mathcal{D}_X),$$

where the source is a full subcategory of $D^b_{\mathrm{cons}}(X, \mathbf{Q}_p)$ spanned by what we call "weakly Hodge–Tate sheaves" (including all sheaves of geometric origin). This functor commutes with proper pushforward, intertwines Verdier and Grothendieck duality, and interacts well with the perverse t-structure.

Theorem 6.4 represents ongoing work in progress with Lurie, and the statement above is not quite optimal (e.g., there is a variant for singular X).

Remark 6.5 (The case of local systems). The functor $\mathrm{RH}_{\mathcal{D}}$ from Theorem 6.4 is not really new for local systems: up to a certain nilpotent operator encoding that a weakly Hodge–Tate local system is not quite de Rham, it coincides with the one appearing in [165, THEOREM 1.5] (and is thus related to constructions from [203]; see also [1, 74, 87]). However, for geometric applications such as Example 6.7 below, it is critical to apply $\mathrm{RH}_{\mathcal{D}}$ to constructible complexes that are not local systems.

Remark 6.6 (Why is the Hodge filtration automatic?). Theorem 6.4 implies that constructible $\mathbf{Q}_p$-sheaves F of geometric origin on a variety X/K as above have a functorially attached filtered $\mathcal{D}_X$-module $\mathcal{M} := \mathrm{RH}_{\mathcal{D}}(F)$, i.e., the Hodge filtration on the $\mathcal{D}_X$-module $\mathcal{M}$ is actually determined by F, unlike in the correspondence ($\widetilde{\mathrm{RH}}$). This discrepancy is ultimately because the constructible sheaves in Theorem 6.4 carry Galois symmetries as they are defined over K. Moreover, this is perfectly consistent with known phenomena in p-adic Hodge theory that stem ultimately from the richness of the absolute Galois group G_K of K. For instance, when $X = \mathrm{Spec}(K)$ and $F = Rf_*\mathbf{Q}_p$ for a smooth proper map $f : Y \to X$, we are simply observing that the G_K-representation $H^*(Y_{\overline{K}}, \mathbf{Q}_p)$ knows the de Rham cohomology of $H^*_{\mathrm{dR}}(Y/K)$ as a filtered vector space (and in particular knows the Hodge numbers of X) via the de Rham comparison; see [137] for a purely geometric application of this fact.

As Theorem 6.4 gives an honest functor, one can now directly apply $\mathrm{RH}_{\mathcal{D}}$ to deep theorems on the constructible side, such as the BBDG decomposition theorem [16], to obtain highly nontrivial results on the coherent side. This mechanism appears robust enough to yield some results in birational geometry that are traditionally best understood via mixed Hodge module theory, e.g., Kollár's vanishing theorems [148,149] (see [201, §25] for the Hodge module proof); we sketch the argument for vanishing next to illustrate this idea.

Example 6.7 (Recovering Kollár vanishing, p-adically). Fix a finite extension $K/\mathbf{Q}_p$. Suppose $f : Y \to X$ is a projective surjective morphism of proper K-varieties of dimensions d_Y and d_X, respectively, with Y smooth. Consider the functor

$$\mathrm{RH} : D^b_{\mathrm{wHT}}(Y, \mathbf{Q}_p) \to D^b_{\mathrm{coh,gr}}(Y)$$

obtained by composing the functor $\mathrm{RH}_{\mathcal{D}}$ from Theorem 6.4 with $\Omega^*(-) \circ \mathrm{gr}^*(-)$, as in ($\widetilde{\mathrm{RH}}$). Essentially by the local Hodge–Tate decomposition of [203], we have

$$\mathrm{RH}\big(\mathbf{Q}_p[d_Y]\big) = \bigoplus_i \Omega^i_{Y/K}[d_Y - i]$$

with its natural grading, so i-forms have weight i. (If Y were singular, one would have a similar formula with the Deligne–Du Bois variants $\underline{\Omega}^i_{Y/K}$ of differential forms, as in [79] and [191, §7.3], on the right by [110].) Pushing forward along f, using the proper pushforward compatibility of RH, and extracting the weight d_Y summand gives

$$\mathrm{RH}\big(Rf_*\mathbf{Q}_p[d_Y]\big)_{\mathrm{wt}=d_Y} = Rf_*\omega_Y.$$

On the other hand, the decomposition theorem [16,72] shows that

$$Rf_*\mathbf{Q}_p[d_Y] \simeq \left(\bigoplus_{i=-(d_Y-d_X)}^{d_Y-d_X} {}^p\mathcal{H}^i[-i] \right) \oplus N$$

where each ${}^p\mathcal{H}^i$ is perverse and N is a summand of $Rg_*\mathbf{Q}_p[d_Y]$ with $g : Y_Z \to Z \subset X$ being the restriction of f over the closed subvariety $Z \subsetneq X$ where f is not smooth. The singular variant of the reasoning just used for f applied to g then shows that

$$\mathrm{RH}\big(Rg_*\mathbf{Q}_p[d_Y]\big)_{\mathrm{wt}=d_Y} = Rg_*\underline{\Omega}^{d_Y}_{Y_Z} \simeq 0,$$

where the last vanishing follows as $\underline{\Omega}^{d_Y}_{Y_Z} = 0$ since $d_Y > \dim(Y_Z)$ (see [110] for a purely p-adic proof of this property of Deligne–Du Bois complexes). But then the same vanishing is also true for the summand N of $Rg_*\mathbf{Q}_p[d_Y]$, so we learn that

$$Rf_*\omega_Y = \mathrm{RH}\big(Rf_*\mathbf{Q}_p[d_Y]\big)_{\mathrm{wt}=d_Y} = \bigoplus_{i=-(d_Y-d_X)}^{d_Y-d_X} \mathrm{RH}\big(\mathcal{H}^i[-i]\big)_{\mathrm{wt}=d_Y}.$$

The perverse exactness properties of RH now imply that the ith summand on the right lies in $D^{\leq i}$ whence $Rf_*\omega_Y \in D^{\leq d_Y - d_X}$ as $i \leq d_Y - d_X$, i.e.,

$$R^j f_*\omega_Y = 0 \quad \text{for } j > d_Y - d_X,$$

proving the Kollár vanishing theorem [148, THEOREM 2.1]. From this perspective, one answer to Kollár's question "Why is ω_Y better behaved than $\mathcal{O}_Y$?" [149] could be the following: as ω_Y is the highest Hodge–Tate weight summand of $\mathrm{RH}(\mathbf{Q}_p[d_Y])$, it does not see interference from smaller dimensional varieties when moved around via operations such as Rf_*.

ACKNOWLEDGMENTS

The landscape of algebraic geometry in mixed characteristic has changed substantially in the last decade. I feel extremely fortunate to have had the opportunity to witness many aspects of this metamorphosis up close. I am thus eternally indebted to my mentors, collaborators, and friends for their support, wisdom and generosity (with time and ideas) over the years. Thanks also to Johan de Jong, Valia Gazaki, Toby Gee, Lars Hesselholt, Jacob Lurie, Linquan Ma, Akhil Mathew, Davesh Maulik, Mircea Mustata, Arthur Ogus, Alex Perry, Peter Scholze, Karl Schwede, Kevin Tucker, and especially Wei Ho, Luc Illusie, Matthew Morrow, and Anurag Singh for numerous helpful comments during the preparation of this survey.

FUNDING

This work was partially supported by the NSF (#1801689, #1952399, #1840234), the Packard Foundation, and the Simons Foundation (#622511).

REFERENCES

[1] A. Abbes, M. Gros, and T. Tsuji, *The p-adic Simpson correspondence*. Ann. of Math. Stud., Princeton University Press, Princeton, 2016.

[2] P. Achinger and J. Suh, Some refinements of the Deligne–Illusie theorem. 2021, arXiv:2003.09857.

[3] Y. André, La conjecture du facteur direct. *Publ. Math. IHÉS* **127** (2018), no. 1, 71–93.

[4] Y. André, Le lemme d'Abhyankar perfectoide. *Publ. Math. IHÉS* **127** (2018), no. 1, 1–70.

[5] Y. André, Perfectoid spaces and the homological conjectures. In *Proc. Int. Cong. of Math.—2018, Rio de Janeiro, Vol. 1*, pp. 249–260, World Scientific, 2018.

[6] Y. André, Weak functoriality of Cohen–Macaulay algebras. *J. Amer. Math. Soc.* **33** (2020), no. 2, 363–380.

[7] Y. André and L. Fiorot, On the canonical, fpqc, and finite topologies on affine schemes. The state of the art. 2019, arXiv:1912.04957.

[8] J. Anschütz and A.-C. L. Bras, Prismatic Dieudonné theory. 2019, arXiv:1907.10525.

[9] B. Antieau, Periodic cyclic homology and derived de Rham cohomology. *Ann. K-Theory* **4** (2019), no. 3, 505–519.

[10] B. Antieau and R. Datta, Valuation rings are derived splinters. *Math. Z.* (2021), 1–25.

[11] B. Antieau, A. Mathew, M. Morrow, and T. Nikolaus, On the Beilinson fibre square. 2020, arXiv:2003.12541.

[12] K. Aomoto and Y. Kato, A q-analogue of de Rham cohomology. In *ICM-90 satellite conference proceedings*, pp. 30–62, Springer, 1991.

[13] A. Beilinson, Height pairing between algebraic cycles. In *K-theory, arithmetic and geometry*, pp. 1–26, Springer, 1987.

[14] A. Beilinson, p-adic periods and derived de Rham cohomology. *J. Amer. Math. Soc.* **25** (2012), no. 3, 715–738.

[15] A. Beilinson, On the crystalline period map. *Camb. J. Math.* **1** (2013), no. 1, 1–51.

[16] A. Beilinson, J. Bernstein, P. Deligne, and O. Gabber, Faisceaux pervers. *Astérisque* **100** (2018).

[17] P. Berthelot, *Cohomologie cristalline des schemas de caracteristique* $p > 0$. Lecture Notes in Math. 407, Springer, 2006.

[18] P. Berthelot and A. Ogus, *Notes on crystalline cohomology*. Princeton University Press, 1978.

[19] B. Bhatt, p-adic derived de Rham cohomology. 2012, arXiv:1204.6560.

[20] B. Bhatt, Completions and derived de Rham cohomology. 2012, arXiv:1207.6193.

[21] B. Bhatt, Derived splinters in positive characteristic. *Compos. Math.* **148** (2012), no. 6, 1757–1786.

[22] B. Bhatt, p-divisibility for coherent cohomology. *Forum Math. Sigma* **3** (2015).

[23] B. Bhatt, On the direct summand conjecture and its derived variant. *Invent. Math.* **212** (2018), no. 2, 297–317.

[24] B. Bhatt, Specializing varieties and their cohomology from characteristic 0 to characteristic p. In *Algebraic geometry: Salt Lake City 2015*, pp. 43–88, Proc. Sympos. Pure Math. 97, American Mathematical Society, 2018.

[25] B. Bhatt, Cohen–Macaulayness of absolute integral closures. 2020, arXiv:2008.08070.

[26] B. Bhatt, M. Blickle, G. Lyubeznik, A. K. Singh, and W. Zhang, Stabilization of the cohomology of thickenings. *Amer. J. Math.* **141** (2019), no. 2, 531–561.

[27] B. Bhatt, D. Clausen, and A. Mathew, Remarks on K(1)-local K-theory. *Selecta Math. (N.S.)* **26** (2020), no. 3, 1–16.

[28] B. Bhatt and S. Li, Totaro's inequality for classifying spaces. 2021, arXiv:2107.04111.

[29] B. Bhatt and J. Lurie, A p-adic Riemann–Hilbert functor I: torsion coefficients. Preprint.

[30] B. Bhatt and J. Lurie, A p-adic Riemann–Hilbert functor II: $\mathbf{Q}_p$-coefficients (in preparation).

[31] B. Bhatt and J. Lurie, A Riemann–Hilbert correspondence in positive characteristic. *Camb. J. Math.* **7** (2019), no. 1, 71–217.

[32] B. Bhatt and J. Lurie, Absolute prismatic cohomology. Preprint.

[33] B. Bhatt and J. Lurie, Absolute prismatic gauges (in preparation).

[34] B. Bhatt, J. Lurie, and A. Mathew, Revisiting the de Rham–Witt complex. *Astérisque* **424** (2021).

[35] B. Bhatt and A. Mathew (in preparation).

[36] B. Bhatt and A. Mathew, The arc-topology. *Duke Math. J.* **170** (2021), no. 9, 1899–1988.

[37] B. Bhatt, M. Morrow, and P. Scholze, Integral p-adic Hodge theory. *Publ. Math. IHÉS* **128** (2018), no. 1, 219–397.

[38] B. Bhatt, M. Morrow, and P. Scholze, Topological Hochschild homology and integral p-adic Hodge theory. *Publ. Math. IHÉS* **129** (2019), no. 1, 199–310.

[39] B. Bhatt and P. Scholze, The pro-étale topology for schemes. *Astérisque* **369** (2015), 99–201.

[40] B. Bhatt and P. Scholze, Projectivity of the Witt vector affine Grassmannian. *Invent. Math.* **209** (2017), no. 2, 329–423.

[41] B. Bhatt and P. Scholze, Prisms and prismatic cohomology. 2019, arXiv:1905.08229.

[42] B. Bhatt and P. Scholze, Prismatic F-crystals and crystalline Galois representations. 2021, arXiv:2106.14735.

[43] B. Bhatt, L. Ma, Z. Patakfalvi, K. Schwede, K. Tucker, J. Waldron, and J. Witaszek, Globally $+$-regular varieties and the minimal model program for threefolds in mixed characteristic. 2020, arXiv:2012.15801.

[44] C. Birkar, Existence of flips and minimal models for 3-folds in char p. *Ann. Sci. Éc. Norm. Supér. (4)* **49** (2016), no. 1, 169–212.

[45] C. Birkar and J. Waldron, Existence of Mori fibre spaces for 3-folds in characteristic p. *Adv. Math.* **313** (2017), 62–101.

[46] M. Blickle, K. Schwede, and K. Tucker, F-singularities via alterations. *Amer. J. Math.* **137** (2015), no. 1, 61–109.

[47] S. Bloch, Algebraic K-theory and crystalline cohomology. *Publ. Math. IHÉS* **47** (1977), 187–268.

[48] S. Bloch, Algebraic cycles and higher K-theory. *Adv. Math.* **61** (1986), no. 3, 267–304.

[49] S. Bloch, H. Esnault, and M. Kerz, p-adic deformation of algebraic cycle classes. *Invent. Math.* **195** (2014), no. 3, 673–722.

[50] S. Bloch and K. Kato, p-adic étale cohomology. *Publ. Math. IHÉS* **63** (1986), no. 1, 107–152.

[51] S. Bloch and K. Kato, L-functions and Tamagawa numbers of motives. In *The Grothendieck Festschrift*, pp. 333–400, Springer, 2007.

[52] S. Bloch and S. Lichtenbaum, A spectral sequence for motivic cohomology. Preprint, 1995.

[53] M. Bökstedt, Topological Hochschild homology. Preprint, Bielefeld, 1985.

[54] M. Bökstedt, Topological Hochschild homology of $\mathbf{Z}$ and $\mathbf{Z}/p$. Preprint, Bielefeld, 1985.

[55] M. Bökstedt, W. C. Hsiang, and I. Madsen, The cyclotomic trace and algebraic K-theory of spaces. *Invent. Math.* **111** (1993), no. 1, 465–539.

[56] A. Borel, P.-P. Grivel, B. Kaup, A. Haefliger, B. Malgrange, and F. Ehlers, *Algebraic D-modules. Vol. 2*. Academic Press, 1987.

[57] J. Borger, The basic geometry of Witt vectors, I: the affine case. *Algebra Number Theory* **5** (2011), no. 2, 231–285.

[58] D. Bragg and M. Olsson, Representability of cohomology of finite flat abelian group schemes. 2021, arXiv:2107.11492.

[59] C. Breuil, Représentations semi-stables et modules fortement divisibles. *Invent. Math.* **136** (1999), no. 1, 89–122.

[60] C. Breuil and W. Messing, Torsion étale and crystalline cohomologies. *Astérisque* **279** (2002), 81–124.

[61] A. Buium, Arithmetic analogues of derivations. *J. Algebra* **198** (1997), no. 1, 290–299.

[62] K. Česnavičius and T. Koshikawa, The A_{inf}-cohomology in the semistable case. *Compos. Math.* **155** (2019), no. 11, 2039–2128.

[63] K. Česnavičius and P. Scholze, Purity for flat cohomology. 2019, arXiv:1912.10932.

[64] A. Chatzistamatiou, q-crystals and q-connections. 2020, arXiv:2010.02504.

[65] D. Clausen and A. Mathew, Hyperdescent and étale K-theory. *Invent. Math.* (2021), 1–96.

[66] D. Clausen, A. Mathew, and M. Morrow, K-theory and topological cyclic homology of henselian pairs. *J. Amer. Math. Soc.* **34** (2021), no. 2, 411–473.

[67] P. Colmez, G. Dospinescu, and W. Nizioł, Integral p-adic étale cohomology of Drinfeld symmetric spaces. *Duke Math. J.* **170** (2021), no. 3, 575–613.

[68] P. Colmez and W. Nizioł, Syntomic complexes and p-adic nearby cycles. *Invent. Math.* **208** (2017), no. 1, 1–108.

[69] O. Das and J. Waldron, On the log minimal model program for 3-folds over imperfect fields of characteristic $p > 5$. 2019, arXiv:1911.04394.

[70] R. Datta and K. Tucker, Openness of splinter loci in prime characteristic. 2021, arXiv:2103.10525.

[71] P. Deligne, *Équations différentielles à points singuliers réguliers*. Lecture Notes in Math. 163, Springer, 1970.

[72] P. Deligne, Décompositions dans la catégorie dérivée. In *Motives, Proc. Symposia in Pure Math.*, pp. 115–128, American Mathematical Society, 1994.

[73] P. Deligne and L. Illusie, Relevements modulo p^2 et décomposition du complexe de de Rham. *Invent. Math.* **89** (1987), no. 2, 247–270.

[74] H. Diao, K.-W. Lan, R. Liu, and X. Zhu, Logarithmic Riemann–Hilbert correspondences for rigid varieties. 2018, arXiv:1803.05786.

[75] V. Drinfeld, A stacky approach to crystals. 2018, arXiv:1810.11853.

[76] V. Drinfeld, Prismatization. 2020, arXiv:2005.04746.

[77] V. Drinfeld, A 1-dimensional formal group over the prismatization of $\mathrm{Spf}(\mathbf{Z}_p)$. 2021, arXiv:2107.11466.

[78] H. Du and T. Liu, A prismatic approach to $(\varphi, \hat{G})$-modules and F-crystals. 2021, arXiv:2107.12240.

[79] P. Du Bois, Complexe de de Rham filtré d'une variété singulière. *Bull. Soc. Math. France* **109** (1981), 41–81.

[80] T. Dupuy, E. Katz, J. Rabinoff, and D. Zureick-Brown, Total p-differentials on schemes over $\mathbf{Z}/p^2$. *J. Algebra* **524** (2019), 110–123.

[81] M. Emerton and T. Gee, Moduli stacks of étale (ϕ, Γ)-modules and the existence of crystalline lifts. 2019, arXiv:1908.07185.

[82] M. Emerton and M. Kisin, The Riemann–Hilbert correspondence for unit F-crystals. *Astérisque* (2004).

[83] G. Faltings, p-adic Hodge theory. *J. Amer. Math. Soc.* **1** (1988), no. 1, 255–299.

[84] G. Faltings, Does there exist an arithmetic Kodaira–Spencer class? *Contemp. Math.* **241** (1999), 141–146.

[85] G. Faltings, Integral crystalline cohomology over very ramified valuation rings. *J. Amer. Math. Soc.* **12** (1999), no. 1, 117–144.

[86] G. Faltings, Almost étale extensions. *Astérisque* **279** (2002), 185–270.

[87] G. Faltings, A p-adic Simpson correspondence. *Adv. Math.* **198** (2005), no. 2, 847–862.

[88] G. Faltings, A p-adic Simpson correspondence II: small representations. *Pure Appl. Math. Q.* **7** (2011), no. 4, 1241–1264.

[89] B. Farb, M. Kisin, and J. Wolfson, Essential dimension via prismatic cohomology. 2021, arXiv:2110.05534.

[90] L. Fargues and P. Scholze, Geometrization of the local Langlands correspondence. 2021, arXiv:2102.13459.

[91] A. Fonarev and D. Kaledin, Bokstedt periodicity generator via K-theory. 2021, arXiv:2107.03753.

[92] J.-M. Fontaine, Groupes p-divisibles sur les corps locaux. *Astérisque* **47–48** (1977), 401.

[93] J.-M. Fontaine, Sur certains types de représentations p-adiques du groupe de Galois d'un corps local; construction d'un anneau de Barsotti–Tate. *Ann. of Math.* **115** (1982), no. 3, 529–577.

[94] J.-M. Fontaine and U. Jannsen, Frobenius gauges and a new theory of p-torsion sheaves in characteristic p. I. 2013, arXiv:1304.3740.

[95] J.-M. Fontaine and W. Messing, p-adic periods and p-adic étale cohomology. In *Current trends in arithmetical algebraic geometry*, pp. 179–207, Amer. Math. Soc., 1987.

[96] D. S. Freed, Classical Chern–Simons theory, part 1. 1992, arXiv:hep-th/9206021.

[97] E. M. Friedlander and A. Suslin, The spectral sequence relating algebraic k-theory to motivic cohomology. *Ann. Sci. Éc. Norm. Supér.* **35** (2002), 773–875.

[98] O. Gabber, K-theory of henselian local rings and henselian pairs. *Contemp. Math.* **126** (1992), 59–70.

[99] O. Gabber, Observations made after the MSRI workshop on homological conjectures. 2018, https://docs.google.com/viewer?url=https://www.msri.org/workshops/842/schedules/23854/documents/3322/assets/31362.

[100] O. Gabber and L. Ramero, *Almost ring theory*. Springer, 2003.

[101] O. Gabber and L. Ramero, Foundations for almost ring theory—release 7.5. 2004, arXiv:math/0409584.

[102] E. Gazaki, A finer Tate duality theorem for local Galois symbols. *J. Algebra* **509** (2018), 337–385.

[103] T. Geisser, Motivic cohomology over Dedekind rings. *Math. Z.* **248** (2004), no. 4, 773–794.

[104] T. Geisser and L. Hesselholt, The de Rham–Witt complex and p-adic vanishing cycles. *J. Amer. Math. Soc.* **19** (2006), no. 1, 1–36.

[105] T. Geisser and L. Hesselholt, On relative and bi-relative algebraic K-theory of rings of finite characteristic. *J. Amer. Math. Soc.* **24** (2011), no. 1, 29–49.

[106] T. Geisser and M. Levine, The K-theory of fields in characteristic p. *Invent. Math.* **139** (2000), no. 3, 459–493.

[107] M. Gros, B. Le Stum, and A. Quirós, Twisted divided powers and applications. *J. Number Theory* (2019).

[108] M. Gros, B. L. Stum, and A. Quirós, Twisted differential operators and q-crystals. 2020, arXiv:2004.14320.

[109] A. Grothendieck and M. Raynaud, Cohomologie locale des faisceaux coh'erents et théorémes de Lefschetz locaux et globaux. 2005, arXiv:math/0511279.

[110] H. Guo, Hodge–Tate decomposition for non-smooth spaces. 2019, arXiv:1909.09917. To appear in *J. Eur. Math. Soc. (JEMS)*.

[111] H. Guo and S. Li, Period sheaves via derived de Rham cohomology. 2020, arXiv:2008.06143. To appear in *Compositio*.

[112] C. Hacon, A. Lamarche, and K. Schwede, Global generation of test ideals in mixed characteristic and applications. 2021, arXiv:2106.14329.

[113] C. Hacon and C. Xu, On the three dimensional minimal model program in positive characteristic. *J. Amer. Math. Soc.* **28** (2015), no. 3, 711–744.

[114] R. C. Heitmann, The direct summand conjecture in dimension three. *Ann. of Math.* **156** (2002), no. 2, 695–712.

[115] R. Heitmann and L. Ma, Big Cohen–Macaulay algebras and the vanishing conjecture for maps of tor in mixed characteristic. *Algebra Number Theory* **12** (2018), no. 7, 1659–1674.

[116] L. Hesselholt, On the topological cyclic homology of the algebraic closure of a local field. 2005, arXiv:math/0508309.

[117] L. Hesselholt and I. Madsen, Cyclic polytopes and the K-theory of truncated polynomial algebras. *Invent. Math.* **130** (1997), no. 1, 73–97.

[118] L. Hesselholt and I. Madsen, On the K-theory of finite algebras over Witt vectors of perfect fields. *Topology* **36** (1997), no. 1, 29–101.

[119] L. Hesselholt and I. Madsen, On the K-theory of local fields. *Ann. of Math.* (2003), 1–113.

[120] L. Hesselholt and T. Nikolaus, Topological cyclic homology. 2019, arXiv:1905.08984.

[121] L. Hesselholt and P. Scholze, Arbeitsgemeinschaft: topological cyclic homology. *Oberwolfach Rep.* **15** (2019), no. 2, 805–940.

[122] M. Hochster, Big Cohen–Macaulay algebras in dimension three via Heitmann's theorem. *J. Algebra* **254** (2002), no. 2, 395–408.

[123] M. Hochster, The homological conjectures: past, present, and future. Notes for a talk at MSRI in 2018. Available at https://docs.google.com/viewer?url=https://www.msri.org/workshops/842/schedules/23828/documents/3276/assets/31030.

[124] M. Hochster, et al., Homological conjectures, old and new. *Illinois J. Math.* **51** (2007), no. 1, 151–169.

[125] M. Hochster and C. Huneke, Tight closure, invariant theory, and the Briançon–Skoda theorem. *J. Amer. Math. Soc.* (1990), 31–116.

[126] M. Hochster and C. Huneke, Infinite integral extensions and big Cohen–Macaulay algebras. *Ann. of Math.* (1992), 53–89.

[127] M. Hochster and C. Huneke, Tight closure of parameter ideals and splitting in module-finite extensions. *J. Algebraic Geom.* **3** (1994), no. 4, 599–670.

[128] M. Hochster and J. Jeffries, A Jacobian criterion for nonsingularity in mixed characteristic. 2021, arXiv:2106.01996.

[129] M. Hoyois, From algebraic cobordism to motivic cohomology. *J. Reine Angew. Math.* **2015** (2015), no. 702, 173–226.

[130] C. Huneke, Absolute integral closure. In *Commutative algebra and its connections to geometry*, pp. 119–135, AMS, 2011.

[131] C. Huneke and G. Lyubeznik, Absolute integral closure in positive characteristic. *Adv. Math.* **210** (2007), no. 2, 498–504.

[132] L. Illusie, *Complexe cotangent et déformations I.* Lecture Notes in Math. 239, Springer, 1971.

[133] L. Illusie, *Complexe cotangent et déformations II* Lecture Notes in Math. 283, Springer, 1972.

[134] L. Illusie, Complexe de de Rham–Witt et cohomologie cristalline. *Ann. Sci. Éc. Norm. Supér.* **12** (1979), 501–661.

[135] L. Illusie, A new approach to de Rham–Witt complexes, after Bhatt–Lurie–Mathew. *Thirty Years of Berkovich Spaces. IHP*. 2018.

[136] L. Illusie, Partial degeneration of Hodge to de Rham spectral sequences and Kodaira type vanishing theorems for locally complete intersections in positive characteristic. Available at https://www.imo.universite-paris-saclay.fr/~illusie/DI-revisited2a.pdf.

[137] T. Ito, Birational smooth minimal models have equal Hodge numbers in all dimensions. In *Calabi–Yau varieties and mirror symmetry*, pp. 183–194, Fields Institute Communications 38, American Mathematical Society, 2003.

[138] A. Joyal, δ-anneaux et vecteurs de Witt. *C. R. Math. Acad. Sci. Can.* **7** (1985), no. 3, 177–182.

[139] K. Kato, Logarithmic structures of Fontaine–Illusie. In *Algebraic analysis, geometry, and number theory*, pp. 191–224, 1989.

[140] K. Kato, Syntomic complexes of F-crystals and Tamagawa number conjecture in characteristic p, with appendix by Arthur Ogus. 2019, arXiv:1909.13181.

[141] K. Kato, et al., On p-adic vanishing cycles (application of ideas of Fontaine–Messing). In *Algebraic geometry, Sendai, 1985*, pp. 207–251, Mathematical Society of Japan, 1987.

[142] Y. Kawamata, Semistable minimal models of threefolds in positive or mixed characteristic. *J. Algebraic Geom.* **3** (1994), no. 3, 463–491.

[143] K. S. Kedlaya and R. Liu, Relative p-adic Hodge theory: foundations. 2013, arXiv:1301.0792.

[144] S. Keel, Basepoint freeness for nef and big line bundles in positive characteristic. *Ann. of Math.* (1999), 253–286.

[145] M. Kerz, Milnor K-theory of local rings with finite residue fields. *J. Algebraic Geom.* **19** (2010), no. 1, 173–191.

[146] M. Kisin, Crystalline representations and F-crystals. In *Algebraic geometry and number theory*, pp. 459–496, Springer, 2006.

[147] M. Kisin, Potentially semi-stable deformation rings. *J. Amer. Math. Soc.* **21** (2008), no. 2, 513–546.

[148] J. Kollár, Higher direct images of dualizing sheaves I. *Ann. of Math.* **123** (1986), no. 1, 11–42.

[149] J. Kollár, Higher direct images of dualizing sheaves II. *Ann. of Math.* **124** (1986), no. 1, 171–202.

[150] T. Koshikawa, Logarithmic prismatic cohomology I. 2020, arXiv:2007.14037.

[151] T. Koshikawa and Z. Yao, Logarithmic prismatic cohomology II. In preparation.

[152] S. J. Kovács, Rational singularities. 2017, arXiv:1703.02269.

[153] A. Krause and T. Nikolaus, Bökstedt periodicity and quotients of dvrs. 2019, arXiv:1907.03477.

[154] D. Kubrak and A. Prikhodko, p-adic Hodge theory for Artin stacks. 2021, arXiv:2105.05319.

[155] E. Kunz, Characterizations of regular local rings of characteristic p. *Amer. J. Math.* **91** (1969), no. 3, 772–784.

[156] M. Land, A. Mathew, L. Meier, and G. Tamme, Purity in chromatically localized algebraic K-theory. 2020, arXiv:2001.10425.

[157] R. K. Lazarsfeld, *Positivity in algebraic geometry I: Classical setting: line bundles and linear series*. Ergeb. Math. Grenzgeb. 48, Springer, 2017.

[158] L. Lebruyn, Who dreamed up the primes = knots analogy? Blog post, 2011, http://www.neverendingbooks.org/who-dreamed-up-the-primesknots-analogy.

[159] M. Levine, The homotopy coniveau tower. *J. Topol.* **1** (2008), no. 1, 217–267.

[160] S. Li, Integral p-adic Hodge filtrations in low dimension and ramification. 2020, arXiv:2002.00431. To appear in *J. Eur. Math. Soc. (JEMS)*.

[161] S. Li and T. Liu, Comparison of prismatic cohomology and derived de Rham cohomology. 2020, arXiv:2012.14064.

[162] S. Li and S. Mondal, On endomorphisms of the de Rham cohomology functor. 2021, arXiv:2109.04303.

[163] A. Lindenstrauss and I. Madsen, Topological Hochschild homology of number rings. *Trans. Amer. Math. Soc.* **352** (2000), no. 5, 2179–2204.

[164] R. Liu and G. Wang, Topological cyclic homology of local fields. 2020, arXiv:2012.15014.

[165] R. Liu and X. Zhu, Rigidity and a Riemann–Hilbert correspondence for p-adic local systems. *Invent. Math.* **207** (2017), no. 1, 291–343.

[166] T. Liu, A note on lattices in semi-stable representations. *Math. Ann.* **346** (2010), no. 1, 117–138.

[167] J.-L. Loday, *Cyclic homology*. Grundlehren Math. Wiss. 301, Springer, 2013.

[168] M. Lüders and M. Morrow, Milnor K-theory of p-adic rings. 2021, arXiv:2101.01092.

[169] J. Lurie, Higher topos theory. *Ann. of Math. Stud.* (2012).

[170] J. Lurie, Higher algebra. Preprint, 2017. Available at https://www.math.ias.edu/~lurie/.

[171] J. Lurie, Spectral algebraic geometry. Preprint, 2018. Available at https://www.math.ias.edu/~lurie/.

[172] F. Ma, Splitting in integral extensions, Cohen–Macaulay modules and algebras. *J. Algebra* **116** (1988), no. 1, 176–195.

[173] L. Ma, The vanishing conjecture for maps of tor and derived splinters. *J. Eur. Math. Soc. (JEMS)* **20** (2018), no. 2, 315–338.

[174] L. Ma and T. Polstra, F-singularities: a commutative algebra approach. 2021, https://www.math.purdue.edu/~ma326/F-singularitiesBook.pdf.

[175] L. Ma and K. Schwede, Perfectoid multiplier/test ideals in regular rings and bounds on symbolic powers. *Invent. Math.* **214** (2018), no. 2, 913–955.

[176] L. Ma and K. Schwede, Singularities in mixed characteristic via perfectoid big Cohen–Macaulay algebras. 2018, arXiv:1806.09567. To appear in *Duke*.

[177] L. Ma and K. Schwede, Recent applications of p-adic methods to commutative algebra. *Not. Amer. Math. Soc.* **66** (2019), no. 6.

[178] L. Ma, K. Schwede, K. Tucker, J. Waldron, and J. Witaszek, An analog of adjoint ideals and plt singularities in mixed characteristic. 2019, arXiv:1910.14665.

[179] A. Mathew, Some recent advances in topological Hochschild homology. 2021, arXiv:2101.00668.

[180] B. Mazur, Remarks on the Alexander polynomial. 1963, https://people.math.harvard.edu/~mazur/papers/alexander_polynomial.pdf.

[181] J. S. Milne, Duality in the flat cohomology of a surface. *Ann. Sci. Éc. Norm. Supér.* **9**, (1976), 171–201.

[182] Y. Min, Integral p-adic Hodge theory of formal schemes in low ramification. *Algebra Number Theory* **15** (2021), no. 4, 1043–1076.

[183] M. Morrow, Notes on the A_{inf}-cohomology of integral p-adic Hodge theory. In *p-adic Hodge theory (Simons Symposia)*, pp. 1–69, Springer, 2020.

[184] M. Morrow and T. Tsuji, Generalised representations as q-connections in integral p-adic Hodge theory. 2020, arXiv:2010.04059.

[185] Y. P. Nesterenko and A. A. Suslin, Homology of the full linear group over a local ring, and Milnor's K-theory. *Izvestiya* **34** (1989), no. 1, 121–146.

[186] T. Nikolaus and P. Scholze, On topological cyclic homology. *Acta Math.* **221** (2018), no. 2, 203–409.

[187] W. Nizioł, Crystalline conjecture via K-theory. *Ann. Sci. Éc. Norm. Supér.* **31** (1998), 659–681.

[188] A. Ogus, The saturated de Rham–Witt complex for schemes with toroidal singularities. 2020. Preprint, available at https://math.berkeley.edu/~ogus/preprints/drwtoric6.pdf.

[189] A. Ogus and V. Vologodsky, Nonabelian Hodge theory in characteristic p. *Publ. Math. IHÉS* **106** (2007), no. 1, 1–138.

[190] M. C. Olsson, Logarithmic geometry and algebraic stacks. *Ann. Sci. Éc. Norm. Supér.* **36** (2003), 747–791.

[191] C. A. Peters and J. H. Steenbrink, *Mixed Hodge structures*. Ergeb. Math. Grenzgeb. 52, Springer, 2008.

[192] J. P. Pridham, On q-de Rham cohomology via λ-rings. *Math. Ann.* **375** (2019), no. 1, 425–452.

[193] D. Quillen, Higher algebraic K-theory: I. In *Higher K-theories*, pp. 85–147, Springer, 1973.

[194] D. G. Quillen, *Homotopical algebra*. Lecture Notes in Math. 43, Springer, 2006.

[195] M. Raynaud, Contre-exemple au "Vanishing Theorem" en caractéristique $p > 0$. In *CP Ramanujam—A Tribute*, pp. 273–278, TIFR Stud. Math. 8, Springer, 1978.

[196] E. Reinecke, The cohomology of the moduli space of curves at infinite level. 2019, arXiv:1911.07392.

[197] M. Saito, Mixed Hodge modules. *Publ. Res. Inst. Math. Sci.* **26** (1990), no. 2, 221–333.

[198] T. Saito, Frobenius–Witt differentials and regularity. 2020, arXiv:2008.04728.

[199] K. Sato, p-adic étale Tate twists and arithmetic duality. *Ann. Sci. Éc. Norm. Supér.* **40** (2007), 519–588.

[200] P. Schneider, p-adic points of motives. In *Motives*, Proc. Sympos. Pure Math., pp. 225–249, American Mathematical Society, 1994.

[201] C. Schnell, An overview of Morihiko Saito's theory of mixed Hodge modules. 2014, arXiv:1405.3096.

[202] P. Scholze, Perfectoid spaces. *Publ. Math. IHÉS* **116** (2012), no. 1, 245–313.

[203] P. Scholze, p-adic Hodge theory for rigid-analytic varieties. *Forum Math. Pi* **1** (2013).

[204] P. Scholze, Canonical q-deformations in arithmetic geometry. *Ann. Fac. Sci. Univ. Toulouse Math.* **26** (2017), 1163–1192.

[205] P. Scholze, Etale cohomology of diamonds. 2017, arXiv:1709.07343.

[206] P. Scholze, p-adic geometry. In *Proceedings of the International Congress of Mathematicians: Rio de Janeiro 2018*, pp. 899–933, World Scientific, 2018.

[207] P. Scholze and J. Weinstein, Berkeley lectures on p-adic geometry. *Ann. of Math. Stud.* (2018).

[208] K. Schwede, A canonical linear system associated to adjoint divisors in characteristic p> 0. *J. Reine Angew. Math.* **2014** (2014), no. 696, 69–87.

[209] S. Sen, Continuous cohomology and p-adic Galois representations. *Invent. Math.* **62** (1980), no. 1, 89–116.

[210] A. Shiho, Notes on generalizations of local Ogus–Vologodsky correspondence. 2012, arXiv:1206.5907.

[211] K. Shimomoto, Integral perfectoid big Cohen–Macaulay algebras via Andrés theorem. *Math. Ann.* **372** (2018), no. 3, 1167–1188.

[212] C. T. Simpson, Nonabelian Hodge theory. In *Proceedings of the International Congress of Mathematicians, Vol. 1*, pp. 747–756, 1990.

[213] C. T. Simpson, Higgs bundles and local systems. *Publ. Math. IHÉS* **75** (1992), 5–95.

[214] A. K. Singh, **Q**-Gorenstein splinter rings of characteristic p are F-regular. *Math. Proc. Cambridge Philos. Soc.* **127** (1999), 201–205.

[215] K. E. Smith, Vanishing, singularities and effective bounds via prime characteristic local algebra. In *Algebraic geometry—Santa Cruz 1995*, Proc. Sympos. Pure Math. 62, pp. 289–328, American Mathematical Society, 1997.

[216] L. Stigant, Mori fibrations in mixed characteristic. 2021, arXiv:2110.06067.

[217] A. Suslin, On the K-theory of algebraically closed fields. *Invent. Math.* **73** (1983), no. 2, 241–245.

[218] S. Takagi and K. Watanabe, F-singularities: applications of characteristic p methods to singularity theory. *Sugaku Expositions* **31** (2018), no. 1, 1–42.

[219] T. Takamatsu and S. Yoshikawa, Minimal model program for semi-stable threefolds in mixed characteristic. 2020, arXiv:2012.07324.

[220] H. Tanaka, Minimal model program for excellent surfaces. *Ann. Inst. Fourier (Grenoble)* **68** (2018), no. 1, 345–376.

[221] R. W. Thomason, Algebraic K-theory and etale cohomology. *Ann. Sci. Éc. Norm. Supér.* **18** (1985), 437–552.

[222] Y. Tian, Finiteness and duality for the cohomology of prismatic crystals. 2021, arXiv:2109.00801.

[223] B. Toën and G. Vezzosi, Homotopical algebraic geometry I: topos theory. *Adv. Math.* **193** (2005), no. 2, 257–372.

[224] B. Toën and G. Vezzosi, Homotopical algebraic geometry II: Geometric stacks and applications. *Mem. Amer. Math. Soc.* **193** (2008), x+224 pp.

[225] B. Totaro, Milnor K-theory is the simplest part of algebraic K-theory. *K-Theory* **6** (1992), no. 2, 177–189.

[226] T. Tsuji, p-adic étale cohomology and crystalline cohomology in the semi-stable reduction case. *Invent. Math.* **137** (1999), no. 2, 233–411.

[227] V. Voevodsky, Open problems in the motivic stable homotopy theory. I. In *Motives, polylogarithms and Hodge theory, Part I (Irvine, CA, 1998), Vol. 3*, pp. 3–34, International Press, 2002.

[228] C. Weibel, The Hodge filtration and cyclic homology. *K-Theory* **12** (1997), no. 2, 145–164.

[229] J. Witaszek, Keel's base point free theorem and quotients in mixed characteristic. 2020, arXiv:2002.11915. To appear in *Annals*.

[230] Z. Wu, Galois representations, (φ, Γ)-modules and prismatic F-crystals. 2021, arXiv:2104.12105.

[231] L. Xie and Q. Xue, On the termination of the mmp for semi-stable fourfolds in mixed characteristic. 2021, arXiv:2110.03115.

[232] B. Zavyalov, Mod p Poincaré duality in p-adic analytic geometry. 2021, arXiv:2111.01830.

BHARGAV BHATT

University of Michigan, Ann Arbor, USA, bhargav.bhatt@gmail.com

This Page Intentionally Left Blank

DYNAMICS OF DILUTE GASES: A STATISTICAL APPROACH

THIERRY BODINEAU, ISABELLE GALLAGHER, LAURE SAINT-RAYMOND, AND SERGIO SIMONELLA

ABSTRACT

The evolution of a gas can be described by different models depending on the observation scale. A natural question, raised by Hilbert in his sixth problem, is whether these models provide consistent predictions. In particular, for rarefied gases, it is expected that continuum laws of kinetic theory can be obtained directly from molecular dynamics governed by the fundamental principles of mechanics.

In the case of hard sphere gases, Lanford [46] showed that the Boltzmann equation emerges as the law of large numbers in the low density limit, at least for very short times. The goal of this survey is to present recent progress in the understanding of this limiting process, providing a complete statistical description.

MATHEMATICS SUBJECT CLASSIFICATION 2020

Primary 82C40; Secondary 82C21, 76P05, 60F10

KEYWORDS

Hard sphere gas, Boltzmann equation, Boltzmann-Grad limit, fluctuations, large deviations, cluster expansion

ICM INTERNATIONAL CONGRESS OF MATHEMATICIANS 2022 JULY 6–14

Proc. Int. Cong. Math. 2022, Vol. 2, pp. 750–795

DOI 10.4171/ICM2022/130

Published by EMS Press

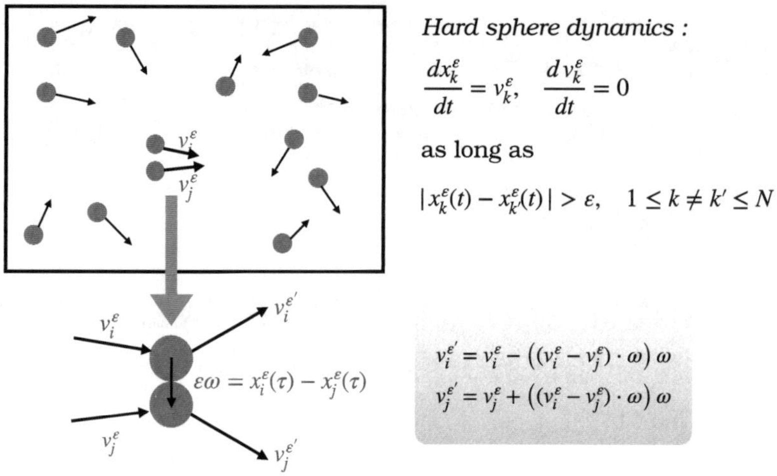

FIGURE 1
At time t, the hard-sphere system is described by the positions $(x_k^\varepsilon(t))_{k\le N}$ and the velocities $(v_k^\varepsilon(t))_{k\le N}$ of the N particles. Particles move in straight lines and when two particles touch each other at distance $\varepsilon > 0$ (the diameter of the spheres), they are scattered according to elastic reflection laws. The scattering rules, mapping the precollisional velocities $(v_i^\varepsilon, v_j^\varepsilon)$ to the postcollisional velocities $(v_i^{\varepsilon'}, v_j^{\varepsilon'})$, are determined in terms of the relative position $\omega = (x_i^\varepsilon(\tau) - x_j^\varepsilon(\tau))/\varepsilon$ of the particles at the collision time τ. The collisions preserve the total momentum $v_i^\varepsilon + v_j^\varepsilon = v_i^{\varepsilon'} + v_j^{\varepsilon'}$ and the kinetic energy $\frac{1}{2}(|v_i^\varepsilon|^2 + |v_j^\varepsilon|^2) = \frac{1}{2}(|v_i^{\varepsilon'}|^2 + |v_j^{\varepsilon'}|^2)$.

1. AIM: PROVIDING A STATISTICAL PICTURE OF DILUTE GAS DYNAMICS

1.1. A very simple physical model

Even though at the time Boltzmann published his famous paper [17], the atomistic theory was still dismissed by some scientists, it is now well established that matter is composed of atoms, which are the elementary constituents of all solid, liquid, and gaseous substances. The particularity of dilute gases is that their atoms are very weakly bound and almost independent. In other words, there are very few constraints on their geometric arrangement because their volume is negligible compared to the total volume occupied by the gas.

If we neglect the internal structure of atoms (consisting of a nucleus and electrons) and their possible organization into molecules, we can represent a gas as a large system of correlated interacting particles. We will also neglect the effect of long range interactions and assume strong interatomic forces at very short distance. Each particle moves freely most of the time and occasionally collides with some other particle leading to an almost instantaneous scattering. The simplest example of such a model consists in assuming that the particles are identical tiny balls of unit mass interacting only by contact (see Figure 1). We then speak of a *gas of hard spheres*. All the results we will present should nevertheless extend to isotropic, compactly supported stable interaction potentials [57, 63].

This microscopic description of a gas is daunting because the number of particles involved is extremely large, the individual size of these particles is tiny (of diameter $\varepsilon \ll 1$) and therefore positions are very sensitive to small spatial shifts (see Figure 2). In practice,

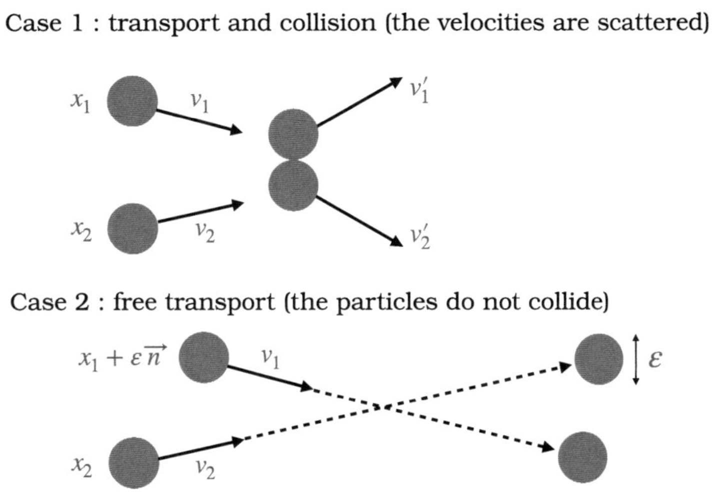

FIGURE 2
Particles are very small (of diameter $\varepsilon \ll 1$) and therefore the dynamics is very sensitive to small spatial shifts. In the first case depicted above, two particles with initial positions x_1, x_2 and velocities v_1, v_2 collide and are scattered. In the second case, by shifting the first particle by a distance ε in the direction $\vec{n}$, the two particles no longer collide and they move in straight lines. Thus a perturbation of order ε of the initial conditions can lead to very different trajectories.

this model is not efficient for making theoretical predictions, and numerical methods are often in favor of Monte Carlo simulations. The question we would like to address here is a more fundamental one, namely the consistency of this (simplified) atomic description with the kinetic or fluid models used in applications. This question was formalized by Hilbert at the ICM in 1900, in his sixth problem: "Boltzmann's work on the principles of mechanics suggests the problem of developing mathematically the limiting processes, there merely indicated, which lead from the atomistic view to the laws of motion of continua."

The Boltzmann equation, mentioned by Hilbert and which we will present in more detail later, expresses that the distribution of particles evolves under the combined effect of free transport and collisions. For these two effects to be of the same order of magnitude, a simple calculation shows that, in dimension $d \geq 2$, the number of particles N and their diameter size ε must satisfy the scaling relation $N\varepsilon^{d-1} = O(1)$, the so-called *Boltzmann–Grad scaling* [40]. Indeed, the regime described by the Boltzmann equation is such that the mean free path, namely the average distance covered by a particle traveling in straight line between two collisions, is of order 1. Thus a typical particle trajectory should span a tube of volume $1 \times \varepsilon^{d-1}$ between two collisions. This means that, on average, this tube should intersect the position of one of the other $(N-1)$ particles (see Figure 3). Note that in this regime the total volume occupied by the particles at a given time is proportional to $N\varepsilon^d$ and therefore is negligible compared to the total volume occupied by the gas. We speak then of a *dilute gas*.

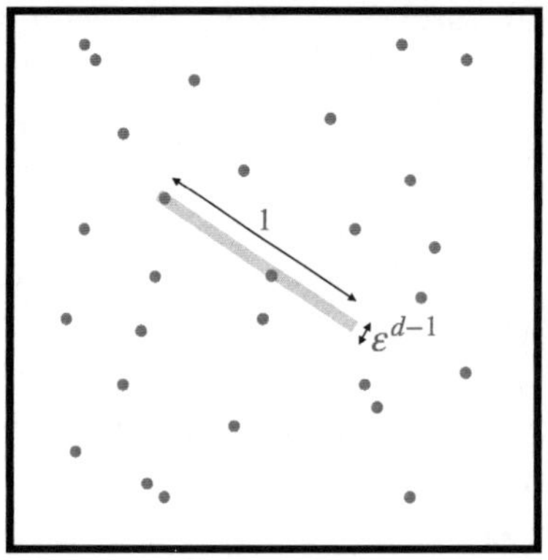

FIGURE 3

Consider N spheres of diameter ε uniformly distributed in a box. If the mean free path is equal to 1, then the grey tube of length 1 and section area of order ε^{d-1} represents the volume spanned by a typical particle between two collisions. The Boltzmann–Grad scaling $N\varepsilon^{d-1} = 1$ is tuned such that on average this tube intersects one particle.

1.2. Three levels of averaging

As already shown in the previous scaling argument, the equations that we want to derive describe the behavior of "typical particles." We therefore have to introduce several averaging processes, and then to describe the average dynamics.

For a statistical description of a monoatomic gas, all particles are considered identical (same geometry, same mass, same interaction law, ...). This is referred to as the exchangeability assumption. The *empirical distribution* of particles is defined as

$$\pi_t^N(x,v) = \frac{1}{N}\sum_{i=1}^{N} \delta_{x-x_i^\varepsilon(t)}\delta_{v-v_i^\varepsilon(t)}, \tag{1.1}$$

where $(x_i^\varepsilon(t), v_i^\varepsilon(t))_{i\le N}$ stands for the positions and velocities of the N particles at time t and δ_x stands for the Dirac mass at $x = 0$. This measure is completely symmetric (i.e., invariant under any permutation of the particle labels) due to the exchangeability assumption. However, this first averaging is not enough to obtain a simple description of the dynamics when N is large because of the instabilities mentioned in the previous section (see Figure 2) which lead to a strong dependency in ε of the particle trajectories. We will therefore introduce a second averaging with respect to initial configurations.

From the physical point of view, this averaging is natural as only fragmentary information on the initial configuration is available. A natural starting point is the particle distribution $f^0 = f^0(x,v)$ which prescribes the probability for a particle to be at position x with velocity v. As N is large, we assume that the initial data $(X_N, V_N) = (x_i, v_i)_{1\le i\le N}$ are independent random variables identically distributed according to f^0. This assumption has, however, to be slightly corrected in order to take into account the exclusion between particles $|x_i - x_j| > \varepsilon$ for $i \neq j$. This statistical framework is referred to as the *canonical* ensemble [63]. This is a simple framework to derive rigorous foundations for the kinetic theory, i.e., to characterize, in the large N asymptotics, the average dynamics and more precisely the evolution equation governing the distribution $f(t,x,v)$ at time t of a typical particle.

In this paper, our goal is actually to go beyond this average dynamics, and to understand in a fine way the correlations arising dynamically inside the gas. Fixing a priori the number N of particles induces additional correlations and thus technical difficulties. To bypass them, we introduce a third level of averaging, by assuming that the number N of particles is also a random variable, and that only its average tuned by $\mu_\varepsilon = \varepsilon^{-(d-1)}$ is determined according to the Boltzmann–Grad scaling. Roughly speaking, N is chosen according to a Poisson law of mean close to μ_ε, and then for any fixed N, the variables (X_N, V_N) are identically distributed, and independent up to the spatial exclusion. More precisely, the variables (N, X_N, V_N) are chosen jointly under the so-called *grand canonical* measure which will be introduced later in (2.3). This is referred to as the *grand canonical ensemble* and from now on, we will use this setting.

We therefore seek to understand the statistical behavior of the empirical measure

$$\pi_t^\varepsilon(x,v) = \frac{1}{\mu_\varepsilon}\sum_{i=1}^{N} \delta_{x-x_i^\varepsilon(t)}\delta_{v-v_i^\varepsilon(t)}, \tag{1.2}$$

where the initial configuration $(N, (X_N^{\varepsilon 0}, V_N^{\varepsilon 0}))$ is a random variable, but the microscopic dynamics is completely deterministic (governed by the hard sphere equations represented in Figure 1).

1.3. A probabilistic approach

The first question is to determine the law of large numbers, that is, the limiting distribution of a typical particle when $\mu_\varepsilon \to \infty$. In the case of N independent identically distributed variables $(\eta_i)_{1\le i\le N}$, the law of large numbers implies in particular that the average converges in probability to its expectation

$$\frac{1}{N}\sum_{i=1}^{N}\eta_i \xrightarrow[N\to\infty]{} \mathbb{E}(\eta).$$

For the interacting particle system, two difficulties arise. The first is that, even at time 0, the variables $(x_i, v_i)_{1\le i\le N}$ are weakly correlated due to the exclusion. In the low density regime, this problem is well understood by classical methods of equilibrium statistical mechanics (see, e.g., **[63]**). In particular, denoting the average of any continuous test function h under the initial empirical measure by

$$\langle \pi_0^\varepsilon, h\rangle = \frac{1}{\mu_\varepsilon}\sum_{i=1}^{N} h(x_i^{\varepsilon 0}, v_i^{\varepsilon 0}),$$

the following convergence in probability holds:

$$\langle \pi_0^\varepsilon, h\rangle - \int f^0 h(x,v)\,dx dv \xrightarrow[\mu_\varepsilon\to\infty]{} 0 \quad \text{under the grand-canonical measure.}$$

We stress the fact that, throughout this paper, the limit $\mu_\varepsilon \to \infty$ implies that the sphere diameter ε tends also to 0 as both parameters are linked by the Boltzmann–Grad scaling

$\mu_\varepsilon \varepsilon^{d-1} = 1$. The second difficulty, which is the main challenge, is to understand whether the initial quasiindependence is propagated in time so that there exists a function $f(t,x,v)$ such that the following convergence in probability holds:

$$\langle \pi_t^\varepsilon, h \rangle - \int f(t) h \, dx dv \xrightarrow[\mu_\varepsilon \to \infty]{} 0 \quad \text{under the grand-canonical measure on initial configurations,} \tag{1.3}$$

and whether $f(t)$ evolves according to a deterministic equation, namely the Boltzmann equation. As we will see, this question is particularly delicate since the Boltzmann equation obtained in the limit is singular (see (2.1)). The major result proving this convergence goes back to Lanford [46] and will be explained in Section 2.2.

The approximation (1.3) of the empirical measure neglects two types of errors. The first is the fact that there are corrector terms which converge to 0 as $\mu_\varepsilon \to +\infty$. The second is related to the vanishing probability of the initial configurations for which the convergence does not hold. A classical question in statistical physics is to quantify more precisely these errors, by studying fluctuations, i.e., deviations between the empirical measure and its expectation. In the case of N independent and identically distributed random variables $(\eta_i)_{1 \le i \le N}$, the central limit theorem implies that the fluctuations are of order $O(1/\sqrt{N})$ and the following convergence in law holds:

$$\sqrt{N}\left(\frac{1}{N} \sum_{i=1}^{N} \eta_i - \mathbb{E}(\eta) \right) \xrightarrow[N \to \infty]{\text{(law)}} \mathcal{N}\big(0, \operatorname{Var}(\eta)\big),$$

where $\mathcal{N}(0, \operatorname{Var}(\eta))$ is the normal law of variance $\operatorname{Var}(\eta) = \mathbb{E}((\eta - \mathbb{E}(\eta))^2)$. In particular, at this scale, some randomness is retrieved. Investigating the same fluctuation regime for the dynamics of hard sphere gases consists in considering the scaled fluctuation field ζ_t^ε defined by duality

$$\langle \zeta_t^\varepsilon, h \rangle = \sqrt{\mu_\varepsilon} \big(\langle \pi_t^\varepsilon, h \rangle - \mathbb{E}_\varepsilon(\langle \pi_t^\varepsilon, h \rangle) \big), \tag{1.4}$$

where h is a continuous test function, and $\mathbb{E}_\varepsilon$ denotes the expectation on initial configurations under the grand-canonical measure. A series of recent works [13–16] has allowed to characterize these dynamical fluctuations, and to derive a stochastic evolution equation governing the limiting process. These results will be presented in Sections 3.4 and 4.2.

The last question generally studied in a classical statistical approach is that of quantifying rare events, i.e., of estimating the probability of observing an atypical behavior (which deviates macroscopically from the average). For independent and identically distributed random variables, this probability is exponentially small, and it is therefore natural to study the asymptotics

$$I(m) := \lim_{\delta \to 0} \lim_{N \to \infty} -\frac{1}{N} \log \mathbb{P}\left(\left| \frac{1}{N} \sum_{i=1}^{N} \eta_i - m \right| < \delta \right) \quad \text{with } m \neq \mathbb{E}(\eta). \tag{1.5}$$

The limit $I(m)$ is called the large deviation function and it can be expressed as the Legendre transform of the log-Laplace transform of a single variable $u : \mathbb{R} \mapsto \log \mathbb{E}(\exp(u\eta))$ [23]. To generalize this statement to correlated variables, it is necessary to compute a more global

Laplace transform and this requires a control on the correlations with exponential accuracy. The methods of dynamical cumulants introduced in [13,14] are a key tool to compute exponential moments of the hard sphere distribution and, in this way, to control the measure of events up to scales which are vanishing exponentially fast. We will give a flavor of those techniques in Section 3.4.

Note that precise conjectures regarding those three questions are formulated by Rezakhanlou in [62].

2. TYPICAL DYNAMICAL BEHAVIOR

2.1. Boltzmann's great intuition

The equation which rules the typical evolution of a hard sphere gas was proposed heuristically by Boltzmann [17] about one century before its rigorous derivation by Lanford [46] as the "limit" of the particle system when $\mu_\varepsilon \to +\infty$. The revolutionary idea of Boltzmann was to write an evolution equation for the probability density $f = f(t, x, v)$ giving the proportion of particles at position x with velocity v at time t. In the absence of collisions and in a domain without boundary, this density f would be exactly transported along the physical trajectories $x(t) = x(0) + vt$, meaning that $f(t, x, v) = f^0(x - vt, v)$. The difficulty consists then in taking into account the statistical effect of collisions. Insofar as the size of the particles is negligible, one can consider that these collisions are pointwise both in t and x. Boltzmann proposed therefore a rather intuitive counting:

- the number of particles with velocity v is increased when a particle of velocity v' collides with a particle of velocity v'_1, and jumps to velocity v (see (2.2)). Notice that here, the pair (v', v'_1) plays the role of precollisional velocities, while instead in Figure 1 this notation was used for the postcollisional velocities in the particle system;
- the number of particles with velocity v is decreased when a particle of velocity v collides with a particle of velocity v_1, and is deflected into another velocity.

The probability of these jumps is described by a transition rate, referred to as the *collision cross-section* b. The function $b(v, v_1, \omega)$ is nonnegative, depends only on the relative velocity $|v - v_1|$ and on the angle between $(v - v_1)$ and ω, a scattering vector which is distributed uniformly in the unit sphere $\mathbb{S}^{d-1} \subset \mathbb{R}^d$. For the hard sphere interactions, we shall see that ω keeps track of the way two hard spheres collide (see Figure 1) and that $b(v - v_1, \omega) = ((v - v_1) \cdot \omega)_+$. In particular, it is invariant under $(v, v_1) \mapsto (v_1, v)$ (exchangeability) and under $(v, v_1, \omega) \mapsto (v', v'_1, \omega)$ (microscopic reversibility).

The fundamental assumption in Boltzmann's theory is that, in a rarefied gas, the correlations between two particles about to collide should be very weak. Therefore the joint probability to have both precollisional particles of velocities v and v_1 at position x at time t should be well approximated by $f(t, x, v) f(t, x, v_1)$. This independence property is called the molecular *chaos assumption*. The equation then states

$$\begin{cases} \partial_t f + \underbrace{v \cdot \nabla_x f}_{\text{transport}} = \underbrace{C(f,f)}_{\text{collision}} \\ C(f,f)(t,x,v) \\ \quad = \iint \big[\underbrace{f(t,x,v')f(t,x,v_1')}_{\text{gain term}} - \underbrace{f(t,x,v)f(t,x,v_1)}_{\text{loss term}}\big] \underbrace{b(v-v_1,\omega)}_{\text{cross section}}\, dv_1 d\omega, \end{cases} \tag{2.1}$$

where the scattering rules

$$v' = v - \big((v-v_1)\cdot\omega\big)\omega, \quad v_1' = v_1 + \big((v-v_1)\cdot\omega\big)\omega \tag{2.2}$$

are analogous to the microscopic collision rules introduced in Figure 1, with the important difference that ω is now a random vector chosen uniformly in the unit sphere $\mathbb{S}^{d-1} \subset \mathbb{R}^d$. Indeed, the relative position of the colliding particles has been forgotten in the limit $\varepsilon \to 0$. As a consequence, the Boltzmann equation is singular as it involves a product of densities at the same point x.

Boltzmann's idea of reducing to a kinetic equation the Hamiltonian dynamics describing the atomistic behavior was revolutionary and opened the way to the description of nonequilibrium phenomena by mesoscopic equations. However, the Boltzmann equation was first heavily criticized as it seems to violate some basic physical principles. Indeed, what made Boltzmann's theory such a breakthrough, but also made it unacceptable by many of his contemporaries, is that it predicts a time irreversible evolution, providing actually a quantitative formulation of the second principle of thermodynamics. The Boltzmann equation (2.1) has indeed a Lyapunov functional defined by $S(t) = -\iint f \log f(t,x,v)dxdv$ and referred to as the entropy, which can only increase along the evolution $\frac{d}{dt}S(t) \geq 0$, with equality if and only if the gas is at thermal equilibrium. At first sight, this irreversibility does not seem to be compatible with the fact that the hard sphere dynamics is governed by a Hamiltonian system, i.e., a system of ordinary differential equations which is completely time reversible. Soon after Boltzmann postulated his equation, these two different behaviors were considered by Loschmidt as a paradox and an obstruction to Boltzmann's theory. A fully satisfactory mathematical explanation of this issue remained open during almost one century, until the role of probability was precisely identified: the underlying dynamics is reversible, but the description which is given of this dynamics is only partial (obtained by averaging or looking at the most probable path) and therefore is not reversible.

2.2. Lanford's theorem

Lanford's result [46] shows in which sense the Boltzmann equation (2.1) is a good approximation of the hard sphere dynamics. Let us first define the initial distribution.

Initial data. *Consider $\mathbb{T}^d = [0,1]^d$ the unit domain with periodic boundary conditions and $f^0 = f^0(x,v)$ a Lipschitz probability density in $\mathbb{T}^d \times \mathbb{R}^d$, with Gaussian tails at large velocities. To define a system of hard spheres which are initially independent (up to the exclusion) and identically distributed according to f^0, we introduce the grand canonical measure:*

the probability density of finding N particles with coordinates $Z_N = (x_i, v_i)_{i \leq N}$ is given by

$$\frac{1}{N!} W_N^\varepsilon(Z_N) = \frac{1}{\mathcal{Z}^\varepsilon} \frac{\mu_\varepsilon^N}{N!} \prod_{i=1}^N f^0(x_i, v_i) \prod_{i \neq j} \mathbf{1}_{|x_i - x_j| > \varepsilon}, \quad \text{for } N = 0, 1, 2, \ldots, \tag{2.3}$$

where the constant $\mathcal{Z}^\varepsilon$ is the normalization factor of the probability measure. Once the random initial configuration is chosen, the hard sphere dynamics evolve deterministically and the corresponding probability and expectation on the particle trajectories will be denoted by $\mathbb{P}_\varepsilon$ and $\mathbb{E}_\varepsilon$.

Lanford's result can be stated as follows (this is not exactly the original formulation, see in particular Section 2.5 below for comments).

Theorem 2.1 (Lanford). *In the Boltzmann–Grad limit ($\mu_\varepsilon \to \infty$ with $\mu_\varepsilon \varepsilon^{d-1} = 1$), the empirical measure π_t^ε of the hard sphere system defined by* (1.2) *concentrates on the solution of the Boltzmann equation* (2.1)*, i.e., for any bounded and continuous function h,*

$$\forall \delta > 0, \quad \lim_{\mu_\varepsilon \to \infty} \mathbb{P}_\varepsilon \left(\left| \langle \pi_t^\varepsilon, h \rangle - \int f(t) h \, dx \, dv \right| \geq \delta \right) = 0,$$

on a time interval $[0, T_L]$ depending only on the initial distribution f^0.

Let us comment on the time of validity T_L of the approximation. This time depends on the initial data f^0 and turns out to be of the order of a fraction of the mean time between two successive collisions for a typical particle. This time is large enough for the microscopic system to undergo a large number of collisions (of the order $O(\mu_\varepsilon)$), and in particular irreversibility already shows up at this scale. But this time is (far) too small to see phenomena such as relaxation towards (local) thermodynamic equilibrium, and a fortiori hydrodynamic regimes. Physically we do not expect this time to be critical, in the sense that the dynamics would change nature afterwards. Actually, in practice the Boltzmann equation is used in many applications (such as calculations for the reentrance of spatial vehicles in the atmosphere) without time restriction. However, it is important to note that a time restriction may not be only technical: from the mathematical point of view, one cannot exclude that the Boltzmann equation exhibits singularities (typically, spatial concentrations which would prevent making sense of the collision term, and which would also contradict locally the low density assumption). In order to construct global in time solutions for the Boltzmann equation, one actually has either to consider small fluctuations around some equilibrium, or to introduce a renormalization procedure [28]. These two approaches rely strongly on entropy production estimates, which do not have any counterpart at the microscopic level (i.e., for fixed $\mu_\varepsilon, \varepsilon$). In the current state of our knowledge, the problem of extending Lanford's convergence result to longer times faces serious obstructions, even to the time of existence and uniqueness of the solution to the Boltzmann equation. This will be discussed later on in Section 4.1 (see also Section 5). In Section 4, we will also present some recent results in this direction, providing a global in time convergence for the fluctuation field at equilibrium.

2.3. Heuristics of the proof

Let us now explain informally how the Boltzmann equation (2.1) can be guessed from the particle dynamics. The goal is to transport the initial grand-canonical measure, defined in (2.3), along the dynamics and then to project this measure at time t on the 1-point particle phase space. We therefore define by duality $F_1^\varepsilon(t,z)$ the density of a typical particle with respect to the test function h as

$$\int F_1^\varepsilon(t,z)h(z)dz = \mathbb{E}_\varepsilon\big(\langle \pi_t^\varepsilon, h\rangle\big), \tag{2.4}$$

where the empirical measure π_t^ε was introduced in (1.2). More generally, we are going to introduce $\pi_{k,t}^\varepsilon$, the natural extension of the empirical measure π_t^ε to k distinct particles. For simplicity, the particle coordinates $(x_i^\varepsilon(t), v_i^\varepsilon(t))$ at time t will be denoted by $z_i^\varepsilon(t)$. For any test function h_k of k variables, we define

$$\langle \pi_{k,t}^\varepsilon, h_k\rangle = \frac{1}{\mu_\varepsilon^k} \sum_{(i_1,\dots,i_k)} h_k\big(z_{i_1}^\varepsilon(t), \dots, z_{i_k}^\varepsilon(t)\big) \tag{2.5}$$

and the sum is over the k-tuples of indices among all the particles at time t. We stress the fact that $\pi_{k,t}^\varepsilon$ differs from $(\pi_t^\varepsilon)^{\otimes k}$ as the variables are never repeated. We will study the k-particle correlation functions F_k^ε which are symmetric finite dimensional projections of the probability measure

$$\int F_k^\varepsilon(t, Z_k)h_k(Z_k)dZ_k = \mathbb{E}_\varepsilon\big(\langle \pi_{k,t}^\varepsilon, h_k\rangle\big), \tag{2.6}$$

denoting $Z_k = (x_i, v_i)_{1\le i\le k}$. The correlation functions are key to describe the kinetic limit. In particular, Theorem 2.1 shows that $F_1^\varepsilon(t,z)$ converges to the solution of the Boltzmann equation $f(t)$ in the Boltzmann–Grad limit ($\mu_\varepsilon \to \infty$ with $\mu_\varepsilon \varepsilon^{d-1} = 1$). Let us explain briefly why this holds.

Let h be a bounded smooth test function on $\mathbb{T}^d \times \mathbb{R}^d$. Consider the evolution of the empirical measure during a short time interval $[t, t+\delta]$ and split the different contributions according to the number of collisions for each particle

$$\begin{aligned}
&\mathbb{E}_\varepsilon\big[\langle \pi_{t+\delta}^\varepsilon, h\rangle\big] - \mathbb{E}_\varepsilon\big[\langle \pi_t^\varepsilon, h\rangle\big] \\
&\quad = \mathbb{E}_\varepsilon\Bigg[\frac{1}{\mu_\varepsilon} \sum_{\substack{j \\ \text{no collision}}} \big(h\big(z_j^\varepsilon(t+\delta)\big) - h\big(z_j^\varepsilon(t)\big)\big)\Bigg] \\
&\qquad + \mathbb{E}_\varepsilon\Bigg[\frac{1}{2\mu_\varepsilon} \sum_{\substack{(i,j) \\ \text{with 1 collision}}} \big(h\big(z_i^\varepsilon(t+\delta)\big) + h\big(z_j^\varepsilon(t+\delta)\big) - h\big(z_i^\varepsilon(t)\big) - h\big(z_j^\varepsilon(t)\big)\big)\Bigg] + O(\delta^2),
\end{aligned} \tag{2.7}$$

and we are going to argue that the error term δ^2 takes into account all the groups of particles undergoing at least 2 collisions in the short time interval δ.

The asymptotic behavior when δ tends to 0 will be analyzed now for each term in (2.7). The transport contribution arises from the particles moving in straight line without

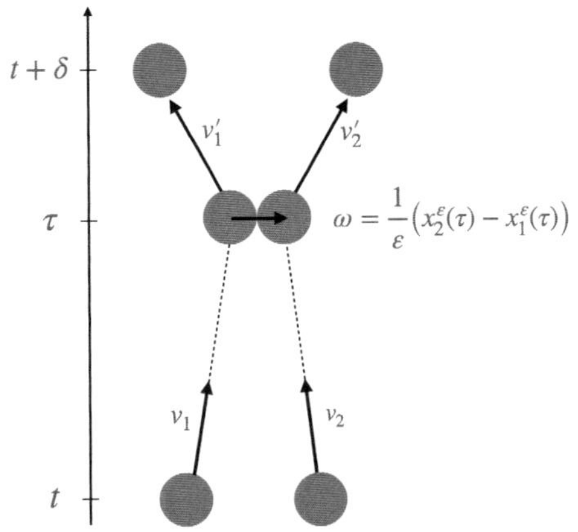

FIGURE 4
Two particles collide in the time interval $[t, t+\delta]$ according to the scattering rules of Figure 1. The collision occurs at time τ if $x_1 - x_2 + (\tau - t)(v_1 - v_2) = -\varepsilon\omega$. Therefore x_2 has to be in a tube with axis $v_1 - v_2$ and the coordinates z_1, z_2 at time t can be parametrized by $(x_1, v_1, v_2, \tau, \omega)$. This change of variables has a Jacobian $dz_1 dz_2 = \varepsilon^{d-1}((v_1 - v_2)\cdot\omega)_+ d\omega d\tau dx_1 dv_1 dv_2$.

collisions; indeed, if the distribution F_1^ε is smooth enough, one gets

$$\mathbb{E}_\varepsilon\Bigg[\frac{1}{\mu_\varepsilon}\sum_{\substack{j\\ \text{no collision}}}\Big(h\big(z_j^\varepsilon(t+\delta)\big) - h\big(z_j^\varepsilon(t)\big)\Big)\Bigg] = \delta\int dz_1 F_1^\varepsilon(t, z_1) v_1 \cdot \nabla_x h(z_1) + o(\delta).$$

We turn next to the term involving one collision. Note first that two particles starting at (x_1, v_1) and (x_2, v_2) at time t collide at a later time $\tau \le t + \delta$ if the following geometric condition holds (see Figure 4):

$$x_1 - x_2 + (\tau - t)(v_1 - v_2) = -\varepsilon\omega. \tag{2.8}$$

This implies that their relative position must belong to a tube oriented in the direction $v_1 - v_2$ with length $\delta|v_1 - v_2|$ and width ε. This set has a size proportional to $\delta\varepsilon^{d-1}|v_2 - v_1|$ with respect to the Lebesgue measure. More generally, a series of $k-1$ collisions between k particles imposes $k-1$ constraints of the previous form. Using the Boltzmann–Grad scaling $\mu_\varepsilon\varepsilon^{d-1} = 1$ and neglecting the velocity contribution, one can show that this event has a vanishing probability bounded from above by

$$\left(\frac{\delta}{\mu_\varepsilon}\right)^{k-1}. \tag{2.9}$$

Since there are, on average, μ_ε^k ways of choosing these k colliding particles, we deduce that the occurrence of $k-1$ collisions in (2.7) has a probability of order $\delta^{k-1}\mu_\varepsilon$. This explains why in (2.7) the probability of the terms involving more than 1 collision, i.e., involving $k \ge 3$ colliding particles, has been estimated by $O(\delta^2)$.

This crude estimate is not sufficient to recover the collision operator $C(f, f)$ of the Boltzmann equation (2.1). We are going now to analyze more carefully the term with one collision in (2.7) in order to identify $C(f, f)$. As the collision term involves 2 particles, it is no longer a function of the empirical measure. The correlation function F_2^ε defined in (2.6) will be needed to rewrite it :

$$\begin{aligned}\text{Coll} &= \mathbb{E}_\varepsilon\Bigg[\frac{1}{2\mu_\varepsilon}\sum_{\substack{(i,j)\\ \text{with 1 collision}}}\Big(h\big(z_i^\varepsilon(t+\delta)\big)+h\big(z_j^\varepsilon(t+\delta)\big)-h\big(z_i^\varepsilon(t)\big)-h\big(z_j^\varepsilon(t)\big)\Big)\Bigg]\\ &= \frac{\mu_\varepsilon}{2}\int dz_1dz_2F_2^\varepsilon(t,z_1,z_2)\mathbf{1}_{1\text{ and }2\text{ collide}}\big[h\big(z_1(\delta)\big)+h\big(z_2(\delta)\big)-h(z_1)-h(z_2)\big]\\ &\quad+o(\delta), \qquad (2.10)\end{aligned}$$

where $z_1(\delta), z_2(\delta)$ stands for the particle coordinates after a time δ. After the collision, the velocities are scattered to v_1', v_2' according to the deflection parameter ω (see Figure 4), but the positions are almost unchanged as $\delta \ll \varepsilon$. Since the function h is smooth, the last term in (2.10) can be approximated by the velocity jump

$$\Delta h(z_1,z_2,\omega)=h\big(x_1,v_1'\big)+h\big(x_2,v_2'\big)-h(z_1)-h(z_2). \qquad (2.11)$$

By condition (2.8), it is equivalent to parametrize two colliding particles either by their coordinates z_1, z_2 at time t or by their coordinates at the collision time τ which are determined by $x_1, v_1, \tau, \omega, v_2$ (see Figure 4). This change of variables has a Jacobian $\varepsilon^{d-1}((v_1-v_2)\cdot\omega)_+$. Since $\varepsilon^{d-1}=1/\mu_\varepsilon$ and $\delta \ll \varepsilon$, we deduce from (2.11) that

$$\text{Coll}=\frac{1}{2}\int_t^{t+\delta}d\tau\int dz_1dv_2d\omega F_2^\varepsilon(\tau,z_1,z_2)\big((v_1-v_2)\cdot\omega\big)_+\Delta h(z_1,z_2,\omega)+o(\delta), \qquad (2.12)$$

with $z_2=(x_1+\varepsilon\omega, v_2)$, as both particles are next to each other at the collision time. The cross-section $b(v_1-v_2,\omega)=((v_1-v_2)\cdot\omega)_+$ in the Boltzmann equation can be identified from the equation above. From the previous heuristics, the relation (2.7) provides "almost" a weak formulation of the collision operator in (2.1) in the limit $\delta\to 0$,

$$\begin{aligned}\partial_t&\int dz_1F_1^\varepsilon(t,z_1)h(z_1)\\ &=\int dz_1F_1^\varepsilon(t,z_1)v_1\cdot\nabla h(z_1)\\ &\quad+\frac{1}{2}\int dz_1\,d\omega\,dv_2\delta_{x_2-x_1-\varepsilon\omega}F_2^\varepsilon(t,z_1,z_2)\big((v_1-v_2)\cdot\omega\big)_+\Delta h(z_1,z_2,\omega), \qquad (2.13)\end{aligned}$$

where we used the Dirac notation to stress that $z_2=(x_1+\varepsilon\omega, v_2)$. The key step to close the equation is the *molecular chaos assumption* postulated by Boltzmann which asserts that the precollisional particles remain independently distributed at any time so that

$$F_2^\varepsilon(t,z_1,z_2)\simeq F_1^\varepsilon(t,z_1)F_1^\varepsilon(t,z_2). \qquad (2.14)$$

When the diameter of the spheres ε tends to 0, the coordinates x_1 and x_2 coincide and the scattering parameter ω becomes a random parameter. Assuming that F_1^ε converges, its limit has to satisfy the Boltzmann equation (2.1). Establishing rigorously the factorization (2.14) requires implementing a different and more involved strategy which will be presented in Section 2.4.

2.4. Some elements of proof

Lanford's proof [46] has been completed and improved over the years; we refer to the monographs [21, 22, 67] for accounts of the related results. In the more recent years, several quantitative convergence results were established, and the proofs extended to the case of compactly supported potentials [37, 57, 58]. In the following, we sketch the main steps of the proof for the hard sphere dynamics.

The proof of Lanford's theorem relies on the study of the correlation functions F_k^ε defined in (2.6), characterizing joint probabilities of k particles. In particular, we do not consider directly the empirical measure, but only its average F_1^ε under the grand-canonical probability $\mathbb{P}_\varepsilon$. The starting point is the system of ordinary differential equations for the hard sphere positions and velocities (see Figure 1), which provides, by applying Green's formula to the Liouville equation, the following equation on the first correlation function:

$$\begin{cases} \partial_t F_1^\varepsilon + \underbrace{v \cdot \nabla_x F_1^\varepsilon}_{\text{transport}} = \underbrace{C^\varepsilon(F_2^\varepsilon)}_{\text{collision at distance } \varepsilon}, \\ C^\varepsilon(F_2^\varepsilon)(t,x,v) \\ \quad = \iint \big[\underbrace{F_2^\varepsilon(t,x,v',x+\varepsilon\omega,v_1')}_{\text{gain term}} - \underbrace{F_2^\varepsilon(t,x,v,x-\varepsilon\omega,v_1)}_{\text{loss term}}\big] \underbrace{\big((v-v_1)\cdot\omega\big)_+}_{\text{cross-section}} dv_1 d\omega. \end{cases} \tag{2.15}$$

A weak form of this equation has been stated in (2.13). In the limit $\mu_\varepsilon \to \infty$, we expect that it can be closed by the factorization $F_2^\varepsilon \sim F_1^\varepsilon \otimes F_1^\varepsilon$, called the propagation of chaos (2.14). We are unable to prove it directly, nor will it be shown directly from (2.15) that the limit F_1 of F_1^ε satisfies an infinitesimal evolution equation of the previous form. We will rather obtain a series expansion of F_1, which will be identified with the solution of the Boltzmann equation by a uniqueness argument. The proof is therefore very different from the heuristics presented in Section 2.3.

The proof can be divided into three steps. The first is to rewrite $F_1^\varepsilon(t,x,v)$ as an "average" (weighted with the initial correlation functions $F_k^{\varepsilon,0}$) of all possible dynamics such that at time t, a particle stands at position x with velocity v. The analytical way of doing so is to derive evolution equations similar to (2.15) for all correlation functions F_k^ε, and then to write the iterated Duhamel formula for this hierarchy of equations, called the BBGKY hierarchy after Bogoliubov–Born–Green–Kirkwood–Yvon (see [22] for an account and references). We will not give the details of these technical computations here, but will retrieve the final series expansion (formally) using a more probabilistic perspective based on geometric representations in terms of pseudotrajectories.

The idea is to track back the history of the particle sitting at position x with velocity v at time t, referred to as particle $*$, in order to characterize all initial configurations which contribute to $F_1^\varepsilon(t,x,v)$. We start by following (backward in time) this particle, which has a uniform rectilinear motion $x(t') = x - v(t-t')$ until it collides with another particle, called particle 1, say at time t_1. Note that this collision can actually be either a physical collision (with scattering) or a mathematical artefact coming from the loss term of equation (2.15)

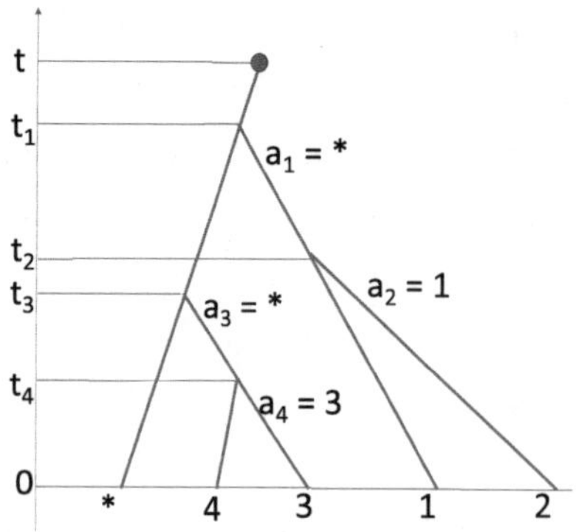

FIGURE 5

The sequence of collisions in the backward history can be encoded in a tree with the root indexed by the particle $*$ and n branchings (here $n = 4$). At each creation time, the label of the particle colliding with the fresh particle is indicated. For example, at time t_3, the particle $*$ collides with particle 3 so that $a_3 = *$.

(particles touch each other but are not deflected). Thus in order to understand the history of particle $*$, we need to track back the history of both particles $*$ and 1 before time t_1. From time t_1, both particles are then transported by the 2-particle backward flow until the next collision, say with particle 2 at time t_2, etc., and we iterate this procedure until time 0. Notice that in between the creations of new particles, the particles may collide between themselves as they are transported by the backward hard sphere flow: this will be called *recollision*. The history of the particle $*$ can be reconstructed (see Figure 5) by prescribing

- the total number of collisions n;
- the combinatorics of collisions, encoded in a tree $a \in \mathcal{A}_{1,n}$ with root indexed by the label $*$ and n branchings ($a_i \in \{*, 1, \dots, i-1\}$ for $1 \le i \le n$);
- the collision parameters $(T_n, V_n, \Omega_n) = (t_i, v_i, \omega_i)_{1 \le i \le n}$ with $0 < t_n < \cdots < t_1 < t$.

We then define the pseudotrajectory $\Psi^{\varepsilon}_{1,n}$ starting from $z = (x, v)$ at time t as follows:

- on $]t_i, t_{i-1}[$, the group of i particles is transported by the backward flow;
- at time t_i, particle i is added at position $x_{a_i}(t_i) + \varepsilon\omega_i$, with velocity v_i;
- if the velocities $(v_i, v_{a_i}(t_i^+))$ are postcollisional, meaning that $(v_{a_i}(t_i^+) - v_i) \cdot \omega_i > 0$, then they are instantaneously scattered as in Figure 1 (with deflection angle ω_i).

We stress the fact that pseudotrajectories are not particle trajectories of the physical system, but a geometric interpretation of an iterated Duhamel expansion. In particular, pseudotrajectories do not involve a fixed number of particles, they are coded in terms of random trees (with creation of particles at random times as in Figure 5) and of signs associated with the gain and loss terms of the collision operator.

Note that not all collision parameters (T_n, V_n, Ω_n) are admissible since particles should never overlap. We denote by $\mathcal{G}^\varepsilon$ the set of admissible parameters. With these notations, we obtain the following representation of F_1^ε:

$$F_1^\varepsilon(t,x,v) = \sum_{n=0}^{+\infty} \sum_{a \in \mathcal{A}_{1,n}} \int_{\mathcal{G}^\varepsilon} dT_n dV_n d\Omega_n \mathcal{C}(\Psi_{1,n}^\varepsilon) F_{1+n}^{\varepsilon,0}(\Psi_{1,n}^\varepsilon(0)), \tag{2.16}$$

where $\Psi_{1,n}^\varepsilon(0)$ stands for the particle configuration at time 0 of the pseudotrajectory and the term $\mathcal{C}(\Psi_{1,n}^\varepsilon)$ comes from the collision cross-sections

$$\mathcal{C}(\Psi_{1,n}^\varepsilon) = \prod_{i=1}^{n} \big((v_i - v_{a_i}(t_i^+)) \cdot \omega_i \big).$$

The elementary factor indexed by i is positive if the addition of particle i corresponds to a physical collision (with scattering), and negative if not.

Remark 2.2. A similar formula holds for the k point correlation function F_k^ε, except that collision trees $a \in \mathcal{A}_{k,n}$ have k roots and n branchings.

Formula (2.16) for the first correlation function has been obtained in a rather formal way. In order to study the convergence as μ_ε tends to infinity, we need to establish the uniform convergence of the series (2.16). We actually use very rough estimates (forgetting in particular the signs of the gain and loss terms in (2.15), although the cancelations between these different contributions should improve the estimates) and prove that the series is absolutely convergent for short times uniformly with respect to ε. Note that this is the only argument in the proof which requires a restriction on short kinetic times.

Let us now estimate the size of the term in (2.16) corresponding to n branchings. The different contributions are:

- a combinatorial factor taking into account all the branching choices $|\mathcal{A}_{1,n}| = n!$;
- the volume $t^n/n!$ of the simplex in time $\{t_n < \cdots < t_1 < t\}$;
- the L^∞-norm of $F_{1+n}^{\varepsilon,0}$ which grows like $\|f^0\|_\infty^n$.

This leads to an upper bound of the form $(C\|f^0\|_\infty t)^n$ which implies that the series is absolutely convergent uniformly in ε on a small time interval depending only on a (weighted) L^∞-norm of f^0.

Remark 2.3. For the sake of simplicity, we do not discuss here the problem of large velocities which create a divergence in the collision cross-section $\mathcal{C}(\Psi_{1,n}^\varepsilon)$. It can be dealt with similar, but more technical arguments, introducing weighted functional spaces encoding the exponential decay of correlation functions $F_{1+n}^{\varepsilon,0}$ at large energies.

The convergence of F_1^ε, as μ_ε tends to infinity, will then follow termwise. In this third step of the proof, we therefore fix the number n of branchings, as well as the collision

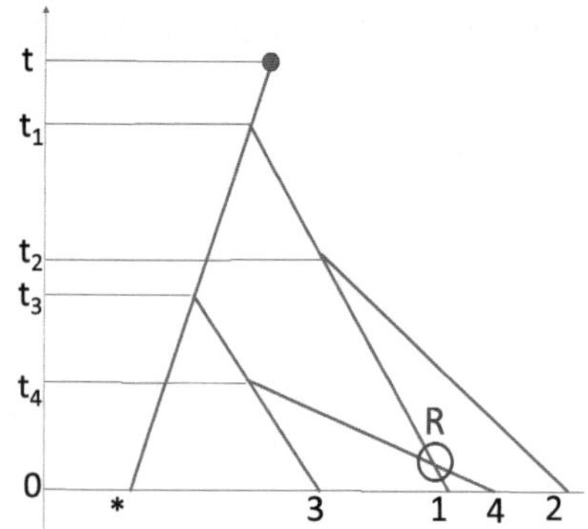

FIGURE 6

When two particles recollide in the backward flow for fixed ε, their velocities are scattered and the resulting pseudodynamics is quite different from the Boltzmann pseudodynamics. The sets $\hat{B}_n^\varepsilon$ are the sets of integration parameters leading to at least a recollision within a pseudotrajectory (as on the picture with $n = 4$).

tree $a \in \mathcal{A}_{1,n}$. One goal is to understand the asymptotic behavior of the pseudotrajectories $\Psi_{1,n}^\varepsilon$. Going back to their definition, we see that it is natural to define limit pseudotrajectories $\Psi_{1,n}$ (when μ_ε tends to ∞) as follows:

- on $]t_i, t_{i-1}[$, the group of i particles is transported by the backward free flow (since the particles become pointwise in the limit, they cannot see each other);
- at time t_i, particle i is added at position $x_{a_i}(t_i^+)$, with velocity v_i (the spatial shift at the creation time disappears);
- if the velocities $(v_i, v_{a_i}(t_i^+))$ are postcollisional, then they are scattered (with deflection angle ω_i).

Note that in the limit, all collision parameters are admissible (since the non overlap condition disappears). With this definition of $\Psi_{1,n}$, we see that there is a very natural coupling between $\Psi_{1,n}^\varepsilon$ and $\Psi_{1,n}$: in most cases, the velocities are exactly equal and the positions differ at most by $n\varepsilon$. The only problem is when two particles of size ε recollide (see Figure 6) in the backward flow on some interval $]t_i, t_{i-1}[$: in this case they are deflected, and the pseudotrajectory $\Psi_{1,n}^\varepsilon$ is no longer close to $\Psi_{1,n}$ on $[0, t_{i-1}]$. We therefore split the set of collision parameters (T_n, V_n, Ω_n) into two parts (and correspondingly split each term in (2.16) into two integrals): the first subset corresponds to admissible integration parameters such that there is no recollision in $\Psi_{1,n}^\varepsilon$, and the second subset, denoted by $\hat{B}_n^\varepsilon$ corresponds either to nonadmissible integration parameters (leading to some overlap) or to integration parameters for which $\Psi_{1,n}^\varepsilon$ has at least one recollision. Using the coupling between $\Psi_{1,n}^\varepsilon$ and $\Psi_{1,n}$ and the regularity of the initial limiting correlation functions (which are nothing else than $(f^0)^{\otimes(1+n)}$), we easily obtain the convergence of the first integral. It remains then to prove that the set $\hat{B}_n^\varepsilon$ has vanishing measure so that the corresponding integral has a negligible contribution. The recollision (or overlap) condition implies that the relative velocity between the two recolliding particles

j_1 and j_2 has to be in a small cone, which imposes strong constraints on the last creation involving either j_1 or j_2. We do not detail these geometric estimates here, but they are quite explicit and provide the following rate of convergence for t sufficiently small (independently of ε)

$$\left\| F_1^\varepsilon(t) - F_1(t) \right\|_\infty \leq C\varepsilon^\alpha \quad \text{for any } \alpha < 1,$$

provided that f^0 is Lipschitz. This concludes the proof, as the series expansion defining F_1 turns out to be the (unique) solution of the Boltzmann equation with initial data f^0. Note that the convergence still holds if f^0 is only continuous, but, in that case, we lose the explicit rate of convergence.

Remark 2.4. Actually one can prove (see [12]) the following quantitative propagation of chaos, where the sets $\mathcal{B}_k^\varepsilon$ have vanishing measure:

$$\sup_{t \leq T_L} \sup_{Z_k \notin \mathcal{B}_k^\varepsilon} \left| F_k^\varepsilon(t, Z_k) - \prod_{i=1}^k f(t, z_i) \right| \leq C^k \varepsilon^\alpha, \tag{2.17}$$

for some $\alpha > 0$ and a constant C depending on the initial measure f^0. This is a much stronger notion of convergence than the one stated in Theorem 2.1.

2.5. On the irreversibility

In this paragraph, we are going to argue that the answer to the irreversibility paradox is hidden in the chaos assumption (2.14) which holds only for specific configurations. Understanding the range of validity of the chaos assumption will be the key to derive not only the Boltzmann equation, but also the stochastic corrections.

Actually, the notion of convergence which appears in the statement of Theorem 2.1 differs slightly from the one used in the proof (see Section 2.4): Theorem 2.1 states the convergence of observables $\langle \pi_t^\varepsilon, h \rangle$, that is, a convergence in the sense of measures since the test function h has to be continuous. This convergence is rather weak and is actually not enough to ensure the stability of the collision term in the Boltzmann equation since this term involves traces. In the proof of Lanford's theorem, one actually considers all the correlation functions F_k^ε introduced in (2.6), and one shows that each one of these correlation functions converges uniformly outside a set $\mathcal{B}_k^\varepsilon$ of vanishing measure when μ_ε tends to infinity (see Remark 2.4). Moreover, the set $\mathcal{B}_k^\varepsilon$ of bad microscopic configurations (t, Z_k) (on which F_k^ε is not converging) is somehow transverse to the set of precollisional configurations (as can be seen in Figure 7, two particles in $\mathcal{B}_2^\varepsilon$ tend to move far apart so that they are unlikely to collide). The convergence defect is therefore not an obstacle to taking limits in the collision term, however, these singular sets $\mathcal{B}_k^\varepsilon$ carry important information on the time correlations: in particular, they encode the memory of the evolution and by neglecting them it is no longer possible to reverse time and to retrace the dynamics backwards. Thus by discarding the microscopic information encoded in $\mathcal{B}_k^\varepsilon$, one can only recover an irreversible kinetic description which is far from describing the complete microscopic dynamics. The singular sets $\mathcal{B}_k^\varepsilon$ have been described in [12,24,68] and their complex structure has been made more precise in [14] by means of the cumulants which will be introduced in Section 3.3.

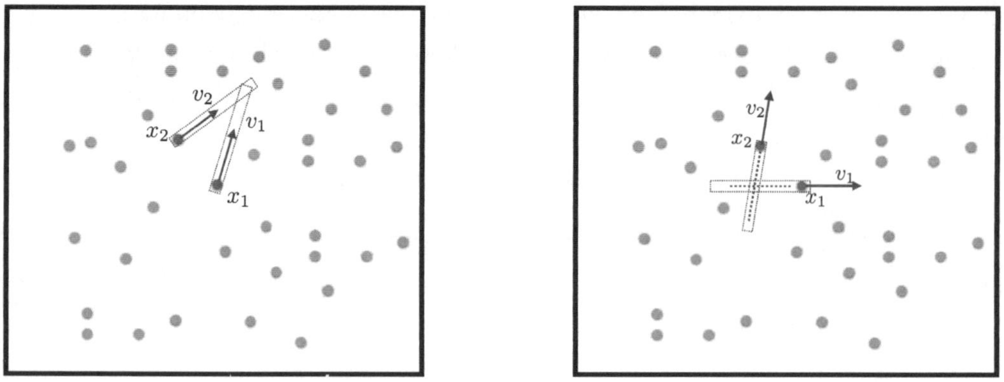

$F_2^\varepsilon(t,x_1,v_1,x_2,v_2) \simeq F_1^\varepsilon(t,x_1,v_1)\,F_1^\varepsilon(t,x_2,v_2)$ $\qquad$ $F_2^\varepsilon(t,x_1,v_1,x_2,v_2) \not\simeq F_1^\varepsilon(t,x_1,v_1)F_1^\varepsilon(t,x_2,v_2)$

FIGURE 7

(Left) Particles 1 and 2 will encounter in the future so they are likely not to have collided in the past, and we expect that the correlation function F_2^ε factorizes in the limit $\mu_\varepsilon \to \infty$. (Right) The particle coordinates belong to the bad set $\mathcal{B}_2^\varepsilon$, meaning that they have met in the past. In this case, microscopic correlations have been built dynamically and the factorization (2.14) should not be valid.

The sets leading to a forward or a backward collision have a similar geometric structure and a similar size which vanishes with respect to the Lebesgue measure when ε tends to 0. However, they play different roles: the memory of the system is encoded in the sets $\mathcal{B}_2^\varepsilon$; on the other hand, the forward sets are the only ones relevant for the chaos assumption. The sets $\mathcal{B}_k^\varepsilon$ are built similarly in terms of the backward flow of k particles (see **[12]**).

3. CORRELATIONS AND FLUCTUATIONS

3.1. From instability to stochasticity

In order to understand the specific features of the hard sphere dynamics in the low density regime (dilute Boltzmann–Grad limit), it is worthwhile to compare its behavior to the mean field dynamics. For this, let us consider more general microscopic dynamics interpolating between the short range and the mean field regimes. For a given number N of particles, we set

$$\forall i \le N, \quad \frac{d}{dt}x_i = v_i, \quad \frac{d}{dt}v_i = -\frac{1}{N\lambda^d}\sum_j \nabla\Phi\Big(\frac{x_i - x_j}{\lambda}\Big),$$

for some smooth repulsive (radial decreasing) potential $\Phi : [0,1]^d \to \mathbb{R}^+$ and a fixed parameter $\lambda \in (0,1]$. This dynamics is Hamiltonian and by choosing $\lambda = \varepsilon$ (with $N\varepsilon^{d-1} = 1$), one recovers dynamics with a short range potential which behaves qualitatively as the hard sphere gas and which follows a Boltzmann equation in the limit **[37,57]**. For fixed λ, however, say $\lambda = 1$, the limiting behavior is mean field like and the typical density follows the Vlasov equation **[20]**

$$\partial_t f(t,x,v) + v \cdot \nabla_x f(t,x,v) = \Big(\int dy\,dw\, f(t,y,w)\nabla\Phi(x-y)\Big) \cdot \nabla_v f(t,x,v).$$

The Vlasov equation has very different properties from the Boltzmann equation, in particular it is reversible, as the microscopic dynamics. Furthermore, contrary to the hard sphere dynamics, the precise structure of the initial data plays no role in the limiting behavior and it has even been shown in [20] that the fluctuations of the initial data are simply transported by the linearized Vlasov equation. Finally, we stress the fact that the chaos assumption (2.14) is known to be propagated in a very strong sense for the mean field dynamics [38, 42].

A drastic difference between the two regimes comes from the fact that the mean field dynamics is not sensitive to a small shift of the coordinates, as the function Φ is smooth for fixed λ. This is not the case for the choice $\lambda = \varepsilon$ in the Boltzmann–Grad limit. Indeed, in the latter situation the scattering behaves qualitatively as in Figure 2, where asymptotically for ε small the deflection parameter decouples completely from the positions and becomes random (cf. Section 2.3). This gives a probabilistic flavor to the surface integral in Boltzmann's collision operator. As we shall see in Theorem 3.4, the corrections to the limiting Boltzmann equation are driven by a stochastic noise which is also generated by the dynamical instabilities. Thus the limiting structure of the hard sphere dynamics behaves qualitatively as a stochastic process, combining free transport and a random jump process in the velocity space. Notice that in the mean field regime, some instability remains for large times $O(\mu_\varepsilon)$ and this is expected to lead to the Lenard–Balescu stochastic correction [30, 52].

The crucial role of randomness in the low density limit was understood by Mark Kac. He devised a purely stochastic process [43] whose limiting distribution is a solution to the homogeneous Boltzmann equation. Mathematically, at the microscopic level, this model has a very different structure from the Hamiltonian dynamics previously mentioned. Indeed, it is a Markov chain restricted only to particle velocities and the collisions are modeled by a jump process with a random deflection parameter. For Kac's model, the chaos assumption has been derived in a very strong sense [51].

In the following sections, we are going to argue that the hard sphere dynamics shares, however, many similarities with Kac's model, not only at the typical level, but also at the level of the fluctuations and of the large deviations. In this respect, random modeling is an excellent approximation of the hard sphere dynamics. The key step to accessing this refined statistical information will be to understand more precisely the chaos assumption (2.14).

3.2. Defects in the chaos assumption

Going back to the equation (2.15) on F_1^ε, one can see that up to the small spatial shifts in the collision term (known as Enskog corrections to the Boltzmann equation), deviations from the Boltzmann dynamics are due to the defect of factorization $F_2^\varepsilon - F_1^\varepsilon \otimes F_1^\varepsilon$, the so-called second order cumulant. In terms of our geometric interpretation, this corresponds to pseudotrajectories which are correlated. Recall that F_2^ε can be described by interacting collision trees with two roots, say labeled by 1^* and 2^*, and $n_1 + n_2$ branchings (see Remark 2.2), while the tensor product is described by two independent collision trees each with one root, and n_1, n_2 branchings, respectively. The main difference when building the pseudodynamics corresponding to F_2^ε is that particles from tree 1^* and 2^* may (or may not) interact. We start by extracting the pseudotrajectories of F_2^ε having at least one interaction between the two trees, which will be called an *external recollision* (see Figure 8) in contrast with a recollision inside a collision tree which will be called *internal*.

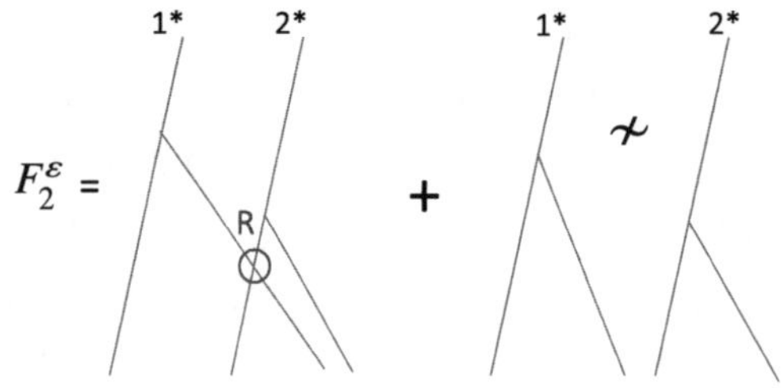

FIGURE 8
Among the pseudodynamics describing F_2^ε, we separate those having a recollision between trees 1^* and 2^*, and those where particles from tree 1^* and particles from tree 2^* remain at a distance greater than ε, which will be denoted by $\not\sim$. In this picture, $n_1 = n_2 = 1$.

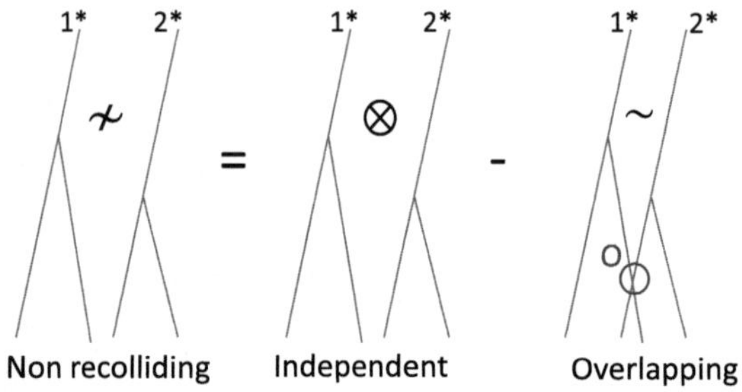

FIGURE 9
Expanding the dynamical exclusion condition leads to the definition of overlaps.

We stress that pseudodynamics without external recollision are not independent since they satisfy a dynamical exclusion condition. We therefore decompose the exclusion condition $\mathbf{1}_{1^* \not\sim 2^*} = 1 - \mathbf{1}_{1^* \sim 2^*}$ (see Figure 9).

Note that this decomposition is a pure mathematical artefact to compare pseudodynamics without external recollision with independent pseudodynamics. In particular, the overlapping condition $1^* \sim 2^*$ does not affect the dynamics itself (overlapping particles are not scattered!). If we ignore the correlation encoded in the initial data, we then end up with a representation of the second order cumulant by trees which are coupled by external recollisions or overlaps (see Figure 10).

Remark 3.1. Recall that the initial measure does not factorize exactly $F_2^{\varepsilon,0} \neq F_1^{\varepsilon,0} \otimes F_1^{\varepsilon,0}$ due to the exclusion condition. Thus the initial data induces also a small correlation which is actually much smaller than the dynamical correlations (by a factor ε), so we will neglect it in the following.

Recolliding and overlapping pseudotrajectories should provide a contribution of order 1 in L^∞ to $F_2^\varepsilon - F_1^\varepsilon \otimes F_1^\varepsilon$. For $n_1 = n_2 = 0$, i.e., for collision trees without branchings, this defines the bad set of configurations $\mathcal{B}_2^\varepsilon$ (mentioned in Sections 2.4–2.5) encoding the collisions between two particles in the backward flow (see Figure 7). In particular, by choosing z_{1^*} and z_{2^*} at time t such that $|x_{1^*} - x_{2^*} - (v_{1^*} - v_{2^*})(t-s)| \leq \varepsilon$ for some $s \leq t$,

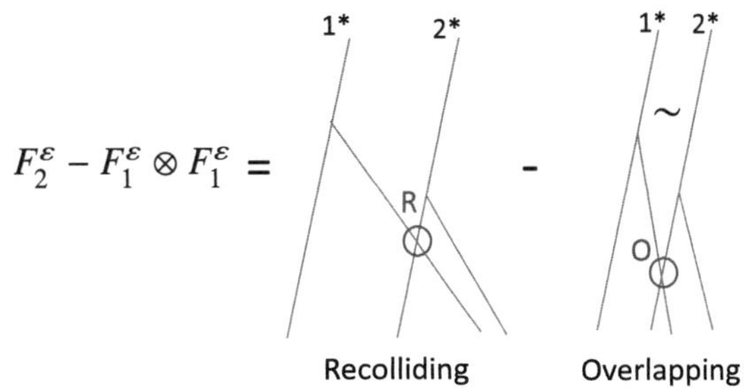

FIGURE 10
The second order cumulant corresponds to pseudotrajectories with at least one external recollision or overlap.

the contribution to the cumulant of the pseudodynamics with $n_1 = n_2 = 0$ is expected to be nonzero (except at equilibrium when recollisions and overlaps almost compensate). The smallness of the second cumulant $F_2^\varepsilon - F_1^\varepsilon \otimes F_1^\varepsilon$ actually comes from the size of its support. The right norm to measure the smallness of correlations is thus the L^1-norm and the quantity to be studied asymptotically is the rescaled second-order cumulant

$$f_2^\varepsilon = \mu_\varepsilon \big(F_2^\varepsilon - F_1^\varepsilon \otimes F_1^\varepsilon \big). \tag{3.1}$$

With this scaling, we expect that f_2^ε has a limit f_2 in the sense of measures. The set supporting the function f_2^ε records the correlation between two pseudotrajectories (rooted in 1^* and 2^*) via a recollision or an overlap. On the other hand, once the two pseudotrajectories are correlated by a recollision or an overlap then any additional recollision, overlap or internal recollision will impose stronger geometric constraints and they can be discarded in the limit as in Lanford's proof (see Figure 6). Therefore the limit f_2 corresponds to pseudotrajectories with exactly one (external) recollision or overlap on $[0, t]$.

In order to understand fluctuations with respect to the Boltzmann dynamics, we also need to understand time correlations. To characterize these time correlations, one can proceed exactly in the same way, using a kind of duality method with weighted pseudotrajectories. Recall that F_2^ε is by definition

$$\int F_2^\varepsilon(t, z_{1^*}, z_{2^*}) h_1(z_{1^*}) h_2(z_{2^*}) dz_{1^*} dz_{2^*} = \mathbb{E}_\varepsilon \bigg(\frac{1}{\mu_\varepsilon^2} \sum_{(i_1, i_2)} h_1\big(z_{i_1}^\varepsilon(t)\big) h_2\big(z_{i_2}^\varepsilon(t)\big) \bigg),$$

meaning that there is a weight $h_1(z_{1^*}) h_2(z_{2^*})$ at time t in the geometric representation. The counterpart for the time correlations

$$F_2^\varepsilon\big[(h_i, \theta_i)_{i \le 2}\big] = \mathbb{E}_\varepsilon \bigg(\frac{1}{\mu_\varepsilon^2} \sum_{(i_1, i_2)} h_1\big(z_{i_1}^\varepsilon(\theta_1)\big) h_2\big(z_{i_2}^\varepsilon(\theta_2)\big) \bigg) \tag{3.2}$$

is to construct the same pseudotrajectories $\Psi_{2,n}^\varepsilon$ starting from some $\theta_2 > \theta_1$, and to evaluate the weight h_1 on the resulting configuration of particle 1^* at time θ_1 and the weight h_2 on the resulting configuration of particle 2^* at time θ_2 (see Figure 11).

We then define the rescaled weighted second order cumulant

$$f_2^\varepsilon\big[(h_i, \theta_i)_{i \le 2}\big] = \mu_\varepsilon \big(F_2^\varepsilon\big[(h_i, \theta_i)_{i \le 2}\big] - F_1^\varepsilon[h_1, \theta_1] F_1^\varepsilon[h_2, \theta_2] \big), \tag{3.3}$$

$$F_2^\varepsilon[(h_i, \theta_i)_{1\le i\le 2}] =$$

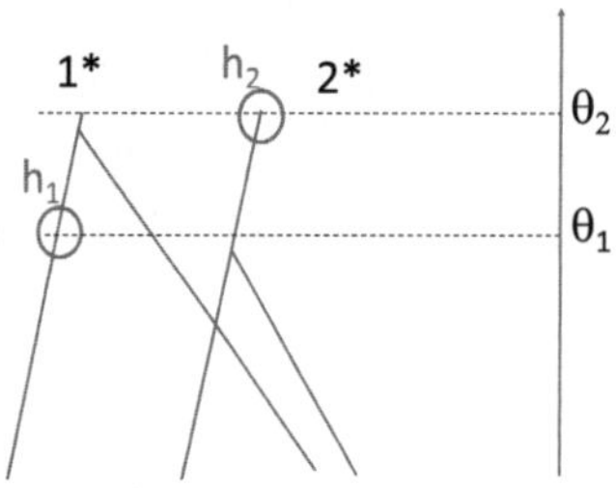

FIGURE 11

Time correlations (3.2) can be computed by introducing weights along the pseudotrajectories.

and performing the same geometric analysis as before, the cumulant $f_2^\varepsilon[(h_i, \theta_i)_{i\le 2}]$ at different times converges also to a limit $f_2[(h_i, \theta_i)_{i\le 2}]$ as μ_ε diverges.

3.3. Higher-order correlations and exponential moments

For a Gaussian process, the first two correlation functions F_1^ε, F_2^ε determine completely all other correlation functions F_k^ε, but in general part of the information is encoded in the (scaled) cumulants of higher order defined by (restricting here for simplicity to only one time)

$$f_k^\varepsilon(t, Z_k) = \mu_\varepsilon^{k-1} \sum_{\ell=1}^{k} \sum_{\sigma\in\mathcal{P}_k^\ell} (-1)^{\ell-1}(\ell-1)! \prod_{i=1}^{\ell} F_{|\sigma_i|}^\varepsilon(t, Z_{\sigma_i}),$$

where $\mathcal{P}_k^\ell$ is the set of partitions of $\{1, \dots, k\}$ in ℓ parts with $\sigma = \{\sigma_1, \dots, \sigma_\ell\}$, $|\sigma_i|$ stands for the cardinality of the set σ_i and $Z_{\sigma_i} = (z_j)_{j\in\sigma_i}$. Each cumulant encodes finer and finer correlations. Contrary to correlation functions F_k^ε, they do not duplicate the information which is already encoded at lower orders.

From the geometric point of view, one can extend the analysis of the previous paragraph and show that the cumulant of order k can be represented by k pseudotrajectories which are completely connected either by external recollisions or by overlaps (see Figure 12).

One can classify these completely connected pseudotrajectories by associating them with a dynamical graph G with k vertices representing the different trees encoding the external recollisions (edge with a $+$ sign) and the overlaps (edges with a $-$ sign). Furthermore, one can define a systematic procedure to extract from this connected graph G a minimally connected graph T by identifying $k-1$ "clustering recollisions" or "clustering overlaps" (see Figure 13). Here we use a cluster expansion reminiscent of the method originally developed by Penrose to deal with correlations in the grand canonical Gibbs measure **[54, 55]**.

We then expect the scaled cumulant f_k^ε to decompose in a sum of $2^{k-1}k^{k-2}$ terms obtained by grouping all pseudotrajectories compatible with each one of the signed minimally connected graphs T (recall that k^{k-2} is the number of trees on k labeled vertices, known as Cayley's formula). For each given signed minimally connected graph, the recollision/overlap conditions can be written as $k-1$ "independent" constraints on the config-

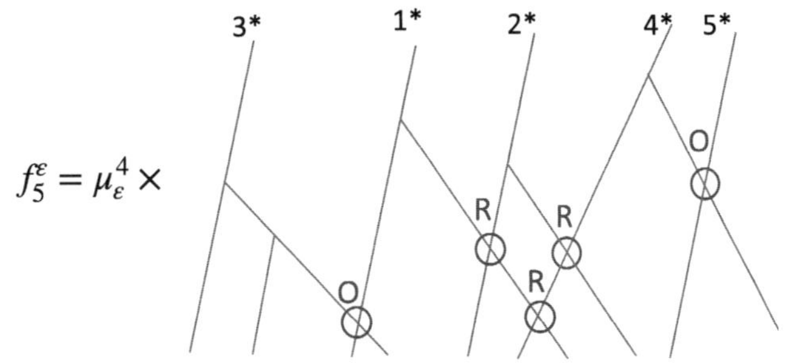

FIGURE 12

The cumulant of order k corresponds to pseudotrajectories issued from $z_{1^*}, \dots, z_{k^*}$ completely connected by external recollisions or overlaps.

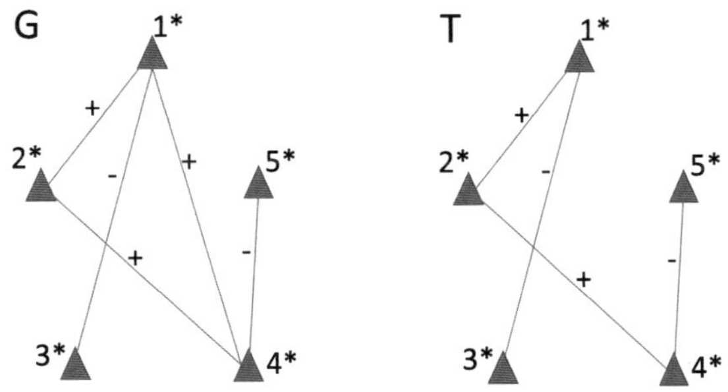

FIGURE 13

All recollisions and overlaps from the pseudotrajectories depicted in Figure 12 are encoded in the graph G. Only recollisions/overlaps which do not create a cycle (going backward in time) are kept in the tree T.

uration $z_{1^*}, \dots, z_{k^*}$ at time t. Therefore, neglecting the velocity dependence as in (2.9), this contribution to the cumulant f_k^ε has a support of size $O((t/\mu_\varepsilon)^{k-1})$ with respect to Lebesgue measure and from this we deduce the expected L^1 estimate

$$\| f_k^\varepsilon \|_{L^1} \leq \underbrace{\mu_\varepsilon^{k-1}}_{\text{scaling}} \times \underbrace{2^{k-1} k^{k-2}}_{\text{number of signed trees}} \times \underbrace{\left(\frac{Ct}{\mu_\varepsilon} \right)^{k-1}}_{\text{support size}} \leq k!(Ct)^{k-1}. \tag{3.4}$$

Furthermore, a geometric argument similar to the one developed in Lanford's proof (see Section 2.4) and already used in the study of the second order cumulant allows showing that f_k^ε converges to some limiting cumulant f_k and that only the pseudotrajectories having exactly $k-1$ recollisions or overlaps (and no cycle) contribute in the limit.

This geometric approach allows characterizing *all corrections to the chaos assumption, up to exponential order*, at least for times of the same order as T_L [13, 14]. Actually, a classical and rather straightforward computation (based on the series expansions of the exponential and logarithm) shows that cumulants are nothing else than the coefficients of the series expansion of the exponential moment

$$\mathcal{J}_t^\varepsilon(h) = \frac{1}{\mu_\varepsilon}\log \mathbb{E}_\varepsilon\big[\exp\big(\mu_\varepsilon\langle\pi_t^\varepsilon, h\rangle\big)\big] = \frac{1}{\mu_\varepsilon}\log \mathbb{E}_\varepsilon\Big[\exp\Big(\sum_i h\big(z_i^\varepsilon(t)\big)\Big)\Big] \tag{3.5}$$
$$= \sum_{k=1}^{\infty}\frac{1}{k!}\int f_k^\varepsilon(t, Z_k)\prod_{i=1}^{k}\big(e^{h(z_i)}-1\big)dZ_k.$$

The quantity $\mathcal{J}_t^\varepsilon(h)$ is referred to as the *cumulant generating function*. Estimate (3.4) provides the analyticity of $\mathcal{J}_t^\varepsilon(h)$ as a functional of e^h, and this uniformly with respect to ε (small enough). The limit $\mathcal{J}_t$ of $\mathcal{J}_t^\varepsilon$ can then be determined as a series in terms of the limiting cumulants f_k.

Instead of using the cumulant expansion, we present a heuristic approach to characterize the limit $\mathcal{J}_t$ as the solution of the *Hamilton–Jacobi equation* (3.8). At first reading, this formal derivation can be skipped and the reading can be resumed at Equation (3.8). We proceed as in Section 2.3 for the Boltzmann equation (2.1) and write the formal equation satisfied by $\mathcal{J}_t^\varepsilon(h)$ for fixed ε. Considering an evolution for a short time δ as in (2.7) and then taking a formal limit $\delta \to 0$, we get

$$\mathbb{E}_\varepsilon\Big[\exp\Big(\sum_i h(z_i^\varepsilon(t))\Big)\Big]\,\partial_t\mathcal{J}_t^\varepsilon(h)$$
$$= \mathbb{E}_\varepsilon\Big[\Big(\frac{1}{\mu_\varepsilon}\sum_j\frac{dx_j^\varepsilon}{dt}\cdot\nabla_x h\big(z_j^\varepsilon(t)\big)\Big)\exp\Big(\sum_i h\big(z_i^\varepsilon(t)\big)\Big)\Big]$$
$$+\int d\omega\,\mathbb{E}_\varepsilon\Bigg[\frac{1}{\mu_\varepsilon^2}\sum_{j_1\neq j_2}\delta_{x_{j_2}^\varepsilon(t)-x_{j_1}^\varepsilon(t)-\varepsilon\omega}\big(\big(v_{j_2}^\varepsilon(t)-v_{j_1}^\varepsilon(t)\big)\cdot\omega\big)_+$$
$$\times\Big(e^{h(z_{j_1}^\varepsilon(t^+))+h(z_{j_2}^\varepsilon(t^+))}-e^{h(z_{j_1}^\varepsilon(t^-))+h(z_{j_2}^\varepsilon(t^-))}\Big)\exp\Big(\sum_{i\neq j_1,j_2}h\big(z_i^\varepsilon(t)\big)\Big)\Bigg],$$

where ω becomes a random parameter after changing variables at the collision time as in (2.12). We used the Dirac notation as in (2.13) to stress that $x_{j_2}^\varepsilon(t) = x_{j_1}^\varepsilon(t)+\varepsilon\omega$ at the collision. Denoting by $\pi_{2,t}^\varepsilon$ the generalized empirical measure depending on 2 arguments (see (2.6)), we get

$$\mathbb{E}_\varepsilon\Big[\exp\big(\mu_\varepsilon\langle\pi_t^\varepsilon, h\rangle\big)\Big]\,\partial_t\mathcal{J}_t^\varepsilon(h) = \mathbb{E}_\varepsilon\big[\pi_t^\varepsilon\{v\cdot\nabla_x h\}\exp\big(\mu_\varepsilon\langle\pi_t^\varepsilon, h\rangle\big)\big]$$
$$+\frac{1}{2}\int d\omega\,\mathbb{E}_\varepsilon\big[\pi_{2,t}^\varepsilon\{\delta_{x_2-x_1-\varepsilon\omega}\big(e^{\Delta h(z_1,z_2,\omega)}-1\big)\}\exp\big(\mu_\varepsilon\langle\pi_t^\varepsilon, h\rangle\big)\big], \tag{3.6}$$

where $\Delta h(z_1, z_2, \omega) = h(x_1, v_1') + h(x_2, v_2') - h(z_1) - h(z_2)$ was already introduced in (2.11). To obtain a closed equation, it remains to find the counterparts of the correlation functions F_1^ε and F_2^ε which describe the distribution under the measure tilted by the exponential weight $\langle\pi_t^\varepsilon, h\rangle$.

Differentiating the exponential moment (3.5) at h in the direction φ, we recover the quantity $\langle \pi_t^\varepsilon, \varphi \rangle$

$$\left\langle \frac{\partial \mathcal{J}_t^\varepsilon}{\partial h}(h), \varphi \right\rangle = \lim_{\delta \to 0} \frac{1}{\delta} \big(\mathcal{J}_t^\varepsilon(h + \delta\varphi) - \mathcal{J}_t^\varepsilon(h) \big)$$
$$= \frac{1}{\mathbb{E}_\varepsilon[\exp(\mu_\varepsilon \langle \pi_t^\varepsilon, h \rangle)]} \mathbb{E}_\varepsilon\big[\langle \pi_t^\varepsilon, \varphi \rangle \exp\big(\mu_\varepsilon \langle \pi_t^\varepsilon, h \rangle\big)\big].$$

Thus the transport term has the form $\langle \frac{\partial \mathcal{J}_t^\varepsilon}{\partial h}(h), v \cdot \nabla_x h \rangle$. By taking a second derivative, the tilted distribution of the two-point correlations can be identified in terms of

$$\frac{1}{\mu_\varepsilon} \frac{\partial^2 \mathcal{J}_t^\varepsilon}{\partial h^2}(h) + \frac{\partial \mathcal{J}_t^\varepsilon}{\partial h}(h) \otimes \frac{\partial \mathcal{J}_t^\varepsilon}{\partial h}(h).$$

The collision term is singular, but formally the right-hand side of (3.6) can be rewritten as

$$\begin{aligned} \partial_t \mathcal{J}_t^\varepsilon(h) = {} & \frac{1}{2} \left\langle \frac{\partial \mathcal{J}_t^\varepsilon}{\partial h}(h) \otimes \frac{\partial \mathcal{J}_t^\varepsilon}{\partial h}(h), \int d\omega \big((v_2 - v_1) \cdot \omega \big)_+ \delta_{x_2 - x_1 - \varepsilon\omega} \big(e^{\Delta h(z_1, z_2, \omega)} - 1 \big) \right\rangle \\ & + \frac{1}{2\mu_\varepsilon} \left\langle \frac{\partial^2 \mathcal{J}_t^\varepsilon}{\partial h^2}(h), \int d\omega \big((v_2 - v_1) \cdot \omega \big)_+ \delta_{x_2 - x_1 - \varepsilon\omega} \big(e^{\Delta h(z_1, z_2, \omega)} - 1 \big) \right\rangle \\ & + \left\langle \frac{\partial \mathcal{J}_t^\varepsilon}{\partial h}(h), v \cdot \nabla_x h \right\rangle. \end{aligned} \tag{3.7}$$

We recognize here a kind of Hamilton–Jacobi equation, with a small "viscous" term (involving derivatives of order 2 with respect to h, but without a definite sign). Thus the limiting functional $\mathcal{J}_t$ has to satisfy the following Hamilton–Jacobi equation obtained by formally taking the limit $\mu_\varepsilon \to \infty$,

$$\begin{aligned} \partial_t \mathcal{J}_t(h) = {} & \frac{1}{2} \left\langle \frac{\partial \mathcal{J}_t}{\partial h}(h) \otimes \frac{\partial \mathcal{J}_t}{\partial h}(h), \int d\omega \big((v_2 - v_1) \cdot \omega \big)_+ \delta_{x_2 - x_1} \big(e^{\Delta h(z_1, z_2, \omega)} - 1 \big) \right\rangle \\ & + \left\langle \frac{\partial}{\partial h} \mathcal{J}_t(h), v \cdot \nabla_x h \right\rangle. \end{aligned} \tag{3.8}$$

The structure of this Hamilton–Jacobi equation is reminiscent of the Boltzmann equation (3.8), with a collision term and a transport term. However, it encodes a much more complete description of the hard sphere dynamics, including in particular the structure of the exponentially small correlations and of the large deviations (see Theorem 3.5).

As in (3.2), further information on the correlations in a time interval $[0, t]$ can be obtained by generalizing (3.5)

$$\mathcal{J}_{[0,t]}^\varepsilon(H) = \frac{1}{\mu_\varepsilon} \log \mathbb{E}_\varepsilon \left[\exp\left(\sum_i H\big(z_i^\varepsilon([0, t])\big) \right) \right], \tag{3.9}$$

for functions H depending on the trajectory of a particle in $[0, t]$. For example, a sampling at different times $\theta_1 < \theta_2 < \cdots < \theta_k \leq t$ by test functions $(h_\ell)_{\ell \leq k}$ is obtained by considering

$$H\big(z([0, t])\big) = \sum_{\ell=1}^k h_\ell\big(z(\theta_\ell)\big). \tag{3.10}$$

Remark 3.2. The procedure described here allows to obtain easily the limiting equation (3.8) without having to guess how to combine the different cumulant terms (which happens to be quite technical). However, the weak understanding we have on this equation does not allow to use it to justify the limit as $\mu_\varepsilon \to \infty$ (without going through the cumulant analysis of [14]).

Remark 3.3. In the absence of spatial inhomogeneities, one can discard the transport term and retrieve asymptotically the same cumulant generating function as for the Kac model, i.e., the dynamics in which collisions are given by a random jump process [4,41,47,60]. This indicates that in the limit $\mu_\varepsilon \to \infty$, both models are indistinguishable (up to exponentially small corrections). In other words, the Hamilton–Jacobi equation (3.8) conserves the stochastic reversibility, but not the deterministic reversibility: one cannot hope for any strong convergence result.

3.4. A complete statistical picture for short times

As mentioned in the previous paragraph, the cumulant generating function provides a complete statistical picture of the hard sphere dynamics. We now explain how it can be used to answer the main questions raised in Section 1.3 (on a short time $T^\star$, of the same order as Lanford's time T_L in Theorem 2.1).

As a first consequence of the uniform estimates on the cumulant generating function $\mathcal{J}^\varepsilon_{[0,t]}$, the convergence of the fluctuation field, defined by (1.4) and recalled below

$$\langle \zeta^\varepsilon_t, h\rangle = \sqrt{\mu_\varepsilon}\big(\langle \pi^\varepsilon_t, h\rangle - \mathbb{E}_\varepsilon(\langle \pi^\varepsilon_t, h\rangle)\big),$$

can be obtained.

At time 0, it is known that, under the grand-canonical measure introduced on page 757, the fluctuation field ζ^ε_0 converges in the Boltzmann–Grad limit to a Gaussian field ζ_0 with covariance

$$\mathbb{E}\big(\zeta_0(h)\zeta_0(g)\big) = \int dz f^0(z)h(z)g(z). \tag{3.11}$$

The following theorem controls the dynamical fluctuations.

Theorem 3.4 (Bodineau, Gallagher, Saint-Raymond, Simonella [15]). *Under the assumptions on the initial data stated on page 757, the fluctuation field ζ^ε_t of the hard sphere system converges, in the Boltzmann–Grad limit ($\mu_\varepsilon \to \infty$ with $\mu_\varepsilon \varepsilon^{d-1} = 1$), on a time interval $[0, T^\star]$ towards a process ζ_t, solution to the fluctuating Boltzmann equation*

$$\begin{cases} d\zeta_t = \underbrace{\mathcal{L}_t \zeta_t dt}_{\text{linearized Boltzmann operator}} + \underbrace{d\eta_t}_{\text{Gaussian noise}} \\ \mathcal{L}_t h = \underbrace{-v\cdot\nabla_x h}_{\text{transport}} + \underbrace{C(f_t, h) + C(h, f_t)}_{\text{linearized collision operator}} \end{cases} \tag{3.12}$$

where f_t denotes the solution at time t to the Boltzmann equation (2.1) with initial data f^0, and $d\eta_t$ is a centered Gaussian noise delta-correlated in t, x with covariance

$$\mathrm{Cov}_t(h_1, h_2) = \frac{1}{2}\int dz_1 dz_2 d\omega \big((v_2 - v_1)\cdot\omega\big)_+ \delta_{x_2 - x_1} f(t, z_1) f(t, z_2)\Delta h_1 \Delta h_2(z_1, z_2, \omega)$$

with $\Delta h(z_1, z_2, \omega) = h(z_1') + h(z_2') - h(z_1) - h(z_2)$ as in (2.11).

As hinted in Section 3.2, the limiting noise is a consequence of the asymptotically unstable structure of the microscopic dynamics (see Figure 2). The randomness of the initial configuration is transported deterministically by the dynamics and generates a white noise in space and time through a particular class of collisions. The velocity scattering mechanism is coded in the covariance of the noise.

If the system starts initially from an equilibrium measure, i.e., with particle positions spatially independent (up to the exclusion) and velocities identically distributed according to the Maxwell–Boltzmann equilibrium distribution

$$f^0(x, v) = M(v) = \frac{1}{(2\pi)^{d/2}} \exp\left(-\frac{|v|^2}{2}\right), \tag{3.13}$$

then $f_t = f^0$ so that the linearized operator is time independent and it will be denoted by $\mathcal{L}_{\mathrm{eq}}$. The limiting stochastic partial differential equation $d\zeta_t = \mathcal{L}_{\mathrm{eq}}\zeta_t + d\eta_t$ satisfies the fluctuation/dissipation relation: the dissipation from the linearized operator $\mathcal{L}_{\mathrm{eq}}$ is exactly compensated by the noise η_t. As the equilibrium measure is time invariant, it was expected on physical grounds that a stochastic correction should emerge in order to keep this invariance in time. In fact, the equation governing the covariance of the limiting process $\mathrm{Cov}(\zeta_t)$ away from equilibrium was obtained, and the full fluctuating equation for $(\zeta_t)_{t\geq 0}$ conjectured, in the pioneering works by Spohn **[65–67]**. In particular, it was already understood in **[65]** that out of equilibrium, a nontrivial contribution to $\mathrm{Cov}(\zeta_t)$ is provided by the second-order cumulant (3.1). Note that the predictions on the stochastic corrections from the Kac model **[49, 50, 59]** fully agree with the stochastic equation emerging from the deterministic hard sphere dynamics. Thus from a phenomenological point of view, it is equivalent to consider a stochastic model (including as well the positions as in **[59]**) or a deterministic evolution. We refer also to the work by Ernst and Cohen **[34]** for further discussion on the time correlations and the fluctuations.

Note that equilibrium fluctuations for a microscopic evolution with spatial coordinates and stochastic collisions have been derived in **[59]** for arbitrary long times. We will see in Theorem 4.2 that the convergence time of the previous theorem can be greatly improved at equilibrium.

Out of equilibrium, although the solution f to the Boltzmann equation (describing the averaged dynamics) is very smooth on $[0, T^\star]$, the fluctuating Boltzmann equation is quite singular: the linearized operator $\mathcal{L}_t$ is nonautonomous, non-self-adjoint, and the corresponding semigroup is not a contraction. Thus we consider a very weak notion of solution of (3.12), requiring only that

- the process ζ_t is Gaussian;
- its covariance defined, for test functions h_1, h_2 and times θ_1, θ_2, as

$$\mathcal{C}(\theta_1, h_1, \theta_2, h_2) = \lim_{\varepsilon\to 0} \mathbb{E}_\varepsilon\left(\langle\zeta^\varepsilon_{\theta_1}, h_1\rangle\langle\zeta^\varepsilon_{\theta_2}, h_2\rangle\right) \tag{3.14}$$

  satisfies a set of equations governed by the linearized Boltzmann equation.

The convergence of the process $(\zeta^\varepsilon_t)_{t\leq T^\star}$ can be derived in 3 steps:

- *The convergence of the time marginals to a Gaussian process*
The characteristic function of the process tested at times $\theta_1 < \cdots < \theta_k \leq T^\star$ by functions $(h_\ell)_{\ell \leq k}$ is encoded by the exponential moment (3.9) by choosing $H(z([0,T^\star])) = \frac{i}{\sqrt{\mu_\varepsilon}} \sum_{\ell=1}^k h_\ell(z(\theta_\ell))$ as in (3.10)

$$\log \mathbb{E}_\varepsilon \left[\exp\left(i \sum_{\ell=1}^k \left(\langle \zeta^\varepsilon_{\theta_\ell}, h_\ell \rangle + \sqrt{\mu_\varepsilon} \mathbb{E}_\varepsilon \left(\langle \pi^\varepsilon_{\theta_\ell}, h_\ell \rangle \right) \right) \right) \right] = \mu_\varepsilon \mathcal{J}^\varepsilon_{[0,T^\star]}(H). \quad (3.15)$$

The cumulant expansion (3.5) combined with sharp controls on the cumulants ensure that $\mathcal{J}^\varepsilon_{[0,T^\star]}(H)$ is an analytic function of H in a neighborhood of 0 so that complex values can also be handled. Furthermore, in the scaling considered for the fluctuations, H is of order $\frac{1}{\sqrt{\mu_\varepsilon}}$. Thus in the cumulant expansion (3.5), the term of order n scales as

$$f^\varepsilon_n\left((e^H - 1)^{\otimes n} \right) \simeq \frac{1}{\mu_\varepsilon^{n/2}},$$

so that the asymptotics of the characteristic function (3.15) is only determined by the cumulants of order less than 2. This implies that the Wick rule holds and therefore the limiting variables are Gaussian.

- *The characterization of the limit covariance*
The evolution equation of the covariance $\mathcal{C}(\theta_1, h_1, \theta_2, h_2)$ can be recovered from the equations satisfied by the first two cumulants. As already pointed out in **[65]**, we stress that the behavior of the covariance $\mathcal{C}(\theta_1, h_1, \theta_2, h_2)$ is determined by means of a careful analysis of the second cumulant $f^\varepsilon_2[(h_\ell, \theta_\ell)_{\ell \leq 2}]$ introduced in (3.3). Out of equilibrium, the cumulant of order 2 takes into account the contribution of one external recollision or of one overlap (as explained in Section 3.2). Even though the contribution of the recollisions vanishes when deriving the Boltzmann equation (recall the chaos assumption (2.14)), it plays an important role in the stochastic corrections.

- *The tightness of the sequence $(\zeta^\varepsilon_t)_{\varepsilon > 0}$*
This is the most technical part of the proof as it requires to control uniform estimates in time for a wide class of test functions h,

$$\mathbb{E}_\varepsilon \Big[\sup_{|s-s'| \leq \delta} \left| \langle \zeta^\varepsilon_s, h \rangle - \langle \zeta^\varepsilon_{s'}, h \rangle \right| \Big].$$

We will not discuss further this point and refer to **[14]** for details.

Note that Theorem 3.4, which is a kind of central limit theorem, does not use the fine structure of cumulants: a sufficient decay of the correlations is enough to control the typical fluctuations (which are of size $O(1/\sqrt{\mu_\varepsilon})$).

The strength of the cumulant generating function appears at the level of large deviations, i.e., for very unlikely trajectories which are at a "distance" $O(1)$ from the averaged dynamics. The counterpart of the large deviation statement (1.5) for independent variables can be rephrased, in a loose way, as follows: observing an empirical particle distribution close to the density $\varphi(t,x,v)$ during the time interval $[0,T^\star]$ decays exponentially fast with a rate quantified by the large deviation functional $\mathcal{F}$,

$$\mathbb{P}_\varepsilon\big(\pi_t^\varepsilon \simeq \varphi_t,\ \forall t \le T^\star\big) \sim \exp\big(-\mu_\varepsilon \mathcal{F}(\varphi)\big).$$

Notice that at time 0, under the grand-canonical measure introduced on page 757, it is known that the large deviations around a density φ^0 can be informally stated as follows:

$$\mathbb{P}_\varepsilon\big(\pi_0^\varepsilon \simeq \varphi^0\big) \sim \exp\big(-\mu_\varepsilon H(\varphi^0|f^0)\big),$$

with a static large deviation functional given by the relative entropy

$$H(\varphi^0|f^0) = \int \Big(\varphi^0 \log\frac{\varphi^0}{f^0} - (\varphi^0 - f^0)\Big)dz.$$

More precisely, the distance between π^ε and φ is measured with respect to a weak topology on the Skorokhod space of measure valued functions. This topology is used in the theorem below.

Theorem 3.5 (Bodineau, Gallagher, Saint-Raymond, Simonella [14]). *Under the assumptions on the initial data stated on page 757, there is a time $T^\star > 0$ such that the empirical measure $(\pi_t^\varepsilon)_{t\le T^\star}$ satisfies, in the Boltzmann–Grad limit $\mu_\varepsilon \to \infty$ ($\mu_\varepsilon\varepsilon^{d-1} = 1$), the following large deviation estimates:*

$$\begin{cases} \limsup\limits_{\mu_\varepsilon\to\infty} \dfrac{1}{\mu_\varepsilon}\log \mathbb{P}_\varepsilon\big[\pi^\varepsilon \in K \textit{ compact}\big] \le -\inf\limits_{\varphi\in K}\mathcal{F}(\varphi), \\ \liminf\limits_{\mu_\varepsilon\to\infty} \dfrac{1}{\mu_\varepsilon}\log \mathbb{P}_\varepsilon\big[\pi^\varepsilon \in O \textit{ open}\big] \ge -\inf\limits_{\varphi\in O\cap\mathcal{R}}\mathcal{F}(\varphi), \end{cases}$$

for some (nontrivial) restricted set $\mathcal{R}$.

The large deviation functional $\mathcal{F}$ is defined by convex duality from the cumulant generating function $\mathcal{J}_{[0,T^\star]}$ (obtained as the limit of (3.9)). It coincides on the restricted set $\mathcal{R}$ with

$$\begin{cases} \tilde{\mathcal{F}}(\varphi) = \underbrace{H(\varphi^0|f^0)}_{\text{relative entropy of the initial data}} + \underbrace{\sup_p \int_0^{T^\star} \big(\langle p, (\partial_t + v\cdot\nabla_x)\varphi\rangle - \mathcal{H}(\varphi,p)\big)}_{\text{Legendre transform of the Hamiltonian}}, \\ \mathcal{H}(\varphi,p) = \dfrac{1}{2}\displaystyle\int dz_1 dz_2\, d\omega\big((v_2-v_1)\cdot\omega\big)_+ \delta_{x_2-x_1}\varphi(z_1)\varphi(z_2)\big(e^{\Delta p(z_1,z_2,\omega)} - 1\big), \end{cases} \tag{3.16}$$

with $\Delta p(z_1,z_2,\omega) = p(z_1') + p(z_2') - p(z_1) - p(z_2)$ as in (2.11).

All the functionals appearing in the above statement are quite singular (notice that the Hamiltonian is defined by an integral over a manifold of codimension d with a weight growing for large velocities) and our method is restricted to considering very smooth and sufficiently decaying test functions. These restrictions on the functional spaces are the reason why we are not able to obtain a more precise large deviation principle, or to identify clearly the large deviation functional. We refer to [14] for the proof which follows a quite standard path, once the limiting cumulant generating function $\mathcal{I}_{[0,T^\star]}$ has been constructed. The identification between $\mathcal{F}$ and $\tilde{\mathcal{F}}$ relies on the limiting Hamilton–Jacobi equation (3.8).

Remark 3.6. Note that the large deviation functional $\tilde{\mathcal{F}}$ defined by (3.16) was conjectured in [62] and [19]. As already mentioned, it actually corresponds to the large deviation functional for stochastic microscopic processes, such as the Kac model (in the absence of transport) [41,47], or intermediate models (with transport and stochastic collisions) introduced by Rezakhanlou [68].

4. BEYOND LANFORD'S TIME

Up to a short time, Theorems 3.4 and 3.5 provide a good statistical description of the hard sphere dynamics in the Boltzmann–Grad limit ($\mu_\varepsilon \to \infty$ with $\mu_\varepsilon \varepsilon^{d-1} = 1$). The stochastic corrections to the Boltzmann equation emerge from the complex interplay between the random initial data and the asymptotic instability of the dynamics.

However, these results are still far from being satisfactory as the time restriction is not expected from physics: it does not allow understanding the relaxation toward equilibrium (and the corresponding entropy cascades between cumulants), or deriving fluid limits. This question remains quite open, and the goal of this last section is to discuss theoretical obstructions and methodological difficulties, as well as some recent progress close to equilibrium.

4.1. Main difficulties

A natural way to address this problem is trying to understand what kind of convergence one can hope for beyond Lanford's time T_L. Recall that Lanford's theorem describes the approximation of a reversible system by an irreversible system, where a macroscopic part of the information is missing. This excludes any kind of "strong" convergence in terms of relative entropy. This implies in particular that one will hardly use the fine knowledge one might have on the solution to the Boltzmann equation to obtain a robust notion of stability which would be as well compatible with the microscopic system.

Remark 4.1. In the framework of fluid limits, modulated energy or modulated entropy methods are among the most powerful to prove convergence theorems [39,64,71] since they require very few properties on the original system, typically

- an energy/entropy inequality satisfied by weak solutions;
- the consistency of the approximation (meaning that the limiting equations are those inferred from the formal asymptotics);
- some bootstrap estimates controlling (nonlinear) fluxes in terms of the modulated energy/entropy.

An alternative would be to establish some weak convergence $F_1^\varepsilon \rightharpoonup f$, which paradoxically requires better compactness estimates on the sequence $(F_1^\varepsilon)_\varepsilon$. In this framework, the best one can do in general is to retrieve the structure of the limiting equation and its good (weak) stability properties from the solutions F_1^ε for fixed ε, and this uniformly in ε. The problem here, as mentioned in Section 2.2, is that the Boltzmann equation does not have such a weak stability. Two ingredients are necessary to construct solutions satisfying only physical bounds (mass, energy, and entropy estimates):

- a renormalization procedure to tame the possible singularity (concentration in x) in the loss collision term $f(t,x,v) \times \int f(t,x,v_1) b(v-v_1,\omega) d\omega dv_1$;
- a bound on the entropy dissipation to control the gain term by the loss term.

These ingredients have been used in [61] to recover the Boltzmann equation from a microscopic dynamics with stochastic collisions, but they do not seem to have a clear counterpart for a deterministic microscopic evolution.

The Hamilton–Jacobi equation (3.7) retains much more information on the system, thus the convergence of $\mathcal{J}_t^\varepsilon$ to $\mathcal{J}_t$, in a sense to be understood, could provide a more stable framework to study the kinetic limit for large times. This would then imply the convergence to the Boltzmann equation.

4.2. Close to equilibrium

An easier setting to control the long-time evolution is to consider a perturbation of an equilibrium measure. Here the stationarity of the equilibrium becomes a key tool in order to provide uniform estimates in time and to control the pathological behaviors previously mentioned. In a series of recent works [15,16], we took advantage of the equilibrium structure to extend Theorem 3.4 to arbitrarily long kinetic times, and even slowly diffusive times.

Theorem 4.2 (Bodineau, Gallagher, Saint-Raymond, Simonella [15,16]). *Consider a system of hard spheres initially at equilibrium, i.e., with a spatially uniform distribution and with a Maxwell–Boltzmann distribution M in velocities as in* (3.13) *(Gibbs grand-canonical ensemble, $f^0 = M$ in* (2.3)*).*

Then, in the Boltzmann–Grad limit $\mu_\varepsilon \to \infty$ ($\mu_\varepsilon \varepsilon^{d-1} = 1$), the fluctuation field $(\zeta_t^\varepsilon)_{t\geq 0}$ of the hard sphere system converges on any time interval $[0, T_\varepsilon]$, with $T_\varepsilon = O(\log\log\log \mu_\varepsilon)$, towards the process $(\zeta_t)_{t\geq 0}$, solution to the fluctuating Boltzmann equation

$$\begin{cases} d\zeta_t = \underbrace{\mathcal{L}_{\text{eq}} \zeta_t \, dt}_{\text{linearized Boltzmann operator}} + \underbrace{d\eta_t}_{\text{Gaussian noise}} \\ \mathcal{L}_{\text{eq}} h = \underbrace{-v \cdot \nabla_x h}_{\text{transport}} + \underbrace{C(h,M) + C(M,h)}_{\text{linearized collision operator}} \end{cases} \tag{4.1}$$

where the linearized operator $\mathcal{L}_{\text{eq}}$ is time independent and η is a Gaussian noise delta-correlated in t, x with a time independent covariance

$$\operatorname{Cov}(h_1,h_2) = \frac{1}{2}\int dz_1 dz_2 d\omega \big((v_2-v_1)\cdot\omega\big)_+ \delta_{x_2-x_1} M(v_1) M(v_2) \Delta h_1 \Delta h_2(z_1,z_2,\omega),$$

with $\Delta h(z_1,z_2,\omega) = h(z_1') + h(z_2') - h(z_1) - h(z_2)$ as in (2.11).

Since the approximation holds true for very long times compared to the mean free time (diverging to infinity as $\log\log\log\mu_\varepsilon$), it makes sense to look at fluid limits, i.e., at regimes when the collision process is much faster than the transport (density is still low but makes the collisions a bit more likely) $\mu_\varepsilon\varepsilon^{d-1} = \alpha^{-1}$ with $\alpha \gg \varepsilon$, $\alpha \to 0$. Starting from the scaled linearized Boltzmann equation

$$\partial_t h + v \cdot \nabla_x h = \frac{1}{\alpha}\big(C(h, M) + C(M, h)\big),$$

it is well known [3] that, in the limit $\alpha \to 0$, the gas will be close to a local thermodynamic equilibrium, with density, bulk velocity, and temperature satisfying the acoustic equations. Zooming out on longer times $O(1/\alpha)$, these acoustic waves become fast oscillating and thus converge weakly to 0, but the incompressible component has a diffusive behavior, satisfying the Stokes–Fourier equations. This by now classical asymptotic analysis can be actually combined with Theorem 4.2 to derive directly the Stokes–Fourier equations from the dynamics of hard spheres as in [11]. In a work in progress, we also take into account the noise, and get the corresponding fluctuating hydrodynamics (satisfying the fluctuation-dissipation principle).

4.3. Some elements of the proof of Theorem 4.2

As in the previous sections, we will not enter into the technicalities of the proof, which is actually quite involved. We will just focus here on some key arguments, providing a better understanding of large time asymptotics. We work directly on moments of the fluctuation field, defined for any collection of times $\theta_1 < \cdots < \theta_p$ by

$$\mathbb{E}_\varepsilon\big[\langle\zeta^\varepsilon_{\theta_1}, h_1\rangle \cdots \langle\zeta^\varepsilon_{\theta_p}, h_p\rangle\big], \tag{4.2}$$

and we are going to prove their convergence to the moments of the field in the stochastic equation $d\zeta_t = \mathcal{L}_{\mathrm{eq}}\zeta_t dt + d\eta_t$. Combined with the tightness results from [14], this fully characterizes the convergence of the microscopic fluctuation field.

Let us start with $p = 2$ and compute the covariance $\mathbb{E}_\varepsilon[\zeta^\varepsilon_{\theta_1}(h_1)\zeta^\varepsilon_{\theta_2}(h_2)]$. The idea is to pull back the observable h_2 from time θ_2 to θ_1 in order to reduce the estimates at a single time θ_1. A similar strategy was presented in Sections 2.4 and 3.2 to transport the correlation up to time 0 for which the distribution was known. In particular, we have seen that the correlation functions at a time θ_2 can be represented by backward pseudotrajectories involving collision trees with a number m of additional particles encoding the dynamical history during the time interval $[\theta_1, \theta_2]$. The time restriction T_L for the convergence to Boltzmann equation in Theorem 2.1 was due to the lack of control on the growth of the tree sizes m at large times. Indeed, dynamical correlations may develop and form giant components of correlated particles for very pathological trajectories. In order to reach larger time scales, one has to show that the contribution of these bad trajectories with large m remains negligible. For this we perform a *time sampling*. The idea is to build the pseudotrajectories iteratively from θ_2 to θ_1 on time steps of length $\tau \ll 1$ and to neglect the collision trees with a fast (superexponential) growth during a time τ (see Figure 14). The large collision trees are therefore discarded before they reach the time θ_1, i.e., before their sizes become uncontrollable. This can be achieved by using the time invariance property of the equilibrium measure which provides

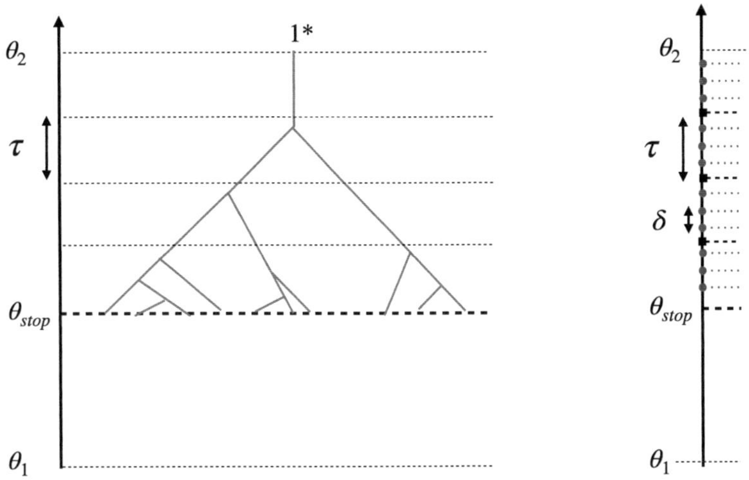

FIGURE 14
Pseudotrajectories are build iteratively on short time intervals of length τ starting from θ_2. The procedure stops before reaching time θ_1 if superexponential branchings occur in a time interval of length τ. The corresponding pseudotrajectories stop at time θ_{stop} and are then discarded. A double sampling at scales $\delta \ll \tau \ll 1$, depicted on the right figure, is implemented to control the recollisions.

a priori controls on the statistics. This kind of sampling was introduced for the first time in the context of the Boltzmann–Grad limit in [10, 11], but it is also an important ingredient in the weak coupling limit for quantum systems leading to quantum diffusion [32, 33].

Another key ingredient, to derive the convergence to the Boltzmann equation, is the procedure to neglect the "bad" trajectories involving recollisions (see Section 2.4). Controlling the growth of the collision trees is also essential to discard recollisions. The idea is to introduce a double sampling in time (with time scales $\delta \ll \tau \ll 1$, see Figure 14) which takes care simultaneously of the recollisions and of the collision tree growth. The backward iteration is stopped and the corresponding pseudotrajectories are discarded as soon as one of the following conditions is violated:

- there is at least one recollision on the last very small interval of size $\delta = O(\varepsilon^{1-\frac{1}{2d}})$;
- on the last small interval of size $\tau = (\log\log \mu_\varepsilon)^{-1/2}$, the number of particles has been multiplied at least by 2.

Note that both conditions are entangled. On the one hand, the bigger the size of the system, the easier for recollisions to occur. On the other hand, it is rather difficult to control the growth of the system if there are recollisions.

Assuming that the pseudotrajectories can be controlled by the previous time sampling, let us now explain the *weak convergence method* for computing the covariance. The two-time correlation $\mathbb{E}_\varepsilon[\langle \zeta^\varepsilon_{\theta_1}, h_1\rangle\langle \zeta^\varepsilon_{\theta_2}, h_2\rangle]$ can be rephrased as the expectation of two fluc-

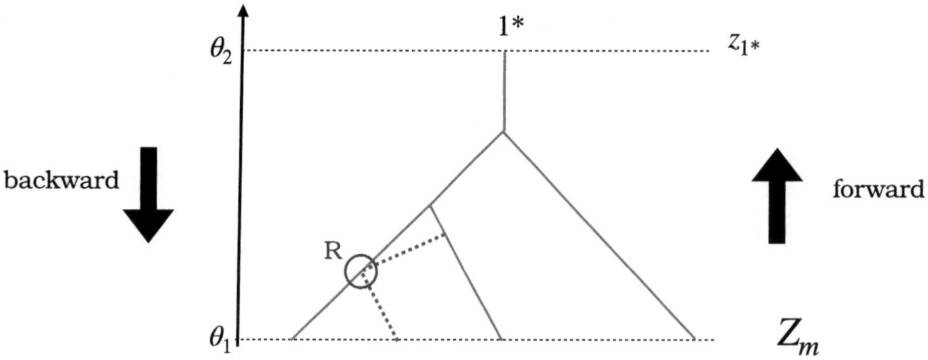

FIGURE 15
Starting from z_{1*} at time θ_2, the blue pseudotrajectory is built backward and leads to a configuration Z_m at time θ_1 (with $m = 3$ on the picture). The dual procedure goes forward, starting from Z_m in order to reconstruct z_{1*} as a function of Z_m at time θ_2. Following the forward flow, a tree is built by removing one of the particles at each encounter between two particles. Notice that one has to choose which particle will be removed and if a scattering occurs. Thus there are potentially several ways to build forward trajectories, but their combinatorics is well under control. This is no longer the case when recollisions can occur. Indeed, this adds the possibility that when two particles encounter in the forward flow, none of them disappears (see the dotted path on the figure) so when the number of recollisions is not bounded the combinatorics diverges.

tuation fields at the same time θ_1

$$\mathbb{E}_\varepsilon\big[\langle\zeta^\varepsilon_{\theta_1}, h_1\rangle\langle\zeta^\varepsilon_{\theta_2}, h_2\rangle\big] \text{ ``=''} \sum_m \mathbb{E}_\varepsilon\big[\langle\zeta^\varepsilon_{\theta_1}, h_1\rangle\langle\zeta^\varepsilon_{m,\theta_1}, \phi_{\theta_2-\theta_1}\rangle\big], \tag{4.3}$$

where the new test function $\phi_{\theta_2-\theta_1}(Z_m)$ is obtained from h_2 by considering all possible forward flows starting from Z_m at time θ_1 and having only one particle left at time θ_2 (see Figure 15). In this sense, (4.3) is dual to the backward representation of the correlation functions (2.16). The price to pay, to reduce the expectation at a single time, is that the new test function $\phi_{\theta_2-\theta_1}$ depends on m particles (a parameter related to the size of the collision trees in the time interval $[\theta_1, \theta_2]$) so that the fluctuation field $\zeta^\varepsilon_{m,\theta_1}$ has the form

$$\langle\zeta^\varepsilon_{m,\theta_1}, \phi_{\theta_2-\theta_1}\rangle = \sqrt{\mu_\varepsilon}\left(\frac{1}{\mu_\varepsilon^m}\sum_{(i_1,\dots,i_m)} \phi_{\theta_2-\theta_1}\big(z^\varepsilon_{i_1}(\theta_1), \dots, z^\varepsilon_{i_m}(\theta_1)\big) - \mathbb{E}_\varepsilon(\phi_{\theta_2-\theta_1})\right),$$

which is related to the generalized empirical measure defined in (2.6), with the abbreviation

$$\mathbb{E}_\varepsilon(\phi_{\theta_2-\theta_1}) = \mathbb{E}_\varepsilon\big(\langle\pi^\varepsilon_{m,\theta_1}, \phi_{\theta_2-\theta_1}\rangle\big).$$

In the following, we will abusively forget the subscript m.

The difficulty to make sense of the pullback in (4.3) is that the forward flow is not a priori well defined. Indeed, different backward pseudotrajectories may end up at time θ_1 with the same particle configuration Z_m. Thus starting from Z_m, there are many possibilities to build the forward flow from θ_1 to θ_2: when two particles touch each other, we need to prescribe whether one of them will be deleted (corresponding to a creation in the backward flow) or not (corresponding to a recollision), and in the case of deletion whether there is scattering of the remaining particle (see Figure 15). The combinatorics of these choices is diverging very fast if the number of recollisions is not under control. The very short time sampling δ is introduced so that the number of recollisions during a time δ is controlled with high probability under the equilibrium measure.

Then the pullback relation (4.3) is obtained by successive iterations of the sampling time δ. After the first elementary time step in the time interval $[\theta_2 - \delta, \theta_2]$, the pathological events are discarded and then the elementary pullback can be iterated. This means that, at each time $\theta_2 - r\delta$, remainder terms due to recollisions are neglected, and that, at each time $\theta_2 - k\tau$, remainder terms due to superexponential growth can also be discarded. Let $\theta_{\text{stop}} \in [\theta_1, \theta_2]$ be the first time at which a pseudotrajectory becomes pathological (see Figure 14). The corresponding terms obtained by forward transport from the time θ_{stop} are generically denoted by $\phi^{\text{bad}}_{\theta_{\text{stop}}}$ and are proved to be small by using the time invariance of the equilibirum measure. Indeed, the *time decoupling* follows from a Cauchy–Schwarz estimate

$$\left|\mathbb{E}_\varepsilon\left[\langle\zeta^\varepsilon_{\theta_1}, h_1\rangle\langle\zeta^\varepsilon_{\theta_{\text{stop}}}, \phi^{\text{bad}}_{\theta_{\text{stop}}}\rangle\right]\right| \le \mathbb{E}_\varepsilon\left[\langle\zeta^\varepsilon_{\theta_1}, h_1\rangle^2\right]^{1/2}\mathbb{E}_\varepsilon\left[\langle\zeta^\varepsilon_{\theta_{\text{stop}}}, \phi^{\text{bad}}_{\theta_{\text{stop}}}\rangle^2\right]^{1/2}, \tag{4.4}$$

and from the strong geometric constraints on the corresponding pathological pseudotrajectories which can be estimated under the equilibrium measure on can deduce that

$$\mathbb{E}_\varepsilon\left[\langle\zeta^\varepsilon_{\theta_{\text{stop}}}, \phi^{\text{bad}}_{\theta_{\text{stop}}}\rangle^2\right] \to 0 \quad \text{as } \mu_\varepsilon \to \infty.$$

The last important step to prove that the limiting process is Gaussian boils down to showing that, asymptotically when $\mu_\varepsilon \to \infty$, the moments, defined in (4.2), are determined by the covariances according to *Wick's rule*

$$\lim_{\mu_\varepsilon \to \infty}\left|\mathbb{E}_\varepsilon\left[\langle\zeta^\varepsilon_{\theta_1}, h_1\rangle\cdots\langle\zeta^\varepsilon_{\theta_p}, h_p\rangle\right] - \sum_{\eta\in\mathfrak{S}^{\text{pairs}}_p}\prod_{\{i,j\}\in\eta}\mathbb{E}_\varepsilon\left[\langle\zeta^\varepsilon_{\theta_i}, h_i\rangle\langle\zeta^\varepsilon_{\theta_j}, h_j\rangle\right]\right| = 0, \tag{4.5}$$

where $\mathfrak{S}^{\text{pairs}}_p$ is the set of partitions of $\{1, \dots, p\}$ made only of pairs. Notice that if p is odd then $\mathfrak{S}^{\text{pairs}}_p$ is empty and the product of the moments is asymptotically 0.

To understand this pairing mechanism, let us start with a simpler example for which explicit computations can be achieved. Consider the moments of the fluctuation field at time 0, under the equilibrium measure with independently distributed particles. This reduces to the case $\varepsilon = 0$ and $\theta_1 = \cdots = \theta_p = 0$. Assuming furthermore that the test functions are of mean $\mathbb{E}_0(h_i) = 0$ (we abusively write here $\mathbb{E}_0$ for this *iid* case, not to be confused with $\mathbb{E}_\varepsilon$ for $\varepsilon = 0$), we get

$$\mathbb{E}_0\left[\prod_{\ell=1}^p\langle\zeta^\varepsilon_0, h_\ell\rangle\right] = \frac{1}{\mu_\varepsilon^{p/2}}\mathbb{E}_0\left[\prod_{\ell=1}^p\Big(\sum_{i_\ell} h_\ell(z_{i_\ell})\Big)\right] = \frac{1}{\mu_\varepsilon^{p/2}}\mathbb{E}_0\left[\sum_{i_1,\dots,i_p}\prod_{\ell=1}^p h_\ell(z_{i_\ell})\right], \tag{4.6}$$

where the sum is over all the possible choices (with repetition) among N particles (with $N \simeq \mu_\varepsilon$ under the grand-canonical measure). As the mean of the test functions is assumed to be 0, each particle has to be chosen at least twice, otherwise by the independence of the variables the expectation is equal to 0. Thus in the sum over $i_1, \dots, i_p$, the number k of different particles is such that $k \le p/2$. Choosing k different particles gives a combinatorial factor μ_ε^k so that only the pairings with $k = p/2$ and p even contribute to the limiting moment. In this way, one recovers the Wick decomposition (4.5) in terms of pairings. Note that for $\varepsilon > 0$, a similar result holds (at time zero) in the Boltzmann–Grad limit, but a cluster expansion of the equilibrium measure is necessary to control the (weak) correlations of the Gibbs measure.

For time-dependent fluctuation fields, the pairing cannot be achieved in one step as in the previous example. One has instead to proceed iteratively. Let us revisit the computation above to explain the idea first in this simple setting. We start by focusing on the product of two fields and decompose it as follows:

$$\langle \zeta_0^\varepsilon, h_p \rangle \langle \zeta_0^\varepsilon, h_{p-1} \rangle = \underbrace{\frac{1}{\mu_\varepsilon} \sum_i h_p(z_i) h_{p-1}(z_i)}_{=\Psi} + \underbrace{\frac{1}{\mu_\varepsilon} \sum_{i \neq j} h_p(z_i) h_{p-1}(z_j)}_{=\langle \zeta_0^\varepsilon, h_p \rangle \otimes \langle \zeta_0^\varepsilon, h_{p-1} \rangle} . \tag{4.7}$$

The pairing between $\langle \zeta_0^\varepsilon, h_p \rangle$ and $\langle \zeta_0^\varepsilon, h_{p-1} \rangle$ is coded by the function Ψ which is called a *contracted product* as the variables are repeated. As the variables are independent, the covariance between h_p and h_{p-1} is given by

$$\mathbb{E}_0\big[\langle \zeta_0^\varepsilon, h_p \rangle \langle \zeta_0^\varepsilon, h_{p-1} \rangle\big] = \mathbb{E}_0[\Psi]. \tag{4.8}$$

From the central limit theorem, Ψ can be interpreted as a small fluctuation around the covariance

$$\Psi = \mathbb{E}_0[\Psi] + \frac{1}{\sqrt{\mu_\varepsilon}} \widehat{\Psi} \quad \text{with } \widehat{\Psi} = \frac{1}{\sqrt{\mu_\varepsilon}} \Big(\sum_i h_p(z_i) h_{p-1}(z_i) - \mu_\varepsilon \mathbb{E}_0[h_p h_{p-1}] \Big), \tag{4.9}$$

where $\widehat{\Psi}$ behaves as a random variable with finite covariance (uniformly in ε). The second term in (4.7) will be called a $\otimes$-product and denoted by $\langle \zeta_0^\varepsilon, h_p \rangle \otimes \langle \zeta_0^\varepsilon, h_{p-1} \rangle$. It behaves qualitatively as a fluctuation field as the variables are not repeated.

Returning to (4.6), to extract the pairing between $\langle \zeta_0^\varepsilon, h_p \rangle$ and $\langle \zeta_0^\varepsilon, h_{p-1} \rangle$, we write

$$\mathbb{E}_0\Big[\prod_{\ell=1}^{p} \langle \zeta_0^\varepsilon, h_\ell \rangle\Big] = \underbrace{\mathbb{E}_0\Big[\Big(\prod_{\ell=1}^{p-2} \langle \zeta_0^\varepsilon, h_\ell \rangle\Big) \Psi\Big]}_{\text{pairing of } h_p, h_{p-1}} + \underbrace{\mathbb{E}_0\Big[\Big(\prod_{\ell=1}^{p-2} \langle \zeta_0^\varepsilon, h_\ell \rangle\Big) \big(\langle \zeta_0^\varepsilon, h_p \rangle \otimes \langle \zeta_0^\varepsilon, h_{p-1} \rangle\big)\Big]}_{\text{product of } p-1 \text{ fields}} . \tag{4.10}$$

The second term can be seen as a product of $p-1$ fields which will be treated recursively at the next step. The pairing between $\langle \zeta_0^\varepsilon, h_p \rangle$ and $\langle \zeta_0^\varepsilon, h_{p-1} \rangle$ can be extracted from the first term as follows. Using the decomposition (4.9), we get

$$\begin{aligned}
\mathbb{E}_0\Big[\Big(\prod_{\ell=1}^{p-2} \langle \zeta_0^\varepsilon, h_\ell \rangle\Big) \Psi\Big] &= \mathbb{E}_0\Big[\prod_{\ell=1}^{p-2} \langle \zeta_0^\varepsilon, h_\ell \rangle\Big] \mathbb{E}_0[\Psi] + \frac{1}{\sqrt{\mu_\varepsilon}} \mathbb{E}_0\Big[\Big(\prod_{\ell=1}^{p-2} \langle \zeta_0^\varepsilon, h_\ell \rangle\Big) \widehat{\Psi}\Big] \\
&= \mathbb{E}_0\Big[\prod_{\ell=1}^{p-2} \langle \zeta_0^\varepsilon, h_\ell \rangle\Big] \mathbb{E}_0\big[\langle \zeta_0^\varepsilon, h_p \rangle \langle \zeta_0^\varepsilon, h_{p-1} \rangle\big] + O\Big(\frac{1}{\sqrt{\mu_\varepsilon}}\Big),
\end{aligned}$$

where the smallness of the last term follows from Hölder's inequality

$$\left| \mathbb{E}_0\Big[\Big(\prod_{\ell=1}^{p-2} \langle \zeta_0^\varepsilon, h_\ell \rangle\Big) \widehat{\Psi}\Big] \right| \leq \mathbb{E}_0\big[\widehat{\Psi}^2\big]^{\frac{1}{2}} \prod_{\ell=1}^{p-2} \mathbb{E}_0\big[\langle \zeta_0^\varepsilon, h_\ell \rangle^{2(p-2)}\big]^{\frac{1}{2(p-2)}}, \tag{4.11}$$

provided bounds on the moments of single fields can be obtained. For independent variables, this procedure is far from optimal; however, it will be extremely useful to decouple fields at different times. In this way, the pairing between $\langle \zeta_0^\varepsilon, h_p \rangle$ and $\langle \zeta_0^\varepsilon, h_{p-1} \rangle$ can be extracted without investigating the correlations between these two fields and the $p-2$ other fields. Note that a time decoupling inequality similar to (4.11) was used in the computation of the covariance (4.4) to neglect bad pseudotrajectories. Finally, it remains to iterate this procedure with $\mathbb{E}_0[\prod_{\ell=1}^{p-2} \langle \zeta_0^\varepsilon, h_\ell \rangle]$ and the second term in (4.10) which involves a product of at most $p-1$ fluctuation fields.

We turn now to the time dependent case (4.5) and proceed backward in time to achieve the pairing step by step. First, the fluctuation at time θ_p is pulled back at time θ_{p-1} as a sum of (more complicated) fluctuations by the same duality method as for the covariance (4.3). Using analogous notation as in (4.3), the test function h_p is transformed into a function $\phi^{(p)}_{\theta_p - \theta_{p-1}}$ with m variables. Forgetting for a moment the product $\prod_{\ell=1}^{p-2} \langle \zeta_{\theta_\ell}^\varepsilon, h_\ell \rangle$, we focus on the product of the fields at time θ_{p-1},

$$\big\langle \zeta_{\theta_{p-1}}^\varepsilon, h_{p-1} \big\rangle \big\langle \zeta_{m,\theta_{p-1}}^\varepsilon, \phi^{(p)}_{\theta_p - \theta_{p-1}} \big\rangle \tag{4.12}$$

and decompose it as in (4.7) according to the repeated indices in the spirit of the example above. This leads to two types of contribution:

- a "contracted product" (by analogy with the function Ψ) which records all the repeated indices in the product (4.12) at time θ_{p-1}. By Hölder's inequality as in (4.11), this term can be decoupled from the rest of the weight formed by the moments $\prod_{\ell=1}^{p-2} \langle \zeta_{\theta_\ell}^\varepsilon, h_\ell \rangle$. This strategy is particularly relevant for time-dependent fields as it reduces the estimates to computing moments of fields at a single time. In an equilibrium regime, the moments of the field at a single time can be easily analyzed as the distribution is time invariant. In this way, the moments at θ_p and θ_{p-1} are paired and their covariance $\mathbb{E}_\varepsilon[\langle \zeta_{\theta_{p-1}}^\varepsilon, h_{p-1} \rangle \langle \zeta_{\theta_p}^\varepsilon, h_p \rangle]$ is recovered. It remains then to study the remaining moments $\mathbb{E}_\varepsilon[\prod_{\ell=1}^{p-2} \langle \zeta_{\theta_\ell}^\varepsilon, h_\ell \rangle]$.

- a "$\otimes$-product", which by definition takes into account the nonrepeated indices, and which can be interpreted as a product of two independent fluctuations at time θ_{p-1}. In a very loose way, we have to evaluate now the following structure:

$$\mathbb{E}_\varepsilon \left[\left(\prod_{\ell=1}^{p-2} \big\langle \zeta_0^\varepsilon, h_\ell \big\rangle \right) \big\langle \zeta_{\theta_{p-1}}^\varepsilon, h_{p-1} \big\rangle \otimes \big\langle \zeta_{m,\theta_{p-1}}^\varepsilon, \phi^{(p)}_{\theta_p - \theta_{p-1}} \big\rangle \right],$$

  with a more complicated fluctuation field at time θ_{p-1}.

The key point here is that using the cumulant techniques introduced in Section 3.3, one can then prove that the tensorized structure $\otimes$ is essentially preserved by the pullback of test functions: the configurations for which the $\otimes$-product breaks can be neglected. Thus with high probability the fields $\langle \zeta_{\theta_{p-1}}^\varepsilon, h_{p-1} \rangle \otimes \langle \zeta_{m,\theta_{p-1}}^\varepsilon, \phi^{(p)}_{\theta_p - \theta_{p-1}} \rangle$ can be pulled back up to time θ_{p-2} as if they were independent. Then we apply the pairing procedure at time θ_{p-2}. This leads to new pairings between $\langle \zeta_{\theta_{p-2}}^\varepsilon, h_{p-2} \rangle$ and the pulled-back fields. In particular, the covariances $\mathbb{E}_\varepsilon[\langle \zeta^\varepsilon_{\theta_{p-2}}, h_{p-2}\rangle \langle \zeta^\varepsilon_{\theta_p}, h_p\rangle]$ and $\mathbb{E}_\varepsilon[\langle \zeta^\varepsilon_{\theta_{p-2}}, h_{p-2}\rangle \langle \zeta^\varepsilon_{\theta_{p-1}}, h_{p-1}\rangle]$ can be identified. The nonrepeated variables at time θ_{p-2} build new $\otimes$-products involving the fluctuation fields (or their pullbacks) from times θ_{p-2}, θ_{p-1} and θ_p.

Iterating this procedure up to time θ_1, all the pairings can be recovered and the Wick's decomposition (4.5) is obtained in the limit $\mu_\varepsilon \to \infty$. This shows that the limiting process is Gaussian, thus achieving the proof of Theorem 4.2.

5. OPEN PROBLEMS AND PERSPECTIVES

The research program that we conducted during this last decade and which is presented in this survey has led to two important breakthroughs compared to the state-of-the-art after Lanford's theorem:

- an extended statistical picture of the dynamics of hard-sphere gases for short times, including fluctuations and large deviations;
- a complete answer to Hilbert's sixth problem connecting the three levels of modeling (atomistic, kinetic, and fluid) for linear equations of dilute hard-sphere gases close to equilibrium.

Nevertheless, the problem of the axiomatization of gas dynamics remains largely open, even in dilute regimes. We propose in this final section to review some important directions to be explored in the future. We choose to discuss here only kinetic limits, involving a separation of scales, for which an enterprise in the spirit of those discussed above is conceivable (albeit possibly hard).

5.1. Long time behavior for dilute gases

The only case in which we have a complete picture of the transition from the atomistic description to fluid models is the *equilibrium* case. Nevertheless, the diffusive scaling considered in these linear regimes is sublogarithmic (see, e.g., [10, 15]). It would be interesting to reach more relevant physical scales, for which we expect the limiting picture to remain unchanged.

The law of large numbers in the equilibrium case is trivial, and the fluctuations are governed by linear models. In order to extend this analysis to gases which are initially *out of equilibrium*, a major obstruction is to define a good notion of stability for the nonlinear Boltzmann equation, which plays the role of pivot between the microscopic and macroscopic scales. In other words, this requires designing a good notion of convergence. The weak convergence method developed in the equilibrium case uses a topology which is a priori too weak to make sense of the nonlinear collision operator. Based on our analysis, we believe that stronger convergence methods require a rather precise understanding of the mechanisms responsible for the entropy cascade through the cumulants, retaining enough information in the limiting system. Note that this information is encoded in the supports of the cumulants, which have a finer and finer structure as the order of the cumulant increases. This structure might well be a key ingredient, as entropy and entropy dissipation play a crucial role in the stability of the Boltzmann equation.

Beyond the law of large numbers, it would be also natural to extend the analysis of fluctuations and large deviations for long kinetic times, and even diffusive times. This would allow deriving the fluctuating hydrodynamics (typically, the fluctuating Navier–Stokes–Fourier equations). A fine understanding of the Hamilton–Jacobi equations and of the associated gradient structure would be certainly a major step in this direction.

5.2. The role of microscopic interactions

Our study is focused on the case of hard-sphere gases, for which the interaction is pointwise in time and the scattering law is very simple. The papers [37, 45, 57] have shown that, despite technical complications, the same average behavior, in the low density limit, is obtained for compactly-supported potentials satisfying some suitable lower bound (thermodynamic stability). Only the collision cross-section (i.e., the transition rate of the jump process in the velocity space) and, consequently, the hydrodynamic transport coefficients are modified. One expects, and can prove for short times [45], that multiple collisions (three or more particles simultaneously interacting at a given time) are a correlation of higher order with respect to the dynamical correlations determining the fluctuation theory. It is then very likely that the description of fluctuations and large deviations for short times can be also extended to this *short-range* case. Notice that the absence of monotonicity of the potential would require a more delicate treatment, as some trajectories can be trapped for a very long time [57].

A problem of a much higher level of difficulty is to deal with *long-range* interactions. We know that, as soon as the potential is not compactly supported, the collision cross-section (which can be computed by solving the two-body problem) has a nonintegrable divergence at grazing angles. It is therefore impossible to define solutions of the Boltzmann equation without taking into account the cancelations between the gain and loss terms in the collision operator, which would imply to find new ideas (in our methods dealing with microscopic systems, such cancelations are never used). Close to equilibrium, using a sampling to discard superexponential growth (as in Section 4 above), N. Ayi [2] has proved a convergence result for very fast decaying potentials, but the method does not seem robust enough to deal with weaker decays or systems out of equilibrium.

A natural idea, often used by physicists, would be to decouple the short range part (acting as "collisions"), and the long range part of the interaction potential (to be dealt with by mean field methods). However, from the mathematical point of view, this leads to a major issue: no analysis method is available so far, as the techniques used for the low density limit and for the mean field limit are completely different and apparently incompatible. This problem is investigated in [27], where a linear Boltzmann–Vlasov equation is derived rigorously for a simple (Lorentz gas) model system (see also [26]).

A related issue is how to precisely identify and separate the long range and the collisional part for a given potential law, capturing the good scaling for both parts. There are some delicate aspects here involving the details of the potential and the dimension of the problem

[52,53]. Formal considerations as in [7] indicate that, in case of power law potentials $1/x^s$, the low density scaling should lead to a Boltzmann equation for $s > d - 1$, to a Boltzmann–Vlasov equation for $s = d - 1$, and to a Vlasov equation (with Boltzmann's operator still describing the collisions as a long time correction) for $s \in (d - 2, d - 1)$. For the Coulomb potential (and for smaller values of s), the Boltzmann operator has to be replaced by a diffusive variant of it (Landau, or Lenard–Balescu operator; see also Section 5.4). We refer to [52] for details.

We remark that the combination of mean-field and collisions has an interest in connection with the problem of binary mixtures exhibiting phase segregation [5] (see also [1] on a derivation result for mixtures).

5.3. Nonequilibrium stationary states

For short times, Lanford's theorem allows considering particle systems which are initially put out of equilibrium, provided that their distribution is controlled in some sense by an equilibrium state. This assumption is a key argument to get uniform bounds (even for short times when the relaxation phenomenon cannot be observed). In this situation, one can use a comparison principle because nothing forces the system to stay out of equilibrium, and the invariant measure is well known.

A natural extension is to deal with a gas evolving in a domain with *boundary conditions*, rather than the whole space or the periodic setting as considered previously. In the case of boundary conditions ensuring conservation of energy, we still have a control by the invariant measure, and the main extra difficulty caused by the presence of boundaries lies in the geometric analysis of recollisions. This has been discussed so far in the case of simple geometries [29,48] (see also [35] for the case of external forces).

A much more delicate situation is when the system of interacting particles is maintained out of equilibrium by a forcing or a boundary condition (reservoir, thermostat, ...). One would like to derive, in this nonequilibrium framework, the Boltzmann equation and more generally the properties of the steady states. As exposed in [18], this question is a "challenge to theorists," and few quantitative results are known either for gas dynamics or for other mechanical systems such as chains of anharmonic oscillators. Even though, under reasonable assumptions on the nonequilibrium forces, the existence of a *stationary measure* of the microscopic dynamics is expected, one does not know how to construct such a measure or any exact solution which would play the role of supersolution for the actual distribution of particles. In particular, a good starting point for the analysis of the low density limit seems to be missing at present. Finally, it is worth mentioning that the theory of stationary solutions for the Boltzmann equation with thermal reservoirs is still far from mature, see [36] for a recent review.

Beyond the derivation of the Boltzmann equation for boundary driven systems, it would be interesting to investigate the large deviations as they can provide some knowledge on the invariant measure [6,25]. Also it is conjectured [9,18] that the Fourier law should be valid for a dilute gas maintained out of equilibrium by reservoirs. To prove its validity would require an analysis beyond the kinetic time scale in order to derive fluid equations out of equilibrium.

5.4. A realm of kinetic limits

Besides the low density (Boltzmann–Grad) scaling discussed so far, there is a variety of interacting particle models admitting a kinetic limit and sharing many similarities with the classical Boltzmann gas [67]. We shall only mention here the two main obvious modifications of our assumptions (which are reviewed in detail in [56]): (i) start from a microscopic description based on *quantum mechanics* instead of classical mechanics, namely replace the Newton equations by the N-body Schrödinger equation, including additional symmetry/antisymmetry constraints which take into account the specificity of bosons/fermions; (ii) perform a high-density, *weak-coupling* scaling with potential $\varepsilon^{\alpha}\phi(x/\varepsilon)$, where $\alpha \in [0, 1]$ and the particle density is correspondingly tuned as $-d + 1 - 2\alpha$. For $\alpha \in (0, 1)$, the latter scaling should lead to the diffusive Landau equation in the case of classical systems, and is suited to a description of collisions in plasmas. The diffusion emerges from a central limit type effect on an accumulation of many weak collisions. The limiting point $\alpha = 1$ is expected to capture the famous Lenard–Balescu correction. Conversely, in the case of quantum systems, each value of α should lead to a quantum version of the Boltzmann equation. The amount and quality of quantum features surviving in the limit depends on the particular value of α. For $\alpha = 0$, the collision operator contains the full quantum cross-section. On the other hand, for $\alpha = 1/2$ (when only the first term of the Born series survives), one expects to get additional cubic terms in the collision operator, expressing the inclination of particles to aggregate (Bose–Einstein condensation) or to repel each other (Pauli's exclusion principle).

For such a variety of situations, no rigorous full derivation result is available at present, not even for short kinetic times; see however [8,56,69,70] for consistency results and attempts in this direction (full results are instead available for Lorentz type (linear) models, see [31,44] for the classical case and [32] for a review in the quantum case). When trying to reproduce Lanford's strategy, one stumbles indeed upon many difficulties. The construction of the equilibrium measure is delicate, and it is not completely clear how to identify the suitable functional spaces for the study of the limit. The Wigner transform, which allows computing observables, is nonpositive and quadratic with respect to the wave function: this implies that the combinatorics associated with the Duhamel series, which can be represented by Feynman diagrams is much worse than the combinatorics of collision trees. In general, these formal series are never absolutely convergent.

All the open questions regarding the long-time behavior, the structure of correlations and the deviations from the average dynamics, the role of microscopic interactions or the stationary nonequilibrium case remain, also in these different settings, as challenges for the future.

ACKNOWLEDGMENTS

We thank P. Dario, C. Garban, E. Ghys, F. Golse, and J. Marklof for their very useful comments on a preliminary version of this manuscript.

FUNDING

This work was partially supported by ANR-15-CE40-0020-01 grant LSD.

REFERENCES

[1] I. Ampatzoglou, J. K. Miller, and N. Pavlović, A rigorous derivation of a Boltzmann system for a mixture of hard-sphere gases. 2021, arXiv:2104.14480.

[2] N. Ayi, From Newton's law to the linear Boltzmann equation without cut-off. *Comm. Math. Phys.* **350** (2017), no. 3, 1219–1274.

[3] C. Bardos, F. Golse, and C. D. Levermore, Fluid dynamic limits of kinetic equations. II: Convergence proofs for the Boltzmann equation. *Comm. Pure Appl. Math.* **46** (1993), no. 5, 667–753.

[4] G. Basile, D. Benedetto, L. Bertini, and C. Orrieri, Large deviations for Kac-like walks. *J. Stat. Phys.* **184** (2021), no. 1, 27.

[5] S. Bastea, R. Esposito, J. L. Lebowitz, and R. Marra, Binary fluids with long range segregating interaction. I: Derivation of kinetic and hydrodynamic equations. *J. Stat. Phys.* **101** (2000), no. 5–6, 1087–1136.

[6] L. Bertini, A. De Sole, D. Gabrielli, G. Jona-Lasinio, and C. Landim, Macroscopic fluctuation theory. *Rev. Modern Phys.* **87** (2015), 593–636.

[7] A. V. Bobylev, P. Dukes, R. Illner, and H. D. jun. Victory, On Vlasov–Manev equations. I: Foundations, properties, and nonglobal existence. *J. Stat. Phys.* **88** (1997), no. 3–4, 885–911.

[8] A. Bobylev, M. Pulvirenti, and C. Saffirio, From particle systems to the Landau equation: a consistency result. *Comm. Math. Phys.* **319** (2013), no. 3, 683–702.

[9] T. Bodineau, I. Gallagher, and L. Saint-Raymond, De la dynamique des sphères dures aux équations de Stokes–Fourier: une analyse L^2 de la limite de Boltzmann-Grad. *C. R. Math. Acad. Sci. Paris* **353** (2015), no. 7, 623–627.

[10] T. Bodineau, I. Gallagher, and L. Saint-Raymond, The Brownian motion as the limit of a deterministic system of hard-spheres. *Invent. Math.* **203** (2016), no. 2, 493–553.

[11] T. Bodineau, I. Gallagher, and L. Saint-Raymond, From hard sphere dynamics to the Stokes–Fourier equations: An analysis of the Boltzmann–Grad limit. *Ann. PDE* **3** (2017), no. 1, 2.

[12] T. Bodineau, I. Gallagher, L. Saint-Raymond, and S. Simonella, One-sided convergence in the Boltzmann–Grad limit. *Ann. Fac. Sci. Univ. Toulouse Math.* **27** (2018), no. 5, 985–1022.

[13] T. Bodineau, I. Gallagher, L. Saint-Raymond, and S. Simonella, Fluctuation theory in the Boltzmann–Grad limit. *J. Stat. Phys.* **180** (2020), no. 1, 873–895.

[14] T. Bodineau, I. Gallagher, L. Saint-Raymond, and S. Simonella, Statistical dynamics of a hard sphere gas: fluctuating Boltzmann equation and large deviations. 2020, arXiv:2008.10403.

[15] T. Bodineau, I. Gallagher, L. Saint-Raymond, and S. Simonella, Long-time correlations for a hard-sphere gas at equilibrium. 2020, arXiv:2012.03813. To appear in *Comm. Pure Appl. Math.*

[16] T. Bodineau, I. Gallagher, L. Saint-Raymond, and S. Simonella, Long-time derivation at equilibrium of the fluctuating Boltzmann equation. 2022, arXiv:2201.04514.

[17] L. Boltzmann, Weitere Studien über das Wärmegleichgewicht unter Gasmolecülen. *Wien. Ber.* **66** (1872), 275–370.

[18] F. Bonetto, J. L. Lebowitz, and L. Rey-Bellet, Fourier's law: a challenge to theorists. In *Mathematical physics 2000*, pp. 128–150, Imperial College Press, London, 2000.

[19] F. Bouchet, Is the Boltzmann equation reversible? A large deviation perspective on the irreversibility paradox. *J. Stat. Phys.* **181** (2020), no. 2, 515–550.

[20] W. Braun and K. Hepp, The Vlasov dynamics and its fluctuations in the $1/N$ limit of interacting classical particles. *Comm. Math. Phys.* **56** (1977), 101–113.

[21] C. Cercignani, V. I. Gerasimenko, and D. Y. Petrina, *Many-particle dynamics and kinetic equations*. Math. Appl. 420, Kluwer Academic Publishers Group, Dordrecht, 1997.

[22] C. Cercignani, R. Illner, and M. Pulvirenti, *The mathematical theory of dilute gases*. Appl. Math. Sci. 106, Springer, New York, 1994.

[23] A. Dembo and O. Zeitouni, *Large deviations techniques and applications*. Stoch. Model. Appl. Probab. 38, Springer, Berlin, 2010.

[24] R. Denlinger, The propagation of chaos for a rarefied gas of hard spheres in the whole space. *Arch. Ration. Mech. Anal.* **229** (2018), no. 2, 885–952.

[25] B. Derrida, Microscopic versus macroscopic approaches to non-equilibrium systems. *J. Stat. Mech.* **2011** (2011), no. 01, P01030.

[26] L. Desvillettes and M. Pulvirenti, The linear Boltzmann equation for long-range forces: a derivation from particle systems. *Math. Models Methods Appl. Sci.* **9** (1999), no. 8, 1123–1145.

[27] L. Desvillettes, C. Saffirio, and S. Simonella, Collisions in a mean-field : kinetic limit for the Lorentz gas with long-range forces (in preparation).

[28] R. J. DiPerna and P. L. Lions, On the Cauchy problem for Boltzmann equations: Global existence and weak stability. *Ann. of Math. (2)* **130** (1989), no. 2, 321–366.

[29] T. Dolmaire, About Lanford's theorem in the half-space with specular reflection. 2021, arXiv:2102.05513.

[30] M. Duerinckx and L. Saint-Raymond, Lenard–Balescu correction to mean-field theory. *Probab. Math. Phys.* **2** (2021), no. 1, 27–69.

[31] D. Dürr, S. Goldstein, and J. L. Lebowitz, Asymptotic motion of a classical particle in a random potential in two dimensions: Landau model. *Comm. Math. Phys.* **113** (1987), 209–230.

[32] L. Erdős, In *Lecture notes on quantum Brownian motion*, pp. 3–98, Oxford University Press, 2012.

[33] L. Erdős, M. Salmhofer, and H.-T. Yau, Quantum diffusion of the random Schrödinger evolution in the scaling limit. *Acta Math.* **200** (2008), no. 2, 211–277.

[34] M. Ernst and E. Cohen, Nonequilibrium fluctuations in μ space. *J. Stat. Phys.* **25** (1981), no. 1, 153–180.

[35] R. Esposito, Y. Guo and R. Marra, Validity of the Boltzmann equation with an external force. *Kinet. Relat. Models* **4** (2011), no. 2, 499–515.

[36] R. Esposito and R. Marra, Stationary non equilibrium states in kinetic theory. *J. Stat. Phys.* **180** (2020), no. 1–6, 773–809.

[37] I. Gallagher, L. Saint-Raymond, and B. Texier, *From Newton to Boltzmann: hard spheres and short-range potentials*. European Mathematical Society (EMS), Zürich, 2013.

[38] F. Golse, In *On the dynamics of large particle systems in the mean field limit*, pp. 1–144, Springer, Cham, 2016.

[39] F. Golse, C. D. Levermore, and L. Saint-Raymond, La méthode de l'entropie relative pour les limites hydrodynamiques de modèles cinétiques. In *Séminaire: Équations aux Dérivées Partielles, 1999–2000*, Exp. No. XIX, Sémin. Équ. Dériv. Partielles 23, École Polytech, Palaiseau, 2000.

[40] H. Grad, In *Principles of the kinetic theory of gases*, pp. 205–294, Springer, Berlin–Göttingen–Heidelberg, 1958.

[41] D. Heydecker, Large deviations of Kac's conservative particle system and energy non-conserving solutions to the Boltzmann equation: A counterexample to the predicted rate function. 2021, arXiv:2103.14550.

[42] P.-E. Jabin and Z. Wang, Mean field limit and propagation of chaos for Vlasov systems with bounded forces. *J. Funct. Anal.* **271** (2016), no. 12, 3588–3627.

[43] M. Kac, In *Foundations of kinetic theory*, pp. 171–197, University of California Press, Berkeley and Los Angeles, 1956.

[44] H. Kesten and G. C. Papanicolaou, A limit theorem for stochastic acceleration. *Comm. Math. Phys.* **78** (1980), 19–63.

[45] F. G. King, *BBGKY hierarchy for positive potentials*. University of California, Berkeley, 1975.

[46] O. E. Lanford III, Time evolution of large classical systems. *Lecture Notes in Phys.* **38** (1975), 1–111.

[47] C. Léonard, On large deviations for particle systems associated with spatially homogeneous Boltzmann type equations. *Probab. Theory Related Fields* **101** (1995), no. 1, 1–44.

[48] C. Le Bihan, Boltzmann–Grad limit of a hard sphere system in a box with diffusive boundary conditions. 2021, arXiv:2104.04354. To appear in *Discrete Contin. Dyn. Syst.*

[49] J. Logan and M. Kac, Fluctuations and the Boltzmann equation. *Phys. Rev. A* **13** (1976), no. 1, 458.

[50] S. Méléard, Convergence of the fluctuations for interacting diffusions with jumps associated with Boltzmann equations. *Stoch. Stoch. Rep.* **63** (1998), no. 3–4, 195–225.

[51] S. Mischler and C. Mouhot, Kac's program in kinetic theory. *Invent. Math.* **193** (2013), no. 1, 1–147.

[52] A. Nota, J. J. L. Velázquez, and R. Winter, Interacting particle systems with long-range interactions: scaling limits and kinetic equations. *Atti Accad. Naz. Lincei Rend. Lincei Mat. Appl.* **32** (2021), no. 2, 335–377.

[53] A. Nota, J. J. L. Velázquez, and R. Winter, Interacting particle systems with long range interactions: approximation by tagged particles in random fields. 2021, arXiv:2103.09740.

[54] O. Penrose, In *Convergence of fugacity expansions for classical systems*, p. 101, Benjamin, New York, 1967.

[55] S. Poghosyan and D. Ueltschi, Abstract cluster expansion with applications to statistical mechanical systems. *J. Math. Phys.* **50** (2009), no. 5, 053509.

[56] M. Pulvirenti, The weak-coupling limit of large classical and quantum systems. In *Proceedings of the international congress of mathematicians (ICM), Madrid, Spain*, pp. 229–256, European Mathematical Society (EMS), Zürich, 2006.

[57] M. Pulvirenti, C. Saffirio, and S. Simonella, On the validity of the Boltzmann equation for short range potentials. *Rev. Math. Phys.* **26** (2014), no. 2, 1450001.

[58] M. Pulvirenti and S. Simonella, The Boltzmann–Grad limit of a hard sphere system: analysis of the correlation error. *Invent. Math.* **207** (2017), no. 3, 1135–1237.

[59] F. Rezakhanlou, Equilibrium fluctuations for the discrete Boltzmann equation. *Duke Math. J.* **93** (1998), no. 2, 257–288.

[60] F. Rezakhanlou, Large deviations from a kinetic limit. *Ann. Probab.* **26** (1998), no. 3, 1259–1340.

[61] F. Rezakhanlou, Boltzmann–Grad limits for stochastic hard sphere models. *Comm. Math. Phys.* **248** (2004), no. 3, 553–637.

[62] F. Rezakhanlou, Kinetic limits for interacting particle systems. In *Entropy methods for the Boltzmann equation. Lectures from a special semester at the Centre Émil Borel, Institut H. Poincaré, Paris 2001*, Lecture Notes in Math. 1916, Springer, Berlin, 2008, xii, 107 pp. ISBN 978-3-540-73704-9/pbk.

[63] D. Ruelle, *Statistical mechanics*. World Scientific Publishing Co., Inc., River Edge, NJ; Imperial College Press, London, 1999.

[64] L. Saint-Raymond, *Hydrodynamic limits of the Boltzmann equation*. Lecture Notes in Math. 1971, Springer, Berlin, 2009.

[65] H. Spohn, Fluctuations around the Boltzmann equation. *J. Stat. Phys.* **26** (1981), no. 2, 285–305.

[66] H. Spohn, Fluctuation theory for the Boltzmann equation. In *Nonequilibrium phenomena, I*, pp. 225–251, Stud. Stat. Mech. 10, North-Holland, Amsterdam, 1983.

[67] H. Spohn, *Large scale dynamics of interacting particles*. Springer, 2012.

[68] H. van Beijeren, O. E. Lanford III, J. L. Lebowitz, and H. Spohn, Equilibrium time correlation functions in the low-density limit. *J. Stat. Phys.* **22** (1980), no. 2, 237–257.

[69] J. Velázquez and R. Winter, From a Non-Markovian system to the Landau equation. *Comm. Math. Phys.* **361** (2018), no. 1, 1–49.

[70] R. Winter, Convergence to the Landau equation from the truncated BBGKY hierarchy in the weak-coupling limit. *J. Differential Equations* **283** (2021), 1–36.

[71] H.-T. Yau, Relative entropy and hydrodynamics of Ginzburg–Landau models. *Lett. Math. Phys.* **22** (1991), no. 1, 63–80.

THIERRY BODINEAU

CNRS, Ecole Polytechnique, Institut Polytechnique de Paris, Palaiseau, France, thierry.bodineau@polytechnique.edu

ISABELLE GALLAGHER

Ecole Normale Supérieure, Paris, France, Isabelle.Gallagher@ens.fr

LAURE SAINT-RAYMOND

IHES, Bures sur Yvette, France, laure@ihes.fr

SERGIO SIMONELLA

Ecole Normale Supérieure de Lyon, CNRS, Lyon, France, sergio.simonella@ens-lyon.fr

AUTOMORPHIC FUNCTIONS ON MODULI SPACES OF BUNDLES ON CURVES OVER LOCAL FIELDS: A SURVEY

ALEXANDER BRAVERMAN AND DAVID KAZHDAN

ABSTRACT

This paper is the written version of D. Kazhdan's plenary talk at ICM 2022. It is dedicated to an exposition of recent results and (mostly) conjectures attempting to construct an analog of the theory of automorphic functions on moduli spaces of bundles on curves over local fields (both archimedean and non-archimedean). The talk is based on joint works of D. Kazhdan with A. Braverman, P. Etingof, E. Frenkel, and A. Polishchuk.

MATHEMATICS SUBJECT CLASSIFICATION 2020

Primary 11F70; Secondary 22E57, 14H70 , 14D21

KEYWORDS

Automorphic forms, Langlands program, Hecke operators

ICM INTERNATIONAL CONGRESS OF MATHEMATICIANS 2022 JULY 6–14

Proc. Int. Cong. Math. 2022, Vol. 2, pp. 796–824
DOI 10.4171/ICM2022/167

1. INTRODUCTION

1.1. Langlands correspondence over functional fields

Let $\mathcal{C}$ be a smooth projective irreducible curve over a finite field $\mathbb{F}_q$. One can consider the global field $\mathbb{F} = \mathbb{F}_q(\mathcal{C})$ of rational functions on $\mathcal{C}$ and its adele ring $\mathbb{A}$. Given a split semisimple group G one can study automorphic forms on the adelic group $G(\mathbb{A})$ – these are (by definition) irreducible representations of $G(\mathbb{A})$ which appear in the space of $\mathbb{C}$-valued functions on $G(\mathbb{A})/G(\mathbb{F})$. For many purposes, it is important to consider discrete automorphic representations – these are automorphic representations appearing in $L^2(G(\mathbb{A}_F/G(\mathbb{F}))$.

In this introduction we restrict our attention to unramified automorphic representations, i.e., those which have a $G(\mathcal{O}_{\mathbb{F}})$-invariant vector where $\mathcal{O}_{\mathbb{F}} \subset \mathbb{A}$ is the ring of integral adeles. In other words, we consider functions on $G(\mathcal{O}_{\mathbb{F}})\backslash G(\mathbb{A})/G(\mathbb{F})$ which are eigenfunctions of certain commuting family of linear operators, called Hecke operators; for every place c of $\mathbb{F}$ (which is the same as a point of $\mathcal{C}(\overline{\mathbb{F}_q})$ up to the action of Frobenius), one constructs the algebra of Hecke operators which is isomorphic to the complexified Grothendieck ring of finite-dimensional representations of the Langlands dual group $G^\vee$ (and for different c these algebras commute with each other). The weak form of the Langlands conjecture (now proved by V. Lafforgue for global fields of positive characteristic) asserts that (after the replacement of the coefficient field $\mathbb{C}$ by $\overline{\mathbb{Q}_\ell}$) the common eigenvalues of all the Hecke operators come from ℓ-adic $G^\vee$-local systems on $\mathcal{C}$.

The quotient $G(\mathcal{O}_{\mathbb{F}})\backslash G(\mathbb{A})/G(\mathbb{F})$ is canonically isomorphic to the set of $\mathbb{F}_q$-points of the moduli stack $\mathrm{Bun}_G(\mathcal{C})$ of principal G-bundles on $\mathcal{C}$. Thus Hecke eigenfunctions are functions on $\mathrm{Bun}_G(\mathcal{C})(\mathbb{F}_q)$ and unramified discrete automorphic forms correspond to Hecke eigenfunction lying in $L^2(\mathrm{Bun}_G(\mathcal{C})(\mathbb{F}_q))$ (with respect to the Tamagawa measure).

We fix a curve $\mathcal{C}$ and a group G, and will write Bun instead of $\mathrm{Bun}_G(\mathcal{C})$ when it does not lead to a confusion.

1.2. Hecke eigenfunctions on moduli spaces of bundles over local fields

This survey reports on an attempt to extend the above constructions and results to the case when instead of a curve over $\mathbb{F}_q$ we start with a curve over a local field F. The idea to consider Hecke eigenfunctions in this case was first formulated by Langlands in the case $F = \mathbb{C}$ (cf. [25] and also [18]) several years ago. A systematic study of this question was started in [9] in a slightly different framework. To simplify notations we often assume that G is semisimple and the genus g of $\mathcal{C}$ is ≥ 2.

Here several difficulties are present. First, since Bun is a stack, it is not clear what space of functions on $\mathrm{Bun}(F)$ to consider. In fact, a big part of this paper is devoted to a discussion of three different spaces with actions of Hecke algebras one can attach to stacks over local fields and the relation between them (cf. Sections 2 and 3). In the first approach (which follows the papers of P. Etingof, E. Frenkel, and D. Kazhdan), the action of the Hecke algebra is defined on the Hilbert space $L^2(\mathrm{Bun}_{\mathrm{st}}(F))$ of half-measures where $\mathrm{Bun}_{\mathrm{st}} \subset \mathrm{Bun}$ is the open Deligne–Mumford substack of stable bundles. In this case the space is familiar, but one has to justify the convergence of the integrals defining Hecke operators. In the second and third approaches, the action of Hecke operators is well defined, but it is not easy to describe spaces on which they act. Some of our conjectures are on the relation between these different realizations.

In all three approaches the definition of Hecke operators, in fact, comes from [6] where some version of Satake isomorphism for Hecke algebras over a local field F is studied (formally, [6] only deals with non-archimedian fields, but the extension to archimedian case is straightforward).

Remark 1.1. In [9] (which deals with the case $F = \mathbb{C}$), the role of Hecke operators is played by the algebra $\mathcal{D}$ of global differential operators on $\mathrm{Bun}(F)$ (and their complex conjugate). In fact, as was observed in [3] there is no nontrivial regular differential operators acting on functions, but there is a large algebra of differential operators on half-forms. This algebra $\mathcal{D}$ is commutative and is equal to algebra of functions on the moduli space of certain special $G^\vee$-local systems on $\mathcal{C}$ called opers. This is another reason why half-forms are better suited for this problem. One of the main purposes of [9] is a formulation of a conjectural description of eigenvalues of the algebra $\mathcal{A} = \mathcal{D} \otimes \overline{\mathcal{D}}$ in terms of certain $G^\vee$-local systems on $\mathcal{C}$ (opers with real monodromy). For $G = \mathrm{SL}(2)$, a very close conjecture was formulated by J. Teschner in [27].

A systematic study of Hecke operators as self-adjoint operators acting on a Hilbert space started in [10] (in the case $F = \mathbb{C}$). As was mentioned above, the definition of Hecke operators is based on [6], and it again follows from [6] that in order to define Hecke operators one must work with half-forms; in this case Hecke operators are given by certain integrals (which are not guaranteed to converge). In [10] the authors conjectured that these integrals, in fact, define compact self-adjoint operators on $L^2(\mathrm{Bun})(F)$ for any local field F (in particular, contrary to the case of finite fields, their common spectrum on $L^2(\mathrm{Bun}(F))$ is discrete); in the case $F = \mathbb{C}$, it is expected that their eigenvectors are essentially the same as the eigenvectors for the algebra $\mathcal{A}$ (we shall give a precise formulation in Section 6). It is also explained in [10] (in the case $F = \mathbb{C}$) how to produce Hecke eigenvalues from opers with real monodromy (again, this is reviewed in Section 6). For non-archimedian fields F and $G = \mathrm{SL}_2$, analogous conjectures were formulated earlier by M. Kontsevich in [22].

In Section 5 we propose two other constructions of modules over the Hecke algebra – the last one only in the non-archimedian case. As was mentioned before, the space $L^2(\mathrm{Bun})$ is not the only choice of functional space one can work with. One can define another functional space (still having to do with half-forms) on which the Hecke operators will automatically act. The relationship between this space and $L^2(\mathrm{Bun})$, in the case when $G = \mathrm{SL}_2$, is the subject of a forthcoming paper by A. Braverman, D. Kazhdan, and A. Polishchuk. We review the relevant definitions and statements in Sections 2 and 3.

As a byproduct, when F is non-archimedian and the curve $\mathcal{C}$ is defined over its ring of integers $\mathcal{O}_F$ (and has good reduction), we give a conjectural construction of finite-dimensional spaces of *cuspidal* functions with an action of Hecke operators generalizing the space of cuspidal functions on $\mathrm{Bun}(\mathbb{F}_q)$, where $\mathbb{F}_q$ is the residue field of F (but in this way one gets only a very small portion of Hecke eigen-functions).[1] This is reviewed in Section 5.

In Sections 6 and 7, we formulate in the archimedian case a precise conjecture on the interpretation of the spectrum of Hecke operators on $L^2(\mathrm{Bun}(F))$ in terms of some kind of Galois data (involving the dual group $G^\vee$). It would be extremely interesting to find an interpretation of the spectrum of Hecke operators to for the non-archimedian case.

1.3. Relation of the archimedian case to geometric Langlands correspondence and conformal field theory

In the case when the field F is archimedian, our program is related to the quantum gauge theory (see [19]).

In this case Beilinson and Drinfeld associate to every $G^\vee$-oper o a certain algebraic $\mathbb{D}$-module M_o on Bun which is a *Hecke eigenmodule* which is equipped with a canonical generator (here $\mathbb{D}$ stands for the sheaf of differential operators on Bun acting on half-forms). This is an important part of a general geometric Langlands conjecture. The $\mathbb{D}$-module M_o can be thought of as a system of linear differential equations on $\mathrm{Bun}_{\mathrm{st}}$. The corresponding Hecke eigen-half-form (in the case when o has real monodromy) is a solution of both this system of equations and its complex conjugate.

The difference between the traditional categorical Langlands correspondence and the analytic Langlands correspondence for complex curves can be illustrated by an analogy with the two-dimensional conformal field theory (CFT). In CFT, there are two types of correlation functions. The first is chiral correlation functions, also known as conformal blocks. They form a vector space for fixed values of the parameters of the CFT, so we obtain a vector bundle of conformal blocks on the space of parameters, equipped with a projectively flat connection (or more generally, a twisted $\mathbb{D}$-module). Conformal blocks are its *multivalued* horizontal sections. The second type is the "physical" correlation functions. They can be expressed as sesquilinear combinations of conformal blocks and their complex conjugates (anticonformal blocks), which is a *single-valued* function of the parameters.

The Hecke eigensheaves on Bun constructed in the categorical Langlands correspondence may be viewed as sheaves of conformal blocks of a certain CFT. They are parametrized by all $G^\vee$-opers on the curve. It turns out that for special $G^\vee$-opers (namely, the real ones) there exists a sesquilinear linear combinations of these conformal blocks and their complex conjugates which are single-valued functions (more precisely, $1/2$-measures) on Bun. These are the automorphic forms of the analytic theory. Thus, the objects of the analytic theory of automorphic forms on Bun can be constructed from the objects of the categorical theory in roughly the same way as the correlation functions of CFT are obtained from conformal blocks (see [16] and the references therein for more details). An important difference with traditional CFT is that while usually in CFT the monodromy of conformal blocks is typically unitary, here the monodromy is expected to be in a split real group.

1 The construction itself is, in fact, not conjectural – we can do it rigorously. But at the moment, we cannot prove that the resulting eigenfunctions are not equal to 0.

1.4. Notations

We shall use the letter k to denote an arbitrary field (which could be finite) and the letter F for local fields. For a variety (or stack) X over k, we denote by $X(k)$ the set of k-points (for a stack we consider isomorphism classes of points). If F is non-archimedian, we denote by $\mathcal{O}_F$ its ring of integers. We shall also consider the field $\mathcal{K} = k((t))$ (or $F((t))$) with ring of integers which we denote just by $\mathcal{O}$.

For a split semisimple group G, we denote by $G^\vee$ the Langlands dual group of G considered as a group over $\mathbb{C}$. We fix a Borel subgroup $B = TU$ of G, where T is a maximal torus and U is a maximal unipotent subgroup; similarly we have a Borel subgroup $B^\vee = T^\vee U^\vee \subset G^\vee$.

We denote by Λ and $\Lambda^\vee$ the lattices of coweights and of weights of T (so Λ is also the lattice of weights of $T^\vee$) and by $\Lambda^+ \subset \Lambda$ the subset of dominant coweights.

1.5. Organization of the paper

In Section 2 we review some basic information about varieties and stacks over local fields and various spaces of functions on them. In Section 3 we begin the discussion of the moduli stack Bun of G-bundles on a curve $\mathcal{C}$ over a local field F and formulate some conjectures about the relation between various function spaces one attaches to Bun. In Section 4 we review the definition of Hecke operators and the formulation of the unramified Langlands correspondence for curves over $\mathbb{F}_q$. In Section 5 we explain the definition of Hecke operators in the case of local fields, formulate our main conjectures and also discuss some constructions specific for the non-archimedian case. Section 6 is dedicated to the case $F = \mathbb{C}$ and Section 7 to the case $F = \mathbb{R}$.

2. SMOOTH SECTIONS OF LINE BUNDLES ON VARIETIES AND STACKS

2.1. Smooth sections on varieties

If X is an algebraic variety over a local field F (archimedian or not), the set $X(F)$ is endowed with a natural topology.

Definition 2.1. A function $f : X(F) \to \mathbb{C}$ is smooth if

(a) F is non-archimedian and f is locally constant;

(b) F is archimedian and (locally) there exists a closed embedding $X \hookrightarrow Y$ where Y is a smooth variety over F and a C^∞-function $\bar{f} : Y(F) \to \mathbb{C}$ such that $f = \bar{f}|_{X(F)}$.

We denote by $C^\infty(X)$ the space of smooth functions on $X(F)$ and by $\mathcal{S}(X)$ its subspace of functions with compact support.

For a line bundle $\mathcal{L}$ over X, we denote by $\mathcal{L}^0 := \mathcal{L}\backslash X$ the corresponding $\mathbb{G}_m$-torsor over X and set

$$|\mathcal{L}|^\kappa = \mathcal{L}^0(F) \underset{F^*}{\times} \mathbb{C}_\kappa,$$

where $\mathbb{C}_\kappa$ denotes the 1-dimensional space $\mathbb{C}$ on which F^* acts by $|\cdot|^\kappa$. Then $|\mathcal{L}|^\kappa$ is a complex line bundle over $X(F)$. Since the bundle $|\mathcal{L}|^\kappa$ is locally trivial with respect to the natural topology, we can define its space of smooth sections which we denote by $C^\infty(X, |\mathcal{L}|^\kappa)$. Similarly, we denote by $\mathcal{S}(X, |\mathcal{L}|^\kappa) \subset C^\infty(X, |\mathcal{L}|^\kappa)$ the subspace of sections with compact support.

In the case when X is smooth we shall often take $\mathcal{L} = \omega_X$, where ω_X is the line bundle of differential forms of top degree and write $\mathcal{S}_\kappa(X)$ instead of $\mathcal{S}(X, |\omega|^\kappa)$. The case $\kappa = 1/2$ is of special interest since the space $\mathcal{S}_{1/2}(X)$ is endowed with a natural Hermitian product. We denote by $L^2(X)$ its Hilbert space completion.

Remark 2.2. (1) If $U \subset X$ is an open subset and $Z = X\backslash U$ then we have a short exact sequence

$$0 \to \mathcal{S}\big(U, |\mathcal{L}|_U^\kappa\big) \to \mathcal{S}\big(X, |\mathcal{L}|^\kappa\big) \to \mathcal{S}\big(Z, |\mathcal{L}|_Z^\kappa\big) \to 0.$$

(2) More generally, instead of choosing $\mathcal{L} \in \mathrm{Pic}(X)$ and $\kappa \in \mathbb{C}$ we can start with any element of $\mathrm{Pic}(X) \otimes \mathbb{C}$ – all the above definitions make sense in this context.

2.2. Smooth sections on stacks

In this subsection we extend the above definitions to a class of algebraic stacks.

Definition 2.3. An algebraic stack $\mathcal{Y}$ is *admissible* if locally there exists a presentation of $\mathcal{Y}$ as a quotient stack X/G where X is a smooth variety and G is an affine algebraic group. We denote by $p : X \to \mathcal{Y}$ the projection.

A presentation of $\mathcal{Y}$ as a quotient $\mathcal{Y} = X/\mathrm{GL}_n$ is called an *admissible presentation.*[2]

Remark 2.4. (1) Any smooth admissible stack of finite type can be presented as a quotient X/GL_n for a smooth variety X (see **[21]**). As follows from the Hilbert's 90, we have $\mathcal{Y}(F) = X(F)/\mathrm{GL}_n(F)$.

(2) Any admissible stack is automatically locally of finite type.

(3) A line bundle on a quotient X/G is a G-equivariant line bundle on X.

Definition 2.5. (1) Assume that F is non-archimedian, $\mathcal{Y}$ is an admissible stack of finite type over F. Choose an admissible presentation $\mathcal{Y} = X/\mathrm{GL}_n$ for some variety X and set

$$\mathcal{S}\big(\mathcal{Y}, |\mathcal{L}|^\kappa\big) = \mathcal{S}\big(X, |\mathcal{L}_X|^\kappa\big)_{\mathrm{GL}(n,F)},$$

2 The definition of admissibility that we use here is close to the one introduced in **[21]** but slightly different. It is easy to see that every admissible stack locally has an admissible presentation.

where the latter space stands for the space of $\mathrm{GL}(n, F)$-coinvariants on $\mathcal{S}(X, |\mathcal{L}_X|^\kappa)$.

(2) If F is non-archimedian and $\mathcal{Y}$ is only locally of finite type, then we can write $\mathcal{Y}$ as a direct limit of open substacks $\mathcal{Y}_i$ of finite type over F and define $\mathcal{S}(\mathcal{Y}, |\mathcal{L}|^\kappa) := \lim\limits_{\rightarrow} \mathcal{S}(\mathcal{Y}_i, |\mathcal{L}|^\kappa)$.

(3) In the case when F is archimedian we make an analogous definition but take coinvariants $\mathcal{S}(X, |\mathcal{L}_X|^\kappa)_{\mathrm{GL}(n,F)}$ in the category of topological spaces where $\mathcal{S}(X, |\mathcal{L}_X|^\kappa)$ is endowed with Fréchet topology.[3]

The above definition makes sense because of the following

Claim 2.6. *If $\mathcal{Y}$ is an admissible stack of finite type then the space $\mathcal{S}(\mathcal{Y}, |\mathcal{L}|^\kappa)$ does not depend on a choice of an admissible presentation $\mathcal{Y} = X/\mathrm{GL}_n$.*

Remark 2.7. In the case when F is non-archimedian, $\mathcal{L} = \omega_X$, and $\kappa = 1$, this claim is proven in [21, SECTION 6]. The same arguments work in the general case.

2.3. Functoriality

If $\mathcal{Y}$ is an admissible stack and $\mathcal{U}$ is an open substack, we have a natural map $\mathcal{S}(\mathcal{U}, |\mathcal{L}|^\kappa) \to \mathcal{S}(\mathcal{Y}, |\mathcal{L}|^\kappa)$, which is not injective in general.

More generally, let $f : \mathcal{Z} \to \mathcal{Y}$ be a smooth representable map of admissible stacks and $\omega_{\mathcal{Z}/\mathcal{Y}}$ be the relative canonical bundle. Then we have a natural ("integration over the fibers") map

$$\mathcal{S}\big(\mathcal{Z}, |\mathcal{L}|^\kappa \otimes |\omega_{\mathcal{Z}/\mathcal{Y}}|\big) \to \mathcal{S}\big(\mathcal{Y}, |\mathcal{L}|^\kappa\big).$$

2.4. An example: stacks over $\mathcal{O}_F$

In this subsection we consider the case when the field F is non-archimedian (with residue field k) and construct some explicit elements in $\mathcal{S}(\mathcal{Y}, |\mathcal{L}|^\kappa)$. Assume that $\mathcal{Y} = X/G$ where both X and G are defined over $\mathcal{O}_F$ and that $X_{\mathcal{O}_F}$ is a regular scheme over $\mathcal{O}_F$ such that $\mathcal{Y}(F) = X(F)/G(F)$. Assume also that the line bundle $\mathcal{L}$ is defined over $\mathcal{O}_F$. Then in the same way as before we can define $\mathcal{S}(\mathcal{Y}_{\mathcal{O}_F}, |\mathcal{L}|^\kappa)$ with an obvious map $\mathcal{S}(\mathcal{Y}_{\mathcal{O}_F}, |\mathcal{L}|^\kappa) \to \mathcal{S}(\mathcal{Y}, |\mathcal{L}|^\kappa)$.

Consider now the case when $\mathcal{L} = \omega_\mathcal{Y}$. Then the complex line bundle $|\mathcal{L}|$ has a canonical trivialization on $\mathcal{Y}(\mathcal{O}_F)$. Let $\mathcal{S}(\mathcal{Y}(k))$ denote the space of $\mathbb{C}$-valued functions with finite support on $\mathcal{Y}(k)$. Then the above trivialization gives rise to a map $\mathcal{S}(\mathcal{Y}(k)) \to \mathcal{S}(\mathcal{Y}_{\mathcal{O}_F}, |\mathcal{L}|^\kappa)$. Composing it with the map $\mathcal{S}(\mathcal{Y}_{\mathcal{O}_F}, |\mathcal{L}|^\kappa) \to \mathcal{S}(\mathcal{Y}, |\mathcal{L}|^\kappa)$, we get a map $E_{\mathcal{Y},\kappa} : \mathcal{S}(\mathcal{Y}(k)) \to \mathcal{S}_\kappa(\mathcal{Y})$.

Remark 2.8. (1) This map is often not injective.

---

3 We define the space $\mathcal{S}(X, |\mathcal{L}_X|^\kappa)_{\mathrm{GL}(n,F)}$ as the quotient of $\mathcal{S}(X, |\mathcal{L}_X|^\kappa)$ by the closure of the subset generated by elements of the form $g(s) - s$ where $g \in \mathrm{GL}_n(F)$ and $s \in \mathcal{S}(X, |\mathcal{L}_X|^\kappa)$.

(2) We will be mostly interested in the space $\mathcal{S}_{1/2}(\mathcal{Y})$ (for a particular choice of $\mathcal{Y}$). In the case when $\mathcal{Y}$ was a smooth scheme, this space had a canonical Hermitian product. We do not expect to see a Hermitian product on $\mathcal{S}_{1/2}(\mathcal{Y})$ for general admissible stacks $\mathcal{Y}$ but we define a class of *excellent stacks* when such a product exists.

(3) We write $\mathcal{M}(\mathcal{Y}) := \mathcal{S}_{1/2}(\mathcal{Y})$.

2.5. Nice and excellent stacks

In this subsection we assume that $\mathcal{Y}$ is an admissible stack which contains an open substack $\mathcal{Y}_{\mathrm{vs}} \subset \mathcal{Y}$ such that $\mathcal{Y}_{\mathrm{vs}} = Y_{\mathrm{vs}}/Z$ where Y_{vs} is a smooth scheme and Z is a finite group acting trivially on Y_{vs}.[4]

Remark 2.9. (1) To simplify the notations, let us assume that $Z = \{e\}$ (but generalization to arbitrary Z is straightforward).

(2) A choice of this open substack is not unique, and some of the definitions below depend on this choice.

(3) Let $L^2(\mathcal{Y}_{\mathrm{vs}})$ be the Hilbert space completion of the space of smooth half-measures on $\mathcal{Y}_{\mathrm{vs}}(F)$ with compact support. It is easy to see that this space is in fact independent of the choice of $\mathcal{Y}_{\mathrm{vs}}$.

If $\mathcal{Y}$ is of finite type over F, we choose a presentation $\mathcal{Y} = X/\operatorname{GL}_n$, denote by U the preimage of $\mathcal{Y}_{\mathrm{vs}}$ in X and by $p : U \to \mathcal{Y}_{\mathrm{vs}}$ the quotient map.

Let s be a smooth section with compact support of the complex line bundle $p^*|\omega_{\mathcal{Y}_{\mathrm{vs}}}|^\kappa \otimes |\omega_{X/\mathcal{Y}}|$. Then $s|_U$ is a section of $p^*|\omega_{\mathcal{Y}_{\mathrm{vs}}}|^\kappa \otimes |\omega_{U/\mathcal{Y}}|$. We can try to integrate it over the fibers of p to get a section of $|\omega_{\mathcal{Y}_{\mathrm{vs}}}|^\kappa$ on $\mathcal{Y}_{\mathrm{vs}}$. The problem is that these integrals might not converge since the intersection of the support of s with the fibers of the map p might not be compact.

Definition 2.10. (1) The stack $\mathcal{Y}$ is κ-bounded if there exists an open substack of finite type $\mathcal{Y}_0 \subset \mathcal{Y}$ such that the map $\mathcal{S}_\kappa(\mathcal{Y}_0) \to \mathcal{S}_\kappa(\mathcal{Y})$ is an isomorphism.

(2) A pair $(\mathcal{Y}, \mathcal{Y}_{\mathrm{vs}})$ is κ-*nice* if $\mathcal{Y}$ is κ-bounded and for every s as above supported on the preimage of $\mathcal{Y}_{\mathrm{vs}}$ the push-forward $p_*(s)$ is well-defined (i.e., it is absolutely convergent) and defines a smooth section of $|\omega_{\mathcal{Y}_{\mathrm{vs}}}|^\kappa$ on $\mathcal{Y}_{\mathrm{vs}}$.

(3) A pair as above is *excellent* if it is nice for all $\kappa \geq 1/2$ and for $\kappa = 1/2$ we have $p_*(s) \in L^2(\mathcal{Y}_{\mathrm{vs}})$ for every smooth section s with compact support.

When the substack $\mathcal{Y}_{\mathrm{vs}} \subset \mathcal{Y}$ is fixed we refer to the stack $\mathcal{Y}$ as "nice" or "excellent."

---

4 The subscript vs stands for "very stable." The reason for this notation is that later when we work with the stack Bun of G-bundles on a curve, we define $\mathrm{Bun}_{vs} \subset \mathrm{Bun}$ as the open subset of very stable bundles.

Remark 2.11. The convergence in the definition of κ-niceness is automatically true for $\kappa \geq 1$.

If $\mathcal{Y}$ is κ-nice, then the map $s \mapsto p_*(s)$ descends to a map $\mathcal{S}_\kappa(\mathcal{Y}) \to C^\infty(\mathcal{Y}_{\mathrm{vs}})$. If $\mathcal{Y}$ is excellent we get a map $\mathcal{M}(\mathcal{Y}) = \mathcal{S}_{1/2}(\mathcal{Y}) \to L^2(\mathcal{Y}_{\mathrm{vs}}) = L^2(\mathcal{Y})$.

Example 2.12. Let $X = (\mathbb{P}^1)^3$, $G = \mathrm{PGL}_2$, and $\mathcal{Y} = X/G$, where G acts diagonally; we take U to be the complement to all diagonals in $(\mathbb{P}^1)^3$. Then G acts freely on U and we set $\mathcal{Y}_{\mathrm{vs}} = U/G$ (note that $\mathcal{Y}_{\mathrm{vs}}$ is just Spec F). In this case one can check that $\mathcal{Y}$ is nice for $\kappa > 1/3$ and the stack $\mathcal{Y}$ is excellent.

3. THE CASE OF Bun_G: PRELIMINARIES

We fix a split connected semisimple group G and denote by Z it center.

Let $\mathcal{C}$ be a smooth complete irreducible curve over a field k.

Definition 3.1.

(1) Bun_G is the stack of the principal G-bundles on $\mathcal{C}$ and $\mathrm{Bun}_{G,\mathrm{st}} \subset \mathrm{Bun}_G$ is the open substack of stable bundles.

(2) For a G-bundle $\mathcal{F}$ on $\mathcal{C}$ we denote by $\mathrm{Ad}_{\mathcal{F}}$ the adjoint bundle to $\mathcal{F}$ associated with the adjoint action of G on $\mathfrak{g}$.

(3) A G-bundle $\mathcal{F}$ is *very stable* if there is no nonzero section of $\Gamma(\mathcal{C}, \mathrm{Ad}_{\mathcal{F}}) \otimes \omega_{\mathcal{C}}$ whose values at all points of $\mathcal{C}$ are nilpotent.

(4) We denote by $\mathrm{Bun}_{G,\mathrm{vs}} \subset \mathrm{Bun}_G$ the substack of very stable bundles.

Remark 3.2. If $\mathcal{C}$ is of genus ≥ 2 then:

(1) Every very stable bundle is stable.

(2) $\mathrm{Bun}_{\mathrm{st}}$ is a dense open subset of Bun of the form Y/Z where Y is a smooth scheme of finite type over F and Z acts trivially on Y.

(3) $\mathrm{Bun}_{\mathrm{vs}}$ is a dense open subset of $\mathrm{Bun}_{\mathrm{st}}$.

(4) When it does not lead to a confusion we shall drop the subscript G from the notation (e.g., we shall write Bun for Bun_G).

Claim 3.3. *The stack* Bun *is κ-bounded for all κ.*

Remark 3.4. This statement is inspired by the proof of the main result of [7].

Conjecture 3.5. Assume that the genus g of $\mathcal{C}$ is ≥ 2.

(1) Bun is κ-nice for $\mathrm{Re}(\kappa) \geq 1/2$. In particular, for $\kappa \geq 1/2$, we get a map $\iota_\kappa : \mathcal{S}_\kappa(\mathrm{Bun}) \to C^\infty_\kappa(\mathrm{Bun}_{\mathrm{vs}})$.

(2) For $\kappa \geq 1/2$ any section in the image of the map ι_κ extends to a continuous section of $|\omega_{\mathrm{Bun}}|^\kappa$ on $\mathrm{Bun}_{\mathrm{st}}$.

(3) Bun is excellent.

For $G = \mathrm{PGL}_2$, the first assertion of Conjecture 3.5 (as well as some special cases of the second and third assertions) will appear in a forthcoming paper of A. Braverman, D. Kazhdan, and A. Polishchuk. Let us note that (again for $G = \mathrm{PGL}_2$) the second assertion can be reduced to the following purely algebro-geometric statement using [2] (we can prove the conjecture for curves of genus 2 and 3).

Conjecture 3.6. Let $\mathcal{E}$ be a stable bundle on $\mathcal{C}$ of degree $2g-1$. Let $F_{\mathcal{E}}$ denote the scheme of pairs $(\mathcal{L}, s)$ where $\mathcal{L} \in \mathrm{Pic}^0(\mathcal{C})$ and $s \in \mathbb{P}(H^0(\mathcal{C}, \mathcal{L} \otimes \mathcal{E}))$. Then

(1) $F_{\mathcal{E}}$ is irreducible.

(2) $\dim F_{\mathcal{E}} = g$.

(3) $F_{\mathcal{E}}$ has rational singularities.

4. AFFINE GRASSMANNIAN AND HECKE OPERATORS: THE CASE OF FINITE FIELD

In this section we collect some facts about the canonical class of certain Schubert varieties that we shall need in the future. All the results of this section follow easily from [12] and [3]. In what follows we a ground field k and set denote by $\mathcal{O}$ the ring functions on the formal one-dimensional disc D over k and by $\mathcal{K}$ the field of functions on the punctured disc $D^{\star}$. So $\mathcal{O} \sim k[[t]]$ and $\mathcal{K} \sim k((t))$. We denote by ω_D the canonical bundle on D and fix a square root $\omega_D^{1/2}$ (unique up to an isomorphism; the isomorphism is unique up to ± 1).

4.1. The affine Grassmannian

Let G be a split semisimple group over k and $\mathrm{Gr}_G := G(\mathcal{K})/G(\mathcal{O})$. It is known that Gr_G has a natural structure of a proper ind-scheme over k and the orbits of the group $G(\mathcal{O})$ on Gr_G are parameterized by the elements of Λ_+.

For each $\lambda \in \Lambda^+$, we shall denote by Gr_G^{λ} the corresponding orbit and by $\overline{\mathrm{Gr}}_G^{\lambda}$ the closure of Gr_G^{λ}.

4.2. Satake isomorphism

In the rest of this section we assume that k is a finite field.

Let $\mathcal{H}(G,k)$ be the algebra of compactly supported $G(\mathcal{O})$-biinvariant distributions on $G(\mathcal{K})$ (by choosing a Haar measure on $G(\mathcal{K})$ such that $G(\mathcal{O})$ has volume 1, we can identify these distributions with functions). Let $G^{\vee}$ be the Langlands dual group, considered as a group over $\mathbb{C}$. The Satake isomorphism identifies $\mathcal{H}(G,k)$ with the complexified Grothendieck ring of the category $\mathrm{Rep}(G^{\vee})$ of finite-dimensional representations of $G^{\vee}$. It can also be identified with the algebra $\mathbb{C}[T^{\vee}]^W$ of W-invariant polynomial functions on $T^{\vee}$.

4.3. Hecke operators

Let now $\mathcal{C}$ be a smooth projective irreducible curve over k. As before we consider the stack $\mathrm{Bun} := \mathrm{Bun}_G$ of principal G-bundles on $\mathcal{C}$. Let $c \in \mathcal{C}$ be a closed point with residue field k' which is a finite extension of k. Choose of a local parameter near c,[5] (in the end nothing will depend on this choice) and consider the stack Hecke_c classifying triples $(\mathcal{E}_1, \mathcal{E}_2, \eta)$ where every $\mathcal{E}_i$ is a principal G-bundle on $\mathcal{C}$ and η is an isomorphism between $\mathcal{E}_1$ and $\mathcal{E}_2$ on $\mathcal{C}\backslash\{c\}$. We have canonical projections

$$\begin{array}{ccc} \mathrm{Hecke}_c & \xrightarrow{\mathrm{pr}_2} & \mathrm{Bun} \\ {\scriptstyle \mathrm{pr}_1}\downarrow & & \\ \mathrm{Bun}\,. & & \end{array} \tag{1}$$

Every fiber of the map pr_2 is isomorphic to Gr_G and this isomorphism is canonical up to the action of $G(\mathcal{O})$. Thus every $h \in \mathcal{H}(G, k')$ defines a canonical function $\tilde{h}$ on Hecke_c. We can use it as a correspondence, and set

$$T_{h,c}(f) = \mathrm{pr}_{2,*}\big(\mathrm{pr}_1^*(f) \cdot \tilde{h}\big)$$

for any $f : \mathrm{Bun}(\mathbb{F}_q) \to \mathbb{C}$. This construction defines an action of the algebra $\mathcal{H}(G, k')$ on the space of all functions on $\mathrm{Bun}(k)$ (given a choice of c as above). For different choices of c, these operators commute.

Claim 4.1. (1) *The operators $T_{h,c}$ preserve the space $\mathcal{S}(\mathrm{Bun})$ of functions with finite support on* $\mathrm{Bun}(k)$.

(2) *Let $L^2(\mathrm{Bun}(k))$ be the L^2-completion of the space $\mathcal{S}(\mathrm{Bun}(k))$ with respect to the standard L^2-norm given by the measure on the (discrete) set* $\mathrm{Bun}(k)$ *where the volume of every $\mathcal{E}$ is equal to $\frac{1}{\#\mathrm{Aut}(\mathcal{E})}$. Then for every c the action of $\mathcal{H}(G, k)$ extends to an action on $L^2(\mathrm{Bun}(k))$ by bounded operators. If h is real-valued, the operator $T_{h,c}$ is self-adjoint.*

4.4. Langlands conjectures

In the theory of automorphic forms, we are usually interested in eigenfunctions of all the operators $T_{h,c}$. Let us replace the field of coefficients $\mathbb{C}$ by $\overline{\mathbb{Q}}_\ell$ where ℓ is a prime number different from the characteristic of $\mathbb{F}_q$. Then (the weak form) of the Langlands conjecture states that if f is such an eigenfunction, then the eigenvalues of all the operators $T_{h,c}$ come from a homomorphism $\rho : \mathbb{W}(\mathcal{C}) \to G^\vee(\overline{\mathbb{Q}}_\ell)$ where $\mathbb{W}(\mathcal{C})$ is the Weil group of $\mathcal{C}$ (a close cousin of the fundamental group of $\mathcal{C}$). In fact, in this form the Langlands conjecture has been proved by V. Lafforgue (cf. [24]).

Let us recall the connection between Hecke-eigenvalues and homomorphisms ρ as above. First of all, any c defines a conjugacy class $\mathrm{Fr}_c \subset \mathbb{W}(\mathcal{C})$. For any $V \in \mathrm{Rep}(G^\vee)$ the by Satake isomorphism associates to V an element in $\mathcal{H}(G, k)$, which we denote by h_V. We denote $T_{V,c}$ the corresponding Hecke operator. We say that the eigenvalue of an eigenfunction f comes from ρ if

$$T_{V,c}(f) = \mathrm{Tr}\big(\rho(\mathrm{Fr}_c), V\big) \cdot f \tag{2}$$

for all c and V.

---

5 That is an identification of the formal neighborhood of c with $\mathrm{Spec}\, k'[[t]]$.

In general, Hecke eigenfunctions lie neither in $\mathcal{S}(\mathrm{Bun})$ nor in $L^2(\mathrm{Bun})$ (here we come back to considering $\mathbb{C}$-coefficients). Those which lie in the former are called *cuspidal*, and those which lie in the latter are called *discrete*. The fact that not all eigenfunctions are discrete is related to the fact that the operators T_h have both discrete and continuous spectrum.

Remark 4.2. Note that the operators $T_{h,c}$ would be compact, if the stack Bun were of finite type over k (in fact, $L^2(\mathrm{Bun}(k))$ would be finite-dimensional in this case), and so in that their common spectrum would be discrete. So, the existence of continuous spectrum of Hecke operators is related to the fact that Bun is not of finite type over k.

5. THE AFFINE GRASSMANNIAN AND HECKE OPERATORS: THE CASE OF LOCAL FIELD

5.1. More on formal discs

We are going to make a very mild change of notation (compared to the previous section). Namely, let F be a field (very soon we shall assume that F is a local field). In what follows we denote by $\mathcal{O}$ some discrete valuation ring over F which (as a discrete valuation ring) is isomorphic to $F[[t]]$ (the point is that we do not want to fix this isomorphism). We let $\mathcal{K}$ be the field of fractions of $\mathcal{O}$. We set $D = \mathrm{Spec}(\mathcal{O})$, $D^* = \mathrm{Spec}(\mathcal{K})$. We shall denote by 0 the canonical F-point of D.

We let ω_D be canonical sheaf of D and let $\omega_{D,0}$ be its fiber at 0. This is a vector space over F.

5.2. Line bundles on Gr_G

It is well-known (cf. [3] and [12]) that every finite-dimensional representation V of G gives rise to a (determinant) line bundle $\mathcal{L}_V$ on Gr_G; the fiber of this bundle over a point $g \in G(\mathcal{K})/G(\mathcal{O})$ is equal to the determinant of the vector space $g(V(\mathcal{O}))/g(V(\mathcal{O})) \cap V(\mathcal{O})$. In particular, we let $\mathcal{L}_{\mathfrak{g}}$ denote the line bundle corresponding to the adjoint representation of G. The line bundle $\mathcal{L}_{\mathfrak{g}}^{-1}$ has a square root (unique up to isomorphism) which we denote by $\mathcal{L}_{\mathrm{crit}}$.

The following result from [3] is crucial for us:

Theorem 5.1. *For every $\lambda \in \Lambda_+$, there is a canonical isomorphism*

$$\mathcal{L}_{\mathrm{crit}}|_{\mathrm{Gr}_G^\lambda} \simeq \omega_{\mathrm{Gr}_G^\lambda} \otimes \omega_{D,0}^{-\langle\lambda,\rho^\vee\rangle}.$$

(Here, as before, $\omega_{\mathrm{Gr}_G^\lambda}$ denotes the canonical bundle of on Gr_G^λ*).*[6]

6 The formulation of the theorem requires a clarification when G is not simply connected, since in this case $\langle\lambda,\rho^\vee\rangle$ might be a half-integer (and not an integer). It is sufficient for our purposes to say that we choose a square root of $\omega_{D,0}$ and that the isomorphism above is canonical up to ± 1. This potential sign will disappear when we apply $|\cdot|$ to both sides which we shall do in applications.

We need more information about the structure of the varieties $\overline{\mathrm{Gr}}_G^{\lambda}$. The following result is proved in [12] (cf. also [23] and [26] for the corresponding result in characteristic 0).

Theorem 5.2. (1) *Each $\overline{\mathrm{Gr}}_G^{\lambda}$ is a normal and Cohen–Macaulay projective variety over F.*

(2) *Each $\overline{\mathrm{Gr}}_G^{\lambda}$ has a resolution of singularities*[7] *and, for every such resolution $\pi^{\lambda} : \widetilde{\mathrm{Gr}}_G^{\lambda} \to \overline{\mathrm{Gr}}_G^{\lambda}$, one has*

$$R\pi_*^{\lambda}(\mathcal{O}_{\widetilde{\mathrm{Gr}}_G^{\lambda}}) = \mathcal{O}_{\overline{\mathrm{Gr}}_G^{\lambda}}.$$

(in other words, $\overline{\mathrm{Gr}}_G^{\lambda}$ has rational singularities).

The next result is an easy corollary of Theorems 5.2 and 5.1 (cf. [6] for a proof):

Theorem 5.3. (1) *For every $\lambda \in \Lambda_+$, the variety $\overline{\mathrm{Gr}}_G^{\lambda}$ is Gorenstein. Moreover, the canonical sheaf of $\overline{\mathrm{Gr}}_G^{\lambda}$ is isomorphic to $\mathcal{L}_{\mathrm{crit}}|_{\overline{\mathrm{Gr}}_G^{\lambda}} \otimes \omega_{D,0}^{\langle\lambda,\rho^{\vee}\rangle}$. Abusing the notation, we shall denote this sheaf by $\omega_{\overline{\mathrm{Gr}}_G^{\lambda}}$.*

(2) *For any $\lambda \in \Lambda_+$, let $\pi^{\lambda} : \widetilde{\mathrm{Gr}}_G^{\lambda} \to \overline{\mathrm{Gr}}_G^{\lambda}$ be any resolution of singularities. Then the identification between $(\pi^{\lambda})^*\omega_{\overline{\mathrm{Gr}}_G^{\lambda}}$ and $\omega_{\widetilde{\mathrm{Gr}}_G^{\lambda}}$ that one has at the generic point of $\widetilde{\mathrm{Gr}}_G^{\lambda}$ comes from an embedding*

$$(\pi^{\lambda})^*\omega_{\overline{\mathrm{Gr}}_G^{\lambda}} \hookrightarrow \omega_{\widetilde{\mathrm{Gr}}_G^{\lambda}}.$$

(In the case $\mathrm{char}\, k = 0$, *this implies that $\overline{\mathrm{Gr}}_G^{\lambda}$ has canonical singularities).*

5.3. Hecke algebra over local field

In this subsection F can be any local field.

Let us now work over a local field F instead of k with the corresponding ring $\mathcal{O}$ and its field of fractions $\mathcal{K}$.[8] Then we would like to define the Hecke algebra $\mathcal{H}(G, F)$. First we consider the space

$$C^{\infty}_{1/2}(\mathrm{Gr}_G) = \varprojlim \mathcal{S}(\overline{\mathrm{Gr}}^{\lambda}(F), |\mathcal{L}_{\mathrm{crit}}|).$$

Assume first that F is non-archimedian. Then we define $\mathcal{H}(G, F)$ to be the space of all $G(\mathcal{O})$-invariant linear functionals on $C^{\infty}_{1/2}(\mathrm{Gr}_G)$ with compact support. The latter condition means that we consider functionals $\delta : C^{\infty}_{1/2}(\mathrm{Gr}_G) \to \mathbb{C}$ which factorize through a map $C^{\infty}_{1/2}(\mathrm{Gr}_G) \to \mathcal{S}(\overline{\mathrm{Gr}}^{\lambda}(F), |\mathcal{L}_{\mathrm{crit}}|)$ for some λ. It is easy to see that $\mathcal{H}(G, F)$ is an algebra with respect to convolution.

It turns our that Theorems 5.1, 5.2, and 5.3 allow one to construct a lot of elements in $\mathcal{H}(G, F)$ (what follows is essentially equivalent to the main result of [6]). Namely, let λ be as above and let $\phi \in C^{\infty}_{1/2}(\mathrm{Gr}_G)$. Let us first trivialize the space $\omega_{D,0}$. Then $\phi|_{\mathrm{Gr}_G^{\lambda}}$

---

7 Of course, this statement is not a priori clear only if char $F > 0$.

8 In the case when F is non-archimedian the reader should not confuse $\mathcal{O} = F[[t]]$ with $\mathcal{O}_F$ which is the ring of integers of F.

is a distribution on $\mathrm{Gr}^{\lambda}(F)$ and we can try to consider its integral. A priori it might not be well defined since $\mathrm{Gr}_G^{\lambda}(F)$ is not compact, but it is explained in [6] that Theorems 5.2 and 5.3 imply that in fact this integral is absolutely convergent and thus defines an element $h_\lambda \in \mathcal{H}(G, F)$. These elements have the property that for any dominant λ and μ we have

$$h_\lambda \star h_\mu = h_{\lambda+\mu}.$$

In other words, we get an embedding $\mathbb{C}[\Lambda_+] \hookrightarrow \mathcal{H}(G, F)$. It is easy to see that it is actually an isomorphism.

If we do not want to trivialize the space $\omega_{D,0}$ then canonically h_λ is a map from $|\omega_{D,0}|^{-\langle\lambda,\rho^\vee\rangle} \to \mathcal{H}(G, F)$, and we get an isomorphism

$$\bigoplus_{\lambda\in\Lambda_+} |\omega_{D,0}|^{\langle\lambda,\rho^\vee\rangle} \simeq \mathcal{H}(G, F)$$

(the left-hand side has an obvious algebra structure).

5.4. Hecke operators for curves over local fields: the first approach

We would like to define Hecke operators in some space of actual functions on Bun (or, rather, sections of $|\omega_{\mathrm{Bun}}|^{1/2}$), or maybe some open subset of it. Let us assume that the genus of $\mathcal{C}$ is ≥ 2. Then, as we have discussed before, Bun contains a dense open substack $\mathrm{Bun}_{\mathrm{st}}$ of stable bundles which is a Deligne–Mumford stack. So, one can try to start with a smooth section ϕ of $|\omega_{\mathrm{Bun}}|^{1/2}$ on $\mathrm{Bun}_{\mathrm{st}}(F)$ and apply the operator $\mathbb{T}_{\lambda,c}$ using the diagram (1).[9]

In this case the definition will involve integration over Gr_G^{λ}, and we are not guaranteed that the corresponding integral is convergent. The trouble is caused by the following: take some $\mathcal{E} \in \mathrm{Bun}(F)$ (which one can assume to be stable or even very stable) and consider $\mathrm{pr}_2^{-1}(\mathcal{E})$. Let us identify it with Gr_G and consider the corresponding $G(\mathcal{O})$-invariant subset $\overline{\mathrm{Gr}}_G^{\lambda}$ in it. Let S be a compact subset of $\mathrm{Bun}_{\mathrm{st}}(F)$. Then typically $\mathrm{pr}_1^{-1}(S) \cap \overline{\mathrm{Gr}}_G^{\lambda}$ is not compact.

We say that $\phi \in C^{\infty}_{1/2}(\mathrm{Bun}_{\mathrm{vs}})$ is good if the integral defining $\mathbb{T}_{\lambda,c}(\phi)$ is absolutely convergent and the result is again an element of $C^{\infty}_{1/2}(\mathrm{Bun}_{\mathrm{vs}})$. The following result is easy:

Claim 5.4. *Assume the validity of Conjecture 3.5(1). Then the image of the map $\iota_{1/2}$ consists of good sections and the map $\iota_{1/2}$ commutes with the operator $\mathbb{T}_{\lambda,c}$.*

Note that the image of $\iota_{1/2}$ obviously contains $\mathcal{S}_{1/2}(\mathrm{Bun}_{\mathrm{vs}})$. Thus Conjecture 3.5(1) implies that any $\phi \in \mathcal{S}_{1/2}(\mathrm{Bun}_{\mathrm{vs}})$ is good. On the other hand, without assuming Conjecture 3.5(1) we cannot a priori construct any good element of $C^{\infty}_{1/2}(\mathrm{Bun}_{\mathrm{vs}})$.

We now proceed to the discussion of the action of the Hecke operators on $L^2(\mathrm{Bun})$. The main expectation is the following:

---

9 We are slightly abusing the notation here: namely, we are going to denote by $\mathbb{T}_{\lambda,c}$ both the operator on $\mathcal{M}(\mathrm{Bun})$ and on some space of sections of $|\omega_{\mathrm{Bun}}|^{1/2}$ which we are going to discuss below. We hope that it does not lead to a confusion.

Conjecture 5.5. The operators $\mathbb{T}_{\lambda,c}$ on $L^2(\mathrm{Bun})$ are bounded, compact, and self-adjoint. In particular, their common spectrum is discrete.

Philosophically, the reason for the fact that in the case of local fields the operators $\mathbb{T}_{\lambda,c}$ have discrete spectrum (as opposed to the case of finite fields) is that in the case of local fields we always work only with some open subset of Bun of finite type (cf. also Claim 3.3), and as was noted in Remark 4.2, the source for noncompactness of the Hecke operators in the case of finite fields has to do with the fact that the stack Bun is not globally of finite type (and in particular, not quasicompact).

5.5. Hecke operators for curves over local fields: the second approach

We now go back to the setup of Section 3. We would like to define Hecke operators in this context. First, we need to decide on what space they are going to act. The first (and the easiest) choice is to work with the space $\mathcal{M}(\mathrm{Bun}) = \mathcal{S}_{1/2}(\mathrm{Bun})$ (another choice is discussed in the next subsection). In what follows it will be convenient (but not necessary) to choose a particular square root $\omega_{\mathrm{Bun}}^{1/2}$ of ω_{Bun} (this is always possible, but the choice is slightly not canonical).

Let us also choose a closed point c of the scheme $\mathcal{C}$ with residue field F' which is a finite unramified extension of F; we shall take $\mathcal{O}$ to be the local ring of c (so, it is a discrete valuation ring over F' noncanonically isomorphic to $F'[[t]]$). To emphasize the dependence on c, we denote the corresponding Hecke algebra by $\mathcal{H}_c(G)$ (instead of $\mathcal{H}(G, F')$).

Then we again can consider the diagram (1) as in Section 4.3. Then since the line bundle $\mathcal{L}_{\mathrm{crit}}$ on Gr_G is $G(\mathcal{O})$-equivariant, we can define a line bundle $\widetilde{\mathcal{L}}_{\mathrm{crit}}$ on Hecke_c whose restriction to every fiber of pr_2 is canonically isomorphic to $\mathcal{L}_{\mathrm{crit}}$ (this property makes sense since every fiber is canonically isomorphic to Gr_G up to the action of $G(\mathcal{O})$).

Lemma 5.6. *We have*

$$\mathrm{pr}_1^* \, \omega_{\mathrm{Bun}}^{1/2} \simeq \mathrm{pr}_2^* \, \omega_{\mathrm{Bun}}^{1/2} \otimes \widetilde{\mathcal{L}}_{\mathrm{crit}}. \tag{3}$$

The isomorphism (3) easily allows one to define action of $\mathcal{H}(G, F') \simeq \mathbb{C}[\Lambda_+]$ on $\mathcal{M}(\mathrm{Bun})$. We denote by $\mathbb{T}_{\lambda,c}$ the operator corresponding to $h_{\lambda,c}$ (more generally, we denote by $\mathbb{T}_{h,c}$ the operator corresponding to any $h \in \mathcal{H}(G, F')$). For different choices of c, these actions commute. Therefore, one can try to study eigenvectors of all these operators in $\mathcal{M}(\mathrm{Bun})$.

Remark 5.7. Recall that the operators $\mathbb{T}_{\lambda,c}$ are canonically defined only up to a scalar; canonically each $\mathbb{T}_{\lambda,c}$ is an operator from $\mathcal{M}(\mathrm{Bun})$ to $\mathcal{M}(\mathrm{Bun}) \otimes |\omega_{\mathcal{C},c}|^{\langle \lambda, \rho^\vee \rangle}$. Therefore when we vary c each eigenvalue gives rise to a section of $|\omega_{\mathcal{C}}|^{-\langle \lambda, \rho^\vee \rangle}$. This will not be important for us until the end of Section 6 (where it will in fact become quite crucial).

Note that $\mathcal{M}(\mathrm{Bun})$ is an analog of the space of functions with finite support on $\mathrm{Bun}(k)$ (where k is a finite field). But unlike in the case of finite fields, we expect the following (some philosophical reasons for this difference are discussed in the next subsection):

Conjecture 5.8. Assume that F is non-archimedian. Then the space $\mathcal{M}(\mathrm{Bun})$ has a basis of Hecke eigenvectors. Similarly, in the archimedian case, the space $\mathcal{M}(\mathrm{Bun})$ has a topological basis of Hecke eigenvectors.

Before we try to say something about the eigenvalues, let us discuss a slightly different version of Hecke operators.

5.6. Example

We now want to explain how to produce some Hecke eigenfunction using the construction of Section 2.4.

In the case when F is non-archimedian and that $\mathcal{C}$ is defined over $\mathcal{O}_F$, i.e., we choose a model $\mathcal{C}_{\mathcal{O}_F}$ of $\mathcal{C}$ over $\mathcal{O}_F$. We assume that $\mathcal{C}_{\mathcal{O}_F}$ is a regular scheme and we denote by $\mathcal{C}_k$ the corresponding curve over k. Then the stack Bun is canonically defined over $\mathcal{O}_F$, and we have the map $E_{\mathrm{Bun},1/2} : \mathcal{S}(\mathrm{Bun}(k)) \to \mathcal{M}(\mathrm{Bun})$ (see Section 2.4).

We claim that this map commutes with the Hecke operators in the appropriate sense. Namely, let F' be a finite Galois extension of F with ring of integers $\mathcal{O}_{F'}$ and residue field k'. Then one can construct a homomorphism $\gamma_{F'} : \mathcal{H}(G, F') \to \mathcal{H}(G, k')$ with the following property. Let c be a closed point of $\mathcal{C}$ whose residue field is F'. Note that $\mathcal{C}(F') = \mathcal{C}(\mathcal{O}_{F'})$, so c has canonical reduction $\bar{c}$ which is a closed point of $\mathcal{C}_k$ with residue field k'. Then for any $h \in \mathcal{H}(F, F')$ and for any $\phi \in \mathcal{S}(\mathrm{Bun}(k))$ we have

$$E_{\mathrm{Bun},1/2}(T_{\gamma_{F'}(h)}(\phi)) = \mathbb{T}_h\big(E_{\mathrm{Bun},1/2}(\phi)\big). \tag{4}$$

Remark 5.9. We do not know how to describe the map $\gamma_{F'}$ in general. It is easy to see that $\gamma_{F'}(h_\lambda)$ is supported on $\overline{\mathrm{Gr}}^{\lambda}_G(k')$ (when viewed as a function on $\mathrm{Gr}_G(k')$). But this information is sufficient only in the case when $G = \mathrm{PGL}_n$ when minuscule coweights generate Λ.

Equation (4) implies that $E_{\mathrm{Bun},1/2}$ sends Hecke eigenfunctions to Hecke eigenfunctions. This operator is certainly not injective, but we expect it to be injective on cuspidal functions. More precisely (assuming the validity of Conjecture 3.5), we formulate the following

Conjecture 5.10. Assume the validity of Conjecture 3.5. Then the composition of $\iota_{1/2} \circ E_{\mathrm{Bun},1/2}$ is unitary on cuspidal functions.

Conjecture 5.10 implies that we can attach a nonzero Hecke eigenvector in $L^2(\mathrm{Bun})$ to any *cuspidal* Hecke eigenfunction in $\mathcal{S}(\mathrm{Bun}(k))$. On the other hand, we expect that the map $E_{\mathrm{Bun},1/2}$ is highly noninjective on noncuspidal functions. For example, let $G = \mathrm{PGL}_2$ and let $\mathcal{S}(\mathrm{Bun}(k))^{\perp}_{\mathrm{cusp}}$ denote the space of functions with finite support which are orthogonal to all cuspidal functions (with respect to the standard Hermitian product). This space is infinite-dimensional, but we expect that

$$\dim E_{\mathrm{Bun},1/2}\big(\mathcal{S}\big(\mathrm{Bun}(k)\big)^{\perp}_{\mathrm{cusp}}\big) = 1.$$

Note that equation (4) implies that the action of any $\mathbb{T}_{\lambda,c}$ on any section in the image of $E_{\mathrm{Bun},1/2}$ depends only on $\bar{c}$ (and not on c). This is certainly a very restrictive condition.

Also, one should think about $E_{\text{Bun},1/2}$ as some kind of Eisenstein series operator between the group $G(k)$ and the group $G(F)$ (with $G(\mathcal{O}_F)$ playing the role of a parabolic subgroup). This is in fact the source for our notation.

5.7. Parabolic bundles

We would like to introduce a generalization of the above setup, which allows in particular, to consider the case of curves of genus ≤ 1 when we may analyze some explicit nontrivial examples.

Definition 5.11. (1) Let us denote by Fl the variety of Borel subgroups of G.

(2) For a G-bundles $\mathcal{F}$ on $\mathcal{C}$, we denote by $\text{Fl}_{\mathcal{F}}$ the associated Fl-bundle over $\mathcal{C}$.

(3) For a divisor $D \subset \mathcal{C}$ defined over k, we denote by Bun^D the stack of G-bundles $\mathcal{F}$ on $\mathcal{C}$ with a section a of $\text{Fl}_{\mathcal{F}}$ over D.

It is easy to extend the definition of the Hecke operators $\mathbb{T}_{\lambda,c}$ for $c \notin D$. All our constructions and conjectures can be extended to this case. As was noted above, considering parabolic points allows one to consider explicit examples. For example, in the case when $\mathcal{C} = \mathbb{P}^1$, D consists of at least 3 points and G is of rank 1, Conjecture 5.5 is proved in [8] (Proposition 3.13).

5.8. More spaces with Hecke action

5.8.1. The map $E_{\mathcal{Y},\kappa,n}$

Here we would like to discuss how to generalize the construction of Sections 2.4 and 5.6. Namely, let $\mathcal{Y}$ be as in Section 2.4. Let $A_n = \mathcal{O}_F/\mathfrak{m}_F^n$. Let $\mathcal{Y}_n$ denote the reduction of $\mathcal{Y}_{\mathcal{O}_F}$ modulo $\mathfrak{m}_F^n$. This is a regular stack over A_n. We consider the set $\mathcal{Y}_n(A_n)$ and we set $\mathcal{S}(\mathcal{Y}_n(A_n))$ to be the vector space of $\mathbb{C}$-valued functions on $\mathcal{Y}_n(A_n)$ with finite support. Then for any $\kappa \in \mathbb{C}$, we have the obvious map

$$\mathcal{S}\big(\mathcal{Y}_n(A_n)\big) \to \mathcal{S}(\mathcal{Y}_{\mathcal{O}_F}) \simeq \mathcal{S}\big(\mathcal{Y}_{\mathcal{O}_F}, |\omega_{\mathcal{Y}}^{\kappa}|\big).$$

Composing it with the natural map $\mathcal{S}(\mathcal{Y}_{\mathcal{O}_F}, |\omega_{\mathcal{Y}}^{\kappa}|) \to \mathcal{S}_\kappa(\mathcal{Y})$, we get a map

$$E_{\mathcal{Y},\kappa,n} : \mathcal{S}\big(\mathcal{Y}_n(A_n)\big) \to \mathcal{S}_\kappa(\mathcal{Y}).$$

It is easy to see that this map is surjective if the map $\mathcal{Y}(\mathcal{O}) \to \mathcal{Y}(F)$ is surjective; in particular, this is true for $\mathcal{Y} = \text{Bun}_G$ for a reductive group G.

In fact, when G is a reductive but not semisimple group, we also need the following variant of the definition of $\mathcal{S}_\kappa(Y)$ for $Y = \text{Bun}_G$.

Definition 5.12. Let G be a reductive group and Z the connected component of the center of G (so Z is a torus). For a character $\chi_n : \text{Bun}_{Z,n}(A_n) \to \mathbb{C}^\times$, we denote by $\mathcal{S}_{\chi_n}(\text{Bun}_{G,n}(A_n))$ the vector the space of $(\text{Bun}_{Z,n}(A_n), \chi_n)$-coinvariants in $\mathcal{S}(\text{Bun}_{G,n}(A_n))$. Similarly, for $\chi : \text{Bun}_Z(F) = \text{Bun}_Z(\mathcal{O}) \to \mathbb{C}^\times$, we denote by $\mathcal{S}_\kappa(\text{Bun}_G)_\chi$ the space of $(\text{Bun}_Z(F), \chi)$-coinvariants in $\mathcal{S}_\kappa(\text{Bun}_G)$.

As before we have a map $E_{\mathrm{Bun}_G,\kappa,n,\chi} : \mathcal{S}_{\chi_n}(\mathrm{Bun}_{G,n}(A_n)) \to \mathcal{S}_{\kappa,\chi}(\mathrm{Bun}_G)$ provided that χ is equal to the pullback of κ_n under the natural map $\mathrm{Bun}_Z(F) = \mathrm{Bun}_Z(\mathcal{O}) \to \mathrm{Bun}_{Z,n}(A_n)$.

5.8.2. Commutation with Hecke operators

We now want to specialize to the case $\mathcal{Y} = \mathrm{Bun}_G$ and $\kappa = 1/2$. We claim that in this case the map $E_{\mathrm{Bun},1/2,n}$ commutes with the Hecke operators in the sense similar to (4). To explain the formulation we first need to discuss an analog of the homomorphism γ_F; this is a local question.

Namely, let us consider the ring $\mathcal{K}_n = A_n((t))$. This is a locally compact topological ring; its subring $\mathcal{O}_n = A_n[[t]]$ is open and compact. Thus the group $G(\mathcal{K}_n)$ is a totally disconnected locally compact topological group with an open compact subgroup $G(\mathcal{O}_n)$. Hence, we may consider the corresponding Hecke algebra

$$\mathcal{H}_n(G) = \mathcal{H}\big(G(\mathcal{K}_n), G(\mathcal{O}_n)\big).$$

Here is a variant. Let $\mathcal{C}$ be a smooth projective curve over $\mathcal{O}_F$. We denote by $\mathcal{C}_n$ its reduction mod $\mathfrak{m}_F^n$. Let F' be a finite unramified extension of F and let c be an F'-point of $\mathcal{C}$. As before we can also view it as an $\mathcal{O}_{F'}$-point of $\mathcal{C}$ and we denote by c_n its reduction modulo $\mathfrak{m}_{F'}^n$. This is an A'_n-point of $\mathcal{C}_n$. Then we might consider the corresponding Hecke algebra $\mathcal{H}_{c_n}(G)$. It is noncanonically isomorphic to $\mathcal{H}_n(G)$.

This Hecke algebra is quite bad: it is not commutative for $n > 1$ and apparently it does not have any reasonable description. However, it has the following two important features:

(1) Let $\mathcal{C}$ above and let c be a point of $\mathcal{C}$ defined over a finite unramified extension F' of F. Then the (noncommutative) algebra $\mathcal{H}_{c_n}(G)$ acts on $\mathcal{S}(\mathrm{Bun}_n)$. Given $h \in \mathcal{H}_{c_n}(G)$, we denote by $T_{h,n}$ the corresponding operator on $\mathcal{S}(\mathrm{Bun}_n)$.

(2) We have a canonical homomorphism $\gamma_{c,n} : \mathcal{H}_c(G) \to \mathcal{H}_{c_n}(G)$.

(3) For any $h \in \mathcal{H}_c(G)$ and $\phi \in \mathcal{S}(\mathrm{Bun}_n)$, we have

$$E_{\mathrm{Bun},1/2,n}\big(\gamma_{c,n}(\mathbb{T}_{h,c})(\phi)\big) = \mathbb{T}_{h,c}\big(E_{\mathrm{Bun},1/2,n}(\phi)\big). \tag{5}$$

5.8.3. Eigenfunctions and cuspidal functions: the idea

Definition 5.13. Let χ be a unitary character of $\mathrm{Bun}_Z(A_n)$.

(1) A function $\phi \in \mathcal{S}(\mathrm{Bun}_{G,n})_\chi$ is cuspidal if the span of $\{T_{h,n}(\phi)\}, h \in \mathcal{H}_c(G)$, $c \in \mathcal{C}_n$ is finite dimensional.

(2) $\mathcal{S}_{\mathrm{cusp}}(\mathrm{Bun}_{G,n})_\chi \subset \mathcal{S}(\mathrm{Bun}_n)_\chi$ is the subspace of cuspidal functions.

Conjecture 5.14. $\mathcal{S}_{\mathrm{cusp}}(\mathrm{Bun}_n)_\chi$ is finite dimensional for any n and $\dim(\mathcal{S}_{\mathrm{cusp}}(\mathrm{Bun}_n)_\chi) \sim q^{n \dim(\mathrm{Bun}_{G/Z})}$ for $q \gg 1$.

Remark 5.15. Such that Hecke operators $\gamma_{c,n}(\mathbb{T}_{h,c})$ are self-adjoint with respect to the natural Hermitian form on $\mathcal{S}_{\text{cusp}}(\text{Bun}_n)_\chi$.

Since $\text{Bun}_G(\mathcal{O})$ maps surjectively to $\text{Bun}_G(F)$, it follows from the statement at the end of Section 5.8.1 that

$$\mathcal{S}_\kappa(\text{Bun}_G) = \bigcup_n E_{\text{Bun}_G,\kappa,n}(\mathcal{S}(\text{Bun}_{G,n}(A_n)))$$

(and a similar statement holds for the space $\mathcal{S}_{\kappa,\chi}(\text{Bun}_G)$). We can now define

$$\mathcal{S}_{\kappa,\text{cusp}}(\text{Bun}_G) = \bigcup_n E_{\text{Bun}_G,\kappa,n}(\mathcal{S}_{\text{cusp}}(\text{Bun}_{G,n}(A_n)))$$

(and again similarly for $\mathcal{S}_{\kappa,\chi,\text{cusp}}(\text{Bun}_G)$). Note that for $\kappa = 1/2$ this space is locally finite dimensional with respect to the Hecke operators.

5.9. The case of $G = \text{GL}_2$

The proof of Conjecture 5.14 in the case $G = \text{GL}_2$ will appear in a forthcoming publication by A. Braverman, D. Kazhdan, and A. Polishchuk. In this subsection we outline a notion of the *constant term* used in our proof Conjecture 5.14 (again in the case in the case when $G = \text{GL}_2$; for simplicity, we shall also restrict ourselves to the case $n = 2$). This notion is used for a different (but equivalent) definition of cuspidality.

5.9.1. The constant term in the usual case

Recall that the usual constant term operator (for $n = 1$) is defined as follows. Let P be a parabolic subgroup of G; it has a natural homomorphism to M – the Levi factor. Consider the diagram

$$\begin{array}{ccc} \text{Bun}_P(k) & \xrightarrow{p} & \text{Bun}_G(k) \\ {\scriptstyle q}\big\downarrow & & \\ \text{Bun}_M(k). & & \end{array} \tag{6}$$

Then the constant term $\mathbf{c}_{G,P}$ is equal to $q_! \circ p^*$ (when k is a finite field).

Claim 5.16. *A function ϕ on $\text{Bun}_G(k)$ is cuspidal in the sense of Definition 5.13 if and only if $\mathbf{c}_{G,P}(\phi) = 0$ for all parabolic subgroups P of G; let $\mathcal{S}_{\text{cusp}}(\text{Bun}_G(k))$ be the space of all cuspidal functions.*

The following facts are well known and are easy to prove:

(1) $\mathcal{S}_{\text{cusp}}(\text{Bun}_G(k))$ is invariant under the Hecke operators.

(2) $\mathcal{S}_{\text{cusp}}(\text{Bun}_G(k))$ consists of functions with finite support.

(3) $\dim \mathcal{S}_{\text{cusp}}(\text{Bun}_G(k)) < \infty$ if G is semisimple. More generally, $\dim \mathcal{S}_{\text{cusp},\chi}(\text{Bun}_G(k)) < \infty$ if G is reductive and χ is a character of Bun_{Z^0} where Z^0 is the connected componenent of identity of the center of G.

We expect that for a proof of Conjecture 5.14, one has to extend the definition of a *constant term* onto the space $\mathcal{S}(\mathrm{Bun}_n)_\chi$.

The definition is not completely straightforward; as was mentioned above, we shall only discuss the case $G = \mathrm{GL}_2$ and $n = 2$. So, we shall now assume that $G = \mathrm{GL}_2$ and again just write Bun instead of $\mathrm{Bun}_{\mathrm{GL}_2}$. Also in this case the only proper parabolic subgroup up to conjugacy is the Borel subgroup; we shall also denote the corresponding constant term operator (that we are going to define) simply by $\mathbf{c}^{(2)}$.

In this case $T = \mathbb{G}_m \times \mathbb{G}_m$, so $\mathrm{Bun}_T(\mathcal{C}_2) = \mathrm{Pic}(\mathcal{C}_2) \times \mathrm{Pic}(\mathcal{C}_2)$, so it would be natural to expect that our constant term operator $\mathbf{c}^{(2)}$ maps functions on $\mathrm{Bun}(\mathcal{C}_2)$ to functions on $\mathrm{Pic}(\mathcal{C}_2) \times \mathrm{Pic}(\mathcal{C}_2)$. However, we do not know how to define such an operator if we want it to commute with the Hecke operators in some reasonable sense. Instead, let us do the following. Consider the semigroup Pic_2' (which contains the Picard group Pic_2 of $\mathcal{C}_2$). By definition, Pic_2' consists of coherent sheaves $\mathcal{M}$ on $\mathcal{C}$ such that $t\mathcal{M} \neq 0$ and there exists an imbedding $\mathcal{M} \hookrightarrow \mathcal{L}$ where $\mathcal{L}$ is a line bundle on $\mathcal{C}$. The tensor product defines the semigroup structure on Pic_2'.

Example 5.17. Let $\mathcal{C} = \mathrm{Spec}(A_2[x])$, $J \subset A_2[x]$ be the maximal ideal generated by (x, t), and $\mathcal{J}$ the corresponding sheaf on $\mathcal{C}$. Then $\mathcal{J} \otimes \mathcal{J} = \mathcal{I}$ where $I \subset xA_2[x]$ is generated by (x^2, tx).

We would like now to define an analog of the diagram (6). Namely, we consider the diagram

$$\begin{array}{ccc} \mathrm{Bun}_B'(\mathcal{C}_2) & \xrightarrow{p_2} & \mathrm{Bun}_2 \\ {\scriptstyle q_2}\big\downarrow & & \\ \mathrm{Pic}_2' \times \mathrm{Pic}_2', & & \end{array} \tag{7}$$

where $\mathrm{Bun}_B'(\mathcal{C}_2)$ consists of all short exact sequences

$$0 \to \mathcal{L}_1 \to \mathcal{F} \to \mathcal{L}_2 \to 0,$$

where $\mathcal{F} \in \mathrm{Bun}_2$, $\mathcal{L}_1$, $\mathcal{L}_2 \in \mathrm{Pic}_2'$. It is easy to see that in this case we have $\mathcal{L}_i = \underline{\mathrm{Hom}}(\mathcal{L}_j, \det(\mathcal{F}))$ for $i, j = 1, 2$ and $i \neq j$. So if we fix $\det(\mathcal{F})$ and one of the bundles $\mathcal{L}_1$ of $\mathcal{L}_2$, this determines the isomorphism class of the other.

Theorem 5.18. (1) *The space* $\mathcal{S}_{\mathrm{cusp}}(\mathrm{Bun}_2)$ *of Conjecture* 5.14 *is the space of functions* $\phi \in \mathcal{S}(\mathrm{Bun}_2)$ *such that* $(q_2)_! p_2^*(\phi) = 0$.

(2) $\dim \mathcal{S}_{\mathrm{cusp},\chi}(\mathrm{Bun}_2) < \infty$ *for any unitary character* $\chi : \mathrm{Pic}_2 \to \mathbb{C}^*$. *In fact, Conjecture* 5.14 *holds in this case.*

The proof will be discussed in another publication. Let us note that it is not difficult to deduce the second assertion of Theorem 5.18 from the first.

5.10. Main question

Assuming the above conjectures, one can ask how to describe the Hecke eigenvalues. It would be extremely interesting to relate them to some kind of Galois data (involving the dual group $G^\vee$). At the moment, we do not know how to do it in the non-archimedian case even for $G = \mathrm{GL}_2$ when we defined an action of these operators of finite-dimensional spaces $E_{\mathrm{Bun},1/2,n}(\mathcal{S}_{\mathrm{cusp},\chi}(\mathrm{Bun}_n))$.

In the archimedian case, a precise conjecture of this sort is formulated in [9] and [10]. We discuss it in the next section.

6. THE CASE $F = \mathbb{C}$

6.1. From Hecke operators to differential operators: the idea

In this section we specialize to the case $F = \mathbb{C}$. In this case, in addition to Hecke operators, one can introduce another player, namely the algebra of twisted (polynomial) differential operators on Bun, which will, roughly speaking, act on the same space as the Hecke operators and the two actions will commute. This will allow us to formulate a variant of Langlands conjecture in this case. More precisely, we are going to relate the Hecke eigenvalues to some particular $G^\vee$-local systems on $\mathcal{C}$ – opers with real monodromy. Let us begin by recalling basic information about opers and differential operators on Bun.

6.2. Opers

For a principal $G^\vee$-bundle $\mathcal{G}$ on $\mathcal{C}$, we denote by $\mathrm{Fl}_{\mathcal{G}}$ the associated Fl-bundle on $\mathcal{C}$ where Fl is the variety of Borel subgroups of $G^\vee$.

Definition 6.1. (1) A $G^\vee$-oper on $\mathcal{C}$ is a triple $(\mathcal{G}, \nabla, s)$, where $\mathcal{G}$ is a principal $G^\vee$-bundle on $\mathcal{C}$, ∇ is a connection on $\mathcal{G}$, and s is a section of $\mathrm{Fl}_{\mathcal{G}}$ satisfying an analog of the Griffiths-type condition with respect to ∇ (see [4]). We denote by $\mathrm{Oper}_{G^\vee}(\mathcal{C})$ the variety of opers.[10]

(2) For an oper $o = (\mathcal{F}, \nabla, s)$, we denote by $\zeta_o : \pi_1(C) \to G^\vee(\mathbb{C})$ the morphism defined by the connection ∇. We denote by $\mathrm{Oper}^\vee_G(\mathcal{C})^{\mathbb{R}} \subset \mathrm{Oper}^\vee_G(\mathcal{C})$ the subset of opers o such that the homomorphisms ζ_o and $\overline{\zeta_o}$ are conjugate, where $^- : G^\vee(\mathbb{C}) \to G^\vee(\mathbb{C})$ is the complex conjugation corresponding to a choice of a split real form of $G^\vee$.

Let us make several comments. First, it is known (cf. [3]) that given just a pair $(\mathcal{G}, \nabla)$, the B structure s is unique if it exists. Thus $\mathrm{Oper}_{G^\vee}(\mathcal{C})$ is actually a closed subset of the moduli stack of $G^\vee$-bundles with a connection (in other words, for such a local system to be an oper is a property rather than a structure). Second, let us comment on the reality condition in (2). Obviously, one way to guarantee this condition is to require that the monodromy representation of $\pi_1(\mathcal{C})$ corresponding to $(\mathcal{G}, \nabla)$ is conjugate to a homomorphism going

10 If $G^\vee$ is adjoint then the moduli stack of opers is, in fact, an algebraic variety (which is isomorphic to an affine space of dimension $\mathrm{rk}(G)$). If $G^\vee$ is not adjoint then formally one needs to consider the coarse moduli space here, since the center Z of $G^\vee$ is equal to the group of automorphisms of every oper. We shall ignore this subtlety for the rest of this section.

into $G^\vee(\mathbb{R})$ for a real split form of $G^\vee$. We expect that the converse is also true, and this is proved for $G^\vee = \mathrm{SL}(2)$ in [10] (Remark 1.8), but we do not know how to prove this in general. However, it is not hard to see (cf. again [10]) that up to conjugation the image of the monodromy homomorphism $\pi_1(\mathcal{C}) \to G^\vee$ lies in some inner form of the split real form of $G^\vee$. When we are in the setup of Section 5.7 and $|D| \geq 1$ it is also shown in [10] that the monodromy is lies in the split real form of $G^\vee$.

6.3. Opers and differential operators

Let $\mathcal{D}$ be the algebra of global sections of the sheaf $\mathbb{D}_{1/2}(\mathrm{Bun})$ of regular differential operators on $\omega_{\mathrm{Bun}}^{1/2}$. We denote by $\tau : \mathcal{D} \to \mathcal{D}$ the involution on $\mathcal{D}$ induced by the Cartan involution of G.

The following statement is one of the main results of [3] (a local version of this result appears in [13]).

Theorem 6.2.

(1) *The algebra $\mathcal{D}$ is commutative.*

(2) $\mathrm{Spec}(\mathcal{D}) = \mathrm{Oper}_{G^\vee}(\mathcal{C})$.

(3) *Let $o \in \mathrm{Oper}_{G^\vee}(\mathcal{C})$ and let $\chi_o : \mathcal{D} \to \mathbb{C}$ be the corresponding homomorphism. Let also $I_o \subset \mathbb{D}_{1/2}(\mathrm{Bun})$ be the sheaf of ideals of $\mathbb{D}_{1/2}(\mathrm{Bun})$ generated elements of the form $d - \chi_o(d)$ where $d \in \mathcal{D}$. Then the $\mathbb{D}_{1/2}(\mathrm{Bun})$-module $M_o := \mathbb{D}_{1/2}(\mathrm{Bun})/I_o$ is $\mathcal{O}_{\mathrm{Bun}}$-coherent when restricted to $\mathrm{Bun}_{\mathrm{vs}}$.*[11]

6.4. Differential operators and Hecke operators

Recall that we denote by $C^\infty_{1/2}(\mathrm{Bun}_{\mathrm{vs}})$ the space of smooth $1/2$-forms on $\mathrm{Bun}_{\mathrm{vs}}$. The algebra $\mathcal{A} := \mathcal{D} \otimes \overline{\mathcal{D}}$ acts naturally on $C^\infty_{1/2}(\mathrm{Bun}_{\mathrm{vs}})$. We denote by $\hat{\tau}$ the involution on $\mathcal{A}$ such that $\hat{\tau}(d_1 \otimes \bar{d}_2) = d_2^\tau \otimes \overline{d_1^\tau}$ and define $\mathcal{A}^{\mathbb{R}} \subset \mathcal{A}$ as the subalgebra of $\hat{\tau}$-fixed points.

We would like to claim that the action of the algebra $\mathcal{A}$ on $1/2$-forms commutes with the action of the Hecke operators. Here we must be careful, as a priori it is not clear how to construct one vector space on which both algebras will act. For this, we need to formulate one more definition.

Let us define a space $\mathcal{S}\mathrm{ch}(\mathrm{Bun})$ – "the Schwartz space of Bun." Namely, we set

$$\mathcal{S}\mathrm{ch}(\mathrm{Bun}) = \left\{\phi \in C^\infty_{1/2}(\mathrm{Bun}_{\mathrm{vs}}) \mid a(\phi) \in L^2(\mathrm{Bun}) \text{ for any } a \in \mathcal{A}\right\}. \tag{8}$$

For $a \in \mathcal{A}$, we denote by $\hat{a}$ the induced endomorphism of $\mathcal{S}\mathrm{ch}(\mathrm{Bun})$.

Note that by definition $\mathcal{S}\mathrm{ch}(\mathrm{Bun}) \subset L^2(\mathrm{Bun})$ and also $\mathcal{S}(\mathrm{Bun}_{\mathrm{vs}}) \subset \mathcal{S}\mathrm{ch}(\mathrm{Bun})$. The reader might ask why we start with C^∞-forms on $\mathrm{Bun}_{\mathrm{vs}}$ rather than on $\mathrm{Bun}_{\mathrm{st}}$. The reason is that below we want to study eigenvectors of $\mathcal{A}$ on $\mathcal{S}\mathrm{ch}(\mathrm{Bun})$, and it follows easily from Theorem 6.2(3) that any such eigenvector is automatically smooth on $\mathrm{Bun}_{\mathrm{vs}}$ (but there is no reason for it to be smooth on $\mathrm{Bun}_{\mathrm{st}}$).

---

11 Part (3) of this theorem explains the reason for our belief in Conjecture 3.5(2).

Conjecture 6.3. (1) Any $a \in \mathcal{A}^{\mathbb{R}}$ extends to an (unbounded) self-adjoint operator on $L^2(\mathrm{Bun})$.

(2) The space $\mathcal{S}\mathrm{ch}(\mathrm{Bun})$ is stable under the action of all Hecke operators.

(3) $\mathcal{S}\mathrm{ch}(\mathrm{Bun}) = \iota_{1/2}(\mathcal{S}_{1/2}(\mathrm{Bun}))$.[12]

(4) The action of $\mathcal{A}$ on $\mathcal{S}\mathrm{ch}(\mathrm{Bun})$ commutes with the action of Hecke operators.

(5) There exists a dense (in the L^2-sense) subspace $\mathcal{S}\mathrm{ch}(\mathrm{Bun})_0$ of $\mathcal{S}\mathrm{ch}(\mathrm{Bun})$ which is stable under $\mathcal{A}$ and the Hecke operators and such that $\mathcal{S}\mathrm{ch}(\mathrm{Bun})_0$ is a direct sum of 1-dimensional eigenspaces for $\mathcal{A}$ (in other words, the space $\mathcal{S}\mathrm{ch}(\mathrm{Bun})_0$ is locally finite dimensional for $\mathcal{A}$ and every generalized eigenvalue has multiplicity 1).

Let us comment on the multiplicity 1 statement. A 1/2-form is actually an eigenvector if it satisfies a certain system of linear differential equations. Locally on $\mathrm{Bun}_{\mathrm{vs}}(\mathbb{C})$, the space of solutions is finite dimensional but certainly not one dimensional (this has to do with the fact the $\mathbb{D}$-module M_o has high rank on $\mathrm{Bun}_{\mathrm{vs}}$; for example, for SL_2 this rank is 2^{3g-3}). However, globally most of these solutions become multivalued, so the multiplicity-one conjecture says that only one-dimensional space of solutions is single-valued globally. This, in fact, would follow if we knew that the $\mathbb{D}$-module M_o was irreducible and had regular singularities. For $G = \mathrm{PGL}_2$, this can be deduced from [20] (and probably similar analysis can be carried over for PGL_n).

Conjecture 6.3 implies that $L^2(\mathrm{Bun})$ is a (completed) direct sum of eigenspaces for $\mathcal{A}$ and eigenvalues have multiplicity 1. A priori any such eigenvalue is given by a pair of opers (o, o'), but part (1) of Conjecture 6.3 implies that $o' = \overline{o^{\tau}}$, so we are supposed to attach an eigenspace to a single oper o. It is also not difficult to see that $o \in \mathrm{Oper}_{G^\vee}(\mathcal{C})^{\mathbb{R}}$. We denote the corresponding eigenspace by $L^2(\mathrm{Bun})_o$. Note that Conjecture 6.3 implies that $L^2(\mathrm{Bun})_o \subset \mathcal{S}\mathrm{ch}(\mathrm{Bun})$.

Conjecture 6.4. We have $L^2(\mathrm{Bun})_o \neq 0$ if and only if $o \in \mathrm{Oper}_{G^\vee}(\mathcal{C})^{\mathbb{R}}$.

Remark 6.5. As was remarked above, the "only if" direction is easy. What is hard is to prove existence of eigenvectors for $\mathcal{A}$ which lie in L^2.

Note that Conjectures 5.5, 6.3, and 6.4 together imply the following

Corollary 6.6. *Let $\mathcal{W}$ denote the set of Hecke eigenvalues on $L^2(\mathrm{Bun})$. Then there exists a surjective map $\eta : \mathrm{Oper}_{G^\vee}(\mathcal{C})^{\mathbb{R}} \to \mathcal{W}$ such that for any $c \in \mathcal{C}$ and any $h \in \mathcal{H}(G, \mathbb{C})$ the operator $\mathbb{T}_{h,c}$ acts on $L^2(\mathrm{Bun})_0$ by $\eta(o)(h)$.*

Let us comment on the connection between Corollary 6.6 and Conjecture 5.5. We actually expect the map η to be finite-to-one (and in many cases it should be an isomorphism), so Conjecture 5.5 should imply that $\mathrm{Oper}_{G^\vee}(\mathcal{C})^{\mathbb{R}}$ should be a discrete subset of $\mathrm{Oper}_{G^\vee}(\mathcal{C})$.

---

12 Note that if we assume the validity of Conjecture 3.5 for $F = \mathbb{C}$, then (3) implies (2).

This assertion is not obvious, and at the moment we do not know how to prove it in general, but let us note that for $G^\vee = \mathrm{PGL}_2$ it was proven by G. Faltings in [11].

6.5. Eigenvalues of Hecke operators

We conclude this section by describing a conjectural formula for the map η (the contents of this subsection is described in more detail in [10]). More precisely, we are going to do the following. We would like to understand the scalar by which the operator $\mathbb{T}_{\lambda,c}$ acts in $L^2(\mathrm{Bun})_o$. We can actually regard c as a variable here. In view of Remark 5.7, this eigenvalue is, in fact, a section $\Phi_{\lambda,o}$ of $|\omega_{\mathcal{C},c}|^{-\langle\lambda,\rho^\vee\rangle}$ (recall that $\rho^\vee$ denotes the half-sum of positive coroots of G).

For $\lambda \in \Lambda^+$, let V_λ be the corresponding irreducible finite-dimensional representation of $G^\vee$. Choose an $o = (\mathcal{F}, \nabla, s) \in \mathrm{Oper}_{G^\vee}(\mathcal{C})$. Moreover, the Griffiths transversality condition implies that the $T^\vee$-bundle induced from the $B^\vee$-structure s by means of the homomorphism $B^\vee \to T^\vee$ is induced from $\omega_\mathcal{C}$ by means of the cocharacter $\rho^\vee : \mathbb{G}_m \to T^\vee$.[13] Therefore if we denote by $(\mathcal{V}_{o,\lambda}, \nabla_{o,\lambda})$ the vector bundle on $\mathcal{C}$ associated to $\mathcal{F}$ via the representation V_λ (with the corresponding flat connection), then s defines an embedding

$$\omega_C^{\langle\lambda,\rho^\vee\rangle} \hookrightarrow \mathcal{V}_{o,\lambda}$$

and hence a morphism

$$\mathcal{O}_\mathcal{C} \hookrightarrow \omega_C^{-\langle\lambda,\rho^\vee\rangle} \otimes \mathcal{V}_{o,\lambda}.$$

We let σ_λ be the image of 1 under this morphism.

Let now $o \in \mathrm{Oper}_{G^\vee}(C)^\mathbb{R}$. Then we have isomorphism of $\mathcal{V}_{o,\lambda}$ and $\overline{\mathcal{V}}_{o,\lambda}$ as flat C^∞-bundles (and this isomorphism is canonical up to the action of the center of $G^\vee$). Since $V_\lambda^* \simeq V_{-w_0(\lambda)}$, we get a pairing $(\cdot,\cdot)_\lambda$ between C^∞-sections of $\mathcal{V}_{o,\lambda}$ and of $\overline{\mathcal{V}}_{o,\lambda}$. Since $\langle -w_0(\lambda), \rho^\vee\rangle = \langle\lambda,\rho^\vee\rangle$, we can regard $\overline{\sigma}_{-w_0(\lambda)}$ as a section of $\overline{\omega}_\mathcal{C}^{-\langle\lambda,\rho^\vee\rangle} \otimes \overline{\mathcal{V}}_\lambda^*$. Since $\omega_\mathcal{C}^{-\langle\lambda,\rho^\vee\rangle} \otimes \overline{\omega}_\mathcal{C}^{-\langle\lambda,\rho^\vee\rangle} = |\omega_\mathcal{C}|^{-\langle\lambda,\rho^\vee\rangle}$, we can formulate the following Conjecture (cf. [10]):

Conjecture 6.7.

$$\Phi_{\lambda,o} = (\sigma_\lambda, \overline{\sigma}_{-w_0(\lambda)})_\lambda \in C^\infty\big(\mathcal{C}, |\omega_\mathcal{C}|^{-\langle\lambda,\rho^\vee\rangle}\big).$$

6.6. Parabolic bundles: results

All the conjectures of this section can be easily generalized to the setup of Section 5.7. In the case when $\mathcal{C}$ is $\mathbb{P}^1$ and the cardinality of the divisor D is 3, 4, or 5, they are proven in [8] (and most of them are proven in [8] even for $|D| > 5$).

13 Strictly speaking, this makes sense only if $G^\vee$ is simply connected since $\rho^\vee$ is a well-defined cocharacter of $T^\vee$ only in that case. For general G, the corresponding $T^\vee$-bundle is induced from $\omega_\mathcal{C}^{1/2}$ by the character $2\rho^\vee$ for some choice of $\omega_\mathcal{C}^{1/2}$. To simplify the notation, we are going to write the answer in the case when $G^\vee$ is simply connected – the generalization to any G is straightforward.

7. THE CASE $F = \mathbb{R}$

In this section we would like to describe the conjectural picture of the analytic Langlands correspondence in the case $F = \mathbb{R}$. This picture has been developed by P. Etingof, E. Frenkel, D. Gaiotto, D. Kazhdan, and E. Witten, and is discussed in [19, SECTION 6].

Warning. Some of the letters used in the previous section (such as σ or τ) will have a different meaning in this section.

7.1. Real groups, L-groups, and all that

Let G be a connected complex semisimple group. Recall that a *real structure* on G is defined by an antiholomorphic involution $\sigma : G \to G$. The corresponding *group of real points* is G^σ (it may be disconnected). The inner class of σ gives rise to a based root datum involution $s = s_\sigma$ for G which is also one for $G^\vee$. If G is semisimple, this is just a Dynkin diagram automorphism.

Recall [1] that to G, s we may attach the *Langlands L-group* ${}^LG = {}^LG_s$, the semidirect product of $\mathbb{Z}/2 = \mathrm{Gal}(\mathbb{C}/\mathbb{R})$ by $G^\vee$, with the action of $\mathbb{Z}/2$ defined by $\gamma \circ s$, where γ is the Cartan involution.

7.2. L-systems

Let $\mathcal{C}$ be a compact complex Riemann surface of genus $g \geq 2$. Let $\tau : \mathcal{C} \to \mathcal{C}$ be an antiholomorphic involution. Given a holomorphic principal G-bundle $\mathcal{E}$ on $\mathcal{C}$, we can define the antiholomorphic bundle $\tau(\mathcal{E})$, hence a holomorphic bundle $\sigma\tau(\mathcal{E})$. Let us say that $\mathcal{E}$ is *real* under σ if there exists an isomorphism $A : \mathcal{E} \to \sigma\tau(\mathcal{E})$ such that

$$\sigma\tau(A) \circ A = 1. \tag{9}$$

This isomorphism A is unique if it exists, and (9) is automatic if $\mathrm{Aut}(\mathcal{E}) = 1$, which happens generically for stable bundles if G is adjoint. In this case, $gA : \mathcal{E} \to g\sigma\tau(\mathcal{E})$ has the same property for $\sigma' = g\sigma$, where $g \in G$ and $g\sigma(g) = 1$. Thus the moduli space of such stable bundles depends only on s [5, PROPOSITION 3.8]. We will denote it by $\mathrm{Bun}_{G,s}$.

Consider the simplest case when τ has no fixed points, i.e., $\mathcal{C}(\mathbb{R}) = \emptyset$. Let ζ be a local system on the nonorientable surface $\mathcal{C}/\tau$ with structure group ${}^LG$. Let us say that ζ is an *L-system* if it attaches to every orientation-reversing path in $\mathcal{C}/\tau$ a conjugacy class in ${}^LG$ that maps to the nontrivial element in $\mathbb{Z}/2$. The following conjecture is formulated in [19, SECTION 6] (in the case of the compact inner class).

Conjecture 7.1. The spectrum of Hecke operators on $L^2(\mathrm{Bun})$ is parametrized by L-systems on $\mathcal{C}/\tau$ with values in ${}^LG = {}^LG_s$ whose pullback to $\mathcal{C}$ has a structure of a G-oper.

Example 7.2. Let $s = \gamma$. Then ${}^LG = \mathbb{Z}/2 \times G^\vee$, so an L-system is the same thing as a $G^\vee$-local system on $\mathcal{C}/\tau$. So in this case the condition on the $G^\vee$-local system on $\mathcal{C}$ to occur in the spectrum is (conjecturally) that it extends to the 3-manifold $M := (\mathcal{C} \times [-1, 1])/(\tau, -\mathrm{Id})$ whose boundary is $\mathcal{C}$ (and this extension is a part of the data).

Namely, in this case the spectral local systems are ζ which are isomorphic to ζ^τ and such that ζ is an oper (hence also an anti-oper), so ζ is a real oper "with real coefficients." But among these we should only choose those that extend to $\mathcal{C}/\tau$ (and then the multiplicity of eigenvalue may be related to the number of such extensions). This agrees with the picture [19, SECTION 6.2] coming from 4-dimensional supersymmetric gauge theory.[14] More precisely, recall that by a result of Beilinson and Drinfeld [3], opers for adjoint groups have no nontrivial automorphisms. So for any connected semisimple G, we get an obstruction for such ζ to extend to $\mathcal{C}/\tau$ which lies in $Z/Z^2 = H^2(\mathbb{Z}/2, Z)$, where Z is the center of $G^\vee$.[15] Moreover, if this obstruction vanishes then the freedom for choosing the extension is in a torsor over $H^1(\mathbb{Z}/2, Z) = Z_2$, the 2-torsion subgroup in Z.

Example 7.3. Let $G = K \times K$ for some complex group K, and s be the permutation of components (the only real form in this inner class is K regarded as a real group). This is equivalent to the case $F = \mathbb{C}$ considered above (for $\mathcal{C}$ defined over $\mathbb{R}$). Then ${}^LG_s = {}^LG_{\gamma\circ s} = \mathbb{Z}/2 \ltimes (K^\vee \times K^\vee)$, where $\mathbb{Z}/2$ acts by permutation. So an L-system is a $K^\vee \times K^\vee$ local system on $\mathcal{C}$ of the form (ρ, ρ^τ). Thus the spectrum is parametrized by ζ such that both ζ and ζ^τ are opers, i.e., ζ is both an oper and an anti-oper, i.e. a real oper, which agrees with the conjecture for $F = \mathbb{C}$. (Note that in this case $H^i(\mathbb{Z}/2, Z) = 1$ so there is no obstructions or freedom for extensions).

Remark 7.4. If $\mathcal{C}(\mathbb{R}) \neq \emptyset$, the story gets more complicated, and we will not discuss the details here. Let us just indicate that, as explained in [19, SECTION 6], to define the appropriate moduli space and the spectral problem on it, we need to fix a real form G_i of G in the inner class s for each component (oval) C_i of $\mathcal{C}(\mathbb{R})$, and the eigenvalues of Hecke operators are conjecturally parametrized by a certain kind of "real" opers corresponding to this data, i.e., opers with real coefficients satisfying appropriate reality conditions on the monodromy of the corresponding $G^\vee$-connection. Furthermore, in the tamely ramified case, when we also have a collection of marked points D on $\mathcal{C}$ defined over $\mathbb{R}$, to define the most general version of our spectral problem, we need to fix a unitary representation π_i of the real group G_i for every marked point $c \in D$ on C_i and a unitary representation of the complex group $G_\mathbb{C}$ for every pair of complex conjugate marked points $c, \bar{c} \in D$ not belonging to $\mathcal{C}(\mathbb{R})$. For example, the case of parabolic structures corresponds to taking s to be the split inner class, G_i the split forms, and π_i the unitary principal series representations. In the genus zero case,

---

14 More precisely, as was explained to us by E. Witten, what comes from ordinary gauge theory is this picture for the compact inner class s. To obtain other inner classes, one needs to consider twisted gauge theory where the twisting is by a Dynkin diagram automorphism of G. Namely, gauge fields in this theory are invariant under complex conjugation τ up to such an automorphism.

15 Indeed, $\pi_1(\mathcal{C}/\tau)$ is generated by $\pi_1(\mathcal{C})$ and an element t such that $tbt^{-1} = \beta(b)$ for some automorphism β of $\pi_1(\mathcal{C})$, and $t^2 = c \in \pi_1(\mathcal{C})$, so that $\beta^2(b) = cbc^{-1}$. So given a representation $\zeta : \pi_1(\mathcal{C}) \to G^\vee$, an L-system would be given by an assignment $\zeta(t) = T \in G^\vee$ such that (1) $T^2 = \zeta(c)$ and (2) $T\zeta(a)T^{-1} = \zeta(\beta(a))$. If $\zeta \cong \rho \circ \beta$ then T satisfying (2) is unique up to multiplying by $u \in Z$, and $T^2 = \zeta(c)z$, $z \in Z$. Moreover, if T is replaced by Tu then z is replaced by zu^2, hence the obstruction to satisfying (1) lies in Z/Z^2.

this was discussed in detail in [8], and it was shown that this problem leads to appearance of T-systems.

7.3. Connection to Gaudin model

Recall that the *Gaudin model* for a simple complex Lie algebra $\mathfrak{g}$ is the problem of diagonalization of the *Gaudin hamiltonians*

$$H_i := \sum_{1 \leq j \leq N, j \neq i} \frac{\Omega_{ij}}{z_i - z_j}$$

on the space $(V_1 \otimes \cdots \otimes V_N)^{\mathfrak{g}}$, where V_i are finite-dimensional $\mathfrak{g}$-modules, $z_i \in \mathbb{C}$ are distinct points, $\Omega \in (S^2\mathfrak{g})^{\mathfrak{g}}$ is the Casimir tensor dual to the Killing form, and Ω_{ij} denotes the action of Ω in the ith and jth factor. These operators commute, and if $\mathfrak{g} \neq \mathfrak{sl}_2$ then there are also higher Gaudin hamiltonians associated to the Feigin–Frenkel higher Sugawara central elements at the critical level (see [14]), which commute with each other and with H_i, and the problem is to simultaneously diagonalize all these operators.

It turns out that this problem (for real z_i) is a special case of the spectral problem considered in this paper, in the case $F = \mathbb{R}$. Namely, let us take $\mathscr{C} = \mathbb{P}^1$ with the usual real structure and fix the compact inner class s of the complex simply connected group G with $\mathrm{Lie}(G) = \mathfrak{g}$. As explained in the previous remark, on the real locus $\mathbb{P}^1(\mathbb{R})$, we are supposed to fix a real form of G in this inner class, and we fix the compact form G_c. Further, consider marked points $z_1, \ldots, z_N$ on the real locus (the tamely ramified case). Then we are supposed to fix a unitary representation of G_c at every z_i, and we take it to be V_i. Then the Hilbert space of the analytic Langlands theory is $\mathscr{H} = (V_1 \otimes \cdots \otimes V_n)^{G_c}$ (so in this case it is finite dimensional), and the quantum Hitchin system comprises the Gaudin hamiltonians (including the higher ones), cf. [14].

As is explained in [15, 17], the Bethe ansatz method shows that the eigenvectors of the Gaudin hamiltonians are labeled by monodromy-free $G^\vee$-opers on $\mathbb{P}^1$ with first-order poles at z_i and residues in the conjugacy class of $-\lambda_i - \rho$, where λ_i is the highest weight of V_i. These are exactly the "real opers" for this situation. Thus the results of [15, 17] may be considered as a finite-dimensional instance of the tamely ramified analytic Langlands correspondence for genus zero and $F = \mathbb{R}$.

ACKNOWLEDGMENTS

The first author was partially supported by NSERC. The second named was partially supported by the ERC grant No 669655. We would like to thank P. Etingof, E. Frenkel, and A. Polishchuk for the help with the creation of this text and D. Gaiotto and E. Witten for a very useful discussion.

REFERENCES

[1] J. Adams, D. Barbasch, and D. Vogan, *The Langlands classification and irreducible characters for real reductive groups*, Progr. Math. 104, Birkhäuser, Boston, 1992.

[2] A. Aizenbud and N. Avni, Representation growth and rational singularities of the moduli space of local systems. *Invent. Math.* **204** (2016), 245–316.

[3] A. Beilinson and V. Drinfeld, Quantization of Hitchin's integrable system and Hecke eigen-sheaves, preprint.

[4] A. Beilinson and V. Drinfeld, Opers. 2005, arXiv:math/0501398.

[5] I. Biswas, O. Garcia-Prada, and J. Hurtubise, Pseudo-real principal G-bundles over a real curve. 2015, arXiv:1502.00563.

[6] A. Braverman and D. Kazhdan, Some examples of Hecke algebras for two-dimensional local fields. *Nagoya Math. J.* **184** (2006), 57–84.

[7] V. Drinfeld and D. Gaitsgory, Compact generation of the category of D-modules on the stack of G-bundles on a curve. *Camb. J. Math.* **3** (2015), no. 1–2, 19–125.

[8] P. Etingof, E. Frenkel, and D. Kazhdan, Analytic Langlands correspondence for PGL(2) on $\mathbb{P}^1$ with parabolic structures over local fields. 2021, arXiv:2106.05243.

[9] P. Etingof, E. Frenkel, and D. Kazhdan, An analytic version of the Langlands correspondence for complex curves. In *Integrability, quantization, and geometry. II*, pp. 137–202, Proc. Sympos. Pure Math. 103.2, Amer. Math. Soc., Providence, RI, 2021.

[10] P. Etingof, E. Frenkel, and D. Kazhdan, Hecke operators and analytic Langlands correspondence for curves over local fields. 2021, arXiv:2103.01509.

[11] G. Faltings, Real projective structures on Riemann surfaces. *Compos. Math.* **48** (1983), 223–269.

[12] G. Faltings, Algebraic loop groups and moduli spaces of bundles. *J. Eur. Math. Soc.* **5** (2003), 41–68.

[13] B. Feigin and E. Frenkel, Affine Kac–Moody algebras at the critical level and Gelfand–Dikii algebras. In *Infinite analysis (Kyoto, 1991)*, pp. 197–215, Adv. Ser. Math. Phys. 16, World Sci. Publ., River Edge, NJ, 1992.

[14] B. Feigin, E. Frenkel, and N. Reshetikhin, Gaudin model, Bethe Ansatz and critical level. *Comm. Math. Phys.* **166** (1994), 27–62.

[15] E. Frenkel, Opers on the projective line, flag manifolds and Bethe ansatz. *Mosc. Math. J.* **4** (2004), no. 3, 655–705.

[16] E. Frenkel, Is there an analytic theory of automorphic functions for complex algebraic curves? *SIGMA* **16** (2020), arXiv:1812.08160.

[17] E. Frenkel, Gaudin model and opers. In *Infinite dimensional algebras and quantum integrable systems*, edited by P. P. Kulish, N. Manojlovich, and H. Samtleben, Progr. Math. 237, Birkhäuser, Basel, 2005.

[18] E. Frenkel, Gauge theory and Langlands duality talk given at MSRI, Berkeley, in September 2014, available at https://youtu.be/NQfeBeRKMiw?t=4190.

[19] D. Gaiotto and E. Witten, Gauge theory and the analytic form of the geometric Langlands program. 2021, arXiv:2107.01732.

[20] D. Gaitsgory, Outline of the proof of the geometric Langlands conjecture for GL(2). 2014, arXiv:1302.2506.

[21] D. Gaitsgory and D. Kazhdan, Algebraic groups over a 2-dimensional local field: some further constructions. In *Studies in Lie theory*, pp. 97–130, Progr. Math. 243, Birkhäuser, 2006.

[22] M. Kontsevich, Notes on motives in finite characteristic. In *Algebra, arithmetic, and geometry, in honor of Yu. I. Manin, vol. II*, pp. 213–247, Progr. Math. 270, Springer, 2010.

[23] S. Kumar, Demazure character formula in arbitrary Kac–Moody setting. *Invent. Math.* **89** (1987), no. 2, 395–423.

[24] V. Lafforgue, Shtukas for reductive groups and Langlands correspondence for function field. *J. Amer. Math. Soc.* **31** (2018), no. 3, 719–891.

[25] R. P. Langlands, On analytic form of geometric theory of automorphic forms (in Russian). Preprint, http://publications.ias.edu/rpl/paper/2678.

[26] O. Mathieu, Formules de caractères pour les algèbres de Kac–Moody générales. *Astérisque* **159–160** (1988), 267.

[27] J. Teschner, Quantisation conditions of the quantum Hitchin system and the real geometric Langlands correspondence. In *Geometry and physics, Festschrift in honour of Nigel Hitchin, Vol. I*, edited by J. E. Andersen, A. Dancer, and O. Garcia-Prada, pp. 347–375, Oxford University Press, 2018, arXiv:1707.07873.

ALEXANDER BRAVERMAN

Department of Mathematics, University of Toronto, Toronto, Canada, Perimeter Institute of Theoretical Physics, Waterloo, Canada, braval@math.toronto.edu

DAVID KAZHDAN

Department of Mathematics, Hebrew University of Jerusalem, Jerusalem, Israel, kazhdan@math.huji.ac.il

This Page Intentionally Left Blank

EVOLUTION OF FORM AND SHAPE

TOBIAS HOLCK COLDING

ABSTRACT

The evolution of form and shape can be described by differential equations. Many of these equations originate in various branches of science and engineering. They are fundamental and in a sense canonical. The fact that they make sense geometrically means that they are relevant everywhere and have fundamental properties that appear over and over in many settings. Understanding them requires simultaneous insight into analysis and geometry and the interplay between these.

MATHEMATICS SUBJECT CLASSIFICATION 2020

Primary 53C44; Secondary 35K55, 53A10, 53E20, 49Q05, 35J15

KEYWORDS

Geometric flows, function theory on manifolds, minimal surfaces, arrival time, degenerate elliptic and parabolic PDEs, optimal regularity and growth, uniqueness of blowups

ICM INTERNATIONAL CONGRESS OF MATHEMATICIANS 2022 JULY 6–14

Proc. Int. Cong. Math. 2022, Vol. 2, pp. 826–871
DOI 10.4171/ICM2022/142

Published by EMS Press

1. INTRODUCTION

The evolution of form and shape can be described by differential equations. These equations are classical, and those we will consider are variants of the heat equation that governs how heat distributes over time. The questions and equations, many of which originate in various branches of science and engineering, are fundamental and in a sense canonical, and as a consequence come up in many areas. The Laplace equation, for example, is the canonical linear second order partial differential equation once we have a metric structure. The Laplace operator appears classically in the physics of gravity, electricity and magnetism, fluid mechanics, and quantum mechanics, it has played a central role in many areas of mathematics, and its study in increasing generality played a central role in the development of the theory of PDEs. The fact that the equations make sense geometrically means that they are relevant everywhere in physical settings, and they have certain fundamental properties that appear over and over. Understanding them requires simultaneous insight into analysis and geometry, and the interplay between these. The new ideas and techniques to deal with these questions apply to many different situations. Recent years have seen dramatic progress on many of these questions thanks to the combined efforts of many people with different points of views and techniques. The goal here is to give a flavor of some of these results.

The first equation we will consider is mean curvature flow of hypersurfaces. Surface tension is the tendency of fluid surfaces to shrink into the smallest surface area possible. Mathematically, the force of surface tension is described by the mean curvature. In equilibrium the mean curvature is zero and one gets minimal surfaces. Minimal surfaces date back to Euler and Lagrange and the beginning of the calculus of variations. Many of the techniques developed have played key roles in geometry and partial differential equations. Examples include monotonicity and tangent cone analysis originating in the regularity theory for minimal surfaces, estimates for nonlinear equations based on the maximum principle arising in Bernstein's classical work, and even Lebesgue's definition of the integral that he developed in his thesis on the Plateau problem for minimal surfaces.

Under mean curvature flow, the surface moves to decrease surface area as fast as possible. If we think of the hypersurface as the level set of a function and insist that all level sets move by mean curvature flow, then this gives rise to a nonlinear degenerate parabolic PDE on a Euclidean space. This is the level set formulation of the equation. The level set method has been intensively studied in many pure and applied fields over the last 35 years. One of the first questions that comes up is the regularity of solutions. The equation is degenerate and a priori solutions are only defined weakly. We will see that the regularity of solutions is equivalent to a question that has been widely studied in geometry over the last 40 years, namely, the question of uniqueness of blowups. This is very much in the spirit of the simple fact that a function is differentiable at a point if, at all sufficiently small scales, it not only looks like a linear function but the same linear function independent of scale.

As growth of solutions to PDEs plays an important role in many different areas, we will discuss the growth of some classical and basic equations on manifolds. These include harmonic and caloric functions. That is, functions that are either solutions to the Laplace equation or the heat equation. We will also discuss more general eigenfunctions of drift equations. Drift Laplacians are ubiquitous in many areas, including quantum field theory, stochastic PDE, and anywhere the heat equation or Gaussian appears, such as functional inequalities, parabolic PDEs, geometric flows, and probability. The drift term arises in two different ways. One is whenever there is a natural scaling or, more generally, a gradient flow. A second way it arises is when there is a natural measure, in which case the drift operator is the canonical self-adjoint second order operator. There is a long history of studying the growth of solutions to differential equations, inequalities, and systems. These new growth estimates have direct application to longstanding open questions.

Analysis of noncompact manifolds almost always requires some controlled behavior at infinity. Without such, one can neither show nor expect strong properties. On the other hand, such assumptions restrict the possible applications and often too severely. In a wide range of areas, noncompact spaces come with a Gaussian weight and a drift Laplacian. Eigenfunctions are L^2 in the weighted space allowing for extremely rapid growth. Rapid growth would be disastrous for many applications. Surprisingly, for very general tensors, manifolds, and weights, we will show the same polynomial growth bounds that Laplace and Hermite observed for functions on a Euclidean space for the standard Gaussian. This covers all shrinkers for Ricci and mean curvature flows.

These new growth estimates for the PDEs open a door to study delicate analytical questions on a wide class of non-compact manifolds without assuming any asymptotic decay at infinity. They provide an analytic framework for investigating nonlinear PDE on Gaussian spaces where previously the Gaussian weight allowed wild growth that made it impossible to approximate nonlinear by linear. They are key to bound the growth of diffeomorphisms of noncompact manifolds and to solving the "gauge problem." Many key problems are defined intrinsically without a canonical coordinate system. In those problems, the infinite-dimensional diffeomorphism group (gauge group) becomes a major issue and dealing with it a major obstacle. Ricci flow is such an example. There are many problems where this degeneracy under diffeomorphisms plays a central role, but most techniques rely on compactness or rapid decay which we do not have in the situations we consider.

Another common feature for all of these problems is that they are dynamical and can be thought of as infinite-dimensional dynamical systems. Classical results from dynamics do not apply directly, but they do give some guiding principles, [85, 88, 92]. In mathematics, structural stability is a fundamental property of a dynamical system, which means that the qualitative behavior of the trajectories is unaffected by small perturbations. Given a smooth function f on a finite-dimensional space, the gradient ∇f points in the direction of the steepest ascent. The critical points of f are the points where ∇f vanishes. If p is a local minimum of f, then the second derivative test tells us that the Hessian matrix of f at p is nonnegative. More generally, the number of negative eigenvalues of the Hessian is called the index of the critical point. A fundamental method to find the minimum of f is the method of gradient descent. Here, we make an initial guess p_0 and then iteratively move in the negative gradient direction, the direction of the steepest descent, by setting $p_{i+1} = p_i - \nabla f(p_i)$. The function $f(x(t))$ decreases as efficiently as possible as $x(t)$ heads towards the minimum.

The dynamics near a nondegenerate critical point are determined by the index. If the index is zero, then the critical point is attracting and the entire neighborhood flows towards the critical point. However, when the index is positive, a generic point will flow out of the neighborhood, missing the critical point. In the final part we will discuss stable structures in geometry.

Part 1. Optimal regularity of PDEs. In mean curvature flow, the velocity vector field is the mean curvature vector and the evolving front is the level set of a function that satisfies a nonlinear degenerate parabolic equation. Solutions are defined in a weak, so-called "viscosity" sense; in general, they may not even be differentiable (let alone twice differentiable). However, it turns out that for a monotonically advancing front viscosity solutions are in fact twice differentiable and satisfy the equation in the classical sense. Moreover, the situation becomes very rigid when the second derivative is continuous.

Suppose $\Sigma \subset \mathbf{R}^{n+1}$ is an embedded hypersurface and $\mathbf{n}$ is the unit normal of Σ. The *mean curvature* is given by $H = \mathrm{div}_\Sigma(\mathbf{n})$. Here

$$\mathrm{div}_\Sigma(\mathbf{n}) = \sum_{i=1}^{n} \langle \nabla_{e_i} \mathbf{n}, e_i \rangle,$$

where e_i is an orthonormal basis for the tangent space of Σ. For example, at a point where $\mathbf{n}$ points in the x_{n+1} direction and the principal directions are in the other axis directions,

$$\mathrm{div}_\Sigma(\mathbf{n}) = \sum_{i=1}^{n} \frac{\partial \mathbf{n}_i}{\partial x_i}$$

is the sum (n times the mean) of the principal curvatures. If $\Sigma = u^{-1}(s)$ is the level set of a function u on $\mathbf{R}^{n+1}$ and s is a regular value, then $\mathbf{n} = \frac{\nabla u}{|\nabla u|}$ and

$$H = \sum_{i=1}^{n} \langle \nabla_{e_i} \mathbf{n}, e_i \rangle = \mathrm{div}_{\mathbf{R}^{n+1}} \left(\frac{\nabla u}{|\nabla u|} \right).$$

The last equality used that $\langle \nabla_{\mathbf{n}} \mathbf{n}, \mathbf{n} \rangle$ is automatically 0 because $\mathbf{n}$ is a unit vector.

A one-parameter family of smooth hypersurfaces $M_t \subset \mathbf{R}^{n+1}$ flows by the *mean curvature flow* if the speed is equal to the mean curvature and points inward:

$$x_t = -H\mathbf{n},$$

where H and $\mathbf{n}$ are the mean curvature and unit normal of M_t at the point x. Our flows will always start at a smooth embedded connected hypersurface, even if it becomes disconnected and nonsmooth at later times. The earliest reference to the mean curvature flow we know of is in the work of Birkhoff from the 1910s, where he used a discrete version of this, and independently in the material science literature of the 1920s.

Two key properties.

- H is the gradient of area, so the mean curvature flow is the negative gradient flow for volume (Vol M_t decreases most efficiently).
- (Avoidance property) If M_0 and N_0 are disjoint, then M_t and N_t remain disjoint.

The avoidance principle is simply a geometric formulation of the maximum principle. An application of it shows that if one closed hypersurface encloses another, then the outer one can never catch up with the inner. The reason for this is that if it did there would be a first point of contact, and right before that the inner one would contract faster than the outer, contradicting that the outer was catching up.

Curve shortening flow. When $n = 1$ and the hypersurface is a curve, the flow is the curve shortening flow. Under the curve shortening flow, a round circle shrinks through round circles to a point in finite time. A remarkable result of Grayson [103] from 1987 (using earlier work of Gage and Hamilton [100]) shows that any simple closed curve in the plane remains smooth under the flow until it disappears in finite time in a point. Right before it disappears, the curve will be an almost round circle.

Level set flow. The analytical formulation of the flow is the level set equation that can be deduced as follows. Given a closed embedded hypersurface $\Sigma \subset \mathbf{R}^{n+1}$, choose a function $v_0 : \mathbf{R}^{n+1} \to \mathbf{R}$ that is zero on Σ, positive inside the domain bounded by Σ, and negative outside. (Alternatively, choose a function that is negative inside and positive outside.)

- If we simultaneously flow $\{v_0 = s_1\}$ and $\{v_0 = s_2\}$ for $s_1 \neq s_2$, then avoidance implies they stay disjoint.
- In the level set flow, we look for $v : \mathbf{R}^{n+1} \times [0, \infty) \to \mathbf{R}$ so that each level set $t \to \{v(\cdot, t) = s\}$ flows by mean curvature and $v(\cdot, 0) = v_0$.
- If $\nabla v \neq 0$ and the level sets of v flow by mean curvature, then

$$v_t = |\nabla v| \operatorname{div}\left(\frac{\nabla v}{|\nabla v|}\right).$$

This is degenerate parabolic and undefined when $\nabla v = 0$. It may not have classical solutions.

In a paper from 1988, Osher and Sethian [159] studied this equation numerically. The analytical foundation was provided by Evans and Spruck [98] in a series of four papers in the early 1990s and, independently and at the same time, by Chen, Giga, and Goto [41]; see also [5]. Both of these two groups constructed (continuous) viscosity solutions and showed uniqueness. The notion of viscosity solutions had been developed by Lions and Crandall in the early 1980s. The work of these two groups on the level set flow was one of the significant applications of this theory.

Examples of singularities. Under mean curvature flow, a round sphere remains round but shrinks and eventually becomes extinct in a point. A round cylinder remains round and eventually becomes extinct in a line. The marriage ring is the example of a thin torus of revolution in $\mathbf{R}^3$. Under the flow, the marriage ring shrinks to a circle then disappears.

Dumbbell. If the neck is sufficiently thin, then under the evolution the neck of a rotationally symmetric mean convex dumbbell in $\mathbf{R}^3$ pinches off first and the surface disconnects into two components. Later each component (bell) shrinks to a round point. This example falls into a larger category of surfaces that are rotationally symmetric around an axis. Because of the symmetry, then the solution reduces to a one-dimensional heat equation. This was analyzed already in the early 1990s by Angenent, Altschuler, and Giga [4]; cf. also the work of Soner and Souganidis from around the same time. A key tool in the arguments of Angenent–Altschuler–Giga was a parabolic Sturm–Liouville theorem of Angenent that holds in one spatial dimension.

Singular set. Under mean curvature flow, closed hypersurfaces contract, develop singularities, and eventually become extinct. The *singular set* $\mathcal{S}$ is the set of points in space and time where the flow is not smooth.

In the first three examples—the sphere, cylinder, and marriage ring—$\mathcal{S}$ is a point, line, and closed curve, respectively. In each case, the singularities occur only at a single time. In contrast, the dumbbell has two singular times with one singular point at the first time and two at the second.

Mean convex flows. A hypersurface is convex if every principal curvature is positive. It is mean convex if $H > 0$, i.e., if the sum of the principal curvatures is positive at every point. Under the mean curvature flow, a mean convex hypersurface moves inward and, since mean convexity is preserved, it will continue to move inward and eventually sweep out the entire compact domain bounded by the initial hypersurface.

Monotone movement can be modeled particularly efficiently numerically by the Fast Marching Method of Sethian.

Level set flow for mean convex hypersurfaces. When the hypersurfaces are mean convex, the equation can be rewritten as a degenerate elliptic equation for a function u defined by

$$u(x) = \{t \mid x \in M_t\}.$$

We say that u is the *arrival time* since it is the time the hypersurfaces M_t arrive at x as the front sweeps through the compact domain bounded by the initial hypersurface. Kohn and Serfaty [131] provided a game theoretic interpretation of the arrival time. It follows easily that if we set $v(x,t) = u(x) - t$, then v satisfies the level set flow. Now the level set equation $v_t = |\nabla v| \operatorname{div}(\nabla v / |\nabla v|)$ becomes

$$-1 = |\nabla u| \operatorname{div}\left(\frac{\nabla u}{|\nabla u|}\right).$$

This is a degenerate elliptic equation that is undefined when $\nabla u = 0$. Note that if u satisfies this equation, then so does u plus a constant. This just corresponds to shifting the time when the flow arrives by a constant. A particular example of a solution to this equation is the function $u = -\frac{1}{2}(x_1^2 + x_2^2)$, that is, the arrival time for shrinking round cylinders in $\mathbf{R}^3$. In general, Evans–Spruck (cf. Chen–Giga–Goto) constructed Lipschitz solutions to this equation.

Singular set of mean convex level set flow. The singular set of the flow is the critical set of u. Namely, $(x, u(x))$ is singular if and only if $\nabla_x u = 0$. For instance, in the example of the shrinking round cylinders in $\mathbf{R}^3$, the arrival time is given by $u = -\frac{1}{2}(x_1^2 + x_2^2)$ and the flow is singular in the line $x_1 = x_2 = 0$; that is, exactly where $\nabla u = 0$.

We will next see that even though the arrival time was only a solution to the level set equation in a weak sense, it always turns out to be a twice differentiable classical solution.

Differentiability [79, 80].

- u is twice differentiable everywhere, with bounded second derivatives, and smooth away from the critical set.
- u satisfies the equation everywhere in the classical sense.
- At each critical point, the Hessian is symmetric and has only two eigenvalues 0 and $-\frac{1}{k}$; $-\frac{1}{k}$ has multiplicity $k+1$.

This result is equivalent to saying that at a critical point, say $x = 0$ and $u(x) = 0$, the function u is (after possibly a rotation of $\mathbf{R}^{n+1}$) up to higher order terms equal to the quadratic polynomial

$$-\frac{1}{k}\left(x_1^2 + \cdots + x_{k+1}^2\right).$$

This second-order approximation is simply the arrival time of the shrinking round cylinders. It suggests that the level sets of u right before the critical value and near the origin should be approximately cylinders (with an $(n-k)$-dimensional axis). This has indeed been known for a long time and is due to Huisken [114–116], White [182–184], Huisken–Sinestrari [117, 118], Andrews [8], and Haslhofer–Kleiner [109]. It also suggests that those cylinders should be nearly the same (after rescaling to unit size). That is, the axis of the cylinders should not depend on the value of the level set. This last property, however, was only very recently established in [78] (cf. [90]) and is the key to proving that the function is twice differentiable.[1] The proof that the axis is unique, independent on the level set, relies on a key new inequality that draws its inspiration from real algebraic geometry although the proof is entirely new. This kind of uniqueness is a famously difficult problem in geometric analysis and no general case had previously been known.

Regularity of solutions. We have seen that the arrival time is always twice differentiable, and one may wonder whether there is even more regularity. Huisken [116] showed already in 1990 that the arrival time is C^2 for *convex* M_0. However, in 1992 Ilmanen gave an example of a rotationally symmetric *mean convex* M_0 in $\mathbf{R}^3$ where u is not C^2. This result of Ilmanen [120] shows that the above theorem about differentiability cannot be improved to C^2. We will see later that in fact one can entirely characterize when the arrival time is C^2. In the plane, Kohn and Serfaty [131] showed that u is C^3, and for $n > 1$ Sesum [168] gave an example of a *convex* M_0 where u is not C^3. Thus Huisken's result is optimal for $n > 1$.

The next result shows that one can entirely characterize when the arrival time is C^2.

Continuous differentiability [82]. u is C^2 if and only if:

---

1 Uniqueness of the axis is parallel to the fact that a function is differentiable at a point precisely if on all sufficiently small scales at that point it looks like the *same* linear function.

- There is exactly one singular time (where the flow becomes extinct).
- The singular set $\mathcal{S}$ is a k-dimensional, closed, connected, embedded, C^1 submanifold of cylindrical singularities.

Moreover, the axis of each cylinder is the tangent plane to $\mathcal{S}$.

When u is C^2 in $\mathbf{R}^3$, the singular set $\mathcal{S}$ is either:

(1) A single point with a spherical singularity, or

(2) A simple closed C^1 curve of cylindrical singularities.

The examples of the sphere and marriage ring show that each of these phenomena can happen, whereas the example of the dumbbell does not fall into either, showing that in that case the arrival time is not C^2.

We can restate this result for $\mathbf{R}^3$ in terms of the structure of the critical set and Hessian: u is C^2 if and only if u has exactly one critical value and the critical set is either:

(1) A single point where Hess_u is $-\frac{1}{2}$ times the identity, or

(2) A simple closed C^1 curve where Hess_u has eigenvalues 0 and -1 with multiplicities 1 and 2, respectively.

In case (2), the kernel of Hess_u is tangent to the curve, in fact, more is true, see [84].

2. UNIQUENESS OF BLOWUPS IN GEOMETRY

We saw that the key for optimal regularity for the level set equation was to show that the second-order approximation to a solution is independent of scale. The level sets of the second-order approximation are cylinders, and the key was that the axis of the cylinders was independent of scales.

This, independence of scale, is part of a larger question about uniqueness of blowups that has been widely studied whenever singularities occur. Indeed, once singularities occur, one naturally wonders what the singularities are like. A standard technique for analyzing singularities is to magnify around them. Unfortunately, singularities in many of the interesting problems in geometric PDEs looked at under a microscope will resemble one blowup, but under higher magnification, it might (as far as anyone knows) resemble a completely different blowup. Whether this ever happens is perhaps the most fundamental question about singularities; see, e.g., [171] and [108]. By general principles, the set of blowups is connected and, thus, the difficulty for uniqueness is when the blowups are not isolated in the space of blowups.

One of the first major results on uniqueness was by Allard–Almgren in 1981 [3], where uniqueness of tangent cones with smooth cross-section for minimal varieties is proven under an additional integrability assumption on the cross-section. The integrability condition applies in a number of important cases, but it is difficult to check and is not satisfied in many other important cases.

The next breakthrough on uniqueness was inspired by some old results in real algebraic geometry. Perhaps surprisingly, blowups for a number of important geometric PDEs can essentially be reformulated as infinite-dimensional gradient flows of analytic functionals. Thus, the uniqueness question would follow from an infinite-dimensional version of Lojasiewicz's theorem for gradient flows of analytic functionals. In real algebraic geometry, Lojasiewicz's theorem asserts that any integral curve of the gradient flow of an analytic function that has an accumulation point has a unique limit. Lojasiewicz proved this result in the early 1960s as a consequence of his gradient inequality. Infinite-dimensional versions of Lojasiewicz's theorem and the underlying Lojasiewicz inequalities were proven in a celebrated work of Simon [170] for the area, energy, and related functionals, and used, in particular, to prove a fundamental result about uniqueness of tangent cones with smooth cross-section of minimal surfaces. This holds, for instance, at all singular points of an area-minimizing hypersurface in $\mathbf{R}^8$. It also holds for singularities with smooth compact tangent flows for mean curvature flow by Schulze [174].

These method are very powerful and have had a major impact, but they do not apply when the blowups are noncompact. Indeed, in the most important examples, for essentially all of the natural flows the most common singularities are products with nontrivial Euclidean factors and thus are noncompact.

We will say that a singular point is *cylindrical* if at least one tangent flow is a multiplicity-one cylinder $\mathbf{S}^k \times \mathbf{R}^{n-k}$. We will later see that these are the most common and most important singularities. In [78] we showed that at each cylindrical singular point of a mean curvature flow the blowup is unique, that is, it does not depend on the sequence of rescalings.

Theorem 2.1. *Let M_t be an MCF in $\mathbf{R}^{n+1}$. At each cylindrical singular point, the tangent flow is unique. That is, any other tangent flow is also a cylinder with the same $\mathbf{R}^k$ factor that points in the same direction.*

This settled a major open problem that was open even in the case of mean convex hypersurfaces where it was known that all singularities are cylindrical. Moreover, this was the first general uniqueness theorem for blowups to a geometric PDE at a noncompact singularity.

To prove our uniqueness result, we established two completely new infinite-dimensional Lojasiewicz-type inequalities. Infinite-dimensional Lojasiewicz inequalities were pioneered 30 years ago by Simon [170]. However, unlike all other infinite-dimensional Lojasiewicz inequalities we know of, ours do not follow from a reduction to the classical finite-dimensional Lojasiewicz inequalities from the 1960s from algebraic geometry, rather we prove our inequalities directly and do not rely on Lojasiewicz's arguments or results.

This is only a brief introduction to a very central and active area, see [37, 39, 47, 52, 74, 76, 78, 95, 101, 112, 154, 155, 174].

3. REGULARITY OF SINGULAR SET

A major theme in PDEs over the last 50 years has been understanding singularities and the set where singularities occur. In the presence of a scale-invariant monotone quantity, blowup arguments can often be used to bound the dimension of the singular set; see, e.g., [3]. Unfortunately, these dimension bounds say little about the structure of the set. The key to get more structure is uniqueness of blowups. Uniqueness of tangents has important applications to regularity of the singular set; see, e.g., [171]. We will see in this section that the results of the previous sections lead to a rather complete description of the singular set for MCF with cylindrical singularities:

Theorem 3.1 ([81]). *Let $M_t \subset \mathbf{R}^{n+1}$ be an MCF of closed embedded hypersurfaces with only cylindrical singularities, then the space-time singular set is contained in finitely many (compact) embedded C^1 submanifolds each of dimension at most $(n-1)$ together with a set of dimension at most $(n-2)$.*

In fact, [81] proves considerably more than what is stated in Theorem 3.1; see Theorem 4.18 there. For instance, instead of just proving the first claim of the theorem, the entire stratification of the space-time singular set is Lipschitz of the appropriate dimension. Moreover, this holds without ever discarding *any* subset of measure zero of any dimension as is always implicit in any definition of rectifiable. To illustrate the much stronger version, consider the case of evolution of surfaces in $\mathbf{R}^3$. In that case, this gives that the space-time singular set is contained in finitely many (compact) embedded Lipschitz curves with cylinder singularities together with a countable set of spherical singularities. In higher dimensions, the direct generalization of this is proven.

Theorem 3.1 has the following corollaries:

Corollary 3.2 ([81]). *Let $M_t \subset \mathbf{R}^{n+1}$ be an MCF of closed embedded mean convex hypersurfaces or an MCF with only cylindrical singularities, then the conclusion of Theorem 3.1 holds.*

More can be said in dimensions three and four:

Corollary 3.3 ([81]). *If M_t is as in Theorem 3.1 and $n = 2$ or 3, then the evolving hypersurface is completely smooth (i.e., has no singularities) at almost all times. In particular, any connected subset of the space-time singular set is completely contained in a time-slice.*

A key technical point in [81] is to prove a strong parabolic Reifenberg property for MCF with generic singularities. In fact, the space-time singular set is proven to be (parabolically) Reifenberg vanishing. In analysis, a subset of a Euclidean space is said to be Reifenberg (or Reifenberg flat) if on all sufficiently small scales it is, after rescaling to unit size close, to a k-dimensional plane. The dimension of the plane is always the same but the plane itself may change from scale to scale. Many snowflakes, like the Koch snowflake, are Reifenberg with Hausdorff dimension strictly larger than one. A set is said to be Reifenberg vanishing if the closeness to a k-plane goes to zero as the scale goes to zero. It is said to have the strong

Reifenberg property if the k-dimensional plane depends only on the point but not on the scale.

Using the uniqueness of tangent flows, [81] shows that the singular set in space-time is strong (half) Reifenberg vanishing with respect to the parabolic Hausdorff distance. This is done in two steps, showing first that nearby singularities sit inside a parabolic cone (i.e., between two oppositely oriented space-time paraboloids that are tangent to the time-slice through the singularity). In fact, this parabolic cone property holds with vanishing constant. Next, in the complementary region of the parabolic cone in space-time (that is essentially space-like), the parabolic Reifenberg essentially follows from the space Reifenberg that the uniqueness of tangent flows implies.

An immediate consequence, of independent interest, of the parabolic cone property with vanishing constant is that nearby a generic singularity in space-time (nearby is with respect to the parabolic distance) all other singularities happen at almost the same time.

These results should be contrasted with a result of Altschuler–Angenent–Giga [4] showing that in $\mathbf{R}^3$ the evolution of any rotationally symmetric surface obtained by rotating the graph of a function $r = u(x)$, $a < x < b$ around the x-axis is smooth except at finitely many singular times where either a cylindrical or spherical singularity forms. For more general rotationally symmetric surfaces (even mean convex), the singularities can consist of nontrivial curves. For instance, consider a torus of revolution bounding a region Ω. If the torus is thin enough, it will be mean convex. Since the symmetry is preserved and because the surface always remains in Ω, it can only collapse to a circle. Thus at the time of collapse, the singular set is a simple closed curve.

White showed that a mean convex surface evolving by MCF in $\mathbf{R}^3$ must be smooth at almost all times, and at no time can the singular set be more than 1-dimensional. In fact, White's general dimension reducing argument [180, 181] gives that the singular set of any MCF with only cylindrical singularities has dimension at most $(n-1)$.

These results motivate the following conjecture:

Conjecture 3.4 ([81]). *Let M_t be an MCF of closed embedded hypersurfaces in $\mathbf{R}^{n+1}$ with only cylindrical singularities. Then the space-time singular set has only finitely many components.*

If this conjecture was true, then it would follow that in $\mathbf{R}^3$ and $\mathbf{R}^4$ MCF with only generic singularities is smooth except at finitely many times; cf. the three-dimensional conjecture at the end of Section 5 in [183].

Part 2. Growth of solutions to differential equations. On a Riemannian manifold M with metric $\langle \cdot, \cdot \rangle$ and Levi-Civita connection ∇, the gradient of a function f is defined by

$$V(f) = \langle \nabla f, V \rangle \quad \text{for all vectors fields } V. \tag{3.5}$$

The Laplacian of f is the trace of the Hessian. That is, if e_i is an orthonormal frame for M, then

$$\Delta f = \operatorname{Tr} \operatorname{Hess}_f = \sum_i \operatorname{Hess}_f(e_i, e_i) = \sum_i \langle \nabla_{e_i} V, e_i \rangle. \tag{3.6}$$

The Laplace operator is the canonical linear second order partial differential equation once we have a metric structure.

4. HARMONIC FUNCTIONS WITH POLYNOMIAL GROWTH

The classical Liouville theorem, named after Joseph Liouville (1809–1882), states that a bounded (or even just positive) harmonic function on all of $\mathbf{R}^n$ must be constant. There is a very short proof of this for bounded functions using the mean value property:

> *Given two points, choose two balls with the given points as centers and of equal radius. If the radius is large enough, the two balls will coincide except for an arbitrarily small proportion of their volume. Since the function is bounded, the averages of it over the two balls are arbitrarily close, and so the function assumes the same value at any two points.*

The Liouville theorem has had a huge impact across many fields, such as complex analysis, partial differential equations, geometry, probability, discrete mathematics, and complex and algebraic geometry, as well as many applied areas. The impact of the Liouville theorem has been even larger as the starting point of many further developments.

On manifolds with nonnegative Ricci curvature, mean values inequalities hold, but are no longer equalities, and the above proof does not give a Liouville type property. However, in the 1970s, S. T. Yau **[187]** showed that the Liouville theorem holds for such manifolds. Later, in the mid 1970s, Yau together with S. Y. Cheng **[42]** showed a gradient estimate on these manifolds giving an effective version of the Liouville theorem; see also Schoen **[165]**.

The situation is very different for negatively curved manifolds such as hyperbolic space. This is easiest seen in two dimensions where being harmonic is conformally invariant, so each harmonic function on the Euclidean disk is also harmonic in the hyperbolic metric. In particular, each continuous function on the circle extends to a harmonic function on the disk and the space of bounded harmonic functions is infinite dimensional; cf. Anderson **[6]**, Sullivan **[173]**, and Anderson–Schoen **[7]**.

On a Euclidean space, as soon as one allows a polynomial rate of growth, there are lots of harmonic functions. In fact, on a Euclidean space the harmonic functions with polynomial growth are the harmonic polynomials which play a central role in analysis. On a general manifold, the situation is much more complicated, and one does not expect an explicit representation. Given a manifold M and a constant d, $\mathcal{H}_d(M)$ is the linear space of harmonic functions of polynomial growth at most d. Namely, $u \in \mathcal{H}_d(M)$ if $\Delta u = 0$ and for some $p \in M$ and a constant C_u depending on u

$$\sup_{B_R(p)} |u| \leq C_u(1+R)^d \quad \text{for all } R. \tag{4.1}$$

In 1974, S. T. Yau conjectured that manifolds with nonnegative Ricci curvature should have a strong Liouville property, namely that $\mathcal{H}_d(M)$ is finite dimensional for each d when $\mathrm{Ric}_M \geq 0$. The conjecture was settled in **[59]**; see **[86]** for more results.[2] In fact, **[59,62,63]** proved finite dimensionality under much weaker assumptions of:

(1) A volume doubling bound,

(2) A scale-invariant Poincaré inequality or mean value inequality.

Both (1) and (2) hold for $\mathrm{Ric} \geq 0$ by the Bishop–Gromov volume comparison and work of Buser. However, these properties do not require much regularity of the space and are quite flexible. In particular, they make sense for more general metric-measure spaces and are preserved by bi-Lipschitz changes of the metric. Moreover, properties (1) and (2) make sense also for discrete spaces, vastly extending the theory and methods out of the continuous world. This extension opens up applications to geometric group theory and discrete mathematics, some of which we will touch upon later.

An interesting feature of these dimension estimates is that they follow from "rough" properties of M and are therefore surprisingly stable under perturbation. Unlike a Ricci curvature bound, these properties are stable under bi-Lipschitz transformations, cf. **[134]**. Moreover, these properties make sense also for discrete spaces, vastly extending the theory and methods out of the continuous world. Kleiner **[128]** (see also Shalom–Tao **[169,175,176]**) used, in part, this in his new proof of an important and foundational result in geometric group theory, originally due to Gromov **[104]**. Harmonic functions also play a central role in complex geometry, **[136,142,157]**.

5. ANCIENT CALORIC FUNCTIONS WITH POLYNOMIAL GROWTH

Harmonic functions are functions that are in equilibrium for the Laplace equation. For the heat equation, equilibrium is reached when solutions have existed for all prior times. This naturally leads to the question of whether there is a generalization of the results in the previous section to ancient solutions of the heat equation with polynomial growth. Ancient solutions are those that are defined for all negative t. Many solutions of the heat equation, including the fundamental solution, cannot be extended to all negative t. Given $d > 0$, $u \in \mathcal{P}_d(M)$ if u is ancient (defined for all negative t), $\partial_t u = \Delta u$ and for some $p \in M$ and a constant C_u,

$$\sup_{B_R(p)\times[-R^2,0]} |u| \leq C_u(1+R)^d \quad \text{for all } R. \tag{5.1}$$

2 For Yau's 1974 conjecture, see: page 117 in **[188]**, problem 48 in **[189]**, Conjecture 2.5 in **[97,124–126,165]**, Conjecture 1 in **[137]**, and problem (1) in **[138]**, amongst others.

On $\mathbf{R}^n$, these functions are the classical caloric polynomials that include the spherical harmonics and generalize the Hermite polynomials.

A manifold has polynomial volume growth if there are constants C and d_V so that $\mathrm{Vol}(B_R(p)) \leq C(1+R)^{d_V}$ for some $p \in M$, and all $R > 0$.[3] In [89] the following sharp inequality, which is an equality on $\mathbf{R}^n$, was shown:

Theorem 5.2. *If M has polynomial volume growth and k is a nonnegative integer, then*

$$\dim \mathcal{P}_{2k}(M) \leq \sum_{i=0}^{k} \dim \mathcal{H}_{2i}(M). \tag{5.3}$$

Since $\mathcal{H}_{d_1} \subset \mathcal{H}_{d_2}$ for $d_1 \leq d_2$, Theorem 5.2 implies:

Corollary 5.4. *If M has polynomial volume growth, then for all $k \geq 1$,*

$$\dim \mathcal{P}_{2k}(M) \leq (k+1) \dim \mathcal{H}_{2k}(M). \tag{5.5}$$

Combining this with the bound $\dim \mathcal{H}_d(M) \leq C d^{n-1}$ when $\mathrm{Ric}_{M^n} \geq 0$ from [59] gives:

Corollary 5.6. *There exists $C = C(n)$ so that if $\mathrm{Ric}_{M^n} \geq 0$, then for $d \geq 1$,*

$$\dim \mathcal{P}_d(M) \leq C d^n. \tag{5.7}$$

The exponent n in (5.7) is sharp: There is a constant c depending on n so that for $d \geq 1$,

$$c^{-1} d^n \leq \dim \mathcal{P}_d\left(\mathbf{R}^n\right) \leq c\, d^n. \tag{5.8}$$

Recently, Lin and Zhang [141] proved very interesting related results, adapting the methods of [59, 62, 63] to get the bound d^{n+1}.

An immediate corollary of the parabolic gradient estimate of Li and Yau [139] is that if $d < 2$ and $\mathrm{Ric} \geq 0$, then $\mathcal{P}_d(M) = \mathcal{H}_d(M)$ consists only of harmonic functions of polynomial growth. In particular, $\mathcal{P}_d(M) = \{\text{constant functions}\}$ for $d < 1$ and, moreover, $\dim \mathcal{P}_1(M) \leq n+1$, by Li and Tam [138], with equality if and only if $M = \mathbf{R}^n$ by [38].

The exponent $n-1$ is also sharp in the bound for $\dim \mathcal{H}_d$ when $\mathrm{Ric}_{M^n} \geq 0$. However, as in Weyl's asymptotic formula, the coefficient of d^{n-1} can be related to the volume [63]:

$$\dim \mathcal{H}_d(M) \leq C_n \, \mathrm{V}_M \, d^{n-1} + o\left(d^{n-1}\right), \tag{5.9}$$

where

- V_M is the "asymptotic volume ratio" $\lim_{r \to \infty} \mathrm{Vol}(B_r)/r^n$.
- $o(d^{n-1})$ is a function of d with $\lim_{d \to \infty} o(d^{n-1})/d^{n-1} = 0$.

Combining (5.9) with Corollary 5.4 gives $\dim \mathcal{P}_d(M) \leq C_n \, \mathrm{V}_M \, d^n + o(d^n)$ when $\mathrm{Ric}_{M^n} \geq 0$.

---

3 A volume-doubling space with doubling constant C_D has polynomial volume growth of degree $\log_2 C_D$.

6. GROWTH OF DRIFT EQUATIONS

The Laplacian Δ is self-adjoint with respect to the ordinary L^2 inner product. However, if we instead use a weighted L^2 inner product, then the Laplacian may not be self-adjoint but there is a natural self-adjoint elliptic operator known as the drift Laplacian. Drift Laplacians are ubiquitous in many areas, including quantum field theory, stochastic PDEs, and anywhere the heat equation or Gaussian appear, such as functional inequalities, parabolic PDEs, geometric flows, and probability. The drift term arises whenever there is natural measure or a natural scaling or, more generally, a gradient flow.

To make the drift Laplacian precise, fix a function ϕ and define the weighted L^2-norm $\|\cdot\|_\phi$ by

$$\|u\|_\phi^2 \equiv \int_M u^2 \, \mathrm{e}^{-\phi} . \tag{6.1}$$

Similarly, we will define the weighted inner product by

$$\langle u, v\rangle_\phi \equiv \int_M uv \, \mathrm{e}^{-\phi} . \tag{6.2}$$

The drift Laplacian $\mathcal{L}_\phi$ is defined by

$$\mathcal{L}_\phi u = \Delta u - \langle \nabla\phi, \nabla u\rangle = \mathrm{e}^{\phi} \operatorname{div}\big(\mathrm{e}^{-\phi} \, \nabla u\big) \tag{6.3}$$

and

$$\langle \mathcal{L}_\phi u, v\rangle_\phi = -\int_M \langle \nabla u, \nabla v\rangle \, \mathrm{e}^{-\phi} = \langle u, \mathcal{L}_\phi v\rangle_\phi . \tag{6.4}$$

The operator is self adjoint and under reasonable hypothesis has discrete eigenvalues going to infinity, see, for instance, [11,43,111,144]. The best-known example is the Ornstein–Uhlenbeck operator on $\mathbf{R}^n$,

$$\mathcal{L} = \Delta - \frac{1}{2}\nabla_x , \tag{6.5}$$

where $\phi = \frac{|x|^2}{4}$ and $\|\cdot\|_\phi$ is the Gaussian L^2-norm.

Drift Laplacians were considered very early on. Laplace discovered that on the line eigenfunctions of $\mathcal{L}u = u'' - \frac{x}{2}u'$ in the Gaussian L^2 space are polynomials whose degree is exactly twice the eigenvalue. These polynomials were later rediscovered twice. First by Chebyshev and a few years later by Hermite. They are now known as the Hermite polynomials and the eigenvalue equation as the Hermite equation. The first few eigenfunctions are: constants with eigenvalue 0, the linear function x with eigenvalue $\frac{1}{2}$, and the quadratic polynomial $x^2 - 2$ with eigenvalue 1. The Hermite polynomials and their higher-dimensional analogues play an important role in diverse fields. We will describe a vast generalization of these results that has many applications.

6.1. Growth of drift equations

We will next describe optimal polynomial growth bounds for eigenfunctions of drift Laplacians in a general setting that includes all shrinking solitons for both Ricci and mean curvature flows (or MCF). These bounds are sharp for the Ornstein–Uhlenbeck operator on Euclidean space.

There is a long history of studying the growth of solutions to differential equations, inequalities, and systems. At a very rough level, there are two main techniques. The first, exemplified in the work of Carleman and Hörmander, is to consider weighted L^2-norms with growing weights. The second, seen, for instance, in the work of Hadamard and Almgren, is to study the growth of spherical maxima or averages. The second is an extreme version of the first where the weight is a measure concentrated on a lower-dimensional set. As such, the second method typically gives stronger information and requires greater structure, such as invariance under dilations. However, general manifolds do not come with any dilation structure.

The growth estimates that we describe here hold in remarkable generality and without any assumptions on asymptotic decay. This is surprising and in contrast to most other situations, like unique continuation, that require very strong geometric assumptions on the space. A typical starting point for growth estimates is a Pohozaev identity or commutator estimate that come from a dilation, or approximate dilation, structure. We have none of these here in this general setting. In contrast, we rely on a miraculous cancelation for just the right quantity. A consequence of the generality is that the growth estimates hold for all singularities which is key for applications.

In many settings, one has an n-dimensional Riemannian manifold (M, g), that could even be flat Euclidean space, with two nonnegative functions f and S satisfying

$$\Delta f + S = \frac{n}{2}, \tag{6.6}$$

$$|\nabla f|^2 + S = f, \tag{6.7}$$

and where f is proper and C^n. The weight e^{-f} gives a drift Laplacian $\mathcal{L}$ on tensors u

$$\mathcal{L}u = e^f \operatorname{div}\big(e^{-f}\,\nabla u\big) = \Delta u - \nabla_{\nabla f} u \tag{6.8}$$

that is self-adjoint with respect to the L^2-norm $\|u\|_{L^2}^2 = \int |u|^2 e^{-f}$. Using the function f, we can define a very natural exhaustion function b that will share many of the same properties that the distance function has on a Euclidean space with the standard Gaussian measure. Since $|\nabla \sqrt{f}| \leq \frac{1}{2}$ by (6.7), $b = 2\sqrt{f}$ satisfies $|\nabla b| \leq 1$ as in [35]. On $\mathbf{R}^n$, $f = \frac{|x|^2}{4}$ and $S = 0$ satisfy (6.6), (6.7) with $\mathcal{L} = \Delta - \frac{1}{2}\nabla_x$ the Ornstein–Uhlenbeck operator and $b = |x|$. In a Ricci flow, singularities are gradient shrinking solitons, f is the potential, and S is scalar curvature.[4] In an MCF, singularities are shrinkers $\Sigma \subset \mathbf{R}^N$, $f = \frac{|x|^2}{4}$, and $S = |\mathbf{H}|^2$, where $\mathbf{H}$ is the mean curvature vector.[5]

Throughout, $\lambda > 0$ is a constant and u is a tensor on M. We will often assume that

$$\langle \mathcal{L}u, u\rangle \geq -\lambda |u|^2; \tag{6.9}$$

4 See [32, 40, 49–51, 107, 129, 160, 178].

5 See, e.g., [72, 78, 115].

this includes eigentensors with $\mathcal{L}u = -\lambda u$. To understand the growth of u, we will study a weighted average of $|u|^2$ on level sets of b,

$$I(r) = r^{1-n} \int_{b=r} |u|^2 |\nabla b|. \tag{6.10}$$

This is defined at regular values of b, but extends continuously to all values to be differentiable a.e. and absolutely continuous. The weight $|\nabla b|$ will play a crucial role (cf. [1,53,60,61,75,105]). The growth of I will be bounded above in terms of the solid integral

$$D(r) = r^{2-n} \mathrm{e}^{\frac{r^2}{4}} \int_{b<r} \left(|\nabla u|^2 + \langle \mathcal{L}u, u\rangle\right) \mathrm{e}^{-f} . \tag{6.11}$$

The frequency $U = \frac{D}{I}$ is defined when I is positive and will measure the growth of $\log I$.

The frequency U describes the rate of growth of the function u. To illustrate this, when u is a degree m Hermite polynomial, so $\lambda = \frac{m}{2}$, it is easy to see that

$$U(r) = m\left(1 + O\left(r^{-2}\right)\right) = 2\lambda\left(1 + O\left(r^{-2}\right)\right). \tag{6.12}$$

The next theorem from [91] shows that an L^2 tensor satisfying (6.9) has frequency bounded by 2λ and, accordingly, it grows at most polynomially at this rate. This may seem surprising since the weight e^{-f} decays rapidly, so the L^2 condition a priori allows extremely rapid growth.

Theorem 6.13. *Suppose $u, \mathcal{L}u \in L^2$, (6.6), (6.7), (6.9) hold, and u does not vanish identically outside a compact set. Given $\varepsilon > 0$, there exists $R = R(n, \lambda, \varepsilon)$ such that if $r > R$, then*

$$U(r) \leq 2\lambda + \varepsilon, \tag{6.14}$$

and for all $r_2 > r_1 > R$,

$$I(r_2) \leq I(r_1)\left(\frac{r_2}{r_1}\right)^{2(2\lambda+\varepsilon)} . \tag{6.15}$$

This is sharp for the Ornstein–Uhlenbeck operator on $\mathbf{R}^n$ where the L^2 eigenfunctions are Hermite polynomials with degree twice the eigenvalue. Note that u cannot vanish on an open set if u has unique continuation, e.g., if $\mathcal{L}u = -\lambda u$.

Our results give that polynomially growing "special functions" are dense in L^2. This gives manifold versions of some very classical problems in analysis. Whereas Weierstrass's approximation theorem shows that polynomials are dense among continuous functions on any compact interval, the classical Bernstein problem [145], dating back to 1924, asks if polynomials are dense on $\mathbf{R}$ in the weighted $L^p(\mathrm{e}^{-f}\, dx)$ space if f is assumed to grow sufficiently fast at infinity. On the line, the Hermite polynomials are dense in $L^2(\mathrm{e}^{-\frac{|x|^2}{4}}\, dx)$ and Lennart Carleson (and implicitly Izumi–Kawata) showed that polynomials are dense in $L^p(\mathrm{e}^{-|x|^\alpha}\, dx)$ if and only if $\alpha \geq 1$. A similar problem in several complex variables is the *completeness problem*, going back to Carleman in 1923, about the density of polynomials in weighted L^2 spaces of holomorphic functions [22].

Almgren's frequency has been used to show unique continuation [102] and structure of the nodal sets [143]; prior to this, the main tool in unique continuation was Carleman estimates that still is the primary technique. Almgren's frequency bounds relied on scaling for $\mathbf{R}^n$; cf. [60, 61]. The papers [18] (cf. [179]), [83] developed frequencies for conical and cylindrical MCF shrinkers and did not involve a weight like $|\nabla b|$. Theorem 6.13, in contrast, holds very generally, including for all shrinkers in both Ricci flow and MCF. A much weaker version of Theorem 6.13, that was not relative, was proven in [83] in the special case of MCF.

Part 3. Stable structures. In mathematics, structural stability is a fundamental property of a dynamical system which means that the qualitative behavior of the trajectories is unaffected by small perturbations. Given a smooth function f on a finite-dimensional space, the gradient ∇f points in the direction of the steepest ascent. The critical points of f are the points where ∇f vanishes. If p is a local minimum of f, then the second derivative test tells us that the Hessian matrix of f at p is nonnegative. More generally, the number of negative eigenvalues of the Hessian is called the index of the critical point. A fundamental method to find the minimum of f is the method of gradient descent. Here, we make an initial guess p_0 and then iteratively move in the negative gradient direction, the direction of the steepest descent, by setting $p_{i+1} = p_i - \nabla f(p_i)$. This can also be done continuously by defining a negative gradient flow

$$\frac{dx}{dt} = -\nabla f\big(x(t)\big). \tag{6.16}$$

The function $f(x(t))$ decreases as efficiently as possible as $x(t)$ heads towards the minimum. The dynamics near a nondegenerate critical point are determined by the index. If the index is zero, then the critical point is attracting and the entire neighborhood flows towards the critical point. However, when the index is positive, a generic point will flow out of the neighborhood, missing the critical point.

Many of the fundamental problems in geometry can be understood as problems about dynamical systems on an infinite-dimensional space. Sometimes this is immediate. For instance, in the case of geodesics or minimal surfaces. Geodesics are critical points for energy, whereas minimal surfaces are critical points for area. Another example where the connection to dynamical systems is immediate is the mean curvature flow that is the negative gradient flow for area. In other cases the connection is hidden, but no less fundamental. An example of this is uniqueness of blowups, that we discussed earlier. Uniqueness can be thought of as the question of whether a related recurrent flow has a limit or is wandering. One of the most basic and fundamental questions about a dynamical system is the question of equilibria: which equilibria are stable (generic) and which are not. For a nongeneric equilibrium, a nearby flow line passes by the equilibria and thus the nongeneric ones can typically be ignored.

We will look for stable structures in four situations and discuss what is known and unknown, see [58]. Those four are: (1) minimal hypersurfaces; (2) minimal submanifolds of higher codimension; (3) singularities that are stable or generic, and cannot be perturbed away, for motion by mean curvature of hypersurfaces; and, finally, (4) singularities for motion by mean curvature in higher codimension.

7. MINIMAL SURFACES

Let $\Sigma^n \subset \mathbf{R}^N$ be a smooth submanifold (possibly with boundary). Given an infinitely differentiable (i.e., smooth), compactly supported, normal (orthogonal to Σ) vector field V on Σ, consider the one-parameter variation

$$\Sigma_{s,V} = \{x + sV(x) \mid x \in \Sigma\}. \tag{7.1}$$

This gives a path $s \to \Sigma_{s,V}$ in the space of submanifolds with $\Sigma_{0,V} = \Sigma$. The so-called first variation formula of area or volume is the equation (integration is with respect to d Vol)

$$\frac{d}{ds}_{s=0} \mathrm{Vol}(\Sigma_{s,V}) = \int_\Sigma \langle V, \mathbf{H} \rangle, \tag{7.2}$$

where $\mathbf{H}$ is the mean curvature vector. When Σ is a hypersurface, $\mathbf{H}$ is the unit normal times the sum of the principal curvatures. In general, $\mathbf{H} = -\sum_i A(e_i, e_i)$ where A is the second fundamental form and e_i is an orthonormal frame for the tangent space of Σ; $A(e_i, e_j) = A_{ij} = \nabla^{\perp}_{e_i} e_j$ where ∇ is the Euclidean derivative and "$\perp$" is the component orthogonal to the submanifold. When Σ is noncompact, $\Sigma_{s,V}$ is replaced by $\Gamma_{s,V} = \{x + sV(x) \mid x \in \Gamma\}$ where Γ is a compact subset of Σ containing the support of V.

The submanifold Σ is said to be a *minimal* if

$$\frac{d}{ds}_{s=0} \mathrm{Vol}(\Sigma_{s,V}) = 0 \quad \text{for all } V, \tag{7.3}$$

or, equivalently, by (7.2), if $\mathbf{H}$ is identically zero. Thus Σ is minimal if and only if it is a critical point for the volume functional. Since a critical point is not necessarily a minimum, the term minimal is misleading but time-honored. It is easy to see that being minimal is equivalent to all the coordinate functions of $\mathbf{R}^N$ restricted to the submanifold are harmonic with respect to the Laplacian, Δ_Σ, on the submanifold. In higher codimension, the minimal surface equation is a complicated system.

A computation shows that if Σ is minimal, then the second derivative of volume is

$$\frac{d^2}{ds^2}_{s=0} \mathrm{Vol}(\Sigma_{s,V}) = -\int_\Sigma \langle V, LV \rangle, \tag{7.4}$$

where $LV = \Delta_\Sigma V + \langle A_{ij}, \mathrm{V} \rangle A_{ij}$ is the so-called second variational (or Jacobi) operator. This is an operator on the normal bundle of Σ and is the Laplacian plus a zeroth-order term. When the submanifold is a hypersurface, this simplifies and becomes $LV = \Delta_\Sigma V + |A|^2 V$, where $|A|^2$ is the sum of the squares of the principal curvatures. It simplifies further if one identifies V with its projection $\phi = \langle V, \mathbf{n} \rangle$ onto the unit normal $\mathbf{n}$. Then $L\phi = \Delta_\Sigma \phi + |A|^2 \phi$.

A minimal submanifold is *stable* if it passes the second derivative test

$$\frac{d^2}{ds^2} \mathrm{Vol}(\Sigma_{s,V}) \geq 0 \quad \text{for all } V. \tag{7.5}$$

Obviously, if a minimal surface is area or volume minimizing among competitors with the same boundary, then it is stable as well. However, stability is much more general than being minimizing. Stability becomes a question about whether the Jacobi operator L is nonnegative or not. The operator L is much simpler for hypersurfaces and, in particular, it is easy to see that a minimal graph is stable. In higher codimension, the question of stability becomes much more complicated because of the vector-valued nature of L and the curvature of the normal bundle. For example, minimal graphs are not necessarily stable in higher codimension.[6]

A classical theorem of Bernstein from 1916 shows that entire (that is, where the domain of definition is all of $\mathbf{R}^2$) minimal graphs in $\mathbf{R}^3$ are planes. Whether this is true in higher dimensions became known as the Bernstein problem. This problem played an important role in the field for decades and is closely related to regularity for area minimizers. In 1965 and 1966, De Giorgi and Almgren proved the Bernstein theorem for graphs in $\mathbf{R}^4$ and $\mathbf{R}^5$. In 1968, Simons extended the Bernstein theorem to $\mathbf{R}^6$, $\mathbf{R}^7$, and $\mathbf{R}^8$, which was shown to be sharp the next year by Bombieri, De Giorgi, and Giusti. Simons' influential paper introduced the second variation operator and stability to minimal surface theory. Stability of hypersurfaces was studied by Schoen–Simon–Yau [166], who showed that, as long as the dimension of the hypersurface is at most six and the volumes of balls are up to a constant the same as Euclidean balls of the same radius and dimension, all stable minimal hypersurfaces are planes, cf. [186] and references there. In $\mathbf{R}^3$ Fischer-Colbrie and Schoen [99] showed the same, but without assuming area bounds. This was also proved independently by Do Carmo and Peng. Schoen [164] (see also [46, 57]) later showed a local version of this that has had a huge influence on the development of minimal surfaces in three dimensions. Stable minimal surfaces can be constructed variationally, see, for instance, [152]. These estimates can also be applied to low index minimal surfaces, [146, 147, 172]. See [64–71] and [161] for more about minimal surfaces.

The situation is much more complicated in higher codimension where there is no analog of the Bernstein theorem, cf. [96,163]. A simple argument of Wirtinger from the 1930s, using Stokes' formula, shows that any complex submanifold of $\mathbf{C}^N$ is volume minimizing among things with the same boundary and, thus, a stable minimal submanifold. This gives a plethora of area-minimizing, and thus also stable, minimal submanifolds once the codimension is at least two. Moreover, these examples can have arbitrarily large areas. Remarkably, Micallef [153] proved a converse in $\mathbf{R}^4$. Namely, he showed that a stable oriented, parabolic minimal surface in $\mathbf{R}^4$ is complex for some orthogonal complex structure. Being parabolic is a conformal property that holds, for instance, if the volume of balls grows at most quadratically. Examples of Arezzo and Micallef show that this converse does not hold for surfaces in codimension larger than two.

6 By [153], Osserman's minimal graph $x_3 = \frac{1}{2} \cos \frac{x_2}{2} (e^{x_1} - 3\, e^{-x_1})$ and $x_4 = -\frac{1}{2} \sin \frac{x_2}{2} (e^{x_1} - 3\, e^{-x_1})$ in $\mathbf{R}^4$ is not stable.

8. MOTION BY MEAN CURVATURE

Surface tension is the tendency of fluid surfaces to shrink into the minimum surface area possible. Mathematically, the force of surface tension is described by the mean curvature.

A one-parameter family of n-dimensional submanifolds $M_t \subset \mathbf{R}^N$ is said to move by motion by mean curvature, see, for instance, [9, 92], if the time derivative of the position vector x moves by minus the mean curvature. That is,

$$\frac{\partial x}{\partial t} = -\mathbf{H}. \tag{8.1}$$

It follows from the first variation formula that the mean curvature flow is the negative gradient flow for area. That is, the mean curvature flow moves the submanifold in the direction where the area or volume decreases as fast as possible.

We can view the mean curvature flow as a type of heat equation. This is exemplified by that the coordinate functions of the ambient Euclidean space restricted to the evolving submanifolds satisfy the heat equation

$$\frac{\partial x}{\partial t} = \Delta_{M_t} x. \tag{8.2}$$

This equation is nonlinear since the Laplacian Δ_{M_t} depends on M_t. Moreover, since the submanifolds are evolving, the induced metric is time-varying so the Laplacian Δ_{M_t} is also time-varying. From the first variation formula (7.2), it follows easily that the mean curvature flow moves in the direction where the volume decreases as fast as possible; thus, the mean curvature flow is the negative gradient flow of volume. The motion is by surface tension. In higher codimension, (8.1) and (8.2) are complicated parabolic systems where much less is known.

Since the coordinate functions on the evolving submanifolds satisfy the heat equation, it follows from the parabolic maximum principle that the evolving submanifolds remain inside the convex hull of the initial submanifold. A straightforward computation shows that also the function $|x|^2 - 2nt$ satisfies the heat equation on the evolving submanifolds. At the initial time $t = 0$, this is nonnegative and therefore, by the parabolic maximum principle, it remains nonnegative as long as the flow exists. Since we have already seen that $\max_{M_t} |x|^2$ remains bounded under the evolution, it follows that the flow must become extinct in finite time and, thus, singularities occur. There are two approaches either considering a weak flow through singularities or considering flow with surgery through singularites; see, [17, 30, 110, 119, 130] for surgery.

For a fixed constant $c > 0$, rescaling the flow parabolically

$$t \to c M_{c^{-2}} = M_{c,t} \tag{8.3}$$

gives a new solution to motion by mean curvature that has the effect that the submanifolds are magnified by the constant c. If we simultaneously with rescaling also reparametrize time, then we get a rescaled mean curvature flow. It is easy to see that such a one-parameter family satisfies the rescaled mean curvature flow equation

$$\frac{\partial x}{\partial t} = \frac{x^{\perp}}{2} - \mathbf{H}. \tag{8.4}$$

The rescaled mean curvature flow, which is so critical for understanding the mean curvature flow, can itself be interpreted as the negative gradient flow of a functional that we call the Gaussian surface area.

8.1. Gaussian surface area and entropy

The Gaussian surface area F of an n-dimensional submanifold $\Sigma^n \subset \mathbf{R}^N$ is

$$F(\Sigma) = (4\pi)^{-\frac{n}{2}} \int_\Sigma \mathrm{e}^{-\frac{|x|^2}{4}} . \tag{8.5}$$

The constant $(4\pi)^{-\frac{n}{2}}$ is a normalization that makes the Gaussian area equal to one for an n-plane through the origin. Following [72], the entropy λ is the supremum of F over all translations and dilations

$$\lambda(\Sigma) = \sup_{c,x_0} F(c\Sigma + x_0). \tag{8.6}$$

By considering all centers and scales and taking the supremum over these, we get some rough low-regularity measure of the complexity of the submanifold. In particular, it is easy to see that the entropy is always at least 1 and achieved only on a n-dimensional plane.

It follows easily from Huisken's monotonicity formula that the entropy is monotone under mean curvature flow and, moreover, the entropy at the initial time gives an upper bound for the entropy of any future singularity; see [72].

Prior to the entropy, many results focused on either convexity conditions or graphical restrictions as these were preserved under the flow by the maximum principle. These properties, however, are pretty strong and heavily restrict the types of singularities that can occur. The entropy now plays a central role in mean curvature flow and a great deal is now known about low entropy flows, [20, 21, 45, 48, 55].

If V is a normal vector field and $\Sigma_{s,V}$, as before, is the variation $\Sigma_{s,V} = \{x + sV(x) \mid x \in \Sigma\}$, then an easy computation shows that

$$\frac{d}{ds}_{s=0} F(\Sigma_{s,V}) = (4\pi)^{-\frac{n}{2}} \int_\Sigma \left\langle V, \mathbf{H} - \frac{x^\perp}{2} \right\rangle \mathrm{e}^{-\frac{|x|^2}{4}} . \tag{8.7}$$

It follows that the Gaussian surface area F is monotone nonincreasing under the rescaled mean curvature flow and constant if and only if

$$\mathbf{H} = \frac{x^\perp}{2}. \tag{8.8}$$

This equation is the shrinker equation and is equivalent to the rescaled flow is static. Or, equivalently, the evolution under the mean curvature flow is by rescaling. That is, a later time slice is exactly like an earlier, just scaled down. That Gaussian surface area is monotone under the rescaled flow corresponds to Huisken's celebrated monotonicity formula [115]. From this, it follows also that the entropy is a Lyapunov function for both the mean curvature flow and the rescaled mean curvature flow.

From Huisken's monotonicity [115], as well as work of Ilmanen [121] and White [180], one knows that every sequence $c_i \to \infty$ has a subsequence (also denoted by c_i) such that $M_{c_i,t}$ converges to a shrinker $M_{\infty,t}$ (so $M_{\infty,t} = \sqrt{-t}\, M_{\infty,-1}$) with $\sup_t \lambda(M_{\infty,t}) \leq$

$\sup_t \lambda(M_t)$. Such a limit is said to be a tangent flow at the origin. Similarly, one can magnify (blow up) around any other point in space time. If one does not fix the point around where one blows up, but still looks at limits of a sequence of blowups, then the limiting flows are not shrinkers, but even then the limiting flows will exist for all negative times and are said to be ancient flows.

The shrinker equation (8.8) is a second order nonlinear elliptic equation that is closely related to the classical minimal surface equation. In fact, shrinkers are minimal surfaces for a conformally changed metric that is not particularly well-behaved: it is not complete and the curvature is unbounded. This perspective has limited utility for global questions, but it is very useful for local regularity (e.g., any tangent cone is a minimal cone); cf. [55,72,73].

8.2. Second variation and stability

We have already seen that shrinkers are critical points for the Gaussian area. The critical points for the Gaussian surface area are the fixed points for the rescaled flow. To understand the dynamics of the flow, we would like to understand which fixed points can be avoided and, more generally, the dynamics near any fixed point.

When Σ is a shrinker, we therefore look at the second derivative. A calculation (see [72]) gives

$$\frac{d^2}{ds^2}_{s=0} F(\Sigma_{s,V}) = -(4\pi)^{-\frac{n}{2}} \int_\Sigma \langle V, LV\rangle \, \mathrm{e}^{-\frac{|x|^2}{4}} . \tag{8.9}$$

Here $LV = \mathcal{L}V + \langle A_{ij}, V\rangle A_{ij} + \frac{1}{2}V$ is the second variation operator, and $\mathcal{L}V = \Delta_\Sigma V - \frac{1}{2}\nabla^\perp_{x^T} V$ is the Ornstein–Uhlenbeck operator on the normal bundle. For hypersurfaces, there is a similar simplification of the operator L, as we saw for the second derivative of volume; cf. [10,14,135] for higher codimension.

For any shrinker, translations and scaling give directions where the Gaussian area decreases [72], so there are no stable shrinkers in the usual sense. Translation of a submanifold in the direction $E \in \mathbf{R}^N$ is infinitesimally given by the normal part $E^\perp$ of E. Similarly, rescaling is given by the normal vector field $\frac{x^\perp}{2}$. This corresponds to $E^\perp$ (with $E \in \mathbf{R}^N$) and $\mathbf{H} = \frac{x^\perp}{2}$ being eigenvectors of L with eigenvalues $-\frac{1}{2}$ and -1, respectively. Perturbing by either translation or scaling has the effect of moving the same singularity to a different point in space or time. However, the singularity is not avoided; it just occurs at another time or place for the flow. For this reason, we say [72] that a shrinker is *F-stable* if

$$\frac{d^2}{ds^2}_{s=0} F(\Sigma_{s,V}) \geq 0 \quad \text{for all } V \text{ orthogonal to } \mathbf{H} \text{ and to all } E^\perp. \tag{8.10}$$

Here orthogonal means with respect to the Gaussian inner product on the space of normal vector fields. It is easy to see that spheres and planes are F-stable in any codimension. In [72] the F-stable hypersurfaces were classified.

Theorem 8.11. *The only F-stable hypersurfaces are the planes and the round sphere.*

At first it may seem surprising that round cylinders are not F-stable. Indeed, for nonompact shrinkers, it turns out that the right notion of stability is that of entropy stability, however, for compact singularities those two notions of stability are the same [72]. A shrinker is *entropy-stable* if it is a local minimum for the entropy λ. Entropy-unstable shrinkers are singularities that can be perturbed away, whereas entropy-stable ones cannot; see [72].

Even for hypersurfaces, examples show that singularities of the mean curvature flow are too numerous to classify. The hope is that the generic ones that cannot be perturbed away are much simpler. Indeed, in all dimensions, generic singularities (that is, entropy-stable shrinkers) of hypersurfaces moving by mean curvature flow have been classified in [72].

Theorem 8.12. *In all dimensions, generic singularities (that is, entropy-stable shrinkers) of hypersurfaces are round generalized cylinders* $\mathbf{S}^k_{\sqrt{2k}} \times \mathbf{R}^{n-k}$.

The generic singularities in $\mathbf{R}^3$ are the sphere $\mathbf{S}^2_2$, cylinder $\mathbf{S}^1_{\sqrt{2}} \times \mathbf{R}$, and plane $\mathbf{R}^2$. In contrast to the Bernstein theorems for minimal hypersurfaces, this classification of generic singularities holds in every dimension.

The paper [55] showed that for hypersurfaces round spheres are the shrinkers with the smallest entropy. The authors of [55] conjectured further that round spheres had the least entropy for any closed hypersurface; this was proven by Bernstein–Wang [20] up to dimension 7 and extended by Zhu [190] to higher dimensions; cf. also [21, 24, 127, 185]. For surfaces embedded shrinkers with genus zero has been classified by Brendle, [28].

8.3. Higher codimension

For the mean curvature flow in higher codimension, we search again for the stable singularities. Recall that stable singularities are those that are entropy stable, which is equivalent to being F-stable for closed shrinkers. In higher codimension, [87] gave the following bound for the entropy:

Theorem 8.13. *If* $\Sigma^2 \subset \mathbf{R}^N$ *is an* F*-stable shrinker diffeomorphic to a two-sphere, then*

$$\lambda(\Sigma) < 4 = \mathrm{e}\,\lambda\big(\mathbf{S}^2_2\big). \tag{8.14}$$

The sharp constant is unknown, but (8.14) is at most off by a factor of e. By [87], similar bounds also hold for other closed shrinking surfaces of any finite index where the entropy bound depends on the genus and index. This implies that any such F-stable shrinker, that, a priori, lies in a high-dimensional Euclidean space, in fact, lies in a linear subspace of some fixed small dimension. The sharp bound for the dimension of the linear space is unknown, though [87] provides sharp dimension bounds in various other important situations.

There is no analog of (8.14) for minimal surfaces in $\mathbf{R}^4$. Namely, viewing $\mathbf{R}^4$ as $\mathbf{C}^2$, one sees that the parametrized complex submanifold $z \to (z, z^m)$ is a stable minimal variety that is topologically a plane for each integer m. It has $\mathrm{Area}(B_r \cap \Sigma) \geq Cmr^2$ for $r \geq 1$. In contrast, [87] implies that $\mathrm{Area}(B_r \cap \Sigma) \leq C(1 + \gamma)r^2$ for a closed stable 2-dimensional shrinker Σ of genus γ. Similarly, there is no analog of the codimension bound for minimal surfaces. Indeed, for each m, the parametrized surface $z \to (z, z^2, z^3, \dots, z^{m+1})$ is a stable minimal variety that is topologically a plane. Its real codimension is $2m$ and it is not contained in a proper subspace.

Once one has the entropy bound in (8.14), to conclude that stable singularities have low codimension, one needs a result about the number of linearly independent coordinate functions. The coordinate functions on a mean curvature flow produce a linear space of caloric functions, i.e., solutions of the heat equation, that grow at most linearly. The bound on the codimension is a consequence of a much more general result about polynomial growth caloric functions on an ancient mean curvature flow that has a variety of other useful applications.

Let $M_t^n \subset \mathbf{R}^N$ be an ancient mean curvature flow of n-dimensional submanifolds with entropies $\lambda(M_t) \leq \lambda_0 < \infty$. Recall that ancient flows are solutions that exist for all negative times. The space $\mathcal{P}_d$ of polynomial growth caloric functions consists of $u(x,t)$ on $\bigcup_t M_t \times \{t\}$ so that $(\partial_t - \Delta_{M_t})u = 0$ and there exists C depending on u with

$$\left|u(x,t)\right| \leq C\left(1 + |x|^d + |t|^{\frac{d}{2}}\right) \quad \text{for all } (x,t) \text{ with } x \in M_t,\ t < 0. \tag{8.15}$$

The simplest example is when the flow consists of a static (constant in time) hyperplane $\mathbf{R}^n$. In this case, $\mathcal{P}_d(\mathbf{R}^n)$ consists of polynomials in $(t, x_1, \dots, x_n)$ known as the caloric polynomials and, using the special structure in this case, it is easy to see that $\dim \mathcal{P}_d(\mathbf{R}^n) \approx c_n d^n$. The paper [87] showed sharp bounds for $\dim \mathcal{P}_d$ for all $d \geq 1$ for an ancient flow with $\lambda(M_t) \leq \lambda_0$,

$$\dim \mathcal{P}_d \leq C_n \lambda_0 d^n. \tag{8.16}$$

One remarkable consequence when $d = 1$ is a bound for the codimension. Namely, the flow sits inside a linear subspace of dimension at most $\dim \mathcal{P}_1$, since a linear relation for coordinate functions specifies a hyperplane containing the flow.

The next result we will describe gives sharp bounds for codimension in arguably some of the most important situations for ancient flows. The bounds mentioned above were sharp in the exponent of d and, thus, asymptotically sharp as $d \to \infty$. The next result is more delicate and obtains sharp constants for d fixed.

Suppose that $M_t^n \subset \mathbf{R}^N$ is an ancient MCF with $\sup_t \lambda(M_t) < \infty$. For each constant $c > 0$ define the flow $M_{c,t}$ by $M_{c,t} = \frac{1}{c} M_{c^2 t}$. It follows that $M_{c,t}$ is an ancient MCF as well. Since $\sup_t \lambda(M_t) < \infty$, it follows from Huisken's monotonicity [115], as well as work of Ilmanen [121] and White [180], that every sequence $c_i \to \infty$ has a subsequence (also denoted by c_i) such that $M_{c_i,t}$ converges to a shrinker $M_{\infty,t}$ (so $M_{\infty,t} = \sqrt{-t}\, M_{\infty,-1}$) with $\sup_t \lambda(M_{\infty,t}) \leq \sup_t \lambda(M_t)$. We will say that such an $M_{\infty,t}$ is a tangent flow at $-\infty$ of the original flow. In [87] the following sharp bound for the codimension was shown:

Theorem 8.17. *If $M_t^n \subset \mathbf{R}^N$ is an ancient MCF and one tangent flow at $-\infty$ is a cylinder $\mathbf{S}^k_{\sqrt{2k}} \times \mathbf{R}^{n-k}$, then M_t is a flow of hypersurfaces in a Euclidean subspace.*

Combining this result with results of Angenent–Daskalopoulos–Sesum [12, 13], Brendle–Choi [29], and Choi–Haslhofer–Hershkovits [48] gives uniqueness for ancient flows of surfaces in higher codimension.

Part 4. The gauge group. Comparing and recognizing metrics can be extraordinarily difficult because of the group of diffeomorphisms. Two metrics, which could even be the same, could look completely different in different coordinates. Many key problems are defined intrinsically without a canonical coordinate system. In those problems, the infinite-dimensional diffeomorphism group (gauge group) becomes a major issue and dealing with it a major obstacle. Ricci flow is such an example.

> *"Gauge theory is a term which has connotations of being a fearsomely complicated part of mathematics—for instance, playing an important role in quantum field theory, general relativity, geometric PDEs, and so forth. But the underlying concept is really quite simple: a gauge is nothing more than a coordinate system that varies depending on location ...By fixing a gauge (thus breaking or spending the gauge symmetry), the model becomes something easier to analyse mathematically ...Deciding exactly how to fix a gauge (or whether one should spend the gauge symmetry at all) is a key question in the analysis of gauge theories, and one that often requires the input of geometric ideas and intuition into that analysis."*
> [177]

One of the most interesting results of transformation groups is the existence of slices. A slice for the action of a group on a manifold is a submanifold which is transverse to the orbits near a given point.[7] Ebin and Palais proved the existence of a slice for the infinite-dimensional diffeomorphism group of a *compact* manifold acting on the space of all Riemannian metrics. However, here we will be interested in when the manifolds are not compact.

8.4. A new approach to dealing with the gauge group

We describe a new way of dealing with the diffeomorphism group from [91] that should be useful in a broad range of applications, and explain how it can be used to solve a well-known problem in Ricci flow. A key new tool is a detailed analysis of a natural second-order system operator $\mathcal{P}$. The operator will be used to "fix the gauge." The analysis applies to all noncompact singularities. This makes it particularly useful, but also delicate. At each scale, a diffeomorphism is applied to fix the gauge, requiring precise and delicate estimates for the growth of the diffeomorphism. The gauge-fixing diffeomorphism satisfies a nonlinear system of PDEs, where $\mathcal{P}$ is the linearization. We will need, and show, strong bounds for the displacement function of the gauge-fixing diffeomorphism.

Suppose we have two weighted manifolds. Assume that on a large, but compact set, the manifolds, metrics, and weights almost agree after identification by a diffeomorphism.

---

7 If the group is compact and Lie and the space is completely regular, Mostow proved, as a generalization of works of Gleason, Koszul, Montgomery, Yang, and others, that there is a slice through every point. If the group is not compact but Lie and if the space is a Cartan space, then Palais proves the same result.

On this set, in these coordinates, we write the metric on one as g and on the other as $g + h$, where h is small, and the weights as e^{-f} and e^{-f-k}, where k is small. We would like to mod out by the diffeomorphism group, by adjusting by a diffeomorphism to put the equation in an appropriate gauge so that the difference h in the metrics is orthogonal to the action of the group. Orthogonality corresponds to making $\operatorname{div}_f h = 0$,[8] which means finding a diffeomorphism Φ so that

$$\operatorname{div}_f\big(\Phi^*(g+h) - g\big) = 0. \tag{8.18}$$

The pullback metric is quadratic in the differential of Φ, so this is a second-order nonlinear system of PDEs for Φ. This is the PDE that is in the spirit of the slice theorem for group actions and a solution Φ gives the desired "gauge-fixing." Terms involving $\operatorname{div}_f h$ come up again and again, so many quantities simplify in this gauge and things become easier.

In **[91]** we construct the diffeomorphism solving (8.18) using an iteration scheme for the linearized operator $\mathcal{P}$ on vector fields Y. We show first sharp polynomial bounds on $\mathcal{P}$ and then use them to show sharp polynomial bounds for the displacement function of Φ

$$x \to \operatorname{dist}_g\big(x, \Phi(x)\big). \tag{8.19}$$

The bounds are relative, meaning that better initial bounds give better bounds further out. These optimal bounds hold on all singularities and give a key new tool for dealing with the gauge group of all noncompact singularities.

The linearization of (8.18) is to find a vector field whose Lie derivative of the metric has div_f equal to $-\operatorname{div}_f h$. The Lie derivative in a direction Y can also be written as $-2\operatorname{div}_f^* Y$, where div_f^* is the operator adjoint of div_f with respect to the weighted measure. Therefore, the linearization of (8.18) is $\mathcal{P}Y = \frac{1}{2}\operatorname{div}_f h$, where

$$\mathcal{P}Y = \operatorname{div}_f \circ \operatorname{div}_f^* Y. \tag{8.20}$$

Solutions of $\mathcal{P}Y = \frac{1}{2}\operatorname{div}_f h$ are unique once we require that Y is orthogonal to the kernel of $\mathcal{P}$. The kernel is the Killing fields. We will solve $\mathcal{P}Y = \frac{1}{2}\operatorname{div}_f h$ on any shrinker and show via L^2-methods that $\|Y\|_{W^{1,2}} \leq \|\operatorname{div}_f h\|_{L^2}$. Given the noncompactness, the L^2-estimates are not sufficient to implement the iteration scheme, and we need stronger polynomial estimates. The problems are magnified by that initial closeness is only on a given compact set. As one builds out to get closeness on larger sets, one needs at each step to adjust the entire diffeomorphism so that the normalization is zero on larger and larger sets. Understanding $\mathcal{P}$ and proving growth estimates is a major point.

The L^2-theory for $\mathcal{P}$ shares formal similarities with Hörmander's influential L^2 $\bar\partial$-method in several complex variables. In the L^2 $\bar\partial$-method, one solves the Poisson equation $\bar\partial u = F$, with estimates, where $\bar\partial F = 0$. To do so, one introduces the adjoint of $\bar\partial$ with respect to a weight. Hörmander's idea for the weight came from Carleman's method for proving unique continuation of a PDE. Here we solve $\mathcal{P}Y = F$, where $F = \frac{1}{2}\operatorname{div}_f h$ is

8 For a symmetric two-tensor h, the f-divergence is $\operatorname{div}_f(h) = e^f \operatorname{div}(e^{-f} h) = \operatorname{div}(h) - h(\nabla f, \cdot)$.

orthogonal to the kernel of div_f^*. Hörmander's method gives weighted L^2-bounds for $\bar{\partial}$ similar to our weighted bounds for $\mathcal{P}$. To introduce a second weight to capture the growth à la Carleman and Hörmander is less natural here. Instead, we go a different route to prove stronger bounds.

8.5. Bounding the growth of gauge transformations

We need to control the growth of Y to control the metric in the new coordinates, but Y will be constructed using weighted L^2-methods and, thus, a priori could grow rapidly. The next theorem from [91] shows that an L^2 eigenvector field with eigenvalue λ for $\mathcal{P}$ grows polynomially of degree at most $4\lambda + 1$. A Poisson version is used to control Y with $\mathcal{P}Y = \frac{1}{2}\operatorname{div}_f h$. We set $b = 2\sqrt{f}$ and measure the growth of Y by the weighted average

$$I_Y(r) = r^{1-n}\int_{b=r} |Y|^2|\nabla b|. \tag{8.21}$$

A one-parameter family of smooth manifolds, [15–17, 25, 106, 129, 130], is said to flow by the Ricci flow if

$$g_t = -2\operatorname{Ric}.$$

The triple (M, g, f) is a gradient shrinking soliton, or shrinker for short, if

$$\operatorname{Ric} + \operatorname{Hess}_f = \frac{1}{2}g;$$

shrinkers are the singularities in Ricci flow, [33, 36, 107, 122, 158, 162].

Theorem 8.22. *For any shrinker (M, g, f), if $Y \in L^2$, $\mathcal{P}Y = \lambda Y$ and*

$$Z = Y + \frac{2}{2\lambda+1}\nabla\operatorname{div}_f(Y),$$

then $\operatorname{div}_f(Z) = 0$ and for any $\delta > 0$ and $r_2 > r_1 > R = R(\lambda, n, \delta)$,

$$I_{\nabla\operatorname{div}_f(Y)}(r_2) \leq \left(\frac{r_2}{r_1}\right)^{4\lambda+\delta} I_{\nabla\operatorname{div}_f(Y)}(r_1), \tag{8.23}$$

$$I_Z(r_2) \leq \left(\frac{r_2}{r_1}\right)^{8\lambda+2+\delta} I_Z(r_1). \tag{8.24}$$

Each of these growth bounds is sharp and so is the requirement that $Y \in L^2$. Combining them bounds Y. As a corollary, L^2 Killing fields on a shrinker grow at most linearly.

Corollary 8.25. *On any shrinker, for any L^2 Killing field Y, $\nabla\operatorname{div}_f(Y)$ is parallel and if $Z = Y + 2\nabla\operatorname{div}_f(Y)$, then $\operatorname{div}_f(Z) = 0$ and for any $\delta > 0$ and $r_2 > r_1 > R = R(n, \delta)$,*

$$I_Z(r_2) \leq \left(\frac{r_2}{r_1}\right)^{2+\delta} I_Z(r_1). \tag{8.26}$$

It is easy to see that this is sharp; on the two-dimensional Gaussian soliton, $Y = x_2e_1 - x_1e_2$ is a Killing field with $\operatorname{div}_f(Y) = 0$ that grows linearly.

On a shrinker, the operator $\mathcal{P}$ relates to the manifold version of the much studied Ornstein–Uhlenbeck operator $\mathcal{L}$ on vector fields Y by the formula

$$-2\mathcal{P}Y = \nabla\operatorname{div}_f Y + \mathcal{L}Y + \frac{1}{2}Y. \tag{8.27}$$

Whereas $\mathcal{P}$ is a true system operator, $\mathcal{L}$ is not, and for that reason, $\mathcal{P}$ is more complicated. On the other hand, on solitons, $\mathcal{P}$ has many nice properties: it commutes with $\mathcal{L}$ and if Y is an eigenvector field of $\mathcal{P}$ with eigenvalue $-\lambda$, then $\nabla \operatorname{div}_f(Y)$ is an eigenvector field of $\mathcal{L}$ with eigenvalue λ. The unweighted version of $\mathcal{P}$ was used implicitly by Bochner to show that closed manifolds with negative Ricci curvature have no Killing fields. Building on this the unweighted operator was later used by Bochner and Yano to show that the isometry group of such manifolds is finite. The unweighted operator also arises in general relativity. The relationship between $\mathcal{P}$ and the unweighted version, used by Bochner, mirrors the relationship between the Ornstein–Uhlenbeck operator and the Laplacian.

8.6. Applications

This new understanding of the "gauge group" can be used to settle a well-known problem in Ricci flow. Namely, using it one can show, see [91], a strong rigidity for cylinders, quotients of cylinders, and more general shrinking solitons; [23, 34], cf. [140].

Theorem 8.28. *Let Σ be the round cylinder $\mathbf{S}^\ell \times \mathbf{R}^{n-\ell}$ (or quotient of such) as a shrinker with potential $f_\Sigma = \frac{|x|^2}{4} + \frac{\ell}{2}$. There exists an $R = R(n)$ such that if (M^n, g, f) is another shrinker and $\{f_\Sigma \leq R\} \cap \Sigma$ is close to $\{f \leq R\} \subset M$ in the smooth topology and f_Σ and f are close on this set, then (M, g, f) is a round shrinking cylinder (or quotient of such).*

Since blowups only converge on compact subsets, rather than globally, the most useful characterizations involve only a compact subset as in Theorem 8.28. An important difficulty is that there are nontrivial infinitesimal variations, i.e., variations in the kernel of the linearized operator (not generated by diffeomorphisms). One consequence of Theorem 8.28 is that these infinitesimal variations are not integrable; cf. also [54].

The principle behind Theorem 8.28 is that closeness to a large enough piece of Σ propagates outwards, becoming even closer on larger scales. We will explain some of the ideas behind this shortly. A much weaker extension will follow from pseudolocality [160], which says that flatness propagates forward in time; accordingly, flatness propagates outward in space for shrinkers. This gives a priori curvature estimates on a slightly larger scale. However, it gives little control over the metric itself because of the gauge invariance and, second, there is a loss in the estimates that makes it impossible to iterate. There are three major ingredients in the proof of Theorem 8.28; we loosely refer to these as propagation of almost splitting, gauge fixing, and quadratic rigidity in the right gauge. These are of independent interest and will described in order next.

"Propagation of almost splitting" shows that if a shrinker is close to a product $N \times \mathbf{R}^{n-\ell}$ on a large scale, then it remains close on a fixed larger scale. The closeness on the first scale is used to get $n - \ell$ eigenvalues that are exponentially close to $\frac{1}{2}$, which is a lower bound for any shrinker that is only achieved by linear functions on products. The corresponding eigenfunctions will have exponentially small L^2-bounds for their Hessians, which forces the gradients to be virtually parallel on small sets but says little on large balls because of the Gaussian weight. It is here that the growth bounds from [91] first play a crucial role, showing that the Hessian bounds can only grow polynomially so the initial exponential smallness gives control on larger scales. These almost parallel vector fields are then used to construct a diffeomorphism to Σ on the larger scale, giving vastly more control than what followed from pseudolocality. This is very much a Ricci flow fact that does not have an MCF analogue where we do not have a corresponding description of the bottom of the spectrum.

The almost splitting gives considerable control on the larger scale, but does not fix the gauge—the difference in metrics is small, but is not orthogonal to the action of the gauge group. Moreover, even when the two metrics are the same, the difference between the potentials could be a linear function, corresponding to a translation along the axis.

There are many other important uniqueness results in Ricci flow, see, for instance, [16, 26, 27, 132, 133].

Part 5. Minimal surfaces. Surfaces that locally minimize area have been extensively used to model physical phenomena, including soap films, black holes, compound polymers, protein folding, etc. The mathematical field dates to the 1740s.

Minimal surfaces with uniform curvature or area bounds are well understood, yet essentially nothing was known without such bounds. We discuss here the theory of embedded (i.e., without self-intersections) minimal surfaces in Euclidean space $\mathbf{R}^3$ without a priori bounds; see [64–70, 77, 161] for more. The study is divided into three cases, depending on the topology of the surface. In case one the surface is a disk, in case two the surface is a planar domain (genus zero), and the third case is that of finite (nonzero) genus. The complete understanding of the disk case is applied in both cases two and three. In all three cases the surface is allowed to have a boundary. This is an essential point and makes the results particularly useful. For instance, given any minimal surface, independent of its topology, if a component of the intersection of the surface with a Euclidean ball is a disk, then case one applies and gives a good description of that component. Similarly, for cases two and three. The surface itself may then be thought of as built out of these snapshots (or building blocks). We will here mostly only discuss the case of disks.

The helicoid, which is a double spiral staircase, was discovered to be a minimal surface by Meusnier in 1776. As we will see, the helicoid is the most important example of an embedded minimal disk. In fact, we will see that every such disk is either a graph of a function or part of a double spiral staircase. For planar domains the fundamental examples are the catenoid, also discovered by Meusnier in 1776, and the Riemann examples discovered by Riemann in the beginning of the 1860s.[9] Finally, for general fixed genus, an important example is the recent example by Hoffman–Weber–Wolf of a genus-one helicoid. The genus-one helicoid is a complete minimal surface that on a large scale, away from the genus, looks essentially like an ordinary helicoid. This illustrates that the helicoid is one of the basic building blocks of general minimal surfaces. This is also true for the Riemann examples. The Riemann examples are a two-parameter family of complete minimal surfaces. As the

9 Riemann worked on minimal surfaces in the period 1860–1861. He died in 1866. The Riemann example was published post-mortem in 1867 in an article edited by Poggendorf.

parameters degenerate, the Riemann examples looks like either a collection of catenoids stacked on top of each other or two oppositely oriented helicoids (with parallel axes) glued together.

In the last section we discuss why (complete) embedded minimal surfaces are automatically proper (i.e., why divergent sequences of points on the surface diverge in Euclidean space). This question is known as the Calabi–Yau conjectures for embedded surfaces. For immersed (but not embedded) surfaces, there are counterexamples by Jorge-Xavier and Nadirashvili.

8.7. Minimal graphs and the helicoid

The derivation of the equation for a minimal graph goes back to Lagrange's 1762 memoir. There are questions of existence of solutions, uniqueness of equilibria, and the global structure of the space (or spaces) of examples. At the intersection of all of these questions is the question of what the (shape of the) natural building blocks are. In a broad sense, graphs and helicoids are in a fundamental way the key building blocks of embedded minimal surfaces.

There are two local models for embedded minimal *disks*. One model is the plane (or, more generally, a minimal graph) and the other is a piece of a helicoid.

Minimal graphs over proper simply connected domains in $\mathbf{R}^2$ gives a large class of embedded minimal disks, however, by a classical theorem of Bernstein from 1916 entire (i.e., where $\Omega = \mathbf{R}^2$) minimal graphs are planes.

The second model comes from the helicoid which was discovered by Meusnier in 1776.[10] The helicoid is a "double spiral staircase" given by sweeping out a horizontal line rotating at a constant rate as it moves up a vertical axis at a constant rate. Each half-line traces out a spiral staircase and together the two half-lines trace out (up to scaling) the double spiral staircase $(s \cos t, s \sin t, t)$, where $s, t \in \mathbf{R}$.

For the results about embedded minimal disks, it will be important to understand a sequence of helicoids obtained from a single helicoid by rescaling as follows:

Consider the sequence $\Sigma_i = a_i \Sigma$ of rescaled helicoids where $a_i \to 0$. (That is, rescale $\mathbf{R}^3$ by a_i, so points that used to be distance d apart will in the rescaled $\mathbf{R}^3$ be distance $a_i d$ apart.) The curvatures of this sequence of rescaled helicoids are blowing up (i.e., the curvatures go to infinity) along the vertical axis. The sequence converges (away from the vertical axis) to a foliation by flat parallel planes; that is, it converges to the collection of planes $x_3 =$ constant. The singular set (the axis) then consists of removable singularities.

10 Meusnier had been a student of Monge. He also discovered that the catenoid is minimal in the sense of Lagrange, and he was the first to characterize a minimal surface as a surface with vanishing mean curvature. Unlike the helicoid, the catenoid is not topologically a plane but rather a cylinder.

8.8. Multivalued graphs, spiral staircases, double spiral staircases

To be able to give a precise meaning to the statement that the helicoid is a double spiral staircase, we will need the notion of a multivalued graph, each staircase will be a multivalued graph. Intuitively, a multivalued graph is a surface covering an annulus, such that over a neighborhood of each point of the annulus, the surface consists of N graphs. To make this notion precise, let D_r be the disk in the plane centered at the origin and of radius r and let $\mathcal{P}$ be the universal cover of the punctured plane $\mathbf{C} \setminus \{0\}$ with global polar coordinates (ρ, θ) so $\rho > 0$ and $\theta \in \mathbf{R}$. An *N-valued graph* on the annulus $D_s \setminus D_r$ is a single valued graph of a function u over $\{(\rho, \theta) \mid r < \rho \leq s, |\theta| \leq N\pi\}$. For working purposes, we generally think of the intuitive picture of a multisheeted surface in $\mathbf{R}^3$, and we identify the single-valued graph over the universal cover with its multivalued image in $\mathbf{R}^3$.

The multivalued graphs that we will consider will all be embedded, which corresponds to a nonvanishing separation between the sheets (or the floors). If Σ is the helicoid, then $\Sigma \setminus \{x_3 - \text{axis}\} = \Sigma_1 \cup \Sigma_2$, where Σ_1, Σ_2 are ∞-valued graphs on $\mathbf{C} \setminus \{0\}$; Σ_1 is the graph of the function $u_1(\rho, \theta) = \theta$ and Σ_2 is the graph of the function $u_2(\rho, \theta) = \theta + \pi$. (Further, Σ_1 is the subset where $s > 0$ in the parametrization of the helicoid and Σ_2 the subset where $s < 0$.) In either case the separation between the sheets is constant, equal to 2π. A *multivalued minimal graph*, see chapter 1 in [71], is a multivalued graph of a function u satisfying the minimal surface equation.

8.9. Structure of embedded minimal disks

All of our results for disks, as well as for other topological types, require only a piece of a minimal surface. In particular, the surfaces may well have boundaries and when we, for instance, say in the next theorem "Any embedded minimal disk in $\mathbf{R}^3$ is *either* a graph of a function *or* part of a double spiral staircase", then we mean that if the surface is contained in a Euclidean ball of radius r_0 and the boundary is contained in the boundary of that ball, then in a concentric Euclidean ball with radius a fixed (small) fraction of r_0, any component of the surface is *either* a graph of a function *or* part of a double spiral staircase. That the surfaces are allowed to have boundaries is a major point and makes the results particularly useful. Note also that as the conclusion is for a "fixed fraction of the surface" this is an interior estimate.

The following is the main structure theorem for embedded minimal disks:

Theorem 8.29. *Any embedded minimal disk in* $\mathbf{R}^3$ *is* either *a graph of a function* or *part of a double spiral staircase. In particular, if for some point the curvature is sufficiently large, then the surface is part of a double spiral staircase (it can be approximated by a piece of a rescaled helicoid). On the other hand, if the curvature is below a certain threshold everywhere, then the surface is a graph of a function.*

As a consequence of this structure theorem we get the following compactness result:

Corollary 8.30. *A sequence of embedded minimal disks with curvatures blowing up (i.e., going to infinity*[11]*) at a point mimics the behavior of a sequence of rescaled helicoids with curvature going to infinity.*

8.10. Two key ideas behind the proof of the structure theorem for disks

The first of these key ideas says that if the curvature of such a disk Σ is large at some point $x \in \Sigma$, then near x a multivalued graph forms (in Σ), and this extends (in Σ) almost all the way to the boundary[12] of Σ. Moreover, the inner radius, r_x, of the annulus where the multivalued graph is defined is inversely proportional to $|A|(x)$, and the initial separation between the sheets is bounded by a constant times the inner radius.

An important ingredient in the proof of Theorem 8.29 is that general embedded minimal disks with large curvature at some interior point can be built out of N-valued graphs. In other words, any embedded minimal disk can be divided into pieces each of which is an N-valued graph. Thus the disk itself should be thought of as being obtained by stacking these pieces (graphs) on top of each other.

The second key result (Theorem 8.31) is a curvature estimate for embedded minimal disks in a half-space (in this theorem r_0 is a scaling factor, which after rescaling can be taken to be one):

Theorem 8.31. *There exists $\varepsilon > 0$ such that for all $r_0 > 0$, if $\Sigma \subset B_{2r_0} \cap \{x_3 > 0\} \subset \mathbf{R}^3$ is an embedded minimal disk with $\partial\Sigma \subset \partial B_{2r_0}$, then for all components Σ' of $B_{r_0} \cap \Sigma$ which intersect $B_{\varepsilon r_0}$,*

$$\sup_{x \in \Sigma'} \left|A_\Sigma(x)\right|^2 \leq r_0^{-2}. \tag{8.32}$$

This theorem has an equivalent formulation that may be easier to appreciate. Namely, for $\varepsilon > 0$ sufficiently small, (8.32) is equivalent to the statement that Σ' is a graph over (a domain in) the plane $\{x_3 = 0\}$.

Theorem 8.31 is an interior estimate where the curvature bound, (8.32), is on the ball B_{r_0} of one-half of the radius of the ball B_{2r_0} containing Σ. This is just like a gradient estimate for a harmonic function where the gradient bound is on one-half of the ball where the function is defined. Theorem 8.31 is often referred to as *the one-sided curvature estimate* (since Σ is assumed to lie on one side of a plane). The assumption in Theorem 8.31 that Σ is simply connected (i.e., that Σ is a disk) is crucial, as can be seen from the example of a rescaled catenoid. Rescaled catenoids converge (with multiplicity two) to the flat plane. Likewise, by considering the universal cover of the catenoid, one sees that Theorem 8.31 requires the disk to be embedded, and not just immersed.

The one-sided curvature estimate has strong implications for embedded minimal surfaces. We will return to some of these applications later, but note here that it can be

---

11 A minimal surface in $\mathbf{R}^3$ the curvature $K = -\frac{1}{2}|A|^2$ is nonpositive; so that by the curvatures of a sequence is going to infinity we mean that $K \to -\infty$ or, equivalently, $|A|^2 \to \infty$.

12 Our results require only that we have a piece of a minimal surface and thus it may have boundary.

applied even to ends of embedded minimal surfaces with finite topology to give a different of a conjecture of Nitsche, see [56, 93].

8.11. Uniqueness theorems

There is a long history of uniqueness theorems for properly embedded minimal surfaces, but all of those made very strong assumptions. A typical example is Catalan's theorem. Catalan proved in 1842 that any complete ruled minimal surface is either a plane or a helicoid. A surface is said to be *ruled* if it has the parametrization $X(s,t) = \beta(t) + s\delta(t)$, where $s, t \in \mathbf{R}$, and β and δ are curves in $\mathbf{R}^3$. The curve $\beta(t)$ is called the *directrix* of the surface, and a line having $\delta(t)$ as direction vector is called a *ruling*. For the helicoid, the x_3-axis is a directrix, and for each fixed t the line $s \to (s\cos t, s\sin t, t)$ is a ruling. More recent uniqueness results (for instance, by Lopez, Meeks, Nirenberg, Nitsche, Osserman, Perez, Ros, Schoen, Shiffman, and Simon) assumed either finite total curvature or periodicity. The structure theorems in [65–68] opened up the possibility of showing uniqueness theorems in complete generality.

To give a flavor of some of the results that led to spetacular development in the theory of minimal surfaces, we will mention just a few highlights. Using the above structure theorem for disks, Meeks–Rosenberg [150] proved, cf. [19], that the plane and the helicoid are the only complete properly embedded simply-connected minimal surfaces in $\mathbf{R}^3$. The Riemann examples were shown to be unique by Meeks–Perez–Ros [148]. In addition to the structure theory for disks, they also used the structure theory of all finite-genus embedded minimal surfaces from [70]. The paper [148] also introduced two very interesting new techniques into the subject: the KdV equation and a careful analysis of the Shiffman function.

9. EMBEDDED MINIMAL SURFACES ARE AUTOMATICALLY PROPER

Implicit in all of the results mentioned above was an assumption that the minimal surfaces were proper. However, as we will see next, it turns out that embedded minimal surfaces are, in fact, automatically proper. This was the content of the Calabi–Yau conjectures which were proven to be true for embedded surfaces in [66].

9.1. Proper embeddings

An immersed surface in $\mathbf{R}^3$ is *proper* if the preimage of any compact subset of $\mathbf{R}^3$ is compact in the surface. For instance, a line is proper whereas a curve that spiral infinitely into a circle is not.

9.2. The Calabi–Yau conjectures; the statements and examples

The Calabi–Yau conjectures about surfaces date back to the 1960s. Their original form was given in 1965 where Calabi [31] made the following two conjectures about minimal surfaces[13]:

Conjecture 9.1. *Prove that a complete minimal surface in* $\mathbf{R}^3$ *must be unbounded.*

Calabi continued: "It is known that there are no compact minimal surfaces in $\mathbf{R}^3$ (or of any simply connected complete Riemannian 3-dimensional manifold with sectional curvature ≤ 0). A more ambitious conjecture is":

Conjecture 9.2. *A complete [non-flat] minimal surface in* $\mathbf{R}^3$ *has an unbounded projection in every line.*

The *immersed* versions of these conjectures turned out to be false. Namely, Jorge and Xavier [123] constructed non-flat minimal immersions contained between two parallel planes in 1980, giving a counterexample to the immersed version of the more ambitious Conjecture 9.2. Another significant development came in 1996, when Nadirashvili [156] constructed a complete immersion of a minimal disk into the unit ball in $\mathbf{R}^3$, showing that Conjecture 9.1 also failed for immersed surfaces; cf. [2].

The main result in [70] is an effective version of properness for disks, giving a chord–arc bound.[14] Obviously, intrinsic distances are larger than extrinsic distances, so the significance of a chord–arc bound is the reverse inequality, i.e., a bound on intrinsic distances from above by extrinsic distances. Given such a chord–arc bound, one has that as intrinsic distances go to infinity, so do extrinsic distances. Thus as an immediate consequence:

Theorem 9.3. *A complete embedded minimal disk in* $\mathbf{R}^3$ *must be proper.*

Theorem 9.3 gives immediately that the first of Calabi's conjectures is true for *embedded* minimal disks. Another immediate consequence of the chord–arc bound together with the one-sided curvature estimate (i.e., Theorem 8.31) is a version of that estimate for intrinsic balls. As a corollary of this intrinsic one-sided curvature estimate, we get that the second, and more ambitious, of Calabi's conjectures is also true for *embedded* minimal disks. The second Calabi conjecture (for embedded disks) is an immediate consequence of the following half-space theorem:

Theorem 9.4. *The plane is the only complete embedded minimal disk in* $\mathbf{R}^3$ *in a half-space.*

Theorem 9.4 is a byproduct of the proof of Theorem 9.3. However, given Theorem 9.3, Theorem 9.4 follows from the half-space theorem of [113].

The results for disks imply both of Calabi's conjectures and properness also for embedded surfaces with finite topology. A surface Σ is said to have finite topology if it is homeomorphic to a closed Riemann surface with a finite set of points removed or "punctures." Each puncture corresponds to an end of Σ.

---

13 S. S. Chern [44] also promoted these conjectures at roughly the same time and they were revisited several times by S. T. Yau.

14 A chord–arc bound is a bound above and below for the ratio of intrinsic to extrinsic distances.

See [94,149,151] for related results and further references.

FUNDING

The author was partially supported by NSF Grant DMS 2104349.

REFERENCES

[1] V. Agostiniani, M. Fogagnolo, and L. Mazzieri, Sharp geometric inequalities for closed hypersurfaces in manifolds with nonnegative Ricci curvature. *Invent. Math.* **222** (2020), no. 3, 1033–1101.

[2] A. Alarcon and F. Forstneric, New complex analytic methods in the theory of minimal surfaces: a survey. *J. Aust. Math. Soc.* **106** (2019), no. 3, 287–341.

[3] W. K. Allard and F. J. Almgren Jr., On the radial behavior of minimal surfaces and the uniqueness of their tangent cones. *Ann. of Math. (2)* **113** (1981), no. 2, 215–265.

[4] S. Altschuler, S. B. Angenent, and Y. Giga, Mean curvature flow through singularities for surfaces of rotation. *J. Geom. Anal.* **5** (1995), no. 3, 293–358.

[5] L. Ambrosio and H. M. Soner, Level set approach to mean curvature flow in arbitrary codimension. *J. Differential Geom.* **43** (1996), no. 4, 693–737.

[6] M. T. Anderson, The Dirichlet problem at infinity for manifolds of negative curvature. *J. Differential Geom.* **18** (1983), no. 4, 701–721.

[7] M. T. Anderson and R. Schoen, Positive harmonic functions on complete manifolds of negative curvature. *Ann. of Math. (2)* **121** (1985), no. 2, 429–446.

[8] B. Andrews, Noncollapsing in mean-convex mean curvature flow. *Geom. Topol.* **16** (2012), no. 3, 1413–1418.

[9] B. Andrews, B. Chow, C. Guenther, and M. Langford, *Extrinsic geometric flows*. Grad. Stud. Math. 206, American Mathematical Society, Providence, RI, 2020.

[10] B. Andrews, H. Li, and Y. Wei, $\mathcal{F}$-stability for self-shrinking solutions to mean curvature flow. *Asian J. Math.* **18** (2014), no. 5, 757–777.

[11] B. Andrews and L. Ni, Eigenvalue comparison on Bakry–Emery manifolds. *Comm. Partial Differential Equations* **37** (2012), no. 11, 2081–2092.

[12] S. Angenent, P. Daskalopoulos, and N. Sesum, Unique asymptotics of ancient convex mean curvature flow solutions. *J. Differential Geom.* **111** (2019), no. 3, 381–455.

[13] S. Angenent, P. Daskalopoulos, and N. Sesum, Uniqueness of two-convex closed ancient solutions to the mean curvature flow. *Ann. of Math. (2)* **192** (2020), no. 2, 353–436.

[14] C. Arezzo and J. Sun, Self-shrinkers for the mean curvature flow in arbitrary codimension. *Math. Z.* **274** (2013), no. 3–4, 993–1027.

[15] R. Bamler, Structure theory of singular spaces. *J. Funct. Anal.* **272** (2017), no. 6, 2504–2627.

[16] R. Bamler, Recent developments in Ricci flows. *Not. Amer. Math. Soc.* **68**, (2021), no. 9, 1486–1498.

[17] R. Bamler and B. Kleiner, Uniqueness and stability of Ricci flow through singularities. *Acta Math.*, to appear.

[18] J. Bernstein, Asymptotic structure of almost eigenfunctions of drift Laplacians on conical ends. *Amer. J. Math.* **142** (2020), no. 6, 1897–1929.

[19] J. Bernstein and C. Breiner, Conformal Structure of Minimal Surfaces with Finite Topology. *J. Reine Angew. Math.* **655** (2011), 129–146.

[20] J. Bernstein and L. Wang, A sharp lower bound for the entropy of closed hypersurfaces up to dimension six. *Invent. Math.* **206** (2016), no. 3, 601–627.

[21] J. Bernstein and L. Wang, A topological property of asymptotically conical self-shrinkers of small entropy. *Duke Math. J.* **166** (2017), no. 3, 403–435.

[22] S. Biard, J. Fornaess, and J. Wu, Weighted L^2 version of Mergelyan and Carleman approximation. *J. Geom. Anal.* **31** (2021), 3889–3914.

[23] C. Böhm, Inhomogeneous Einstein metrics on low-dimensional spheres and other low-dimensional spaces. *Invent. Math.* **134** (1998), no. 1, 145–176.

[24] K. Brakke, *The motion of a surface by its mean curvature*. Math. Notes 20, Princeton University Press, Princeton, 1978.

[25] S. Brendle, *Ricci flow and the sphere theorem*. Grad. Stud. Math. 111, American Mathematical Society, Providence, RI, 2010.

[26] S. Brendle, Rotational symmetry of self-similar solutions to the Ricci flow. *Invent. Math.* **194** (2013), no. 3, 731–764.

[27] S. Brendle, Rotational symmetry of Ricci solitons in higher dimensions. *J. Differential Geom.* **97** (2014), no. 2, 191–214.

[28] S. Brendle, Embedded self-similar shrinkers of genus 0. *Ann. of Math. (2)* **183** (2016), no. 2, 715–728.

[29] S. Brendle and K. Choi, Uniqueness of convex ancient solutions to mean curvature flow in $\mathbf{R}^3$. *Invent. Math.* **217** (2019), no. 1, 35–76.

[30] S. Brendle and G. Huisken, Mean curvature flow with surgery of mean convex surfaces in $\mathbf{R}^3$. *Invent. Math.* **203** (2016), no. 2, 615–654.

[31] E. Calabi, Final chapter in Problems in differential geometry. In *Proceedings of the United States-Japan seminar in differential geometry, Kyoto, Japan, 1965*, edited by S. Kobayashi and J. Eells Jr. Nippon Hyoronsha Co., Ltd., Tokyo, 1966.

[32] H. D. Cao, Recent progress on Ricci solitons. Recent advances in geometric analysis. In *Recent advances in geometric analysis*, pp. 1–38, Adv. Lect. Math. (ALM) 11, Int. Press, Somerville, MA, 2010.

[33] H. D. Cao, R. Hamilton, and T. Ilmanen, Gaussian densities and stability for some Ricci solitons. 2004, arXiv:math/0404165v1.

[34] H. D. Cao and C. He, Linear stability of Perelman's ν-entropy on symmetric spaces of compact type. *J. Reine Angew. Math.* **709** (2015), 229–246.

[35] H. D. Cao and D. Zhou, On complete gradient shrinking Ricci solitons. *J. Differential Geom.* **85** (2010), 175–186.

[36] J. Carrillo and L. Ni, Sharp logarithmic Sobolev inequalities on gradient solitons and applications. *Comm. Anal. Geom.* **17** (2009), no. 4, 721–753.

[37] J. Cheeger and T. H. Colding, Lower bounds on Ricci curvature and the almost rigidity of warped products. *Ann. of Math. (2)* **144** (1996), no. 1, 189–237.

[38] J. Cheeger, T. H. Colding, and W. P. Minicozzi II, Linear growth harmonic functions on complete manifolds with nonnegative Ricci curvature. *Geom. Funct. Anal.* **5** (1995), no. 6, 948–954.

[39] J. Cheeger and G. Tian, On the cone structure at infinity of Ricci flat manifolds with Euclidean volume growth and quadratic curvature decay. *Invent. Math.* **118** (1994), no. 3, 493–571.

[40] B.-L. Chen, Strong uniqueness of the Ricci flow. *J. Differential Geom.* **82** (2009), no. 2, 363–382.

[41] Y. G. Chen, Y. Giga, and S. Goto, Uniqueness and existence of viscosity solutions of generalized mean curvature flow equations. *J. Differential Geom.* **33** (1991), no. 3, 749–786.

[42] S. Y. Cheng and S. T. Yau, Differential equations on Riemannian manifolds and their geometric applications. *Comm. Pure Appl. Math.* **28** (1975), 333–354.

[43] X. Cheng and D. Zhou, Eigenvalues of the drifted Laplacian on complete metric measure spaces. *Commun. Contemp. Math.* **19** (2017), 1650001.

[44] S. S. Chern, The geometry of G-structures. *Bull. Amer. Math. Soc.* **72** (1966), 167–219.

[45] O. Chodosh, K. Choi, C. Mantoulidis, and F. Schulze, Mean curvature flow with generic low-entropy initial data. Preprint.

[46] O. Chodosh and C. Li, Stable minimal hypersurfaces in $\mathbf{R}^4$. Preprint.

[47] O. Chodosh and F. Schulze, Uniqueness of asymptotically conical tangent flows. *Duke Math. J.* 170 (2021), no. 16, 3601–3657.

[48] K. Choi, R. Haslhofer, and O. Hershkovits, Ancient low entropy flows, mean convex neighborhoods, and uniqueness. *Acta Math.* (to appear).

[49] B. Chow and P. Lu, On κ-noncollapsed complete noncompact shrinking gradient Ricci solitons which split at infinity. *Math. Ann.* **366** (2016), no. 3–4, 1195–1206.

[50] B. Chow, P. Lu, and L. Ni, *Hamilton's Ricci flow*. Grad. Stud. Math. 77, AMS, Providence, RI, 2006.

[51] B. Chow, S.-C. Chu, D. Glickenstein, C. Guenther, J. Isenberg, T. Ivey, D. Knopf, P. Lu, F. Luo, and L. Ni, *The Ricci flow: techniques and applications. Part I. Geometric aspects*. Math. Surveys Monogr. 135, American Mathematical Society, Providence, RI, 2007.

[52] T. H. Colding, Spaces with Ricci curvature bounds. In *Proceedings of the International Congress of Mathematicians*, Vol. II (Berlin, 1998). Doc. Math. 1998, Extra Vol. II, 299–308.

[53] T. H. Colding, New monotonicity formulas for Ricci curvature and applications, I. *Acta Math.* **209** (2012), no. 2, 229–263.

[54] T. H. Colding, T. Ilmanen, and W. P. Minicozzi II, Rigidity of generic singularities of mean curvature flow. *Publ. Math. Inst. Hautes Études Sci.* **121** (2015), 363–382.

[55] T. H. Colding, T. Ilmanen, W. P. Minicozzi II, and B. White, The round sphere minimizes entropy among closed self-shrinkers. *J. Differential Geom.* **95** (2013), 53–69.

[56] T. H. Colding and W. P. Minicozzi II, Complete properly embedded minimal surfaces in $\mathbf{R}^3$. *Duke Math. J.* 107 (2001), no. 2, 421–426.

[57] T. H. Colding and W. P. Minicozzi II, Estimates for parametric elliptic integrands. *Int. Math. Res. Not.* **6** (2002), 291–297.

[58] T. H. Colding and W. P. Minicozzi II, In search of stable geometric structures. *Notices Amer. Math. Soc.* **66** (2019), no. 11, 1785–1791.

[59] T. H. Colding and W. P. Minicozzi II, Harmonic functions on manifolds. *Ann. of Math. (2)* **146** (1997), no. 3, 725–747.

[60] T. H. Colding and W. P. Minicozzi II, Harmonic functions with polynomial growth. *J. Differential Geom.* **46** (1997), no. 1, 1–77.

[61] T. H. Colding and W. P. Minicozzi II, Large scale behavior of kernels of Schrödinger operators. *Amer. J. Math.* **119** (1997), no. 6, 1355–1398.

[62] T. H. Colding and W. P. Minicozzi II, Liouville theorems for harmonic sections and applications. *Comm. Pure Appl. Math.* **51** (1998), no. 2, 113–138.

[63] T. H. Colding and W. P. Minicozzi II, Weyl type bounds for harmonic functions. *Invent. Math.* **131** (1998), no. 2, 257–298.

[64] T. H. Colding and W. P. Minicozzi II, Disks that are double spiral staircases. *Notices Amer. Math. Soc.* **50** (2003), no. 3, 327–339.

[65] T. H. Colding and W. P. Minicozzi II, The space of embedded minimal surfaces of fixed genus in a 3-manifold I; Estimates off the axis for disks. *Ann. of Math. (2)* **160** (2004), no. 1, 27–68.

[66] T. H. Colding and W. P. Minicozzi II, The space of embedded minimal surfaces of fixed genus in a 3-manifold II; Multi-valued graphs in disks. *Ann. of Math. (2)* **160** (2004), no. 1, 69–92.

[67] T. H. Colding and W. P. Minicozzi II, The space of embedded minimal surfaces of fixed genus in a 3-manifold III; Planar domains. *Ann. of Math. (2)* **160** (2004), no. 2, 523–572.

[68] T. H. Colding and W. P. Minicozzi II, The space of embedded minimal surfaces of fixed genus in a 3-manifold IV; Locally simply connected. *Ann. of Math. (2)* **160** (2004), no. 2, 573–615.

[69] T. H. Colding and W. P. Minicozzi II, Shapes of embedded minimal surfaces. *Proc. Natl. Acad. Sci. USA* **103** (2006), no. 30, 11106–11111.

[70] T. H. Colding and W. P. Minicozzi II, The Calabi–Yau conjectures for embedded surfaces. *Ann. of Math. (2)* **167** (2008), no. 1, 211–243.

[71] T. H. Colding and W. P. Minicozzi II, *A course in minimal surfaces*. Grad. Stud. Math. 121, AMS, Providence, RI, 2011.

[72] T. H. Colding and W. P. Minicozzi II, Generic mean curvature flow I; generic singularities. *Ann. of Math.* **175** (2012), no. 2, 755–833.

[73] T. H. Colding and W. P. Minicozzi II, Smooth compactness of self-shrinkers. *Comment. Math. Helv.* **87** (2012), no. 2, 463–475.

[74] T. H. Colding and W. P. Minicozzi II, On uniqueness of tangent cones for Einstein manifolds. *Invent. Math.* **196** (2014), no. 3, 515–588.

[75] T. H. Colding and W. P. Minicozzi II, Ricci curvature and monotonicity for harmonic functions. *Calc. Var. Partial Differential Equations* **49** (2014), no. 3–4, 1045–1059.

[76] T. H. Colding and W. P. Minicozzi II, Łojasiewicz inequalities and applications. In *Regularity and evolution of nonlinear equations, Essays dedicated to Richard Hamilton, Leon Simon, and Karen Uhlenbeck*, pp. 63–82, Surv. Differ. Geom. 19, International Press, 2015.

[77] T. H. Colding and W. P. Minicozzi II, The space of embedded minimal surfaces of fixed genus in a 3-manifold V; fixed genus. *Ann. of Math. (2)* **181** (2015), no. 1, 1–153.

[78] T. H. Colding and W. P. Minicozzi II, Uniqueness of blowups and Łojasiewicz inequalities. *Ann. of Math.* **182** (2015), no. 1, 221–285.

[79] T. H. Colding and W. P. Minicozzi II, Differentiability of the arrival time. *Comm. Pure Appl. Math.* **LXIX** (2016), 2349–2363.

[80] T. H. Colding and W. P. Minicozzi II, Level set method for motion by mean curvature. *Notices Amer. Math. Soc.* **63** (2016), no. 10, 1148–1153.

[81] T. H. Colding and W. P. Minicozzi II, The singular set of mean curvature flow with generic singularities. *Invent. Math.* **204** (2016), no. 2, 443–471.

[82] T. H. Colding and W. P. Minicozzi II, Regularity of the level set flow. *Comm. Pure Appl. Math.* **71** (2018), no. 4, 814–824.

[83] T. H. Colding and W. P. Minicozzi II, Sharp frequency bounds for eigenfunctions of the Ornstein-Uhlenbeck operator. *Calc. Var. Partial Differential Equations* **57** (2018), no. 5, 138.

[84] T. H. Colding and W. P. Minicozzi II, Arnold–Thom gradient conjecture for the arrival time. *Comm. Pure Appl. Math.* **72** (2019), no. 7, 1548–1577.

[85] T. H. Colding and W. P. Minicozzi II, Dynamics of closed singularities. *Ann. Inst. Fourier (Grenoble)* **69** (2019), no. 7, 2973–3016.

[86] T. H. Colding and W. P. Minicozzi II, Liouville properties. *Notices ICCM* **7** (2019), no. 1, 16–26.

[87] T. H. Colding and W. P. Minicozzi II, Complexity of parabolic systems. *Publ. Math. Inst. Hautes Études Sci.* (2020), 83–135.

[88] T. H. Colding and W. P. Minicozzi II, Wandering singularities. *J. Differential Geom.* **119** (2021), 403–420.

[89] T. H. Colding and W. P. Minicozzi II, Optimal bounds for ancient caloric functions. *Duke Math. J.* 170 (2021), no. 18, 4171-4182.

[90] T. H. Colding and W. P. Minicozzi II, Regularity of elliptic and parabolic systems. Preprint.

[91] T. H. Colding and W. P. Minicozzi II, Singularities of Ricci flow and diffeomorphisms. Preprint.

[92] T. H. Colding, W. P. Minicozzi II, and E. K. Pedersen, Mean curvature flow. *Bull. Amer. Math. Soc. (N.S.)* **52** (2015), no. 2, 297–333.

[93] P. Collin, Topologie et courbure des surfaces minimales proprement plongees de $\mathbf{R}^3$. *Ann. of Math. (2)* **145** (1997), no. 1, 1–31.

[94] B. Coskunuzer, W. H. Meeks III, and G. Tinaglia, Non-properly embedded H-planes in $\mathbf{H}$. *J. Differential Geom.* **105** (2017), no. 3, 405–425.

[95] C. De Lellis, The regularity theory for the area functional (in geometric measure theory). Preprint.

[96] C. De Lellis, E. Spadaro, and L. Spolaor, Uniqueness of tangent cones for two-dimensional almost-minimizing currents. *Comm. Pure Appl. Math.* **70** (2017), no. 7, 1402–1421.

[97] H. Donnelly and C. Fefferman, Nodal domains and growth of harmonic functions on noncompact manifolds. *J. Geom. Anal.* **2** (1992), 79–93.

[98] L. C. Evans and J. Spruck, Motion of level sets by mean curvature I. *J. Differential Geom.* **33** (1991), 635–681.

[99] D. Fischer-Colbrie and R. Schoen, The structure of complete stable minimal surfaces in 3-manifolds of nonnegative scalar curvature. *Comm. Pure Appl. Math.* **33** (1980), no. 2, 199–211.

[100] M. Gage and R. S. Hamilton, The heat equation shrinking convex plane curves. *J. Differential Geom.* **23** (1986), no. 1, 69–96.

[101] Z. Gang and D. Knopf, Universality in mean curvature flow neckpinches. *Duke Math. J.* **164** (2015), no. 12, 2341–2406.

[102] N. Garofalo and F. H. Lin, Monotonicity properties of variational integrals, A_p weights and unique continuation. *Indiana Univ. Math. J.* **35** (1986), no. 2, 245–268.

[103] M. A. Grayson, The heat equation shrinks embedded plane curves to round points. *J. Differential Geom.* **26** (1987), no. 2, 285–314.

[104] M. Gromov, Groups of polynomial growth and expanding maps. *Publ. Math. Inst. Hautes Études Sci.* **53** (1981), 53–73.

[105] M. Gursky and J. Viaclovsky, Rigidity and stability of Einstein metrics for quadratic curvature functionals. *J. Reine Angew. Math.* **700** (2015), 37–91.

[106] R. Hamilton, Three-manifolds with positive Ricci curvature. *J. Differential Geom.* **17** (1982), 255–306.

[107] R. Hamilton, The formation of singularities in the Ricci flow. In *Surveys in differential geometry, Vol. II (Cambridge, MA, 1993)*, pp. 7–136, Int. Press, Cambridge, MA, 1993.

[108] R. M. Hardt, Singularities of harmonic maps. *Bull. Amer. Math. Soc. (N.S.)* **34** (1997), no. 1, 15–34.

[109] R. Haslhofer and B. Kleiner, Mean curvature flow of mean convex hypersurfaces. *Comm. Pure Appl. Math.* **70** (2017), no. 3, 511–546.

[110] R. Haslhofer and B. Kleiner, Mean curvature flow with surgery. *Duke Math. J.* **166** (2017), no. 9, 1591–1626.

[111] H.-J. Hein and A. Naber, New logarithmic Sobolev inequalities and an ϵ-regularity theorem for the Ricci flow. *Comm. Pure Appl. Math.* **67** (2014), no. 9, 1543–1561.

[112] H. J. Hein and S. Sun, Calabi-Yau manifolds with isolated conical singularities. *Publ. Math. Inst. Hautes Études Sci.* 126 (2017), 73–130.

[113] D. Hoffman and W. H. Meeks III, The strong halfspace theorem for minimal surfaces. *Invent. Math.* **101** (1990), no. 2, 373–377.

[114] G. Huisken, Flow by mean curvature of convex surfaces into spheres. *J. Differential Geom.* **20** (1984), no. 1, 237–266.

[115] G. Huisken, Asymptotic behavior for singularities of the mean curvature flow. *J. Differential Geom.* **31** (1990), no. 1, 285–299.

[116] G. Huisken, Local and global behaviour of hypersurfaces moving by mean curvature. In *Differential geometry: partial differential equations on manifolds (Los Angeles, CA, 1990). Part 1*, pp. 175–191, Proc. Sympos. Pure Math. 54, Amer. Math. Soc., Providence, RI, 1993.

[117] G. Huisken and C. Sinestrari, Convexity estimates for mean curvature flow and singularities of mean convex surfaces. *Acta Math.* **183** (1999), no. 1, 45–70.

[118] G. Huisken and C. Sinestrari, Mean curvature flow singularities for mean convex surfaces. *Calc. Var. Partial Differ. Equ.* **8** (1999), 1–14.

[119] G. Huisken and C. Sinestrari, Mean curvature flow with surgeries of two-convex hypersurfaces. *Invent. Math.* **175** (2009), no. 1, 137–221.

[120] T. Ilmanen, Generalized flow of sets by mean curvature on a manifold. *Indiana Univ. Math. J.* **41** (1992), no. 3, 671–705.

[121] T. Ilmanen, Singularities of mean curvature flow of surfaces. Preprint, 1995.

[122] T. Ivey, Ricci solitons on compact three-manifolds. *Differential Geom. Appl.* **3** (1993), no. 4, 301–307.

[123] L. Jorge and F. Xavier, A complete minimal surface in $\mathbf{R}^3$ between two parallel planes. *Ann. of Math. (2)* **112** (1980), 203–206.

[124] A. Kasue, Harmonic functions of polynomial growth on complete manifolds. In *Differential geometry: partial differential equations on manifolds (Los Angeles, CA, 1990). Part 1*, pp. 281–290, Proc. Sympos. Pure Math. 54, Amer. Math. Soc., Providence, RI, 1993.

[125] A. Kasue, Harmonic functions of polynomial growth on complete manifolds II. *J. Math. Soc. Japan* **47** (1995), 37–65.

[126] J. Kazdan, Parabolicity and the Liouville property on complete Riemannian manifolds. In *Aspects of Math.*, pp. 153–166, Vieweg, Braunschweig, 1987.

[127] D. Ketover and X. Zhou, Entropy of closed surfaces and min–max theory. *J. Differential Geom.* **110** (2018), no. 1, 31–71.

[128] B. Kleiner, A new proof of Gromov's theorem on groups of polynomial growth. *J. Amer. Math. Soc.* **23** (2010), no. 3, 815–829.

[129] B. Kleiner and J. Lott, Notes on Perelman's papers. *Geom. Topol.* **12** (2008), no. 5, 2587–2855.

[130] B. Kleiner and J. Lott, Singular Ricci flows. *Acta Math.* (2017), 65–134.

[131] R. V. Kohn and S. Serfaty, A deterministic-control-based approach to motion by curvature. *Comm. Pure Appl. Math.* **59** (2006), no. 3, 344–407.

[132] B. Kotschwar and L. Wang, Rigidity of asymptotically conical shrinking gradient Ricci solitons. *J. Differential Geom.* **100** (2015), no. 1, 55–108.

[133] B. Kotschwar and L. Wang, A uniqueness theorem for asymptotically cylindrical shrinking Ricci solitons. *J. Differential Geom.* (to appear).

[134] P. Kuchment, An overview of periodic elliptic operators. *Bull. Amer. Math. Soc. (N.S.)* **53** (2016), no. 3, 343–414.

[135] Y.-I. Lee and Y.-K. Lue, The stability of self-shrinkers of mean curvature flow in higher co-dimension. *Trans. Amer. Math. Soc.* **367** (2015), no. 4, 2411–2435.

[136] P. Li, Linear growth harmonic functions on Kähler manifolds with non–negative Ricci curvature. *Math. Res. Lett.* **2** (1995), 79–94.

[137] P. Li, The theory of harmonic functions and its relation to geometry. In *Differential geometry: partial differential equations on manifolds (Los Angeles, CA, 1990)*, pp. 307–315, Proc. Sympos. Pure Math. 54, Part 1, Amer. Math. Soc., Providence, RI, 1993.

[138] P. Li and L. F. Tam, Linear growth harmonic functions on a complete manifold. *J. Differential Geom.* **29** (1989), 421–425.

[139] P. Li and S. T. Yau, On the parabolic kernel of the Schrödinger operator. *Acta Math.* **156** (1986), no. 3–4, 153–201.

[140] Y. Li and B. Wang, Rigidity of the round cylinders in Ricci shrinkers. Preprint.

[141] F. H. Lin and Q. S. Zhang, On ancient solutions of the heat equation. *Comm. Pure Appl. Math.* **72** (2019), no. 9, 2006–2028.

[142] G. Liu, Three-circle theorem and dimension estimate for holomorphic functions on Kähler manifolds. *Duke Math. J.* **165** (2016), no. 15, 2899–2919.

[143] A. Logunov, Nodal sets of Laplace eigenfunctions: polynomial upper estimates of the Hausdorff measure. *Ann. of Math. (2)* **187** (2018), no. 1, 221–239.

[144] J. Lott, Some geometric properties of the Bakry–Emery–Ricci tensor. *Comment. Math. Helv.* **78** (2003), no. 4, 865–883.

[145] D. S. Lubinsky, A survey of weighted polynomial approximation with exponential weights. *Surv. Approx. Theory* **3** (2007), 1–105.

[146] F. C. Marques and A. Neves, Existence of infinitely many minimal hypersurfaces in positive Ricci curvature. *Invent. Math.* **209** (2017), no. 2, 577–616.

[147] F. C. Marques and A. Neves, The space of cycles, a Weyl law for minimal hypersurfaces and Morse index estimates. In *Surveys in differential geometry 2017. Celebrating the 50th anniversary of the Journal of Differential Geometry*, pp. 319–329, Surv. Differ. Geom. 22, Int. Press, Somerville, MA, 2018.

[148] W. Meeks III, J. Perez, and A. Ros, Properly embedded minimal planar domains. *Ann. of Math. (2)* **181** (2015), no. 2, 473–546.

[149] W. Meeks III, J. Perez, and A. Ros, The embedded Calabi-Yau conjecture for finite genus. *Duke Math. J.* 170 (2021), no. 13, 2891–2956.

[150] W. Meeks III and H. Rosenberg, The uniqueness of the helicoid. *Ann. of Math. (2)* **161** (2005), no. 2, 727–758.

[151] W. Meeks III and G. Tinaglia, Limit lamination theorem for H-disks. *Invent. Math.* **226** (2021), 393–420.

[152] W. Meeks and S.-T. Yau, Topology of three-dimensional manifolds and the embedding problems in minimal surface theory. *Ann. of Math. (2)* **112** (1980), no. 3, 441–484.

[153] M. Micallef, Stable minimal surfaces in Euclidean space. *J. Differential Geom.* **19** (1984), no. 1, 57–84.

[154] A. Naber, The geometry of Ricci curvature. In *Proceedings of the International Congress of Mathematicians–Seoul 2014. Vol. II*, pp. 911–937, Kyung Moon Sa, Seoul, 2014.

[155] A. Naber and D. Valtorta, The singular structure and regularity of stationary varifolds. *J. Eur. Math. Soc. (JEMS)* **22** (2020), no. 10, 3305–3382.

[156] N. Nadirashvili, Hadamard's and Calabi–Yau's conjectures on negatively curved and minimal surfaces. *Invent. Math.* **126** (1996), no. 3, 457–465.

[157] L. Ni, A monotonicity formula on complete Kähler manifolds with nonnegative bisectional curvature. *J. Amer. Math. Soc.* **17** (2004), no. 4, 909–946.

[158] L. Ni and N. Wallach, On a classification of gradient shrinking solitons. *Math. Res. Lett.* **15** (2008), no. 5, 941–955.

[159] S. Osher and J. A. Sethian, Fronts propagating with curvature-dependent speed: algorithms based on Hamilton–Jacobi formulations. *J. Comput. Phys.* **79** (1988), no. 1, 12–49.

[160] G. Perelman, The entropy formula for the Ricci flow and its geometric applications. 2002, arXiv:math/0211159.

[161] J. Perez, A new golden age of minimal surfaces. Notices Amer. Math. Soc. **64** (2017), no. 4, 347–358.

[162] P. Petersen and W. Wylie, Rigidity of gradient Ricci solitons. *Pacific J. Math.* **241** (2009), 329–345.

[163] T. Riviere and G. Tian, The singular set of 1–1 integral currents. *Ann. of Math. (2)* **169** (2009), no. 3, 741–794.

[164] R. Schoen, Estimates for stable minimal surfaces in three-dimensional manifolds. Seminar on minimal submanifolds. In *Ann. of Math. Stud.*, pp. 111–126, 103, Princeton Univ. Press, Princeton, NJ, 1983.

[165] R. Schoen, The effect of curvature on the behavior of harmonic functions and mappings. In *Nonlinear partial differential equations in differential geometry (Park City, UT, 1992)*, pp. 127–184, IAS/Park City Math. Ser. 2, 1992, Amer. Math. Soc., Providence, RI, 1996.

[166] R. Schoen L, Simon, and S. T. Yau, Curvature estimates for minimal hypersurfaces. *Acta Math.* **134** (1975), no. 3–4, 275–288.

[167] F. Schulze, Uniqueness of compact tangent flows in mean curvature flow. *J. Reine Angew. Math.* **690** (2014), 163–172.

[168] N. Sesum, Rate of convergence of the mean curvature flow. *Comm. Pure Appl. Math.* **61** (2008), no. 4, 464–485.

[169] Y. Shalom and T. Tao, A finitary version of Gromov's polynomial growth theorem. *Geom. Funct. Anal.* **20** (2010), no. 6, 1502–1547.

[170] L. Simon, Asymptotics for a class of evolution equations, with applications to geometric problems. *Ann. of Math.* **118** (1983), 525–571.

[171] L. Simon, Rectifiability of the singular sets of multiplicity 1 minimal surfaces and energy minimizing maps. In *Surveys in differential geometry, Vol. II*, pp. 246–305, Int. Press, Cambridge, MA, 1995.

[172] A. Song, Existence of infinitely many minimal hypersurfaces in closed manifolds. Preprint.

[173] D. Sullivan, The Dirichlet problem at infinity for a negatively curved manifold. *J. Differential Geom.* **18** (1983), no. 4, 723–732.

[174] G. Székelyhidi, Uniqueness of certain cylindrical tangent cones, preprint.

[175] T. Tao, Kleiner's proof of Gromov's theorem. https://terrytao.wordpress.com/2008/02/14/kleiners-proof-of-gromovs-theorem/

[176] T. Tao, A proof of Gromov's theorem. https://terrytao.wordpress.com/2010/02/18/a-proof-of-gromovs-theorem/

[177] T. Tao, "What is a gauge". https://terrytao.wordpress.com/2008/09/27/what-is-a-gauge/.

[178] P. Topping, *Lectures on the Ricci flow*. London Math. Soc. Lecture Note Ser. 325, Cambridge University Press, Cambridge, 2006.

[179] L. Wang, Uniqueness of self-similar shrinkers with asymptotically conical ends. *J. Amer. Math. Soc.* **27** (2014), 613–638.

[180] B. White, Partial regularity of mean-convex hypersurfaces flowing by mean curvature. *Int. Math. Res. Not. IMRN* **4** (1994), 185–192.

[181] B. White, Stratification of minimal surfaces, mean curvature flows, and harmonic maps. *J. Reine Angew. Math.* **488** (1997), 1–35.

[182] B. White, The size of the singular set in mean curvature flow of mean-convex sets. *J. Amer. Math. Soc.* **13** (2000), no. 3, 665–695.

[183] B. White, Evolution of curves and surfaces by mean curvature. In *Proceedings of the international congress of mathematicians, Vol. I (Beijing, 2002)*, pp. 525–538, Higher Ed. Press, 2002.

[184] B. White, The nature of singularities in mean curvature flow of mean-convex sets. *J. Amer. Math. Soc.* **16** (2003), no. 1, 123–138.

[185] B. White, A local regularity theorem for mean curvature flow. *Ann. of Math. (2)* **161** (2005), no. 3, 1487–1519.

[186] N. Wickramasekera, Regularity and compactness for stable codimension 1 CMC varifolds. In *Current developments in mathematics 2017*, pp. 87–174, Int. Press, Somerville, MA, 2019.

[187] S. T. Yau, Harmonic functions on complete Riemannian manifolds. *Comm. Pure Appl. Math.* **28** (1975), 201–228.

[188] S. T. Yau, Some function-theoretic properties of complete Riemannian manifold and their applications to geometry. *Indiana Univ. Math. J.* **25** (1976), no. 7, 659–670.

[189] S. T. Yau, Nonlinear analysis in geometry. *Enseign. Math. (2)* **33** (1987), 109–158.

[190] J. Zhu, On the entropy of closed hypersurfaces and singular self-shrinkers. *J. Differential Geom.* **114** (2020), no. 3, 551–593.

TOBIAS HOLCK COLDING

MIT, Dept. of Math., 77 Massachusetts Avenue, Cambridge, MA 02139-4307, USA, colding@math.mit.edu

THE REGULARITY THEORY FOR THE AREA FUNCTIONAL (IN GEOMETRIC MEASURE THEORY)

CAMILLO DE LELLIS

ABSTRACT

The aim of this article is to give a rather extensive, and yet nontechnical, account of the birth of the regularity theory for generalized minimal surfaces, of its various ramifications along the decades, of the most recent developments, and of some of the remaining challenges.

MATHEMATICS SUBJECT CLASSIFICATION 2020

Primary 28A75; Secondary 49Q15, 35B65, 35J47

KEYWORDS

Geometric measure theory, Plateau problem, regularity theory, area-minimizing currents, stationary varifolds

INTERNATIONAL CONGRESS OF MATHEMATICIANS 2022 JULY 6–14

Proc. Int. Cong. Math. 2022, Vol. 2, pp. 872–913
DOI 10.4171/ICM2022/122

1. INTRODUCTION

Let U be a bounded open subset of the Euclidean space $\mathbb{R}^{m+n}$ and let $\Sigma \subset U$ be an m-dimensional surface (e.g., a C^1 m-dimensional submanifold, but we will allow more general concept of surfaces for most of this note). Then Σ is said to be a "critical point of the area functional," or more commonly a "minimal surface," if

$$\left.\frac{d}{dt}\right|_{t=0} \mathrm{Vol}^m\big(\Phi_t(\Sigma)\big) = 0 \tag{1.1}$$

for every smooth one-parameter family of diffeomorphisms $[-\delta, \delta] \ni t \mapsto \Phi_t$ of $\overline{U}$ such that:

(a) $\Phi_t(x) = x$ for every $x \in \partial U$ and every t;

(b) $\Phi_0(x) = x$ for every $x \in \overline{U}$.

Here Vol^m denotes a suitable concept of m-dimensional volume: in the case of classical submanifolds, we can take the usual one from differential geometry.

Notable examples of minimal surfaces are those that minimize the volume in some suitable class $\mathscr{C}$. It suffices to assume that $\mathscr{C}$ is closed under deformations satisfying (a) and (b) above to conclude that any minimizer in $\mathscr{C}$ is necessarily a critical point of the area functional.

We call the attention of the reader to condition (a): the deformations fix the boundary of the open set U. Thus an example of a class $\mathscr{C}$ is that formed by those surfaces Σ's whose boundary (in a suitable sense, for instance, we can take the usual one of differential topology, if we are dealing with smooth surfaces) is a fixed Γ contained in ∂U. Such minimizer is then a surface of "least area spanning the contour Γ." However, we ultimately have to agree on the very definition of an "admissible surface" (Must it be embedded or do we allow self-intersections? Do we allow any topological type? In fact, must it be smooth or do we allow singularities? If we allow singularities, which type should we allow?), on what it means to span Γ, and how we define its volume.

Having assumed that we have answered all the above questions, i.e., that we have selected a suitable class $\mathscr{C}$ and a related concept of m-dimensional volume, a minimizer in $\mathscr{C}$ can be regarded as one possible solution to a celebrated problem in the calculus of variations, which goes under the name of *Plateau problem*. Indeed, the Belgian physicist Joseph Plateau investigated it in the early 19th century with the intention of finding a good description of soap films. However, the problem had already appeared in the mathematical literature decades before the investigations of Plateau, and can be found in the works of Lagrange, Meusnier, Monge, and Légendre. In particular, Lagrange considered minimizers of the area as early as the 1760s and he used his newly established method (which leads to what nowadays are called "Euler–Lagrange" conditions for minima of integral energies (cf. [79])) to describe 2-dimensional minimal graphs in $\mathbb{R}^3$ through a suitable partial differential equation.

As it is well known, if Σ is of class C^2, minimality in the sense of (1.1) is equivalent to the vanishing of the mean curvature vector. The latter is a condition which can be explained without any knowledge of differential geometry and it is, in fact, fairly easy to describe to anybody with a basic knowledge of multivariate calculus. Having fixed a point

$p_0 \in \Sigma$, choose first an orthonormal system of coordinates so that Σ is the graph of a map $\psi : \mathbb{R}^m \supset \Omega \to \mathbb{R}^n$ with the properties that

- $p_0 = (0, \psi(0))$ (i.e., p_0 is the origin of the coordinate system);
- and $\nabla\psi(0) = 0$ (i.e., the tangent to Σ at the origin is horizontal).

Then the mean curvature vector of Σ vanishes at p_0 if and only if $\Delta\psi(0) = 0$. One way to think about minimal surfaces is thus to understand them as solutions to a (somewhat complicated) nonlinear elliptic system of partial differential equations which linearize to the Laplace equation, namely $\Delta\psi = 0$, when we rotate the coordinates so that the tangent to the graph is horizontal.

The Laplace equation is universally considered as the prototypical elliptic partial differential equation of second order, and its solutions, i.e., the harmonic functions, are the prototype to understand the behavior of solutions to more general second order elliptic PDEs. The Laplace equation as well has a variational flavor, since it characterizes critical points of the Dirichlet energy $\int |\nabla\psi|^2$. But one could argue that minimal surfaces are even more natural objects than harmonic functions: indeed, a surface is minimal independently of the system of coordinates used to describe the ambient Euclidean space, while rigid motions of graphs which "mix domain and target" do not preserve the harmonicity: the latter is a concept which depends strongly on the selection of the dependent and independent variables used to describe the surface as a graph.

Critical points of the area functional have fascinated (and have been the object of study of) generations of mathematicians throughout at least two centuries and a half. One very interesting aspect of minimal surface theory is that it is relatively easy to produce singular examples. A particularly simple instance is given by holomorphic subvarieties in $\mathbb{C}^n$: if we identify $\mathbb{C}^n$ with $\mathbb{R}^{2n}$ and we understand holomorphic subvarieties of dimension k as $2k$-dimensional surfaces (with singularities), then the latter are always minimal. In fact, they are much more than just minimal: they minimize the area among a vast class of possible deformations. So an object as seemingly innocent as the complex algebraic curve $\Sigma = \{(z, w) \in \mathbb{C}^2 : z^2 = w^3\}$ is a 2-dimensional minimal surface in $\mathbb{R}^4$ and, in fact, it is minimal according to one of the most restrictive meanings that we can give. However, the origin is a point where Σ is not a regular submanifold: in particular, there is no neighborhood of 0 in which it can be described as the graph of a function (at least if we understand our functions "classically" and do not allow them to take more than one value at each fixed point of their domain).

Another simple example is given by the connected set E of least length which contains three noncolinear points $p_1, p_2, p_3 \in \mathbb{R}^2$. Such a set is the union of three distinct segments σ_i's which:

- have p_i as one endpoint,
- have all a common q as the other endpoint,
- meet at q forming angles of 120° degrees.

Again q is a "singular point" in the sense of differential topology: E is not a submanifold of $\mathbb{R}^2$ in any neighborhood of q.

If we accept that a good theory of minimal surfaces must include singular objects, we then open a Pandora box, in particular the following seemingly innocent questions immediately come to mind:

- How do we define the admissible surfaces, or otherwise put, which kind of singularities do we allow?
- Which kind of deformations do we take into account?
- What is the m-dimensional volume of a singular surface?

All these questions can be studied from different points of view and can be given very different answers depending on which goals one has in mind. For instance, answers which deal efficiently with the problem of minimizing the m-dimensional volume of surfaces in some fixed homology class of a given Riemannian manifold do not seem to give a satisfactory description of the complexity of soap films in real life. On the other hand, even though real life soap films display singularities, it can be proved that any 2-dimensional integral homology class in a closed smooth Riemannian 3-manifold has a smooth representative which minimizes the area. At any rate, whichever the goal, a rather large number of answers to these questions can be given in a subject of modern mathematics called geometric measure theory.

Geometric measure theory provides powerful tools to study various variational questions linked to the theory of minimal surfaces and has produced, in more than half a century, several notions of "singular minimal surfaces." In what follows, I will address several of them and, for lack of a better term, I will call all of them "generalized minimal surfaces." A subtopic of geometric measure theory, which is commonly called "regularity theory," studies natural questions like:

- Under which conditions singularities can be ruled out, i.e., the generalized minimal surfaces of a particular class end up being classical minimal submanifolds?
- How large can the set of singularities be when its existence cannot be completely ruled out?
- Which structural properties can the singularities have?

The ambition of this article is to give a rather extensive, and yet nontechnical, account of the birth of this topic, of its various ramifications along the decades, of the most recent developments, and of some of the remaining challenges. Since the topic is vast and complicated, I will probably not do a good service to much of the existing literature and I emphasize from the start that I consider my views quite biased.

2. PLATEAU'S PROBLEM, CRITICALITY, AND STABILITY

Before coming to a description of the "regularity theory," I will first introduce, in this section, some of the most common notions of "generalized minimal surfaces" considered in geometric measure theory.

2.1. Plateau's problem: two general approaches

As already mentioned, the Plateau problem can be loosely described as "looking for the surfaces Σ of least volume spanning a given contour Γ." If the surfaces Σ in question are C^1 submanifolds, it is commonly understood that the m-dimensional volume is the usual one from calculus books. If the given contour is as well a submanifold, a seemingly natural possibility to give a rigorous definition of "Σ spans Γ" is to say that Γ is the boundary of Σ in the usual sense of differential topology. From this point of view, it is also natural to consider ambient spaces more general than the Euclidean one, and a common natural choice is to have a general complete and smooth Riemannian ambient manifold. Practically all the "positive results" which we will discuss in this note have a generalization to smooth Riemannian ambient manifolds, but in order to be as nontechnical as possible, I will refrain to state the general theorems and always assume that the ambient is Euclidean. There are, however, some counterexamples which have been thus far found only in ambient Riemannian manifolds, and given their relevance in some of the problems examined below, I felt that they should be discussed.

If we want to enlarge the class of surfaces (and, in particular, allow minima with singularities in our class $\mathscr{C}$), we then have to specify at the same time what we mean by a surface, its volume, and the fact that it spans a given contour Γ. First of all, we will fix the convention that the dimension of the surfaces in $\mathscr{C}$ is m, the dimension of Γ is $m-1$, and the ambient Euclidean space (or Riemannian manifold, when the general case will be discussed) has dimension $m+n$. Secondly, we will restrict our attention to regular contours Γ: even though this can be relaxed considerably and one can fix certain type of nonsmooth boundaries (depending on the framework), clearly, when dealing with "boundary regularity" theorems, it is natural to assume that Γ itself has some regularity to start with. Once we have defined our generalized class of surfaces, their generalized volume, and what it means for them to span Γ, we will say that we have a "variational framework" for the Plateau problem.

It is possible to subdivide the various variational frameworks proposed in geometric measure theory in two large classes, which follow two rather different philosophical approaches. I will loosely describe them as:

- *Set-theoretic*. We insist in this case that our generalized surfaces Σ are just merely closed subsets of the ambient space which include Γ as a subset. The fact that they "span" Γ will then be encoded in some topological condition which Σ must satisfy, while the m-dimensional volume is defined by a suitable "measure" which satisfies the usual requirements of measure theory and coincides with the classical m-dimensional volume when Σ is a subset of a C^1 surface (or a countable union of subsets of C^1 surfaces).

- *Functional-analytic*. In this case we focus first on some nice, sufficiently regular, class of surfaces Σ, and we prescribe that their boundary is Γ in some suitable convenient sense coming from algebraic topology: let us denote this privileged class by $\mathscr{R}$. On $\mathscr{R}$ the concept of volume will be also given in terms of classical differential geometry and algebraic topology, and this will give us a functional $\mathcal{A}$ (the area functional) on $\mathscr{R}$. We then introduce some topology on $\mathscr{R}$, for instance, a distance, and the class $\mathscr{C}$ will be a suitable completion of this topological space, while the functional $\mathcal{A}$ will be extended to $\mathscr{C}$ in some natural way (for instance, we can take its lower semicontinuous envelope).

Observe that the "classical" parametric approach of Douglas and Rado does not fit in any of these two broad descriptions. The fact that I am not including it in the scope of these notes does not reflect any judgment on its mathematical interest: the "classical parametric theory" is a beautiful piece of mathematics, but it has a rather different flavor compared to the results and problems which will be discussed here.

In all the variational frameworks which we will examine, no matter whether they fall in one class or the other, there is a common and recurrent use of two important objects from geometric measure theory: the Hausdorff m-dimensional measure and rectifiable sets. The Hausdorff m-dimensional measure, which we will denote by $\mathcal{H}^m$, is a very natural way of extending the classical notion of m-dimensional volume to *any* subset of the Euclidean space (or, more generally, of a metric space). In fact, it is just one possibility, while a general theory of such extensions can be given in terms of the so-called "Caratheodory construction" and we refer the reader to some of the several textbook in the literature which treats it (cf. [59,62,88]). Rectifiable m-dimensional sets are a very natural class of sets which contain C^1 surfaces but is closed under many more operations which are natural from the point of view of measure theory: they consist of countable unions of closed subsets of C^1 m-dimensional submanifold plus a set of zero $\mathcal{H}^m$ measure (some people, for instance, Misha Gromov, consider the latter a somewhat very unpleasant technical addition, and the author agrees that there might be an efficient theory which works without the annoying technicality of adding null sets; however, most people in analysis grew accustomed to it, as "completing" a σ-algebra by adding sets of measure zero is a fairly common operation).

It was a major discovery of Besicovitch in the first half of the 20th century that any set of finite $\mathcal{H}^1$ measure can be decomposed into the union of a "rectifiable portion" and a "purely unrectifiable portion" (cf. [14–16]). The latter is somewhat "orthogonal" to any C^1 submanifold, in the sense that it intersects any C^1 curve in a set of $\mathcal{H}^1$ measure zero, even though it might have positive $\mathcal{H}^1$ measure. The more general theory of m-dimensional rectifiable sets was developed later Federer, cf. [61] (and see also [92]). While the rectifiable part has much of the features of C^1 submanifolds, and can be considered as a weak version of the latter, the purely unrectifiable part behaves in a rather counterintuitive way and in what follows we will discount it: it is, however, one of the major early developments of geometric measure theory that one can, without loss of generality, discard unrectifiable sets pretty much in all variational theories for the area functional and I am hiding quite deep and beautiful theorems here.

2.2. Examples of set-theoretic approaches

The first to pioneer what I dubbed "set-theoretic approach" was Reifenberg in [96]. In his variational framework, the definition of "E spans Γ" is that Γ is trivial in the relative Cech homology of E (cf. [96] for the precise definition). More recently Harrison (cf. [71, 72]) suggested another, very elegant, possible definition of "E spans Γ" which for simplicity we describe in the easiest case of $(m-1)$-dimensional Γ's in $\mathbb{R}^{m+1}$: any closed curve $\gamma \subset \mathbb{R}^{m+1} \setminus \Gamma$ which is not contractible in $\mathbb{R}^{m+1} \setminus \Gamma$ must intersect E.

Another point of view is that taken by Almgren in his theory of (M, ε, δ)-minimal sets, cf. [9]: rather than giving a precise notion of "spanning," we focus on which deformations are allowed and assume that our class $\mathscr{C}$ is closed under the latter deformations. In his work, Almgren gave a far-reaching existence and regularity theory, and the existence part was recently revisited and extended in [60]. Concerning deformations, a very interesting point raised only recently by David is that in practically all the works in the literature thus far the authors used deformations which completely "fix" the boundary Γ, while it would be much more natural to impose that they, in fact, map Γ onto itself in some controlled way (for instance, they are isotopic to the identity within the class of diffeomorphisms of Γ): this idea is at the base of his recent theory of "sliding minimizers", cf. [25, 27, 28].

In all these variational frameworks, for any given sequence E_k of compact sets there is a natural notion of convergence, that of Hausdorff, for which we can extract a converging subsequence. However, the Hausdorff measure $\mathcal{H}^m$ does not behave well in terms of the latter convergence, in the sense that it is not lower semicontinuous. On the other hand, one can suitably adjust *minimizing sequences* so to achieve the lower semicontinuity of $\mathcal{H}^m$: it thus suffices to prove that the limit is in the considered class $\mathscr{C}$ to achieve a minimizer. The author, in a joint work with F. Ghiraldin and F. Maggi in [37], pointed out that there is, in fact, no need to adjust the minimizing sequence and that a suitable compactness and lower-semicontinuity statement is valid for any minimizing sequence in a class $\mathscr{C}$ as soon as it allows a rather limited number of basic competitors. In particular, this gives a unified framework which treats all known examples of set-theoretic approaches put forward thus far, cf. also [36, 57, 58].

From the point of view of differential and algebraic topology, all the set-theoretic approaches have some very undesirable properties. Typical set-theoretic minimizers of the Plateau problem will always have singularities: if the boundary Γ is complicated, it is energetically convenient to form "triple junctions" along a singularity of codimension 1. On the other hand, one of the biggest achievements of the functional-analytic approach is that for every smooth closed embedded curve Γ in $\mathbb{R}^3$ there is always a smooth oriented 2-dimensional submanifold with boundary Γ which minimizes the 2-dimensional area among all smooth oriented 2-dimensional submanifolds with boundary Γ. Likewise, it is possible to show that every 2-dimensional integral homology class in a closed Riemannian 3-manifold has a smooth representative which minimizes the area. The set-theoretic approaches are not able to detect these two beautiful phenomena.

On the other hand, actual soap films do form triple junction singularities (and even more complicated ones) in real life, and these phenomena do not seem to be efficiently captured by functional-analytic frameworks (even though such singularities do occur in some specific situations, see below). Much of the research in the set-theoretic frameworks is thus motivated by the original intention of Plateau of finding a good variational description of soap films. In that respect the recent paper [85] by Maggi, Scardicchio, and Stuvard pointed out that much of the investigations in the mathematical literature have thus far ignored some very relevant physical attributes of real-life soap films. Combining some of the aspects of the set-theoretic approaches with other modern techniques, like Γ-convergence, and with more accurate considerations from mathematical physics, the papers [76–78] propose a new variational theory which promises to provide a much more accurate description of real-life soap films.

2.3. Functional-analytic frameworks

The pioneer of functional-analytic frameworks seems to be Renato Caccioppoli. In his works [18,19], Caccioppoli proposed the following definition of "perimeter" of a general (Lebesgue measurable) set of $\mathbb{R}^{m+1}$ (I will actually describe a slight variation of Caccioppoli's approach, but the actual differences are just of technical nature and for the purposes of this discussion I will ignore them). First of all, if the set has a C^1 boundary, its perimeter is defined to be the usual m-dimensional volume of the boundary. Next, given a general Lebesgue measurable set $E \subset \mathbb{R}^{m+1}$, we consider all possible sequences E_k of sets with C^1 boundaries with the property that the Lebesgue measure of the symmetric difference $E_k \Delta E$ goes to 0. We then consider

$$\liminf_{k\to\infty} \mathcal{H}^m(E_k)$$

and we further take the infimum of all such numbers among all approximating sequences $\{E_k\}$. The latter is defined to be the perimeter of E. If it is finite, E is commonly called a *set of finite perimeter* or (especially if you are Italian!) *Caccioppoli set*.

Caccioppoli's approach is very natural in the calculus of variations. We start from a class of "good" objects, the open sets with smooth boundary, over which the energy we are interested in, i.e., their perimeter, is classically defined. However, a sequence of smooth sets with uniformly controlled perimeter might converge to nonsmooth sets (for instance, one can easily form a corner, cusp, or other type of singularity) and we therefore would like to enlarge this class. We then take a much larger class, that of all measurable sets, with a topology in which the good objects are dense and we extend the energy to be the lower semicontinuous envelope. Interestingly Caccioppoli's approach was initially dismissed by his contemporaries (cf. the reviews by L. C. Young of the aforementioned papers) because he was not able to relate his abstract definition to any concrete notion of perimeter in a measure-theoretic sense. In the early 1950s, De Giorgi took up Caccioppoli's approach and proved, in his celebrated works on the isoperimetric property of the sphere (cf. [29–31]), that:

- the class of sets with finite perimeter in the sense of Caccioppoli is compact, under a uniform bound on their perimeter;
- the perimeter has a precise measure-theoretic interpretation, i.e., if the set Ω has a finite perimeter, one can introduce a suitable notion of (oriented) measure-theoretic boundary which turns out to be rectifiable and whose Hausdorff measure is indeed the perimeter of Ω.

De Giorgi also reformulated the theory of sets of finite perimeters through a useful duality: if correctly interpreted, the usual divergence theorem holds for them, and the boundary integral in the formulation is, in fact, a classical integral, in the sense of measure theory, over the measure-theoretic boundary. For open sets with smooth boundaries, the "measure-theoretic" one coincides with the topological one. An interesting byproduct (not at all obvious from the definition) is that the perimeter as defined by Caccioppoli is, in fact, the classical surface area of the topological boundary when the latter is smooth.

Thus, the oriented "generalized" boundaries of Caccioppoli and De Giorgi act as linear functionals on vector fields. In the celebrated theory developed later by Federer and Fleming (cf. [64]), these are particular instances of "integral currents," which act on general forms (and hence can have arbitrary codimension). Like De Giorgi's theory of Caccioppoli sets, the theory of integral currents of Federer and Fleming can also be seen as a suitable variational completion: after introducing an appropriate class of good objects (in this case integral smooth chains, which are formal linear combinations, with integer coefficients, of smooth oriented submanifolds with smooth boundaries), the more general objects, namely the integral currents, can be characterized as the limits, in an appropriate weak topology, of sequences of those good objects, under uniform bound on their volume and on the volume of their boundaries. Like in the case of De Giorgi's theory of Caccioppoli sets, integral currents can be represented, in a suitable measure-theoretic sense, as integration over "oriented" rectifiable sets.

While the duality with differential forms limits the choice of coefficient groups in the formal linear combinations to integer and real coefficients (or anyway to subgroups of the reals), the "completion point of view" allows choosing other "coefficient groups" (endowed with an appropriate norm, so that we can make sense of the notion of "mass"), cf. the foundational paper of [66] for the case of finite groups. Notable choices are the so-called "flat chains mod p" (which, with a slight abuse of terminology, we will call currents mod p). In the latter case, p is a positive integer larger than 1 and the coefficient group is $\mathbb{Z}_p = \mathbb{Z}/(p\mathbb{Z})$ (for an element $[q] \in \mathbb{Z}_p$, endowed with the usual norm

$$\big|[q]\big| \min\big\{|q - kp| : k \in \mathbb{Z}\big\}.$$

In this note the coefficient group will always be either $\mathbb{Z}$, or $\mathbb{Z}_p$. Note in particular that in both cases the norm will always take integer values, a fact which will play a fundamental role in our discussions.

In all these instances, we have a framework where we can apply the direct methods of the calculus of variations. In particular,

- the concept of a boundary comes naturally either from the duality with differential forms, or from the closure procedure;
- the underlying space of generalized objects is closed;
- the generalized area functional (often called the mass) is lower semicontinuous and its sublevel sets are compact (if we assume that the boundary of our generalized surfaces is a fixed given one).

In particular, the Plateau problem in the above frameworks has a very elegant existence theory.

2.4. Varifolds and the calculus variations "in the large"

One notable drawback of the functional-analytic frameworks outlined above is that the mass is not continuous for the natural convergence in the underlying spaces. Continuity along a sequence might be lost because of two mechanisms:

- High frequency oscillations: for instance, the graphs of the functions $\frac{1}{k}\sin kx$ in the two-dimensional plane have locally bounded length and they converge, in the sense of integral currents, to the straight line. It is, however, easy to see that the total length of any segment in the limiting line is strictly less than the limit of the corresponding approximations.
- Cancelation: a line in $\mathbb{R}^2$ can be given two distinct orientations, thereby defining two different integral currents. However, their sum is 0. If we approximate the two different oriented lines with a sequence of two shifted oriented lines with disjoint supports, then we get a sequence of integral currents with masses uniformly bounded from below which converges to the trivial current.

We are concerned mostly with the second, since the approximating sequence is a sequence of minimal surfaces. Indeed, we can reasonably expect that a sequence of critical surfaces will not exhibit the oscillatory behavior of the first example (this fact was, in fact, proved by Allard as a byproduct his famous regularity theory, see below). On the other hand, the criticality assumption does not rule out the second example, which is therefore "particularly bad" because it shows that in the space of currents we cannot expect any reasonable type of "Palais–Smale" property.

A way to remedy this loss of continuity is to introduce the notion of varifold, which is just a positive measure on the Grassman space of m-dimensional unoriented m-planes in the tangent bundle of the Euclidean space (or, more generally, of a Riemannian manifold). For a smooth (not necessarily oriented) surface Σ, the corresponding varifold is given by

$$\delta_{T_x\Sigma} \otimes d\,\mathrm{Vol}_\Sigma\,. \tag{2.1}$$

General varifolds were introduced by L. C. Young (cf. [126]), while in the context of minimal surfaces Almgren introduced them precisely in order to tackle general existence problems for *critical* points of the area functional, cf. [7]. A particularly useful subclass of varifolds is that of integral varifolds, which satisfy a structure as in (2.1) where $d\,\mathrm{Vol}_\Sigma$ is substituted by the Hausdorff k-dimensional measure restricted on a general m-dimensional rectifiable set R (with an integer-valued weight), and $\delta_{T_x\Sigma}$ is substituted by $\delta_{T_x R}$, where $T_x R$ is a natural measure-theoretic generalization, to rectifiable sets, of the tangent to a smooth surface.

In his notable monograph [95], based on some groundbreaking ideas of Almgren [7], Pitts developed a quite powerful variational theory for finding generalized critical points of the area functional. In codimension 1, i.e., in the case of hypersurfaces, the theory of Almgren and Pitts has found striking geometric applications in the works of A. Neves and F. Coda Marques, cf. [81, 86, 87]. These results have spurred a number of interesting works in the area, and Pitts' existence theory has been revisited in several different ways, see, for instance, [21, 22, 56, 75, 127].

Varifolds can be naturally deformed using one-parameter family of diffeomorphisms, and this allows introducing a rather natural notion of the kth variation of the varifold along smooth vector fields. Of particular relevance are then

- *stationary* varifolds, i.e., varifolds for which the first variation vanishes along any vector field,
- and *stable* varifolds, i.e., stationary varifolds for which the second variation is nonnegative along any vector field.

Since all the objects encountered above in the existence theories for the Plateau problem naturally induce corresponding varifolds, all the minimizers in the various senses given above are, in fact, stable varifolds.

3. MONOTONICITY FORMULA AND TANGENT CONES

One simple and very powerful tool in the regularity theory for minimal submanifolds is the monotonicity formula. In order to gain an intuition about it, consider a smooth m-dimensional surface $\Sigma \subset \mathbb{R}^{m+n}$ which minimizes the volume in some suitable class of comparison surfaces and fix an "interior" point $p \in \Sigma$. Consider then a ball $\mathbf{B}_r(p)$ which does not intersect $\partial\Sigma$. We wish to compare the volume of $\Sigma \cap \mathbf{B}_r(p)$ to the volume of the cone Σ_c with vertex p and base $\Sigma \cap \partial\mathbf{B}_r(p)$. By Sard's lemma, we can assume that Σ intersects $\partial\mathbf{B}_r(p)$ transversally. Note that our comparison surface is somewhat singular because of the vertex singularity of the cone and the discontinuity in the tangents that might be introduced by cutting $\Sigma \cap \mathbf{B}_r(p)$ out and replacing it with Γ. On the other hand, it is also simple to see that $\Sigma_c \cup (\Sigma \setminus \mathbf{B}_r(p))$ can be obtained as limit of deformations of Σ by smooth isotopies of the ambient space: in particular, it is a good comparison surface in pretty much all the variational frameworks considered so far.

The minimizing property of Σ implies then that

$$\mathrm{Vol}^m\big(\Sigma \cap \mathbf{B}_r(p)\big) \le \mathrm{Vol}^m(\Sigma_c) = \frac{r}{m}\,\mathrm{Vol}^{m-1}\big(\Sigma \cap \partial\mathbf{B}_r(p)\big),$$

which in turn (considering that $\mathrm{Vol}^{m-1}(\Sigma \cap \partial \mathbf{B}_r(p)) \leq \frac{d}{dr} \mathrm{Vol}^m(\Sigma \cap \mathbf{B}_r(p)))$ gives

$$\frac{d}{dr} \frac{\mathrm{Vol}^m(\Sigma \cap \mathbf{B}_r(p))}{r^m} \geq 0. \tag{3.1}$$

The latter is the classical monotonicity formula for minimal submanifolds. It is very robust, in the sense that

(a) It can be derived for critical points by using stationarity with respect to some specific radial deformations. In particular, it holds for stationary varifolds (see [4]).

(b) Allowing for suitable multiplicating factors like e^{Cr}, the formula holds for much more general objects, in particular for stationary varifolds in smooth Riemannian manifolds (cf. again [4]).

(c) A suitable version of the formula can be derived at boundary points too, under the assumption that the boundary $\partial\Sigma$ is smooth enough (cf. [5]). An intuition for this can be gained through the following observation: if the boundary $\partial\Sigma$ were an affine subspace passing through p, then the competitor surface $\Sigma_c \cup (\Sigma \setminus \mathbf{B}_r(p))$ would have the same boundary, namely $\partial\Sigma$.

(d) Perhaps most importantly, a more refined version of the arguments leading to (3.1) shows that the equality case in holds if and only if Σ_c coincides with Γ, i.e., if Σ itself is a cone with vertex p.

For further reference, we will call *density of Σ at p* (denoted by $\Theta(\Sigma, p)$ the limit of the "mass ratio"

$$\lim_{r \downarrow 0} \frac{\mathrm{Vol}^m(\Sigma \cap \mathbf{B}_r(p))}{\omega_m r^m},$$

where ω_m is the volume of the m-dimensional disk. Obviously, the density is not particularly interesting for smooth Σ's as it will be 1 at every interior point and $\frac{1}{2}$ at every boundary point (or k at every interior point and $\frac{k}{2}$ at every boundary point, if we entertain the possibility of allowing for multiplicities in the normed groups $\mathbb{Z}$ and $\mathbb{Z}_p$, or if we consider integral varifolds). However, due to (a), the density exists for any "generalized minimal surface" encountered in the previous section and this is a very nontrivial information, given that the latter might be singular. Another interesting byproduct of the monotonicity formula is that the density is nowhere smaller than 1 at interior points, which in turn implies that some suitable definition of "support" of the generalized minimal surface is a closed rectifiable set of locally finite Hausdorff measure.

3.1. Tangent cones

Fact (d) above is maybe the most relevant, as it is the starting point for a fruitful fundamental concept in minimal surface theory. Let us fix a point p in (the support of) our generalized minimal surface Σ (and, in light of (c) above, we might even fix it at the boundary as long as the latter is sufficiently smooth). For every radius r, consider then the translated and rescaled surface

$$\Sigma_{p,r} := \frac{\Sigma - p}{r} = \{y : p + ry \in \Sigma\}.$$

The volume of the surface in $\mathbf{B}_R(0)$ is then uniformly bounded for every fixed R, independently of the parameter r. Again, while this is not particularly exciting for a smooth Σ, it is a highly nontrivial fact for the generalized minimal surfaces, which are potentially singular at p. Given the uniform bound and the compactness properties available for all the generalized minimal surfaces introduced thus far, up to subsequences we can assume that $\Sigma_{p,r}$ converges to a generalized minimal surface in the same class, which for convenience we will denote by Σ_c.

Assuming convergence of the volume (which is, in fact, correct for objects like varifolds because of their definition, while it is a property shared by minimizers out of variational arguments, in any of the classes described thus far) the mass ratio $R^{-m} \operatorname{Vol}^m(\Sigma_c \cap \mathbf{B}_R)$ is constant in R, and hence by point (d) it is a cone. In the literature, Σ_c is called *a tangent cone to Σ at p*. Note that we are speaking about *a* tangent cone: the uniqueness of this object, i.e., its independence of the subsequence $r_k \downarrow 0$, is a widely open problem, even though several fundamental result have been proved in the past (see Section 15 below).

At a regular point p, i.e., a point in a neighborhood of which the generalized minimal surface Σ is smooth, the tangent cone Σ_c is, of course, unique and it is given by the tangent space to Σ at p (counted with the appropriate multiplicity, depending upon the chosen variational framework), or half of the tangent space if p is a boundary point. A later section will examine under which assumption the latter conclusion is correct.

At any rate, even in the possible presence of singularities, we have gained a great deal of new information about Σ_c compared to Σ: Σ_c is a "global minimal surface" (it has no boundary if p is in the interior, or its boundary is affine if $p \in \partial\Sigma$) and, moreover, it is conical. In particular, its spherical cross-section carries all the information about Σ_c, even when Σ_c is singular: at all effects, Σ_c must be less complex than Σ, i.e., Σ_c has "lost one dimension."

4. INVARIANT SPACES AND STRATA

For simplicity, in what follows we will focus on tangent cones Σ_c at interior points p, even though a variant of the following discussion applies to boundary tangent cones as well. Since p is not a boundary point, Σ_c has no boundary. Moreover, two simple corollaries of the monotonicity formula are that:

- $\Theta(\Sigma_c, 0) \geq \Theta(\Sigma_c, q)$ for every q,
- and if the quality holds at some $q \neq 0$, then Σ_c "splits off a line," i.e., it is invariant under translations in the direction q.

The latter property is the starting point of Federer in his celebrated "dimension reduction argument" (cf. **[62, 63]**), which we will illustrate below. Here we want to present Almgren's stratification theory, which is a far-reaching generalization of Federer's original idea.

First of all, it follows from the above consideration that the set

$$V = \{q \in \Sigma_c : \Theta(\Sigma_c, q) = \Theta(\Sigma_c, 0)\}$$

is a linear subspace of $\mathbb{R}^{m+n}$. If V has the same dimension as Σ_c, then, in fact, Σ_c coincides with V (counted with the correct multiplicity and, in some cases, given the correct orientation). Otherwise, assuming that $k = \dim(V)$, Σ_c is the product of V and a minimal cone Σ_0 in the orthogonal complement of $V^\perp$, which is not invariant by any translation. This is a great deal of information, and in several cases implies severe restrictions upon k. For instance, for area-minimizing integral currents, it can be easily checked that $k \leq m - 2$ in general, while in the particular case of codimension 1 the celebrated paper of Simons on stable minimal hypercones (cf. [62]) implies that $k \leq m - 7$ (again this will be discussed further below)! In [11] Almgren coined the term *building dimension of the cone* Σ_c to identify the nonnegative integer k and introduced a stratification of the interior points $p \in \Sigma$ according to the maximal building dimension of its tangent cones. In particular, the stratum $\mathscr{S}_k$ is the set of (interior) points $p \in \Sigma$ such that the building dimension of any tangent cone to Σ at p is at most k. Almgren's fundamental discovery is the following

Theorem 4.1. *For a stationary integral varifold Σ, the stratum $\mathscr{S}_k$ is a closed set of Hausdorff dimension at most k.*

Almgren's approach is very general and can be applied to a variety of different context. For a framework which is very flexible and covers a wide range of applications, see [123]. Almost four decades after the work of Almgren, groundbreaking ideas allowed Naber and Valtorta to improve massively upon Almgren's original theorem, showing (cf. [93,94])

Theorem 4.2. *For a stationary integral varifold Σ, the stratum $\mathscr{S}_k$ is k-rectifiable, i.e., it can be covered, up to a set of $\mathcal{H}^k$-measure zero, with countably many C^1 submanifolds of dimension k.*

Theorem 4.2, which was predated by pioneering works of Simon (cf. [108,109]) covering some particular cases (most notably the stratum $\mathscr{S}_{m-7}$ for area-minimizing integral currents, see below for more details), builds upon a new sophisticated version of Reifenberg's topological disk theorem combined with a clever use of the remainder in the monotonicity formula. The ideas are quite general and can be applied to other contexts.

5. INTERIOR ε-REGULARITY AT MULTIPLICITY 1 POINTS

Following the above terminology, two things are obvious: the stratum $\mathscr{S}_m$ coincides with the whole support of the m-dimensional generalized minimal surface and the stratum $\mathscr{S}_{m-1}$ consists necessarily of singular points. A point $p \in \mathscr{S}_m \setminus \mathscr{S}_{m-1}$ is clearly a good candidate for being a regular point, since we know that at least one tangent cone to Σ at p is, in fact, a plane (counted with its multiplicity). However, a famous theorem by Federer shows that the existence of a "flat tangent" does not guarantee the regularity of the point. Indeed, based on a classical theorem of Wirtinger in Kähler geometry, Federer proved (cf. [62])

Theorem 5.1. *Any holomoprhic subvariety Σ of complex dimension k in $\mathbb{C}^n$ induces an area-minimizing integral current of dimension $2k$ in $\mathbb{R}^{2n}$.*

It can be readily checked that the holomorphic curve

$$\Sigma = \{z^2 = w^3 : (z, w) \in \mathbb{C}^2\} \tag{5.1}$$

gives then an example of an area-minimizing integral current of dimension 2 in $\mathbb{R}^4$ for which $0 \in \mathscr{S}^2 \setminus \mathscr{S}^1$ is a singular point. One crucial fact is, however, that the flat tangent at 0 is a 2-dimensional plane (i.e., the complex line $\{z = 0\}$) but counted with *multiplicity* 2. A celebrated theorem of Allard (cf. [4]), extensively used in the literature, shows that the naive expectation "flat tangent cone $\Longleftrightarrow$ regular point" is indeed correct if the flat tangent cone has multiplicity 1.

Theorem 5.2. *If a stationary integral varifold Σ is sufficiently close in $\mathbf{B}_{2r}(p)$ to a plane (counted with multiplicity 1) in the weak topology, then in $\mathbf{B}_r(p)$ it is a smooth graph over that plane. Moreover, at any interior point p where the density of Σ is 1, such a plane always exists for a sufficiently small r.*

Among the various objects examined in this note, there are three situations where it is relatively simple to see a priori that our generalized minimal surface Σ will not "pick higher multiplicity" at flat points:

(a) Σ is a portion of the boundary of some Caccioppoli set;

(b) Σ is a solution of the Plateau problem in one of the set-theoretic senses described in Section 2.2;

(c) Σ is an area-minimizing current mod 2 or an area-minimizing current mod 3.

In fact, Theorem 5.2 was realized independently by De Giorgi and Reifenberg, in [32] and [97,98], respectively in the particular cases of (a) and (b) (this is literally correct for De Giorgi, while in reality Reifenberg in [97,98] dealt with the only set-theoretic solutions of the Plateau problem known at his time, which were those he himself introduced in [96]; it must also be noticed that De Giorgi's monograph appeared three years before Reifenberg's paper, but it was probably not yet widely known when Reifenberg wrote his papers [97,98]). The two pioneering approaches are rather different, but they both rely on the fact that the "linearization" of a minimal surface, understood as a graph over his tangent plane, is harmonic (in fact, it would be more correct to say that the linearization of the minimal surface equation is the Laplace equation, or that, at the level of the energies, the Dirichlet energy is the second order Taylor expansion of the area functional).

Reifenberg used harmonic competitors to estimate how much an area-minimizing surface deviates from being conical if it is close to a plane, and derived his famous "epiperimetric inequality," which can be thought as a quantitative improvement of the cone-comparison outlined above to prove the monotonicity formula. De Giorgi used a linearization technique which has a more PDE flavor, and which was generalized afterwards by Almgren in any codimension and for much more general energy functionals, cf. [7]. Both approaches exploit in a substantial way the minimizing property of the surfaces in question. Allard's proof of Theorem 5.2, while still based on the intuition that harmonic functions provide a good approximation for minimal graphs, deviates drastically from both of them, having to deal with stationary objects. But ultimately it is fair to say that Allard's approach borrows much more substantially from the works of De Giorgi and Almgren, than from that of Reifenberg.

It is worth spending some words on why all the approaches mentioned above for the ε-regularity theory fail at the origin in the example (5.1): no matter how small is the scale that we look at, it is not possible to approximate efficiently (5.1) around the origin with the graph of a single-valued function. Of course, before knowing Theorem 5.2 we also do not know that, under the corresponding assumptions, a generalized surface is graphical over the approximating plane: however, a crucial point in Allard's proof of Theorem 5.2 is that, before proving any regularity, he was able to produce a graphical approximation which covers *most* of the support of the generalized minimal surface. In contrast, no matter how small the r is, a single-valued graph will cover no more than half of $\Sigma \cap \mathbf{B}_r(p)$ when Σ is given by (5.1).

The assumption on the multiplicity of the varifold severely limits the effectiveness of Theorem 5.2 in bounding the size of the singular set for stationary integral varifolds. In fact, it would be natural to expect that singular points with a flat tangent cone form anyway a set of relatively modest size: according to the known examples, its dimension is likely $m-2$. The latter is less than the dimension of $\mathscr{S}_{m-1}$ and so one could reasonably conjecture that the singular set of a stationary integral varifold has dimension at most $m-1$. On the other hand, so far the best that we can conclude is still a corollary of Theorem 5.2 noted by Allard in [4] almost 50 years ago.

Corollary 5.3. *Let Σ be a stationary m-dimensional varifold in $U \subset \mathbb{R}^{m+n}$. Then the singular set of Σ is a closed subset which has empty (relative) interior.*

6. BOUNDARY ε-REGULARITY AT MULTIPLICITY $\frac{1}{2}$ POINTS

In his second groundbreaking work [5], Allard proved a statement parallel to Theorem 5.2 at boundary points. The following is an informal description of his main "boundary regularity" theorem.

Theorem 6.1. *Assume Σ is an m-dimensional integral varifold in some open set $U \subset \mathbb{R}^{m+n}$, which is stationary for variations which keep fixed a smooth $(m-1)$-dimensional submanifold Γ. Then the following conclusions hold:*

(a) *If $p \in \Gamma$ belongs to the support of the varifold, then $\Theta(p, \Sigma) \geq \frac{1}{2}$.*

(b) *If in $\mathbf{B}_{2r}(p)$ the varifold is sufficiently close, in the weak topology, to a single copy of half of an m-dimensional plane π, then in $\mathbf{B}_r(p)$ it is a C^1 graph over a suitable portion of π.*

(c) *If* $\Theta(p, \Sigma) = \frac{1}{2}$, *then the assumption of (b) (and hence the corresponding conclusion) holds for a sufficiently small* r.

For this boundary version as well, the overall intuition is that V is, in first approximation, very well approximated by the graph of a (single-valued) harmonic function.

While in the rest of this note I will touch upon interior regularity results for many different notions of generalized minimal surfaces, concerning boundary regularity I will only focus on the case of area-minimizing integral currents. This is also due to the fact that there are not many other cases studied in the literature. Aside from Allard's general theorem (i.e., Theorem 6.1 stated above), the author is only aware of:

- the work [80] (cf. also [91]), which contains a conjectural list of boundary tangent cones for set-theoretic 2-dimensional solutions of the Plateau problem;
- the recent work of David [26], which, for 2-dimensional sliding minimizers, generalizes the conclusion of Theorem 6.1 to the union of two half-planes, and possible additional transverse cones as in the classical theorem of Taylor in the interior (cf. Theorem 8.1);
- an argument by White which shows how to gain curvature estimates for stable minimal hypercurrents at the boundary, under some convexity assumption (cf. [45, SECTION 6.4]).

7. INTERIOR REGULARITY THEORY: MINIMIZING INTEGRAL HYPERCURRENTS

Even though Allard's Theorem 5.2 needs the multiplicity 1 assumption, the latter might be dropped in the case of integral area-minimizing currents of codimension 1 (which for simplicity we will call hypercurrents from now on). The key point is that area-minimizing integral hypercurrents Σ can be locally decomposed into the sum of area-minimizing boundaries of Caccioppoli sets (this is a consequence of the Coarea formula, see, for instance, [104]). If in $\mathbf{B}_{2r}(p)$ the original current Σ is close to a multiple Q of a hyperplane π, each of these boundaries is then close to a multiplicity 1 copy of π. We can then apply Allard's theorem to prove that each of them is a C^1 graph in $\mathbf{B}_r(p)$, obtaining what can be called (cf. for instance [102]) a "sheeting theorem" for $\Sigma \cap \mathbf{B}_r(p)$. However, each of these sheets must be ordered (for minimizing reasons they cannot cross) and they must touch at the point p: the maximum principle (each of these graphs is a solution of the minimal surface equation) then implies that they collapse all into a single smooth surface counted with the appropriate multiplicity, which must be Q.

This argument rules out that an example like (5.1) could exist for integral area-minimizing *hyper*currents. We are therefore in the luckiest of situations where we can infer that a single flat tangent cone at p is indeed a necessary and sufficient condition for regularity at p. If we introduce the notation $\mathrm{Sing}_i(\Sigma)$ for the interior singularities of Σ, when Σ is an m-dimensional area-minimizing integral current in $\mathbb{R}^{m+1}$ (or, more generally, in a complete smooth Riemannian manifold of dimension $m+1$), we infer $\mathrm{Sing}_i(\Sigma) \subset \mathscr{S}_{m-1}$. Consider, however, that the existence of a point $p \in \mathscr{S}_{m-1} \setminus \mathscr{S}_{m-2}$ implies the existence of a singular 1-dimensional area-minimizing cone in $\mathbb{R}^2$, and it is rather elementary to see that the latter cones do not exist, namely that $\mathrm{Sing}_i(\Sigma) \subset \mathscr{S}^{m-2}$. Of course, we can now wonder whether for some Σ the set $\mathscr{S}_{m-2} \setminus \mathscr{S}_{m-3}$ is nonempty, which is equivalent to the existence of an area-minimizing 2-dimensional cone in $\mathbb{R}^3$ which is not a plane (i.e., it is singular at the origin). In [63] Federer introduced his well-known reduction argument, which could be formalized as follows.

Theorem 7.1. *Let m be the smallest integer with the property that there is an m-dimensional area-minimizing integral current Σ_0 in $\mathbb{R}^{m+1}$ which is a nonplanar cone with vertex at the origin. Then Σ_0 is everywhere regular except at the origin.*

It was also realized by De Giorgi in [33] that the well-known Bernstein problem, i.e., whether a complete minimal graph over $\mathbb{R}^{m+1}$ must be affine, would also be implied by the nonexistence of nonplanar area-minimizing oriented hypercones in $\mathbb{R}^{m+1}$. After progress by Fleming, De Giorgi, and Almgren (cf. [8, 33, 65]), Simons in [112] proved his famous result about stable minimal hypercones, namely

Theorem 7.2. *If $m \le 6$ and $\Gamma_0 \subset \partial \mathbf{B}_1 \subset \mathbb{R}^{m+1}$ is a smooth connected submanifold of dimension $m-1$, such that the cone Σ_0 with base Γ_0 and vertex 0 is a stable varifold, then Γ_0 is a great sphere (i.e., Σ_0 is planar). On the other hand,*

$$\Sigma_s := \left\{x_1^2 + x_2^2 + x_3^2 + x_4^2 = x_5^2 + x_6^2 + x_7^2 + x_8^2\right\} \subset \mathbb{R}^8 \tag{7.1}$$

is a nonplanar, oriented, stable singular cone of dimension 7.

Since area-minimizing currents are automatically stable varifolds, in combination with Federer's reduction argument, the first part of Theorem 7.2 implies that $\mathrm{Sing}_i(\Sigma) \subset \mathscr{S}_{m-7}$ for any m-dimensional area-minimizing integral hypercurrents. In particular, for $m \le 6$, Σ is a regular hypersurface (in the interior), while, for $m \ge 7$, $\mathrm{Sing}_i(\Sigma)$ has dimension at most $m-7$. In fact, by Theorem 4.2 we can conclude that $\mathrm{Sing}_i(\Sigma)$ is $(m-7)$-rectifiable. The latter conclusion was first reached by Simon in his pioneering work [108]. However, compared to Simon's techniques, the approach by Naber and Valtorta (cf. [94]) allows proving the stronger conclusion, namely

Theorem 7.3. *Let Σ be an m-dimensional area-minimizing integral current in $\mathbb{R}^{m+1}$. Then $\mathrm{Sing}_i(\Sigma)$ has locally finite Hausdorff $(m-7)$-dimensional measure, and it is $(m-7)$-rectifiable.*

In their famous work [17], Bombieri, De Giorgi, and Giusti completed the solution of the Bernstein problem showing that indeed the Simons cone (7.1) is an area-minimizing integral current of dimension 7, and that in addition there is a nonaffine global solution $u : \mathbb{R}^8 \to \mathbb{R}$ of the minimal surface equation.

8. INTERIOR REGULARITY THEORY: MINIMAL SETS

As already mentioned, the phenomenon of "picking higher multiplicity at flat points" is absent in the solutions of the Plateau problem that fall in the "set-theoretic" approach. This was pioneered by Reifenberg in [97, 98], who proved that his m-dimensional solutions of the Plateau problem are always real analytic except for a closed $\mathcal{H}^m$-null set. A much more general statement, valid in a variety of contexts and also for a vast class of elliptic energies was proved by Almgren in [9].

Following the Remarks of Section 4, we conclude that the singular set of an m-dimensional set-theoretic solution of the Plateau's problem is necessarily contained in $\mathcal{S}^{m-1}$. While Theorem 4.2 implies that $\mathcal{S}^{m-1}$ is rectifiable, much more can actually be said in the codimension 1 case. First of all, for 2-dimensional minimizing sets in $\mathbb{R}^3$ Taylor in [117] proved the following complete structure theorem.

Theorem 8.1. *Let Σ be a 2-dimensional set which minimizes the area in the sense of Almgren. Then:*

(a) *$\mathcal{S}_1(\Sigma) \setminus \mathcal{S}_2(\Sigma)$ is the (locally finite) union of $C^{1,\alpha}$ arcs and for each $p \in \mathcal{S}_1(\Sigma)$ there is a neighborhood U of p in which Σ is the union of three classical minimal surfaces meeting in $\mathcal{S}_1(\Sigma) \cap U$ at 120 degrees;*

(b) *$\mathcal{S}_0(\Sigma)$ consists of isolated points and for each $p \in \mathcal{S}_0(\Sigma)$ there is a neighborhood U in which Σ is diffeomorphic to the cone over a regular tetrahedron.*

The same conclusion as in part (a) of the above remarkable theorem is, in fact, valid for the stratum $\mathcal{S}_{m-1} \setminus \mathcal{S}_{m-2}$ of m-dimensional area-minimizing sets in $\mathbb{R}^{m+1}$ and can be inferred from Simon's theory on the uniqueness of multiplicity 1 cylindrical cones, cf. [106]. Part (b) can also be generalized to a similar statement for m-dimensional area-minimizing sets of $\mathbb{R}^{m+1}$, implying in particular that $\mathcal{S}_2 \setminus \mathcal{S}_{m-3}$ is an $(m-2)$-dimensional submanifold. This generalization was announced by White in [120] and a proof has been recently published by Colombo, Edelen, and Spolaor in [23], as a corollary of a more general result. The main Theorem in [23] also implies that the stratum $\mathcal{S}_{m-3}$ has finite $\mathcal{H}^{m-3}$ measure.

9. INTERIOR REGULARITY THEORY: STABLE HYPERSURFACES AND STABLE HYPERVARIFOLDS

In [103] Schoen, Simon, and Yau realized that Simons' theorem on stable minimal hypercones could be recast in a suitable a priori estimate for the curvature of stable minimal surfaces. More precisely, combining Simons' inequality with techniques from elliptic PDEs they were able to prove the following groundbreaking theorem.

Theorem 9.1. *Let Σ be a smooth minimal hypersurface in $U \subset \mathbb{R}^{m+1}$ with $m \leq 5$. Then for every $V \subset\subset U$ there is a constant C which depends on U, V, and $\mathcal{H}^m(\Sigma)$ such that the Hilbert–Schmidt norm of the second fundamental form A of Σ is bounded by C at every point of $\Sigma \cap V$.*

In their subsequent work [102], Schoen and Simon were able to cover the case $m = 6$ of the above statement and also to give a "GMT regularity theory" counterpart of the Schoen–Simon–Yau estimates. More precisely, they were able to prove

Theorem 9.2. *Assume Σ is a stable m-dimensional varifold in $U \subset \mathbb{R}^{m+1}$ with the property that*[1] $\mathcal{H}^{m-2}(\mathrm{Sing}_i(\Sigma)) < \infty$. *Then* $\mathrm{Sing}_i(\Sigma) \subset \mathcal{S}_{m-7}$.

It has been recently shown by Simon, see [110, 111], that the subsequent conclusion that $\mathrm{Sing}_i(\Sigma)$ is $(m-7)$-rectifiable is optimal, in the sense that there are stable m-dimensional varifolds in $m+1$ smooth Riemannian manifolds whose singular sets are closed sets of arbitrary Hausdorff dimension $\alpha \leq m-7$. On the other hand, the assumption $\mathcal{H}^{m-2}(\mathrm{Sing}_i(\Sigma)) = 0$ is not at all optimal. Based on the examples known thus far, one could expect that for a general stable hypervarifold the top stratum $\mathcal{S}_{m-1} \setminus \mathcal{S}_{m-2}$ is a $C^{1,\alpha}$ $(m-1)$-dimensional submanifold and that if the latter is empty, then $\mathrm{Sing}_i(\Sigma)$ is contained in $\mathcal{S}_{m-7}$. A notable theorem in this direction, in particular covering the second conclusion, has been achieved by Wickramasekera in his deep regularity theory of stable hypervarifolds. The main conclusion of his paper [124] is the following

Theorem 9.3. *Assume Σ is a stable m-dimensional varifold in a connected set $U \subset \mathbb{R}^{m+1}$. Then:*

- *either* $\mathrm{Sing}_i(\Sigma)$ *contains a point p in a neighborhood of which Σ consists of a finite number of smooth minimal hypersurfaces meeting at a common $C^{1,\alpha}$ $(m-1)$-dimensional boundary (which in particular is a nonempty subset of $\mathcal{S}_{m-1} \setminus \mathcal{S}_{m-2}$),*
- *or otherwise* $\mathrm{Sing}_i(\Sigma) \subset \mathcal{S}_{m-7}$.

While the latter is a remarkable achievement, for general stable hypervarifolds the best unconditional regularity result is still that which can be concluded from the sole condition of stationarity through Allard's work, namely Corollary 5.3.

10. INTERIOR REGULARITY THEORY: MINIMIZING INTEGRAL CURRENTS IN HIGHER CODIMENSION

As already witnessed in Example (5.1), the regularity theory for area-minimizing integral currents in codimension larger than 1 differs dramatically from the regularity theory for hypercurrents, since there are singular points which belong to $\mathcal{S}_m \setminus \mathcal{S}_{m-1}$, and which from now on we will call "flat singular points." The major problem of how to give a suitable dimension bound for "flat singular points" was finally conquered by Almgren in a titanic effort, which resulted in a famous 1728 page preprint in the early 1980s (cf. [10]), published

1 In their paper, Schoen and Simon assume the stronger property that $\mathcal{H}^{m-2}(\mathrm{Sing}_i(\Sigma)) = 0$, but it is well known by the experts that their arguments apply if the Hausdorff measure is, in fact, finite.

pusthumously thanks to the editorial work of Scheffer and Taylor in [11]. Almgren's monograph achieves the optimal dimension bound for area-minimizing integral currents in any dimension and codimension.

Theorem 10.1. *Let Σ be an area-minimizing integral current of dimension m in $\mathbb{R}^{m+n}$. Then the Hausdorff dimension of the set of interior flat singular points is at most $m-2$, while the stratum $\mathscr{S}_{m-2} \setminus \mathscr{S}_{m-1}$ is empty. In particular,* $\dim_H(\mathrm{Sing}_i(\Sigma)) \le m-2$.

Almgren invented several tools to prove Theorem 10.1. In particular,

(i) he introduced an entire new concept of "multivalued functions minimizing the Dirichlet energy" in order to find the "appropriate linearization" of area-minimizing integral currents at flat singular points, and he developed a subsequent existence and regularity theory for these new objects;

(ii) he introduced several flexible techniques to approximate currents with Lipschitz multivalued graphs;

(iii) he developed a very intricate regularization technique to find a sufficiently smooth "central sheet" at possible branching singularities (the so-called "center manifold");

(iv) he discovered a new monotonicity formula for a harmonic function (the monotonicity of the "frequency") which has meanwhile been used in a variety of different contexts in elliptic and parabolic partial differential equations (see, e.g., [67,70,84]).

Almgren's theory has been revisited by the author and Emanuele Spadaro in the series of works [47–51]. Besides making the proof of Theorem 10.1 shorter, these works improve upon Almgren's monograph in several aspects, and, moreover, they have been the starting point of several further developments, which will be detailed in the next sections

Shortly after Almgren completed his 3-volume preprint, White proved that in the case of 2-dimensional area-minimizing currents the stratum $\mathscr{S}_0$ consists, in fact, of isolated points, cf. [119] (this is indeed a corollary of a more precise theorem which shows the uniqueness of tangent cones in that particular case and which will be discussed further in Section 15. The program of understanding the singularities of 2-dimensional area-minimizing currents was then completed by Chang in [20].

Theorem 10.2. *Let Σ be a 2-dimensional area-minimizing current in $\mathbb{R}^{2+m}$. Then* $\mathrm{Sing}_i(\Sigma)$ *consists of isolated points. Moreover, for each $p \in \mathrm{Sing}_i(\Sigma)$ there is a neighborhood U in which Σ can be decomposed as the union of a finite number N of branched minimal immersed disks D_i with the following properties:*

- *each D_i is an embedding, except for the point p;*
- *$D_i \cap D_j$ is either the empty set or consists only of the point p.*

However, the proof given in [20] is, strictly speaking, incomplete, as Chang needs the existence of a suitable generalization of Almgren's center manifold to a "branched version." For the latter, he just gives a 4-page sketch (cf. the appendix of [20]), invoking suitable modifications of Almgren's statements (it must be noted that the construction of the center manifold occupies more than half of Almgren's monograph [11]). Based on the works [47–51], the author, Spadaro, and Luca Spolaor gave a complete independent proof of the existence of a branched center manifold in [52]. We also developed a suitable more general counterpart of Chang's theory in the papers [53–55], proving in particular the same regularity result for spherical cross-sections of area-minimizing 3-dimensional cones and for semicalibrated 2-dimensional currents (previous theorems in [12, 13] proved some cases of particular interest, based on the works of Rivière and Tian, see [99–101]). Almgren's dimension bound in Theorem 10.1 has also been extended to semicalibrated currents by Spolaor in [113].

11. INTERIOR REGULARITY THEORY: MINIMIZING CURRENTS MOD p

The regularity theory for area-minimizing currents mod p started around the same time as the regularity theory for integral currents. As a consequence of Almgren's generalization of De Giorgi's ε regularity theorem, the cases $p = 2, 3$ were already rather well understood in the 1960s. In particular, the absence of flat singular points allowed inferring the following theorem (the case $p = 2$ is due to Federer, cf. his pioneering work on the reduction argument [63]).

Theorem 11.1. *If Σ is an m-dimensional area-minimizing current mod 2 in $\mathbb{R}^{m+n}$, then $\operatorname{Sing}_i(\Sigma) \subset \mathscr{S}_{m-2}$. If Σ is an m-dimensional area-minimizing current mod 3, then $\operatorname{Sing}_i(\Sigma) \subset \mathscr{S}_{m-1}$.*

The case $p = 2$ in codimension 1 allows even more restrictive results (the same regularity as for integral area-minimizing currents holds and, in fact, locally any area-minimizing integral hypercurrent mod 2 is the boundary of a Caccioppoli set), while in higher codimension there are indeed area-minimizing 2-dimensional currents mod 2 with point singularities.

For $p = 3$, the union of three half-planes in $\mathbb{R}^3$ meeting at a common line at 120 degrees gives an obvious example for which $\mathscr{S}_{m-1} \neq \emptyset$. The beautiful result of Taylor [116] gave a complete description of the interior singular set for area-minimizing 2-dimensional currents mod 3 in $\mathbb{R}^3$: locally the singular set is always diffeomorphic to the above example. The subsequent work of Simon [106] on the uniqueness of cylindrical tangent cones allowed giving a suitable generalization of Taylor's result in any dimension and codimension. The final outcome is the following

Theorem 11.2. *If Σ is an m-dimensional area-minimizing current mod 3 in $\mathbb{R}^{m+n}$, then $\mathscr{S}_{m-1} \setminus \mathscr{S}_{m-2}$ is an $(m-1)$-dimensional submanifold and at every $p \in \mathscr{S}_{m-1} \setminus \mathscr{S}_{m-2}$ there is a neighborhood U in which Σ consists of 3 smooth minimal surfaces meeting at $\mathscr{S}_{m-1} \cap U$ at 120 degrees. If Σ is in addition an hypercurrent (i.e., $n = 1$), then $\mathscr{S}_{m-2} \setminus \mathscr{S}_{m-3}$ is an empty set. In particular, if $m = 2$ and $n = 1$, then $\operatorname{Sing}_i(\Sigma)$ consists of pairwise disjoint* *closed simple curves and pairwise disjoint simple arcs with endpoints lying in the support of the boundary of* Σ.

In order to progress beyond Corollary 5.3 for higher moduli, it is necessary to either rule out flat singular points or bound their dimension. In the special case of mod 4 hypercurrents, White in [118] discovered a beautiful fact which allowed him to derive the following structural result

Theorem 11.3. *If* Σ *is an* m*-dimensional area-minimizing current mod* 4 *in* $\mathbb{R}^{m+1}$*, then it can be locally decomposed, away from its boundary, into the union of two* m*-dimensional area-minimizing currents mod* 2.

He also showed a converse to Theorem 11.3. In particular, his results imply the existence of flat singular points even for hypercurrents mod $2k$. More precisely, consider a function $u : \mathbb{R}^2 \supset B_1 \to \mathbb{R}$ which solves the minimal surface equation and, after applying a suitable translation and rotation, assume that $u(0) = 0$, $\nabla u(0) = 0$, and $D^2u(0) \neq 0$. Since $\Delta u(0) = 0$, it follows that the zero set of u in a neighborhood of 0 consists of 2 arcs crossing orthogonally in 0. We can thus assume that the disk $B_r(0) \subset \mathbb{R}^2$ is subdivided by $\{u = 0\}$ into 4 sectors S_1, S_2, S_3, S_4. We then consider in $\mathbb{C}_r := B_r(0) \times \mathbb{R} \subset \mathbb{R}^3$ the union of the four sectors $S_i \times \{0\}$ and of the four portions G_i of the graph of u lying over the respective sector S_i. We give to S_i opposite alternating orientations and sum them to construct an integral current S in $\mathbb{C}_r(0)$. Clearly, ∂S is formed by the four arcs which describe $\{u = 0\} \times \{0\}$, suitably oriented and counted with multiplicity 2. In particular, S is a cycle mod 2. We then perform an analogous operation with the 4 portions G_i of the graph of u and construct a corresponding integral current T. By choosing the orientations correctly, we can achieve that $\partial T = \partial S$. Therefore the current $\Sigma = T + S$ is a cycle mod 4 and, according to the results in [118], it is area-minimizing (as a cycle mod 4). In particular, 0 is a flat singular point for Σ.

This phenomenon is typical of even moduli, and indeed in his subsequent work [121] White proved that area-minimizing hypercurrents mod $2k + 1$ cannot have singular flat points.

Theorem 11.4. *If* Σ *is an* m*-dimensional area-minimizing current mod* p *in* $\mathbb{R}^{m+1}$ *and* p *is odd, then* $\mathrm{Sing}_i(\Sigma) \subset \mathscr{S}_{m-1}$.

In the papers [41, 42], the author, Jonas Hirsch, Andrea Marchese, and Salvatore Stuvard developed a theory to bound the dimension of flat singular points of a general area-minimizing current Σ mod p (i.e., in any dimension and codimension), which implies that the Hausdorff dimension of the set of flat singular points of Σ is at most $m - 1$.

Theorem 11.5. *If* Σ *is an* m*-dimensional area-minimizing current mod* p *in* $\mathbb{R}^{m+n}$*, then* $\dim_H(\mathrm{Sing}_i(\Sigma)) \leq m - 1$.

While the latter theorem is a considerable improvement compared to what was known before (aside from the cases covered by Theorems 11.1, 11.3, and 11.4, in all others the best known result was that the singular set is meager, thanks to Corollary 5.3). Indeed, the known examples would suggest that the set of flat singular points of any area-minimizing current mod p is at most $m-2$. The work [40] and the forthcoming one [38], by the author, Hirsch, Marchese, Spolaor, and Stuvard, give a first step towards the latter picture in codimension 1.

Theorem 11.6. *Let Σ be an m-dimensional area-minimizing current mod p in $\mathbb{R}^{m+1}$. Then:*

(a) *$\mathcal{S}_{m-1} \setminus \mathcal{S}_{m-2}$ is a $C^{1,\alpha}$ submanifold and for every $q \in \mathcal{S}_{m-1} \setminus \mathcal{S}_{m-2}$ there is a neighborhood U in which Σ consists of p minimal hypersurface meeting at $\mathcal{S}_{m-1} \cap U$;*

(b) *At every flat singular point there is a unique tangent cone, which is a flat plane with multiplicity $\frac{p}{2}$ (in particular, p must be even).*

In fact, after the appearance of [40] Minder and Wickramasekera (cf. [90]) pointed out to the authors that it is possible to derive Theorem 11.6 directly from the theory developed in [124], starting from one observation in [40] concerning tangent cones in the top stratum $\mathcal{S}_{m-1} \setminus \mathcal{S}_{m-2}$ and the verification of Simon's no hole condition. In [39], Theorem 11.6 will be further used to confirm the conjectural picture in codimension 1, namely to prove

Theorem 11.7. *Let Σ be an m-dimensional area-minimizing current mod p in $\mathbb{R}^{m+1}$. Then* $\mathrm{Sing}_i(\Sigma) \cap \mathcal{S}_m$ *is empty for p odd (as implied by Theorem* 11.4*), while* $\mathrm{Sing}_i(\Sigma) \cap \mathcal{S}_m$ *has dimension at most $m-2$ for even p.*

12. BOUNDARY REGULARITY THEORY: MINIMIZING INTEGRAL HYPERCURRENTS

The first boundary regularity theorem for area-minimizing integral currents Σ was proved by Allard in his PhD thesis [3] in codimension 1. More precisely, we have

Theorem 12.1. *Assume Σ is an area-minimizing integral current of dimension m in $\mathbb{R}^{m+1}$ and assume that*

(a) *$\partial\Sigma$ is a smooth (more precisely C^2) $(m-1)$-dimensional surface Γ with multiplicity* 1*;*

(b) *there is a uniformly convex smooth (more precisely C^2) bounded open set U such that $\Gamma \subset \partial U$.*

Then Σ is smooth in a neighborhood of Γ; more precisely, there is an open set $V \supset \Gamma$ such that $V \cap \Sigma$ is a smooth minimal hypersurface (with boundary) and its boundary (in V) is precisely Γ (in the classical sense of differential topology).

In fact, the proof in [3] contains an ε-regularity result which is the precursor of Theorem 6.1, while assumption (a) is combined with a suitable classification of boundary tangent cones to prove that any point $p \in \Gamma$ has density $\frac{1}{2}$. In order to remove the "convex barrier" of assumption (a), one needs to handle situations in which p might be a "2-sided" boundary point.

To illustrate the latter point, consider a 2-dimensional plane V in $\mathbb{R}^3$ and the two circles $\gamma_1 = \partial \mathbf{B}_1(0) \cap V$ and $\gamma_2 = \partial \mathbf{B}_2(0) \cap V$. Give to γ_1 and γ_2 the "same orientation," so that they bound the disks $D_1 = \mathbf{B}_1(0) \cap V$ and $D_2 = \mathbf{B}_2(0) \cap V$, taken with the same orientation where they overlap. It can be easily shown that, if $\Gamma = \gamma_1 + \gamma_2$, then $\Sigma = D_1 + D_2$ is the unique area-minimizing integral current bounded by Γ. Moreover, Σ can be described as the sum of the corona $D_2 \setminus D_1$, counted with multiplicity 1, and the disk D_1, counted with multiplicity 2. A point $p \in \gamma_1$ is what can be naturally called a "2-sided" boundary point, and note that its density is $\frac{3}{2}$ (for a more rigorous definition, cf. [35]). The regularity theory at such points is rather subtle, and (in codimension 1) it was handled in the famous work [69] by Hardt and Simon.

Theorem 12.2. *Let Γ be a smooth oriented closed $(m-1)$-dimensional submanifold of $\mathbb{R}^{m+1}$ and let Σ be an area-minimizing integral current whose boundary (in the sense of currents) is given by Γ counted with multiplicity* 1. *Then every point $p \in \Sigma$ is regular, namely one of the following two mutually exclusive possibilities holds:*

(i) *either the density of Σ at p is $\frac{1}{2}$ and hence the conclusion of Theorem* 6.1 *applies in a neighborhood U of p;*

(ii) *or the density of Σ at p is $k + \frac{1}{2}$ for some positive integer k; in this case there is a neighborhood U of p and a minimal hypersurface Λ of U without boundary such that:*

- *Λ contains Γ;*
- *Γ subdivides Λ in two regions Λ^+ and Λ^-;*
- *Σ in U is given by Λ^+ counted with multiplicity $k+1$ and Λ^- counted with multiplicity k.*

Among the many ideas introduced in [69], one has been highly influential in several other problems in minimal surface theory, and it is the so-called Hardt–Simon inequality. In a nutshell, the Hardt–Simon inequality makes clever use of the remainder in the monotonicity formula (namely the precise expression for the quantity $\frac{d}{dr}\frac{\mathrm{Vol}^m(\Sigma \cap \mathbf{B}_r)}{r^m}$) in order to infer nontrivial information on the graphical approximation of Σ at small scales.

While we have stated Theorems 12.1 and 12.2 as "global theorems," suitable local versions of them are also valid, and, in fact, the very nature of the main arguments is completely local.

13. BOUNDARY REGULARITY THEORY: MINIMIZING INTEGRAL CURRENTS WITH SMOOTH BOUNDARIES OF MULTIPLICITY 1

In his fundamental boundary regularity paper [5], Allard noticed that Theorem 6.1 can be used to generalize the conclusion of Theorem 12.1 to all codimensions.

Theorem 13.1. *Let Γ be a smooth $(m-1)$-dimensional closed oriented submanifold of $\mathbb{R}^{m+n}$ and let U be a bounded smooth uniformly convex set such that $\Gamma \subset \partial U$. Then any area-minimizing integral current Σ whose boundary is given by Γ (counted with multiplicity 1) is smooth in a neighborhood of Γ, in the sense of the conclusion of Theorem 12.1.*

Again, a local version of the above theorem holds as well; in fact, in order to conclude that a boundary point p is regular and one-sided in the sense of Theorem 12.2(i), it suffices to find a uniformly convex "barrier" which touches Γ at p and so that Σ lies (locally) on one side of it, cf. [68]. A simple argument furnishes such a barrier for any smooth $\Gamma \subset \mathbb{R}^{m+n}$: for instance, one could consider the smallest closed ball containing Γ. It then follows that under the mere assumption that Γ is sufficiently smooth, an area-minimizing current bounding Γ (taken with multiplicity 1) has always at least one boundary regular point.

Up until recently nothing more was known, except that in codimension higher than 1 singular boundary points are certainly possible. A simple example is given by the union of a smooth simple curve $\gamma_1 \subset \{x_1 = x_2 = 0\} \subset \mathbb{R}^4$ containing the origin and a smooth simple curve $\gamma_2 \subset \{x_3 = x_4 = 0\} \subset \mathbb{R}^4$ which *does not* contain the origin. This union bounds an area-minimizing 2-dimensional integral current for which 0 is a boundary singular point. Moreover, since in a general Riemannian manifold the barrier argument outlined in the previous paragraph is not available, even in the simplest case of a smooth simple closed curve in a closed smooth Riemannian 4-manifold M, the results outlined so far could not exclude the possibility that *all* boundary points of an area-minimizing current $\Sigma \subset M$ bounding Γ are singular.

As in the case of Theorem 12.2, the main difficulty in removing the convex barrier assumption is the possibility that boundary points have density larger than $\frac{1}{2}$. And as in the case of Theorem 10.1, the most problematic issue is that, unfortunately, the existence of a flat tangent cone at the boundary does not guarantee regularity: flat boundary singular points exist as soon as the codimension is larger than 1, cf. [35]. In [35], the author, Guido De Philippis, Hirsch, and Annalisa Massaccesi were able to develop a suitable "Almgren-type" regularity theory for boundary points, building on a previous important step of Hirsch [73]. In particular, we proved the following

Theorem 13.2. *Let Γ be a smooth closed oriented $(m-1)$-dimensional submanifold of $\mathbb{R}^{m+n}$ and let Σ be an area-minimizing integral current whose boundary is given by Γ taken with multiplicity 1. Then the set of boundary regular points, understood as points where one of the two alternatives (i) and (ii) of Theorem 12.2 hold, is a dense relatively open subset of Γ.*

While Theorem 13.2 might look very far from optimal, it turns out that a naive counterpart of the bound of the dimension of the interior singular set is, in fact, false. In [35] we prove also the following

Theorem 13.3. *There is a smooth* 1*-dimensional embedded submanifold of* $\mathbb{R}^4$ *which bounds an area-minimizing current* Σ *of* $\mathbb{R}^4$ *whose boundary singular set has Hausdorff dimension* 1.

Theorem 13.3 leaves open the possibility that at least the set of boundary singular points has zero $(m-1)$-dimensional Hausdorff measure and that it has dimension $m-2$ if the boundary is real analytic. We also caution the reader that a less restrictive definition of a boundary regular point might restrict the size of boundary singularities even in the C^∞ case. For a more detailed discussion of all these possibilities we refer the reader to Section 16. However, that boundary regularity is subtle is also witnessed by the following example of the author, De Philippis, and Hirsch (cf. [34]).

Theorem 13.4. *There is a smooth closed* 4*-dimensional Riemannian manifold* M *and a smooth simple closed curve* $\Gamma \subset M$ *which bounds a unique area-minimizing* 2*-dimensional current* Σ *which is smooth in* $M \setminus \Gamma$ *and whose first homology group is infinite-dimensional. In fact,* Σ *is smooth except at a single point* $p \in \Gamma$.

14. BOUNDARY REGULARITY THEORY: MINIMIZING INTEGRAL CURRENTS WITH SMOOTH BOUNDARIES OF HIGHER MULTIPLICITY

In the previous sections we examined the boundary regularity of area-minimizing integral currents under the assumption that the multiplicity of the boundary is 1. A rather intriguing and widely open question, already raised by Allard in his PhD thesis [3], is what happens when the multiplicity is an integer larger than 1 (the fact that it *must* be an integer is, of course, a consequence of the integrality assumption, but we also remind the reader that when T is integer rectifiable and ∂T has finite mass, ∂T is necessarily integer rectifiable, cf. [62]).

The problem raised by Allard in [3] is highlighted again by White in [1]. In the same reference, White observes also that, thanks to the decomposition theorem for area-minimizing hypercurrents, if Σ is an m-dimensional area-minimizing integral current in $\mathbb{R}^{m+1}$ whose boundary is a smooth submanifold Γ counted with multiplicity $Q > 1$, then Σ can be decomposed into the sum of Q area-minimizing integral currents whose boundary is Γ counted with multiplicity 1 also. In particular, we are in the position of applying the Hardt–Simon Theorem 12.2 to each element of the decomposition. While this is the same "codimension 1 phenomenon" that rules out flat singular points in the interior for area-minimizing integral hypercurrents, we pause a moment to make one important remark. It is well known that there are smooth $(m-1)$-dimensional oriented closed submanifolds of $\mathbb{R}^{m+1}$ that bound more than one area-minimizing integral current. This is already the case for smooth simple closed curves Γ in $\partial \mathbf{B}_1 \subset \mathbb{R}^3$. Consider in particular one such Γ and let Σ_1

and Σ_2 be two area-minimizing integral 2-dimensional currents which bound Γ (with multiplicity 1). Thanks to the above decomposition theorem, $\Sigma = \Sigma_1 + \Sigma_2$ is an area-minimizing current which bounds a double copy of Γ[2]. By the interior regularity theory, Σ_1 and Σ_2 have no interior point in common. Therefore, by the Hopf boundary lemma, there is no boundary point at Γ in which Σ_1 and Σ_2 have the same tangent: Σ_1 and Σ_2 meet at every point of Γ *transversally*.

In light of the above example, it seems sensible to give the following definition of a boundary regular point.

Definition 14.1. Assume that $\Gamma \subset \mathbb{R}^{m+n}$ is a smooth oriented $(m-1)$-dimensional submanifold and that Q is a positive integer. Let Σ be an area-minimizing integral current in $\mathbb{R}^{m+n}$ whose boundary is given by Q copies of Γ. Then $p \in \Gamma$ is a regular boundary point if one of the following two alternatives occur in some neighborhood U of p:

(i) There are N positive integers k_i with $\sum_i k_i = Q$ and N smooth minimal surfaces Λ_i in U with boundary Γ such that $\Sigma \cap U = \sum_i k_i \Lambda_i$, and each distinct pair Λ_i and Λ_j meet transversally at p;

(ii) There is a minimal surface Λ in U without boundary, which contains Γ: the latter subdivides Λ in two regions Λ^+ and Λ^- and

$$\Sigma \cap U = (Q+k)\Lambda^+ + Q\Lambda^-$$

for some positive integer k.

In particular, the discussion above reduces the following statement to a mere corollary of Theorem 12.2:

Corollary 14.2. *Let Γ be a smooth oriented closed $(m-1)$-dimensional submanifold of $\mathbb{R}^{m+1}$, Q be a positive integer, and Σ an area-minimizing integral current with $\partial \Sigma = Q\Gamma$. Then every boundary point $p \in \Gamma$ is regular in the sense of Definition* 14.1.

The boundary regularity theory for $Q > 1$ and in codimension larger than 1 is widely open. A very first preliminary result, which is a counterpart of Theorem 12.1 for 2-dimensional area-minimizing currents, has been proved very recently by the author, Stefano Nardulli, and Simone Steinbrüchel in **[43,44]**, building in part upon the theory developed in **[35]** and the paper **[74]**.

Theorem 14.3. *Consider a smooth 1-dimensional closed submanifold Γ of $\mathbb{R}^{2+n}$ and assume that there is a bounded smooth uniformly convex open set U such that $\Gamma \subset \partial U$.*

---

2 It is one of the most beautiful discoveries of geometric measure theory that this conclusion is in general false in higher codimension. In particular, following the pioneering work of L. C. Young **[125]**, there are several constructions of smooth simple curves Γ in $\mathbb{R}^4$ with the following remarkable property. If we let $m(\Gamma)$ be the mass of an area-minimizing 2-dimensional current which bounds one copy of Γ and $m(2\Gamma)$ the mass of an area-minimizing integral current which bounds two copies of Γ, then $m(2\Gamma) < 2m(\Gamma)$.

Let Σ be an area-minimizing 2-dimensional integral current such that $\partial\Sigma = Q\Gamma$ for some integer Q. Then every point $p \in \Gamma$ is a boundary regular point and moreover alternative (i) in Definition 14.1 *holds at every such point.*

15. UNIQUENESS OF TANGENT CONES

One major open question in the regularity theory of minimal submanifolds, which has attracted the attention of a large number of researchers since the dawn of geometric measure theory, is the uniqueness of tangent cones. This amounts to the question of whether there is at every point p a unique limit for the rescalings $\frac{\Sigma-p}{r}$ of the minimal submanifold Σ. In some situations the question is intimately connected to the understanding of the regularity properties of the various strata $\mathscr{S}_k \setminus \mathscr{S}_{k-1}$. For instance, the pioneering works of Taylor [116, 117] leading to the Theorems 8.1 and 11.2 can be reduced to suitable uniqueness statements for the relevant tangent cones.

The most striking result in the area is the celebrated theorem of Simon in [105].

Theorem 15.1. *Let Σ be a stationary integral varifold and assume that the spherical cross-section of one tangent cone Σ_0 at an interior point p of Σ is a regular submanifold of $\partial\mathbf{B}_1$ with multiplicity* 1. *Then Σ_0 is the unique tangent cone to Σ at p.*

Once again a similar uniqueness theorem is widely open when the multiplicity of the cross-section is allowed to take multiplicity higher than 1, except for some lucky situations in which the case of higher multiplicity can be reduced to that of multiplicity 1. Two notable examples are that of area-minimizing hypercurrents and that of area-minimizing currents mod 2.

Corollary 15.2. *Let Σ be an area-minimizing m-dimensional integral current in $\mathbb{R}^{m+1}$ or an area-minimizing m-dimensional current mod 2 in $\mathbb{R}^{m+n}$ and assume that p is an interior point at which one tangent cone Σ_0 has smooth spherical cross-section. Then Σ_0 is the unique tangent cone at p.*

Even in the case of multiplicity 1, the uniqueness of tangent cones whose spherical cross-section is not smooth is a much more subtle issue. Before discussing it, we wish to introduce a suitable concept which has played a pivotal role in many contexts. Let Σ_0 be a stationary varifold which is a cone with smooth cross-section Γ_0, taken with multiplicity 1. It is then well known that Γ_0 is a minimal submanifold of the sphere $\partial\mathbf{B}_1$. If Σ_k is a sequence of cones converging to Σ_0, with cross-section Γ_k, up to extraction of a subsequence Γ_k is the graph of a solution of a suitable linear elliptic PDE over Γ_0 plus higher order terms. Such solutions are called Jacobi fields in the literature, and they are the higher-dimensional counterpart of the classical Jacobi fields on geodesics. A Jacobi field u is called "integrable" if there is a sequence Γ_k of minimal submanifolds of $\partial\mathbf{B}_1$ converging to Γ_0 which generates u as outlined above. Prior to Theorem 15.1, Allard and Almgren in [6] proved the following important result:

Theorem 15.3. *Let Σ be a stationary integral varifold, let p be an interior point and assume that a tangent cone Σ_0 to Σ at p satisfies the following two properties:*

(i) *The spherical cross-section Γ_0 of Σ_0 is smooth and is taken with multiplicity* 1*;*

(ii) *Every Jacobi field of Γ_0 is integrable.*

Then Σ_0 is the unique tangent cone to Σ at p and, moreover, the rescalings $\frac{\Sigma-p}{r}$ converge to Σ_0 with a polynomial rate.

The integrability condition (ii) of Theorem 15.3 has several drawbacks. In order to verify it, one must know rather explicitly the cross-section Γ_0. But even in the cases in which Γ_0 is known and has a rather simple formula, verifying the condition is in general quite hard (in particular, it requires a classification result for all the solutions of some particular elliptic PDE). Last but not least, there are examples in which it does not hold, see **[2]**, and in which the convergence rate of $\frac{\Sigma-p}{r}$ to Σ_0 is just logarithmic. The powerful approach of Simon to Theorem 15.1 avoids any discussion of the integrability of the Jacobi vector fields thanks to his realization that the convergence of $\frac{\Sigma-p}{r}$ to Σ_0 can be reduced to an infinite-dimensional version of a classical result of Lojasiewicz for finite-dimensional gradient flows. The corresponding "Lojasiewicz–Simon inequality" has been widely used to study the convergence of parabolic PDEs to a unique steady state and the uniqueness of model singularities in other geometric variational problems.

Coming back to cones whose spherical cross-sections are not smooth, a particularly simple subclass are called "cylindrical tangent cones." In his notable investigation **[186]**, Simon has been able to prove a useful generalization of the Allard–Almgren Theorem 15.3.

Theorem 15.4. *Let Σ be a stationary integral varifold, let p be an interior point, and assume that a tangent cone Σ_0 to Σ at p satisfies the following structural properties:*

(i) $\Sigma_0 = V \times \Lambda_0$ *for some minimal cone Λ_0 and some linear subspace V;*

(ii) *The spherical cross-section Γ_0 of Λ_0 is smooth and is taken with multiplicity* 1*;*

(iii) *Every Jacobi field of Γ_0 is integrable;*

(iv) *The following "no hole condition" holds for a sufficiently small $\delta(\Gamma_0)$: provided $\frac{\Sigma-p}{r}$ is sufficiently close to Σ_0, every $\mathbf{B}_\delta(q)$ with $q \in \mathbf{B}_1 \cap V$ contains a point x of density $\Theta(\Sigma, x)$ larger than $\Theta(\Sigma, p) - \delta$.*

Then Σ_0 is the unique tangent cone to Σ at p and, moreover, the rescalings $\frac{\Sigma-p}{r}$ converge to Σ_0 with a polynomial rate.

Quite a few of the structural results for singular strata mentioned in the previous sections depend heavily on the above result (or can be deduced from it). A notable exception is the uniqueness theorem of Taylor which underlines the second conclusion of Theorem 8.1 (and the higher-dimensional counterpart in **[23]**). The latter is, in fact, derived through a direct epiperimetric inequality à la Reifenberg.

One major drawback of the approaches to Theorems 15.1, 15.3, and 15.4 is that the underlying PDE arguments rely heavily on the ε-regularity result of Allard, namely Theorem 5.2 (or on some other analogous results). For instance, in Theorem 15.1 the multiplicity 1 assumption and the regularity of Γ_0 allow us to conclude that $\partial \mathbf{B}_1 \cap \frac{\Sigma - p}{r}$ is a smooth graph over Γ_0. On the contrary, the epiperimetric inequality, which is based on exhibiting a suitable competitor, can be applied in situations where the cross-section is irregular or taken with higher multiplicity. On the other hand, its applicability is limited to minimizers. Up until recently, another obvious objection to a wider plausibility of an epiperimetric inequality à la Reifenberg was that it immediately implies a polynomial decay rate, which is known to be false in general cf. [2]. However, the recent paper of Colombo, Spolaor, and Velichkov [24] shows that Theorem 15.1 can be recovered (and, in fact, generalized to suitable "quasi-minima") through a suitable generalization of Reifenberg's epiperimetric (it must be noted that the proof of the latter is nonetheless achieved using the Lojasiewicz–Simon inequality).

An important case in which an epiperimetric inequality can be proved and used effectively to prove uniqueness of tangent cones (while a "PDE-approach" has not been given yet) is that of 2-dimensional area-minimizing currents at interior points. In particular, White in [119] proved

Theorem 15.5. *Let Σ be a 2-dimensional area-minimizing integral current in $\mathbb{R}^{2+n}$. Then the tangent cone to Σ is unique at every interior point p.*

A counterpart to Theorem 15.5 has been shown by Hirsch and Marini in [74] at smooth boundaries taken with multiplicity 1. However, as noticed in [44], the proof of Hirsch and Marini can easily be adapated to the case of smooth boundaries with arbitrary multiplicities, thus giving a complete result for 2-dimensional area-minimizing integral currents in any dimension and codimension.

16. OPEN PROBLEMS

In this section we will collect some open questions. I wish to emphasize that the selection given here by no means exhausts the interesting open problems in the area, it rather reflects a personal choice of the author.

16.1. Stationary and stable varifolds

Perhaps the most intriguing question is whether it is possible to improve Corollary 5.3 in any situation which is not the trivial one of 1-dimensional stationary varifolds. The most modest goal would be to show that the singular set of stationary 2-dimensional integral varifolds in $\mathbb{R}^3$ has zero 2-dimensional Hausdorff measure. In general, there is no example of singular stationary m-dimensional varifolds (in any codimension) for which the singular set has dimension larger than $m-1$.

In the case of stable varifolds of codimension 1, the deep theory of Wickramasekera developed in [124] (see also [89] for some further progress) gives one hope that in the future some final unconditional structural result might be at hand. A coronation of the efforts in the area would be a theorem which proves that $\mathscr{S}_{m-1} \setminus \mathscr{S}_{m-2}$ is a $C^{1,\alpha}$ $(m-1)$-dimensional submanifold, while the set of flat singularities is $(m-2)$-rectifiable. The latter statement seems to be reachable in the very particular case of area-minimizing hypercurrents mod p.

A widely open problem is whether stability allows going beyond Allard's conclusion in codimension higher than 1. It is quite baffling that no further regularity information has been concluded thus far for stable varifolds as soon as the codimension is larger than 1. A particularly intriguing case would be that of 2-dimensional stable varifolds, already in $\mathbb{R}^4$. A first question in that direction is whether some counterpart of the Schoen–Simon–Yau estimates and hence a corresponding compactness theorem hold for classical (possibly branched) minimal 2-dimensional surfaces in, say, $\mathbb{R}^4$. In other words, assume that Σ_k is such a sequence in some bounded open set $U \subset \mathbb{R}^4$, that the area of Σ_k is uniformly bounded, and that each Σ_k is stable. Is it possible to extract a subsequence which is converging (in the varifold sense) to a classical stable (possibly branched) immersed minimal surface? What if we restrict further Σ_k and ask that they are embedded except for a finite number of branching singularities? Note that Theorem 10.2 does imply the desired conclusion if each Σ_k can be oriented so to give an area-minimizing integral 2-dimensional current.

16.2. Singularities of area-minimizing integral hypercurrents

Area-minimizing integral currents of dimension m in $\mathbb{R}^{m+1}$ are the objects for which we have the strongest regularity theory. Is it possible to prove more facts about the structure of the singular set? In particular, is it true that $\mathscr{S}_{m-7} \setminus \mathscr{S}_{m-8}$ is a $C^{1,\alpha}$ submanifold, or rather are there examples (as the recent stable minimal hypersurfaces in some Riemannian manifolds given by Simon in [110]) for which $\mathscr{S}_{m-7}$ has a fractal Hausdorff dimension $m-8 < \alpha < m-7$? Does it make a difference if the ambient is a smooth Riemannian manifold rather than Euclidean space?

A closely related question is whether the no-hole condition of (iv) in Theorem 15.4 can be violated at some point p of an area-minimizing hypercurrent (in the Euclidean space or in a general smooth Riemannian ambient). This is indeed the case for some points in the examples of stable minimal hypersurfaces constructed in [110], while in the Euclidean space a completely different example has been given by Gabor Székelyhidi in [115].

16.3. Singularities of area-minimizing integral currents in codimension higher than 1

It is very tempting to conjecture that for $m \geq 3$ Almgren's partial regularity theorem can improved to say that the singular set of any area-minimizing integral m-dimensional current in $\mathbb{R}^{m+n}$ is $(m-2)$-rectifiable. This problem seems intimately linked to the "simplest" open case of uniqueness of tangent cones for area-minimizing currents in codimension $n \geq 2$:

- Consider an area-minimizing integral current Σ of dimension m in $\mathbb{R}^{m+n}$ and let $p \in \mathrm{Sing}_i(\Sigma)$ be a point where one tangent cone is flat. Is the latter the *unique* tangent cone to Σ at p?

The forthcoming work [46] seems to suggest that a positive answer to the latter question, together with the additional information that the convergence rate is polynomial, would imply $(m-2)$-rectifiability of $\mathrm{Sing}_i(\Sigma)$.

On the other hand, the works [82, 83] suggest that further structural results cannot be expected, at least not in general smooth ambient manifolds, and instead there are 3-dimensional area-minimizing integral currents in closed smooth Riemannian manifolds whose singular sets have any preassigned Hausdorff dimension $\alpha \in (0, 1)$.

16.4. Singularities of area-minimizing currents mod p

As already mentioned, in the works [38–40] (see also [90, 124]) we plan to show that, for an m-dimensional area-minimizing current mod p in $\mathbb{R}^{m+1}$, the stratum $\mathscr{S}^{m-1} \setminus \mathscr{S}^{m-2}$ is a $C^{1,\alpha}$ $(m-1)$-dimensional submanifold, while the set of flat singular points has dimension at most $m-2$. In fact, it is expected that the latter is $(m-2)$-rectifiable. The same properties could be expected in higher codimension, but the problem poses considerable difficulties. Moreover, the author does not know examples in which the stratum $\mathscr{S}^{m-2} \setminus \mathscr{S}^{m-3}$ is nonempty. To that respect, the most basic question is whether there is any counterpart of Taylor's theorem for the case $p = 3$: is there any 2-dimensional area-minimizing cone mod p in $\mathbb{R}^3$ which is not invariant under some translation? The works [116] and [118] imply that the answer is no for $p = 3$ and 4 (while it is a simple exercise to see that it is no for $p = 2$ as well, since it reduces to the case of integral currents).

16.5. Boundary regularity of area-minimizing integral currents at multiplicity 1 boundaries

Is it possible to improve Theorem 13.2 and show that for general smooth Γ the set of boundary singular points has zero Hausdorff $(m-1)$-dimensional measure? It must also be noted that, in the examples of Theorem 13.3 given by the argument of [35], most of the boundary singular points p's are of "crossing type," i.e., in some neighborhood U of such p's the area-minimizing current can be decomposed in one area-minimizing current which takes the boundary Γ smoothly and a second one which is area-minimizing and has no boundary (but includes p in its support). In particular, the following two conjectures seem likely:

- Boundary singularities of noncrossing type have a much lower dimension (according to the examples, the best we can hope is $m-2$).
- Since crossing-type singularities have necessarily dimension $m-2$ when Γ (and the ambient Riemannian manifold) is real analytic, the whole boundary singular set has dimension at most $m-2$ under the latter assumption.

In fact, the following elegant conjecture is due to White in [122].

Conjecture 16.1. *Let $\Gamma \subset \mathbb{R}^{2+n}$ be a simple closed real-analytic curve and Σ an area-minimizing integral current such that $\partial\Sigma = \Gamma$. Then the union of the boundary and interior singular points of Σ is discrete. In particular,*

- *the "overall singular set" is finite,*
- *Σ has finite genus g,*
- *and it is a classical Douglas–Rado solution of the Plateau problem among surfaces of genus g.*

16.6. Boundary regularity of area-minimizing integral currents at boundaries with higher multiplicity

It is tempting to conjecture that Theorem 14.3 holds for m-dimensional integral currents for $m \geq 2$, but in reality the situation might be more complicated. Otherwise, a more modest expectation is that for general m, under the assumptions of Theorem 14.3, the boundary singular set has dimension at most $m - 3$. Nothing is known in the case of a general integral multiplicity Q and a general boundary Γ, i.e., without the assumption that there is a "convex barrier" at (a portion of) Γ. One might expect that the counterpart of Theorem 13.2 holds for general multiplicities $Q \geq 1$.

16.7. Uniqueness of tangent cones

The uniqueness of interior tangent cones when the multiplicity of the cross-section is larger than 1 is widely open. As already mentioned, the most striking case is that of flat singular points, i.e., points at which at least one tangent cone is a plane with higher multiplicity, but the generalized minimal surface is not regular. This problem is open for integral area-minimizing currents of dimension $m \geq 3$ in codimension larger than $n \geq 2$, but it is also open for stationary and stable varifolds in dimension $m \geq 2$ and codimension 1.

It is also widely open whether Simon's Theorem 15.4 can be improved. In particular, can one drop the "no-hole condition" (iv) or the integrability condition (iii), at least for some suitable subclass of stationary varifolds? Some situations in which the "no-hole condition" can be dropped are given in [107], while the recent work [114] is the first, to the best of author's knowledge, in which the uniqueness of the cylindrical cone is proved for one example in which both conditions (iii) and (iv) in Theorem 15.4 can be dropped.

We finish this survey by mentioning that the following very innocent question is still open (even in the case $m = 3$ and $n + 2 = 5$):

- Consider an m-dimensional area-minimizing integral current Σ in $\mathbb{R}^{m+n}$ with $m \geq 3$ and $n \geq 2$. Assume that one tangent cone Σ_0 at some point $p \in \mathrm{Sing}_i(\Sigma)$ is the union of two distinct linear planes counted both with multiplicity 1. Is Σ_0 the unique tangent cone to Σ at p?

ACKNOWLEDGMENTS

I am very grateful to Guido De Philippis and Luca Spolaor for carefully reading a very preliminary version of this manuscript, suggesting several precious improvements, and reminding me of a few pertinent results in the literature.

FUNDING

This work was partially supported by the National Science Foundation through the grant FRG-1854147.

REFERENCES

[1] Some open problems in geometric measure theory and its applications suggested by participants of the 1984 AMS summer institute. In *Geometric measure theory and the calculus of variations (Arcata, Calif., 1984)*, edited by J. E. Brothers, pp. 441–464, Proc. Sympos. Pure Math. 44, Amer. Mathd. Soc., Providence, RI, 1986.

[2] D. Adams and L. Simon, Rates of asymptotic convergence near isolated singularities of geometric extrema. *Indiana Univ. Math. J.* **37** (1988), 225–254.

[3] W. K. Allard, *On boundary regularity for the Plateau problem*. Ph.D. thesis, Brown University, Providence, 1968.

[4] W. K. Allard, On the first variation of a varifold. *Ann. of Math. (2)* **95** (1972), 417–491.

[5] W. K. Allard, On the first variation of a varifold: boundary behavior. *Ann. of Math. (2)* **101** (1975), 418–446.

[6] W. K. Allard and F. J. Jr. Almgren, On the radial behavior of minimal surfaces and the uniqueness of their tangent cones. *Ann. of Math. (2)* **113** (1981), 215–265.

[7] F. J. Jr. Almgren, *The theory of varifolds: A variational calculus in the large for the k-dimensional area integrand*. Mimeographed notes, Princeton Univ. Math. Library, 1965.

[8] F. J. Jr. Almgren, Some interior regularity theorems for minimal surfaces and an extension of Bernstein's theorem. *Ann. of Math. (2)* **84** (1966), 277–292.

[9] F. J. Jr. Almgren, Existence and regularity almost everywhere of solutions to elliptic variational problems with constraints. *Mem. Amer. Math. Soc.* **4** (1976), no. 165, viii+199 pp.

[10] F. J. Jr. Almgren, Q-valued functions minimizing Dirichlet's integral and the regularity of area minimizing rectifiable currents up to codimension two. *Bull. Amer. Math. Soc. (N.S.)* **8** (1983), 327–328.

[11] F. J. Jr. Almgren, *Almgren's big regularity paper. Q-valued functions minimizing Dirichlet's integral and the regularity of area-minimizing rectifiable currents up to codimension* 2. *With a preface by Jean E. Taylor and Vladimir Scheffer*. Monogr. Ser. Math. 1, World Scientific, River Edge, NJ, 2000.

[12] C. Bellettini, Almost complex structures and calibrated integral cycles in contact 5-manifolds. *Adv. Calc. Var.* **6** (2013), 339–374.

[13] C. Bellettini and T. Rivière, The regularity of special Legendrian integral cycles. *Ann. Sc. Norm. Super. Pisa Cl. Sci. (5)* **11** (2012), 61–142.

[14] A. S. Besicovitch, On the fundamental geometrical properties of linearly measurable plane sets of points. *Math. Ann.* **98** (1928), 422–464.

[15] A. S. Besicovitch, On the fundamental geometrical properties of linearly measurable plane sets of points II. *Math. Ann.* **115** (1938), 296–329.

[16] A. S. Besicovitch, On the fundamental geometrical properties of linearly measurable plane sets of points III. *Math. Ann.* **116** (1939), 349–357.

[17] E. Bombieri, E. De Giorgi, and E. Giusti, Minimal cones and the Bernstein problem. *Invent. Math.* **7** (1969), 243–268.

[18] R. Caccioppoli, Misura e integrazione sugli insiemi dimensionalmente orientati. *Atti Accad. Naz. Lincei Rend. Cl. Sci. Fis. Mat. Natur. (8)* **12** (1952), 3–11.

[19] R. Caccioppoli, Misura e integrazione sugli insiemi dimensionalmente orientati II. *Atti Accad. Naz. Lincei Rend. Cl. Sci. Fis. Mat. Natur. (8)* **12** (1952), 137–146.

[20] S. X. Chang, Two-dimensional area minimizing integral currents are classical minimal surfaces. *J. Amer. Math. Soc.* **1** (1988), 699–778.

[21] O. Chodosh and C. Mantoulidis, Minimal surfaces and the Allen-Cahn equation on 3-manifolds: index, multiplicity, and curvature estimates. *Ann. of Math. (2)* **191** (2020), 213–328.

[22] T. H. Colding and C. De Lellis, The min–max construction of minimal surfaces. In *Surveys in differential geometry, Vol. VIII*, Int. Press, Somerville, MA, 2003.

[23] M. Colombo, N. Edelen, and L. Spolaor, The singular set of minimal surfaces near polyhedral cones. 2017, arXiv:1709.09957.

[24] M. Colombo, L. Spolaor, and B. Velichkov, (Log)-epiperimetric inequality and regularity over smooth cones for almost area-minimizing currents. 2018, arXiv:1802.00418.

[25] G. David, Should we solve Plateau's problem again? In *Advances in analysis: the legacy of Elias M. Stein*, pp. 108–145, Princeton Math. Ser. 50, Princeton Univ. Press, Princeton, NJ, 2014.

[26] G. David, A local description of 2-dimensional almost minimal sets bounded by a curve. 2019, arXiv:1901.10701.

[27] G. David, Local regularity properties of almost- and quasiminimal sets with a sliding boundary condition. *Astérisque* **411** (2019), ix+377 pp.

[28] G. David, Sliding almost minimal sets and the Plateau problem. In *Harmonic analysis and applications*, pp. 199–256, IAS/Park City Math. Ser. 27, Amer. Math. Soc., Providence, RI, 2020.

[29] E. De Giorgi, Su una teoria generale della misura $(r-1)$-dimensionale in uno spazio ad r dimensioni. *Ann. Mat. Pura Appl. (4)* **36** (1954), 191–213.

[30] E. De Giorgi, Nuovi teoremi relativi alle misure $(r-1)$-dimensionali in uno spazio ad r dimensioni. *Ric. Mat.* **4** (1955), 95–113.

[31] E. De Giorgi, Sulla proprietà isoperimetrica dell'ipersfera, nella classe degli insiemi aventi frontiera orientata di misura finita. *Atti Accad. Naz. Lincei, Mem. Cl. Sci. Fis. Mat. Nat., Sez. Ia (8)* **5** (1958), 33–44.

[32] E. De Giorgi, *Frontiere orientate di misura minima*. Semin. Mat. Sc. Norm. Super. Pisa, Editrice Tecnico Scientifica, Pisa, 1961.

[33] E. De Giorgi, Una estensione del teorema di Bernstein. *Ann. Sc. Norm. Super. Pisa Cl. Sci. (3)* **19** (1965), 79–85.

[34] C. De Lellis, G. De Philippis, and J. Hirsch, Nonclassical minimizing surfaces with smooth boundary. 2019, arXiv:1906.09488. To appear in *J. Differential Geom.*

[35] C. De Lellis, G. De Philippis, J. Hirsch, and A. Massaccesi, On the boundary behavior of mass-minimizing integral currents. 2018, arXiv:1809.09457. To appear in *Mem. Amer. Math. Soc.*

[36] C. De Lellis, A. De Rosa, and F. Ghiraldin, A direct approach to the anisotropic Plateau problem. *Adv. Calc. Var.* **12** (2019), 211–223.

[37] C. De Lellis, F. Ghiraldin, and F. Maggi, A direct approach to Plateau's problem. *J. Eur. Math. Soc. (JEMS)* **19** (2017), 2219–2240.

[38] C. De Lellis, J. Hirsch, A. Marchese, L. Spolaor, and S. Stuvard, in preparation.

[39] C. De Lellis, J. Hirsch, A. Marchese, L. Spolaor, and S. Stuvard, in preparation.

[40] C. De Lellis, J. Hirsch, A. Marchese, L. Spolaor, and S. Stuvard, Area minimizing hypersurfaces modulo p: a geometric free-boundary problem. 2021, arXiv:2105.08135.

[41] C. De Lellis, J. Hirsch, A. Marchese, and S. Stuvard, Area-minimizing currents mod $2Q$: linear regularity theory. 2019, arXiv:1909.03305. To appear in *Comm. Pure Appl. Math.*

[42] C. De Lellis, J. Hirsch, A. Marchese, and S. Stuvard, Regularity of area minimizing currents mod p. *Geom. Funct. Anal.* **30** (2020), 1224–1336.

[43] C. De Lellis, S. Nardulli, and S. Steinbrüchel, An Allard-type boundary regularity theorem for 2d minimizing currents at smooth curves with arbitrary multiplicity. 2021, arXiv:2111.02991

[44] C. De Lellis, S. Nardulli, and S. Steinbrüchel, Uniqueness of tangent cones at the boundary for 2-dimensional area-minimizing currents. 2021, arXiv:2111.02981

[45] C. De Lellis and J. Ramic, Min–max theory for minimal hypersurfaces with boundary. *Ann. Inst. Fourier (Grenoble)* **68** (2018), 1909–1986.

[46] C. De Lellis and A. Skorobogatova, in preparation.

[47] C. De Lellis and E. N. Spadaro, Q-valued functions revisited. *Mem. Amer. Math. Soc.* **211** (2011), no. 991, vi+79 pp.

[48] C. De Lellis and E. N. Spadaro, Regularity of area minimizing currents I: gradient L^p estimates. *Geom. Funct. Anal.* **24** (2014), 1831–1884.

[49] C. De Lellis and E. N. Spadaro, Multiple valued functions and integral currents. *Ann. Sc. Norm. Super. Pisa Cl. Sci. (5)* **14** (2015), 1239–1269.

[50] C. De Lellis and E. N. Spadaro, Regularity of area minimizing currents II: center manifold. *Ann. of Math. (2)* **183** (2016), 499–575.

[51] C. De Lellis and E. N. Spadaro, Regularity of area minimizing currents III: blow-up. *Ann. of Math. (2)* **183** (2016), 577–617.

[52] C. De Lellis, E. N. Spadaro, and L. Spolaor, Regularity theory for 2-dimensional almost minimal currents II: branched center manifold. *Ann. PDE* **3** (2017), no. 2, 18.

[53] C. De Lellis, E. N. Spadaro, and L. Spolaor, Uniqueness of tangent cones for two-dimensional almost-minimizing currents. *Comm. Pure Appl. Math.* **70** (2017), 1402–1421.

[54] C. De Lellis, E. N. Spadaro, and L. Spolaor, Regularity theory for 2-dimensional almost minimal currents I: Lipschitz approximation. *Trans. Amer. Math. Soc.* **370** (2018), 1783–1801.

[55] C. De Lellis, E. N. Spadaro, and L. Spolaor, Regularity theory for 2-dimensional almost minimal currents III: blowup. *J. Differential Geom.* **116** (2020), 125–185.

[56] C. De Lellis and D. Tasnady, The existence of embedded minimal hypersurfaces. *J. Differential Geom.* **95** (2013), 355–388.

[57] G. De Philippis, A. De Rosa, and F. Ghiraldin, A direct approach to Plateau's problem in any codimension. *Adv. Math.* **288** (2016), 59–80.

[58] G. De Philippis, A. De Rosa, and F. Ghiraldin, Existence results for minimizers of parametric elliptic functionals. *J. Geom. Anal.* **30** (2020), 1450–1465.

[59] L. C. Evans and R. F. Gariepy, *Measure theory and fine properties of functions. Revised edition*. Textb. Math., CRC Press, Boca Raton, 2015.

[60] Y. Fang and S. Kolasinski, Existence of solutions to a general geometric elliptic variational problem. *Calc. Var. Partial Differential Equations* **57** (2018), no. 3, 91.

[61] H. Federer, The (φ, k) rectifiable subsets of n-space. *Trans. Amer. Math. Soc.* **62** (1947), 114–192.

[62] H. Federer, *Geometric measure theory*. Grundlehren Math. Wiss. 153, Springer, New York, 1969.

[63] H. Federer, The singular sets of area minimizing rectifiable currents with codimension one and of area minimizing flat chains modulo two with arbitrary codimension. *Bull. Amer. Math. Soc.* **76** (1970), 767–771.

[64] H. Federer and W. H. Fleming, Normal and integral currents. *Ann. of Math. (2)* **72** (1960), 458–520.

[65] W. H. Fleming, On the oriented Plateau problem. *Rend. Circ. Mat. Palermo (2)* **11** (1962), 69–90.

[66] W. H. Fleming, Flat chains over a finite coefficient group. *Trans. Amer. Math. Soc.* **121** (1966), 160–186.

[67] N. Garofalo and F.-H. Lin, Monotonicity properties of variational integrals, A_p weights and unique continuation. *Indiana Univ. Math. J.* **35** (1986), 245–268.

[68] R. M. Hardt, On boundary regularity for integral currents or flat chains modulo two minimizing the integral of an elliptic integrand. *Comm. Partial Differential Equations* **2** (1977), 1163–1232.

[69] R. Hardt and L. Simon, Boundary regularity and embedded solutions for the oriented Plateau problem. *Ann. of Math. (2)* **110** (1979), 439–486.

[70] R. Hardt and L. Simon, Nodal sets for solutions of elliptic equations. *J. Differential Geom.* **30** (1989), 505–522.

[71] J. Harrison, Soap film solutions of Plateau's problem. *J. Geom. Anal.* **24** (2014), 271–297.

[72] J. Harrison and H. Pugh, Existence and soap film regularity of solutions to Plateau's problem. 2013, arXiv:1310.0508.

[73] J. Hirsch, Boundary regularity of Dirichlet minimizing Q-valued functions. *Ann. Sc. Norm. Super. Pisa Cl. Sci. (5)* **16** (2016), 1353–1407.

[74] J. Hirsch and M. Marini, Uniqueness of tangent cones to boundary points of two-dimensional almost-minimizing currents. 2019, arXiv:1909.13383.

[75] D. Ketover, Genus bounds for min–max minimal surfaces. *J. Differential Geom.* **112** (2019), 555–590.

[76] D. King, F. Maggi, and S. Stuvard, Plateau's problem as a singular limit of capillarity problems. 2019, arXiv:1907.00551.

[77] D. King, F. Maggi, and S. Stuvard, Collapsing and the convex hull property in a soap film capillarity model. 2020, arXiv:2002.06273.

[78] D. King, F. Maggi, and S. Stuvard, Smoothness of collapsed regions in a capillarity model for soap films. 2020, arXiv:2007.14868.

[79] J. L. Lagrange, Essai d'une nouvelle méthode pour determiner les maxima et les minima des formules integrales indefinies. *Misc. Taurinensis* **2** (1760), no. 325, 173–199.

[80] G. Lawlor and F. Morgan, Curvy slicing proves that triple junctions locally minimize area. *J. Differential Geom.* **44** (1996), 514–528.

[81] Y. Liokumovich, F. C. Marques, C. Fernando, and A. Neves, Weyl law for the volume spectrum. *Ann. of Math. (2)* **187** (2018), 933–961.

[82] Z. Liu, Calibrated area-minimizing surfaces with a fractal singular set. 2021, arXiv:2110.13137

[83] Z. Liu, Every finite graph arises as the singular set of a compact 3-d calibrated area minimizing surface. 2021, arXiv:2106.03199.

[84] A. Logunov, Nodal sets of Laplace eigenfunctions: proof of Nadirashvili's conjecture and of the lower bound in Yau's conjecture. *Ann. of Math. (2)* **187** (2018), 241–262.

[85] F. Maggi, S. Stuvard, and A. Scardicchio, Soap films with gravity and almost-minimal surfaces. *Discrete Contin. Dyn. Syst.* **39** (2019), 6877–6912.

[86] F. C. Marques and A. Neves, Min–max theory and the Willmore conjecture. *Ann. of Math. (2)* **179** (2014), 683–782.

[87] F. C. Marques, A. Neves, and A. Song, Equidistribution of minimal hypersurfaces for generic metrics. *Invent. Math.* **216** (2019), 421–443.

[88] P. Mattila, *Geometry of sets and measures in Euclidean spaces. Fractals and rectifiability*. Cambridge Stud. Adv. Math. 44, Cambridge University Press, Cambridge, 1995.

[89] P. Minter, The structure of stable codimension one integral varifolds near classical cones of density $\frac{5}{2}$. 2021, arXiv:2108.02614.

[90] P. Minter and N. Wickramasekra, A Structure Theory for Stable Codimension 1 Integral Varifolds with Applications to Area Minimising Hypersurfaces mod p. 2021, arXiv:2111.11202

[91] F. Morgan, *Geometric measure theory. A beginner's guide. Fifth edn.* Elsevier/Academic Press, Amsterdam, 2016.

[92] A. P. Morse and J. F. Randolph, The ϕ-rectifiable subsets of the plane. *Trans. Amer. Math. Soc.* **55** (1944), 236–305.

[93] A. Naber and D. Valtorta, Rectifiable-Reifenberg and the regularity of stationary and minimizing harmonic maps. *Ann. of Math. (2)* **185** (2017), 131–227.

[94] A. Naber and D. Valtorta, The singular structure and regularity of stationary varifolds. *J. Eur. Math. Soc. (JEMS)* **22** (2020), 3305–3382.

[95] J. T. Pitts, *Existence and regularity of minimal surfaces on Riemannian manifolds.* Math. Notes 27, Princeton University Press, Princeton, NJ, 1981.

[96] E. R. Reifenberg, Solution of the Plateau Problem for m-dimensional surfaces of varying topological type. *Acta Math.* **104** (1960), 1–92.

[97] E. R. Reifenberg, An epiperimetric inequality related to the analyticity of minimal surfaces. *Ann. of Math. (2)* **80** (1964), 1–14.

[98] E. R. Reifenberg, On the analyticity of minimal surfaces. *Ann. of Math. (2)* **80** (1964), 15–21.

[99] T. Rivière, A lower-epiperimetric inequality for area-minimizing surfaces. *Comm. Pure Appl. Math.* **57** (2004), 1673–1685.

[100] T. Rivière and G. Tian, The singular set of J-holomorphic maps into projective algebraic varieties. *J. Reine Angew. Math.* **570** (2004), 47–87.

[101] T. Rivière and G. Tian, The singular set of 1–1 integral currents. *Ann. of Math. (2)* **169** (2009), 741–794.

[102] R. Schoen and L. Simon, Regularity of stable minimal hypersurfaces. *Comm. Pure Appl. Math.* **34** (1981), 741–797.

[103] R. Schoen, L. Simon, and S. T. Yau, Curvature estimates for minimal hypersurfaces. *Acta Math.* **134** (1975), 275–288.

[104] L. Simon, *Lectures on geometric measure theory*. Proc. Centre Math. Anal. Austral. Nat. Univ. 3, The Australian National University, Canberra, 1983.

[105] L. Simon, Asymptotics for a class of nonlinear evolution equations, with applications to geometric problems. *Ann. of Math. (2)* **118** (1983), 525–571.

[106] L. Simon, Cylindrical tangent cones and the singular set of minimal submanifolds. *J. Differential Geom.* **38** (1993), 585–652.

[107] L. Simon, Uniqueness of some cylindrical tangent cones. *Comm. Anal. Geom.* **2** (1994), 1–33.

[108] L. Simon, Rectifiability of the singular sets of multiplicity 1 minimal surfaces and energy minimizing maps. In *Surveys in differential geometry, Vol. II*, pp. 246–305, Int. Press, Cambridge, MA, 1995.

[109] L. Simon, *Theorems on regularity and singularity of energy minimizing maps. Based on lecture notes by Norbert Hungerbühler*. Lectures Math. ETH Zürich, Birkhäuser, Basel, 1996.

[110] L. Simon, A Liouville-type theorem for stable minimal hypersurfaces. 2021, arXiv:2101.06404.

[111] L. Simon, Stable minimal hypersurfaces in $\mathbb{R}^{N+1+\ell}$ with singular set an arbitrary closed K in $\{0\} \times \mathbb{R}^{\ell}$. 2021, arXiv:2101.06401.

[112] J. Simons, Minimal varieties in Riemannian manifolds. *Ann. of Math. (2)* **88** (1968), 62–105.

[113] L. Spolaor, Almgren's type regularity for semicalibrated currents. *Adv. Math.* **350** (2019), 757–815.

[114] G. Székelyhidi, Uniqueness of certain cylindrical tangent cones. 2020, arXiv:2012.02065.

[115] G. Székelyhidi, Minimal hypersurfaces with cylindrical tangent cones. 2021, arXiv:2107.14786.

[116] J. E. Taylor, Regularity of the singular sets of two-dimensional area-minimizing flat chains modulo 3 in $\mathbb{R}^3$. *Invent. Math.* **22** (1973), 119–159.

[117] J. E. Taylor, The structure of singularities in soap-bubble-like and soap-film-like minimal surfaces. *Ann. of Math. (2)* **103** (1976), 489–539.

[118] B. White, The structure of minimizing hypersurfaces mod 4. *Invent. Math.* **53** (1979), 45–58.

[119] B. White, Tangent cones to two-dimensional area-minimizing integral currents are unique. *Duke Math. J.* **50** (1983), 143–160.

[120] B. White, Regularity of the singular sets in immiscible fluid interfaces and solutions to other Plateau-type problems. In *Miniconference on geometry and partial differential equations (Canberra, 1985)*, pp. 244–249, Proc. Centre Math. Anal. Austral. Nat. Univ. 10, Austral. Nat. Univ., Canberra, 1986.

[121] B. White, A regularity theorem for minimizing hypersurfaces modulo p. In *Geometric measure theory and the calculus of variations (Arcata, CA, 1984)*, pp. 413–427, Proc. Sympos. Pure Math. 44, Amer. Math. Soc., Providence, RI, 1986.

[122] B. White, Classical area minimizing surfaces with real-analytic boundaries. *Acta Math.* **179** (1997), 295–305.

[123] B. White, Stratification of minimal surfaces, mean curvature flows, and harmonic maps. *J. Reine Angew. Math.* **488** (1997), 1–35.

[124] N. Wickramasekera, A general regularity theory for stable codimension 1 integral varifolds. *Ann. of Math. (2)* **179** (2014), 843–1007.

[125] L. C. Young, Some extremal questions for simplicial complexes. V. The relative area of a Klein bottle. *Rend. Circ. Mat. Palermo (2)* **12** (1963), 257–274.

[126] L. C. Young, *Lectures on the calculus of variations and optimal control theory*. W. B. Saunders Co., Philadelphia–London–Toronto, ON, 1969.

[127] X. Zhou and J. Zhu, Min–max theory for constant mean curvature hypersurfaces. *Invent. Math.* **218** (2019), 441–490.

CAMILLO DE LELLIS

School of Mathematics, Institute for Advanded Study, 1 Einstein Drive, Princeton, NJ 08540, USA, camillo.delellis@ias.edu

A MATHEMATICAL PERSPECTIVE OF MACHINE LEARNING

WEINAN E

ABSTRACT

What lies at the heart of modern neural network-based machine learning is the ability to approximate very high dimensional functions with good accuracy. This opens up two major avenues of research. The first is to develop machine learning-based algorithms for scientific problems that suffer from the *curse of dimensionality*. The second is to build a theoretical framework that helps us to form a better foundation for machine learning. For the latter, the most important questions that need to be addressed include: Why do neural network models work so well in high dimension? Why does their performance depend so sensitively on the choice of the hyperparameters? Can we develop more robust and equally accurate new machine learning models and algorithms? In this article, we review some of the major progresses made in these directions.

MATHEMATICS SUBJECT CLASSIFICATION 2020

Primary 65; Secondary 68, 41, 60, 91, 93

KEYWORDS

Machine learning, scientific computing, curse of dimensionality, approximation theory, neural networks, error analysis, continuous formulation, integral differential equations

INTERNATIONAL CONGRESS OF MATHEMATICIANS 2022 JULY 6–14

Proc. Int. Cong. Math. 2022, Vol. 2, pp. 914–954
DOI 10.4171/ICM2022/155

1. INTRODUCTION

Supervised learning. We begin with the simplest task in machine learning (ML), supervised learning. The goal is to approximate an unknown target function from a finite training dataset. Denote by $f^* : X = [0, 1]^d \rightarrow \mathbb{R}$ the target function. Let $S = \{(\boldsymbol{x}_j, y_j = f^*(\boldsymbol{x}_j)), j \in [n] = \{1, 2, \ldots, n\}\}$ be the available dataset. Our objective is to approximate f^* as accurately as we can. This usually means that we would like to minimize the *population risk* in a given function class:

$$\mathcal{R}(f) = \mathbb{E}\big(f(\boldsymbol{x}) - f^*(\boldsymbol{x})\big)^2 = \int_X \big(f(\boldsymbol{x}) - f^*(\boldsymbol{x})\big)^2 d\mu,$$

where μ is a given probability distribution on X.

A typical supervised learning algorithm consists of the following three major components:

- Defining a *hypothesis space*. This is a set of functions that we use to approximate f^*. It is the analog of the finite element trial function space, except that in modern ML, we typically use neural network functions as the trial functions. We will use $\mathcal{H}_m$ to denote the hypothesis space where m is roughly the dimension of $\mathcal{H}_m$. We denote the functions in $\mathcal{H}_m$ generically as $f(\cdot, \theta)$ and we use θ to parametrize the functions in $\mathcal{H}_m$.

- Setting up an optimization problem for finding the optimal parameters. Though we are interested in minimizing the population risk, in practice we have to work with the *empirical risk* (or its variants):

  $$\mathcal{R}_n(\theta) = \frac{1}{n}\sum_j \big(f(\boldsymbol{x}_j, \theta) - y_j\big)^2 = \frac{1}{n}\sum_j \big(f(\boldsymbol{x}_j, \theta) - f^*(\boldsymbol{x}_j)\big)^2$$

  or, more generally,

  $$\mathcal{R}_n(\theta) = \frac{1}{n}\sum_j \ell_j(\theta),$$

  where the term ℓ_j is the loss for the jth data point. Regularization terms are sometimes added to this expression. The population and empirical risks are more commonly referred to as the training and testing errors, respectively.
  The difference between the true objective, the population risk, and the objective function that we work with in practice, the empirical risk, is an important issue that differentiates the optimization problems in ML from those in other settings.

- Solving this optimization problem. The simplest idea is to use the gradient descent algorithm (GD),

  $$\theta_{k+1} = \theta_k - \eta \nabla \mathcal{R}_n(\theta_k) = \theta_k - \eta \frac{1}{n} \sum_j \nabla \ell_j(\theta_k),$$

  where η is called the learning rate. Since the full gradient is an average over all training samples and is costly to evaluate, in practice, one often randomly selects one term in that average and uses it instead of the full gradient. This leads to the stochastic gradient descent algorithm (SGD),

$$\theta_{k+1} = \theta_k - \eta \nabla \ell_{j_k}(\theta_k),$$

where $j_1, j_2, \ldots$ are i.i.d. random variables uniformly drawn from $\{1, 2, \ldots, n\}$. One of the main mysteries in ML is that SGD is not only more efficient than GD, it often leads to a smaller test error.

How do we choose the hypothesis space? In classical numerical algorithms, we choose polynomials, piecewise polynomials, wavelets, and the like. In linear regression, we choose functions of the form $f(\boldsymbol{x}) = \beta \cdot \boldsymbol{x} + \beta_0$, where β and β_0 are the parameters to be found. Neural network models are the most popular choice in modern ML. A simple neural network model takes the form $f(\boldsymbol{x}) = \sum_k a_k \sigma(\boldsymbol{w}_k \cdot \boldsymbol{x} + c_k)$, where σ is some scalar nonlinear function, called the activation function. Popular choices of σ include $\sigma(x) = \max(x, 0)$, the ReLU (rectified linear units) function, and $\sigma(x) = (1 + e^{-x})^{-1}$, the sigmoid function. This is called a two-layer neural network model since there are two affine transformations (represented by the parameters $\{a_k\}$ and $\{\boldsymbol{w}_k\}$, respectively) involved. As is usually the case in ML, we have neglected the constant terms in the affine transformations. To include them, one can think of $\boldsymbol{x}$ as being $(\boldsymbol{x}^T, 1)^T$ and change the dimensionality accordingly. We will adopt this convention throughout this report. Multilayer neural network models, or deep neural networks (DNN), are formed by compositions of functions of the form above:

$$f(\boldsymbol{x}, \theta) = \boldsymbol{W}_L \sigma \circ \big(\boldsymbol{W}_{L-1} \sigma \circ \big(\cdots \sigma \circ (\boldsymbol{W}_0 \boldsymbol{x})\big)\big), \quad \theta = (\boldsymbol{W}_0, \boldsymbol{W}_1, \ldots, \boldsymbol{W}_L).$$

Here the $\boldsymbol{W}$'s are vectors or matrices, "$\circ$" means that the scalar function is applied to each component of the vector. In practice, it has been found that training such networks is often quite hard when L is large due to the *exploding or vanishing gradient problem* [50]: The gradient with respect to the parameters either grows or diminishes fast as L, the number of layers or the depth, increases. This problem is very much alleviated if one switches to a residual form:

$$\begin{aligned} \boldsymbol{z}_0(\boldsymbol{x}) &= \boldsymbol{V}\boldsymbol{x}, \\ \boldsymbol{z}_{l+1}(\boldsymbol{x}) &= \boldsymbol{z}_l(\boldsymbol{x}) + \boldsymbol{U}_l \sigma \circ \big(\boldsymbol{W}_l \boldsymbol{z}_l(\boldsymbol{x})\big), \quad l = 0, 1, \ldots, L-1, \end{aligned} \tag{1.1}$$

and $f(\boldsymbol{x}, \theta) = \alpha \cdot \boldsymbol{z}_L(\boldsymbol{x})$ for some vector α. This is the *residual neural network* model, or the *ResNet* model [48]. The issue of exploding or vanishing gradients has been analyzed for DNNs in [47], but we still lack a rigorous mathematical analysis for ResNets.

In addition to supervised learning, there are two other major subjects in classical machine learning:

- unsupervised learning, which is mainly concerned with finding some aspects of an underlying probability distribution using a finite sample;
- reinforcement learning, which is about finding the optimal strategy for a *Markov decision process* [93].

Deep neural network-based ML is commonly referred to as *deep learning* [62,85].

Deep learning is a very powerful tool. In the last ten years or so, deep learning has achieved tremendous success for a wide variety of problems. The most representative example is in computer vision, e.g., the classification of images. Typically, the images are labeled into several different categories according to the content of each image. Our task is to predict the correct category for images of the same kind. This is a supervised learning problem where the target function is the mapping from each image to its content, i.e., the category of that image.[1]

Another example is generating extremely real-looking pictures of fake human faces.[2] Using pictures of real human faces as samples, generative ML models can produce new samples which are pictures of fake human faces. This is an example of unsupervised learning. We can view pictures of human faces as being a random variable in the spaces of images. The probability distribution of that random variable is unknown to us. But we do know some samples of that probability distribution, namely the pictures of real human faces. From that sample, one can approximate the underlying probability distribution sufficiently accurately that one can produce new samples. These new samples are the pictures of fake human faces.

The best known example of reinforcement learning is AlphaGo [90]. Given the strategy of the opponent, the Go game can be formulated as a Markov decision process whose optimal strategy satisfies the underlying Bellman equation. What AlphaGo did was to solve that Bellman equation approximately for an increasingly better opponent.

Approximating functions, probability distributions, and solutions of difference or differential equations are among the most common tasks in computational mathematics. One is naturally led to ask: What is different in the tasks described above from those that are commonly done in mathematics? One most important difference is the dimensionality of the problems. Take the CIFAR-10 dataset as an example. We can view each image as being a point in a $d = 32 \times 32 \times 3 = 3072$-dimensional space, counting the number of pixels and the dimensionality of the color space. Classical algorithms in computational mathematics are not able to handle problems in such high dimension.

The curse of dimensionality. To see this more clearly, let us take a look at a typical result in classical approximation theory, the approximation by piecewise linear functions over a regular mesh. Let h be the typical size of the mesh. Then we have

$$\inf_{f \in \mathcal{H}_m} \| f^* - f \|_{L^2(X)} \leq C_d h^2 \| f^* \|_{H^2(X)} \sim C_d m^{-2/d} \| f^* \|_{H^2(X)},$$

where $\| f^* \|_{H^2(X)}$ is the Sobolev norm of f^*. If we want to reduce the error by a factor of 10, we need to reduce h by a factor of $\sqrt{10}$ and increase m by a factor of $10^{d/2}$. For $d = 3072$, this is truly a huge number.

1 See, for example, https://www.cs.toronto.edu/~kriz/cifar.html.

2 See, for example, https://machinelearningmastery.com/resources-for-getting-started-with-generative-adversarial-networks/.

The problem described here is referred to as the *curse of dimensionality* (CoD): As dimensionality grows, computational cost grows exponentially. This phenomenon is common to all classical algorithms, such as algorithms based on fixed meshes and wavelets.

CoD has been a major obstacle for many problems in science and engineering, including quantum and classical many-body problems, dynamic programming and control problems, and nonparametric statistics. Before deep learning, many approximate algorithms and models have been developed to bypass the CoD problems. The most well-known ones include the Hartree and Hartree–Fock approximation in quantum mechanics, the generalized linear models in statistics, and approximate dynamic programming models. Although these models are heavily used in practice, we lack systematic ways to improve their accuracy. It is fair to say that deep learning seems to be the first general methodology that is capable of handling a large class of such problems with satisfactory accuracy.

2. DEEP LEARNING-BASED ALGORITHMS FOR PROBLEMS IN SCIENTIFIC COMPUTING

Deep learning has been very successful for many high-dimensional problems in computer vision and natural language processing [62]. It is natural to ask whether it can be used to solve high-dimensional problems in other areas such as scientific computing and computational science. This has indeed been a very active research area since 2016. Below we briefly review some representative progresses in this direction.

2.1. Control problems

The first successful application of deep learning to problems in scientific computing was presented in [43] for stochastic control problems. Consider the stochastic dynamic model

$$z_{l+1} = z_l + \mathbf{g}_l(z_l, \boldsymbol{u}_l) + \xi_l, \tag{2.1}$$

where z_l, $\boldsymbol{u}_l$, ξ_l denotes the state of the system, the control, and the noise at step l, respectively. Our objective is

$$\min_{\{\boldsymbol{u}_l\}_{l=0}^{T-1}} \mathbb{E}_{\{\xi_l\}} \left\{ \sum_{l=0}^{T-1} c_l\big(z_l, \boldsymbol{u}_l(z_l)\big) + c_T(z_T) \right\}. \tag{2.2}$$

We are interested in looking for the feedback control (or closed-loop control)

$$\boldsymbol{u}_l = \boldsymbol{u}_l(z),$$

and we will approximate this function by some neural network model (the details of the network model is not important for this discussion)

$$\boldsymbol{u}_l(z) \approx \tilde{\boldsymbol{u}}_l(z|\theta_l), \quad l = 0, \dots, T-1.$$

With this approximation, the optimization problem becomes

$$\min_{\{\theta_l\}_{l=0}^{T-1}} \mathbb{E}_{\{\xi_l\}} \left\{ \sum_{l=0}^{T-1} c_l\big(z_l, \tilde{\boldsymbol{u}}_l(z_l|\theta_l)\big) + c_T(z_T) \right\}. \tag{2.3}$$

This was the first example on developing deep learning-based algorithms for problems in scientific computing. The motivation for using this as the first example was the close similarity between stochastic control problems and ResNet-based deep learning: the dynamic model (2.1) in the control problem plays the role of the ResNet, the objective function (2.2) plays the role of the empirical risk and the random noise in (2.1) plays the role of the training data. Using this analogy, Han and E developed an SGD and neural network-based algorithm for the stochastic control problem and demonstrated that it can readily handle very high dimensional problems [43]. The neural network model used was a composite network, with the control at each step represented by a subnetwork.

Subsequently, there have been many developments on deep learning-based algorithms for control problems. For a survey of the activities in this area, we refer to the UCLA IPAM workshop in the Spring of 2020. We mention in particular the extension to deterministic control problems in [77]. These developments have demonstrated adequately the potential of deep learning-based algorithms for solving real world control problems. Yet there are still serious work to be done to fully realize that potential in practice. There are two main obstacles. The first is that we often lack reliable dynamic models for the practical problems we are interested in. The second is the robustness of the deep learning-based algorithms in realistic settings.

2.2. High-dimensional partial differential equations

Motivated by the success for control problems, E, Han, and Jentzen developed deep learning-based algorithms for nonlinear parabolic partial differential equations (PDEs). The idea is to use backward stochastic differential equations (BSDEs) to reformulate the nonlinear PDE as a control-like problem, and then follow similar strategies for stochastic control problems [26, 44].

Consider the initial value problem

$$\frac{\partial v}{\partial t} = \frac{1}{2}\sigma\sigma^T : \nabla^2 v + \mu \cdot \nabla v + f(\sigma^{\mathrm{T}}\nabla v), \quad v(0, \boldsymbol{x}) = g(\boldsymbol{x}).$$

It is better to turn this into a terminal value problem by reversing the direction of time. Let $u(t, \cdot) = v(T - t, \cdot)$. Then the problem above becomes

$$\frac{\partial u}{\partial t} + \frac{1}{2}\sigma\sigma^T : \nabla^2 u + \mu \cdot \nabla u + f(\sigma^{\mathrm{T}}\nabla u) = 0, \quad u(T, \boldsymbol{x}) = g(\boldsymbol{x}).$$

One can reformulate this as a stochastic optimization problem using BSDEs [78]:

$$\inf_{Y_0, \{Z_t\}_{0\le t\le T}} \mathbb{E}\left|g(X_T) - Y_T\right|^2, \tag{2.4}$$

$$\text{such that} \quad X_t = X_0 + \int_0^t \mu(s, X_s)\, ds + \int_0^t \sigma(s, X_s)\, dW_s, \tag{2.5}$$

$$Y_t = Y_0 - \int_0^t f(Z_s)\, ds + \int_0^t (Z_s)^{\mathrm{T}}\, dW_s. \tag{2.6}$$

It can be shown that both problems have unique solutions and these solutions are related to each other by [79]

$$Y_t = u(t, X_t) \quad \text{and} \quad Z_t = \sigma^T(t, X_t)\, \nabla u(t, X_t). \tag{2.7}$$

Problem (2.4) is very much like a stochastic control problem and one can then develop algorithms using ideas similar to those described above. The resulted algorithm, the *Deep BSDE method*, has turned out to be an elegant and powerful tool for solving (nonlinear) Black–Scholes equations in finance, Hamilton–Jacobi–Bellman equations, as well as BSDEs. See [27] for a review.

In the Deep BSDE method, much effort has gone into the reformulation of the PDE problem as a control-like problem, in order to explore the intrinsic structure of the underlying problem. In the opposite direction, [82,91] developed strategies that are "foolproof." The idea is to use least squares and formulate the PDE and boundary condition as an optimization problem, and then more or less blindly apply ML to that optimization problem [82,91]. This has become quite popular in applied mathematics since it offers applied mathematicians a way to gain experience in deep learning by playing with the problems they are familiar with.

2.3. Parametrizing solutions of differential equations

Another idea is to explore the representative power of deep neural network models and parametrize solutions of PDEs as a functional of the coefficients and boundary data. This was first demonstrated by Khoo, Lu, and Ying for the Schrödinger equation with random potential [59]. For a more systematic development along this direction, we refer to [66].

In contrast to most other applications in which the object of interest is a function (though maybe in high dimension), in this setting, the object of interest is an operator on an infinite-dimensional space. This raises new mathematical issues beyond those discussed below.

2.4. Molecular dynamics

In molecular dynamics, we model the dynamic trajectory of each atom in a material or a molecule by solving the Newton's equation

$$m_i \frac{d^2 \boldsymbol{x}_i}{dt^2} = -\nabla_{\boldsymbol{x}_i} V, \quad V = V(\boldsymbol{x}_1, \boldsymbol{x}_2, \ldots, \boldsymbol{x}_i, \ldots, \boldsymbol{x}_N),$$

where m_i, $\boldsymbol{x}_i$ are the mass and position of the ith atom, respectively. The key question is how to model the potential energy function V that describes the interaction between the atoms. Traditionally, this has been modeled either empirically or by solving quantum mechanics-based models, such as density functional theory, on the fly computing the forces between the atoms [12,22]. Neither is satisfactory: the empirical approach is unreliable; the on-the-fly quantum mechanics-based approach is expensive and limited to systems with only hundreds or thousands of atoms.

With the advent of ML, we can contemplate a new paradigm in which quantum mechanics models are used to provide data, from which one can learn a highly accurate potential energy function, which can then be used to perform molecular dynamics. Such a paradigm was first proposed in [9]. One of the most successful examples of such a model is the Deep Potential models developed in [45,110] (see Figure 1). Using high performance computing resources, one can perform molecular dynamics calculation with *ab initio* accuracy for systems with hundreds of millions atoms [54].

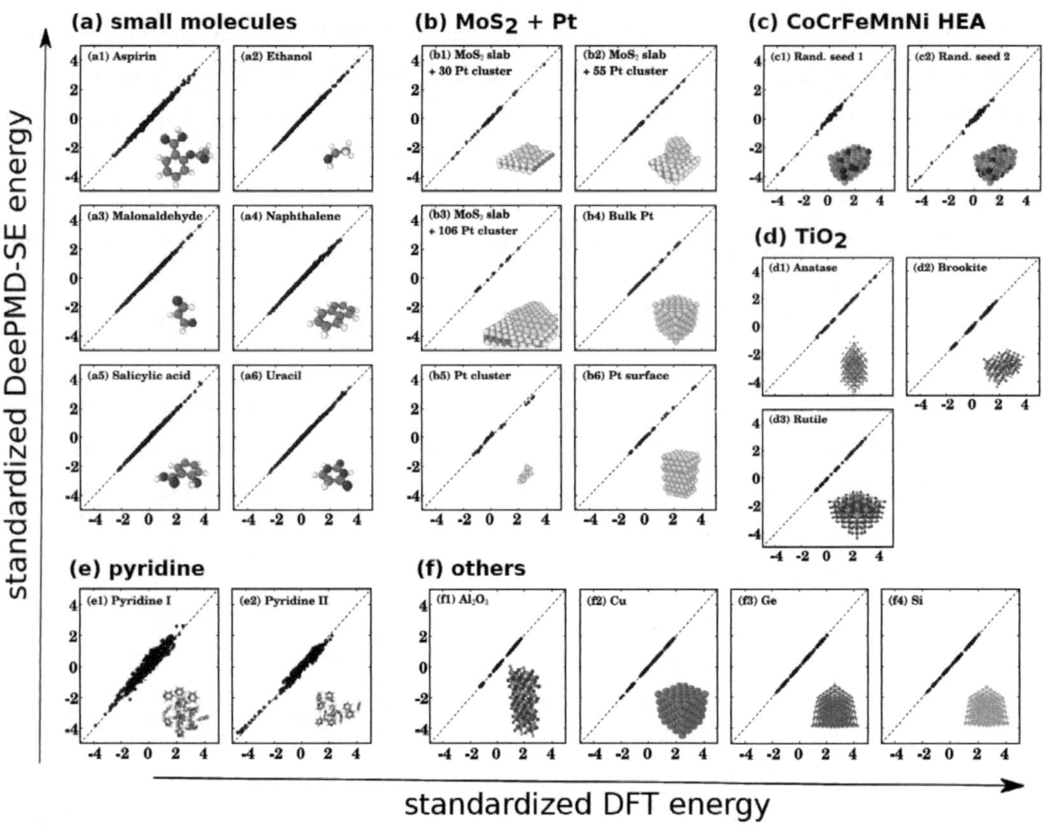

FIGURE 1

Comparison of the accuracy of the energies predicted by the Deep Potential model and density functional theory for different kinds of systems [110].

With the Deep Potential model, one can do many things that were either impossible or very difficult before. Examples include complex reaction processes in combustion [109], crystal nucleation of liquid silicon [10], liquid–liquid phase transition of water [38], one-dimensional cooperative diffusion in a three-dimensional crystal [98], structural order in quasicrystal growth [42], and the phase diagram of water [112].

2.5. Multiscale modeling

It has long been recognized that multiscale modeling can be a very effective tool in computational science and engineering (see Figure 2). However, its practical usage has been hampered by our inadequate ability to analyze the data obtained from the underlying microscopic model [22]. This is exactly where ML can help. Indeed ML-based *ab initio* molecular dynamics is an example of the application of ML to multiscale modeling. Besides molecular dynamics, ML-based multiscale models have been developed for density functional theory, coarse-grained molecular dynamics, moment closure models for the kinetic equations, hydrodynamic models for non-Newtonian fluids, etc. There is no doubt that this will continue to be a very fruitful line of research.

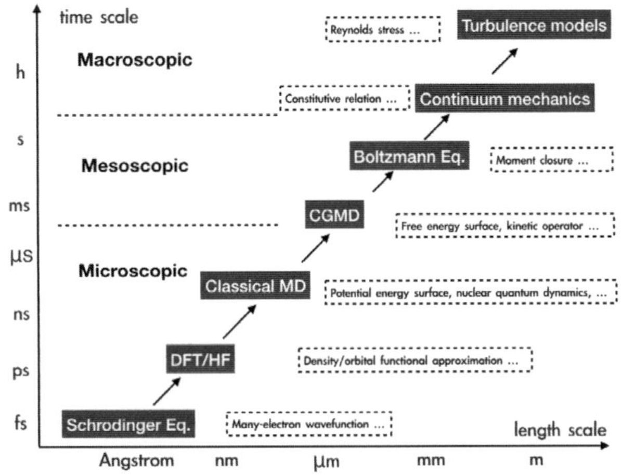

FIGURE 2
The hierarchy of multiscale physical models.

Most if not all existing applications of ML to multiscale modeling belong to the class of *sequential multiscale modeling* [22], i.e., ML algorithms are used at the pre-computing stage to obtain accurate coarse-grained models. One is naturally led to ask: How can we develop reliable and interpretable new physical models using ML? There are three most important issues involved here [28]:

- The first is how to collect the training data. The training dataset needs to be representative enough of all the practical situations that the model is intended for, yet at the same time, it needs to be as small as possible since each data point usually involves solving the microscale model. For this purpose, Zhang et al. developed the ELT (exploration–labeling–training) algorithm and it has been successfully applied to molecular dynamics and coarse-grained molecular dynamics [28,111,113].

- The second issue is the starting point of the new physical model. To be interpretable, it usually helps to formulate the new physical model as some kind of projection of the underlying microscale model. An example is the moment-closure model for the kinetic equation. The projection scheme, or the coarse-grained model, should not violate the physical conservation laws in the problem.
To formulate this projection scheme, one needs to know the set of coarse-grained variables. In principle, ML can also be a very powerful tool for this purpose. In practice, this is still a relatively unexplored area.

- The projected model is usually not closed and involves terms that need to be modeled. These terms are analogous to the constitutive relations in classical models such as the Navier–Stokes equation. This situation is very similar to that of the heterogeneous multiscale method [22,25]. The third issue is therefore to formulate

ML models for the unclosed terms in the coarse-grained model. To do so, one has to take into account physical constraints such as the symmetries in the system.

For a discussion of these issues, we refer to [28].

2.6. The many-electron Schrödinger equation

The many-electron Schrödinger equation in quantum mechanics is a notoriously hard problem not only due to its high dimensionality but also the fact that its solution must satisfy the Pauli exclusion principle, i.e., the wave-function must be antisymmetric. It is also arguably the most fundamental problem in computational science since it represents the true first principle. This latter feature is becoming increasingly more clear due to the advance of ML-based algorithms and the Schrödinger equation is the ultimate provider of the data that we use to train more coarse-grained models, particularly density functional theory models.

For the spin Schrödinger equation, Carleo and Troyer developed an algorithm using the restricted Boltzmann machine and the least squares formulation [13]. Deep learning-based algorithm for the many-electron Schrödinger equation was first developed in [46]. More sophisticated ansatz for the antisymmetric part of the wave-function was developed in [49,80]. A spectral projection algorithm was proposed in [102] to fully take advantage of the linear character of the Schrödinger equation. It is fair to say that at this stage, deep learning-based algorithm still remains an experimental effort and has not outperformed traditional quantum chemistry methods.

2.7. Purely data-driven methods

The most remarkable example of the purely data-driven method is AlphaFold2 [58]. By using only the structures in the protein data bank and protein sequence data, AlphaFold2 is able to predict the native structure of proteins to experimental accuracy. This was quite unexpected and has changed the way things are done in structural biology.

As a structural optimization problem, protein folding can be considered as a (classical) many-body problem. This is NOT how AlphaFold2 solved the problem. AlphaFold2 did not try to find the native structure by exploring the high-dimensional configuration space of the protein of some energy function. Instead, it took an interpolation viewpoint: Given the structures we know in the protein data bank, try to find the unknown structures by exploring the similarity between the given protein sequence and the sequences in the protein data bank. For this purpose, one needs to explore the structure of the sequence space. This is done by pushing multiple sequence alignment to a limit [58].

3. MATHEMATICAL THEORY OF NEURAL NETWORK-BASED MACHINE LEARNING MODELS

At this point, the objective of a mathematical theory for deep learning is not to explain in detail everything we see in practice, but rather to formulate general principles that can help organize our thoughts and guide future work.

The two most important puzzles in deep learning are:

- Why do deep learning models work so well on such seemingly very complicated tasks?
- Why does the performance of deep learning models depend so sensitively on the choice of the hyper-parameters, such as the network size, architecture, and the learning rate in the optimization algorithm?

A more advanced question is whether we can come up with new formulations of ML models that are both accurate and robust.

There are many different ways of looking at these issues, ranging from classical learning theory [94], statistical physics perspective [108], to information theory perspective [1]. We will take the viewpoint of classical numerical analysis (approximation theory, convergence of training algorithms, convergence rates, etc.) but put emphasis on the feature of high dimensionality. For a review along this line of thoughts, we refer to [34].

Before proceeding further, let us note that we will use the terminology "norm" in a loose way, in the sense that the triangle inequality is not necessarily satisfied.

3.1. An overview of approximation theory

Approximation theory is concerned with the question whether a given hypothesis space can efficiently approximate the target functions we are interested in. In this direction, there are three kinds of results.

The first is the so-called *Universal Approximation Theorem (UAT)*, which, roughly speaking, asserts that under mild conditions, one can use neural network functions to approximate arbitrary continuous functions uniformly on compact domains [18]. Such results are of course important, without them the whole foundation of neural network models would be in doubt, but they do not explain why neural network models are so much better than classical polynomial approximations. After all, as we know from the Weierstrass theorem, UAT also holds for polynomial approximations, which we know is a bad idea in high dimension. To see the difference between the two kinds of approximations, we must study the rate of convergence.

The second kind of results are convergence rates of neural network approximations for functions with certain regularity conditions. A typical result states that if a function has derivatives of order up to k, then it can be approximated by neural networks with an error of $O(m^{-k/d})$ where m is the total number of parameters. The first systematic result of this type can be found in [107]. The most recent and sharpest results can be found in [70]. These results do suffer from CoD. But they are useful for analyzing neural network-based algorithms for low dimensional problems.

The third kind of results are convergence rates for neural network approximations that do not suffer from CoD. This line of research began with the pioneering work of Barron [7,8,11,57]. We will focus on this type of results.

3.2. General remarks about high-dimensional problems

Before continuing, let us recap the important parameters that we have: m is the dimensionality of the hypothesis space; n is the size of the training sample; d is the dimensionality of the input variable to the ML model. We are interested in the case when $d \gg 1$.

The one high-dimensional problem that has been very well studied is high dimensional numerical integration. We are interested in approximating the following integral:

$$I(g) = \int_X g(\boldsymbol{x}) d\boldsymbol{x}$$

by a sum $I_m(g) = \frac{1}{m} \sum_j g(\boldsymbol{x}_j)$. If we use grid-based quadrature rules such as the Trapezoidal Rule, then the error behaves like

$$I(g) - I_m(g) \sim \frac{C(g)}{m^{\alpha/d}}$$

for some fixed constant α, indicating CoD. If instead we use Monte Carlo integration, say by taking $\{\boldsymbol{x}_j, j \in [m]\}$ to be independent, uniformly distributed in X, then we have

$$\mathbb{E}\big(I(g) - I_m(g)\big)^2 = \frac{\operatorname{var}(g)}{m}, \quad \operatorname{var}(g) = \int_X g^2(\boldsymbol{x}) d\boldsymbol{x} - \left(\int_X g(\boldsymbol{x}) d\boldsymbol{x}\right)^2.$$

The $O(1/\sqrt{m})$ rate is (almost) the best we can hope for, and is independent of d: Improvements on the convergence rate, say using quasi-Monte Carlo or other lattices, diminish quickly as d becomes large [20].

The variance $\operatorname{var}(g)$ can be very large in high dimension. For this reason many variance-reduction algorithms have been developed. These ideas allow physicists to study statistical physical models in very high dimension.

Function approximation is a harder problem than numerical integration. In light of the discussion above, the best we can hope for function approximation in high dimension are results of the following type:

$$\inf_{f \in \mathcal{H}_m} \mathcal{R}(f) = \inf_{f \in \mathcal{H}_m} \|f - f^*\|^2_{L^2(d\mu)} \lesssim \frac{\Gamma(f^*)}{m}.$$

The questions that we need we address are: Can this be true? Given a neural network model, say two-layer neural networks or ResNets, for what class of functions is this true? If true, what should the quantity $\Gamma(f^*)$ be?

3.3. Approximation theory for the random feature model

To explain the general philosophy, we will use the random feature model [81] as an illustration. Let $\phi(\cdot; \boldsymbol{w})$ denote some feature function parametrized by $\boldsymbol{w}$, e.g., $\phi(\boldsymbol{x}, \boldsymbol{w}) = \sigma(\boldsymbol{w}^T \boldsymbol{x})$. A random feature model is defined by

$$f_m(\boldsymbol{x}; \boldsymbol{a}) = \frac{1}{m} \sum_{j=1}^{m} a_j \phi(\boldsymbol{x}; \boldsymbol{w}_j^0), \tag{3.1}$$

where $\{\boldsymbol{w}_j^0\}_{j=1}^m$ are i.i.d. samples drawn from a prefixed distribution π_0. Once drawn, $\{\boldsymbol{w}_j^0\}_{j=1}^m$ are fixed; $\boldsymbol{a} = (a_1, \ldots, a_m)^T \in \mathbb{R}^m$ are the trainable parameters. For simplicity, we assume $\Omega := \operatorname{supp}(\pi_0)$ is compact. Denote $\boldsymbol{W}^0 = (\boldsymbol{w}_1^0, \ldots, \boldsymbol{w}_m^0)^T \in \mathbb{R}^{m \times d}$. Note that random feature models are linear models.

If the inner parameters $\{\boldsymbol{w}_j^0\}$ are also allowed to change, then this becomes a (generalized) two-layer neural network model. For this reason, the random feature model can be considered as a simplified two-layer neural network model in which the inner parameters are frozen at some random initial value. This connection has proven to be important for understanding the two-layer neural network model.

We are interested in identifying the function class and the functional $\Gamma(\cdot)$ for this model. To this end, consider the reproducing kernel Hilbert space (RKHS) [2] induced by the kernel $k(\boldsymbol{x}, \boldsymbol{x}') = \mathbb{E}_{\boldsymbol{w} \sim \pi_0}[\phi(\boldsymbol{x}; \boldsymbol{w})\phi(\boldsymbol{x}'; \boldsymbol{w})]$. Denote by $\mathcal{H}_k$ this RKHS. Then for any $f \in \mathcal{H}_k$, there exists $a(\cdot) \in L^2(\pi_0)$ such that

$$f(\boldsymbol{x}) = \int a(\boldsymbol{w})\phi(\boldsymbol{x}; \boldsymbol{w})d\pi_0(\boldsymbol{w}) \tag{3.2}$$

and

$$\|f\|_{\mathcal{H}_k}^2 = \int a^2(\boldsymbol{w})d\pi_0(\boldsymbol{w}). \tag{3.3}$$

Theorem 1 (Direct Approximation Theorem). *For any $f^* \in \mathcal{H}_k$, let*

$$f^*(\boldsymbol{x}) = \int a^*(\boldsymbol{w})\phi(\boldsymbol{x}; \boldsymbol{w})d\pi_0(\boldsymbol{w}). \tag{3.4}$$

Then we have

$$\mathbb{E}_{\boldsymbol{W}^0}\| f_m(\cdot; a^*(\boldsymbol{W}^0)) - f^*\|_{L^2}^2 \leq \frac{\|f^*\|_{\mathcal{H}_k}^2}{m},$$

where $a^*(\boldsymbol{W}^0) = (a^*(\boldsymbol{w}_1^0), \ldots, a^*(\boldsymbol{w}_m^0))^T$.

Theorem 2 (Inverse Approximation Theorem). *Let $(\boldsymbol{w}_j^0)_{j=0}^\infty$ be a realization of the sequence of i.i.d. random samples drawn from π_0. Let f^* be a continuous function on $X = [0, 1]^d$. Assume that there exists a constant C and a sequence $(a_j)_{j=0}^\infty$ satisfying $\sup_j |a_j| \leq C$, such that*

$$\lim_{m\to\infty} \frac{1}{m}\sum_{j=1}^m a_j\phi(\boldsymbol{x}; \boldsymbol{w}_j^0) = f^*(\boldsymbol{x}), \tag{3.5}$$

for all $\boldsymbol{x} \in X$. Then with probability 1, there exists a function $a^*(\cdot) : \Omega \mapsto \mathbb{R}$ such that

$$f^*(\boldsymbol{x}) = \int_\Omega a^*(\boldsymbol{w})\phi(\boldsymbol{x}; \boldsymbol{w})d\pi_0(\boldsymbol{w}).$$

Moreover, $\|a^*\|_\infty \leq C$.

This pair of direct and inverse theorems are not exactly converses of each other since different norms (L^2 and L^∞) are used. But they do tell us that the associated RKHS is the appropriate function space to study in connection with the random feature model. These results are not new, but it seems difficult to identify the origin of these results.

3.4. Approximation theory for two-layer neural networks

We will restrict our attention to the case when ReLU is used as the activation function. The hypothesis space for two-layer neural networks is defined by

$$\mathcal{H}_m = \left\{ f_m(\boldsymbol{x}) = \frac{1}{m}\sum_j a_j \sigma\big(\boldsymbol{w}_j^T \boldsymbol{x}\big) \right\}$$

A good candidate for the associated function space for this model is the *Barron space* [30,33] (see also [6,7,35,60]). To define the Barron space, consider functions $f : X = [0,1]^d \mapsto \mathbb{R}$ of the following form:

$$f(\boldsymbol{x}) = \int_\Omega a\sigma(\boldsymbol{w}^T\boldsymbol{x})\rho(da, d\boldsymbol{w}) = \mathbb{E}_\rho\big[a\sigma(\boldsymbol{w}^T\boldsymbol{x})\big], \quad \boldsymbol{x} \in X,$$

where $\Omega = \mathbb{R}^1 \times \mathbb{R}^{d+1}$ and ρ is a probability distribution on Ω. The "Barron norm" is defined by

$$\|f\|_{\mathcal{B}_p} = \inf_{\rho \in P_f} \big(\mathbb{E}_\rho\big[a^p \|\boldsymbol{w}\|_1^p\big]\big)^{1/p},$$

where $P_f := \{\rho : f(\boldsymbol{x}) = \mathbb{E}_\rho[a\sigma(\boldsymbol{w}^T\boldsymbol{x})]\}$. Let $\mathcal{B}_p = \{f \in C^0 : \|f\|_{\mathcal{B}_p} < \infty\}$. Functions in $\mathcal{B}_p$ are called Barron functions. As was shown in [33], we actually have $\|\cdot\|_{\mathcal{B}_p} = \|\cdot\|_{\mathcal{B}_q}$ for any $1 \le p \le q \le \infty$. Hence, we will use $\|\cdot\|_{\mathcal{B}}$ and $\mathcal{B}$ denote the Barron norm and Barron space.

One immediate question is: What kinds of function are Barron functions? In this direction, a general result is given by:

Theorem 3 ([60]). *Let $\gamma_2(f) := \int_{\mathbb{R}^d} \|\omega\|_1^2 |\tilde{f}(\omega)| d\omega < \infty$, where $\tilde{f}$ is the Fourier transform of f, then f can be represented as*

$$f(\boldsymbol{x}) = \int_\Omega a\sigma(\boldsymbol{w}^T\boldsymbol{x})\rho(da, d\boldsymbol{w}).$$

Moreover, $\|f\|_{\mathcal{B}} \le 2\gamma_2(f) + 2\|\nabla f(0)\|_1 + 2|f(0)|$.

Remark 4. One should note the difference between the Barron norm and the quantities like γ_2 which were originally introduced by Barron [7]. The Barron norm is defined using a probabilistic setting. The quantity $\gamma_2(f)$ and the like are defined using the Fourier transform which is related to the regularity property of f. We believe that the probabilistic setup is the right direction to go and this is partly confirmed by subsequent results on continuous ResNets (see below) and multilayer neural networks. To avoid further confusion, we propose to call quantities γ_2 and the like *Barron's spectral norm*. For further results in this direction, as well as some interesting analysis on the relationship between these spaces, we refer to [88,89].

An interesting structural theorem about Barron functions is proved in [35].

Theorem 5. *Let f be in Barron space. Then $f = \sum_{i=1}^\infty f_i$ where $f_i \in C^1(X \setminus V_i)$ where V_i is a k-dimensional affine subspace of X for some $0 \le k \le d-1$.*

As an immediate corollary, we see that the distance to the unit sphere is not a Barron function.

The claim that Barron space is the natural space associated with two-layer networks is justified by the following series of results [30].

Theorem 6 (Direct Approximation Theorem). *For any* $f \in \mathcal{B}$ *and* $m \in \mathbb{N}^+$, *there exists a two-layer neural network* f_m *with* m *neurons* $\{(a_i, \boldsymbol{w}_i)\}$ *such that*

$$\|f - f_m\|_{L^2(P)} \lesssim \frac{\|f\|_{\mathcal{B}}}{\sqrt{m}}.$$

Theorem 7 (Inverse Approximation Theorem). *Let*

$$\mathcal{N}_C \stackrel{\text{def}}{=} \left\{ \frac{1}{m} \sum_{k=1}^{m} a_k \sigma(\boldsymbol{w}_k^T \boldsymbol{x}) : \frac{1}{m} \sum_{k=1}^{m} |a_k| \|\boldsymbol{w}_k\|_1 \leq C,\ m \in \mathbb{N}^+ \right\}.$$

Let f^* *be a continuous function. Assume there exist a constant* C *and a sequence of functions* $\{f_m\} \subset \mathcal{N}_C$ *such that*

$$f_m(\boldsymbol{x}) \to f^*(\boldsymbol{x})$$

for all $\boldsymbol{x} \in X$, *then there exists a probability distribution* ρ^* *on* Ω *such that*

$$f^*(\boldsymbol{x}) = \int a\sigma(\boldsymbol{w}^T \boldsymbol{x}) \rho^*(da, d\boldsymbol{w}),$$

for all $\boldsymbol{x} \in X$. *Moreover,* $\|f^*\|_{\mathcal{B}} \leq C$.

3.5. Approximation theory for residual neural networks

Consider a residual network model

$$\begin{aligned}
z_0(\boldsymbol{x}) &= \boldsymbol{V}\boldsymbol{x}, \\
z_{l+1}(\boldsymbol{x}) &= z_l(\boldsymbol{x}) + \frac{1}{L} \boldsymbol{U}_l \sigma \circ \big(\boldsymbol{W}_l z_l(\boldsymbol{x})\big), \quad l = 0, 1, \ldots, L-1, \\
f(\boldsymbol{x}, \theta) &= \alpha \cdot z_L(\boldsymbol{x}),
\end{aligned}$$

where $\boldsymbol{x} \in \mathbb{R}^d$ is the input, $\boldsymbol{V} \in \mathbb{R}^{D \times d}$, $D \geq d$, $\boldsymbol{W}_l \in \mathbb{R}^{m \times D}$, $\boldsymbol{U}_l \in \mathbb{R}^{D \times m}$, $\alpha \in \mathbb{R}^D$ and we use $\Theta := \{\boldsymbol{V}, \boldsymbol{U}_1, \ldots, \boldsymbol{U}_L, \boldsymbol{W}_l, \ldots, \boldsymbol{W}_L, \alpha\}$ to denote all the parameters to be learned from data. Without loss of generality, we will fix $\boldsymbol{V}$ to be

$$\boldsymbol{V} = \begin{bmatrix} \boldsymbol{I}_{d \times d} \\ \boldsymbol{0}_{(D-d) \times d} \end{bmatrix}. \tag{3.6}$$

To look for the appropriate associated function space, let us consider the following flow-based representation of functions (see next section):

$$\begin{aligned}
z(\boldsymbol{x}, 0) &= \boldsymbol{V}\boldsymbol{x}, \\
\dot{z}(\boldsymbol{x}, t) &= \mathbb{E}_{(\boldsymbol{U}, \boldsymbol{W}) \sim \rho_t} \boldsymbol{U} \sigma\big(\boldsymbol{W} z(\boldsymbol{x}, t)\big), \\
f_{\alpha, \{\rho_t\}}(\boldsymbol{x}) &= \alpha^T z(\boldsymbol{x}, 1).
\end{aligned}$$

For $p \geq 1$, consider the following linear ODEs associated with the representation above:

$$\begin{aligned}
N_p(0) &= \mathbf{e}, \\
\dot{N}_p(t) &= \left(\mathbb{E}_{\rho_t} (|\boldsymbol{U}||\boldsymbol{W}|)^p\right)^{1/p} N_p(t),
\end{aligned}$$

where $\boldsymbol{e}$ is a column vector with every component equal to 1, $|\mathbf{A}|$ and $\mathbf{A}^q$ are elementwise operations for the matrix $\mathbf{A}$ and a positive number q. The following function spaces and "norms" were introduced in [33].

Definition 8. Let f be a function that satisfies $f = f_{\alpha,\{\rho_t\}}$ for a pair of $(\alpha, \{\rho_t\})$. Define

$$\|f\|_{\mathcal{D}_p(\alpha,\{\rho_t\})} = |\alpha|^T N_p(1) \tag{3.7}$$

to be the $\mathcal{D}_p$ norm of f with respect to the pair $(\alpha, \{\rho_t\})$, where $|\alpha|$ is a vector obtained from α by taking elementwise absolute values. We define

$$\|f\|_{\mathcal{D}_p} = \inf_{f=f_{\alpha,\{\rho_t\}}} |\alpha|^T N_p(1) \tag{3.8}$$

to be the $\mathcal{D}_p$ norm of f, and let $\mathcal{D}_p = \{f : \|f\|_{\mathcal{D}_p} < \infty\}$.

Definition 9. Let f be a function that satisfies $f = f_{\alpha,\{\rho_t\}}$ for a pair of $(\alpha, \{\rho_t\})$. Define

$$\|f\|_{\tilde{\mathcal{D}}_p(\alpha,\{\rho_t\})} = |\alpha|^T N_p(1) + \left\|N_p(1)\right\|_1 - D \tag{3.9}$$

to be the $\tilde{\mathcal{D}}_p$ norm of f with respect to the pair $(\alpha, \{\rho_t\})$. We define

$$\|f\|_{\tilde{\mathcal{D}}_p} = \inf_{f=f_{\alpha,\{\rho_t\}}} |\alpha|^T N_p(1) + \left\|N_p(1)\right\|_1 - D \tag{3.10}$$

to be the $\tilde{\mathcal{D}}_p$ norm of f. The space $\tilde{\mathcal{D}}_p$ is defined as the set of functions that admit the representation $f_{\alpha,\{\rho_t\}}$ with finite $\tilde{\mathcal{D}}_p$ norm.

These two kinds of "norms" appear to be similar but different. These function spaces were introduced in [33] and are named *flow-induced function spaces*.

For the approximation theorems, we will make use of the following "Lipschitz continuity" condition for $\{\rho_t\}$.

Definition 10. Given a family of probability distributions $\{\rho_t,\ t \in [0,1]\}$, the "Lipschitz coefficient" of $\{\rho_t\}$, denoted by $\mathrm{Lip}_{\{\rho_t\}}$, is defined as the infimum of all the numbers that satisfy

$$\left|\mathbb{E}_{\rho_t}\boldsymbol{U}\sigma(\boldsymbol{W}\boldsymbol{z}) - \mathbb{E}_{\rho_s}\boldsymbol{U}\sigma(\boldsymbol{W}\boldsymbol{z})\right| \le \mathrm{Lip}_{\{\rho_t\}}|t-s||\boldsymbol{z}| \tag{3.11}$$

and

$$\left|\left\|\mathbb{E}_{\rho_t}|\boldsymbol{U}||\boldsymbol{W}|\right\|_{1,1} - \left\|\mathbb{E}_{\rho_s}|\boldsymbol{U}||\boldsymbol{W}|\right\|_{1,1}\right| \le \mathrm{Lip}_{\{\rho_t\}}|t-s|, \tag{3.12}$$

for any $t, s \in [0,1]$, where $\|\cdot\|_{1,1}$ is the sum of the absolute values of all the entries in a matrix. The "Lipschitz norm" of $\{\rho_t\}$ is defined as

$$\left\|\{\rho_t\}\right\|_{\mathrm{Lip}} = \left\|\mathbb{E}_{\rho_0}|\boldsymbol{U}||\boldsymbol{W}|\right\|_{1,1} + \mathrm{Lip}_{\{\rho_t\}}. \tag{3.13}$$

Finally, we define a discrete "path norm" for residual networks.

Definition 11. For a residual network defined by (3.6) with parameters $\Theta = \{\alpha, \boldsymbol{U}_l, \boldsymbol{W}_l, l = 0, 1, \ldots, L-1\}$, we define the l_1 path norm of Θ to be

$$\|\Theta\|_{\mathcal{P}} = |\alpha|^T \prod_{l=1}^{L}\left(I + \frac{1}{L}|\boldsymbol{U}_l||\boldsymbol{W}_l|\right)\boldsymbol{e}. \tag{3.14}$$

With the definitions above, we are ready to state the direct and inverse approximation theorems in the flow-induced function spaces [33].

Theorem 12 (Direct Approximation Theorem). *Let $f \in \tilde{\mathcal{D}}_2$, $\delta \in (0,1)$. Assume there exists a constant l_0 such that, for any $\varepsilon > 0$, there exists $(\alpha, \{\rho_t\})$ that satisfies $f = f_{\alpha,\{\rho_t\}}$ and $\|f\|_{\tilde{\mathcal{D}}_1(\alpha,\{\rho_t\})} < \|f\|_{\tilde{\mathcal{D}}_1} + \varepsilon$, $\|\{\rho_t\}\|_{\mathrm{Lip}} \le l_0$. Then there exists an L_0, depending polynomially on D, m, l_0, and $\|f\|_{\tilde{\mathcal{D}}_1}$, such that for any $L \ge L_0$, there exists an L-layer residual network $f_L(\cdot;\Theta)$ that satisfies*

$$\left\|f - f_L(\cdot;\Theta)\right\|^2 \le \frac{3\|f\|^2_{\tilde{\mathcal{D}}_1}}{L^{1-\delta}} \tag{3.15}$$

and

$$\|\Theta\|_{\mathcal{P}} \le 9\|f\|_{\tilde{\mathcal{D}}_1}. \tag{3.16}$$

Theorem 13 (Inverse Approximation Theorem). *Let f^* be a function defined on $X = [0,1]^d$. Assume that there exists a sequence of residual networks $\{f_L(\cdot;\Theta_L)\}_{L=1}^{\infty}$ such that $\|f^*(\boldsymbol{x}) - f_L(\boldsymbol{x};\Theta)\| \to 0$ as $L \to \infty$ for all $\boldsymbol{x} \in X$. Assume further that the parameters in $\{f_L(\cdot;\Theta)\}_{L=1}^{\infty}$ are (entrywise) bounded by c_0. Then, we have $f^* \in \mathcal{D}_\infty$ and $\|f^*\|_{\mathcal{D}_\infty} \le \frac{2e^{m(c_0^2+1)}D^2 c_0}{m}$. Moreover, if there exists a constant c_1 such that $\|f_L\|_{\mathcal{D}_1} \le c_1$ holds for any $L > 0$, then we have $\|f^*\|_{\mathcal{D}_1} \le c_1$.*

A natural question is how big the flow-induced norms are compared with the Barron norm. In this direction, we have [33]

Theorem 14. *For any function $f \in \mathcal{B}$, and $D \ge d + 2$, $m \ge 1$, we have $f \in \tilde{\mathcal{D}}_1$ and*

$$\|f\|_{\tilde{\mathcal{D}}_1} \le 2\|f\|_{\mathcal{B}}. \tag{3.17}$$

In this sense, going from two-layer neural networks to ResNets is like variance reduction in Monte Carlo methods.

3.6. The generalization gap

The second main issue in theoretical machine learning is the difference between training and test accuracy, in other words, the difference between the empirical and population risk. Estimating the difference between these two quantities is complicated by the fact that the parameters obtained from the training process are highly correlated with the data. There are many ways to bypass this difficulty. The simplest idea is to use the trivial bound

$$\left|\mathcal{R}(\hat{f}) - \mathcal{R}_n(\hat{f})\right| \le \sup_{f\in\mathcal{H}_m} \left|\mathcal{R}(f) - \mathcal{R}_n(f)\right| = \sup_{f\in\mathcal{H}_m} \left|I(g) - I_n(g)\right|, \tag{3.18}$$

where $\hat{f} \in \mathcal{H}_m$ is the function in the hypothesis space obtained from training, $g = (f - f^*)^2$. One of the most effective ways of estimating the right-hand side is to use the notion of Rademacher complexity.

Definition 15. Let $\mathcal{H}$ be a set of functions, and $S = (\boldsymbol{x}_1, \boldsymbol{x}_2, \ldots, \boldsymbol{x}_n)$ be a set of data points. The Rademacher complexity of $\mathcal{H}$ with respect to S is defined as

$$\mathrm{Rad}_S(\mathcal{H}) = \frac{1}{n}\mathbb{E}_\xi\left[\sup_{h\in\mathcal{H}}\sum_{i=1}^{n}\xi_i h(\boldsymbol{x}_i)\right], \tag{3.19}$$

where $\{\xi_i\}_{i=1}^n$ are i.i.d. random variables taking values ± 1 with equal probability.

Rademacher complexity is useful since it bounds the quantity of interest, $\sup_{h\in\mathcal{H}} |I(h) - I_n(h)|$, from above and below.

Theorem 16 ([86, THEOREM 26.5]). *For any $\delta \in (0, 1)$, with probability at least $1-\delta$ over the random samples $S = (\boldsymbol{x}_1, \ldots, \boldsymbol{x}_n)$, we have*

$$\sup_{h\in\mathcal{H}}\left|\mathbb{E}_{\boldsymbol{x}}\left[h(\boldsymbol{x})\right] - \frac{1}{n}\sum_{i=1}^{n}h(\boldsymbol{x}_i)\right| \le 2\,\mathrm{Rad}_S(\mathcal{H}) + \sup_{h\in\mathcal{H}}\|h\|_\infty\sqrt{\frac{\log(2/\delta)}{2n}},$$

$$\sup_{h\in\mathcal{H}}\left|\mathbb{E}_{\boldsymbol{x}}\left[h(\boldsymbol{x})\right] - \frac{1}{n}\sum_{i=1}^{n}h(\boldsymbol{x}_i)\right| \ge \frac{1}{2}\,\mathrm{Rad}_S(\mathcal{H}) - \sup_{h\in\mathcal{H}}\|h\|_\infty\sqrt{\frac{\log(2/\delta)}{2n}}.$$

Roughly speaking, Rademacher complexity quantifies the degree to which functions in the hypothesis space can approximate random noise on a given dataset. The larger the hypothesis space, the larger the Rademacher complexity.

As example, if $\mathcal{H}$ is the unit ball in the space of continuous functions, then we obviously have $\mathrm{Rad}_S(\mathcal{H}) = O(1)$. If if $\mathcal{H}$ is the unit ball in the space of Lipschitz continuous functions, then it can be shown that [96]

$$\mathrm{Rad}_S(\mathcal{H}) = O(n^{-1/d}).$$

This signals another potential source of CoD, namely that the training sample size needed grows exponentially as d grows.

Fortunately, for the function spaces we identified earlier, their Rademacher complexity has roughly the optimal scaling. For Barron space, we have

Theorem 17 ([6]). *Let $\mathcal{F}_Q = \{f \in \mathcal{B}, \|f\|_{\mathcal{B}} \le Q\}$ and let $S = (\boldsymbol{x}_1, \ldots, \boldsymbol{x}_n)$. Then we have*

$$\mathrm{Rad}_S(\mathcal{F}_Q) \le 2Q\sqrt{\frac{2\ln(2d)}{n}}.$$

The $n^{-1/2}$ scaling at the right-hand side is consistent with the Monte Carlo scaling that one would expect at a first sight.

The Rademacher complexity estimate is only established for a family of modified flow-induced function norms $\|\cdot\|_{\hat{\mathcal{D}}_p}$ (note the factor 2 in the definition below). It is not clear at this stage whether this is only a technical issue.

Let

$$\|f\|_{\hat{\mathcal{D}}_p} = \inf_{f=f_{\alpha,\{\rho_t\}}} |\alpha|^T \hat{N}_p(1) + \left\|\hat{N}_p(1)\right\|_1 - D + \left\|\{\rho_t\}\right\|_{\mathrm{Lip}}, \tag{3.20}$$

where $\hat{N}_p(t)$ is given by

$$\hat{N}_p(0) = 2\boldsymbol{e},$$
$$\dot{\hat{N}}_p(t) = 2\big(\mathbb{E}_{\rho_t}\big(|\boldsymbol{U}||\boldsymbol{W}|\big)^p\big)^{1/p}\hat{N}_p(t).$$

Denote by $\hat{\mathcal{D}}_p$ the space of functions with finite $\hat{\mathcal{D}}_p$ norm. Then, we have

Theorem 18 ([33]). *Let* $\hat{\mathcal{D}}_p^Q = \{f \in \hat{\mathcal{D}}_p : \|f\|_{\hat{\mathcal{D}}_p} \le Q\}$ *and let* $S = (\boldsymbol{x}_1, \dots, \boldsymbol{x}_n)$*. Then we have*

$$\mathrm{Rad}_S\big(\hat{\mathcal{D}}_2^Q\big) \le 18Q\sqrt{\frac{2\log(2d)}{n}}. \tag{3.21}$$

3.7. A priori estimates of the population risk for regularized models

Our objective is to show that one can find accurate approximations of the target function using a finite training sample. Ideally, we would like to have the following kind of results:

$$\mathcal{R}(\hat{f}) \lesssim \frac{\Gamma_1(f^*)}{m} + \frac{\Gamma_2(f^*)}{\sqrt{n}}. \tag{3.22}$$

For appropriately regularized models, results of this kind have been established for the random feature model, the two-layer neural network model, and ResNets.

For the random feature model, consider the regularized model

$$\mathcal{L}_{n,\lambda}(\boldsymbol{a}) = \mathcal{R}_n(\boldsymbol{a}) + \frac{\lambda}{\sqrt{n}}\frac{\|\boldsymbol{a}\|}{\sqrt{m}},$$

and define the regularized estimator

$$\hat{\boldsymbol{a}}_{n,\lambda} = \operatorname{argmin} \mathcal{L}_{n,\lambda}(\boldsymbol{a}).$$

Theorem 19. *Fix any* $\lambda > 0$*. For any* $\delta \in (0, 1)$*, with probability* $1 - \delta$*, we have*

$$\mathcal{R}(\hat{\boldsymbol{a}}_{n,\lambda}) \le \frac{1}{m}\left(\log(n/\delta)\|f^*\|_{\mathcal{H}_k}^2 + \frac{\log^2(n/\delta)}{m}\|a^*\|_\infty^2\right) + \frac{1}{\sqrt{n}}\left(\|f^*\|_{\mathcal{H}_k} + \left(\frac{\log(1/\delta)}{m}\right)^{1/4}\|a^*\|_\infty + \sqrt{\log(2/\delta)}\right). \tag{3.23}$$

Such results should be standard. But a complete proof seems to be only found in [34].

In the same way, for the two-layer neural network model, one can consider the regularized model

$$\mathcal{L}_n(\theta) = \mathcal{R}_n(\theta) + \lambda\sqrt{\frac{\log(2d)}{n}}\|\theta\|_{\mathcal{P}}, \quad \hat{\theta}_{n,\lambda} = \operatorname{argmin} \mathcal{L}_n(\theta),$$

where the path norm is defined by

$$\|\theta\|_{\mathcal{P}} = \frac{1}{m}\sum_{j=1}^{m} |a_j|\|\boldsymbol{w}_j\|_1,$$

and let $\hat{\theta}_n = \operatorname{argmin} \mathcal{L}_n(\theta)$.

Theorem 20 ([30]). *Assume $f^*: X \mapsto [0,1] \in \mathcal{B}$. There exists an absolute constant C_0 such that if $\lambda \geq C_0$ then for any $\delta > 0$, with probability at least $1-\delta$ over the choice of the training set, we have*

$$\mathcal{R}(\hat{\theta}_n) \lesssim \frac{\|f^*\|_{\mathcal{B}}^2}{m} + \lambda \|f^*\|_{\mathcal{B}} \sqrt{\frac{\log(2d)}{n}} + \sqrt{\frac{\log(n/\delta)}{n}}.$$

For ResNets, instead of the path norm (3.14), we have to consider a weighted path norm

$$\|\Theta\|_{WP} = |\alpha|^T \prod_{l=1}^{L} \left(I + \frac{2}{L} |\boldsymbol{U}_l||\boldsymbol{W}_l| \right) \boldsymbol{e}, \tag{3.24}$$

which assigns larger weights to paths that pass through more nonlinearities. Consider the regularized model

$$\min_{\Theta} \mathcal{J}(\Theta) = \hat{\mathcal{R}}(\Theta) + 3\lambda \|\Theta\|_{WP} \sqrt{\frac{2\log(2d)}{n}}. \tag{3.25}$$

Theorem 21 ([29]). *Let $f^*: X \to [0,1]$. Assume that $\widehat{\Theta}$ is an optimal solution of the regularized model* (3.25). *Let $\lambda \geq 4 + 2/[3\sqrt{2\log(2d)}]$. Then for any $\delta \in (0,1)$, with probability at least $1-\delta$ over the random training samples, the population risk satisfies*

$$\mathcal{R}(\widehat{\Theta}) \leq \frac{3\|f\|_{\mathcal{B}}^2}{LD} + (4\|f\|_{\mathcal{B}} + 1)\frac{3(4+\lambda)\sqrt{2\log(2d)}+2}{\sqrt{n}} + 4\sqrt{\frac{2\log(14/\delta)}{n}}. \tag{3.26}$$

One unsatisfactory aspect of this result is that it is proved for a Barron function, not functions in the flow-induced space.

3.8. The loss function and the loss landscape

The a priori estimates for the regularized models establish the existence of accurate approximations to the target function in the hypothesis space. The next question is how to find them. At this point, there is a vast amount of experience suggesting that one can find accurate solutions using simple gradient-based algorithms, without any explicit regularization, but the result may depend sensitively on the choice of the hyperparameters, such as the network parameters, the initialization of the training algorithms, the learning rate, etc. Sensitive dependence on the network parameters suggests that the landscape of the loss function changes qualitatively as these parameters change.

At a first sight, it is quite surprising that simple gradient-based algorithms such as the gradient descent can work at all. After all, the loss function, say the empirical risk, is a nonconvex function of many variables with potentially very complicated landscape. In the case of molecular structural optimization such as protein folding, gradient descent would get stuck very quickly at a bad local minima. In the case of training neural network models, one can often avoid this by tuning the hyperparameters in the training algorithm. Obviously, this means that the landscape of the molecular structural optimization and the landscape for training neural network models are qualitatively very different. Therefore one first issue might be to understand how the landscape looks. In this direction, one important result is that of Cooper who considered overparametrized neural networks with a smooth activation function and characterized the structure of the set of global minima [17]. Cooper proved that the locus of the global minima is generically (i.e., possibly after an arbitrarily small change to the data set) a smooth $(m-n)$-dimensional submanifold of $\mathbb{R}^m$ where m is the number of free parameters in the neural network model and n is the training data size.

3.9. Training dynamics

Two-layer neural networks with mean-field scaling. "Mean-field" is a notion in statistical physics that describes a particular form of the interaction between particles. In the mean-field situation, particles interact with each other only through a mean-field formed through the collective effort of all the particles. The most elegant mean-field picture in machine learning is found in the case of two-layer neural networks: If one views the neurons as interacting particles, then these particles only interact with each other through the function represented by the neural network, which is the mean-field in this case. This observation was first made in [15,75,84,92]. By taking the hydrodynamic limit for the gradient flow of finite neuron systems, these authors obtained a continuous integral differential equation that describes the evolution of the probability measure for the weights associated with the neurons.

Given the two-layer neural network model

$$f_m(\boldsymbol{x}) = \frac{1}{m}\sum_j a_j\sigma\left(\boldsymbol{w}_j^T\boldsymbol{x}\right),$$

let

$$I(\boldsymbol{u}_1,\ldots,\boldsymbol{u}_m) = \mathcal{R}(f_m), \quad \boldsymbol{u}_j = (a_j, \boldsymbol{w}_j).$$

Consider the gradient descent dynamics

$$\frac{d\boldsymbol{u}_j}{dt} = -m\nabla_{\boldsymbol{u}_j} I(\boldsymbol{u}_1,\ldots,\boldsymbol{u}_m), \quad \boldsymbol{u}_j(0) = \boldsymbol{u}_j^0, \quad j \in [m]. \tag{3.27}$$

Lemma 22. *Let*

$$\rho(d\boldsymbol{u}, t) = \frac{1}{m}\sum_j \delta_{\boldsymbol{u}_j(t)}.$$

Then the gradient descent dynamics (3.27) *can be expressed equivalently as*

$$\partial_t\rho = \nabla(\rho\nabla V), \quad V = \frac{\delta\mathcal{R}_n}{\delta\rho}. \tag{3.28}$$

Equation (3.28) is the mean-field equation that describes the evolution of the probability distribution for the weights associated with each neuron. The lemma above simply states that (3.28) is satisfied for the finite neuron system without the need to take the infinite particle limit.

It is well known that (3.28) is the gradient flow of $\mathcal{R}$ under the Wasserstein metric. This brings the hope that the mathematical tools developed in the theory of optimal transport can be brought to bear for the analysis of (3.28) [95]. In particular, we would like to use these tools to study the qualitative behavior of the solutions of (3.28) as $t \to \infty$. Unfortunately, straightforward application of the results from optimal transport theory requires that the risk functional be displacement convex [74], a property that rarely holds in ML. As a result, less than expected has been achieved using the optimal transport theory.

The one important result, due originally to Chizat and Bach [15], is the following. We will state the result for the population risk. Again we consider the ReLU activation function.

Theorem 23 ([15, 16, 99]). *Let $\{\rho_t\}$ be a solution of the Wasserstein gradient flow such that*

- *ρ_0 is a probability distribution on the cone $\Theta := \{|a|^2 \leq |\boldsymbol{w}|^2\}$.*
- *Every open cone in Θ has positive measure with respect to ρ_0.*

Then the following are equivalent:

- *The velocity potentials $\frac{\delta \mathcal{R}}{\delta \rho}(\rho_t, \cdot)$ converge to a unique limit as $t \to \infty$.*
- *$\mathcal{R}(\rho_t)$ decays to the global infimum value as $t \to \infty$.*

If either condition is met, the unique limit of $\mathcal{R}(\rho_t)$ is zero. If ρ_t also converges in the Wasserstein metric, then the limit ρ_∞ is a minimizer.

A few remarks are in order:

- There are further technical conditions for the theorem to hold.
- Convergence of subsequences of $\frac{\delta \mathcal{R}}{\delta \rho}(\rho_t, \cdot)$ is guaranteed by compactness.
- The first assumption on ρ_0 is a smoothness assumption needed for the existence of the gradient flow.
- The second assumption on ρ_0 is called *omnidirectionality*. It ensures that ρ can shift mass in any direction which reduces risk. The requirement that the support of the initial distribution be sufficiently large seems to be confirmed by practical experience.

Two-layer neural networks with conventional scaling. In practice, people often use the scaling (instead of the mean-field scaling)

$$f_m(\boldsymbol{x}; \boldsymbol{a}, \boldsymbol{W}) = \sum_{j=1}^{m} a_j \sigma(\boldsymbol{w}_j^T \boldsymbol{x}) = \boldsymbol{a}^T \sigma(\boldsymbol{W}\boldsymbol{x}).$$

A popular initialization [48, 63] is as follows:

$$a_j(0) \sim \mathcal{N}(0, \beta^2), \quad \boldsymbol{w}_j(0) \sim \mathcal{N}(0, I/d),$$

where $\beta = 0$ or $1/\sqrt{m}$. We define the Gram matrix $K = (K_{ij}) \in \mathbb{R}^{n \times n}$ as

$$K_{i,j} = \frac{1}{n} \mathbb{E}_{\boldsymbol{w} \sim \pi_0} \big[\sigma(\boldsymbol{w}^T \boldsymbol{x}_i) \sigma(\boldsymbol{w}^T \boldsymbol{x}_j)\big].$$

In this case, a lot is known in the so-called *highly overparametrized regime*. In this part, for simplicity, we will assume that the domain of interest is the unit ball S^{d-1} instead of the unit cube.

There is both good and bad news. The good news is that one can prove exponential convergence to global minima of the empirical risk.

Theorem 24 ([21]). *Let $\lambda_n = \lambda_{\min}(K)$ and assume $\beta = 0$. For any $\delta \in (0, 1)$, assume that $m \gtrsim n^2 \lambda_n^{-4} \delta^{-1} \ln(n^2 \delta^{-1})$. Then with probability at least $1 - 6\delta$, we have*

$$\mathcal{R}_n\big(\boldsymbol{a}(t), \boldsymbol{W}(t)\big) \le e^{-m\lambda_n t} \mathcal{R}_n\big(\boldsymbol{a}(0), \boldsymbol{W}(0)\big). \tag{3.29}$$

Now the bad news: the generalization property of the converged solution is no better than that of the associated random feature model, defined by freezing $\{\boldsymbol{w}_j\} = \{\boldsymbol{w}_j(0)\}$ and only training $\{a_i\}$.

The first piece of insight that the underlying dynamics in this regime is effectively linear is given in [19]. Jacot et al. [53] termed the effective kernel the "*neural tangent kernel*" and this terminology has got a lot of popularity. Later it was proved rigorously that in this regime, the entire gradient descent path for the two-layer neural network model is uniformly close to that of the associated random feature model [3, 31].

Theorem 25 ([31]). *Let $\boldsymbol{W}_0 = \boldsymbol{W}(0)$. Denote by $f_m(\cdot; \tilde{\boldsymbol{a}}, \boldsymbol{W}_0)$ the solution of the gradient descent dynamics for the random feature model. Under the same setting as in Theorem 24, we have*

$$\sup_{\boldsymbol{x} \in S^{d-1}} \big| f_m\big(\boldsymbol{x}; \boldsymbol{a}(t), \boldsymbol{W}(t)\big) - f_m\big(\boldsymbol{x}; \tilde{\boldsymbol{a}}(t), \boldsymbol{W}_0\big)\big| \lesssim \frac{(1 + \sqrt{\ln(1/\delta)})^2 \lambda_n^{-1}}{\sqrt{m}}. \tag{3.30}$$

This can also be seen from the (m, n) hyperparameter space. Shown in Figure 3 are the heat maps of the test errors under the conventional and mean-field scaling, respectively. We see that the test error changes smoothly as m changes for the mean-field scaling. In contrast, there is a clear "phase transition" in the heat map for the conventional scaling where we see the coexistence of a good (darker region) phase with small test error and a bad (lighter

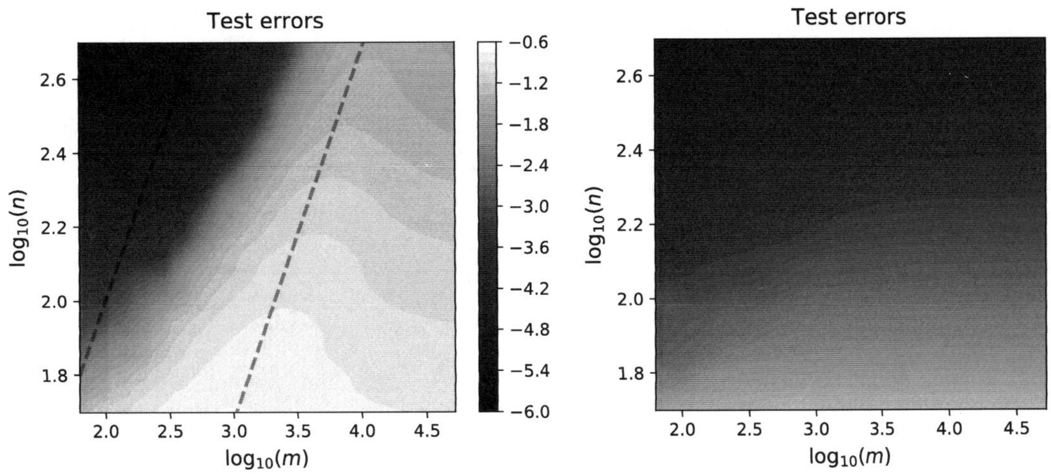

FIGURE 3
How the network width affects the test error of gradient descent solutions. The test errors are given in logarithmic scale. These experiments are conducted for the single-neuron target function with $d = 20$ and learning rate $\eta = 0.0005$. The two dashed lines correspond to $m = n/(d+1)$ (left) and $m = n$ (right), respectively. (Left) Conventional scaling. (Right) Mean-field scaling. For more details, see [72].

region) phase where the test error is much larger. This means that, in practice, one has to tune the network parameters so that they fall into the good phase. For the details of this study, we refer to [72].

From this, it is natural to speculate that the sensitive dependence of the performance on the hyperparameters is a consequence of this kinds of phase transition. For two layer neural networks, the phase diagram in the hyperparameter space is relatively simple. For more complicated neural network models, the phase diagram should be more complicated and tuning the parameters becomes a much harder task.

Hardness of training. In high dimensions, a "dynamic curse of dimensionality" may affect gradient descent training if the target function is not in Barron space.

Theorem 26 ([100]). *There exists f^* with Lipschitz constant and L^∞-norm bounded by 1 such that the parameter measures $\{\rho_t\}$ defined by the 2-Wasserstein gradient flow of either $\mathcal{R}_n$ or $\mathcal{R}$ satisfy*

$$\limsup_{t\to\infty}\left[t^\gamma\, \mathcal{R}(\rho_t)\right] = \infty$$

for all $\gamma > \frac{4}{d-2}$.

What makes matters worse is that even for functions in the Barron space, a dynamic CoD might also happen. Livni et al. [67] show that learning Barron functions is equivalent to solving some well-known hard problems in cryptography. This means that learning Barron functions is computationally as hard as breaking a cryptosystem. Such results are powerful but abstract. In the following, we provide an explicit understanding from the perspective of learning orthonormal classes, which is a reinterpretation of the results in [73, 87].

Consider a subset of the Barron space over $X = [0,1]^d$: $\mathcal{F} = \{f_{\boldsymbol{w}} = 2\sin(2\pi \boldsymbol{w}^T\cdot) : \sum_{i=1}^d w_i \le d, w_i \in \mathbb{N}_+\}$. Note that the following statements hold: (1) $|\mathcal{F}| \ge \exp(d)$; (2) $\langle f_{\boldsymbol{w}}, f_{\boldsymbol{w}'}\rangle = \delta_{\boldsymbol{w},\boldsymbol{w}'}$; (3) $\|f\|_{\mathcal{B}} \le Cd^2$, $\forall f \in \mathcal{F}$. Statements (1) and (2) are quite obvious and a proof can be found in [7]. Statement (3) directly follows from Theorem 3. Consider learning the function in $\mathcal{F}$ using the parametric model $h(\cdot;\theta)$ that includes, but is not limited to, the two-layer neural network model. Let $\mathcal{R}^f(\theta) = \mathbb{E}_{\boldsymbol{x}}[(h(\boldsymbol{x};\theta) - f(\boldsymbol{x}))^2]$. Notice that $\nabla_\theta \mathcal{R}^f(\theta) = 2\mathbb{E}_{\boldsymbol{x}}[(h(\boldsymbol{x};\theta) - f(\boldsymbol{x}))\nabla_\theta h(\boldsymbol{x};\theta)] = C(\theta) - 2\langle \nabla_\theta h(\cdot;\theta), f\rangle$. Let ν be the uniform distribution over $\mathcal{F}$. Then,

$$\begin{aligned} \operatorname{var}_{f\sim\nu}\big(\nabla \mathcal{R}^f(\theta)\big) &\le 4\mathbb{E}_{f\sim\nu}\langle \nabla_\theta h(\cdot;\theta), f\rangle^2 = \frac{4}{|\mathcal{F}|}\sum_{f\in\mathcal{F}}\langle \nabla_\theta h(\cdot;\theta), f\rangle^2 \\ &\le \frac{4\mathbb{E}_{\boldsymbol{x}}\|\nabla_\theta h(\boldsymbol{x};\theta)\|^2}{|\mathcal{F}|}, \end{aligned} \tag{3.31}$$

where the last inequality uses the fact that the functions in $\mathcal{F}$ are orthonormal.

Since $|\mathcal{F}| = \exp(d)$, the variance of the gradient with respect to different target function is exponentially small as d increases. This means that the gradient can barely distinguish different target functions. As a result, gradient-based optimizations algorithms are unlikely to succeed.

These hardness results suggest that the Barron space is likely to be too large for studying the training of two-layer neural networks. It is an important open problem to identify the right function space, in which the functions can be learned in polynomial time by two-layer neural networks.

3.10. Other results

Classification problems. A binary classification problem can be approached as a regression problem with the additional knowledge that the target function only takes the values ± 1. This a priori knowledge gives us different mathematical means, and we commonly choose to interpret $\{f > 0\}$ as $\{f \approx 1\}$ and similarly for negative values. It is therefore not necessary that f should take a particular value, but only that f have the correct sign (and possibly be bounded away from zero). This is encoded in the common hinge loss and logistic loss functions

$$\ell_{\text{hinge}}(h, y) = \max\{0, 1 - hy\}, \quad \ell_{\log}(h, y) = \frac{\log(1 + \exp(-hy))}{\log 2},$$

which primarily force alignment between the classifier h and the label $y \in \{-1, 1\}$. Both the hinge loss and the logistic loss differ from the ℓ^2-loss geometrically from the optimization perspective: While the ℓ^2-loss vanishes at exactly one point, hinge-loss vanishes whenever the classifier has the correct sign and magnitude ≥ 1, whereas the logistic loss never vanishes. The risk functional therefore has a much larger set of minimizers or none at all in typical classification problems.

Another key difference is the fact that we are minimizing a *surrogate loss*. While our goal is to minimize the measure of the misclassified set $\mathbb{E}_{(\boldsymbol{x},y)\sim\mu}[1_{\{f(\boldsymbol{x})\cdot y \leq 0\}}]$, we use convex loss functions ℓ which bound the zero–one loss function from above:

$$\ell(h, y) \geq 1_{\{h \cdot y \leq 0\}}.$$

The bounds on the true risk functional to minimize are therefore coarser by nature.

The nonexistence of minimizers for logistic loss has interesting implications from the optimization perspective. It was shown in [16] that as the risk decays to zero along a gradient flow trajectory, the geometry of a two-layer neural network adapts not only to correct classification, but also a higher-order optimality condition (maximum margin classification), where the "confidence" (or margin)

$$\min_{(\boldsymbol{x},y)\in \operatorname{spt}(\mu)} h(\boldsymbol{x}) \cdot y$$

becomes as large as possible. The notion of margin is easiest to interpret for linear classifiers, where it corresponds to the distance to the decision boundary, and harder to interpret in classes of nonlinear functions such as neural networks.

In multiclass classification, a similar philosophy holds, but the classifier has to align with the vectors $\boldsymbol{e}_1, \ldots, \boldsymbol{e}_k$ corresponding to the k classes, rather than the directions ± 1. The most popular loss functional in this case is the cross-entropy loss, which generalizes the logistic loss. Just as the logistic loss, the cross-entropy loss function does not admit minimizers either.

The frequency principle. One of the most interesting observations for training dynamics is the so-called "frequency principle": During training, the low frequency components in the target function tend to be recovered earlier than the high ones [103]. This is opposite to the situation we usually see in numerical analysis, for example, the numerical solution of elliptic PDEs. It is well known that when using iterative algorithms to solve the algebraic equations obtained from the numerical discretization of elliptic PDEs, it takes longer to remove the low frequency errors than the high frequency errors. This is why multi-grid methods are useful there. In ML, the opposite seems to be true.

The reason behind this is as follows. Roughly speaking, when solving PDEs, the high frequency components correspond roughly to large eigenvalues of the underlying algebraic system. As we have seen earlier, the training dynamics in machine learning is more like solving some integral equation, therefore the high frequency components correspond roughly to small eigenvalues. This should be an important avenue for understanding the training dynamics.

Generative models. Generative models are ways of approximating probability distributions using finite samples. One of the most well-known generative models is the generative adversarial network, or GAN. Given a sample set S, the empirical distribution formed from S, δ_S, can be considered as an approximation to the underlying probability distribution. This approximation is unsatisfactory since it cannot provide any new samples. However, it can be shown, or at least argued, that without any explicit regularization, generative models will always converge to the empirical distribution (see, for example, [39,104]). Therefore the merit of a generative model must be that during training, it can produce better approximations to the underlying probability distribution before ultimately converges to the empirical distribution. Theoretically, this means that one has to study the situation with "early stopping," not the ultimate convergence, and analyze whether the statement above really holds.

Results of this type have been proved for the so-called "bias potential" model [104]. However, it is fair to say that there are more open questions for the theoretical understanding of generative models than for the case of supervised learning. One of the difficulty is that generative models are not variational problems but rather game theory problems.

Machine learning of dynamical systems. Given a sample of time series, one would like to learn the underlying dynamical system that produced the time series. For these problems, besides CoD, there is also the issue of *curse of memory* [65]: The cost increases exponentially as memory increases. For linear dynamical systems, this issue has been analyzed quite thoroughly in [65].

Reinforcement learning. Reinforcement learning (RL) is an area of machine learning that is concerned with how an agent should interact with an unknown environment in order to maximize the expected cumulative reward [93]. To deal with practical problems that involve a large number of states or in high dimensions, one needs to introduce function approximation for the value or policy functions. Indeed, RL has had remarkable success in Atari games [76], Go [90], and robotics [61] using deep neural network approximations. Despite the practical success of RL with function approximation, most existing theoretical results is only applicable to the tabular setting [4, 5, 55], in which both the state and action spaces are finite and no function approximation is involved. Relatively simple function approximation methods, such as the linear model [56, 105], have been studied recently. RL with kernel function approximation has been studied in [37, 68, 69, 106], and RL with neural network approximation has been studied in [36, 69, 97]. These results require the existence of a reference distribution such that all possible state–action distributions under the admissible policies are close to in a certain sense, and [68] shows the necessity of this type of assumptions.

In a way, the mathematical issues for RL are a lot like those for supervised learning, but more complicated. Specifically,

- For the approximation error: Analyze the conditions on the reward function and transition probability under which the value or policy functions can be efficiently approximated by neural networks. Since the value and policy functions are obtained from the Bellman equation, this question is related to the new regularity theory for PDEs discussed below.
- For the generalization error: This is the sample complexity problem. It is similar in spirit to the Rademacher complexity except that there is an additional dynamic component.
- For the optimization error: Again the dynamic component complicates things. This is dynamics within dynamics: The optimization algorithm involves dynamics. Within that there is the dynamics of the underlying problem such as the dynamics of the Go game.

New regularity theory of PDEs. The approximation theory discussed earlier suggests that in high dimension, the classical smoothness-based function spaces such as Sobolev spaces should be replaced by new spaces such as the Barron space or Barron's spectral space. It is natural to ask whether the solutions of prototypical PDEs lies in these new spaces or whether one can develop a regularity theory for the relevant PDEs in these new spaces. This issue is of practical significance because of the success of ML-based algorithms for solving PDEs in high dimension (see Section 2). In this direction, [101] considered the simplest PDEs such as the Poisson equation, heat equation, and Hamilton–Jacobi equation, and studied whether the solutions lie in Barron space if the data does. Lu et al. [71] carried out a more thorough analysis and developed a regularity theory for Barron's spectral space. Also of relevance is the work in [40, 52].

4. MACHINE LEARNING FROM A CONTINUOUS VIEWPOINT

To define a machine learning model, one only needs two things: A way of representing functions and a way of finding the parameters in the representation. The latter is usually formulated as an optimization problem. Neural networks are special classes of functions. In contrast to piecewise polynomials, they can be defined without using a mesh. This "continuous" nature is quite helpful when constructing numerical algorithms. One is naturally led to ask: Can we push this continuous nature further?

One interesting idea proposed in [23,41] is to use the solutions of ordinary differential equations (ODEs) to represent trial functions. This proposal is now made popular through the name of *neural ODEs* [14]. ResNets can be viewed as discretizations of neural ODEs, and the back-propagation algorithm can be viewed as a particular way of solving the adjoint equation for the gradients of the solutions [14,23].

A more systematic presentation of the continuous formulation is found in [32]. The framework suggested there consists of the following main components:

- representation of functions;
- the variational problem for minimizing the population risk;
- gradient flow for the variational problem.

Since we are aiming at high-dimensional problems, the function representation should be of a probabilistic nature. E et al. [32] suggested two classes of representations, integral transform-based and flow-based.

Once we have a continuous formulation, one can then discretize the continuous formulation to obtain concrete algorithms. Again, since we are aiming at high dimension, the particle method is the most natural algorithm for discretizing the dynamic equations. We will see that some of the most popular neural network-based ML algorithms can be derived this way. At the same time, one can also come up with new algorithms.

4.1. Integral transform-based representation

Consider the (parametric) representation

$$f(\boldsymbol{x}; a, \pi) = \int_{\mathbb{R}^d} a(\boldsymbol{w})\sigma(\boldsymbol{w}^T\boldsymbol{x})\pi(d\boldsymbol{w}) = \mathbb{E}_{\boldsymbol{w}\sim\pi} a(\boldsymbol{w})\sigma(\boldsymbol{w}^T\boldsymbol{x}), \tag{4.1}$$

or

$$f(\boldsymbol{x}; \rho) = \mathbb{E}_{(a,\boldsymbol{w})\sim\rho} a\sigma(\boldsymbol{w}^T\boldsymbol{x}). \tag{4.2}$$

Given a target function f^*, the variational problem for minimizing the population risk is given by

$$\min_{\theta} \mathcal{R}(\theta), \quad \mathcal{R}(\theta) = \mathbb{E}_{\boldsymbol{x}\sim\mu}\big(f(\boldsymbol{x};\theta) - f^*(\boldsymbol{x})\big)^2,$$

where θ denotes abstract parameters. For (4.1), $\theta = (a, \pi)$; for (4.2), $\theta = \rho$. These two ingredients together form the variational problem we are interested in. We can either discretize this variational problem and then solve the resulted discretized optimization problem, or formulate an optimization problem at the continuous level and then discretize. We will discuss the latter approach.

To define the gradient flow, we borrow ideas from nonequilibrium statistical physics [51]. We regard the population risk as the free energy, and the parameter θ as the "order parameter". As in [51], we can divide the order parameters into two different classes, the conserved and nonconserved ones. For example, as a probability measure, π and ρ are conserved order parameters. In contrast, the coefficient a is nonconserved.

For nonconserved order parameters, we use the so-called "model A" gradient flow, e.g.,

$$\frac{\partial a}{\partial t} = -\frac{\delta \mathcal{R}}{\delta a}.$$

For conserved order parameters, we use "model B" gradient flow, e.g.,

$$\frac{\partial \pi}{\partial t} + \nabla \cdot \mathbf{J} = 0,$$

where

$$\mathbf{J} = \pi \boldsymbol{v}, \quad \boldsymbol{v} = -\nabla V, \quad V = \frac{\delta \mathcal{R}}{\delta \pi}.$$

It is instructive to look at some specific examples. For the first example, we use the representation (4.1). We fix π and optimize over a. The gradient flow in this case is given by

$$\partial_t a(\boldsymbol{w}, t) = -\frac{\delta \mathcal{R}}{\delta a}(\boldsymbol{w}, t) = -\int a(\tilde{\boldsymbol{w}}, t) K(\boldsymbol{w}, \tilde{\boldsymbol{w}}) \pi(d\tilde{\boldsymbol{w}}) + \tilde{f}(\boldsymbol{w}), \tag{4.3}$$

where

$$K(\boldsymbol{w}, \tilde{\boldsymbol{w}}) = \mathbb{E}_{\boldsymbol{x}}\big[\sigma(\boldsymbol{w}^T \boldsymbol{x})\sigma(\tilde{\boldsymbol{w}}^T \boldsymbol{x})\big], \quad \tilde{f}(\boldsymbol{w}) = \mathbb{E}_{\boldsymbol{x}}\big[f^*(\boldsymbol{x})\sigma(\boldsymbol{w}^T \boldsymbol{x})\big].$$

This is an integral equation with a symmetric positive definite kernel.

As an example of the conservative gradient flow, let us consider the representation

$$f(\boldsymbol{x}) = \mathbb{E}_{\boldsymbol{u} \sim \rho} \phi(\boldsymbol{x}, \boldsymbol{u}),$$

where ϕ is some general feature function. One example is $\boldsymbol{u} = (a, \boldsymbol{w})$, $\phi(\boldsymbol{x}, \boldsymbol{u}) = a\sigma(\boldsymbol{w}^T \boldsymbol{x})$. Let

$$V(\boldsymbol{u}) = \frac{\delta \mathcal{R}}{\delta \rho}(\boldsymbol{u}) = \mathbb{E}_{\boldsymbol{x}}\big[\big(f(\boldsymbol{x}) - f^*(\boldsymbol{x})\big)\phi(\boldsymbol{x}, \boldsymbol{u})\big] = \int \tilde{K}(\boldsymbol{u}, \tilde{\boldsymbol{u}})\rho(d\tilde{\boldsymbol{u}}) - \tilde{f}(\boldsymbol{u})$$

be the potential with a kernel $\tilde{K}$ defined similarly to K, then the model B gradient flow dynamics is given by

$$\partial_t \rho = \nabla(\rho \nabla V). \tag{4.4}$$

This is the same as the mean-field equation derived in [75, 84, 92]. For an interesting modification of this model to improve convergence, we refer to [83].

Next, we turn to the discretization of the continuous formulation. There are several levels of discretization to consider:

- Discretizing the variational problem. In the current setting, this is straightforward: By using data, we discretize the population risk into the empirical risk.
- Discretizing the function representation and the gradient flow. This can be done using a number of different numerical methods. In low dimensions, the spectral method can be a powerful tool. In high dimensions, the most obvious choice is the particle method since this is the dynamic version of Monte Carlo. One can

also use the smoothed particle method which has shown better performance at least on low dimensional problems.

As an example, consider (4.1). If we use

$$\pi(d\boldsymbol{w}) \sim \frac{1}{m}\sum_j \delta_{\boldsymbol{w}_j}, \quad a(\boldsymbol{w}_j, t) \sim a_j(t)$$

to discretize (4.3), we obtain

$$\frac{d}{dt}a_j(t) = -\frac{1}{m}\sum_k K(\boldsymbol{w}_j, \boldsymbol{w}_k)a_k(t) + \tilde{f}(\boldsymbol{w}_j).$$

This is exactly the gradient descent dynamics for the random feature model.

Next, consider the integral differential equation (4.4). If we use the particle method discretization, we have

$$\rho(da, d\boldsymbol{w}, t) \sim \frac{1}{m}\sum_j \delta_{(a_j(t), \boldsymbol{w}_j(t))} = \frac{1}{m}\sum_j \delta_{\boldsymbol{u}_j(t)}$$

and the discretized problem becomes

$$\frac{d\boldsymbol{u}_j}{dt} = -\nabla_{\boldsymbol{u}_j} I(\boldsymbol{u}_1, \ldots, \boldsymbol{u}_m),$$

where

$$I(\boldsymbol{u}_1, \ldots, \boldsymbol{u}_m) = \mathcal{R}(f_m), \quad \boldsymbol{u}_j = (a_j, \boldsymbol{w}_j), \quad f_m(\boldsymbol{x}) = \frac{1}{m}\sum_j a_j\sigma\big(\boldsymbol{w}_j^T \boldsymbol{x}\big).$$

This is exactly the gradient descent dynamics for two-layer neural networks under the mean-field scaling.

4.2. Flow-based representation

Consider the following flow-based representation:

$$\begin{aligned} z_0^{\boldsymbol{x}} &= V\boldsymbol{x}, \\ \frac{dz_\tau^{\boldsymbol{x}}}{d\tau} &= \mathbb{E}_{\boldsymbol{w}\sim\rho_\tau}\boldsymbol{\phi}(z, \boldsymbol{w}), \quad \forall \tau \in [0, 1], \\ f(\boldsymbol{x}; \theta) &= \mathbf{1}^T z_1^{\boldsymbol{x}}, \end{aligned} \tag{4.5}$$

where $\boldsymbol{w} \in \Omega$ and $\boldsymbol{\phi} : \mathbb{R}^D \times \Omega \mapsto \mathbb{R}^D$, V is a $D \times d$ matrix. For simplicity, we will fix V. The parameters are then $\theta = \rho = (\rho_\tau)_{\tau\in[0,1]}$, a sequence of probability measures.

The form of the right-hand side in (4.5) is chosen because of the following two considerations. The first is that it is an integral transformation-based representation that we just discussed. More importantly, it arises as the natural continuum limit of a suitable ResNet model with random parameters [33].

Minimizing the population risk using this flow-based representation is then a control problem where the population risk serves as the objective function, and the parameters $\rho = (\rho_\tau)_{\tau\in[0,1]}$ serve as the control. One useful tool from the control perspective is Pontryagin's maximum principle. To state this maximum principle, denote by $\Sigma := \{\rho : [0, 1] \mapsto \mathcal{P}_2(\Omega)\}$, the space of all feasible controls, and define the Hamiltonian $H : \mathbb{R}^d \times \mathbb{R}^d \times \mathcal{P}_2(\Omega) :\mapsto \mathbb{R}$ as

$$H(\boldsymbol{z}, \boldsymbol{p}, \mu) = \mathbb{E}_{\boldsymbol{u}\sim\mu}\big[\boldsymbol{p}^T \boldsymbol{\phi}(\boldsymbol{z}, \boldsymbol{u})\big].$$

Here $\boldsymbol{z}, \boldsymbol{p} \in \mathbb{R}^D$ and $\mu \in \mathcal{P}_2(\Omega)$. $\boldsymbol{p}$ is the costate that corresponds to the state $\boldsymbol{z}$.

Pontryagin's maximum principle states that the solutions of this control problem must satisfy

$$\rho_\tau^* = \operatorname{argmax}_\rho \mathbb{E}_{\boldsymbol{x}}\big[H\big(\boldsymbol{z}_\tau^{\boldsymbol{x}}, \boldsymbol{p}_\tau^{\boldsymbol{x}}, \rho\big)\big], \quad \forall \tau \in [0, 1], \tag{4.6}$$

and for each $\boldsymbol{x}$, $(\boldsymbol{z}_\tau^{\boldsymbol{x}}, \boldsymbol{p}_\tau^{\boldsymbol{x}})$ are defined by the forward/backward equations:

$$\begin{aligned} \frac{d\boldsymbol{z}_\tau^{\boldsymbol{x}}}{d\tau} &= \nabla_{\boldsymbol{p}} H = \mathbb{E}_{\boldsymbol{u}\sim\rho_\tau^*}\big[\boldsymbol{\phi}\big(\boldsymbol{z}_\tau^{\boldsymbol{x}}, \boldsymbol{u}\big)\big], \\ \frac{d\boldsymbol{p}_\tau^{\boldsymbol{x}}}{d\tau} &= -\nabla_{\boldsymbol{z}} H = \mathbb{E}_{\boldsymbol{u}\sim\rho_\tau^*}\big[\nabla_{\boldsymbol{z}}^T \boldsymbol{\phi}\big(\boldsymbol{z}_\tau^{\boldsymbol{x}}, \boldsymbol{u}\big)\boldsymbol{p}_\tau^{\boldsymbol{x}}\big], \end{aligned} \tag{4.7}$$

with the boundary conditions:

$$\boldsymbol{z}_0^{\boldsymbol{x}} = \boldsymbol{x}, \tag{4.8}$$

$$\boldsymbol{p}_1^{\boldsymbol{x}} = 2f\big(\boldsymbol{x}; \rho_1^* - f^*(\boldsymbol{x})\big)\mathbf{1}. \tag{4.9}$$

With this, one can then construct maximum principle-based algorithms. This was first done in [64] and it was based on an extension of the *method of successive approximation (MSA)*. This is an iterative algorithm that alternates between solving the Hamiltonian system for the states and costates and finding the optimal parameters at each step. Symbolically, one can write it as, at the step k:

- Solve
$$\frac{d\boldsymbol{z}_\tau^k}{d\tau} = \nabla_{\boldsymbol{p}} H\big(\boldsymbol{z}_\tau^k, \boldsymbol{p}_\tau^k, \theta_\tau^k\big), \quad \boldsymbol{z}_0^k = V\boldsymbol{x}.$$
- Solve
$$\frac{d\boldsymbol{p}_\tau^k}{d\tau} = -\nabla_{\boldsymbol{z}} H\big(\boldsymbol{z}_\tau^k, \boldsymbol{p}_\tau^k, \theta_\tau^k\big), \quad \boldsymbol{p}_1^k = 2\big(f(\boldsymbol{x}; \theta^k) - f^*(\boldsymbol{x})\big)\mathbf{1}.$$
- Set $\theta_\tau^{k+1} = \operatorname{argmax}_{\theta\in\Theta} H(\boldsymbol{z}_\tau^k, \boldsymbol{p}_\tau^k, \theta)$, for each $\tau \in [0, 1]$.

Compared with the usual gradient descent-based algorithms, the advantage is that the optimization problems are decoupled for different values of τ. Li et al. [64] presented numerical evidence which suggests that an extended version of this algorithm is quite competitive, compared with several different versions of SGD.

The gradient flow for this model was derived in [32]. For any $\rho^1, \rho^2 \in X$, consider the following metric:

$$\mathcal{D}^2(\rho^1, \rho^2) := \int_0^1 W_2^2\big(\rho_\tau^1, \rho_\tau^2\big)d\tau,$$

where $W_2(\cdot, \cdot)$ is the 2-Wasserstein distance.

Proposition 27. *The gradient flow in the metric space* $(\Sigma, \mathcal{D})$ *for the population risk is given by*

$$\partial_t \rho_\tau(\boldsymbol{w}, t) = \nabla_{\boldsymbol{w}} \cdot \left(\rho_\tau \mathbb{E}_{\boldsymbol{x}}\left[\boldsymbol{v}(\boldsymbol{z}_\tau^{\boldsymbol{x}}, \boldsymbol{p}_\tau^{\boldsymbol{x}}, \boldsymbol{w})\right]\right), \quad \forall \tau \in [0, 1], \tag{4.10}$$

where

$$\boldsymbol{v}(\boldsymbol{z}, \boldsymbol{p}, \boldsymbol{w}) = \nabla_{\boldsymbol{w}} \frac{\delta H}{\delta \mu} = \nabla_{\boldsymbol{w}}^T \boldsymbol{\phi}(\boldsymbol{z}, \boldsymbol{w}) \boldsymbol{p},$$

and for each $\boldsymbol{x}$, $(\boldsymbol{z}_\tau^{\boldsymbol{x}}, \boldsymbol{p}_\tau^{\boldsymbol{x}})$ *satisfies* (4.7) *and* (4.8) *with* ρ^* *replaced by* ρ *at time* t.

This is a one parameter family of coupled flows. For this flow, the energy dissipation relation is given by

$$\frac{d\mathcal{R}}{dt} = -\int_0^1 \mathbb{E}_{\boldsymbol{w} \sim \rho_\tau(\cdot;t)}\left[\left\|\mathbb{E}_{\boldsymbol{x}} \nabla_{\boldsymbol{w}}^T \boldsymbol{\phi}(\boldsymbol{z}_\tau^{\boldsymbol{x}}, \boldsymbol{w}) \boldsymbol{p}_\tau^{\boldsymbol{x}}\right\|^2\right] d\tau. \tag{4.11}$$

Consider now the discretization of the flow using the particle method. Letting $\rho_\tau(\cdot, t) = \frac{1}{m} \sum_{j=1}^m \delta(\boldsymbol{w}_\tau^j(t) - \cdot)$, the discretized gradient flow is given by

$$\begin{aligned} \frac{d\boldsymbol{z}_\tau^{\boldsymbol{x}}}{d\tau} &= \frac{1}{m} \sum_{j=1}^m \phi(\boldsymbol{z}_\tau^{\boldsymbol{x}}, \boldsymbol{w}_\tau^j), \quad \tau \in [0, 1], \\ \frac{d\boldsymbol{p}_\tau^{\boldsymbol{x}}}{d\tau} &= -\frac{1}{m} \sum_{j=1}^m \nabla_z \phi(\boldsymbol{z}_\tau^{\boldsymbol{x}}, \boldsymbol{w}_\tau^j) \boldsymbol{p}_\tau^{\boldsymbol{x}}, \quad \tau \in [0, 1], \\ \frac{d\boldsymbol{w}_\tau^j}{dt} &= -\mathbb{E}_{\boldsymbol{x}}\left[\nabla_{\boldsymbol{w}}^T \phi(\boldsymbol{z}_\tau^{\boldsymbol{x}}, \boldsymbol{w}_\tau^j)^T \boldsymbol{p}_\tau^{\boldsymbol{x}}\right], \quad j = 1, \ldots, m. \end{aligned} \tag{4.12}$$

Upon further discretizing the flow-based representation, one essentially recovers the gradient descent algorithm for ResNets together with back-propagation (for more details, see [32]).

The gradient descent-based algorithm and the maximum principle-based algorithms are two representative classes of training algorithms for deep neural networks. There are two major components in these algorithms: the propagation and back-propagation of the states and the costates, and the optimization of the parameters. In gradient-descent algorithms, for each iteration of the gradient descent, one performs a full cycle of forward and backward propagation. In maximum principle-based algorithms, for each cycle of the forward and backward propagation, one performs the full optimization. These two classes of algorithms stand at the opposite extreme as far as the balance of these two components are concerned. Obviously, the most efficient algorithm should lie somewhere in-between.

Another interesting question is the comparison between the mean field and the continuous philosophies. In the simplest setting, the mean-field and the continuous formulation give rise to the same continuous model. However, one should note that their starting point is quite different:

- For the mean field approach: discrete $\to$ continuous by taking the hydrodynamic limit, as in the study of interacting particle systems in statistical physics.
- For the continuous formulation: continuous $\to$ discrete by discretization. This viewpoint is more like the one in classical numerical analysis where one starts from continuous problems and then discretize.

Continuous formulation allows us to think about machine learning "outside the box" of neural network models. It also seems to be a "double-sided sword": While the continuous formulation seems to be quite attractive, it is difficult to initialize. In practice, it seems that initializing using i.i.d. random samples under the conventional scaling leads to better performance. This is still a puzzle that needs to be resolved.

5. SOME PERSPECTIVES AND CONCLUDING REMARKS

What have we really learned? Perhaps the single most important thing is that neural network-based machine learning is a very powerful tool for overcoming the CoD, or for discrete problems, the *combinatorial explosion*. This should be the most important guiding principle for designing new algorithms, trying new applications, or developing the theory. Neural networks might also be useful for problems with very few degrees of freedom, but at the moment, we still lack convincing evidence in this regard.

One of the most exciting recent development is the application of machine learning to science. AlphaFold2 and DeePMP are two of the most representative examples. The former is a powerful solution of a fundamental problem in science using data-driven methods. The latter is a powerful extension of a classical theoretical tool, namely molecular dynamics, that substantially advanced its realm of applicability. As we discussed in Section 2, machine learning seems to provide the missing tool for realizing the goals put forward in the multiscale modeling program. In addition, using machine learning to improve the efficiency of experimental work is also an area with a lot of promise. Indeed, one can argue that *AI for Science* has been the most exciting development in AI or science during the last couple of years, and it is changing the paradigm with which we do science.

On the theoretical side, even though we are still quite far from having a satisfactory theory for neural network-based machine learning, the roadmap to such a theory is emerging. This roadmap includes understanding the approximation theory, generalization gap, the landscape for the training problem, dynamical path during training, the difference between the landscape for the empirical and population risks, and so on. Perhaps more importantly, a consensus is starting to emerge regarding what the right questions are. One such consensus is that what is important is not the specific values of the neural network parameters, but rather their probability distribution. This underlies most of the theoretical advances discussed here.

Besides these abstract studies, there is also the need to study in more detail the structure of practical datasets. The fact that one can perform classification of images using neural network models suggests that the task itself is not so complicated, at least when represented using multilayer neural networks. It is worthwhile to look into the details of the structures of such a representation.

In addition to supervised learning, there is also the need to build some theoretical understanding of unsupervised learning, learning dynamical systems, reinforcement learning, as well as the new tasks that have emerged in the application of machine learning to scientific computing. The efforts to develop such an understanding is likely going to lead us to a new subject in mathematics, namely high-dimensional analysis.

Regarding the impact that machine learning will have on applied mathematics as a whole, we refer the readers to the article [24].

Finally, machine learning is not the ultimate solution of AI. It has a lot of problems, including the difficulties with interpretability, the need for a large training dataset, the vulnerability to adversarial attacks, and so on. Traditional rule-based methods are much better on these issues. Naturally one should ask whether it is possible to combine rule-based and learning-based approaches to build better AI algorithms.

ACKNOWLEDGMENTS

I would like to express my sincere gratitude to my collaborators, particularly Roberto Car, Jiequn Han, Qianxiao Li, Jihao Long, Chao Ma, Cheng Tai, Han Wang, Stephan Wojtowytsch, Lei Wu, Hongkang Yang, and Linfeng Zhang, for their work and for the discussions we have had that helped to shape this report.

There is a vast amount of literature in machine learning and it grows very fast. We feel that we have included the most important literature that is relevant to the topics discussed here, but we apologize for any of the missing items. For additional literature, the reader is referred to the review articles [27, 34].

FUNDING

This work was partially supported by a gift to the Princeton University from iFlytek. Part of the work is also supported by an ONR grant N00014-13-1-0338.

REFERENCES

[1] E. Abbe and C. Sandon, Poly-time universality and limitations of deep learning. 2020, arXiv:2001.02992.

[2] N. Aronszajn, Theory of reproducing kernels. *Trans. Amer. Math. Soc.* **68** (1950), no. 3, 337–404.

[3] S. Arora, S. S. Du, W. Hu, Z. Li, R. Salakhutdinov, and R. Wang, On exact computation with an infinitely wide neural net. 2019, arXiv:1904.11955.

[4] M. G. Azar, R. Munos, and H. Kappen, On the sample complexity of reinforcement learning with a generative model. In *International conference on machine learning*, 2012.

[5] M. G. Azar, I. Osband, and R. Munos, Minimax regret bounds for reinforcement learning. In *International conference on machine learning*, pp. 263–272, PMLR, 2017.

[6] F. Bach, Breaking the curse of dimensionality with convex neural networks. *J. Mach. Learn. Res.* **18** (2017), no. 19, 1–53.

[7] A. R. Barron, Universal approximation bounds for superpositions of a sigmoidal function. *IEEE Trans. Inf. Theory* **39** (1993), no. 3, 930–945.

[8] A. R. Barron, Approximation and estimation bounds for artificial neural networks. *Mach. Learn.* **14** (1994), no. 1, 115–133.

[9] J. Behler and M. Parrinello, Generalized neural-network representation of high-dimensional potential-energy surfaces. *Phys. Rev. Lett.* **98** (2007), no. 14, 146401.

[10] L. Bonati and M. Parrinello, Silicon liquid structure and crystal nucleation from ab initio deep metadynamics. *Phys. Rev. Lett.* **121** (2018), no. 26, 265701.

[11] L. Breiman, Hinging hyperplanes for regression, classification, and function approximation. *IEEE Trans. Inf. Theory* **39** (1993), no. 3, 999–1013.

[12] R. Car and M. Parrinello, Unified approach for molecular dynamics and density-functional theory. *Phys. Rev. Lett.* **55** (1985), no. 22, 2471.

[13] G. Carleo and M. Troyer, Solving the quantum many-body problem with artificial neural networks. *Science* **355** (2017), no. 6325, 602–606.

[14] R. T. Q. Chen, Y. Rubanova, J. Bettencourt, and D. K. Duvenaud, Neural ordinary differential equations. In *Advances in neural information processing systems*, pp. 6571–6583, 2018.

[15] L. Chizat and F. Bach, On the global convergence of gradient descent for over-parameterized models using optimal transport. In *Advances in neural information processing systems*, pp. 3036–3046, 2018.

[16] L. Chizat and F. Bach, Implicit bias of gradient descent for wide two-layer neural networks trained with the logistic loss. In *Conference on learning theory*, pp. 1305–1338, PMLR, 2020.

[17] Y. Cooper, The loss landscape of overparameterized neural networks. 2018, arXiv:1804.10200.

[18] G. Cybenko, Approximation by superpositions of a sigmoidal function. *Math. Control Signals Systems* **2** (1989), no. 4, 303–314.

[19] A. Daniely, SGD learns the conjugate kernel class of the network. In *Advances in neural information processing systems*, pp. 2422–2430, 2017.

[20] J. Dick, F. Y. Kuo, and I. H. Sloan, High-dimensional integration: the quasi-Monte Carlo way. *Acta Numer.* **22** (2013), 133–288.

[21] S. S. Du, X. Zhai, B. Poczos, and A. Singh, Gradient descent provably optimizes over-parameterized neural networks. In *International conference on learning representations*, 2019.

[22] W. E, *Principles of multiscale modeling*. Cambridge University Press, 2011.

[23] W. E, A proposal on machine learning via dynamical systems. *Commun. Math. Stat.* **5** (2017), no. 1, 1–11.

[24] W. E, The dawning of a new era in applied mathematics. *Not. Amer. Math. Soc.* **68** (2021), no. 4, 565–571.

[25] W. E and B. Engquist, The heterogenous multiscale methods. *Commun. Math. Sci.* **1** (2003), no. 1, 87–132.

[26] W. E, J. Han, and A. Jentzen, Deep learning-based numerical methods for high-dimensional parabolic partial differential equations and backward stochastic differential equations. *Commun. Math. Stat.* **5** (2017), no. 4, 349–380.

[27] W. E, J. Han, and A. Jentzen, Algorithms for solving high dimensional PDEs: from nonlinear Monte Carlo to machine learning. 2020, arXiv:2008.13333.

[28] W. E, J. Han, and L. Zhang, Machine-learning-assisted modeling. *Physics Today* **74** (2021), no. 7, 36–41.

[29] W. E, C. Ma, and Q. Wang, Rademacher complexity and the generalization error of residual networks. *Commun. Math. Sci.* **18** (2020), no. 6, 1755–1774.

[30] W. E, C. Ma, and L. Wu, A priori estimates of the population risk for two-layer neural networks. *Commun. Math. Sci.* **17** (2019), no. 5, 1407–1425. 2018, arXiv:1810.06397.

[31] W. E, C. Ma, and L. Wu, A comparative analysis of the optimization and generalization property of two-layer neural network and random feature models under gradient descent dynamics. *Sci. China Math.* (2020), 1–24. 2019, arXiv:1904.04326.

[32] W. E, C. Ma, and L. Wu, Machine learning from a continuous viewpoint, I. *Sci. China Math.* **63** (2020), no. 11, 2233–2266.

[33] W. E, C. Ma, and L. Wu, The Barron space and the flow-induced function spaces for neural network models. *Constr. Approx.* (2021), 1–38.

[34] W. E, C. Ma, L. Wu, and S. Wojtowytsch, Towards a mathematical understanding of neural network-based machine learning: what we know and what we don't. *CSIAM Trans. Appl. Math.* **1** (2020), no. 4, 561–615.

[35] W. E and S. Wojtowytsch, Representation formulas and pointwise properties for Barron functions. 2020, arXiv:2006.05982.

[36] J. Fan, Z. Wang, Y. Xie, and Z. Yang, A theoretical analysis of deep Q-learning. In *Learning for dynamics and control*, pp. 486–489, PMLR, 2020.

[37] A.-m. Farahmand, M. Ghavamzadeh, C. Szepesvári, and S. Mannor, Regularized policy iteration with nonparametric function spaces. *J. Mach. Learn. Res.* **17** (2016), no. 1, 4809–4874.

[38] T. E. Gartner, L. Zhang, P. M. Piaggi, R. Car, A. Z. Panagiotopoulos, and P. G. Debenedetti, Signatures of a liquid–liquid transition in an ab initio deep neural network model for water. *Proc. Natl. Acad. Sci.* **117** (2020), no. 42, 26040–26046.

[39] I. Goodfellow, J. Pouget-Abadie, M. Mirza, B. Xu, D. Warde-Farley, S. Ozair, A. Courville, and Y. Bengio, Generative adversarial nets. In *Advances in neural information processing systems*, pp. 2672–2680, 2014.

[40] P. Grohs, F. Hornung, A. Jentzen, and P. Von Wurstemberger, A proof that artificial neural networks overcome the curse of dimensionality in the numerical approximation of Black–Scholes partial differential equations. 2018, arXiv:1809.02362.

[41] E. Haber and L. Ruthotto, Stable architectures for deep neural networks. *Inverse Probl.* **34** (2017), no. 1, 014004.

[42] I. Han, J. T. McKeown, L. Tang, C.-Z. Wang, H. Parsamehr, Z. Xi, Y.-R. Lu, M. J. Kramer, and A. J. Shahani, Dynamic observation of dendritic quasicrystal growth upon laser-induced solid-state transformation. *Phys. Rev. Lett.* **125** (2020), no. 19, 195503.

[43] J. Han and W. E, Deep learning approximation for stochastic control problems. In *NIPS2016, Deep Reinforcement Learning Workshop*, 2016.

[44] J. Han, A. Jentzen, and W. E, Solving high-dimensional partial differential equations using deep learning. *Proc. Natl. Acad. Sci.* **115** (2018), no. 34, 8505–8510.

[45] J. Han, L. Zhang, R. Car, and W. E, Deep potential: a general representation of a many-body potential energy surface. *Commun. Comput. Phys.* **23** (2018), no. 3, 629–639.

[46] J. Han, L. Zhang, and W. E, Solving many-electron Schrödinger equation using deep neural networks. *J. Comput. Phys.* **399** (2019), 108929.

[47] B. Hanin and D. Rolnick, How to start training: the effect of initialization and architecture. In *Advances in neural information processing systems*, pp. 571–581, 2018.

[48] K. He, X. Zhang, S. Ren, and J. Sun, Delving deep into rectifiers: surpassing human-level performance on ImageNet classification. In *Proceedings of the IEEE International Conference on Computer Vision*, pp. 1026–1034, 2015.

[49] J. Hermann, Z. Schätzle, and F. Noé, Deep-neural-network solution of the electronic Schrödinger equation. *Nat. Chem.* **12** (2020), no. 10, 891–897.

[50] S. Hochreiter, The vanishing gradient problem during learning recurrent neural nets and problem solutions. *Internat. J. Uncertain. Fuzziness Knowledge-Based Systems* **6** (1998), no. 02, 107–116.

[51] P. C. Hohenberg and B. I. Halperin, Theory of dynamic critical phenomena. *Rev. Modern Phys.* **49** (1977), no. 3, 435.

[52] M. Hutzenthaler, A. Jentzen, T. Kruse, T. Anh Nguyen, and P. von Wurstemberger, Overcoming the curse of dimensionality in the numerical approximation of semilinear parabolic partial differential equations. *Proc. R. Soc. A* **476** (2020), no. 2244, 20190630.

[53] A. Jacot, F. Gabriel, and C. Hongler, Neural tangent kernel: convergence and generalization in neural networks. In *Advances in neural information processing systems*, pp. 8580–8589, 2018.

[54] W. Jia, H. Wang, M. Chen, D. Lu, L. Lin, R. Car, W. E, and L. Zhang, Pushing the limit of molecular dynamics with ab initio accuracy to 100 million atoms with machine learning. In *Proceedings of the international conference for high performance computing, networking, storage and analysis, SC'20*, pp. 1–14, IEEE Press, 2020.

[55] C. Jin, Z. Allen-Zhu, S. Bubeck, and M. I. Jordan, Is Q-learning provably efficient? In *Proceedings of the 32nd International Conference on Neural Information Processing Systems*, pp. 4868–4878, 2018.

[56] C. Jin, Z. Yang, Z. Wang, and M. I. Jordan, Provably efficient reinforcement learning with linear function approximation. In *Conference on learning theory*, pp. 2137–2143, PMLR, 2020.

[57] L. K. Jones, A simple lemma on greedy approximation in Hilbert space and convergence rates for projection pursuit regression and neural network training. *Ann. Statist.* (1992), 608–613.

[58] J. Jumper, R. Evans, A. Pritzel, T. Green, M. Figurnov, O. Ronneberger, K. Tunyasuvunakool, R. Bates, A. Žídek, A. Potapenko, et al., Highly accurate protein structure prediction with AlphaFold. *Nature* **596** (2021), no. 7873, 583–589.

[59] Y. Khoo, J. Lu, and L. Ying, Solving parametric PDE problems with artificial neural networks. *European J. Appl. Math.* **32** (2021), no. 3, 421–435.

[60] J. M. Klusowski and A. R. Barron, Risk bounds for high-dimensional ridge function combinations including neural networks. 2016, arXiv:1607.01434.

[61] J. Kober, J. A. Bagnell, and J. Peters, Reinforcement learning in robotics: a survey. *Internat. J. Robot. Res.* **32** (2013), no. 11, 1238–1274.

[62] Y. LeCun, Y. Bengio, and G. Hinton, Deep learning. *Nature* **521** (2015), no. 7553, 436–444.

[63] Y. LeCun, L. Bottou, G. B. Orr, and K.-R. Müller, Efficient backprop. In *Neural networks: tricks of the trade*, pp. 9–48, Springer, 2012.

[64] Q. Li, L. Chen, C. Tai, and W. E, Maximum principle based algorithms for deep learning. *J. Mach. Learn. Res.* **18** (2017), no. 1, 5998–6026.

[65] Z. Li, J. Han, W. E, and Q. Li, On the curse of memory in recurrent neural networks: approximation and optimization analysis. In *International conference on learning representations*, 2020.

[66] Z. Li, N. Kovachki, K. Azizzadenesheli, B. Liu, K. Bhattacharya, A. Stuart, and A. Anandkumar, Fourier neural operator for parametric partial differential equations. 2020, arXiv:2010.08895.

[67] R. Livni, S. Shalev-Shwartz, and O. Shamir, On the computational efficiency of training neural networks. In *Advances in neural information processing systems*, pp. 855–863, 2014.

[68] J. Long and J. Han, Perturbational complexity by distribution mismatch: a systematic analysis of reinforcement learning in reproducing kernel Hilbert space. 2021, arXiv:2111.03469.

[69] J. Long, J. Han, and W. E, An L^2 analysis of reinforcement learning in high dimensions with kernel and neural network approximation. 2021, arXiv:2104.07794.

[70] J. Lu, Z. Shen, H. Yang, and S. Zhang, Deep network approximation for smooth functions. *SIAM J. Math. Anal.* **53** (2021), no. 5, 5465–5506.

[71] Y. Lu, J. Lu, and M. Wang, A priori generalization analysis of the deep Ritz method for solving high dimensional elliptic partial differential equations. In *Conference on learning theory*, pp. 3196–3241, PMLR, 2021.

[72] C. Ma, L. Wu, and W. E, The quenching-activation behavior of the gradient descent dynamics for two-layer neural network models. 2020, arXiv:2006.14450.

[73] E. Malach and S. Shalev-Shwartz, When hardness of approximation meets hardness of learning. 2020, arXiv preprint arXiv:2008.08059.

[74] R. J. McCann, A convexity principle for interacting gases. *Adv. Math.* **128** (1997), no. 1, 153–179.

[75] S. Mei, A. Montanari, and P.-M. Nguyen, A mean field view of the landscape of two-layer neural networks. *Proc. Natl. Acad. Sci.* **115** (2018), no. 33, E7665–E7671.

[76] V. Mnih, K. Kavukcuoglu, D. Silver, A. Graves, I. Antonoglou, D. Wierstra, and M. Riedmiller, Playing Atari with deep reinforcement learning. 2013, arXiv:1312.5602.

[77] T. Nakamura-Zimmerer, Q. Gong, and W. Kang, Adaptive deep learning for high-dimensional Hamilton–Jacobi–Bellman equations. *SIAM J. Sci. Comput.* **43** (2021), no. 2, A1221–A1247.

[78] E. Pardoux and S. Peng, Adapted solution of a backward stochastic differential equation. *Systems Control Lett.* **14** (1990), no. 1, 55–61.

[79] E. Pardoux and S. Peng, Backward stochastic differential equations and quasi-linear parabolic partial differential equations. In *Stochastic partial differential equations and their applications*, pp. 200–217, Springer, 1992.

[80] D. Pfau, J. S. Spencer, A. G. Matthews, and W. M. C. Foulkes, Ab initio solution of the many-electron Schrödinger equation with deep neural networks. *Phys. Rev. Res.* **2** (2020), no. 3, 033429.

[81] A. Rahimi and B. Recht, Random features for large-scale kernel machines. In *Proceedings of the 20th international conference on neural information processing systems*, pp. 1177–1184, 2007.

[82] M. Raissi, P. Perdikaris, and G. E. Karniadakis, Physics-informed neural networks: a deep learning framework for solving forward and inverse problems involving nonlinear partial differential equations. *J. Comput. Phys.* **378** (2019), 686–707.

[83] G. Rotskoff, S. Jelassi, J. Bruna, and E. Vanden-Eijnden, Neuron birth-death dynamics accelerates gradient descent and converges asymptotically. In *International conference on machine learning*, pp. 5508–5517, 2019.

[84] G. Rotskoff and E. Vanden-Eijnden, Parameters as interacting particles: long time convergence and asymptotic error scaling of neural networks. In *Advances in neural information processing systems*, pp. 7146–7155, 2018.

[85] J. Schmidhuber, Deep learning in neural networks: an overview. *Neural Netw.* **61** (2015), 85–117.

[86] S. Shalev-Shwartz and S. Ben-David, *Understanding machine learning: from theory to algorithms*. Cambridge university press, 2014.

[87] O. Shamir, Distribution-specific hardness of learning neural networks. *J. Mach. Learn. Res.* **19** (2018), no. 1, 1135–1163.

[88] J. W. Siegel and J. Xu, Approximation rates for neural networks with general activation functions. *Neural Netw.* **128** (2020), 313–321.

[89] J. W. Siegel and J. Xu, Characterization of the variation spaces corresponding to shallow neural networks. 2021, arXiv:2106.15002.

[90] D. Silver, A. Huang, C. J. Maddison, A. Guez, L. Sifre, G. Van Den Driessche, J. Schrittwieser, I. Antonoglou, V. Panneershelvam, M. Lanctot, S. Dieleman, D. Grewe, J. Nham, N. Kalchbrenner, I. Sutskever, T. Lillicrap, M. Leach, K. Kavukcuoglu, and D. Hassabis, Mastering the game of Go with deep neural networks and tree search. *Nature* **529** (2016), no. 7587, 484–489.

[91] J. Sirignano and K. Spiliopoulos, DGM: a deep learning algorithm for solving partial differential equations. *J. Comput. Phys.* **375** (2018), 1339–1364.

[92] J. Sirignano and K. Spiliopoulos, Mean field analysis of neural networks: a central limit theorem. 2018, arXiv:1808.09372.

[93] R. S. Sutton and A. G. Barto, *Reinforcement learning: an introduction.* MIT Press, 2018.

[94] V. N. Vapnik and A. Y. Chervonenkis, On the uniform convergence of relative frequencies of events to their probabilities. In *Measures of complexity*, pp. 11–30, Springer, 2015.

[95] C. Villani, *Optimal transport: old and new*. 338, Springer, 2008.

[96] U. Von Luxburg and O. Bousquet, Distance-based classification with Lipschitz functions. *J. Mach. Learn. Res.* **5** (2004), 669–695.

[97] L. Wang, Q. Cai, Z. Yang, and Z. Wang, Neural policy gradient methods: global optimality and rates of convergence. In *International conference on learning representations*, 2019.

[98] Y. Wang, J. Wang, A. Hermann, C. Liu, H. Gao, E. Tosatti, H.-T. Wang, D. Xing, and J. Sun, Electronically driven 1D cooperative diffusion in a simple cubic crystal. *Phys. Rev. X* **11** (2021), no. 1, 011006.

[99] S. Wojtowytsch, On the global convergence of gradient descent training for two-layer ReLU networks in the mean field regime. 2020, arXiv:2005.13530.

[100] S. Wojtowytsch and W. E, Can shallow neural networks beat the curse of dimensionality? a mean field training perspective. *IEEE Trans. Artif. Intell.* **1** (2020), no. 2, 121–129.

[101] S. Wojtowytsch and W. E, Some observations on partial differential equations in Barron and multi-layer spaces. 2020, arXiv:2012.01484.

[102] P. Xie and W. E, Coarse-grained spectral projection: a deep learning assisted approach to quantum unitary dynamics. *Phys. Rev. B* **103** (2021), no. 2, 024304.

[103] Z.-Q. J. Xu, Y. Zhang, T. Luo, Y. Xiao, and Z. Ma, Frequency principle: Fourier analysis sheds light on deep neural networks. 2019, arXiv:1901.06523.

[104] H. Yang and W. E, Generalization and memorization: the bias potential model. 2020, arXiv:2011.14269.

[105] L. Yang and M. Wang, Sample-optimal parametric Q-learning using linearly additive features. In *International conference on machine learning*, pp. 6995–7004, PMLR, 2019.

[106] Z. Yang, C. Jin, Z. Wang, M. Wang, and M. Jordan, Provably efficient reinforcement learning with kernel and neural function approximations. *Adv. Neural Inf. Process. Syst.* **33** (2020).

[107] D. Yarotsky, Error bounds for approximations with deep ReLU networks. *Neural Netw.* **94** (2017), 103–114.

[108] L. Zdeborová and F. Krzakala, Statistical physics of inference: thresholds and algorithms. *Adv. Phys.* **65** (2016), no. 5, 453–552.

[109] J. Zeng, L. Cao, M. Xu, T. Zhu, and J. Z. Zhang, Complex reaction processes in combustion unraveled by neural network-based molecular dynamics simulation. *Nat. Commun.* **11** (2020), no. 1, 1–9.

[110] L. Zhang, J. Han, H. Wang, W. Saidi, R. Car, and W. E, End-to-end symmetry preserving inter-atomic potential energy model for finite and extended systems. In *Advances in neural information processing systems 31*, pp. 4441–4451, Curran Associates, Inc., 2018.

[111] L. Zhang, D.-Y. Lin, H. Wang, R. Car, and W. E, Active learning of uniformly accurate interatomic potentials for materials simulation. *Phys. Rev. Mater.* **3** (2019), no. 2, 023804.

[112] L. Zhang, H. Wang, R. Car, and W. E, Phase diagram of a deep potential water model. *Phys. Rev. Lett.* **126** (2021), no. 23, 236001.

[113] L. Zhang, H. Wang, and W. E, Reinforced dynamics for enhanced sampling in large atomic and molecular systems. *J. Chem. Phys.* **148** (2018), no. 12, 124113.

WEINAN E

Center for Machine Learning Research and School of Mathematical Sciences, Peking University, Beijing, China, weinan@math.pku.edu.cn

Beijing Institute for Big Data Research, Beijing, China, weinan@bibdr.org

Department of Mathematics and Program in Applied and Computational Mathematics, Princeton University, Princeton, USA, weinan@math.princeton.edu

This Page Intentionally Left Blank

HOMOMORPHIC ENCRYPTION: A MATHEMATICAL SURVEY

CRAIG GENTRY

ABSTRACT

If the first thing that comes to mind when you hear the word "encryption" is the Enigma machine, you might think that encryption is complicated and mathematically uninteresting. In fact, many modern encryption systems are quite simple from a mathematical point of view, especially encryption systems that are *homomorphic*. In these systems, the starting point is a homomorphism that respects some binary operation(s), such as $+$ or $\times$. Depicting this homomorphism with a rectangular commutative diagram, the objects on the top level of the diagram are called ciphertexts, and the objects on the bottom level are called messages or plaintexts. The downward arrows in the diagram are the homomorphism, which we call decryption. The rightward arrows are the operation(s). Decryption commutes with the operations. To the commutative diagram we add one extra ingredient, computational complexity. Specifically, we need for it to be *easy* (in the sense of polynomial-time) for anyone to compute the rightward arrows in the diagram, but *hard* to learn how to compute the downward (decryption) arrows except with some special information that we will call a "secret key." In short, homomorphic encryption is simply a homomorphism that has been "hardened" in the complexity-theoretic sense.

Homomorphic encryption allows anyone to compute on encrypted data, without needing (or being able) to decrypt, has many exciting applications. *Fully homomorphic encryption* (FHE) systems, which allow a rich (functionally complete) set of operations, were finally discovered in 2009. But all of the FHE systems that we have discovered so far follow the same blueprint, and we still wonder whether there are other ways to build FHE.

This survey presents homomorphic encryption from a mathematical point of view, illustrating with several examples how to start from a homomorphism and harden it to make it suitable for cryptography, pointing out pitfalls and attacks to avoid, laying out the current blueprint for FHE, and (I hope) serving as an inspiration and useful guide in the development of new approaches to FHE.

MATHEMATICS SUBJECT CLASSIFICATION 2020

Primary 68P25; Secondary 68Q17, 14G50

KEYWORDS

Homomorphism, cryptography, encryption, complexity theory

ICM INTERNATIONAL CONGRESS OF MATHEMATICIANS 2022 JULY 6–14

Proc. Int. Cong. Math. 2022, Vol. 2, pp. 956–1006
DOI 10.4171/ICM2022/165

1. INTRODUCTION

Let me sketch *homomorphic encryption* in two different ways, a cryptographic way and a mathematical way.

Cryptographically, homomorphic encryption has the usual 3 algorithms of encryption—namely, key generation K, encryption E, and decryption D. Key generation K generates a random key pair (ek, sk), an encryption key and a secret decryption key. In a "symmetric" encryption system, $ek = sk$; in an "asymmetric" encryption system, ek is public (not secret) and does not equal sk. Encryption E is a randomized algorithm that maps a message $m \in \mathcal{M}$ and some randomness to a ciphertext, $c = \mathsf{E}(ek, m, r)$. Decryption D is a deterministic algorithm that recovers a message from a ciphertext, $m = \mathsf{D}(sk, c)$. It should hold that $m = \mathsf{D}(sk, \mathsf{E}(ek, m, r))$ for all key pairs in the image of K, all $m \in \mathcal{M}$, and all randomness r. For the encryption system to be secure, for any two messages $m_0, m_1 \in \mathcal{M}$ chosen by an adversary, it should be computationally "hard" for the adversary to distinguish encryptions of m_0 from encryptions of m_1, even after seeing many encryptions of these and other values. Homomorphic encryption also has a fourth procedure, V, for evaluation. This procedure uses an additional evaluation key evk, which is public. Evaluation V allows anyone to process encrypted data while it remains encrypted, without using the secret decryption key. For example, one might be able to apply some binary operation $\boxplus$ to two ciphertexts to produce a new ciphertext that encrypts the sum of the original two messages. Formally, for some set of binary operations $\mathcal{F}$ associated to the system, the following is true: for any correctly generated key tuple (ek, sk, evk), and for any ciphertexts c_1, c_2 in the image of E under the key tuple, as well as for any $f \in \mathcal{F}$, we have $\mathsf{D}(\mathsf{V}(evk, f, c_1, c_2)) = f(\mathsf{D}(sk, c_1), \mathsf{D}(sk, c_2))$, that is, running V with the function f on two ciphertexts that happen to decrypt to m_1 and m_2 produces a new ciphertext that decrypts to $f(m_1, m_2)$. We say the system is "unbounded" if operations can be applied repeatedly, indefinitely, not only on ciphertexts in the image of E but also ciphertexts in the image of V. The set of ciphertexts should be finite. (Actually, ciphertexts should be compactly expressible, for efficiency). Thus, homomorphic encryption, via the algorithm V, allows the *processing* of data without giving away *access* to the data. The applications are numerous. For example, if the system is unbounded for operations $+$ and $\times$ (i.e., is *fully homomorphic*), you could give your encrypted financial information to an online service, which could prepare a (encrypted) completed tax form for you (which you could then decrypt), without the online service learning any of your private information.

The cryptographic way also emphasizes an approach called *provable security*. In this approach, one invokes a well-established computational assumption, such as the assumption that it is hard to factor large integers.[1] Then, one constructs a cryptosystem, and *proves* that it is secure if your computational assumption is true. Specifically, one shows that if there is an efficient adversary that violates the security of the cryptosystem, then from that adversary one

---

1 We will discuss computational complexity and security in more detail later, but as a first approximation one can view the notion of computational "hardness" here as requiring at least that $P \neq NP$, i.e., that there exist problems for which one can *verify* a solution efficiently if given a *witness*, but not *find* a solution efficiently.

can build an efficient algorithm to solve the assumedly hard problem. Provable security is an elegant and necessary approach that puts cryptography on a firm and rigorous foundation: anyway, about as firm as possible, given that we are not even certain that P $\neq$ NP. The provable security approach performs the essential function of discouraging and weeding out unproven cryptosystems that might look hard to break at first glance, but are usually broken eventually.

Mathematically, the essence of unbounded homomorphic encryption is captured by a commutative diagram

$$\begin{array}{ccc} \mathcal{C}^2 & \xrightarrow{\boxplus,\boxtimes} & \mathcal{C} \\ \Big\downarrow{\scriptstyle \mathsf{D}(sk,\cdot,\cdot)} & & \Big\downarrow{\scriptstyle \mathsf{D}(sk,\cdot)} \\ \mathcal{M}^2 & \xrightarrow{+,\times} & \mathcal{M}. \end{array}$$

In the diagram, $\mathcal{M}$ is the set of valid messages and $\mathcal{C}$ is the set of valid ciphertexts. The downward arrows are decryption, which I have drawn as dashed since the arrows should be "hard" to compute without the secret decryption key sk. The rightward arrows are binary operations over $\mathcal{M}$ and $\mathcal{C}$, which anyone can compute easily. To make the diagram cleaner, I simply assumed that the binary operations over $\mathcal{M}$ are $+$ and $\times$ (though other possibilities are interesting), and I used $\boxplus$ and $\boxtimes$, instead of the more ponderous $\mathsf{V}(evk, +, \cdot, \cdot)$ and $\mathsf{V}(evk, \times, \cdot, \cdot)$. The diagram displays how decryption commutes with the binary operations: starting from 2 ciphertexts in the upper-left corner, applying componentwise decryption and then $+$ (for example) produces the same result as first applying $\boxplus$ and then decryption. With the dashed arrows, the diagram depicts homomorphic encryption as a rather straightforward marriage of homomorphism and computational complexity.

The mathematical way does not avoid provable security (nor would we wish it to). Also, the commutative diagram does not refer explicitly to K (key generation) or E (encryption). But the diagram, implicit in the dashed arrow, has a lot to say about provable security, K and E. Ciphertexts are preimages of messages under the decryption map. For the system to be secure, for any two messages $m_0, m_1 \in \mathcal{M}$ chosen by an adversary, it should be computationally "hard" for the adversary to distinguish preimages of m_0 from preimages of m_1, even after seeing many (image, preimage) pairs. In particular, it should be hard to distinguish samples from the *kernel* of the decryption map versus samples from all of $\mathcal{C}$. Typically, one proves the security of a homomorphic encryption system by reducing security to precisely that assumption, namely that $\mathcal{C}$ and $\ker(\mathsf{D}(sk, \cdot))$ are hard to distinguish from samples. Similarly, encryption of m, that is, sampling a random preimage of m, often proceeds by picking *some* preimage c_1 of m and then randomizing it by sampling random $c_2 \leftarrow \ker(\mathsf{D}(sk, \cdot))$ and setting $c = c_1 \boxplus c_2$.

There are already many surveys of homomorphic encryption that follow the cryptographic way **[1, 2, 7, 11, 22, 42, 43, 48, 62, 72, 74]**.[2] This survey is aimed at mathematicians. So,

---

2 Silverberg's survey **[72]** is aimed at mathematicians, but in a different way than I intend here.

our journey will follow the mathematical way, starting always with a homomorphism (rather than a well-established cryptographic assumption), and then seeking ways to "harden" the homomorphism to make it suitable for cryptography. My ulterior motive for following this way is that I want to encourage mathematicians to be creative, to try to introduce new useful algebraic structures into cryptography's limited repertoire, and to invent new homomorphic encryption systems (subject, eventually, to the constraints of provable security).

Accordingly, the plan of the survey is to be maximally accessible, useful and inspiring to mathematicians, by presenting:

- Several examples of simple homomorphic encryption systems, starting from their homomorphisms, showing how their homomorphisms are "hardened," and giving their proofs of security (after defining security);
- General results about homomorphic encryption—including "fully" homomorphic encryption (FHE), which allows arbitrary computations to be performed on data while it remains encrypted—most of which follow directly from the commutative diagram defining the system's correctness;
- Some discussion of why ring homomorphisms do not seem to give secure FHE systems;
- A clear exposition of an actual FHE system, including how we start with a ring homomorphism, how we harden the ring homomorphism by adding "noise," and how to base the security of the system on a "hard" problem over integer lattices;
- Some discussion of failed attempts to use different algebraic structures to build fully homomorphic encryption systems in a way that falls outside of the current blueprint;
- A mercifully concise discussion of practical matters, such as the performance characteristics of FHE.

By the end, we will see that the algebraic structures underlying current fully homomorphic encryption (FHE) systems are rather bizarre. In known FHE systems, the set of messages $\mathcal{M}$ is a ring with natural $+$ and $\times$. The set of ciphertexts $\mathcal{C}$ has analogous binary operations $\boxplus$ and $\boxtimes$, but is not a ring, but rather a commutative "double magma"—in particular, the binary operations are not even associative. As an algebraic structure, the ciphertexts are very unstructured. It is an intriguing question whether FHE can be built with a set of ciphertexts that is more structured, e.g., a nonsolvable group.

In the next section, we review some simple early homomorphic encryption systems, their commutative diagrams, and their proofs of security. After these examples, we present some general definitions and results about homomorphic encryption in Section 3, most notably the bootstrapping theorems, which demonstrate that to get a homomorphic encryption system capable of correctly evaluating *any function* on encrypted data (that is, an FHE system), it is enough to get a homomorphic encryption system that can correctly evaluate *a single special function*. In Section 4, we describe in detail the construction of an FHE system. The construction starts with a homomorphism that respects a rich set of operations—such as a ring homomorphism—and hardens it by adding "noise" to it. The noise turns the unbounded homomorphism into a bounded one, but the bounded homomorphism is "bootstrappable," as needed to obtain FHE. We show how to base the security of different versions of the FHE system on different versions of the learning with errors (LWE) problem, whose hardness in turn can be based on hard problems over integer lattices. Finally, in Section 5, we suggest directions for future research.

2. SOME SIMPLE HOMOMORPHIC ENCRYPTION SYSTEMS

Here, as a (safely skippable) warm-up, we present some simple homomorphic encryption systems, starting from their homomorphisms, showing how their homomorphisms are "hardened," and giving their proofs of security (after defining security).

First, some history. Rivest, Adleman, and Dertouzos [67] proposed the notion of homomorphic encryption in 1978—calling it a "privacy homomorphism." They were inspired by a homomorphic property of the RSA encryption system, which Rivest, Shamir, and Adleman [68] had proposed the previous year—namely, that if you multiply two ciphertexts encrypted under the same key, it has the effect of multiplying the messages encrypted inside. They wondered whether it was possible to take this further: to construct a privacy homomorphism capable of general computation on encrypted data, not just multiplications modulo an integer. In [67], they proposed several systems allowing general computation. They knew these systems were insecure against realistic attacks—for example, in some of the systems, if you obtain a few encryptions of 0, it becomes trivial to recover the secret key. These systems were inspiring to later researchers, who eventually found ways to modify them to make them secure—in particular, with "noise"—to construct the fully homomorphic encryption systems that we have today.

Fortunately, for the purposes of this survey, we have some simple homomorphic encryption systems that are also provably secure, based on natural computational assumptions, under the "right" model of security for an encryption system. For these examples, we can start with a homomorphism, show how to "harden" it, and provide a proper proof of security in the "right" model of security. In these examples, the proof of security in this model makes heavy use of the homomorphism. In fact, the assumption used in the proof of security is simply that it is computationally hard to distinguish samples from the kernel of the homomorphism from random samples.

Our first example is the Goldwasser–Micali encryption system, described in 1982 [47]. Goldwasser and Micali were the first to prove an encryption system secure under a natural computational assumption using the "right" model of security. Granted, they had an advantage here, because they also *defined* the model of security. But, to their credit, this model has stood the test of time and is still considered the right one.

2.1. Goldwasser–Micali: HE starting from the Legendre symbol

For a fixed prime p, the Legendre symbol $\left(\frac{\cdot}{p}\right) : (\mathbb{Z}/p\mathbb{Z})^\times \to \{\pm 1\}$ is a group homomorphism, mapping an element of $(\mathbb{Z}/p\mathbb{Z})^\times$ to 1 if it is a quadratic residue (square) modulo p, and to -1 if it is a nonresidue. We have the following commutative diagram:

$$\begin{array}{ccc} (\mathbb{Z}/p\mathbb{Z})^\times \times (\mathbb{Z}/p\mathbb{Z})^\times & \xrightarrow{\quad\boxtimes\quad} & (\mathbb{Z}/p\mathbb{Z})^\times \\ \Big\downarrow{\scriptstyle \left(\frac{\cdot}{p}\right),\left(\frac{\cdot}{p}\right)} & & \Big\downarrow{\scriptstyle \left(\frac{\cdot}{p}\right)} \\ \{\pm 1\} \times \{\pm 1\} & \xrightarrow{\quad\times\quad} & \{\pm 1\}, \end{array}$$

where $\times$ denotes multiplication in $\{\pm 1\}$, and $\boxtimes$ denotes multiplication in $(\mathbb{Z}/p\mathbb{Z})^\times$.

How can we "harden" the Legendre symbol homomorphism to build a homomorphic encryption system? The downward arrow, which will eventually become decryption, currently requires only knowledge of p, so we must hide p in some way. A natural way to hide p is to reveal only a composite integer $N = p \cdot q$, where p and q are both large prime integers; N hides p only if it is "hard" to recover p from N via factorization, so we will at least need to assume that factorization is hard. We now have the following commutative diagram:

$$\begin{array}{ccc} (\mathbb{Z}/N\mathbb{Z})^\times \times (\mathbb{Z}/N\mathbb{Z})^\times & \xrightarrow{\quad\boxtimes\quad} & (\mathbb{Z}/N\mathbb{Z})^\times \\ \vdots\,{\scriptstyle \left(\frac{\cdot}{p}\right),\left(\frac{\cdot}{p}\right)} & & \vdots\,{\scriptstyle \left(\frac{\cdot}{p}\right)} \\ \downarrow & & \downarrow \\ \{\pm 1\} \times \{\pm 1\} & \xrightarrow{\quad\times\quad} & \{\pm 1\}, \end{array}$$

where now $\boxtimes$ is multiplication modulo N, and the downward arrows are dashed because (we hope) it is hard to learn how to compute the Legendre symbol $\left(\frac{\cdot}{p}\right)$ without the secret p, even after seeing many (image, preimage pairs).

For several reasons, it makes sense to use the subgroup of $(\mathbb{Z}/N\mathbb{Z})^\times$, which we will denote by J_N, of elements with Jacobi symbol 1. First, the fact that the Jacobi symbol $\left(\frac{\cdot}{N}\right)$ is efficiently computable even without the factorization of N makes cryptographers nervous. We can make the Jacobi symbol useless to an attacker by using only elements that have the same Jacobi symbol. Second, restricting to J_N makes the system cleaner by removing unneeded cosets from $(\mathbb{Z}/N\mathbb{Z})^\times$. Third, using J_N will make the computational assumption easier to state. We now have the following commutative diagram:

$$\begin{array}{ccc} J_N \times J_N & \xrightarrow{\quad\boxtimes\quad} & J_N \\ \vdots\,{\scriptstyle \left(\frac{\cdot}{p}\right),\left(\frac{\cdot}{p}\right)} & & \vdots\,{\scriptstyle \left(\frac{\cdot}{p}\right)} \\ \downarrow & & \downarrow \\ \{\pm 1\} \times \{\pm 1\} & \xrightarrow{\quad\times\quad} & \{\pm 1\}. \end{array}$$

The downward arrows are still surjective. Half of the elements of J_N are *squares* in $(\mathbb{Z}/N\mathbb{Z})^*$ with Legendre symbol 1 for both p and q, and half are *nonsquares* with Legendre symbol -1 for both p and q.

Now, let us build a homomorphic encryption system from the commutative diagram. Our diagram indicates that our set of ciphertexts is J_N, and that we decrypt a ciphertext c by computing $\left(\frac{c}{p}\right)$. The diagram also depicts the homomorphism of our system—namely, that by multiplying ciphertexts (modulo N), we implicitly multiply the underlying messages (in $\{\pm 1\}$). So, we have already built the decryption function D (which maps a ciphertext to a message) and the evaluation function V (which uses the binary relation(s) over ciphertexts to implicitly apply the analogous binary relation(s) to the messages that are encrypted).

All what remains is to build key generation K (which generates a random key pair (ek, sk), an encryption key and a secret decryption key) and encryption E (which maps a message from $\mathcal{M}$ and some randomness to a ciphertext using the encryption key ek, i.e., $c = \mathsf{E}(ek, m, r)$.) What do we need to put in the public encryption key to allow a user to generate a random encryption of either $\{\pm 1\}$? Notice that the encryptions of 1, i.e., the subset of J_N that has Legendre symbol 1 for p, are precisely the quadratic residues (squares) in $(\mathbb{Z}/N\mathbb{Z})^\times$. So, given N, anybody can generate a random encryption of 1 easily by taking a random element of $(\mathbb{Z}/N\mathbb{Z})^\times$ and squaring it. To generate a random encryption of -1, the user needs *some* encryption u of -1, namely, a nonsquare in J_N, which it can then randomize via multiplication with a random square. Hence, it suffices to provide (N, u) as the public encryption key.[3]

Below is a cleaner presentation of the Goldwasser–Micali encryption system. Let $\mathsf{CompositeGen}(\lambda, r)$ be a function that takes a security parameter λ and some randomness r as input, and which outputs integer primes p, q of size determined by λ (they have a number of bits polynomial in λ) and their product $N = p \cdot q$.

Goldwasser–Micali encryption system.

- Key Generation: $\mathsf{K}(\lambda, r)$ takes a security parameter λ and some randomness r as input. It outputs $(p, q, N) \leftarrow \mathsf{CompositeGen}(\lambda, r)$. Also, it uses the randomness to generate random $u \in J_N$ that is a nonsquare. The secret key sk is p. The public encryption ek is (N, u). The message set $\mathcal{M}$ is $\{-1, 1\}$. The ciphertext set $\mathcal{C}$ is J_N.

- Encryption: $\mathsf{E}(ek, m, r)$ takes the encryption key ek, a message $m \in \mathcal{M}$ and some randomness r as input. It generates random $t \in (\mathbb{Z}/N\mathbb{Z})^*$. If $m = 1$, it outputs ciphertext $c \leftarrow t^2 \bmod N$, else it outputs $c \leftarrow u \cdot t^2 \bmod N$.

- Decryption: $\mathsf{D}(sk, c)$ takes the secret key sk and a ciphertext $c \in \mathcal{C}$ as input. It outputs $m \leftarrow \left(\frac{c}{p}\right) \in \mathcal{M}$.

- Homomorphic multiplication: it takes two ciphertexts $c_1, c_2 \in \mathcal{C}$ and outputs $c \leftarrow c_1 \boxtimes c_2$.

---

3 See Section 3.2 for a more generally applicable approach to key generation and encryption, in which key generation involves populating the public key with encryptions (preimages under the decryption map) of several known values, and encryption involves applying the binary relation to the ciphertexts (preimages) in the public key to generate a random encryption (preimage) of the desired value.

Now, let us turn to security. Goldwasser and Micali defined the security of an encryption system using the following game [15, 47].

Definition 1 (IND-CPA game). The IND-CPA game between a "challenger" and an "adversary" is as follows:

- Key Generation: The challenger uses K and λ to generate a key pair (sk, ek). If the encryption system is asymmetric (public-key), it sends ek to the adversary. It keeps sk secret. The challenger samples a random bit $b \in \{0, 1\}$.
- Training and Challenges: Repeatedly, the adversary selects some messages $m_{i,0}, m_{i,1} \in \mathcal{M}$ that it sends to the challenger. The challenger generates randomness r_i and sends $c_i \leftarrow \mathsf{E}(ek, m_{i,b}, r_i)$ to the adversary. (If the adversary is free to set $m_{i,0} = m_{i,1}$ if it wants an encryption of a known message.)
- Guess: The adversary guesses a bit $b' \in \{0, 1\}$. It wins if $b' = b$.

Definition 2 (Adversary's advantage). In a game against system $\mathcal{E}$ with security parameter λ in which an adversary is trying to guess a random bit $b \in \{0, 1\}$, we define the adversary's advantage $\mathrm{Adv}_{\mathcal{A}}^{\mathcal{E}}(\lambda)$ to be $|\Pr[\mathcal{A}$ guesses b correctly$] - \frac{1}{2}|$.

Definition 3 (IND-CPA security of encryption). We say that an encryption system $\mathcal{E}$ is IND-CPA-secure if, for all probabilistic polynomial time adversaries $\mathcal{A}$ (i.e., that run in time polynomial in λ), the adversary's advantage $\mathrm{Adv}_{\mathcal{A}}^{\mathcal{E}}(\lambda)$ in the IND-CPA game is negligible (i.e., $o(1/\lambda^c)$ for all constants c).

See Appendix A for a discussion of why, for most settings, IND-CPA is a minimal viable notion of security for encryption; here, we just make a few comments about it. In our presentation of the IND-CPA game, we consider only fixed-length messages from the set $\mathcal{M}$; if queries with variable-length messages are allowed, the game requires $m_{i,0}, m_{i,1}$ to be the same length.

In its attack, the adversary is, of course, free to use $\mathsf{E}(ek, \cdot)$ and/or V to produce ciphertexts on its own if the system is asymmetric and/or homomorphic.

An encryption system can be secure under the IND-CPA game only if it is *probabilistic*—that is, there are many ciphertexts for each message. If the system were deterministic, the adversary could easily win by obtaining an encryption of a known message m, and then querying $(m_{i,0} = m, m_{i,1})$ for some $m_{i,1} \neq m$.

The IND-CPA game and our commutative diagram imply that, in a secure encryption system, it is hard to distinguish samples of $\ker(\mathsf{D}(sk, \cdot))$ from samples of $\mathcal{C}$. Making the kernel of the decryption homomorphism indistinguishable from the entire set of ciphertexts is the essence, and hardest part, of hardening a homomorphism to make it suitable for cryptography.

It only remains to define a clear *computational assumption*, and *prove* the security of Goldwasser–Micali based on the assumption. The computational assumption is that it is hard to distinguish whether a randomly sampled element of J_N is a square modulo N. This assumption is formalized as follows:

Definition 4 (Quadratic Residuosity (QR) assumption). For security parameter λ and randomness r, compute $(p, q, N) \leftarrow \mathsf{CompositeGen}(\lambda, r)$. Sample v uniformly from J_N. The QR assumption is that given (N, v) (but not p, q, r), all probabilistic polynomial time (in λ) adversaries $\mathcal{A}$ have negligible advantage in guessing whether v is a quadratic residue modulo N. (The probability in the assumption is taken not just over the sampling of J_N, but also over the choice of N.)

In other words, the QR assumption is that samples from J_N are indistinguishable from samples from the subset of J_N that is in $\ker\left(\left(\frac{\cdot}{p}\right)\right)$.

Note that the assumption (like all of the computational assumptions that we will make) depends on how elements are presented. For example, if our presentation of an element $j \in J_N$ is "too revealing" in that we give j not just as element of $(\mathbb{Z}/N\mathbb{Z})^*$ but also give the value $\left(\frac{j}{p}\right)$, then clearly the assumption becomes false. Generally speaking, it will be clear what "hardened" presentation the assumption is using.

How hard is the QR problem? We do not know of any algorithm for the QR problem that is faster than factoring N. The fastest algorithm for integer factorization is currently the number field sieve [57], which runs in time $\exp(O(\log N)^{\frac{1}{3}}(\log\log N)^{\frac{2}{3}})$, i.e., subexponential (but superpolynomial) time.

For fixed N, Goldwasser and Micali show how to amplify the success probability of an QR algorithm—given an algorithm that guesses correctly with probability $\frac{1}{2} + \varepsilon$, one can construct an algorithm that uses about $O(\varepsilon^{-1})$ times the computation and guesses with probability very close to 1. This follows from the fact that the QR problem is *random self-reducible*, given a *particular* $j \in J_N$, we can run the initial algorithm on many *random* $j \cdot r$, where r is a random square, and aggregate the results. In other words, random self-reducibility exploits the homomorphism to generate many samples that have the same preimage by multiplying an initial sample with many elements of the kernel.

Now, let us prove the security of the Goldwasser–Micali encryption system. The proof is quite simple. Nonetheless, it is useful because it reduces the security of a complex system (that allows the IND-CPA adversary to "train" and adapt by making interactive dynamically-chosen queries) to a crisp and concise computational assumption.

Theorem 1. *The Goldwasser–Micali encryption system is IND-CPA-secure based on the QR assumption.*

Proof. Suppose that there exists an efficient adversary $\mathcal{A}$ that wins the IND-CPA game with probability $\frac{1}{2} + \varepsilon$. Then, we claim that there exists an efficient algorithm $\mathcal{B}$, running in about the same time as $\mathcal{A}$, that solves the QR problem with probability $\frac{1}{2} + \frac{\varepsilon}{2}$. The theorem follows from this claim.

Here is how algorithm $\mathcal{B}$ works: $\mathcal{B}$ is given an instance of the QR problem, namely (N, v) such that N is a composite number chosen according to the specified distribution and v is sampled uniformly from J_N. Here $\mathcal{B}$'s task is to distinguish whether v is a quadratic residue modulo N. To solve its task, $\mathcal{B}$ assumes the role of the challenger in the IND-CPA game with $\mathcal{A}$. Then $\mathcal{B}$ gives $ek \leftarrow (N, v)$ to $\mathcal{A}$ as the public encryption key, and $\mathcal{B}$ samples a random $b \in \{0, 1\}$. When $\mathcal{A}$ sends query $(m_{i,0}, m_{i,1})$, $\mathcal{B}$ samples randomness r_i for encryption, sets $c_i \leftarrow \mathsf{E}(ek, m_{i,b}, r_i)$, and sends c_i to $\mathcal{A}$. Also $\mathcal{A}$ outputs a guess $b' \in \{0, 1\}$. If $b' = b$, then $\mathcal{B}$ guesses that v is a quadratic nonresidue; otherwise, it guesses that u is quadratic residue.

Now, we have two cases: either v is a nonresidue (as it should be in the real system), or it is a residue. Each case happens with probability $\frac{1}{2}$. In the former case, the public key ek and all of the ciphertexts generated by $\mathcal{B}$ have the same distribution as in the IND-CPA game. Therefore, in this case, the adversary guesses correctly ($b' = b$) with probability $\frac{1}{2} + \varepsilon$ by assumption. In the latter case (if v is a quadratic residue), then v and all of the ciphertexts generated by $\mathcal{B}$ are uniformly random quadratic residues. In particular, the ciphertexts are independent of the messages they are supposed to encrypt, and hence independent of b. In this case, the adversary's guess b' is also independent of b. Thus, $\mathcal{A}$'s success (or lack of it) gives $\mathcal{B}$ a clue about whether or not v is a quadratic nonresidue (or residue). In detail, using QR and QNR to denote the events that v is a quadratic residue or nonresidue, respectively, we have:

$$
\begin{aligned}
\Pr[\mathcal{B}\text{ correct}] = {} & \Pr[\mathcal{B}\text{ correct}|\text{QR and }\mathcal{A}\text{ correct}] \cdot \Pr[\text{QR and }\mathcal{A}\text{ correct}] \\
& + \Pr[\mathcal{B}\text{ correct}|\text{QR and }\mathcal{A}\text{ incorrect}] \cdot \Pr[\text{QR and }\mathcal{A}\text{ incorrect}] \\
& + \Pr[\mathcal{B}\text{ correct}|\text{QNR and }\mathcal{A}\text{ correct}] \cdot \Pr[\text{QNR and }\mathcal{A}\text{ correct}] \\
& + \Pr[\mathcal{B}\text{ correct}|\text{QNR and }\mathcal{A}\text{ incorrect}] \cdot \Pr[\text{QNR and }\mathcal{A}\text{ incorrect}] \\
= {} & 0 + 1 \cdot \frac{1}{2} \cdot \frac{1}{2} + 1 \cdot \frac{1}{2}\left(\frac{1}{2} + \varepsilon\right) + 0 \\
= {} & \frac{1}{2} + \frac{\varepsilon}{2}.
\end{aligned}
$$

■

2.2. ElGamal: HE starting from a linear homomorphism

We provide one more example of a simple homomorphic encryption system. The ElGamal cryptosystem [35] was probably directly inspired by the Diffie–Hellman protocol [31], but it *could* have been invented by starting from a linear homomorphism, and then hardening the homomorphism, as follows.

Let q be a prime integer. For $\vec{s} \in (\mathbb{Z}/q\mathbb{Z})^n$, the inner product $\langle \vec{s}, \cdot \rangle : (\mathbb{Z}/q\mathbb{Z})^n \to \mathbb{Z}/q\mathbb{Z}$ is a linear homomorphism. We have the following commutative diagram:

$$
\begin{array}{ccc}
(\mathbb{Z}/q\mathbb{Z})^n \times (\mathbb{Z}/q\mathbb{Z})^n & \xrightarrow{\quad\boxplus\quad} & (\mathbb{Z}/q\mathbb{Z})^n \\
\Big\downarrow {\scriptstyle \langle \vec{s},\cdot\rangle, \langle \vec{s},\cdot\rangle} & & \Big\downarrow {\scriptstyle \langle \vec{s},\cdot\rangle} \\
\mathbb{Z}/q\mathbb{Z} \times \mathbb{Z}/q\mathbb{Z} & \xrightarrow{\quad + \quad} & \mathbb{Z}/q\mathbb{Z},
\end{array}
$$

where $\boxplus$ is vector addition.

How can we "harden" the linear homomorphism to build a homomorphic encryption system? The IND-CPA game (see Definition 1) allows an adversary to obtain many (message, ciphertext) pairs. If such pairs have the form $(m, \vec{c}) \in (\mathbb{Z}/q\mathbb{Z})^{n+1}$ such that $m = \langle \vec{s}, \vec{c} \rangle$, the adversary can efficiently solve for $\vec{s}$ using linear algebra. One countermeasure to this linear algebra attack is to put the elements of $\mathbb{Z}/q\mathbb{Z}$ "in the exponent." That is, let G be a cyclic group of prime order q whose binary operation we denote multiplicatively. For example, G could be a subgroup of the multiplicative group of a finite field, or of the group of points of an elliptic curve group over a finite field. Let g be a generator of G. In such groups, given G and g, recovering $a \in \mathbb{Z}/q\mathbb{Z}$ from g^a is called the *discrete logarithm* (DL) problem, which is believed to be hard for appropriate groups.[4] We now have the following commutative diagram:

$$\begin{array}{ccc} G^n \times G^n & \xrightarrow{\boxtimes} & G^n \\ \Big\downarrow{\scriptstyle \langle \vec{s},\cdot\rangle, \langle \vec{s},\cdot\rangle} & & \Big\downarrow{\scriptstyle \langle \vec{s},\cdot\rangle} \\ G \times G & \xrightarrow{\times} & G, \end{array}$$

where now $\langle \vec{s}, \cdot \rangle$ is applied "in the exponent," i.e., for $(g_1, \ldots, g_n) \in G^n$, we have

$$\langle \vec{s}, (g_1, \ldots, g_n) \rangle = \prod_{i=1}^{n} g_i^{s_i},$$

and $\boxtimes$ is componentwise multiplication.

Now, let us build the Elgamal encryption system from the commutative diagram. Elgamal uses a 2-dimensional secret key $\vec{s}$ with the special form $(-s, 1)$. So, ciphertexts and messages live in G^2 and G, respectively. As depicted by the commutative diagram, decryption involves applying $\langle \vec{s}, \cdot \rangle$ "in the exponent." The diagram also specifies the multiplicative homomorphism. Now, what do we need to put in the public key to allow anyone to generate a random encryption of any element of G? A random encryption of $m \in G$ is simply *some* encryption of m multiplied (via $\boxtimes$) by a random encryption of g^0. Anybody can easily compute *some* encryption of m as $\vec{c} = (g^0, m)$, since $\langle \vec{s}, \vec{c} \rangle = (g^0)^{-s} \cdot m^1 = m$. A random encryption of g^0 has the form $(g^r, g^{r \cdot s})$ for r sampled uniformly from $\mathbb{Z}/q\mathbb{Z}$. To enable generation of a random encryption of g^0, the public encryption key needs only a some nontrivial encryption of g^0, in particular, (g, g^s) suffices.

ElGamal encryption system.

- Key Generation: $\mathsf{K}(\lambda, r)$ takes a security parameter λ and some randomness r as input. It uses λ and the randomness to generate (G, g), a group and generator of order $q = q(\lambda)$. (Alternatively, the group may be preset and common to many users.) It generates a random $s \in \mathbb{Z}/q\mathbb{Z}$ and sets $h \leftarrow g^s$. The secret key sk is s. The public encryption ek is (G, g, h). The message set $\mathcal{M}$ is G. The ciphertext set $\mathcal{C}$ is G^2.

4 The reverse problem of computing g^a from g for $a \in \mathbb{Z}/q\mathbb{Z}$ can be solved efficiently using only $O(\log q)$ multiplications in G using the technique of "repeated squaring."

- Encryption: $\mathsf{E}(ek, m, r)$ takes the encryption key ek, a message $m \in \mathcal{M}$ and some randomness as input. It generates random $t \in (\mathbb{Z}/q\mathbb{Z})$. It outputs $c \leftarrow (g^t, m \times h^t)$.
- Decryption: $\mathsf{D}(sk, c)$ takes the secret key sk and a ciphertext $c = (c_0, c_1) \in \mathcal{C}$ as input. It outputs $m \leftarrow c_0^{-s} \times c_1 \in \mathcal{M}$.
- Homomorphic multiplication: it takes two ciphertexts $c^{(1)}, c^{(2)} \in \mathcal{C}$ and outputs $c \leftarrow c_1 \boxtimes c_2$.

The computational assumption underlying ElGamal is called the Diffie–Hellman assumption.

Definition 5 (Diffie–Hellman (DH) assumption). Let G be a fixed group of order q (determined by security parameter λ) with generator g. Sample a random bit $\beta \in \{0, 1\}$. If $\beta = 0$, sample x and y randomly from $\mathbb{Z}/q\mathbb{Z}$ and set $z = x \cdot y$. If $\beta = 1$, sample x, y, and z randomly from $\mathbb{Z}/q\mathbb{Z}$. Output (G, g, g^x, g^y, g^z). The DH assumption is that all probabilistic polynomial time (in λ) adversaries $\mathcal{A}$ have negligible advantage in guessing the bit β.

Note that is easy to determine whether the discrete logarithms of a tuple satisfy a given linear equation. The DH assumption is basically that it is hard to distinguish whether the discrete logarithms satisfy a degree-2 equation. For some elliptic curve groups over finite fields, the fastest algorithm for distinguishing Diffie–Hellman is to solve the discrete logarithm problem by using the "baby-step giant-step" method, which takes roughly $\sqrt{q}$ computational steps.

Now, we prove the security of ElGamal based on the Diffie–Hellman (DH) assumption.

Theorem 2. *The Elgamal encryption system is IND-CPA secure under the DH assumption.*

Proof. Let $\mathcal{B}$ be an algorithm that is given an instance of the DH problem, namely, (G, g, g^x, g^y, g^z) such that if $\beta = 0$ then $z = x \cdot y$, but if $\beta = 1$ then z is sampled uniformly and independently modulo q. Here $\mathcal{B}$'s task is to distinguish the bit β while $\mathcal{B}$ and $\mathcal{A}$ play the roles of the challenger and adversary in the IND-CPA game. Algorithm $\mathcal{B}$ gives (G, g, g^x) to $\mathcal{A}$ as the public encryption key. Then $\mathcal{B}$ chooses a random bit $b \in \{0, 1\}$. When $\mathcal{A}$ queries messages $(m_{i,0}, m_{i,1})$, $\mathcal{B}$ samples randomness $r_i \in \mathbb{Z}/q\mathbb{Z}$, and sends the ciphertext $(g^0, m_{i,b}) \boxtimes ((g^y)^{r_i}, (g^z)^{r_i})$. Adversary $\mathcal{A}$ outputs a guess $b' \in \{0, 1\}$. If $b' = b$, then $\mathcal{B}$ guesses that $\beta = 0$; otherwise it guesses that $\beta = 1$.

One can check that $\mathcal{B}$'s advantage in the DH game is $\frac{\varepsilon}{2}$, where ε is $\mathcal{A}$'s advantage in the IND-CPA game. The idea, as in the security proof for Goldwasser–Micali system, is that everything—namely the public key and ciphertexts—is distributed properly when $\beta = 0$ and $z = x \cdot y$, and so $\mathcal{A}$ should have advantage ε in that case. In particular, $((g^y)^{r_i}, (g^z)^{r_i})$ is a random encryption of g^0, and so the ith ciphertext is indeed a random encryption of $m_{i,b}$. However, when $\beta = 1$, with high probability $((g^y)^{r_i}, (g^z)^{r_i})$ is an encryption of a random value. Therefore the ciphertexts generated by $\mathcal{B}$ encrypt random values independent of b, and $\mathcal{A}$ has no advantage in guessing b. So, $\mathcal{A}$'s success (or lack of it) gives $\mathcal{B}$ a clue about the value of β. ■

3. GENERAL RESULTS ABOUT HOMOMORPHIC ENCRYPTION

Now that we have in mind some simple examples of homomorphic encryption systems, let us provide a general definition of homomorphic encryption and some general results.

3.1. Formal definition of HE

A homomorphic encryption system is, first of all, an encryption system:

Definition 6 (Encryption (syntax)). An encryption system consists of 3 functions: key generation K, encryption E, and decryption D:

- $(sk, ek, \mathsf{params}) \leftarrow \mathsf{K}(\lambda, r)$ takes the security parameter λ and some randomness r and outputs a secret decryption key sk, an encryption key ek, some parameters params of the system, such as the message set $\mathcal{M}$. In a symmetric system, the encryption key ek equals sk and is kept secret. In an asymmetric (or public-key) system, ek is public and does not equal sk. We omit mentioning params as an input to the other functions.

- $c \leftarrow \mathsf{E}(ek, m, r)$ takes the encryption key ek, a message $m \in \mathcal{M}$, and some randomness r and outputs a ciphertext. Encryption is probabilistic: new randomness r is sampled for each encryption.

- $m \leftarrow \mathsf{D}(sk, c)$ takes a secret key and ciphertext and returns a message $m \in \mathcal{M}$.

We write $\vec{c} \leftarrow \mathsf{E}(ek, \vec{m}, \vec{r})$ and $\vec{m} \leftarrow \mathsf{D}(sk, \vec{c})$ for vectors of messages and ciphertexts; K, E, and D all can be computed in time polynomial in the security parameter λ.

Definition 7 (Correctness of encryption). It must hold that $m = \mathsf{D}(sk, \mathsf{E}(ek, m, r))$ for all key tuples (sk, ek) in the image of K, all $m \in \mathcal{M}$, and all randomness r.

The IND-CPA security of encryption system is as described in Definition 3.

A homomorphic encryption system also has a fourth function V (evaluation). The homomorphic property requires some tweaks to K as well. (The functions E and D are as before.)

Definition 8 (Homomorphic encryption (syntax)). A homomorphic encryption system consists of 4 functions: key generation K, encryption E, decryption D, and evaluation V:

- K: As in an encryption system, except that K also outputs a public evaluation key evk, and params includes some description of a set $\mathcal{F}$ of functions, with input and output over $\mathcal{M}$, that the homomorphic encryption system is capable of evaluating correctly (see below).

- E: As in an encryption system.
- D: As in an encryption system.
- $c \leftarrow \mathsf{V}(evk, f, c_1, \ldots, c_t)$ takes evk, a function $f \in \mathcal{F}$, and t ciphertexts $c_1, \ldots, c_t$, where t is the number of inputs to f. It outputs a ciphertext c.

The above K, E, D, and V all can be computed in time polynomial in the security parameter λ, though V's complexity necessarily also depends (polynomially) on the complexity of the function f being evaluated.

The security notion of homomorphic encryption remains IND-CPA security, without reference to V. This is because V is a public function with no secrets. The adversary is, of course, free to try to use V in its attack.

A homomorphic encryption system must satisfy not only the basic correctness of encryption, but also correctness of evaluation. We will define the correctness of evaluation with commutative diagrams. First, note that the images of E and of V need not be the same in general (though they were the same for the simple homomorphic encryption systems we presented in Section 2).

Definition 9 (Fresh and evaluated ciphertexts). We differentiate between two types of ciphertexts:

- "Fresh ciphertexts" (denoted by $\mathcal{C}_{\mathsf{E}}$): ciphertexts in the image of E,
- "Evaluated ciphertexts" (denoted by $\mathcal{C}_{\mathsf{V}}$): a superset of $\mathcal{C}_{\mathsf{E}}$ that also includes ciphertexts in the image of V when evaluated on a function $f \in \mathcal{F}$ and ciphertexts from $\mathcal{C}_{\mathsf{E}}$.

Though the notation suppresses it, these sets depend on the particular encryption key ek and evaluation evk being used.

Definition 10 (Correctness of evaluation (bounded homomorphic encryption)). A homomorphic encryption system correctly evaluates a set of functions $\mathcal{F}$, and is called $\mathcal{F}$-homomorphic, if

$$\mathsf{D}\big(sk, \mathsf{V}(evk, f, c_1, \ldots, c_t)\big) = f\big(\mathsf{D}(sk, c_1), \ldots, \mathsf{D}(sk, c_t)\big)$$

for all (sk, ek, evk) in the image of K, all fresh ciphertexts $\{c_i\}$ in $\mathcal{C}_{\mathsf{E}}$ (for key ek), and all $f \in \mathcal{F}$. This correctness requirement is depicted by the commutative diagram for bounded homomorphic encryption in Figure 1a.

Unless otherwise specified, a homomorphic encryption system is *bounded*, as depicted in Figure 1a, since correctness of evaluation is *a priori* guaranteed to hold only when the inputs are fresh ciphertexts come from $\mathcal{C}_{\mathsf{E}}$, not necessarily from $\mathcal{C}_{\mathsf{V}}$. Bounded homomorphic encryption systems can be trivial and uninteresting—for example, when V does nothing but output f and its input ciphertexts $c_1, \ldots, c_t$, leaving it to the decryption function D to decrypt the c_i's and apply f to the messages.

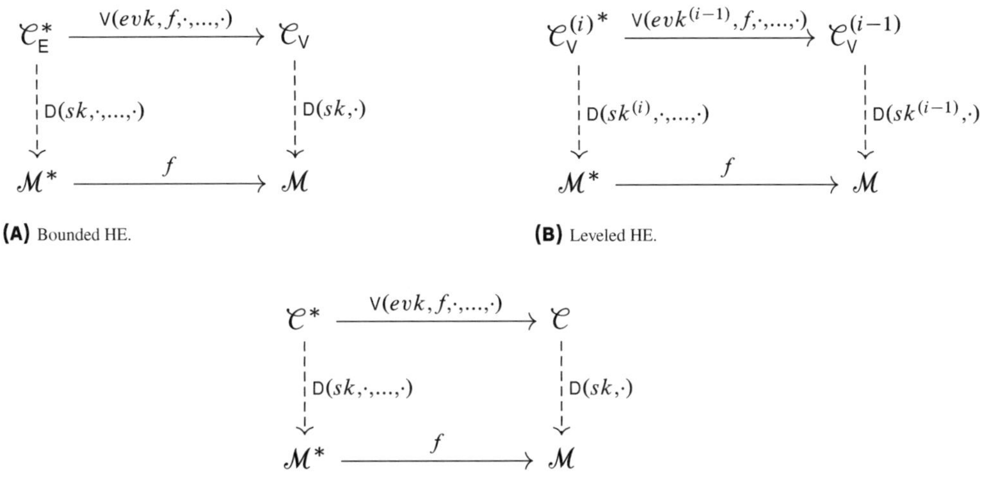

(A) Bounded HE.

(B) Leveled HE.

(C) Unbounded HE.

FIGURE 1

Bounded, leveled, and unbounded systems with $f \in \mathcal{F}$ for function set $\mathcal{F}$. The notation $\mathcal{M}^*$ indicates that the number of copies of $\mathcal{M}$ depends on the number of inputs to f.

One property that can make a homomorphic encryption system nontrivial is if it is *unbounded.*

Definition 11 (Unbounded homomorphic encryption). We say a homomorphic encryption system is unbounded $\mathcal{F}$-homomorphic if, for some set of ciphertexts $\mathcal{C}$, the commutative diagram in Figure 1c holds.

For an unbounded system, the functions in $\mathcal{F}$ can be applied repeatedly, indefinitely. The size of the ciphertexts, and the cost of their decryption, does not depend on how many homomorphic operations were performed. The simple homomorphic encryption systems discussed in Section 2 are unbounded with respect to a single binary operation.

In-between bounded and unbounded, we have a notion of leveled homomorphic encryption.

Definition 12 (Leveled homomorphic encryption). We say a homomorphic encryption system is leveled $\mathcal{F}$-homomorphic if, for any number $n \in \mathbb{N}$ (a parameter to be included in params that indicates the number of levels), the commutative diagram in Figure 1b holds for all $i \in \{1, \ldots, n\}$. The functions K, E, D, V are required to be independent of n, aside from the fact that K generates a key-tuple $(sk^{(i)}, ek^{(i)}, evk^{(i)})$ for each level.

Our definition of leveled homomorphic encryption here is strict; sometimes the definition allows the complexity of the functions to grow polynomially in n. Note that, in the leveled system, V converts ciphertexts under key $ek^{(i)}$ to the next key $ek^{(i-1)}$.

Now what about the set $\mathcal{F}$? The set $\mathcal{F}$ can be limited or powerful. The systems considered in Section 2 are unbounded, but can perform only a single binary operation that is insufficient to perform arbitrary computations. In contrast, we say that $\mathcal{F}$ is *functionally complete* if it contains a functionally complete (or *universal*) set of operations or "gates." Examples of a universal set of gates are $\{\mathsf{AND}, \mathsf{NOT}\}$ and $\{\mathsf{NAND}\}$. Boolean circuits composed of a universal set of Boolean gates are very powerful: any problem that can be computed in polynomial time by a deterministic Turing machine can also be computed by a polynomial-size Boolean circuit family. That is, Boolean circuits can efficiently perform general efficient computation (as classically defined with respect to Turing machines). Our ultimate goal is a homomorphic encryption system that is unbounded for universal gates.

Definition 13 (Fully homomorphic encryption (FHE)). A homomorphic encryption system is called fully homomorphic (resp. leveled fully homomorphic) if it is unbounded (resp. leveled) and its $\mathcal{F}$ contains a functionally complete set of gates.

With a fully homomorphic encryption (FHE) system, we can do arbitrary computations on data while it remains encrypted. Most of the rest of this survey focuses on building FHE.

3.2. General approach to key generation and encryption

Decryption D and evaluation V are the stars of our commutative diagrams, but let us focus on key generation K and encryption E for a moment.

Encryption E is simply the inverse of D. Given $m \in \mathcal{M}$, we encrypt m by sampling from the preimage of m under the decryption map $\mathsf{D}(sk, \cdot)$. In an asymmetric (public-key) system, such sampling must be possible only with the public encryption key ek, without sk. Also, for the system to be IND-CPA secure, this sampling must be probabilistic. Can we devise a simple, natural, secure way to encrypt, that is, to sample randomly from the decryption-map preimage of a message?

For a homomorphic encryption system, a possible answer immediately presents itself. Populate the encryption key ek with (image, preimage) pairs $\{(m_i, c_i)\}$. Then, the encrypter *uses the homomorphism of the system* to generate a random preimage of its m from the given preimages of $\{m_i\}$ in ek. That is, the encrypter finds a random function $f \in \mathcal{F}$ such that $f(m_1, \dots, m_t) = m$, and then outputs the ciphertext $c = \mathsf{V}(ek, f, c_1, \dots, c_t)$ as its encryption of m.

But we need to be careful here. The ciphertext c certainly decrypts to the correct message m. But how can we be sure that c does not retain some detectable residue of its history? A homomorphic encryption system may be IND-CPA secure, yet still allow an adversary to distinguish that a ciphertext c was produced by evaluating a particular function f on $c_1, \dots, c_t$, which in this case would allow the adversary to recover m. In fact, the "trivial" bounded HE system mentioned after Definition 10 (in Section 3.1) has exactly this property. For the encryption procedure to be secure, we need to ensure that the encrypter's ciphertext c "forgets" how it was made.

Suppose that an evaluated ciphertext c is at most ℓ bits, i.e., c only "remembers" ℓ bits. And also suppose that we generate c as (say) a random linear combination of the c_i's in ek, subject to the same random linear combination of the m_i's equaling m. If t (the number of ciphertexts in ek) is very large, such that the entropy of the random linear combination is much greater than ℓ, then indeed (whp) c's precise history will be very ambiguous given only c itself.

Rothblum [69] formalizes this intuition for homomorphic encryption systems that have sufficiently compact ciphertexts and a $\boxplus$ operation that is additively homomorphic over $\mathcal{M} = \mathbb{Z}/2\mathbb{Z}$. Precisely, he proves:

Theorem 3 (Rothblum). *Let $\mathcal{E}_{\text{sym}}$ be an IND-CPA secure symmetric encryption system that is homomorphic with respect to addition modulo 2. Suppose that there is a polynomial t such that homomorphically adding $t(\lambda)$ fresh ciphertexts results in a ciphertext of at most $t(\lambda)/10$ bits. Then from $\mathcal{E}_{\text{sym}}$ one can build an IND-CPA secure asymmetric system $\mathcal{E}_{\text{asym}}$.*

Rothblum presents the result somewhat differently as constructing asymmetric encryption from homomorphic symmetric encryption. But the bottom line is the same: the encryption algorithm practically writes itself as long as there is an efficient procedure (using sk) to sample a handful of random (image, preimage) pairs of the decryption map to put in ek, and we can evaluate homomorphic addition modulo 2 compactly.

In detail, Rothblum builds the asymmetric system as follows:

Rothblum's asymmetric system $\mathcal{E}_{\text{asym}}$.

- $\mathcal{E}_{\text{asym}}.\mathsf{K}$: Compute $\mathcal{E}_{\text{sym}}.\mathsf{K}$ to obtain sk. Use $\mathcal{E}_{\text{sym}}.\mathsf{E}(sk,\cdot,\cdot)$ to generate $X^{(0)} = (X_1^{(0)}, \ldots, X_t^{(0)})$ and $X^{(1)} = (X_1^{(1)}, \ldots, X_t^{(1)})$, which are $t = t(\lambda)$ encryptions of 0 and t encryptions of 1, respectively. The secret decryption key is sk. The public encryption key ek is $(X^{(0)}, X^{(1)})$.

- $\mathcal{E}_{\text{asym}}.\mathsf{E}$: Let $\boxplus$ denote $\mathcal{E}_{\text{sym}}$'s homomorphic addition mod 2 operation. To encrypt $m \in \{0, 1\}$, sample a random vector $r \in \{0, 1\}^t$ such that $r_1 + \cdots + r_t = m \bmod 2$ and output the ciphertext $c = X_1^{(r_1)} \boxplus \cdots \boxplus X_\ell^{(r_t)}$.

- $\mathcal{E}_{\text{asym}}.\mathsf{D}$: Identical to $\mathcal{E}_{\text{sym}}.\mathsf{D}$.

Correctness follows easily from the properties of $\mathcal{E}_{\text{sym}}$.

Rothblum's proof of IND-CPA security comes in two parts. First, he proves that if $\mathcal{E}_{\text{sym}}$ is indeed IND-CPA secure, a polynomial-time adversary will not notice if $X^{(1)}$ is replaced by t more encryptions of 0. This follows immediately from the definition of IND-CPA security (see Definition 3).

Second, assuming now that $X^{(0)}$ and $X^{(1)}$ are now $2t$ i.i.d. encryptions of 0, Rothblum shows that a ciphertext generated as $c = X_1^{(r_1)} \boxplus \cdots \boxplus X_t^{(r_t)}$ "forgets" the value $r_1 + \cdots + r_t \bmod 2$ (the value that is supposed to be encrypted). Specifically, the possible preimages $(r_1, \ldots, r_t)$ for c satisfy $r_1 + \cdots + r_t = 0 \bmod 2$ with probability at most $\frac{1}{2} + 2^{-0.2t+\ell+1}$, where ℓ is the number of bits in c (and similarly for the case of 1 mod 2). As $\ell < t/10$, this probability is negligibly close to $\frac{1}{2}$.

3.3. Getting to functional completeness

Now, let us return to our main goal, which is constructing FHE. Suppose we have an unbounded system $\mathcal{E}$ that is correct for a non-functionally-complete set $\mathcal{F}$. Can we use $\mathcal{E}$ to get an FHE system $\mathcal{E}_{\text{FHE}}$?

In some cases, yes. Here is a silly example. The arithmetic gates $\{+, \times\}$ are not functionally complete over GF(2). In particular, any circuit composed of $\{+, \times\}$ gates can only output 0 when the inputs are all 0 (and so such circuits cannot express functions that output 1 when the inputs are all 0). But this technicality is not a real obstacle to constructing FHE. As long as $\mathcal{E}$ is capable of producing a single encryption of 1 (e.g., via encryption) it can evaluate $\mathsf{NOT}(x)$ as $1 + x$. (And it can emulate $\mathsf{AND}(x, y)$ as $x \times y$.)

Here is a less trivial example. The gates $\{\mathsf{AND}, \mathsf{OR}\}$ are not functionally complete. Circuits composed of $\{\mathsf{AND}, \mathsf{OR}\}$ gates can only compute *monotone* functions (not general functions), where a function f is called monotone if $f(x) \leq f(y)$ whenever $x_i \leq y_i$ for all i. However, via De Morgan's law, we can reexpress any Boolean circuit as a circuit that is monotone except at the input level, which is allowed to have NOT gates. Applying De Morgan's law, given an unbounded $\{\mathsf{AND}, \mathsf{OR}\}$-homomorphic system $\mathcal{E}$, we can construct an FHE system $\mathcal{E}_{\text{FHE}}$ as follows. An $\mathcal{E}_{\text{FHE}}$ ciphertext encrypting "1" consists of an ordered pair of two $\mathcal{E}$ ciphertexts encrypting 1 and 0, respectively. An $\mathcal{E}_{\text{FHE}}$ ciphertext encrypting "0" consists of an ordered pair of two $\mathcal{E}$ ciphertexts encrypting 0 and 1, respectively. Performing a NOT gate in $\mathcal{E}_{\text{FHE}}$ is simple: just swap the $\mathcal{E}$ ciphertexts in the ordered pair. To perform an AND gate in $\mathcal{E}_{\text{FHE}}$, take the $\mathcal{E}$-AND of the first $\mathcal{E}$ ciphertexts in each pair, and the $\mathcal{E}$-OR of the second $\mathcal{E}$ ciphertexts in each pair.

3.4. Homomorphic encryption unbound: Recryption and bootstrapping

Suppose we have a bounded system $\mathcal{E}$ that is $\mathcal{F}$-homomorphic. Can we use $\mathcal{E}$ to get an FHE system $\mathcal{E}_{\text{FHE}}$? Is there some "special" function f such that, if $f \in \mathcal{F}$, we automatically get FHE?

Here is a crazy idea for the "special" function f: *the system's own decryption function* D*!* D is a function, expressible as a circuit, that takes a secret key sk and ciphertext c as input, and outputs a message m. So, can D be in $\mathcal{F}$? Does this sort of self-embedding lead to impossibilities, as in Gödel's Incompleteness Theorem and Turing's Halting Problem? Or, does the self-embedding actually work, and what are the consequences? We will see that, if a homomorphic encryption system can evaluate its own decryption function, plus "a little bit more," we can *bootstrap* the system to obtain a fully homomorphic system.

First, let us work out what happens when $\mathsf{D} \in \mathcal{F}$, and we evaluate D "inside" V. We start with the commutative diagram in Figure 1a for a bounded homomorphic encryption system. In the diagram, $\mathcal{C}_{\mathsf{E}}$ denotes the image of the encryption algorithm E—i.e., "fresh" ciphertexts—and $\mathcal{C}_{\mathsf{V}}$ denotes the superset of $\mathcal{C}_{\mathsf{E}}$ of "evaluated" ciphertexts. The commutative diagram captures the correctness requirement on V with respect to functions from $\mathcal{F}$. Assuming $\mathsf{D} \in \mathcal{F}$, we are interested in what happens when we start with some value in the upper-left corner, and apply D homomorphically (inside V). Since the diagram is commuta-

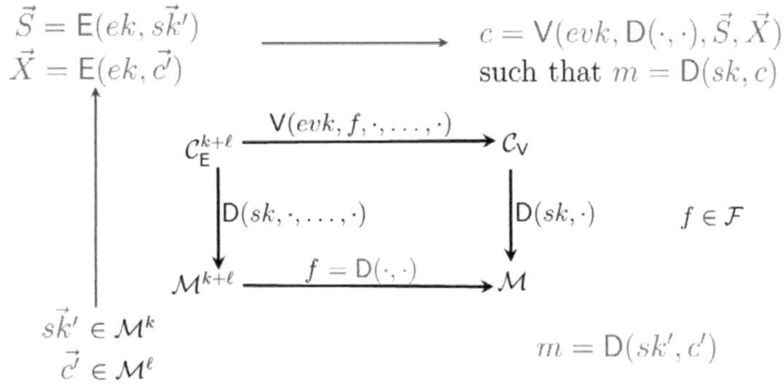

FIGURE 2

Recryption: Evaluating decryption homomorphically.

tive, we can gain useful information about what happens by also considering the lower path through the diagram.

Accordingly, now let us assign the bottom rightward arrow f to be $\mathsf{D}(\cdot,\cdot)$ and see what happens as we complete the diagram. Follow along in Figure 2. Though the input to $f = \mathsf{D}(\cdot,\cdot)$ could be arbitrary, the natural input to D is a pair (sk', c') such that sk' is a secret key and c' is a ciphertext with $m = \mathsf{D}(sk', c')$. (Note that sk' might, or might not, equal the key sk that is already in the diagram.) We put those values in the lower-left and lower-right. For this to make sense in the diagram, sk' and c' must be expressible as strings over the message set $\mathcal{M}$. So, suppose that they can be expressed as strings in $\mathcal{M}^k$ and $\mathcal{M}^\ell$, respectively, and let us write them as vectors, $\vec{sk'}$ and $\vec{c'}$. (This reexpression is especially straightforward when $\mathcal{M}$ is simply $\mathbb{Z}/2\mathbb{Z}$, i.e., when messages are bits.) We abuse notation by using D even when the domain is $\mathcal{M}^{k+\ell}$. Continuing to complete the diagram, for the left downward arrow to hold, we need the upper-left to be fresh encryptions of the secret key and ciphertext "bits." So, we will assume for now that ek is public, and we set $\vec{S} = \mathsf{E}(ek, \vec{sk'})$ and $\vec{X} = \mathsf{E}(ek, \vec{c'})$. (We have omitted the randomness of encryption.) Finally, in the top rightward arrow, we apply V with $f = \mathsf{D}(\cdot,\cdot)$ to the freshly encrypted bits of the secret key sk' and ciphertext c', to obtain a ciphertext c in the upper-right corner. Recall that $\mathsf{D} \in \mathcal{F}$ by assumption, so V is guaranteed to satisfy the correctness requirement. So, what does c encrypt? By the commutativity of the diagram, it must encrypt m, just like the original ciphertext c'! If $\mathsf{D} \in \mathcal{F}$, then given an initial ciphertext c' that encrypts m under sk', we can produce a new ciphertext c that encrypts m under sk, by running decryption homomorphically (inside V) using encryptions of bits of the secret key sk' and the ciphertext c'. This process is called *recryption*.

Interestingly, $\vec{X} = \mathsf{E}(ek, \vec{c'})$ is a "double encryption" of m, with the inner encryption under ek', and the outer encryption under ek. Starting with $\vec{X}$ and then taking the down-then-right path through the diagram, we remove the outer encryption first (as you would expect), then the inner encryption, to recover m. Taking the right-then-down path, we remove the inner encryption under ek' first, then the outer encryption!

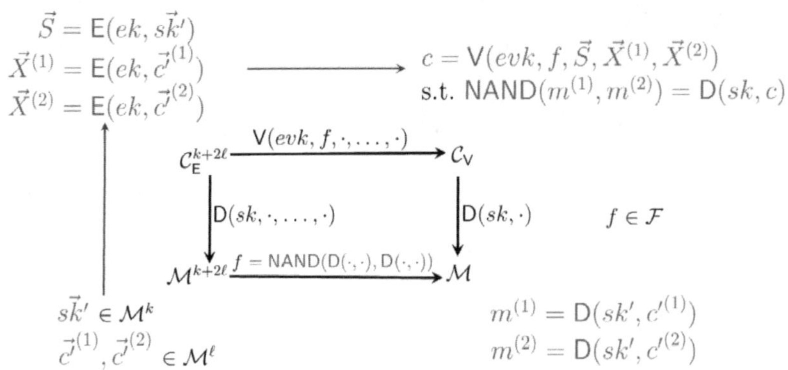

FIGURE 3

Recryption-then-NAND and Bootstrappable Homomorphic Encryption.

You may be underwhelmed. What have we gained by recryption? The homomorphic encryption system probably already provides us with much simpler ways to obtain an encryption of the same message, such as applying homomorphic addition $\boxplus$ with an encryption of 0, or homomorphic multiplication $\boxtimes$ with an encryption of 1. Why bother with the more complex process?

First, notice that we do not have any guarantee that c' can participate correctly in even simple operations such as $\boxplus$ and $\boxtimes$. The ciphertext c' is not necessarily a fresh encryption. On the other hand, recryption does not perform additional operations on c'. Instead, it performs operations on *fresh* encryptions from $\mathcal{C}_{\mathsf{E}}$ that encrypt the bits of sk' and c'. Recryption is guaranteed to work correctly assuming $\mathsf{D} \in \mathcal{F}$.

Second, as hinted above, a good motivation for getting a new encryption of the same message would be if the new ciphertext is "refreshed," i.e., it can participate correctly in more operations, guaranteed. Trivially, we could refresh c' by decrypting it then applying E to obtain a fresh encryption of the same message. But this process requires sk', which we want to keep secret. So we arrive at the real motivation for considering recryption: maybe it can refresh a ciphertext by decrypting it *homomorphically*, requiring only an *encrypted* secret key.

If a recrypted ciphertext c is indeed refreshed, it means we should be able to apply some additional operation after recryption, such as a NAND gate (if the messages are bits). So, assume that the messages are bits and that the function $f = \mathsf{NAND}(\mathsf{D}(\cdot,\cdot), \mathsf{D}(\cdot,\cdot))$ is in $\mathcal{F}$. This function f takes as input a secret key sk' and two ciphertexts $c'^{(1)}, c'^{(2)}$, decrypts the two ciphertexts with the secret key, and applies the NAND gate to the two messages. Since this f is in $\mathcal{F}$ by assumption, we have the guarantee that if $m^{(1)} = \mathsf{D}(sk', c'^{(1)})$ and $m^{(2)} = \mathsf{D}(sk', c'^{(2)})$, then going clockwise through the commutative diagram in Figure 3 gives us a ciphertext c such that $\mathsf{NAND}(m^{(1)}, m^{(2)}) = \mathsf{D}(sk, c)$. By using this process for every NAND gate, we can evaluate an arbitrary circuit of NAND gates. Recall that NAND is, by itself, a functionally complete gate, enabling general computation. Therefore, we obtain a fully homomorphic encryption system.

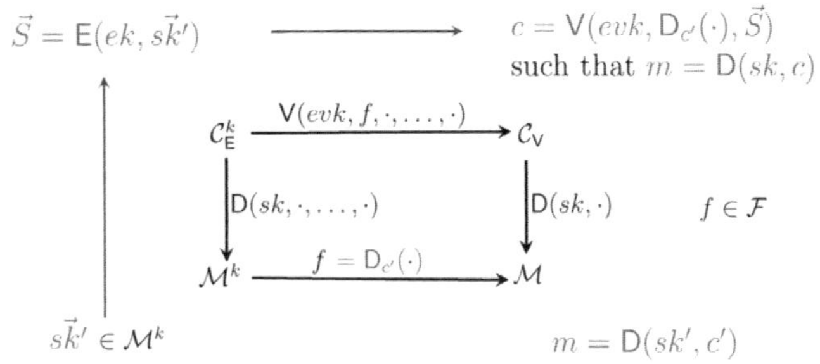

FIGURE 4

Recryption variant: Evaluating decryption homomorphically with ciphertext prewired.

To get FHE, all we need is a bounded homomorphic encryption system such that this weird function $\mathsf{NAND}(\mathsf{D}(\cdot,\cdot), \mathsf{D}(\cdot,\cdot))$ is in $\mathcal{F}$! We call such a bounded homomorphic encryption system *bootstrappable*, and call this process *bootstrapping*.

In retrospect, bootstrapping seems like the almost inevitable answer for how to refresh a ciphertext generically. Generically, if all we are given is the commutative diagram of a bounded homomorphic encryption system (with its function set $\mathcal{F}$) and a ciphertext c in $\mathcal{C}_\mathsf{V}$ but not $\mathcal{C}_\mathsf{E}$ that we want to refresh, the only possible way to refresh it is to somehow use V. We cannot input c as a ciphertext directly into V, since it is not in $\mathcal{C}_\mathsf{E}$. Yet, we must give V some inputs that retain the information in c, and that V can operate on correctly. We have only two choices: either c must be embedded in *fresh* ciphertexts input to V (e.g., encrypted bitwise, as above) or c must be embedded in the function f that we give to V to evaluate. In either case, since the only useful thing we know about c is that it is in $\mathcal{C}_\mathsf{V}$ and encrypts some m, it seems that the only meaningful functions we can evaluate on c are functions that first decrypt c (and thereafter perform some operations on m). And so we arrive at bootstrapping. (This reasoning does not preclude nongeneric techniques for refreshing ciphertexts.)

Embedding the ciphertext(s) to be refreshed in the function to be evaluated—as opposed to encrypting the bits of these ciphertexts—is actually preferable. If we do not encrypt the bits of ciphertext(s), we do not need ek to be public (the encryption system can be symmetric). Also, we do not need to reexpress ciphertexts in terms of message "bits." In detail, in this approach, instead of evaluating $\mathsf{D}(\cdot,\cdot)$, we can evaluate the function $\mathsf{D}_{c'} = \mathsf{D}(\cdot, c')$, where c' comes "prewired." Similarly, we can replace $\mathsf{NAND}(\mathsf{D}(\cdot,\cdot), \mathsf{D}(\cdot,\cdot))$ with $\mathsf{NAND} \circ \mathsf{D}_{c'^{(1)}, c'^{(2)}}(\cdot)$, a function that when given sk' as input, decrypts ciphertexts $c'^{(1)}$ and $c'^{(2)}$ and then NANDs their respective messages. We provide revised versions of recryption and recrypt-then-NAND in Figures 4 and 5.

Now, let us state our FHE result a bit more formally.

Definition 14 (Bootstrappable homomorphic encryption). A (possibly bounded) homomorphic encryption system $\mathcal{E}$ with function set $\mathcal{F}$ is *bootstrappable* if there is a functionally complete set of binary gates Γ, such that for all $g \in \Gamma$, and all ciphertexts $c^{(1)}, c^{(2)} \in \mathcal{C}_\mathsf{V}$, $g \circ \mathsf{D}_{c^{(1)}, c^{(2)}}(\cdot) \in \mathcal{F}$.

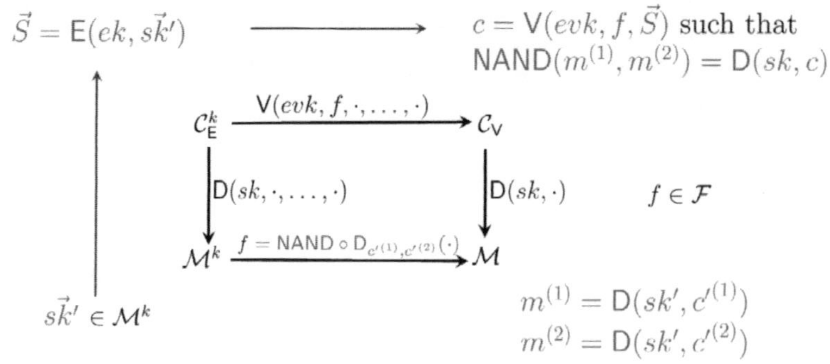

FIGURE 5

Recryption-then-NAND and Bootstrappable Homomorphic Encryption, with two ciphertexts prewired; $\mathsf{NAND} \circ \mathsf{D}_{c'^{(1)}, c'^{(2)}}(\cdot)$ is a function that, when given sk' as input, decrypts ciphertexts $c'^{(1)}$ and $c'^{(2)}$ and then NANDs their respective messages.

We limited the above definition to binary gates for convenience, and because it usually seems to suffice in practice. Figure 5 depicts a bootstrappable homomorphic encryption system where Γ includes NAND, which is functionally complete by itself.

Theorem 4 (Bootstrapping to leveled FHE [40, 41]). *If $\mathcal{E}$ is a bootstrappable system that is IND-CPA secure, then it can be transformed into a leveled FHE system $\mathcal{E}_{LFHE}$ that is IND-CPA secure.*

Theorem 4 gives us leveled FHE, not unbounded FHE. For unbounded FHE, there is an issue in the proof of IND-CPA security. Specifically, to obtain unbounded FHE, we need for it to be secure to encrypt $\mathcal{E}$'s secret key sk under its companion encryption key ek. We call this property *circular security*. (For leveled FHE, we can avoid the circular security issue by encrypting each $sk^{(i)}$ under the *next* encryption key $ek^{(i-1)}$, so that the encrypted secret keys form an acyclic chain.)

Definition 15 (Circular security). A encryption system $\mathcal{E}$ is circular secure if it is IND-CPA secure even when $\{\vec{S} = \mathsf{E}(ek, \vec{sk})\}$ is public, where $\vec{sk} \in \mathcal{M}^k$ are the "bits" of the secret key sk.

Theorem 5 (Bootstrapping to unbounded FHE [40, 41]). *If $\mathcal{E}$ is a bootstrappable system that is circular secure, then it can be transformed into an unbounded FHE system $\mathcal{E}_{FHE}$ that is circular secure.*

Let us prove Theorem 5 first, because it is simpler.

Proof of Theorem 5. The construction of $\mathcal{E}_{\text{FHE}}$ is given below; $\mathcal{E}_{\text{FHE}}$ is unbounded, since $\mathcal{E}_{\text{FHE}}.\mathsf{V}$ outputs a ciphertext in $\mathcal{C}_\mathsf{V}$ (the set of evaluated ciphertexts of $\mathcal{E}$) whenever the input ciphertexts are in $\mathcal{C}_\mathsf{V}$. It correctly evaluates any gate g in the functionally complete set Γ (in time polynomial in the security parameter). Therefore, it is fully homomorphic.

The circular security of $\mathcal{E}_{\text{FHE}}$ follows directly from that of $\mathcal{E}$, since the K, E, and D functions of the two systems are the same, the IND-CPA game makes no reference to the V function, and the encrypted secret key $\vec{S}$ is the same in the two systems. ■

Unbounded FHE construction. Let $\mathcal{E} = (\mathcal{E}.\mathsf{K}, \mathcal{E}.\mathsf{E}, \mathcal{E}.\mathsf{D}, \mathcal{E}.\mathsf{V})$ be a circular-secure bootstrappable homomorphic encryption system. We construct a circular-secure FHE system $\mathcal{E}_{\text{FHE}} = (\mathcal{E}_{\text{FHE}}.\mathsf{K}, \mathcal{E}_{\text{FHE}}.\mathsf{E}, \mathcal{E}_{\text{FHE}}.\mathsf{D}, \mathcal{E}_{\text{FHE}}.\mathsf{V})$ as follows. (Below, we will omit the randomness of the encryptions.)

- $\mathcal{E}_{\text{FHE}}.\mathsf{K}$: Identical to $\mathcal{E}.\mathsf{K}$. Note that $\mathcal{E}$ publishes $\vec{S} = \mathsf{E}(ek, \vec{sk})$, which we call part of the evaluation key evk.
- $\mathcal{E}_{\text{FHE}}.\mathsf{E}$: Identical to $\mathcal{E}.\mathsf{E}$.
- $\mathcal{E}_{\text{FHE}}.\mathsf{D}$: Identical to $\mathcal{E}.\mathsf{D}$.
- $\mathcal{E}_{\text{FHE}}.\mathsf{V}(evk, g, c^{(1)}, c^{(2)})$: output $\mathcal{E}.\mathsf{V}(evk, g \circ \mathsf{D}_{c^{(1)},c^{(2)}}(\cdot), \vec{S})$.

So, we have a very clean construction of unbounded circular-secure FHE from circular-secure bootstrappable encryption. Unfortunately, we do not understand circular security very well. Indeed, we can construct encryption systems that are IND-CPA secure (under natural assumptions), but which break completely when an encryption of the secret key is published. These systems tend to be contrived (specifically designed to break), but still these counterexamples are sobering. Also, encryption systems that are *provably* circular secure (based on natural assumptions) are rare compared to provably IND-CPA secure systems. More to the point, and looking ahead, we know how to build IND-CPA secure bootstrappable encryption based on well-studied assumptions about the hardness of computational problems over integer lattices, while the assumptions underlying current unbounded FHE systems are less well-understood.

With that motivation, let us now prove Theorem 4, which says that we can get IND-CPA secure *leveled* FHE from IND-CPA secure bootstrappable encryption.

The proof uses a common technique in cryptographic security proofs called a *hybrid argument*. In a hybrid argument, in each step we change what one ciphertext encrypts—e.g., from encrypting a secret key to encrypting 0—and prove that if $\mathcal{E}$ is IND-CPA secure then an adversary should not notice the difference. By the end of the hybrid argument, all purported encrypted secret keys are in fact encryptions of 0, and are therefore useless to the adversary. The hybrid argument works for a leveled system because the encrypted secret keys $\vec{S}^{(i)}$ form an acyclic chain: the secret key at level i is encrypted under the encryption key at level $i-1$. If the encrypted secret keys form a loop, the hybrid argument does not go through. In particular, the first time we replace an encrypted secret key with an encryption of 0, we break the key loop, and this change may be efficiently distinguishable if the system is not circular secure.

Proof of Theorem 4. The construction of $\mathcal{E}_{\text{LFHE}}$ is given below. By Lemma 1, $\mathcal{E}_{\text{LFHE}}$ is a leveled FHE system. The IND-CPA security of $\mathcal{E}_{\text{LFHE}}$ follows from Lemma 7, which says that if there is an adversary $\mathcal{A}$ in the IND-CPA game against $\mathcal{E}_{\text{LFHE}}$ with n levels that has advantage ε, then there is an adversary $\mathcal{B}$ in the IND-CPA game against $\mathcal{E}$ that has advantage at least $\varepsilon/2(n+1)$, and runs in about the same time as $\mathcal{A}$. ■

Leveled FHE construction. Let $\mathcal{E} = (\mathcal{E}.\mathsf{K}, \mathcal{E}.\mathsf{E}, \mathcal{E}.\mathsf{D}, \mathcal{E}.\mathsf{V})$ be the bootstrappable system with function set $\mathcal{F}$ including functionally complete gates Γ. We construct a leveled FHE system

$$\mathcal{E}_{\text{LFHE}} = (\mathcal{E}_{\text{LFHE}}.\mathsf{K}, \mathcal{E}_{\text{LFHE}}.\mathsf{E}, \mathcal{E}_{\text{LFHE}}.\mathsf{D}, \mathcal{E}_{\text{LFHE}}.\mathsf{V})$$

as follows. (Below, we will omit the randomness of the key generations and encryptions.)

- $\mathcal{E}_{\text{LFHE}}.\mathsf{K}$: Let $n \in \mathbb{N}$ be the number of levels specified in params. For $i \in \{0, \ldots, n\}$, compute key tuples $(sk^{(i)}, ek^{(i)}, evk^{(i)}) \leftarrow \mathcal{E}.\mathsf{K}(\lambda, i)$. For $i \in \{1, \ldots, n\}$, set $\vec{S}^{(i)} = \mathsf{E}(ek^{(i-1)}, \vec{sk}^{(i)})$. For $i \in \{1, \ldots, n\}$, set $evk'^{(i-1)} = (evk^{(i-1)}, \vec{S}^{(i)})$.
- $\mathcal{E}_{\text{LFHE}}.\mathsf{E}$: Identical to $\mathcal{E}.\mathsf{E}$ using $ek^{(n)}$ and attaching the label n to the ciphertext.
- $\mathcal{E}_{\text{LFHE}}.\mathsf{D}$: The ciphertext c comes with a label in $i \in \{0, \ldots, n\}$. Output $\mathcal{E}.\mathsf{D}(sk^{(i)}, c)$.
- $\mathcal{E}_{\text{LFHE}}.\mathsf{V}$: Given gate g and two ciphertexts $c^{(1)}, c^{(2)}$ with label i, output

  $$\mathcal{E}_{\text{LFHE}}.\mathsf{V}(evk'^{(i-1)}, g, c^{(1)}, c^{(2)}) = \mathcal{E}.\mathsf{V}\big(evk^{(i-1)}, g \circ \mathsf{D}_{c^{(1)},c^{(2)}}(\cdot), \vec{S}^{(i)}\big),$$

  and set the label of the ciphertext to $i-1$.

Lemma 1. *If $\mathcal{E}$ is bootstrappable, the system $\mathcal{E}_{LFHE}$ described above is leveled fully homomorphic.*

Proof of Lemma 1. Recall that $\mathcal{C}_{\mathsf{E}}^{(i)}$ is the image of $\mathcal{E}.\mathsf{E}$ with encryption key $ek^{(i)}$ and $\mathcal{C}_{\mathsf{V}}^{(i)}$ is the superset of $\mathcal{C}_{\mathsf{E}}^{(i)}$ that also includes ciphertexts in the image of $\mathcal{E}.\mathsf{V}$ under $evk^{(i)}$ with functions from $\mathcal{F}$ and ciphertexts from $\mathcal{C}_{\mathsf{E}}^{(i)}$. $\mathcal{E}_{\text{LFHE}}$ is leveled, as depicted in Figure 1b, since $\mathcal{E}_{\text{LFHE}}.\mathsf{V}$ outputs a ciphertext in $\mathcal{C}_{\mathsf{V}}^{(i-1)}$ whenever the input ciphertexts are in $\mathcal{C}_{\mathsf{V}}^{(i)}$.

Moreover, $\mathcal{E}_{\text{LFHE}}.\mathsf{V}$ correctly evaluates any gate g in the functionally complete set Γ (in time polynomial in the security parameter). So, it is a leveled FHE system. (Note that the system must perform some bookkeeping relating to the labels of the ciphertexts, but this is not part of the actual functions $\mathcal{E}_{\text{LFHE}}.\mathsf{K}$, $\mathcal{E}_{\text{LFHE}}.\mathsf{E}$, $\mathcal{E}_{\text{LFHE}}.\mathsf{D}$, and $\mathcal{E}_{\text{LFHE}}.\mathsf{V}$, and does not contribute to their complexity.) ■

We provide Lemma 7 and its proof in Appendix B.

3.5. Computational hardness, cryptanalysis, and learning

The aspect of homomorphic encryption (and cryptography in general) that is probably hardest to understand is *computational hardness*.

Computational hardness is often described in terms of the P vs. NP question. Roughly speaking, P consists of problems that can be solved (on a Turing machine) in time polynomial in the size of the problem instance. For example, two λ bit numbers can be multiplied together in time $O(\lambda^2)$ using grade-school multiplication, or even time $O(\lambda \cdot \log \lambda)$ using more sophisticated techniques [49]. NP consists of problems for which a solution (together with a "witness" or proof) can be verified in polynomial time. For example, integer factorization is in NP, since given a nontrivial factorization (p, q) of integer N, one can verify in polynomial time that $N = p \cdot q$ via multiplication (which is polynomial time). On the other hand, integer factorization is not known to be in P, since there are no known polynomial-time algorithms for *finding* p and q.

We have no proof that P $\neq$ NP, and therefore nobody knows whether computational hardness (of the type needed for the security of public-key encryption systems) even exists. Certainly if P $=$ NP, public-key encryption systems are insecure. In this case, one could efficiently *find* the randomness r used in key generation, since one can efficiently *verify* running K with randomness r indeed generates the targeted key pair. Even if we assume P $\neq$ NP, this assumption provides little support for public-key encryption systems, which rely on the hardness of problems that are unlikely to be NP-complete.

In the absence of helpful lower bounds, we are forced to turn to upper bounds. That is, we consider well-studied problems—such as integer factorization, the discrete logarithm problem, and finding short vectors in integer lattices—for which the fastest known algorithms run in time superpolynomial (preferably exponential) in the size of the problem instance. Then, we assume that the best known algorithms are not much worse than the best possible algorithms, and base the security of our cryptosystem on the assumed hardness of the well-studied problem. Even for well-studied problems, this approach can be precarious. For example, although the integer factorization problem has been considered for centuries, a dramatic algorithmic improvement came in 1990 with the invention of the number field sieve [57], which factors λ-bit integers in subexponential time $\exp(O(\log N)^{\frac{1}{3}}(\log\log N)^{\frac{2}{3}})$, considerably faster than the previous quadratic sieve algorithm, which takes time $\exp(O(\log N)^{\frac{1}{2}}(\log\log N)^{\frac{1}{2}})$.

Another surprise has been quantum computation. In 1994, Shor [71] made quantum computation famous by discovering an efficient quantum algorithm for problems including integer factorization and discrete logarithm. More generally, we now have the following result by Watrous [78]:

Theorem 6 (Watrous). *Let G be a solvable (e.g., abelian) group, given by generators. There is a polynomial-time quantum algorithm to compute $|G|$ (with small error probability).*

Corollary 1 (Armknecht et al. [10]). *Group homomorphic encryption systems in which the ciphertext set is a finite solvable group and decryption is a group homomorphism cannot be IND-CPA secure against a quantum adversary.*

The corollary follows because Watrous' algorithm allows one to distinguish between the entire group G of ciphertexts and the proper subgroup H of encryptions of 1 (the kernel of the decryption map). The attack also applies when the ciphertext set is a finite ring and decryption is a ring homomorphism, since the encryptions of 0 form an ideal that is an abelian additive subgroup of the ciphertext set.

There are also efficient quantum attacks **[77]** (and subexponential classical attacks **[18]**) on FHE systems that have a zero oracle—that is, an efficient method to distinguish when a ciphertext encrypts 0. But we are anyway only interested in IND-CPA secure systems, for which no such oracle can exist.

Currently there are no general attacks on homomorphic encryption systems based on nonsolvable groups, but there are also no known plausibly-secure constructions. Again, for the construction to be secure, one must ensure that the ciphertext group G and the proper subgroup H of encryptions of 1 are hard to distinguish. A homomorphic encryption system using nonsolvable groups must avoid at least the following attacks:

- Solvability of H: If H is solvable and G is not, they can be distinguished easily by computing their respective derived series.
- Watrous' order finding algorithm: Even if G and H are both nonsolvable, Watrous' algorithm can distinguish them if the cyclic subgroups generated by randomly sampled elements $g \leftarrow G$ and $h \leftarrow H$ have distinguishably different distributions of orders.
- Linear representations: If one can efficiently compute (or one is given) linear representations of G and H, one may be able to distinguish G and H using linear algebra.

It may seem obvious, but it is an essential point: in a secure encryption system, the decryption function cannot be linear, since linear decryption leads to a trivial linear algebra attack (e.g., Gaussian elimination).[5]

More generally, the decryption function cannot be *learnable*, in the sense of Valiant's "probabilistically approximately correct" (PAC) learning model **[75]**.[6] In the PAC learning model, one is given samples $(x, f(x))$ with x coming from a training set X, and the goal is to learn f well-enough to output $f(x)$ with high probability on a new sample x. This model is nearly identical to the IND-CPA game, where we use $f = \mathsf{D}(sk, \cdot)$.

Since the models are so similar, we can look to learning theory to help us design a decryption function for an IND-CPA secure system **[16]**. For example, Linial, Mansour, and Nisan **[58]** give an algorithm to learn a function expressible as an AC circuit of size s, depth d, and n inputs with accuracy parameter ε in time $n^{O(\log s/\varepsilon)^d}$. (AC circuits are Boolean circuits that have AND and OR gates with arbitrarily many inputs, as well as NOT gates.) So, we cannot have decryption be a constant depth AC circuit if we want (as we usually do) it to take time exponential (or at least subexponential) in the security parameter λ for an adversary to break our system.

---

5 There are a surprising number of FHE proposals without proofs of security, and they are almost always insecure for the simple reason that decryption is linear.

6 Interestingly, the decryption function of a secure bootstrappable encryption system must satisfy an interesting dichotomy: it must be simultaneously *unlearnable* (complex) and *evaluatable* (not too complex).

On the other hand, learning theory suggests that it can be difficult to learn functions from samples that are *noisy*, i.e., from samples $(x, f(x) + e)$, where e is some *error* or *noise*. Accordingly, two learning problems that have become useful to cryptography are the *learning with errors (LWE)* problem [65], and the *learning parity with noise (LPN)* problem, where one is tasked with distinguishing whether given samples are completely random or have the form $(\vec{a}_i, \langle \vec{a}_i, \vec{s} \rangle + e_i)$, where $\vec{s}$ is a secret vector and the e_i's are random noise values. In LWE the vectors are over $\mathbb{Z}/q\mathbb{Z}$ and $|e_i| \ll q$, while in LPN the vectors are over $\mathbb{Z}/2\mathbb{Z}$ and the e_i's are Bernoulli random variables that are usually 0 but sometimes 1. While the inner product is linear, the noise introduces nonlinearity, in particular, it defeats linear algebra. As far as we know, these learning problems are hard even for quantum adversaries. The security of all current FHE constructions relies on the hardness of such learning with noise problems.

4. NOISY CONSTRUCTIONS OF FULLY HOMOMORPHIC ENCRYPTION

The simple FHE systems in Section 2 are unbounded, but only for a single operation that is not functionally complete. How can we construct an unbounded system capable of evaluating, say, $\{+, \times\}$?

For evaluating $\{+, \times\}$, a natural approach to try is a ring homomorphism. However, as we saw in Section 3.5, a system in which decryption is a ring homomorphism can be broken efficiently by a quantum adversary. Moreover, ring homomorphisms have a linearity that may be exploited even by classical adversaries.

How can we defeat linear algebra attacks? As we saw in Section 2.2, one answer is to put values "in the exponent." Unfortunately, in groups for which the Diffie–Hellman assumption holds, values in the exponent can be added efficiently but not multiplied.[7]

As we saw in Section 3.5, another way to defeat linear algebra is to add "noise" or "errors" to the linear equations. Linear algebra is notoriously brittle against noise. As we will see, while adding noise "hardens" the homomorphism, this security comes at a cost: the noise turns our unbounded ring homomorphism into a bounded one. Fortunately, by calibrating the noise level, we can make the bounded homomorphism bootstrappable to achieve FHE while basing security on reasonable hardness assumptions.

4.1. Overview of the noisy approach

Virtually all known constructions of FHE follow essentially the same blueprint:[8] (1) construct bootstrappable encryption by (perhaps implicitly) starting from an insecure

---

7 Some groups for which the discrete logarithm is hard feature a bilinear map—such as a Weil or Tate pairing—that effectively allows one multiplication "in the exponent" (see [17, 54]). Cryptographically-secure multilinear maps are an ongoing area of research [19, 38].

8 Some constructions avoid this blueprint by constructing FHE from *cryptographic program obfuscation* [14, 26, 31, 39, 53]. While much progress has been made on basing obfuscation systems on well-established computational assumptions [53], all current constructions of obfuscation still rely on the hardness of learning with noise problems, and are less efficient and more complicated than more "direct" constructions of FHE.

unbounded homomorphism that respects some functionally complete set of gates and then "hardening" the homomorphism with noise, and (2) invoke the bootstrapping theorems (Theorems 4 and 5) to get FHE from bootstrappable encryption. Here we sketch a fairly general technique for hardening a homomorphism with noise.

For convenience, let us fix our message space to $\mathcal{M} = \mathbb{Z}/2\mathbb{Z}$ and our gates to $\{+, \times\}$. We start with an unbounded homomorphism that respects $\{+, \times\}$:

$$\begin{array}{ccc} \mathcal{C} \times \mathcal{C} & \xrightarrow{\boxplus,\boxtimes} & \mathcal{C} \\ \Big\downarrow {\scriptstyle \delta_{sk}(\cdot),\delta_{sk}(\cdot)} & & \Big\downarrow {\scriptstyle \delta_{sk}(\cdot)} \\ \mathbb{Z}/2\mathbb{Z} \times \mathbb{Z}/2\mathbb{Z} & \xrightarrow{+,\times} & \mathbb{Z}/2\mathbb{Z}. \end{array}$$

For example, $\mathcal{C}$ might be $\mathbb{Z}[x]$, and $\delta_{sk}(\cdot)$ could be evaluation of the ciphertext polynomial at the secret key $sk \in \{0, 1\}^n$ modulo 2. As another example, $\mathcal{C}$ could be a set of integer matrices that all have sk as an eigenvector with integer eigenvalue, $\boxplus$ and $\boxtimes$ could be matrix addition and multiplication, and decryption could be recovering the eigenvalue modulo 2.

In these examples, the homomorphism is unbounded but unsuitable for cryptography. For the polynomial evaluation homomorphism, one problem is that as we apply $\boxtimes$, the degree increases and the number of monomials can increase exponentially.[9] To control the number of monomials, one approach is to publish a Gröbner basis G for the ideal $I(sk) = \{p(x) : p(sk) = 0\}$ of polynomials that evaluate to 0 at sk. But while keeping the monomial basis small helps efficiency, it opens up a trivial linear algebra attack to recover sk. The matrix-based system is also breakable via linear algebra.

To harden these homomorphisms, we add *noise*. Here is one way to do it. First, we make a trivial observation: if we replace δ_{sk} with $\lfloor \delta_{sk} \rceil$ in the above diagram, nothing changes since δ_{sk} is already integral over $\mathcal{C}$. But now let us expand the set of ciphertexts so that $\delta_{sk}(c)$ is *not necessarily integral*. Rather we let ciphertexts be *noisy*. We refer to the value $\delta_{sk}(c) - \lfloor \delta_{sk}(c) \rceil$ as the *noise* of the ciphertext c. Now decryption involves applying $\delta_{sk}(c)$, removing the noise to obtain $\lfloor \delta_{sk}(c) \rceil$, and then reducing modulo 2. But now we must ask: do the $\boxplus$, $\boxtimes$ operations "play well" with the noise? Starting from fresh ciphertexts in $\mathcal{C}_{\mathsf{E}}$, which presumably have a small amount of noise, how many (possibly modified) $\boxplus$, $\boxtimes$ operations can we apply with the guarantee that the following diagram commutes?

$$\begin{array}{ccc} \mathcal{C}_{\mathsf{E}}^{*} & \xrightarrow{\text{bounded number of tweaked } \boxplus,\boxtimes} & \mathcal{C}_{\mathsf{V}} \\ \Big\downarrow {\scriptstyle \lfloor \delta_{sk}(\cdot) \rceil, \lfloor \delta_{sk}(\cdot) \rceil} & & \Big\downarrow {\scriptstyle \lfloor \delta_{sk}(\cdot) \rceil} \\ \mathbb{Z}/2\mathbb{Z} \times \mathbb{Z}/2\mathbb{Z} & \xrightarrow{\text{bounded number of } +,\times} & \mathbb{Z}/2\mathbb{Z} \end{array}$$

---

9 Fellows and Koblitz **[37]** described an encryption system where decryption involves evaluating a ciphertext polynomial at a secret point $\vec{s}$. However, it is not practically usable as a homomorphic encryption system due to this explosion of monomials.

The diagram will commute for function f if evaluating f on fresh ciphertexts—say, ciphertexts of noise at most some ε_0—always results in ciphertexts with noise bounded comfortably below 1/2 that is, the noise is guaranteed not to wrap modulo 1 and result in a possible decryption error. The hope is that $\boxplus$, $\boxtimes$, or tweaked versions of them, play well with the noise and do not amplify it too much, so that if we choose ε_0 well, we can get the diagram to commute for (say) the recrypt-then-NAND function while achieving IND-CPA security based on a reasonable hardness assumption.

Below, we will discuss an instantiation of this framework in detail. Before we describe this construction, we introduce some hardness assumptions related to learning with noise, and show how to construct a symmetric encryption system from one of these assumptions.

4.2. Learning with noise problems

Suppose that p is a large integer. Whenever you press a button, you get an approximate multiple of p, that is, a random integer of the form $x_i = q_i \cdot p + r_i$. Can you recover p? If r_i is always 0, then you can recover p efficiently using the Euclidean algorithm (as soon as you have samples for which the q_i's are relatively prime). But if the "noise" r_i (and other values) are sampled from well-chosen distributions, this problem, called the approximate GCD problem, appears hard.

Definition 16 (Approximate GCD problem [52,76]). Let λ be a security parameter. Let $\alpha = \alpha(\lambda)$, $\beta = \beta(\lambda)$, and $\gamma = \gamma(\lambda)$ be parameters. Fix integer p, sampled as a random integer of β bits. Given arbitrarily many samples $x_i = q_i \cdot p + r_i$, sampled as random γ-bit integers subject to the constraint that $x_i - p\lfloor x_i/p \rceil$ is at most α bits, output p.

Another learning with noise problem is learning parity with noise.

Definition 17 (Learning Parity with Noise (LPN) problem). Let λ be a security parameter. Let $n = n(\lambda)$ be an integer, and $\chi = \chi(\lambda)$ a Bernoulli distribution. Fix a vector $\vec{s} \in (\mathbb{Z}/2\mathbb{Z})^n$ sampled according to χ^n. Given arbitrarily many samples $(\vec{a}_i, b_i)$ where $\vec{a}_i$ is sampled uniformly from $(\mathbb{Z}/2\mathbb{Z})^n$, e_i is sampled from χ, and $b_i \leftarrow \langle \vec{a}_i, \vec{s} \rangle + e_i \bmod 2$, output $\vec{s}$.

(There is some abuse of terms in this description—with mixing of $\mathbb{Z}$, $\mathbb{Z}/2\mathbb{Z}$, and mod 2—but it will be understood that we are really working over $\mathbb{Z}$ and then reducing modulo 2 to representatives of $\mathbb{Z}/2\mathbb{Z}$.)

The LPN problem is easy if the noise e_i is always 0, in which case we can solve for $\vec{s}$ using linear algebra, but the problem appears to be hard for an appropriate choice of parameters. In the normal formulation of LPN, $\vec{s}$ is sampled uniformly from $\{0, 1\}^n$, but Applebaum et al. [9] showed that the problem is just as hard when $\vec{s}$ is sampled from the noise distribution χ.

Finally, we come to the learning with errors problem.

Definition 18 (Learning with Errors (LWE) problem [65]). Let λ be a security parameter. Let $n = n(\lambda)$ and $q = q(\lambda)$ be integers, and $\chi = \chi(\lambda)$ a distribution over $\mathbb{Z}$. Fix a vector

$\vec{s} \in \mathbb{Z}^n$ sampled according to χ^n. Given arbitrarily many samples $(\vec{a}_i, b_i)$ where $\vec{a}_i$ is sampled uniformly from $(\mathbb{Z}/q\mathbb{Z})^n$, e_i is sampled from χ, and $b_i \leftarrow \langle \vec{a}_i, \vec{s} \rangle + e_i \bmod q$, output $\vec{s}$.

Again, if the noise e_i is always 0, we can solve for $\vec{s}$ using linear algebra given enough samples, but the presence of appropriate noise seems to make the problem hard. Typically, χ is taken to be a discrete Gaussian distribution over $\mathbb{Z}$, with deviation $\sigma \ll q$. Note that $\vec{s}$ is chosen according to χ, as Applebaum et al.'s result [9] (mentioned above) applies in this context as well. Rather than referring explicitly to the noise distribution χ, sometimes it is convenient to refer to a bound ε on the size of the noise.

Definition 19 (ε-bounded distributions). A distribution ensemble $\{\chi_n\}_{n \in \mathbb{N}}$, supported over the integers, is called ε-bounded if $\Pr_{e \leftarrow \chi_n}[|e| > \varepsilon]$ is negligible in n.

All of these learning with noise problems are useful for cryptography. As far as we know, they are hard even for quantum adversaries. We have constructions of leveled FHE based on approximate GCD [28, 46, 76] and LWE [23, 24, 46], while constructing FHE based on LPN appears to be more difficult (see [21, 53]). In this paper, we will focus mainly on LWE, and describe a construction of leveled FHE based on LWE.

Since IND-CPA security is about distinguishing between two distributions, it is helpful to define a decision version of LWE which is also about distinguishing between two distributions.

Definition 20 (Decision LWE). As in Definition 18, except that the challenger sets a random bit $\beta \in \{0, 1\}$, and outputs samples according to one of two distributions:

(1) If $\beta = 0$, it outputs $(\vec{a}_i, b_i)$ as uniform samples from $(\mathbb{Z}/q\mathbb{Z})^{n+1}$.

(2) If $\beta = 1$, it samples them according to the distribution in Definition 18.

The problem is to guess β (with nonnegligible advantage). The $\mathrm{LWE}_{n,q,\chi}$ assumption is that this decision $\mathrm{LWE}_{n,q,\chi}$ problem is hard.

What do we know about the hardness of LWE? For n and q that are polynomial in λ, Regev gave a polynomial-time reduction from search LWE to decision LWE. When the noise is extremely small or has some structure, there are subexponential algorithms to solve LWE [12]. For example, when $e_i \in \{0, 1\}$ for all i, solving LWE is easy given $O(n^2)$ samples, since one can compute $\vec{s} \times \vec{s}$ as the solution to the $O(n^2)$-dimensional system of linear equations given by the equalities $\langle \vec{a}_i, \vec{s} \rangle \cdot (\langle \vec{a}_i, \vec{s} \rangle - 1)$. However, for discrete Gaussian error distributions with σ polynomial in n, the hardness of LWE appears to depend solely on the ratio q/ε.

In particular, the LWE problem has been shown to be as hard on average (for random instances) as certain lattice problems in the worst-case (the hardest instances) [3, 4, 65]. An n-dimensional lattice is a (full-rank) additive subgroup of $\mathbb{R}_n$. For lattice dimension parameter n and number d, the shortest vector problem GapSVP_γ is the problem of distinguishing whether an n-dimensional lattice has a nonzero vector of Euclidean norm less than d or no nonzero vector shorter than $\gamma(n) \cdot d$. The gist of the theorem below is that if there is a quantum algorithm for average-case n-dimensional LWE for ratio q/ε, then there is a quantum algorithm for worst-case n-dimensional GAPSVP_γ for γ just a little larger than q/ε.

Theorem 7 (Regev [65]). *Let n, q be integers and $\alpha \in (0, q)$ be such that $\alpha > 2\sqrt{n}$. Let χ be a discrete Gaussian distribution over $\mathbb{Z}$ with deviation α. If there exists an efficient algorithm that solves $\mathrm{LWE}_{n,q,\chi}$, then there exists an efficient quantum algorithm that approximates the decision version of the shortest vector problem (GAPSVP) and the shortest independent vectors problem (SIVP) to within $\tilde{O}(n \cdot \frac{q}{\alpha})$ in the worst case.*

A discrete Gaussian with deviation α will be ε-bounded for $\varepsilon = \alpha \cdot \tilde{O}(\sqrt{n})$. Again, as long as the deviation of discrete Gaussian noise exceeds a certain lower bound, the hardness of LWE appears to depend only on the ratio between q and the size of the noise.

GAPSVP_γ is NP-hard for any constant γ, but, unfortunately, in cryptography we need γ to be larger (at least n in the theorem above). For $\gamma = \mathrm{poly}(n)$, the fastest algorithm to solve GAPSVP_γ takes time $2^{O(n)}$. As a crude rule of thumb, the fastest known algorithm to solve GAPSVP_{2^k} takes roughly $2^{n/k}$ time [70]. Interestingly, there are no quantum algorithms for GAPSVP that perform significantly better than classical algorithms.

Looking ahead to the construction of FHE, some reformulations and variants of LWE will be useful. Sometimes, we prefer to view LWE samples as polynomials.

Fact 1. *A sample $(\vec{a}, b)$ such that $b = \langle \vec{a}, \vec{s} \rangle + e$ can be viewed as a degree-1 polynomial $c(\vec{x}) = b - \sum_j a_j \cdot x_j$ such that $c(\vec{s}) = e$.*

Homomorphic multiplication of ciphertexts related to LWE samples becomes more natural when the ciphertexts are viewed as polynomials. The reformulation in Fact 1 suggests a generalization of LWE to higher-degree polynomials.

Definition 21 (LWE with higher degree polynomials). The values $\lambda, n, \chi, \vec{s}$ are as in Definition 18. There is also a $\mathrm{poly}(\lambda)$-size set of polynomials $P \subset \mathbb{Z}[x_1, \ldots, x_n]$ that is fixed independently of $\vec{s}$. As in LWE (as reformulated in Fact 1), we are given an arbitrary number of degree-1 polynomials $c(\vec{x})$ such that $c(\vec{s}) \bmod q$ has distribution χ. In addition, for each $p(\vec{x}) \in P$, we are given a degree-1 polynomial $c(\vec{x})$ such that $c(\vec{s}) - p(\vec{s}) \bmod q$ has distribution χ.

That is, instead of getting just degree-1 polynomials $c(\vec{x})$ that always evaluate to something small at $\vec{s}$, we can now also obtain potentially higher-degree polynomials $c(\vec{x}) - p(\vec{x})$ that always evaluate to something small at $\vec{s}$. This problem is related to the circular security of some LWE-based systems.

4.3. Encryption based on LWE

Regev [65] built an IND-CPA asymmetric encryption system based on decision LWE. Below, we describe a symmetric version of Regev's encryption system. Since the symmetric system is additively homomorphic modulo 2 (and satisfies other conditions), we can apply Rothblum's theorem (Theorem 3) to get an asymmetric system that looks similar to Regev's.

The idea of the system is simple. Suppose that we want to encrypt $m \in \{0, 1\}$ under secret $\vec{s} \in (\mathbb{Z}/q\mathbb{Z})^n$. Generate an LWE sample $(\vec{a}, b)$ such that $b = \langle \vec{a}, \vec{s} \rangle + e$. The LWE assumption says that it is hard, without $\vec{s}$, to tell whether $b - \langle \vec{a}, \vec{s} \rangle$ is actually distributed according to χ (with small deviation) or uniformly. In the latter case, b is like a one-time pad, even given $\vec{a}$. Accordingly, to encrypt m, we mask it with b—specifically, we encrypt m as $(\vec{a}, b + m \cdot \lfloor q/2 \rceil)$. The key-holder knows a good approximation of b—namely, $\langle \vec{a}, \vec{s} \rangle$—and therefore can remove b, up to small "noise." Thereafter, it can recover m, whose value is preserved (despite the small noise) in the most significant bit by multiplying it $\lfloor q/2 \rceil$.

More formally, the LWE-based encryption system is as follows.

Symmetric encryption system $\mathcal{E}_{\mathrm{LWE}}$.

- K: takes security parameter λ and randomness r and generates parameters $n = n(\lambda)$, $q = q(\lambda)$, and $\chi = \chi(\lambda)$. It generates secret key $sk = ek = \vec{s} \leftarrow \chi^n$.
- E: takes ek, a message $m \in \{0, 1\}$, and randomness r. It generates a random LWE sample $(\vec{a}, b)$. (That is, it generates random $\vec{a} \in (\mathbb{Z}/q\mathbb{Z})^n$, $e \leftarrow \chi$ and sets $b = \langle \vec{a}, \vec{s} \rangle + e$.) It outputs ciphertext $c = (\vec{a}, u)$, where $u = b + m \cdot \lfloor q/2 \rceil \bmod q$.
- D: takes sk and a ciphertext c. It computes $m' \leftarrow u - \langle \vec{a}, \vec{s} \rangle \bmod q$. Depending on whether m' is close to 0 or $\lfloor q/2 \rceil$, output $m = 0$ or $m = 1$.

Regarding correctness, for a well-formed ciphertext we have $m' = b + m \cdot \lfloor q/2 \rceil - \langle \vec{a}, \vec{s} \rangle = e + m \cdot \lfloor q/2 \rceil$, which is close to 0 when $m = 0$, and to $\lfloor q/2 \rceil$ when $m = 1$.

Theorem 8. *$\mathcal{E}_{\mathrm{LWE}}$ is IND-CPA secure based on LWE.*

The proof of security follows the usual format of having an IND-CPA adversary $\mathcal{A}$, an LWE adversary $\mathcal{B}$ who plays the role of the challenger in the IND-CPA game, and an LWE challenger, with $\mathcal{B}$ mostly forwarding slightly modified transmissions between $\mathcal{A}$ and the challenger. If the LWE samples are uniform, the ciphertexts that $\mathcal{B}$ sends to $\mathcal{A}$ will also be uniform, and $\mathcal{A}$ can have no advantage guessing which bit is encrypted. If the LWE samples are well formed, the ciphertexts that $\mathcal{B}$ sends to $\mathcal{A}$ are well formed and $\mathcal{A}$ should have its assumed advantage ε. $\mathcal{B}$ guesses that the LWE samples are well formed if $\mathcal{A}$ guesses correctly, and $\mathcal{B}$ wins with advantage at least $\varepsilon/2$. (The calculation is as in the proof of Theorem 1.)

4.4. Bootstrappable encryption construction

Here we present a bootstrappable encryption system. Historically, the first bootstrappable system was rather complex [41]. But after a sequence of works [6, 20, 23, 24], bootstrappable encryption became simple enough to describe in a blog post [13]. We mostly follow Barak and Brakerski's excellent exposition [13] here.

We start with the LWE-based encryption system $\mathcal{E}_{\mathrm{LWE}}$ of Section 4.3, but we make some cosmetic changes. First, we view ciphertexts as degree-1 polynomials (see Fact 1). This viewpoint will make multiplication of ciphertexts somewhat more natural than if we viewed them as vectors. Second, we basically divide ciphertexts by $q/2$, allowing them to be fractional. Recall that in $\mathcal{E}_{\mathrm{LWE}}$, a ciphertext c (which we will view now as a polynomial) has the property that $c(\vec{s})$ is close to $m \cdot \lfloor q/2 \rfloor$ modulo q. If we divide the ciphertext by $q/2$, we get something more natural—namely, $c(\vec{s})$ is close to m modulo 2, or in other words $\lfloor c(\vec{s}) \rceil = m \bmod 2$. This change allows for a simple description of decryption. It also allows for a simple definition of the "noise" of a ciphertext—namely $c(\vec{s}) - \lfloor c(\vec{s}) \rceil$, the distance of $c(\vec{s})$ to the nearest integer. We write

$$c(\vec{s}) =_{\varepsilon} m \bmod 2$$

to indicate that ciphertext c has noise of magnitude at most $\varepsilon \geq 0$ and $\lfloor c(\vec{s}) \rceil = m \bmod 2$. This notation will simplify the tracking and bounding of ciphertext noise as we apply homomorphic operations $\boxplus, \boxtimes$. We will use ε_0 to denote the noise bound on fresh ciphertexts output by E. We write the system $\mathcal{E}_{\mathrm{boot}}$ below.

Bootstrappable encryption system $\mathcal{E}_{\mathrm{boot}}$.

- K: As in $\mathcal{E}_{\mathrm{LWE}}$, except that we calibrate the parameters $q = q(\lambda)$ and $\chi = \chi(\lambda)$ to achieve bootstrapping, and we set $q = 2^{\kappa+1}$ for some κ. Also, we set evaluation key evk as described below.
- E: Generate c as in $\mathcal{E}_{\mathrm{LWE}}$. Write c as a polynomial $c(\vec{x}) \in \mathbb{Z}[x_1, \ldots, x_n]$ (as described in Fact 1). Divide c by 2^{κ}, i.e., $c(\vec{x}) \leftarrow c(\vec{x})/2^{\kappa}$, a polynomial in $\mathbb{R}[x_1, \ldots, x_n]$. Output $c(\vec{x})$.
- D: Remove as much precision as possible in the coefficients of $c(\vec{x})$ while maintaining correctness before "actual" decryption. Then output $\lfloor c(\vec{s}) \rceil \bmod 2$.
- V: Apply homomorphic operations $\{\boxplus, \boxtimes\}$ for $\{+, \times\}$ modulo 2, as described below.

Notice that we have split decryption into two phases, namely a preprocessing phase where we drop unneeded precision, and a second phase where we do the "actual" decryption. The purpose of the preprocessing is to facilitate bootstrapping: it is important to minimize the complexity of "actual" decryption as much as possible to get recrypt-then-NAND function inside the function set $\mathcal{F}$ that the system can correctly evaluate.

Before we consider $\boxplus, \boxtimes$, it will be useful to specify the *proper form* of ciphertext polynomials in this system, an invariant that we will maintain while performing homomorphic operations.

Definition 22 (Proper form of ciphertexts in $\mathcal{E}_{boot}$). A ciphertext $c(\vec{x})$ is in *proper form* if it is a degree-1 polynomial with coefficients that are in $(-1, 1]$ with κ bits of precision.

Toward maintaining the proper form invariant, observe that reducing a ciphertext polynomial c modulo 2, that is, adjusting c's coefficients by even integers into the range $(-1, 1]$, does not change the ciphertext's noise or the bit that it decrypts to, since $\vec{s}$ is integral.

So, after we add or multiply ciphertext polynomials, we will always reduce the coefficients modulo 2 back into the range $(-1, 1]$, perhaps without mentioning this explicitly.

Now, let us add and multiply ciphertext polynomials! Suppose we add ciphertexts:

$$c = \sum_i c^{(i)}.$$

Then,

$$c^{(i)}(\vec{s}) =_{\varepsilon_i} m^{(i)} \bmod 2 \Longrightarrow c(\vec{s}) =_{\sum_i \varepsilon_i} m \bmod 2,$$

where $m = \sum_i m^{(i)}$. If the original ciphertexts are in the proper form, then so is c. The ciphertext c has noise bounded by the sum of the noises of the original ciphertexts. It will decrypt to the "right" value—in particular, the noise will not "wrap" and overwhelm the "signal"—as long as $\sum_i \varepsilon_i < 1/2$.

Lemma 2 (Addition $\boxplus$). *Let* $c^{(i)}(\vec{s}) =_{\varepsilon_i} m^{(i)} \bmod 2$ *for all* i. *Let* $c = \sum_i c^{(i)} \bmod 2$ *and* $m = \sum_i m^{(i)} \bmod 2$. *Then* $c(\vec{s}) =_{\sum_i \varepsilon_i} m \bmod 2$.

Suppose we multiply two ciphertexts over $\mathbb{R}[\vec{x}]$:

$$c = c^{(1)} \cdot c^{(2)}.$$

Then, c is a degree-2 polynomial. Regarding the noise, we have:

$$\begin{aligned} c^{(i)}(\vec{s}) &=_{\varepsilon_i} m^{(i)} \bmod 2 \\ &= m^{(i)} + 2k^{(i)} + e^{(i)}, \quad m^{(i)} \in \{0, 1\},\ k^{(i)} \in \mathbb{Z},\ \left|e^{(i)}\right| \le \varepsilon_i. \end{aligned}$$

And so

$$\begin{aligned} &c^{(1)}(\vec{s}) \cdot c^{(2)}(\vec{s}) \\ &= (m^{(1)} + 2k^{(1)} + e^{(1)}) \cdot (m^{(2)} + 2k^{(2)} + e^{(2)}) \\ &= m^{(1)} \cdot m^{(2)} + 2k + e^{(1)}(m^{(2)} + 2k^{(2)}) + e^{(2)}(m^{(1)} + 2k^{(1)}) + e^{(1)} \cdot e^{(2)}, \quad k \in \mathbb{Z} \\ &=_{\varepsilon} m^{(1)} \cdot m^{(2)} \bmod 2, \quad \varepsilon = (\varepsilon_1 + \varepsilon_2) \cdot (|\vec{s}|_1 + 1), \end{aligned}$$

where $|\vec{s}|_1$ is the ℓ_1 norm of $\vec{s}$. The new noise is bounded by ε, because if we let B be an upper bound on $|c^{(i)}(\vec{s})|$, then ε is at most $(\varepsilon_1 + \varepsilon_2) \cdot B$, and B is at most $|\vec{s}|_1 + 1$ since the coefficients of $c^{(i)}$ are in $(-1, 1]$.

But $c = c^{(1)} \cdot c^{(2)}$ is not in the proper form. We can easily reduce its coefficients modulo 2 into the range $(-1, 1]$ and drop precision beyond κ bits. Dropping precision costs us an addition noise term of at most $2^{-\kappa} \cdot |\vec{s}|_1^2$. But the biggest issue is that c is degree-2. We need to somehow *relinearize* c so that it is degree-1, as required.

To relinearize, we publish some polynomials in the evaluation key evk that will help us reduce the degree. You can think of these relinearization polynomials as a "noisy Gröbner basis" (using the term loosely): they allow us to reduce degree-2 polynomials to degree-1, but this reduction introduces additional noise.

Definition 23 (Evaluation key for $\mathcal{E}_{\text{boot}}$ (version 1)). As evk, publish polynomials $P = \{p_{i,j,k}(\vec{x})\}$ that are:

- "Pseudoencryptions" of 0: We have $p_{i,j,k}(\vec{s}) =_{\varepsilon_0} 0 \bmod 2$, where ε_0 is the noise of fresh ciphertexts,
- Slightly quadratic: We have $p_{i,j,k}(\vec{x}) = 2^{-k} \cdot x_i \cdot x_j + \ell_{i,j,k}(\vec{x})$ where $k \in \{0, \ldots, \kappa\}$ for precision parameter κ, and $\ell_{i,j,k}(\vec{x})$ is a degree-1 polynomial.

With this evaluation key evk in hand to facilitate relinearization, here is the entire ⊠ procedure.

⊠ for $\mathcal{E}_{\text{boot}}$.

- Compute $c = c^{(1)} \cdot c^{(2)}$ over $\mathbb{R}[x]$.
- Reduce c modulo 2 into the range $(-1, 1]$.
- Drop precision beyond κ bits.
- Relinearize using polynomials P: Call the polynomial so far $c(\vec{x})$. Write each coefficient $c_{i,j}$ (of monomial $x_i x_j$) in terms of its binary decomposition: $c_{i,j} = \sum_{k=0}^{\kappa} c_{i,j,k} 2^{-k}$ with each $c_{i,j,k} \in \{0, 1\}$. Next, subtract off a subset sum of the $p_{i,j,k}$'s to obtain a linear polynomial
$$\text{relinearize}_P\left(c(\vec{x})\right) = c(\vec{x}) - \sum_{i \leq j,k} c_{i,j,k} \cdot p_{i,j,k}(\vec{x}).$$
- Reduce the resulting polynomial modulo 2 into the range $(-1, 1]$.

The ciphertext polynomial output by ⊠ is in the proper form. Relinearization introduces noise of magnitude at most $n^2 \cdot (\kappa + 1) \cdot \varepsilon_0$. Now, let us bound how ⊠ affects the noise overall.

Lemma 3 (Multiplication ⊠). *Let $c^{(1)}(\vec{s}) =_{\varepsilon_1} m^{(1)} \bmod 2$ for $i \in \{1, 2\}$. Let $c = c^{(1)} \boxtimes c^{(2)}$ and $m = m^{(1)} \times m^{(2)}$. Then $c(\vec{s}) =_{\varepsilon_\boxtimes} m \bmod 2$, for $\varepsilon_\boxtimes = (\varepsilon_1 + \varepsilon_2) \cdot (|\vec{s}|_1 + 1) + 2^{-\kappa} \cdot |\vec{s}|_1^2 + n^2 \cdot (\kappa + 1) \cdot \varepsilon_0$. For reasonable parameter settings, ⊠ multiplies the noise by an $O(poly(n))$ factor.*

Proof. The exact expression for the noise comes from the bounds above on the noise added by individual steps of ⊠. Now, take ε to be the maximum of $\varepsilon_1, \varepsilon_2$, both of which are at least ε_0, the latter being the noise of fresh ciphertexts. Recall that we can choose $\vec{s}$ from the noise distribution (except that, unlike the noise and ciphertexts, we have not divided $\vec{s}$ by 2^κ). So, the coefficients of $\vec{s}$ are bounded by $2^\kappa \cdot \varepsilon_0$, and the middle term $2^{-\kappa} \cdot |\vec{s}|_1^2$ is at most $n \cdot |\vec{s}|_1 \cdot \varepsilon_0$. We satisfy the conditions of Theorem 7 as long as the noise distribution (according to which the coefficients of $\vec{s}$ are also chosen) has deviation $2\sqrt{n}$, so we can take $|\vec{s}|_1 = \text{poly}(n)$. We will also take $\kappa = \text{poly}(n)$. Then, we have that the new noise is bounded by $\varepsilon \cdot \text{poly}(n)$. ■

We have established that (for reasonable parameter settings), a single $\boxplus$ or $\boxtimes$ operation increases the noise by at most a factor of $p(n)$ for some polynomial p. (See Lemmas 2 and 3.) This result gives us the following commutative diagram:

$$\begin{array}{ccc} \mathcal{C}_\varepsilon \times \mathcal{C}_\varepsilon & \xrightarrow{\;\boxplus,\boxtimes\;} & \mathcal{C}_{\varepsilon\cdot p(n)} \\ \Big\downarrow {\scriptstyle \lfloor \mathrm{ev}_{\vec{s}}(\cdot)\rceil \bmod 2} & & \Big\downarrow {\scriptstyle \lfloor \mathrm{ev}_{\vec{s}}(\cdot)\rceil \bmod 2} \\ \mathbb{Z}/2\mathbb{Z}\times\mathbb{Z}/2\mathbb{Z} & \xrightarrow{\;+,\times\;} & \mathbb{Z}/2\mathbb{Z}, \end{array}$$

where $\mathcal{C}_\varepsilon$ denotes ciphertexts with noise bounded by ε, and $\mathrm{ev}_{\vec{s}}(\cdot)$ denotes evaluation of a polynomial at $\vec{s}$.

Now, let us extend this observation to evaluation of an arithmetic circuit C of $\{+,\times\}$ gates. Recall that the depth d of the circuit C is the length of the longest path from an input gate to the output gate, considering the circuit as a directed acyclic graph.

Lemma 4. *Let $p(x)$ be a polynomial such that $\boxplus$ and $\boxtimes$ multiply the noise of input ciphertexts by at most $p(n)$. Given input ciphertexts of noise at most ε_0, the above system can evaluate any arithmetic circuit of depth d with noise at most $\varepsilon_0 \cdot p(n)^d$.*

For example, if we want to evaluate a d-depth circuit so that the noise of the final ciphertext is at most (say) $1/4$, it suffices to take $\varepsilon_0 \leq (1/4)\cdot p(n)^{-d}$.

As we are aiming for a bootstrappable encryption system, we are especially interested in the depth of the decryption circuit (and the recrypt-then-NAND circuit, which has depth 2 more than the decryption circuit). Recall that we drop unneeded precision before "actual" decryption. Importantly, after dropping precision, the complexity of decryption does *not* depend on ε_0 at all. Indeed, ε_0 becomes a free parameter that we can eventually set as small as needed to allow us to evaluate the decryption circuit.

Lemma 5. *Let c be a ciphertext such that $c(\vec{s}) =_{1/4} m \bmod 2$. For any κ' such that $2^{\kappa'} > 4\cdot(|\vec{s}|_1+1)$, we can drop to κ' bits of precision in c while preserving correctness of decryption (to the message m).*

Proof. Dropping to κ' bits of precision adds at most $2^{-\kappa'}\cdot(|\vec{s}|_1+1)$ to the noise. By assumption, $2^{-\kappa'}\cdot(|\vec{s}|_1+1) < 1/4$, so that the total noise remains $< 1/2$, and correctness of decryption is preserved. ∎

Now we bound the arithmetic circuit depth of "actual" decryption (after dropping to $\lceil \log(4\cdot(|\vec{s}|_1+1))\rceil$ bits of precision in c).

Lemma 6. *For reasonable parameters, the decryption circuit has depth*

$$O(\log n + \log\log|\vec{s}|_1) = O(\log n).$$

Proof. The computation of the inner product of two vectors—represented in the natural way, with coefficients in binary representation—is in NC^1 (see Section 4 of **[56]**), meaning that it can be computed in circuit depth proportional to the logarithm of the description length of the vectors. Decryption is an inner product of vectors of dimension n with coefficients of $O(\log \|\vec{s}_1\|)$ bits. Therefore, it can be computed in depth $O(\log n + \log\log |\vec{s}|_1)$. Regarding $\|\vec{s}_1\|$, we can satisfy the conditions of Theorem 7 as long as the noise distribution (according to which the coefficients of $\vec{s}$ are also chosen) has deviation $2\sqrt{n}$, and so we can take $|\vec{s}|_1 = \text{poly}(n)$. The result follows. ■

Theorem 9. *$\mathcal{E}_{\text{boot}}$ is bootstrappable for some ε_0 (the noise bound of fresh ciphertexts) that is $n^{-O(\log n)}$.*

Proof. By Lemma 6, the "actual" decryption circuit (hence recrypt-then-NAND) applied to ciphertexts with dropped precision has depth $O(\log n)$. By Lemmas 4 and 5, for some $\varepsilon_0 = n^{-O(\log n)}$, we can apply recrypt-then-NAND and drop precision while keeping the noise below $1/2$. ■

Theorem 10. *$\mathcal{E}_{\text{boot}}$ is IND-CPA secure based on the LWE with higher degree polynomials assumption (Definition 21) for some $q = n^{O(\log n)}$ and some poly(n)-bounded distribution χ.*

Proof. The proof is similar to the proof of IND-CPA security for the basic symmetric encryption system based on LWE (see Theorem 8), except that we use LWE with higher degree polynomials (versus regular LWE) to sample the "slightly quadratic" polynomials P used for relinearization in ⊠. ■

By Theorem 4, we get leveled FHE based on this assumption. In fact, we can get circular-secure FHE based on variants of the LWE with higher degree polynomials assumption, e.g., where we still only use linear and "slightly quadratic" polynomials, but allow the secret $\vec{s}$ to come from $\{0,1\}^n$ rather than χ^n. We elaborate on this observation in Section 4.5.

But our underlying encryption system $\mathcal{E}_{\text{LWE}}$ is based on the LWE assumption. How can we get a bootstrappable encryption system also based on LWE assumption, rather than the less well-established LWE with higher degree polynomials assumption? Clearly, the issue originates with the current version of $\mathcal{E}_{\text{boot}}$, which requires publication of "slightly quadratic" polynomials that are used in relinearization. Can we publish a different set of polynomials to facilitate relinearization, and somehow base security on LWE?

The trick to get security based on LWE is similar to the trick that we used to get IND-CPA secure leveled FHE from IND-CPA secure bootstrappable encryption—namely, to avoid a circular security issue, we use an acyclic chain of encrypted secret keys. The slightly quadratic relinearization polynomials used in the above version of $\mathcal{E}_{\text{boot}}$ can be viewed as an encryption of the secret key under itself. Specifically, we have:

$$p^{(\ell)}_{i,j,k}(\vec{s}) = 2^{-k} \cdot x_i \cdot x_j + \ell_{i,j,k}(\vec{s}) =_{\varepsilon_0} 0 \bmod 2$$
$$\implies \ell_{i,j,k}(\vec{s}) =_{\varepsilon_0} -2^{-k} \cdot s_i \cdot s_j.$$

That is, the linear polynomial $\ell_{i,j,k}(\vec{x})$, which has the proper form of a ciphertext, encrypts (in some sense) the value $-2^{-k} \cdot s_i \cdot s_j$, a quadratic monomial of the secret key itself. In version 2 of $\mathcal{E}_{\text{boot}}$, we modify the relinearization polynomials so that each one is, in effect, an encryption of a quadratic monomial over some $\vec{s}^{(\ell)}$ under the *next* secret $\vec{s}^{(\ell-1)}$, so that we have an acyclic chain of encrypted secret keys. In detail:

Definition 24 (Evaluation key for $\mathcal{E}_{boot}$ (version 2)). Publish polynomials $P = \{p_{i,j,k}^{(\ell)}(\vec{x}, \vec{y})\}$ that are:

- "Pseudoencryptions" of 0: We have $p_{i,j,k}^{(\ell)}(\vec{s}^{(\ell)}, \vec{s}^{(\ell-1)}) =_{\varepsilon_0} 0 \bmod 2$, where ε_0 is the noise of fresh ciphertexts,
- Slightly quadratic (and some linear): We have

  $$p_{i,j,k}^{(\ell)}(\vec{x}, \vec{y}) = 2^{-k} \cdot x_i \cdot x_j + \ell_{i,j,k}(\vec{y})$$

  where $k \in \{0, \ldots, \kappa\}$ for precision parameter κ, and $\ell_{i,j,k}(\vec{y})$ is a degree-1 polynomial. For $i = 0$, we have linear polynomials $p_{i,j,k}^{(\ell)}(\vec{x}, \vec{y}) = 2^{-k} \cdot x_j + \ell_{i,j,k}(\vec{y})$.

Unlike version 1, we need linear polynomials in addition to the slightly quadratic ones, because we are using the polynomials not only to relinearize but simultaneously to transfer the ciphertexts from one key ($\vec{s}^{(\ell)}$) to the next ($\vec{s}^{(\ell-1)}$). In the system, we can apply $\boxplus$ or $\boxtimes$ to two ciphertexts under $\vec{s}^{(\ell)}$, with the result under $\boxtimes$ being under the next key $\vec{s}^{(\ell-1)}$. The noise analysis is identical to version 1.

Theorem 11. *$\mathcal{E}_{\text{boot}}$ (version 2) is IND-CPA secure based on the LWE assumption (Definition 18) for some $q = n^{O(\log n)}$ and some poly(n)-bounded distribution χ.*

Proof. (Sketch) The proof is similar to that of Lemma 7, where we proved the IND-CPA security of a leveled FHE system that uses an acyclic chain of encrypted secret keys, using a so-called hybrid argument where in a sequence of steps we replace encrypted secret keys with encryptions of 0. The main idea in this proof is that each relinearization polynomial $p_{i,j,k}^{(\ell)}(\vec{x}, \vec{y})$, whose special property is that it evaluates to a small value at $(\vec{s}^{(\ell)}, \vec{s}^{(\ell-1)})$, looks indistinguishable from random to an adversary that does not know $\vec{s}^{(\ell-1)}$, even if it knows $\vec{s}^{(\ell)}$. ■

Recall from Section 4.2 that, as far as we know, LWE is hard even if the ratio of q to the noise is subexponential in n, so Theorem 11 provides a strong security guarantee. It is possible to base the security of leveled FHE on LWE even for factors that are polynomial in n [25].

4.5. Reflections on the overall FHE system

Now that we have completed our modular description of the FHE system, it is interesting to demodularize the system to see what is happening overall. To simplify the overall picture for the moment, let us imagine that the secret $\vec{s}$ is in $\{0, 1\}^n$, so that the "bits" of $\vec{s}$ are in fact the coefficients of $\vec{s}$. We operate on linear ciphertext polynomials $c(\vec{x})$ that have small noise when evaluated at $\vec{s}$. When we multiply polynomials, we use a noisy partial Gröbner basis to reduce the resulting polynomial back to linear while changing the evaluation at $\vec{s}$ by only small noise. As we apply $\boxplus$ and $\boxtimes$, the noisiness increases until we have a $c(\vec{x})$

whose noise at $\vec{s}$ is nearly 1/2, so that it can no longer participate safely in operations. But now imagine that the underlying polynomial that we are evaluating with these $\boxplus$'s and $\boxtimes$'s is $f(\vec{x}) = \mathsf{D}_c(\vec{x})$. It holds that $f(\vec{s}) = m$, the message encrypted by c, but we do not know $\vec{s}$. Instead, we start evaluating $f(\vec{x})$ as a formal polynomial, except that we reduce the degree using our noisy Gröbner basis. This noisy basis allows us to reduce the degree to linear, while preserving (up to small noise) the evaluation at $\vec{s}$ (which is what we care about). At the end, we get a linear polynomial whose evaluation at $\vec{s}$ equals (up to small noise) the value $f(\vec{s}) = m$. Hence, this linear polynomial is a new encryption of m. This new encryption has noise that depends only on the complexity of D_c and the noisiness of our noisy Gröbner basis, not on the noisiness of c as a ciphertext, and therefore it can (if we calibrate our noise appropriately) participate in more operations.

It is also interesting to consider what our ciphertext set looks like as an algebraic structure. Our unbounded system has a clean commutative diagram

$$\begin{array}{ccc} \mathcal{C} \times \mathcal{C} & \xrightarrow{\boxplus,\boxtimes} & \mathcal{C} \\ \Big\downarrow {\scriptstyle \lfloor \mathrm{ev}_{\vec{s}}(\cdot) \rceil \bmod 2} & & \Big\downarrow {\scriptstyle \lfloor \mathrm{ev}_{\vec{s}}(\cdot) \rceil \bmod 2} \\ \mathbb{Z}/2\mathbb{Z} \times \mathbb{Z}/2\mathbb{Z} & \xrightarrow{+,\times} & \mathbb{Z}/2\mathbb{Z}, \end{array}$$

but now $\boxplus$ and $\boxtimes$ hide a lot of complexity, e.g., $\boxtimes$ is in fact $\times \circ \mathsf{D}_{c^{(1)},c^{(2)}}(\cdot)$. The operations $\boxplus, \boxtimes$ are not even associative: $\mathcal{C}$ is a magma under both operations. In fact, these operations are so unstructured that it is almost as if they select a pseudorandom ciphertext that encrypts the appropriate value. While this intuition may not be entirely accurate, Ducas and Stehlé **[34]** show that starting with any c that encrypts m, successive alternations of recryption and small injections of noise converge to a *canonical* distribution over ciphertexts that encrypt m, effectively erasing the ciphertext's history.

5. NEW DIRECTIONS IN HOMOMORPHIC ENCRYPTION

Many questions about FHE remain unresolved. Below, we discuss a few of these questions—in particular, questions about improving efficiency, handling the circular security issue, avoiding bootstrapping, building noise-free FHE, and exploring quantum FHE.

Some of these questions can be answered (not entirely satisfactorily) with the ultimate cryptographic hammer: cryptographic program obfuscation **[14, 31, 39, 53]**. Informally, program obfuscation takes a program P and produces an obfuscated program $O(P)$ that has the same input/output functionality, but where $O(P)$ is otherwise "unintelligible." Program obfuscation was proven impossible to achieve for a certain "virtual black box" notion of unintelligibility **[14]**. However, there are now constructions of program obfuscation **[39]**, even based on well-established computational assumptions **[53]**, for a notion of unintelligibility based on indistinguishability.[10] Namely, an indistinguishability obfuscator iO offers the fol-

10 Say it 10 times fast!

lowing guarantee: if there are two circuits C_0, C_1 of the same size and equivalent input/output functionality and iO obfuscates one of them (C_b for random $b \in \{0, 1\}$) to produce $O(C_b)$, it is computationally hard to distinguish b. (Notice the similarity to IND-CPA security for encryption.) Obfuscation is even more powerful than FHE—in particular, you can build FHE from versions of it [26]—but it is currently also computationally much more expensive than FHE. A crucial difference between an obfuscated program and an FHE-encrypted program is that the obfuscated program has unencrypted output.

How practical can we make the current noisy FHE systems? Since the first construction of FHE in 2009 [40, 41] there have been four generations of FHE systems, with the second, third, and fourth generations each offering significant performance gains. Currently, efforts are underway to standardize homomorphic encryption systems [51].

The second generation of FHE systems simultaneously improved security and efficiency by basing security on LWE [24], and then using variants of LWE over polynomial rings, such as ring-LWE [60] and the NTRU problem [58, 59]. Polynomial rings naturally facilitate *batching* or *SIMD* operations on encrypted data [73]—that is, via the Chinese Remainder Theorem, they allow many individual messages to be packed in single ciphertexts, and for many messages to be (implicitly) operated on through a single ciphertext operation. Automorphisms of the polynomial rings allow message slots to be permuted within a ciphertext. These optimizations, together with better techniques to control the growth of the noise in ciphertexts [23], reduced the overhead of FHE—that is, multiplicative factor of how much processing it takes to operated on FHE-encrypted data versus unencrypted data—to only polylogarithmic in the security parameter λ [45].

The third generation [8, 25, 29, 30, 33, 46] introduced techniques for fast bootstrapping, reducing bootstrapping time to tens of milliseconds. However, so far, these techniques are incompatible with the batching techniques of the second generation.

The fourth generation systems [27] are optimized for operating on floating-point data, making it more friendly for real-world applications such as logistic regression and neural nets. Current estimates are that, depending on the type of computation, the overhead of fourth generation systems is as low as $10000\times$, and this overhead can be further reduced with customized hardware [36, 55, 64].

An inherent limitation of FHE systems is that they do not support certain types of computation, such as RAM (random access model) computations. FHE computations are inherently input-oblivious, i.e., the structure of the computation cannot depend on the input, since IND-CPA security implies no information about the input is disclosed. This limitation can be overcome by using heavier machinery—in particular, cryptographic program obfuscation [44]—since obfuscated programs can disclose unencrypted information.

Can we base unbounded FHE on a well-established computational hardness assumption? Current unbounded (versus leveled) FHE systems require a circular security assumption—namely, that it is secure to encrypt the secret decryption key under its companion encryption key. For the noisy FHE system presented in Section 4, the encryption of the secret key manifests as a collection of nonlinear polynomials that evaluate to some small (noise) value at the secret key $\vec{s}$. Can we reduce the LWE with higher degree polynomials assumption to a more well-established assumption, such as the hardness of problems over integer lattices? Can we somehow avoid nonlinear polynomials altogether, and build unbounded FHE based on LWE?

Can we get unbounded FHE without bootstrapping or recryption? In Section 3.4, we saw that bootstrapping seems unavoidable as a generic technique to convert a bounded homomorphic encryption system to an unbounded one. The one FHE system based on well-established assumptions that does not use bootstrapping [53] instead uses program obfuscation, which currently is less practical than more direct constructions of FHE. Moreover, the technique to build FHE [26] from obfuscation uses obfuscation to decrypt ciphertexts, perform an operation on them, and then encrypt the result—a process that (like bootstrapping) still involves computing the decryption function. Can we get unbounded FHE while avoiding expensive repeated computation of the decryption function? Note that we can get leveled FHE without bootstrapping [23] for a relaxed definition of leveled that allows the parameters to grow with the number of levels. (Our Definition 12 for leveled systems is not relaxed.)

Can we build "noise-free" FHE?. So far, all constructions of FHE use "noise," even the obfuscation-based solution [53]. For building bootstrappable encryption, noise has the nice feature that we can calibrate the noise level, decreasing it until the system becomes bootstrappable. For the system in Section 4.4, this calibration was especially easy because adjusting the noise level did not affect decryption complexity at all (though it affects the computational assumption). While noise has these nice features, it also introduces complexities, and one wonders whether it is possible to construct noise-free FHE.

Mathematically, perhaps the cleanest approach to noise-free FHE would be to construct a multilinear map with suitable cryptographic properties [19]. We have cryptographic *bi*linear maps from Weil and Tate pairings over abelian varieties, which have proven to be enormously useful. We can obtain FHE (and obfuscation) from cryptographic multilinear maps of higher degree, but so far we have no viable noise-free constructions.

Nuida [63] proposed a construction of noise-free FHE using nonsolvable groups with certain properties, but groups with these properties are not known to exist. As discussed in Section 3.5, there are many obstacles to constructing a secure homomorphic encryption based on nonsolvable groups.

Are fundamentally new constructions possible in the quantum setting? The FHE system presented here works for a computation expressible as a polynomial-size circuit with classical gates, like $\{+, \times\}$. What if we want to privately delegate a computation not known to be in P, but which is easy for a quantum computer, like factoring an encrypted integer? For that, we need an FHE system capable of evaluating quantum gates. Mahadev [61] resolved the question of quantum (leveled) FHE, showing that a classical client can privately outsource a quantum computation to a quantum server, under the surprisingly minimal assumption that LWE is hard for quantum computers. One wonders whether this result, and the suggestive similarities between quantum error correction and managing ciphertext noise in FHE systems, will lead to new techniques even for classical FHE systems. The question of quantum obfuscation is not resolved. While preliminary results [5] indicate that virtual black box obfuscation of quantum circuits is impossible, they leave open the question of indistinguishability obfuscation for quantum circuits.

ACKNOWLEDGMENTS

I would like to give special thanks to Dan Boneh for his encouragement to work on this problem, and to Shai Halevi for such fruitful collaboration. I would also like to thank the many others who have made this area so exciting to work in.

A. WHAT DOES IT MEAN FOR AN ENCRYPTION SYSTEM TO BE SECURE?

This section provides informal philosophical discussion about what it means for an encryption system to be secure—in particular, about why IND-CPA security (see Definitions 1 through 3) is the "right" minimal notion of security for an encryption system.

To see why, let us try to reinvent the security model ourselves. We have an "adversary" that is trying to "break" the encryption system. In general, we can model the security of encryption as a game between a "challenger" and the adversary that the adversary is trying to "win." To specify the game, there are 3 aspects to consider:

(1) *Adversary's power inside the system*: How can the adversary interact with the encryption system? Does the adversary know how the algorithms (key generation, encryption, decryption) work, is it allowed to see many ciphertexts, on messages of its choice, can it ask for ciphertexts to be decrypted, can it see how transmitted and decrypted messages affect peoples' "behavior," can it ask for bits of the secret key or functions of the secret key?

(2) *Adversary's power outside the system*: Is the adversary limited to running polynomial time algorithms, polynomial time quantum computation, polynomial time nondeterministic computation, is its computational power unbounded?

(3) *Object of the game*: Is the object to recover the secret key, to recover the message encrypted by a ciphertext, to merely distinguish which of two messages a ciphertext encrypts, to produce a new ciphertext that encrypts a message related to a message encrypted by a given ciphertext?

Now, let us start to prune the numerous possibilities given above.

First, as a theoretical matter, we can assume without loss of generality that adversary knows how the system works. We simply label what the adversary does not know as the secret key. The secret key may include hidden aspects of how key generation, encryption, and decryption operate, but really this is just a matter of semantics, and these algorithms can always be redefined so that secret information is localized to the secret key, and the algorithms themselves are public. Also, as a practical matter, we have Kerckhoff's principle, which (as reformulated by Claude Shannon) simply states that "the enemy knows the system." The practical justification for this principle is that "security by obscurity" rarely works. Rather, empirically, one is more likely to obtain a secure system by making it comprehensible to friendly cryptanalysts. (Provable security, à la Goldwasser and Micali, is an extreme version of Kerckhoff's principle, in which we proudly display a concise clearly-specified mathematical problem on which the cryptosystem's security is based.) So, we take the algorithms of encryption—key generation K, encryption E, decryption D, and evaluation V (if applicable)—as known.

Object of the game. The purpose of an encryption system is to hide a message. Clearly, we should not require the adversary to recover the secret key, since it might be able to recover information about an encrypted message without it. From a ciphertext, the adversary will trivially know an upper bound on the message length, but it should not be able to determine *anything* else. (Lacking a general way to characterize what is *important* in a message, we should require that the adversary cannot distinguish *anything* nontrivial.) Goldwasser and Micali capture this intuition with their definition of semantic security: "Informally, a system is semantically secure if whatever an eavesdropper can compute about the cleartext given the cyphertext, he can also compute without the cyphertext" [47]. In particular, given a ciphertext encrypting a message, the *a priori* and *a posteriori* distributions of the message should be identical (up to a negligible factor) from the perspective of the adversary. Goldwasser and Micali proved that this notion of semantic security is modeled well by the IND-CPA game, which allows the adversary to choose message pairs $\{m_{i,0}, m_{i,1}\}$ whose encryptions it thinks it is most able to distinguish.

To make things even easier on the adversary, we could require only that it produce a new ciphertext (not given to it by the challenger) that encrypts a message related to (e.g., the same as) in the challenge ciphertext. A system that prevents this attack is called *nonmalleable* [32]. We do not consider nonmalleability to be part of a *minimal* viable notion of security for encryption for two reasons. First, we are considering homomorphic encryption, which is inherently malleable; the whole point is to produce new ciphertexts that encrypt values meaningfully related to those of previous ciphertexts. Second, there are general techniques (that we will not review here) that prevent malleability. To a large extent, nonmalleability can be "added on" to an IND-CPA secure encryption system after the fact.

Adversary's power outside the system. Aside from interacting with the system, the adversary's power (outside of the system) comes down to its computational power. Claude Shannon resolved the case of a computationally unbounded adversary. He showed that one can perfectly hide a message (except for an upper bound on its length) by encrypting it with a one-time pad (a perfectly random key as long as the message). In some settings, such as military or diplomatic settings that demand absolute eternal secrecy, a one-time pad might be the right solution. However, we are also interested in many other (most) settings, where distributing a one-time pad is not practical. Accordingly, to allow more practical systems, we permit computational assumptions—that is, we only require our encryption systems to be secure against probabilistic polynomial-time adversaries, and assume that some problems in NP are hard to solve in probabilistic polynomial time. (Of course, even this seemingly minimal assumption may turn out to be false, as currently we are not even certain that $P \neq NP$.)

Adversary's power inside the system. As mentioned before, we must allow the adversary to "know the system." Moreover, as we are not in the setting of the one-time-pad, the adversary should be able to see many ciphertexts encrypted under the same key. Furthermore, since the adversary in real life might be able to influence what the encrypter encrypts, the game should allow the adversary to choose what messages are encrypted, or even have all of the messages depend on some bit that it is trying to guess. The IND-CPA game gives the adversary this power.

But why not give the adversary even more power in our minimal notion of security? For example, why not allow the adversary to choose ciphertexts to be decrypted by the challenger, or to receive some (not-completely-revealing) function of the secret key? Indeed, these forms of security are important. The IND-CCA (indistinguishability of ciphertexts against a chosen ciphertext attack) game allows the adversary to query the decryption of ciphertexts, a model that is actually quite realistic in real life, because adversaries can potentially break cryptosystems by observing how keyholders react after decrypting their ciphertexts, even (or perhaps especially) if those ciphertexts are malformed. Key "leakage" and "side channel" attacks, in which the adversary obtains some limited information about the secret key, are also quite realistic, because (unless special care is taken) even the amount of time the keyholder takes to decrypt can leak information about the secret key.

The reason why we consider IND-CPA still to be acceptable minimal notion of security is that there are techniques for achieving IND-CCA and security against side channel attacks that are mostly orthogonal to IND-CPA security—that is, they can, to a large extent, be applied to an IND-CPA secure system after the fact. Even for homomorphic encryption, which is inherently malleable, one can use so-called noninteractive zero-knowledge arguments to ensure that the keyholder decrypts only after verifying a cryptographic proof that the ciphertext is well formed and resulted from a "permitted" evaluation over valid ciphertexts. In the real world, combining homomorphic encryption systems with proof systems in this way is actually important for preventing devastating attacks. But, again, these considerations are largely orthogonal to the security of the underlying homomorphic system, and we therefore do not consider them to be part of the minimal notion of security.

Conclusion. Out of many possible security notions, we pruned the possibilities to land on IND-CPA security as the "right" minimal notion of security for a homomorphic encryption system. Weaker notions may not provide much security at all in realistic contexts, and stronger notions typically can be achieved by using an IND-CPA-secure system in combination with orthogonal techniques.

B. HYBRID ARGUMENT FOR LEVELED FHE

Lemma 7. *Let $\mathcal{E}$ be an IND-CPA secure encryption system such that any secret key sk can be expressed as a vector $\vec{sk} \in \mathcal{M}^k$, where $\mathcal{M}$ is the message set of the system. Let $\mathcal{E}_{LFHE}$ be a system in which we publish encrypted secret keys $\vec{S}^{(i)} = \mathcal{E}.\mathsf{E}(ek^{(i-1)}, \vec{sk}^{(i)})$, where $(sk^{(i)}, ek^{(i)})$ for $i \in \{0, \ldots, n\}$ is an $\mathcal{E}$ key pair. Suppose that $\mathcal{E}_{LFHE}.\mathsf{E}$ is the same as $\mathcal{E}.\mathsf{E}$, using encryption key $ek^{(n)}$. Then $\mathcal{E}_{LFHE}$ is also IND-CPA secure in the following sense. Suppose that there is an adversary $\mathcal{A}$ in the IND-CPA game against $\mathcal{E}_{LFHE}$ that has advantage ε. Then there is an adversary $\mathcal{B}$ in the IND-CPA game against $\mathcal{E}$ that has advantage at least $\varepsilon/2(n+1)$, and that runs in about the same time as $\mathcal{A}$.*

Proof of Lemma 7. We consider $\mathcal{A}$'s behavior in a sequence of games: Game 0, Game 1, ..., Game n. Game 0 is identical to the IND-CPA game for $\mathcal{E}_{\text{LFHE}}$. Game i is identical, except that the values $\vec{S}^{(i+1)}, \ldots, \vec{S}^{(n)}$ are constructed correctly (as in the system), but the values $\vec{S}^{(1)}, \ldots, \vec{S}^{(i)}$ are all encryptions of 0. Whatever game we are in, the $\mathcal{E}_{\text{LFHE}}$-IND-CPA-challenger samples a random bit $b \in \{0, 1\}$. When $\mathcal{A}$ queries messages $(m_{j,0}, m_{j,1})$, the challenger encrypts $m_{j,b}$ under $ek^{(n)}$.

Let ε_i be $\mathcal{A}$'s advantage in guessing b in Game i. Since Game 0 is the true IND-CPA game for $\mathcal{E}_{\text{LFHE}}$ as given in Definition 3, we have $\varepsilon_0 = \varepsilon$. Therefore either ε_n or $\varepsilon_i - \varepsilon_{i+1}$ for some $i \in \{0, \ldots, n-1\}$ must exceed $\varepsilon/(n+1)$ in magnitude. Set i^* so that the magnitude of $\varepsilon_{i^*} - \varepsilon_{i^*+1}$ is maximized (or set $i^* = n$ if ε_n it is the biggest contributor).

Then $\mathcal{B}$ attacks $\mathcal{E}$ by using $\mathcal{A}$ as follows: $\mathcal{B}$ participates in an IND-CPA game with an $\mathcal{E}$-challenger who flips a bit $\beta \in \{0, 1\}$. This game is associated to some key pair, which $\mathcal{B}$ will label formally as $(sk^{(i^*)}, ek^{(i^*)})$. If the system is asymmetric, it will receive $ek^{(i^*)}$ from the challenger. Also $\mathcal{B}$ assumes the role of the challenger in the $\mathcal{E}_{\text{LFHE}}$ game and flips a bit $b \in \{0, 1\}$.

Then $\mathcal{B}$ uses $\mathcal{E}.\mathsf{K}$ to generate key tuples $(sk^{(i)}, ek^{(i)})$ for all $i \neq i^*$. Here is how $\mathcal{B}$ generates the $\vec{S}^{(i)}$ values for $i \in \{1, \ldots, n\}$. For all $i \geq i^* + 2$, it generates each $\vec{S}^{(i)}$ correctly (as in the system) as an encryption of $\vec{sk}^{(i)}$ under $ek^{(i-1)}$. For $i \leq i^*$, it generates each $\vec{S}^{(i)}$ as an encryption of 0 under $ek^{(i-1)}$.

If $i^* = n$, then this is a complete set of $\vec{S}^{(i)}$'s, and $\mathcal{B}$ sends the complete $\mathcal{E}_{\text{LFHE}}$ public key to $\mathcal{A}$. When $\mathcal{A}$ queries messages $(m_{j,0}, m_{j,1})$, $\mathcal{B}$ forwards these messages to the $\mathcal{E}$-challenger as queries. The $\mathcal{E}$-challenger encrypts $m_{j,\beta}$ under $ek^{(n)}$, and sends the ciphertext to $\mathcal{B}$, which forwards the ciphertext to $\mathcal{A}$. Then $\mathcal{A}$ submits a guess and $\mathcal{B}$ forwards that guess to the $\mathcal{E}$-challenger. Now $\mathcal{B}$'s advantage is the same as $\mathcal{A}$'s. Since the distribution seen by $\mathcal{A}$ is precisely as in Game n, $\mathcal{A}$'s advantage is ε_n.

If $i^* \neq n$, then $\mathcal{B}$ generates $\vec{S}^{(i^*+1)}$ as follows. It submits $\vec{0}$ and $\vec{sk}^{(i^*+1)}$ to the $\mathcal{E}$-challenger. If $\beta = 0$, the challenger sends to $\mathcal{B}$ the ciphertexts $\mathcal{E}.\mathsf{E}(ek^{(i^*)}, \vec{0})$, else it sends the ciphertexts $\mathcal{E}.\mathsf{E}(ek^{(i^*)}, \vec{sk}^{(i^*+1)})$. Then $\mathcal{B}$ labels the ciphertext from the $\mathcal{E}$-challenger as $\vec{S}^{(i^*+1)}$ and sends the complete $\mathcal{E}_{\text{LFHE}}$ public key to $\mathcal{A}$. When $\mathcal{A}$ queries messages $(m_{j,0}, m_{j,1})$, $\mathcal{B}$ encrypts $m_{j,b}$ under $ek^{(n)}$. Notice that from $\mathcal{A}$'s perspective, if $\beta = 0$ then its view is as in Game $i^* + 1$, and if $\beta = 1$ its view is as in Game i^*. Therefore, $\mathcal{A}$'s advantage is ε_{i^*+1} if $\beta = 0$ and ε_{i^*} if $\beta = 1$. Also $\mathcal{B}$ guesses that $\beta = 1$ if $\mathcal{A}$ guesses b

correctly, otherwise it guesses that $\beta = 0$. Now $\mathcal{B}$'s success probability is

$$\begin{aligned}\Pr[\mathcal{B}\text{ correct}] &= \Pr[\mathcal{B}\text{ correct}|\beta = 0\text{ and }\mathcal{A}\text{ correct}]\cdot\Pr[\beta = 0\text{ and }\mathcal{A}\text{ correct}]\\ &\quad+\Pr[\mathcal{B}\text{ correct}|\beta = 0\text{ and }\mathcal{A}\text{ incorrect}]\cdot\Pr[\beta = 0\text{ and }\mathcal{A}\text{ incorrect}]\\ &\quad+\Pr[\mathcal{B}\text{ correct}|\beta = 1\text{ and }\mathcal{A}\text{ correct}]\cdot\Pr[\beta = 1\text{ and }\mathcal{A}\text{ correct}]\\ &\quad+\Pr[\mathcal{B}\text{ correct}|\beta = 1\text{ and }\mathcal{A}\text{ incorrect}]\cdot\Pr[\beta = 1\text{ and }\mathcal{A}\text{ incorrect}]\\ &= 0\cdot\big[(1/2)(1/2+\varepsilon_{i^*+1})\big]+1\cdot\big[(1/2)(1/2-\varepsilon_{i^*+1})\big]\\ &\quad+1\cdot\big[(1/2)(1/2+\varepsilon_{i^*})\big]+0\cdot\big[(1/2)(1/2-\varepsilon_{i^*})\big]\\ &= 1/2+(\varepsilon_{i^*}-\varepsilon_{i^*+1})/2. \qquad\blacksquare\end{aligned}$$

REFERENCES

[1] A. Acar, H. Aksu, A. S. Uluagac, and M. Conti, A survey on homomorphic encryption schemes: Theory and implementation. *ACM Comput. Surv.* **51** (2018), no. 4, 1–35.

[2] C. Aguilar-Melchor, S. Fau, C. Fontaine, G. Gogniat, and R. Sirdey, Recent advances in homomorphic encryption: a possible future for signal processing in the encrypted domain. *IEEE Signal Process. Mag.* **30** (2013), no. 2, 108–117.

[3] M. Ajtai, Generating hard instances of lattice problems. In *Proceedings of the twenty-eighth annual ACM symposium on theory of computing*, pp. 99–108, ACM, 1996.

[4] M. Ajtai and C. Dwork, A public-key cryptosystem with worst-case/average-case equivalence. In *Proceedings of the twenty-ninth annual ACM symposium on theory of computing*, pp. 284–293, ACM, 1997.

[5] G. Alagic, Z. Brakerski, Y. Dulek, and C. Schaffner, Impossibility of quantum virtual black-box obfuscation of classical circuits. 2020, arXiv:2005.06432.

[6] M. R. Albrecht, P. Farshim, J.-C. Faugere, and L. Perret, Polly cracker, revisited. In *International conference on the theory and application of cryptology and information security*, pp. 179–196, Springer, 2011.

[7] M. Albrecht, M. Chase, H. Chen, J. Ding, S. Goldwasser, S. Gorbunov, S. Halevi, J. Hoffstein, K. Laine, K. Lauter, S. Lokam, D. Micciancio, D. M. T. Morrison, A. Sahai, and V. Vaikuntanathan, Homomorphic encryption standard. Available at http://homomorphicencryption.org/, accessed February 2019, November 2018.

[8] J. Alperin-Sheriff and C. Peikert, Faster bootstrapping with polynomial error. In *Advances in cryptology—CRYPTO 2014, Part I,* edited by J. A. Garay and R. Gennaro, pp. 297–314, Springer, 2014.

[9] B. Applebaum, D. Cash, C. Peikert, and A. Sahai, Fast cryptographic primitives and circular-secure encryption based on hard learning problems. In *Advances in cryptology—CRYPTO 2009, 29th annual international cryptology conference, Santa Barbara, CA, USA, August 16–20, 2009. Proceedings*, pp. 595–618, Springer, 2009.

[10] F. Armknecht, T. Gagliardoni, S. Katzenbeisser, and A. Peter, General impossibility of group homomorphic encryption in the quantum world. In *International workshop on public key cryptography*, pp. 556–573, Springer, 2014.

[11] F. Armknecht, C. Boyd, C. Carr, K. Gjøsteen, A. Jäschke, C. A. Reuter, and M. Strand, A guide to fully homomorphic encryption. *IACR Cryptol. ePrint Arch.* **2015** (2015), 1192.

[12] S. Arora and R. Ge, New algorithms for learning in presence of errors. In *ICALP (1)*, pp. 403–415, Lecture Notes in Comput. Sci. 6755, Springer, 2011.

[13] B. Barak and Z. Brakerski, Building the Swiss Army Knife. https://windowsontheory.org/2012/05/02/building-the-swiss-army-knife/, May 2, 2012.

[14] B. Barak, O. Goldreich, R. Impagliazzo, S. Rudich, A. Sahai, S. Vadhan, and K. Yang, On the (im) possibility of obfuscating programs. In *Annual international cryptology conference*, pp. 1–18, Springer, 2001.

[15] M. Bellare, A. Desai, E. Jokipii, and P. Rogaway, A concrete security treatment of symmetric encryption. In *Proceedings 38th annual symposium on foundations of computer science*, pp. 394–403, IEEE, 1997.

[16] A. Blum, M. Furst, M. Kearns, and R. J. Lipton, Cryptographic primitives based on hard learning problems. In *Annual international cryptology conference*, pp. 278–291, Springer, 1993.

[17] D. Boneh and M. Franklin, Identity-based encryption from the weil pairing. In *Annual international cryptology conference*, pp. 213–229, Springer, 2001.

[18] D. Boneh and R. J. Lipton, Algorithms for black-box fields and their application to cryptography. In *Annual international cryptology conference*, pp. 283–297, Springer, 1996.

[19] D. Boneh and A. Silverberg, Applications of multilinear forms to cryptography. *Contemp. Math.* **324** (2003), no. 1, 71–90.

[20] Z. Brakerski, Fully homomorphic encryption without modulus switching from classical GapSVP. In *Annual cryptology conference*, pp. 868–886, Springer, 2012.

[21] Z. Brakerski, When homomorphism becomes a liability. In *Theory of cryptography conference*, pp. 143–161, Springer, 2013.

[22] Z. Brakerski, Fundamentals of fully homomorphic encryption. In *Providing sound foundations for cryptography: on the work of Shafi Goldwasser and Silvio Micali*, pp. 543–563, ACM, 2019.

[23] Z. Brakerski, C. Gentry, and V. Vaikuntanathan, (Leveled) fully homomorphic encryption without bootstrapping. *ACM Trans. Comput. Theory* **6** (2014), no. 3, 13.

[24] Z. Brakerski and V. Vaikuntanathan, Efficient fully homomorphic encryption from (standard) lwe. *SIAM J. Comput.* **43** (2014), no. 2, 831–871.

[25] Z. Brakerski and V. Vaikuntanathan, Lattice-based FHE as secure as PKE. In *Innovations in theoretical computer science, ITCS'14*, edited by M. Naor, pp. 1–12, ACM, 2014.

[26] R. Canetti, H. Lin, S. Tessaro, and V. Vaikuntanathan, Obfuscation of probabilistic circuits and applications. In *Theory of cryptography conference*, pp. 468–497, Springer, 2015.

[27] J. H. Cheon, A. Kim, M. Kim, and Y. S. Song, Homomorphic encryption for arithmetic of approximate numbers. In *ASIACRYPT (1)*, pp. 409–437, Lecture Notes in Comput. Sci. 10624, Springer, 2017.

[28] J. H. Cheon and D. Stehlé, Fully homomophic encryption over the integers revisited. In *Annual international conference on the theory and applications of cryptographic techniques*, pp. 513–536, Springer, 2015.

[29] I. Chillotti, N. Gama, M. Georgieva, and M. Izabachène, Faster fully homomorphic encryption: bootstrapping in less than 0.1 seconds. In *Advances in cryptology–ASIACRYPT 2016. ASIACRYPT 2016*, pp. 3–33, Lecture Notes in Comput. Sci. 10031, Springer, 2016.

[30] I. Chillotti, N. Gama, M. Georgieva, and M. Izabachène, Faster packed homomorphic operations and efficient circuit bootstrapping for TFHE. In *ASIACRYPT (1)*, pp. 377–408, Lecture Notes in Comput. Sci. 10624, Springer, 2017.

[31] W. Diffie and M. Hellman, New directions in cryptography. *IEEE Trans. Inf. Theory* **22** (1976), no. 6, 644–654.

[32] D. Dolev, C. Dwork, and M. Naor, Nonmalleable cryptography. *SIAM Rev.* **45** (2003), no. 4, 727–784.

[33] L. Ducas and D. M. FHEW, bootstrapping homomorphic encryption in less than a second. In *EUROCRYPT (1)*, pp. 617–640, Lecture Notes in Comput. Sci. 9056, Springer, 2015.

[34] L. Ducas and D. Stehlé, Sanitization of fhe ciphertexts. In *Proceedings, Part I, of the 35th annual international conference on advances in cryptology—EUROCRYPT 2016*, pp. 294–310, Lecture Notes in Comput. Sci. 9665, Springer, 2016.

[35] T. ElGamal, A public key cryptosystem and a signature scheme based on discrete logarithms. *IEEE Trans. Inf. Theory* **31** (1985), no. 4, 469–472.

[36] A. Feldmann, N. Samardzic, A. Krastev, S. Devadas, R. Dreslinski, K. Eldefrawy, N. Genise, C. Peikert, and D. Sanchez, F1: A fast and programmable accelerator for fully homomorphic encryption (extended version). 2021, arXiv:2109.05371.

[37] M. Fellows and N. Koblitz, Combinatorial cryptosystems galore! *Contemp. Math.* **168** (1994), 51–51.

[38] S. Garg, C. Gentry, and S. Halevi, Candidate multilinear maps from ideal lattices. In *Annual international conference on the theory and applications of cryptographic techniques*, pp. 1–17, Springer, 2013.

[39] S. Garg, C. Gentry, S. Halevi, M. Raykova, A. Sahai, and B. Waters, Candidate indistinguishability obfuscation and functional encryption for all circuits. *SIAM J. Comput.* **45** (2016), no. 3, 882–929.

[40] C. Gentry, *A fully homomorphic encryption scheme*. Stanford university, 2009.

[41] C. Gentry, Fully homomorphic encryption using ideal lattices. In *Proceedings of the 41st ACM symposium on theory of computing—STOC 2009*, pp. 169–178, ACM, 2009.

[42] C. Gentry, Computing arbitrary functions of encrypted data. *Commun. ACM* **53** (2010), no. 3, 97–105.

[43] C. Gentry, Computing on the edge of chaos: Structure and randomness in encrypted computation. In *Proceedings of the international congress of mathematicians (ICM)*, pp. 609–632, Kyung Moon SA, Seoul, 2014.

[44] C. Gentry, S. Halevi, M. Raykova, and D. Wichs, Outsourcing private RAM computation. In *2014 IEEE 55th annual symposium on foundations of computer science*, pp. 404–413, IEEE, 2014.

[45] C. Gentry, S. Halevi, and N. Smart, Fully homomorphic encryption with polylog overhead. In *Advances in cryptology—EUROCRYPT 2012*, pp. 465–482, Lecture Notes in Comput. Sci. 7237, Springer, 2012. Full version at http://eprint.iacr.org/2011/566.

[46] C. Gentry, A. Sahai, and B. Waters, Homomorphic encryption from learning with errors: conceptually-simpler, asymptotically-faster. In *Advances in cryptology—CRYPTO 2013, Part I*, edited by R. Canetti, and J. A. Garay, pp. 75–92, Springer, 2013.

[47] S. Goldwasser and S. Micali, Probabilistic encryption. *J. Comput. System Sci.* **28** (1984), no. 2, 270–299.

[48] S. Halevi, Homomorphic encryption. In *Tutorials on the foundations of cryptography*, pp. 219–276, Springer, 2017.

[49] D. Harvey and J. Van Der Hoeven, Integer multiplication in time $o(n \log n)$. *Ann. of Math.* **193** (2021), no. 2, 563–617.

[50] J. Hoffstein, J. Pipher, and J. H. Silverman NTRU, A ring-based public key cryptosystem. In *ANTS*, edited by J. Buhler, pp. 267–288, Lecture Notes in Comput. Sci. 1423, Springer, 1998.

[51] Homomorphic Encryption Standardization. 2021, https://homomorphicencryption.org.

[52] N. Howgrave-Graham, Approximate integer common divisors. In *International cryptography and lattices conference*, pp. 51–66, Springer, 2001.

[53] A. Jain, H. Lin, and A. Sahai, Indistinguishability obfuscation from well-founded assumptions. In *Proceedings of the 53rd annual ACM SIGACT symposium on theory of computing*, pp. 60–73, ACM, 2021.

[54] A. Joux, A one round protocol for tripartite Diffie–Hellman. In *International algorithmic number theory symposium*, pp. 385–393, Springer, 2000.

[55] W. Jung, S. Kim, J. H. Ahn, J. H. Cheon, and Y. Lee, Over 100× faster bootstrapping in fully homomorphic encryption through memory-centric optimization with gpus. *IACR Cryptol. ePrint Arch.* **2021** (2021), 508.

[56] R. M. Karp and V. Ramachandran, *A survey of parallel algorithms for shared-memory machines*. United States: N. p., 1989.

[57] A. K. Lenstra, H. W. Jr. Lenstra, M. S. Manasse, and J. M. Pollard, The number field sieve. In *Proceedings of the twenty-second annual ACM symposium on theory of computing*, pp. 564–572, ACM, 1990.

[58] N. Linial, Y. Mansour, and N. Nisan, Constant depth circuits, Fourier transform, and learnability. *J. ACM* **40** (1993), no. 3, 607–620.

[59] A. López-Alt, E. Tromer, and V. Vaikuntanathan, On-the-fly multiparty computation on the cloud via multikey fully homomorphic encryption. In *STOC*, pp. 1219–1234, ACM, 2012.

[60] V. Lyubashevsky, C. Peikert, and O. Regev, On ideal lattices and learning with errors over rings. In *Advances in cryptology—EUROCRYPT'10*, edited by H. Gilbert, pp. 1–23, Lecture Notes in Comput. Sci. 6110, Springer, 2010.

[61] U. Mahadev, Classical homomorphic encryption for quantum circuits. *SIAM J. Comput.* **(0):FOCS18–189** (2020).

[62] P. Martins, L. Sousa, and A. Mariano, A survey on fully homomorphic encryption: an engineering perspective. *ACM Comput. Surv.* **50** (2017), no. 6, 1–33.

[63] K. Nuida, Towards constructing fully homomorphic encryption without ciphertext noise from group theory. In *International symposium on mathematics, quantum theory, and cryptography*, pp. 57–78, Springer, Singapore, 2021.

[64] B. Reagen, W.-S. Choi, Y. Ko, V. T. Lee, H.-H. S. Lee, G.-Y. Wei, and D. B. Cheetah, Optimizing and accelerating homomorphic encryption for private inference. In *IEEE international symposium on high-performance computer architecture (HPCA)*, pp. 26–39, IEEE, 2021.

[65] O. Regev, On lattices, learning with errors, random linear codes, and cryptography. In *Proceedings of the thirty-seventh annual ACM symposium on theory of computing*, pp. 84–93, ACM, 2005. Full version in [66].

[66] O. Regev, On lattices, learning with errors, random linear codes, and cryptography. *J. ACM* **56** (2009), no. 6, 34:1–34:40.

[67] R. Rivest, L. Adleman, and M. Dertouzos, On data banks and privacy homomorphisms. In *Foundations of secure computation*, pp. 169–177, Academic Press, 1978.

[68] R. L. Rivest, A. Shamir, and L. Adleman, A method for obtaining digital signatures and public-key cryptosystems. *Commun. ACM* **21** (1978), no. 2, 120–126.

[69] R. Rothblum, Homomorphic encryption: from private-key to public-key. In *Theory of cryptography conference*, pp. 219–234, Springer, 2011.

[70] C.-P. Schnorr, A hierarchy of polynomial time lattice basis reduction algorithms. *Theoret. Comput. Sci.* **53** (1987), no. 2–3, 201–224.

[71] P. W. Shor, Polynomial-time algorithms for prime factorization and discrete logarithms on a quantum computer. *SIAM Rev.* **41** (1999), no. 2, 303–332.

[72] A. Silverberg, Fully homomorphic encryption for mathematicians. In *Women in numbers 2: research directions in number theory*, p. 111, Contemp. Math. 606, AMS, 2013.

[73] N. P. Smart and F. Vercauteren, Fully homomorphic SIMD operations. *Des. Codes Cryptogr.* **71** (2014), no. 1, 57–81. Early version at http://eprint.iacr.org/2011/133.

[74] V. Vaikuntanathan, Computing blindfolded: new developments in fully homomorphic encryption. In *2011 IEEE 52nd annual symposium on foundations of computer science*, pp. 5–16, IEEE, 2011.

[75] L. G. Valiant, A theory of the learnable. *Commun. ACM* **27** (1984), no. 11, 1134–1142.

[76] M. van Dijk, C. Gentry, S. Halevi, and V. Vaikuntanathan, Fully homomorphic encryption over the integers. In *Advances in cryptology—EUROCRYPT 2010, 29th annual international conference on the theory and applications of cryptographic techniques, French Riviera, May 30—June 3, 2010. Proceedings*, pp. 24–43, Springer, 2010.

[77] W. Van Dam, S. Hallgren, and L. Ip, Quantum algorithms for some hidden shift problems. *SIAM J. Comput.* **36** (2006), no. 3, 763–778.

[78] J. Watrous, Quantum algorithms for solvable groups. In *Proceedings of the thirty-third annual ACM symposium on theory of computing*, pp. 60–67, ACM, 2001.

CRAIG GENTRY

Algorand Foundation, New York, NY, USA, craigbgentry@gmail.com

This Page Intentionally Left Blank

RARE EVENTS IN RANDOM MATRIX THEORY

ALICE GUIONNET

ABSTRACT

The uses of random matrix models have spread in many domains of mathematics, physics and computer science. As a consequence, the theory of large random matrices has grown into a diverse and mature field during the last 40 years, yielding answers to increasingly sophisticated questions. In these proceedings, we discuss the applications of large deviations techniques in random matrix theory.

MATHEMATICS SUBJECT CLASSIFICATION 2020

Primary 60B20; Secondary 60F10, 46L54

KEYWORDS

Large random matrices, large deviations

INTERNATIONAL CONGRESS OF MATHEMATICIANS 2022 JULY 6–14

Proc. Int. Cong. Math. 2022, Vol. 2, pp. 1008–1052
DOI 10.4171/ICM2022/174

Published by EMS Press

1. INTRODUCTION

Large random matrices appear in a wide variety of domains. They were first introduced in statistics in the work of Wishart [183] to analyze a large array of noisy data, a point of view that turns out to be particularly relevant and useful nowadays in principal component analysis and statistical learning. Goldstine and von Neumann considered random matrices to model the inevitable errors made in measurements [180]. Wigner [182] and Dyson [79] later conjectured that the statistics of their eigenvalues model very well those of high energy levels in heavy nuclei. Even more surprisingly, Montgomery [154] showed that random matrices are intimately related to the zeroes of Riemann Zeta function, a conjecture which nowadays provides a great intuition for many mathematical results, see, e.g., [5,135]. Random matrices also play a central role in operator algebra theory since Voiculescu [175,177] proved that they are asymptotically free. Random matrices are, moreover, intimately related to integrable systems to which they furnish key examples. The computation of the joint law of the eigenvalues of invariant matrices goes back to Weyl [181] and Cartan [55]. They showed that this distribution is characterized by a density proportional to a power of the Vandermonde determinant of the eigenvalues. As a consequence, the eigenvalues of random matrices furnish an example of a strongly interacting particles system, in connection with many other models such as Coulomb gases or random tilings. For all these reasons, the study of large random matrices (LRM) has grown into a diverse and mature field during the last 40 years, yielding answers to increasingly sophisticated questions. The most basic questions often involve the distribution of the eigenvalues as the size of the matrix goes to infinity. Such a question was first tackled in the breakthrough paper of Wigner [182] who showed that the distribution of the spectrum of a self-adjoint matrix with independent entries (modulo the symmetry constraint) is described by the semicircle law when the dimension goes to infinity. This article discusses how to estimate the probability that the spectrum follows a different distribution in large dimensions. More generally, we will investigate the probability of rare events, that is, of large deviations, in the context of random matrices. In this introduction, we will first outline some of the main results of random matrix theory for the famous Gaussian ensembles, placing the questions on large deviations in the wider context of this theory. We will then motivate the study of large deviations for large random matrices. An important aspect of random matrix theory lies in its connection with the so-called Beta-ensembles, and we will sketch a few applications of large deviations for Beta-ensembles beyond random matrix theory. This introduction is short and therefore unfortunately bypasses many beautiful aspects of large random matrix theory: we refer the interested reader to the introductory books [2,3,14,93,94,153,157] for more.

1.1. Introduction to random matrix theory

1.1.1. The Gaussian ensembles

The most famous model of random matrices is given by the Gaussian ensembles, the Gaussian orthogonal ensemble (GOE) and the Gaussian unitary ensemble (GUE). We say that $\mathbf{G}^n$ follows the law of the GUE (resp. the GOE) if it is an $n \times n$ self-adjoint matrix with independent centered complex (resp. real) Gaussian entries above the diagonal with independent real and imaginary parts with variance $1/(2n)$ (resp. with variance $1/n$), the entries on the diagonal being centered real Gaussians with variance $1/n$ (resp. $2/n$). Their distribution is given by

$$d\mathbb{P}_\beta^n(\mathbf{G}^n) = \frac{1}{\mathbb{Z}_\beta^n} e^{-\frac{\beta n}{4}\operatorname{Tr}((\mathbf{G}^n)^2)} d\mathbf{G}^n, \tag{1.1}$$

where $\beta = 1$ for the GOE and $\beta = 2$ for the GUE. The measure $d\mathbf{G}^n$ denotes the Lebesgue measure over the corresponding set of matrices (symmetric if $\beta = 1$, Hermitian if $\beta = 2$), which is simply the product of the Lebesgue measure on the entries $d\mathbf{G}^n = \prod_{i\le j} dG_{ij}^n$ if $\beta = 1$ and $d\mathbf{G}^n = \prod_{i\le j} d\Re(G_{ij}^n)\prod_{i<j} d\Im(G_{ij}^n)$ if $\beta = 2$. The constant $\mathbb{Z}_\beta^n$ is the normalizing constant such that $\mathbb{P}_\beta^n$ is a probability measure. These ensembles have a remarkable property: their distribution is invariant under conjugation $\mathbf{G}^n \to \mathbf{U}\mathbf{G}^n\mathbf{U}^*$ by unitary (resp. orthogonal) matrices if $\beta = 2$ (resp. $\beta = 1$). Because of this invariance, the eigenvectors of $\mathbf{G}^n$ are uniformly distributed on the sphere and hence delocalized in the sense that their entries are typically of order of the inverse of the square root of the dimension. Moreover, a change of variables shows that the eigenvalues of $\mathbf{G}^n$, $\vec{\lambda} = (\lambda_1, \dots, \lambda_n)$, $\lambda_1 \ge \lambda_2 \dots \ge \lambda_n$, are distributed according to

$$dP_\beta^n(\vec{\lambda}) = \frac{1}{Z_\beta^n}\Delta(\vec{\lambda})^\beta e^{-\frac{\beta n}{4}\sum_{i=1}^n(\lambda_i)^2}\prod d\lambda_i, \tag{1.2}$$

where $\Delta(\vec{\lambda}) = \prod_{i<j}|\lambda_i - \lambda_j|$ is the modulus of the Vandermonde determinant. There exists a third Gaussian ensemble, the Gaussian symplectic ensemble (GSE), with quaternionic entries and which is invariant under conjugation by symplectic matrices. Its eigenvalues are distributed according to P_4^n. However, we shall not highlight this case in the sequel. The Gaussian ensembles are also called the $G\beta E$'s with $\beta = 1, 2, 4$ for the GOE, GUE, and GSE, respectively. Remarkably, for any $\beta > 0$, P_β^n was shown **[77]** to describe the law of the eigenvalues of the $n \times n$ self-adjoint tridiagonal matrix $\sqrt{\beta n}^{-1}\mathbf{X}_\beta^n$ where the diagonal entries $\{X_\beta^n(i,i), 1 \le i \le n\}$ are independent centered Gaussian variables with variance 2, independent of the off-diagonal entries $\{X_\beta^n(i,i+1), 1 \le i \le n-1\}$ which are independent and such that $X_\beta^n(i,i+1)$ is a chi-distributed variable with $\beta(n-i)$ degrees of freedom for $i \in \{1, \dots, n-1\}$. Thanks to formula (1.2), the Gaussian ensembles were studied in detail. We next review a few classical results involving these random matrices. We will see in the core of the text that some of these results generalize to other random matrices, for instance, the Wigner matrices which are similar to the Gaussian ensembles but with entries that are not necessarily Gaussian, namely symmetric or Hermitian matrices with independent centered entries and with variance $1/n$.

1.1.2. Typical events

The celebrated law of large numbers states that the sum of independent identically distributed variables, once properly renormalized, converges almost surely towards its mean. More precisely, if $\mathbf{x} = (x_1, \dots, x_n, \dots)$ is a sequence of independent real random variables

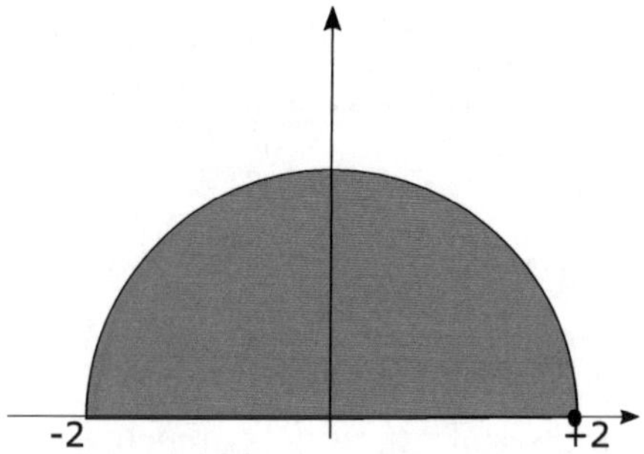

FIGURE 1
The semicircle law and the asymptotic distribution of the spectrum

with the same distribution μ such that $\int |x| d\mu(x)$ is finite, the empirical mean

$$m_n(\mathbf{x}) := \frac{1}{n} \sum_{i=1}^{n} x_i \tag{1.3}$$

converges almost surely towards the mean when n goes to infinity:

$$\lim_{n \to \infty} m_n(\mathbf{x}) = \int x d\mu(x) \quad \text{a.s.}$$

Historically the first, and particularly simple, application of this theorem applied to coin tossing. The distribution of one toss can be modeled by the Bernoulli law $\mu = \mu_p = p\delta_1 + (1-p)\delta_0$ if the coin has probability p to show heads, which is represented by the value $\{1\}$. The law of large numbers shows that if one flips a coin many times independently, one should see heads approximately a proportion p of the times. There are many proofs of the law of large numbers, and in simple cases like coin tossing, it follows from counting the number of ways to see a given number of heads out of n flips.

The emergence of an almost sure deterministic phenomenon from many independent random events is a usual feature in probability theory or statistical mechanics. In the latter, many random particles collaborate to give a deterministic macroscopic behavior. In random matrix theory (RMT), Wigner [182] showed that the distribution of the eigenvalues of Gaussian ensembles converges almost surely towards a deterministic limit given by the semicircle law.

Theorem 1.1 ([182]). *Let $\lambda_1 \geq \lambda_2 \geq \cdots \geq \lambda_n$ be the eigenvalues of the GβE for $\beta = 1, 2$, or 4. Then, for any $a < b$,*

$$\lim_{n \to \infty} \frac{1}{n} \#\{i : \lambda_i \in [a, b]\} = \sigma([a, b]) \quad \textit{almost surely}, \tag{1.4}$$

where σ is the semicircle law,

$$\sigma(dx) = \frac{1}{2\pi} \sqrt{4 - x^2} 1_{|x| \leq 2} dx. \tag{1.5}$$

Equation (1.4) can be seen as the almost sure weak convergence of the empirical measure $\hat{\mu}_n = \frac{1}{n} \sum_{i=1}^{n} \delta_{\lambda_i}$ of the eigenvalues in the sense that it is equivalent to the convergence, for any bounded continuous function f, of

$$\lim_{n\to\infty}\frac{1}{n}\sum_{i=1}^{n} f(\lambda_i) = \int f(x)d\sigma(x) \quad \text{a.s.} \tag{1.6}$$

This result was proved by Wigner for matrices $\mathbf{X}^n$ with independent centered entries (modulo the symmetry constraint) with variance $1/n$ and finite moments and not only for Gaussian entries. However, the proof of Wigner's theorem is much less obvious than that of the classical law of large numbers because the spectrum is a complicated function of the entries of the matrix. The key point of Wigner was to observe that moments of the empirical measure of the eigenvalues are more explicit functions of the entries than indicator functions since, for any integer number k,

$$\frac{1}{n}\sum_{i=1}^{n}\lambda_i^k = \frac{1}{n}\mathrm{Tr}\big((\mathbf{G}^n)^k\big) = \frac{1}{n}\sum_{i_1,\dots,i_k=1}^{n}\mathbf{G}^n_{i_1i_2}\cdots\mathbf{G}^n_{i_ki_1}. \tag{1.7}$$

The expectation and variance of the right-hand side of (1.7) can be estimated, yielding by Borel–Cantelli's lemma the almost sure convergence of traces of moments. Moreover, by the Weierstrass approximation theorem, the almost sure convergence of the moments (1.7) implies (1.6) and then (1.4) because the semicircle law is compactly supported and has no atoms.

We are also interested in more detailed convergence of the spectrum, for instance, convergence of the largest eigenvalue λ_1. Fűredi and Komlós [95] show that it sticks to the bulk in the sense that the largest eigenvalue converges almost surely towards 2, the boundary of the support of the semicircle law (strictly speaking, [95] assumes that the entries are bounded, but the proof easily generalizes to sub-Gaussian entries, see, e.g., [3]). This is analogous to the statement from classical probability theory that the supremum of independent variables with law μ converges almost surely towards the upper boundary of the support of μ, except that this is infinite if the variables are unbounded like the Gaussians.

1.1.3. Fluctuations

The probability to make a small error in the law of large numbers is specified by the well-known central limit theorem. It asserts that errors are of the order of the square root of the dimension and fluctuations are Gaussian. More precisely, coming back to the example of the empirical mean (1.3) of independent variables, it states that, if $\int |x|^2 d\mu(x)$ is finite and we set $\sigma(\mu) = (\int x^2 d\mu(x) - (\int x d\mu(x))^2)^{1/2}$, then $\sqrt{n}(m_n(\mathbf{x}) - \int x d\mu(x))$ converges in distribution towards a centered Gaussian variable with variance $\sigma(\mu)$, so that for every real number t,

$$\lim_{n\to\infty}\mathbb{P}\left(\sqrt{n}\left(m_n(\mathbf{x}) - \int x d\mu(x)\right) \le t\sigma(\mu)\right) = \int_{-\infty}^{t} e^{-\frac{x^2}{2}}\frac{dx}{\sqrt{2\pi}}.$$

In the context of random matrices, the fluctuations of the eigenvalues are much smaller, see Figure 2. The fluctuations of the empirical measure were first studied in [134, 137]. We describe below the result obtained by Johansson [131] in the case of the Gaussian ensembles.

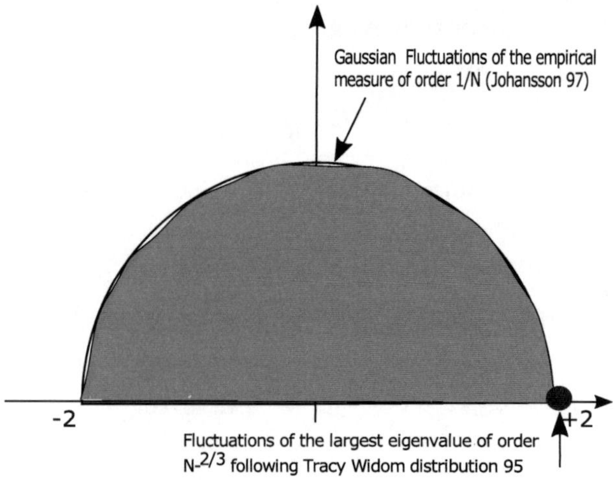

FIGURE 2
Fluctuations of the spectrum. Courtesy of D. Coulette

He showed that for every sufficiently smooth test function f,

$$\sum_{i=1}^{n} f(\lambda_i) - n \int f(x)d\sigma(x) \tag{1.8}$$

converges in distribution towards a Gaussian variable. This Gaussian variable is not centered in general when $\beta \neq 2$, but both its mean and variance are explicit. The original proof [131] relies on the explicit joint law of the eigenvalues (1.2) and is far from obvious because of the strong correlations between the eigenvalues due to the Vandermonde determinant. This result was generalized to the case of Wigner matrices by using moment estimates [4,146], resulting in the universality of the fluctuations within the class of entries with four first moments equal to the Gaussian ones.

Remarkably, the fluctuations are of order 1 over the dimension (since the convergence of (1.8) holds without any normalization): this indicates that the eigenvalues fluctuate much less than independent variables. This phenomenon was quantified by the so-called local law [84,85] which asserts that the convergence in Wigner's theorem (1.4) can be refined into a quantitative estimate to showing that the number of eigenvalues in a set $[a, b] \subset (-2, 2)$ such that $b - a \gg 1/n$ is still of the order of $n\sigma([a, b])$. This can often be improved to get the rigidity property [88], namely that the eigenvalues in the bulk stay at a distance of order $n^{-1+o(1)}$ from their deterministic limit.

Fluctuations are not always described by the Gaussian distribution: for instance, the maximum of independent variables with fast decaying tails follows a limiting Gumbel distribution. In a breakthrough paper [170], the largest eigenvalue of Gaussian ensembles was shown to fluctuate on the scale $n^{-2/3}$ and the fluctuations to be distributed according to the Tracy–Widom laws. The fluctuations of the eigenvalues inside the bulk are also known and are in the scale n^{-1} (see [151] for $\beta = 1, 2$). These remarkable results were derived thanks to the explicit joint distribution of the eigenvalues (1.2). In particular, the case where $\beta = 2$ was analyzed thanks to the fact that the density is the square of a determinant, allowing for the use of orthogonal polynomials and integrable system theory. In a series of major contributions, these results were shown to hold for Wigner matrices with entries with finite second and fourth moments, respectively [86, 87, 169]. The proofs of these results are sophisticated and build on comparison with the Gaussian case.

1.1.4. Rare events

The interest in estimating the probability of rare events goes back to Boltzmann, Gibbs, and Shannon who defined the entropy as the logarithm of the volume of configurations (or microstates) achieving a given macrostate. Going back to the coin tossing example, with a probability p to show heads, a macrostate was defined as the set of configurations such that n tosses give approximately ρn heads, namely the event that $m_n(\mathbf{x})$ is approximately equal to ρ for independent equidistributed x_i with law μ_p. The volume, or probability, of such a macrostate is easily seen to be given by

$$\lim_{\delta\downarrow 0}\lim_{n\to\infty}\frac{1}{n}\ln \mathbb{P}\big(\mathbf{x} : |m_n(\mathbf{x})-\rho| \leq \delta\big) = -S_p(\rho) = -\rho\ln(\rho/p)-(1-\rho)\ln\left(\frac{1-\rho}{1-p}\right), \tag{1.9}$$

where $-S_p(\rho)$ is the entropy of ρ. This result can be inferred from counting the configurations and using Stirling's formula. Large deviations theory is the art of estimating such rare events in a general framework [73, 75, 78, 174] by proving large deviation principles (LDPs) that we now define. We will hereafter consider a sequence of probability measures $(\mu_n)_{n\geq 0}$ on a Polish space E. In this article, we will mainly consider the case where E is the real line or the set of probability measures on the real line equipped with its weak topology. Let $(a_n)_{n\in\mathbb{N}}$ be a sequence of nonnegative real numbers going to infinity as n goes to infinity. We say that $(\mu_n)_{n\geq 0}$ satisfies an LDP with speed a_n and good rate function I, denoted in short by LDP(a_n, I), if and only if

- $I : E \to \mathbb{R}^+$ has compact level sets $\{x \in E : I(x) \leq M\}$ for every $M \in \mathbb{R}^+$,
- For each Borel measurable set $B \subset E$,

$$-\inf_{\mathring{B}} I \leq \liminf_{n\to\infty}\frac{1}{a_n}\ln \mu_n(B) \leq \limsup_{n\to\infty}\frac{1}{a_n}\ln\mu_n(B) \leq -\inf_{\bar{B}} I. \tag{1.10}$$

Taking B to be a small ball $B = B(\rho,\delta)$ for some $\rho \in E$ and $\delta > 0$ as small as wished (but independent of n) shows that the LDP allows estimating the probability of small balls as

$$\mu_n\big(B(\rho,\delta)\big) \simeq e^{-a_n I(\rho)}$$

in the sense that for any $\rho \in E$,

$$\lim_{\delta\downarrow 0}\liminf_{n\to\infty}\frac{1}{a_n}\ln\mu_n\big(B(\rho,\delta)\big) = \lim_{\delta\downarrow 0}\limsup_{n\to\infty}\frac{1}{a_n}\ln\mu_n\big(B(\rho,\delta)\big) = -I(\rho). \tag{1.11}$$

Such an estimate is called a weak large deviation principle. By a covering argument, (1.11) can be shown to be equivalent to the LDP if E is compact or if μ_n satisfies a property called exponential tightness [73, (1.2.17)]. An important consequence of the LDP(a_n, I) is that if the rate function I vanishes at a single point $x^* \in E$, then μ_n converges weakly towards a Dirac mass at this point.

The two most well-known results from the large deviations theory are Cramèr's and Sanov's theorems. Cramèr's theorem [73] asserts that the distribution of the empirical mean $m_n(\mathbf{x})$ satisfies an LDP with speed n when the $(x_i)_{i\geq 0}$ are independent equidistributed real-valued random variables with distribution μ with a finite Laplace transform in a vicinity of the origin, see (1.9) in the case $\mu = \mu_p$. Sanov's theorem shows that the law of the empirical measure $\frac{1}{n}\sum_{i=1}^n \delta_{x_i}$ satisfies as well an LDP$(n, H(\cdot|\mu))$, so that for any probability measure ν,

$$\mathbb{P}\left(d\left(\frac{1}{n}\sum_{i=1}^{n}\delta_{x_i}, \nu\right) < \delta\right) \simeq e^{-nH(\nu|\mu)} \tag{1.12}$$

if d is the distance on the set $\mathcal{P}(\mathbb{R})$ of probability measures on the real line defined by

$$d(\mu,\nu) = \sup_{\|f\|_L\leq 1} \left|\int f(x)d\mu(x) - \int f(x)d\nu(x)\right|,$$

where $\|f\|_L = \sup_{x\neq y}|x-y|^{-1}|f(x)-f(y)| + \sup_x |f(x)|$. Here, $H(\nu|\mu)$ is the relative entropy: it is infinite unless ν is absolutely continuous with respect to μ and then equals $\int \ln \frac{d\nu}{d\mu} d\nu$. The proofs of such theorems are more sophisticated than in the coin tossing example since they cannot rely on direct combinatorial arguments. They often rather follow from clever changes of measures (also called tilts) that reveal how the distributions should be changed to make a given rare event typical. These arguments are very much based on the independence of the variables $(x_i)_{i\geq 0}$. Large deviations theory was mainly developed to tackle the distribution of sums of independent random variables, or of "weakly" dependent variables such as Markov chains, or probability measures obtained either by a push-forward or a nice density from the latter, see the work of Cramèr, Varadhan, and many others [73, 75, 81]. This classical theory does not apply to large random matrices in general. Indeed, even if the random matrices are chosen with independent entries, the spectrum or the eigenvectors are complicated functions of these entries. We can take the example of the trace of a power of a matrix, see (1.7): as soon as the power k is higher or equal to 3, it cannot be written as a sum of independent entries and understanding the large deviations of such functionals for Wigner matrices is still open in general, see [9, 10] for entries with sharp sub-Gaussian tails or without Gaussian tails. The case of the Gaussian ensembles is simpler because of the explicit law of the eigenvalues (1.2). Even if the classical large deviations theory does not apply to the distribution of the eigenvalues (1.2) because of the strong interaction due to the Vandermonde determinant term in its density, LDPs were derived in this case to estimate the probability that the empirical measure of the eigenvalues or the largest eigenvalue deviates from their typical behavior, see Figure 3.

Theorem 1.2. *Let $\lambda_1 \geq \lambda_2 \cdots \geq \lambda_n$ be distributed according to* (1.2) *for some $\beta > 0$. Then*

- ([25]) *For $\mu \in \mathcal{P}(\mathbb{R})$, set*

$$E(\mu) = \frac{1}{2}\int\int\left(\frac{x^2}{4} + \frac{y^2}{4} - \ln|x-y|\right)d\mu(x)d\mu(y)$$

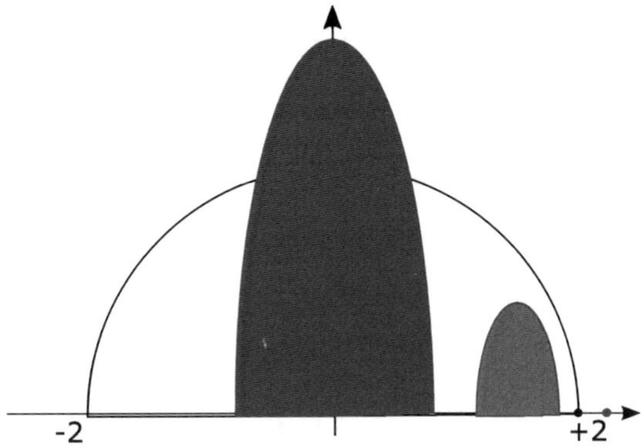

FIGURE 3
Large deviations of the spectrum with exponentially small probability. Courtesy of D. Coulette

and $\mathcal{E}(\mu) = E - \inf E$. *Then* $\mathcal{E}$ *is a good rate function. The distribution of the empirical measure of the eigenvalues* $\hat{\mu}_n = \frac{1}{n}\sum_{i=1}^n \delta_{\lambda_i}$ *under* P_β^n *satisfies an* $\mathrm{LDP}(\beta n^2, \mathcal{E})$, *that is, for every closed set* F,

$$\limsup_{n\to\infty} \frac{1}{\beta n^2} \ln P_\beta^n(\hat{\mu}_n \in F) \le -\inf_F \mathcal{E},$$

whereas for any open set O,

$$\limsup_{n\to\infty} \frac{1}{\beta n^2} \ln P_\beta^n(\hat{\mu}_n \in O) \ge -\inf_O \mathcal{E}.$$

- (**[24, THEOREM 6.2]**) *Let* $I_{\mathrm{GOE}}(x) = \frac{1}{2}\int_2^x \sqrt{y^2-4}\,dy$ *for* $x \ge 2$ *and* $I_{\mathrm{GOE}}(x) = +\infty$ *for* $x < 2$. *Then the distribution of* λ_1 *satisfies an* $\mathrm{LDP}(\beta n, I_{\mathrm{GOE}})$.

Notice that the speed of the LDP for the empirical measure is n^2, in contrast with the speed n in Sanov's theorem, showing again that the eigenvalues of Gaussian ensembles are much less random than independent variables. Moreover, it can be seen that $\mathcal{E}$ vanishes only at the semicircle law, implying Theorem 1.1 (see Section 2.1 for more detail). Similarly, I_{GOE} vanishes at 2 only, ensuring the convergence of the largest eigenvalue towards 2. The proof of this theorem relies on Laplace's principle. Indeed, the distribution of the empirical measure of the eigenvalues and of the largest eigenvalue can be seen to have approximately the density $e^{-\beta n^2 E(\hat{\mu}_n)}/Z_n$ and $e^{-\beta n I_{\mathrm{GOE}}(\lambda_1)}/z_n$, respectively, where Z_n, z_n are appropriate normalizing constants. The theorem would follow by the Laplace principle if E and I_{GOE} were continuous. The main point is to make the above approximations precise and to show that, even though E is not continuous (because the logarithm is not bounded), the result is still valid.

One of the main goals of this article is to discuss how to generalize this theorem. For instance, how can it be extended to general Wigner matrices? In this case, no explicit formula for the law of the eigenvalues such as (1.2) is known. On the other hand, the LDP a priori depends on the whole distribution of the entries as in the case of Sanov's theorem, contrarily to fluctuations which often depend mainly on a finite number of moments. Such universal classes are not expected in large deviations theory. Even conjecturing the rate functions for such LDPs is not clear. LDPs were also obtained for other invariant models such as Wishart or unitary matrices [125], or non-Hermitian Gaussian matrices [28], but the distributions of their eigenvalues all enjoy a rather explicit form. LDPs for Gaussian random matrices with independent centered entries but variance different from those of the Gaussian ensembles are also still open, see [105] for large deviation upper bounds. We will see that other invariant matrix models, such as models involving several matrices, remain very challenging as well.

Large deviations theory is key to study laws of dependent variables such as Boltzmann–Gibbs distributions in statistical mechanics. They are probability measures of the form

$$d\mu_\beta^n(\mathbf{x}) = \frac{1}{Z_\beta^n} e^{-\beta n E_n(\mathbf{x})} d\mu_n^0(\mathbf{x}), \tag{1.13}$$

where E_n is a function from the space of states (for instance, $\mathbb{R}^n$) into $\mathbb{R}$, often called the energy or the Hamiltonian, β is a real parameter proportional to the inverse of the temperature, $\mu_n^0(\mathbf{x})$ is some reference probability measure, and Z_β^n is the so-called partition function, namely the constant which turns μ_β^n into a probability measure. The properties of such measures when the dimension n goes to infinity are better understood when the distribution of $E_n(\mathbf{x})$ under $\mu_n^0(\mathbf{x})$ satisfies an LDP(n, I). The typical values of the energy can then be inferred from the fact that for every $y \in \mathbb{R}$,

$$\mu_\beta^n\left(\mathbf{x} : \left|E_n(\mathbf{x}) - y\right| < \delta\right) \simeq \frac{1}{Z_\beta^n} e^{-\beta n y - nI(y) + nO(\delta)},$$

from which it is clear that $E_n(\mathbf{x})$ concentrates in a neighborhood of the minimizers of $I_\beta(y) = \beta y + I(y)$ with large probability when n goes to infinity. Varadhan's integral lemma [73, THEOREM 4.3.1] states more precisely that the distribution of E_n under μ_β^n satisfies an LDP$(n, I_\beta - \inf I_\beta)$. This type of analysis often holds for the so-called mean field interacting systems that are distributions such that all variables interact in the same way, for instance, where the energy $E_n(\mathbf{x})$ is a function of the empirical mean m_n or of the empirical measure. A celebrated example is the Curie–Weiss model, where E_n is a quadratic polynomial in the empirical mean m_n and $d\mu_n^0 = \prod_{i=1}^n d\mu_p$. The LDP for this model can be proven as above, as well as the convergence of the empirical mean towards the minimizers of the rate function. It can be shown that this minimizer is unique, equals zero for small enough β, but takes a nonzero value after some critical β_c. This provides a simple example of a phase transition known as spontaneous magnetization. Such applications are also important in RMT when studying matrix models, see Section 1.2.4.

We present in the rest of this introduction a few additional motivations for the study of large deviations for large random matrices, as well as extensions to related fields. We will then review the main results of this emerging field, focusing first on large deviations for the spectrum of one random matrix, and then on multimatrix models where noncommutativity raises new challenges. Along the way, we highlight a few open problems.

1.2. Motivations

In this section, we discuss a few additional motivations to establish large deviation principles in random matrix theory.

1.2.1. Bernoulli matrices

Matrices with entries equal to zero or one can be interpreted as the adjacency matrices of random graphs where the entry at (ij) is equal to one iff there is an edge between the vertices i and j. In particular, random matrices with independent Bernoulli entries are the adjacency matrices of Erdős–Rényi graphs. The spectrum of the adjacency matrix of a graph is intimately related to the graph's geometric properties, such as being an expander. Moreover, traces of moments count particular subgraphs, for instance, the trace of the adjacency matrix to the third power counts the number of triangles in the graph. Understanding how a random graph looks like when a rare event happens is a natural question [60]. As we will see, studying the large deviations for the spectrum of matrices with non-Gaussian entries such as Bernoulli's is far more difficult, basically because the law of the eigenvalues is not given by an explicit distribution as in (1.2). In particular, one needs to understand more precisely the best scheme to perform a given large deviation event.

1.2.2. The BBP transition

The largest eigenvalue is often used to test whether an array of data contains information, just by comparing it with the largest eigenvalue of an array taken at random. Even though such applications involve usually asymmetric matrices and their singular values, the famous Wishart matrices in RMT [183], we stick to Wigner matrices in this article for consistency. The renowned BBP transition [15] asserts that the largest eigenvalue of a random matrix perturbed by a finite-rank signal pops out of the bulk at a critical value of the intensity of the signal (more precisely, of its largest eigenvalue), above which the weak recovery of the signal u from the observation of the perturbed signal is possible [30]. The large deviations for the largest eigenvalue have then been used in statistics to assert the risk of statistical tests [34]. In the related problem of estimating a low-rank tensor in Gaussian noise [27] requires large deviations for the largest eigenvalue of a rank-one perturbation of a Gaussian matrix, which were derived in [111,147].

1.2.3. The complexity of random functions

The interest in optimizing random functions grew in the last ten years from its relevance to deep learning, building on its importance in spin glass theory. However, random functions in high dimensions are complex in the sense that they have many local minima and finding their global minima may be a complicated task, in fact, an NP-hard problem. In the last ten years, the study of the complexity of random functions grew into a field on its own, for instance, allowing to estimate the expectation of the number of local minima of a random function with a given index and level. Such estimates are based on Kac–Rice for-

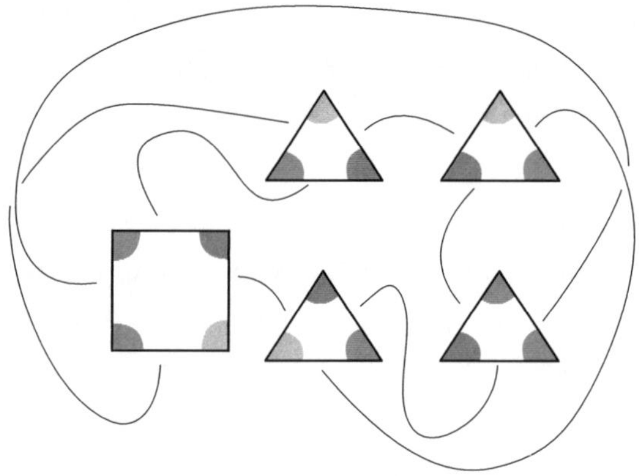

FIGURE 4
Gluing of 4 triangles and 1 square. Corners of the same color belong to the same vertex of the map in the final surface. Courtesy of G. Miermont

mula. Because the Hessian of a random function can be seen as a random matrix, the large deviations for the latter are crucial to getting such estimates **[6,7,23,27,61,96,167]**.

1.2.4. Random matrices and the enumeration of maps

The relation between random matrices and the enumeration of maps goes back to **[50,123,168]** where it was proved that if $\mathbf{G}^n$ follows the GUE, then for every integer k,

$$\mathbb{E}\left[\frac{1}{n}\operatorname{Tr}\left((\mathbf{G}^n)^{2k}\right)\right] = \sum_{g\geq 0}\frac{1}{n^{2g}}M_g(k),$$

where $M_g(k)$ is the number of ways to glue the sides of a $2k$-polygon in pairs such that the resulting surface has genus g. The counting is made after labeling the sides clockwise, or, equivalently, after drawing the polygon on an orientable surface with a distinguished root side. Maps are the same only if all the matchings occur between sides with the same labels.

This relation between maps and random matrices extends to several polygons, see Figure 4, if one considers the distribution

$$d\mathbb{P}^n_{V,2}(\mathbf{X}^n) = \frac{1}{\mathbb{Z}^n_{V,2}}e^{-n\operatorname{Tr}(V(\mathbf{X}^n))}d\mathbb{P}^n_2(\mathbf{X}^n) \tag{1.14}$$

for some potential V. Here, $V(\mathbf{X}^n)$ is defined as the matrix with the same eigenvectors as $\mathbf{X}^n$ and eigenvalues given by the image by V of the eigenvalues of $\mathbf{X}^n$. The measure $\mathbb{P}^n_2$ denotes the law of the GUE (1.1) and the constant $\mathbb{Z}^n_{V,2}$ is the normalizing constant so that $\mathbb{P}^n_{V,2}$ is a probability measure. We will assume that V is a polynomial, $V(x) = -\sum_{i=3}^{p} t_i x^i$, with p even and $t_p < 0$, so that (1.14) makes sense. It was shown **[50,168]** that

$$F^n_{V,2} = \frac{1}{n^2}\ln \mathbb{Z}^n_{V,2} = \sum_{g\geq 0}\frac{1}{n^{2g}}\sum_{k_1,\ldots,k_p=0}^{\infty}\prod_{1\leq i\leq p}\frac{t_i^{k_i}}{k_i!}M_g\big((k_i,i)_{1\leq i\leq p}\big), \tag{1.15}$$

where $M_g((k_i,i)_{1\leq i\leq p})$ denotes the number of ways to glue pairwise the sides of k_i polygons with i sides, $1 \leq i \leq p$, and get a connected two-dimensional surface of genus g. The counting is done with labeled sides. Equivalently, we can think of a polygon with i sides as a vertex with i half-edges drawn on an orientable surface. The number $M_g((k_i, i)_{1\leq i\leq p})$ then counts the maps, that is, the connected graphs drawn on a surface, built by matching the half-edges of k_i vertices with i half-edges, $1 \leq i \leq p$. Half-edges are labeled. The genus of the map is the genus of the surface in which the graph can be properly embedded, which is such that the faces, obtained by cutting the surface along the edges of the map, are homeomorphic to disks. It can be computed from the fact that the Euler characteristic $2 - 2g$ is equal to the number of vertices minus the number of edges plus the number of faces.

Observe also that making a small change $V \to V + \delta x^\ell$ in (1.14), and identifying the linear term in δ, shows that

$$\int \frac{1}{n} \operatorname{Tr}((\mathbf{X}^n)^\ell) d\mathbb{P}^n_{V,2}(\mathbf{X}^n) = \sum_{g\geq 0} \frac{1}{n^{2g}} \sum_{k_1,\dots,k_p=0}^{\infty} \prod \frac{t_i^{k_i}}{k_i!} M_g\big((k_i, i)_{1\leq i\leq p}, (1, \ell)\big), \quad (1.16)$$

where $(1, \ell)$ means that the maps contain an additional polygon with ℓ sides, also called external face. A priori, (1.15) and (1.16) are equalities of formal series. They are obtained by expanding all the terms depending on V and using Wick formula (or, equivalently, Feynman diagrams) to compute the resulting Gaussian expectations. These equalities can be turned into an asymptotic expansion up to errors of order n^{-2k} for any integer number k as soon as the parameters t_i, $1 \leq i \leq p$, are small enough, p is even and with $t_p > 0$ [83]. Therefore, computing the large n limit of the free energy $F^n_{V,2}$ or the limit of the empirical measure of the spectral measure of the eigenvalues allows effectively enumerating planar maps. This route was followed in [50] where random triangulations and quadrangulations were studied, corresponding to cubic and quartic polynomials. Note that in the first case p is odd and $\mathbb{Z}^n_{V,2}$ a priori infinite, but the above relations can be generalized by restricting the integration to matrices with spectral radius bounded by a large enough constant. Such computations can be done more generally by using large deviations theory [106,113].

1.2.5. Beta-ensembles

A change of variables shows that the eigenvalues $\vec{\lambda} = (\lambda_1, \dots, \lambda_n)$ of $\mathbf{X}^n$ following $\mathbb{P}^n_{V,2}$ of (1.14) are distributed according to the distribution $P^n_{\frac{1}{2}x^2+V,\beta}$ where

$$dP^n_{V,\beta}(\vec{\lambda}) = \frac{1}{Z^n_{V,\beta}} \Delta(\vec{\lambda})^\beta e^{-\frac{\beta n}{2}\sum_{i=1}^n V(\lambda_i)} \prod d\lambda_i, \quad (1.17)$$

and $\beta = 2$. The case $\beta = 1$ corresponds to symmetric matrices and $\beta = 4$ to quaternionic entries. We only considered the case $\beta = 2$ in the previous section because the combinatorial interpretation of the other cases is less clear in general, see, e.g., [56,104,141] for $\beta = 1$. In fact, $P^n_{V,\beta}$ makes sense for any $\beta > 0$ and is called a Beta-ensemble. Equation (1.17) furnishes a classical example of particles in strong interaction belonging to the family of Coulomb gases in dimension 1, see, e.g., [28,161] for higher dimensions. Large deviations are useful in analyzing the limiting distribution of the particles.

Equation (1.17) also provides another route to estimate the asymptotics of the free energy $F^n_{V,2}$ or of the empirical measure of the matrix models (1.14) and hence study the enumeration of maps, as proposed in [50] to complement Tutte's combinatorial approach [172].

1.2.6. Multimatrix models and the enumeration of maps

Equation (1.15) generalizes to colored maps for matrix models of the form

$$d\mathbb{P}^n_{V,2}(\mathbf{X}^n_1,\ldots,\mathbf{X}^n_d) = \frac{1}{\mathbb{Z}^n_{V,2}} e^{-n\,\mathrm{Tr}(V(\mathbf{X}^n_1,\ldots,\mathbf{X}^n_d)) - \frac{n}{2}\mathrm{Tr}(\sum(\mathbf{X}^n_i)^2)} d\mathbf{X}^n_1 \cdots d\mathbf{X}^n_d, \tag{1.18}$$

where V is a self-adjoint polynomial going to infinity fast enough. If $V(a_1,\ldots,a_d) = -\sum_{i=1}^p t_j q_j(a_1,\ldots,a_d)$ with monomials q_j, then the authors of [58,168] show that

$$\frac{1}{n^2}\log \mathbb{Z}^n_{V,2} = \sum_{g\geq 0} \frac{1}{n^{2g}} \sum_{k_1,\ldots,k_p=0}^{\infty} \prod_{1\leq i\leq p} \frac{(t_i)^{k_i}}{k_i!} M_g\big((k_i,q_i)_{1\leq i\leq p}\big) \tag{1.19}$$

where $M_g((k_i,q_i)_{1\leq i\leq p})$ counts maps with genus g built over k_i colored polygons of type q_i. A colored polygon of type $q = a_{i_1}\cdots a_{i_k}$ is a polygon drawn on an orientable surface so that its first side has color $i_1 \in \{1,\ldots,m\}$ (the root), second has color i_2, and so on until the last one which has color i_k. Maps are constructed by matching sides with the same color and counting is done with labeled sides. Note that a colored polygon is in bijection with a rooted vertex with ordered colored half-edges and maps are then obtained by matching half-edges of the same color. Even though this equality holds a priori at the level of formal power series, it can be turned into an asymptotic expansion [113]. This equality allows representing many physical models in terms of random matrices, such as the Ising model or the Potts model on random maps [45,89]. Multimatrix integrals turn out to be much more difficult to estimate than one matrix integrals, basically because noncommutativity kicks in. This fact is not surprising given the complicated combinatorial questions that they eventually represent. We will see in Section 3 that the case of the so-called AB interaction is better understood than the general case discussed in Section 4.

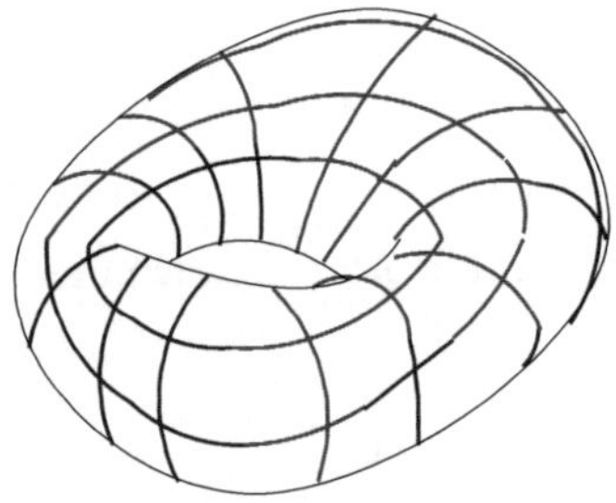

Ising model on the lattice

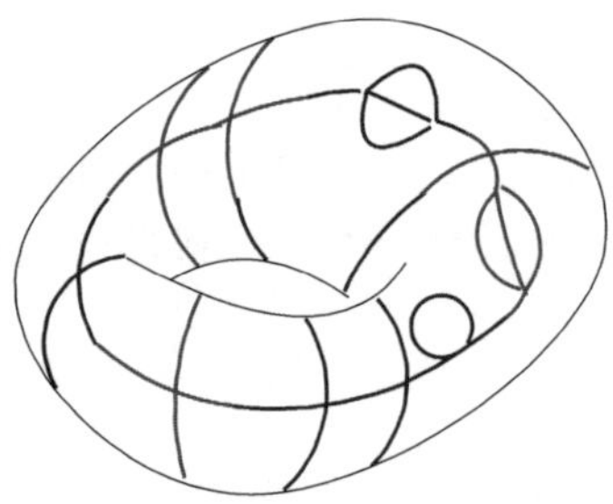

Ising model on random graphs

1.2.7. Multimatrix models and Voiculescu's entropy

One of the most challenging goals in studying large deviations for random matrices was provided by Voiculescu [176,178] in the 1990s when he defined notions of entropy in the context of free probability. Free probability is a probability theory where random variables do not commute and the notion of independence is replaced by freeness. A central point in free probability theory is that Gaussian random matrices are free variables in the limit where their size goes to infinity [175]. Free probability is intimately related to von Neumann algebras, and Voiculescu's hope was to define an invariant for von Neumann algebras to classify them. His ideas were inspired by Minkowski content and entropy in classical probability theory. Voiculescu microstates entropy can be seen as a generalization of Shannon's entropy as it measures the volume of matrices which approximate in a weak sense a given set of noncommutative random variables. In the case of a single variable, the noncommutative entropy is roughly speaking given by the rate function of the large deviation principle for the law of the empirical measure of the eigenvalues of Gaussian ensembles in Theorem 1.2 [178]. Understanding better Voiculescu's entropies would have groundbreaking applications in the theory of von Neumann algebras. Moreover, random matrices can serve to construct interesting noncommutative laws, see, e.g., [110]. We discuss these issues in Section 4.

1.3. Extensions

Beta-ensembles and random matrices are connected with many other fields, of which we describe briefly a few below, see, e.g., [2, 93] for more.

1.3.1. Beta-ensembles and quantum physics

Beta-ensembles and Coulomb gases arise in many domains of physics, including condensed matter physics, statistical physics, and quantum mechanics, we refer to [161] for a survey including higher dimensional generalization. Variants of Beta-ensembles involving hyperbolic Vandermonde determinants appear in quantum integrable models solvable by the quantum separation of variables method, such as the Toda chain [136] or the lattice Sinh-Gordon model [144]. Such integrals then correspond to normalizations of the n-particles wave functions and, more generally, to matrix elements of local operators. Some of their large-n properties were investigated in [42]. Furthermore, integrals similar to Beta-ensembles but having more general interactions with the same singularity arise in the form factor expansions of Wightman functions in massive integrable quantum field theories in $1+1$ dimension [164]. The large deviation techniques discussed in this article allow estimating such integrals.

1.3.2. Random tilings

Beta-ensembles extend to the discrete case. They then model the distribution of horizontal lozenge tiles in a lozenge tiling taken at random. Indeed, consider discrete ensembles given for a weight function w by

$$P_w^n(\vec{\ell}) = \frac{1}{Z_n^\omega} \prod_{i<j} |\ell_j - \ell_i|^2 \prod_i w(\ell_i, n). \tag{1.20}$$

The coordinates $\ell_1, \ldots, \ell_n$ are discrete and such that $\ell_{i+1} - \ell_i \in \mathbb{N}^*$. This probability measure arises in the setting of lozenge tilings of domains such as the hexagon. In fact, considering an hexagon with sides of size A, B, C, along the vertical line at distance t of the vertical side of size A (see Figure 5), the distribution of horizontal lozenges corresponds to

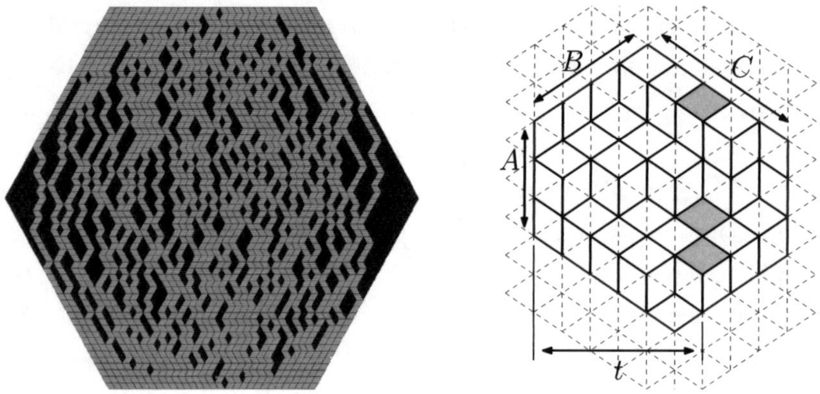

FIGURE 5
Tiling of the hexagon. Courtesy of L. Petrov and V. Gorin

a potential of the form

$$w(\ell, n) = \left[(A + B + C + 1 - t - \ell)_{t-B}(\ell)_{t-C}\right], \tag{1.21}$$

where $(a)_k = a(a+1)\cdots(a+k-1)$ is the Pochhammer symbol, and n is the total number of horizontal lozenges. Large deviations can be used to describe the limiting surface of the tiling when n goes to infinity, for instance, recovering the limiting well-known arctic circle, see, e.g., [62,162] for large deviations of the whole surface. The measure in (1.20) corresponds to $\beta = 2$ ensembles, but can be generalized to all $\beta > 0$, see [40].

1.3.3. Zeroes of random polynomials

The distribution of zeroes of random polynomials also follows a kind of Beta-ensembles distribution: this connection was used in [185] to study large deviations for the distribution of such zeroes. In the same direction, [102] studies the topology of a random real hypersurface in a given smooth real projective manifold by estimating the mean of their Betti numbers thanks to large deviation principles. Such questions are closely related to the study of the complexity of random functions discussed in Section 1.2.3.

1.3.4. Longest increasing subsequence and discrete polynuclear growth

Beta-ensembles also describe the distribution of the discrete polynuclear growth and the length of the longest increasing subsequence of a permutation taken at random, a relation which allowed to study precisely the fluctuations and the large deviations of these models. It was shown in [130] that the distribution of the length of the longest increasing subsequence of a permutation of n elements taken uniformly at random is closely related with Beta-ensembles. This formed the basis for the evaluation of the fluctuations of the longest increasing subsequence in [16]. In [132], the distribution of the discrete polynuclear growth given by

$$G(M, N) = \max_{\pi} \sum_{(i,j)\in\pi} w(i, j),$$

where π is a up-right path from $(0, 1)$ to (M, N), was shown to be intimately related with a discrete Beta-ensemble when the w are independent equidistributed geometric variables. These connections with random matrices allowed studying large deviations [18,19,74,133,168].

1.3.5. Sum rules

Gamboa et al. [97, 98] found out that equating large deviations rate functions in random matrix theory was also fruitful in getting a deep understanding of the sum rules of Killip and Simon [138], also called GEM relations in spectral theory. The latter states highly nontrivial equalities between different functionals on the space of measures. [97, 98] interpreted both sides of the equalities as rate functions for the large deviations for the spectral measure given by $\hat{\mu}_e^n(f) = \langle e, f(\mathbf{G}^n)e\rangle$ for a deterministic unit vector e (and a GOE/GUE matrix $\mathbf{G}^n$). Indeed, one can take two different routes to compute the probability of deviations of this spectral measure: either by relating it to the spectrum of $\mathbf{G}^n$ or to the recursion relations of the associated orthogonal polynomials. Equating the resulting rate functions allows recovering the sum rules of [138] and proving new sum rules. Even the fact that both sides of these equalities are finite at the same time is surprising, see [48] for a pedagogical introduction.

1.3.6. Gibbs ensembles for Toda lattice

Recently, the interest in tridiagonal matrices was revived by Spohn [165, 166] who related them with the Toda lattice. The latter is described by the evolution of n particles with position q_j and momentum p_j satisfying

$$\partial_t q_j = p_j, \quad \partial_t p_j = e^{-r_j} - e^{-r_{j-1}},$$

where $r_j = q_j - q_{j-1}$ and the periodic boundary conditions $q_{j+n} = q_j + cn$. We consider the Lax matrix L_n which is the self-adjoint tridiagonal matrix with entries p_j on the diagonal and $L_n(j, j+1) = L_n(j+1, j) = e^{-r_j/2}$ with periodic boundary condition. It is easy to see that for any function V, $\mathrm{Tr}(V(L_n))$ and $\sum r_j$ are left invariant under the dynamics so that natural invariant measures, called generalized Gibbs measures for the Toda lattice, are given by

$$d\mathbb{T}^n_{V,P}(p,r) = \frac{1}{\mathbb{Z}^{n,\mathbb{T}}_{V,P}} \exp\{-\mathrm{Tr}(V(L_n))\} \prod_{i=1}^n e^{-Pr_i} dr_i dp_i, \tag{1.22}$$

where $\mathbb{Z}^{n,\mathbb{T}}_{V,P}$ is the partition function of the Toda Gibbs measure,

$$\mathbb{Z}^{n,\mathbb{T}}_{V,P} = \int \exp\{-\mathrm{Tr}(V(L_n))\} \prod_{i=1}^n e^{-Pr_i} dr_i dp_i. \tag{1.23}$$

The goal is then to characterize the limiting spectrum of the Lax matrix under $\mathbb{T}^n_{V,P}$. Spohn related this problem with the Beta-ensembles, hence allowing to describe rather explicitly the equilibrium measure of this model. When $V(x) = x^2$, we see that L_n is a tridiagonal matrix with standard independent Gaussian variables on the diagonal and independent chi-distributed variables with a fixed degree on the off-diagonal, allowing comparisons with the

Beta-ensembles thanks to [77]. This also led to LDPs [113] and convergence for a wider set of potentials V.

2. ONE MATRIX MODELS

In this section, we discuss the main large deviations results encompassing only one matrix. We start with the invariant ensembles and more generally Beta-ensembles. We then discuss Wigner matrices.

2.1. Beta-ensembles

The Beta-ensembles are defined in (1.17). As in the Gaussian case of Theorem 1.2, they are amenable to a large deviation analysis and we have the following more general statement.

Theorem 2.1 ([25]). *Let V be a continuous function going to infinity at infinity faster than $\ln|x|$. For a probability measure μ on $\mathbb{R}$, set*

$$E_V(\mu) = \frac{1}{2}\int\int \big(V(x) + V(y) - \ln|x-y|\big)d\mu(x)d\mu(y)$$

and $\mathcal{E}_V(\mu) = E_V(\mu) - \inf E_V$. Then $\mathcal{E}_V$ is a good rate function and the distribution of the empirical measure of the eigenvalues under $P^n_{V,\beta}$ satisfies an LDP *with rate function $\mathcal{E}_V$ and speed βn^2. In particular, the free energy $\frac{1}{\beta n^2}\ln Z^n_{V,\beta}$ converges towards $-\inf E_V$.*

This theorem implies the almost sure convergence of the empirical measure of the eigenvalues as $\mathcal{E}_V$ vanishes at a unique probability measure μ_V. Indeed, E_V is strictly convex on the space of probability measures [158] because it is equal to the sum of a linear functional $\mu \to \int V d\mu$ and a strictly convex function since, for any probability measures μ, μ' on the real line,

$$-\int \ln|x-y| d(\mu-\mu')(x)d(\mu-\mu')(y) = \int_0^\infty \frac{1}{t}\left|\int e^{itx} d(\mu-\mu')(x)\right|^2 dt \geq 0.$$

This ensures the uniqueness of the minimizers of $\mathcal{E}_V$ and hence the following corollary.

Corollary 2.2 ([25, 158]). *Let V be a continuous function going to infinity at infinity faster than $\ln|x|$. Then, $\hat{\mu}_n$ converges almost surely towards a distribution μ_V which is the unique probability measure μ such that there exists a constant C such that for every $x \in \mathbb{R}$,*

$$V_{\text{eff}}(x) := V(x) - \int \ln|x-y| d\mu(y) - C \geq 0$$

with equality μ almost surely.

It is easy to see that V_{eff} goes to infinity under our assumptions and hence μ_V has compact support. The case when the potential satisfies a weaker growth assumption is different [122]. An LDP can also be proven for the extreme eigenvalues in the sense that the probability that some eigenvalue goes away from the support of the equilibrium measure decays exponentially fast if V_{eff} is positive there [3, 24, 39, 41]. It was shown [91] that, conversely, if the effective potential is not strictly positive outside of the support of the limiting measure, eigenvalues may deviate towards the points where it vanishes.

Theorem 2.3. *Let S be the support of μ_V. Assume that V_{eff} is positive outside S and V is C^2. Then, for any closed set F in S^c,*

$$\limsup_{n\to\infty} \frac{1}{\beta n} \ln P^n_{V,\beta}\big(\exists i \in \{1,n\} : \lambda_i \in F\big) \le -\inf_F V_{\text{eff}},$$

whereas for any open set $O \subset S^c$,

$$\liminf_{n\to\infty} \frac{1}{\beta n} \ln P^n_{V,\beta}\big(\exists i \in \{1,n\} : \lambda_i \in O\big) \ge -\inf_O V_{\text{eff}}.$$

An important question, both in physics and for the applications to map enumerations, is to understand the phase transitions for these models. It can be seen that this often occurs when the support of the equilibrium measure changes (or its density vanishes).

Remark. Theorem 2.1 can be extended to the case where β goes to zero with n [101]. If βn goes to a finite constant $P > 0$, the speed of the LDP is n and the rate function contains a new entropy term coming from Sanov's theorem.

But what can we say about the large deviations for the traces of moments? Because polynomials are unbounded functions, this is not implied by Theorem 2.1. In fact, such large deviations are mainly due to the deviations of the extreme eigenvalues [97, 98] and their speed depends on the moment. The following result was obtained in [10].

Theorem 2.4. *Let $V(x) = c|x|^\alpha + v(x)$ where $\alpha \ge 2$, $c > 0$, v is convex and $v(x)/|x|^\alpha$ goes to zero at infinity. Then, for any $\beta > 0$, any $p > \alpha$, the law of $n^{-1}\sum_{i=1}^n |\lambda_i|^p$ under $P^n_{V,\beta}$ satisfies an* LDP$(n^{1+\frac{\alpha}{p}}, I_{p,\alpha})$ *where $I_{p,\alpha}$ is infinite if $x < \mu_V(y^p)$ and otherwise is given by*

$$I_{p,\alpha}(x) = \frac{\beta}{2} c\big(x - \mu_V(y^p)\big)^\alpha.$$

In Section 1.3.5, we have seen that LDPs for the spectral measure, given for a deterministic vector e as the probability measure $\hat{\mu}^e_n$ such that

$$\hat{\mu}^e_n(f) = \langle e, f(\mathbf{G}^n)e\rangle = \sum f(\lambda_i)\langle e, v_i\rangle^2,$$

are also interesting. They depend a priori on the large deviations of the whole spectrum and of the scalar products $(\langle e, v_i\rangle^2)_{1\le i\le n}$, while the empirical measure of the eigenvalues stays close to the semicircle law with overwhelming probability. Because $\mathbf{G}^n$ follows the Gaussian ensembles, the distribution of by the spectral measure does not depend on e. Interestingly, the rate function depends on the "reverse relative entropy," see [100, 145] for related works. This yields the following result, see [99] for general Beta-ensembles.

Theorem 2.5 ([97]). *The distribution of $\hat{\mu}^e_n$ satisfies an* LDP$(\beta n, \mathcal{J})$ *where $\mathcal{J}(\mu)$ is infinite unless there exists a nonnegative measure ν and countably many atoms $\{E_i\}_{i\in\mathbb{N}}$ such that*

$\mu = \nu + \sum_{i>0} \alpha_i \delta_{E_i}$, $\alpha_i > 0$, *and then*

$$\mathcal{J}(\mu) = H(\sigma|\nu) + \sum_{i>0} I_{\mathrm{GOE}}(|E_i|)$$

where $H(\sigma|\nu)$ *is the relative entropy of the semicircle law* σ *with respect to* ν *and* I_{GOE} *is the rate function for the largest eigenvalue of the GOE, see Theorem* 1.2.

We have also seen in Sections 1.2.3 and 1.2.2 that large deviations for rank-one perturbations of Gaussian matrices appear naturally in statistics. It is not hard to see that the law of the eigenvalues of the perturbed matrix $\mathbf{Y}^n = \mathbf{G}^n + \theta e e^T$ is absolutely continuous with respect to the law of $\mathbf{G}^n$ and with density given by the spherical integral. The spherical integral evaluated at an $n \times n$ self-adjoint matrix $\mathbf{A}^n$ and a real parameter θ is given by

$$I_\beta^n(\mathbf{A}^n, \theta) := \mathbb{E}_e\big[e^{\frac{n\beta}{2}\theta\langle e, \mathbf{A}^n e\rangle}\big], \tag{2.1}$$

where the expectation holds over the vector e which follows the uniform measure on the sphere in $\mathbb{C}^n$ if $\beta = 2$ and $\mathbb{R}^n$ if $\beta = 1$. The spherical integral $\mathbf{A}^n \to I_\beta^n(\mathbf{A}^n, \theta)$ is an eigenfunction of the Laplacian which only depends on the eigenvalues of $\mathbf{A}^n$. It appears as a natural Laplace transform in RMT and, as such, plays a key role in many large deviations questions. In particular, large deviations for the extreme eigenvalues of $\mathbf{Y}^n$ are based on asymptotic estimates for these integrals. We discuss spherical integrals for matrices with higher rank in Section 3.

Theorem 2.6.

- ([111]) *Let* $\mathbf{A}^n$ *be a sequence of* $n \times n$ *self-adjoint deterministic matrices whose largest eigenvalues converge towards* ρ *whereas the empirical measures of their eigenvalues converge weakly towards* μ_A*. Then, for any* $\theta \geq 0$*, there exists a finite constant* $J(\mu_A, \rho, \theta)$ *such that*

$$\lim_{n\to\infty} \frac{1}{\beta n} \ln I_\beta^n(\mathbf{A}^n, \theta) = J(\mu_A, \rho, \theta). \tag{2.2}$$

- ([147]) *For any unit vector* u *and if* $\mathbf{G}^n$ *follows the GUE or GOE, the law of the largest eigenvalue of* $\mathbf{G}^n + \theta u u^*$ *satisfies an LDP with speed* βn *and rate function* $x \to I_{\mathrm{GOE}}(x) - J(\sigma, x, \theta) - \inf\{I_{\mathrm{GOE}} - J(\sigma, \cdot, \theta)\}$.

Idea of proof 2.1. Again, the density of the eigenvalues of a rank-one deformation of a Gaussian matrix is given by the spherical integral in (2.2) so that Laplace's principle and (2.2) gives the result. The estimation of spherical integrals can itself use the representation of the uniform law on the sphere by Gaussian variables [111], or in terms of Dirichlet laws [109] or in terms of Schur functions [103]. The limit $J(\mu_A, \rho, \theta)$ is explicit and depends on ρ only for θ large enough.

Open Problems 2.7. Theorems 2.6 and 2.5 are restricted to invariant ensembles: generalize them to noninvariant matrix ensembles such as random matrices with bounded entries.

In the last part of this section we outline the relation of LDPs with the local fluctuations of the spectrum. As we stressed in the introduction, fluctuations and large deviations are a priori different concepts. However, they were shown to be associated in RMT in two different ways. First, the tails of Tracy–Widom laws were demonstrated to be intimately related to the rate function of the largest eigenvalue [43,72] (where the probability that the largest eigenvalue takes a value strictly smaller than two is given by the large deviations for the empirical measure, which then cannot converge to the semicircle law). The fluctuations of the eigenvalues inside the bulk could also be described by an LDP in [140]. To do so, the authors considered the finite configuration around E given by the nonnegative measure on $\mathbb{R}$,

$$\vec{X}_n(E) = \sum_{i=1}^{n} \delta_{n(\lambda_i - E)}.$$

From [173], we know that the finite configuration converges vaguely almost surely inside the bulk when V is quadratic (see [20,21,46,47] for extensions to general V). In other words, for any integer number p and any compactly supported bounded continuous function f,

$$\frac{1}{2s}\int_{-s}^{s} du \int f(x_1,\dots,x_p)d\vec{X}_n(E+u)(x_1)d\vec{X}_n(E+u)(x_2)\cdots d\vec{X}_n(E+u)(x_p)$$

converges almost surely as n goes to infinity and s goes to zero with n slowly enough for any $E \subset (-2,2)$. To state large deviations, [140] considers the tagged empirical field given for $\Sigma \subset \mathbb{R}$ by the following probability measure on the space of nonnegative measures,

$$\mathrm{Emp}_n(\vec{X}_n)(\Sigma) := \frac{1}{|\Sigma|}\int_{\Sigma} \delta_{E,\vec{X}_n(E)} dE.$$

Such $\mathrm{Emp}_n(\vec{X}_n)(\Sigma)$ converges vaguely almost surely towards the so-called Sine-Beta process if Σ has size going to zero, but $|\Sigma|$ is much bigger than $1/n$. Leblé and Serfaty [140] prove the following LDP.

Theorem 2.8 ([140]). *The distribution of* $\mathrm{Emp}_n(\vec{X}_n)$ *satisfies a large deviation principle with speed n for the vague topology.*

The rate function is the sum of the relative entropy with respect to the Poisson law and a complicated term coming from the Coulomb interaction. Even though it is not very explicit, it was proved in [82] that it achieves its minimal value at a unique point for every $\beta > 0$, hence providing another characterization of the Sine-Beta process.

Open Problems 2.9.

- In higher dimensions, Theorem 2.8 also holds for Coulomb and Riesz gases [140], but the uniqueness of the minimizers of the rate function is still unknown.
- It would be interesting to characterize as well the Airy process describing the fluctuations at the boundary by an LDP, for which one should first understand how to generalize the notion of tagged empirical field. It would also be interesting to relate the large deviations for the KPZ equation [143,171] with large deviations of the eigenvalues, see [69] for heuristics.

2.2. Wigner matrices

We recall that a Wigner matrix $\mathbf{X}^n$ is an $n \times n$ matrix with independent centered entries above the diagonal with variance $1/n$. Wigner's theorem [182] and Kómlos–Füredi's theorem [95] apply in great generality.

Theorem 2.10 ([13, 142]). *Assume that the family $((\sqrt{n}X_{ij}^n)^2)_{i\leq j}$ is uniformly integrable. Then, almost surely, for any $a < b$,*

$$\lim_{n\to\infty} \frac{1}{n}\#\{i : \lambda_i \in [a,b]\} = \sigma([a,b])$$

Moreover, if there exists $\epsilon > 0$ such that

$$B_\epsilon := \sup_{n\in\mathbb{N}} \sup_{(i,j)\in\{1,\dots,n\}^2} \mathbb{E}[|\sqrt{n}\mathbf{X}_{ij}^n|^{4+\epsilon}] < \infty, \tag{2.3}$$

then the largest eigenvalue of $\mathbf{X}^n$ converges to 2 almost surely.

When the entries do not have a finite variance, for instance, have α-stable distribution, the limiting distribution of the spectrum differs [26, 29, 44, 184] and the extreme eigenvalues go to infinity because of the presence of large entries in the matrix [8].

The large deviations of the spectrum of Wigner matrices are still poorly understood in many cases, for instance, when the entries $\sqrt{n}\mathbf{X}_{ij}^n$ of the matrix are bounded. In this case we expect the large deviations for the empirical measure to have the same speed n^2 as for Gaussian matrices because of concentration results [118], but no LDP was derived. These large deviations question are related to a new large deviations theory called nonlinear large deviations [11,59,67,80] which allows one to analyze large deviations for functions of independent variables whose gradients have low complexity (in a certain sense). Understanding large deviations for Wigner matrices remains a challenge because, as we will see, large deviations are often created both by events that have low entropy (like a few large entries in the matrix) coupled with high entropy events (like changing all entries a little), a combination that so far resisted a systematic approach. We start our journey in the LDPs for Wigner matrices by mentioning the breakthrough paper [36] which tackled the case when the tail of the entries decays slower than the Gaussian. Assume that for some $\alpha \in (0,2)$, there exists $a > 0$ so that for every i, j,

$$\lim_{t\to\infty} 2^{-1_{i=j}} t^{-\alpha} \ln \mathbb{P}(|\sqrt{n}X_{ij}^n| \geq t) = -a.$$

Theorem 2.11.

- ([36]) *The law of the empirical measure satisfies an* LDP *with speed $n^{1+\frac{\alpha}{2}}$ and rate function $\mathcal{E}_\alpha$ which is infinite except at probability measures given by the free convolution $\sigma \boxplus \nu$ of the semicircle law and a probability measure ν. It is then equal to $a\int |x|^\alpha d\nu(x)$.*
- ([9]) *The law of the largest eigenvalue satisfies an* $\mathrm{LDP}(n^{\frac{\alpha}{2}}, \mathrm{C}(\alpha)(\int \frac{d\sigma(y)}{\cdot - y})^{-\alpha})$.

Above, $\mu \boxplus \nu$ denotes the free convolution of μ and ν, see Section 4.3.

Idea of proof 2.2. Large deviations are here created by making a few large entries of order one to create a large eigenvalue and $O(n)$ large entries to change the empirical measure, the rest of the matrix behaving like a typical Wigner matrix.

The large deviations for sparse matrices are also partly understood, in particular if one considers the eigenvalues of the adjacency matrix of Erdős–Rényi graphs where an entry is equal to one with probability p/n, and to zero otherwise. In this case, [37] gives an LDP for the empirical measure with speed n. Moreover, the largest eigenvalues go to infinity. When $\ln(1/np) \ll \ln n$ and $np \ll \sqrt{\ln n / \ln\ln n}$, [33] proves an LDP with respect to the typical behavior.

Open Problems 2.12. Prove LDPs for Wigner matrices with heavy tails (such as α-stable laws). We expect the LDPs for the empirical measure to have speed n, following the concentration of measures estimates of [38].

Recently, there was some progress in understanding the large deviations properties of the largest eigenvalue of Wigner matrices with compactly supported or sub-Gaussian entries. Surprisingly, it turns out that they are universal for the so-called sharp sub-Gaussian entries, that is, entries whose laws P_{ij} satisfy, for every real number t,

$$\ln \int e^{tx} dP_{ij}(x) \le \frac{t^2}{2} \tag{2.4}$$

if the entries are real (and if they are complex, we assume the real and imaginary parts independent and the bound (2.4) holds for both real and imaginary parts. This is the case of Rademacher entries $P_{ij} = \frac{1}{2}\delta_{-1} + \frac{1}{2}\delta_{+1}$ and the uniform measure on $[-\sqrt{3}, \sqrt{3}]$. We tune the variances of the entries so that they are the same as in the Gaussian ensembles. We then have, see [108]:

Theorem 2.13. *Let $\mathbf{X}^n$ be a Wigner matrix with sharp sub-Gaussian entries. Then the law of the largest eigenvalue satisfies an* LDP *with speed βn and the same rate function I_{GOE} than in the Gaussian case.*

More generally, assume that the entries are sub-Gaussian,

$$A := \sup_{ij} \sup_{t \in \mathbb{R}} \frac{2}{t^2} \ln \int e^{tx} dP_{ij}(x) \in [1, +\infty).$$

Then there is a transition in the LDP if $A > 1$:

Theorem 2.14 ([12]). *Under some technical hypothesis, there exist $2 \le x_1 \le x_2 < \infty$ and a good rate function I_μ such that for $x \in [2, x_1] \cup [x_2, \infty)$,*

$$\lim_{\delta \downarrow 0} \liminf_{n \to \infty} \frac{1}{n} \ln \mathbb{P}(|\lambda_1 - x| \le \delta) = \lim_{\delta \downarrow 0} \limsup_{n \to \infty} \frac{1}{n} \ln \mathbb{P}(|\lambda_1 - x| \le \delta) = -\beta I_\mu(x).$$

Moreover, $I_\mu(x) \simeq \frac{x^2}{4A}$ when x goes to infinity, whereas $I_\mu(x) = I_{\mathrm{GOE}}(x)$ when $x \le x_1$. Furthermore, for $A \in (1,2)$ we can take $x_1 = (A-1)^{1/2} + (A-1)^{-1/2} > 2$.

This result shows a transition where the "heavy tails" created by $A > 1$ kick in. It is related to the optimal way to create these large deviations: for small enough values, the best way to create large deviations is delocalized, meaning that one better changes a bit all the entries of the matrix, whereas for very large deviations one better changes one or $o(n)$ entries. This is also related to a transition between a localized or a delocalized eigenvector.

Unfortunately, the same kind of universality does not hold for the empirical measure and we do not expect to have a universal rate function. For instance, the probability that the empirical measure of the eigenvalues of a Wigner matrix with Rademacher entries is close to a Dirac mass at 0 is bounded below by $(1/2)^{n^2}$, the probability that all entries equal $+1$, whereas the Gaussian rate function is infinite at any Dirac mass. This nonuniversal behavior persists also in examples with entries possessing a density, and thus contrasts with the large deviations for the empirical measure of the zeroes of random polynomials [51].

Open Problems 2.15.

- Prove an LDP for the empirical measure of the eigenvalues of a Wigner matrix with Rademacher entries, or, more generally, any Wigner matrix with sub-Gaussian tails (which is not Gaussian).
- Complete the LDP for the extreme eigenvalues of Wigner matrices with sub-Gaussian entries and understand the localization of the eigenvectors for the extreme eigenvalues conditionally to their large deviations.

Large deviations for traces of moments are also interesting, see [11] for LDPs of traces of moments of Wigner matrices with sharp sub-Gaussian tails such as Rademachers. It can also be relevant in combinatorics to consider traces of moments of random matrices with Bernoulli entries. If one considers the matrix $\mathbf{B}^n$ with Bernoulli entries of mean p, $\operatorname{Tr}((\mathbf{B}^n)^3)$ is the number $T_{n,p}$ of triangles in the Erdős–Rényi graph. Observe that its expectation is of order p^3n^3. In [60], the following theorem was proved:

Theorem 2.16. *Let*

$$I_p(f) = \sup_h \left\{ \int_0^1 \int_0^1 f(x,y)h(x,y)dxdy - \frac{1}{2} \int \int \log\big(pe^{2h(x,y)} + (1-p)\big)dxdy \right\}$$

and set $\varphi(p,t) = \inf\{I_p(f), \int f(x,y)f(y,v)f(v,x)dxdydv \geq 6t\}$. *Then for each* $p \in (0,1)$,

$$\lim_{n\to\infty} \frac{1}{n^2} \log \mathbb{P}(T_{n,p} \geq tn^3) = -\varphi(p,t).$$

Wigner matrices assume that all entries are taken at random, but it is in many cases more relevant to consider band matrices, for instance, to reflect the notion of neighbors and the geometry of the underlying space. The most common model under consideration is that of matrices with independent centered entries but with nontrivial variance profile $(\sigma_{i,j})_{1\leq i,j\leq n}$, for instance, $\sigma_{ij} = 1_{|i-j|\leq W} W^{-1}$ with W going to infinity with the dimension. In this setting, the convergence of the empirical measure [163] and of the largest eigenvalue towards the boundary of the support (when W goes to infinity fast enough with the dimension) are also known [1,4]. But very little is known about large deviations even when the entries are Gaussian because the law of the eigenvalues is not explicit. There are, however, LDPs proved for the largest eigenvalue for nice variance profile [126] and a large deviation upper bound for the empirical measure [105].

Open Problems 2.17.

- Obtain LDPs for the empirical measure of Wigner matrices with a variance profile.

- Obtain the optimal assumptions on the profile to prove an LDP for the law of the largest eigenvalue.
- Derive a local LDP similar to Theorem 2.8 for matrices with a variance profile.

A more tractable setting for large deviations is a band matrix with finite width W, independent of the dimension. Indeed, in this case, we can see that the trace of polynomials in the matrix is a sum of functions on the entries which only depend on $2W$ entries of the matrix, hence making the use of Markov chains' approach or the so-called $2W$ dependent large deviations applicable [186]. However, even in this case the rate function is not very explicit and the analysis of associated Boltzmann distributions quite difficult in general. A remarkable special case is when $W = 1$ and the entries are chosen independent centered Gaussian variables with variance β on the diagonal and independent chi distributed variables with $(n-i)$ degrees of freedom for $i \in \{1, \dots, n\}$. Indeed, it was then shown [77] that the eigenvalues of such a matrix follows the Beta-ensemble (1.17) and therefore large deviations can be derived with an explicit good rate function, see Section 2.1.

3. MATRIX MODELS WITH AN EXTERNAL FIELD

In this section we shall start our journey towards noncommutative matrix models by considering $n \times n$ self-adjoint random matrices following the distribution

$$d\mathbb{P}^n_{V,\Lambda,\beta}(\mathbf{X}^n) = \frac{1}{\mathbb{Z}^n_{V,\Lambda,\beta}} e^{n\frac{\beta}{2}\operatorname{Tr}(\mathbf{X}^n\Lambda) - n\operatorname{Tr}(V(\mathbf{X}^n))} d\mathbf{X}^n,$$

where Λ is a deterministic self-adjoint matrix. We can integrate either on Hermitian ($\beta = 2$) or symmetric ($\beta = 1$) matrices. We could also consider Λ random and study two random matrices with AB interaction such as

$$d\mathbb{P}^n_{V_1,V_2,\beta}(\mathbf{X}^n, \mathbf{Y}^n) = \frac{1}{\mathbb{Z}^n_{V_1,V_2,\beta}} e^{cn\operatorname{Tr}(\mathbf{X}^n\mathbf{Y}^n) - n\operatorname{Tr}(V_1(\mathbf{X}^n)) - n\operatorname{Tr}(V_2(\mathbf{Y}^n))} d\mathbf{X}^n d\mathbf{Y}^n.$$

The latter includes the Ising model on random graphs as it is intimately connected with their combinatorics, see (1.19). If one takes, for instance, $V_i(x) = \frac{1}{2}x^2 + t_i x^4$ and $\beta = 2$, then the limiting free energy was computed [152], hence providing the first formula for the enumeration of the Ising model on planar maps (see, e.g., [90] for generalizations). We refer to [49] for numerous other motivations. Clearly, diagonalizing the matrices $\mathbf{X}^n$ and $\mathbf{Y}^n$, we see that the main new ingredient to analyze such probability measures is again a spherical integral, the famous Harish-Chandra–Itzykson–Zuber integral given by

$$I^\beta_n(\mathbf{A}^n, \mathbf{B}^n) = \int e^{\frac{\beta}{2}n\operatorname{Tr}(\mathbf{A}^n\mathbf{U}^n\mathbf{B}^n(\mathbf{U}^n)^*)} d\mathbf{U}^n,$$

where $d\mathbf{U}^n$ denotes the Haar measure over the orthogonal (resp. unitary and symplectic) group when $\beta = 1$ (resp. 2 and 4). When $\beta = 2$, this integral was shown by Harish-Chandra [124] and then Itzykson and Zuber [127] to be equal to a determinant

$$I^2_n(\mathbf{A}^n, \mathbf{B}^n) = c_n \frac{\det[e^{n a_i b_j}]_{1\le i,j\le n}}{\prod_{i<j}(a_i - a_j)(b_i - b_j)}, \tag{3.1}$$

where $\mathbf{a}=(a_1,a_2,\dots,a_n)$ and $\mathbf{b}=(b_1,b_2,\dots,b_n)$ are the eigenvalues of $\mathbf{A}^n$ and $\mathbf{B}^n$, respectively. This formula allows showing that Schur functions are intimately related to spherical integrals. Note, however, that Harish-Chandra–Itzykson–Zuber formula does not help to estimate it asymptotically as it expresses the integral as a large signed sum of terms with modulus going to infinity. These asymptotics were first studied in [149], and made rigorous in [106,119] for $\beta = 1,2$, and finally for $\beta = 4$ in [107], where the result is also extended to rectangular spherical integrals, by computing the Laplace transform of the real part of $\mathrm{Tr}(AUBV)$ for rectangular matrices A, B and independent unitary matrices U, V.

Theorem 3.1. *Let $\mathbf{A}^n, \mathbf{B}^n \in \mathbb{R}^{n\times n}$ (resp. $\mathbf{A}^n, \mathbf{B}^n \in \mathbb{C}^{n\times n}$) be self-adjoint and $\mathbf{U}^n \in O(n)$ (resp. $U(n)$) following the Haar distribution over orthogonal group (resp. unitary group) for $\beta = 1$ (resp. $\beta = 2$). We assume that the empirical measure of the eigenvalues $\hat{\mu}_A^n$ and $\hat{\mu}_B^n$ of $\mathbf{A}^n$ and $\mathbf{B}^n$ converge weakly to μ_A and μ_B, respectively. We, moreover, assume that for $C = A$ or B, we have $\sup_n \hat{\mu}_C^n(x^2) < \infty$ and $\Sigma(\mu_C) := \int \ln|x-y| d\mu_C(x)d\mu_C(y) > -\infty$. Then, the following limit of spherical integral exists:*

$$\lim_{n\to\infty} \frac{1}{n^2} \log I_n(\mathbf{A}^n, \mathbf{B}^n) = \frac{\beta}{2} I(\mu_A, \mu_B).$$

It is given explicitly by

$$\begin{aligned} I(\mu_A, \mu_B) = &- \inf_{\{\rho_t\}_{0\le t\le 1}} \left\{ \int_0^1 \int u_s^2\, \rho_s \mathrm{d}x\mathrm{d}s + \frac{\pi^2}{3} \int_0^1 \int \rho_s^3 \mathrm{d}x\mathrm{d}s \right\} \\ &+ \mu_A(x^2) + \mu_B(x^2) - \big(\Sigma(\mu_A) + \Sigma(\mu_B)\big) + c, \end{aligned} \tag{3.2}$$

where c is a constant. The infimum is taken over continuous measure-valued processes $(\rho_t(x)\mathrm{d}x)_{0<t<1}$ such that

$$\lim_{t\to 0} \rho_t(x)\mathrm{d}x = \mu_A, \quad \lim_{t\to 1} \rho_t(x)\mathrm{d}x = \mu_B. \tag{3.3}$$

Moreover, u is given as the weak solution of the following conservation of mass equation:

$$\partial_s \rho_s + \partial_x(\rho_s u_s) = 0.$$

Idea of proof 3.1. The proof follows from the fact that the density of the law of the matrix $\mathbf{G}^n + \mathbf{A}^n$ is given by the spherical integral. As a consequence, it is enough to prove an LDP for the empirical measure of the eigenvalues of $\mathbf{G}^n + \mathbf{A}^n$ to derive the limit of the spherical integral. On the other hand, we can think of $\mathbf{G}^n + \mathbf{A}^n$ as $\mathbf{H}_1^n + \mathbf{A}^n$ where $\mathbf{H}^n$ is a symmetric or an Hermitian Brownian motion, that is, a Wigner matrix whose Gaussian entries are replaced by Brownian motions, see Figure 6. The interest of this point of view is that the eigenvalues of $\mathbf{H}_t^n + \mathbf{A}^n$ follow a Dyson Brownian motion: $\lambda_0^i = a_i$ and for every $t \ge 0$,

$$d\lambda_t^i = \frac{\sqrt{2}}{\sqrt{\beta n}} dW_t^i + \frac{1}{n} \sum_{j:j\neq i} \frac{1}{\lambda_t^i - \lambda_t^j} dt, \quad 1 \le i \le n. \tag{3.4}$$

The large deviations for the empirical measure-valued process of the $(\lambda_t^i)_{1\le i\le n}$ would then be standard to derive if the drift was not singular, as (3.4) shows that the eigenvalues of the Hermitian (or symmetric) Brownian motion are simply particles in mean-field interaction. The whole point is again to show that this singularity does not matter.

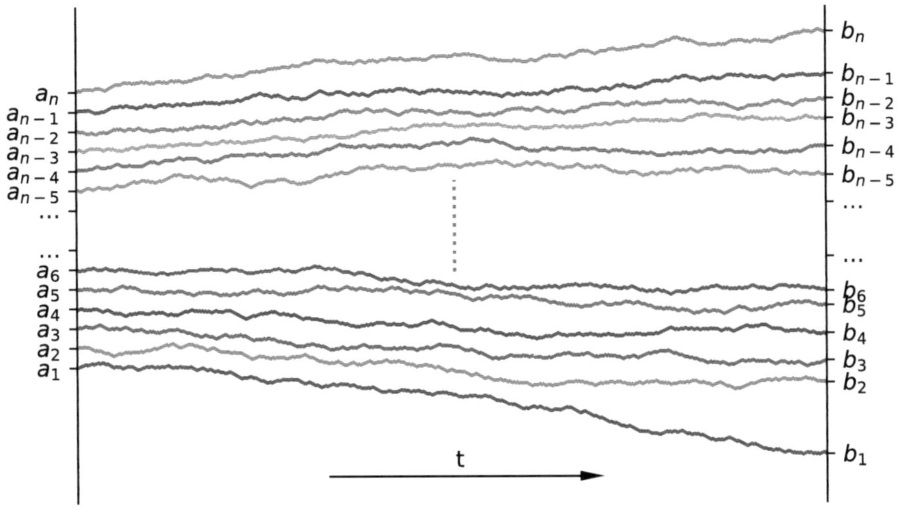

FIGURE 6
The Dyson Brownian motion between $(a_1, \ldots, a_n)$ and $(b_1, \ldots, b_n)$. Courtesy of D. Coulette

As a consequence, we find again by Laplace's principle that the two matrix models with AB interaction converge [106] in the following sense.

Corollary 3.2. *Assume that V_1 and V_2 are polynomials going to $+\infty$ at infinity. Then, the law of the empirical measure of* **X** *or* **Y** *under $\mathbb{P}^n_{V_1,V_2,\beta}$ satisfies an LDP with speed βn^2. Its rate function has a unique minimizer towards which the empirical measure converges almost surely.*

Similar statements hold for the matrix model with an external field, provided the empirical measure of the eigenvalues of Λ converges.

Open Problems 3.3.

- Theorem 3.1 describes the asymptotic of the spherical integral when $\mathbf{B}^n$ has full rank, where Theorem 2.6 deals with the case where it has rank one. As long as the rank does not go to infinity too fast with n, it can be seen that spherical integrals factorize [66,109]. It would be interesting to understand the transition from this factorization phenomenon at low rank and the full rank case.
- Study the corrections to the large n limit of spherical integrals in nonperturbative situations (see [116] for the perturbative case).
- Study the LDP for Brownian motions interacting via more singular potentials such as Riesz's which corresponds to an interaction of the form $\sum h(\lambda_i - \lambda_j)$ with h blowing up at the origin like $x/|x|^{s+2}$ for some $s > 0$.
- Study the LDP for the law of the largest particle $(\lambda_1(t), t \in [0,1])$ with general initial condition, hence generalizing [76].

4. MULTIMATRIX MODELS

4.1. Setup

We next study the asymptotic of traces of words in several matrices. More precisely, let $(\mathbf{A}_1^n, \ldots, \mathbf{A}_d^n)$ be a family of d self-adjoint matrices of size $n \times n$. Their empirical distribution generalizes the empirical measure of the eigenvalues as follows. We consider the set of polynomials $\mathbb{C}\langle X_1, \ldots, X_d\rangle$ in d noncommutative variables given by the complex linear span of words in $X_1, \ldots, X_d$ and equip it with the involution

$$(z X_{i_1} X_{i_2} \cdots X_{i_k})^* = \bar{z} X_{i_k} \cdots X_{i_1}.$$

The empirical distribution of $\mathbf{A}_1^n, \ldots, \mathbf{A}_d^n$ is defined as the linear form on $\mathbb{C}\langle X_1, \ldots, X_d\rangle$ such that, for every $P \in \mathbb{C}\langle X_1, \ldots, X_d\rangle$,

$$\hat{\mu}^n_{A_1,\ldots,A_d}(P) = \frac{1}{n}\operatorname{Tr}\big(P(\mathbf{A}_1^n, \ldots, \mathbf{A}_d^n)\big).$$

We let $\mathcal{M}_d$ be the set of linear functionals τ on the set of polynomials in d noncommutative variables such that

$$\tau(PP^*) \geq 0, \quad \tau(1) = 1, \quad \tau(PQ) = \tau(QP).$$

Clearly, $\hat{\mu}^n_{A_1,\ldots,A_d}$ belongs to $\mathcal{M}_d$. We will say that the empirical distribution $\hat{\mu}^n_{A_1,\ldots,A_d}$ converges weakly as n goes to infinity towards τ iff for every $P \in \mathbb{C}\langle X_1, \ldots, X_d\rangle$,

$$\lim_{n\to\infty} \hat{\mu}^n_{A_1,\ldots,A_d}(P) = \tau(P).$$

If the empirical distribution of $\mathbf{A}_1^n, \ldots, \mathbf{A}_d^n$ converges weakly towards τ, for any self-adjoint polynomial P, $P = P^*$, the empirical measure of the eigenvalues of the $n \times n$ self-adjoint matrix $P(\mathbf{A}_1^n, \ldots, \mathbf{A}_d^n)$ converges towards τ_P, the probability measure on the real line such that

$$\int x^k \, d\tau_P(x) = \tau(P^k), \quad \forall k \in \mathbb{N}. \tag{4.1}$$

Also τ_P is unique as soon as the moments do not grow too fast. Strong convergence requires additionally that the operator norm of $P(\mathbf{A}_1^n, \ldots, \mathbf{A}_d^n)$ converges to the largest point in the support of $\tau_{|P|}$ for any polynomial $P \in \mathbb{C}\langle X_1, \ldots, X_d\rangle$:

$$\lim_{n\to\infty} \big\| P(\mathbf{A}_1^n, \ldots, \mathbf{A}_d^n)\big\|_\infty = \lim_{n\to\infty}\lim_{k\to\infty} \hat{\mu}^n_{A_1,\ldots,A_d}\big((PP^*)^k\big)^{\frac{1}{2k}} = \lim_{k\to\infty} \tau\big((PP^*)^k\big)^{\frac{1}{2k}}.$$

We will denote by $\mathcal{M}_d^R$ the elements of $\mathcal{M}_d$ bounded by R (that is, $|\tau(X_{i_1} \cdots X_{i_k})| \leq R^k$ for all choices of indices $i_l \in \{1, \ldots, d\}$).

Another important feature of random matrices is their role in free probability, as a toy example of matrices whose large dimension limit is free. Free probability is a theory of noncommutative variables equipped with a notion of freeness. Freeness is a condition on the joint distribution of noncommutative variables. We say that $X_1, \ldots, X_d$ are free under τ iff

$$\tau\big(P_1(X_{i_1}) \cdots P_\ell(X_{i_\ell})\big) = 0 \tag{4.2}$$

as soon as $\tau(P_j(X_{i_j})) = 0$ for all j and $i_j \neq i_{j+1}$, $1 \leq j \leq \ell - 1$. The latter property was introduced by Voiculescu and named freeness, as it is related to the usual notion of free generators of a group. He also proved the key result **[175]**:

Theorem 4.1 ([3, 175]). *Let* $(\mathbf{X}_1^n, \dots, \mathbf{X}_d^n)$ *be n independent Wigner matrices with entries with finite moments. Then, for any choice of* $i_1, \dots, i_k \in \{1, \dots, d\}^k$,

$$\lim_{n\to\infty} \frac{1}{n} \operatorname{Tr}(\mathbf{X}_{i_1}^n \cdots \mathbf{X}_{i_k}^n) = \sigma^d(X_{i_1} \cdots X_{i_k}) \quad a.s.,$$

where σ^d *is the law of d free semicircular variables. It is uniquely described by the facts that the moments of a single* X_i *are given by the Catalan numbers, and their joint moments satisfy* (4.2).

Voiculescu also showed that matrices $\mathbf{Y}_j = \mathbf{U}_j \mathbf{D}_j \mathbf{U}_j^*$ with deterministic matrices $\mathbf{D}_j$ and independent Haar distributed orthogonal or unitary matrices are asymptotically free in the sense that their joint moments satisfy in the large n limit the freeness property (4.2). Hence, matrices become asymptotically free if the position of their eigenvectors are "sufficiently" independent.

In the groundbreaking article [121], it was shown that independent Gaussian matrices are not only asymptotically free, but also strongly asymptotically free in the sense that they converge strongly to free semicircular variables.

Theorem 4.2. *Let* $(\mathbf{X}_1^n, \dots, \mathbf{X}_d^n)$ *be n independent GUE matrices, then for any polynomial P,*

$$\lim_{n\to\infty} \left\| P(\mathbf{X}_1^n, \dots, \mathbf{X}_d^n) \right\|_\infty = \lim_{k\to\infty} \sigma^d\left((PP^*)^k\right)^{\frac{1}{2k}} \quad a.s.$$

This result was generalized to the GOE and GSE [159], to Wigner matrices with entries satisfying Poincaré inequality [54], to polynomials in GUE matrices and deterministic matrices in [148], to polynomials in deterministic matrices and Haar distributed unitary matrices in [65]. These results are based on the linearization trick that allows comparing the spectrum of a polynomial in matrices with the spectrum of a larger matrix obtained by sums of tensor products of the original matrices. The main drawback of this approach is that the estimates for this convergence are far from optimal: to remedy this point, an interpolation trick was introduced [17, 64].

4.2. Large deviations and Voiculescu's entropies

Free entropy was defined by Voiculescu as a generalization of classical entropy to the noncommutative context. There are several definitions of free entropy; we shall concentrate on two of them. The first is the so-called microstates' entropy that measures a volume of matrices with empirical distribution approximating a given law. The second, called the microstates-free entropy, is defined via a noncommutative version of Fisher information. The classical analog of these definitions is, on the one hand, the definition of the entropy of a measure μ as the volume of points whose empirical distribution approximates μ, and, on the other hand, the well-known entropy $-\int \frac{d\mu}{dx} \log \frac{d\mu}{dx} dx$. In this classical setting, Sanov's theorem shows that these two entropies are equal. The free analog statement is still open but we shall give in this section bounds to compare the microstates and the microstates-free entropies [35, 52].

Definition 4.3. Let $R \in \mathbb{R}^+$ and $\tau \in \mathcal{M}_d^R$. For $\varepsilon > 0$ and $k, N \in \mathbb{N}$, we define the microstate as the following subset of the set $\mathcal{H}_n^d$ of d Hermitian matrices of size $n \times n$:

$$\Gamma_n(\tau; \varepsilon, k, R) = \{\mathbf{A}_1^n, \ldots, \mathbf{A}_d^n \in \mathcal{H}_n^d : \max_{1 \le i \le d} \|A_i\|_\infty \le R,$$
$$|\hat{\mu}^n_{A_1,\ldots,A_d}(X_{i_1} \cdots X_{i_p}) - \tau(X_{i_1} \cdots X_{i_p})| \le \varepsilon$$
$$\text{for all } i_j \in \{1, \ldots, d\}, \text{all } j \in \{1, \ldots, p\}, p \le k\}$$

We then define the microstates entropy of τ by

$$\chi(\tau) = \limsup_{\substack{\varepsilon \to 0, L \to \infty \\ k \to \infty}} \limsup_{n \to \infty} \frac{1}{n^2} \log(\mathbb{P}_2^n)^{\otimes d}\big(\Gamma_n(\tau; \varepsilon, k, L)\big). \tag{4.3}$$

Remark 4.4.

- The classical analogue is Sanov's theorem (1.12) which computes the volume of small balls for the weak topology. Besides noncommutativity, it differs from the above definition by using bounded continuous test functions, instead of polynomials, and so do not need the cut-off $\bigcap_i \{\|\mathbf{A}_i^n\|_\infty \le R\}$.
- It was shown that noncommutative laws with finite entropy have nice properties. For instance, if P is a self-adjoint noncommutative polynomial, the law τ_P of $P(a_1, \ldots, a_d)$ as defined in (4.1) has no atoms **[57]**.

We denote by ∂_i the noncommutative derivative given on monomials by

$$\partial_i(X_{i_1} \cdots X_{i_k}) = \sum_{j : i_j = i} X_{i_1} \cdots X_{i_{j-1}} \otimes X_{i_{j+1}} \cdots X_{i_k}$$

and $\mathcal{D}_i = m \circ \partial_i$ the cyclic derivative, where $m(P \otimes Q) = QP$. Let us now introduce the microstates-free entropy. Its definition is based on the notion of free Fisher information which is given, for a tracial state τ, by

$$\Phi^*(\tau) = 2 \sum_{i=1}^{d} \sup_{P \in \mathbb{C}\langle X_1, \ldots, X_d \rangle} \left\{\tau \otimes \tau(\partial_i P) - \frac{1}{2}\tau(P^2)\right\}.$$

Then, we define the microstates-free entropy χ^* by

$$\chi^*(\tau) = -\frac{1}{2} \int_0^1 \Phi^*(\tau_{tX + \sqrt{t(1-t)}S}) dt$$

with $S = (S_1, \ldots, S_d)$ being a d-dimensional free semicircular vector, free from $X = (X_1, \ldots, X_d)$ with law τ. An equivalent definition of χ^* is given by optimizing the entropy of the distribution of the noncommutative law $(\hat{\mu}_{\mathbf{H}_s^1, \ldots, \mathbf{H}_s^d}, s \in [01])$ of independent Hermitian Brownian motions $(\mathbf{H}^1, \ldots, \mathbf{H}^d)$. We let $(\tau_t)_{t \in [0,1]}$ be a continuous process with values in $\mathcal{M}_d^R$. Then we define the dynamical entropy $\Xi : C([0,1], \mathcal{M}_d^R) \to [0, \infty]$ to be infinite if τ_0 is not the distribution of d operators equal to 0 and to be otherwise given by

$$\Xi(\tau_.) = \sup_F \left\{\tau_1(F_1) - \tau_0(F_0) - \int_0^1 \left[\tau_s(\partial_s F_s) + \frac{1}{2} \sum_{i=1}^{d} \tau_s \otimes \tau_s(\partial_i \mathcal{D}_i F_s)\right] ds \right.$$
$$\left. - \frac{1}{2} \sum_{i=1}^{d} \int_0^1 \tau_s(|\mathcal{D}_i F_s|^2) ds \right\}$$

where the supremum is taken over smooth noncommutative self-adjoint test functions F. Such Ξ is the candidate rate function for the large deviation of $s \to \hat{\mu}_{\mathbf{H}_s^1,\ldots,\mathbf{H}_s^d}$, generalizing to the noncommutative setting the large deviations of Theorem 3.1. It is easily seen by Riesz's theorem that the supremum over F is achieved at K such that $\sum_{i=1}^d \int_0^1 \tau_s(|\mathcal{D}_i K|^2)ds$ is finite and such that for every F,

$$\tau_1(F_1) - \tau_0(F_0) - \int_0^1 \tau_s(\partial_s F) - \frac{1}{2}\int_0^1 \sum_{i=1}^d \tau_s \otimes \tau_s(\partial_i \mathcal{D}_i F_s)ds$$
$$= \sum_{i=1}^d \int_0^1 \tau_s(\mathcal{D}_i F_s \cdot \mathcal{D}_i K_s)ds. \tag{4.4}$$

The entropy is infinite if such a K does not exist. Then taking $\tau_0 = \delta_0$, $\chi^*(\mu) = \inf_{\tau_1=\mu}\{\Xi(\tau_.)\}$. We define as well χ^{**} in the same way, but by taking the infimum only over processes such that the associated field K is smooth (the entropy is $-\infty$ if there is no such process ending near τ). Then [35, 52, 53] proved that

Theorem 4.5. *For every $\tau \in \mathcal{M}_d^R$,*

$$\chi^{**}(\tau) \leq \chi(\tau) \leq \chi^*(\tau).$$

Open Problems 4.6.

- Show that the limsup in the definition (4.3) of χ can be replaced by a liminf. The two bounds above still hold if we perform this change.
- Prove that $\chi = \chi^*$ at least whenever $\chi < \infty$. In [70, 128], it was proven that $\chi(\tau_V) = \chi^*(\tau_V)$ when τ_V is the equilibrium measure of matrix models with convex potentials, see Section 4.4.
- Prove that $\chi^{**} = \chi^*$ in general. This is already true if μ is close to some τ_1 obtained as the value at time 1 of a process satisfying (4.4) with K smooth. In particular, τ_1 can be smoothly constructed from the increments of an Hermitian Brownian motions by a smooth differential equation. In a breakthrough series of papers, it was recently shown that there exist tracial states that cannot be approximated by a sequence of noncommutative empirical distributions of d matrices [129]. Hence, the question of estimating noncommutative laws by differential equations is far from trivial, in particular because the weak closure of the set of noncommutative empirical distributions of d matrices is not very well understood.
- Prove an LDP for the operator norm of polynomials in independent GUE matrices, in the line of the topological entropy introduced by Voiculescu [179].

4.3. Free convolution

A long-standing question posed by Weyl was to describe the spectrum of the sum of two Hermitian matrices. A complete description was conjectured by Horn, and proved by Knutson and Tao [139]. But what should be the spectrum of the sum of two matrices taken at random? This question was tackled [31,32] when the two matrices are asymptotically free. It was characterized by an analog of the Fourier transform, the so-called R-transform. It is defined as follows: Let G_μ be the Stieltjes transform of a probability measure μ given for complex number z by

$$G_\mu(z) = \int \frac{1}{z-x} d\mu(x).$$

Then G_μ is invertible in a neighborhood of infinity, with inverse K_μ equivalent to $1/z$ in a neighborhood of the origin. The R-transform R_μ is given in a neighborhood of the origin by

$$R_\mu(z) = K_\mu(z) - \frac{1}{z}.$$

It is not hard to see that R_μ defines uniquely μ as it defines uniquely G_μ.

Theorem 4.7 ([31,32,156]). *If the empirical measures $\hat{\mu}^n_{X_1}$ and $\hat{\mu}^n_{X_2}$ of $\mathbf{X}^n_1$ and $\mathbf{X}^n_2$ converge respectively towards μ_1 and μ_2, and $\mathbf{X}_1$ and $\mathbf{X}_2$ are asymptotically free, then the empirical measure $\hat{\mu}^n_{X_1+X_2}$ of the eigenvalues of $\mathbf{X}^n_1 + \mathbf{X}^n_2$ converges weakly in L^1 towards the unique probability measure $\mu_1 \boxplus \mu_2$ defined by*

$$R_{\mu_1 \boxplus \mu_2}(z) = R_{\mu_1}(z) + R_{\mu_2}(z).$$

The above result holds in particular for $\mathbf{X}^n_1 + \mathbf{U}^n \mathbf{X}^n_2 (\mathbf{U}^n)^*$ if $\mathbf{X}^n_1, \mathbf{X}^n_2$ are two deterministic Hermitian matrices whose spectral measures converge, independent of $\mathbf{U}$ which follows the Haar measure on the unitary or orthogonal group. Theorem 4.7 was shown then to be a direct consequence of the asymptotics of spherical integrals [111]. But what can we say about the large deviations of the empirical measure and the largest eigenvalue of $\mathbf{X}_1 + \mathbf{U}^n \mathbf{X}_2 (\mathbf{U}^n)^*$? The description of the spectrum of the sum of two self-adjoint matrices is complicated and depicted by Horn's problem [139]. Understanding which of these possible spectrum has a finite entropy is a natural question which was attacked in [68,187] by noticing that the Fourier transform of the density of the spectrum can be written in terms of Harish-Chandra–Itzykson–Zuber integrals. Unfortunately, this formula so far has resisted asymptotic analysis as they require complex matrices and hence oscillatory integrals. We now, however, have a quite complete series of results on the large deviations for the sum of two random Hermitian matrices.

Theorem 4.8. *Let $\mathbf{X}^n_1, \mathbf{X}^n_2$ be two Hermitian matrices whose empirical measures of the eigenvalues $\hat{\mu}^n_{X_1}$ and $\hat{\mu}^n_{X_2}$ of $\mathbf{X}^n_1$ and $\mathbf{X}^n_2$ converge respectively towards μ_1 and μ_2. Let $\mathbf{U}^n$ follow the Haar measure on the orthogonal or unitary group.*

- ([112]) *Assume that the largest eigenvalues of $\mathbf{X}^n_1$ and $\mathbf{X}^n_2$ stick to the bulk. Then the largest eigenvalue of $\mathbf{X}^n_1 + \mathbf{U}^n \mathbf{X}^n_2 (\mathbf{U}^n)^*$ satisfies an LDP in the scale βn.*
- ([22]) *The law of $N^{-1} \sum_{i=1}^N \delta_{(\mathbf{U}^n \mathbf{X}^n_1 (\mathbf{U}^n)^*)_{ii}}$ satisfies an LDP in the scale βn^2 and good rate function $I^D(\mu) = \sup_\nu \{\frac{1}{2} \int_0^1 T_\nu(x) T_\mu(x) - I(\nu, \mu_1)\}$ where T_μ is the inverse of $F_\mu(x) = \mu((-\infty, x])$.*
- ([22]) *The law of $\hat{\mu}^n_{X_1+UX_2U^*}$ satisfies a weak large deviation estimate* (1.11) *in the scale βn^2 and good rate function $I^{X_1+X_2}(\mu) = \sup_\nu \{I(\mu,\nu) - I(\nu,\mu_1) -$*

$I(\nu, \mu_2)\}$ *at any* μ *so that* $\operatorname{argmax}(I^{X_1+X_2}(\mu)) \neq \operatorname{argmax}(I^{X_1+X_2}(\mu'))$ *for all* $\mu' \neq \mu$.

- ([155]) *Assume that for* $j = 1$ *and* 2, *the eigenvalues* $(\lambda_i^j)_{1 \le i \le n}$ *of* $\mathbf{X}_j^n$ *are such that* $\lambda_i^j = f_j(\frac{i}{n})$ *with strictly increasing functions* f_j. *Then the law of* $\hat{\mu}^n_{X_1+UX_2U^*}$ *satisfies a weak large deviation principle in the scale* βn^2.

It would be interesting to understand how the two last results relate. The first three results above were obtained by tilting the laws by spherical integrals, and using their limit $I(\cdot,\cdot)$ from Theorem 3.1, the last is derived by using large deviations on an interesting object called random hives, closer to [139].

4.4. Multimatrix models

Recall the definition (1.18) of the multimatrix model, which can be extended to $\beta = 1$:

$$d\mathbb{P}^n_{V,\beta}(\mathbf{X}_1^n, \dots, \mathbf{X}_d^n) = \frac{1}{\mathbb{Z}^n_{V,\beta}} e^{-\beta n \operatorname{Tr}(V(\mathbf{X}_1^n,\dots,\mathbf{X}_d^n)) - \frac{\beta n}{4} \operatorname{Tr}(\sum (\mathbf{X}_i^n)^2)} d\mathbf{X}_1^n \cdots d\mathbf{X}_d^n. \tag{4.5}$$

Here V is a self-adjoint polynomial that decomposes as $V = -\sum t_i q_i$ with words (or monomials) q_i in d noncommutative letters. We assume either that V is bounded from below uniformly, or we restrict the integration over $\bigcap_i \{\|\mathbf{X}_i^n\| \le M\}$ for some $M > 2$.

Theorem 4.9 ([113, 150]). *Let* $\beta = 1$ *or* 2. *For all* $g \in \mathbb{N}$, *there exists* $\varepsilon_g > 0$ *such that for every* $|\varepsilon| \le \varepsilon_g$, *every monomial* q,

$$\begin{aligned} &\int \hat{\mu}^n_{X_1,\dots,X_d}(q) d\mathbb{P}^n_{\varepsilon V,\beta}(\mathbf{X}_1^n, \dots, \mathbf{X}_d^n) \\ &\quad = \sum_{\ell=0}^{g} \frac{1}{n^\ell} \sum_{k_1,\dots,k_p} \prod \frac{(\varepsilon t_i)^{k_i}}{k_i!} M_\ell^\beta\big((k_i, q_i)_{1\le i\le p}, (1,q)\big) + o\Big(\frac{1}{n^g}\Big). \end{aligned}$$

Moreover, for every monomial q, $\hat{\mu}^n_{X_1,\dots,X_d}(q)$ *converges almost surely towards*

$$\tau_{\varepsilon V}(q) = \sum_{k_1,\dots,k_p \in \mathbb{N}} \prod \frac{(\varepsilon t_i)^{k_i}}{k_i!} M_0^2\big((k_i, q_i)_{1\le i\le p}, (1,q)\big).$$

Note that when $\beta = 1$, the expansion is in $1/n$ rather than $1/n^2$. The first-order expansion is the same, $M_0^1((k_i, q_i)_{1\le i\le p}, (1,q)) = M_0^2((k_i, q_i)_{1\le i\le p}, (1,q))$, but the higher orders differ. $M_\ell^1((k_i, q_i)_{1\le i\le p}, (1,q))$ can also be seen to enumerate certain maps, but in locally orientable surfaces, see, e.g., [104, 141].

The proof of this theorem follows by showing that $\hat{\mu}^n_{X_1,\dots,X_d}(q)$ is tight and its moments satisfy the so-called Dyson–Schwinger equations as a consequence of integration by parts. Showing the uniqueness of the solutions to the limiting Dyson–Schwinger equation gives the result for $g = 0$. A more detailed study of the solution of Dyson–Schwinger equations allows obtaining the higher-order corrections [114, 150].

Remark 4.10. • Theorem 4.9 was extended to the case where one integrates over the Haar measure on the unitary or orthogonal groups [63,116] and to $SO(n)$ lattice gauge theory [58].

- Distribution $\tau_{\varepsilon V}$ extends by linearity to polynomials. It is a priori unclear that τ_V is a noncommutative law, in particular that $\tau_{\varepsilon V}(PP^*) \geq 0$ for all polynomials P. This is part of the result.

- The noncommutative distribution $\tau_{\varepsilon V}$ has finite entropy, and hence the spectral distribution of polynomials has no atoms by [57]. Much more was proved in [120]: there exist noncommutative functions given by absolutely converging series such that $\tau_{\varepsilon V}$ is the push-forward of $\tau_0 = \sigma^d$ by these functions (and vice versa). This implies that the C^* and von Neumann algebras associated with $\tau_{\varepsilon V}$ by the so-called GNS construction are isomorphic to those of d free semicircular variables.

- The central limit theorem for the empirical distribution can be proven by analyzing the asymptotic of more general moments of the empirical distribution [114], allowing to derive the next order expansion of the free energy related to maps with higher genus. The fact that the eigenvalues fluctuate locally like independent GUE was proven in [92] by constructing approximate transport maps.

It should be expected that the convergence in Theorem 4.9 (which amounts to taking $g = 0$) holds for large ε, at least till a certain phase transition. In the one-matrix case, this phase transition is usually related to the point where the support of the equilibrium measure splits, which is the case, for instance, when the potential has several wells that become deeper when the parameters vary. This, in particular, does not happen when V is convex. The same is true for several matrices. Of course, for potentials in several matrices the notion of convexity itself needs to be clarified, see [70,71,117,128]. The most handy one, in the sense that it is easier to check, relies on matrices and simply states that, in any dimension n, the map $\mathbf{X}_1^n, \dots, \mathbf{X}_d^n \to \mathrm{Tr}V(\mathbf{X}_1^n, \dots, \mathbf{X}_d^n)$ is a convex function of the entries of the self-adjoint matrices $(\mathbf{X}_1^n, \dots, \mathbf{X}_d^n)$.

Theorem 4.11 ([70,128]). *Assume that the noncommutative function $V(X_1, \dots, X_d) + (1 - \delta)\beta \sum X_i^2$ is convex for some $\delta > 0$. Then the empirical distribution $\hat{\mu}^n_{X_1,\dots,X_d}$ converges $\mathbb{P}^n_{V,2}$-almost surely towards τ_V. Moreover, $\chi^*(\tau_V) = \chi(\tau_V)$ is the limit of the classical entropy of $\mathbb{P}^n_{V,2}$.*

This result uses again the dynamics of the Hermitian Brownian motions and the fact that they converge uniformly to their invariant measures $\mathbb{P}^n_{V,2}$ thanks to convexity. In this case it is also seen that $\chi^*(\tau_V) = \chi^{**}(\tau_V)$. Unfortunately, except for multimatrix models whose interaction is related to spherical integrals, even the convergence of the matrix models is unknown in general (such a convergence will result in the possibility of changing the lim sup by a lim inf in the definition of χ which would have important consequences). Recently, [115] undertook the study of matrix models at "low temperature" in the sense that the constant ϵ in Theorem 4.9 is now very large. In this case, we can give sufficient conditions on the potential V so that the matrices stay bounded in norm with high probability. The limit point of the empirical distribution then satisfies the Dyson–Schwinger equations. Unfortunately, the uniqueness of the solutions to these equations is in general not true and convergence is unclear. We can, however, study more in detail special situations when the case $\epsilon = \infty$ is simple. We detail below a few results that hold under $\mathbb{P}^n_{V,\beta}$ for T small enough.

- Assume $V = \frac{1}{T} V_0 + W$ where V_0 is uniformly strictly convex. Let $(\alpha_i)_{1 \le i \le d}$ be the unique minimizer of V_0 in $\mathbb{R}^d$. Then, the matrices will concentrate near $(\alpha_i I)_{1 \le i \le d}$ when n goes to infinity and then T to zero. Moreover, the empirical distribution $\hat{\mu}^n_{X_1,\ldots,X_d}$ converges almost surely towards a noncommutative law which can be obtained as a smooth push-forward of d free semicircular variables.
- Assume $V(X_1, \ldots, X_d) = \frac{1}{T} V_1(X_1) + V_1(X_1) W(X_1, \ldots, X_d)$ with V_1 nonnegative and vanishing at $(\alpha_i)_{1 \le j \le m}$. Then, the spectrum of X_1 will asymptotically belong to a neighborhood of the minimizers of V_1. Moreover, the empirical distribution $\hat{\mu}_{X_1,\ldots,X_d}$ converges almost surely towards a noncommutative law which can be obtained as a smooth push-forward of free semicircular variables and a projection.
- If $V(X_1, X_2) = -\frac{1}{T}[X_1, X_2]^2 + W_1(X_1) + W_2(X_2)$, then the matrices will asymptotically commute and their respective spectrum will converge towards the minimizers of W_1 and W_2 with nontrivial masses.

The last result is interesting because we see that the matrices asymptotically commute but are not a multiple of the identity in general. Indeed, the case where we have 3 matrices and the strong interaction presents two commutators $[X_1, X_2]^2 + [X_1, X_3]^2$, it is easy to see by an entropy argument that X_1 will be forced to be a multiple of the identity, regardless of the rest of the potential of order one. It was therefore tempting to think that all such limit laws would asymptotically commute because they are trivial, which is not the case. This is only the beginning of the journey towards the understanding of multimatrix models at low temperature and large dimension.

ACKNOWLEDGMENTS

I thank G. Ben Arous, C. Bordenave, G. Borot, C. Garban, J. Huang, K. Kozlowski, G. Miermont, and O. Zeitouni for their numerous comments on a preliminary version of these proceedings.

FUNDING

This work was partially supported by ERC Project LDRAM: ERC-2019-ADG Project 884584.

REFERENCES

[1] O. H. Ajanki, L. Erdős, and T. Krüger, Universality for general Wigner-type matrices. *Probab. Theory Related Fields* **169** (2017), no. 3–4, 667–727.

[2] G. Akemann, J. Baik, and P. Di Francesco, *The Oxford handbook of random matrix theory*. Oxford University Press, Oxford, 2015.

[3] G. W. Anderson, A. Guionnet, and O. Zeitouni, *An introduction to random matrices*. Cambridge Stud. Adv. Math. 118, Cambridge University Press, Cambridge, 2010.

[4] G. W. Anderson and O. Zeitouni, A CLT for a band matrix model. *Probab. Theory Related Fields* **134** (2005), 283–338.

[5] L.-P. Arguin, D. Belius, P. Bourgade, M. Radziwill, and K. Soundararajan, Maximum of the Riemann zeta function on a short interval of the critical line. *Comm. Pure Appl. Math.* **72** (2019), no. 3, 500–535.

[6] G. Ben Arous, Y. V. Fyodorov, and B. Khoruzhenko, Counting equilibria of large complex systems by instability index. *Proc. Natl. Acad. Sci. USA* **118** (2021), no. 34.

[7] A. Auffinger and G. Ben, Arous, Complexity of random smooth functions on the high-dimensional sphere. *Ann. Probab.* **41** (2013), no. 6, 4214–4247.

[8] A. Auffinger, G. Ben Arous, and S. Péché, Poisson convergence for the largest eigenvalues of heavy tailed random matrices. *Ann. Inst. Henri Poincaré Probab. Stat.* **45** (2009), no. 3, 589–610.

[9] F. Augeri, Large deviations principle for the largest eigenvalue of Wigner matrices without Gaussian tails. *Electron. J. Probab.* **21** (2016), 32, 49 pp.

[10] F. Augeri, On the large deviations of traces of random matrices. *Ann. Inst. Henri Poincaré Probab. Stat.* **54** (2018), no. 4, 2239–2285.

[11] F. Augeri, Nonlinear large deviation bounds with applications to Wigner matrices and sparse Erdős–Rényi graphs. *Ann. Probab.* **48** (2020), no. 5, 2404–2448.

[12] F. Augeri, A. Guionnet, and J. Husson, Large deviations for the largest eigenvalue of sub-Gaussian matrices. *Comm. Math. Phys.* **383** (2021), no. 2, 997–1050.

[13] Z. D. Bai, Methodologies in spectral analysis of large-dimensional random matrices, a review. *Statist. Sinica* **9** (1999), 611–677.

[14] Z. Bai and J. W. Silverstein, *Spectral analysis of large dimensional random matrices. Second edn*. Springer Ser. Statist., Springer, New York, 2010.

[15] J. Baik, G. Ben Arous, and S. Péché, Phase transition of the largest eigenvalue for nonnull complex sample covariance matrices. *Ann. Probab.* **33** (2005), no. 5, 1643–1697.

[16] J. Baik, P. Deift, and K. Johansson, On the distribution of the length of the longest increasing subsequence of random permutations. *J. Amer. Math. Soc.* **12** (1999), 1119–1178.

[17] A. Bandeira, M. Boedihardjo, and R. van Handel, Matrix concentration inequalities and free probability. 2021, arXiv:2108.06312.

[18] R. Basu and S. Ganguly, Connecting eigenvalue rigidity with polymer geometry: Diffusive transversal fluctuations under large deviation. 2019, arXiv:1902.09510.

[19] R. Basu, S. Ganguly, and A. Sly, Delocalization of polymers in lower tail large deviation. *Comm. Math. Phys.* **370** (2019), no. 3, 781–806.

[20] F. Bekerman, Transport maps for β-matrix models in the multi-cut regime. *Random Matrices Theory Appl.* **7** (2018), no. 1, 1750013, 36 pp.

[21] F. Bekerman, A. Figalli, and A. Guionnet, Transport maps for β-matrix models and universality. *Comm. Math. Phys.* **338** (2015), no. 2, 589–619.

[22] S. Belinschi, A. Guionnet, and J. Huang, Large deviation principles via spherical integrals. 2020, arXiv:2004.07117.

[23] G. Ben Arous, P. Bourgade, and B. McKenna, Exponential growth of random determinants beyond invariance. 2020, arXiv:2105.05000.

[24] G. Ben Arous, A. Dembo, and A. Guionnet, Aging of spherical spin glasses. *Probab. Theory Related Fields* **120** (2001), no. 1, 1–67.

[25] G. Ben Arous and A. Guionnet, Large deviations for Wigner's law and Voiculescu's non-commutative entropy. *Probab. Theory Related Fields* **108** (1997), no. 4, 517–542.

[26] G. Ben Arous and A. Guionnet, The spectrum of heavy tailed random matrices. *Comm. Math. Phys.* **278** (2008), no. 3, 715–751.

[27] G. Ben Arous, S. Mei, A. Montanari, and M. Nica, The landscape of the spiked tensor model. *Comm. Pure Appl. Math.* (2019), 2282–2330.

[28] G. Ben Arous and O. Zeitouni, Large deviations from the circular law. *ESAIM Probab. Stat.* **2** (1998), 123–134.

[29] F. Benaych-Georges, A. Guionnet, and C. Male, Central limit theorems for linear statistics of heavy tailed random matrices. *Comm. Math. Phys.* **329** (2014), no. 2, 641–686.

[30] F. Benaych-Georges and R. R. Nadakuditi, The eigenvalues and eigenvectors of finite, low rank perturbations of large random matrices. *Adv. Math.* **227** (2011), no. 1, 494–521.

[31] H. Bercovici and D. Voiculescu, Lévy–Hinčin type theorems for multiplicative and additive free convolution. *Pacific J. Math.* **153** (1992), 217–248.

[32] H. Bercovici and D. Voiculescu, Free convolution of measures with unbounded support. *Indiana Univ. Math. J.* **42** (1993), 733–773.

[33] B. B. Bhattacharya, S. Bhattacharya, and S. Ganguly, Spectral edge in sparse random graphs: Upper and lower tail large deviations. *Ann. Probab.* **49** (2021), no. 4, 1847–1885.

[34] P. Bianchi, M. Debbah, M. Maida, and J. Najim, Performance of statistical tests for single-source detection using random matrix theory. *IEEE Trans. Inf. Theory* **57** (2011), no. 4, 2400–2419.

[35] P. Biane, M. Capitaine, and A. Guionnet, Large deviation bounds for matrix Brownian motion. *Invent. Math.* **152** (2003), no. 2, 433–459.

[36] C. Bordenave and P. Caputo, A large deviation principle for Wigner matrices without Gaussian tails. *Ann. Probab.* **42** (2014), no. 6, 2454–2496.

[37] C. Bordenave and P. Caputo, Large deviations of empirical neighborhood distribution in sparse random graphs. *Probab. Theory Related Fields* **163** (2015), no. 1–2, 149–222.

[38] C. Bordenave, P. Caputo, and D. Chafaï, Spectrum of non-Hermitian heavy tailed random matrices. *Comm. Math. Phys.* **307** (2011), no. 2, 513–560.

[39] C. Bordenave and A. Guionnet, Localization and delocalization of eigenvectors for heavy-tailed random matrices. *Probab. Theory Related Fields* **157** (2013), no. 3–4, 885–953.

[40] A. Borodin, V. Gorin, and A. Guionnet, Gaussian asymptotics of discrete β-ensembles. *Publ. Math. IHÉS* (2016), 1–78.

[41] G. Borot, A. Guionnet, and K. K. Kozlowski, Large-N asymptotic expansion for mean field models with Coulomb gas interaction. *Int. Math. Res. Not. IMRN* **20** (2015), 10451–10524.

[42] G. Borot, A. Guionnet, and K. K. Kozlowski, *Asymptotic expansion of a partition function related to the sinh-model*. Math. Phys. Stud., Springer, Cham, 2016.

[43] G. Borot and C. Nadal, Right tail asymptotic expansion of Tracy–Widom beta laws. *Random Matrices Theory Appl.* **1** (2012), no. 3, 1250006, 23.

[44] J.-P. Bouchaud and P. Cizeau, Theory of Lévy matrices. *Phys. Rev. E* **3** (1994), 1810–1822.

[45] D. V. Boulatov and V. A. Kazakov, The Ising model on a random planar lattice: the structure of the phase transition and the exact critical exponents. *Phys. Lett. B* **186** (1987), no. 3–4, 379–384.

[46] P. Bourgade, L. Erdős, and H.-T. Yau, Edge universality of beta ensembles. *Comm. Math. Phys.* **332** (2014), no. 1, 261–353.

[47] P. Bourgade, L. Erdős, and H.-T. Yau, Universality of general β-ensembles. *Duke Math. J.* **163** (2014), no. 6, 1127–1190.

[48] J. Breuer, B. Simon, and O. Zeitouni, Large deviations and the Lukic conjecture. *Duke Math. J.* **167** (2018), no. 15, 2857–2902.

[49] E. Brézin and S. Hikami, Universal singularity at the closure of a gap in a random matrix theory. *Phys. Rev. E (3)* **57** (1998), 4140–4149.

[50] G. P. E. Brézin, C. Itzykson, and J. B. Zuber, Planar diagrams. *Comm. Math. Phys.* **59** (1978), 35–51.

[51] R. Butez and O. Zeitouni, Universal large deviations for Kac polynomials. *Electron. Commun. Probab.* **22** (2017), 10.

[52] T. Cabanal-Duvillard and A. Guionnet, Large deviations upper bounds for the laws of matrix-valued processes and non-communicative entropies. *Ann. Probab.* **29** (2001), 1205–1261.

[53] T. Cabanal-Duvillard and A. Guionnet, Discussions around Voiculescu's free entropies. *Adv. Math.* **174** (2003), no. 2, 167–226.

[54] M. Capitaine and C. Donati-Martin, Strong asymptotic freeness for Wigner and Wishart matrices. *Indiana Univ. Math. J.* **56** (2007), no. 2, 767–803.

[55] E. Cartan, Sur la détermination d'un système orthogonal complet dans un espace de Riemann symétrique clos. *Rend. Circ. Mat. Palermo* **53** (1929), 217–252.

[56] G. Chapuy and M. Dołęga, A bijection for rooted maps on general surfaces. *J. Combin. Theory Ser. A* **145** (2017), 252–307.

[57] I. Charlesworth and D. Shlyakhtenko, Free entropy dimension and regularity of non-commutative polynomials. *J. Funct. Anal.* **271** (2016), no. 8, 2274–2292.

[58] S. Chatterjee, Rigorous solution of strongly coupled $SO(N)$ lattice gauge theory in the large N limit. *Comm. Math. Phys.* **366** (2019), no. 1, 203–268.

[59] S. Chatterjee and A. Dembo, Nonlinear large deviations. *Adv. Math.* **299** (2016), 396–450.

[60] S. Chatterjee and S. R. S. Varadhan, The large deviation principle for the Erdős–Rényi random graph. *European J. Combin.* **32** (2011), no. 7, 1000–1017.

[61] W.-K. Chen, D. Panchenko, and E. Subag, The generalized TAP free energy. II. *Comm. Math. Phys.* **381** (2021), no. 1, 257–291.

[62] H. Cohn, R. Kenyon, and J. Propp, A variational principle for domino tilings. *J. Amer. Math. Soc.* **14** (2001), no. 2, 297–346.

[63] B. Collins, A. Guionnet, and E. Maurel-Segala, Asymptotics of unitary and orthogonal matrix integrals. *Adv. Math.* **222** (2009), no. 1, 172–215.

[64] B. Collins, A. Guionnet, and F. Parraud, On the operator norm of non-commutative polynomials in deterministic matrices and iid GUE matrices. 2019, arXiv:math/0608192v1.

[65] B. Collins and C. Male, The strong asymptotic freeness of Haar and deterministic matrices. *Ann. Sci. Éc. Norm. Supér. (4)* **47** (2014), no. 1, 147–163.

[66] B. Collins and P. Śniady, New scaling of Itzykson–Zuber integrals. *Ann. Inst. Henri Poincaré Probab. Stat.* **43** (2007), no. 2, 139–146.

[67] N. Cook and A. Dembo, Large deviations of subgraph counts for sparse Erdős–Rényi graphs. *Adv. Math.* **373** (2020), 107289, 53.

[68] R. Coquereaux, C. McSwiggen, and J.-B. Zuber, On Horn's problem and its volume function. *Comm. Math. Phys.* **376** (2020), no. 3, 2409–2439.

[69] I. Corwin, P. Ghosal, A. Krajenbrink, P. Le Doussal, and L.-C. Tsai, Coulomb-gas electrostatics controls large fluctuations of the KPZ equation. 2018, arXiv:1803.05887.

[70] Y. Dabrowski, A Laplace principle for Hermitian Brownian motion and free entropy I: the convex functional case. 2016, arXiv:1604.06420.

[71] Y. Dabrowski, A. Guionnet, and D. Shlyakhtenko, Free transport for convex potentials. *New Zealand J. Math.* **52** (2021), 259–359.

[72] D. Dean and S. Majumdar, Large deviations of extreme eigenvalues of random matrices. *Phys. Rev. Lett.* **97** (2006), no. 16, 160201, 4 pp.

[73] A. Dembo and O. Zeitouni, *Large deviations techniques and applications. Second edn.* Appl. Math. (N. Y.) 38, Springer, New York, 1998.

[74] J.-D. Deuschel and O. Zeitouni, On increasing subsequences of i.i.d. samples. *Combin. Probab. Comput.* **8** (1999), no. 3, 247–263.

[75] J. D. Deuschel and D. W. Stroock, *Large deviations*. Academic Press, Boston, MA, 1989.

[76] C. Donati-Martin and M. Maïda, Large deviations for the largest eigenvalue of an Hermitian Brownian motion. *ALEA Lat. Am. J. Probab. Math. Stat.* **9** (2012), no. 2, 501–530.

[77] I. Dumitriu and A. Edelman, Matrix models for beta ensembles. *J. Math. Phys.* **43** (2002), 5830–5847.

[78] P. Dupuis and R. S. Ellis, *A weak convergence approach to the theory of large deviations*. John Wiley & Sons, Chichester, 1997.

[79] F. J. Dyson, A Brownian-motion model for the eigenvalues of a random matrix. *J. Math. Phys.* **3** (1962), 1191–1198.

[80] R. Eldan, Gaussian-width gradient complexity, reverse log-Sobolev inequalities and nonlinear large deviations. *Geom. Funct. Anal.* **28** (2018), no. 6, 1548–1596.

[81] R. S. Ellis, *Entropy, large deviations, and statistical mechanics*. Springer, Berlin, 2006.

[82] M. Erbar, M. Huesmann, and T. Leblé, The one-dimensional log-gas free energy has a unique minimizer. *Comm. Pure Appl. Math.* **74** (2021), no. 3, 615–675.

[83] N. M. Ercolani and K. D. T.-R. McLaughlin, Asymptotics of the partition function for random matrices via Riemann–Hilbert techniques and applications to graphical enumeration. *Int. Math. Res. Not.* **14** (2003), 755–820.

[84] L. Erdős, A. Knowles, H.-T. Yau, and J. Yin, The local semicircle law for a general class of random matrices. *Electron. J. Probab.* **18** (2013), no. 59, 58.

[85] L. Erdős, B. Schlein, and H.-T. Yau, Local semicircle law and complete delocalization for Wigner random matrices. *Comm. Math. Phys.* **287** (2009), no. 2, 641–655.

[86] L. Erdős, B. Schlein, and H.-T. Yau, Universality of random matrices and local relaxation flow. *Invent. Math.* **185** (2011), no. 1, 75–119.

[87] L. Erdős and H.-T. Yau, Gap universality of generalized Wigner and β-ensembles. *J. Eur. Math. Soc. (JEMS)* **17** (2015), no. 8, 1927–2036.

[88] L. Erdős, H.-T. Yau, and J. Yin, Rigidity of eigenvalues of generalized Wigner matrices. *Adv. Math.* **229** (2012), no. 3, 1435–1515.

[89] B. Eynard, *Counting surfaces*. Prog. Math. Phys. 70, Springer, Cham, 2016.

[90] B. Eynard and M. L. Mehta, Matrices coupled in a chain. I. Eigenvalue correlations. *J. Phys. A* **31** (1998), no. 19, 4449–4456.

[91] C. Fan, A. Guionnet, Y. Song, and A. Wang, Convergence of eigenvalues to the support of the limiting measure in critical β matrix models. *Random Matrices Theory Appl.* **4** (2015), no. 3, 1550013, 22 pp.

[92] A. Figalli and A. Guionnet, Transport maps for several-matrix models and universality. *Acta Math.* **217** (2016), 81–176.

[93] P. J. Forrester, *Log-gases and random matrices* 34. Princeton University Press, Princeton, NJ, 2010.

[94] P. J. Forrester and E. M. Rains, Interrelationships between orthogonal, unitary and symplectic matrix ensembles. In *Random matrix models and their applications*, pp. 171–207, Math. Sci. Res. Inst. Publ. 40, Cambridge Univ. Press, Cambridge, 2001.

[95] Z. Füredi and J. Komlós, The eigenvalues of random symmetric matrices. *Combinatorica* **1** (1981), no. 3, 233–241.

[96] Y. V. Fyodorov, Complexity of random energy landscapes, glass transition, and absolute value of the spectral determinant of random matrices. *Phys. Rev. Lett.* **92** (2004), no. 24, 4.

[97] F. Gamboa, J. Nagel, and A. Rouault, Sum rules via large deviations. *J. Funct. Anal.* **270** (2016), no. 2, 509–559.

[98] F. Gamboa, J. Nagel, and A. Rouault, Sum rules and large deviations for spectral matrix measures. *Bernoulli* **25** (2019), no. 1, 712–741.

[99] F. Gamboa, J. Nagel, and A. Rouault, Sum rules via large deviations: extension to polynomial potentials and the multi-cut regime. *J. Funct. Anal.* **282** (2022), no. 3.

[100] H. Ganesh and O'Connell, A large-deviation principle for Dirichlet posteriors. *Bernoulli* **6** (2000), 1021–1034.

[101] D. García-Zelada, A large deviation principle for empirical measures on Polish spaces: application to singular Gibbs measures on manifolds. *Ann. Inst. Henri Poincaré Probab. Stat.* **55** (2019), no. 3, 1377–1401.

[102] D. Gayet and J.-Y. Welschinger, Betti numbers of random real hypersurfaces and determinants of random symmetric matrices. *J. Eur. Math. Soc. (JEMS)* **18** (2016), no. 4, 733–772.

[103] V. Gorin and G. Panova, Asymptotics of symmetric polynomials with applications to statistical mechanics and representation theory. *Ann. Probab.* **43** (2015), no. 6, 3052–3132.

[104] I. P. Goulden and D. M. Jackson, Maps in locally orientable surfaces and integrals over real symmetric surfaces. *Canad. J. Math.* **49** (1997), no. 5, 865–882.

[105] A. Guionnet, Large deviation upper bounds and central limit theorems for band matrices. *Ann. Inst. Henri Poincaré Probab. Stat.* **38** (2002), 341–384.

[106] A. Guionnet, First order asymptotics of matrix integrals; a rigorous approach towards the understanding of matrix models. *Comm. Math. Phys.* **244** (2004), no. 3, 527–569.

[107] A. Guionnet and J. Huang, Large deviations asymptotics of rectangular spherical integral. 2021, arXiv:2106.07146.

[108] A. Guionnet and J. Husson, Large deviations for the largest eigenvalue of Rademacher matrices. *Ann. Probab.* **48** (2020), no. 3, 1436–1465.

[109] A. Guionnet and J. Husson, Asymptotics of k dimensional spherical integrals and applications. 2021, arXiv:2101.01983.

[110] A. Guionnet, V. F. R. Jones, and D. Shlyakhtenko, Random matrices, free probability, planar algebras and subfactors. In *Quanta of maths*, pp. 201–239, Clay Math. Proc. 11, Amer. Math. Soc., Providence, RI, 2010.

[111] A. Guionnet and M. Maïda, A Fourier view on the R-transform and related asymptotics of spherical integrals. *J. Funct. Anal.* **222** (2005), no. 2, 435–490.

[112] A. Guionnet and M. Maïda, Large deviations for the largest eigenvalue of the sum of two random matrices. *Electron. J. Probab.* **25** (2020), 24.

[113] A. Guionnet and E. Maurel Segala, Combinatorial aspects of matrix models. *ALEA Lat. Am. J. Probab. Math. Stat.* **1** (2006), 241–279.

[114] A. Guionnet and E. Maurel Segala, Second order asymptotics for matrix models. *Ann. Probab.* **35** (2007), no. 6, 2160–2212.

[115] A. Guionnet and E. Maurel Segala, Low temperature expansion for matrix models. 2022, work in progress.

[116] A. Guionnet and J. Novak, Asymptotics of unitary multimatrix models: the Schwinger–Dyson lattice and topological recursion. *J. Funct. Anal.* **268** (2015), no. 10, 2851–2905.

[117] A. Guionnet and D. Shlyakhtenko, Free diffusions and matrix models with strictly convex interaction. *Geom. Funct. Anal.* **18** (2008), no. 6, 1875–1916.

[118] A. Guionnet and O. Zeitouni, Concentration of the spectral measure for large matrices. *Electron. Commun. Probab.* **5** (2000), 119–136 (electronic).

[119] A. Guionnet and O. Zeitouni, Large deviations asymptotics for spherical integrals. *J. Funct. Anal.* **188** (2002), 461–515.

[120] D. Guionnet and A. Shlyakhtenko, Free monotone transport. *Invent. Math.* **197** (2014), 613–661.

[121] U. Haagerup and S. Thorbjørnsen, Random matrices and k-theory for exact C^*-algebras. *Doc. Math.* **4** (1999), 341–450 (electronic).

[122] A. Hardy, A note on large deviations for 2D Coulomb gas with weakly confining potential. *Electron. Commun. Probab.* **17** (2012), no. 19, 12.

[123] J. Harer and D. Zagier, The Euler characteristic of the moduli space of curves. *Invent. Math.* **85** (1986), 457–485.

[124] Harish-Chandra, Invariant differential operators on a semisimple Lie algebra. *Proc. Natl. Acad. Sci. USA* **42** (1956), 252–253.

[125] F. Hiai and D. Petz, *The semicircle law, free random variables and entropy*. Math. Surveys Monogr. 77, American Mathematical Society, Providence, RI, 2000.

[126] J. Husson, Large deviations for the largest eigenvalue of matrices with variance profiles. 2020, arXiv:2002.01010.

[127] C. Itzykson and J. B. Zuber, The planar approximation. II. *J. Math. Phys.* **21** (1980), 411–421.

[128] D. Jekel, An elementary approach to free entropy theory for convex potentials. *Anal. PDE* **13** (2020), no. 8, 2289–2374.

[129] Z. Ji, A. Natarajan, T. Vidick, J. Wright, and H. Yuen MIP* $=$ RE. 2020, arXiv:2001.04383.

[130] K. Johansson, The longest increasing subsequence in a random permutation and a unitary random matrix model. *Math. Res. Lett.* **5** (1998), no. 1–2, 63–82.

[131] K. Johansson, On fluctuations of eigenvalues of random Hermitian matrices. *Duke Math. J.* **91** (1998), 151–204.

[132] K. Johansson, Shape fluctuations and random matrices. *Comm. Math. Phys.* **209** (2000), no. 2, 437–476.

[133] K. Johansson, Shape fluctuations and random matrices. *Comm. Math. Phys.* **209** (2000), 437–476.

[134] D. Jonsson, Some limit theorems for the eigenvalues of a sample covariance matrix. *J. Multivariate Anal.* **12** (1982), 1–38.

[135] J. P. Keating and N. C. Snaith, Random matrix theory and $\zeta(1/2 + it)$. *Comm. Math. Phys.* **214** (2000), no. 1, 57–89.

[136] S. Kharchev and D. Lebedev, Integral representation for the eigenfunctions of quantum periodic Toda chain. *Lett. Math. Phys.* **50** (1999), 53–77.

[137] A. M. Khorunzhy, B. A. Khoruzhenko, and L. A. Pastur, Asymptotic properties of large random matrices with independent entries. *J. Math. Phys.* **37** (1996), 5033–5060.

[138] R. Killip and B. Simon, Sum rules for Jacobi matrices and their applications to spectral theory. *Ann. of Math. (2)* **158** (2003), no. 1, 253–321.

[139] A. Knutson and T. Tao, Honeycombs and sums of Hermitian matrices. *Notices Amer. Math. Soc.* **48** (2001), no. 2, 175–186.

[140] T. Leblé and S. Serfaty, Large deviation principle for empirical fields of Log and Riesz gases. *Invent. Math.* **210** (2017), no. 3, 645–757.

[141] M. Ledoux, A recursion formula for the moments of the Gaussian orthogonal ensemble. *Ann. Inst. Henri Poincaré Probab. Stat.* **45** (2009), no. 3, 754–769.

[142] J. O. Lee and J. Yin, A necessary and sufficient condition for edge universality of Wigner matrices. *Duke Math. J.* **163** (2014), no. 1, 117–173.

[143] Y. Lin and L.-C. Tsai, Short time large deviations of the KPZ equation. *Comm. Math. Phys.* **386** (2021), no. 1, 359–393.

[144] S. Lukyanov, Form-factors of exponential fields in the sine-Gordon model. *Modern Phys. Lett. A* **12** (1997), 2543–2550.

[145] J. Lynch and J. Sethuraman, Large deviations for processes with independent increments. *Ann. Probab.* **15** (1987), 610–627.

[146] A. Lytova and L. Pastur, Central limit theorem for linear eigenvalue statistics of random matrices with independent entries. *Ann. Probab.* **37** (2009), no. 5, 1778–1840.

[147] M. Maïda, Large deviations for the largest eigenvalue of rank one deformations of Gaussian ensembles. *Electron. J. Probab.* **12** (2007), 1131–1150.

[148] C. Male, The norm of polynomials in large random and deterministic matrices. *Probab. Theory Related Fields* **154** (2012), no. 3–4, 477–532.

[149] A. Matytsin, On the large-N limit of the Itzykson–Zuber integral. *Nuclear Phys. B* **411** (1994), no. 2–3, 805–820.

[150] E. Maurel Segala, High order asymptotics of matrix models and enumeration of maps. 2006, arXiv:1912.04588.

[151] M. Mehta, *Random matrices*. Academic Press, 1991.

[152] M. L. Mehta, A method of integration over matrix variables. *Comm. Math. Phys.* **79** (1981), no. 3, 327–340.

[153] M. L. Mehta, *Random matrices. Third edn*. Pure Appl. Math. (Amst.) 142, Elsevier/Academic Press, Amsterdam, 2004.

[154] H. L. Montgomery, The pair correlation of zeros of the zeta function. In *Analytic number theory (Proc. Sympos. Pure Math., Vol. XXIV, St. Louis Univ., St. Louis, MO, 1972)*, pp. 181–193, Amer. Math. Soc., Providence, R.I., 1972.

[155] H. Narayanan and S. Sheffield, Large deviations for random hives and the spectrum of the sum of two random matrices. 2021, arXiv:2111.00421.

[156] L. A. Pastur, On the spectrum of random matrices. *Teor. Math. Phys.* **10** (1972), 67–74.

[157] L. Pastur and M. Shcherbina, *Eigenvalue distribution of large random matrices*. Math. Surveys Monogr. 171, American Mathematical Society, Providence, RI, 2011.

[158] E. B. Saff and V. Totik, *Logarithmic potentials with external fields*. Grundlehren Math. Wiss. 316, Springer, Berlin, 1997.

[159] H. Schultz, Non-commutative polynomials of independent Gaussian random matrices. The real and symplectic cases. *Probab. Theory Related Fields* **131** (2005), no. 2, 261–309.

[160] T. Seppäläinen, Large deviations for increasing sequences on the plane. *Probab. Theory Related Fields* **112** (1998), no. 2, 221–244.

[161] S. Serfaty, Systems of points with Coulomb interactions. In *Proceedings of the International Congress of Mathematicians—Rio de Janeiro 2018. Vol. I. Plenary lectures*, pp. 935–977, World Sci. Publ., Hackensack, NJ, 2018.

[162] S. Sheffield, *Random surfaces*. Astérisque, (2005), no. 304, vi+175 pp.

[163] D. Shlyakhtenko, Gaussian random band matrices and operator-valued free probability theory. In *Quantum probability (Gdańsk, 1997)*, pp. 359–368, Banach Center Publ. 43, Polish Acad. Sci, Warsaw, 1998.

[164] F. Smirnov, *Form factors in completely integrable models of quantum field theory*. Adv. Ser. Math. Phys. 14, World Scientific, 1992.

[165] H. Spohn, Ballistic space-time correlators of the classical Toda lattice. *J. Phys. A* **53** (2020), 265004.

[166] H. Spohn, Generalized Gibbs ensembles of the classical Toda chain. *J. Stat. Phys.* **180** (2020), no. 1–6, 4–22.

[167] E. Subag, The geometry of the Gibbs measure of pure spherical spin glasses. *Invent. Math.* **210** (2017), no. 1, 135–209.

[168] G. t'Hooft, Magnetic monopoles in unified gauge theories. *Nuclear Phys. B* **79** (1974), 276–284.

[169] T. Tao and V. Vu, Random matrices: universality of local eigenvalue statistics. *Acta Math.* **206** (2011), no. 1, 127–204.

[170] C. A. Tracy and H. Widom, Level spacing distributions and the Airy kernel. *Comm. Math. Phys.* **159** (1994), 151–174.

[171] L.-C. Tsai, Exact lower tail large deviations of the KPZ equation. 2018, arXiv:1809.03410.

[172] W. T. Tutte, A census of planar maps. *Canad. J. Math.* **15** (1963), 249–271.

[173] B. Valkó and B. Virág, Continuum limits of random matrices and the Brownian carousel. *Invent. Math.* **177** (2009), no. 3, 463–508.

[174] S. R. S. Varadhan, Asymptotic probabilities and differential equations. *Comm. Pure Appl. Math.* **19** (1966), 261–286.

[175] D. Voiculescu, Limit laws for random matrices and free products. *Invent. Math.* **104** (1991), 201–220.

[176] D. Voiculescu, The analogues of entropy and of Fisher's information measure in free probability theory. I. *Comm. Math. Phys.* **155** (1993), no. 1, 71–92.

[177] D. Voiculescu, In *Lectures on probability theory and statistics: Ecole D'Été de Probabilités de Saint-Flour XXVIII – 1998*, pp. 283–349, Lecture Notes in Math. 1738, Springer, New York, NY, 2000.

[178] D. Voiculescu, Free entropy. *Bull. Lond. Math. Soc.* **34** (2002), 257–278.

[179] D. Voiculescu, The topological version of free entropy. *Lett. Math. Phys.* **62** (2002), no. 1, 71–82.

[180] J. von Neumann and H. H. Goldstine, Numerical inverting of matrices of high order. *Bull. Amer. Math. Soc.* **53** (1947), 1021–1099.

[181] H. Weyl, *The classical groups: their invariants and representations*. Princeton University Press, Princeton, NJ, 1939.

[182] E. P. Wigner, Characteristic vectors of bordered matrices with infinite dimensions. *Ann. of Math.* **62** (1955), 548–564.

[183] J. Wishart, The generalized product moment distribution in samples from a normal multivariate population. *Biometrika* **20A** (1928), 32–52.

[184] I. Zakharevich, A generalization of Wigner's law. *Comm. Math. Phys.* **268** (2006), 403–414.

[185] O. Zeitouni and S. Zelditch, Large deviations of empirical measures of zeros of random polynomials. *Int. Math. Res. Not. IMRN* **20** (2010), 3935–3992.

[186] D. Zhang, Tridiagonal random matrix: Gaussian fluctuations and deviations. *J. Theoret. Probab.* **30** (2017), no. 3, 1076–1103.

[187] J.-B. Zuber, Horn's problem and Harish-Chandra's integrals. Probability density functions. *Ann. Inst. Henri Poincaré D* **5** (2018), no. 3, 309–338.

ALICE GUIONNET

ENS de Lyon, CNRS, France, Alice.Guionnet@ens-lyon.fr

This Page Intentionally Left Blank

DECOUPLING ESTIMATES IN FOURIER ANALYSIS

LARRY GUTH

ABSTRACT

Decoupling is a recent development in Fourier analysis, which has applications in harmonic analysis, PDEs, and number theory. We survey some applications of decoupling and some of the ideas in the proof.

MATHEMATICS SUBJECT CLASSIFICATION 2020

Primary 42B15; Secondary 42B20, 11L15

KEYWORDS

Restriction theory, circle method

INTERNATIONAL CONGRESS OF MATHEMATICIANS 2022 JULY 6–14

Proc. Int. Cong. Math. 2022, Vol. 2, pp. 1054–1089
DOI 10.4171/ICM2022/172

1. INTRODUCTION

Decoupling is a recent development in Fourier analysis, which has applications in harmonic analysis, PDEs, and number theory. To put it in context, let us start by recalling some basic ideas of Fourier analysis. In Fourier analysis, we represent a function as a Fourier series or Fourier integral. For instance, if $f : \mathbb{R}^n \to \mathbb{C}$ is a reasonably nice function, then we can write it as a Fourier integral

$$f(x) = \int_{\mathbb{R}}^{n} \hat{f}(\omega) e^{2\pi i \omega \cdot x} dx. \tag{1.1}$$

Here $\omega \cdot x$ is the dot product of ω and x, which we will also abbreviate as just ωx.

Here are a couple reasons that it is useful to represent a function f using a Fourier series or integral. First, the functions $e^{2\pi i \omega x}$ are eigenfunctions for the partial derivative operators ∂x_j. This makes the Fourier representation interact well with partial derivatives, and it helps to study PDEs. Second, the functions $e^{2\pi i \omega \cdot x}$ are eigenfunctions of the translation operator T_v defined by $T_v f(x) = f(x + v)$. This makes the Fourier representation useful in problems involving the translation structure of $\mathbb{R}^n$, including problems in additive number theory.

But there is also a serious downside to representing a function f as a Fourier series/integral. To evaluate $f(x)$, we have to compute an integral or a sum with many terms. It often happens that the terms have various phases in the complex plane, and it is difficult to tell what happens when we add them all up. In general, given some information about $\hat{f}$, it can be difficult to determine what that information it has to say about f. We will see some longstanding open questions of this flavor below.

Decoupling is helpful for estimating $\|f\|_{L^p}$ in terms of information about $\hat{f}$. Now $\|f\|_{L^2}$ is directly related to $\hat{f}$ because of orthogonality: Plancherel's theorem states that

$$\|f\|_{L^2} = \|\hat{f}\|_{L^2}. \tag{1.2}$$

But for other values of p, it is much harder to connect $\|f\|_{L^p}$ with information about $\hat{f}$.

Estimates for $\|f\|_{L^p}$ for $p \neq 2$ occur often in harmonic analysis, PDEs, and analytic number theory. You may wonder, if we have a good understanding of $\|f\|_{L^2}$, what more do we learn by understanding $\|f\|_{L^p}$ for other values of p. I like to think of this question in terms of superlevel sets. Define the superlevel set $U_\lambda(f)$ by

$$U_\lambda(f) := \{x : |f(x)| > \lambda\}. \tag{1.3}$$

We denote the volume of a set U by $|U|$. If we know $\|f\|_{L^p}$ for every p, we typically get accurate estimates for $|U_\lambda(f)|$ for every λ, which gives us basically all the possible information about how "big" the function f is. But if we only know $\|f\|_{L^2}$, we get only limited information about $|U_\lambda(f)|$.

Other motivations for studying $\|f\|_{L^p}$ come from applications in PDE and analytic number theory. In nonlinear PDEs, bounds involving $\|f\|_{L^p}$ are important for understanding how close a solution to a nonlinear PDE is to a solution of a corresponding linear PDE. In analytic number theory, the number of solutions to certain diophantine systems is equal to

$\int |f|^p$ for a well-chosen function f and exponent p. These are just a couple samples among many applications for estimating $\|f\|_{L^p}$.

Decoupling is a new tool for estimating $\|f\|_{L^p}$ in terms of Fourier-analytic information about f. It was first formulated by Wolff in [56], where he was able to prove sharp estimates for large values of p. In [14], Bourgain and Demeter proved sharp decoupling estimates for all p. This breakthrough has led to solutions to problems in harmonic analysis that had seemed far out of reach a decade ago.

In the next two subsections, we will introduce two main areas where decoupling has had an impact. We will give examples of hard open problems and also examples of problems that were solved using decoupling.

1.1. Restriction theory

The Fourier representation of a function $f : \mathbb{R}^n \to \mathbb{C}$ is

$$f(x) = \int_{\mathbb{R}^n} \hat{f}(\omega) e^{2\pi i \omega \cdot x} \, d\omega.$$

There are two basic estimates connecting the L^p-norms of f and the L^p-norms of $\hat{f}$:

- (Orthogonality) $\|f\|_{L^2} = \|\hat{f}\|_{L^2}$;
- (Triangle inequality) $\|f\|_{L^\infty} \leq \|\hat{f}\|_{L^1}$.

Interpolating between these gives the Hausdorff–Young inequality

$$\|f\|_{L^p} \leq \|\hat{f}\|_{L^q} \quad \text{if } 1 \leq p \leq 2 \text{ and } \frac{1}{q} = 1 - \frac{1}{p}. \tag{1.4}$$

These are all of the L^p-type estimates for the Fourier transform operator.

If $\hat{f}$ is supported in a subset $\Omega \subset \mathbb{R}^n$, we can write

$$f(x) = \int_{\Omega} \hat{f}(\omega) e^{2\pi i \omega \cdot x} \, d\omega.$$

Restriction theory studies how the geometry of Ω relates to properties of f such as $\|f\|_{L^p}$. One of the most interesting cases is when $\hat{f}$ is supported in a compact submanifold $S \subset \mathbb{R}^n$. In this case, the Fourier representation of f has the form

$$f(x) = \int_S a(\omega) e^{2\pi i \omega x} \, d\mu_S(\omega), \tag{1.5}$$

where $d\mu_S$ is the surface area measure of S.

Stein proposed studying L^p-estimates of the form

$$\|f\|_{L^p(\mathbb{R}^n)} \leq C \|a\|_{L^q(S)} \tag{1.6}$$

and made the remarkable discovery that the estimates for the operator E_S depend on the geometry of S. If S is a flat disk, then the only estimate of the form (1.6) is the triangle inequality $\|f\|_{L^\infty} \leq \|a\|_{L^1(S)}$. But if S is a curved surface, then there are more inequalities.

One central problem in the field is to understand all the L^p-inequalities of form (1.6) when S is a curved hypersurface, like a paraboloid. Let us write P for the truncated paraboloid

$$P := \left\{ \omega \in \mathbb{R}^n \mid \omega_n = \sum_{j=1}^{n-1} \omega_j^2 \text{ and } \sum_{j=1}^{n-1} \omega_j^2 \leq 1 \right\}. \tag{1.7}$$

In this case, the Fourier representation of f is

$$f(x) = \int_P a(\omega) e^{2\pi i \omega x} d\mu_P(\omega). \tag{1.8}$$

Example 1.1. Suppose $a(\omega) = 1$ on P, and f is given by (1.8).

First note that $f(0) = \int_P d\mu_P$ is equal to the area of P, which is ~ 1. When x is large, there is a lot of cancellation in the integral (1.8) coming from rapid oscillation of the function $e^{2\pi i \omega x}$ as ω varies over P. This effect can be estimated accurately using stationary phase, and one finds that

$$\left|f(x)\right| \lesssim |x|^{-\frac{n-1}{2}}.$$

This bound is sharp for most x. Therefore $\|f\|_{L^p(\mathbb{R}^n)} < \infty$ if and only if $p > \frac{2n}{n-1}$.

Stein conjectured that the same L^p-bounds hold whenever $|a(\omega)| \leq 1$ for all ω.

Conjecture 1.2 (Restriction conjecture **[49]**). *Suppose that f has the form* (1.8) *and that $|a(\omega)| \leq 1$ for all $\omega \in P$. If $p > \frac{2n}{n-1}$, then*

$$\|f\|_{L^p(\mathbb{R}^n)} \leq C(p,n).$$

Notice that the hypothesis that f has the form (1.8) with $|a(\omega)| \leq 1$ for all ω is a hypothesis about $\hat{f}$. The restriction conjecture asks what this information about $\hat{f}$ tells us about $\|f\|_{L^p}$. The 2-dimensional case of Conjecture 1.2 was proven by Fefferman in **[26]**. But for dimension $n \geq 3$, the conjecture remains open after intensive work by many people. In Section 5, we will some reasons the problem is so difficult.

In Conjecture 1.2, we considered the bound $\|a(\omega)\|_{L^\infty} \leq 1$. Bounds of the form $\|a\|_{L^q(P)}$ are also interesting for other q. The case $q = 2$ is the most important, and it was completely worked out by Strichartz **[51]** following work by Tomas and Stein. It has turned out to be important in PDEs. It reads as follows.

Theorem 1.3 (Strichartz inequality **[51]**). *Suppose that f has the form* (1.8). *If $p \geq \frac{2(n+1)}{n-1}$, then*

$$\|f\|_{L^p(\mathbb{R}^n)} \leq C(n) \left\|a(\omega)\right\|_{L^2(P)}.$$

This theorem plays an important role in the study of the Schrödinger equation. Recall that the linear Schödinger equation for a function $u(x,t)$ with $x \in \mathbb{R}^d$ and $t \in \mathbb{R}$ is

$$\partial_t u = i \sum_{j=1}^{d} \partial_{x_j}^2 u. \tag{1.9}$$

If u obeys the linear Schrödinger equation, then $\hat{u}$ is a distribution supported on the paraboloid, and so the Strichartz estimate can be used to understand $\|u\|_{L^p}$. Theorem 1.3 tells us that for any solution of the linear Schrödinger equation (1.9) with initial data $u(x,0) = u_0(x)$,

$$\|u\|_{L^{\frac{2(d+2)}{d}}(\mathbb{R}^d \times \mathbb{R})} \leq C \|u_0\|_{L^2(\mathbb{R}^d)}. \tag{1.10}$$

This theorem has played a central role in PDEs, especially in nonlinear PDEs. The L^2-norm on the right-hand side is important in PDEs because $\|u_0\|_{L^2} = \|\hat{u}_0\|_{L^2}$ and also $\|u_0\|_{L^2} = \|u(y,t)\|_{L^2_y}$ for every t. In nonlinear PDEs, it leads to sharp estimates about when the solution to a nonlinear PDE is close to the solution of the corresponding linear PDE.

The Strichartz estimate describes a spreading-out effect. To get a sense of it, first suppose that u_0 is a smooth bump concentrated on a ball in spacetime. As t increases, the function $u(x,t)$ spreads out and gets smaller. As it does so, $\int_{\mathbb{R}^d} |u(x,t)|^2 dx$ remains constant, and $\int_{\mathbb{R}^d} |u(x,t)|^p dx$ gets smaller for any $p > 2$. Because of this spreading out effect, $\int_{\mathbb{R}^d \times \mathbb{R}} |u(x,t)|^{\frac{2(d+2)}{d}} dx dt$ is finite.

The exponent $\frac{2(d+2)}{d}$ is the only exponent for which (1.10) holds. To see what is special about this exponent, it helps me to translate the Strichartz estimate into an estimate for superlevel sets. Let $U_\lambda(u) := \{(x,t) \in \mathbb{R}^d \times \mathbb{R} : |u(x,t)| > \lambda\}$. The Strichartz inequality implies that if $\|u_0\|_{L^2(\mathbb{R}^d)} = 1$, then

$$\left|U_\lambda(u)\right| \leq C \lambda^{-\frac{2(d+2)}{d}}.$$

This estimate is sharp: for any choice of λ, we can find initial data u_0 with $\|u_0\|_{L^2(\mathbb{R}^d)} = 1$ so that the solution of the Schrödinger equation has $|U_\lambda(u)| \geq c\lambda^{-\frac{2(d+2)}{d}}$.

It is also worth mentioning that the choice of the paraboloid in this discussion is just one interesting example. There are similar theorems and conjectures for other surfaces, such as the sphere and the cone, and these help to study other PDEs, such as the Laplace eigenfunction equation $\Delta u = \lambda u$ and the wave equation.

One striking application of decoupling involves Strichartz estimates on flat tori. The Schrödinger equation makes sense on any Riemannian manifold, and for each manifold we can ask for the best inequality in the spirit of (1.10). Understanding the Strichartz estimates on closed manifolds is extremely difficult. It is known that different closed manifolds behave quite differently from each other—for example, round spheres behave differently from flat tori. But very few examples are understood. Before decoupling, sharp Strichartz estimates were only known for S^1 and $S^1 \times S^1$ (by Bourgain in the 1990s [8]) and S^3 (by Burq–Gerard–Tzvetkov [19]). In all these examples, the value of the exponent p is an even integer, and we will discuss in Section 1.2 why this is important.

The simplest flat torus in the unit cube torus $\mathbb{R}^d/\mathbb{Z}^d$. A solution to the Schrödinger equation on the unit cube torus is just a solution $u(x,t)$ on $\mathbb{R}^d \times \mathbb{R}$ which is $\mathbb{Z}^d$-periodic in the x variable. Any such solution can be written in the form

$$u(x,t) = \sum_{n \in \mathbb{Z}^d} a_n e^{2\pi i (n \cdot x + |n|^2 t)}. \tag{1.11}$$

Notice that this Fourier representation is analogous to (1.8), except that the integral in (1.8) is replaced by a sum. We say that u has "frequency at most N" if the coefficients a_n are supported in the cube $Q_N := \{(n_1, \dots, n_d) \in \mathbb{Z}^d : |n_j| \le N \text{ for all } j\}$.

Example 1.4. Suppose that u is given by (1.11) where $a_n = 1$ if $n \in Q_N$ and $a_n = 0$ otherwise. In other words,

$$u(x,t) = \sum_{n \in Q_N} e^{2\pi i (n \cdot x + |n|^2 t)}.$$

First note that $u(0,0) = |Q_N| \sim N^d$. We have $|u(x,t)| \sim N^d$ when $|x| \le \frac{1}{10dN}$ and $|t| \le \frac{1}{10dN^2}$, because then each term in the sum is almost 1. As x and t increase, we get cancellation in the sum coming from oscillations in $e^{2\pi i(n\cdot x+|n|^2 t)}$. So far this behavior is similar to Example 1.1.

However, in the torus case, $|u(x,t)|$ is also large when (x,t) lies near to a rational point of the form $(\frac{p_1}{q}, \dots, \frac{p_d}{q}, \frac{p_t}{q})$. Taking account of all these peaks near rational points, it turns out that $U_\lambda(u) \cap [0,1]^{d+1}$ has volume $\sim N^{d+2} \lambda^{-\frac{2(d+2)}{d}}$ for all λ in the range $N^{d/2} \le \lambda \le N^d$. (This range includes all interesting values of λ.)

A natural analogue of the restriction conjecture in the periodic setting would say

Conjecture 1.5. *Suppose that u is given by* (1.11) *and that $|a_n| \le 1$ for all $n \in Q_N$ and $a_n = 0$ for $n \notin Q_N$. Then $|U_\lambda(u) \cap [0,1]^{d+1}| \le C(d,\varepsilon) N^{d+2+\varepsilon} \lambda^{-\frac{2(d+2)}{d}}$ for all λ in the range $N^{d/2} \le \lambda \le N^d$.*

This conjecture used to sound to me just as hard as the restriction conjecture or maybe harder. The setup is similar. And Example 1.4 in this periodic setting is more intricate and complex than Example 1.1 in the setting of the original restriction conjecture. However, Bourgain and Demeter proved this conjecture as a corollary of their sharp Strichartz estimate on tori. This theorem is one of the first applications of decoupling.

Theorem 1.6 (Bourgain and Demeter [14]). *Suppose that u is given by* (1.11) *and that a_n is supported in Q_N. Then*

$$\|u\|_{L^{\frac{2(d+2)}{d}}([0,1]^{d+1})} \le C(d,\varepsilon) N^{\varepsilon} \|a_n\|_{\ell^2}. \tag{1.12}$$

Notice that if $u_0(x) = u(x,0)$, then $\|a_n\|_{\ell^2} = \|u_0\|_{L^2([0,1]^d)}$, so this inequality is very similar to the Strichartz inequality for the Schrödinger equation on $\mathbb{R}^d$ recorded in (1.10).

To finish this section, let us try to roughly indicate why the Strichartz inequality on the torus is much harder than the Strichartz inequality on $\mathbb{R}^d$. Recall that the Strichartz inequality encodes a spreading-out effect. First, imagine a solution $u(x,t)$ on a Euclidean space, and suppose that the initial data u_0 is concentrated in a very small ball. As time increases, the solution $u(x,t)$ spreads out. At a small time t_0, the solution is spread over a unit ball. In the Euclidean space, it can continue to spread out in all directions indefinitely. The proof of Strichartz estimates this effect in a quantitative way.

Now let u_P be the solution on the torus with the same initial data u_0. The function u_P is given by periodizing u:

$$u_P(x,t) = \sum_{z\in\mathbb{Z}^d} u(x+z,t). \tag{1.13}$$

For times up to t_0, $u(x,t)$ is supported on a unit ball in the x variable, and so $u_P(x,t) = u(x,t)$. But beyond this time, $u(x,t)$ is spread over a much bigger ball, and there are many nonzero terms in the sum (1.13). If we visualize $u_P(x,t)$, the solution starts to wrap around the torus. Different pieces of the solution, which have traveled around the torus in different ways, get added up, and we have to prove that there is a lot of cancellation in that sum.

Before decoupling, Theorem 1.6 was known for $d = 1, 2$ only because of a connection with number theory. In the next section, we describe some connections between Fourier analysis and number theory, and we will flesh this out.

1.2. Analytic number theory

When p is an even integer, L^p-estimates have a special interpretation which connects them with problems in additive number theory.

Suppose that $A \subset \mathbb{Z}^d$ is a finite set. We define $E_s(A)$ (the additive s-energy of A) by

$$E_s(A) := \#\{(a_1,\dots,a_s,b_1,\dots,b_s) \in A^{2s} : a_1+\dots+a_s = b_1+\dots+b_s\}. \tag{1.14}$$

For each A, we can also define a function $f_A(x)$ with Fourier series

$$f_A(x) = \sum_{a\in A} e^{2\pi i a\cdot x}. \tag{1.15}$$

The function $f_A : \mathbb{R}^d \to \mathbb{C}$ is $\mathbb{Z}^d$-periodic because $A \subset \mathbb{Z}^d$, and so each function $e^{2\pi i a\cdot x}$ is $\mathbb{Z}^d$-periodic.

Lemma 1.7. *For any finite set* $A \subset \mathbb{Z}^d$,

$$\int_{[0,1]^d} |f_A(x)|^{2s}\,dx = E_s(A).$$

Proof sketch. We expand the integral on the left-hand side:

$$\begin{aligned}\int_{[0,1]^d} |f_A(x)|^{2s}\,dx &= \int_{[0,1]^d} f_A^s \bar{f}_A^s\,dx \\ &= \int_{[0,1]^d} \sum_{a_1,\dots,a_s,b_1,\dots,b_s\in A} e^{2\pi i(a_1+\dots+a_s-b_1-\dots-b_s)x}\,dx.\end{aligned}$$

Now if $m \in \mathbb{Z}^d$, then $\int_{[0,1]^d} e^{2\pi i m\cdot x}\,dx$ is 1 if $m = 0$ and 0 otherwise. And so the only terms that contribute to the integral above are terms where $a_1+\dots+a_s-b_1-\dots-b_s = 0$. So the last integral is $E_s(A)$. ■

For instance, if $A_{k,N} := \{1^k, 2^k, \dots, N^k\} \subset \mathbb{Z}$ then

$$\begin{aligned} E_s(A_{k,N}) &= \#\text{ of solutions to } a_1^k + \cdots + a_s^k = b_1^k + \cdots + b_s^k, \\ &\text{with } a_j, b_j \in \mathbb{Z}, 1 \le a_j, b_j \le N. \end{aligned} \tag{1.16}$$

In this case, the relevant function f is

$$f_{k,N}(x) = \sum_{a=1}^{N} e^{2\pi i a^k x}, \tag{1.17}$$

and Lemma 1.7 tells us that

$$\int_0^1 |f_{k,N}(x)|^{2s} dx = E_s(A_{k,N}). \tag{1.18}$$

Lemma 1.7 tells us that a certain L^p-norm is equal to the number of solutions to a certain diophantine equation. The lemma is useful in both directions. If we know something about the number of solutions to the diophantine equation, then we can get information about the L^p-norm. If we know something about the L^p-norm, then we can get information about the number of solutions to the diophantine equation.

For instance, consider the diophantine equation $a_1^2 + a_2^2 = b_1^2 + b_2^2$, with a_i, b_i between 1 and N. First let us estimate the number of solution directly. Rearranging we get $a_1^2 - b_1^2 = b_2^2 - a_2^2$, and factoring one side we see that

$$(a_1 + b_1)(a_1 - b_1) = b_2^2 - a_2^2.$$

If we fix a_2, b_2, then the number of (a_1, b_1) solving this equation depends on the number of factors of $b_2^2 - a_2^2$. Because of unique factorization, the number of different factors of an integer M is fairly small, at most $C_\varepsilon M^\varepsilon$ for any $\varepsilon > 0$. Using this, we see that the number of integer solutions to $a_1^2 + a_2^2 = b_1^2 + b_2^2$ with $1 \le a_j, b_j \le N$ is at most $C_\varepsilon N^{2+\varepsilon}$. Lemma 1.7 tells us that the number of solutions is equal to $\int_0^1 |f_{2,N}(x)|^4 dx$, and so we conclude that this integral is bounded by $C_\varepsilon N^{2+\varepsilon}$.

On the other hand, Weyl used the differencing method to give pointwise estimates for the function $f_{2,N}$. These estimates imply that $\int_0^1 |f_{2,N}(x)|^4 dx \le C_\varepsilon N^{2+\varepsilon}$ which then gives an analytic proof that the number of integer solutions to $a_1^2 + a_2^2 = b_1^2 + b_2^2$ with $1 \le a_j, b_j \le N$ is at most $C_\varepsilon N^{2+\varepsilon}$.

Hardy and Littlewood made a conjecture that generalizes these estimates from squares to higher powers.

Conjecture 1.8 (Hardy and Littlewood). *For any $k \ge 2$, $E_k(A_{k,N}) \le C_\varepsilon N^{k+\varepsilon}$. Equivalently,*

$$\int_0^1 |f_{k,N}(x)|^{2k} dx \le C_\varepsilon N^{k+\varepsilon}.$$

This conjecture is open for all $k \ge 3$. The Fourier series of $f_{3,N}$ is fairly simple to write down. But it is very difficult to determine good bounds for the L^p-norms of $f_{3,N}$, or for the size of superlevel sets $U_\lambda(f_{3,N})$. This is a classical and striking example of how difficult it is to read off information about $f(x)$ from information about its Fourier series.

On the other hand, there are cases when we can use Fourier analysis to estimate an L^p-norm and then use Lemma 1.7 to get a new estimate for the number of solutions to a diophantine equation. One of the most interesting examples of this kind concerns Vinogradov's mean value theorem, which is a multivariable generalization of the functions we just considered.

Define

$$F_{k,N}(x_1,\dots,x_k) = \sum_{a=1}^{N} e^{2\pi i(ax_1+a^2x_2+\dots+a^kx_k)}.$$

By Lemma 1.7, $\int_{[0,1]^k} |F_{k,N}(x)|^{2s}dx$ is equal to the number of solutions to the following diophantine system of equations:

$$a_1^j+\dots+a_s^j = b_1^j+\dots+b_s^j \quad \text{for all } 1\le j\le k, \text{ with } a_i,b_i\in\mathbb{Z}, 1\le a_i,b_i\le N.$$

Vinogradov [52] studied the L^p-norms of $F_{k,N}$ in the 1930s. He was able to prove sharp estimates for $\|F_{k,N}\|_{L^p}$ for sufficiently large p. He used these bounds to greatly improve the estimates for Weyl sums and Waring's problem in large degree, and also to improve the bounds on the zero-free region of the Riemann zeta function. Vinogradov's argument cleverly exploited both sides of equation (1.7): some parts of the argument directly count the number of solutions to some diophantine systems in the variables a_i, b_i, and other parts of the argument estimate integrals in the x variable. Some important ideas in the proof of decoupling are related to Vinogradov's argument, and we will discuss this more in Section 4.5.

In the last decade, mathematicians have proven estimates for $\|F_{k,N}\|_{L^p}$ that are sharp up to factors of $C(k,\varepsilon)N^\varepsilon$ for every k and p. As a corollary, we get estimates for the number of solutions to the Vinogradov system that are sharp up to a factor $C(k,\varepsilon)N^\varepsilon$.

Theorem 1.9 ([16, 58, 59]).

$$\|F_{k,N}\|_{L^p([0,1]^k)} \le C(k,\varepsilon)N^\varepsilon\big(N^{1/2}+N^{1-\frac{k(k+1)}{2p}}\big).$$

The proof in [16] uses decoupling and the proof in [59] uses the method of efficient congruencing. (Historically, Wooley developed efficient congruencing starting in the 1990s, cf. [57]. He improved Vinogradov's estimates and gave sharp estimates for $k=3$ in [58]. Then [16] used decoupling to prove Theorem 1.9 and immediately afterwards, [59] used efficient congruencing to give a different proof of Theorem 1.9.)

Both [16] and [59] are quite technical. Recently, Guo–Li–Yung–Zorin-Kranich [28] gave a dramatically simpler proof of Theorem 1.9, combining some of the features of [16] and [59] with some new clarifying ideas. Their paper is ten pages long and essentially self-contained.

Lemma 1.7 is a special trick for understanding L^p-norms when p is an even integer. This even integer trick also plays an important role in the problems we discussed in Section 1.1. In [26], Fefferman used a version of the even integer trick to prove Conjecture 1.2 in dimension $n=2$. The L^p-exponent in Conjecture 1.2 is $p=\frac{2n}{n-1}$, which is an even integer when $n=2$ but not for any $n\ge 3$. In the early 1990s, in [8], Bourgain used the even integer trick to prove sharp periodic Strichartz estimates when $d = 1, 2$ (the cases $d = 1, 2$ in Theorem 1.6). The exponent in the Strichartz estimate is $\frac{2(d+2)}{d}$, which is an even integer when $d = 1, 2$, but not for any $d > 2$. Another important problem in this circle is Montgomery's conjecture about the L^p-norms of Dirichlet polynomials. When p is an even integer, Montgomery gave sharp estimates for the relevant L^p-norms in just a page (cf. [42]). But giving a sharp estimate for any other value of p is a major open problem. This might help explain why, even though Theorem 1.6 was already known in dimensions $d = 1, 2$, it still seemed far out of reach to prove it for any other d.

Before decoupling, the situation concerning periodic Strichartz estimates in dimensions 1, 2 was rather curious. The periodic Strichartz estimate can be considered as a result in PDEs, resolving a problem of mathematical physics. But the proof depended on number theory facts, such as unique factorization. The decoupling proof of Theorem 1.6 is purely analytic—with no input from number theory. The argument can then recover some of the number theory that went into the original proof. The relevant number theory estimates are not that difficult, but proving them by analysis is still interesting. Building on this, Bourgain and Demeter began to work on Vinogradov's mean value theorem in [15], eventually leading to Theorem 1.9 and new results in number theory.

Theorem 1.9 leads to improved bounds for Waring's problem on the number of ways to write an integer as a sum of kth powers and the related problem of Weyl sums. Other applications of decoupling have led to incremental improvements in very classical problems of analytic number theory such as the Lindelof hypothesis [13] and the Gauss circle problem and [18]. Guo–Zhang [29] and Guo–Zorin-Kranich [30] have extended Theorem 1.9 to more complex systems of diophantine equations, introduced in number theory by Arkhipov–Chubarikov–Karatsuba [1].

1.3. Influence of the proof

Besides the new results, the method of proof of decoupling has had a big influence on the field. There is a classical toolbox in harmonic analysis with tools like orthogonality, integration by parts, and Hölder's inequality. For hard problems in this area, such as the restriction conjecture, people who have worked a lot on them generally feel that this set of classical tools is not sufficient to understand the problem. Over the last 25 years, mathematicians have brought into play ideas from other areas in order to attack some of these hard problems. For instance, Wolff ([54] and [56]) brought in ideas from combinatorial geometry and topology, Bourgain [9] brought in ideas from combinatorial number theory, and Dvir [24] brought in ideas from error-correcting codes and algebraic geometry. In contrast to these developments, the proof of decoupling is based on the classical toolbox. The most important idea in the proof is to take advantage of estimates at many different scales. Using many different scales is also a classical idea in harmonic analysis. But it is really striking how powerful it turns out to be in the context of decoupling. I personally was shocked that it is possible to prove Theorem 1.6 using only these tools. The main goal of the article is to explore how combining information at many scales helps to prove theorems like Theorems 1.6 and 1.9.

1.4. Outline of the rest of the article

In Section 2, we will introduce the statement of decoupling. In Section 3, we will begin to discuss multiscale arguments, and we will see how the statement of decoupling was carefully crafted to work well in such arguments. In Section 4, we will discuss some ideas of the proof of decoupling.

In Section 5, we will discuss the connection between the restriction problem and the Kakeya problem, and try to explain why the restriction problem seems to be so difficult. Then we will discuss why decoupling turns out to be easier than restriction.

In Section 6, we will survey some other applications of decoupling in harmonic analysis.

In Section 7, we will discuss some limitations of the method, some frustrating aspects of the proof, and some open problems.

2. THE STATEMENT OF DECOUPLING

Now that we have seen some applications of decoupling, we turn to the actual statement of decoupling. The statement of decoupling was crafted carefully, and after we state it we will spend two sections digesting it and discussing some of the choices involved in the statement.

Suppose that $\Omega \subset \mathbb{R}^n$ and that Ω is a disjoint union of subsets θ, $\Omega = \sqcup\theta$. If $f : \mathbb{R}^n \to \mathbb{C}$ is a function, and $\hat{f}$ is supported in Ω, then we can decompose $f = \sum_\theta f_\theta$ where f_θ is defined by

$$f_\theta = \int_\theta \hat{f}(\omega) e^{2\pi i \omega x} d\omega.$$

Decoupling has to do with the relationship between L^p-norm of f and the L^p-norms of f_θ for the different θ in the decomposition $\Omega = \sqcup\theta$.

Definition 2.1. Suppose that $\Omega \subset \mathbb{R}^n$ and Ω is a disjoint union of subsets θ, $\Omega = \sqcup\theta$. For each exponent p, we define the decoupling constant $D_p(\Omega = \sqcup\theta)$ to be the smallest constant so that for every function f with $\hat{f}$ supported in Ω,

$$\|f\|_{L^p(\mathbb{R}^n)}^2 \le D_p(\Omega = \sqcup\theta)^2 \sum_\theta \|f_\theta\|_{L^p(\mathbb{R}^n)}^2. \tag{2.1}$$

If $p = 2$, then orthogonality gives $\|f\|_{L^2}^2 = \sum_\theta \|f_\theta\|_{L^2}^2$, and so $D_2(\Omega = \sqcup\theta) = 1$ for any decomposition $\Omega = \sqcup\theta$. Decoupling theorems for higher p are a kind of strengthening of orthogonality. For $p > 2$, the value of $D_p(\Omega = \sqcup\theta)$ depends on the geometry of the decomposition.

As an example of a decomposition, first let P denote the truncated parabola:

$$P = \{(\omega_1, \omega_2) \in \mathbb{R}^2 : \omega_2 = \omega_1^2, -1 \le \omega_1 \le 1\}.$$

Definition 2.2. For a large parameter N, we let Ω be the N^{-2}-neighborhood of P. For $j = -N, \dots, N$, we define

$$\theta_j := \Omega \cap \left\{ \frac{j}{N} - \frac{1}{2N} \le \omega_1 \le \frac{j}{N} + \frac{1}{2N} \right\}.$$

Each θ_j is approximately a rectangular box of dimensions $N^{-2} \times N^{-1}$.

We have $\Omega = \bigsqcup_{j=1}^{N} \theta_j$, and we abbreviate this whole decomposition as P_N.

We can now state our first decoupling theorem.

Theorem 2.3 ([14]). *For each $\varepsilon > 0$, for each $2 \le p \le 6$, $D_p(P_N) \le C_\varepsilon N^\varepsilon$.*

In other words, if $2 \le p \le 6$, and if $\hat{f}$ is supported in the N^{-2}-neighborhood of P, then

$$\|f\|^2_{L^p(\mathbb{R}^2)} \le C_\varepsilon N^\varepsilon \sum_{j=1}^{N} \|f_{\theta_j}\|^2_{L^p(\mathbb{R}^2)}. \tag{2.2}$$

This decoupling theorem can be applied to exponential sums, and it implies Theorem 1.6 in the case $d = 1$ and Theorem 1.9 in the case $k = 2$. Theorem 1.6 for a d-dimensional torus follows from a decoupling theorem for the paraboloid in $\mathbb{R}^{d+1}$, and Theorem 1.9 for higher k follows from a decoupling theorem for the moment curve in $\mathbb{R}^k$.

Let us see how this decoupling theorem leads to L^p-estimates for exponential sums. This will help a little to digest the definition of D_p. Suppose we start with an exponential sum using frequencies on the truncated parabola. For $j = -N, \dots, N$, we define the frequency $\omega_j = (\frac{j}{N}, \frac{j^2}{N^2}) \in P$, and we let f be the exponential sum

$$f(x) = \sum_{j=-N}^{N} a_j e^{2\pi i \omega_j x}.$$

If the a_j were chosen randomly, then with high probability, we would have $|f(x)| \sim (\sum_j |a_j|^2)^{1/2}$ for most x. In this random case, we would have $\|f\|_{L^p(B_R)} \sim (\sum_j |a_j|^2)^{1/2} |B_R|^{1/p}$. So the best possible bound we could hope for has the form

$$\|f\|_{L^p(B_R)} \sim \left(\sum_j |a_j|^2\right)^{1/2} |B_R|^{1/p}.$$

Decoupling achieves such a bound up to a factor of N^ε when $2 \le p \le 6$ and R is large enough. This bound in turn implies Theorem 1.6 for $d = 1$ and Theorem 1.9 for $k = 2$.

Here is how to apply decoupling. Note that the frequency ω_j lies in θ_j. In fact, if we write $f = \sum_j f_{\theta_j}$, then $f_{\theta_j} = a_j e^{2\pi i \omega_j x}$. Directly applying Theorem 2.3 does not tell us anything because $\|f_{\theta_j}\|_{L^p(\mathbb{R}^2)}$ is infinite. But with a little technical work, one can prove that a similar estimate holds with L^p-norms on large balls instead of L^p-norms on the whole plane. In particular, if $R \ge N^2$, then

$$\|f\|^2_{L^p(B_R)} \le 100 D_p(P_N)^2 \sum_{j=-N}^{N} \left\|a_j e^{2\pi i \omega_j x}\right\|^2_{L^p(B_R)}.$$

(The extra factor 100 comes from the technical work of passing from $\mathbb{R}^2$ to B_R.) If $p = 6$, then we can plug in $D_6(P_N) \le C_\varepsilon N^\varepsilon$ and simplify everything to get

$$\|f\|_{L^6(B_R)} \le C_\varepsilon N^\varepsilon \left(\sum_{j=1}^{N} |a_j|^2\right)^{1/2} |B_R|^{1/6}.$$

This bound matches the random example above up to the factor $C_\varepsilon N^\varepsilon$, and so in particular it is tight up to this factor. This estimate is the periodic Strichartz estimate for $d = 1$ and the Vinogradov mean value theorem for $k = 2$.

The definition of the decoupling constant D_p was crafted partly to make this computation work. This explains the squares in Definition 2.1.

3. INDUCTION ON SCALES

The definition of decoupling was crafted by Thomas Wolff in his work on local smoothing [56]. He noticed that this definition is well suited for combining information from many scales. The whole field of decoupling leans on this observation. The first example of combining scales is the following lemma, which essentially appears in [56].

Lemma 3.1. *$D_p(P_{N_1 N_2}) \le D_p(P_{N_1}) D_p(P_{N_2})$.*

Let us first discuss why this is significant, and then we will sketch the proof. If we iterate this lemma k times, we get

$$D_p(P_{N_1^k}) \le D_p(P_{N_1})^k. \tag{3.1}$$

Suppose that we are able to find a single number N_1 for which we can prove $D_p(P_{N_1}) \le N_1^{\frac{1}{1000}}$. Then equation (3.1) implies that $D_p(P_N) \le N^{\frac{1}{1000}}$ when N is any power of N_1. This implies the decoupling theorem, Theorem 2.3, with $\varepsilon = \frac{1}{1000}$. For any particular N_1, the decoupling constant $D_p(N_1)$ can be approximated to a given accuracy by a finite computation. This is not immediately obvious from the definition, but it is not that difficult to show. So, in principle, there exists a brute force proof of Theorem 2.3 with $p = 6$ (the most interesting p) and $\varepsilon = \frac{1}{1000}$, where the proof is a giant finite computation to check that $D_6(P_{N_1}) \le N_1^{\frac{1}{1000}}$ for a particular N_1 together with Lemma 3.1.

This situation is very different from the periodic Strichartz estimate, Theorem 1.6, or Vinogradov's mean value theorem, Theorem 1.9. For instance, suppose we somehow knew that Theorem 1.6 holds when $d = 3$ and $N = 10^{10}$. Recall that Theorem 1.6 is an L^p-estimate for periodic solutions to the Schrödinger equation with frequencies at most N. If we somehow knew optimal bounds for periodic solutions with frequency at most 10^{10}, I do not see how we could use that information to say anything about solutions with much larger frequencies, like 10^{1000}.

By switching our point of view from the original problem of periodic Strichartz estimates to the decoupling problem, we make it easier to combine information from different scales. The real proof of the decoupling theorem does not involve a giant brute force computation like we described above. It combines the multiscale idea from Lemma 3.1 with other ideas from the field, and we will discuss it more in the next section.

Next let us talk about the proof of Lemma 3.1. The proof is very short, and it illustrates how the statement of decoupling was crafted to combine information from different scales.

The first observation is that decoupling behaves in a nice way under translations and under linear changes of variable. Suppose that $L : \mathbb{R}^n \to \mathbb{R}^n$ is a linear change of variables, or a translation, or a composition of those. If we start with a decomposition $\Omega = \sqcup\theta$, then we get a new decomposition $L\Omega = \sqcup L\theta$. The first observation is that the new decomposition has the same decoupling constant as the original one:

$$D_p(L\Omega = \sqcup L\theta) = D_p(\Omega = \sqcup\theta). \tag{3.2}$$

If g has Fourier support in Ω and $g = \sum g_\theta$, then we can perform a change of a variables to get a new function $\tilde{g}$ with Fourier support in $L\theta$. Since the Fourier transform behaves in a nice way with respect to linear changes of variables and to translations, it is easy to track how the decoupling constant behaves and check (3.2).

Now we start the proof sketch of Lemma 3.1. Suppose that $\hat{f}$ is supported in Ω, the $(N_1 N_2)^{-2}$-neighborhood of P. This neighborhood is divided into blocks θ of length $(N_1 N_2)^{-1}$, and we need to prove that

$$\|f\|_{L^p}^2 \le D_p(P_{N_1})^2 D_p(P_{N_2})^2 \sum_\theta \|f_\theta\|_{L^p}^2.$$

We prove this bound in two steps. Note that Ω is contained in the N_1^{-2}-neighborhood of P, which we can divide into blocks τ of length N_1^{-1}. By definition of $D_p(P_{N_1})$, we have

$$\|f\|_{L^p}^2 \le D_p(P_{N_1})^2 \sum_\tau \|f_\tau\|_{L^p}^2. \tag{Step 1}$$

The support of $\hat{f}_\tau$ is contained in $\Omega \cap \tau$, which we can decompose as $\Omega \cap \tau = \bigsqcup_{\theta\subset\tau} \theta$.

By the definition of D_p,

$$\|f_\tau\|_{L^p}^2 \le D_p\left(\Omega \cap \tau = \bigsqcup_{\theta\subset\tau} \theta\right)^2 \sum_{\theta\subset\tau} \|f_\theta\|_{L^p}^2.$$

Notice that there are N_2 different θ in each τ. In fact, there is a linear change of variables that takes $\Omega \cap \tau$ to the N_2^{-1}-neighborhood of P and takes each θ to a block of length N_2^{-1}. Therefore, $D_p(\Omega \cap \tau = \bigsqcup_{\theta\subset\tau} \theta) = D_p(P_{N_2})$. Plugging in to the last indented equation, we get

$$\|f_\tau\|_{L^p}^2 \le D_p(N_2)^2 \sum_{\theta\subset\tau} \|f_\theta\|_{L^p}^2. \tag{Step 2}$$

Now if we combine Steps 1 and 2, we get the desired inequality:

$$\|f\|_{L^p}^2 \le D_p(P_{N_1})^2 \sum_\tau \|f_\tau\|_{L^p}^2 \le D_p(N_1)^2 D_p(N_2)^2 \sum_\theta \|f_\theta\|_{L^p}^2.$$

4. IDEAS OF THE PROOF

In this section, we discuss some of the ideas in the proof of the decoupling theorem for the parabola, Theorem 2.3. By now there are actually several proofs of Theorem 2.3 (cf.

[14, 32, 38]). Each proof has some advantages. We will focus on the original proof in [14], but as we go we will try to highlight certain ideas that appear in all of the proofs.

Recall that P is the truncated parabola in $\mathbb{R}^2$. We let Ω be the N^{-2}-neighborhood of P, and we decompose Ω into N pieces θ, which are each approximately rectangles of dimensions $N^{-2} \times N^{-1}$. Suppose $\hat{f}$ is supported on Ω and decompose $f = \sum_\theta f_\theta$. To help illustrate the ideas, we focus on the following corollary of Theorem 2.3.

Corollary 4.1. *If $f = \sum_\theta f_\theta$ as in the last paragraph, and $\|f_\theta\|_{L^\infty(\mathbb{R}^2)} \leq 1$ for every θ, then*

$$\left|U_{N/10}(f) \cap B_{N^2}\right| \leq C_\varepsilon N^{1+\varepsilon}.$$

First, let us give a little context for the numbers that appear in this bound. By the triangle inequality, $|f(x)| \leq \sum_\theta |f_\theta(x)| \leq N$. So $U_{N/10}(f)$ is the region where $|f(x)|$ is biggest. The bound in Corollary 4.1 is sharp, as we can see from the following example.

Example 4.2. Let $f(x)$ be the exponential sum

$$f(x) = \sum_{n=1}^{N} e^{2\pi i(\frac{n}{N}x_1 + \frac{n^2}{N^2}x_2)}.$$

Each f_θ is a single term in the sum, and so $\|f_\theta\|_{L^\infty} = 1$.

We can check directly that $f(mN, 0) = N$ for any integer m because each term in the sum is 1. Also if x lies in a ball of radius $1/100$ around $(mN, 0)$, then each term in the sum has real part more than $1/2$, and so $|f(x)| \geq N/2$. Therefore, $U_{N/10}(f) \cap B_{N^2}$ contains $\sim N$ balls of measure ~ 1 and so itself has measure $\gtrsim N$.

We will give a rough sketch of the proof of Corollary 4.1. The proof of Corollary 4.1 is simpler than the whole proof of Theorem 2.3, but it shows most of the main ideas.

4.1. Orthogonality

Under the hypotheses of Corollary 4.1, it may well happen that $|f_\theta(x)| \sim 1$ for every x and θ. To prove Corollary 4.1, we need to show that for most points $x \in B_{N^2}$, there is a lot of cancellation in the sum $f(x) = \sum_\theta f_\theta(x)$. The most fundamental tool for proving cancellation in Fourier analysis is orthogonality. Since the sets θ are disjoint, the functions f_θ are orthogonal, and so

$$\int_{\mathbb{R}^2} |f|^2 = \sum_\theta \int_{\mathbb{R}^2} |f_\theta|^2.$$

The functions f_θ are exactly orthogonal on $\mathbb{R}^2$. They are also approximately orthogonal over any sufficiently large set. Since the distance between any two (nonadjacent) θ's is at least $1/N$, the functions f_θ are morally orthogonal on any ball of radius N. The rough reason for this approximate orthogonality is the following. Suppose $\omega_1 \in \theta_1$ and $\omega_2 \in \theta_2$. We have to check that the functions $e^{2\pi i \omega_1 x}$ and $e^{2\pi i \omega_2 x}$ are approximately orthogonal on a ball $B_N(x_0)$. The inner product of $e^{2\pi i \omega_1 x}$ and $e^{2\pi i \omega_2 x}$ on $B_N(x_0)$ is

$$\int_{B_N(x_0)} e^{2\pi i \omega_1 x} \overline{e^{2\pi i \omega_2 x}} \, dx = \int_{B_N(x_0)} e^{2\pi i(\omega_1 - \omega_2)x} \, dx.$$

Since $|\omega_1 - \omega_2| \geq 1/N$, the function $e^{2\pi i(\omega_1-\omega_2)x}$ oscillates significantly on $B_N(x_0)$, which causes some cancellation in that integral. This approximate argument suggests the following heuristic.

Heuristic 4.3 (Approximate orthogonality). If B is a square box of side length at least N, then

$$\int_B |f|^2 dx \approx \sum_\theta \int_B |f_\theta|^2 dx.$$

As written, this heuristic is not quite true, but there are more technical substitutes for it. It is morally true, and it helps to imagine it in our proof sketch.

By approximate orthogonality,

$$\int_{B_{N^2}} |f|^2 dx \approx \sum_\theta \int_{B_{N^2}} |f_\theta|^2 dx \leq CN|B_{N^2}| = CN^5.$$

This gives an upper bound

$$\left|U_{N/10}(f) \cap B_{N^2}\right| \leq CN^3. \tag{4.1}$$

To prove Corollary 4.1, we will have to improve the bound N^3 to $N^{1+\varepsilon}$.

So far, we have only used that the rectangles θ are disjoint (and separated by at least $1/N$). We will have to use more information about the θ in order to do better. In fact, if the rectangles θ were laid out along a straight line, then bound (4.1) would be best possible. (We can see that by considering the exponential sum $f(x) = \sum_{n=1}^N e^{2\pi i \frac{n}{N} x_1}$.) To do better, we will have to take advantage of the way the rectangles θ follow the curve of the parabola. In the next two subsections we set up some basic tools that will allow us to take advantage of the curvature of the parabola.

4.2. Multiple scales

We want to study $f = \sum_\theta f_\theta$. We can divide this sum into pieces in various ways. If $M < N$, then we can cover Ω by M rectangles τ of dimensions $M^{-1} \times M^{-2}$. Imagine that M divides N so that each θ is contained in exactly one τ. Then we can write

$$f_\tau = \sum_{\theta \subset \tau} f_\theta \quad \text{and} \quad f = \sum_\tau f_\tau.$$

In order to get better bounds for f, we will consider the functions f_τ at many different intermediate scales (many different choices of M).

The number of $\theta \subset \tau$ is $\frac{N}{M}$, and so $|f_\tau(x)| \leq \frac{N}{M}$. We define $N_\tau = \frac{N}{M}$, which is the number of θ in τ.

Since $|f(x)| \leq \sum_\tau |f_\tau(x)|$, we see that if $|f(x)| \geq N/10$, then $|f_\tau(x)| \geq \frac{N_\tau}{20}$ for at least $M/20$ different τ. This suggests studying $U_{N_\tau/20}(f_\tau)$ for each τ.

We can use orthogonality (Lemma 4.3) to bound $|U_\lambda(f_\tau)|$. By itself, this will not lead to any new bounds. In addition to that, we will study the shape of $U_\lambda(f_\tau)$. Because of their shapes, the sets $U_{N_\tau/20}(f_\tau)$ cannot overlap too much. This geometric input will lead to improvement on the bound for $U_{N/10}(f)$.

4.3. Wave packets

Suppose that $\tau \subset \mathbb{R}^2$ is a rectangle. Suppose that $\hat{f}_\tau$ is supported in τ. Then f_τ itself has a special geometric structure, which is called a wave packet decomposition.

This wave packet decomposition is based on a tiling of $\mathbb{R}^2$ which is in some sense dual to τ. First, let τ^* be the dual rectangle. If τ has dimensions $M^{-1} \times M^{-2}$, then τ^* would have dimensions $M \times M^2$. The axis of τ^* with length M^2 corresponds to the axis of τ with length M^{-2}. Next, let $\mathbb{T}_\tau$ be a tiling of $\mathbb{R}^2$ by rectangles congruent to τ^*.

Heuristic 4.4 (Locally constant heuristic). If $\hat{f}_\tau$ is supported on a rectangle τ with center ω_τ, then for each rectangle $T \in \mathbb{T}_\tau$,

$$f_\tau(x) \approx a_T e^{2\pi i \omega_\tau x},$$

where $a_T \in \mathbb{C}$ is a constant. In particular,

$$|f_\tau(x)| \quad \text{is approximately constant on each rectangle } T \in \mathbb{T}_\tau.$$

According to this heuristic, we can describe f_τ on all of $\mathbb{R}^2$ in the form

$$f_\tau(x) \approx \sum_{T \in \mathbb{T}_\tau} a_T e^{2\pi i \omega_\tau x} \chi_T. \tag{4.2}$$

Here χ_T is the characteristic function of T (or a smoothed out version of it). Each term on the right-hand side is called a wave packet, and equation (4.2) is called the wave packet decomposition of f_τ.

This heuristic is again not quite literally true, but it can be replaced by more technical statements that are true. It is morally true.

One origin of wave packet decompositions is particle–wave duality in quantum mechanics. If $\hat{f}$ is supported in the parabola P, then f satisfies the Schrödinger equation, which describes a quantum mechanical particle moving in a vacuum. (Here we have $f(x_1, x_2)$, and we think of x_2 as the time variable t.) Quantum mechanical particles can behave almost like classical particles for significant time periods. A classical particle in a vacuum moves with constant velocity, tracing out a straight line in space time. A single wave packet describes a quantum mechanical particle behaving almost classically.

Let us try to give some idea why the locally constant heuristic makes sense. The Fourier transformation behaves in a nice way with respect to linear changes of variables and translations. Because of this, it actually suffices to understand the wave packet decomposition when τ is the square $[-1, 1]^2$. Also the wave packet decomposition makes sense in any dimension, and the proofs are basically the same. For simplicity, let us consider dimension 1. Now we have a function $f : \mathbb{R} \to \mathbb{C}$ with $\hat{f}$ supported in $[-1, 1]$. The wave packet decomposition, equation (4.2), says that $f(x)$ is roughly constant on each unit interval. This vague statement is closely related to the the Whittaker–Shannon–Nyquist interpolation theorem, which says that if $\hat{f}$ is supported in $[-1, 1]$, then the whole function $f(x)$ can be recovered from the values $f(n/2)$, with $n \in \mathbb{Z}$. Informally, this suggests that "nothing significant is happening on length scales smaller than 1/2." Here is another way to think about it. Since $\hat{f}$

is supported in $[-1, 1]$,

$$f(x) = \int_{-1}^{1} \hat{f}(\omega) e^{2\pi i \omega x} dx.$$

For $|\omega| \leq 1$, each function $e^{2\pi i \omega x}$ varies slowly, and looks roughly constant on any scale significantly smaller than 1. The function f itself is a linear combination of these slowly varying functions, and so we may hope that f also looks roughly constant at scales smaller than 1.

4.4. Transversality

We are now ready to return to the proof sketch of Corollary 4.1. By bringing into play the wave packet structure of f_τ, we will see how to improve on the bound from Section 4.1, which only used orthogonality. At this point, the curvature of the parabola will come into play.

Recall that each θ is an $N^{-2} \times N^{-1}$ rectangle in the N^{-2}-neighborhood of the truncated parabola P. By the hypotheses of Corollary 4.1, we know that $\|f_\theta\|_{L^\infty(\mathbb{R}^2)} \leq 1$. We want to bound $U_{N/10}(f) \cap B_{N^2}$.

As in Section 4.2, set $M = N^{1/2}$, and cover the parabola with M rectangles τ of dimensions $M^{-1} \times M^{-2}$. Set $N_\tau = N/M = N^{1/2}$, and let us try to understand $U_{N_\tau/20}(f_\tau)$ for each τ. We know that $|f_\tau|$ is locally constant on translates of τ^*, which have dimensions $M \times M^2 = N^{1/2} \times N$. We can also use orthogonality to estimate $\int_{B_N} |f_\tau|^2 dx$ for each ball B_N of radius N. Putting together this information, we conclude that for each B_N, $U_{N_\tau/10}(f_\tau) \cap B_N$ is contained in $\lesssim 1$ translates of τ^*. In other words, on each B_N, each f_τ has only around 1 wave packet of amplitude $\sim N_\tau$.

Now we are ready to take advantage of the curvature of the parabola. Because of the curvature of P, the rectangles τ are oriented in different directions, and so the dual rectangles τ^* point in different directions. On each B_N, $U_{N_\tau/10}(f_\tau)$ is essentially one translate of τ^*. Because all these rectangles point in different directions, they do not overlap very much. The set $U_{N/10}(f)$ should lie in $U_{N_\tau/20}(f_\tau)$ for most τ, and so $U_{N/10}(f) \cap B_N$ has to lie in a constant number of balls of radius $N^{1/2}$. This geometric observation allows us to improve the bound for $U_{N/10}(f)$ beyond what we got from orthogonality alone.

The most effective way to study $U_{N/10}(f)$ on each of these balls of radius $N^{1/2}$ is to repeat the same method, using larger τ's with $M = N^{1/4}$. Continuing in this way through many scales, we eventually see that $U_{N/10}(f) \cap B_N$ has to lie in at most N^ε balls of radius 1. This gives an upper bound

$$\left| U_{N/10}(f) \cap B_{N^2} \right| \leq C_\varepsilon N^{2+\varepsilon}. \tag{4.3}$$

We will call the argument in this section the orthogonality/transversality method, because those are the two main tools that go into it. This argument is essentially due to Bennett–Carbery–Tao [3]. We will discuss their work more in Section 5.1 below. The orthogonality/transversality method improves on just orthogonality, but to prove Corollary 4.1, we will have to improve the bound $N^{2+\varepsilon}$ to $N^{1+\varepsilon}$.

4.5. Induction on scales and transversality together

To get the sharp bound in Corollary 4.1, Bourgain and Demeter combined the ideas from the last subsection with induction on scales (as in Section 3). As in the last subsection, we set $M = N^{1/2}$ and cover the parabola with M rectangles τ of dimensions $M^{-1} \times M^{-2}$. In the orthogonality/transversality argument, we had to understand $U_{N_\tau/20}(f_\tau)$, and we controlled it with the following observation:

(1) Local orthogonality gives an upper bound on $|U_{N_\tau/20}(f_\tau) \cap B_N|$ for each box B_N of side length N.

We can also bring into play induction on scales. After a change of variables, estimating $|U_{N_\tau}(f_\tau)|$ is equivalent to our original problem, Corollary 4.1, but with $N^{1/2}$ rectangular tiles instead of N tiles. So we can also use induction on scales to bound $|U_{N_\tau}(f_\tau)|$.

(2) Induction on scales gives an upper bound on $|U_{N_\tau}(f_\tau) \cap B_{N^2}|$.

The proof of decoupling in **[14]** uses (1) and (2) together. Combining them leads to the sharp bound in Corollary 4.1

When I was first reading the proof of decoupling, I was surprised and even troubled that combining induction on scales with orthogonality/transversality is so powerful. The orthogonality/transversality method gives an interesting but suboptimal bound. Induction on scales by itself does not give any bound. Why do these ingredients become so much stronger when we mix them together?

Initially, the argument even felt fishy to me. Let us look back at points (1) and (2) above. Why should we combine them? If (2) is stronger than (1), then why not just use (2)? If (1) is stronger than (2), then why not just use (1)? I gradually realized that (1) and (2) give different types of information about $U_{N_\tau/20}(f_\tau)$. Neither one is stronger than the other. They are different and give complementary information.

Induction on scales gives information about the total measure of $U_{N_\tau/20}(f_\tau)$ in B_{N^2}. Local orthogonality also implies a bound on the total measure of $U_{N_\tau/20}(f_\tau)$ in B_{N^2}. The bound on the total measure coming from induction is stronger than the bound coming from orthogonality. But (1) is a local bound: it bounds $|U_{N_\tau/20}(f_\tau) \cap B_N|$ for each box of side N. For a small box B_N of side length N, the bound on $|U_{N_\tau/20}(f_\tau) \cap B_N|$ coming from (1) is stronger than the bound coming from (2). Induction on scales controls the total measure of $U_{N_\tau/20}(f_\tau)$, and local orthogonality forces $U_{N_\tau/20}(f_\tau)$ to be rather spread out.

To summarize, the bound (2) from induction gives the best information about the measure of $U_{N_\tau/20}(f_\tau)$. But the bound (1) from local orthogonality gives us additional information about the shape of $U_{N_\tau/20}(f_\tau)$: in particular, for any box B_N of side length N, $U_{N_\tau/20}(f_\tau) \cap B_N$ consists of at most a constant number of $N^{1/2} \times N$ rectangles.

Now we have digested the information that (1) and (2) give us about f_τ, for each τ. The reader may wonder why information about the shape of $U_{N_\tau/20}(f_\tau)$ helps bound the measure of $U_{N/10}(f)$. The point is that it is difficult for different functions f_{τ_1} and f_{τ_2} to be large in the same place. Notice that if $|f(x)| = |\sum_\tau f_\tau(x)|$ is large, then we must have

$|f_\tau(x)|$ large for many different τ at the same point x. If we knew (2) but not (1), it would be possible for $U_{N_\tau/20}(f_{\tau_1})$ and $U_{N_\tau/20}(f_{\tau_2})$ to be equal to each other. But if we use (1) and (2) together, then we get a much stronger estimate for the measure of the intersection $U_{N_\tau/20}(f_{\tau_1}) \cap U_{N_\tau/20}(f_{\tau_2})$.

Here is another way to think about the leverage we get by adding induction on scales to the transversality/orthogonality argument from Section 4.4. Recall that we covered our original tiles θ with M rectangles τ with dimensions $M^{-1} \times M^{-2}$, and we considered f_τ. In the argument from Section 4.4, we started by picking $M = N^{1/2}$. Continuing through the argument, we then used $M = N^{1/4}$, then $M = N^{1/8}$, and so on. At each of these scales, we used the wave packet structure of the f_τ and we took advantage of transversality between the wave packets of the different f_τ's.

When we add in induction on scales, we are implicitly considering many different scales. We started as before by using the scale $M = N^{1/2}$. When we apply induction to a given f_τ, and we unwind the induction, then we are really applying the same argument to f_τ. When we apply the argument to f_τ, it gets decomposed as $f_\tau = \sum_\gamma f_\gamma$, where each γ contains $N^{1/4}$ of the θ in τ. The total number of γ covering all the different τ's is $M = N^{3/4}$, a scale that we never used in Section 4.4. If we fully unwind the inductive argument, it brings into play wave packets at every scale. And it takes advantage of transversality between wave packets at every scale. In some sense, the extra power comes from using transversality at every scale instead of just the special scales $M = N^{1/2}, N^{1/4}, N^{1/8}, \dots$, which were used in Section 4.4.

4.6. Final comments

As we mentioned earlier, there are a number of different proofs of decoupling. In [38], Zane Li gave a new proof of Theorem 2.3 based on Wooley's method of efficient congruencing (cf. [57–59]). In [32], Maldague, Wang, and I gave a new proof of Theorem 2.3 based on ideas from projection theory in geometric measure theory such as Orponen's work [43]. One common feature of all these proofs is to bring into play f_τ with τ at every scale, and to take advantage of some type of transversality at every scale.

Vinogradov's work on the mean value conjecture [52] already has this key feature: it uses f_τ for τ of every scale (after unwinding the induction) and takes advantage of some type of transversality at every scale. Vinogradov's work [52] is the first work I am aware of to take advantage of many scales of τ in estimating an exponential sum. Within harmonic analysis, Wolff's work on local smoothing [56] used this key feature. Bourgain's work on the restriction problem [6] took advantage of the transversality of wave packets of f_τ for a single scale of τ. Wolff's work [56] introduced a version of induction on scales which allowed him to take advantage of transversality of wave packets at every scale. Using this method, he proved a decoupling theorem (for the cone) for large exponents p.

The papers [52] and [56] prove estimates for $|U_\lambda(f)|$, which are sharp when λ takes the largest possible value, but not sharp for smaller λ. For instance, the methods of [52] or [56] could prove Corollary 4.1. The advantage of Theorem 2.3 is to give sharp estimates for $|U_\lambda(f)|$ for every λ. For simplicity, we illustrated the method with $\lambda = N/10$, the largest possible value. The same general method works for every value of λ, although there are some extra wrinkles in the argument.

5. THE KAKEYA CONJECTURE

In this section, we discuss why the restriction conjecture, Conjecture 1.2, remains out of reach in dimension $n \geq 3$. As we saw in the last section, Fourier-analytic estimates in restriction theory are related to understanding how much rectangles pointing in different directions can overlap each other. The Kakeya conjecture is a precise question about how much rectangles pointing in different directions can overlap each other. (Actually, there are several related conjectures.)

Let us formulate the Kakeya conjecture in a way that connects with our discussion of wave packets. Recall that $P \subset \mathbb{R}^n$ denotes the truncated paraboloid:

$$P = \left\{ \omega \in \mathbb{R}^n : \omega_n = \sum_{j=1}^{n-1} \omega_j^2 \text{ and } 0 \leq \omega_n \leq 1 \right\}.$$

Cover P with N^{n-1} rectangular boxes θ of dimensions $\frac{1}{N} \times \cdots \times \frac{1}{N} \times \frac{1}{N^2}$. For each θ, let θ^* denote the dual box with dimensions $N \times \cdots \times N \times N^2$. The long direction of θ^* is equal to the short direction of θ. For each θ, let T_θ denote a translate of θ^*.

The tubes T_θ are related to wave packets that occur in the restriction problem. In the restriction problem, we consider a function f of the form

$$f(x) = \int_P a(\omega) e^{2\pi i \omega x} d\mu_P(\omega). \tag{5.1}$$

The restriction problem asks to estimate $\|f\|_{L^p(\mathbb{R}^n)}$ assuming that $|a(\omega)| \leq 1$ for every ω. We can decompose f as $f = \sum_\theta f_\theta$ where

$$f_\theta(x) = \int_{P \cap \theta} a(\omega) e^{2\pi i \omega x} d\mu_P(\omega). \tag{5.2}$$

Heuristically, each function f_θ is organized into wave packets, and in particular $|f_\theta|$ is locally constant on translates of θ^*. So the tubes T_θ correspond to wave packets of f. Understanding how much the wave packets overlap helps estimate $\|f\|_{L^p}$.

Now we are ready to formulate one version of the Kakeya conjecture.

Conjecture 5.1 (Kakeya conjecture for volume). *Suppose $n \geq 2$. For each θ in the covering of $P \subset \mathbb{R}^n$, let T_θ be a translate of θ^*. Then for each $\varepsilon > 0$,*

$$\left| \bigcup_\theta T_\theta \right| \geq C(n, \varepsilon) N^{-\varepsilon} \sum_\theta |T_\theta|.$$

An argument of Fefferman [25] shows that the restriction conjecture implies the Kakeya conjecture. If a set of tubes $\{T_\theta\}$ is a counterexample to the Kakeya conjecture, we could build a counterexample to the restriction conjecture by choosing f_θ to concentrate on a single wave packet supported on T_θ.

Around 1920, Besicovitch constructed a remarkable example in 2 dimensions where $|\bigcup_\theta T_\theta| \sim \frac{1}{\log N} \sum_\theta |T_\theta|$. Fefferman used this construction in [25] to give a counterexample to a cousin of the restriction conjecture called the ball multiplier problem.

When $n = 2$, Besicovitch's construction turns out to be tight: Davies proved that $|\bigcup_\theta T_\theta| \geq \frac{c}{\log N} \sum_\theta |T_\theta|$. If $n \geq 3$, Besicovitch's construction still works, but we do not know good bounds in the other direction. For example, if $n = 3$, then Davies's method gives only

$$\left|\bigcup_\theta T_\theta\right| \geq \frac{c}{N} \sum_\theta |T_\theta|.$$

Bourgain [6] improved the $\frac{c}{N}$ to $\frac{c}{N^{2/3}}$ and Wolff [53] improved it further to $\frac{c}{N^{1/2}}$. At this point, it becomes very difficult to go further. The best current bound is

$$\left|\bigcup_\theta T_\theta\right| \geq \frac{c}{N^{1/2-\varepsilon_0}} \sum_\theta |T_\theta|,$$

where ε_0 is a small positive constant. The proofs do not make ε_0 explicit, but the best value given by current techniques is probably around $1/1000$. This estimate was proven under an extra assumption by Katz–Laba–Tao [35] and then proven in full generality by Katz–Zahl [36]. The arguments of [6] and [53] are fairly short, about five pages each, but the arguments of [35] and [36] are much more complex, about 50 pages each.

The reason that it is very difficult to improve on $\frac{c}{N^{1/2}}$ has to do with an "almost counterexample" which takes place in $\mathbb{C}^3$. This almost counterexample was first described in [35]. Consider the set

$$H = \{(z_1, z_2, z_3) \in \mathbb{C}^3 : |z_1|^2 + |z_2|^2 - |z_3|^2 = 1\}.$$

This set is a 5-dimensional real manifold in $\mathbb{C}^3$. Its key feature is that it contains many complex lines. Each point of H lies in infinitely many complex lines contained in H. Using this set H as a guide, [35] constructed a set of "complex tubes" T_j with "dimensions" $N \times N \times N^2$, where $|\bigcup_j T_j| = \frac{c}{N^{1/2}} \sum_j |T_j|$. These tubes overlap each other in a very intricate way. They are complex tubes instead of real tubes, and they do not actually all point in different directions, but Wolff's argument from [53] does apply to them. To beat the Kakeya estimate from [53], one has to introduce into the argument some tool that rules out this "almost counterexample." The papers [35] and [36] succeed in doing this, but the tools are much more complex and the quantitative bounds are rather weak. It would be major progress in the field to give a good quantitative improvement to the Kakeya bound in [53], let alone proving the Kakeya conjecture in full.

There is also a stronger version of the Kakeya conjecture which involves L^p-norms. This version is important for the coming subsection.

Conjecture 5.2 (Kakeya conjecture for L^p-norms). *Suppose $n \geq 2$. For each θ in the covering of $P \subset \mathbb{R}^n$, let T_θ be the characteristic function of translate of θ^*, and let $T_{\theta,0}$ be the characteristic function of θ^* itself. The difference is that θ^* is centered at 0, but T_θ could* *have any center. Then for any $\varepsilon > 0$ and any p,*

$$\Big\| \sum_\theta T_\theta \Big\|_{L^p(\mathbb{R}^n)} \le C(n,\varepsilon) N^\varepsilon \Big\| \sum_\theta T_{\theta,0} \Big\|_{L^p(\mathbb{R}^n)}.$$

To digest this formula, notice that $\sum_\theta T_\theta(x)$ is the number of tubes through x. The pth power of the left-hand side is $\int_{\mathbb{R}}^n |\sum_\theta T_\theta(x)|^p dx$. This is large if many points x lie in many tubes from our set of tubes. So the L^p Kakeya conjecture says that not too many points x can lie in many different tubes.

The restriction conjecture implies this stronger version of the Kakeya conjecture, which in turn implies the Kakeya conjecture for volumes, Conjecture 5.1.

Bourgain and Demeter proved a sharp decoupling theorem for the paraboloid $P \subset \mathbb{R}^n$ for all n, which they used to give a sharp Strichartz estimate for tori in all dimensions, Theorem 1.6. One reason this result came as a big surprise has to do with the Kakeya conjecture. The proof of decoupling for the paraboloid involves estimating how much tubes pointing in different directions overlap. When $n = 2$, we know a great deal about how rectangles in different directions overlap, including the Kakeya conjecture for $n = 2$. But when $n \ge 3$, we do not know the Kakeya conjecture. Although there was no formal connection between Kakeya and decoupling for the paraboloid, the Kakeya conjecture still made a sharp decoupling theorem in high dimensions seem out of reach, especially for an approach which is heavily based on estimating the overlaps of tubes pointing in different directions.

5.1. Multilinear Kakeya

The Kakeya-type input into the proof of decoupling is called multilinear Kakeya. It was formulated and proven by Bennett–Carbery–Tao [3]. Multilinear Kakeya is a cousin of Kakeya. The setup is a little different, and we will explain it below, but it still gets at the idea that tubes pointing in different directions cannot overlap too much. Remarkably, Bennett–Carbery–Tao proved sharp multilinear Kakeya estimates in all dimensions. Their proof was simplified in [31] down to a few pages.

The multilinear Kakeya estimate in $\mathbb{R}^n$ is an L^p-type estimate. Suppose that $\ell_{j,a} \subset \mathbb{R}^n$ is a line that makes a small angle with the x_j axis (an angle at most $\frac{1}{100n}$ will do). Let $T_{j,a}$ be the characteristic function of the unit neighborhood of $\ell_{j,a}$—the characteristic function of a tube. Let $B_R \subset \mathbb{R}^n$ denote a cube of side length R.

Theorem 5.3 (Multilinear Kakeya [3]).

$$\int_{B_R} \prod_{j=1}^n \left(\sum_{a=1}^{A_j} T_{j,a}(x) \right)^{\frac{1}{n-1}} dx \le C(n,\varepsilon) R^\varepsilon \prod_{j=1}^n A_j^{\frac{1}{n-1}}.$$

Let us take a moment to digest this estimate. For a fixed j, think of the tubes $\{T_{j,a}\}_{a=1}^{A_j}$ as tubes "in direction j." Now $\sum_{a=1}^{A_j} T_{j,a}(x)$ is the number of tubes in direction j going through x. The integrand is $\prod_{j=1}^n (\sum_{a=1}^{A_j} T_{j,a}(x))^{\frac{1}{n-1}}$, which is big if x lies in many tubes from each direction. So the integral on the left-hand side measures how many points x lie in many tubes from each direction. The multilinear Kakeya inequality says that there cannot be too many points which lie in many tubes from each direction.

The exponent $\frac{1}{n-1}$ makes the inequality sharp in two natural examples: the example when all the tubes go through the origin and an example when the tubes are arranged in a rectangular grid. The exponent $\frac{1}{n-1}$ is the most important, and this bound implies sharp estimates with any other exponent.

It makes sense to compare Theorem 5.3 with the L^p Kakeya conjecture, Conjecture 5.2. The main difference between them is that in the multilinear Kakeya theorem, the integrand is a product of n factors, and we assume that the n factors are transverse to each other in a strong sense. The word "multilinear" refers to this product structure.

Theorem 5.3 is also proven by induction on scales. In the case that the tubes $T_{j,a}$ are exactly parallel to the x_j axis (for all j and a), Theorem 5.3 reduces to the Loomis–Whitney inequality [39], which we will recall a moment. The general case of multilinear Kakeya is proven by applying Loomis–Whitney at many scales (cf. [31]). The multilinear Kakeya inequality grew out of work by Bennett–Carbery–Wright on nonlinear versions of the Loomis–Whitney inequality [4].

For completeness let us recall the statement of the Loomis–Whitney inequality. One version is an inequality for integrals that looks reminiscent of Hölder's inequality. Suppose that $\pi_j : \mathbb{R}^n \to \mathbb{R}^{n-1}$ are projections onto the coordinate hyperplanes. Then the Loomis–Whitney inequality says

$$\int_{\mathbb{R}^n} \prod_{j=1}^{n} f_j(\pi_j(x))^{\frac{1}{n-1}} dx \le \prod_{j=1}^{n} \|f_j\|_{L^1(\mathbb{R}^{n-1})}^{\frac{1}{n-1}}.$$

There is a geometric corollary of this inequality which may feel more intuitive. Suppose that $U \subset \mathbb{R}^n$ is an open set, and that the projection of U onto every coordinate hyperplane has $(n-1)$-volume at most A. Then U has n-volume at most $A^{\frac{n}{n-1}}$. The case $n = 2$ is straightforward, but the case $n = 3$ is quite subtle. It is one of my favorite problems to think through with students studying analysis.

When multilinear Kakeya was first proven, it seemed natural and remarkable, but it was not clear just how much impact it would have in restriction theory. In [3], Bennett, Carbery, and Tao [3] formulated and proved an interesting multilinear restriction conjecture. They proved multilinear restriction by using multilinear Kakeya at many scales. But it was not clear whether these multilinear estimates would lead to bounds on problems that were not multilinear, such as the original restriction conjecture.

The paper [17] used these multilinear estimates to prove new partial results about the restriction problem. It introduced a technique called the broad/narrow method which can sometimes reduce linear estimates to multilinear estimates.

Remarkably, sharp decoupling theorems follow from multilinear Kakeya, even though there is nothing obviously multilinear about the statement of decoupling. This was one of the big surprises in the development of the field. The original Kakeya problem is much harder than multilinear Kakeya. The original restriction problem is much harder than multilinear restriction. There is also a multilinear version of decoupling. A key fact that makes decoupling accessible is that the original decoupling problem is EQUIVALENT to multilinear decoupling. This equivalence was noticed implicitly by Bourgain in [10], and explicitly by Bourgain and Demeter in [14]. Because of this connection between decoupling and multilinear decoupling, we can prove sharp estimates for the original decoupling problem using multilinear Kakeya, even though we do not know sharp estimates for the original Kakeya problem.

The connection between decoupling and multilinear decoupling is another important application of induction on scales. It is based on the broad/narrow method. Because of considerations of space, we do not give a detailed description here.

When multilinear Kakeya first appeared, it seemed like it might not have very many applications in harmonic analysis compared with the original Kakeya conjecture. But now the situation has reversed: multilinear Kakeya currently has more applications in harmonic analysis than the original Kakeya conjecture would have even if we knew it.

6. APPLICATIONS OF DECOUPLING IN HARMONIC ANALYSIS

Decoupling theory has led to the solutions of several longstanding problems in harmonic analysis. We give three examples here. Each of these problems seemed out of reach a decade ago.

6.1. The helical maximal function

Hardy and Littlewood introduced their maximal function in the early 20th century. The Hardy–Littlewood maximal function is based on averages over balls. If $f : \mathbb{R}^n \to \mathbb{R}$, then the average value of f on the ball of radius r around x can be written as

$$\frac{1}{|B_r|} \int_{B_r} f(x+y)dy.$$

The Hardy–Littlewood maximal function is defined by taking the supremum over r,

$$Mf(x) = \sup_r \frac{1}{|B_r|} \int_{B_r} |f(x+y)| dy.$$

Hardy and Littlewood proved that $\|Mf\|_{L^p} \le C(p,n)\|f\|_{L^p}$ for all $p > 1$ but not for $p = 1$.

In the 1960s, Stein introduced a spherical maximal function [48]. Suppose $f : \mathbb{R}^n \to \mathbb{R}$. The average value of f on the sphere of radius r around x can be written as

$$\frac{1}{|S^{n-1}|} \int_{S^{n-1}} f(x + r\theta)d\theta.$$

The spherical maximal function is defined by taking the supremum over r,

$$M_S f(x) := \sup_{r>0} \frac{1}{|S^{n-1}|} \int_{S^{n-1}} |f(x + r\theta)| d\theta. \tag{6.1}$$

For $n \ge 3$, Stein proved that in $\mathbb{R}^n$, $\|M_S f\|_{L^p} \le C(n,p)\|f\|_{L^p}$ for all $p > \frac{n}{n-1}$, but not for $p \le \frac{n}{n-1}$. He conjectured that the same was true for $n = 2$. The case $n = 2$ was proven by Bourgain in [5].

Stein's result was striking for the following reason. A function $f \in L^p$ need only be defined almost everywhere. It may be undefined or infinite on a lower-dimensional submanifold like a sphere. So for a particular x and r, the integral on the right-hand side of

(6.1) may be infinite or undefined. Nevertheless, if $f \in L^p$ for $p > \frac{n}{n-1}$, Stein showed that the spherical maximal function is actually defined for almost every x. The curvature of the sphere is crucial in this estimate. The spherical maximal function and the restriction conjecture were two fundamental connections between curvature and harmonic analysis that Stein investigated.

The spherical maximal function can be generalized by replacing the sphere by other curved submanifolds. Many of the corresponding problems are still open. After the sphere and circle, the next most fundamental case to look at is the case of the moment curve in $\mathbb{R}^n$. Here is the definition. Consider the moment curve parametrized by $\gamma(t) = (t, t^2, t^3, \dots, t^n)$. We can build an averaging operator based on the moment curve as follows. Suppose $f : \mathbb{R}^n \to \mathbb{R}$ and define

$$Af(x) = \int_0^1 f\big(x + \gamma(t)\big)dt.$$

Geometrically, $Af(x)$ is the average value of f on the translate of the moment curve starting at x. Next we can consider different scalings of the moment curve. Define

$$A_r f(x) = \int_0^1 f\big(x + r\gamma(t)\big)dt.$$

Geometrically, $A_r f(x)$ is the average value of f on a moment curve which has been scaled by a factor of r and then translated to start at x. Finally, we can define the helical maximal function by taking the maximum of these averages over different choices of r,

$$M_{\text{hel}} f(x) := \sup_{r>0} A_r f(x).$$

In analogy with the work of Stein and Bourgain on the circular maximal function, it is natural to ask when $\|M_{\text{hel}} f\|_{L^p(\mathbb{R}^n)} \lesssim \|f\|_{L^p(\mathbb{R}^n)}$. In [45], Pramanik and Seeger connected this problem (when $n = 3$) to the decoupling problem for the cone, which Wolff had recently introduced in [56]. In [14], Bourgain and Demeter gave sharp estimates for the decoupling for the cone, but that by itself is not enough to give sharp estimates for the helical maximal function. Recently, Ko–Lee–Oh [37] and Beltran–Guo–Hickman–Seeger [2] independently proved the sharp L^p-estimate for the helical maximal function when $n = 3$.

Theorem 6.1 ([37] and [2]). *For $p > 3$ $\|M_{\text{hel}} f\|_{L^p(\mathbb{R}^3)} \le C(p)\|f\|_{L^p(\mathbb{R}^3)}$.*
If $p \le 3$, this estimate does not hold.

The case of higher dimensions remains open, although both groups have proven interesting estimates on helical averages in other dimensions as well.

6.2. Pointwise convergence for the Schrödinger equation

Consider the initial value problem for the linear Schrödinger equation in $\mathbb{R}^d \times \mathbb{R}$,

$$\partial_t u(x,t) = i\Delta u(x,t), u(x,0) = u_0(x).$$

We can write down the solution u with the help of the Fourier transform. If the initial data u_0 is rough, then the solution $u(x,t)$ will be rough also. In this situation, $u(x,t)$ will solve the differential equation in a distributional sense, even if $u(x,t)$ is discontinuous.

Carleson [20] raised the following problem.

Question 6.2. What is the smallest s so that whenever $u_0 \in H^s(\mathbb{R}^d)$ and $u(x,t)$ is a distributional solution to the Schrödinger equation on $\mathbb{R}^d \times \mathbb{R}$ with initial data $u_0(x)$, then $\lim_{t\to 0} u(x,t) = u_0(x)$ for almost every $x \in \mathbb{R}^d$?

This question helps describe how regular distributional solutions to the Schrodinger equation are. This question is actually a cousin of the restriction problem and the Strichartz estimate, although we will have to rewrite it a little bit to see how they are connected.

Because u solves the Schrödinger equation, the spacetime Fourier transform $\hat{u}$ is supported on the infinite paraboloid. One has to prove some estimates about how badly $u(x,t)$ oscillates for small t. After some standard arguments (scaling and Littlewood–Paley), one can reduce these estimates to the case that $\hat{u}$ is supported on the truncated paraboloid P and normalize so that $\|u_0\|_{L^2(\mathbb{R}^d)} = 1$. Now consider $U_\lambda(u) \subset \mathbb{R}^d \times \mathbb{R}$. The Strichartz estimates give sharp bounds for $|U_\lambda(u)|$ in terms of λ. A small variation gives sharp estimates for $|U_\lambda(u) \cap [0,R]^{d+1}|$ in terms of λ and R. Now let $\Pi_{\mathbb{R}^d}(x,t) = x$ be the projection from spacetime to space. Carleson's pointwise convergence problem is related to the following question about the size of $\Pi_{\mathbb{R}^d}(U_\lambda(u))$:

Question 6.3. Suppose that $\hat{u}$ is supported on the truncated paraboloid P. Let $u_0(x) = u(x,0)$, and suppose that $\|u_0\|_{L^2(\mathbb{R}^d)} = 1$. For any given λ, R, estimate the maximum possible size of $|\Pi_{\mathbb{R}^d}(U_\lambda u \cap [0,R]^{d+1})|$.

The key difference between this problem and the Strichartz inequality is we have to estimate the d-volume of the projection of $U_\lambda(u)$ instead of the $(d+1)$-volume of $U_\lambda(u)$ itself. This general problem is still open. However, we do understand a special case, which is sufficient to resolve the pointwise convergence problem. Here is the special case:

Question 6.4. Suppose that $\hat{u}$ is supported on the truncated paraboloid P. Let $u_0(x) = u(x,0)$, and suppose that $\|u_0\|_{L^2(\mathbb{R}^d)} = 1$. Suppose that $|\Pi_{\mathbb{R}^d}(U_\lambda u \cap [0,R]^{d+1})| \geq cR^d$. How big can λ be?

As a first example, suppose that u_0 is a smooth bump function approximating a constant function on $[0,R]^d$. Because $\|u_0\|_{L^2} = 1$, we have $|u_0(x)| \sim R^{-d/2}$ on most of $[0,R]^d$. In this case, $u(x,t)$ is roughly constant on $[0,R]^{d+1}$, and so λ is also $\sim R^{-d/2}$.

This first example is not the worst case. In case $d = 1$, the worst case example was found by Dahlberg–Kenig [21]. It is given when $u(x,t)$ is a single wave packet, essentially supported on a tilted rectangle with dimensions $R^{1/2} \times R$.

In this case, $u_0(x)$ is essentially supported on an interval of length $R^{1/2}$, and so $|u_0(x)| \sim R^{-1/4}$ on this interval. Then $|u(x,t)| \sim R^{-1/4}$ on the whole wave packet, and we get $\lambda \sim R^{-1/4}$. Carleson [20] had showed previously that this value of λ is optimal. This settles Carleson's problem in the case $d = 1$, but the case of higher dimensions was open for 30+ years.

In higher dimensions, we can adapt the Dahlberg–Kenig example by taking many parallel wave packets with disjoint projections onto $\mathbb{R}^d$. This gives $\lambda = R^{-\frac{d}{2}+\frac{1}{4}}$. For a long time, it seemed plausible that this construction was sharp in any dimension. In the last decade, mathematicians found other much more intricate examples. The first was given by Bourgain [11] and there were several improvements leading up to [12] (cf. also [40]). The last example gives $\lambda = R^{-\frac{d}{2}+\frac{d}{2d+2}}$.

This last example turns out to be sharp. The case $d = 2$ was proven in [22] and the case of all d was proven in [23]. Even for $d = 2$, the proof in [23] is simpler. The key ingredient in these proofs is decoupling. Decoupling is applied in a somewhat indirect way. In particular, the proofs use decoupling many times at different scales.

Theorem 6.5 ([12, 23]). *Suppose that $s > \frac{d}{2d+2}$. If $u_0 \in H^s(\mathbb{R}^d)$, and $u(x,t)$ is a (distributional) solution to the linear Schrödinger equation with initial data u_0. Then* $\lim_{t\to 0} u(x,t) = u_0(x)$ *for almost every x.*

Suppose that $s < \frac{d}{2d+2}$. There exists a function $u_0 \in H^s(\mathbb{R}^d)$ with the following bad behavior. Let $u(x,t)$ be the (distributional) solution to the linear Schrödinger equation with initial data u_0. For this function, $\limsup_{t\to 0} |u(x,t)| = +\infty$ *for almost every $x \in \mathbb{R}^d$.*

6.3. The local smoothing problem

Wolff introduced decoupling in his work on the local smoothing problem [56]. This problem is an estimate about solutions to the wave equation.

Suppose that $u(x,t)$ solves the wave equation $\partial_t^2 u = \Delta u$, with $x \in \mathbb{R}^d$ and $t \in \mathbb{R}$, and with initial data $u(x,0) = u_0(x)$ and $\partial_t u(x,0) = u_1(x)$. The local smoothing problem concerns Sobolev-type bounds for the wave equation: Given bounds on some Sobolev norms of u_0 and u_1, what bounds can we prove on the Sobolev norms of u?

To make things simple and concrete, let us suppose that $u_1 = 0$ and that $\hat{u}_0$ is supported in a ball of radius N in frequency space. Then we would like to find all bounds of the form

$$\left\|u(x,t)\right\|_{L^p(\mathbb{R}^d\times[0,1])} \le C N^\alpha \left\|u_0(x)\right\|_{L^p(\mathbb{R}^d)}.$$

The word "local" in "local smoothing" refers to the time interval $[0,1]$. A global estimate would give a bound on $\mathbb{R}^d \times \mathbb{R}$, whereas a fixed-time estimate would give a bound for $\mathbb{R}^d \times \{t_0\}$ for some fixed t_0 (such as $t_0 = 1$). Global in time estimates, local in time estimates, and fixed time estimates are all interesting. Sharp fixed time estimates were established by Peral [44] and Miyachi [41] around 1980. The word "smoothing" in "local smoothing" is because the power of α in the local in time estimates is smaller than the power in a fixed time estimate.

In [47], Sogge formulated the local smoothing conjecture, and he proved the first local smoothing estimates improving upon the α given by the fixed time estimates.

Conjecture 6.6 ([47]). *Suppose $d \ge 2$. Suppose that $u(x,t)$ solves the wave equation in $\mathbb{R}^d \times \mathbb{R}$, with initial data $u(x,0) = u_0(x)$ and $\partial_t u(x,0) = 0$. Suppose that $\hat{u}_0$ is supported in the ball of radius N. Then, if $2 \le p \le \frac{2d}{d-1}$, then*

$$\left\|u(x,t)\right\|_{L^p(\mathbb{R}^d\times[0,1])} \le C(d,\varepsilon) N^\varepsilon \|u_0\|_{L^p(\mathbb{R}^d)}.$$

If $p > \frac{2d}{d-1}$, *then*

$$\left\|u(x,t)\right\|_{L^p(\mathbb{R}^d\times[0,1])} \le C(d,\varepsilon)N^{\frac{d-1}{2}-\frac{d}{p}+\varepsilon}\|u_0\|_{L^p(\mathbb{R}^d)}.$$

The case $p = \frac{2d}{d-1}$ is the critical exponent, and it implies all the other estimates for a given dimension d. In [56], Wolff introduced decoupling and used it to show that Conjecture 6.6 holds when $d = 2$ and $p > 74$. Wolff also observed that the local smoothing conjecture in dimension d implies the Kakeya conjecture in dimension d, by adapting Fefferman's argument from [25]. Therefore, the full conjecture remains out of reach for all $d \geq 3$.

In [14], Bourgain and Demeter proved a complete decoupling theorem for the cone. This implies that Conjecture 6.6 holds in $\mathbb{R}^d$ for all $p > \frac{2(d+1)}{d-1}$. In particular, when $d = 2$, local smoothing holds for all $p > 6$. When $d = 2$, the critical exponent for local smoothing is $p = 4$.

In [33], Wang, Zhang, and I proved the local smoothing conjecture when $d = 2$ for $p = 4$ (and hence for all p). The proof of local smoothing does not use decoupling per se, but it is strongly influenced by the ideas in the proof of decoupling, including induction on scales.

7. FRUSTRATIONS, LIMITATIONS, AND OPEN PROBLEMS

Decoupling and the ideas in the proof of decoupling have led to solutions of many problems that seemed out of reach a decade ago. The proof is elegant in some ways. In some ways, it feels like a proof "from the book." It is essentially self-contained and it is not that long. But in other ways the proof is frustrating. (Actually, there are now several proofs, and they have various advantages and disadvantages. The community is actively trying to understand decoupling from different angles, and in five or ten years, we may have a different sense of the essential ingredients.)

In this section, I discuss some of my frustrations with the proof of decoupling, some limitations of the method, and some open problems.

7.1. Too much induction

On the one hand, induction on scales is the central idea in the proof of decoupling. On the other hand, the heavy reliance on induction makes the proof difficult to read. A lot of important stuff is happening inside the induction.

For example, as we discussed in Section 4.5, I think that the leverage in the proof of decoupling comes from taking advantage of the transversality of wave packets of every scale, not just at a few scales. For instance, suppose we cut the parabola P into M rectangles τ with $M = N^{5/16}$. The proof of decoupling takes advantage of the transversality between the wave packets at this scale, but it is not easy to locate the place in the argument where this transversality is used because it is a little bit buried in the induction. Even though I have thought through the proof many times, it took me a good while to locate where wave packets at this particular scale are used.

Reading through the full proof of decoupling for the paraboloid, we see many different tricks for taking advantage of induction on scales. Loomis–Whitney is used at many scales to prove multilinear Kakeya. Multilinear Kakeya is used at many scales in the argument in Section 4.4. The key induction on scales is described in Section 4.5. Induction on scales is also used in a different way to go back and forth between multilinear estimates and the original linear estimates, as we discussed in Section 5.1. Finally, many applications of decoupling actually use decoupling many times at different scales, as in Section 6.2.

We might look at this and feel that using multiple scales is a craft with many aspects. But we might also start to get the feeling that this is too many different tricks, and that we should try to take advantage of many scales in a more systematic way.

7.2. What does decoupling say about the shapes of superlevel sets?

Decoupling gives an estimate for $\|f\|_{L^p}$ or for the measure of the superlevel sets $U_\lambda(f)$. Besides the measure of the sets $U_\lambda(f)$, decoupling also seems to be connected to the shape of the superlevel sets $U_\lambda(f)$. Looking back through our discussion in Sections 4.4 and 4.5, the shape of $U_\lambda(f)$ plays an important role, even though the final estimate only concerns the measure of $U_\lambda(f)$. In particular, during the argument, we make use of some information about $|U_\lambda(f_\tau) \cap B|$ for various balls B and for various τ. This information roughly describes how much the set $U_\lambda(f)$ can concentrate in balls. The shape of $U_\lambda(f)$ is also connected to some applications of decoupling, such as the work on Carleson's pointwise convergence problem discussed in Section 6.2.

Perhaps the shape of $U_\lambda(f)$ should be a more central character in decoupling. What is the full information about the shape of $U_\lambda(f)$ which the proof method of decoupling gives? Unfortunately this question is quite vague. There are many possible ways we could describe the shape of $U_\lambda(f)$, and it is not clear which language to use. But it is possible that discussing the shape of $U_\lambda(f)$ systematically throughout the whole story might make the arguments clearer or even stronger…

Here is one question from the harmonic analysis literature that has to do with the shape of $U_\lambda(f)$. We consider a measure μ supported on a large ball $B_R \subset \mathbb{R}^n$ which obeys the Frostman condition

$$\mu\big(B_r(x)\big) \le r^\alpha. \tag{7.1}$$

Here $0 < \alpha < n$ is fixed.

Question 7.1. As in the restriction problem or the Strichartz inequality, suppose that $f : \mathbb{R}^n \to \mathbb{C}$ is given by

$$f(x) = \int_P a(\omega) e^{2\pi i \omega x} d\mu_P(\omega).$$

For a given n and α, what is the best exponent γ in the inequality

$$\|f\|_{L^2(d\mu)} \le C R^\gamma \|a\|_{L^2(P)},$$

among all functions f as above and all measures μ obeying the Frostman condition (7.1) with exponent α.

In dimension $n = 2$, this question is well understood for all α by work of Mattila and Wolff, cf. [55]. But for $n \geq 3$, the problem is far from understood. In [23], Du and Zhang gave a sharp answer for $\alpha = n - 1$. No other cases are fully understood. The Du–Zhang estimate for $\alpha = n - 1$ is closely related to the solution of Carleson's problem on pointwise convergence for solutions of the Schrödinger equation. Decoupling and multilinear restriction are the essential tools in their approach, and they use decoupling at many different scales.

How much can the method of decoupling tell us about other values of α? Is there anything fundamentally special about $\alpha = n - 1$? Also the Frostman condition (7.1) can be replaced by other conditions, by replacing the function r^α by other functions of r. This would lead to other kinds of estimates about the shape of $U_\lambda(f)$.

7.3. Limitations of the information used in the proof

In the statement of decoupling, we assume that $\hat{f}$ is supported in Ω, and we try to bound $\|f\|_{L^p}$ in terms of some information about $\|f_\theta\|_{L^p}$ for all the θ in the decomposition of Ω. If we look through the proof and check where the hypothesis $\operatorname{supp}(\hat{f}) \subset \Omega$ is used, we find that it is used only in fairly simple ways.

In the course of the proof, we consider f_τ for many different rectangles τ. The proof relies crucially on two facts. The first is the locally constant heuristic:

$$\text{For each } \tau,\ |f_\tau| \text{ is approximately constant on each translate of } \tau*. \tag{7.2}$$

The second is the local orthogonality heuristic. If τ is a rectangle, and γ are smaller rectangles contained in τ, and if nonadjacent γ are separated by at least s, then

$$\int_B |f_\tau|^2 \approx \sum_{\gamma \subset \tau} \int_B |f_\gamma|^2, \tag{7.3}$$

whenever B is a cube whose side length is longer than s^{-1}.

The Fourier support properties of the different functions f, f_τ, f_θ are only really used to justify these two heuristics. These two heuristics are consequences of the Fourier support hypotheses, but they do not encode all the information given by the Fourier support hypotheses.

This raises the question: Which theorems of restriction theory can we prove only using the locally constant heuristic and local orthogonality? Which theorems require us to use the Fourier support hypothesis in some other way?

The proofs of the different decoupling theorems essentially only use these two properties. (I say essentially because some of the proofs also involve some pigeonholing of wave packets.) Also, the strongest current work on the restriction conjecture only uses these two properties. It is possible that the full restriction conjecture might follow only using these two properties.

In restriction theory there are currently very few examples of techniques for exploiting the Fourier support of f that use Fourier support information in some other way. (One example is to use the even integer trick, Lemma 1.7, together with number-theoretic input. An interesting recent example of this approach is the work on Strichartz-type estimates for the periodic Airy equation by Hughes–Wooley [34].)

However, there are a number of problems in restriction theory where I strongly doubt that these two properties are sufficient to give full answers. One example is the the problem of estimating the L^p-norms of the functions

$$f_{k,N}(x) = \sum_{a=1}^{N} e^{2\pi i a^k x}.$$

As we discussed in Section 1.2, the L^p-norms of $f_{k,N}$ are well understood for $k = 2$ and wide open for $k \geq 3$. When $k = 2$, the different proofs all use some information besides the locally constant heuristic and local orthogonality. I believe the sharp estimates for $k = 2$ cannot be proven by an argument using only those two properties.

There is an interesting generalization of this L^p-problem which I think is a good test case for going beyond the locally constant property and local orthogonality. As we mentioned in Section 1.2, $\|f_{2,N}\|_{L^4([0,1])} \leq C_\varepsilon N^{1/2+\varepsilon}$.

Question 7.2. We consider a sequence of frequencies ω_a, with $a = 1, \dots, N$, which behave approximately like the squares a^2, in the sense that

$$\omega_{a+1} - \omega_a \sim a \quad \text{and} \quad (\omega_{a+1} - \omega_a) - (\omega_a - \omega_{a-1}) \sim 1.$$

For such a choice of frequencies ω_a, define

$$f(x) = \sum_{a=1}^{N} e^{2\pi i \omega_a x}.$$

Estimate $\|f\|_{L^4([0,1])}$. Is it true that $\|f\|_{L^4([0,1])} \leq C_\varepsilon N^{1/2+\varepsilon}$?

As far as I know, it is possible that $\|f\|_{L^4([0,1])} \leq C_\varepsilon N^{1/2+\varepsilon}$ in this much more general setting. However, the proofs that work for $f_{2,N}$ do not generalize to this setting. And the method of decoupling can prove only limited things. In [27], Fu, Maldague, and I explored how much we can say about this question using ideas of decoupling theory. As part of that investigation, we explain the version of the locally constant property which appears in this setting, which goes back to Bourgain's work [7] on Montgomery's conjecture. The main theorems of [27] give sharp L^p-estimates for much shorter sums, namely sums of length $\sim N^{1/2}$. For these shorter sums, the locally constant property and the methods of decoupling are effective. But for longer sums, they seem much less effective, and I believe that some different tools are needed.

Question 7.2 is also related to a question of Erdős about sumsets of convex sets. A sequence ω_a is called convex if $(\omega_{a+1} - \omega_a) - (\omega_a - \omega_{a-1}) > 0$ for all a. Notice that the set of frequencies in Question 7.2 is a convex sequence. If A is a convex sequence, then Erdős conjectured that $|A + A| \geq c_\varepsilon |A|^{2-\varepsilon}$. Here $A + A$ denotes all sums of two elements of A. This conjecture is open. There is interesting recent work on it by Schoen and Shkredov [46], who proved that $|A + A| \geq c_\varepsilon |A|^{1.6-\varepsilon}$. This beats the previous best estimate $|A|^{1.5}$, which had stood for a long time. If A denotes the frequencies in Question 7.2, and if indeed $\|f\|_{L^4([0,1])} \leq C_\varepsilon N^{1/2+\varepsilon}$, then it would follow that $|A + A| \geq c_\varepsilon |A|^{2-\varepsilon}$. The best bound I could prove using the methods of decoupling gives $|A + A| \geq c|A|^{1.5}$. Work in combinatorics such as [46] may give clues on how to go further in problems like Question 7.2.

ACKNOWLEDGMENTS

The author is supported by a Simons Investigator award.

REFERENCES

[1] G. I. Arkhipov, V. N. Chubarikov, and A. A. Karatsuba, *Trigonometric sums in number theory and analysis*. de Gruyter Exp. Math. 39, Walter de Gruyter GmbH & Co. KG, Berlin, 2004. Translated from the 1987 Russian original.

[2] D. Beltran, J. Hickman, S. Guo, and A. Seeger, Sharp L^p bounds for the helical maximal function. 2021, arXiv:2102.08272.

[3] J. Bennett, A. Carbery, and T. Tao, On the multilinear restriction and Kakeya conjectures. *Acta Math.* **196** (2006), no. 2, 261–302.

[4] J. Bennett, A. Carbery, and J. Wright, A non-linear generalisation of the Loomis–Whitney inequality and applications. *Math. Res. Lett.* **12** (2005), no. 4, 443–457.

[5] J. Bourgain, Averages in the plane over convex curves and maximal operators. *J. Anal. Math.* **47** (1986), 69–85.

[6] J. Bourgain, Besicovitch type maximal operators and applications to Fourier analysis. *Geom. Funct. Anal.* **1** (1991), no. 2, 147–187.

[7] J. Bourgain, Remarks on Montgomery's conjectures on Dirichlet sums. In *Geometric aspects of functional analysis (1989–1990)*, pp. 153–165, Lecture Notes in Math. 1469, Springer, Berlin, 1991.

[8] J. Bourgain, Fourier transform restriction phenomena for certain lattice subsets and applications to nonlinear evolution equations. I. Schrödinger equations. *Geom. Funct. Anal.* **3** (1993), no. 2, 107–156.

[9] J. Bourgain, On the dimension of Kakeya sets and related maximal inequalities. *Geom. Funct. Anal.* **9** (1999), no. 2, 256–282.

[10] J. Bourgain, Moment inequalities for trigonometric polynomials with spectrum in curved hypersurfaces. *Israel J. Math.* **193** (2013), no. 1, 441–458.

[11] J. Bourgain, On the Schrödinger maximal function in higher dimension. *Proc. Steklov Inst. Math.* **280** (2013), no. 1, 46–60.

[12] J. Bourgain, A note on the Schrödinger maximal function. *J. Anal. Math.* **130** (2016), 393–396.

[13] J. Bourgain, Decoupling, exponential sums and the Riemann zeta function. *J. Amer. Math. Soc.* **30** (2017), no. 1, 205–224.

[14] J. Bourgain and C. Demeter, The proof of the l^2 decoupling conjecture. *Ann. of Math. (2)* **182** (2015), no. 1, 351–389.

[15] J. Bourgain and C. Demeter, Decouplings for curves and hypersurfaces with nonzero Gaussian curvature. *J. Anal. Math.* **133** (2017), 279–311.

[16] J. Bourgain, C. Demeter, and L. Guth, Proof of the main conjecture in Vinogradov's mean value theorem for degrees higher than three. *Ann. of Math. (2)* **184** (2016), no. 2, 633–682.

[17] J. Bourgain and L. Guth, Bounds on oscillatory integral operators based on multilinear estimates. *Geom. Funct. Anal.* **21** (2011), no. 6, 1239–1295.

[18] J. Bourgain and N. Watt, Mean square of zeta function, circle problem and divisor problem revisited. 2017, arXiv:1709.04340.

[19] N. Burq, P. Gérard, and N. Tzvetkov, Strichartz inequalities and the nonlinear Schrödinger equation on compact manifolds. *Amer. J. Math.* **126** (2004), no. 3, 569–605.

[20] L. Carleson, Some analytic problems related to statistical mechanics. In *Euclidean harmonic analysis (Proc. Sem., Univ. Maryland, College Park, Md, 1979)*, pp. 5–45, Lecture Notes in Math. 779, 1979.

[21] B. Dahlberg and C. Kenig, A note on the almost everywhere behavior of solutions to the Schrödinger equation. In *Harmonic analysis (Minneapolis, MN, 1981)*, pp. 205–209, Lecture Notes in Math. 908, Springer, Berlin–New York, 1982.

[22] X. Du, L. Guth, and X. Li, A sharp Schrödinger maximal estimate in $\mathbb{R}^2$. *Ann. of Math. (2)* **186** (2017), no. 2, 607–640.

[23] X. Du and R. Zhang, Sharp L^2 estimates of the Schrödinger maximal function in higher dimensions. *Ann. of Math. (2)* **189** (2019), no. 3, 837–861.

[24] Z. Dvir, On the size of Kakeya sets in finite fields. *J. Amer. Math. Soc.* **22** (2009), no. 4, 1093–1097.

[25] C. Fefferman, The multiplier problem for the ball. *Ann. of Math. (2)* **94** (1971), 330–336.

[26] C. Fefferman, A note on spherical summation multipliers. *Israel J. Math.* **15** (1973), 44–52.

[27] Y. Fu, L. Guth, and D. Maldague, Decoupling inequalities for short generalized Dirichlet sequences. 2021, arXiv:2104.00856.

[28] S. Guo, Z. Li, P.-L. Yung, and P. Zorin-Kranich, A short proof of ℓ^2 decoupling for the moment curve. *Amer. J. Math.* **143** (2021), no. 6, 1983–1998.

[29] S. Guo and R. Zhang, On integer solutions of Parsell–Vinogradov systems. *Invent. Math.* **218** (2019), no. 1, 1–81.

[30] S. Guo and P. Zorin-Kranich, Decoupling for moment manifolds associated to Arkhipov–Chubarikov–Karatsuba systems. *Adv. Math.* **360** (2020).

[31] L. Guth, A short proof of the multilinear Kakeya inequality. *Math. Proc. Cambridge Philos. Soc.* **158** (2015), no. 1, 147–153.

[32] L. Guth, D. Maldague, and H. Wang, Improved decoupling for the parabola. 2020, arXiv:2009.07953.

[33] L. Guth, H. Wang, and R. Zhang, A sharp square function estimate for the cone in $\mathbb{R}^3$. *Ann. of Math. (2)* **192** (2020), no. 2, 551–581.

[34] K. Hughes and T. Wooley, Discrete restriction for (x, x^3) and related topics. 2019, arXiv:1911.12262.

[35] N. Katz, I. Laba, and T. Tao, An improved bound on the Minkowski dimension of Besicovitch sets in $\mathbb{R}^3$. *Ann. of Math. (2)* **152** (2000), no. 2, 383–446.

[36] N. Katz and J. Zahl, An improved bound on the Hausdorff dimension of Besicovitch sets in $\mathbb{R}^3$. *J. Amer. Math. Soc.* **32** (2019), no. 1, 195–259.

[37] H. Ko, S. Lee, and S. Oh, Maximal estimates for averages over space curves. 2021, arXiv:2102.07175.

[38] Z. Li, An ℓ^2 decoupling interpretation of efficient congruencing: the parabola. *Rev. Mat. Iberoam.* **37** (2021), no. 5, 1761–1802.

[39] L. Loomis and H. Whitney, An inequality related to the isoperimetric inequality. *Bull. Amer. Math. Soc.* **55** (1949), 961–962.

[40] R. Luca and K. Rogers, A note on pointwise convergence for the Schrödinger equation. *Math. Proc. Cambridge Philos. Soc.* **166** (2019), no. 2, 209–218.

[41] A. Miyachi, On some estimates for the wave equation in L^p and H^p. *J. Fac. Sci., Univ. Tokyo, Sect. 1A, Math.* **27** (1980), no. 2, 331–354.

[42] H. Montgomery, Mean and large values of Dirichlet polynomials. *Invent. Math.* **8** (1969), 334–345.

[43] T. Orponen, On the dimension and smoothness of radial projections. *Anal. PDE* **12** (2019), no. 5, 1273–1294.

[44] J. C. Peral, L^p estimates for the wave equation. *J. Funct. Anal.* **36** (1980), no. 1, 114–145.

[45] M. Pramanik and A. Seeger, L^p regularity of averages over curves and bounds for associated maximal operators. *Amer. J. Math.* **129** (2007), no. 1, 61–103.

[46] T. Schoen and I. Shkredov, On sumsets of convex sets. *Combin. Probab. Comput.* **20** (2011), no. 5, 793–798.

[47] C. Sogge, Propagation of singularities and maximal functions in the plane. *Invent. Math.* **104** (1991), no. 2, 349–376.

[48] E. Stein, Maximal functions. I. Spherical means. *Proc. Natl. Acad. Sci. USA* **73** (1976), no. 7, 2174–2175.

[49] E. Stein, Some problems in harmonic analysis. In *Harmonic analysis in Euclidean spaces (Proc. Sympos. Pure Math., Williams Coll., Williamstown, MA, 1978), Part 1*, pp. 3–20, Proc. Sympos. Pure Math. XXXV, Amer. Math. Soc., Providence, RI 1978.

[50] E. Stein, *Oscillatory integrals in Fourier analysis, Beijing lectures in harmonic analysis*. Ann. Math. Stat. 112, Princeton University Press, 1986.

[51] R. Strichartz, A priori estimates for the wave equation and some applications. *J. Funct. Anal.* **5** (1970), 218–235.

[52] I. Vinogradov, The method of trigonometrical sums in the theory of numbers (in Russian). *Trav. Inst. Math. Stekloff* **23** (1947), 109 pp.

[53] T. Wolff, An improved bound for Kakeya type maximal functions. *Rev. Mat. Iberoam.* **11** (1995), no. 3, 651–674.

[54] T. Wolff, A Kakeya-type problem for circles. *Amer. J. Math.* **119** (1997), no. 5, 985–1026.

[55] T. Wolff, Decay of circular means of Fourier transforms of measures. *Int. Math. Res. Not.* **10** (1999), 547–567.

[56] T. Wolff, Local smoothing type estimates on L^p for large p. *Geom. Funct. Anal.* **10** (2000), no. 5, 1237–1288.

[57] T. Wooley, Large improvements in Waring's problem. *Ann. of Math. (2)* **135** (1992), no. 1, 131–164.

[58] T. Wooley, The cubic case of the main conjecture in Vinogradov's mean value theorem. *Adv. Math.* **294** (2016), 532–561.

[59] T. Wooley, Nested efficient congruencing and relatives of Vinogradov's mean value theorem. *Proc. Lond. Math. Soc. (3)* **118** (2019), no. 4, 942–1016.

LARRY GUTH

Department of Mathematics, MIT, 77 Massachusetts Ave, Cambridge, MA 02139, USA, larryg@mit.edu

ONE-DIMENSIONAL QUASIPERIODIC OPERATORS: GLOBAL THEORY, DUALITY, AND SHARP ANALYSIS OF SMALL DENOMINATORS

SVETLANA JITOMIRSKAYA

ABSTRACT

Spectral theory of one-dimensional discrete one-frequency Schrödinger operators is a field with the origins in and strong ongoing ties to physics. It features a fascinating competition between randomness (ergodicity) and order (periodicity), which is often resolved on a deep arithmetic level. This leads to an especially rich spectrum of phenomena, many of which we are only beginning to understand. The corresponding analysis involves, in particular, dealing with small denominator problems. It has led to the development of non-KAM methods in this traditionally KAM domain, and to results completely unattainable by the old techniques, also in a number of other settings. This article accompanies the author's lecture at the International Congress of Mathematicians 2022. It covers several related recent developments.

MATHEMATICS SUBJECT CLASSIFICATION 2020

Primary 47B36; Secondary 37C55, 82B26, 37D25

KEYWORDS

Quasiperiodic operators, small denominators, Lyapunov exponents, spectrual theory, localization

INTERNATIONAL CONGRESS OF MATHEMATICIANS 2022 JULY 6–14

Proc. Int. Cong. Math. 2022, Vol. 2, pp. 1090–1120
DOI 10.4171/ICM2022/175

One-dimensional discrete one-frequency Schrödinger operators

$$(H_{V,\alpha,x}u)_n := u_{n-1} + u_{n+1} + V(x + n\alpha)u_n,$$
$$u \in \ell^2(\mathbb{Z}), \quad \alpha \in \mathbb{T} := \mathbb{R}\backslash\mathbb{Q},\ x \in \mathbb{T},\ V : \mathbb{T} \to \mathbb{R}, \tag{0.1}$$

and related questions of the dynamics of quasiperiodic cocycles have not been underrepresented at the ICMs. As I remember, roughly within the last 25 years, there were sectional lectures by H. Eliasson in 1998, myself in 2002, B. Fayad, R. Krikorian, and J. You in 2018, as well as plenary lectures by A. Avila in 2010 and 2014, devoted either in part or in full to this topic.

The field itself is not at all new. It may be seen as having been originated in physics when Peierls [103] and later his student Harper [61] studied the tight-binding two-dimensional electron in a uniform perpendicular magnetic field (also known as the Harper model) and derived the by now iconic family $H_{2\lambda\cos,\alpha,x}$ that we now, following Barry Simon [105], call the almost Mathieu operator. It remains hugely popular in physics, being directly linked to several remarkable experimental discoveries and Nobel prizes, providing, in particular, the theoretical underpinning of the Quantum Hall Effect, as proposed by D. J. Thouless in 1983 (see, e.g. [18,19]). A Google search for "Harper's model physics" leads to many thousands of hits.

The field may also be seen as having been originated in a numerical experiment, as the interest was picked after Douglas Hofstadter came up with what we now call the Hofstadter's butterfly [64]—a beautiful numerically produced fractal (Figure 1), discovered even before the word "fractal" was coined by Benoit Mandelbrot. Finally, the field may be seen as having been originated from the first application of KAM in the spectral theory—a pioneering work of Dinaburg and Sinai [37], that preceded Hofstadter. The field has consistently attracted top mathematical physicists (e.g., Bellissard, Deift, Simon, Sinai, Spencer), dynamicists (e.g., Avila, Eliasson, Herman, Krikorian, You), and analysts (e.g., Bourgain, Eliott, Sarnak, Schlag). Indeed, it turned out to be a fantastic ever-expanding playground for the analysts and dynamicists alike, leading to strong cross-fertilization of ideas that have a tendency to later expand to other subjects. Jean Bourgain wrote a book [28] devoted to analytic, mostly one-dimensional, quasiperiodic operators that summarized significant new understanding achieved around the turn of the century, where the work of Jean and collaborators was central.

It is therefore all the more surprising that as of the time of this writing it seems that the field is on the verge of further significant breakthroughs, with our current understanding covering just the tip of an exciting iceberg. Given the remarkable current momentum, we will refrain from making an attempt at an overview of the vast past literature, neither even very recent nor a number of important milestones, and will concentrate instead only on two selected topics that enjoyed significant recent advances and hold a particular promise to shape some of the future discourse.

For the review up to about five years ago, see [82], and for various fine issues related to continuity of the Lyapunov exponents, featuring, in particular, very important work by M. Goldstein and W. Schlag, see the recent book by P. Duarte and S. Klein [38]. The 2018

FIGURE 1
Hofstadter's butterfly.

ICM proceedings by J. You [117] summarize, among other things, the quantitative reducibility breakthrough developed in his group, that has led to a number of powerful consequences. There are also recent expositions [68, 80] that include some further remarkable results of roughly the last decade that could not make it into this article.

1. SPECTRAL THEORY MEETS (DUAL) DYNAMICS

Quasiperiodic operators (0.1) are, of course, a particular case of one-dimensional discrete ergodic Schrödinger operators

$$(H_x u)_n := u_{n-1} + u_{n+1} + V(T^n x)u_n, \quad u \in \ell^2(\mathbb{Z}), \tag{1.1}$$

where $x \in X$, and (X, μ, T) is an ergodic dynamical system. Operators with ergodic potentials (also in the continuum or in a more general multidimensional/covariant setting) always have spectra and closures of the other spectral components constant for μ-a.e. x [95, 102]. In case of the minimal underlying dynamics, such as, e.g., the irrational rotation of the circle in (0.1), the spectra [21] and absolutely continuous spectra in the one-dimensional case [97] are constant for *all* x. In contrast, the point and singular continuous parts (that are constant a.e.) can depend sensitively on x. It is an interesting problem, usually attributed to B. Simon, and open even in the setting of (0.1) whether this still holds when they are combined together (see Problem 6 in [67]).

The spectral theory of one-dimensional ergodic Schrödinger operators (1.1) is deeply connected to the study of linear cocycles over corresponding underlying dynamics. By an SL$(2, \mathbb{R})$ cocycle, we mean a pair (T, A), where $T : X \to X$ is ergodic, A is a measurable 2×2 matrix-valued function on X and $\det A = 1$.

We can regard it as a dynamical system on $X \times \mathbb{R}^2$ with

$$(T, A) : (x, f) \mapsto \big(Tx, A(x)f\big), \quad (x, f) \in X \times \mathbb{R}^2.$$

A one-parameter family of Schrödinger cocycles over (X, μ, T), indexed by the energy $E \in \mathbb{C}$, is given by $(T, A) : (X, \mathbb{R}^2) \mapsto (X, \mathbb{R}^2)$ where $(T, A) : (x, y) \mapsto (Tx, A(x, E)y)$, and $A \in \mathrm{SL}(2, \mathbb{C})$ is the transfer-matrix

$$A(x, E) := \begin{pmatrix} E - v(x) & -1 \\ 1 & 0 \end{pmatrix},$$

with $x \in X$, $y \in \mathbb{R}^2$, and $E \in \mathbb{C}$. The eigenvalue equation $Hu = Eu$ can be rewritten dynamically as

$$\begin{pmatrix} u_{n+1} \\ u_n \end{pmatrix} = A\big(T^n x, E\big) \begin{pmatrix} u_n \\ u_{n-1} \end{pmatrix}.$$

The (top) Lyapunov exponent is then defined as $L(E) := \lim_{n\to\infty} \int \frac{1}{n} \ln \|A_n(x, E)\| \, d\mu$, where

$$A_n(x, E) := \prod_{i=n-1}^{0} A(T^i x, E). \tag{1.2}$$

Two classical results link dynamics/Lyapunov exponents to the spectral theory of ergodic operators:

- (Johnson's theorem [91]) For minimal (X, μ, T), the spectrum $\sigma(H)$ (which is constant in $x \in X$) is given by the set of $E \in \mathbb{R}$ such that the Schrödinger cocycle $(T, A(\cdot, E))$ is not uniformly hyperbolic.
- (Kotani theory [94]) The absolutely continuous spectrum $\sigma_{ac}(H)$ (μ- a.e. constant for any ergodic (X, μ, T) and constant for minimal systems [97]) is given by the essential closure of the set $\{E : L(E) = 0\}$.

Therefore, for minimal, and in particular quasiperiodic, underlying dynamics, spectrum and absolutely continuous spectrum of H_x are encoded by the dynamics of the one-parameter family $A(x, E)$ of transfer-matrix cocycles, indexed by the energy E, but, for the spectrum, not by any explicit quantity. One recent surprising development is that for analytic one-frequency quasiperiodic Schrodinger operators, the spectrum (and therefore absence of uniform hyperbolicity of the corresponding cocycles) can be characterized more directly. In [47] we introduce a new object, dual Lyapunov exponent $\hat{L}(E)$, and prove

Theorem 1.1 ([47]). *For quasiperiodic operators* (0.1) *with analytic* V,

$$\sigma(H) = \big\{E : L(E)\hat{L}(E) = 0\big\}. \tag{1.3}$$

Exponent $\hat{L}(E)$ is defined as the limit of lowest Lyapunov exponents of dual high-dimensional cocycles (see Sections 2 and 4) which is proved to exist. There are interesting questions of varying levels of difficulty on whether this can be appropriately extended to higher-dimensional analytic one-frequency quasiperiodic Schrodinger cocycles, corresponding to operators on the strips, to multifrequency analytic cocycles, to nonanalytic potentials, or even other underlying dynamics. Perhaps the most natural question is whether one can find an analytic characterization of the absence of uniform hyperbolicity for all analytic one-frequency quasiperiodic cocycles. For the latter, there is a topological obstruction, but one can reduce the question, say, to cocycles homotopic to the identity.

2. AUBRY DUALITY AND HIGHER-DIMENSIONAL COCYCLES

The early work of Dinaburg–Sinai [37] notwithstanding, it is fair to say that the study of the spectral theory of quasiperiodic operators has been largely shaped around and driven by several explicit models, all coming from physics. The most prominent of those is the almost Mathieu family $H_{2\lambda\cos,\alpha,x}$, which can be argued to be the tight-binding analogue of a harmonic oscillator. Besides being the main model in the related physics studies and that featured in the Hofstadter's butterfly, it is also the simplest, in many ways, analytic case, yet it seems to represent most of the nontrivial properties expected to be encountered in the more general situation. In some sense, it plays the same role in the theory of quasiperiodic operators that the Ising model plays in statistical mechanics, and similarly to the latter, it does have an important additional symmetry.

Namely, we define the Aubry dual of the one-frequency Schrödinger operator (0.1) as

$$(\hat{H}_{V,\alpha,\theta}u)_n = \sum_{k=-\infty}^{\infty} V_k u_{n+k} + 2\cos 2\pi(\theta + n\alpha)u_n, \quad n \in \mathbb{Z}, \tag{2.1}$$

where V_k is the kth Fourier coefficient of V.[1] It can be useful to view this as a transformation of the entire family indexed by x for fixed V, α. In this regard, this transform can be viewed as a unitary conjugation on $\mathcal{H} = L^2(\mathbb{T} \times \mathbb{Z})$, via

$$U\psi(x,n) = \hat{\psi}(n, x + \alpha n), \tag{2.2}$$

where $\hat{\psi} : L^2(\mathbb{Z} \times \mathbb{T}) \to L^2(\mathbb{T} \times \mathbb{Z})$ is the Fourier transform. The almost Mathieu family is self-dual with respect to this transformation $\hat{H}_{2\lambda\cos,\alpha,x} = H_{\frac{2}{\lambda}\cos,\alpha,\theta}$, and, in particular, $H_{2\cos,\alpha,x}$, that is, $H_{2\lambda\cos,\alpha,x}$ with $\lambda = 1$, is the self-dual (also called critical) point.

Aubry duality can be explained by the magnetic nature and corresponding gauge invariance of two-dimensional magnetic Laplacians that lead to $H_{V,\alpha,x}$ [101]. In particular, spectra and integrated densities of states of $H_{V,\alpha,x}$ and $\hat{H}_{V,\alpha,x}$ coincide. However, it is not the case for the spectral type, and indeed it is natural to expect that a Fourier-type transform would take localized eigenfunctions (point spectrum!) into extended ones (absolutely continuous spectrum!), and vice versa. That was the basis for several predictions by physicists Aubry and Andre [1] about the almost Mathieu family with irrational α, namely that the spectrum of $H_{2\lambda\cos,\alpha,x}$ is absolutely continuous for $\lambda < 1$ (called subcritical) and pure point for $\lambda > 1$ (called supercritical). This was described in the paper where transformation (2.1)

---

1 There is a more general, multidimensional definition, but we stick to the one-dimensional case for this exposition.

was introduced in the context of the almost Mathieu family, leading to the name Aubry duality. This problem, along with a few others related to this family, was heavily popularized by Barry Simon in [106, 108], fueling an increased interest in the mathematics community.

Aubry duality has been formulated and explored on different levels, e.g., [10, 55, 101]. It has consistently played a central role in the analysis of quasiperiodic operators, in proving absolutely continuous spectrum and reducibility [10, 31], point spectrum [17, 24, 50, 57, 70],[2] or its absence [11, 69].

In general, operator (2.1) is long-range. If V is a trigonometric polynomial of degree d, the transfer-matrix $A(x, E)$ of the eigenvalue equation $\hat{H}_{V,\alpha,x}\Psi = E\Psi$ gives rise to a $2d$-dimensional cocycle, which has a complex-symplectic structure [60], so we will view it as an $\mathrm{Sp}(2d, \mathbb{C})$ cocycle (α, A), $A \in \mathrm{Sp}(2d, \mathbb{C})$, a linear skew product

$$(\alpha, A) : \left\{ \begin{array}{ccc} \mathbb{T} \times \mathbb{C}^{2d} & \to & \mathbb{T} \times \mathbb{C}^{2d} \\ (x, v) & \mapsto & (x + \alpha, A(x, E) \cdot v) \end{array} \right\}.$$

The Lyapunov exponents $L_1(\alpha, A) \geq L_2(\alpha, A) \geq \cdots \geq L_{2d}(\alpha, A)$, repeated according to their multiplicity, are defined by

$$L_k(\alpha, A) = \lim_{n \to \infty} \frac{1}{n} \int_{\mathbb{T}} \ln\big(\sigma_k\big(A_n(x)\big)\big) dx,$$

where for a matrix $B \in M_m(\mathbb{C})$, $\sigma_1(B) \geq \cdots \geq \sigma_m(B)$ denote its singular values (eigenvalues of $\sqrt{B^* B}$). Since for real E the transfer-matrix $A(x, E)$ of the eigenvalue equation $\hat{H}_{V,\alpha,x}\Psi = E\Psi$ is symplectic, its Lyapunov exponents come in the opposite pairs $\{\pm L_i(\alpha, A)\}_{i=1}^{d}$. We will now denote

$$\hat{L}_i = L_{d-i}(\alpha, A), \tag{2.3}$$

so that $0 \leq \hat{L}_1 \leq \hat{L}_2 \leq \cdots \leq \hat{L}_d$.

In general, Lyapunov exponents are not nicely behaved with respect to parameter changes. They can be (and most likely, typically are) discontinuous in α at $\alpha \in \mathbb{Q}$ (the almost Mathieu cocycle is one example), are generally discontinuous in A in C^0, and can be discontinuous in A even in C^∞ [35, 81, 113, 114]. It is a remarkable fact, enabling much of the related theory, that Lyapunov exponents are continuous in the analytic category.

Theorem 2.1 ([12, 29, 31, 73]). *The functions $\mathbb{R} \times C^\omega(\mathbb{T}, M_m(\mathbb{C})) \ni (\alpha, A) \mapsto L_k(\alpha, A) \in [-\infty, \infty)$ are continuous at any (α', A') with $\alpha' \in \mathbb{R} \backslash \mathbb{Q}$.*[3]

For the almost Mathieu operator, it leads to the exact formula for the Lyapunov exponent for energies E in the spectrum of $H_{2\lambda \cos,\alpha,x}$. We have $L_{\lambda,\alpha}(E) = \max\{\ln|\lambda|, 0\}$ [30].

For Diophantine α, this continuity extends to sufficiently smooth Gevrey spaces [35, 92], and it is a remarkable recent result [48] that for certain α the transition in the topology

---

2 Made possible with the development of recent powerful methods [7, 14, 65, 118] to establish nonperturbative reducibility directly and independently of localization for the dual model.

3 In dimension one, it extends to the Lyapunov exponents of multifrequency cocycles $\mathbb{R} \times C^\omega(\mathbb{T}^b, \mathrm{SL}_2(\mathbb{C})) \ni (\alpha, A) \mapsto L(\alpha, A) \in [0, \infty)$.

for continuity of L occurs sharply at the Gevrey space G^2. It should be noted that both the original spectacular counterexample [113] and its refinements [48, 114] require α to be a fixed irrational of bounded type, i.e., having a continued fraction expansion with bounded coefficients. This set includes the golden mean but forms a set of zero Lebesgue measure. The authors of all these papers also vary the cocycle, i.e., the potential. This still leaves open the question whether continuous behavior of the Lyapunov exponents at least for Schrödinger cocycles with regularity lower than G^2 is possible if α is not of bounded type. Another open question is whether it is true that for a fixed potential of lower than G^2 regularity, the Lyapunov exponent is necessarily a continuous function of energy.

3. AVILA'S GLOBAL THEORY AND CLASSIFICATION OF ANALYTIC ONE-FREQUENCY COCYCLES

While many results exist in lower regularity, the analyticity of V in (0.1) brings on board powerful ideas related to subharmonicity (leading, in particular, to the crucially important for other developments continuity results) and the technique of semialgebraic sets introduced to the field by J. Bourgain [28]. As a result, a lot more can be said about analytic quasiperiodic operators. Particularly, while Kotani theory based its characterization of the absolutely continuous spectrum on compexifying the energy, for analytic quasiperiodic operators there is one more natural parameter to complexify, namely the phase. This idea goes back to M. Herman [63], and has been fruitfully used to prove positivity (and later continuity) of the Lyapunov exponent in [29, 63, 110]. Avila [5] discovered a remarkable related structure that has served as a foundation of his global theory (later extended to the high-dimensional cocycles in [12]). Define

$$L_\epsilon(E) := \lim_{n\to\infty} \int \frac{1}{n} \ln \left\| \prod_{j=n-1}^{0} A_j(x + j\alpha + i\epsilon, E) \right\| d\mu.$$

Avila observed that, for a given cocycle, L_ϵ is a convex function of ϵ, and proved that it has quantized derivative in ϵ.

Theorem 3.1 ([5]). *For any complex-analytic one-frequency cocycle,*

$$\omega(A) = \lim_{\epsilon\to 0^+} \frac{L_\epsilon(A) - L_0(A)}{2\pi\epsilon} \in \mathbb{Z}.$$

This was enabled through approximation by the rationals due to the continuity of the Lyapunov exponent in the analytic category [32]. The fact that such continuity does not hold even for higher Gevrey cocycles [48, 113, 114] complicates potential nonanalytic extensions.

Theorem 3.1 already enables full analytic computation of the Lyapunov exponents for E in the spectrum, as well as of their complexifications L_ϵ and further analysis for several models originating and relevant in physics: the almost Mathieu operator [5], the extended Harper's model [81], recently discovered models with mobility edges [112] and unitary almost Mathieu operator [34], models arising in the study of the quantum graph graphene [23], and others.

Avila classified analytic cocycles $A(x)$ depending on the behavior of the Lyapunov exponent L_ϵ of the complexified cocycle $A(x+i\epsilon)$. Namely, he distinguishes three cases, with the terminology inspired by the almost Mathieu family:

(Subcritical) $L_\epsilon = 0, \epsilon < \delta,\ \delta > 0$, or, alternatively, $L_0 = \omega(A) = 0$.

(Critical) $L_0 = 0, L_\epsilon > 0, \epsilon > 0$, or, alternatively, $L_0 = 0, \omega(A) > 0$.

(Supercritical) $L_0 > 0$.

For the almost Mathieu family, these three regimes are uniform over the spectrum, corresponding to the supercritical ($\lambda > 1$), subcritical ($\lambda < 1$), and critical ($\lambda = 1$) values of the coupling constant. Spectrally, there is purely absolutely continuous spectrum for all x and all $\alpha \in \mathbb{R}\backslash\mathbb{Q}$ in the subcritical case [3], purely singular continuous spectrum for all x and all $\alpha \in \mathbb{R}\backslash\mathbb{Q}$ in the critical case [69], and pure point spectrum for a.e. x, α with sharp spectral transitions depending on the arithmetics of both α and x between pure point spectrum and singular continuous spectrum in the supercritical case (see Section 5). Remarkably, the critical almost Mathieu operators appear at the boundary of the two other regimes.

For general quasiperiodic operators, this classification leads to the corresponding division of energies in the spectrum, depending on (sub/super)criticality of the cocycle $A(\cdot, E)$. For convenience we will call the energy in the spectrum (super/sub)critical according to whether the corresponding transfer-matrix cocycle is such. It is expected that the key spectral properties of spectra in the three above regimes follow those of the corresponding almost Mathieu operators.

Indeed, pure point spectrum for a.e. x, α holds through the *supercritical* set of energies, for any analytic potential [30]. It is an important open problem to make this result arithmetic, and it is expected that certain universal features of the transitions and structure of the eigenfunctions discovered in [77,78] will hold globally, throughout the supercritical regime, see Section 6.3.

The *subcritical* regime is subject to the almost reducibility conjecture (ARC) which claims that subcritical cocycles are almost reducible, that is, have constant cocycles in the closure of their analytic conjugacy class (note that since almost reducibility implies subexponential growth of the iterates of the cocycle that is uniform in the (complexified) phase, the converse is obviously true). The idea of reducing nonperturbative (global) to perturbative (local) results originated from an earlier work by Avila and Krikorian [14]. ARC was first formulated in [10], and first established for the almost Mathieu operator [3,10]. It was solved by Avila for the Liouville case in [4], and the solution for the complementary Diophantine case has been announced [5] to appear in [2]. Also, L. Ge has recently found a different proof [46].

Almost reducible (and therefore subcritical) cocycles enjoy all the dynamical and spectral consequences of the Eliasson's perturbative regime [39]. In particular, there is purely absolutely continuous spectrum throughout the subcritical regime. Moreover, reducibility can be made quantitative [117], and even arithmetically so [50], allowing for a wealth of conclusions. However, it remains true that the absolutely continuous spectrum is fully characterized by the subcritical regime, with no delicate dependence, as far as the spectral decomposition goes, on any other parameters.

The *critical* regime is expected (see [11, 82]) to support only singular continuous spectrum (again, no dependence on the other parameters, as long as α is irrational) but fully establishing it even for the critical almost Mathieu operator took decades and was only accomplished recently [69].

On the other hand, the key result of Avila's global theory [5] is that operators with critical energies throughout the spectrum, like the critical almost Mathieu operator, are an anomaly, that does not happen typically. In fact, for prevalent (in a certain measure-theoretic sense) potentials, there are no critical energies, and the spectrum is contained in finitely many intervals, with either only subcritical or only supercritical regime within each.[4] Moreover, the set of potentials and energies (V, E) such that E is critical is contained in a countable union of codimension-one analytic submanifolds of $C^{\omega}(\mathbb{T};\mathbb{R}) \times \mathbb{R}$. Another remarkable related fact is that Lyapunov exponent enjoys even much stronger regularity when restricted to potentials and energies with a fixed value of acceleration: it becomes real-analytic on this (typically rather irregular) set, in both the energy E and any parameter λ ranging in a real analytic manifold Λ, if V_λ in $C^{\omega}(\mathbb{T};\mathbb{R})$ is a family real-analytic in parameter λ.

From the point of view of the global theory, it becomes particularly important to study the universal features of the two prevalent regimes, subcritical and supercritical. As mentioned above, the absolutely continuous spectrum is fully characterized by the subcritical regime, with no delicate dependence, as far as the spectral decomposition goes, on any other parameters. The picture for the supercritical regime is a lot more interesting, and is in a certain sense at the beginning of its development.

Going back to the complexified cocycle L_ϵ, quantizatization of acceleration means that as a function of $\epsilon > 0$, L_ϵ is convex, piecewise affine, and thus is fully characterized by $L = L_0$ and monotone increasing sequences of turning points b_i and slopes $n_i \in 2\pi\mathbb{Z}_+$, so that the slope of L_ϵ between b_i and b_{i+1} is n_i. Clearly, sequences b_i and n_i present a very important intrinsic characterization of the cocycle and the corresponding Schrödinger operator. What information do they give us?

4. DUAL LYAPUNOV EXPONENTS OR GLOBAL THEORY DEMYSTIFIED

It turns out that Aubry duality not only provides a new proof of quantization of acceleration, but holds key to the mystery of the global theory. We have

Theorem 4.1 ([47]). *Assume $\alpha \in \mathbb{R}\backslash\mathbb{Q}$ and $V \in C^{\omega}(\mathbb{T}, \mathbb{R})$. Then there exist nonnegative $\{\hat{L}_i(E)\}$ such that for any $E \in \mathbb{R}$,*

$$\hat{L}_i(E) = \lim_{d\to\infty} \hat{L}_i^d(E),$$

4 A part of this picture was previously established in the semiclassical regime in the continuum in [40].

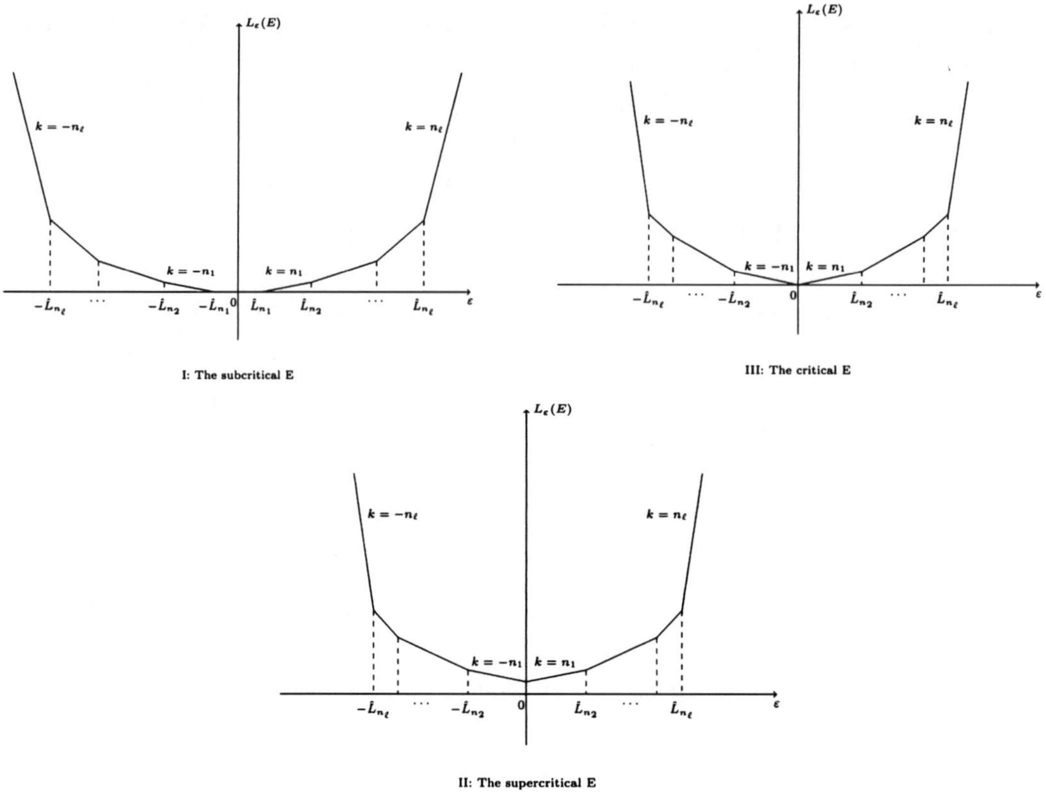

FIGURE 2
The complexified Lyapunov exponent.

where $\hat{L}_i^d(E), i = 1, \ldots, d$, are the Lyapunov exponents, as defined in (2.3), of the $\mathrm{Sp}(2d, \mathbb{C})$ *transfer-matrix cocycle of the dual eigenvalue equation $\hat{H}_{V^d,\alpha,x}\Psi = E\Psi$, with $V^d(x) = D_d \star V$ and D_d being the Dirichlet kernel. Moreover,*

$$L_\epsilon(E) = L_0(E) - \sum_{\{i:\hat{L}_i(E)<2\pi|\epsilon|\}} \hat{L}_i(E) + 2\pi\big(\#\{i : \hat{L}_i(E) < 2\pi|\epsilon|\}\big)|\epsilon|$$

In fact, the theorem also holds for $V \in C_h^\omega(\mathbb{T}, \mathbb{R})$ and $|\epsilon| < h$, where $C_h^\omega(\mathbb{T}, \mathbb{R})$ is the space of bounded analytic functions f defined on a strip $\{|\Im z| < h\}$ with the norm $\|f\|_h = \sup_{|\Im z|<h} |f(z)|$. See Fig. 2 for an illustration of the three possible scenarios.

This means that for the trigonometric polynomials V the turning points b_i are given precisely by the Lyapunov exponents $\hat{L}_i(E)$ of the dual cocycle, and increases in the slopes are given by the 2π times their multiplicities; for analytic V, these objects are given by the limits of those quantities for successive trigonometric polynomial cutoffs of V. We call $\hat{L}_i(E)$ the dual Lyapunov exponents, the objects that play a role similar to that of zeros of an analytic function in the Jensen's formula. In particular, the acceleration $\omega(E)$ turns out to be precisely the number of vanishing dual Lyapunov exponents (an analogue of the winding number for an analytic function on $\mathbb{T}$).

Besides unraveling the mystery of the behavior of complexified Lyapunov exponents, this leads to a new understanding of the key statement of Avila's global theory, namely that for prevalent operators (0.1), almost all pairs of potentials and energies are acritical. Indeed, it immediately follows that

Theorem 4.2 ([47]). *Assume $\alpha \in \mathbb{R}\backslash\mathbb{Q}$ and V is analytic, then the energy $E \in \mathbb{R}$ is*

(1) outside the spectrum *if $L(E) > 0$ and $\hat{L}_1(E) > 0$,*

(2) supercritical *if $L(E) > 0$ and $\hat{L}_1(E) = 0$,*

(3) critical *if $L(E) = 0$ and $\hat{L}_1(E) = 0$,*

(4) subcritical *if $L(E) = 0$ and $\hat{L}_1(E) > 0$.*

Thus, in the regime $L(E) = 0$, criticality is in the locus of vanishing of an additional continuous [12] function $\hat{L}_1(E)$, implying the prevalence of the acriticality claim. Theorem 4.2, of course, also contains the statement of Theorem 1.1, with $\hat{L} := \hat{L}_1$, as well as the fact that Schrödinger cocycle is subcritical if and only if its dual Lyapunov exponents are all positive. It also leads to a number of other powerful spectral corollaries, both for the general analytic case and several particular models [47]. It also has exciting physics applications [100].

5. PRECISE ANALYSIS OF SMALL DENOMINATORS

One of the most fascinating features of the spectral theory of one-frequency quasiperiodic operators in the supercritical regime is its delicate dependence on the arithmetics, that can be analyzed to a remarkable depth, and in some cases completely. There were many exciting recent developments where the arithmetics has played a crucial role (e.g., [9,15,89]) but here we focus only on the analysis of small denominators in the proofs of point spectrum and related study of the eigenfunctions.

The main difficulty in proving point spectrum (or the phenomenon of Anderson localization, that is, pure point spectrum with exponentially decaying eigenfunctions) and analyzing the corresponding eigenfunctions of ergodic operators is in the fact that the eigenvalues are dense in the spectrum. Formal perturbative expansions of eigenfunctions and eigenvalues include the $(V(T^n x) - V(T^m x))^{-1}$ terms that, of course, get arbitrarily large. More generally, when we have *resonances*, that is, restrictions to boxes that are not too far away from each other that have eigenvalues that are too close (something that is bound to happen for ergodic operators), small denominators are created. Thus localization for ergodic and, in particular, quasiperiodic operators can be viewed as a small denominator problem.

Indeed, it has been traditionally approached in a perturbative way: through KAM-type schemes for large couplings [39,44,109], which all required Diophantine conditions on the frequency α. Small denominators are not simply a nuisance, but lead to actual change in the spectral behavior, since in the opposite regime of very Liouville frequencies (*too* small denominators), there is no localization even with the positivity of the Lyapunov exponent; and delocalization (which in this case means singular continuous spectrum) can be proved by perturbation of nearby periodic operators [20,54]. At the same time, for exponentially approximated frequencies that are neither far from nor close enough to rationals, there is nothing left to perturb about or to remove. Tackling those cannot be approached perturbatively, but requires a precise analysis, giving the problem a strong number-theoretic flavor.

It should be noted that the topology of the one-dimensional line is such that even occasional barriers make it difficult to pass through, strongly favoring localization in the presence of even small irregularities. For example, in the one-dimensional random case, localization holds for all couplings λ, when considering a family of potentials λV, and the same is expected but is apparently difficult to prove even for the underlying dynamics (X, μ, T) with very weak chaotic properties, such as a skew shift. It has even been conjectured by Kotani and Last that absolutely continuous spectrum is impossible for one-dimensional operators that are not almost periodic, but it has been disproved [6,111], and with a particularly simple construction in [119]. Those examples notwithstanding, the presence of metal–insulator transitions (that roughly correspond to transitions between the spectral types) as couplings change remains a distinctive feature of quasiperiodic operators.

The transitions in coupling between absolutely continuous and singular spectrum are fully determined by the vanishing/nonvanishing of the Lyapunov exponent. In the supercritical regime, absolutely continuous spectrum is impossible, but whether the spectrum is point or singular continuous is resolved in the competition between the depth of the small denominators—the strength of the resonances—and the Lyapunov growth.

Two types of resonances have played a special role in the spectral theory of quasiperiodic operators. *Frequency* resonances, when $|V(x) - V(x + k\alpha)|$ is small simply because $\|(x + k\alpha) - x)\|_{\mathbb{R}/\mathbb{Z}} = \|k\alpha\|_{\mathbb{R}/\mathbb{Z}}$ is small, where $\|x\|_{\mathbb{R}/\mathbb{Z}} = \inf_{\ell \in \mathbb{Z}} |x - \ell|$, were first exploited in [21] based on [54] to prove the absence of eigenvalues (and therefore singular continuous spectrum in the hyperbolic regime) for quasiperiodic operators with Liouville frequencies. Their strength is measured by the arithmetic parameter

$$\beta(\alpha) = \limsup_{k \to \infty} -\frac{\ln \|k\alpha\|_{\mathbb{R}/\mathbb{Z}}}{|k|} \tag{5.1}$$

that is equal to zero for Diophantine (thus a.e.) α. Frequency resonances are ubiquitous for all quasiperiodic potentials.

Another class of resonances, appearing for all *even* potentials, was discovered in [83], where it was shown that the arithmetic properties of the phase also play a role and may lead to singular continuous spectrum even for the Diophantine frequencies. Indeed, for even potentials, phases with almost symmetries, when $|V(x) - V(x + k\alpha|$ is small because $\|(x + k\alpha) - (-x)\|_{\mathbb{R}/\mathbb{Z}}$ is small, lead to resonances, regardless of the values of other parameters. The strength of *phase* resonances is measured by the arithmetic parameter

$$\delta(\alpha, \theta) = \limsup_{k \to \infty} -\frac{\ln \|2\theta + k\alpha\|_{\mathbb{R}/\mathbb{Z}}}{|k|}. \tag{5.2}$$

Phase resonances are symmetry based and exist for all even functions V.

It was conjectured in [66] that for the almost Mathieu family no other resonances appear and the competition between the Lyapunov growth and combined exponential resonance strength resolves in a sharp way: there is a pure point spectrum for $L(E) >$

$\beta(\alpha) + \delta(\alpha, x)$ and a singular continuous spectrum in the regime $L(E) < \beta(\alpha) + \delta(\alpha, x)$. We note that for the special case of *α-rational x*, that is, such that $2x \in \mathbb{Z}\alpha + \mathbb{Z}$, we have $\delta(\alpha, x) = \beta(\alpha)$ so the resonances "double up" and the conjectured threshold becomes $2\beta(\alpha)$.

An early nonperturbative localization method was first developed in the 1990s for the almost Mathieu operator [84] and represented perhaps the first case of solving a traditionally KAM problem in a direct way, without an inductive procedure. It presented a (simple, but not sharp) technique to treat the *nonresonant* case, $\beta(\alpha) = \delta(\alpha) = 0$. Further breakthroughs came in [85] where the role of the Lyapunov exponents and corresponding deviations was first understood, allowing to achieve the nonresonant result up to the actual Lyapunov transition, and then in the work of Bourgain and collaborators [28,30] where robust nonperturbative methods were developed for general analytic potentials and more, leading to the proofs of localization for a.e. frequency throughout the supercritical regime. The ideas of [85] hold more generally, and have, in particular, led to very simple proofs of localization for the one-dimensional Anderson model [90]. Most importantly, however, their arithmetic nature has been crucial for further developments. For example, the fact that localization holds for α-rational x,[5] enabled Puig's proof [104] of the ten martini problem (that the spectrum is a Cantor set) for Diophantine α. The solution of the full ten martini problem [8,9] required, in particular, dealing with intermediate frequencies that are neither Diophantine nor Liouville, thus with the frequency resonances. A method to treat those has been devised in [9] leading to the proof of localization for $L(E) > \frac{16}{9}\beta$, but failing in the neighborhood of the actual transition. A sharp method to treat pure frequency resonances was developed in [77], and a sharp method to treat pure phase resonances in [78].

Therefore, the sharp arithmetic spectral transition conjecture of [66] has been established for single-type-resonances: for pure frequency resonances (that is, for the so-called α-Diophantine phases for which $\delta(\alpha, x) = 0$ so there are no exponential phase resonances) in [17,52,77],[6] and for pure phase resonances (that is, for Diophantine frequencies for which $\beta(\alpha) = 0$ so there are no exponential frequency resonances) in [78].

The methods to treat pure frequency and phase resonances in [77,78] are robust in a sense that weak exponential resonances of the other type can be added easily, but it is still an open problem to treat *combined* frequency and phase resonances in a sharp way. However, there were two very recent breakthroughs.

Namely, W. Liu has developed a way to sharply treat doubled resonances for the almost Mathieu operator, proving localization up to the conjectured threshold:

---

5 This was, in fact, established in [72].

6 In [17] the pure frequency part of the conjecture of [66] has been proved by a completely different method, namely through quantitative reducibility [117] and duality, but in a measure-theoretic in x sense, i.e., losing the control over the arithmetics of x. A recent breakthrough by Ge–You [50] where an arithmetic version of quantitative reducibility was developed has lead to a way to obtain sharp arithmetic in phase results through duality as well, enabling, in particular, an arithmetic duality-based proof of the frequency part of the conjecture [52], that works also for all Aubry duals (2.1) of operators (0.1).

Theorem 5.1 ([99]). *Operator $H_{2\lambda\cos,\alpha,x}$ with α-rational x has Anderson localization whenever $L(E) > 2\beta(\alpha)$ (or equivalently, $\lambda > e^{2\beta(\alpha)}$).*

In Liu's earlier work, this was established for $L(E) > 3\beta(\alpha)$ [98], but a significant new understanding of treatment of doubled resonances was necessary to go sharp, and it was achieved in [99]. Also α-rational phases x hold special importance for various questions because eigenvalues for such x are located at gap edges [104]. Puig's proof of the ten martini problem for the Diophantine case [104] was based precisely on localization for α-rational x. The original plan to prove the full ten martini problem was to establish localization for α-rational x and $L(E) > \beta(\alpha)$ [8]. Not surprisingly, it failed, prompting the resonance doubling-up conjecture in [9] that is now solved [99]. It should be noted that the singular-continuous part of the conjecture, namely singular-continuous spectrum for α-rational x and $L(E) < 2\beta(\alpha)$, is still open.

In a different direction, R. Han, F. Yang, and I [58] developed a sharp method to treat the third type of resonances: high barriers (that effectively play the role of *antiresonances*), and, moreover, *combinations* of frequency resonances and high barriers, in another popular quasiperiodic family originating in physics, the Maryland model.

Maryland model is a family

$$(M_{\lambda,\alpha,\theta}u)_n = u_{n+1} + u_{n-1} + \lambda\tan\big(\pi(\theta + n\alpha)\big)u_n, \tag{5.3}$$

where $\lambda > 0$ is the coupling constant, irrational $\alpha \in \mathbb{T} = [0,1]$ is the frequency, and $\theta \in \mathbb{T}$ is the phase with $\theta \notin \Theta = \{\frac{1}{2} + \alpha\mathbb{Z} + \mathbb{Z}\}$.

It was originally proposed by Grempel, Fishman, and Prange [56] as a linear version of the quantum kicked rotor and has attracted continuing interest from the physics community, see, e.g., [26, 42, 45], due to its exactly solvable nature. It has explicit expression for the Lyapunov exponent, integrated density of states, and even (a little less explicit) for the eigenvalues and eigenfunctions. In particular, the Lyapunov exponent $L_\lambda(E)$ is an explicit function of λ, E not dependent on α. However, the implicit expressions for the eigenfunctions do not allow for easy conclusions about their behavior, which is expected to be quite interesting, with transfer matrices satisfying certain exact renormalization [41].

Phase resonances do not exist for the Maryland model, and as a result, for Diophantine (i.e., nonresonant) frequencies it has localization for *all* phases [87, 107]. However, it does have barriers, when the trajectory of a given phase approaches the singularity too early. Barriers compensate for the resonances, and therefore serve as what we call in [58] the *antiresonances*, providing the reason why for the Maryland model there are phases with localization even for the most Liouville frequencies [76]. Thus Maryland model features a combination of frequency resonances and phase antiresonances.

Maryland model was the first one where the spectral decomposition has been resolved completely, for *all* values of the parameters [76].[7] Let p_n/q_n be the continued fraction approximants of α. We note that the frequency resonance index $\beta(\alpha)$ defined in (5.1)

7 It also remains the only one with spectral transitions where this could be claimed.

also satisfies $\beta(\alpha) = \limsup_{n\to\infty} \frac{\ln q_{n+1}}{q_n}$. A new index, $\delta^M(\alpha,\theta)$, was introduced in [76] as

$$\delta^M(\alpha,\theta) := \limsup_{n\to\infty} \frac{\ln q_{n+1} + \ln \|q_n(\theta - \frac{1}{2})\|_{\mathbb{T}}}{q_n}. \tag{5.4}$$

We have

Theorem 5.2 ([76]). *$H_{\lambda,\alpha,\theta}$ has purely singular continuous spectrum on $\{E : L_\lambda(E) < \delta^M(\alpha,\theta)\}$, and pure point spectrum on $\{E : L_\lambda(E) > \delta^M(\alpha,\theta)\}$.*[8]

This provides complete spectral analysis, for all α, θ, but was established implicitly: through the combination of Cayley and Fourier transforms and the study of a resulting explicit cohomological equation, making sharp the previous work in [56,107]. The extension of the analysis from a.e. θ in [107] to *all* θ in [76] required accounting for the effect of the barriers, and Cayley transform allowed to do it, albeit in a highly implicit way. In particular, this proof did not allow the analysis of the structure of eigenfunctions.

The method of [85] was adapted to the Maryland model in [87] where the nonresonant situation was treated and localization for Diophantine α was shown, developing the initial framework to study the eigenfunctions in the much more difficult resonant situation.

In [58] we show that $\delta(\alpha,\theta)$ can be interpreted as the exponential strength of frequency resonances, $\beta(\alpha)$, combined with the (negative) exponential strength of phase antiresonances, defined as the positions of exponential smallness of the $\cos(\pi(\theta + k\alpha))$,[9] and develop the approach to sharply treat the "resonance tamed by an antiresonance" situation. In particular, we give a constructive proof of the localization part of Theorem 5.2 and obtain

Theorem 5.3 ([58]). *For any $\alpha \in \mathbb{R}\backslash\mathbb{Q}$ and any θ, the spectrum on $\{E : L_\lambda(E) \ge \delta^M(\alpha,\theta)\}$ is pure point and for any eigenvalue $E \in \{L_\lambda(E) > \delta^M(\alpha,\theta)\}$ and any $\epsilon > 0$, the corresponding eigenfunction ϕ_E satisfies $|\phi_E(k)| < e^{-(L_\lambda(E)-\delta^M(\alpha,\theta)-\epsilon)|k|}$ for sufficiently large $|k|$.*

Theorem 5.3 provides the sharp upper envelope, and develops the key tools to study the fine behavior of the eigenfunctions, see Section 6.2. In fact, such a study is the most exciting outcome of the proofs of localization based on sharp analysis of resonances.

There are several other models where sharp arithmetic spectral transitions have been conjectured and partially established, most notably the extended Harper's model, where for the complete analysis one would need to develop tools to study the simultaneous presence of three different types of resonances: frequency, phase, and singularity-induced antiresonances. However, for a.e. phase we expect the arithmetic frequency transition to be universal in the class of general analytic potentials. As for the arithmetic transitions in phase, we expect the same results to hold for general even analytic potentials for a.e. frequency. We note that the singular continuous part up to the conjectured transition is already established, even in a far greater generality, in [17,71,78].

---

8 It follows from the explicit formula for $L_\lambda(E)$ that the equality can only happen for two values of E.

9 So exponential largeness of the tan.

Finally, there is a question of arithmetic interfaces, e.g., what happens for the almost Mathieu operators with $L(E) = \beta(\alpha) + \delta(\alpha, \theta)$? It turns out that (in the pure resonance situations) both pure point and singular continuous spectra are possible depending on the finer arithmetic properties of parameters [13, 86, 88]. So far we do not even have a good conjecture on where the arithmetic thresholds within the transition lines lie. Making a significant progress on this problem would require a development of polynomial (as contrasted with current exponential) methods to tackle resonances, a very important problem in its own right, as it could lead to universal hierarchical structures (see Section 6) on polynomial scales.

6. EXACT ASYMPTOTICS AND UNIVERSAL HIERARCHICAL STRUCTURE OF EIGENFUNCTIONS

A very captivating question and a longstanding theoretical challenge is to explain the self-similar hierarchical structure visually obvious in the Hofstadter's butterfly, as well as the hierarchical structure of eigenfunctions, as related to the arithmetics of parameters. Such structure was first predicted for the almost Mathieu operator in the work of Azbel in 1964 [22], some 12 years before Hofstadter [64], and before numerical experimentation was possible. Such self-similar behavior is present for spectra and eigenfunctions of all quasiperiodic operators.

While this does not describe or explain the self-similarity, a step in the right direction is to prove that the spectrum is a Cantor set. Mark Kac offered ten martinis in 1982 for the proof of the Cantor set part of Azbel's 1964 conjecture. It was dubbed the Ten Martini problem by Barry Simon, who advertised it in his lists of 15 mathematical physics problems [106] and later, mathematical physics problems for the XXI century [108]. Most substantial partial solutions were made by Bellissard, Simon, Sinai, Helffer, Sjöstrand, Choi, Eliott, Yui, and Last [25, 36, 62, 96, 109], between 1983 and 1993. J. Puig [104] solved it for Diophantine α by noticing that localization at $\theta = 0$ [73, 85] leads to gaps at corresponding (dense) eigenvalues. The final solution was given in [9]. Cantor spectrum is also prevalent for general one-frequency operators with analytic potential: in the subcritical regime [10], and, by very different methods, in the supercritical regime [53] (and it is conjectured [11] also in the critical regime, which is nongeneric in itself [5]). Moreover, even all gaps predicted by the gap labeling are open in the noncritical almost Mathieu case [10, 16], the statement that is also expected to be true in the critical case, and recently claimed in the physics literature [27] to follow directly from [69].

As for the understanding the hierarchical behavior of the eigenfunctions, despite significant numerical studies and even a discovery of Bethe Ansatz solutions [116], it has remained an important open challenge even at the physics level, although some indications existed in the perturbative regime [33, 62, 109, 120].

Sharp analysis of resonances and small denominators has led to the discovery of universal self-similar structures of eigenfunctions defined by the type of resonance. The universal nature of these structures manifests in two ways: there is the same universal function that depends only on the type of the resonance, that governs the behavior around each exponential frequency or phase resonance (upon (possibly) reflection and renormalization), and it is the same structure for all the parameters involved: any (Diophantine) frequency α, (any α-Diophantine phase θ) with $\beta(\alpha) < L$ ($\delta(\alpha,\theta) < L$), and any eigenvalue E. It has been discovered and proved for the almost Mathieu operator **[77,78]** but is expected to be universal also throughout the class of analytic potentials, and more,[10] that is to hold in the regime of pure resonances. For example, the same universal structure for frequency resonances has already been proved for the Maryland model **[59]**, for a.e. phase, namely, phases without the exponential antiresonances, see also a result on the hierarchical structure in the semiclassical regime **[93]**. However, for phases whose trajectories approach the barrier too fast, the hierarchical structure of the eigenfunctions is very different, and the complete analysis is extremely delicate.

Generally, one can identify four types of (anti)resonances that lead to different universal structures:

- frequency
- phase (only even potentials)
- barriers (antiresonance)
- singularity (antiresonance for Jacobi matrices)

We describe the universal structures for phase and frequency resonances **[77,78]** in the following subsections, and the one for the barrier antiresonances will appear in **[59]**.

We expect that when different types of resonances are present, there will be further different self-similar structures, universal for all corresponding parameters and different resonance positions. Describing these structures for different combinations of resonances is very challenging but seems to be potentially within reach. In particular, in **[58]** we developed the tools to fully describe the universal structures for the Maryland model for all parameters, that is for combinations of frequency resonances and barrier antiresonances. We expect it to be done in **[59]**. We also expect the latter structures to be universal in the class of monotone potentials with a simple pole.

To give a glimpse into the universality results, we present two of them in more detail.

6.1. Frequency resonances

In **[77]** we find explicit universal functions $f(k)$ and $g(k)$, depending only on the Lyapunov exponent and the position of k in the hierarchy defined by the denominators q_n of the continued fraction approximants of the flux α, that completely define the exponential behavior of, correspondingly, eigenfunctions and norms of the transfer matrices of the almost Mathieu operators, for all eigenvalues corresponding to α-Diophantine phase, see Theorem 6.1. This result holds for *all* frequency and coupling pairs in the frequency-

10 For example, C^2 cos-type potentials have been a popular object of study **[43, 49, 51, 109, 115]** and there are reasons to believe that they will feature the same structure, at least in the perturbative regime.

resonance localization regime. Since the behavior is fully determined by the frequency and does not depend on the phase, it is the same, eventually, around any starting point, so is also seen unfolding at different scales when magnified around local eigenfunction maxima, thus describing the exponential universality in the hierarchical structure.

Since we are interested in exponential growth/decay, the behavior of f and g becomes most interesting in case of frequencies with exponential rate of approximation by the rationals.

These functions allow describing *precise* asymptotics of *arbitrary* solutions of $H_{\lambda,\alpha,\theta}\varphi = E\varphi$ where E is an eigenvalue. The precise asymptotics of the norms of the transfer-matrices provides the first example of this sort for nonuniformly hyperbolic dynamics. Since those norms sometimes differ significantly from the reciprocals of the eigenfunctions, this leads to further interesting and unusual consequences, for example, exponential tangencies between contracted and expanded directions at the resonant sites.

Given $\alpha \in \mathbb{R}\backslash\mathbb{Q}$, we define functions $f, g : \mathbb{Z}^+ \to \mathbb{R}^+$ in the following way. Let $\frac{p_n}{q_n}$ be the continued fraction approximants to α. For any $\frac{q_n}{2} \le k < \frac{q_{n+1}}{2}$, define $f(k), g(k)$ as follows:

Case 1. $q_{n+1}^{\frac{8}{9}} \ge \frac{q_n}{2}$ or $k \ge q_n$.
If $\ell q_n \le k < (\ell+1)q_n$ with $\ell \ge 1$, set

$$f(k) = e^{-|k-\ell q_n|\ln|\lambda|}\bar{r}_\ell^n + e^{-|k-(\ell+1)q_n|\ln|\lambda|}\bar{r}_{\ell+1}^n, \tag{6.1}$$

and

$$g(k) = e^{-|k-\ell q_n|\ln|\lambda|}\frac{q_{n+1}}{\bar{r}_\ell^n} + e^{-|k-(\ell+1)q_n|\ln|\lambda|}\frac{q_{n+1}}{\bar{r}_{\ell+1}^n}, \tag{6.2}$$

where for $\ell \ge 1$,

$$\bar{r}_\ell^n = e^{-(\ln|\lambda| - \frac{\ln q_{n+1}}{q_n} + \frac{\ln \ell}{q_n})\ell q_n}.$$

Set also $\bar{r}_0^n = 1$ for convenience.
If $\frac{q_n}{2} \le k < q_n$, set

$$f(k) = e^{-k\ln|\lambda|} + e^{-|k-q_n|\ln|\lambda|}\bar{r}_1^n, \tag{6.3}$$

and

$$g(k) = e^{k\ln|\lambda|}. \tag{6.4}$$

Case 2. $q_{n+1}^{\frac{8}{9}} < \frac{q_n}{2}$ and $\frac{q_n}{2} \le k \le \min\{q_n, \frac{q_{n+1}}{2}\}$.
Set

$$f(k) = e^{-k\ln|\lambda|}, \tag{6.5}$$

and

$$g(k) = e^{k\ln|\lambda|}. \tag{6.6}$$

Notice that f, g only depend on α and λ, but not on θ or E; $f(k)$ decays and $g(k)$ grows exponentially, globally, at varying rates that depend on the position of k in the hierarchy defined by the continued fraction expansion of α, see Figures 3 and 4.

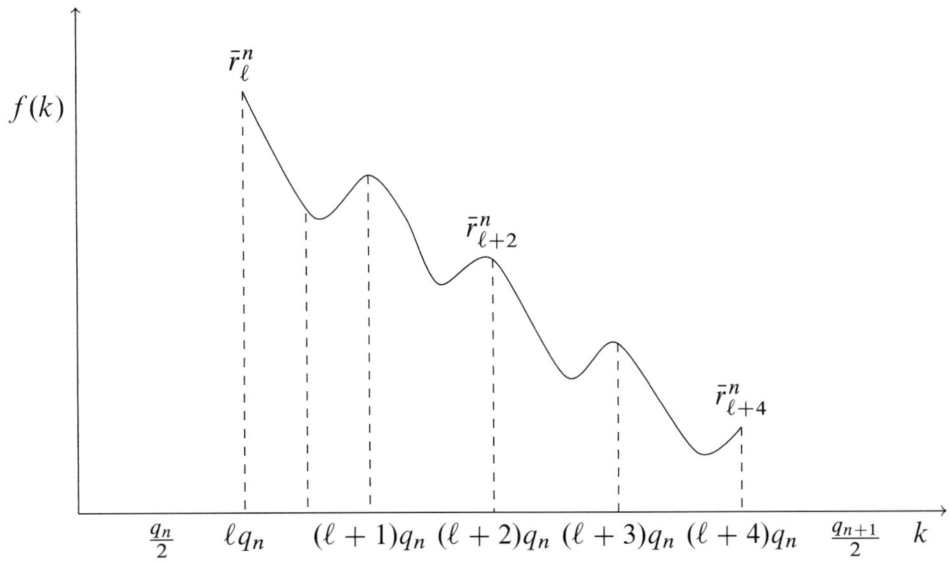

FIGURE 3
The universal behavior of eigenfunctions at scale n.

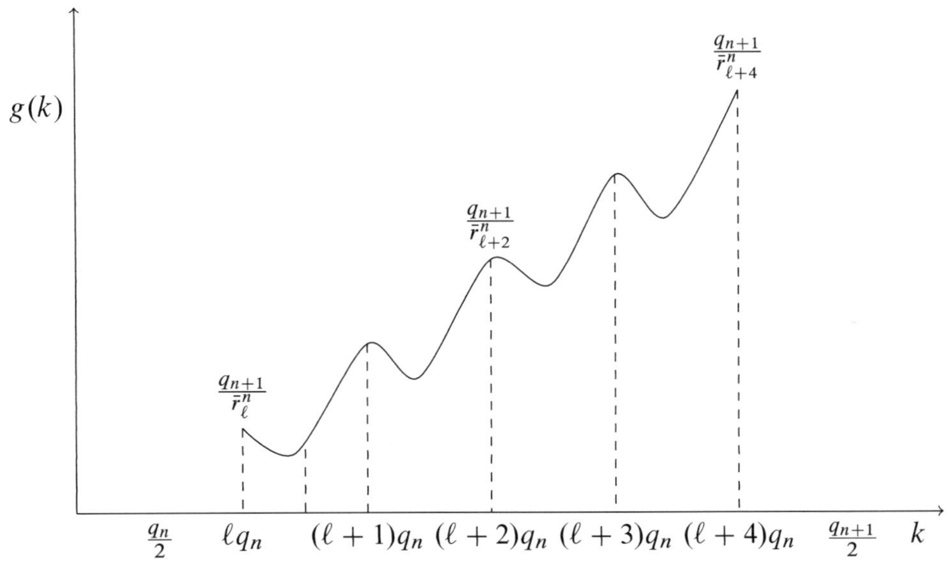

FIGURE 4
The universal behavior of transfer matrix norms at scale n.

It turns out that, in the entire regime $L(E) > \beta$, the exponential asymptotics of the eigenfunctions and norms of transfer matrices at the eigenvalues are completely determined by $f(k), g(k)$.

Theorem 6.1. *Let $\alpha \in \mathbb{R}\backslash\mathbb{Q}$ be such that $|\lambda| > e^{\beta(\alpha)}$. Suppose θ is Diophantine with respect to α, E is an eigenvalue of $H_{\lambda,\alpha,\theta}$, and ϕ is the eigenfunction. Let $U(k) = \binom{\phi(k)}{\phi(k-1)}$. Then for any $\varepsilon > 0$, there exists K (depending on $\lambda, \alpha, \hat{C}, \varepsilon$) such that for any $|k| \geq K$, $U(k)$ and*

A_k[11] *satisfy*

$$f(|k|)e^{-\varepsilon|k|} \leq \|U(k)\| \leq f(|k|)e^{\varepsilon|k|} \tag{6.7}$$

and

$$g(|k|)e^{-\varepsilon|k|} \leq \|A_k\| \leq g(|k|)e^{\varepsilon|k|}. \tag{6.8}$$

In fact, the theorem is formulated in **[77]** for *generalized eigenfunctions*, thus can also be used to establish pure point spectrum throughout the indicated regime. Certainly, there is nothing special about $k = 0$, so the behavior described in Theorem 6.1 happens around an arbitrary point $k = k_0$. This implies the self-similar nature of the eigenfunctions: $U(k)$ behave as described at scale q_n but, when looked at in windows of size $q_k, q_k \leq q_{n-1}$, will demonstrate the same universal behavior around appropriate local maxima/minima.

To further illustrate the above, let ϕ be an eigenfunction and $U(k) = \begin{pmatrix} \phi(k) \\ \phi(k-1) \end{pmatrix}$. An immediate corollary of Theorem 6.1 is the universality of behavior at all appropriately defined nonresonant local maxima. We will say k_0 is a local j-maximum of ϕ if $\|U(k_0)\| \geq \|U(k)\|$ for $|k - k_0| \sim q_j$. Then, with an appropriate notion of nonresonance (see **[77]**), we have

Theorem 6.2 (**[77]**). *Given $\varepsilon > 0$, there exists $j(\varepsilon) < \infty$ such that if k_0 is a nonresonant local j-maximum for $j > j(\epsilon)$, then*

$$f(|s|)e^{-\varepsilon|s|} \leq \frac{\|U(k_0 + s)\|}{\|U(k_0)\|} \leq f(|s|)e^{\varepsilon|s|}, \tag{6.9}$$

for $|s - k_o| \sim q_j$.

In case $\beta(\alpha) > 0$, Theorem 6.1 also guarantees an abundance (and a hierarchical structure) of local maxima of each eigenfunction.

Let k_0 be a global maximum. The self-similar hierarchical structure of local maxima can be described in the following way. We will say that a scale n_{j_0} is exponential if $\ln q_{n_{j_0}+1} > cq_{n_{j_0}}$. Then there is a *constant* scale $\hat{n}_0$, thus a constant $C := q_{\hat{n}_0+1}$, such that for any exponential scale n_j and any eigenfunction there are local n_j-maxima within distance C of $k_0 + sq_{n_{j_0}}$ for each $0 < |s| < e^{cq_{n_{j_0}}}$. Moreover, these are all the local n_{j_0}-maxima in $[k_0 - e^{cq_{n_{j_0}}}, k_0 + e^{cq_{n_{j_0}}}]$.

The exponential behavior of the eigenfunction in the local neighborhood (of size of order $q_{n_{j_0}}$) of each such local maximum, normalized by the value at the local maximum is given by f. Note that only exponential behavior at the corresponding scale is determined by f and fluctuations of much smaller size are invisible.

Now, let $n_{j_1} < n_{j_0}$ be another exponential scale. Denoting "depth 1" local maximum located near $k_0 + a_{n_{j_0}} q_{n_{j_0}}$ by $b_{a_{n_{j_0}}}$, we then have a similar picture around $b_{a_{n_{j_0}}}$: there are local n_{j_1}-maxima in the vicinity of $b_{a_{n_{j_0}}} + sq_{n_{j_1}}$ for each $0 < |s| < e^{cq_{n_{j_1}}}$. Again, this describes all the local $q_{n_{j_1}}$-maxima within an exponentially large interval. And again, the exponential (for the n_{j_1} scale) behavior in the local neighborhood (of size of order $q_{n_{j_1}}$) of each such local maximum, normalized by the value at the local maximum, is given by f.

---

11 Products A_k are defined in (1.2).

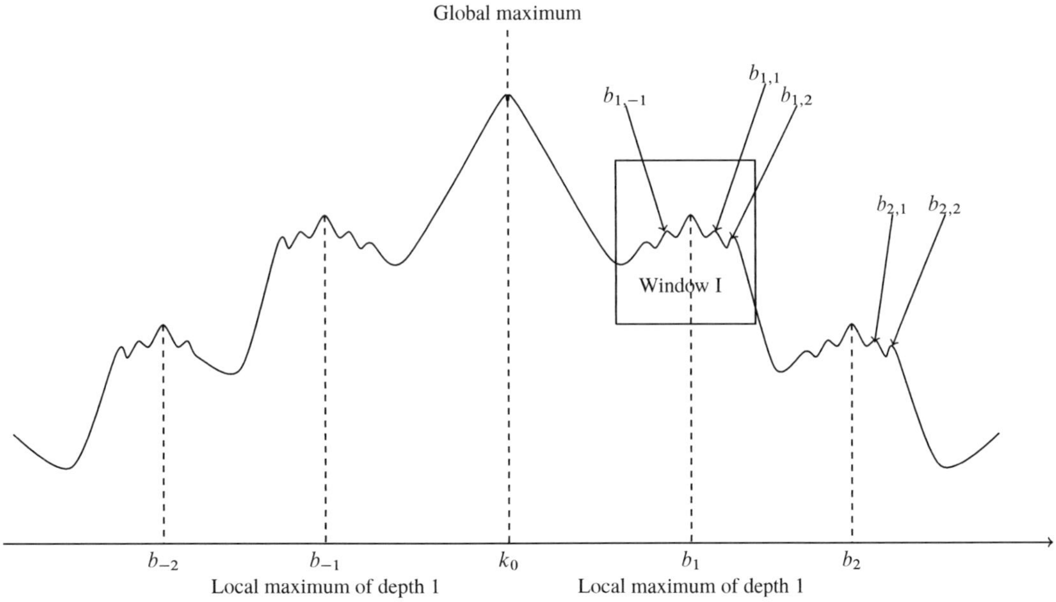

FIGURE 5
Universal self-similar structure of eigenfunctions

Denoting those "depth 2" local maxima located near $b_{a_{n_{j_0}}} + a_{n_{j_1}} q_{n_{j_1}}$ by $b_{a_{n_{j_0}}, a_{n_{j_1}}}$, we then get the same picture taking the magnifying glass another level deeper, and so on. At the end we obtain a complete hierarchical structure of local maxima that we denote by $b_{a_{n_{j_0}}, a_{n_{j_1}}, \dots, a_{n_{j_s}}}$ with each "depth $s+1$" local maximum $b_{a_{n_{j_0}}, a_{n_{j_1}}, \dots, a_{n_{j_s}}}$ being in the corresponding vicinity of the "depth s" local maximum $b_{a_{n_{j_0}}, a_{n_{j_1}}, \dots, a_{n_{j_{s-1}}}}$, and with universal behavior at the corresponding scale around each. The quality of the approximation of the position of the next maximum gets lower with each level of depth, yet the depth of the hierarchy that can be so achieved is at least $j/2 - C$, Figure 5 schematically illustrates the structure of local maxima of depth one and two, and Figure 6 illustrates that the neighborhood of a local maximum appropriately magnified looks like a picture of the global maximum. See [77] for the exact statement.

6.2. Phase resonances

In [78] we found another universal structure, this time for phase resonances. Once again, we found (different) functions f that determine universal asymptotics of the eigenfunctions, also locally around the resonances, which features a self-similar hierarchical structure. In particular, we have Theorem just like Theorem 6.1 but with new f and for $\beta(\alpha) = 0$ and $L > \delta(\alpha, \theta)$ [78]. The behavior described in this theorem happens around an arbitrary point. This, coupled with effective control of parameters at the local maxima, allows uncover-

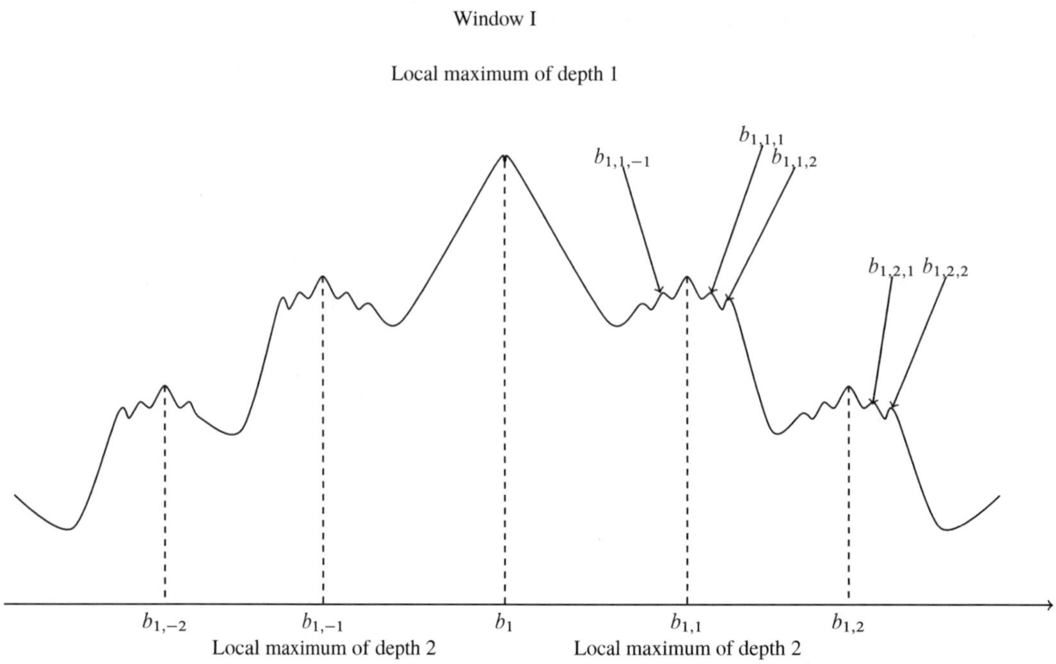

FIGURE 6
Universal self-similar structure of eigenfunctions, zoomed in

ing the self-similar nature of the eigenfunctions, but this time one needs not only the rescaling but also alternating reflections, leading to what we call the *reflective-hierarchical* structure.

Assume phase θ satisfies $0 < \delta(\alpha, \theta) < \ln \lambda$. Fix $0 < \varsigma < \delta(\alpha, \theta)$. Let k_0 be a global maximum of eigenfunction ϕ. Let K_i be the positions of exponential resonances of the phase $\theta' = \theta + k_0\alpha$ defined by

$$\|2\theta + (2k_0 + K_i)\alpha\|_{\mathbb{R}/\mathbb{Z}} \le e^{-\varsigma|K_i|}. \tag{6.10}$$

This means that $|v(\theta' + \ell\alpha) - v(\theta' + (K_i - \ell)\alpha)| \le Ce^{-\varsigma|K_i|}$, uniformly in ℓ, or, in other words, the potential $v_n = v(\theta + n\alpha)$ is $e^{-\varsigma|K_i|}$-almost symmetric with respect to $(k_0 + K_i)/2$.

Since α is Diophantine, we have

$$|K_i| \ge ce^{c|K_{i-1}|}, \tag{6.11}$$

where c depends on ς and α through the Diophantine constants κ, τ. On the other hand, K_i is necessarily an infinite sequence. Let ϕ be an eigenfunction, and $U(k) = \begin{pmatrix} \phi(k) \\ \phi(k-1) \end{pmatrix}$. We say k is a local K-maximum if $\|U(k)\| \ge \|U(k+s)\|$ for all $s - k \in [-K, K]$.

The informal description of the *reflective-hierarchical* structure of local maxima is the following. There exists a *constant* $\hat{K}$ such that there is a local cK_j-maximum b_j within distance $\hat{K}$ of each resonance K_j. The exponential behavior of the eigenfunction in the local cK_j-neighborhood of each such local maximum, normalized by the value at the local maximum, is given by the *reflection* of f.

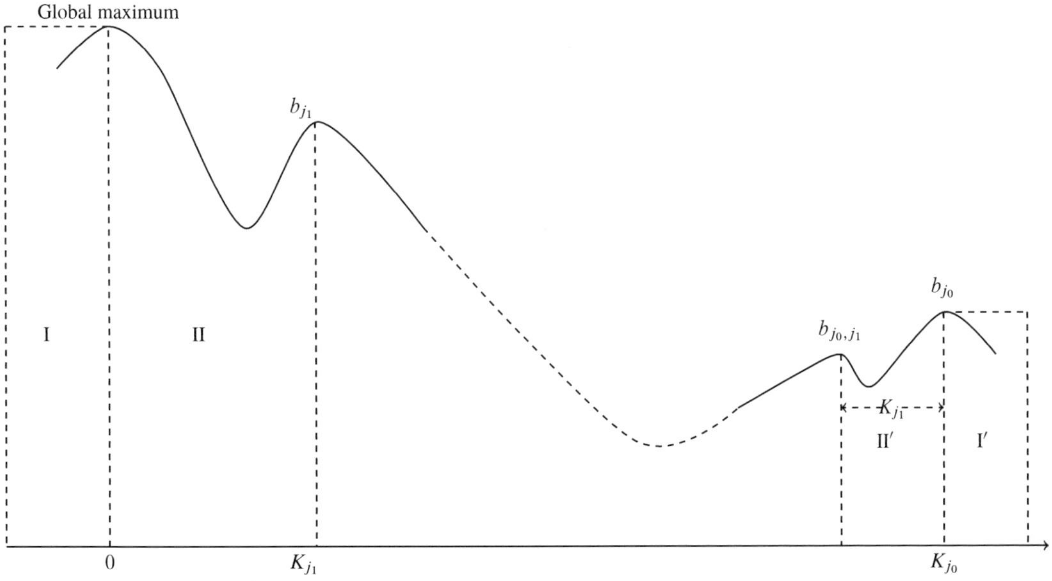

FIGURE 7
Reflective self-similarity of an eigenfunction.

Moreover, this describes the entire collection of local maxima of depth 1, that is, all K such that K is a cK-maximum. Then we have a similar picture in the vicinity of b_j: there are local cK_i-maxima $b_{j,i}, i < j$, within distance $\hat{K}^2$ of each $K_j - K_i$. The exponential (on the K_i scale) behavior of the eigenfunction in the local cK_i-neighborhood of each such local maximum, normalized by the value at the local maximum, is given by f.

Then we get the next level maxima $b_{j,i,s}$, $s < i$ in the $\hat{K}^3$-neighborhood of $K_j - K_i + K_s$ and reflected behavior around each, and so on, with reflections alternating with steps. At the end we obtain a complete hierarchical structure of local maxima that we denote by $b_{j_0,j_1,\ldots,j_s}$, with each "depth $s+1$" local maximum $b_{j_0,j_1,\ldots,j_s}$ being in the corresponding vicinity of the "depth s" local maximum $b_{j_0,j_1,\ldots,j_{s-1}} \approx k_0 + \sum_{i=0}^{s-1}(-1)^i K_{j_i}$ and with universal behavior at the corresponding scale around each. The quality of the approximation of the position of the next maximum gets lower with each level of depth, with $b_{j_0,j_1,\ldots,j_{s-1}}$ determined with $\hat{K}^s$ precision, thus it presents an accurate picture as long as $K_{j_s} \gg \hat{K}^s$.

Thus the behavior of $\phi(x)$ is described by the same universal f in each $\sim K_{j_s}$ window around the corresponding local maximum $b_{j_0,j_1,\ldots,j_s}$ after alternating reflections. The positions of the local maxima in the hierarchy are determined up to errors that at all but possibly the last step are superlogarithmically small in K_{j_s}. We call such a structure *reflective hierarchy*.

Figure 7 depicts reflective self-similarity of an eigenfunction with global maximum at 0. The self-similarity is seen as follows: I′ is obtained from I by scaling the x-axis proportional to the ratio of the heights of the maxima in I and I′; II′ is obtained from II by scaling the x-axis proportional to the ratio of the heights of the maxima in II and II′. The behavior in the regions I′, II′ mirrors the behavior in regions I, II upon reflection and corresponding dilation.

6.3. Universality and extensions

The hierarchical structures of Sections 6.1 and 6.2 are expected to hold universally for most in the appropriate sense (albeit not all, as for the almost Mathieu) local maxima for general analytic potentials. Establishing this fully would require certain new ideas since so far even an arithmetic version of localization for the Diophantine case has not been established for the general analytic family, the current state-of-the-art result by Bourgain–Goldstein [30] being measure-theoretic in α.

The universality of the hierarchical structures of Sections 6.1 and 6.2 is twofold: not only it is the same universal function that governs the behavior around each exponential frequency or phase resonance (upon reflection and renormalization), it is the same structure for all the parameters involved: any (Diophantine) frequency α (any α-Diophantine phase θ) with $\beta(\alpha) < L$ ($\delta(\alpha, \theta) < L$), and any eigenvalue E. The universal reflective-hierarchical structure in Section 6.2 requires the evenness of the function defining the potential and, moreover, resonances of other types may also be present in general. However, we conjectured in [78] that for general even analytic potentials for a.e. frequency only finitely many other exponentially strong resonances will appear, thus the structure described in Section 6.2 will hold for the corresponding class.

The key elements of the technique developed for the treatment of arithmetic resonances are robust and have made it possible to approach other questions and, in particular, study delicate properties of the singular continuous regime. Among other things, it has allowed obtaining upper bounds on fractal dimensions of the spectral measures and quantum dynamics for the singular continuous almost Mathieu operator [79], as well as potentials defined by general trigonometric analytic functions [75], and determining also the *exact* exponent of the exponential decay rate in expectation for the two-point function [74], the first result of this kind for any model. These methods are also expected to be applicable to many other models.

FUNDING

This work was partially supported by NSF DMS-1901462, DMS-2052899, and Simons 681675. S.J. was a 2020–2021 Simons fellow.

REFERENCES

[1] S. Aubry and G. André, Analyticity breaking and Anderson localisation in incommensurate lattices. *Ann. Isr. Phys. Soc.* **3** (1980), 133–164.

[2] A. Avila, in preparation.

[3] A. Avila, The absolutely continuous spectrum of the almost Mathieu operator. 2008, arXiv:0810.2965.
[4] A. Avila, Almost reducibility and absolute continuity I. 2010, arXiv:1006.0704.
[5] A. Avila, Global theory of one-frequency Schrödinger operators. *Acta Math.* **215** (2015), no. 1, 1–54.
[6] A. Avila, On the Kotani–Last and Schrödinger conjectures. *J. Amer. Math. Soc.* **28** (2015), 579–616.
[7] A. Avila, B. Fayad, and R. Krikorian, A KAM Scheme for $SL(2,\mathbb{R})$ Cocycles with Liouvillean Frequencies. *Geom. Funct. Anal.* **21** (2011), no. 5, 1001–1019.
[8] A. Avila and S. Jitomirskaya, Solving the ten martini problem. *Lecture Notes in Phys.* **690** (2006), 5–16.
[9] A. Avila and S. Jitomirskaya, The Ten Martini Problem. *Ann. of Math. (2)* **170** (2009), no. 1, 303–342.
[10] A. Avila and S. Jitomirskaya, Almost localization and almost reducibility. *J. Eur. Math. Soc. (JEMS)* **12** (2010), no. 1, 93–131.
[11] A. Avila, S. Jitomirskaya, and C. Marx, Spectral theory of extended Harper's model and a question by Erdős and Szekeres. *Invent. Math.* **210** (2017), no. 1, 283–339.
[12] A. Avila, S. Jitomirskaya, and C. Sadel, Complex one-frequency cocycles. *J. Eur. Math. Soc. (JEMS)* **16** (2014), no. 9, 1915–1935.
[13] A. Avila, S. Jitomirskaya, and Q. Zhou, Second phase transition line. *Math. Ann.* **370** (2018), 271–285.
[14] A. Avila and R. Krikorian, Reducibility or nonuniform hyperbolicity for quasiperiodic Schrödinger cocycles. *Ann. of Math.* **164** (2006), 911–940.
[15] A. Avila, Y. Last, M. Shamis, and Q. Zhou, On the abominable properties of the almost Mathieu operator with well approximated frequencies. 2021, arXiv:2110.07974.
[16] A. Avila, J. You, and Q. Zhou, in preparation.
[17] A. Avila, J. You, and Q. Zhou, Sharp phase transitions for the almost Mathieu operator. *Duke Math. J.* **166** (2017), no. 14, 2697–2718.
[18] J. E. Avron, Topological quantum states and the 2016 Nobel prize. *Bull. IAMP* (2017), 3–22.
[19] J. Avron, D. Osadchy, and R. Seiler, A Topological Look at the Quantum Hall Effect. *Phys. Today* **56** (2003), 8–38.
[20] J. Avron and B. Simon, Singular continuous spectrum for a class of almost periodic Jacobi matrices. *Bull. Amer. Math. Soc. (N.S.)* **6** (1982), 81–85.
[21] J. Avron and B. Simon, Almost periodic operators, II. The density of states. *Duke Math. J.* **50** (1983), 369–391.
[22] M. Y. Azbel, Energy spectrum of a conduction electron in a magnetic field. *Sov. Phys. JETP* **19** (1964), 634–645.
[23] S. Becker, R. Han, and S. Jitomirskaya, Cantor spectrum of graphene in magnetic fields. *Invent. Math.* **2** (2019), no. 18, 979–1041.

[24] J. Bellissard, R. Lima, and D. Testard, A metal–insulator transition for the almost Mathieu model. *Comm. Math. Phys.* **88** (1983), no. 2, 207–234.

[25] J. Bellissard, B. Simon, Cantor spectrum for the almost Mathieu equation. *J. Funct. Anal.* **48**, (1982), 408–419.

[26] M. Berry, Incommensurability in an exactly-soluble quantal and classical model for a kicked rotator. *Phys. D* **10** (1984), 369–378.

[27] D. S. Borgnia and R.-J. Slagerhttps, The Dry Ten Martini Problem at Criticality. 2021, arXiv:2112.06869.

[28] J. Bourgain, *Green's function estimates for lattice Schrödinger operators and applications*. Ann. of Math. Stud. 158, Princeton University Press, Princeton, NJ, 2005.

[29] J. Bourgain, Positivity and continuity of the Lyapunov exponent for shifts on $\mathbb{T}^d$ with arbitrary frequency vector and real analytic potential. *J. Anal. Math.* **96** (2005), 313–355.

[30] J. Bourgain and M. Goldstein, On nonperturbative localization with quasiperiodic potential. *Ann. of Math.* **152** (2000), 835–879.

[31] J. Bourgain and S. Jitomirskaya, Absolutely continuous spectrum for 1D quasiperiodic operators. *Invent. Math.* **148** (2002), 453–463.

[32] J. Bourgain and S. Jitomirskaya, Continuity of the Lyapunov exponent for quasiperiodic operators with analytic potential. *J. Stat. Phys.* **108** (2002), 1203–1218. Dedicated to David Ruelle and Yasha Sinai on the occasion of their 65th birthdays.

[33] V. Buslaev and A. Fedotov, The monodromization and Harper equation. *Sémin. Équ. Dériv. Partielles* exp. no. **21** (1993–1994), 1–21.

[34] C. Cedzich, J. Fillman, and D. C. Ong, Almost everything about the unitary almost Mathieu operator. 2021, arXiv:2112.03216.

[35] H. Cheng, L. Ge, J. You, and Q. Zhou, Global rigidity for ultra-differentiable quasiperiodic cocycles and its spectral applications. 2021, arXiv:2101.11150.

[36] M. D. Choi, G. A. Elliott, and N. Yui, Gauss polynomials and the rotation algebra. *Invent. Math.* **99** (1990), 225–246.

[37] E. Dinaburg and Ya. Sinai, The one-dimensional Schrödinger equation with a quasi-periodic potential. *Funct. Anal. Appl.* **9** (1975), 279–289.

[38] P. Duarte and S. Klein, *Lyapunov exponents of linear cocycles: continuity via large deviations*. Atlantis Ser. Dyn. Syst. 3, Atlantis Press, 2016.

[39] L. H. Eliasson, Reducibility and point spectrum for linear quasi-periodic skew products. *Doc. Math.* Extra volume ICM 1998 II, 779–787.

[40] A. Fedotov and F. Klopp, Anderson transitions for a family of almost periodic Schrödinger equations in the adiabatic case. *Comm. Math. Phys.* **227** (2002), 1–92.

[41] A. Fedotov and F. Sandomirskiy, An exact renormalization formula for the Maryland model. *Comm. Math. Phys.* **334** (2015), no. 2, 1083–1099.

[42] S. Fishman, Anderson localization and quantum chaos maps. *Scholarpedia* **5** (2010), no. 8, 9816.

[43] Y. Forman and T. VandenBoom, Localization and Cantor spectrum for quasiperiodic discrete Schrödinger operators with asymmetric, smooth, cosine-like sampling functions. 2021, arXiv:2107.05461.

[44] J. Fröhlich, T. Spencer, and P. Wittwer, Localization for a class of one dimensional quasi-periodic Schrödinger operators. *Comm. Math. Phys.* **132** (1990), 5–25.

[45] S. Ganeshan, K. Kechedzhi, and S. Das Sarma, Critical integer quantum hall topology and the integrable Maryland model as a topological quantum critical point. *Phys. Rev. B* **90** (2014), no. 4, 041405.

[46] L. Ge, in preparation.

[47] L. Ge, S. Jitomirskaya, J. You, and Q. Zhou, Multiplicative Jensen's formula and quantitative global theory of one-frequency Schrödinger operators, in preparation.

[48] L. Ge, Y. Wang, J. You, and X. Zhao, Transition space for the continuity of the Lyapunov exponent of quasiperiodic Schrödinger cocycles. 2021, arXiv:2102.05175.

[49] L. Ge, J. Xu, and J. Wang, The Hölder continuity of Lyapunov exponents for a class of cos-type quasiperiodic Schrödinger cocycles. 2020, arXiv:2006.03381.

[50] L. Ge and J. You, Arithmetic version of Anderson localization via reducibility. *Geom. Funct. Anal.* **30** (2020), no. 5, 1370–1401.

[51] L. Ge, J. You, and X. Zhao, Arithmetic version of Anderson localization for quasiperiodic Schrödinger operators with even cosine type potentials. 2021, arXiv:2107.08547.

[52] L. Ge, J. You, and X. Zhao, Arithmetic version of the frequency transition conjecture: new proof and generalization. *Peking J. Math.* (2021). DOI 10.1007/s42543-021-00040-y.

[53] M. Goldstein and W. Schlag, On resonances and the formation of gaps in the spectrum of quasi-periodic Schroedinger equations. *Ann. of Math.* **173** (2011), 337–475.

[54] A. Y. Gordon, The point spectrum of the one-dimensional Schrödinger operator. *Uspekhi Mat. Nauk* **31** (1976), 257–258.

[55] A. Y. Gordon, S. Jitomirskaya, Y. Last, and B. Simon, Duality and singular continuous spectrum in the almost Mathieu equation. *Acta Math.* **178** (1997), 169–183.

[56] D. Grempel, S. Fishman, and R. Prange, Localization in an incommensurate potential: an exactly solvable model. *Phys. Rev. Lett.* **49** (1982), no. 11, 833.

[57] R. Han and S. Jitomirskaya, Full measure reducibility and localization for quasiperiodic Jacobi operators: a topological criterion. *Adv. Math.* **319** (2017), 224–250.

[58] R. Han, S. Jitomirskaya, and F. Yang, Anti-resonances and sharp analysis of Maryland localization for all parameters, 2022, arXiv:2205.04021

[59] R. Han, S. Jitomirskaya, and F. Yang, Universal hierarchical structure of eigenfunctions in the Maryland model, in preparation.

[60] A. Haro and J. Puig, A Thouless formula and Aubry duality for long-range Schrödinger skew-products. *Nonlinearity* **26** (2013), 1163–1187.

[61] P. G. Harper, Single band motion of conducting electrons in a uniform magnetic field. *Proc. Phys. Soc. A* **68** (1955), 874–878.

[62] B. Helffer and J. Sjöstrand, Semi-classical analysis for Harper's equation. III. Cantor structure of the spectrum. *Mém. Soc. Math. Fr. (N.S.)* **39** (1989), 1–139.

[63] M. R. Herman, Une méthode pour minorer les exposants de Lyapounov et quelques exemples montrant le caractere local d'un théoreme d'Arnold et de Moser sur le tore de dimension 2. *Comment. Math. Helv.* **58** (1983), 453–502.

[64] D. R. Hofstadter, Energy levels and wave functions of Bloch electrons in rational and irrational magnetic fields. *Phys. Rev. B* **14** (1976), 22–39.

[65] X. Hou and J. You, Almost reducibility and non-perturbative reducibility of quasi-periodic linear systems. *Invent. Math.* **190** (2012), no. 1, 209–260.

[66] S. Jitomirskaya, Almost everything about the almost Mathieu operator. II. In *XIth International Congress of Mathematical Physics (Paris, 1994)*, pp. 373–382, International Press of Boston, 1994.

[67] S. Jitomirskaya, Ergodic Schrödinger operators (on one foot), Spectral theory and mathematical physics: a Festschrift in honor of Barry Simon's 60th birthday, pp. 613–647, Proc. Sympos. Pure Math. 76, Part 2, Amer. Math. Soc., Providence, RI, 2007.

[68] S. Jitomirskaya, Critical phenomena, arithmetic phase transitions, and universality: some recent results on the almost Mathieu operator. *Curr. Dev. Math.* **2019** (2019), no. 1, 1–42.

[69] S. Jitomirskaya, On point spectrum at critical coupling. *Adv. Math.* **392** (2021), 6 pp.

[70] S. Jitomirskaya and I. Kachkovskiy, $L2$-reducibility and localization for quasiperiodic operators. *Math. Res. Lett.* **23** (2016), no. 2, 431–444.

[71] S. Jitomirskaya and S. Kocic, Spectral theory of Schrödinger operators over circle diffeomorphisms, *Int. Math. Res. Not.* **2** (2021). DOI 10.1093/imrn/rnaa362.

[72] S. Jitomirskaya, D. A. Koslover, and M. S. Schulteis, Localization for a family of one-dimensional quasiperiodic operators of magnetic origin. *Ann. Henri Poincaré* **6** (2005), 103–124.

[73] S. Jitomirskaya, D. A. Koslover, and M. S. Schulteis, Continuity of the Lyapunov Exponent for analytic quasiperiodic cocycles. *Ergodic Theory Dynam. Systems* **29** (2009), 1881–1905.

[74] S. Jitomirskaya, H. Krüeger, and W. Liu, Exact dynamical decay rate for the almost Mathieu operator. *Math. Res. Lett.* **27** (2020), no. 3, 789–808.

[75] S. Jitomirskaya and W. Liu, in preparation.

[76] S. Jitomirskaya and W. Liu, Arithmetic spectral transitions for the Maryland model. *Comm. Pure Appl. Math.* **70** (2017), no. 6, 1025–1051.

[77] S. Jitomirskaya and W. Liu, Universal hierarchical structure of quasiperiodic eigenfunctions. *Ann. of Math. (2)* **187** (2018), 721–776.

[78] S. Jitomirskaya and W. Liu, Universal reflective-hierarchical structure of quasiperiodic eigenfunctions and sharp spectral transition in phase. 2018, arXiv:1802.00781.

[79] S. Jitomirskaya, W. Liu, and S. Tcheremchantsev, Upper bounds on the fractal spectral dimensions and singular continuous spectrum near the arithmetic transition, in preparation.

[80] S. Jitomirskaya, W. Liu, and S. Zhang, Arithmetic spectral transitions: A competition between hyperbolicity and the arithmetics of small denominators IAS/Park City Math. Ser.: Harmon. Anal. Appl. **27**, pp. 35–72, 2020. DOI 10.1090/pcms/027/02.

[81] S. Jitomirskaya and C. Marx, Analytic quasi-periodic cocycles with singularities and the Lyapunov Exponent of Extended Harper's Model. *Comm. Math. Phys.* **316** (2012), 237–267.

[82] S. Jitomirskaya and C. Marx, Dynamics and spectral theory of quasi-periodic Schrödinger-type operators. *Ergodic Theory Dynam. Systems* **37** (2017), 2353–2393.

[83] S. Jitomirskaya and B. Simon, Operators with singular continuous spectrum: III. Almost periodic Schrödinger operators. *Comm. Math. Phys.* **165** (1994), 201–205.

[84] S. Y. Jitomirskaya, Anderson localization for the almost Mathieu equation: a nonperturbative proof. *Comm. Math. Phys.* **165** (1994), 49–57.

[85] S. Y. Jitomirskaya, Metal-insulator transition for the almost Mathieu operator. *Ann. of Math. (2)* **150** (1999), 1159–1175.

[86] S. Jitomirskaya and F. Yang, Phase transition in phase: the interface. 2021, preprint.

[87] S. Jitomirskaya and F. Yang, Pure point spectrum for the Maryland model: a constructive proof. *Ergodic Theory Dynam. Systems* **41** (2021), no. 1, 283–294.

[88] S. Jitomirskaya, F. Yang, and Q. Zhou, Second phase transition line: a constructive proof, in preparation.

[89] S. Jitomirskaya and S. Zhang, Quantitative continuity of singular continuous spectral measures and arithmetic criteria for quasiperiodic Schödinger operators. *J. Eur. Math. Soc. (JEMS)* (2022). DOI 10.4171/jems/1139.

[90] S. Jitomirskaya and X. Zhu, Large deviations of the Lyapunov exponent and localization for the 1D Anderson model. *Comm. Math. Phys.* **370** (2019), 311–324.

[91] R. A. Johnson, Exponential dichotomy, rotation number, and linear differential operators with bounded coefficients. *J. Differential Equations* **61** (1986), 54–78.

[92] S. Klein, Anderson localization for the discrete one-dimensional quasi-periodic Schrödinger operator with potential defined by a Gevrey-class function. *J. Funct. Anal.* **218** (2005), 255–292.

[93] F. Klopp and A. A. Fedotov, On the hierarchical behavior of solutions of the Maryland equation in the semiclassical approximation. *Math. Notes* **108** (2020), no. 5, 906–910.

[94] S. Kotani, Ljapunov indices determine absolutely continuous spectra of stationary one-dimensional Schrödinger operators. In *Stochastic Analysis*, pp. 225–248, N.-Holl. Math. Libr. 32, Elsevier, 1984.

[95] H. Kunz and B. Souillard, Sur le spectre des opérateurs aux différences finies aléatoires. *Comm. Math. Phys.* **78** (1980), 201–246.

[96] Last, Y.: Zero measure spectrum for the almost Mathieu operator. *Commun. Math Phys.* **164**(1994), 421–432 .

[97] Y. Last and B. Simon, Eigenfunctions, transfer matrices, and absolutely continuous spectrum of one-dimensional Schrödinger operators. *Invent. Math.* **135** (1999), 329–367.

[98] W. Liu, Almost Mathieu operators with completely resonant phases. *Ergodic Theory Dynam. Systems* **40** (2020), 1875–1893.

[99] W. Liu, Small denominators and large numerators of quasiperiodic Schrödinger operators. arXiv:2205.04648

[100] Y. Liu, Q. Zhou, and S. Chen, Localization transition, spectrum structure, and winding numbers for one-dimensional non-Hermitian quasicrystals. *Phys. Rev. B* **104** (2021), no. 2, 024201.

[101] V. A. Mandelshtam and S. Ya. Zhitomirskaya, 1D-quasiperiodic operators. Latent symmetries. *Comm. Math. Phys.* **139** (1991), 589–604.

[102] L. A. Pastur, Spectral properties of disordered systems in the one-body approximation. *Comm. Math. Phys.* **75** (1980), 179–196.

[103] R. Peierls, Zur Theorie des Diamagnetismus von Leitungselektronen. *Z. Phys. A: Hadrons Nucl.* **80** (1933), 763–791.

[104] J. Puig, Cantor spectrum for the almost Mathieu operator. *Comm. Math. Phys.* **244** (2004), 297–309.

[105] B. Simon, Almost periodic Schrödinger operators: A review. *Adv. in Appl. Math.* **3** (1982), 463–490.

[106] B. Simon, Fifteen problems in mathematical physics. *Oberwolfach Anniversary Volume* (1984), 423–454.

[107] B. Simon, Almost periodic Schrödinger operators. IV. The Maryland model. *Ann. Physics* **159** (1985), no. 1, 157–183.

[108] B. Simon, In *Schrödinger operators in the twenty-first century*, pp. 283–288, Math. Physics 2000, Imp. Coll. Press, London, 2000.

[109] Ya. Sinai, Anderson localization for one-dimensional difference Schrödinger operator with quasi-periodic potential. *J. Stat. Phys.* **46** (1987), 861–909.

[110] E. Sorets and T. Spencer, Positive Lyapunov exponents for Schrödinger operators with quasi-periodic potential. *CMP* **142** (1991), 543–566.

[111] A. Volberg and P. Yuditskii, Kotani–Last problem and Hardy spaces on surfaces of Widom type. *Invent. Math.* **197** (2014), 683–740.

[112] Y. Wang, X. Xia, J. You, Z. Zheng, and Q. Zhou, Exact mobility edges for 1D quasiperiodic models. 2021, arXiv:2110.00962.

[113] Y. Wang and J. You, Examples of discontinuity of Lyapunov exponent in smooth quasi-periodic cocycles. *Duke Math. J.* **1** (2013), no. 62, 2363–2412.

[114] Y. Wang and J. You, Quasi-Periodic Schrödinger Cocycles with Positive Lyapunov Exponent are not Open in the Smooth Topology. *Comm. Math. Phys.* **362** (2018), no. 9, 801–826.

[115] Y. Wang and Z. Zhang, Cantor spectrum for a class of C^2 quasiperiodic Schrödinger operators. *Int. Math. Res. Not.* **2017** (2017), no. 8, 2300–2366.

[116] P. B. Wiegmann and A. V. Zabrodin, Quantum group and magnetic translations Bethe ansatz for the Azbel–Hofstadter problem. *Nucl. Phys. B* **422** (1994), 495–514.

[117] J. You, Quantitative almost reducibility and applications. In *Proceedings of ICM 2018*, pp. 2131–2154, World Scientific Publishing Co Pte Ltd, Singapore, 2019.

[118] J. You and Q. Zhou, Embedding of analytic quasi-periodic cocycles into analytic quasi-periodic linear systems and its applications. *Comm. Math. Phys.* **323** (2013), no. 3, 975–1005.

[119] J. You and Q. Zhou, Simple Counter-Examples to Kotani–Last Conjecture Via Reducibility. *Int. Math. Res. Not.* **2015** (2015), no. 19, 9450–9455.

[120] S. Zhitomirskaya, Singular spectral properties of a one dimensional discrete Schrödinger operator with quasiperiodic potential. *Adv. Sov. Math.* **3** (1991), 215–254.

SVETLANA JITOMIRSKAYA

Georgia Institute of Technology, Atlanta GA 30332, USA; and UCI Department of Mathematics, Irvine CA 92617, USA, szhitomi@uci.edu

This Page Intentionally Left Blank

ABELIAN POLE SYSTEMS AND RIEMANN–SCHOTTKY-TYPE PROBLEMS

IGOR KRICHEVER

ABSTRACT

In this survey of works on a characterization of Jacobians and Prym varieties among indecomposable principally polarized abelian varieties via the soliton theory, we focus on a certain circle of ideas and methods which show that the characterization of Jacobians as ppav whose Kummer variety admits a trisecant line and the Pryms as ppav whose Kummer variety admits a pair of symmetric quadrisecants can be seen as an abelian version of pole systems arising in the theory of elliptic solutions to the basic soliton hierarchies. We present also recent results in this direction on the characterization of Jacobians of curves with involution, which were motivated by the theory of two-dimensional integrable hierarchies with symmetries.

MATHEMATICS SUBJECT CLASSIFICATION 2020

Primary 14H42; Secondary 35Q51

KEYWORDS

Riemann–Schottky problem, soliton equations, Baker–Akhiezer functioon, Calogero–Moser system

INTERNATIONAL CONGRESS OF MATHEMATICIANS 2022 JULY 6–14

Proc. Int. Cong. Math. 2022, Vol. 2, pp. 1122–1153
DOI 10.4171/ICM2022/120

1. INTRODUCTION

Novikov's conjecture on the Riemann–Schottky problem, namely that *the Jacobians of smooth algebraic curves are precisely those indecomposable principally polarized abelian varieties (ppavs) whose theta-functions provide solutions to the Kadomtsev–Petviashvili (KP) equation*, was the first evidence of nowadays well-established fact: connections between the algebraic geometry and the modern theory of integrable systems is beneficial for both sides. Novikov' conjecture was proved by T. Shiota in [48].

The first goal of this paper is to present the strongest known characterization of a Jacobian variety in this direction: *an indecomposable ppav X is the Jacobian of a curve if and only if its Kummer variety $K(X)$ has a trisecant line*, which was proved in [26,27]. This characterization is called *Welters' (trisecant) conjecture* after the work of Welters [54] which was motivated by Novikov's conjecture and Gunning's celebrated theorem [22]. The approach to its solution, proposed in [26], is general enough to be applicable to a variety of Riemann–Schottky-type problems. In [21,25] it was used for a characterization of principally polarized Prym varieties. The latter problem is almost as old and famous as the Riemann–Schottky problem, but is much harder.

Our second goal is to present recent results on characterization of Jacobians of curves with involution. The curves with involution naturally appear as a part of algebraic-geometrical data defining solutions to integrable system with symmetries. Numerous examples of such systems include the Kadomtsev–Petviashvili hierarchies of type B and C (BKP and CKP hierarhies, respectively) introduced in [11,12] and the Novikov–Veselov hierarchy introduced in [51,52].

The existence of an involution of a curve is central in proving that the constructed solutions have the necessary symmetry. The solutions corresponding to the same curve are usually parameterized by points of its Prym variety. In other words, the existence of involution plus some extra constraints on the divisor of the Baker–Akhiezer function are *sufficient* conditions ensuring required symmetry. The problem of proving that these conditions are *necessary* for *two-dimensional* integrable hierarchies is much harder and that is the problem solved in [28].

The third and to some extent our primary objective is to take this opportunity to elaborate on motivations underlining the proposed solution of the Riemann–Schottky-type problems and to introduce a certain collection of ideas and methods.

Maybe the most important among them is a mysterious *generating property* of two-dimensional linear differential, differential-functional, difference-functional equations. In fact, we will discuss two sources of generating properties. One of them is *local*, and it concerns equations with *meromorphic* coefficients in one of the variables that have *meromorphic solutions*. The other is *global* and concerns equations with elliptic coefficients that have solutions that are meromorphic sections of a line bundle over an elliptic curve [24].

The three main examples are:

(i) the differential equation

$$\big(\partial_t - \partial_x^2 + u(x,t)\big)\psi(x,t) = 0, \quad u = -2\partial_x^2\tau(x,t), \tag{1.1}$$

(ii) the differential-functional equation

$$\partial_t \psi(x,t) = \psi(x+1,t) + w(x,t)\psi(x,t), \quad w(x,t) = \partial_t \ln\left(\frac{\tau(x+1,t)}{\tau(x,t)}\right), \tag{1.2}$$

(iii) the difference-functional equation

$$\psi_{n+1}(x) = \psi_n(x+1) - v_n(x)\psi_n(x), \quad v_n(x) = \frac{\tau_n(x)\tau_{n+1}(x+1)}{\tau_n(x+1)\tau_{n+1}(x)} \tag{1.3}$$

with unknown functions $\psi_n(x)$, $n \in \mathbb{Z}$.

Each of these equations (after change of notations for independent variables) is one of two auxiliary linear problems for the three fundamental equations in the theory of integrable systems: the Kadomtsev–Petviashvili (KP) equation

$$3u_{yy} = (4u_t - 6uu_x + u_{xxx})_x, \tag{1.4}$$

the $2D$ Toda equation

$$\partial_\xi \partial_\eta \varphi_n = e^{\varphi_n - \varphi_n} - e^{\varphi_n - \varphi_{n+1}}, \quad \varphi_n = \varphi(x = n, \xi, \eta), \tag{1.5}$$

and the Bilinear Discrete Hirota equation (BDHE)

$$\begin{aligned} &\tau_n(l+1,m)\tau_n(l,m+1) - \tau_n(l,m)\tau_n(l+1,m+1) \\ &\quad + \tau_{n+1}(l+1,m)\tau_{n-1}(l,m+1) = 0, \end{aligned} \tag{1.6}$$

respectively.

At the first glance, all three nonlinear equations, the KP equation, 2D Toda equation, and BDHE equation, look very different from each other. But in the theory of integrable systems, it is well known that these fundamental soliton equations share an intimate relation: the KP equation is as a continuous limit of the BDHE, and the 2D Toda equation can be obtained in an intermediate step.

Assume that in the first two cases $\tau(x,t)$ is an *entire* function of the variable x and a (local) smooth function of the variable t, and in the third case $\tau_n(x)$ is a sequence of entire functions of x. It turns out that under some generality assumption for each of the above linear equations, the answer to the question *when it has a meromorphic in x solution* is given in terms of equations describing the evolution of zeros of τ in the second variable.

To give an idea of these equations and why I called the very existence of them mysterious, as an instructive example, consider equation (1.1).

Let ψ be a meromorphic solution of (1.1) with $u = -2\partial_x^2 \ln \tau(x,t)$, where τ is an entire function of x and a smooth function of t in some neighborhood of $t = 0$. The generality assumption is that generic zeros of τ are simple. Consider the Laurent expansions of ψ and u in the neighborhood of a simple zero, $\tau(q(t),t) = 0$, $\partial_x \tau(q(t),t) \neq 0$:

$$\begin{aligned} u &= \frac{2}{(x-q)^2} + v + w(x-q) + \cdots; \\ \psi &= \frac{\alpha}{x-q} + \beta + \gamma(x-q) + \delta(x-q)^2 + \cdots. \end{aligned} \tag{1.7}$$

(The coefficients in these expansions $v, w, \ldots; \alpha, \beta, \ldots$ are smooth functions of the variable t). Substituting (1.7) into (1.1) gives an *infinite* system of equations. The first three of them are

$$\begin{aligned} &\alpha\dot{q} + 2\beta = 0; \\ &\dot{\alpha} + \alpha v + 2\gamma = 0; \\ &\dot{\beta} + v\beta - \gamma\dot{q} + \alpha w = 0. \end{aligned} \tag{1.8}$$

Taking the t-derivative of the first equation and using the other two, we get the equation

$$\ddot{q} = 2w, \tag{1.9}$$

derived first in [6].

We would like to emphasize once again that there is no reason for the fact that the system (1.8) can be reduced to equations for the potential u only. Even more unexpected for the author was that, as we will see later, the existence of *one* meromorphic solution of equation (1.1) is sufficient for the existence of a *one-parameter* family of meromorphic solutions.

Formally, if we represent τ as an infinite product,

$$\tau(x,t) = c(t)\prod_i \big(x - q_i(t)\big), \tag{1.10}$$

then equation (1.9) can be written as the infinite system of equations

$$\ddot{q}_i = -4\sum_{j\neq i} \frac{1}{(q_i - q_j)^3}. \tag{1.11}$$

Equations (1.11) are purely formal because, even if τ has simple zeros at $t = 0$, in the general case there is no nontrivial interval of t where the zeros remain simple. One of the reasons to present (1.11) is that it shows that, when τ is a rational, trigonometric, or elliptic polynomial, equations (1.11) coincide with the equations of motion for the rational, trigonometrical, or elliptic Calogero–Moser (CM) system, respectively.

In a similar way, one can get that the existence of a meromorphic solution for equations (1.2) and (1.3) gives equations on zeros of τ which in the case when τ is an elliptic polynomial in x turned out to be the equations of motion of the elliptic Ruijsenaars–Schneider (RS) model and nested Bethe ansatz equations, respectively.

Recall that the elliptic CM system with k particles is a Hamiltonian system with coordinates $q = (q_1, \ldots, q_k)$, momentums $p = (p_1, \ldots, p_k)$, the canonical Poisson brackets $\{q_i, p_j\} = \delta_{ij}$, and the Hamiltonian

$$H = \frac{1}{2}\sum_{i=1}^{k} p_i^2 + \sum_{i\neq j} \wp(q_i - q_j). \tag{1.12}$$

The corresponding equations of motion admit the Lax representation $\dot{L} = [M, L]$ with

$$L_{ij} = p_i\delta_{ij} + 2(1 - \delta_{ij})\Phi(q_i - q_j, z), \tag{1.13}$$

where

$$\Phi(x,z)=\frac{\sigma(z-x)}{\sigma(z)\sigma(z)}e^{x\zeta(z)} \tag{1.14}$$

and $\sigma, \zeta, \wp$ are classical Weierstrass functions.

The ellliptic RS system is a Hamiltonian system with coordinates $q=(q_1,\dots,q_k)$, momentums $p=(p_1,\dots,p_k)$, the canonical Poisson brackets $\{q_i,p_j\}=\delta_{ij}$, and the Hamiltonian

$$H=\sum_{i=1}^{k} f_i, \tag{1.15}$$

where

$$f_i := e^{p_i}\prod_{j\neq i}\left(\frac{\sigma(q_i-q_j-1)\sigma(q_i-q_j+1)}{\sigma(q_i-q_j)^2}\right)^{1/2}. \tag{1.16}$$

It is a completely integrable Hamiltonian system, whose equations of motion admit the Lax representation $\dot{L}=[M,L]$, where

$$L_{ij}=f_i\Phi(q_i-q_j-1,z), \quad i,j=1,\dots,k, \tag{1.17}$$

The elliptic nested Bethe ansatz equations are a system of algebraic equations

$$\prod_j \frac{\sigma(q_i^n-q_j^{n+1})\sigma(q_i^n-1-q_j^n)\sigma(q_i^n-q_j^{n-1}+1)}{\sigma(q_i^n-q_j^{n-1})\sigma(q_i^{n+1}-q_j^n)\sigma(q_i^n-q_j^{n+1}-1)}=-1 \tag{1.18}$$

for k unknown functions $q_i=\{q_i^n\}$, $i=1,\dots,k$, of a discrete time variable $n\in\mathbb{Z}$.

The above systems are usually called elliptic *pole* systems, since they describe the dependence of the poles of the elliptic solutions of the KP, $2D$ Toda, and BDHE equations, respectively. A correspondence between finite-dimensional integrable systems and the pole systems of various soliton equations was considered in [7,29,32,33]. In [2] it was generalized to the case of *field analogues* of CM type systems.

The most general form of the function τ, known to the author so far, for which the equations for its zeros are not formal, is the case of *abelian* functions, that is, when τ has the form

$$\tau=\tau(Ux+z,t), \tag{1.19}$$

where $x,t\in\mathbb{C}$ and $z\in\mathbb{C}^n$ are independent variables, $0\neq U\in\mathbb{C}n$, and for all t the function $\tau(\cdot,t)$ is a holomorphic section of a line bundle $\mathcal{L}=\mathcal{L}(t)$ on an abelian variety $X=\mathbb{C}^n/\Lambda$, i.e., for all $\lambda\in\Lambda$ it satisfies the monodromy relations

$$\tau(z+\lambda,t)=e^{a_\lambda\cdot z+b_\lambda}\tau(z,t), \tag{1.20}$$

for some $a_\lambda\in\mathbb{C}^n$, $b_\lambda=b_\lambda(y,t)\in\mathbb{C}$.

It is tempting to call them *abelian* CM, RS, and nested Bethe ansatz equations. As we shall see below, they are central for the proof of three particular cases of the Welters conjecture.

2. RIEMANN–SCHOTTKY PROBLEM

Let $\mathbb{H}_g := \{B \in M_g(\mathbb{C}) \mid {}^t B = B, \operatorname{Im}(B) > 0\}$ be the Siegel upper half-space. For $B \in \mathbb{H}_g$ let $\Lambda := \Lambda_B := \mathbb{Z}^g + B\mathbb{Z}^g$ and $X := X_B := \mathbb{C}^g/\Lambda_B$. Riemann's theta function

$$\theta(z) := \theta(z, B) := \sum_{m \in \mathbb{Z}^g} e^{2\pi i(m,z) + \pi i(m,Bm)}, \quad (m, z) = m_1 z_1 + \cdots + m_g z_g, \tag{2.1}$$

is holomorphic and Λ-quasiperiodic in $z \in \mathbb{C}^g$.

The factor space $\mathbb{H}_g/Sp(2g, \mathbb{Z}) \simeq \mathcal{A}_g$ is the moduli space of g-dimensional ppavs. A ppav $(X, [\Theta]) \in \mathcal{A}_g$ is said to be *indecomposable* if the zero-divisor Θ of θ is irreducible.

Let $\mathcal{M}_g$ be the moduli space of nonsingular curves of genus g, and let $J : \mathcal{M}_g \to \mathcal{A}_g$ be the Jacobi map defined by the composition of maps $\mathcal{M}_g \to \mathbb{H}_g \to \mathcal{A}_g$. The first one requires a choice of a symplectic basis a_i, b_i $(i = 1, \dots, g)$ of $H_1(\Gamma, \mathbb{Z})$ which defines a basis $\omega_1, \dots, \omega_g$ of the space of holomorphic 1-forms on Γ such that $\int_{a_i} \omega_j = \delta_{ij}$, and then the *period matrix* and the *Jacobian variety* of Γ by

$$B := \left(\int_{b_i} \omega_j \right) \in \mathbb{H}_g \quad \text{and} \quad J(\Gamma) := \big(X_B, [\Theta_B]\big) \in \mathcal{A}_g,$$

respectively.

The above $J(\Gamma)$ is indecomposable and the Jacobi map J is injective (Torelli's theorem). The *(Riemann–)Schottky problem* is the problem of characterizing the Jacobi locus $\mathcal{J}_g := J(\mathcal{M}_g)$ or its closure $\overline{\mathcal{J}_g}$ in $\mathcal{A}_g$. For $g = 2, 3$, the dimensions of $\mathcal{M}_g$ and $\mathcal{A}_g$ coincide, and hence $\overline{\mathcal{J}_g} = \mathcal{A}_g$ by Torelli's theorem. Since $\mathcal{J}_4$ is of codimension 1 in $\mathcal{A}_4$, the case $g = 4$ is the first nontrivial case of the Riemann–Schottky problem.

A nontrivial relation for the Thetanullwerte of a curve of genus 4 was obtained by F. Schottky [45] in 1888, giving a modular form which vanishes on $\mathcal{J}_4$, and hence at least a *local* solution of the Riemann–Schottky problem in $g = 4$, i.e., $\overline{\mathcal{J}_4}$ is *an irreducible component of* the zero locus $\mathcal{S}_4$ of the Schottky relation. The irreducibility of $\mathcal{S}_4$ was proved by Igusa [23] in 1981, establishing $\overline{\mathcal{J}_4} = \mathcal{S}_4$, an effective answer to the Riemann–Schottky problem in genus 4.

A generalization of the Schottky relation to a curve of higher genus, the so-called Schottky–Jung relations, formulated as a conjecture by Schottky and Jung [46] in 1909, was proved by Farkas–Rauch [18] in 1970. Later, van Geemen [50] proved that the Schottky–Jung relations give a local solution of the Riemann–Schottky problem. They do not give a global solution when $g > 4$, since the variety they define has extra components already for $g = 5$ (Donagi [16]).

Over more than 120 year-long history of the Riemann–Schottky problem, quite a few geometric characterizations of the Jacobians have been obtained. None of them provides an explicit system of equations for the image of the Jacobian locus in the projective space under the level-two theta imbedding.

Following Mumford's review with a remark on Fay's trisecant formula [42], and the advent of algebraic geometrical integration scheme in the soliton theory [30, 31, 43] and Novikov's conjecture, significant progress was made in the 1980s in characterizing Jacobians and Pryms using Fay-like formulas and KP-like equations.

Let us first describe the trisecant identity in geometric terms. The Kummer variety $K(X)$ of $X \in \mathcal{A}_g$ is the image of the Kummer map

$$K = K_X : X \ni z \mapsto \big(: \Theta[\varepsilon, 0](z) :\big) \in \mathbb{CP}^{2^g-1}, \tag{2.2}$$

where $\Theta[\varepsilon, 0](z) = \theta[\varepsilon, 0](2z, 2B)$ are the level-two theta-functions with half-integer characteristics $\varepsilon \in ((1/2)\mathbb{Z}/\mathbb{Z})^g$, i.e., they equal $\theta(2(z + B\varepsilon), 2B)$ up to some exponential factor so that we have

$$\theta(z + w)\theta(z - w) = \sum_{\varepsilon \in ((1/2)\mathbb{Z}/\mathbb{Z})^g} \Theta[\varepsilon, 0](z)\Theta[\varepsilon, 0](w). \tag{2.3}$$

We have $K(-z) = K(z)$ and $K(X) \simeq X/\{\pm 1\}$.

A *trisecant* of the Kummer variety is a projective line which meets $K(X)$ at three points. *Fay's trisecant formula* states that if $X = J(\Gamma)$, then $K(X)$ has a family of trisecants parameterized by 4 points A_i, $1 \le i \le 4$, on Γ. Gunning proved in [22] that, under certain nondegeneracy conditions, that the existence of a *one-parametric* family of trisecants characterizes the Jacobians.

Gunning's work was extended by Welters who proved that a Jacobian variety can be characterized by the existence of a formal one-parameter family of flexes of the Kummer variety [53]. A flex of the Kummer variety is a projective line which is tangent to $K(X)$ at some point up to order 2. It is a limiting case of trisecants when the three intersection points come together.

In [5] Arbarello and De Concini showed that the assumption in Welters' characterization is equivalent to an infinite sequence of partial differential equations known as the KP hierarchy, and proved that only a few first equations in the sequence are sufficient, by giving an explicit bound for the number of equations, $N = [(3/2)^g g!]$, based on the degree of $K(X)$.

An algebraic argument based on earlier results of Burchnall, Chaundy, and the author [10, 30, 31] characterizes the Jacobians using a commutative ring R of ordinary differential operators associated to a solution of the KP hierarchy. A simple counting argument then shows that only the first $2g + 1$ time evolutions in the hierarchy are needed to obtain R. The $2g + 1$ KP flows yield a finite number of differential equations for the Riemann theta function θ of X, to characterize a Jacobian. As for the number of equations, an easy estimate shows that $4g^2$ is enough, although a more careful argument should yield a better bound.

Novikov's conjecture, namely that just the first equation of the hierarchy ($N = 1!$) suffices to characterize the Jacobians, i.e.,

an indecomposable symmetric matrix B with positive definite imaginary part is the period matrix of a basis of normalized holomorphic differentials on a smooth algebraic curve Γ if and only if there are vectors $U \neq 0, V, W$, such that the function

$$u(x, y, t) = 2\partial_x^2 \ln \theta(Ux + Vy + Wt + Z|B), \tag{2.4}$$

satisfies the KP equation (1.4),

for quite some time seemed to be the strongest possible characterization within the reach of the soliton theory.

3. WELTER'S CONJECTURE

Novikov's conjecture is equivalent to the statement that the Jacobians are characterized by the existence of length 3 formal jet of flexes.

In [54] Welters formulated the question: *if the Kummer variety $K(X)$ has* one *trisecant, does it follow that X is a Jacobian?* In fact, there are three particular cases of the Welters conjecture, corresponding to three possible configurations of the intersection points (a, b, c) of $K(X)$ and the trisecant:

(i) all three points coincide ($a = b = c$);

(ii) two of them coincide ($a = b \neq c$);

(iii) all three intersection points are distinct ($a \neq b \neq c \neq a$).

Of course, the first two cases can be regarded as degenerations of the general case (iii). However, when the existence of only one trisecant is assumed, all three cases are independent and require their own approaches. The approaches used in [26,27] were based on the theories of three main soliton hierarchies (see details in [39]): the KP hierarchy for (i), the 2D Toda hierarchy for (ii) and the Bilinear Discrete Hirota Equations (BDHE) for (iii). Recently, pure algebraic proofs of the first two cases of the trisecant conjecture were obtained in [4].

Theorem 3.1. *An indecomposable principally polarized abelian variety (X, θ) is the Jacobian variety of a smooth algebraic curve of genus g if and only if there exist g-dimensional vectors $U \neq 0$, V, A, and constants p and E such that one of the following three equivalent conditions is satisfied:*

(A) *equality* (1.1) *with* $\tau = \theta(Ux + Vt + Z)$ *and*

$$\psi = \frac{\theta(A + Ux + Vt + Z)}{\theta(Ux + Vt + Z)} e^{px+Et} \tag{3.1}$$

holds, for an arbitrary vector Z;

(B) *for all theta characteristics $\varepsilon \in (\frac{1}{2}\mathbb{Z}/\mathbb{Z})^g$,*

$$\big(\partial_V - \partial_U^2 - 2p\partial_U + (E - p^2)\big)\Theta[\varepsilon, 0](A/2) = 0$$

(here and below ∂_U, ∂_V are the derivatives along the vectors U and V, respectively);

(C) *on the theta-divisor $\Theta = \{Z \in X \mid \theta(Z) = 0\}$, the equation*

$$\begin{aligned} &\big[(\partial_V\theta)^2 - (\partial_U^2\theta)^2\big]\partial_U^2\theta + 2\big[\partial_U^2\theta\partial_U^3\theta - \partial_V\theta\partial_U\partial_V\theta\big]\partial_U\theta \\ &\quad + \big[\partial_V^2\theta - \partial_U^4\theta\big](\partial_U\theta)^2 = 0 \end{aligned} \tag{3.2}$$

holds.

The direct substitution of expression (3.1) into equation (1.1) and the use of the addition formula for the Riemann theta-functions shows the equivalence of conditions (A) and (B) in the theorem. Condition (B) means that the image of the point $A/2$ under the Kummer map is an inflection point (case (i) of Welters' conjecture).

Condition (C), which we call the abelian CM system, is the relation that is *really used* in the proof of the theorem. Formally, it is weaker than the other two conditions because its derivation does not use an explicit form of the solution ψ of equation (1.1), but requires only that ψ is a *meromorphic solution*. The latter, as we have seen, implies equation (1.9). Expanding the function θ in a neighborhood of a point $z \in \Theta := \{z \mid \theta(z) = 0\}$ such that $\partial_U \theta(z) \neq 0$, and noting that the latter condition holds on a dense subset of Θ since B is indecomposable, it is easy to see that equation (1.9) is equivalent to (3.2).

Equation (1.1) is one of the two auxiliary linear problems for the KP equation. For the author, the motivation to consider not the whole KP equation but just one of its auxiliary linear problems was his earlier work **[32]** on the elliptic Calogero–Moser (CM) system, where it was observed for the first time that equation (1.1) is all what one needs to construct the elliptic solutions of the KP equation.

The proof of Welters' conjecture was completed in **[27]**. First, here is the theorem which treats case (ii) of the conjecture:

Theorem 3.2. *An indecomposable, principally polarized abelian variety (X, θ) is the Jacobian of a smooth curve of genus g if and only if there exist nonzero g-dimensional vectors $U \neq A \pmod{\Lambda}$, V constants p, E, such that one of the following equivalent conditions holds:*

(A) *equation* (1.2) *with $\tau = \theta(Ux + Vt + Z)$ and ψ as in* (3.1) *holds for an arbitrary Z;*

(B) *the equations*

$$\partial_V \Theta[\varepsilon, 0]\big((A - U)/2\big) - e^p \Theta[\varepsilon, 0]\big((A + U)/2\big) + E\Theta[\varepsilon, 0]\big((A - U)/2\big) = 0,$$

are satisfied for all $\varepsilon \in (\frac{1}{2}\mathbb{Z}/\mathbb{Z})^g$. Here and below ∂_V is the constant vector field on $\mathbb{C}^g$ corresponding to the vector V;

(C) *the equation*

$$\partial_V\big[\theta(Z + U)\theta(Z - U)\big]\partial_V \theta(Z) = \big[\theta(Z + U)\theta(Z - U)\big]\partial^2_{VV}\theta(Z) \quad (3.3)$$

is valid on the theta-divisor $\Theta = \{Z \in X \mid \theta(Z) = 0\}$.

Recall, that equation (1.2) is one of the two auxiliary linear problems for the $2D$ Toda lattice equation (1.5). The idea to use it for the characterization of the Jacobians was motivated by **[26]** and the author's earlier work with Zabrodin **[33]**, where a connection of the theory of elliptic solutions of the $2D$ Toda lattice equations and the theory of the elliptic Ruijsenaars–Schneider system was established.

Statement (B) is the second particular case of the trisecant conjecture: the line in $\mathbb{CP}^{2^g - 1}$ passing through the points $K((A - U)/2)$ and $K((A + U)/2)$ of the Kummer variety is tangent to $K(X)$ at the point $K((A - U)/2)$.

Condition (C) is what we call the abelian RS equation.

The affirmative answer to the third particular case, (iii), of Welters' conjecture is given by the following statement.

Theorem 3.3. *An indecomposable, principally polarized abelian variety* (X, θ) *is the Jacobian of a smooth curve of genus g if and only if there exist nonzero g-dimensional vectors* $U \neq V \neq A \neq U \pmod{\Lambda}$ *such that one of the following equivalent conditions holds:*

(A) *equation* (1.3) *with* $\tau_n(x) = \theta(xU + nV + Z)$ *and*

$$\psi_n(x) = \frac{\theta(A + xU + nV + Z)}{\theta(xU + nV + Z)} e^{xp+nE}, \tag{3.4}$$

holds for an arbitrary Z*;*

(B) *the equations*

$$\Theta[\varepsilon, 0]\left(\frac{A - U - V}{2}\right) + e^p \Theta[\varepsilon, 0]\left(\frac{A + U - V}{2}\right) = e^E \Theta[\varepsilon, 0]\left(\frac{A + V - U}{2}\right),$$

are satisfied for all $\varepsilon \in (\frac{1}{2}\mathbb{Z}/\mathbb{Z})^g$*;*

(C) *the equation*

$$\frac{\theta(Z + U)\theta(Z - V)\theta(Z - U + V)}{\theta(Z - U)\theta(Z + V)\theta(Z + U - V)} = -1 \pmod{\theta} \tag{3.5}$$

is valid on the theta-divisor $\Theta = \{Z \in X \mid \theta(Z) = 0\}$.

Under the assumption that the vector U *spans an elliptic curve* in X, Theorem 3.3 was proved in [29], where the connection of the elliptic solutions of BDHE and the so-called elliptic nested Bethe ansatz equations was established. Condition (C) is its abelian generalization.

4. THE PROBLEM OF CHARACTERIZATION OF PRYM VARIETIES

An involution $\sigma : \Gamma \to \Gamma$ on a smooth algebraic curve Γ naturally determines an involution $\sigma^* : J(\Gamma) \mapsto J(\Gamma)$ on its Jacobian. The odd subspace with respect to this involution is a sum of an Abelian subvariety of lower dimension, called the Prym variety, and a finite group. The restriction of the principal polarization of the Jacobian determines a polarization of the Prym variety which is principal if and only if the original involution of the curve has at most two fixed points. The problem of characterizing the locus $\mathcal{P}_g$ of Prym varieties of dimension g in the space $\mathcal{A}_g$ of all principally polarized Abelian varieties is well known and during its history has attracted considerable interest. This problem is much harder than the Riemann–Schottky problem and until relatively recently its solution in terms of a finite system of equations was completely open.

The problem of characterizing Prym varieties in the case of curves with an involution having two fixed points was solved in [25] in terms of the Schrödinger operators integrable with respect to one energy level. The theory of such operators was developed by Novikov and Veselov in [51, 52], where the authors also introduced the corresponding nonlinear equation, the so-called Novikov–Veselov equation. Curves with an involution having a pair of fixed points can be regarded as a limit of unramified covers. A characterization of the Prym varieties in the latter case in terms of the existence of quadrisecants was obtained the author and Grushevsky in [21].

The existence of families of quadrisecants for curves with an involution having at most two fixed points was proved in [9, 20]. An analogue of Gunning's theorem asserting that the existence of a family of secants characterizes Prym varieties was proved by Debarre [13]. We note that the existence of one quadrisecant does not characterize Prym varieties. A counterexample to the naive generalization of Welters' conjecture was constructed by Beauville and Debarre in [9].

It was proved in [21] that the existence of a symmetric pair of quadrisecants is a characteristic property for Prym varieties of unramified covers.

Theorem 4.1 (Geometric characterization of Prym varieties). *An indecomposable principally polarized Abelian variety $(X, \theta) \in \mathcal{A}_g$ is in the closure of the locus of Prym varieties of smooth unramified double covers if and only if there exist four distinct points $p_1, p_2, p_3, p_4 \in X$, none of them of order two, such that the images of the Kummer map of the eight points $p_1 \pm p_2 \pm p_3 \pm p_4$ lie on two quadrisecants (the corresponding quadruples of points are determined by the number of plus signs).*

We should note that the proof of this statement required constructing and developing the theory of a new integrable equation because before that, in contrast with all other cases, no nonlinear equations whose algebro-geometric solutions are associated to unramified double covers were known.

The auxiliary linear equation of the corresponding analogue of the Novikov–Veselov equation is a discrete analogue of the potential Schrödinger equation considered first in [15]. It has the form

$$\psi_{n+1,m+1} - u_{n,m}(\psi_{n+1,m} - \psi_{n,m+1}) - \psi_{n,m} = 0. \tag{4.1}$$

The analog of condition (C) in the previous theorem which can also be thought as the abelian generalization of some discrete time integrable system (which has not been studied so far) is as follows:

(C) *There are constants $c_i^{\pm}, i = 1, 2, 3$ such that two equations (one for the top choice of signs everywhere, and one for the bottom)*

$$\begin{aligned} &c_1^{\mp 2} c_3^2 \theta(Z + U - V)\theta(Z - U \pm W)\theta(Z + V \pm W) \\ &\quad + c_2^{\mp 2} c_3^2 \theta(Z - U + V)\theta(Z + U \pm W)\theta(Z - V \pm W) \\ &\quad = c_1^{\mp 2} c_2^{\mp 2} \theta(Z - U - V)\theta(Z + U \pm W)\theta(Z + V \pm W) \\ &\quad + \theta(Z + U + V)\theta(Z - U \pm W)\theta(Z - V \pm W) \end{aligned} \tag{4.2}$$

are valid on the theta divisor $\{Z \in X : \theta(Z) = 0\}$.

5. ABELIAN SOLUTIONS OF THE SOLITON EQUATIONS

The general concept of *abelian solutions* of soliton equations was introduced by T. Shiota and the author in [37, 38]. It provides a unifying framework for the theory of the elliptic solutions of these equations and algebraic-geometrical solutions of rank 1 expressible in terms of Riemann (or Prym) theta-function. A solution $u(x, y, t)$ of the KP equation is called *abelian* if it is of the form

$$u = -2\partial_x^2 \ln \tau(Ux + z, y, t), \tag{5.1}$$

where $x, y, t \in \mathbb{C}$, and $z \in \mathbb{C}^n$ are independent variables, $0 \neq U \in \mathbb{C}^n$, and for all y, t the function $\tau(\cdot, y, t)$ is a holomorphic section of a line bundle $\mathcal{L} = \mathcal{L}(y, t)$ on an abelian variety $X = \mathbb{C}^n/\Lambda$, i.e., for all $\lambda \in \Lambda$ it satisfies the monodromy relations (1.19).

In the case of sections of the canonical line bundle on a principally polarized Abelian variety the corresponding theta-function is unique up to normalization. Hence the ansatz (5.1) takes the form $u = -2\partial_x^2 \ln \theta(Ux + Z(y, t) + z)$. Since flows commute with each other, the dependence of the vector $Z(y, t)$ must be linear,

$$u = -2\partial_x^2 \ln \theta(Ux + Vy + Wt + z). \tag{5.2}$$

Therefore, the problem of classification of such Abelian solutions is the same problem as posed by Novikov.

In the case of one-dimensional Abelian varieties, the problem of classification of Abelian solutions is the problem of classification of the elliptic solutions. The theory of elliptic solutions of the KP equation goes back to the remarkable work [1], where it was found that the dynamics of poles of the elliptic (rational or trigonometric) solutions of the Korteweg–de Vries equation can be described in terms of the elliptic (rational or trigonometric) Calogero–Moser (CM) system with certain constraints. It was observed in [32] that, when the constraints are removed, this restricted correspondence becomes an isomorphism when the elliptic solutions of the KP equation are considered. The elliptic solutions of the KP equation are distinguished amongst the general algebraic-geometric solutions by the condition that the corresponding vector U spans an elliptic curve embedded into the Jacobian of the curve. Note that, for any vector U, the closure of the group $\{Ux \mid x \in \mathbb{C}\}$, is an Abelian subvariety $X \subset J(\Gamma)$. So when this closure does not coincide with the whole Jacobian, we get nontrivial examples of Abelian solutions. Briefly, the main result on the classification of Abelian solutions of KP obtained in [37] can be formulated as the statement that all the Abelian solutions are obtained in this manner. To avoid some technical complications, we give the formulation of the corresponding theorem in the situation of general position.

Theorem 5.1. *Let $u(x, y, t)$ be an abelian solution of the KP such that the group $\mathbb{C}U \bmod \Lambda$ is dense in X. Then there exists a unique algebraic curve Γ with smooth marked point $P \in \Gamma$, holomorphic imbedding $j_0 : X \to J(\Gamma)$ and a torsion-free rank 1 sheaf $\mathcal{F} \in \overline{\mathrm{Pic}^{g-1}}(\Gamma)$ where $g = g(\Gamma)$ is the arithmetic genus of Γ, such that setting with the notation $j(z) = j_0(z) \otimes \mathcal{F}$*

$$\tau(Ux + z, y, t) = \rho(z, y, t)\hat{\tau}\big(x, y, t, 0, \ldots \mid \Gamma, P, j(z)\big), \tag{5.3}$$

where $\widehat{\tau}(t_1, t_2, t_3, \dots \mid \Gamma, P, \mathcal{F})$ is the KP τ-function corresponding to the data $(\Gamma, P, \mathcal{F})$, and $\rho(z, y, t) \not\equiv 0$ satisfies the condition $\partial_U \rho = 0$.

Note that if Γ is smooth then

$$\widehat{\tau}\big(x, t_2, t_3, \dots \mid \Gamma, P, j(z)\big) = \theta\Big(Ux + \sum V_i t_i + j(z) \Bigm| B(\Gamma)\Big)e^{Q(x,t_2,t_3,\dots)}, \tag{5.4}$$

where $V_i \in \mathbb{C}^n$, Q is a quadratic form, and $B(\Gamma)$ is the period matrix of Γ. A linearization on $J(\Gamma)$ of the nonlinear (y, t)-dynamics for $\tau(z, y, t)$ indicates the possibility of the existence of integrable systems on spaces of theta-functions of higher level. A CM system is an example of such a system for $n = 1$.

6. THE BAKER–AKHIEZER FUNCTIONS—GENERAL SCHEME

The "only if" part of all the theorems above is a corollary of the general algebraic-geometric construction of solutions of soliton equations based on a concept of the Baker–Akhiezer function.

Let Γ be a nonsingular algebraic curve of genus g with N marked points P_α and fixed local parameters $k_\alpha^{-1}(p)$ in neighborhoods of the marked points. The basic scalar *multipoint* and *multivariable* Baker–Akhiezer function $\psi(t, p)$ is a function of external parameters

$$t = (t_{\alpha,i}), \quad \alpha = 1, \dots, N; \quad i = 0, \dots; \quad \sum_\alpha t_{\alpha,0} = 0, \tag{6.1}$$

only finite number of which is nonzero, and a point $p \in \Gamma$. For each set of the external parameters t it is defined by its analytic properties on Γ.

Remark. For the simplicity we will begin with the assumption that the variables $t_{\alpha,0}$ are integers, i.e., $t_{\alpha,0} \in \mathbb{Z}$.

Lemma 6.1. *For any set of g points $\gamma_1, \dots, \gamma_g$ in a general position there exists a unique (up to constant factor $c(t)$) function $\psi(t, p)$, such that:*

(i) *the function ψ (as a function of the variable $p \in \Gamma$) is meromorphic everywhere except for the points P_α and has at most simple poles at the points $\gamma_1, \dots, \gamma_g$ (if all of them are distinct);*

(ii) *in a neighborhood of the point P_α the function ψ has the form*

$$\psi(t, p) = k_\alpha^{t_{\alpha,0}} \exp\left(\sum_{i=1}^{\infty} t_{\alpha,i} k_\alpha^i\right)\left(\sum_{s=0}^{\infty} \xi_{\alpha,s}(t) k_\alpha^{-s}\right), \quad k_\alpha = k_\alpha(p). \tag{6.2}$$

From the uniqueness of the Baker–Akhiezer function, we obtain

Theorem 6.1. *For each pair $(\alpha, n > 0)$, there exists a unique operator $L_{\alpha,n}$ of the form*

$$L_{\alpha,n} = \partial_{\alpha,1}^n + \sum_{j=0}^{n-1} u_j^{(\alpha,n)}(t)\partial_{\alpha,1}^j, \tag{6.3}$$

(*where* $\partial_{\alpha,n} = \partial/\partial t_{\alpha,n}$) *such that*

$$(\partial_{\alpha,n} - L_{\alpha,n})\psi(t, p) = 0. \tag{6.4}$$

The idea of the proof of the theorems of this type proposed in [30, 31] is universal.

For any formal series of the form (6.2), their exists a unique operator $L_{\alpha,n}$ of the form (6.3) such that

$$(\partial_{\alpha,n} - L_{\alpha,n})\psi(t, p) = O(k_\alpha^{-1}) \exp\left(\sum_{i=1}^{\infty} t_{\alpha,i} k_\alpha^i\right). \tag{6.5}$$

The coefficients of $L_{\alpha,n}$ are universal differential polynomials with respect to $\xi_{s,\alpha}$. They can be found after substitution of the series (6.2) into (6.5).

It turns out that if the series (6.2) is not formal, but is an expansion of the Baker–Akhiezer function in the neighborhood of P_α, the *congruence* (6.5) *becomes an equality*. Indeed, let us consider the function ψ_1,

$$\psi_1 = (\partial_{\alpha,n} - L_{\alpha,n})\psi(t, p). \tag{6.6}$$

It has the same analytic properties as ψ except for one. The expansion of this function in the neighborhood of P_α starts from $O(k_\alpha^{-1})$. From the uniqueness of the Baker–Akhiezer function it follows that $\psi_1 = 0$ and the equality (6.4) is proved.

Corollary 6.1. *The operators* $L_{\alpha,n}$ *satisfy the compatibility conditions*

$$[\partial_{\alpha,n} - L_{\alpha,n}, \partial_{\alpha,m} - L_{\alpha,m}] = 0. \tag{6.7}$$

Equations (6.7) are gauge invariant. For any function $c(t)$, operators

$$\tilde{L}_{\alpha,n} = cL_{\alpha,n}c^{-1} + (\partial_{\alpha,n}c)c^{-1} \tag{6.8}$$

have the same form (6.3) and satisfy the same operator equations (6.7). The gauge transformation (6.8) corresponds to the gauge transformation of the Baker–Akhiezer function

$$\widetilde{\psi}(t, p) = c(t)\psi(t, p). \tag{6.9}$$

In addition to differential equations (6.4), the Baker–Akhiezer function satisfies an infinite system of differential-difference equations. Recall that the discrete variables $t_{\alpha,0}$ are subject to the constraint $\sum_\alpha t_{\alpha,0} = 0$. Therefore, only the first $(N-1)$ of them are independent and $t_{N,0} = -\sum_{\alpha=1}^{N-1} t_{\alpha,0}$. Let us denote by T_α, $\alpha = 1, \dots, N-1$, the operator that shifts the arguments $t_{\alpha,0} \to t_{\alpha,0} + 1$ and $t_{N,0} \to t_{N,0} - 1$, respectively. For the sake of brevity, in the formulation of the next theorem we introduce the operator $T_N = T_1^{-1}$.

Theorem 6.2. *For each pair* $(\alpha, n > 0)$, *there exists a unique operator* $\hat{L}_{\alpha,n}$ *of the form*

$$\hat{L}_{\alpha,n} = T_\alpha^n + \sum_{j=0}^{n-1} v_j^{(\alpha,n)}(t) T_\alpha^j, \quad v_0^{(N,n)}(t) = 0 \tag{6.10}$$

such that

$$(\partial_{\alpha,n} - \hat{L}_{\alpha,n})\psi(t, p) = 0. \tag{6.11}$$

The proof is identical to that in the differential case.

Corollary 6.2. *The operators $\hat{L}_{\alpha,n}$ satisfy the compatibility conditions*

$$[\partial_{\alpha,n} - \hat{L}_{\alpha,n}, \partial_{\alpha,m} - \hat{L}_{\alpha,m}] = 0. \tag{6.12}$$

Theta-functional formulae. It should be emphasized that the algebro-geometric construction is not a sort of abstract "existence" and "uniqueness" theorems. It provides the explicit formulae for solutions in terms of the Riemann theta-functions. They are the corollary of the explicit formula for the Baker–Akhiezer function.

Let $a_i, b_i \in H_1(\Gamma, \mathbb{Z})$, $i = 1, \ldots, g$, be a basis of cycles on Γ with the canonical intersection matrix, i.e., $a_i \cdot a_j = b_i \cdot b_j = 0$, $a_i \cdot b_j = \delta_{ij}$, and let ω_i be the basis of holomorphic differentials on Γ normalized by the equations $\oint_{a_j} \omega_j = \delta_{ij}$. The matrix B of their b-periods $B_{ij} = \oint_{b_i} \omega_j$ is indecomposable symmetric matrix with positive definite imaginary part. By formula (2.1), it defines the Riemann theta-function $\theta(z) = \theta(z|B)$.

Theorem 6.3. *The Baker–Akhiezer function is given by the formula*

$$\psi(t, p) = c(t)\exp\Big(\sum t_{\alpha,i}\Omega_{\alpha,i}(p)\Big)\frac{\theta(A(p) + \sum U_{\alpha,i}t_{\alpha,i} + Z)}{\theta(A(p) + Z)}. \tag{6.13}$$

Here the sum is taken over all the indices $(\alpha, i > 0)$ and over the indices $(\alpha, 0)$ with $\alpha = 1, \ldots, N-1$, and

(a) *$\Omega_{\alpha,i}(p)$ is the abelian integral, $\Omega_{\alpha,i}(p) = \int^p d\Omega_{\alpha,i}$, corresponding to the unique normalized, $\oint_{a_k} d\Omega_{\alpha,i} = 0$, meromorphic differential on Γ, which for $i > 0$ has the only pole of the form $d\Omega_{\alpha,i} = d(k_\alpha^i + O(1))$ at the marked point P_α and for $i = 0$ has simple poles at the marked point P_α and P_N with residues ± 1, respectively;*

(b) *$2\pi i U_{\alpha,j}$ is the vector of b-periods of the differential $d\Omega_{\alpha,j}$, i.e.,*

$$U_{\alpha,j}^k = \frac{1}{2\pi i}\oint_{b_k} d\Omega_{\alpha,j};$$

(c) *$A(p)$ is the Abel transform, i.e., a vector with the coordinates $A_i(p) = \int^p d\omega_i$;*

(d) *Z is an arbitrary vector (it corresponds to the divisor of poles of Baker–Akhiezer function).*

Notice that from the bilinear Riemann relations it follows that the expansion of the Abel transform near the marked point has the form

$$A(p) = A(P_\alpha) - \sum_{i=1}^{\infty}\frac{1}{i}U_{\alpha,i}k_\alpha^{-i}. \tag{6.14}$$

Example 1. One-point Baker–Akhiezer function. KP hierarchy. In the one-point case, the Baker–Akhiezer function has an exponential singularity at a single point P_1 and depends on a single set of variables $t_i = t_{1,i}$. Note that in this case there is no discrete variable,

$t_{1,0} \equiv 0$. Let us choose the normalization of the Baker–Akhiezer function with the help of the condition $\xi_{1,0} = 1$, i.e., an expansion of ψ in the neighborhood of P_1 equals

$$\psi(t_1, t_2, \ldots, p) = \exp\left(\sum_{i=1}^{\infty} t_i k^i\right)\left(1 + \sum_{s=1}^{\infty} \xi_s(t) k^{-s}\right). \tag{6.15}$$

Under this normalization (gauge), the corresponding operator L_n has the form

$$L_n = \partial_1^n + \sum_{i=0}^{n-2} u_i^{(n)} \partial_1^i. \tag{6.16}$$

For example, for $n = 2, 3$, after redefinition $x = t_1$ we have

$$L_2 = \partial_x^2 - u, \quad L_3 = \partial_x^3 - \frac{3}{2} u \partial_x - w, \tag{6.17}$$

with $u = 2\partial_x \xi_1$, $w = 3\partial_x \xi_2 + 3\partial_x^2 \xi_1 - \frac{3}{2} u \xi_1$.

If we define $y = t_2$, $t = t_3$, then from (6.7), with $n = 2$ and $m = 3$, it follows that $u(x, y, t, t_4, \ldots)$ satisfies the KP equation (1.4).

The normalization of the leading coefficient in (6.15) defines the function $c(t)$ in (6.13). This gives the following formula for the normalized one-point Baker–Akhiezer function:

$$\psi(t, p) = \exp\left(\sum t_i \Omega_i(p)\right) \frac{\theta(A(p) + \sum U_i t_i + Z)\theta(Z)}{\theta(\sum U_i t_i + Z)\theta(A(p) + Z)}, \tag{6.18}$$

(shifting Z if needed we may assumed that $A(P_1) = 0$). In order to get the explicit theta-functional form of the solution of the KP equation, it is enough to take the derivative of the first coefficient of the expansion at the marked point of the ratio of theta-functions in the formula (6.18).

Using (6.14), we get the final formula for the algebro-geometric solutions of the KP hierarchy [31], namely

$$u(t_1, t_2, \ldots) = -2\partial_1^2 \ln \theta\left(\sum_{i=1}^{\infty} U_i t_i + Z\right) + \text{const}. \tag{6.19}$$

Example 2. Two-point Baker–Akhiezer function. 2*D* Toda hierarchy. In the two-point case, the Baker–Akhiezer function has exponential singularities at two points $P_\alpha, \alpha = 1, 2$, and depends on two sets of continuous variables $t_{\alpha,i>0}$. In addition, it depends on one discrete variable $n = t_{1,0} = -t_{2,0}$. Let us choose the normalization of the Baker–Akhiezer function with the help of the condition $\xi_{1,0} = 1$.

According to Theorem 6.1, the function ψ satisfies two sets of differential equations. The compatibility conditions (6.7) within the each set can be regarded as two copies of the KP hierarchies. In addition the two-point Baker–Akhiezer function satisfies differential-difference equations (6.10). The first two of them have the form

$$(\partial_{1,1} - T + u)\psi = 0, \quad (\partial_{2,1} - wT^{-1})\psi = 0, \tag{6.20}$$

where

$$u = (T - 1)\xi_{1,1}(n, t), \quad w = e^{\varphi_n - \varphi_{n-1}}, \quad e^{\varphi_n(t)} = \xi_{2,0}(n, t). \tag{6.21}$$

The compatibility condition of these equations is equivalent to the $2D$ Toda equation with $\xi = t_{1,1}$ and $\eta = t_{2,1}$. The explicit formula for the solution $\varphi_n(t)$ is a direct corollary of the explicit formula for the Baker–Akhiezer function,

$$\varphi_n(t_{\alpha,i>0}) = \ln \frac{\theta((n+1)U + \sum U_{\alpha,i} t_{\alpha,i} + Z)}{\theta(nU + \sum U_{\alpha,i} t_{\alpha,i} + Z)}, \quad \alpha = 1, 2. \tag{6.22}$$

Example 3. Three-point Baker–Akhiezer function. Starting with three-point case, in which the number of discrete variables is 2, the Baker–Akhiezer function satisfies certain linear difference equations (in addition to the differential and the differential-difference equations (6.4), (6.11)). The origin of these equations is easy to explain. Indeed, if all the continuous variables vanish, $t_{\alpha,i>0} = 0$, then the Baker–Akhiezer function $\psi_{n,m} := \psi(n, m, p)$, where $n = -t_{1,0}$, $m = -t_{2,0}$, is a meromorphic function having a pole of order $n + m$ at P_3 and zeros of order n and m at P_1 and P_2, respectively, i.e.,

$$\psi_{n,m} \in H^0\big(D + n(P_3 - P_1) + m(P_3 - P_2)\big), \quad D = \gamma_1 + \cdots + \gamma_g. \tag{6.23}$$

The functions $\psi_{n+1,m}, \psi_{n,m+1}, \psi_{n,m}$ are all in the linear space $H^0(D + (n + m + 1)P_3 - nP_1 - mP_2)$. By Riemann–Roch theorem, for a generic D the latter space is 2-dimensional. Hence, these functions are linearly dependent, and they can be normalized such their linear dependence takes the form

$$\psi_{m,n+1} = \psi_{m+1,n} + u_{m,n}\psi_{m,n} \tag{6.24}$$

with

$$u_{n,m} = \frac{\tau_{m+1,n+1}\tau_{m,n}}{\tau_{m,n+1}\tau_{m+1,n}}, \quad \tau_{m,n} := \theta(mU + nV + Z). \tag{6.25}$$

At first glance, it seems that everything here is within the framework of classical algebraic-geometry. What might be new and brought to this subject by the soliton theory is understanding that *the discrete variables $t_{\alpha,0}$ can be replaced by continuous ones*. Of course, if in the formula (6.13) the variable $t_{\alpha,0}$ is not an integer, then ψ is not a single-valued function on Γ. Nevertheless, because the monodromy properties of ψ do not change if the shift of the argument by an integer, it satisfies the same type of linear equations with coefficients given by the same type of formulae. It is necessary to emphasize that in such a form the difference equation becomes a *functional* equation.

In the four-point case, there are three discrete variables n, m, and l. For each two of them the Baker–Akhiezer function satisfies a difference equation. The compatibility of these equations is expressed by the BDHE equation.

7. KEY IDEA AND STEPS OF THE PROOFS

As it was mentioned above, the proof of all the particular cases of Welters' trisecant conjecture uses different hierarchies: the KP, the 2D Toda, and the BDHE. In each case, there are some specific difficulties, but the main ideas and structures of the proof are the same. As an instructive example, we present in this section the idea and key steps of the proof of the first particular case of Welters' conjecture, namely, the proof of Theorem 3.1.

As it was mentioned above, the implication (A) → (C) is a direct corollary of (1.9). Now we are going to show that (1.9), which is satisfied when (1.1) has *one* meromorphic solution, is sufficient for the existence of *one-parametric* family of formal wave solutions below.

The wave solution of (1.1) is a solution of the form

$$\psi(x,y,k) = e^{kx+(k^2+b)t}\left(1+\sum_{s=1}^{\infty}\xi_s(x,t)k^{-s}\right). \tag{7.1}$$

Lemma 7.1. *Suppose that equations* (1.9) *for the zeros of* $\tau(x,t)$ *hold. Then there exist meromorphic wave solutions of equation* (1.1) *that have simple poles at zeros* q *of* τ *and are holomorphic everywhere else.*

Proof. Substitution of (7.1) into (1.1) gives a recurrent system of equations

$$2\xi'_{s+1} = \partial_t \xi_s + u\xi_s - \xi''_s. \tag{7.2}$$

We are going to prove by induction that this system has meromorphic solutions with simple poles at all the zeros q of τ.

Let us expand ξ_s at q to get

$$\xi_s = \frac{r_s}{x-q} + r_{s0} + r_{s1}(x-q) + \cdots. \tag{7.3}$$

Suppose that ξ_s is defined and equation (7.2) has a meromorphic solution. Then the right-hand side of (7.2) has the zero residue at $x = q$, i.e.,

$$\operatorname{res}_q(\dot\xi_s + u\xi_s - \xi''_s) = \dot r_s + v_i r_s + 2r_{s1} = 0. \tag{7.4}$$

We need to show that the residue of the next equation vanishes also. From (7.2) it follows that the coefficients of the Laurent expansion for ξ_{s+1} are equal to

$$r_{s+1} = -\dot q r_s - 2r_{s0}, \quad 2r_{s+1,1} = \dot r_{s0} - r_{s1} + wr_s + vr_{s0}. \tag{7.5}$$

These equations imply

$$\dot r_{s+1} + vr_{s+1} + 2r_{s+1,1} = -r_s(\ddot q - 2w) - \dot q(\dot r_s + vr_s + 2r_{s1}) = 0, \tag{7.6}$$

and the lemma is proved. ∎

λ-periodic wave solutions. Our next step in the proof is to fix a *translation-invariant* normalization of ξ_s which defines wave functions uniquely up to an x-independent factor. It is instructive to consider first the case of the periodic potentials $u(x+1,t) = u(x,t)$ (see the details in [36]).

Equations (7.2) are solved recursively by the formulae

$$\xi_{s+1}(x,t) = c_{s+1}(t) + \xi^0_{s+1}(x,t), \tag{7.7}$$

$$\xi^0_{s+1}(x,t) = \frac{1}{2}\int_{x_0}^{x}(\dot\xi_s - \xi''_s + u\xi_s)dx = 0, \tag{7.8}$$

where $c_s(t)$ are *arbitrary* functions of the variable t. Let us show that the periodicity condition $\xi_s(x+1,t) = \xi_s(x,t)$ defines the functions $c_s(t)$ uniquely up to an additive constant.

Assume that ξ_{s-1} is known and satisfies the condition that the corresponding function ξ_s^0 is periodic. The choice of the function $c_s(t)$ does not affect the periodicity property of ξ_s, but it does affect the periodicity in x of the function $\xi_{s+1}^0(x,t)$. In order to make $\xi_{s+1}^0(x,t)$ periodic, the function $c_s(t)$ should satisfy the linear differential equation

$$\partial_t c_s(t) + B(t)c_s(t) + \int_{x_0}^{x_0+1} \big(\dot{\xi}_s^0(x,t) + u(x,t)\xi_s^0(x,y)\big)\, dx, \tag{7.9}$$

where $B(t) = \int_{x_0}^{x_0+1} u\, dx$. This defines c_s uniquely up to a constant.

In the general case, when u is quasiperiodic, the normalization of the wave functions is defined along the same lines.

Let $Y_U = \langle \mathbb{C}U \rangle$ be the Zariski closure of the group $\mathbb{C}U = \{Ux \mid x \in \mathbb{C}\}$ in X. Shifting Y_U if needed, we may assume, without loss of generality, that Y_U is not in the singular locus Σ defined as the ∂_U-invariant subset of the theta-divisor Θ, i.e., $Y_U \not\subset \Sigma$. Then, for a sufficiently small t, we have $Y_U + Vt \notin \Sigma$ as well. Consider the restriction of the theta-function onto the affine subspace $\mathbb{C}^d + Vt$, where $\mathbb{C}^d :=$ (the identity component of $\pi^{-1}(Y_U)$), and $\pi : \mathbb{C}^g \to X = \mathbb{C}^g/\Lambda$ is the universal covering map of X:

$$\tau(z,t) = \theta(z + Vt), \quad z \in \mathbb{C}^d. \tag{7.10}$$

The function $u(z,t) = -2\partial_U^2 \ln \tau$ is periodic with respect to the lattice $\Lambda_U = \Lambda \cap \mathbb{C}^d$ and, for a fixed t, has a double pole along the divisor $\Theta^U(t) = (\Theta - Vt) \cap \mathbb{C}^d$.

Lemma 7.2. *Let equations* (1.9) *for the zeros of* $\tau(Ux + z, t)$ *hold. Then:*

(i) *equation* (1.1) *with the potential* $u(Ux + z, t)$ *has a wave solution of the form* $\psi = e^{kx + k^2 y}\phi(Ux + z, t, k)$ *such that the coefficients* $\xi_s(z, y)$ *of the formal series*

$$\phi(z,t,k) = e^{bt}\Big(1 + \sum_{s=1}^{\infty} \xi_s(z, tk^{-s}\Big) \tag{7.11}$$

are meromorphic functions of the variable $z \in \mathbb{C}^d$ *with a simple pole at the divisor* $\Theta^U(t)$,

$$\xi_s(z + \lambda, t) = \xi_s(z,t) = \frac{\tau_s(z,t)}{\tau(z,t)}; \tag{7.12}$$

(ii) $\phi(z,t,k)$ *is quasiperiodic with respect to* Λ_U, *i.e., for* $\lambda \in \Lambda_U$,

$$\phi(z + \lambda, t, k; z_0) = \phi(z,t,k;z_0)\mu_\lambda(k); \tag{7.13}$$

(iii) $\phi(z,t,k)$ *is unique up to a* ∂_U*-invariant factor which is an exponent of the linear form,*

$$\phi_1(z,t,k) = \phi(z,t,k)e^{(\ell(k),z)}, \quad \big(\ell(k), U\big) = 0. \tag{7.14}$$

The spectral curve. The next goal is to show that λ-periodic wave solutions of equation (1.1) are common eigenfunctions of rings of commuting operators.

Note that a simple shift $z \to z + Z$, where $Z \notin \Sigma$, gives λ-periodic wave solutions with meromorphic coefficients along the affine subspaces $Z + \mathbb{C}^d$. These λ-periodic wave solutions are related to each other by a ∂_U-invariant factor. Therefore choosing, in a neighborhood of any $Z \notin \Sigma$, a hyperplane orthogonal to the vector U and fixing initial data on this hyperplane at $y = 0$, we define the corresponding series $\phi(z + Z, t, k)$ as a *local* meromorphic function of Z and the *global* meromorphic function of z.

Lemma 7.3. *Let the assumptions of Theorem* 3.1 *hold. Then there is a unique pseudo-differential operator*

$$\mathcal{L}(Z, \partial_x) = \partial_x + \sum_{s=1}^{\infty} w_s(Z)\partial_x^{-s} \tag{7.15}$$

such that

$$\mathcal{L}(Ux + Vy + Z, \partial_x)\psi = k\psi, \tag{7.16}$$

where $\psi = e^{kx+k^2y}\phi(Ux + Z, t, k)$ *is a* λ*-periodic solution of* (1.1). *The coefficients* $w_s(Z)$ *of* $\mathcal{L}$ *are meromorphic functions on the abelian variety* X *with poles along the divisor* Θ.

Proof. Let ψ be a λ-periodic wave solution. The substitution of (7.11) into (7.16) gives a system of equations that recursively define $w_s(Z, t)$ as differential polynomials in $\xi_s(Z, t)$. The coefficients of ψ are local meromorphic functions of Z, but the coefficients of $\mathcal{L}$ are well-defined *global meromorphic functions* on $\mathbb{C}^g \setminus \Sigma$ because different λ-periodic wave solutions are related to each other by a ∂_U-invariant factor, which does not affect $\mathcal{L}$. The singular locus is of codimension ≥ 2. Then Hartogs' holomorphic extension theorem implies that $w_s(Z, t)$ can be extended to a global meromorphic function on $\mathbb{C}^g$.

The translational invariance of u implies the translational invariance of the λ-periodic wave solutions. Indeed, for any constant s, the series $\phi(Vs + Z, t - s, k)$ and $\phi(Z, t, k)$ correspond to λ-periodic solutions of the same equation. Therefore, they coincide up to a ∂_U-invariant factor. This factor does not affect $\mathcal{L}$. Hence, $w_s(Z, t) = w_s(Vt + Z)$.

The λ-periodic wave functions corresponding to Z and $Z + \lambda'$ for any $\lambda' \in \Lambda$ are also related to each other by a ∂_U-invariant factor. Hence, w_s are periodic with respect to Λ and therefore are meromorphic functions on the abelian variety X. The lemma is proved. ∎

Consider now the differential parts of the pseudodifferential operators $\mathcal{L}^m$. Let $\mathcal{L}^m_+$ be the differential operator such that $\mathcal{L}^m_- = \mathcal{L}^m - \mathcal{L}^m_+ = F_m\partial^{-1} + O(\partial^{-2})$. The leading coefficient F_m of $\mathcal{L}^m_-$ is the residue of $\mathcal{L}^m$:

$$F_m = \operatorname{res}_\partial \mathcal{L}^m. \tag{7.17}$$

From the definition of $\mathcal{L}$, it follows that $[\partial_t - \partial_x^2 + u, \mathcal{L}^n] = 0$. Hence,

$$\left[\partial_t - \partial_x^2 + u, \mathcal{L}^m_+\right] = -\left[\partial_t - \partial_x^2 + u, \mathcal{L}^m_-\right] = 2\partial_x F_m. \tag{7.18}$$

The functions F_m are differential polynomials in the coefficients w_s of $\mathcal{L}$. Hence, $F_m(Z)$ are meromorphic functions on X. The next statement is crucial for the proof of the existence of commuting differential operators associated with u.

Lemma 7.4 ([26]). *The abelian functions F_m have at most second order poles on the divisor Θ.*

Let $\hat{F}$ be a linear space generated by $\{F_m, m = 0, 1, \ldots\}$, where we set $F_0 = 1$. It is a subspace of the 2^g-dimensional space of the abelian functions that have at most second order poles at Θ. Therefore, for all but $\hat{g} = \dim \hat{F}$ positive integers n, there exist constants $c_{i,n}$ such that

$$F_n(Z) + \sum_{i=0}^{n-1} c_{i,n} F_i(Z) = 0. \tag{7.19}$$

Let I denote the subset of integers n for which there are no such constants. We call this subset the gap sequence.

Lemma 7.5. *Let $\mathcal{L}$ be the pseudodifferential operator corresponding to a λ-periodic wave function ψ constructed above. Then, for the differential operators*

$$L_n = \mathcal{L}_+^n + \sum_{i=0}^{n-1} c_{i,n} \mathcal{L}_+^{n-i} = 0, \quad n \notin I, \tag{7.20}$$

the equations

$$L_n \psi = a_n(k)\psi, \quad a_n(k) = k^n + \sum_{s=1}^{\infty} a_{s,n} k^{n-s}, \tag{7.21}$$

where $a_{s,n}$ are constants, hold.

Proof. First note that from (7.18) it follows that

$$\left[\partial_t - \partial_x^2 + u, L_n\right] = 0. \tag{7.22}$$

Hence, if ψ is a λ-periodic wave solution of (1.1) corresponding to $Z \notin \Sigma$, then $L_n\psi$ is also a formal solution of the same equation. That implies the equation $L_n\psi = a_n(Z,k)\psi$, where a is ∂_U-invariant. The ambiguity in the definition of ψ does not affect a_n. Therefore, the coefficients of a_n are well-defined *global* meromorphic functions on $\mathbb{C}^g \setminus \Sigma$. The ∂_U-invariance of a_n implies that a_n, as a function of Z, is holomorphic outside of the locus. Hence it has an extension to a holomorphic function on $\mathbb{C}^g$. Equations (7.13) imply that a_n is periodic with respect to the lattice Λ. Hence a_n is Z-independent. Note that $a_{s,n} = c_{s,n}$, $s \leq n$. The lemma is proved. ∎

The operator L_m can be regarded as a $Z \notin \Sigma$-parametric family of ordinary differential operators L_m^Z whose coefficients have the form

$$L_m^Z = \partial_x^n + \sum_{i=1}^{m} u_{i,m}(Ux + Z)\partial_x^{m-i}, \quad m \notin I. \tag{7.23}$$

Corollary 7.1. *The operators L^Z_m commute with each other,*

$$[L^Z_n, L^Z_m] = 0, \quad Z \notin \Sigma. \tag{7.24}$$

From (7.21) it follows that $[L^Z_n, L^Z_m]\psi = 0$. The commutator is an ordinary differential operator. Hence, the last equation implies (7.24).

Lemma 7.6. *Let $\mathcal{A}^Z$, $Z \notin \Sigma$, be a commutative ring of ordinary differential operators spanned by the operators L^Z_n. Then there is an irreducible algebraic curve Γ of arithmetic genus $\hat{g} = \dim \hat{F}$ such that $\mathcal{A}^Z$ is isomorphic to the ring $A(\Gamma, P_0)$ of the meromorphic functions on Γ with the only pole at a smooth point P_0. The correspondence $Z \to \mathcal{A}^Z$ defines a holomorphic imbedding of $X \setminus \Sigma$ into the space of torsion-free rank 1 sheaves $\mathcal{F}$ on Γ*

$$j : X \setminus \Sigma \mapsto \overline{\mathrm{Pic}}(\Gamma). \tag{7.25}$$

The statement of the lemma is a corollary of the following fundamental fact from the theory of commuting differential operators

Theorem 7.1 ([10, 30, 31, 43]). *There is a natural correspondence*

$$\mathcal{A} \leftrightarrow \{\Gamma, P_0, [k^{-1}]_1, \mathcal{F}\} \tag{7.26}$$

between regular *at $x = 0$ commutative rings $\mathcal{A}$ of ordinary linear differential operators containing a pair of monic operators of coprime orders, and sets of algebraic-geometrical data $\{\Gamma, P_0, [k^{-1}]_1, \mathcal{F}\}$, where Γ is an algebraic curve with a fixed first jet $[k^{-1}]_1$ of a local coordinate k^{-1} in the neighborhood of a smooth point $P_0 \in \Gamma$ and $\mathcal{F}$ is a torsion-free rank 1 sheaf on Γ such that*

$$H^0(\Gamma, \mathcal{F}) = H^1(\Gamma, \mathcal{F}) = 0. \tag{7.27}$$

The correspondence becomes one-to-one if the rings $\mathcal{A}$ are considered modulo conjugation $\mathcal{A}' = g(x)\mathcal{A}g^{-1}(x)$.

Note that in [10, 30, 31] the main attention was paid to the generic case of the commutative rings corresponding to smooth algebraic curves. The invariant formulation of the correspondence given above is due to Mumford [43].

The algebraic curve Γ is called the spectral curve of $\mathcal{A}$. The ring $\mathcal{A}$ is isomorphic to the ring $A(\Gamma, P_0)$ of meromorphic functions on Γ with the only pole at the point P_0. The isomorphism is defined by the equation

$$L_a \psi_0 = a \psi_0, \quad L_a \in \mathcal{A}, \quad a \in A(\Gamma, P_0). \tag{7.28}$$

Lemma 7.7 ([26]). *The linear space $\hat{F}$ generated by the abelian functions $\{F_0 = 1, F_m = \mathrm{res}_\partial \mathcal{L}^m\}$ is a subspace of the space H generated by F_0 and by the abelian functions $H_i = \partial_U \partial_{z_i} \ln \theta(Z)$.*

The construction of multivariate Baker–Akhiezer functions presented for smooth curves is a manifestation of a general statement valid for singular spectral curves: flows of the KP hierarchy define deformations of the commutative rings $\mathcal{A}$ of ordinary linear differential operators. The spectral curve is invariant under these flows. For a given spectral curve Γ, the orbits of the KP hierarchy are isomorphic to the generalized Jacobian $J(\Gamma) = \mathrm{Pic}^0(\Gamma)$, which is the equivalence classes of zero degree divisors on the spectral curve (see the details in [30, 31, 47, 48]). Hence, for any $Z \notin \Sigma$, the orbit of the KP flows defines a holomorphic imbedding

$$i_Z : J(\Gamma) \mapsto X. \tag{7.29}$$

From (7.29) it follows that $J(\Gamma)$ is *compact*.

The generalized Jacobian of an algebraic curve is compact if and only if the curve is *smooth* [14]. On a smooth algebraic curve, a torsion-free rank 1 sheaf is a line bundle, i.e., $\overline{\mathrm{Pic}}(\Gamma) = J(\Gamma)$. Then (7.25) implies that i_Z is an isomorphism. Note that for the Jacobians of smooth algebraic curves, the bad locus Σ is empty [48], i.e., the imbedding j in (7.25) is defined everywhere on X and is inverse to i_Z. Theorem 3.1 is proved.

8. CHARACTERIZING JACOBIAN OF CURVES WITH INVOLUTION

As it was mentioned in the Introduction, the characterization problem of Jacobians of curves with involution addressed in [28] was motivated by the construction of solutions of two-dimensional integrable systems with symmetries. To the best of our knowledge, from a pure algebraic-geometrical perspective, the characterization problem of curves with involution in terms of their Jacobians has never been considered in its full generality. The only known to the author works in this direction are [8, 17, 44].

Two characterizations which distinguish such Jacobians were obtained in [28] within the framework of cases (i) and (ii) of Welter's conjecture. Both of them are limited to the case of involutions having at least one fixed point, i.e., to two-sheeted *ramified* covers.

In a certain sense, the setup we consider—the Jacobian and the Prym variety in it—resembles the setup arising in the famous Schottky–Yung relations, and it is tempting to find a way to get these relations by means of the soliton theory. Unfortunately, this challenging problem remains open.

The first characterization, related to the KP theory, is limited to the case of ramified cover for the obvious reason—a curve with one marked point is used in constructing its solutions.

Theorem 8.1. *An indecomposable principally polarized abelian variety (X, θ) is the Jacobian variety of a smooth algebraic curve Γ of genus g with involution $\sigma : \Gamma \to \Gamma$ having at least one point fixed if and only if there exist g-dimensional vectors $U \neq 0, V, A, \zeta$ and constants Ω_1, Ω_2, b_1 such that:*

Condition (A) *of Theorem* 3.1 *is satisfied and*

(B) *the intersection of the theta-divisor $\Theta = \{Z \in X \mid \theta(Z) = 0\}$ with a shifted abelian subvariety $Y \subset X$ which is the Zariski closure of $\pi(Ux + \zeta) \subset X$ is reduced and the equation*

$$\partial_V \theta|_{\Theta \cap Y} = 0 \tag{8.1}$$

holds.

Moreover, the locus Π of points $\zeta \in X$ for which equation (8.1) *holds is the locus of points for which the equation $\zeta + \sigma(\zeta) = 2P + K \in X$, where K is the canonical class, holds.*

Condition (B) implies

(C) *there is a constant b_2 such that the equality*

$$\partial_U \partial_V \ln \theta|_{\hat{Y}} = b_2 \tag{8.2}$$

holds on Y.

From the addition theorem (2.3), it follows that (8.2) is equivalent to the condition that the vector $(\partial_U \partial_V K(0) - b_2 K(0))$ is *orthogonal to the image under the Kummer map $K(\Pi)$ of the shifted abelian subvariety $\hat{Y}$*:

$$\sum_{\varepsilon \in ((1/2)\mathbb{Z}/\mathbb{Z})^g} \big(\partial_U \partial_V \Theta[\varepsilon, 0](0) - b_2 \Theta[\varepsilon, 0](0)\big)\Theta[\varepsilon, 0](z) = 0, \quad z \in \hat{Y}, \tag{8.3}$$

whence follows the condition *of a kind of flatness* of the image under the Kummer map of the shifted Prym subvariety $\Pi \subset X$, that is, $K(\Pi)$ lies in a proper (projective) linear subspace.

The explicit meaning (B) is as follows. As shown in **[19, 48]**, the affine line $Ux + Z$ is not contained in Θ for any vector Z. Hence, the function $\tau(x, t) := \theta(Ux + Vt + z)$, $z \in Y$ is a *nontrivial* entire function of x. The statement that $\Theta \cap Y$ is reduced means that the zeros $q(t)$ of τ, considered as a function of x (depending on t), are generically simple, $\tau(q(t), t) = 0$, $\tau_x(q(t), t) \neq 0$. Then (8.2) is the equation

$$\partial_t q|_{t=0} = 0. \tag{8.4}$$

In the case when U spans an elliptic curve in the Jacobian, the statement that from (B) it follows that the corresponding curve Γ admits an involution is obvious. Indeed, in that case the curve is the normalization of the spectral curve of N-point elliptic CM systems. The latter is defined by the characteristic equation

$$\det\big(k \cdot \mathbb{I} - L(z)\big) = 0$$

of the matrix $L(z)$ defined in (1.13) with $q_i = q_i(0)$ and $p_i = \dot{q}_i(0)$, where $q_i(t)$ are roots of the equation $\theta(Ux + Vt + z) = 0$. If equation (8.4) holds, i.e., $p_i = 0$, then it is easy to see that the matrix $L(z)$ satisfies the equation $L^t(z) = -L(-z)$. The latter implies that the curve is invariant under the involution $(k, z) \to (-k, -z)$. That observation made in **[34]** was the main motivation behind **[28]**.

At the heart of the proof in the general case is the statement that if (B) is satisfied then there is a local coordinate k^{-1} such that if $\psi(x, t, k)$ is the wave solution of (1.1) as in Lemma 7.2 then $\psi(x, 0, -k) = \psi^*(x, 0, k)$ where ψ^* is a wave solution of the equation

$$(\partial_t + \partial_x^2 - u)\psi^*(x, t, k) = 0, \tag{8.5}$$

which is formally adjoint to (1.1).

The second characterization of the Jacobians of curves with involution is related to the 2D Toda theory. A priori, unlike in the KP case, there is no obvious reason why it is not applicable to all types of involution, including unramified covers. It turned out that there is an obstacle for the case of unramified covers, and our second theorem also gives a characterization of the Jacobians of curves with involution *with* fixed points.

Theorem 8.2. *An indecomposable, principally polarized abelian variety* (X, θ) *is the Jacobian of a smooth curve of genus g with involution having fixed points if and only if there exist nonzero g-dimensional vectors* $U \neq A \pmod{\Lambda}$, V, ζ, *constants* Ω_0, Ω_1, b_1 *such that:*

Condition (A) *of Theorem* 3.2 *is satisfied and*

(B) (i) *the intersection of the theta-divisor with the shifted Abelian variety* Y, *which is a closure of* $\pi(Ux + \zeta)$, *is reduced and is not invariant under the shift by* U, $\Theta \cap Y \neq (\Theta + U) \cap Y$, *and* (ii) *the equation*

$$\left((\partial_V \theta(z))^2 + \theta(z + U)\theta(z - U) \right)\Big|_{z \in \Theta \cap Y} = 0 \tag{8.6}$$

holds.

Moreover, the locus of the points $\zeta \in X$ *for which equation* (8.6) *holds is the locus of point for which the equation* $\zeta + \zeta^\sigma = K + P_1 + P_2 \in J(\Gamma)$, *where* (P_1, P_2) *are points of the curves permuted by* σ *and such that* $U = A(P_2) - A(P_1)$ *is satisfied.*

Remark 1. In the case when U spans an elliptic curve in the Jacobian, the statement of the theorem was proved first in **[40]**.

The geometric form of the characterization is the condition that *the vector* $(2\partial_V^2 K(0) - b_2 K(U) - b_3 K(0))$ *is orthogonal to the image under the Kummer map of the abelian subvariety* Π:

$$\sum_{\varepsilon \in ((1/2)\mathbb{Z}/\mathbb{Z})^g} \left(2\partial_V^2 \Theta[\varepsilon, 0](0) - b_2 \Theta[\varepsilon, 0](U) - b_3 \Theta[\varepsilon, 0](0)\right)\right)\Theta[\varepsilon, 0](z) = 0, \tag{8.7}$$

where $z \in \Pi$ and b_3 is a constant.

9. NONLOCAL GENERATING PROBLEM

Until now our main focus was on equations that arise from the *local* generating properties of two-dimensional linear operators with meromorphic coefficients. The *nonlocal* generating properties of the same linear operators do not lead directly to equations of motion for zeros of the τ function. To begin with, they generate the Lax representation of these equations. That nonlocal perspective is known for the elliptic case. Its abelian generalization is an open and challenging problem.

Let $\mathcal{D}$ be a linear differential or difference operator in two variables (x, t) with coefficients which are scalar or matrix elliptic functions of the variable x (i.e., meromorphic double-periodic functions with the periods $2\omega_\alpha$, $\alpha = 1, 2$). We do not assume any special dependence of the coefficients with respect to the second variable. Then it is natural to introduce a notion of *double-Bloch* solutions of the equation

$$\mathcal{D}\Psi = 0. \tag{9.1}$$

We call a *meromorphic* vector-function $f(x)$ that satisfies the following monodromy properties:

$$f(x+2\omega_\alpha) = B_\alpha f(x), \quad \alpha = 1, 2, \tag{9.2}$$

a *double-Bloch function.* The complex numbers B_α are called *Bloch multipliers.* (In other words, f is a meromorphic section of a vector bundle over the elliptic curve.)

In the most general form, a problem considered in the framework of elliptic pole systems is to *classify* and to *construct* all the operators L such that equation (9.1) has *sufficiently many* double-Bloch solutions.

It turns out that the existence of the double-Bloch solutions is so restrictive that only in exceptional cases such solutions do exist. A simple and general explanation of that is due to the Riemann–Roch theorem. Let D be a set of points q_i, $i = 1, \dots, m$, on the elliptic curve Γ_0 with multiplicities d_i and let $V = V(D; B_1, B_2)$ be a linear space of the double-Bloch functions with the Bloch multipliers B_α that have poles at q_i of order less than or equal to d_i and holomorphic outside D. Then the dimension of D is equal to

$$\dim D = \deg D = \sum_i d_i.$$

Now let q_i depend on the variable t. Then for $f \in D(t)$, the function $\mathcal{D} f$ is a double-Bloch function with the same Bloch multipliers, but in general with higher orders of poles because taking derivatives and multiplication by the elliptic coefficients increase orders. Therefore, the operator $\mathcal{D}$ defines a linear operator

$$\mathcal{D}|_D : V\big(D(t); B_1, B_2\big) \mapsto V\big(D'(t); B_1, B_2\big), \quad N' = \deg D' > N = \deg D,$$

and (9.1) is *always* equivalent to an *overdetermined* linear system of N' equations in N unknown variables which are the coefficients $c_i = c_i(t)$ of an expansion of $\Psi \in V(t)$ with respect to a basis of functions $f_i(t) \in V(t)$. With some exaggeration, one may say that in the soliton theory the representation of a system in the form of the compatibility condition of an overdetermined system of the linear problems is considered as equivalent to integrability.

In all of known examples, $N' = 2N$ and the overdetermined system of equations has the form

$$LC = kC, \quad \partial_t C = MC, \tag{9.3}$$

where L and M are $N \times N$ matrix functions depending on a point z of the elliptic curve as a parameter. A compatibility condition of (9.3) has the standard Lax form $\partial_t L = [M, L]$, and is equivalent to a finite-dimensional integrable system.

The basis in the space of the double-Bloch functions can be written in terms of the fundamental function $\Phi(x, z)$ defined by the formula (1.14). Note that $\Phi(x, z)$ is a solution of the Lame equation

$$\left(\frac{d^2}{dx^2} - 2\wp(x)\right)\Phi(x, z) = \wp(z)\Phi(x, z). \tag{9.4}$$

From the monodromy properties, it follows that Φ, considered as a function of z, is doubly-periodic,

$$\Phi(x, z + 2\omega_\alpha) = \Phi(x, z),$$

though it is not elliptic in the classical sense due to an essential singularity at $z = 0$ for $x \neq 0$.

As a function of x, the function $\Phi(x, z)$ is a double-Bloch function, i.e.,

$$\Phi(x + 2\omega_\alpha, z) = T_\alpha(z)\Phi(x, z), \quad T_\alpha(z) = \exp\big(2\omega_\alpha\zeta(z) - 2\zeta(\omega_\alpha)z\big).$$

In the fundamental domain of the lattice defined by $2\omega_\alpha$, the function $\Phi(x, z)$ has a unique pole at the point $x = 0$,

$$\Phi(x, z) = x^{-1} + O(x). \tag{9.5}$$

The gauge transformation

$$f(x) \mapsto \tilde{f}(x) = f(x)e^{ax},$$

where a is an arbitrary constant, does not change the poles of any function and transforms a double Bloch-function into a double-Bloch function. If B_α are Bloch multipliers for f, then the Bloch multipliers for $\tilde{f}$ are equal to

$$\tilde{B}_1 = B_1 e^{2a\omega_1}, \quad \tilde{B}_2 = B_2 e^{2a\omega_2}. \tag{9.6}$$

The two pairs of Bloch multipliers that are connected with each other through the relation (9.6) for some a are called equivalent. Note that for all equivalent pairs of Bloch multipliers, the product $B_1^{\omega_2} B_2^{-\omega_1}$ is a constant depending on the equivalence class only.

From (9.5) it follows that a double-Bloch function $f(x)$ with simple poles q_i in the fundamental domain and with Bloch multipliers B_α (such that at least one of them is not equal to 1) may be represented in the form

$$f(x) = \sum_{i=1}^{N} c_i \Phi(x - q_i, z)e^{kx}, \tag{9.7}$$

where c_i is a residue of f at x_i and z, k are parameters related by

$$B_\alpha = T_\alpha(z)e^{2\omega_\alpha k}. \tag{9.8}$$

(Any pair of Bloch multipliers may be represented in the form (9.8) with an appropriate choice of the parameters z and k.)

To prove (9.7), it is enough to note that as a function of x the difference of the left- and right-hand sides is holomorphic in the fundamental domain. It is a double-Bloch function with the same Bloch multipliers as the function f. But a nontrivial double-Bloch function with at least one of the Bloch multipliers that is not equal to 1 has at least one pole in the fundamental domain.

Example: elliptic CM system. Let us consider equation (1.1) with an elliptic (in x) potential $u(x,t)$. Suppose that equation (1.1) has N linearly independent double-Bloch solutions with equivalent Bloch multipliers and N simple poles $q_i(t)$. The assumption that there exist N linearly independent double-Bloch solutions with equivalent Bloch multipliers implies that they can be written in the form

$$\Psi = \sum_{i=1}^{N} c_i(t,k,z)\Phi\big(x - q_i(t), z\big)e^{kx+k^2t}, \tag{9.9}$$

with the same z but different values of the parameter k.

Let us substitute (9.9) into (1.1). Then (1.1) is satisfied if and if we get a function holomorphic in the fundamental domain. First of all, we conclude that u has poles at q_i only. The vanishing of the triple poles $(x - q_i)^{-3}$ implies that $u(x,t)$ has the form

$$u(x,t) = 2\sum_{i=1}^{N} \wp\big(x - q_i(t)\big). \tag{9.10}$$

The vanishing of the double poles $(x - q_i)^{-2}$ gives the equalities that can be written as a matrix equation for the vector $C = (c_i)$,

$$\big(L(t,z) + k\mathbb{I}\big)C = 0, \tag{9.11}$$

where I is the unit matrix and the Lax matrix $L(t,z)$ is defined in (1.13). Finally, the vanishing of the simple poles gives the equations

$$\big(\partial_t - M(t,z)\big)C = 0, \tag{9.12}$$

where

$$M_{ij} = \bigg(\wp(z) - 2\sum_{j\neq i} \wp(q_i - q_j)\bigg)\delta_{ij} - 2(1-\delta_{ij})\Phi'(q_i - q_j, z). \tag{9.13}$$

The existence of N linearly independent solutions for (1.1) with equivalent Bloch multipliers implies that (9.11) and (9.12) have N independent solutions corresponding to different values of k. Hence, as a compatibility condition, we get the Lax equation $\dot{L} = [M, L]$ for the elliptic CM system.

REFERENCES

[1] H. Airault, H. McKean, and J. Moser, Rational and elliptic solutions of the Korteweg–de Vries equation and related many-body problem. *Comm. Pure Appl. Math.* **30** (1977), no. 1, 95–148.

[2] A. Akhmetshin, I. Krichever, and Yu. Volvoskii, Elliptic families of solutions of the Kadomtsev–Petviashvili equation, and the field analogue of the elliptic Calogero–Moser system. *Funct. Anal. Appl.* **36** (2002), no. 4, 253–266.

[3] E. Arbarello, Survey of Work on the Schottky Problem up to 1996. In *Added section to the 2nd edition of Mumford's Red Book*, pp. 287–291, 301–304, Lecture Notes in Math. 1358, Springer, 1999.

[4] E. Arbarello, G. Codogni, and G. Pareschi, Characterizing Jacobians via the KP equation and via flexes and degenerate trisecants to the Kummer variety: an algebro-geometric approach. 2021, arXiv:2009.14324.

[5] E. Arbarello and C. De Concini, On a set of equations characterizing Riemann matrices. *Ann. of Math. (2)* **120** (1984), no. 1, 119–140.

[6] E. Arbarello, I. Krichever, and G. Marini, Characterizing Jacobians via flexes of the Kummer Variety. *Math. Res. Lett.* **13** (2006), no. 1, 109–123.

[7] O. Babelon, E. Billey, I. Krichever, and M. Talon, Spin generalisation of the Calogero–Moser system and the matrix KP equation. In *Topics in topology and mathematical physics*, pp. 83–119, Amer. Math. Soc. Transl. Ser. 2 170, Amer. Math. Soc., Providence, 1995.

[8] A. Beauville, Vanishing thetanulls on curves with involutions. *Rend. Circ. Mat. Palermo (2)* **62** (2013), no. 1, 61–66.

[9] A. Beauville and O. Debarre, Sur le problème de Schottky pour les variétés de Prym. *Ann. Sc. Norm. Super. Pisa Cl. Sci. (4)* **14**, (1987), no. 4, 613–623.

[10] J. L. Burchnall and T. W. Chaundy, Commutative ordinary differential operators. I, II. *Proc. Lond. Math. Soc.* **21** (1922), 420–440. *Proc. R. Soc. Lond.* **118** (1928), 557–583.

[11] E. Date, M. Jimbo, M. Kashiwara, and T. Miwa, KP hierarchy of orthogonal and symplectic type—Transformation groups for soliton equations VI. *J. Phys. Soc. Jpn.* **5** (1981), no. 0, 3813–3818.

[12] E. Date, M. Jimbo, M. Kashiwara, and T. Miwa, Transformation groups for soliton equations. In *Nonlinear integrable systems – classical theory and quantum theory*, edited by M. Jimbo and T. Miwa, pp. 39–119, World Sci, Singapore, 1983.

[13] O. Debarre, Vers une stratification de l'espace des modules des variétés abéliennes principalement polarisées. In *Complex algebraic varieties (Bayreuth, 1990)*, pp. 71–86, Lecture Notes in Math. 1507, Springer, Berlin, 1992.

[14] P. Deligne and D. Mumford, The irreducibility of the space of curves of given genus. *Publ. Math. Inst. Hautes Études Sci.* **36** (1969), 75–109.

[15] A. Doliwa, P. Grinevich, M. Nieszporski, and P. M. Santini, Integrable lattices and their sub-lattices: from the discrete Moutard (discrete Cauchy–Riemann) 4-point equation to the self-adjoint 5-point scheme. 2004, arXiv:nlin/0410046.

[16] R. Donagi, Non-Jacobians in the Schottky loci. *Ann. of Math.* **1** (1987), no. 26, 193–217.

[17] H. Farkas, S. Grushevsky, and R. Salvati Manni, An explicit solution to the weak Schottky problem. *Algebr. Geom.* **8** (2021), no. 3, 358–373.

[18] H. M. Farkas and H. E. Rauch, Period relations of Schottky type on Riemann surfaces. *Ann. of Math. (2)* **92** (1970), 434–461.

[19] J. D. Fay, *Theta functions on Riemann surfaces*. Lecture Notes in Math. 352, Springer, Berlin–New York, 1973.

[20] J. D. Fay, On the even-order vanishing of Jacobian theta functions. *Duke Math. J.* **51** (1984), no. 1, 109–132.

[21] S. Grushevsky and I. Krichever, Integrable discrete Schrödinger equations and a characterization of Prym varieties by a pair of quadrisecants. *Duke Math. J.* **152** (2010), no. 2, 318–371.

[22] R. Gunning, Some curves in abelian varieties. *Invent. Math.* **66** (1982), no. 3, 377–389.

[23] J. Igusa, On the irreducibility of Schottky's divisor. *J. Fac. Sci., Univ. Tokyo, Sect. 1A, Math.* **28** (1982), no. 3, 531–545.

[24] I. Krichever, Elliptic solutions to difference non-linear equations and nested Bethe ansatz equations. In *Calogero–Moser–Sutherland models (Montreal, QC, 1997)*, pp. 249–271, CRM Ser. Math. Phys., Springer, New York, 2000.

[25] I. Krichever, A characterization of Prym varieties. *Int. Math. Res. Not.* (2006), 81476, 36 pp.

[26] I. Krichever, Integrable linear equations and the Riemann–Schottky problem. In *Algebraic geometry and number theory*, Birkhäuser, Boston, 2006.

[27] I. Krichever, Characterizing Jacobians via trisecants of the Kummer variety. *Ann. of Math.* **1** (2010), no. 72, 485–516.

[28] I. Krichever, Characterizing Jacobians of algebraic curves with involution. 2021, arXiv:2109.13161.

[29] I. Krichever, O. Lipan, P. Wiegmann, and A. Zabrodin, Quantum integrable models and discrete classical Hirota equations. *Comm. Math. Phys.* **188** (1997), no. 2, 267–304.

[30] I. M. Krichever, Integration of non-linear equations by methods of algebraic geometry. *Funct. Anal. Appl.* **11** (1977), no. 1, 12–26.

[31] I. M. Krichever, Methods of algebraic geometry in the theory of non-linear equations. *Russian Math. Surveys* **32** (1977), no. 6, 185–213.

[32] I. M. Krichever, Elliptic solutions of the Kadomtsev–Petviashvili equation and integrable systems of particles. *Funct. Anal. Appl.* **14** (1980), no. 4, 282–290.

[33] I. M. Krichever and A. V. Zabrodin, Spin generalization of the Ruijsenaars–Schneider model, non-abelian 2D Toda chain and representations of Sklyanin algebra. *Uspekhi Mat. Nauk* **50** (1995), no. 6, 3–56.

[34] I. Krichever and N. Nekrasov, Novikov–Veselov symmetries of the two dimensional $O(N)$ sigma model. 2021, arXiv:2106.14201.

[35] I. Krichever and S. Novikov, Two-dimensional Toda lattice, commuting difference operators and holomorphic vector bundles. *Uspekhi Mat. Nauk* **58** (2003), no. 3, 51–88.

[36] I. Krichever and D. H. Phong, Symplectic forms in the theory of solitons. In *Surveys in differential geometry IV*, edited by C. L. Terng and K. Uhlenbeck, pp. 239–313, International Press, 1998.

[37] I. Krichever and T. Shiota, Abelian solutions of the KP equation. In *Geometry, topology and mathematical physics*, edited by V. M. Buchstaber and I. M. Krichever, pp. 173–191, Amer. Math. Soc. Transl. Ser. 2 224, AMS, 2008.

[38] I. Krichever and T. Shiota, Abelian solutions of the soliton equations and geometry of abelian varieties. In *Liaison, Schottky problem and invariant theory*, edited by M. E. Alonso, E. Arrondo, R. Mallavibarrena, and I. Sols, pp. 197–222, Progr. Math. 280, Birkhäuser, 2010.

[39] I. Krichever and T. Shiota, Soliton equations and the Riemann–Schottky problem. In *Handbook of moduli. Vol. II*, pp. 205–258, Adv. Lect. Math. (ALM) 25, Int. Press, Somerville, MA, 2013.

[40] I. Krichever and A. Zabrodin, Constrained Toda hierarchy and turning points of the Ruijsenaars–Schneider model. 2021, arXiv:2109.05240.

[41] I. Krichever and A. Zabrodin, Turning Points and CKP Hierarchy. *Comm. Math. Phys.* **386** (2021), no. 3, 1643–1683.

[42] D. Mumford, Curves and their Jacobians. University of Michigan Press, Ann Arbor, 1975; also included in: *The red book of varieties and schemes, 2nd edition*. Lecture Notes in Math. 1358, Springer, 1999.

[43] D. Mumford, An algebro-geometric construction of commuting operators and of solutions to the Toda lattice equation, Korteweg–de Vries equation and related non-linear equations. In *Proceedings int. symp. algebraic geometry, Kyoto, 1977*, edited by M. Nagata, pp. 115–153, Kinokuniya Book Store, Tokyo, 1978.

[44] C. Poor, The hyperelliptic locus. *Duke Math. J.* **76** (1994), no. 3, 809–884.

[45] F. Schottky, Zur Theorie der Abelschen Functionen von vier Variabeln. *J. Reine Angew. Math.* **1** (1888), no. 02, 304–352.

[46] F. Schottky and H. Jung, Neue Sätze über Symmetrralfunktionen und die Abel'schen Funktionen der Riemann'schen Theorie. *Sitz.ber. Preuss. Akad. Wiss. Berl. Phys. Math. Kl.* **1** (1909), 282–297.

[47] G. Segal and G. Wilson, Loop groups and equations of KdV type. *Publ. Math. Inst. Hautes Études Sci.* **61** (1985), 5–65.

[48] T. Shiota, Characterization of Jacobian varieties in terms of soliton equations. *Invent. Math.* **83** (1986), no. 2, 333–382.

[49] G. van der Geer, The Schottky problem. In *Arbeitstagung Bonn 1984*, edited by F. Hirzebruch et al., pp. 385–406, Lecture Notes in Math. 1111, Springer, Berlin, 1985.

[50] B. van Geemen, Siegel modular forms vanishing on the moduli space of curves. *Invent. Math.* **78** (1984), no. 2, 329–349.

[51] A. P. Veselov and S. P. Novikov, Finite-zone, two-dimensional Schrödinger operators. Potential operators. *Dokl. Akad. Nauk SSSR* **279** (1984), no. 4, 784–788.

[52] A. P. Veselov and S. P. Novikov, Finite-zone, two-dimensional, potential Schrödinger operators. Explicit formulas and evolution equations. *Dokl. Akad. Nauk SSSR* **279** (1984), no. 1, 20–24.

[53] G. E. Welters, On flexes of the Kummer variety (note on a theorem of R. C. Gunning). *Indag. Math. (N.S.)* **45** (1983), no. 4, 501–520.

[54] G. E. Welters, A criterion for Jacobi varieties. *Ann. of Math.* **120** (1984), no. 3, 497–504.

IGOR KRICHEVER
Columbia University, 2990 Broadway, New York, NY 10027, USA, and Skolkovo Institute for Science and Technology, Moscow, Russia, krichev@math.columbia.edu

SEMIORTHOGONAL DECOMPOSITIONS IN FAMILIES

ALEXANDER KUZNETSOV

ABSTRACT

We discuss recent developments in the study of semiorthogonal decompositions of algebraic varieties with an emphasis on their behavior in families.
First, we overview new results concerning homological projective duality.
Then we introduce residual categories, discuss their relation to small quantum cohomology, and compute Serre dimensions of residual categories of complete intersections.
After that we define simultaneous resolutions of singularities and describe a construction that works in particular for nodal degenerations of even-dimensional varieties.
Finally, we introduce the concept of absorption of singularities which works under appropriate assumptions for nodal degenerations of odd-dimensional varieties.

MATHEMATICS SUBJECT CLASSIFICATION 2020

Primary 18G80; Secondary 14D06, 14E15, 14N35

KEYWORDS

Derived categories of coherent sheaves, semiorthogonal decompositions, Lefschetz decompositions, homological projective duality, categorical joins, categorical cones, residual categories, mirror symmetry, quantum cohomology, categorical resolutions of singularities, simultaneous categorical resolutions of singularities, absorption of singularities, Fano threefolds

INTERNATIONAL CONGRESS OF MATHEMATICIANS 2022 JULY 6–14

Proc. Int. Cong. Math. 2022, Vol. 2, pp. 1154–1200
DOI 10.4171/ICM2022/135

1. INTRODUCTION

Semiorthogonal decompositions of derived categories of algebraic varieties were introduced into the realm of algebraic geometry at the turn of the millennium by Bondal and Orlov [18,19]; since then the theory of semiorthogonal decompositions has become one of its central topics. Some advances in this theory have been surveyed in [59]; in this sequel paper we discuss the progress obtained after 2014.

An important point of view in algebraic geometry, explained by Grothendieck, is that geometry should be studied in a relative situation. Thus, the central object of algebraic geometry is a morphism $\mathcal{X} \to B$, i.e., a *family of schemes* $\{\mathcal{X}_b\}_{b \in B}$ parameterized by the points $b \in B$ of a base scheme. Translating this point of view into the context of semiorthogonal decompositions, we understand that we should study semiorthogonal decompositions of schemes $\mathcal{X}/B$, especially *B-linear semiorthogonal decompositions*.

The first step in this direction has been made in [51], where the notions of B-linear triangulated categories and semiorthogonal decompositions have been introduced: an enhanced triangulated category $\mathcal{D}$ is B-linear if it is endowed with a monoidal action of the monoidal category $\mathbf{D}^{\mathrm{perf}}(B)$ of perfect complexes on B. For instance, if $f \colon \mathcal{X} \to B$ is a scheme over B, the bounded derived category of coherent sheaves $\mathbf{D}^{\mathrm{b}}(\mathcal{X})$ is B-linear (where $\mathcal{F} \in \mathbf{D}^{\mathrm{perf}}(B)$ acts on $\mathbf{D}^{\mathrm{b}}(\mathcal{X})$ by $\mathcal{G} \mapsto \mathcal{G} \otimes f^*\mathcal{F}$ for $\mathcal{G} \in \mathbf{D}^{\mathrm{b}}(\mathcal{X})$) and a semiorthogonal decomposition

$$\mathbf{D}^{\mathrm{b}}(\mathcal{X}) = \langle \mathcal{D}_1, \dots, \mathcal{D}_n \rangle \tag{1}$$

is B-linear if $\mathcal{D}_i \otimes f^*(\mathbf{D}^{\mathrm{perf}}(B)) \subset \mathcal{D}_i$ for each i.

The next major step in this direction has been performed in [58], where the notion of base change for B-linear semiorthogonal decompositions has been introduced: given a B-linear semiorthogonal decomposition (1) and a morphism $B' \to B$, a B'-linear semiorthogonal decomposition

$$\mathbf{D}^{\mathrm{b}}\big(\mathcal{X} \times_B B'\big) = \langle \mathcal{D}_{1,B'}, \dots, \mathcal{D}_{n,B'} \rangle$$

has been constructed (under appropriate technical assumptions). In particular, for each point $b \in B$ of the base scheme B, the base change categories $\mathcal{D}_{i,b}$ are defined. This allows one to consider a B-linear category $\mathcal{D}_i$ as a *family of triangulated categories* $\{\mathcal{D}_{i,b}\}_{b \in B}$ parameterized by the points $b \in B$ of the base scheme B.

In this survey we discuss several independent topics, all of which, however, correspond to various semiorthogonal decompositions defined in families.

In Sections 2–3 we discuss some standard material, developing and deepening results from [59]. First of all, we describe the main advances in the theory of homological projective duality (HPD) (see [52], [59, §3]), namely *categorical joins* and *categorical cones*. These categorical constructions provide appropriate homological counterparts of classical constructions of projective geometry and are compatible with HPD. This theory itself relies on a "noncommutative" (or rather categorical) version of HPD that was developed by Alex Perry in **[88]**, so we start Section 2 with a short survey of noncommutative HPD; Section 2.1 can also serve as an introduction to HPD.

After that in Section 2.2, we introduce the construction of categorical joins and explain in what sense it is compatible with HPD, then in Section 2.3 we state the nonlinear HPD theorem, a generalization of the fundamental theorem of HPD in which linear sections of varieties are replaced by arbitrary intersections, and then in Section 2.4 we introduce the construction of categorical cones (a particular case of categorical joins), and combining it with HPD for smooth quadrics and the nonlinear HPD theorem, we deduce the quadratic HPD theorem. As an application of these results we deduce the duality conjecture for Gushel–Mukai varieties. Finally, in Section 2.5 we list a number of important developments in HPD not covered in this survey.

In Section 3 we introduce *residual categories*: given a semiorthogonal decomposition

$$\mathbf{D}^{\mathrm{b}}(X) = \langle \mathcal{R}, \mathcal{B}, \mathcal{B}(1), \dots, \mathcal{B}(m-1) \rangle, \tag{2}$$

where the line bundle $\mathcal{O}_X(1)$ is an mth root of the anticanonical line bundle of a Fano variety X, i.e., $\omega_X^{-1} \cong \mathcal{O}_X(m)$, and $\mathcal{B}$ is an admissible subcategory of $\mathbf{D}^{\mathrm{b}}(X)$, the leftmost component $\mathcal{R}$ of (2) is called the residual category. In Section 3.2 we explain a mirror symmetry interpretation of residual categories, which justifies conjectures relating the structure of the semiorthogonal decomposition (2) and residual category $\mathcal{R}$ to the small quantum cohomology ring of X, stated in Section 3.3. Further, in Section 3.4 we specify the predictions of the above conjectures for some homogeneous varieties, namely Grassmannians and adjoint and coadjoint homogeneous varieties; remarkably, in all these cases the residual categories have a combinatorial nature: when nonzero they are generated by completely orthogonal exceptional collections or equivalent to derived categories of Dynkin quivers. Finally, in Sections 3.5–3.6 we discuss the residual categories of hypersurfaces and complete intersections in projective spaces. In these cases the structure of residual categories is much more complicated. In the case of hypersurfaces, the residual categories have a *fractional Calabi–Yau property*, and in the case of complete intersections, they are fractional Calabi–Yau up to an explicit spherical twist; using this we show that they provide interesting examples of categories with distinct *upper* and *lower Serre dimensions*.

In the last part of this survey (Sections 4–5) we discuss two ways that allow us to find a nice categorical replacement for a *degeneration of schemes*, i.e., for a flat proper morphism $f: \mathcal{X} \to B$ to a smooth pointed curve (B, o) such that

- the morphism $f^o := f|_{\mathcal{X}^o}: \mathcal{X}^o \to B^o$ is smooth, and
- the central fiber $X := \mathcal{X}_o$ of f is singular.

Here and below we denote

$$B^o := B \setminus \{o\}, \quad \mathcal{X}^o := \mathcal{X} \times_B B^o, \quad \mathcal{X}_o := \mathcal{X} \times_B \{o\}, \tag{3}$$

so that we have the following commutative diagram with Cartesian squares:

$$\begin{array}{ccccc} X = \mathcal{X}_o & \xhookrightarrow{\;i\;} & \mathcal{X} & \xhookleftarrow{\;j\;} & \mathcal{X}^o \\ \downarrow & & \downarrow{\scriptstyle f} & & \downarrow{\scriptstyle f^o} \\ \{o\} & \hookrightarrow & B & \hookleftarrow & B^o \end{array} \qquad (4)$$

We show that, in the case where f is projective and the central fiber $X = \mathcal{X}_o$ has only ordinary double points as singularities, under appropriate assumptions, one can find

- a smooth and proper B-linear "modification" $\mathcal{D}$ of the derived category $\mathbf{D}^{\mathrm{b}}(\mathcal{X})$ of the total space, and
- a smooth and proper "modification" $\mathcal{D}_o$ of the derived category $\mathbf{D}^{\mathrm{b}}(\mathcal{X}_o)$ of the central fiber

such that $\mathcal{D}_o$ is the base change of $\mathcal{D}$ along the embedding $\{o\} \hookrightarrow B$. The precise meaning of the word "modification" depends on the parity of $\dim(X)$ and is explained in the following two theorems. For simplicity, we assume that the central fiber has a single ordinary double point.

In the case where $\dim(X)$ is even, we construct a *simultaneous categorical resolution of singularities* of $\mathcal{X}$, which is a special case of a categorical resolution of singularities introduced in **[55]** and **[59, §4]**.

Theorem 1.1 (Theorem 4.5). *Assume given a commutative diagram* (4) *with Cartesian squares, where f is a flat projective morphism to a smooth pointed curve such that f^o is smooth and the central fiber X has a single ordinary double point $x_o \in X$.*

If $\dim(X)$ is even and $\mathcal{X}$ has an ordinary double point at x_o, there is a smooth and proper B-linear triangulated category $\mathcal{D}$ and a commutative diagram

$$\begin{array}{ccccc} \mathbf{D}^{\mathrm{perf}}(X) & \underset{i^*}{\overset{i_*}{\rightleftarrows}} & \mathbf{D}^{\mathrm{perf}}(\mathcal{X}) & \xrightarrow{\;j^*\;} & \mathbf{D}^{\mathrm{perf}}(\mathcal{X}^o) \\ {\scriptstyle \pi_o^*}\downarrow & & {\scriptstyle \pi^*}\downarrow & & \| \\ \mathcal{D}_o & \underset{i^*}{\overset{i_*}{\rightleftarrows}} & \mathcal{D} & \xrightarrow{\;j^*\;} & \mathcal{D}_{B^o} \\ {\scriptstyle \pi_{o*}}\downarrow & & {\scriptstyle \pi_*}\downarrow & & \| \\ \mathbf{D}^{\mathrm{b}}(X) & \underset{i^*}{\overset{i_*}{\rightleftarrows}} & \mathbf{D}^{\mathrm{b}}(\mathcal{X}) & \xrightarrow{\;j^*\;} & \mathbf{D}^{\mathrm{b}}(\mathcal{X}^o) \end{array}$$

where $\mathcal{D}_o$ and $\mathcal{D}_{B^o}$ are the base change categories of $\mathcal{D}$, the functor π^* is left adjoint to π_*, π_o^* is left adjoint to π_{o*}, and

$$\pi_* \circ \pi^* \cong \mathrm{id}, \quad \pi_{o*} \circ \pi_o^* \cong \mathrm{id}\,.$$

In particular, $\mathcal{D}$ provides a categorical resolution of singularities for $\mathcal{X}$ and $\mathcal{D}_o$ provides a categorical resolution of singularities for X.

In fact, even more is true—the functors π^* and π_o^* are also right adjoint functors of π_* and π_{o*}, and the categorical resolutions $\mathcal{D}$ and $\mathcal{D}_o$ of $\mathcal{X}$ and X are *weakly crepant* in the sense of [55, DEFINITION 3.4]; this follows easily from the construction and [55, PROPOSITION 4.5].

The construction in the case where $\dim(X)$ is odd is in some sense opposite to the above. Note that in this case the exceptional divisor $E_o \subset \mathrm{Bl}_{x_o}(X)$ of the blowup $\mathrm{Bl}_{x_o}(X)$ is a smooth quadric of even dimension, hence it comes with two *spinor bundles*.

Theorem 1.2 (Corollary 5.19 and Remark 5.20). *Assume given a commutative diagram* (4) *with Cartesian squares, where f is a flat projective morphism to a smooth pointed curve such that f^o is smooth and the central fiber X has a single ordinary double point $x_o \in X$.*

If $\dim(X)$ is odd, $\mathcal{X}$ is smooth at x_o, and there is an exceptional vector bundle $\mathcal{E}$ on $\mathrm{Bl}_{x_o}(X)$ such that the restriction $\mathcal{E}|_{E_o}$ to the smooth quadric E_o is isomorphic to a spinor bundle then there is a smooth and proper B-linear triangulated category $\mathcal{D}$ and a commutative diagram

$$\begin{array}{ccccc}
\langle \mathcal{P}, \mathcal{D}_o \rangle & \underset{i^*}{\overset{i_*}{\rightleftarrows}} & \langle \langle i_* \mathcal{P} \rangle, \mathcal{D} \rangle & \xrightarrow{\quad j^* \quad} & \mathcal{D}_{B^o} \\
\| & & \| & & \| \\
\mathbf{D}^{\mathrm{b}}(X) & \underset{i^*}{\overset{i_*}{\rightleftarrows}} & \mathbf{D}^{\mathrm{b}}(\mathcal{X}) & \xrightarrow{\quad j^* \quad} & \mathbf{D}^{\mathrm{b}}(\mathcal{X}^o)
\end{array}$$

where $\mathcal{P} \subset \mathbf{D}^{\mathrm{b}}(X)$ is an admissible triangulated subcategory such that the triangulated subcategory $\langle i_* \mathcal{P} \rangle \subset \mathbf{D}^{\mathrm{b}}(\mathcal{X})$ generated by the image of $\mathcal{P}$ under i_* is admissible, the categories $\mathcal{D}_o$ and $\mathcal{D}_{B^o}$ are base changes of $\mathcal{D}$, the functors i_* and i^* in the top row are compatible with the semiorthogonal decompositions, while j^* vanishes on $\langle i_* \mathcal{P} \rangle$.

In both cases we obtain a smooth and proper B-linear category $\mathcal{D}$ such that for each point $b \neq o$ in B the fiber $\mathcal{D}_b$ is equivalent to $\mathbf{D}^{\mathrm{b}}(\mathcal{X}_b)$, the derived category of the fiber of the original family of varieties. Thus, the category $\mathcal{D}_o$ provides a *smooth and proper extension* of the family of categories $\mathbf{D}^{\mathrm{b}}(\mathcal{X}_b)$ across the point $o \in B$. Note, however, that this is achieved in two "opposite" ways—in the situation described in Theorem 1.1, the category $\mathcal{D}_o$ is "larger" than $\mathbf{D}^{\mathrm{b}}(X)$ (in particular, $\mathbf{D}^{\mathrm{b}}(X)$ is a Verdier quotient of $\mathcal{D}_o$), while in the situation described in Theorem 1.2, the category $\mathcal{D}_o$ is "smaller" than $\mathbf{D}^{\mathrm{b}}(X)$ (in particular, $\mathcal{D}_o$ is a semiorthogonal component of $\mathbf{D}^{\mathrm{b}}(X)$).

The construction described in Theorem 1.2 does not have a direct geometric analogue; it is called *absorption of singularities*. More precisely, we say that the subcategory $\mathcal{P} \subset \mathbf{D}^{\mathrm{b}}(X)$ *absorbs singularities* of X and, moreover, it provides a *deformation absorption*. We expect the notions of absorption and deformation absorption of singularities (defined in the general categorical context in Section 5) to become as important as the notion of resolution of singularities for the geometry of schemes.

The category $\mathcal{P}$ constructed in Theorem 1.2 has a particularly interesting structure. It is generated by a single object $\mathrm{P} \in \mathbf{D}^{\mathrm{b}}(X)$ such that

$$\mathrm{Ext}^\bullet(\mathrm{P}, \mathrm{P}) \cong \Bbbk[t], \quad \deg(t) = 2.$$

We call such objects $\mathbb{CP}^\infty$-objects. They can be considered as limiting versions of $\mathbb{CP}^n$-objects of Huybrechts–Thomas [35], but while the latter give rise to autoequivalences of categories containing them, the former provide semiorthogonal decompositions with interesting properties. In particular, if $\mathrm{P} \in \mathbf{D}^{\mathrm{b}}(X)$ is a $\mathbb{CP}^{\infty}$-object and $f: \mathcal{X} \to B$ is a *smoothing* of X, the object $i_*\mathrm{P} \in \mathbf{D}^{\mathrm{b}}(\mathcal{X})$ is exceptional.

We finish Sections 4 and 5 by sample applications of Theorems 1.1 and 1.2 to geometry of cubic fourfolds (see Section 4.3) and Fano threefolds (see Section 5.4), respectively. In particular, we show that the nontrivial components of the derived categories of Fano threefolds of index 2 and degree $1 \leq d \leq 5$ (with a minor modification in the case $d = 1$) can be represented as smooth and proper limits of the nontrivial components of the derived categories of Fano threefolds of index 1 and genus $2d + 2$. This gives a corrected version of a conjecture from [56].

Other important results. Of course, this survey could not cover all interesting results related to semiorthogonal decompositions, so we take this opportunity to list here some important achievements not mentioned in the body of the paper.

In a contrast to dimensions 3 and less, Fano varieties of higher dimensions are not yet classified. However, there are several lists of interesting Fano varieties (e.g., see [48]); and, of course, it is interesting to describe their derived categories, especially when they are expected to have interesting semiorthogonal components. Some results in this direction can be found in [14, 16, 60, 61]. There is also some progress extending results about derived categories known over an algebraically closed field to more general fields [4, 7, 63, 64].

An interesting general question, directly related to the subject of this survey, is if it is possible to extend a semiorthogonal decomposition of a special fiber $\mathcal{X}_o$ of a family $\mathcal{X}/B$ to a B-linear semiorthogonal decomposition. In [13] a positive answer is given under the assumption that $\mathcal{X}_o$ is smooth and proper, and after an étale base change.

An intriguing connection between *L-equivalence* of smooth projective varieties (recall that X_1 is L-equivalent to X_2 if the difference of classes $[X_1] - [X_2]$ is annihilated in the Grothendieck ring of varieties by the class $[\mathbb{A}^d]$ of an affine space) and their derived equivalence was discovered, see [26, 75] and references therein.

Two important general results proved recently are Orlov's gluing theorem [84] and Efimov's embeddability theorem [25]. The first says that any *gluing* of derived categories of smooth projective varieties can be realized as an admissible subcategory of another smooth projective variety. The second gives a criterion for realizability of an enhanced triangulated category as an admissible subcategory in a category generated by exceptional collection. It shows in particular that any phantom category admits such a realization, thus providing a negative answer to [59, CONJECTURE 2.10].

Finally, one of the mostly rapidly developing related areas is the study of Bridgeland stability conditions on semiorthogonal components. We refer to [8, 9] for surveys of this area.

Conventions. In this paper all schemes are separated of finite type over a base field $\Bbbk$ and all categories are $\Bbbk$-linear. We write $\mathcal{A} = \langle \mathcal{A}_1, \dots, \mathcal{A}_n \rangle$ for a semiorthogonal decomposition of a category $\mathcal{A}$ with components $\mathcal{A}_i$. We denote by $\mathbf{D}^{\mathrm{b}}(X)$ the bounded derived category of coherent sheaves and by $\mathbf{D}^{\mathrm{perf}}(X)$ the category of perfect complexes on a scheme X. All pushforward, pullback, and tensor product functors are derived, although we use underived notation for them. When we consider enhancements, we usually mean enhancements by differential graded categories as in [66], but one can also use infinity categories as in [88]. We often use the notions of smoothness and properness for enhanced triangulated categories. Recall that a dg-enhanced triangulated category is smooth if the diagonal bimodule over the underlying differential graded category is perfect, and proper if for all objects $\mathcal{F}_1, \mathcal{F}_2$ of the category the graded vector space $\operatorname{Ext}^\bullet(\mathcal{F}_1, \mathcal{F}_2)$ has finite total dimension.

2. NEW RESULTS IN HOMOLOGICAL PROJECTIVE DUALITY

Homological projective duality studies the family of hyperplane sections of a given projective variety. It was the main subject of the survey [59] (see also [94] for an alternative perspective). In this section we review the main advances in HPD obtained after 2014.

2.1. Noncommutative HPD

It has already become standard to consider nice triangulated categories as derived categories of "noncommutative varieties". From this point of view, it was clear from the very beginning that the operation of homological projective duality is very noncommutative in nature—a manifestation of this is the fact that the result of HPD (even when applied to a commutative variety) is typically noncommutative. However, it took some time for a firm foundation [88] for noncommutative HPD to be developed.

The setup of *noncommutative HPD* is the following. Instead of a smooth proper variety endowed with a morphism to a projective space $\mathbb{P}(V)$ and a Lefschetz decomposition, one considers a smooth and proper Lefschetz category $(\mathcal{A}, \mathcal{A}_0)$ over $\mathbb{P}(V)$. By definition this consists of

- a $\mathbb{P}(V)$-linear category $\mathcal{A}$ (in the sense explained in the Introduction), and
- an admissible subcategory $\mathcal{A}_0 \subset \mathcal{A}$ (called the Lefschetz center of $\mathcal{A}$),

such that $\mathcal{A}_0$ extends to right and left Lefschetz (semiorthogonal) decompositions

$$\mathcal{A} = \langle \mathcal{A}_0, \mathcal{A}_1(1), \dots, \mathcal{A}_{m-1}(m-1) \rangle \quad \text{and} \quad \mathcal{A} = \langle \mathcal{A}_{1-m}(1-m), \dots, \mathcal{A}_{-1}(-1), \mathcal{A}_0 \rangle,$$

respectively, where

$$\mathcal{A}_{m-1} \subset \cdots \subset \mathcal{A}_1 \subset \mathcal{A}_0 \quad \text{and} \quad \mathcal{A}_{1-m} \subset \cdots \subset \mathcal{A}_{-1} \subset \mathcal{A}_0$$

are two chains of admissible subcategories (called the Lefschetz components of $\mathcal{A}$). The components $\mathcal{A}_i$ of both Lefschetz decompositions (if they exist) are determined by $\mathcal{A}_0$, and if one of the Lefschetz decompositions exists then the other exists as well, [55, LEMMA 2.18, 2.19] or [88, LEMMA 6.3]. Moreover, the maximal m such that $\mathcal{A}_{m-1} \neq 0$ equals the maximal m such that $\mathcal{A}_{1-m} \neq 0$; it is called the length of the Lefschetz category and is denoted $\operatorname{length}(\mathcal{A})$.

The length of any Lefschetz category $(\mathcal{A}, \mathcal{A}_0)$ over $\mathbb{P}(V)$ satisfies the inequality

$$\operatorname{length}(\mathcal{A}) \le \dim(V),$$

and if the equality holds and if $m = \mathrm{length}(\mathcal{A})$, the category $\mathcal{A}$ contains

$$\mathcal{A}_{m-1} \otimes \mathbf{D}^{\mathrm{perf}}\big(\mathbb{P}(V)\big) = \big\langle \mathcal{A}_{m-1}, \mathcal{A}_{m-1}(1), \dots, \mathcal{A}_{m-1}(m-1)\big\rangle \tag{5}$$

as a rectangular Lefschetz subcategory (see [88, COROLLARY 6.19]) and HPD for $\mathcal{A}$ reduces to HPD for the orthogonal complement of (5) in $\mathcal{A}$ (this is the *residual category* in the sense of Section 3). Thus, without losing generality, one can always reduce HPD to the case where the Lefschetz category $(\mathcal{A}, \mathcal{A}_0)$ is moderate, i.e., $\mathrm{length}(\mathcal{A}) < \dim(V)$.

Given a Lefschetz category $(\mathcal{A}, \mathcal{A}_0)$ over $\mathbb{P}(V)$, one constructs the HPD Lefschetz category $(\mathcal{A}^\natural, \mathcal{A}_0^\natural)$ by adapting the definition [52, DEFINITION 6.1]. Namely, consider the embedding of the universal hyperplane

$$\mathbf{H}(\mathbb{P}(V)) \xhookrightarrow{\delta} \mathbb{P}(V) \times \mathbb{P}(V^\vee)$$

and the base change $\mathcal{A}_{\mathbf{H}(\mathbb{P}(V))}$ of the $\mathbb{P}(V)$-linear category $\mathcal{A}$ along the natural projection $\mathbf{H}(\mathbb{P}(V)) \to \mathbb{P}(V)$. Then $\mathcal{A}^\natural$ is defined (see [88, DEFINITION 7.1]) as

$$\mathcal{A}^\natural := \big\{\mathcal{F} \in \mathcal{A}_{\mathbf{H}(\mathbb{P}(V))} \mid \delta_* \mathcal{F} \in \mathcal{A}_0 \boxtimes \mathbf{D}^{\mathrm{b}}\big(\mathbb{P}\big(V^\vee\big)\big)\big\} \subset \mathcal{A}_{\mathbf{H}(\mathbb{P}(V))},$$

which is $\mathbb{P}(V^\vee)$-linear category with respect to the $\mathbb{P}(V^\vee)$-linear structure of $\mathcal{A}_{\mathbf{H}(\mathbb{P}(V))}$ induced by the morphism $\mathbf{H}(\mathbb{P}(V)) \to \mathbb{P}(V^\vee)$. Furthermore, the Lefschetz center

$$\mathcal{A}_0^\natural \subset \mathcal{A}^\natural$$

is defined (see [88, LEMMA 7.3] and [69, (2.17)]) as an explicit admissible subcategory in $\mathcal{A}^\natural$.

For a linear subspace $L \subset V$, we denote by

$$L^\perp := \mathrm{Ker}\big(V^\vee \to L^\vee\big) \subset V^\vee$$

its orthogonal subspace and by $\mathcal{A}_{\mathbb{P}(L)}$ and $\mathcal{A}^\natural_{\mathbb{P}(L^\perp)}$ the base change of $\mathcal{A}$ and $\mathcal{A}^\natural$ along the embeddings $\mathbb{P}(L) \hookrightarrow \mathbb{P}(V)$ and $\mathbb{P}(L^\perp) \hookrightarrow \mathbb{P}(V^\vee)$, respectively.

The fundamental theorem of noncommutative HPD is stated as follows. We denote by $\mathcal{A}_j$ and $\mathcal{A}_k^\natural$ the Lefschetz components of $(\mathcal{A}, \mathcal{A}_0)$ and $(\mathcal{A}^\natural, \mathcal{A}_0^\natural)$, respectively.

Theorem 2.1 ([88, THEOREM 8.7, 8.9]). *Let $(\mathcal{A}, \mathcal{A}_0)$ be a moderate Lefschetz category over a projective space $\mathbb{P}(V)$. Then the HPD Lefschetz category $(\mathcal{A}^\natural, \mathcal{A}_0^\natural)$ over the dual projective space $\mathbb{P}(V^\vee)$ is also moderate and the HP double dual category is Lefschetz equivalent to the original*

$$\big((\mathcal{A}^\natural)^\natural, (\mathcal{A}^\natural)_0^\natural\big) \simeq (\mathcal{A}, \mathcal{A}_0).$$

Moreover, if $L \subset V$ is a linear subspace and $L^\perp \subset V^\vee$ is its orthogonal complement with $r = \dim(L)$ and $s = \dim(L^\perp)$, and if $m = \mathrm{length}(\mathcal{A})$, $n = \mathrm{length}(\mathcal{A}^\natural)$, then there are semiorthogonal decompositions

$$\mathcal{A}_{\mathbb{P}(L)} = \big\langle \mathcal{K}_L, \mathcal{A}_s(1), \dots, \mathcal{A}_{m-1}(m-s)\big\rangle,$$
$$\mathcal{A}^\natural_{\mathbb{P}(L^\perp)} = \big\langle \mathcal{A}^\natural_{1-n}(r-n), \dots, \mathcal{A}^\natural_{-r}(-1), \mathcal{K}'_{L^\perp}\big\rangle,$$

and an equivalence of triangulated categories $\mathcal{K}_L \simeq \mathcal{K}'_{L^\perp}$.

The simplest example is linear HPD (see [52, §8] for a relative version).

Example 2.2. Let $0 \subsetneq W \subsetneq V$ be a linear subspace; thus $\mathbb{P}(W)$ is a scheme over $\mathbb{P}(V)$. Then $\mathbf{D}^{\mathrm{b}}(\mathbb{P}(W))$ endowed with the Lefschetz center $\langle \mathcal{O}_{\mathbb{P}(W)} \rangle \subset \mathbf{D}^{\mathrm{b}}(\mathbb{P}(W))$ is a moderate Lefschetz category over $\mathbb{P}(V)$, and the HPD of $(\mathbb{P}(W), \langle \mathcal{O}_{\mathbb{P}(W)} \rangle)$ is given by the Lefschetz category $(\mathbb{P}(W^{\perp}), \langle \mathcal{O}_{\mathbb{P}(W^{\perp})} \rangle)$, where $W^{\perp} \subset V^{\vee}$ is the orthogonal complement of W.

See [50–53] for a number of other examples of HPD.

Homological projective duality is related to classical projective duality via critical loci of morphisms [52, THEOREM 7.9] and this connection persists on the noncommutative level: the classical projective dual of a Lefschetz category $(\mathcal{A}, \mathcal{A}_0)$ (defined as the set of all hyperplanes in $\mathbb{P}(V)$ such that the corresponding hyperplane section of $\mathcal{A}$ is singular) coincides with the set of critical values of $\mathcal{A}^{\natural}$ (defined as the set of points in $\mathbb{P}(V^{\vee})$ such that the corresponding fiber of $\mathcal{A}^{\natural}$ is singular) [88, PROPOSITION 7.19]. When both $\mathcal{A}$ and $\mathcal{A}^{\natural}$ are the derived categories of subvarieties $X \subset \mathbb{P}(V)$ and $Y \subset \mathbb{P}(V^{\vee})$, this reduces to classical projective duality $X^{\vee} = Y$ (see Theorem 2.10 for an example).

Noncommutative HPD itself does not provide new examples of homologically projectively dual varieties (or categories) but, as we already pointed above, it provides a firm background for developing the theory and for proving results like that in the next subsection.

2.2. Categorical joins

The categorical join construction described below provides an appropriate homological extension of the classical join construction in projective geometry; it is perfectly compatible with HPD; moreover, it provides new HPD examples and, as a consequence, new interesting results about derived categories of algebraic varieties.

Recall that the join of two projective varieties $X_1 \subset \mathbb{P}(V_1)$ and $X_2 \subset \mathbb{P}(V_2)$ is defined as the subvariety

$$\mathbf{J}(X_1, X_2) \subset \mathbb{P}(V_1 \oplus V_2)$$

swept out by all lines connecting points of X_1 to points of X_2, where we consider both X_1 and X_2 as subvarieties of $\mathbb{P}(V_1 \oplus V_2)$ via the natural embeddings $\mathbb{P}(V_i) \subset \mathbb{P}(V_1 \oplus V_2)$. It is a well-known result in projective geometry that the join construction commutes with projective duality:

$$\mathbf{J}(X_1, X_2)^{\vee} = \mathbf{J}\big(X_1^{\vee}, X_2^{\vee}\big) \subset \mathbb{P}\big(V_1^{\vee} \oplus V_2^{\vee}\big).$$

In [69] we define the categorical join of Lefschetz categories $(\mathcal{A}^1, \mathcal{A}^1_0)$ over $\mathbb{P}(V_1)$ and $(\mathcal{A}^2, \mathcal{A}^2_0)$ over $\mathbb{P}(V_2)$, and prove a similar duality relation on the HPD level, see Theorem 2.4 below. The definition is carried out in three steps.

In the *first step*, the join $\mathbf{J}(X_1, X_2)$ is replaced by the resolved join

$$\tilde{\mathbf{J}}(X_1, X_2) := \mathbb{P}_{X_1 \times X_2}\big(\mathcal{O}(-1, 0) \oplus \mathcal{O}(0, -1)\big).$$

The resolved join is smooth (as soon as X_1 and X_2 are) and provides a natural resolution of singularities for the join $\mathbf{J}(X_1, X_2)$, which is typically very singular. In particular, we have the universal resolved join

$$\tilde{\mathbf{J}}\big(\mathbb{P}(V_1), \mathbb{P}(V_2)\big) = \mathbb{P}_{\mathbb{P}(V_1)\times\mathbb{P}(V_2)}\big(\mathcal{O}(-1, 0) \oplus \mathcal{O}(0, -1)\big) \cong \mathrm{Bl}_{\mathbb{P}(V_1)\sqcup\mathbb{P}(V_2)}\big(\mathbb{P}(V_1 \oplus V_2)\big).$$

We denote by $\varepsilon_i \colon \mathbb{P}(V_1) \times \mathbb{P}(V_2) \hookrightarrow \tilde{\mathbf{J}}(\mathbb{P}(V_1), \mathbb{P}(V_2))$ the exceptional divisor of the blowup lying over $\mathbb{P}(V_i)$, and by $p \colon \tilde{\mathbf{J}}(\mathbb{P}(V_1), \mathbb{P}(V_2)) \to \mathbb{P}(V_1) \times \mathbb{P}(V_2)$ the $\mathbb{P}^1$-bundle, so that we have a commutative diagram

$$\begin{array}{ccccc} \mathbb{P}(V_1) \times \mathbb{P}(V_2) & \xrightarrow{\ \varepsilon_1\ } & \tilde{\mathbf{J}}(\mathbb{P}(V_1), \mathbb{P}(V_2)) & \xleftarrow{\ \varepsilon_2\ } & \mathbb{P}(V_1) \times \mathbb{P}(V_2) \\ & {\scriptstyle \mathrm{id}} \searrow & \downarrow {\scriptstyle p} & \swarrow {\scriptstyle \mathrm{id}} & \\ & & \mathbb{P}(V_1) \times \mathbb{P}(V_2). & & \end{array}$$

Note that the compositions $p \circ \varepsilon_i$ are isomorphisms.

In the *second step* we define the resolved join of $\mathbb{P}(V_i)$-linear categories $\mathcal{A}^i$ as the base change

$$\tilde{\mathbf{J}}(\mathcal{A}^1, \mathcal{A}^2) := (\mathcal{A}^1 \boxtimes \mathcal{A}^2)_{\tilde{\mathbf{J}}(\mathbb{P}(V_1), \mathbb{P}(V_2))}$$

of the $\mathbb{P}(V_1) \times \mathbb{P}(V_2)$-linear category $\mathcal{A}^1 \boxtimes \mathcal{A}^2$ along the $\mathbb{P}^1$-bundle p. The blowup morphism $\tilde{\mathbf{J}}(\mathbb{P}(V_1), \mathbb{P}(V_2)) \to \mathbb{P}(V_1 \oplus V_2)$ endows $\tilde{\mathbf{J}}(\mathcal{A}^1, \mathcal{A}^2)$ with a $\mathbb{P}(V_1 \oplus V_2)$-linear structure. The morphisms ε_i and p defined above induce a commutative diagram of functors

$$\begin{array}{ccccc} \mathcal{A}^1 \boxtimes \mathcal{A}^2 & \xleftarrow{\ \varepsilon_1^*\ } & \tilde{\mathbf{J}}(\mathcal{A}^1, \mathcal{A}^2) & \xrightarrow{\ \varepsilon_2^*\ } & \mathcal{A}^1 \boxtimes \mathcal{A}^2 \\ & {\scriptstyle \mathrm{id}} \nwarrow & \uparrow {\scriptstyle p^*} & \nearrow {\scriptstyle \mathrm{id}} & \\ & & \mathcal{A}^1 \boxtimes \mathcal{A}^2. & & \end{array}$$

Note that the compositions $\varepsilon_i^* \circ p^*$ are equivalences.

So far, the construction uses the $\mathbb{P}(V_i)$-linear structure of the categories $\mathcal{A}^i$, but is independent of their Lefschetz centers $\mathcal{A}_0^1 \subset \mathcal{A}^1$ and $\mathcal{A}_0^2 \subset \mathcal{A}^2$; they come into play in the *third step* of the construction. We define the subcategories of $\tilde{\mathbf{J}}(\mathcal{A}^1, \mathcal{A}^2)$:

$$\begin{aligned} \mathcal{J}(\mathcal{A}^1, \mathcal{A}^2) &:= \{\mathcal{F} \in \tilde{\mathbf{J}}(\mathcal{A}^1, \mathcal{A}^2) \mid \varepsilon_1^*(\mathcal{F}) \in \mathcal{A}^1 \boxtimes \mathcal{A}_0^2 \text{ and } \varepsilon_2^*(\mathcal{F}) \in \mathcal{A}_0^1 \boxtimes \mathcal{A}^2\}, \quad \text{and} \\ \mathcal{J}_0 &:= p^*(\mathcal{A}_0^1 \boxtimes \mathcal{A}_0^2). \end{aligned}$$

The isomorphisms $\varepsilon_i^* \circ p^* \cong \mathrm{id}$ imply the inclusion $\mathcal{J}_0 \subset \mathcal{J}(\mathcal{A}^1, \mathcal{A}^2)$, and one can prove that the subcategory $\mathcal{J}_0$ is a Lefschetz center in $\mathcal{J}(\mathcal{A}^1, \mathcal{A}^2)$.

Theorem 2.3 ([69, THEOREM 3.21]). *If $(\mathcal{A}^1, \mathcal{A}_0^1)$ and $(\mathcal{A}^2, \mathcal{A}_0^2)$ are Lefschetz categories over projective spaces $\mathbb{P}(V_1)$ and $\mathbb{P}(V_2)$ then $(\mathcal{J}(\mathcal{A}^1, \mathcal{A}^2), \mathcal{J}_0)$ is a Lefschetz category over $\mathbb{P}(V_1 \oplus V_2)$ of length* $\mathrm{length}(\mathcal{A}^1) + \mathrm{length}(\mathcal{A}^2)$.

The categorical join $(\mathcal{J}(\mathcal{A}^1, \mathcal{A}^2), \mathcal{J}_0)$ can be thought of as a categorical resolution of the usual join, see [69, PROPOSITION 3.17 AND REMARK 3.18]. The most important property of the categorical join operation is stated in the following

Theorem 2.4 ([69, THEOREM 4.1]). *If $(\mathcal{A}^1, \mathcal{A}_0^1)$ and $(\mathcal{A}^2, \mathcal{A}_0^2)$ are moderate Lefschetz categories over projective spaces $\mathbb{P}(V_1)$ and $\mathbb{P}(V_2)$ then there is a Lefschetz equivalence*

$$\mathcal{J}(\mathcal{A}^1, \mathcal{A}^2)^\natural \simeq \mathcal{J}((\mathcal{A}^1)^\natural, (\mathcal{A}^2)^\natural),$$

where both sides are considered with their natural Lefschetz structures over $\mathbb{P}(V_1^\vee \oplus V_2^\vee)$.

2.3. Nonlinear HPD theorem

Many (but not all, see **[69, §6.2]**) geometric applications of categorical joins rely on a categorical version of the following simple observation about the usual join operation.

As before assume given a pair of projective varieties $X_1 \subset \mathbb{P}(V_1)$ and $X_2 \subset \mathbb{P}(V_2)$, but now assume $\dim(V_1) = \dim(V_2)$. Assume also given linear isomorphisms

$$\xi_1 \colon V_1 \xrightarrow{\sim} V \quad \text{and} \quad \xi_2 \colon V_2 \xrightarrow{\sim} V,$$

so that we can consider both X_1 and X_2 as subvarieties in $\mathbb{P}(V)$. Let

$$L(\xi_1, \xi_2) := \operatorname{Ker}\bigl(V_1 \oplus V_2 \xrightarrow{(\xi_1, -\xi_2)} V\bigr)$$

be the equalizer of ξ_1 and ξ_2. Then it is easy to check that

$$\mathbf{J}(X_1, X_2) \cap \mathbb{P}\bigl(L(\xi_1, \xi_2)\bigr) = X_1 \cap X_2,$$

where we consider both sides as subvarieties in $\mathbb{P}(V)$ using the identifications ξ_1 and ξ_2 and the induced identification $L(\xi_1, \xi_2) \cong V$. Furthermore, $X_1 \cap X_2$ can be thought of as $X_1 \times_{\mathbb{P}(V)} X_2$, and since the fiber product of varieties is categorified by the tensor product of linear categories (see **[88, §2.3]** for a definition), the above isomorphism has a categorical generalization:

Lemma 2.5 (**[69, Lemma 5.1]**). *If $(\mathcal{A}^1, \mathcal{A}^1_0)$ and $(\mathcal{A}^2, \mathcal{A}^2_0)$ are Lefschetz categories over projective spaces $\mathbb{P}(V_1)$ and $\mathbb{P}(V_2)$ and $\xi_i \colon V_i \xrightarrow{\sim} V$, $i = 1, 2$, are isomorphisms, there is an equivalence of categories*

$$\mathcal{J}(\mathcal{A}^1, \mathcal{A}^2)_{\mathbb{P}(L(\xi_1, \xi_2))} \simeq \mathcal{A}^1 \bigotimes_{\mathbf{D}^{\mathrm{b}}(\mathbb{P}(V))} \mathcal{A}^2$$

between the base change of $\mathcal{J}(\mathcal{A}^1, \mathcal{A}^2)$ along the inclusion $\mathbb{P}(L(\xi_1, \xi_2)) \hookrightarrow \mathbb{P}(V_1 \oplus V_2)$ and the tensor product of the $\mathbb{P}(V)$-linear categories $\mathcal{A}^1$ and $\mathcal{A}^2$ over $\mathbf{D}^{\mathrm{b}}(\mathbb{P}(V))$.

Combining this observation with Theorems 2.1 and 2.4, we obtain

Theorem 2.6 (**[69, Theorem 5.5]**). *Let $(\mathcal{A}^1, \mathcal{A}^1_0)$ and $(\mathcal{A}^2, \mathcal{A}^2_0)$ be moderate Lefschetz categories over projective spaces $\mathbb{P}(V_1)$ and $\mathbb{P}(V_2)$ of equal dimensions. Let*

$$N = \dim(V_1) = \dim(V_2), \quad m = \operatorname{length}\bigl(\mathcal{J}(\mathcal{A}^1, \mathcal{A}^2)\bigr), \quad n = \operatorname{length}\bigl(\mathcal{J}((\mathcal{A}^1)^\natural, (\mathcal{A}^2)^\natural)\bigr),$$

and let $\mathcal{J}_j$ and $\mathcal{J}^\natural_k$ denote the Lefschetz components of $\mathcal{J}(\mathcal{A}^1, \mathcal{A}^2)$ and $\mathcal{J}((\mathcal{A}^1)^\natural, (\mathcal{A}^2)^\natural)$. For any isomorphisms $\xi_i \colon V_i \xrightarrow{\sim} V$, $i = 1, 2$, there are semiorthogonal decompositions

$$\mathcal{A}^1 \bigotimes_{\mathbf{D}^{\mathrm{b}}(\mathbb{P}(V))} \mathcal{A}^2 = \bigl\langle \mathcal{K}_{\xi_1, \xi_2}, \mathcal{J}_N(1), \dots, \mathcal{J}_{m-1}(m - N) \bigr\rangle,$$

$$(\mathcal{A}^1)^\natural \bigotimes_{\mathbf{D}^{\mathrm{b}}(\mathbb{P}(V^\vee))} (\mathcal{A}^2)^\natural = \bigl\langle \mathcal{J}^\natural_{1-n}(N - n), \dots, \mathcal{J}^\natural_{-N}(-1), \mathcal{K}'_{(\xi_1^\vee)^{-1}, (\xi_2^\vee)^{-1}} \bigr\rangle,$$

and an equivalence of categories $\mathcal{K}_{\xi_1, \xi_2} \simeq \mathcal{K}'_{(\xi_1^\vee)^{-1}, (\xi_2^\vee)^{-1}}$, where the $\mathbb{P}(V^\vee)$-linear structures of the HPD categories $(\mathcal{A}^i)^\natural$ are induced by the isomorphisms $(\xi_i^\vee)^{-1} \colon V_i^\vee \xrightarrow{\sim} V^\vee$ for $i = 1, 2$.

If we consider the special case where one of the Lefschetz categories, say $\mathcal{A}^2$, is the derived category of a linear subspace $\mathbb{P}(L) \subset \mathbb{P}(V_2) \cong \mathbb{P}(V)$ with its natural Lefschetz structure, then, as it was explained in Example 2.2, $(\mathcal{A}^2)^\natural$ is Lefschetz equivalent to the derived category of the orthogonal subspace $\mathbb{P}(L^\perp) \subset \mathbb{P}(V_2^\vee) \cong \mathbb{P}(V^\vee)$, and there are equivalences

$$\mathcal{A}^1 \bigotimes_{\mathbf{D}^b(\mathbb{P}(V))} \mathbf{D}^b(\mathbb{P}(L)) \simeq \mathcal{A}^1_{\mathbb{P}(L)},$$
$$(\mathcal{A}^1)^\natural \bigotimes_{\mathbf{D}^b(\mathbb{P}(V^\vee))} \mathbf{D}^b(\mathbb{P}(L^\perp)) \simeq (\mathcal{A}^1)^\natural_{\mathbb{P}(L^\perp)},$$

where the right-hand sides are the base change categories. In this case the statement of Theorem 2.6 is equivalent to the statement of Theorem 2.1; therefore Theorem 2.6 can be considered as a *nonlinear HPD theorem*.

Below we give two sample applications of Theorem 2.6. The first is a consequence of HPD for the Grassmannian Gr(2, 5).

Corollary 2.7 ([85, PROPOSITION 1.1], [20, THEOREM 1.1], [69, THEOREM 6.1]). *Let V_1 and V_2 be vector spaces of dimension 5 and let $\xi_i : \wedge^2 V_i \xrightarrow{\sim} V$, $i = 1, 2$, be linear isomorphisms. Set*

$$X := \mathrm{Gr}(2, V_1) \times_{\mathbb{P}(V)} \mathrm{Gr}(2, V_2) \quad \text{and} \quad Y := \mathrm{Gr}(2, V_1^\vee) \times_{\mathbb{P}(V^\vee)} \mathrm{Gr}(2, V_2^\vee).$$

If the fiber products X and Y have the expected dimension 3, then there is an equivalence of triangulated categories $\mathbf{D}^b(X) \simeq \mathbf{D}^b(Y)$.

The varieties X and Y are deformation equivalent Calabi–Yau threefolds, and as the above corollary states they are derived equivalent. However, they are not birational in general [85, THEOREM 1.2], [20, THEOREM 1.2], and thus the pairs (X, Y) provide counterexamples to the so-called *birational Torelli problem*. See [44] for another similar example.

The second application is a similar consequence of HPD for the connected components $\mathrm{OGr}_\pm(5, 10)$ of the orthogonal isotropic Grassmannian $\mathrm{OGr}(5, 10)$; it provides examples of deformation and derived equivalent, but not birational Calabi–Yau fivefolds. Recall that $\mathrm{OGr}_\pm(5, 10)$ are homogeneous varieties of the simple algebraic group Spin(10), and the primitive generators of their Picard groups embed them into the projectivizations of the two mutually dual 16-dimensional half-spinor representations of Spin(10).

Corollary 2.8 ([81, PROPOSITION 4.2], [69, THEOREM 6.3]). *Let V_1 and V_2 be vector spaces of dimension 10 endowed with nondegenerate quadratic forms, let S_1 and S_2 be the 16-dimensional half-spinor representations of the corresponding groups $\mathrm{Spin}(V_i)$, and let $\xi_i : S_i \xrightarrow{\sim} V$ be linear isomorphisms. Set*

$$X := \mathrm{OGr}_+(5, V_1) \times_{\mathbb{P}(V)} \mathrm{OGr}_+(5, V_2) \quad \text{and} \quad Y := \mathrm{OGr}_-(5, V_1) \times_{\mathbb{P}(V^\vee)} \mathrm{OGr}_-(5, V_2).$$

If the fiber products X and Y have the expected dimension 5, then there is an equivalence of triangulated categories $\mathbf{D}^b(X) \simeq \mathbf{D}^b(Y)$.

Quadratic HPD discussed in Section 2.4 is also a special case of nonlinear HPD.

2.4. Categorical cones and quadratic HPD

As we have seen above, examples of geometrically meaningful Lefschetz categories for which the HPD categories are also geometrically meaningful and well understood lead to applications of the nonlinear HPD theorem with interesting geometric consequences. One nice example of such Lefschetz categories can be obtained from smooth quadrics.

Assume the base field $\Bbbk$ is algebraically closed of characteristic not equal to 2. Let Q be a smooth quadric, i.e., a smooth proper variety isomorphic to a hypersurface of degree 2 in a projective space. Let $\mathcal{O}_Q(1)$ be the ample line bundle that embeds Q as a quadric hypersurface. A morphism $f: Q \to \mathbb{P}(V)$ such that $f^*\mathcal{O}_{\mathbb{P}(V)}(1) \cong \mathcal{O}_Q(1)$ is called standard: thus either

- f is a degree 2 embedding into a linear subspace of $\mathbb{P}(V)$, or
- f is a degree 2 covering over a linear subspace of $\mathbb{P}(V)$ ramified over a quadric.

We say that f is nondegenerate if the subspace above (i.e., the linear span of $f(Q)$) is equal to $\mathbb{P}(V)$. In what follows we always consider Q as a $\mathbb{P}(V)$-linear category by means of a standard morphism $Q \to \mathbb{P}(V)$.

Let $\mathcal{S}$ be a spinor bundle on Q (the only one, if $\dim(Q)$ is odd, or one of the two, if $\dim(Q)$ is even).

Lemma 2.9 ([70, LEMMA 2.4]). *The subcategory* $\mathcal{Q}_0 := \langle \mathcal{S}, \mathcal{O} \rangle \subset \mathbf{D}^{\mathrm{b}}(Q)$ *is a Lefschetz center; the length of the corresponding Lefschetz structure on* $\mathbf{D}^{\mathrm{b}}(Q)$ *is equal to* $\dim(Q)$ *and its Lefschetz components are given by*

$$\mathcal{Q}_i = \begin{cases} \langle \mathcal{S}, \mathcal{O} \rangle, & \text{if } |i| \le 1 - p, \\ \langle \mathcal{O} \rangle, & \text{if } 1 - p < |i| \le \dim(Q) - 1, \end{cases}$$

where $p \in \{0, 1\}$ *is the parity of* $\dim(Q)$.

The Lefschetz structure of $\mathbf{D}^{\mathrm{b}}(Q)$ described above is called a standard Lefschetz structure of Q; note that it depends on the choice of the spinor bundle $\mathcal{S}$ (but the Lefschetz structures associated to different choices of $\mathcal{S}$ are noncanonically equivalent).

Recall that if $Q \subset \mathbb{P}(V)$ is a smooth quadric hypersurface, the classical projective dual of Q is also a smooth quadric hypersurface $Q^\vee \subset \mathbb{P}(V^\vee)$. The HPD for a smooth quadric Q with a standard Lefschetz structure is described in similar terms.

Theorem 2.10 ([70, THEOREM 1.1]). *Let* $f: Q \to \mathbb{P}(V)$ *be a standard nondegenerate morphism of a smooth quadric* Q *endowed with a standard Lefschetz structure. Then the HPD of* Q *is given by a standard nondegenerate morphism* $f^\natural: Q^\natural \to \mathbb{P}(V^\vee)$ *of another smooth quadric* $Q^\natural$*, where:*

(1) *If* f *is a divisorial embedding and* $\dim(Q)$ *is even, then* $Q^\natural = Q^\vee$ *is the classical projective dual of* Q *and* $f^\natural: Q^\natural \to \mathbb{P}(V^\vee)$ *is its natural embedding.*

(2) *If* f *is a divisorial embedding and* $\dim(Q)$ *is odd, then* $f^\natural: Q^\natural \to \mathbb{P}(V^\vee)$ *is the double covering branched along the classical projective dual of* Q*.*

(3) *If f is a double covering and* $\dim(Q)$ *is even, then* $Q^\natural$ *is the classical projective dual of the branch locus of* f *and* $f^\natural\colon Q^\natural \to \mathbb{P}(V^\vee)$ *is its natural embedding.*

(4) *If f is a double covering and* $\dim(Q)$ *is odd, then* $f^\natural\colon Q^\natural \to \mathbb{P}(V^\vee)$ *is the double covering branched along the classical projective dual of the branch locus of* f.

In all cases the HPD Lefschetz structure of $Q^\natural$ *is a standard Lefschetz structure.*

This already allows one to apply the nonlinear HPD theorem, but the application becomes much more powerful after an extension to singular quadrics (see [71, §1.4] for an explanation why this is useful). Note that the derived category of a singular quadric is not smooth and proper, so it does not fit into the framework of HPD adopted in this paper. On the other hand, every singular quadric Q can be written as a cone $\mathbf{C}_{V_0}(\bar{Q})$ over a smooth quadric $\bar{Q}$, and since a cone is a special case of a join, one can use the formalism of categorical joins to find a suitable smooth and proper replacement for $\mathbf{D}^{\mathrm{b}}(Q)$. This is achieved by the *categorical cone* construction.

Let $(\mathcal{A}, \mathcal{A}_0)$ be a Lefschetz category over $\mathbb{P}(V)$ and let $V_0 \neq 0$ be a vector space. We define the categorical cone $\mathcal{C}_{V_0}(\mathcal{A})$ as the categorical join

$$\mathcal{C}_{V_0}(\mathcal{A}) := \mathcal{J}(\mathbb{P}(V_0), \mathcal{A}),$$

where $\mathbb{P}(V_0)$ is endowed with the standard Lefschetz structure from Example 2.2. In fact, in [71] we use another definition, but it is equivalent to the above by [71, PROPOSITION 3.15].

The categorical cone $\mathcal{C}_{V_0}(\mathcal{A})$, being the special case of a categorical join, is a Lefschetz category over $\mathbb{P}(V_0 \oplus V)$, which is moderate if $\mathcal{A}$ is, and can be thought of as a categorical resolution of the usual cone. Combining Theorem 2.4 with Example 2.2, we deduce the following

Theorem 2.11 ([71, THEOREM 1.1]). *Let* $V = V_0 \oplus \bar{V} \oplus V_\infty$ *and let* $(\mathcal{A}, \mathcal{A}_0)$ *be a moderate Lefschetz category over* $\mathbb{P}(\bar{V})$. *Then there is a Lefschetz equivalence*

$$\mathcal{C}_{V_0}(\mathcal{A})^\natural \simeq \mathcal{C}_{V_\infty^\vee}(\mathcal{A}^\natural),$$

where both sides are considered as Lefschetz categories over $\mathbb{P}(V^\vee)$.

Here we add the summand V_∞ to V for higher flexibility of the construction. For instance, in the next application to smooth quadrics it allows us to work with possibly degenerate morphisms of singular quadrics, like the morphism $\mathbf{C}_{V_0}(\bar{Q}) \to \mathbb{P}(V)$ below.

Corollary 2.12 ([71, THEOREM 5.20]). *Let* $V = V_0 \oplus \bar{V} \oplus V_\infty$, *let* $\bar{Q} \to \mathbb{P}(\bar{V})$ *be a standard nondegenerate morphism of a smooth quadric* $\bar{Q}$ *and let* $\bar{Q}^\natural \to \mathbb{P}(\bar{V}^\vee)$ *be its HPD morphism. There is a Lefschetz equivalence*

$$\mathcal{C}_{V_0}(\bar{Q})^\natural \simeq \mathcal{C}_{V_\infty^\vee}(\bar{Q}^\natural),$$

where both sides are considered as Lefschetz categories over $\mathbb{P}(V^\vee)$.

This leads to the following quadratic HPD theorem. In the statement the assumption that a $\mathbb{P}(V)$-linear category $\mathcal{A}$ is supported away from $\mathbb{P}(V_0)$ means that the $\mathbb{P}(V)$-linear structure of $\mathcal{A}$ is induced by a $(\mathbb{P}(V) \setminus \mathbb{P}(V_0))$-linear structure; in this case the linear projection $\mathbb{P}(V) \setminus \mathbb{P}(V_0) \to \mathbb{P}(V/V_0)$ provides $\mathcal{A}$ with a $\mathbb{P}(V/V_0)$-linear structure. A similar convention is applied to $\mathcal{A}^\natural$.

Theorem 2.13 ([71, THEOREM 5.21]). *Let $(\mathcal{A}, \mathcal{A}_0)$ be a moderate Lefschetz category over $\mathbb{P}(V)$ and let $(\mathcal{A}^\natural, \mathcal{A}_0^\natural)$ be its HPD. Assume given a direct sum decomposition $V = V_0 \oplus \bar{V} \oplus V_\infty$ such that the $\mathbb{P}(V)$-linear category $\mathcal{A}$ is supported away from $\mathbb{P}(V_0)$ and the $\mathbb{P}(V^\vee)$-linear category $\mathcal{A}^\natural$ is supported away from $\mathbb{P}(V_\infty^\vee)$. Let $\bar{Q} \to \mathbb{P}(\bar{V})$ be a standard nondegenerate morphism from a smooth quadric $\bar{Q}$ and let $\bar{Q}^\natural \to \mathbb{P}(\bar{V}^\vee)$ be its HPD morphism and denote $Q := \mathbf{C}_{V_0}(\bar{Q})$, $Q^\natural := \mathbf{C}_{V_\infty^\vee}(\bar{Q}^\natural)$. Let*

$$N = \dim(V), \quad m = \operatorname{length}(\mathcal{A}), \quad n = \operatorname{length}(\mathcal{A}^\natural), \quad d = \dim(Q), \quad e = \dim(Q^\natural).$$

Then there are semiorthogonal decompositions

$$\begin{aligned} \mathcal{A}_Q = \langle & \mathcal{K}_Q(\mathcal{A}), \mathcal{A}_e(1) \otimes \mathcal{S}, \dots, \mathcal{A}_{m-1}(m-e) \otimes \mathcal{S}, \\ & \mathcal{A}_{N-d}(1) \otimes \mathcal{O}, \dots, \mathcal{A}_{m-1}(m+d-N) \otimes \mathcal{O} \rangle. \\ (\mathcal{A}^\natural)_{Q^\natural} = \langle & \mathcal{A}^\natural_{1-n}(N-e-n) \otimes \mathcal{O}, \dots, \mathcal{A}^\natural_{e-N}(-1) \otimes \mathcal{O}, \\ & \mathcal{A}^\natural_{1-n}(d-n) \otimes \mathcal{S}^\natural, \dots, \mathcal{A}^\natural_{-d}(-1) \otimes \mathcal{S}^\natural, \mathcal{K}'_{Q^\natural}(\mathcal{A}^\natural) \rangle, \end{aligned}$$

and an equivalence of triangulated categories $\mathcal{K}_Q(\mathcal{A}) \simeq \mathcal{K}'_{Q^\natural}(\mathcal{A}^\natural)$, where $\mathcal{S}$ and $\mathcal{S}^\natural$ are spinor bundles on $\bar{Q}$ and $\bar{Q}^\natural$, $\mathcal{A}_Q$ is defined as the base change of $\mathcal{A}$ along the morphism $Q \to \mathbb{P}(V)$ and $(\mathcal{A}^\natural)_{Q^\natural}$ is defined analogously.

Again, here is a sample application of this result. Recall that a Gushel–Mukai variety [23] is either a quadratic section of a linear section of $\operatorname{Gr}(2,5)$, or a double covering of a linear section of $\operatorname{Gr}(2,5)$ branched at a quadratic section. In other words, a Gushel–Mukai variety can be described uniformly as a dimensionally transverse fiber product

$$X = \operatorname{Gr}(2, V) \times_{\mathbb{P}(\wedge^2 V)} Q,$$

where V is a 5-dimensional vector space and $Q \to \mathbb{P}(\wedge^2 V)$ is a standard (possibly degenerate) morphism of a (possibly singular) quadric. Note that for each Q as above there is a direct sum decomposition

$$\wedge^2 V = V_0 \oplus \bar{V} \oplus V_\infty \tag{6}$$

and a standard nondegenerate morphism $\bar{Q} \to \mathbb{P}(\bar{V})$ from a smooth quadric $\bar{Q}$ such that one has $Q = \mathbf{C}_{V_0}(\bar{Q})$.

Theorem 2.14 ([71, THEOREM 6.4]). *Let V be a vector space of dimension 5, let (6) be a direct sum decomposition of $\wedge^2 V$, let $\bar{Q} \to \mathbb{P}(\bar{V})$ be a standard nondegenerate morphism of a smooth quadric, and let $\bar{Q}^\natural \to \mathbb{P}(\bar{V}^\vee)$ be its HPD morphism. Assume the fiber products*

$$X = \operatorname{Gr}(2, V) \times_{\mathbb{P}(\wedge^2 V)} \mathbf{C}_{V_0}(\bar{Q}) \quad \text{and} \quad Y = \operatorname{Gr}(2, V^\vee) \times_{\mathbb{P}(\wedge^2 V^\vee)} \mathbf{C}_{V_\infty^\vee}(\bar{Q}^\natural) \tag{7}$$

are smooth GM varieties of dimensions $d_X \geq 2$ and $d_Y \geq 2$. Let $\mathcal{U}_X$ and $\mathcal{U}'_Y$ denote the pullbacks to X and Y of the rank 2 tautological bundles of the corresponding Grassmannians.

Then there are semiorthogonal decompositions

$$\mathbf{D}^{\mathrm{b}}(X) = \langle \mathcal{K}_X, \mathcal{O}_X(1), \mathcal{U}_X^{\vee}(1), \dots, \mathcal{O}_X(d_X - 2), \mathcal{U}_X^{\vee}(d_X - 2) \rangle, \tag{8}$$

$$\mathbf{D}^{\mathrm{b}}(Y) = \langle \mathcal{U}'_Y(2 - d_Y), \mathcal{O}_Y(2 - d_Y), \dots, \mathcal{U}'_Y(-1), \mathcal{O}_Y(-1), \mathcal{K}'_Y \rangle, \tag{9}$$

and an equivalence of triangulated categories $\mathcal{K}_X \simeq \mathcal{K}'_Y$.

With a bit more work [71, COROLLARY 6.5] this implies the duality conjecture [68, CONJECTURE 3.7] for Gushel–Mukai varieties.

2.5. Other results

To finish this section we list briefly other results developing HPD that appeared after 2014 and have not been mentioned in [59]. First, there are several works establishing HPD for new classes of varieties. The most interesting among these are:

- The work of Rennemo [89], where the HPD for the symmetric square of a projective space $\mathbb{P}^n$ (considered as a stack) is constructed, see also [34] by Hosono–Takagi for a more geometric description of this HPD for small values of n.
- The work of Rennemo–Segal [90], where a construction that allows to deduce some consequences of HPD for $\mathrm{Gr}(2, 2n + 1)$ (without proving the HPD itself) is suggested.

Besides these major advances, the following papers should be mentioned: [15], where the linear HPD is applied to deduce HPD for determinantal varieties; [6], where a differential graded algebra providing the HPD for a degree d (with $d \geq 3$) Veronese embedding of a projective space is described; and [63, §D], where HPD for $\mathbb{P}^1 \times \mathbb{P}^1 \times \mathbb{P}^1$ is established.

There are also some results contributing to general properties of HPD. Among these one should mention [22], where the HPD for a morphism $f: X \to \mathbb{P}(V)$ is related to the HPD of the same variety X (blown up if necessary) with respect to a morphism $f': X \to \mathbb{P}(V')$ obtained from f by composing with a linear projection $\mathbb{P}(V) \dashrightarrow \mathbb{P}(V')$ (see [69, §B.1] for a categorical version of this result). Finally, one should mention the series of papers [36–41], where an alternative approach to categorical joins is developed and many related results are obtained.

3. RESIDUAL CATEGORIES

Let $\mathcal{C}$ be a triangulated category and let $\boldsymbol{\alpha}_{\mathcal{C}}: \mathcal{C} \to \mathcal{C}$ be an autoequivalence. We say that an admissible subcategory $\mathcal{B} \subset \mathcal{C}$ generates a rectangular Lefschetz collection of length m with respect to $\boldsymbol{\alpha}_{\mathcal{C}}$ if the collection of subcategories $\mathcal{B}, \boldsymbol{\alpha}_{\mathcal{C}}(\mathcal{B}), \dots, \boldsymbol{\alpha}_{\mathcal{C}}^{m-1}(\mathcal{B})$ is semiorthogonal in $\mathcal{C}$. Admissibility of $\mathcal{B}$ implies that this collection extends to a semiorthogonal decomposition

$$\mathcal{C} = \langle \mathcal{R}, \mathcal{B}, \boldsymbol{\alpha}_{\mathcal{C}}(\mathcal{B}), \dots, \boldsymbol{\alpha}_{\mathcal{C}}^{m-1}(\mathcal{B}) \rangle \tag{10}$$

and the component $\mathcal{R}$ of this decomposition is called the residual category, see [72,78].

Residual categories often appear in HPD: if the Lefschetz decomposition of a Lefschetz category $\mathcal{A}$ is rectangular, i.e., $\mathcal{A}_0 = \cdots = \mathcal{A}_{m-1}$, the nontrivial components $\mathcal{K}_{\mathbb{P}(L)}$ appearing in Theorem 2.1 are the residual categories of $\mathcal{A}_{\mathbb{P}(L)}$. In particular, residual categories often appear in families of semiorthogonal decompositions.

3.1. Serre compatibility and rotation functors

The residual category $\mathcal{R}$ defined by (10) has especially nice properties if the subcategory $\mathcal{B}$ is Serre compatible in the sense that the condition

$$\mathbf{S}_{\mathcal{C}}\big(\boldsymbol{\alpha}_{\mathcal{C}}^{m}(\mathcal{B})\big) = \mathcal{B} \tag{11}$$

holds. Note that the Serre functor commutes with any autoequivalence, hence (11) implies that the autoequivalence $\mathbf{S}_{\mathcal{C}} \circ \boldsymbol{\alpha}_{\mathcal{C}}^{m}$ preserves all the components of (10).

One nice consequence of Serre compatibility is the following. First of all, $\mathcal{R}$ comes with a natural autoequivalence, induced by the so-called rotation functor $\mathbf{O}_{\mathcal{B}}$. Below we denote by $\mathbf{L}_{\mathcal{B}}$ and $\mathbf{R}_{\mathcal{B}}$ the left and right mutation functors of $\mathcal{C}$ with respect to $\mathcal{B}$.

Proposition 3.1 ([78, THEOREM 2.8], see also [67, THEOREM 7.7] and [62, COROLLARY 3.18]). *If an admissible subcategory $\mathcal{B} \subset \mathcal{C}$ is Serre compatible then the composition*

$$\mathbf{O}_{\mathcal{B}} := \mathbf{L}_{\mathcal{B}} \circ \boldsymbol{\alpha}_{\mathcal{C}}$$

induces an autoequivalence of the residual category $\mathcal{R}$, with the inverse autoequivalence induced by the composition $\boldsymbol{\alpha}_{\mathcal{C}}^{-1} \circ \mathbf{R}_{\mathcal{B}}$.

Second, the relation between the autoequivalence

$$\boldsymbol{\alpha}_{\mathcal{R}} := \mathbf{O}_{\mathcal{B}}|_{\mathcal{R}}$$

and the Serre functor $\mathbf{S}_{\mathcal{R}}$ of $\mathcal{R}$ is analogous to that of $\boldsymbol{\alpha}_{\mathcal{C}}$ and $\mathbf{S}_{\mathcal{C}}$.

Theorem 3.2 ([78, THEOREM 2.8, REMARK 2.9, PROPOSITION 2.10]). *If $\mathcal{B} \subset \mathcal{C}$ is Serre compatible and $\mathcal{R}$ is the residual category then*

$$\mathbf{S}_{\mathcal{R}} \circ \boldsymbol{\alpha}_{\mathcal{R}}^{m} \cong \big(\mathbf{S}_{\mathcal{C}} \circ \boldsymbol{\alpha}_{\mathcal{C}}^{m}\big)|_{\mathcal{R}}.$$

Moreover, there is a bijection between

- *Lefschetz decompositions of $\mathcal{R}$ with respect to $\boldsymbol{\alpha}_{\mathcal{R}}$ and*
- *Lefschetz decompositions of $\mathcal{C}$ with respect to $\boldsymbol{\alpha}_{\mathcal{C}}$ containing $\mathcal{B}$ in every component.*

3.2. Mirror symmetry interpretation

Before discussing further properties of residual categories we sketch their interpretation from the point of view of mirror symmetry. This section is mostly speculative, but it serves as a motivation for precise mathematical conjectures stated in Section 3.3.

In this subsection, we take $\mathcal{C} = \mathbf{D}^{\mathrm{b}}(X)$ to be the derived category of a smooth complex Fano variety X and let

$$\boldsymbol{\alpha}_{\mathcal{C}}(-) := (-) \otimes \mathcal{L} \tag{12}$$

be the twist autoequivalence given by a line bundle $\mathcal{L}$. Assume also that

$$\omega_X^{-1} \cong \mathcal{L}^m, \tag{13}$$

so that for any admissible subcategory $\mathcal{B} \subset \mathbf{D}^{\mathrm{b}}(X)$ generating a rectangular Lefschetz decomposition of length m the Serre compatibility condition (11) holds.

Homological mirror symmetry predicts the existence of a pair $(Y, \mathbf{w})$ (called a Landau–Ginzburg model) consisting of a proper morphism (called the superpotential)

$$\mathbf{w}: Y \to \mathbb{A}^1$$

from a smooth scheme Y endowed with a relative symplectic form, such that there are two equivalences of triangulated categories

$$\mathbf{D}^{\mathrm{b}}(X) \simeq \mathbf{FS}(Y, \mathbf{w}), \tag{14}$$

$$\mathbf{Fuk}(X) \simeq \mathbf{MF}(Y, \mathbf{w}), \tag{15}$$

where $\mathbf{Fuk}(X)$ is the Fukaya category of X, $\mathbf{FS}(Y, \mathbf{w})$ is the Fukaya–Seidel category of $(Y, \mathbf{w})$, and $\mathbf{MF}(Y, \mathbf{w})$ is the category of matrix factorizations for $(Y, \mathbf{w})$.

A Landau–Ginzburg model for X is very far from being canonically defined. On the other hand, the equivalence (14) implies that the groups of autoequivalences of $\mathbf{D}^{\mathrm{b}}(X)$ and $\mathbf{FS}(Y, \mathbf{w})$ coincide, hence the symmetry of $\mathbf{D}^{\mathrm{b}}(X)$ provided by the autoequivalence (12) should correspond to an autoequivalence $\boldsymbol{\alpha}_{\mathbf{FS}}$ of $\mathbf{FS}(Y, \mathbf{w})$, i.e., to a symmetry of $(Y, \mathbf{w})$. Since the (inverse) Serre functor of $\mathbf{FS}(Y, \mathbf{w})$ corresponds to the 2π-rotation around the origin of the target plane $\mathbb{A}^1 = \mathbb{C}$ of the superpotential $\mathbf{w}$, the autoequivalence $\boldsymbol{\alpha}_{\mathbf{FS}}$ should correspond to the $2\pi/m$-rotation. Therefore, we expect that there exists a *$\boldsymbol{\mu}_m$-equivariant* Landau–Ginzburg model for X, i.e., a Landau–Ginzburg model $(Y, \mathbf{w})$, where Y is endowed with a $\boldsymbol{\mu}_m$-action and the morphism $\mathbf{w}$ is $\boldsymbol{\mu}_m$-equivariant for the standard $\boldsymbol{\mu}_m$-action on $\mathbb{A}^1$.

So, from now on we assume that $(Y, \mathbf{w})$ is $\boldsymbol{\mu}_m$-equivariant. The Fukaya–Seidel category $\mathbf{FS}(Y, \mathbf{w})$ is localized over the critical values of the superpotential $\mathbf{w}$. Let

$$\mathrm{Crit}(\mathbf{w}) = \{z_0, z_1, \dots, z_N\} \subset \mathbb{A}^1 \tag{16}$$

be the set of critical values of $\mathbf{w}$. For each $0 \le i \le N$ choose a C^∞-path γ_i in $\mathbb{A}^1$ connecting the point z_i to $\infty = \mathbb{P}^1 \setminus \mathbb{A}^1$ in such a way that the paths do not intersect and their natural cyclic order corresponding to the way they arrive at ∞ is compatible with the linear ordering of the points z_i in (16). By the definition of $\mathbf{FS}(Y, \mathbf{w})$ this gives a semiorthogonal decomposition (depending on the isotopy class of paths γ_i)

$$\mathbf{FS}(Y, \mathbf{w}) = \langle \mathcal{B}_0, \mathcal{B}_1, \dots, \mathcal{B}_N \rangle$$

such that the component $\mathcal{B}_i$ is localized over z_i.

Since $\mathbf{w}$ is $\boldsymbol{\mu}_m$-equivariant, the set (16) is $\boldsymbol{\mu}_m$-invariant. It may or may not contain the point 0; in any case it will be convenient to set

$$z_0 := 0 \in \mathbb{A}^1$$

and if this is not a critical value of $\mathbf{w}$, set $\mathcal{B}_0 := 0$. So, let

$$z_1, \dots, z_N \in \mathbb{A}^1 \setminus \{0\}$$

be the nonzero critical values of $\mathbf{w}$. Since the action of $\boldsymbol{\mu}_m$ on $\mathbb{A}^1 \setminus \{0\}$ is free, m divides N and reordering the points if necessary we can assume that

$$z_{i+N/m} = \zeta \cdot z_i \tag{17}$$

for $1 \le i \le N - N/m$, where $\zeta = \exp(2\pi\sqrt{-1}/m)$. Furthermore, we can choose the paths γ_i in such a way that $\gamma_{i+N/m} = \zeta \cdot \gamma_i$. Then it follows that

$$\mathcal{B}_{i+N/m} = \boldsymbol{\alpha}_{\mathbf{FS}}(\mathcal{B}_i),$$

so gathering the components $\mathcal{B}_1, \dots, \mathcal{B}_{N/m}$ together and setting

$$\mathcal{B} := \langle \mathcal{B}_1, \dots, \mathcal{B}_{N/m} \rangle, \quad \mathcal{R} := \mathcal{B}_0,$$

we see that $\mathcal{B}$ is admissible, Serre compatible of length m, and we have a semiorthogonal decomposition

$$\mathbf{FS}(Y, \mathbf{w}) = \big\langle \mathcal{R}, \mathcal{B}, \boldsymbol{\alpha}_{\mathbf{FS}}(\mathcal{B}), \dots, \boldsymbol{\alpha}_{\mathbf{FS}}^{m-1}(\mathcal{B}) \big\rangle$$

with residual category $\mathcal{R}$. In particular, if 0 is not a critical value for $\mathbf{w}$, the residual category vanishes and $\mathbf{FS}(Y, \mathbf{w})$ acquires a rectangular Lefschetz decomposition. In view of (14), the category $\mathbf{D}^{\mathrm{b}}(X)$ should have a decomposition of the same type.

The above speculation shows the importance of understanding the critical values of the Landau–Ginzburg superpotential for the structure of $\mathbf{D}^{\mathrm{b}}(X)$. It is interesting that one can find them without describing the Landau–Ginzburg model, purely in terms of X, by using the second equivalence (15). Indeed, the matrix factorization category by definition has a direct sum decomposition

$$\mathbf{MF}(Y, \mathbf{w}) = \bigoplus_{i=0}^{N} \mathbf{MF}(Y, \mathbf{w})_{z_i}$$

with components $\mathbf{MF}(Y, \mathbf{w})_{z_i}$ supported over the same points $z_i \in \mathbb{A}^1$, therefore a similar direct sum decomposition holds for its Hochschild cohomology

$$\mathrm{HH}^\bullet\big(\mathbf{MF}(Y, \mathbf{w})\big) = \bigoplus_{i=0}^{N} \mathrm{HH}^\bullet\big(\mathbf{MF}(Y, \mathbf{w})\big)_{z_i}.$$

Thus, $\mathrm{HH}^\bullet(\mathbf{MF}(Y, \mathbf{w}))$ can be thought of as a finite length coherent sheaf on $\mathbb{A}^1$, i.e., a finite-dimensional module over the ring $\mathbb{C}[z]$ of functions on $\mathbb{A}^1$. So, to control the points z_i, it is enough to understand how the generator z of this ring acts on $\mathrm{HH}^\bullet(\mathbf{MF}(Y, \mathbf{w}))$. For this we use the isomorphism

$$\mathrm{HH}^\bullet\big(\mathbf{MF}(Y, \mathbf{w})\big) \cong \mathrm{HH}^\bullet\big(\mathbf{Fuk}(X)\big) \cong \mathrm{QH}_{\mathrm{can}}(X)$$

(the first isomorphism follows from (15) and the second has been conjectured in [47] and is proved in [30, COROLLARY 9]), where $\mathrm{QH}_{\mathrm{can}}(X)$ is the small quantum cohomology ring of X with the quantum parameters specialized to the anticanonical class (see a more detailed discussion in the introduction to [79]). The right-hand side $\mathrm{QH}_{\mathrm{can}}(X)$ is isomorphic to $\mathrm{H}^{\bullet}(X, \mathbb{C})$ as a vector space and is endowed with the supercommutative quantum multiplication; in particular, one can consider the operator of quantum multiplication by the cohomology class $\kappa_X \in \mathrm{QH}_{\mathrm{can}}(X)$ of the anticanonical line bundle. Note that cohomological degree (divided by 2) induces a $\mathbb{Z}/m$-grading on the even part $\mathrm{QH}^{\mathrm{even}}_{\mathrm{can}}(X)$ of $\mathrm{QH}_{\mathrm{can}}(X)$ (which is a commutative ring), i.e., a $\boldsymbol{\mu}_m$-action on its spectrum

$$\mathrm{QS}(X) := \mathrm{Spec}\big(\mathrm{QH}^{\mathrm{even}}_{\mathrm{can}}(X)\big), \tag{18}$$

and the corresponding morphism

$$\kappa_X : \mathrm{QS}(X) \to \mathbb{A}^1$$

is $\boldsymbol{\mu}_m$-equivariant. It is expected that its image coincides with the $\boldsymbol{\mu}_m$-invariant finite subset (16) in $\mathbb{A}^1$, see [5, THEOREM 6.1] for the toric case and a discussion preceeding it.

3.3. The conjectures

Summarizing the above discussion, we suggest the following precise conjectures. We use the notation introduced in Section 3.2. For a point $z \in \mathbb{A}^1$, denote by $\mathrm{QH}^{\mathrm{even}}_{\mathrm{can}}(X)_{\kappa_X^{-1}(z)}$ the quotient ring of $\mathrm{QH}^{\mathrm{even}}_{\mathrm{can}}(X)$ that corresponds to the union of connected components of $\mathrm{QS}(X)$ supported over z.

Conjecture 3.3. *Let X be a complex Fano variety such that* (13) *holds. Assume the $\boldsymbol{\mu}_m$-invariant subset*

$$\kappa_X\big(\mathrm{QS}(X)\big) \cap \big(\mathbb{A}^1 \setminus \{0\}\big) = \{z_1, \dots, z_N\} \subset \mathbb{A}^1 \setminus \{0\},$$

is ordered in such a way that (17) *holds. Then there is an* $\mathrm{Aut}(X)$*-invariant semiorthogonal decomposition*

$$\mathbf{D}^{\mathrm{b}}(X) = \big\langle \mathcal{R}, \mathcal{B}, \mathcal{B} \otimes \mathcal{L}, \dots, \mathcal{B} \otimes \mathcal{L}^{m-1} \big\rangle \tag{19}$$

and the Hochschild homology spaces of its components are given by

$$\mathrm{HH}_{\bullet}(\mathcal{R}) = \mathrm{QH}_{\mathrm{can}}(X) \bigotimes_{\mathrm{QH}^{\mathrm{even}}_{\mathrm{can}}(X)} \mathrm{QH}^{\mathrm{even}}_{\mathrm{can}}(X)_{\kappa_X^{-1}(0)},$$

$$\mathrm{HH}_{\bullet}(\mathcal{B}) = \bigoplus_{i=1}^{N/m} \Big(\mathrm{QH}_{\mathrm{can}}(X) \bigotimes_{\mathrm{QH}^{\mathrm{even}}_{\mathrm{can}}(X)} \mathrm{QH}^{\mathrm{even}}_{\mathrm{can}}(X)_{\kappa_X^{-1}(z_i)} \Big),$$

where in the right-hand sides we identify $\mathrm{QH}_{\mathrm{can}}(X)$ *with the space* $\mathrm{H}^{\bullet}(X, \mathbb{C}) = \mathrm{HH}_{\bullet}(\mathbf{D}^{\mathrm{b}}(X))$ *with the Hochschild homology grading.*

The following example illustrates how this conjecture works.

Example 3.4. Let $X \subset \mathbb{P}^n$ be a smooth Fano complete intersection of type $(d_1, \dots, d_k)$ with $3 \le \dim(X) \le 2(n+1-\sum d_i) - 1$. Then, by [11, THEOREM], the small quantum cohomology ring of X can be written as

$$\mathrm{QH}_{\mathrm{can}}(X) = \mathbb{C}[h, \alpha]_{\alpha \in \mathrm{H}^{n-k}_{\mathrm{prim}}(X,\mathbb{C})} / \langle h^{n-k+1} - h^{\delta},\ h \cdot \alpha,\ \alpha_1 \cdot \alpha_2 - (\alpha_1, \alpha_2)(h^{n-k} - h^{\delta-1}) \rangle,$$

where h is the hyperplane class, $\mathrm{H}^{n-k}_{\mathrm{prim}}(X, \mathbb{C})$ is the primitive cohomology, $(-, -)$ is the intersection pairing, and $\delta = \sum(d_i - 1)$. Since the anticanonical class of X is a multiple of h, the localization of this ring away from $\boldsymbol{\kappa}_X^{-1}(0)$ is obtained by inverting h, and so it is isomorphic to $\mathbb{C}[h]/(h^{n-k+1-\delta} - 1)$. Note that $m := n - k + 1 - \delta = n + 1 - \sum d_i$ is the Fano index of X, hence $N = m$, $z_1, \dots, z_N$ are mth roots of unity, and $\mathrm{QH}_{\mathrm{can}}(X)_{\boldsymbol{\kappa}_X^{-1}(z_i)} \cong \mathbb{C}$ for $1 \le i \le m$. Thus, Conjecture 3.3 predicts the existence of semiorthogonal decomposition (19) with components $\mathcal{B}$ such that $\mathrm{HH}_\bullet(\mathcal{B}) = \mathbb{C}$. Such a decomposition is indeed easy to construct, it is enough to take $\mathcal{B} = \langle \mathcal{O}_X \rangle$, see Section 3.6.

It also makes sense to combine Conjecture 3.3 with Dubrovin's conjecture [24] that predicts that generic semisimplicity of the *big quantum cohomology ring* $\mathrm{BQH}(X)$ is equivalent to the existence of a full exceptional collection in $\mathbf{D}^{\mathrm{b}}(X)$. Note that generic semisimplicity of $\mathrm{BQH}(X)$ implies that $\mathrm{H}^{\mathrm{odd}}(X, \mathbb{C}) = 0$, hence $\mathrm{QH}_{\mathrm{can}}(X) = \mathrm{QH}^{\mathrm{even}}_{\mathrm{can}}(X)$, see [33, THEOREM 1.3]. Recall the finite length scheme $\mathrm{QS}(X)$ defined in (18) and the $\boldsymbol{\mu}_m$-equivariant morphism $\boldsymbol{\kappa}_X$. Furthermore, let

$$\mathrm{QS}^{\times}(X) := \boldsymbol{\kappa}_X^{-1}(\mathbb{A}^1 \setminus \{0\}) \subset \mathrm{QS}(X), \quad \mathrm{QS}^{\circ}(X) := \mathrm{QS}(X) \setminus \mathrm{QS}^{\times}(X) \subset \mathrm{QS}(X).$$

These are finite $\boldsymbol{\mu}_m$-invariant subschemes of $\mathrm{QS}(X)$ and the action of $\boldsymbol{\mu}_m$ on $\mathrm{QS}^{\times}(X)$ is free. Recall the autoequivalence $\boldsymbol{\alpha}_{\mathcal{R}}$ defined in §3.1.

Conjecture 3.5 ([79, CONJECTURE 1.3]). *Let X be a complex Fano variety such that* (13) *holds and the big quantum cohomology* $\mathrm{BQH}(X)$ *is generically semisimple. Let N be the length of the scheme* $\mathrm{QS}^{\times}(X)$.

(i) *There is a semiorthogonal decomposition* (19), *where the component $\mathcal{B}$ is generated by an* $\mathrm{Aut}(X)$*-invariant exceptional collection of length N/m.*

(ii) *The residual category $\mathcal{R}$ of* (19) *has a completely orthogonal* $\mathrm{Aut}(X)$*-invariant decomposition*

$$\mathcal{R} = \bigoplus_{\xi \in \mathrm{QS}^{\circ}(X)} \mathcal{R}_\xi$$

with components indexed by closed points $\xi \in \mathrm{QS}^{\circ}(X)$; moreover, the component $\mathcal{R}_\xi$ of $\mathcal{R}$ is generated by an exceptional collection of length equal to the length of the scheme $\mathrm{QS}^{\circ}(X)$ at ξ.

(iii) *The autoequivalence $\boldsymbol{\alpha}_{\mathcal{R}}$ permutes the components $\mathcal{R}_\xi$ of the residual category; more precisely, for each point $\xi \in \mathrm{QS}^{\circ}(X)$ it induces an equivalence*

$$\boldsymbol{\alpha}_{\mathcal{R}} \colon \mathcal{R}_\xi \xrightarrow{\sim} \mathcal{R}_{g(\xi)},$$

where g is a generator of $\boldsymbol{\mu}_m$.

Most of the predictions in Conjecture 3.5 are specializations of assertions of Conjecture 3.3 to the case of a category with an exceptional collection. The only exception is the complete orthogonality statement in part (ii). A justification for it, based on a comparison with the Fukaya–Seidel category $\mathbf{FS}(Y, \mathbf{w})$, can be found in [78] just before [78, CONJECTURE 1.12].

Example 3.6. Let $X = \mathbb{P}^{m-1} \times \mathbb{P}^{m-1}$. Then by the quantum Künneth formula [46] one has

$$\mathrm{QH}(X) \cong \mathbb{C}[h_1, h_2]/\big(h_1^m - q_1, h_2^m - q_2\big),$$
$$\mathrm{QH}_{\mathrm{can}}(X) \cong \mathbb{C}[h_1, h_2]/\big(h_1^m - 1, h_2^m - 1\big),$$

and the $\mathbb{Z}/m$-grading is defined by $\deg(h_1) = \deg(h_2) = 1$. Therefore,

$$\mathrm{QS}(X) = \boldsymbol{\mu}_m \times \boldsymbol{\mu}_m \subset \mathbb{A}^2,$$

where the embedding is induced by the natural embeddings $\boldsymbol{\mu}_m \subset \mathbb{A}^1 \setminus \{0\} \subset \mathbb{A}^1$, and up to rescaling of $\mathbb{A}^1$ the map $\boldsymbol{\kappa}_X$ is induced by the summation map $\mathbb{A}^2 \to \mathbb{A}^1$.

If m is odd, $\mathrm{QS}^\circ(X)$ is empty, hence Conjecture 3.5 predicts the existence of an $\mathrm{Aut}(X)$-invariant rectangular Lefschetz collection with zero residual category. Several such collections, indeed, have been constructed in [89].

On the other hand, if m is even, $\mathrm{QS}^\circ(X)$ has length m (it consists of all pairs $(\xi, -\xi)$ for $\xi \in \boldsymbol{\mu}_m$); consequently, Conjecture 3.5 predicts the existence of an $\mathrm{Aut}(X)$-invariant rectangular Lefschetz collection with residual category generated by m completely orthogonal objects. Such collection has been found in [79, EXAMPLE 1.4].

3.4. Residual categories of homogeneous varieties

In this subsection, we make the predictions of Conjecture 3.5 more explicit for some homogeneous varieties of reductive algebraic groups and compare them with known results about their derived categories.

According to an old folklore conjecture homogeneous varieties are expected to have full exceptional collections (the conjecture is still not proved, see a discussion of known cases in [73, §1.2] and more recent developments in [12, 29, 31, 79, 93]). Therefore, Conjecture 3.5 should be applicable and we only need to compute the small quantum cohomology ring. This is pretty easy for Grassmannians.

Example 3.7. The small quantum cohomology ring of $X = \mathrm{Gr}(k, n)$ can be presented as

$$\mathbb{C}[c_1, \dots, c_k, s_1, \dots, s_{n-k}]/\big\langle(1 + c_1 + \dots + c_k)(1 + s_1 + \dots + s_{n-k}) = 1 + (-1)^k q\big\rangle,$$

where c_i and s_j should be thought of as Chern classes of the tautological subbundle and quotient bundle respectively, and q is the quantum parameter (see [91, THEOREM 0.1]). Decomposing formally

$$1 + c_1 + \dots + c_k = \prod_{i=1}^{k}(1 - x_i), \quad 1 + s_1 + \dots + s_{n-k} = \prod_{i=k+1}^{n}(1 - x_i),$$

where $x_1, \dots, x_n$ are the corresponding Chern roots, and specializing the quantum parameter q to $(-1)^{k+1}$, we conclude that $\{x_1, \dots, x_n\} = \boldsymbol{\mu}_n$. Since the canonical class of X is proportional to the first Chern class of the tautological bundle, we have up to rescaling

$$\mathrm{QS}(X) = \big((\boldsymbol{\mu}_n)^k \cap (\mathbb{A}^k \setminus \Delta)\big)/\mathfrak{S}_k,$$

where $\Delta \subset \mathbb{A}^k$ is the big diagonal, $\mathfrak{S}_k$ is the permutation group, and the map $\boldsymbol{\kappa}_X$ is induced by summation of coordinates in $\mathbb{A}^k$. Furthermore, the natural action of $\boldsymbol{\mu}_n$ on $\mathrm{QS}(X)$ is given by simultaneous multiplication of all coordinates by a root of unity. Thus, the $\boldsymbol{\mu}_n$-action is free if and only if $\gcd(k, n) = 1$, and otherwise orbits of length d correspond to subsets of cardinality k/d in $\boldsymbol{\mu}_{n/d}$.

Note that some free orbits may be contained in $\mathrm{QS}^\circ(X)$ (by **[92]** this happens if and only if both k and $n - k$ are sums of nontrivial divisors of n, the simplest example with $\gcd(k, n) = 1$ being $n = 12$ and $k = 5$), so the following conjecture is stronger than the prediction of Conjecture 3.5.

Conjecture 3.8 (**[78, CONJECTURE 3.10 AND LEMMA 3.9]**). *If $X = \mathrm{Gr}(k, n)$ there is a rectangular Lefschetz collection with residual category generated by*

$$R_{k,n} = -\sum_{d \mid \gcd(k,n),\ d>1} \mu(d)\binom{n/d}{k/d}$$

completely orthogonal objects, where

$$\mu(d) := \begin{cases} 1, & \text{if } d \text{ is square-free with an even number of prime factors,} \\ -1, & \text{if } d \text{ is square-free with an odd number of prime factors,} \\ 0, & \text{if } d \text{ has a squared prime factor} \end{cases}$$

is the Möbius function.

By now this conjecture is known for the case $\gcd(k, n) = 1$, where the residual category vanishes **[28, THEOREM 4.3 AND PROPOSITION 4.8]**, as well as for the case where k is a prime number **[78, THEOREM 3.13]**.

For more complicated homogeneous varieties, we can also use the available quantum cohomology computations to formulate a number of precise conjectures. We do this below for some interesting classes of homogeneous varieties.

Recall that the adjoint (resp. coadjoint) homogeneous variety of a simple algebraic group G is the orbit of the highest weight vector in the projectivization of an irreducible representation of G, whose highest weight is the highest *long* (resp. *short*) root of G; in particular, if the group G is simply laced, the adjoint and coadjoint varieties coincide.

The following conjecture is motivated by the results of **[87]**. For a group G we denote by $\mathrm{D}_{\mathrm{short}}(\mathrm{G})$ the short roots subdiagram of the Dynkin diagram of G (if the group G is simply laced, it is the entire Dynkin diagram).

Conjecture 3.9 (**[79, CONJECTURE 1.8]**). *Let X be the coadjoint variety of a simple complex algebraic group* G. *Then $\mathbf{D}^{\mathrm{b}}(X)$ has an $\mathrm{Aut}(X)$-invariant rectangular Lefschetz exceptional collection with residual category $\mathcal{R}$, where*

(1) *if the Dynkin type of* G *is* A_n *and* n *is even, then* $\mathcal{R} = 0$*;*

(2) *otherwise,* $\mathcal{R}$ *is equivalent to the derived category of representations of a quiver of type* $\mathrm{D}_{\mathrm{short}}(\mathrm{G})$.

By now this conjecture is known for all Dynkin types except for the exceptional types E_6, E_7, E_8, see a discussion and references in [79].

For adjoint varieties (of non simply laced groups) the prediction of Conjecture 3.5 has been shown to be true with the last step recently accomplished in [93].

Theorem 3.10. *Let* X *be the adjoint variety of a non simply laced simple complex algebraic group* G. *Then* $\mathbf{D}^{\mathrm{b}}(X)$ *has a full* $\mathrm{Aut}(X)$*-invariant rectangular Lefschetz exceptional collection with zero residual category.*

The proof is a combination of [54, THEOREM 7.1] for type B_n, [78, EXAMPLE 1.4] for type C_n, [93, THEOREM 1.1] for type F_4, and [51, §6.4] for type G_2.

3.5. Residual categories of hypersurfaces

Until now we only met examples of residual categories which were of combinatorial nature (either generated by completely orthogonal exceptional collections or equivalent to derived categories of Dynkin quivers). In Section 3.5–3.6 we consider more complicated examples and concentrate on a description of their basic invariants: *Serre functors* and *Serre dimensions*.

Let $X \subset \mathbb{P}^n$ be a Fano hypersurface, i.e., a hypersurface of degree $d \le n$. Note that

$$\omega_X^{-1} \cong \mathcal{O}_X(n+1-d),$$

hence (13) holds for $\mathcal{L} = \mathcal{O}_X(1)$ and

$$m = n + 1 - d.$$

Furthermore, the category $\mathcal{B} := \langle \mathcal{O}_X \rangle$ is admissible in the derived category of perfect complexes $\mathbf{D}^{\mathrm{perf}}(X)$ (we do not assume X to be smooth and therefore consider $\mathbf{D}^{\mathrm{perf}}(X)$ instead of $\mathbf{D}^{\mathrm{b}}(X)$) and induces a rectangular Lefschetz collection of length m

$$\mathbf{D}^{\mathrm{perf}}(X) = \big\langle \mathcal{R}_X, \mathcal{O}_X, \dots, \mathcal{O}_X(m-1) \big\rangle \tag{20}$$

defining the residual category $\mathcal{R}_X \subset \mathbf{D}^{\mathrm{perf}}(X)$. Finally, as X is a Gorenstein scheme, the category $\mathcal{C} = \mathbf{D}^{\mathrm{perf}}(X)$ has a Serre functor given by tensor product with ω_X and shift by $\dim(X) = n-1$, so if $\boldsymbol{\alpha}_{\mathcal{C}}$ is defined by (12) then the autoequivalence

$$\mathbf{S}_{\mathcal{C}} \circ \boldsymbol{\alpha}_{\mathcal{C}}^m \cong [n-1]$$

preserves any triangulated subcategory $\mathcal{B} \subset \mathbf{D}^{\mathrm{perf}}(X)$; in particular, Serre compatibility (11) holds for $\mathcal{B}$ as above.

One of the most surprising properties of the residual category $\mathcal{R}_X$ of a Fano hypersurface is its fractional Calabi–Yau property (note, however, that the residual categories that appeared in Section 3.4 are also fractional Calabi–Yau). This result has already been explained in [59] for smooth X; we restate it here for completeness.

Theorem 3.11 ([49, COROLLARY 4.3]). *Let $X \subset \mathbb{P}^n$ be a smooth hypersurface of degree d with $1 \le d \le n$. Set $c = \gcd(d, n+1)$. Then*

$$\mathbf{S}_{\mathcal{R}_X}^{d/c} \cong \big[(n+1)(d-2)/c\big]. \tag{21}$$

Remark 3.12. The result also holds true (trivially) when $d = n+1$. Indeed, the definition (20) in this case implies $\mathcal{R}_X = \mathbf{D}^{\mathrm{perf}}(X)$ and the formula (21) reads as $\mathbf{S}_{\mathcal{R}_X} \cong [n-1]$, which is true since X is a Calabi–Yau variety of dimension $n-1$.

Actually, the smoothness assumption in the statement of Theorem 3.11 can be removed; this follows immediately from Theorem 3.14 and Remark 3.15 below.

Example 3.13. If X is a cubic fourfold, one has $\mathbf{S}_{\mathcal{R}_X} \cong [2]$. Thus, the category $\mathcal{R}_X$ is a *K3 category*, see [57].

In [62, THEOREM 3.5] Theorem 3.11 was generalized to the situation where X is a smooth divisor in (or a double covering of) a smooth variety M which admits a rectangular Lefschetz decomposition (so that the residual category of M is zero). We do not state this result separately because it is a special case of Theorem 3.14 stated below, see Remark 3.15. The special case of Theorem 3.11 is obtained by taking $M = \mathbb{P}^n$ with the rectangular Lefschetz decomposition given by the Beilinson exceptional collection

$$\mathbf{D}^{\mathrm{b}}(\mathbb{P}^n) = \big\langle \mathcal{O}, \mathcal{O}(1), \dots, \mathcal{O}(n) \big\rangle.$$

There are many other special cases of [62, THEOREM 3.5] (see [62, §4] for a list) which explain most of the currently known examples of fractional Calabi–Yau categories. For instance, it explains the appearance of K3 categories in derived categories of cubic fourfolds, Gushel–Mukai varieties of even dimensions, and Debarre–Voisin 20-folds, see [62, §4.4].

3.6. Residual categories of complete intersections

The results of [62] have been significantly generalized in [72]. To explain this generalization, recall that, for any (enhanced) functor $\Psi\colon \mathcal{C} \to \mathcal{D}$ between (enhanced) triangulated categories with a right adjoint functor $\Psi^!$ (for instance, for a Fourier–Mukai functor between (perfect) derived categories of projective varieties), one can define twist functors $\mathbf{T}_{\Psi,\Psi^!}\colon \mathcal{D} \to \mathcal{D}$ and $\mathbf{T}_{\Psi^!,\Psi}\colon \mathcal{C} \to \mathcal{C}$ by means of distinguished triangles of functors

$$\Psi \circ \Psi^! \xrightarrow{\varepsilon} \mathrm{id}_{\mathcal{D}} \to \mathbf{T}_{\Psi,\Psi^!} \quad \text{and} \quad \mathbf{T}_{\Psi^!,\Psi} \to \mathrm{id}_{\mathcal{C}} \xrightarrow{\eta} \Psi^! \circ \Psi,$$

where η is the unit and ε is the counit of adjunction. The functor Ψ is called spherical if the twist functors $\mathbf{T}_{\Psi,\Psi^!}$ and $\mathbf{T}_{\Psi^!,\Psi}$ are both autoequivalences (for alternative definitions and characterizations of spherical functors see [1, 2, 43, 62]); in this case the twist functors are known as spherical twists. Note that if the functor Ψ is zero (for instance, if the source or target category of Ψ is zero), it is spherical and the corresponding spherical twists are isomorphic to the identity.

The simplest geometric example of a spherical functor is the pullback functor

$$\Psi := i^*\colon \mathbf{D}^{\mathrm{perf}}(M) \to \mathbf{D}^{\mathrm{perf}}(X)$$

for a divisorial embedding $i\colon X \hookrightarrow M$. Its right adjoint is the pushforward i_* and the corresponding spherical twists are given by

$$\mathbf{T}_{i^*,i_*}(\mathcal{F}) = \mathcal{F} \otimes \mathcal{O}_X(-X)[2] \quad \text{and} \quad \mathbf{T}_{i_*,i^*}(\mathcal{G}) = \mathcal{G} \otimes \mathcal{O}_M(-X).$$

Another interesting example is the pullback functor for a flat double covering $f\colon X \to M$ (see [72, LEMMA 2.9] for the description of the corresponding spherical twists).

Now assume M is a projective Gorenstein variety such that $\omega_M \cong \mathcal{L}_M^{-m}$ for a line bundle $\mathcal{L}_M$ and with a semiorthogonal decomposition

$$\mathbf{D}^{\mathrm{perf}}(M) = \langle \mathcal{R}_M, \mathcal{B}_M, \mathcal{B}_M \otimes \mathcal{L}_M, \dots, \mathcal{B}_M \otimes \mathcal{L}_M^{m-1} \rangle.$$

Assume furthermore given another projective Gorenstein variety X and a spherical functor

$$\Psi\colon \mathbf{D}^{\mathrm{perf}}(M) \to \mathbf{D}^{\mathrm{perf}}(X)$$

such that $\mathbf{T}_{\Psi^!,\Psi}(\mathcal{B}_M) = \mathcal{B}_M \otimes \mathcal{L}_M^d$ for some $1 \le d \le m-1$ and $\Psi(- \otimes \mathcal{L}_M) \cong \Psi(-) \otimes \mathcal{L}_X$ for a line bundle $\mathcal{L}_X$ on X. Under these assumptions the following result is proved.

Theorem 3.14 ([72, COROLLARY 4.19]). *Assume M, $\mathcal{B}_M$, $\mathcal{L}_M$, X, $\mathcal{L}_X$, and Ψ are as above.*

(i) *The functor $\Psi|_{\mathcal{B}_M}$ is fully faithful, the subcategory $\mathcal{B}_X := \Psi(\mathcal{B}_M) \subset \mathbf{D}^{\mathrm{perf}}(X)$ is admissible, and there is a semiorthogonal decomposition*

$$\mathbf{D}^{\mathrm{perf}}(X) = \langle \mathcal{R}_X, \mathcal{B}_X, \mathcal{B}_X \otimes \mathcal{L}_X, \dots, \mathcal{B}_X \otimes \mathcal{L}_X^{m-d-1} \rangle,$$

where $\mathcal{R}_X \subset \mathbf{D}^{\mathrm{perf}}(X)$ is the residual category.

(ii) *The restriction $\Psi_{\mathcal{R}} := \Psi|_{\mathcal{R}_M}$ is a spherical functor $\mathcal{R}_M \to \mathcal{R}_X$ between the residual categories.*

(iii) *If $c = \gcd(d, m)$ then*

$$\mathbf{S}_{\mathcal{R}_M}^{d/c} \cong \mathbf{T}_{\Psi^!_{\mathcal{R}},\Psi_{\mathcal{R}}}^{m/c} \circ \left[\frac{d \dim(M)}{c}\right],$$
$$\mathbf{S}_{\mathcal{R}_X}^{d/c} \cong \mathbf{T}_{\Psi_{\mathcal{R}},\Psi^!_{\mathcal{R}}}^{(m-d)/c} \circ \left[\frac{d \dim(X) - 2(m-d)}{c}\right], \tag{22}$$

where $\mathbf{T}_{\Psi^!_{\mathcal{R}},\Psi_{\mathcal{R}}}$ and $\mathbf{T}_{\Psi_{\mathcal{R}},\Psi^!_{\mathcal{R}}}$ are the spherical twists with respect to the spherical functor $\Psi_{\mathcal{R}}$.

Remark 3.15. In the special case, where $\mathcal{R}_M = 0$, the spherical twist $\mathbf{T}_{\Psi_{\mathcal{R}},\Psi^!_{\mathcal{R}}}$ is isomorphic to the identity, and we conclude that $\mathcal{R}_X$ is a fractional Calabi–Yau category of dimension $\dim(X) - 2(m-d)/d$.

Example 3.16. Assume the base field is algebraically closed of characteristic not equal to 2. Let $M \subset \mathbb{P}^5$ be a smooth quadric, $\mathcal{B}_M = \langle \mathcal{O}_M \rangle$ as in (20), and let $X \subset M$ be a smooth intersection of M with a cubic hypersurface. Then the residual category of M is generated by two completely orthogonal spinor bundles of rank 2:

$$\mathcal{R}_M = \langle \mathcal{S}_+, \mathcal{S}_- \rangle,$$

their restrictions $\mathcal{S}_{+X}$ and $\mathcal{S}_{-X}$ to X are contained in $\mathcal{R}_X$ and form a so-called spherical pair (i.e., induce a spherical functor from the derived category of a disjoint union of two points to $\mathcal{R}_X$, see **[72, §2.2]**), and formula (22) gives

$$\mathbf{S}^3_{\mathcal{R}_X} \cong \mathbf{T}_{\mathcal{S}_{+X}, \mathcal{S}_{-X}} \circ [7],$$

where $\mathbf{T}_{\mathcal{S}_{+X}, \mathcal{S}_{-X}} \colon \mathcal{R}_X \to \mathcal{R}_X$ is the spherical twist with respect to the spherical pair, i.e.,

$$\mathbf{T}_{\mathcal{S}_{+X}, \mathcal{S}_{-X}}(\mathcal{F}) \cong \operatorname{Cone}\bigl(\operatorname{Ext}^\bullet(\mathcal{S}_{+X}, \mathcal{F}) \otimes \mathcal{S}_{+X} \oplus \operatorname{Ext}^\bullet(\mathcal{S}_{-X}, \mathcal{F}) \otimes \mathcal{S}_{-X} \to \mathcal{F}\bigr).$$

Example 3.17. In the situation of Example 3.16, a similar result can be proved for the refined residual category $\mathcal{A}_X$ of X defined as the orthogonal complement of one of the spinor bundles in $\mathcal{R}_X$, say $\mathcal{S}_{+X}$, which is exceptional, so that there is a semiorthogonal decomposition

$$\mathcal{R}_X = \langle \mathcal{A}_X, \mathcal{S}_{+X} \rangle.$$

In this case the projection of the other spinor bundle to $\mathcal{A}_X$ is a spherical object $\mathrm{K} \in \mathcal{A}_X$ and it is proved in **[72, PROPOSITION 5.18]** that

$$\mathbf{S}^3_{\mathcal{A}_X} \cong \mathbf{T}_{\mathrm{K}}^{-1} \circ [7]$$

(where $\mathbf{T}_{\mathrm{K}}$ is the spherical twist with respect to K), quite similarly to the case of $\mathcal{R}_X$.

One can apply Theorem 3.14 in order to compute Serre dimensions of residual categories of complete intersections. Recall that the upper and lower Serre dimensions $\overline{\operatorname{Sdim}}(\mathcal{C})$ and $\underline{\operatorname{Sdim}}(\mathcal{C})$ of a category $\mathcal{C}$ admitting a Serre functor $\mathbf{S}_{\mathcal{C}}$ are defined as the rate of growth of upper and lower cohomological amplitude of powers of $\mathbf{S}_{\mathcal{C}}^{-1}$, see **[27]** or **[72, DEFINITION 6.10]** for details. In the case where $\mathcal{C} = \mathbf{D}^{\mathrm{perf}}(X)$ for a Gorenstein variety X, so that $\mathbf{S}_{\mathcal{C}}(\mathcal{F}) \cong \mathcal{F} \otimes \omega_X[\dim X]$, one obtains from **[27, LEMMA 5.6]** the equalities

$$\overline{\operatorname{Sdim}}\bigl(\mathbf{D}^{\mathrm{perf}}(X)\bigr) = \underline{\operatorname{Sdim}}\bigl(\mathbf{D}^{\mathrm{perf}}(X)\bigr) = \dim(X),$$

so Serre dimensions provide a categorical interpretation of the geometric (Krull) dimension of a variety. In this example the upper and lower Serre dimensions coincide, but in general this is not true, and residual categories of Fano complete intersections provide nice examples of this sort.

Theorem 3.18 (**[72, THEOREM 1.7]**). *Let $X \subset \mathbb{P}^n$ be a smooth Fano complete intersection in $\mathbb{P}^n$ of type $(d_1, d_2, \dots, d_k)$, where*

$$d_1 \ge d_2 \ge \cdots \ge d_k \ge 2.$$

Denote by $\operatorname{ind}(X) = n + 1 - \sum_{i=1}^k d_i$ the Fano index of X. Let $\mathcal{R}_X$ be the residual category of X defined by (20) with $m = \operatorname{ind}(X)$. Assume there exists a chain of smooth varieties

$$X = X_k \subset \cdots \subset X_2 \subset X_1 \subset X_0 = \mathbb{P}^n,$$

where X_i is a complete intersection of type $(d_1, d_2, \dots, d_i)$. Then

$$\overline{\mathrm{Sdim}}(\mathcal{R}_X) = \dim(X) - 2\frac{\mathrm{ind}(X)}{d_1}, \quad \underline{\mathrm{Sdim}}(\mathcal{R}_X) = \dim(X) - 2\frac{\mathrm{ind}(X)}{d_k}.$$

In particular, if $d_1 > d_k$, the upper Serre dimension of $\mathcal{R}_X$ is strictly bigger than the lower Serre dimension.

The assumption of the existence of a chain of smooth complete intersections X_i interpolating between $\mathbb{P}^n$ and X is of technical nature; it well may be that the result is also true without this assumption. Note also that this assumption holds when the characteristic of the base field is zero by Bertini's Theorem, see [72, LEMMA 6.11].

In the situation of Examples 3.16 and 3.17, one deduces from Theorem 3.18 that

$$\overline{\mathrm{Sdim}}(\mathcal{A}_X) = \overline{\mathrm{Sdim}}(\mathcal{R}_X) = 7/3, \quad \underline{\mathrm{Sdim}}(\mathcal{A}_X) = \underline{\mathrm{Sdim}}(\mathcal{R}_X) = 2.$$

In fact, it is easy to describe objects of the categories $\mathcal{A}_X$ or $\mathcal{R}_X$ on which the rate of growth of powers of the inverse Serre functor equals $7/3$ and 2, respectively; indeed, the first happens on the orthogonal complements $\mathrm{K}^\perp \subset \mathcal{A}_X$ and $\mathcal{S}_{+X}^\perp \cap \mathcal{S}_{-X}^\perp \subset \mathcal{R}_X$, respectively, while the second holds on the subcategories generated by K in $\mathcal{A}_X$ and $\mathcal{S}_{\pm X}$ in $\mathcal{R}_X$, respectively.

4. SIMULTANEOUS CATEGORICAL RESOLUTIONS OF SINGULARITIES

The goal of this section is to explain the proof of Theorem 1.1 that provides a simultaneous categorical resolution of singularities for a nodal degeneration of even-dimensional varieties. We start by explaining what we mean by a simultaneous categorical resolution; this notion is similar to a relative version of the definition of a categorical resolution from [55,66].

Let $f: \mathcal{X} \to B$ be a flat proper morphism to a pointed scheme (B, o). Recall notation (3) and (4). We usually assume that B is a curve and f is smooth over B^o.

Definition 4.1 ([65, DEFINITION 1.4]). A simultaneous categorical resolution of $(\mathcal{X}, \mathcal{X}_o)$ is a triple $(\mathcal{D}, \pi^*, \pi_*)$, where

- $\mathcal{D}$ is an enhanced B-*linear* triangulated category, and
- $\pi^*: \mathbf{D}^{\mathrm{perf}}(\mathcal{X}) \to \mathcal{D}$ and $\pi_*: \mathcal{D} \to \mathbf{D}^{\mathrm{b}}(\mathcal{X})$ is a pair of B-*linear* triangulated functors,

such that

(i) $\mathcal{D}$ is smooth and proper over B,

(ii) π^* is left adjoint to π_*,

(iii) $\pi_* \circ \pi^* \cong \mathrm{id}$.

More precisely, the condition in part (ii) means that there is a functorial isomorphism

$$\mathrm{Hom}(\pi^* \mathcal{F}, \mathcal{G}) \cong \mathrm{Hom}(\mathcal{F}, \pi_* \mathcal{G})$$

for all $\mathcal{F} \in \mathbf{D}^{\mathrm{perf}}(\mathcal{X})$, $\mathcal{G} \in \mathcal{D}$, and the condition in part (iii) means that the composition $\pi_* \circ \pi^*$ is isomorphic to the canonical inclusion $\mathbf{D}^{\mathrm{perf}}(\mathcal{X}) \hookrightarrow \mathbf{D}^{\mathrm{b}}(\mathcal{X})$. Furthermore, we usually assume that the base change $\mathcal{D}_{B^o}$ of the category $\mathcal{D}$ along the open embedding $B^o \hookrightarrow B$ is equivalent to $\mathbf{D}^{\mathrm{perf}}(\mathcal{X}^o) = \mathbf{D}^{\mathrm{b}}(\mathcal{X}^o)$ via the functors induced by π^* and π_*.

If the scheme $\mathcal{X}$ has rational singularities and the morphism $f: \mathcal{X} \to B$ admits a simultaneous resolution $\pi: \tilde{\mathcal{X}} \to \mathcal{X}$ in the geometric sense (i.e., a resolution of singularities of $\mathcal{X}$ such that its central fiber $\tilde{\mathcal{X}}_o \to \mathcal{X}_o$ is a resolution of singularities of $\mathcal{X}_o$) then the category $\mathcal{D} := \mathbf{D}^{\mathrm{perf}}(\mathcal{X}) = \mathbf{D}^{\mathrm{b}}(\mathcal{X})$ with the derived pullback π^* and pushforward π_* functors is a simultaneous categorical resolution.

Geometric simultaneous resolutions of singularities exist for surface degenerations with rational double points by **[21]** (see also **[95]**), but not in higher dimensions.

4.1. General results

In **[65]** we suggest a construction of a simultaneous categorical resolution, analogous to the construction of a categorical resolution of a variety X given in **[55]**. Recall that the construction of **[55]** assumes that X is resolved by a single blowup with exceptional divisor E and, as an extra input, one needs a Lefschetz decomposition of $\mathbf{D}^{\mathrm{b}}(E)$ with respect to the conormal line bundle of E.

Similarly, to construct a simultaneous categorical resolution we assume that both the total space $\mathcal{X}$ and the central fiber $\mathcal{X}_o$ of f are resolved by blowups with the same center $Z \subset \mathcal{X}_o \subset \mathcal{X}$, such that both have smooth exceptional divisors E and E_o, and that an appropriate Lefschetz decomposition of $\mathbf{D}^{\mathrm{b}}(E)$ is given. The precise statement is as follows:

Theorem 4.2 (**[65, THEOREM 3.11]**). *Let $f: \mathcal{X} \to B$ be a flat projective morphism to a smooth pointed curve (B, o) such that $\mathcal{X} \times_B B^o$ is smooth over B^o and let $Z \subset \mathcal{X}_o$ be a smooth closed subscheme in the central fiber. Assume the scheme $\mathcal{X}$ has rational singularities, the blowups $\tilde{\mathcal{X}} := \mathrm{Bl}_Z(\mathcal{X})$, $\tilde{\mathcal{X}}_o := \mathrm{Bl}_Z(\mathcal{X}_o)$, and their exceptional divisors E and E_o are all smooth, and the central fiber of $\tilde{\mathcal{X}} \to B$ is reduced. Let $\pi: \tilde{\mathcal{X}} \to \mathcal{X}$, $\pi_o: \tilde{\mathcal{X}}_o \to \mathcal{X}_o$, $p: E \to Z$, $p_o: E_o \to Z$, $\varepsilon: E \to \tilde{\mathcal{X}}$, $\varepsilon_o: E_o \to \tilde{\mathcal{X}}_o$, and $i_E: E_o \to E$ be the natural morphisms, shown on the diagram*

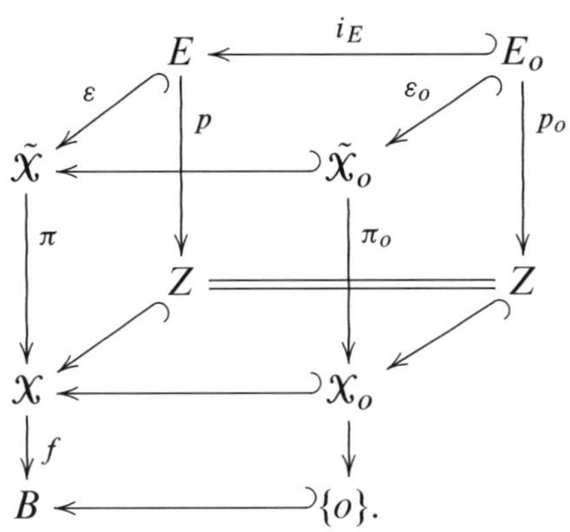

(23)

Furthermore, assume given a Z-linear left Lefschetz decomposition

$$\mathbf{D}^{\mathrm{b}}(E) = \langle \mathcal{A}_{1-m} \otimes \mathcal{O}_E((m-1)E), \dots, \mathcal{A}_{-2} \otimes \mathcal{O}_E(2E), \mathcal{A}_{-1} \otimes \mathcal{O}_E(E), \mathcal{A}_0 \rangle \quad (24)$$

of $\mathbf{D}^{\mathrm{b}}(E)$ *such that* $p^*(\mathbf{D}^{\mathrm{b}}(Z)) \subset \mathcal{A}_0$. *Then the category*

$$\mathcal{D} := \{\mathcal{F} \in \mathbf{D}^{\mathrm{b}}(\tilde{\mathcal{X}}) \mid \varepsilon^*(\mathcal{F}) \in \mathcal{A}_0\} \subset \mathbf{D}^{\mathrm{b}}(\tilde{\mathcal{X}}) \tag{25}$$

provides a categorical resolution of $\mathcal{X}$ *and there is a semiorthogonal decomposition*

$$\mathbf{D}^{\mathrm{b}}(\tilde{\mathcal{X}}) = \langle \varepsilon_*(\mathcal{A}_{1-m} \otimes \mathcal{O}_E((m-1)E)), \dots, \varepsilon_*(\mathcal{A}_{-1} \otimes \mathcal{O}_E(E)), \mathcal{D} \rangle. \tag{26}$$

Moreover, if additionally we have $\mathcal{A}_{-1} = \mathcal{A}_0$, *and the categories* $\mathcal{A}'_k := i_E^*(\mathcal{A}_k)$ *form a semiorthogonal decomposition*

$$\mathbf{D}^{\mathrm{b}}(E_o) = \langle \mathcal{A}'_{1-m} \otimes \mathcal{O}_{E_o}((m-1)E_o), \dots, \mathcal{A}'_{-2} \otimes \mathcal{O}_{E_o}(2E_o), \mathcal{A}'_{-1} \otimes \mathcal{O}_{E_o}(E_o) \rangle \tag{27}$$

of $\mathbf{D}^{\mathrm{b}}(E_o)$ *then:*

(i) *The base change* $\mathcal{D}_o$ *of* $\mathcal{D}$ *along the embedding* $\{o\} \hookrightarrow B$ *is smooth and proper over the residue field of the point o, one has*

$$\mathcal{D}_o \simeq \{\mathcal{F} \in \mathbf{D}^{\mathrm{b}}(\tilde{\mathcal{X}}_o) \mid \varepsilon_o^*(\mathcal{F}) \in \mathcal{A}'_{-1}\}, \tag{28}$$

and there is a semiorthogonal decomposition

$$\mathbf{D}^{\mathrm{b}}(\tilde{\mathcal{X}}_o) = \langle \varepsilon_{o*}(\mathcal{A}'_{1-m} \otimes \mathcal{O}_{E_o}((m-2)E_o)), \dots, \varepsilon_{o*}(\mathcal{A}'_{-2} \otimes \mathcal{O}_{E_o}(E_o)), \mathcal{D}_o \rangle.$$

(ii) *The triple* $(\mathcal{D}, \pi^*, \pi_*)$ *is a simultaneous categorical resolution of* $(\mathcal{X}, \mathcal{X}_o)$*; in particular* $\mathcal{D}$ *is smooth and proper over* B.

The category $\mathcal{D}$ defined by (25) provides a categorical resolution of $\mathcal{X}$ and fits into the semiorthogonal decomposition (26) by [55, THEOREM 4.4 AND PROPOSITION 4.1]. The crucial step in the proof of Theorem 4.2 is the identification (28) of the base change $\mathcal{D}_o$ of the category $\mathcal{D}$, which a priori is a subcategory of the derived category of the central fiber of the morphism $\tilde{\mathcal{X}} \to B$, with a subcategory of $\tilde{\mathcal{X}}_o$. Note that the central fiber is a reduced, but reducible scheme—the union $\tilde{\mathcal{X}}_o \cup E$ of the blowup $\tilde{\mathcal{X}}_o$ of $\mathcal{X}_o$ and the exceptional divisor E of $\tilde{\mathcal{X}}$ with $\tilde{\mathcal{X}}_o \cap E = E_o$. The required identification is achieved in [65, PROPOSITION 3.7], where a more general result of this sort is established. This proves part (i) of the theorem.

On the other hand, by [55, PROPOSITION 4.1] the right-hand side of (28) is an admissible subcategory of $\mathbf{D}^{\mathrm{b}}(\tilde{\mathcal{X}}_o)$; in particular, it is smooth and proper. Therefore, the second part (ii) of the theorem follows from the following general and very useful result.

Theorem 4.3 ([65, THEOREM 2.10]). *Let* $g: \mathcal{Y} \to B$ *be a flat proper morphism of quasiprojective schemes and let*

$$\mathbf{D}^{\mathrm{b}}(\mathcal{Y}) = \langle \mathcal{D}, {}^{\perp}\mathcal{D} \rangle$$

be a B*-linear semiorthogonal decomposition with admissible components and projection functors of finite cohomological amplitude. If for each point* $b \in B$*, the category* $\mathcal{D}_b$ *is smooth and proper over the residue field of b then the category* $\mathcal{D}$ *is smooth and proper over* B.

Remark 4.4. The geometric origin of the category $\mathcal{D}$ in Theorem 4.3 is important for the proof given in [65]. We expect a similar statement for general B-linear categories is not true, although we do not know any example where it fails.

Theorem 4.3 applies immediately to the morphism $g = f \circ \pi\colon \tilde{\mathcal{X}} \to B$. Indeed, for $b \neq o$ the fiber $\mathcal{D}_b$ is equivalent to the category $\mathbf{D}^{\mathrm{b}}(\tilde{\mathcal{X}}_b) = \mathbf{D}^{\mathrm{b}}(\mathcal{X}_b)$ which is smooth and proper because $\mathcal{X} \times_B B^o$ is assumed to be smooth and proper over B^o, while the category $\mathcal{D}_o$ is smooth and proper by part (i) of the theorem.

4.2. Nodal singularities

Theorem 4.2 applies easily to nodal degenerations of even-dimensional varieties under a mild technical assumption that can be satisfied by a simple trick, see Remark 4.6. We use notation introduced in diagram (23).

Theorem 4.5 ([65, THEOREM 3.14]). *Assume the base field is algebraically closed of characteristic not equal to 2. Let $f\colon \mathcal{X} \to B$ be a flat projective morphism of relative dimension $2n$ to a smooth pointed curve (B, o) such that $\mathcal{X} \times_B B^o$ is smooth over B^o. Assume the central fiber $\mathcal{X}_o$ and the total space $\mathcal{X}$ both have an isolated ordinary double point at x_o. Let E and E_o be the exceptional divisors of the blowups $\mathrm{Bl}_{x_o}(\mathcal{X})$ and $\mathrm{Bl}_{x_o}(\mathcal{X}_o)$. Then $(\mathcal{X}, \mathcal{X}_o)$ has a simultaneous categorical resolution of singularities $\mathcal{D}$ fitting into a semiorthogonal decomposition*

$$\mathbf{D}^{\mathrm{b}}(\mathrm{Bl}_{x_o}(\mathcal{X})) = \langle \varepsilon_* \mathcal{O}_E((2n-1)E), \dots, \varepsilon_* \mathcal{O}_E(E), \varepsilon_* \mathcal{S}_E, \mathcal{D} \rangle,$$

where $\mathcal{S}_E$ is a spinor bundle on the smooth quadric E; in particular, $\mathcal{D}$ is smooth and proper over B with $\mathcal{D}_b = \mathbf{D}^{\mathrm{b}}(\mathcal{X}_b)$ for $b \neq o$, and the central fiber $\mathcal{D}_o$ of $\mathcal{D}$ fits into a semiorthogonal decomposition

$$\mathbf{D}^{\mathrm{b}}(\mathrm{Bl}_{x_o}(\mathcal{X}_o)) = \langle \varepsilon_{o*} \mathcal{O}_{E_o}((2n-2)E_o), \dots, \varepsilon_{o*} \mathcal{O}_{E_o}(E_o), \mathcal{D}_o \rangle.$$

Indeed, in this case taking $Z = \{x_0\}$ we see that the exceptional divisor E of the blowup $\tilde{\mathcal{X}} = \mathrm{Bl}_{x_o}(X)$ is a smooth quadric of dimension $2n$, so we can take (24) to be the left Lefschetz decomposition from Lemma 2.9. Then $\mathcal{A}_{-1} = \mathcal{A}_0$ (here it is important that $\dim(\mathcal{X}/B) = \dim(E)$ is even), and since E_o is a smooth quadric of dimension $2n-1$, decomposition (27) holds, again by Lemma 2.9. So, Theorem 4.2 gives the desired results.

Remark 4.6. If $f'\colon \mathcal{X}' \to B'$ is a *smoothing* of a nodal variety X (i.e., $\mathcal{X}'_o \cong X$ and $\mathcal{X}'$ is smooth) applying base change with respect to a double covering $B \to B'$ ramified over o, we obtain a morphism $\mathcal{X} \to B$ such that $\mathcal{X}_o \cong X$ and $\mathcal{X}$ has an ordinary double point at x_o. This double covering trick is quite standard, see [3].

Note that the simultaneous categorical resolution $\mathcal{D}$ constructed in Theorem 4.5 depends on a choice of one of the two spinor bundles $\mathcal{S}_E$; however, two different choices result in equivalent categorical resolutions, and the equivalence can be thought of as an instance of a categorical flop, see [65, PROPOSITION 3.15].

In the case of a degeneration of surfaces (i.e., for $n = 1$), the category $\mathcal{D}$ constructed in Theorem 4.5 is equivalent to the derived category of a small resolution of singularities of $\mathcal{X}$, a choice of one of the two small resolutions corresponds to a choice of one of the two spinor bundles $\mathcal{S}_E$ on the smooth quadric surface E, and the categorical flop mentioned above reduces to the usual Atiyah flop between the small resolutions.

It would be very interesting to find generalizations of Theorem 4.5 to other types of simple singularities.

4.3. Application to nodal degenerations of cubic fourfolds

In this subsection, we provide a simple application of Theorem 4.5 to K3 categories of cubic fourfolds. Recall that for any cubic fourfold $X \subset \mathbb{P}^5$ there is a semiorthogonal decomposition

$$\mathbf{D}^{\mathrm{perf}}(X) = \langle \mathcal{R}_X, \mathcal{O}_X, \mathcal{O}_X(1), \mathcal{O}_X(2) \rangle. \tag{29}$$

In fact, this is just the special case of decomposition (20), so the category $\mathcal{R}_X$ above is the residual category of X. In particular, as we observed in Example 3.13, the Serre functor of $\mathcal{R}_X$ is isomorphic to the shift [2], so $\mathcal{R}_X$ is a *K3 category*.

In the case where X has a single ordinary double point $x_o \in X$, the category $\mathcal{R}_X$ is not smooth, but it admits a categorical resolution by the derived category of a smooth K3 surface. In fact, the linear projection out of x_o identifies the blowup $\mathrm{Bl}_{x_o}(X)$ with the blowup $\mathrm{Bl}_S(\mathbb{P}^4)$ of $\mathbb{P}^4$ along a smooth complete intersection K3 surface $S \subset \mathbb{P}^4$ of type $(2, 3)$ and the derived category of S provides a categorical resolution of singularities for $\mathcal{R}_X$, see [57, THEOREM 5.2]. We write $S(X, x_o)$ for this K3 surface.

In the next theorem we show that the category $\mathbf{D}^{\mathrm{b}}(S(X, x_o))$ can be realized as a limiting category for the K3 categories of smooth cubic fourfolds.

Theorem 4.7 ([65, COROLLARY 1.8]). *Let X be a cubic fourfold with a single ordinary double point $x_o \in X$ over an algebraically closed field $\Bbbk$ of characteristic not equal to 2. There is*

- *a flat proper family $\mathcal{X} \subset \mathbb{P}^5_B \to B$ of cubic fourfolds over a smooth pointed curve (B, o) with central fiber $\mathcal{X}_o \cong X$ such that $\mathcal{X}$ is smooth over B^o and has an ordinary double point at x_o, and*
- *a B-linear category $\mathcal{R}$ smooth and proper over B such that:*
  - (i) *for any point $b \neq o$ in B one has $\mathcal{R}_b \simeq \mathcal{R}_{\mathcal{X}_b}$, i.e., the fiber $\mathcal{R}_b$ is the K3 category of $\mathcal{X}_b$;*
  - (ii) *one has $\mathcal{R}_o \simeq \mathbf{D}^{\mathrm{b}}(S(X, x_o))$.*

In particular, $\mathbf{D}^{\mathrm{b}}(S(X, x_o))$ is a smooth and proper extension of the family of categories $\mathcal{R}_{\mathcal{X}_b}$ across the point $o \in B$.

The construction of the family $\mathcal{X}$ is quite straightforward—we take any smooth cubic fourfold $X' \subset \mathbb{P}^5$ in the ambient projective space of X in such a way that the singular point $x_o \in X$ does not lie on X'. Then, if F and F' are the cubic equations of X and X', we consider the hypersurface in $\mathbb{P}^5 \times \mathbb{A}^1$ given by the equation

$$(1 - t^2)F + t^2 F' = 0,$$

where t is a coordinate on $\mathbb{A}^1$. Throwing away its singular fibers (except for the fiber X over $0 \in \mathbb{A}^1$) we obtain the required family $\mathcal{X} \to B \subset \mathbb{A}^1$.

Next, we consider the smooth and proper B-linear category $\mathcal{D} \subset \mathbf{D}^{\mathrm{b}}(\mathrm{Bl}_{x_o}(\mathcal{X}))$ from Theorem 4.5 and define the subcategory $\mathcal{R} \subset \mathcal{D}$ by the semiorthogonal decomposition

$$\mathcal{D} = \langle \mathcal{R}, \tilde{f}^*\mathbf{D}^{\mathrm{b}}(B), \tilde{f}^*\mathbf{D}^{\mathrm{b}}(B) \otimes \mathcal{O}_{\mathcal{X}/B}(1), \tilde{f}^*\mathbf{D}^{\mathrm{b}}(B) \otimes \mathcal{O}_{\mathcal{X}/B}(2) \rangle, \tag{30}$$

where $\tilde{f}: \mathrm{Bl}_{x_o}(\mathcal{X}) \to B$ is the composition $\mathrm{Bl}_{x_o}(\mathcal{X}) \xrightarrow{\pi} \mathcal{X} \xrightarrow{f} B$ of the blowup morphism and the natural projection and $\mathcal{O}_{\mathcal{X}/B}(i)$ is the pullback to $\mathrm{Bl}_{x_o}(\mathcal{X})$ of the line bundle $\mathcal{O}_{\mathbb{P}^5_B/B}(i)$. Applying the base change of this decomposition to various points of the base B and using **[57, THEOREM 5.2]** for the point o, we obtain the required identifications of the categories $\mathcal{R}_b$ and $\mathcal{R}_o$.

One possible interpretation of Theorem 4.7 is the following. Let $\mathfrak{M}_{\mathrm{cub}}$ be the GIT moduli space of cubic fourfolds **[80]** and let $\mathfrak{M}_{\mathrm{cub}}^{\mathrm{nod}} \subset \mathfrak{M}_{\mathrm{cub}}$ be the divisor of singular cubic fourfolds. Then the family of K3 categories of smooth cubic fourfolds (a priori defined over the open subspace $\mathfrak{M}_{\mathrm{cub}} \setminus \mathfrak{M}_{\mathrm{cub}}^{\mathrm{nod}} \subset \mathfrak{M}_{\mathrm{cub}}$) extends to the general point of the boundary divisor of the root stack $\sqrt{\mathfrak{M}_{\mathrm{cub}}^{\mathrm{nod}}/\mathfrak{M}_{\mathrm{cub}}}$ (the appearance of the root stack corresponds to the necessity of the double covering trick of Remark 4.6).

Remark 4.8. It would be interesting to find analogous extensions of K3 categories of smooth cubic fourfolds to other special loci of the moduli space $\mathfrak{M}_{\mathrm{cub}}$. One particularly interesting point of $\mathfrak{M}_{\mathrm{cub}}$ corresponds to the so-called "chordal cubic" (see **[32, §4.4]** and **[80, §8.2]**), defined as the secant variety of the Veronese surface $v_2(\mathbb{P}^2) \subset \mathbb{P}^5$. It seems likely that to make such an extension possible it is necessary to blowup this point on the moduli space. A general point of the exceptional divisor corresponds to a smooth sextic curve in $\mathbb{P}^2$, and it is natural to expect the derived category of the double covering of $\mathbb{P}^2$ branched at that curve to show up in the extension.

5. ABSORPTION OF SINGULARITIES

The goal of this section is to introduce the notion of absorption of singularities, to explain the proof of Theorem 1.2, and to sketch an application to Fano threefolds.

5.1. Absorption and deformation absorption

We start with the definition of absorption of singularities of a category. For simplicity, we restrict to the case when the category in question is $\mathbf{D}^{\mathrm{b}}(X)$ and X is a proper (singular) variety.

Definition 5.1 (**[76]**). We say that a subcategory $\mathcal{P} \subset \mathbf{D}^{\mathrm{b}}(X)$ absorbs singularities of X if it is admissible and the orthogonal complements $\mathcal{P}^{\perp}$ and ${}^{\perp}\mathcal{P}$ in $\mathbf{D}^{\mathrm{b}}(X)$ are smooth and proper.

Note that, when $\mathcal{P}$ is admissible, the left and right mutation functors with respect to $\mathcal{P}$ induce equivalences of the orthogonal complements $\mathcal{P}^{\perp} \simeq {}^{\perp}\mathcal{P}$, so one of them is smooth and proper if and only if the other is so.

We think of the category $\mathcal{P}^{\perp} \simeq {}^{\perp}\mathcal{P}$ as a smooth and proper "modification" of the category $\mathbf{D}^{\mathrm{b}}(X)$. There is, of course, a trivial example of absorption with $\mathcal{P} = \mathbf{D}^{\mathrm{b}}(X)$

and $\mathcal{P}^\perp = {}^\perp\mathcal{P} = 0$; clearly, this example is not interesting, and it shows that it is desirable to have the absorbing category $\mathcal{P}$ as small as possible (hence its orthogonal complements $\mathcal{P}^\perp$ and ${}^\perp\mathcal{P}$ as big as possible).

The notion of absorption is "opposite" to that of categorical resolution in the sense that in the latter we replace $\mathbf{D}^b(X)$ by a *larger* smooth and proper category, while in the former we replace it by a *smaller* smooth and proper category.

The following is the simplest example of absorption.

Example 5.2. Let $X = X_1 \cup X_2$ be a complete curve with two smooth components intersecting transversely at a point x_0 and with $X_1 \cong \mathbb{P}^1$. Then

$$\mathcal{P} := \langle \mathcal{O}_{X_1}(-1) \rangle \subset \mathbf{D}^b(X)$$

absorbs singularities of X; indeed, the category ${}^\perp\mathcal{P}$ is equivalent to $\mathbf{D}^b(X_2)$ (via the pullback functor for the projection $X \to X_2$ contracting X_1 to the intersection point $x_0 \in X_2$), hence smooth and proper, and $\mathcal{P}^\perp \simeq {}^\perp\mathcal{P}$ because X is Gorenstein.

Other examples of absorption are given by the so-called *Kawamata-type* semiorthogonal decompositions introduced in **[42, DEFINITION 4.1]** (see **[45]** for many decompositions of this type for surfaces).

We will give more examples of absorption in the next subsection, and meanwhile we introduce a stronger notion. Recall that a smoothing of a proper variety X is a Cartesian diagram (4) where f is a flat proper morphism to a smooth pointed curve (B, o) and $\mathcal{X}$ is smooth. Recall that $i: X \to \mathcal{X}$ denotes the embedding of central fiber.

Definition 5.3 (**[76]**). Assume a subcategory $\mathcal{P} \subset \mathbf{D}^b(X)$ absorbs singularities of a proper variety X. We say that $\mathcal{P}$ provides a deformation absorption of singularities of X if for any smoothing $f: \mathcal{X} \to B$ of X the idempotent completion $\langle i_*(\mathcal{P}) \rangle^\oplus \subset \mathbf{D}^b(\mathcal{X})$ of the triangulated subcategory generated in $\mathbf{D}^b(\mathcal{X})$ by the image of $i_*: \mathcal{P} \to \mathbf{D}^b(\mathcal{X})$ is admissible.

Example 5.4. The absorption of singularities described in Example 5.2 is a deformation absorption, because for any smoothing $\mathcal{X} \to B$ of the reducible curve $X = X_1 \cup X_2$ we have

$$\mathcal{N}_{X_1/\mathcal{X}} \cong \mathcal{O}_{X_1}(X_1) \cong \mathcal{O}_{X_1}(-X_2) \cong \mathcal{O}_{X_1}(-x_0) \cong \mathcal{O}_{X_1}(-1),$$

hence $X_1 \subset \mathcal{X}$ is a (-1)-curve on a smooth surface $\mathcal{X}$, so $i_*(\mathcal{O}_{X_1}(-1)) \in \mathbf{D}^b(\mathcal{X})$ is an exceptional object, and hence the subcategory $\langle i_*(\mathcal{P}) \rangle^\oplus = \langle i_*\mathcal{O}_{X_1}(-1) \rangle$ is admissible.

The following result demonstrates how a subcategory providing a deformation absorption can be used to construct a smooth family of categories.

Theorem 5.5 (**[76]**). *Assume a subcategory $\mathcal{P} \subset \mathbf{D}^b(X)$ provides a deformation absorption of singularities of a proper variety X. Let $f: \mathcal{X} \to B$ be a smoothing of X with $\mathcal{X}$ quasiprojective. Define the subcategory $\mathcal{D} \subset \mathbf{D}^b(\mathcal{X})$ from the semiorthogonal decomposition*

$$\mathbf{D}^b(\mathcal{X}) = \langle \langle i_*(\mathcal{P}) \rangle^\oplus, \mathcal{D} \rangle. \tag{31}$$

Then $\mathcal{D}_b = \mathbf{D}^b(\mathcal{X}_b)$ for $b \neq o$, and $\mathcal{D}_o = {}^{\perp}\mathcal{P} \subset \mathbf{D}^b(X)$. In particular, $\mathcal{D}$ is smooth and proper over B.

The main thing to prove here is the equality

$$\langle i_*(\mathcal{P})\rangle_o^{\oplus} = \mathcal{P}$$

of the base change of the category $\langle i_*(\mathcal{P})\rangle^{\oplus}$ with $\mathcal{P}$ (as subcategories of $\mathbf{D}^b(X)$). The inclusion $\mathcal{P} \subset \langle i_*(\mathcal{P})\rangle_o^{\oplus}$ follows from **[58, COROLLARY 5.7]**. To prove the other inclusion, consider the distinguished triangle

$$i^* i_*(\mathcal{F}) \to \mathcal{F} \to \mathcal{F}[2] \tag{32}$$

that exists for any $\mathcal{F} \in \mathcal{P}$. It implies that $i^*(\langle i_*(\mathcal{P})\rangle^{\oplus}) \subset \mathcal{P}$ and using the definition of base change and admissibility of $\mathcal{P}$ the required inclusion easily follows.

Example 5.6. In the setup of Example 5.4, if $\bar{\mathcal{X}}$ is the surface obtained from a smoothing $\mathcal{X}$ of X by contracting the (-1)-curve X_1 then $\bar{\mathcal{X}}$ is smooth, $\mathcal{X}$ is isomorphic to the blowup of $\bar{\mathcal{X}}$ at a point, and the orthogonal complement $\mathcal{D}$ of $\langle i_*(\mathcal{P})\rangle^{\oplus}$ in $\mathbf{D}^b(\mathcal{X})$ is equivalent to $\mathbf{D}^b(\bar{\mathcal{X}})$.

Following this example it is suggestive to think of the category $\mathcal{D}$ from Theorem 5.5 as a categorical contraction of $\mathbf{D}^b(\mathcal{X})$; this point of view is developed in **[76]**.

5.2. $\mathbb{CP}^{\infty}$-objects

In this section we introduce a class of objects that can be used to construct deformation absorptions of singularities.

Definition 5.7 (**[76]**). We say that $\mathrm{P} \in \mathbf{D}^b(X)$ is a $\mathbb{CP}^{\infty}$-object if

$$\mathrm{Ext}^{\bullet}(\mathrm{P}, \mathrm{P}) \cong \Bbbk[t], \quad \text{where } \deg(t) = 2.$$

In other words, the derived endomorphism algebra of P is isomorphic to the cohomology algebra of the topological space $\mathbb{CP}^{\infty}$.

Remark 5.8. In **[76]** we define a more general notion of $\mathbb{P}^{\infty,q}$-objects for an arbitrary positive integer q by assuming that $\mathrm{Ext}^{\bullet}(\mathrm{P}, \mathrm{P}) \cong \Bbbk[t]$ with $\deg(t) = q$. Such objects can be also used to absorb singularities (and there are many geometrically meaningful examples of such absorptions for $q = 1$, see Remark 5.20), but they never provide deformation absorptions unless $q = 2$.

For each $\mathbb{CP}^{\infty}$-object we define the canonical self-extension M of P from the canonical distinguished triangle

$$\mathrm{M} \to \mathrm{P} \xrightarrow{t} \mathrm{P}[2], \tag{33}$$

where the second arrow is given by the generator $t \in \mathrm{Ext}^2(\mathrm{P}, \mathrm{P})$ of $\mathrm{Ext}^{\bullet}(\mathrm{P}, \mathrm{P})$.

The following characterization is useful and easy to prove.

Lemma 5.9 (**[76]**). *If $\mathrm{P} \in \mathbf{D}^b(X)$ is a $\mathbb{CP}^{\infty}$-object and* M *is its canonical self-extension then*

$$\mathrm{Ext}^{\bullet}(\mathrm{M}, \mathrm{P}) \cong \Bbbk. \tag{34}$$

Conversely, if objects $\mathrm{P}, \mathrm{M} \in \mathbf{D}^{\mathrm{b}}(X)$ *satisfy* (34) *and* $\mathrm{Cone}(\mathrm{M} \to \mathrm{P}) \cong \mathrm{P}[2]$, *then* P *is a* $\mathbb{CP}^{\infty}$*-object.*

Example 5.10. Consider the situation described in Example 5.2. Let $\mathcal{L}_0$ be the line bundle on X that restricts to X_1 as $\mathcal{O}_{X_1}(-1)$ and to X_2 as $\mathcal{O}_{X_2}$. Similarly, let $\mathcal{L}_1$ be the line bundle on X that restricts to X_2 as $\mathcal{O}_{X_2}(-x_0)$ and to X_1 as $\mathcal{O}_{X_1}$. Then there is an exact sequence

$$0 \to \mathcal{O}_{X_1}(-1) \to \mathcal{L}_1 \to \mathcal{L}_0 \to \mathcal{O}_{X_1}(-1) \to 0,$$

where the middle arrow is defined as the composition $\mathcal{L}_1 \twoheadrightarrow \mathcal{O}_{X_2}(-x_0) \hookrightarrow \mathcal{L}_0$. It follows that the objects $\mathrm{P} := \mathcal{O}_{X_1}(-1)$ and $\mathrm{M} := \mathrm{Cone}(\mathcal{L}_1 \to \mathcal{L}_0)$ fit into a distinguished triangle of the form (33). It is also easy to check that (34) holds; therefore P is a $\mathbb{CP}^{\infty}$-object.

As the next proposition shows, $\mathbb{CP}^{\infty}$-objects induce semiorthogonal decompositions; it is instructive to compare this with the well-known notion of $\mathbb{CP}^{n}$-objects (see **[35]**), which rather give autoequivalences of derived categories.

Proposition 5.11 (**[76]**). *If X is a proper Gorenstein scheme,* $\mathrm{P} \in \mathbf{D}^{\mathrm{b}}(X)$ *is a* $\mathbb{CP}^{\infty}$*-object, and the canonical self-extension* M *of* P *is perfect, i.e.,* $\mathrm{M} \in \mathbf{D}^{\mathrm{perf}}(X)$, *then the subcategories*

$$\mathcal{P} := \langle \mathrm{P} \rangle \subset \mathbf{D}^{\mathrm{b}}(X) \quad \textit{and} \quad \mathcal{M} := \langle \mathrm{M} \rangle \subset \mathbf{D}^{\mathrm{perf}}(X)$$

are admissible in $\mathbf{D}^{\mathrm{b}}(X)$ *and* $\mathbf{D}^{\mathrm{perf}}(X)$, *respectively. Moreover,*

$$\mathcal{P} \simeq \mathbf{D}^{\mathrm{b}}(\Bbbk[\epsilon]/\epsilon^2) \quad \textit{and} \quad \mathcal{M} \simeq \mathbf{D}^{\mathrm{perf}}(\Bbbk[\epsilon]/\epsilon^2), \quad \textit{where } \deg(\epsilon) = -1,$$

the right-hand sides are the derived categories of DG modules over the DG algebra $\Bbbk[\epsilon]/\epsilon^2$ *which are perfect over* $\Bbbk$ *and* $\Bbbk[\epsilon]/\epsilon^2$, *respectively. Finally,* $\mathcal{P} \cap \mathbf{D}^{\mathrm{perf}}(X) = \mathcal{M}$.

For instance, since in the situation of Example 5.10 the curve X is Gorenstein and the object M is perfect, we conclude that the category $\mathcal{P} \subset \mathbf{D}^{\mathrm{b}}(X)$ generated by $\mathcal{O}_{X_1}(-1)$ is admissible. A similar computation shows that if X is a tree of rational curves, there is a semiorthogonal collection of $\mathbb{CP}^{\infty}$-objects in $\mathbf{D}^{\mathrm{b}}(X)$ absorbing its singularities.

A useful property of $\mathbb{CP}^{\infty}$-objects is given by the following

Theorem 5.12 (**[76]**). *Let* $\mathrm{P}_1, \dots, \mathrm{P}_n$ *be a semiorthogonal collection of* $\mathbb{CP}^{\infty}$*-objects in* $\mathbf{D}^{\mathrm{b}}(X)$. *If the category* $\mathcal{P} = \langle \mathrm{P}_1, \dots, \mathrm{P}_n \rangle$ *absorbs singularities of X, it provides a deformation absorption of singularities.*

To prove the theorem, we consider a smoothing $f : \mathcal{X} \to B$ of X and check that for each $1 \le j \le n$ the object

$$\mathrm{M}_j := i^* i_*(\mathrm{P}_j)$$

is the canonical self-extension of P_j (this follows from the triangle (32) for $\mathcal{F} = \mathrm{P}_j$). Then using the adjunction isomorphisms

$$\mathrm{Ext}^\bullet(i_*\mathrm{P}_j, i_*\mathrm{P}_k) \cong \mathrm{Ext}^\bullet(\mathrm{M}_j, \mathrm{P}_k),$$

Lemma 5.9, and the triangles (33), we deduce exceptionality and semiorthogonality of $i_*\mathrm{P}_j$, which implies admissibility of the subcategory of $\mathbf{D}^{\mathrm{b}}(\mathcal{X})$ generated by $i_*\mathrm{P}_1, \dots, i_*\mathrm{P}_n$.

Combining Theorem 5.12 with Theorem 5.5, we obtain

Corollary 5.13. *Let* $\mathrm{P}_1, \dots, \mathrm{P}_n$ *be a semiorthogonal collection of* $\mathbb{CP}^\infty$*-objects in* $\mathbf{D}^{\mathrm{b}}(X)$. *If the category* $\mathcal{P} = \langle \mathrm{P}_1, \dots, \mathrm{P}_n \rangle$ *generated by the* P_i *absorbs singularities of* X*, then for any smoothing* $f : \mathcal{X} \to B$ *of* X *there is a semiorthogonal decomposition*

$$\mathbf{D}^{\mathrm{b}}(\mathcal{X}) = \langle i_*\mathrm{P}_1, \dots, i_*\mathrm{P}_n, \mathcal{D} \rangle \tag{35}$$

and the subcategory $\mathcal{D}$ *defined by* (35) *is smooth and proper over* B *with* $\mathcal{D}_b = \mathbf{D}^{\mathrm{b}}(\mathcal{X}_b)$ *for* $b \neq o$ *and* $\mathcal{D}_o = {}^{\perp}\mathcal{P} \subset \mathbf{D}^{\mathrm{b}}(X)$.

5.3. Absorption of nodal singularities

In this section under appropriate assumptions we show how to construct collections of $\mathbb{CP}^\infty$-objects absorbing singularities of nodal varieties of odd dimension. We concentrate on the case of threefolds, because this case is technically simpler and the main assumption has clearer geometric meaning.

Definition 5.14 ([42]). A threefold X with isolated singularities is called maximally nonfactorial if the natural morphism from the class group of Weil divisors on X to the sum of local class groups over all singular points of X

$$\mathrm{Cl}(X) \to \bigoplus_{x \in \mathrm{Sing}(X)} \mathrm{Cl}(X, x)$$

is surjective.

For simplicity, consider the case where X has a single ordinary double point $x_0 \in X$. In this case $\mathrm{Cl}(X, x_0) \cong \mathbb{Z}$, and X is nonfactorial if and only if the morphism of the class groups $\mathrm{Cl}(X) \to \mathrm{Cl}(X, x_0)$ is surjective onto a subgroup of finite index. Thus, maximal nonfactoriality is a strengthening of the usual nonfactoriality property.

Further, if $x_0 \in X$ is a nonfactorial ordinary double point of a threefold, there exists a small resolution $\pi : \tilde{X} \to X$, i.e., a smooth threefold $\tilde{X}$ with a projective morphism π such that its exceptional locus

$$L_0 := \pi^{-1}(x_0) \cong \mathbb{P}^1 \tag{36}$$

is a smooth rational curve. In these terms maximal nonfactoriality is equivalent to the existence of a line bundle $\mathcal{L}$ on X such that

$$\mathcal{L}|_{L_0} \cong \mathcal{O}_{L_0}(-1). \tag{37}$$

Theorem 5.15 ([76]). *Let* X *be a maximally nonfactorial proper threefold with a single ordinary double point* $x_0 \in X$ *and assume* $\mathrm{H}^\bullet(X, \mathcal{O}_X) = \Bbbk$. *Let* $\pi : \tilde{X} \to X$ *be a small resolution with exceptional locus* (36). *Then for any line bundle* $\mathcal{L}$ *on* $\tilde{X}$ *for which* (37) *holds the object*

$$\mathrm{P} := \pi_*\mathcal{L} \tag{38}$$

is a $\mathbb{CP}^\infty$*-object in* $\mathbf{D}^{\mathrm{b}}(X)$ *providing a deformation absorption of singularities of* X.

To prove the theorem, we first decompose the derived category $\mathbf{D}^{\mathrm{b}}(\tilde{X})$ into two parts: the first is generated by an exceptional pair and the second is its orthogonal complement. The pair consists of the line bundle $\mathscr{L}$ and the twisted ideal sheaf

$$\mathscr{L}' := \mathscr{I}_{L_0} \otimes \mathscr{L}.$$

Exceptionality of the pair $(\mathscr{L}', \mathscr{L})$ follows from the isomorphism (37) and the fact that L_0 is a $(-1,-1)$-curve. Now we obtain a semiorthogonal decomposition

$$\mathbf{D}^{\mathrm{b}}(\tilde{X}) = \langle \tilde{\mathscr{P}}, \tilde{\mathscr{D}} \rangle, \quad \text{where } \tilde{\mathscr{P}} := \langle \mathscr{L}', \mathscr{L} \rangle \quad \text{and} \quad \tilde{\mathscr{D}} := {}^{\perp}\langle \mathscr{L}', \mathscr{L} \rangle.$$

Since, by **[17, THEOREM 2.14]**, the functor $\pi_*\colon \mathbf{D}^{\mathrm{b}}(\tilde{X}) \to \mathbf{D}^{\mathrm{b}}(X)$ is a Verdier localization with the kernel generated by $\mathscr{O}_{L_0}(-1)$ and since $\mathscr{O}_{L_0}(-1) \cong \mathrm{Cone}(\mathscr{L}' \to \mathscr{L}) \in \tilde{\mathscr{P}}$, it follows that there is a semiorthogonal decomposition

$$\mathbf{D}^{\mathrm{b}}(X) = \langle \mathscr{P}, \mathscr{D} \rangle,$$

such that $\tilde{\mathscr{D}} = \pi^*(\mathscr{D})$ and $\mathscr{P} \simeq \tilde{\mathscr{P}}/\langle \mathscr{O}_{L_0}(-1) \rangle$. The category $\mathscr{D}$ is equivalent to the smooth and proper category $\tilde{\mathscr{D}}$, hence $\mathscr{P}$ absorbs singularities of X. On the other hand, $\mathscr{P}$ is generated by the object $\mathrm{P} := \pi_*(\mathscr{L}) \cong \pi_*(\mathscr{L}')$, and a simple computation shows that there is a distinguished triangle

$$\pi^*(\mathrm{M}) \to \mathscr{L}' \to \mathscr{L}[2]$$

for an object $\mathrm{M} \in \mathbf{D}^{\mathrm{perf}}(X)$ such that (34) holds. Pushing forward this triangle, we obtain (33), hence P is a $\mathbb{CP}^\infty$-object by Lemma 5.9 (and M is its canonical self-extension).

Remark 5.16. The $\mathbb{CP}^\infty$-object P defined in (38) is a reflexive sheaf of rank 1 on X and the image of $[\mathrm{P}] \in \mathrm{Cl}(X)$ in $\mathrm{Cl}(X, x_0)$ is a generator of the local class group. Note that the line bundle $\mathscr{L}$ satisfying (37) is unique up to twist by a line bundle pulled back from X, hence the same is true for the reflexive sheaf P. On the other hand, if $\tilde{X}'$ is the other small resolution of X and P′ is the reflexive generator of $\mathrm{Cl}(X, x_0)$ constructed from it, it follows that P′ is isomorphic to the underived dual $\mathrm{P}^\vee$ of P up to line bundle twist. Thus, the flop $\tilde{X} \dashrightarrow \tilde{X}'$ between the small resolutions corresponds to dualization of the corresponding $\mathbb{CP}^\infty$-object.

Remark 5.17. Theorem 5.15 shows that maximal nonfactoriality is sufficient for the existence of an absorption of singularities of a threefold with a single ordinary double point by a $\mathbb{CP}^\infty$-object. Using **[86, LEMMA 1.11]** and **[42, COROLLARY 3.8]** one can prove that this condition is also necessary, see **[76]**.

Remark 5.18. Let X be a maximally nonfactorial threefold with n ordinary double points and let $\pi\colon \tilde{X} \to X$ be a small resolution of singularities with exceptional curves $L_1, \dots, L_n$. Assuming there is an exceptional collection $(\mathscr{L}_1, \dots, \mathscr{L}_n)$ of line bundles on $\tilde{X}$ such that $\mathscr{L}_j|_{L_k} \cong \mathscr{O}_{L_k}(-\delta_{jk})$ (here δ_{jk} is the Kronecker delta) for all $1 \le j, k \le n$, a similar argument shows that singularities of X are absorbed by the semiorthogonal collection of $\mathbb{CP}^\infty$-objects $\mathrm{P}_j := \pi_*\mathscr{L}_j$, $1 \le j \le n$.

Combining Theorem 5.15 with Theorem 5.12, we obtain

Corollary 5.19 ([76]). *Let X be a maximally nonfactorial projective threefold with a single ordinary double point $x_0 \in X$ and assume $\mathrm{H}^\bullet(X, \mathcal{O}_X) = \Bbbk$. Let $f: \mathcal{X} \to B$ be a smoothing of X. Then there is a semiorthogonal decomposition*

$$\mathbf{D}^{\mathrm{b}}(\mathcal{X}) = \langle i_*\mathrm{P}, \mathcal{D} \rangle,$$

where $\mathrm{P} \in \mathbf{D}^{\mathrm{b}}(X)$ is the $\mathbb{CP}^\infty$-object defined in Theorem 5.15 and the subcategory $\mathcal{D}$ is smooth and proper over B with $\mathcal{D}_b = \mathbf{D}^{\mathrm{b}}(\mathcal{X}_b)$ for $b \neq o$ and $\mathcal{D}_o = {}^\perp\mathrm{P} \subset \mathbf{D}^{\mathrm{b}}(X)$.

Of course, a similar result holds for maximally nonfactorial threefolds with several ordinary double points if the assumption of Remark 5.18 is satisfied.

Remark 5.20. There is a similar construction of deformation absorption that works in higher dimensions. Let X be a variety of *odd* dimension with a single ordinary double point $x_0 \in X$. Let $\tilde{X} = \mathrm{Bl}_{x_0}(X)$; then the exceptional divisor $E \subset \tilde{X}$ is a smooth even-dimensional quadric. Assume there is an exceptional object $\mathcal{E} \in \mathbf{D}^{\mathrm{b}}(\tilde{X})$ such that

$$\mathcal{E}|_E \cong \mathcal{S}_E,$$

where $\mathcal{S}_E$ is a spinor bundle (this condition plays the same role as (37)). Then $\mathrm{P} := \pi_*(\mathcal{E})$ is a $\mathbb{CP}^\infty$-object providing a deformation absorption of singularities of X, see [76]. Of course, there is also a version of this result for several ordinary double points as in Remark 5.18.

An analogous construction for *even-dimensional* varieties produces a $\mathbb{P}^{\infty,1}$-object (as defined in Remark 5.8) which also absorbs singularities of X, but it does not give a deformation absorption.

5.4. Fano threefolds

In this subsection, we apply the above results to clarify and extend the relation between nontrivial components of derived categories of del Pezzo threefolds and prime Fano threefolds that was discovered in [56].

Recall that a prime Fano threefold is a Fano threefold X with $\mathrm{Pic}(X) = \mathbb{Z}K_X$, and its genus $\mathrm{g}(X)$ is defined from the equality

$$(-K_X)^3 = 2\mathrm{g}(X) - 2.$$

It is well known that $2 \leq \mathrm{g}(X) \leq 12$ and $\mathrm{g}(X) \neq 11$.

Mukai proved (see [82], [74, §B.1]) that, for any prime Fano threefold X with

$$\mathrm{g}(X) \in \{6, 8, 10, 12\},$$

there exists a unique exceptional vector bundle $\mathcal{U}_X$ of rank 2 with $\mathrm{c}_1(\mathcal{U}_X) = K_X$ such that $(\mathcal{O}_X, \mathcal{U}_X^\vee)$ is an exceptional pair; it is called the Mukai bundle. Using this observation, the nontrivial part $\mathcal{A}_X \subset \mathbf{D}^{\mathrm{b}}(X)$ was defined in [56] from the semiorthogonal decomposition

$$\mathbf{D}^{\mathrm{b}}(X) = \langle \mathcal{A}_X, \mathcal{O}_X, \mathcal{U}_X^\vee \rangle. \tag{39}$$

Remark 5.21. This definition extends to *general* prime Fano threefolds of genus

$$\mathrm{g}(X) = 4.$$

In fact, any smooth prime Fano threefold of genus 4 is a complete intersection $X \subset \mathbb{P}^5$ of type $(2, 3)$. We will say that X is general if the (unique) quadric passing through $X \subset \mathbb{P}^5$ is smooth; in this case there are two Mukai bundles (the restrictions of the spinor bundles from the quadric) and the corresponding nontrivial parts of $\mathbf{D}^{\mathrm{b}}(X)$ are equal to the refined residual categories from Example 3.17.

Similarly, a del Pezzo threefold is a Fano threefold Y with $-K_Y = 2H$ for a primitive Cartier divisor class H and its degree $\mathrm{d}(Y)$ is defined as

$$\mathrm{d}(Y) = H^3.$$

It is well known that $1 \le \mathrm{d}(Y) \le 5$ for del Pezzo threefolds of Picard rank 1.

If Y is a del Pezzo threefold, the pair of line bundles $(\mathcal{O}_Y, \mathcal{O}_Y(H))$ is exceptional, and this time the nontrivial part $\mathcal{B}_Y \subset \mathbf{D}^{\mathrm{b}}(Y)$ was defined in [56] from the semiorthogonal decomposition

$$\mathbf{D}^{\mathrm{b}}(Y) = \langle \mathcal{B}_Y, \mathcal{O}_Y, \mathcal{O}_Y(H) \rangle \tag{40}$$

(so, in this case these are just the residual categories in the sense of Section 3).

It was observed in [56, PROPOSITION 3.9] that when X and Y are as above and

$$\mathrm{g}(X) = 2\mathrm{d}(Y) + 2,$$

the categories $\mathcal{A}_X$ and $\mathcal{B}_Y$ have isomorphic numerical Grothendieck groups (and their isomorphism is compatible with the Euler pairings). Furthermore, it was proved in [56, THEOREM 3.8] that for each prime Fano threefold X with $\mathrm{g}(X) \in \{8, 10, 12\}$ there is a unique del Pezzo threefold Y (with $\mathrm{d}(Y) = \mathrm{g}(X)/2 - 1 \in \{3, 4, 5\}$) such that

$$\mathcal{A}_X \simeq \mathcal{B}_Y.$$

So, it was expected [56, CONJECTURE 3.7] that the same equivalence takes place for appropriate pairs (X, Y) with $\mathrm{g}(X) \in \{4, 6\}$ and $\mathrm{d}(Y) \in \{1, 2\}$.

However, the conjecture turned out to be false: for $\mathrm{g}(X) = 6$ and $\mathrm{d}(Y) = 2$, it was disproved in [10] or [96, THEOREM 1.2], and for $\mathrm{g}(X) = 4$ and $\mathrm{d}(Y) = 1$, it is false for trivial reasons as in this case $\dim(\mathrm{HH}_1(\mathcal{A}_X)) = 21$ while $\dim(\mathrm{HH}_1(\mathcal{B}_Y)) = 20$. The next theorem clarifies the situation in these two cases.

Theorem 5.22 ([77]). *For $2 \le d \le 5$ let Y be a del Pezzo threefold of degree d and Picard rank 1 and for $d = 1$ let Y be a small resolution of a del Pezzo threefold of degree 1 and Picard rank 1 with a single ordinary double point. Then there exists a flat projective morphism $f : \mathcal{X} \to B$ to a smooth pointed curve (B, o) such that $\mathcal{X}$ is smooth and*

(a) *for any point $b \neq o$ in B the fiber $\mathcal{X}_b$ is a smooth prime Fano threefold of genus $g = 2d + 2$;*

(b) *the central fiber $\mathcal{X}_o$ is a prime Fano threefold of genus $g = 2d + 2$ with a single maximally nonfactorial ordinary double point $x_o \in \mathcal{X}_o$ birational to Y.*

Furthermore, there is a B-linear subcategory $\mathcal{A} \subset \mathbf{D}^{\mathrm{b}}(\mathcal{X})$ which is smooth and proper over B and such that:

(i) *for any point $b \neq o$ in B one has $\mathcal{A}_b = \mathcal{A}_{\mathcal{X}_b} \subset \mathbf{D}^{\mathrm{b}}(\mathcal{X}_b)$;*

(ii) *the central fiber $\mathcal{A}_o$ is equivalent to the component $\mathcal{B}_Y$ of $\mathbf{D}^{\mathrm{b}}(Y)$.*

In particular, the nontrivial part $\mathcal{B}_Y$ of $\mathbf{D}^{\mathrm{b}}(Y)$ is a smooth and proper extension across the point $o \in B$ of the family $\mathcal{A}_{\mathcal{X}_b}$ of the nontrivial parts of $\mathbf{D}^{\mathrm{b}}(\mathcal{X}_b)$.

Remark 5.23. The case $d = 1$ in the theorem is somewhat special; in this case we take Y to be a small resolution $Y \to \bar{Y}$ of a del Pezzo threefold $\bar{Y}$ with a single node (such resolution Y always exists as an algebraic space, but not as a projective variety). Note, however, that the component $\mathcal{B}_Y$ is still defined for this algebraic space Y by the same formula (40), where the line bundle $\mathcal{O}_Y(H)$ is the effective generator of the Picard group of Y. It is remarkable that in this case all prime Fano threefolds $\mathcal{X}_b$ for $b \neq o$ in the constructed family are general in the sense of Remark 5.21.

Let us explain how the family $\mathcal{X}$ is constructed. Let Y be as in the theorem. If $2 \leq d \leq 5$ let $C \subset Y$ be a general smooth rational curve of degree $d - 1$ (with respect to H), and if $d = 1$ let C be the exceptional curve of Y (recall that in this case Y is a small resolution of a nodal del Pezzo threefold $\bar{Y}$). Then one can prove that C has a unique bisecant line $L \subset Y$ and there exists a diagram

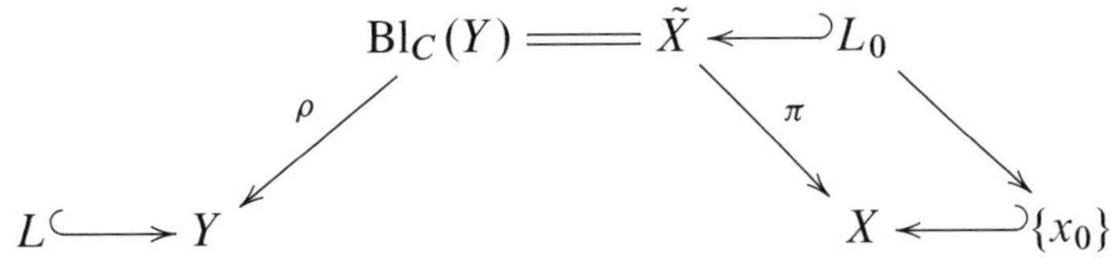

where ρ is the blowup morphism, π is the contraction of the strict transform $L_0 \subset \tilde{X}$ of L, and X is a maximally nonfactorial prime Fano threefold of genus $g = 2d + 2$ with a single ordinary double point $x_0 = \pi(L_0)$. Now we define the family $f \colon \mathcal{X} \to B$ as a smoothing of X (it exists by [83]).

Now we explain how the subcategory $\mathcal{A} \subset \mathbf{D}^{\mathrm{b}}(\mathcal{X})$ is constructed. First, applying Corollary 5.19 to the smoothing $f \colon \mathcal{X} \to B$ constructed above we obtain a B-linear subcategory $\mathcal{D} \subset \mathbf{D}^{\mathrm{b}}(\mathcal{X})$ smooth and proper over B such that

$$\mathcal{D}_b = \mathbf{D}^{\mathrm{b}}(\mathcal{X}_b) \quad \text{for } b \neq o \quad \text{and} \quad \mathcal{D}_o \simeq {}^{\perp}\langle \mathcal{J}_{L_0}(-H), \mathcal{O}_{\tilde{X}}(-H) \rangle \subset \mathbf{D}^{\mathrm{b}}(\tilde{X}),$$

where H is the pullback to $\tilde{X}$ of the hyperplane class of Y. Next, we consider the sheaf

$$\mathcal{U}'_Y := \operatorname{Ker}\big(\mathcal{O}_Y(H) \oplus \mathcal{O}_Y(H) \to \mathcal{O}_C(d)\big),$$

where the morphism is a twist of the evaluation morphism $\mathcal{O}_Y \oplus \mathcal{O}_Y \to \mathcal{O}_C(1)$. We check that there is a vector bundle $\mathcal{U}_X$ on X such that $\mathcal{U}'_Y \cong \rho_*(\pi^* \mathcal{U}_X^{\vee})$, that $(\mathcal{O}_X, \mathcal{U}_X^{\vee})$ is an exceptional pair in $\mathbf{D}^{\mathrm{b}}(X)$, and this pair deforms (possibly after an étale base change) to the nearby fibers of a family $f \colon \mathcal{X} \to B$. Therefore, after a possible étale base change we can assume that the pair is defined on $\mathcal{X}$, and hence we have a B-linear semiorthogonal decomposition

$$\mathcal{D} = \langle \mathcal{A}, f^*\mathbf{D}^{\mathrm{b}}(B), f^*\mathbf{D}^{\mathrm{b}}(B) \otimes \mathcal{U}_{\mathcal{X}}^{\vee} \rangle$$

Finally, we check that $\mathcal{U}_{\mathcal{X}}|_{\mathcal{X}_b}$ is the Mukai bundle of $\mathcal{X}_b$ when $b \neq o$, hence $\mathcal{A}_b = \mathcal{A}_{X_b}$. On the other hand, we find a simple sequence of mutations identifying $\mathcal{A}_o \subset \mathcal{D}_o \subset \mathbf{D}^{\mathrm{b}}(\tilde{X})$ with $\mathcal{B}_Y$. This proves the equivalence $\mathcal{A}_o \cong \mathcal{B}_Y$.

Remark 5.24. When $3 \leq d \leq 5$ the family of threefolds $\mathcal{X}$ in Theorem 5.22 can be chosen in such a way that the family of categories $\mathcal{A}_{\mathcal{X}_b}$ is *isotrivial*, i.e., $\mathcal{A}_{\mathcal{X}_b} \simeq \mathcal{B}_Y$ for all $b \in B$. This is no longer possible for $d \in \{1, 2\}$.

Remark 5.25. There are several interesting examples of del Pezzo threefolds with higher Picard rank: two del Pezzo threefolds of degree 6 (the flag variety $\mathrm{Fl}(1, 2; 3)$ and $(\mathbb{P}^1)^3$) and one del Pezzo threefold of degree 7 (the blowup of $\mathbb{P}^3$ at a point). The construction of Theorem 5.22 works for these threefolds and relates the nontrivial parts of their derived categories (still defined by (40)) to the nontrivial parts (still defined by (39)) of the derived categories of appropriate Fano threefolds with primitive canonical class and genus $g = 14$ and $g = 16$, respectively.

There are other interesting maximally nonfactorial nodal Fano threefolds, e.g., some prime Fano threefolds of odd genus $g \in \{5, 7, 9\}$. They also provide geometrically meaningful extensions of (appropriately defined) nontrivial components of derived categories, see [77] for details.

ACKNOWLEDGMENTS

Most of results described in this survey are obtained in collaboration. I am very grateful to all my coauthors, especially to Alex Perry, Evgeny Shinder, and Maxim Smirnov for their invaluable inputs. I would also like to thank all my colleagues for providing inspiration, sharing important ideas, and answering numerous questions.

Finally, I would like to thank Pieter Belmans, Alex Perry, Evgeny Shinder, and Maxim Smirnov for their comments on the preliminary version of this paper.

FUNDING

This work was performed at the Steklov International Mathematical Center and supported by the Ministry of Science and Higher Education of the Russian Federation (agreement no. 075-15-2019-1614).

REFERENCES

[1] R. Anno, Spherical functors. 2007, arXiv:0711.4409.

[2] R. Anno and T. Logvinenko, Spherical DG-functors. *J. Eur. Math. Soc. (JEMS)* **19** (2017), no. 9, 2577–2656.

[3] M. F. Atiyah, On analytic surfaces with double points. *Proc. R. Soc. Lond. Ser. A* **247** (1958), 237–244.

[4] A. Auel and M. Bernardara, Semiorthogonal decompositions and birational geometry of del Pezzo surfaces over arbitrary fields. *Proc. Lond. Math. Soc. (3)* **117** (2018), no. 1, 1–64.

[5] D. Auroux, Mirror symmetry and T-duality in the complement of an anticanonical divisor. *J. Gökova Geom. Topol. GGT* **1** (2007), 51–91.

[6] M. Ballard, D. Deliu, D. Favero, M. U. Isik, and L. Katzarkov, On the derived categories of degree d hypersurface fibrations. *Math. Ann.* **371** (2018), no. 1–2, 337–370.

[7] M. Ballard, A. Duncan, and P. McFaddin, On derived categories of arithmetic toric varieties. *Ann. K-Theory* **4** (2019), no. 2, 211–242.

[8] A. Bayer, M. Lahoz, E. Macrì, and P. Stellari, Stability conditions on Kuznetsov components. 2017, arXiv:1703.10839.

[9] A. Bayer and E. Macri, The unreasonable effectiveness of wall-crossing in algebraic geometry. In *ICM 2022 Proceedings*. EMS Press, 2022.

[10] A. Bayer and A. Perry, Kuznetsov's Fano threefold conjecture via K3 categories and enhanced group actions. 2022, arXiv:2202.04195.

[11] A. Beauville, Quantum cohomology of complete intersections. *Mat. Fiz. Anal. Geom.* **2** (1995), no. 3–4, 384–398.

[12] P. Belmans, A. Kuznetsov, and M. Smirnov, Derived categories of the Cayley plane and the coadjoint Grassmannian of type F. *Transform. Groups* (2021).

[13] P. Belmans, S. Okawa, and A. T. Ricolfi, Moduli spaces of semiorthogonal decompositions in families. 2020, arXiv:2002.03303.

[14] V. Benedetti and J. Song, Divisors in the moduli space of Debarre–Voisin varieties. 2021, arXiv:2106.06859.

[15] M. Bernardara, M. Bolognesi, and D. Faenzi, Homological projective duality for determinantal varieties. *Adv. Math.* **296** (2016), 181–209.

[16] M. Bernardara, E. Fatighenti, and L. Manivel, Nested varieties of K3 type. *J. Éc. Polytech. Math.* **8** (2021), 733–778.

[17] A. Bondal, M. Kapranov, and V. Schechtman, Perverse schobers and birational geometry. *Selecta Math. (N.S.)* **24** (2018), no. 1, 85–143.

[18] A. Bondal and D. Orlov, Semiorthogonal decomposition for algebraic varieties. 1995, arXiv:alg-geom/9506012.

[19] A. Bondal and D. Orlov, Derived categories of coherent sheaves. In *Proceedings of the International Congress of Mathematicians, Vol. II (Beijing, 2002)*, pp. 47–56, Higher Ed. Press, Beijing, 2002.

[20] L. A. Borisov, A. Căldăraru, and A. Perry, Intersections of two Grassmannians in $\mathbb{P}^9$. *J. Reine Angew. Math.* **760** (2020), 133–162.

[21] E. Brieskorn, Singular elements of semi-simple algebraic groups. In *Actes du Congrès International des Mathématiciens (Nice, 1970), Tome 2*, pp. 279–284, 1971.

[22] F. Carocci and Z. Turčinović, Homological projective duality for linear systems with base locus. *Int. Math. Res. Not. IMRN* **21** (2020), 7829–7856.

[23] O. Debarre and A. Kuznetsov, Gushel–Mukai varieties: classification and birationalities. *Algebr. Geom.* **5** (2018), no. 1, 15–76.

[24] B. Dubrovin, Geometry and analytic theory of Frobenius manifolds. In *Proceedings of the International Congress of Mathematicians, Vol. II (Berlin, 1998)*, pp. 315–326, Extra Vol. II, 1998.

[25] A. Efimov, Wall finiteness obstruction for DG categories. 2020, https://www.youtube.com/watch?v=i3SPxKysBMA.

[26] A. I. Efimov, Some remarks on L-equivalence of algebraic varieties. *Selecta Math. (N.S.)* **24** (2018), no. 4, 3753–3762.

[27] A. Elagin and V. A. Lunts, Three notions of dimension for triangulated categories. *J. Algebra* **569** (2021), 334–376.

[28] A. Fonarëv, Minimal Lefschetz decompositions of the derived categories for Grassmannians. *Izv. Ross. Akad. Nauk Ser. Mat.* **77** (2013), no. 5, 203–224.

[29] A. Fonarëv, Full exceptional collections on Lagrangian Grassmannians. *Int. Math. Res. Not. IMRN* **2** (2022), 1081–1122.

[30] S. Ganatra, Automatically generating Fukaya categories and computing quantum cohomology. 2019, arXiv:1605.07702.

[31] L. A. Guseva, On the derived category of IGr(3, 8). *Sb. Math.* **211** (2020), no. 7, 24–59.

[32] B. Hassett, Special cubic fourfolds. *Compos. Math.* **120** (2000), no. 1, 1–23.

[33] C. Hertling, Y. I. Manin, and C. Teleman, An update on semisimple quantum cohomology and F-manifolds. *Tr. Mat. Inst. Steklova* **264** (2009), 69–76.

[34] S. Hosono and H. Takagi, Towards homological projective duality for $\mathsf{S}^2\mathbb{P}^3$ and $\mathsf{S}^2\mathbb{P}^4$. *Adv. Math.* **317** (2017), 371–409.

[35] D. Huybrechts and R. Thomas, $\mathbb{P}$-objects and autoequivalences of derived categories. *Math. Res. Lett.* **13** (2006), no. 1, 87–98.

[36] Q. Jiang, Derived categories of Quot schemes of locally free quotients, I. 2021, arXiv:2107.09193.

[37] Q. Jiang, Derived category of projectivization and generalized linear duality. 2018, arXiv:1812.05685.

[38] Q. Jiang and N. C. Leung, Blowing up linear categories, refinements, and homological projective duality with base locus. 2019, arXiv:1811.05132.

[39] Q. Jiang and N. C. Leung, Categorical duality between joins and intersections. 2018, arXiv:1811.05135.

[40] Q. Jiang and N. C. Leung, Derived category of projectivization and flops. *Adv. Math.* **396** (2022), 108169.

[41] Q. Jiang, N. C. Leung, and Y. Xie, Categorical Plücker formula and homological projective duality. *J. Eur. Math. Soc. (JEMS)* **23** (2021), no. 6, 1859–1898.

[42] M. Kalck, N. Pavic, and E. Shinder, Obstructions to semiorthogonal decompositions for singular threefolds I: K-theory. *Moscow Math. J.* **21**, no. 3, 567–592.

[43] M. Kapranov, Y. Soibelman, and L. Soukhanov, Perverse schobers and the algebra of the infrared. 2020, arXiv:2011.00845.

[44] M. Kapustka and M. Rampazzo, Torelli problem for Calabi–Yau threefolds with GLSM description. *Commun. Number Theory Phys.* **13** (2019), no. 4, 725–761.

[45] J. Karmazyn, A. Kuznetsov, and E. Shinder, Derived categories of singular surfaces. *J. Eur. Math. Soc. (JEMS)* **24** (2022), no. 2, 461–526.

[46] R. Kaufmann, The intersection form in $\mathscr{H}^*(\overline{\mathscr{M}}_{0n})$ and the explicit Künneth formula in quantum cohomology. *Int. Math. Res. Not.* **19** (1996), 929–952.

[47] M. Kontsevich, Homological algebra of mirror symmetry. In *Proceedings of the International Congress of Mathematicians, Vol. 1, 2 (Zürich, 1994)*, pp. 120–139, Birkhäuser, Basel, 1995.

[48] O. Küchle, On Fano 4-fold of index 1 and homogeneous vector bundles over Grassmannians. *Math. Z.* **218** (1995), no. 4, 563–575.

[49] A. Kuznetsov, Derived category of a cubic threefold and the variety V_{14}. *Tr. Mat. Inst. Steklova* **246** (2004), 183–207.

[50] A. Kuznetsov, Homological projective duality for Grassmannians of lines. 2006, arXiv:math/0610957.

[51] A. Kuznetsov, Hyperplane sections and derived categories. *Izv. Ross. Akad. Nauk Ser. Mat.* **70** (2006), no. 3, 23–128.

[52] A. Kuznetsov, Homological projective duality. *Publ. Math. Inst. Hautes Études Sci.* **105** (2007), 157–220.

[53] A. Kuznetsov, Derived categories of quadric fibrations and intersections of quadrics. *Adv. Math.* **218** (2008), no. 5, 1340–1369.

[54] A. Kuznetsov, Exceptional collections for Grassmannians of isotropic lines. *Proc. Lond. Math. Soc. (3)* **97** (2008), no. 1, 155–182.

[55] A. Kuznetsov, Lefschetz decompositions and categorical resolutions of singularities. *Selecta Math. (N.S.)* **13** (2008), no. 4, 661–696.

[56] A. Kuznetsov, Derived categories of Fano threefolds. *Tr. Mat. Inst. Steklova* **264** (2009), 116–128.

[57] A. Kuznetsov, Derived categories of cubic fourfolds. In *Cohomological and geometric approaches to rationality problems*, pp. 219–243, Progr. Math. 282, Birkhäuser Boston, Boston, MA, 2010.

[58] A. Kuznetsov, Base change for semiorthogonal decompositions. *Compos. Math.* **147** (2011), no. 3, 852–876.

[59] A. Kuznetsov, Semiorthogonal decompositions in algebraic geometry. In *Proceedings of the International Congress of Mathematicians—Seoul 2014. Vol. II*, pp. 635–660, Kyung Moon Sa, Seoul, 2014.

[60] A. Kuznetsov, On Küchle varieties with Picard number greater than 1. *Izv. Ross. Akad. Nauk Ser. Mat.* **79** (2015), no. 4, 57–70.

[61] A. Kuznetsov, Küchle fivefolds of type c5. *Math. Z.* **284** (2016), no. 3–4, 1245–1278.

[62] A. Kuznetsov, Calabi–Yau and fractional Calabi–Yau categories. *J. Reine Angew. Math.* **753** (2019), 239–267.

[63] A. Kuznetsov, Derived categories of families of sextic del Pezzo surfaces. *Int. Math. Res. Not. IMRN* **12** (2021), 9262–9339.

[64] A. Kuznetsov, Derived categories of families of Fano threefolds. 2022, arXiv:2202.12345.

[65] A. Kuznetsov, Simultaneous categorical resolutions. *Math. Z.* (2022).

[66] A. Kuznetsov and V. A. Lunts, Categorical resolutions of irrational singularities. *Int. Math. Res. Not. IMRN* **13** (2015), 4536–4625.

[67] A. Kuznetsov and A. Perry, Derived categories of cyclic covers and their branch divisors. *Selecta Math. (N.S.)* **23** (2017), no. 1, 389–423.

[68] A. Kuznetsov and A. Perry, Derived categories of Gushel–Mukai varieties. *Compos. Math.* **154** (2018), no. 7, 1362–1406.

[69] A. Kuznetsov and A. Perry, Categorical joins. *J. Amer. Math. Soc.* **34** (2021), no. 2, 505–564.

[70] A. Kuznetsov and A. Perry, Homological projective duality for quadrics. *J. Algebraic Geom.* **30** (2021), no. 3, 457–476.

[71] A. Kuznetsov and A. Perry, Categorical cones and quadratic homological projective duality. 2019, arXiv:1902.09824.

[72] A. Kuznetsov and A. Perry, Serre functors and dimensions of residual categories. 2021, arXiv:2109.02026.

[73] A. Kuznetsov and A. Polishchuk, Exceptional collections on isotropic Grassmannians. *J. Eur. Math. Soc. (JEMS)* **18** (2016), no. 3, 507–574.

[74] A. Kuznetsov, Y. Prokhorov, and C. Shramov, Hilbert schemes of lines and conics and automorphism groups of Fano threefolds. *Jpn. J. Math.* **13** (2018), no. 1, 109–185.

[75] A. Kuznetsov and E. Shinder, Grothendieck ring of varieties, D- and L-equivalence, and families of quadrics. *Selecta Math. (N.S.)* **24** (2018), no. 4, 3475–3500.

[76] A. Kuznetsov and E. Shinder, Categorical absorptions of singularities and degenerations (in preparation).

[77] A. Kuznetsov and E. Shinder, Derived categories of Fano threefolds via degenerations (in preparation).

[78] A. Kuznetsov and M. Smirnov, On residual categories for Grassmannians. *Proc. Lond. Math. Soc. (3)* **120** (2020), no. 5, 617–641.

[79] A. Kuznetsov and M. Smirnov, Residual categories for (co)adjoint Grassmannians in classical types. *Compos. Math.* **157** (2021), no. 6, 1172–1206.

[80] R. Laza, The moduli space of cubic fourfolds. *J. Algebraic Geom.* **18** (2009), no. 3, 511–545.

[81] L. Manivel, Double spinor Calabi–Yau varieties. *Épijournal Géom. Algébrique* **3** (2019), 2.

[82] S. Mukai, Fano 3-folds. In *Complex projective geometry (Trieste, 1989/Bergen, 1989)*, pp. 255–263, London Math. Soc. Lecture Note Ser. 179, Cambridge Univ. Press, Cambridge, 1992.

[83] Y. Namikawa, Smoothing Fano 3-folds. *J. Algebraic Geom.* **6** (1997), no. 2, 307–324.

[84] D. Orlov, Smooth and proper noncommutative schemes and gluing of DG categories. *Adv. Math.* **302** (2016), 59–105.

[85] J. C. Ottem and J. V. Rennemo, A counterexample to the birational Torelli problem for Calabi–Yau threefolds. *J. Lond. Math. Soc. (2)* **97** (2018), no. 3, 427–440.

[86] N. Pavic and E. Shinder, K-theory and the singularity category of quotient singularities. *Ann. K-Theory* **6** (2021), no. 3, 381–424.

[87] N. Perrin and M. Smirnov, On the big quantum cohomology of (co)adjoint varieties. 2021, arXiv:2112.12436.

[88] A. Perry, Noncommutative homological projective duality. *Adv. Math.* **350** (2019), 877–972.

[89] J. V. Rennemo, The homological projective dual of $\mathrm{Sym}^2\mathbb{P}(V)$. *Compos. Math.* **156** (2020), no. 3, 476–525.

[90] J. V. Rennemo and E. Segal, Hori-mological projective duality. *Duke Math. J.* **168** (2019), no. 11, 2127–2205.

[91] B. Siebert and G. Tian, On quantum cohomology rings of Fano manifolds and a formula of Vafa and Intriligator. *Asian J. Math.* **1** (1997), no. 4, 679–695.

[92] G. Sivek, On vanishing sums of distinct roots of unity. *Integers* **10** (2010), no. A31, 365–368.

[93] M. Smirnov, On the derived category of the adjoint Grassmannian of type F. 2021, arXiv:2107.07814.

[94] R. P. Thomas, Notes on homological projective duality. In *Algebraic geometry: Salt Lake City 2015*, pp. 585–609, Proc. Sympos. Pure Math. 97, Amer. Math. Soc., Providence, RI, 2018.

[95] G. N. Tjurina, Resolution of singularities of flat deformations of double rational points. *Funktsional. Anal. i Prilozhen.* **4** (1970), no. 1, 77–83.

[96] S. Zhang, Bridgeland moduli spaces for Gushel–Mukai threefolds and Kuznetsov's Fano threefold conjecture. 2020, arXiv:2012.12193.

ALEXANDER KUZNETSOV

Algebraic Geometry Section, Steklov Mathematical Institute of Russian Academy of Sciences, 8 Gubkin str., Moscow 119991, Russia, akuznet@mi-ras.ru

This Page Intentionally Left Blank

WHAT IS A RANDOM SURFACE?

SCOTT SHEFFIELD

ABSTRACT

Given $2n$ unit equilateral triangles, there are finitely many ways to glue each edge to a partner. We obtain a random *sphere-homeomorphic* surface by sampling uniformly from the gluings that produce a topological sphere. As $n \to \infty$, these random surfaces (appropriately scaled) converge in law. The limit is a "canonical" sphere-homeomorphic random surface, much the way Brownian motion is a canonical random path.

Depending on how the surface space and convergence topology are specified, the limit is the *Brownian sphere*, the *peanosphere*, the *pure Liouville quantum gravity sphere*, or a certain *conformal field theory*. All of these objects have concise definitions, and are all in some sense equivalent, but the equivalence is highly nontrivial, building on hundreds of math and physics papers over the past half-century.

More generally, the "continuum random surface embedded in d-dimensional Euclidean space" makes a kind of sense for $d \in (-\infty, 25)$ even when d is not a positive integer; and this can be extended to higher genus surfaces, surfaces with boundary, and surfaces with marked points or other decoration.

These constructions have deep roots in both mathematics and physics, drawing from classical graph theory, complex analysis, probability, and representation theory, as well as string theory, planar statistical physics, random matrix theory, and a simple model for two-dimensional quantum gravity.

We present here an informal, colloquium-level overview of the subject, which we hope will be accessible to both newcomers and experts. We aim to answer, as cleanly as possible, the fundamental question. What is a random surface?

MATHEMATICS SUBJECT CLASSIFICATION 2020

Primary 60G60; Secondary 60G57, 60J67, 60K35

KEYWORDS

Brownian map, peanosphere, Liouville quantum gravity, conformal field theory, Gaussian free field, Schramm–Loewner evolution, continuum random tree, Brownian snake, random planar map, random surface

INTERNATIONAL CONGRESS OF MATHEMATICIANS 2022 JULY 6–14

Proc. Int. Cong. Math. 2022, Vol. 2, pp. 1202–1258
DOI 10.4171/ICM2022/187

Published by EMS Press

As a random path, Brownian motion is *canonical* in the sense that it is uniquely characterized by certain symmetries, and *universal* in the sense that it is a limit of many kinds of discrete random walks. This paper will describe a similarly canonical and universal *random surface*. This random surface has several different formulations. To tell the story in a fanciful way, imagine a dialog, first about random paths, then about random surfaces.

INSTRUCTOR: Consider the simple random walk on $\mathbb{Z}$. At each time step, a coin toss decides whether position goes up or down. If you shrink the graph horizontally by a factor of C and vertically by a factor of $\sqrt{C}$, then the $C \to \infty$ limit is *Brownian motion*.

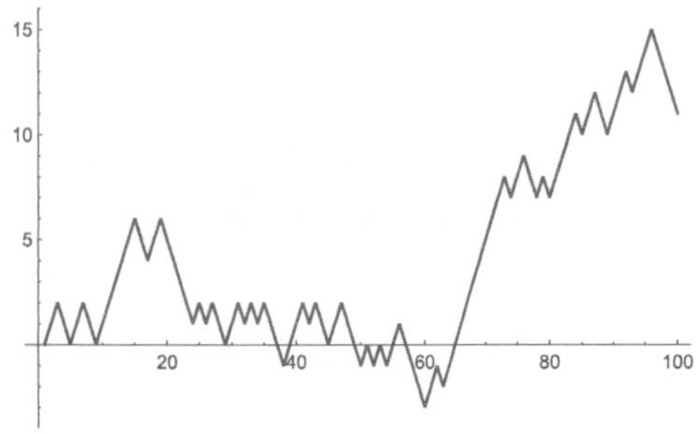

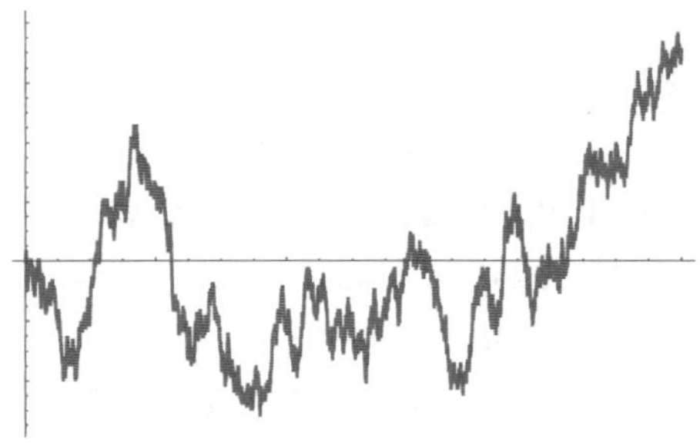

STUDENT: Great! But can you define Brownian motion directly in the continuum?

INSTRUCTOR: Sure! Fix $0 = t_0 < t_1 < \cdots < t_n$. Specify the joint law of $B(t_1), \ldots, B(t_n)$ by making increments $B(t_k) - B(t_{k-1})$ independent normals with mean 0, variance $t_k - t_{k-1}$. Extend to countable dense set (Kolmogorov extension), then all t (Kolomogorov–Čentsov).

STUDENT: Are there other natural ways to characterize Brownian motion?

INSTRUCTOR: Brownian motion is *canonical* in that it is the only random path with certain symmetries (like stationarity/independence of increments). It is *universal* in that (per central limit theorem) it is a limit of many discrete walks. It comes up everywhere.

STUDENT: What if I want a random path embedded in $\mathbb{R}^d$?

INSTRUCTOR: Use a vector $(B_1(t), B_2(t), \ldots, B_d(t))$ of independent Brownian motions.

The student is happy. Now imagine a similar dialog for random surfaces.

INSTRUCTOR: Take a uniformly random triangulation of the sphere with n triangles. The picture below is an example with 30,000 triangles by Budzinski, using some software to embed the surface in three dimensions and give us a view. The $n \to \infty$ limit of this object is a random fractal surface called the *Brownian sphere*. It is also a *peanosphere* and a *pure Liouville quantum gravity sphere*, and a *conformal field theory*.

STUDENT: You just listed four things! Which is the $n \to \infty$ limit of Budzinski's picture?

INSTRUCTOR: They all are! The difference comes down to the topology of convergence and the features of the limit presumed to be measurable. Think of them as different aspects of the same universal object. Four blind mathematicians feel the surface of an elephant and describe four different things:

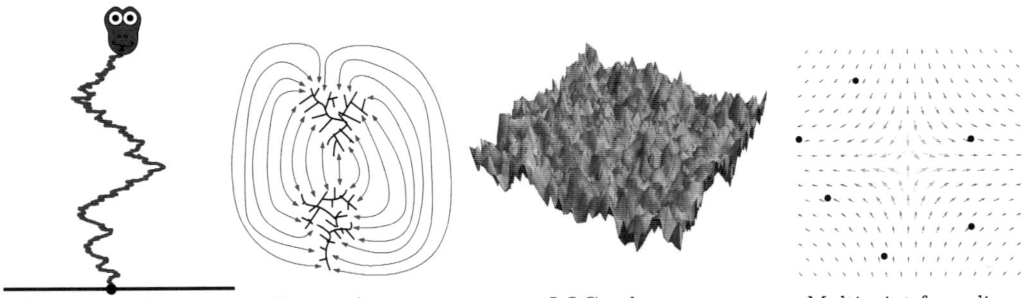

Brownian sphere Peanosphere LQG sphere Multipoint formalism

- *Brownian sphere:* a *random metric measure space* constructed from the so-called *Brownian snake* as explained in surveys by Le Gall, Miermont and Baez [27, 171, 175, 184, 186].
- *Peanosphere:* a *mating of continuum random trees* that encodes both a surface and an *extra tree and/or collection of loops* drawn on top of it, as explained in surveys by Gwynne, Holden, and Sun, and by Biane [51, 128].
- *Liouville quantum gravity sphere:* a *random fractal Riemannian surface* in which areas, lengths, and other measures are given by exponentials of a *Gaussian free field* ϕ, as explained in surveys by Berestycki, Ding, Dubédat, Duplantier, Garban, Gwynne, Miller, Powell, and Werner [33, 37, 91, 101, 113, 125, 188, 259].
- *Conformal field theory:* a *collection of multipoint functions* representing (regularized) integrals of products of the form $\prod e^{\alpha_i \phi(x_i)}$ with respect to a certain infinite measure, as explained mathematically by David, Guillarmou, Kupiainen, Rhodes,

and Vargas [87,120,121,160,254] who also survey the physics literature. The infinite measure is the *Polyakov measure* which is the product of an unrestricted-area measure on LQG spheres (with defining field ϕ) and Haar measure on the Möbius group $\mathrm{PSL}(2,\mathbb{C})$ (to select an embedding in $\mathbb{C}$).

Technical point: The *unit area* Brownian/Peano/LQG-sphere is a sample from a probability measure dS on a space of unit area surfaces. There is a natural *infinite* measure on *unrestricted-area* surfaces (with $k \geq 0$ *marked points*) given by $A^{-7/2+k}dAdS$ where (A, S) is the rescaling of S with area A. This is natural because the number of triangulations (quadrangulations, etc.) with n faces and k marked points scales like $C\beta^n n^{-7/2+k}$ for model-dependent constants C and β. Weighting the counting measure by β^{-n}, we obtain a discrete measure that (appropriately rescaled) converges to the measure above as the area-per-triangle ε goes to zero. If we replace β by the "off-critical" $\beta(1+\varepsilon\mu)$ then the limit is $A^{-7/2+k}e^{-\mu A}dAdS$, which is finite if $\mu > 0$ and $k \geq 3$. Polyakov included the $e^{-\mu A}$ factor and motivated it in a different way, using the *Liouville equation* as we explain later.

STUDENT: Do I really have to learn all four viewpoints?

INSTRUCTOR: A lot of good work has been done by people fluent in only one of the four. But all four have important applications. For example, in the Brownian sphere construction, *distances* are easy to define, and you can show that the surface has fractal dimension 4. So to cover the surface with metric balls of radius $1/n$, you need $\approx n^4$ balls. In the peanosphere construction, one sees the random trees and loops that naturally live on top of a random surface, and most easily makes contact with loop-decorated discrete models. The LQG approach allows one to define *Brownian motion on a random surface* as well as other conformally defined objects (like SLE curves). Conformal field theory is a staple of physics. It ties into quantum field theory (if one analytically continues from Euclidean to Minkowski space and replaces probability measures with quantum wave functions).

STUDENT: Are these surfaces characterized by simple axioms like Brownian motion is?

INSTRUCTOR: Yes. But each viewpoint comes with its own axioms, its own history, its own motivation, its own surveys. The proofs that they agree are long and involved.

STUDENT: Can any of the viewpoints describe a random surface embedded in $\mathbb{R}^d$?

INSTRUCTOR: Sure. You basically weight the law of the surface by the "number of ways" to embed it in $\mathbb{R}^d$. This amounts to weighting by the dth power of a certain "partition function" and this changes the law of the surface itself (even ignoring the embedding). You can apply such a weighting even when d is not a positive integer, so there is actually a one parameter family of random surfaces parameterized by d, with the scaling limit of Budzinski's picture corresponding to $d = 0$. These surfaces are "rougher" when d is large and "smoother" when d is small, converging to the Euclidean sphere as $d \to -\infty$. They are defined as random metric spaces for any $d < 25$, but are only finite-diameter and finite-volume if $d \leq 1$. The family can equivalently be parameterized by related quantities that come up in LQG theory like $Q > 0$ (where $d = 25 - 6Q^2$) or γ (where $Q = 2/\gamma + \gamma/2$ – note that γ is only real if $d \leq 1$ so that $Q \geq 2$) or by the peanosphere correlation coefficient.

STUDENT: I'm getting a bit lost. Can you give me the four definitions you promised?

The remainder of this paper will aim to do just that. We present a narrative-style introduction to each of the four viewpoints above, emphasizing the distinctive intellectual heritage behind each approach, as well as the relationships between them and the relevant recent developments.

Along the way, we will introduce several other natural objects: random trees (like the continuum random tree), random distributions (like the Gaussian free field) and random non-self-crossing curves (like the Schramm–Loewner evolution). We aim to make the exposition accessible to outsiders and newcomers, as well as to researchers who have expertise in one or more of the viewpoints but who might appreciate a high-level overview of the others. Readers interested in a still less technical account might take a look at the recent Quanta Magazine articles on this subject [143–145].

It is ambitious to try to tell four stories in one article, but we will do our best to convey at least the main ideas. The style is informal—no proofs—but we provide accurate definitions and the surveys mentioned above contain more detail. This paper is targeted primarily at a mathematical audience and is meant to be broadly accessible. This subject has deep roots in physics (especially string theory, quantum field theory, and planar statistical physics), but we will do our best to avoid any terminology that would be difficult for non-physicists to understand. Let us also stress that our reference list is long but very far from complete, biased by both the limits of the author's knowledge and the narrative focus of the paper. As such, it is more of a sampling than an exhaustive survey. Many highlights of the subject are not mentioned here.

For concreteness and simplicity, we will focus mostly on *sphere-homeomorphic* random surfaces. But all of the constructions in this paper have analogs that look the same locally but have different global topology. For example, in addition to the Brownian sphere one has a Brownian disk, a Brownian plane, a Brownian torus, Brownian surfaces of arbitrary genus and/or arbitrarily many boundary components, etc. [50,65,84,174,187,222]. Variants like these can be defined within all four viewpoints. One would expect that these definitions are consistent from one viewpoint to another (so that, e.g., a Brownian three-holed torus is somehow equivalent to a corresponding LQG three-holed torus or Peano three-holed torus). Significant work along these lines has been done, but the set of questions one can ask is very large and the program has not been completed yet.

1. BROWNIAN SPHERE: A RANDOM METRIC MEASURE SPACE

Our first construction is a random surface called the *Brownian sphere* (also known as *Brownian map*). The Brownian sphere is a random sphere-homeomorphic metric measure space. It has been described in longer survey articles by Le Gall and Miermont [171, 175, 184, 186, 187] and in a shorter overview by Baez [27].

Both the peanophere and the Brownian sphere can be constructed by "gluing together" a pair of "continuum trees" along their outer boundaries, which produces a sphere decorated by a space-filling curve (that somehow snakes in-between the two trees)—but the law of the pair of trees is different in the two settings. The idea of gluing together trees to obtain a sphere may seem counterintuitive, but we will see that it is well motivated by the discrete models.

Readers familiar with complex dynamics may also recall that the "mating" of two dendritic Julia sets (these are tree-like sets with empty interior) is a topological sphere, see, e.g., [25, 203, 247, 261], or look up the online video animations by Arnaud Chéritat. The Brownian sphere and peanosphere constructions are random versions of this phenomenon. Let us begin by defining *continuum random trees* which are sometimes also called *Brownian trees*.

1.1. Continuum random trees (Brownian trees)

Brownian motion is a random function $B(t)$ defined for $t \in [0, \infty)$. But Brownian motion (like the Brownian sphere) has many *variants* with similar local behavior. For example, the *Brownian bridge* is a Brownian motion on $t \in [0, 1]$ somehow *conditioned* on the (zero probability) event that $B(1) = 0$. The *Brownian excursion* is a Brownian bridge with $B(0) = B(1) = 0$ that is further conditioned to stay positive on $[0, 1]$.

If one starts with the graph of a Brownian excursion B, and identifies two points whenever they are connected by a horizontal chord under the graph (see below), one obtains a random metric space $\mathcal{T}$ called the *continuum random tree*, which was first constructed by Aldous in 1991 [9–11], and which plays a role in the construction of both the Brownian sphere and the peanosphere.

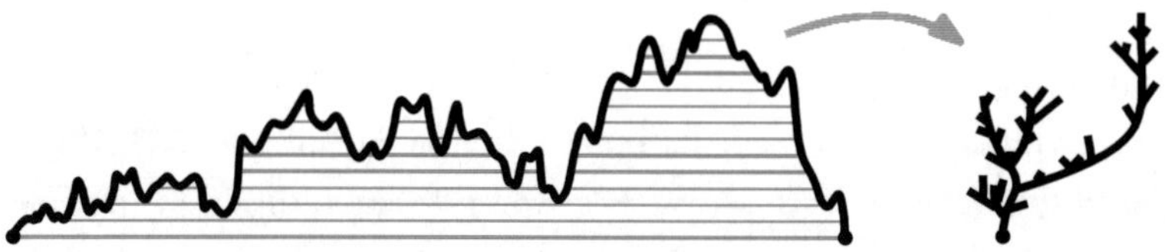

Note that the function taking t to the graph point $(t, B(t))$ induces a function f from $[0, 1]$ to $\mathcal{T}$ that "traces the boundary of the tree clockwise." The pushforward of Lebesgue measure on $[0, 1]$ endows $\mathcal{T}$ with a natural measure. Moreover, one can define the *distance* between $f(s)$ and $f(t)$ to be

$$\Big(B(s) - \inf_{r\in[s,t]} B(r)\Big) + \Big(B(t) - \inf_{r\in[s,t]} B(r)\Big).$$

This measures how far "down and up" one has to travel within $\mathcal{T}$ to get from $f(s)$ and $f(t)$. Since $\mathcal{T}$ comes with a natural measure and a natural metric, we call it a *random metric measure space*.

1.2. Planar map bijections that motivate Brownian sphere and peanosphere

The Brownian sphere has its historical roots in the study of planar maps. A *planar map* is a finite graph together with an embedding in the complex sphere $\mathbb{C} \cup \{\infty\}$, where two embeddings are considered equivalent if there is an orientation-preserving homeomorphism of $\mathbb{C} \cup \{\infty\}$ taking one to the other. For example, the two figures below are isomorphic as graphs but represent different planar maps, as there is no orientation-preserving homeomorphism of $\mathbb{C} \cup \{\infty\}$ mapping one to the other.

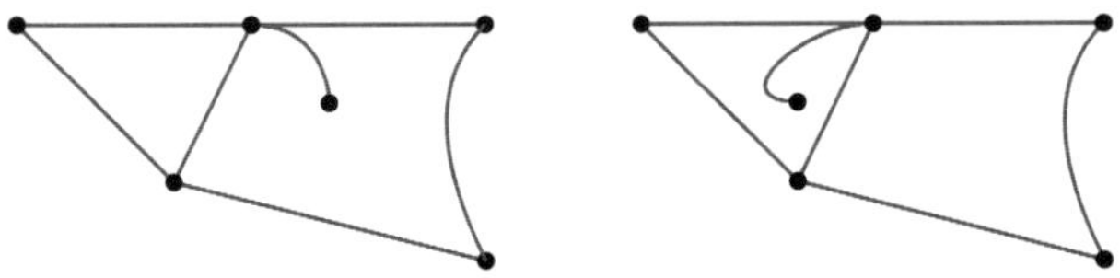

Combinatorially, a planar map is fully determined by the graph together with the "clockwise cyclic ordering" of the edges surrounding each vertex. In particular, the number of distinct planar maps with n edges is finite. To eliminate the ambiguity that comes from having nontrivial automorphisms, it is sometimes helpful to specify a "root" by fixing an oriented edge. If an orientation-preserving homeomorphism of $\mathbb{C} \cup \{\infty\}$ takes a planar map to itself (mapping edges to edges and vertices to vertices) and fixes the oriented edge, one can show that it must induce the identity map on the whole set of vertices and edges. To see why, imagine that your car starts driving along the oriented edge. From there you can describe how to get to any other vertex or edge with a set of directions like "Drive until you reach a vertex, then take the furthest road to your left, then drive until you reach another vertex, then take the third road from the left," and so forth. This method of specifying vertices and edges is preserved by homeomorphisms of $\mathbb{C} \cup \{\infty\}$. In other words, fixing a root gives us a way to uniquely specify all other vertices (just as a Cartesian coordinate system gives us a way to uniquely specify points on a planar lattice). This implies that fixing the root eliminates all non-trivial planar map automorphisms: for example, a triangle has three nontrivial orientation-preserving automorphisms—corresponding to the three rotations—but if the triangle is assigned a root, then only the identity automorphism would fix the root.

Perhaps the most famous planar map problem of all is the "4 color conjecture" (also known as Guthrie's problem) which captivated mathematicians worldwide from its formulation in 1852 to its computer-assisted proof by Appel and Haken in 1976. See, e.g., [81, 112, 146, 223] for historical accounts. The problem is to show, given any finite planar map, that it is possible to color each vertex one of four colors in such a way that no two neighbors have the same color. Note that if one has a map of South America, say, one can create a *planar map* by putting a vertex in the center of each country and drawing an edge connecting two countries wherever they share a nonzero-length border. The four color theorem allows one to color the countries in such a way that no two countries sharing a nonzero-length border have the same color. One of the key contributors to research on this problem, and to graph theory in general, was William Tutte.

By Tutte's own account [253] (see also [211]), he was motivated by the four color theorem when he began thinking about enumerating *all* planar maps. He wrote his famous "census of" series in 1962 and 1963 [249–252] which, among other things, included a remarkable formula for the number of rooted planar maps with n edges, namely $\frac{2}{n+2} \cdot \frac{3^n}{n+1}\binom{2n}{n}$. Using, e.g., Stirling's formula (or the local central limit theorem), one can show that $\binom{2n}{n} = \frac{4^n}{\sqrt{\pi}n^{1/2}}(1 + O(\frac{1}{n}))$ which implies that Tutte's formula grows asymptotically like $C\beta^n n^{-7/2+k}$ (which is the formula from the *technical point* in our introductory dialog) with $C = \frac{2}{\sqrt{\pi}}$, $\beta = 12$, and $k = 1$. This is consistent with the formula from the introduction because the root effectively plays the role of a single marked point, and we recall that k represents the number of marked points.

These papers were followed by another series by Mullin [206–209] which included a formula for the number of rooted planar maps decorated by a distinguished spanning tree, namely $\frac{(2n)!(2n+2)!}{n![(n+1)!]^2(n+2)!}$. This grows asymptotically like $C\beta^n n^{-4+k}$ with $k = 1$ and $\beta = 16$. The addition of the "spanning-tree decoration" somehow changes the growth rate in a fundamental way, changing not only β and C but also the power of n (effectively replacing the $7/2$ with 4).

Both of these formulas have interesting bijective proofs: the former, in its first form due to Cori and Vauquelin in 1981 [80], was later advanced and popularized by Schaeffer [226]. The latter was essentially due to Mullin in 1967 [209], see also the explanation by Bernardi [40], and it reduces the problem of counting spanning-tree-decorated rooted planar maps to a problem about walks in $\mathbb{Z}_+^2$ (which can be separately addressed, e.g., with reflection arguments). These bijections motivate the definition of the Brownian map and the peanosphere, respectively. For now let us describe the bijections side by side, beginning with the Mullin bijection.

Suppose that (X_n, Y_n) is a simple walk in $\mathbb{Z}_+^2$ starting and ending at the origin. Then if we fix a large enough value of C, the graphs of X_n and $C - Y_n$ (linearly interpolated to $\mathbb{R}$) will not intersect; see the figure below. Draw a vertical red line at each time increment. Then declare two points in the graph of X_n (resp. $C - Y_n$) to be equivalent if one can draw a horizontal chord connecting them that does not go above the X_n graph (resp. below the $C - Y_n$ graph). In other words, we "identify" each pair of points connected by a horizontal (blue, black, or green) line segment. We also glue together the leftmost and rightmost red lines. After this gluing is done, the region below X_n collapses to become a tree (shown with black edges and green vertices) as does the region above $C - Y_n$ (shown with black edges and blue vertices) and we are left with a planar map with black/red edges and blue/green vertices. (The left and right red curves are glued together, which makes the topological disk they enclose into a sphere.) This planar map is a *triangulation* in which each triangle contains two red edges and one black edge.

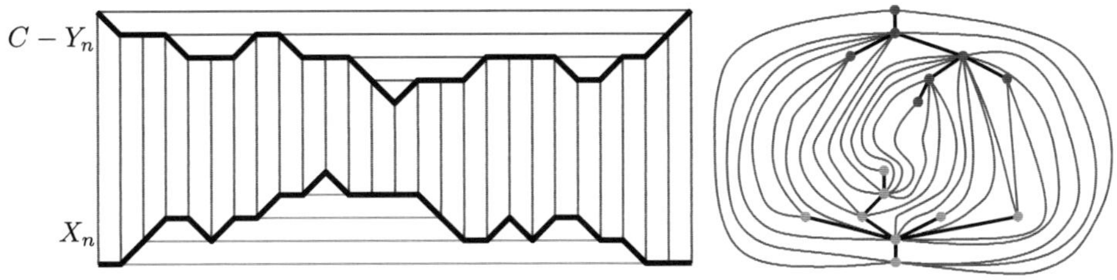

If the black edges are erased, one is left with a red *quadrangulation* $\mathcal{Q}$ which is bipartite: the blue and green vertices are the two partite classes. Each quadrilateral in $\mathcal{Q}$ has one blue-to-blue diagonal. These diagonals together form a planar map M. Conversely, given the planar map M, it is not hard to show that one can recover $\mathcal{Q}$ (by adding a vertex in the center of each face of M and connecting it to all of the boundary vertices of that face). The blue tree T is a spanning tree of M. The green-to-green diagonals of $\mathcal{Q}$ form the dual graph M^*, and the green tree T^* is the dual spanning tree. The story above gives a bijection between:

- Simple walks (X_n, Y_n) in $\mathbb{Z}_+^2$ of length $2N$ that start and end at the origin.
- Pairs (M, T) where M is a rooted planar map and T is a spanning tree of M. Here *rooted* means that a distinguished vertex of M and a distinguished incident vertex of M^* are fixed (to be the roots of T and T^*, i.e., the two vertices attached to the leftmost red line, or the equivalent rightmost red line).

Choosing a uniformly *random* walk (X_n, Y_n) of the type described above produces a uniformly *random* (M, T) pair. The probability of a given map M is proportional to the number of spanning trees M has.

But what if we want to simply count the total number of rooted maps M (or the total number of quadrangulations $\mathcal{Q}$) instead of the number of (M, T) pairs? This is what the Cori–Vauquelin–Schaeffer bijection does. It can be seen as similar to the Mullin bijection but with a few key differences. First, instead of requiring (X_n, Y_n) to traverse lattice edges, we at each step allow Y_n to change by ± 1 and X_n by either 0 or ± 1. Second, instead of perfectly horizontal green chords, we draw chords that are one unit higher on the right than on the left. We draw one such chord leftward starting at each vertex on the graph of X_n, which means that we have to add an extra vertex of minimal height as shown.

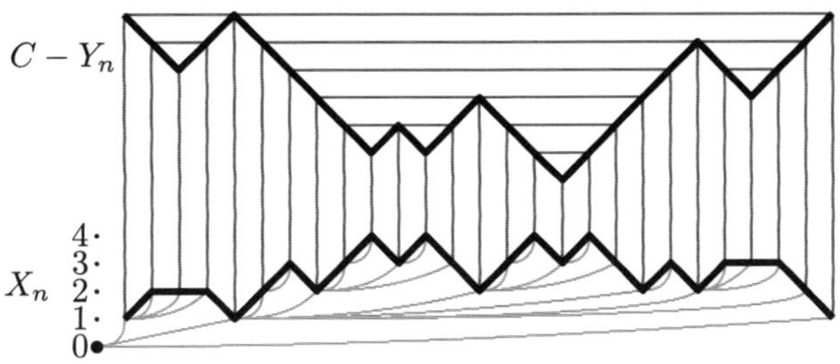

Third, we consider only (X_n, Y_n) pairs for which the above picture has a special property: namely, whenever two red vertical lines are incident to the same blue chord, their lower endpoints have the *same height*.

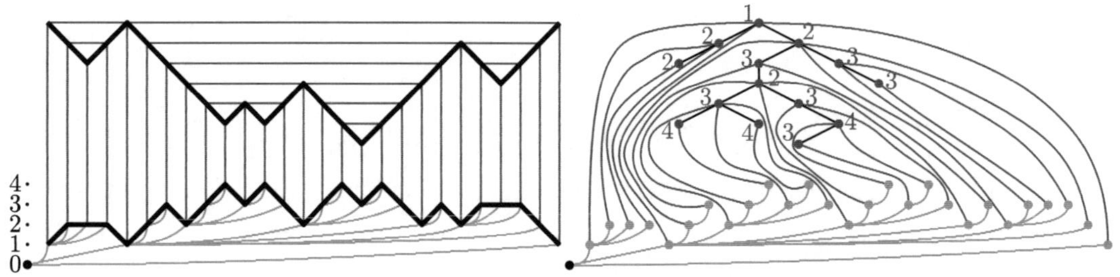

If we glue together the edges on the $C - Y_n$ graph connected by blue chords (the same way as in the Mullin bijection) we obtain a tree, and the third condition above is equivalent to the condition that if two red lines start at the same blue vertex on that tree then they terminate at green vertices of the same height. Thus we can label each vertex in the tree by the height of the green vertex (or vertices) it connects to. We then obtain a planar map by *gluing* the black and green trees on the right to one another: precisely, we glue each green vertex in the tree on the right to the blue vertex it is connected to by a red edge, effectively shrinking each red edge to a point. (Also, the outermost two green edges are understood to be glued/identified with each other, so that the topological disk they surround becomes a sphere.)

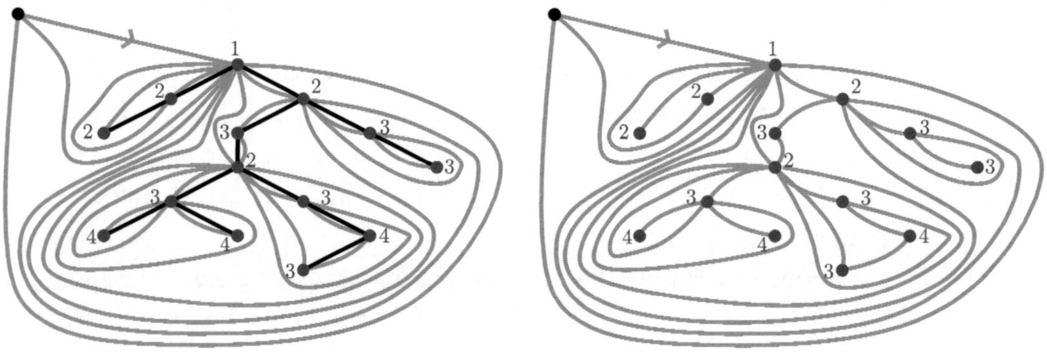

We can think of the graph on the left above as obtained by starting with the region below the X_n graph, then gluing together two black edges of the X_n graph whenever the edges of the $C - Y_n$ graph just above them correspond to the same edge of the upper tree. Whenever two of these black edges are glued together, they are either both horizontal (each part of a triangle with two green edges) or one increasing and one decreasing (respectively part of a two-gon with one green edge and a quadrilateral with three green edges). In each case, once the two black edges are glued together (and erased) we are left with a green quadrilateral.

The construction above yields a bijection between

- Well-labeled rooted planar trees (T, ℓ) (here ℓ maps vertices of T to positive integers, where the root has label one and adjacent vertices differ by 0 or ± 1) and
- Rooted quadrangulations $\mathcal{Q}$.

To prove that the above is a bijection, one has to show that one can reconstruct the green and black trees given just the green-edge quadrangulation Q. Roughly speaking, one reconstructs the original green tree as follows. First, we shorten the green edges slightly so that they are no longer connected to each other at the vertices, but then we reconnect some of them: precisely, we connect each edge (directed "downward" towards its lower-label endpoint) to the successor downward-pointing edge that involves "turning maximally to the left." Once this is done, the green tree is a tree of "leftmost geodesics" built from the directed green edges (each directed in the direction of decreasing distance to the root). Given the green tree, one can construct the X_n in the figure above and it is not too hard to check that one can reconstruct the upper tree as well (since one knows which green vertices have to be equivalent—and hence connected by red–blue–red paths).

We remark that once the black and green trees are glued together, they have vertices (but no edges) in common (this is different from the Mullin bijection) but in a sense the two trees still do not "cross" each other. To try to see why, it is interesting to imagine a path that crosses through the red edges in order from left to right (but does not intersect the green or black trees). Once we take the quotient (with respect to the equivalence class where all points on the same red edge are considered equivalent) this becomes a path that winds between the green and black trees, intersecting the trees at vertices but never crossing a branch of either tree.

This bijection (like the Mullin bijection) constructs a quadrangulation $\mathcal{Q}$ by gluing together a pair of trees. The microscopic details of the gluing are a bit different: in the Mullin bijection the red edges form $\mathcal{Q}$ and in the Cori-Vauquelin-Schaeffer bijection the green edges form $\mathcal{Q}$. There are many variants and analogs of the CVS bijection, see for instance [8, 12, 44, 49, 55, 226] and the references therein. At the global level, the main difference between the CVS bijections (and its variants) and the Mullin bijection (and its variants) is this: in the CVS bijection, one of the trees (the green one) is the so-called *leftmost geodesic tree of edges* and is determined by $\mathcal{Q}$ itself. The second tree (the black one) is in some sense "dual" to the geodesic tree (and is also determined by $\mathcal{Q}$ itself). By contrast, in the Mullin bijection the spanning tree is not determined by $\mathcal{Q}$ and $\mathcal{Q}$ is not uniform among all quadrangulations.

1.3. Unconstrained variant: when $\mathcal{Q}$ is both pointed and rooted

There is a variant of the Mullin bijection in which we relax the restriction that X_n is nonnegative, see below.

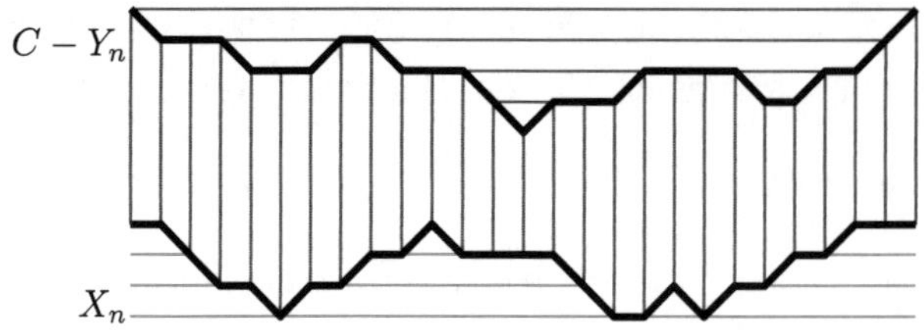

Here we imagine that the left and right sides of the above rectangle are glued to one another (so that both X_n and Y_n then become indexed by a circle). We can then identify points on the same blue chord or the same green (possibly wrapping around) chord, just as before, and we obtain a pair of trees with red edges between them. A key difference from the original construction is that in this construction the root of the lower tree (corresponding to the minimum of X_n) and the root of the upper tree (corresponding to the minimum of Y_n) are no longer required to be adjacent. The minima of X_n and Y_n may occur at different places, and we can think of the vertex corresponding to the X_n minimum as an extra marked point. (In principle, one could allow both X_n and Y_n to take negative values, in which case the two tree roots—and the point described by the leftmost/rightmost red line—would effectively describe three marked points.)

It turns out that the CVS construction also has a similar variant, which corresponds to relaxing the constraint that all of the blue labels are positive. If we relax this constraint, then the corresponding X_n process can be positive and negative, and in particular no longer has to have a minimum at the same place that Y_n does. This means that the root of the upper tree and the root of the geodesic tree are no longer required to be adjacent. With this constraint relaxed, the set of possible labeled upper trees is easier to count: one has a Catalan number $\frac{1}{n+1}\binom{2n}{n}$ of rooted planar trees, and given the tree, there are exactly 3^n ways to choose the labels if we fix the root label to be zero. (We are free to do this since adding a constant to the labels does not affect the map construction, and hence only the labels modulo additive constant are relevant.) The formula $\frac{3^n}{n+1}\binom{2n}{n}$ can also be obtained by starting with Tutte's formula and multiplying by $n+2$ (there are $n+2$ possible to choose an extra marked vertex) and dividing by a factor of 2 (which is related to the choice of rootorientation—this is explained in many places, see e.g. **[226]** or the short explanation in Section 3 of **[68]**).

There is another way to think about the trajectory (X_n, Y_n) obtained from a labeled tree. For every point on a labeled tree, there is a labeled path from that point back to the root. As one traces the tree clockwise, one obtains a corresponding sequence of labeled paths; the labeled path seen at time n can be drawn as a "snake" (a vertical-to-horizontal function defined on $[0, Y_n]$) with the horizontal coordinate indicating the label.

The figures below (read from left to right) are steps in a Markov process on a space of "snakes." Each figure is a sequence of edges, each of which goes up one unit and -1, 0, or 1 units to the right. To move from one figure to the next, one first tosses a coin to decide whether to delete the top edge (with probability $1/2$) or to add an up-left edge (probability $1/6$), up-right edge (probability $1/6$) or up edge (probability $1/6$).

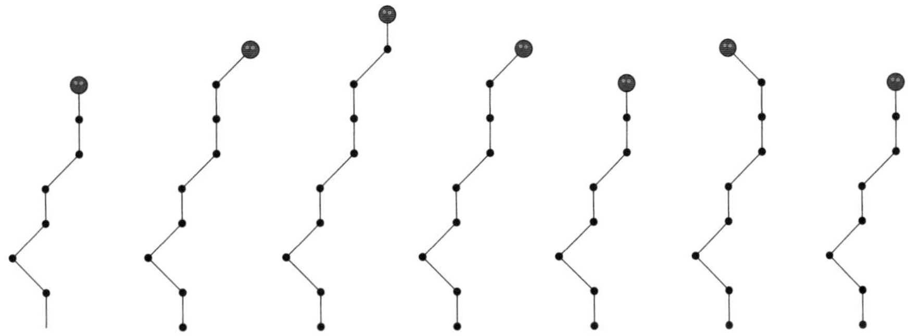

If the head of the snake starts at height zero—and we condition on the head height staying nonnegative and returning to zero at time $2n$—then we can let the head height represent the process Y_n and the head horizontal location represent the process X_n, and it is not hard to see that this corresponds precisely to the (X_n, Y_n) construction from the labeled tree. If one rescales this process vertically by a factor of C and horizontally by $\sqrt{C}$, one obtains a continuum *Brownian snake* model in the limit. The Brownian snake was introduced by Le Gall in the 1990s (see, e.g., [167]) not long after Aldous introduced the continuum random tree (also known as Brownian tree). The terms *Brownian snake* and *Markov snake* were coined by Dynkin and Kuznetsov (who credit Le Gall for the construction) in 1995 [106].

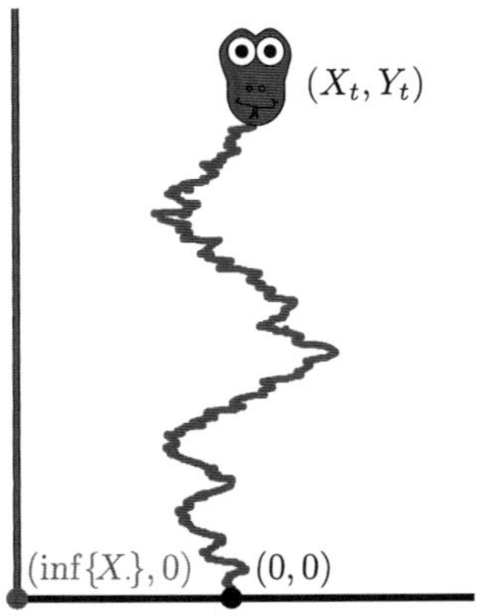

Now the law of the continuum pair (X_t, Y_t) can be described as follows: one first samples Y_t as a Brownian excursion on $[0, 1]$ and then takes a quotient of the graph of Y_t to produce a Brownian tree $\mathcal{T}$, together with the natural map f taking t to $\mathcal{T}$ (which traces around the boundary of $\mathcal{T}$ clockwise as t varies from 0 to 1). One then generates a Brownian motion ζ indexed by $\mathcal{T}$ (taken to be zero at the root of $\mathcal{T}$). This in turn determines a process X_t by $X_t = \zeta(f(t))$. The minimum of this process corresponds to the location of the root of the geodesic tree. The picture of the whole snake at time t (viewed as a vertical-to-horizontal function as above) is described by the restriction of ζ to the path in $\mathcal{T}$ from $f(0)$ and $f(t)$.

1.4. Passing to the continuum

One would expect both the Mullin and the Cori–Vauquelin–Schaeffer bijections to have limits given by the gluing of two fractal trees defined by continuum processes X_t and Y_t, something like the figure below:

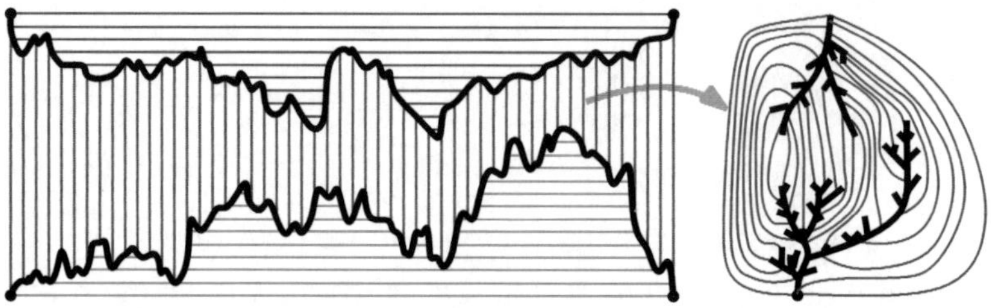

However, we would expect the law of (X_t, Y_t) to be rather different in the two scenarios. In the Mullin limit, X_t and Y_t are independent Brownian excursions, and the picture is a topological quotient of two independent continuum random trees (with respect to the equivalence that identifies two points if they are connected by a red curve). The topological sphere constructed this way is a special case of the *peanosphere* that we will discuss in the next section. In the CVS limit, X_t and Y_t correspond to the horizontal and vertical components of the head location in the Brownian snake process. The *Brownian sphere* (also known as *Brownian map*) is defined to be the metric space obtained by starting with the tree generated by X_t (viewed as a metric space) and then taking the metric quotient with respect to the equivalence that identifies points if they are connected (by the red curves) to the same point on the upper tree defined by Y.

The Brownian map was originally defined a bit differently. Marckert and Mokkadem [180], building on Schaeffer's work [226] (also citing, e.g., the early work of Krikun and of Angel and Schramm [21,157]), argued that the limit of the discrete models could be defined in some sense (since the (X_n, Y_n) process had a limit) and coined the term *Brownian map* to describe the limiting object; however, they did not show convergence in the space of random metric spaces. Le Gall showed that subsequential limits of random quadrangulations exist w.r.t. to the topology obtained from the *Gromov–Hausdorff metric* (a natural "metric on the space of all compact metric spaces") and are a.s. homeomorphic to a metric quotient of the Brownian tree [168]. Le Gall and Paulin subsequently showed that all such subsequential limits are a.s. homeomorphic to the 2-sphere [177], see also the sphericity proof by Miermont [183]. The term *Brownian map* was sometimes used (see e.g. [176]) to describe any one of the subsequential scaling limits of the discrete models, and in this language, the question "Do random quadrangulations have a (non-subsequential) Gromov–Hausdorff scaling limit?" was formulated as the question: "Is the Brownian map unique?" This problem was solved independently by Le Gall and by Miermont [170,185], who also showed that this limit exists in law (without passing to a subsequence) and that the limiting random metric space is indeed the Brownian-snake-based metric quotient defined above.

Since this foundational work, Brownian surfaces have been the subject of a sizable literature. Many different kinds of random planar maps (quadrangulations, triangulations, various variants, etc.) have been shown to have the Brownian sphere as a scaling limit, strengthening the idea that the Brownian map (like Brownian motion) is something "universal" [2–7, 31, 49, 69, 181]. There has also been significant work done on the continuum model. For example, the behavior of the geodesics in the Brownian sphere is quite interesting (geodesics "merge into" one another in ways that are very unlike what one sees for Euclidean spheres) and much has been done to understand their behavior [20, 48, 54, 57, 58, 169, 189].

1.5. An axiomatic approach

Like Brownian motion, the Brownian sphere is uniquely characterized by certain axioms, as shown in a long paper by the author and Miller [196] which draws on related ideas by Bertoin, Budd, Curien, Kortchemski, Krikun, Le Gall, Miermont and others [21, 46, 47, 83, 85, 86, 157, 172, 173]. We won't give a fully precise description here, but let us summarize the rough idea. For Brownian motion, there is a Markov property, which states that *given* $B(t)$ the conditional laws of B restricted to $[0, t]$ and B restricted to $[t, \infty)$ are independent. Now consider an unrestricted-area Brownian sphere with two marked points x and y. The natural Markov property in this setting states that *given* the boundary length of the filled metric ball of radius r centered at x (here the "filled metric ball" is obtained by starting with the ordinary metric ball and adding the components of its complement that don't contain y) the conditional laws of the ball and its complement are independent and depend on the boundary length in a scale invariant way. Furthermore, if we cut a filled metric ball into slices—using geodesics from evenly spaced points around the boundary to the center—then the slices are conditionally independent of each other given the length of their intersections with the metric ball boundary. The main result of [196] (very roughly speaking) is that if a measure on sphere-homeomorphic metric measure spaces satisfies these properties, then it must be the Brownian sphere.

2. PEANOSPHERE: A RANDOM MATING OF TREES

2.1. Basic definition

In Section 1 we saw that the Mullin bijection gives a bijection between lattice walks in $\mathbb{Z}_+^2$ (starting and ending at zero) and pairs (M, T) where M is a rooted planar map and T is a spanning tree on M. The X and Y coordinates of the walk encode, respectively, a tree and a dual tree, which are somehow stitched together to create M. This is actually a special case of a much more general idea. The idea of considering planar maps *together with* an extra structure on the map (a spanning tree, a collection of loops, a distinguished edge subset, a bipolar orientation, etc.) has long been a staple of this subject, both on the mathematics side and the physics side. The extra structure is sometimes called a *decoration* or (on the physics side) a *statistical physics model* or a *matter field*.

It turns out that *many* kinds of planar map decorations can be used to produce a spanning-tree/dual-spanning-tree pair in some way. And the resulting trees are often encoded by *some kind of* lattice walk conditioned to stay in $\mathbb{Z}_+^2$, so that in the fine mesh scaling limit, one obtains *some kind of* Brownian motion conditioned to stay in $\mathbb{R}_+^2$ (starting and ending at the origin). However, in general, the kind of Brownian motion involved (before imposing the quadrant constraint) may be one in which X_t and Y_t are *correlated*. That is, the "diffusion matrix" may be such that $\mathrm{Var}(X_t) = \mathrm{Var}(Y_t) = t$ and $\mathrm{Cov}(X_t, Y_t) = \rho t$ for some possibly nonzero correlation coefficient ρ. (There is no loss in assuming $\mathrm{Var}(X_t) = \mathrm{Var}(Y_t) = t$, since multiplying X or Y by a constant does not change the topological surface construction.)

The limit of the Mullin construction corresponds to $\rho = 0$, but in general one might consider any ρ strictly between -1 (where the Brownian trees are perfectly negatively correlated) and 1 (where the Brownian trees are perfectly correlated). For each ρ there is a natural way to make sense of this Brownian motion *conditioned* to stay in $\mathbb{R}_+^2$ (starting at the origin at time 0 and returning there at time 1). One can then use the corresponding (X_t, Y_t) pair to generate a pair of trees that can be "mated together" in the manner described in the previous section.

Formally, then, the peanosphere is just a pair of random measure-endowed metric planar trees that, when glued together along their boundaries, make a topological sphere, as described in the author's work with Duplantier and Miller [102, 194] and in more recent surveys [51,128]. The sphere comes equipped with a non-self-crossing, space-filling path (also known as *Peano curve*) which in some sense traces the interface between the two continuum trees. (This is the motivation behind the term *peanosphere* which was originally proposed by Richard Kenyon in private communication.) It is also equipped with a measure, which is the pushforward of Lebesgue measure on the parameterizing interval $[0, 1]$. Note that no matter the value of ρ, the law of (X_t, Y_t) is unchanged if we swap the roles of X_t and Y_t, which (by contrast) is very much not the case for the (X_t, Y_t) pair used to define the Brownian map.

Moore in 1925 gave a very general criterion for determining when a topological quotient space is topologically a sphere [205] and as explained in [102, 194] one can verify directly that these criteria are satisfied in our setting, so that the object obtained by "gluing the two trees together along their outer boundaries" is indeed a topological sphere. The *peanosphere* does not *a priori* come with a simple metric space structure like the Brownian sphere does, because neither the tree nor the dual tree is a tree of geodesics. On the other hand, both the tree and the dual tree (as determined by X_t and Y_t) can be viewed as metric measure spaces themselves, so lengths of arcs *within* these trees are well defined.

2.2. Percolation and other simple kinds of decoration

A major reason to study the *peanosphere* is that it helps one understand random (M, T) pairs, where M is a random planar map and T is some kind *extra decoration* on M. Indeed, most of the scaling limit results known for decorated random planar maps make use of the peanosphere construction in some way. But before we discuss that topic, let us say a few words about randomly decorated *deterministic* planar maps, i.e., let us suppose that M is fixed to be a grid or lattice, and we are choosing some way to "decorate" M, e.g., with a coloring or spanning tree.

In statistical physics, it is often interesting to try to find the *simplest possible models* that exhibit behaviors (phase transitions, correlation decays, fractal patterns, etc.) that one might see in complicated real world systems. By understanding these models thoroughly and mathematically, one can hope to get a glimpse of *why* similar behaviors appear in more complex systems. We will describe a few of these very simple models here.

First, let us discuss *lattice percolation* which was introduced to the mathematics literature by Broadbent and Hammersley in **[61]**. In this model, one starts with a set of edges—or vertices or faces—and tosses an independent coin for each one to decide whether it is "open" or "closed." This is easiest to visualize by a computer. For example, the following Mathemetica code generates the picture below it:

```
n=40;Graphics[{Table[If[RandomInteger[1]==1,Line[{{i,j},{i,j+1}}]],{i,0,n},{j,0,n-1}],
  Table[If[RandomInteger[1]==1,Line[{{i,j},{i+1,j}}]],{i,0,n-1},{j,0,n}]}]
```

The above is a 40×40 grid where one tosses a fair coin independently for each edge to decide whether to display it or not. The displayed edges are the open edges; the edges not displayed are closed. A connected component of open edges is called a *cluster*. One can think of this as a simplistic model for a porous medium—with open edges representing pathways that current or fluid can flow through, and closed edges representing obstructions. The figure above is called *bond percolation* (with parameter $p = 1/2$) because it is the edges ("bonds") that are assigned to be open or closed. An alternative is *site percolation* which independently assigns each vertex—or face if one uses the "dual perspective"—to be open with probability p. By way of illustration, the following code generates the figure below it:

```
n=40; Graphics[Table[{If[(i-n)(j-n)==0,Blue, If[i j==0,Yellow,If[RandomInteger[1]==1,
Yellow,Blue]]],RegularPolygon[i{-Sqrt[3],-1}+j{-Sqrt[3],1},{1,0},6]},{i,0,n},{j,0,n}]]
```

This figure represents percolation on the faces of a hexagonal grid. For each interior face, an independent coin toss decides whether it is open (blue) or closed (yellow). Here we have also imposed some deterministic boundary conditions, fixing the colors to be blue along one boundary arc and yellow on the complementary arc. The "blue–yellow" edges (i.e., the edges on the boundary between blue and yellow) form many finite length cycles and one long path, which starts at the lower corner and ends at the upper corner. The long path is the boundary between the outermost yellow cluster (which contains half of the boundary) and the outermost blue cluster (which contains the other half of the boundary). In the fine mesh limit, it converges in law to a random fractal curve called the chordal *Schramm–Loewner evolution with parameter* $\kappa = 6$ (SLE_6) as proved in the breakthrough work of Smirnov **[239]** and Camia and Newman **[67]**, see also the expository paper by Sun **[243]**. We will say more about SLE curves in Section 2.5.

Both bond percolation and site percolation have natural *variants* in which the edge or vertex subsets are sampled from different (nonuniform) probability measures. One of the simplest such models is the *Fortuin–Kasteleyn (FK) random cluster model*, which was introduced by Fortuin and Kasteleyn in 1972 **[109–111]**. It has a "partition function" (defined below) that is equivalent to the famous *Tutte polynomial* introduced by Tutte in 1954 **[248]**. See the historical accounts in **[82,108]** and the overviews by Bernardi and Welsh **[41,256]** of the Tutte polynomial and its (surprisingly numerous) applications.

In the FK random cluster model, one begins with a finite graph $G = (V, E)$ and chooses a random $T \subset E$ (which need not necessarily be a spanning tree). The probability of a given T is proportional to a *weight* given by $W_G(T) := q^{k(T)} w^{|T|}$ where q and w are positive constants, and $k(T)$ represents the number of *clusters* (i.e., connected components) of the graph (V, T). Note that $k(t)$ is maximized and equal to $|V|$ if T is empty. It is minimized and equal to 1 if (V, T) is connected.

If $q = 1$, then the FK random cluster model is just ordinary bond percolation with $p = w/(1+w)$. Generally, a large w gives a bias toward T having more edges, and a small w gives a bias toward T having fewer edges. If q is very large, then the probability measure is biased in favor of (V, T) being highly *disconnected* (i.e., having lots of components). If w is close to zero, then the bias is in favor of T being more *connected* (i.e., having few components). The uniform spanning tree can be obtained as a limiting case of the FK random cluster model. (Taking $w \to 0$ quickly forces T to be a.s. connected; taking $p \to 0$ slowly makes $|T|$ as small as possible, and the connected T that minimize $|T|$ are spanning trees.) The *partition function* of the FK random cluster model is defined to be the sum of the weights, taken over all $T \subset E$,

$$Z_G(q, w) = \sum_{T \subset E} W_G(T),$$

which again is equivalent to the Tutte polynomial (up to a certain coordinate change). An important point to stress is that if $q = 1$, the partition function $W_G(T)$ depends only on the number of edges in G.

For certain parameters, the FK random cluster model is also closely related to the hugely influential *Ising model* which was presented by Ising in 1925 but actually introduced earlier by Lenz in 1920, see, e.g., the historical account at [63]. In the Ising model, not all of the possible face colorings are equally likely; rather one *weights* the probability of each coloring by a constant to the number of adjacent pairs on which the colors disagree.

The decorations we have discussed so far (bond percolation, site percolation, uniform spanning tree, FK random cluster model, Ising model) are some of the very simplest statistical physics models. But there are others. For example, one can decorate a graph G with a *bipolar orientation*, which is a way of *orienting* the edges in G so that only one vertex v (the sink) has all incident edges oriented toward v, and only one vertex w (the source) has all incident edges oriented away from w. The number of bipolar orientations is also encoded in the Tutte polynomial, see, e.g., Bernardi's explanation in [41]. One can also decorate G with a so-called *Schnyder wood* or an instance of the Gaussian free field or a Brownian loop soup (all of which will be mentioned below). The list goes on. We will not properly survey the literature on decorated planar maps here, but a few examples of work in this area include [39, 42, 43, 52, 53, 55, 56, 78].

2.3. Decorated random planar maps

In this section we give some examples of how one can generate a pair of spanning trees if one starts with another type of decoration. The figures below were used in [236] to explain how to generate a pair of trees from an instance of the FK random cluster model. By way of setup, one starts with a planar map M in black, adds a vertex in the center of each face, and draws a green edge connecting it to each vertex on the boundary of the face, thereby producing a green quadrangulation Q. The planar map M^* (which is dual to M) is obtained by connecting red-to-red within each green quadrilateral of Q, and is shown with dotted lines. (Here M itself is recovered from Q by connecting blue to blue within each quadrilateral.)

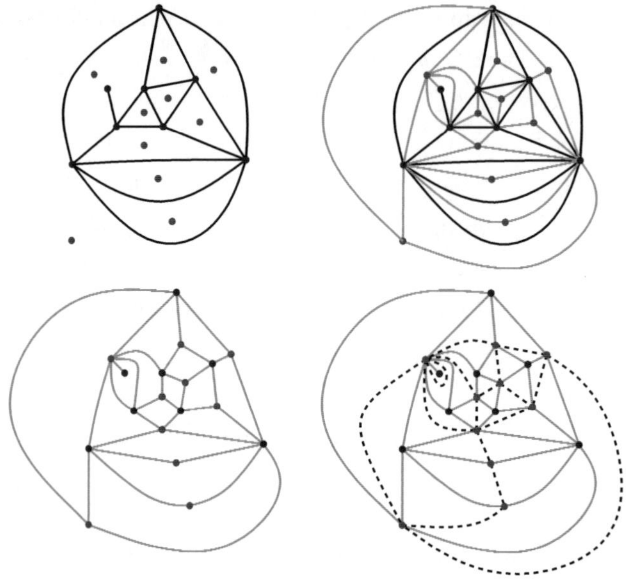

Next, the figures below illustrate an edge subset T in solid black (which happens to form a spanning tree) with the dual T^* shown as dotted black lines. (The dual consists of a red-to-red dotted line contained within each quadrilateral that *does not* have a blue-to-blue solid line.) The interface between the two trees is represented by a red curve that passes through all of the green edges one at a time, and starts at a location between the root and dual root (shown as large blue and red vertices).

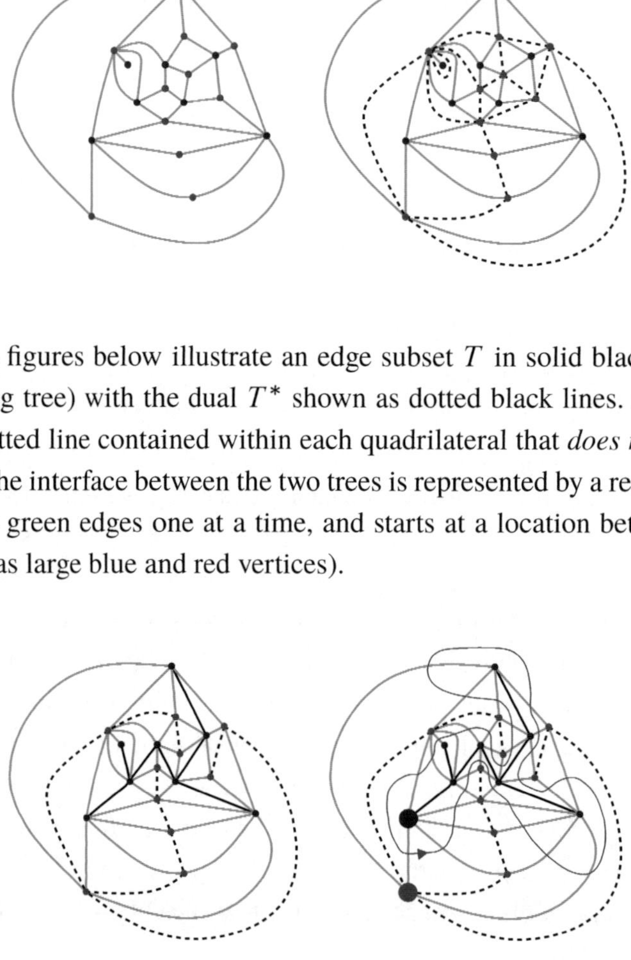

Finally, the figure below on the left illustrates an example of T (solid black lines) that is not a spanning tree (it contains cycles and there is one blue vertex that is not connected to any others). When one draws the red curve (which passes through green lines but never through black solid/dotted lines) one finds that it is not connected—in fact, it has five different components. In the figure on the right, four "swaps" have been made, where the edge in a quadrilateral was changed to a dual edge, or vice versa. The new edges/dual-edges produced by the swaps are colored in yellow. There is a simple algorithm (due to Bernardi in [41], see the author's explanation in [236]) for choosing which edges to swap. (Very roughly speaking, one tries to form a single red loop that crosses all the green edges one at a time, as above—but whenever the path is about to "seal off" a region without traversing it, one replaces a black blue-to-blue edge with a dotted dual red-to-red edge, or vice versa, to prevent this from happening.) Each swap reduces the number of red loops by one—and at the end, one is left with a single red loop, and a tree/spanning-tree pair. This is called the *hamburger–cheeseburger* construction because of its relationship to a simple inventory accumulation model in which two kinds of products ("burgers") are created and consumed [236].

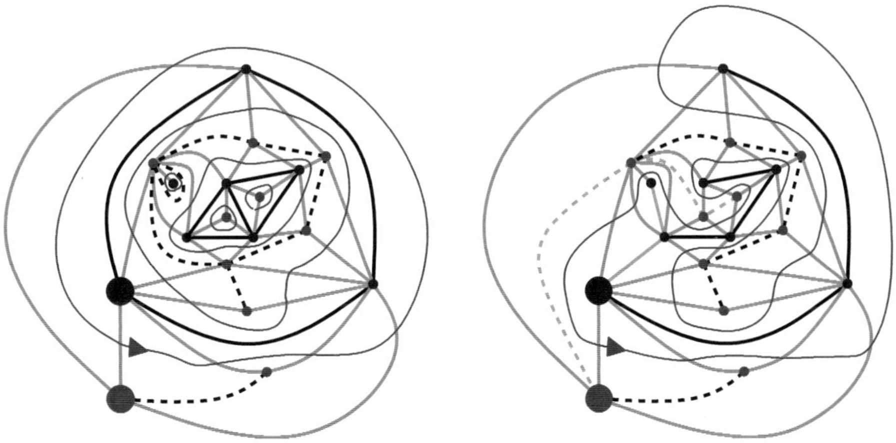

At the end of the day, [236] uses some algebra (which is too detailed to explain here) to show that the lattice path encoding this pair of trees converges to a two-dimensional Brownian motion where the correlation coefficient ρ depends on the FK-cluster parameter in a simple way. While the work in [236] was done for infinite graphs, Gwynne, Mao, and Sun have understood the scaling limits for the FK model on a finite random planar map, and have given much stronger forms of convergence, which also encode the structure of the loops themselves in a direct way [131, 141, 142]. See also [35, 77] for more results about critical FK random maps. Since then, work by Bernardi, Holden, and Sun has explained that a simpler, tailor-made bijection applies for site-percolation-decorated random triangulations, in which case $\rho = 0.5$.

With site or bond percolation, the partition functions depend only on the number of vertices or edges, and once one conditions on this the marginal law of M is the same as in the undecorated model [45]. This suggests that the peanosphere corresponding to $\rho = 0.5$ should be somehow equivalent to the Brownian map (together with extra randomness encoding the percolation structure on the Brownian map). And this is indeed the case—see Section 5.

Other bijections have been found for bipolar orientations and Schnyder woods, and so-called active spanning trees [42, 51, 130, 155, 178]. All of these models have different ρ values. For an extreme case, note that if we naively set $\rho = 1$, then we are effectively gluing two identical Brownian trees to each other, and we simply obtain the same Brownian tree back. On the other hand, there is a way of taking the $\rho \to 1$ limit that produces a nontrivial object, as shown in a remarkable paper by Aru, Holden, Powell, and Sun [22], which shows that there is a nontrivial construction that is far nicer than anyone could have expected, see also [24]. (The $\rho = -1$ limit is expected to correspond to an ordinary Euclidean sphere.)

As a short preview for the experts, let us now note that there is a general relationship between ρ and the κ parameter from SLE theory, a parameter we will say more about in Section 2.5. Using this relationship, one finds for example that the bipolar orientation model corresponds to SLE_{12} and the Schnyder wood model to SLE_{16}. This is because (see Section 5) there is a natural way to "conformally map" the peanosphere onto the Euclidean sphere, and when this is done the space-filling Peano curve becomes a space-filling form of SLE_κ (see Sections 2.5 and 2.6) with $\kappa > 4$, where $\rho = -\cos(4\pi/\kappa)$ [102]. Here κ ranges from 4 to ∞ as ρ ranges from 1 to -1. Let us note that the peanosphere surface itself turns out to be equivalent to an LQG sphere (see Section 3) with relevant parameters satisfying $\gamma^2 = 16/\kappa$ and $Q = 2/\gamma + \gamma/2$ and $d = 25 - 6Q^2$. The special value $\rho = 0.5$ corresponds to $\kappa = 6$ (and $\gamma = \sqrt{8/3}$ and $d = 0$). We will say more about these quantities in subsequent sections.

2.4. Computing the scaling exponent

As mentioned above, it turns out that if we set $\rho = 0.5$ then the peanosphere is a random surface that is equivalent to the Brownian sphere (decorated by a percolation model that does not change the law of the surface itself) although this is far from obvious at first glance. In this case, the two continuum trees are somehow midway between perfectly uncorrelated (as in the Mullin limit) and perfectly correlated. Even before constructing that bijection, it is not hard to argue heuristically that 0.5 is the only correlation coefficient that is consistent with the $C\beta^n n^{-7/2+k}$ formula, just as 0 is the only correlation coefficient consistent with $C\beta^n n^{-4+k}$. In general, one can derive a relationship between ρ and the b in $C\beta^n n^{-b+k}$.

We will not give the full details here, but let us sketch the rough idea, just to show that the relationship between b and ρ is not something mysterious. What is the probability that a simple random walk on $\mathbb{Z}^2$, started at the origin, remains in the wedge of angle θ until time n, and is back at the origin at time n? By way of illustration, part of a wedge of angle $\pi/4$ is shown below on the left (below the generating Mathematica code) along with its conformal image under the map $z \to z^4$.

```
f[z_]=z^4;F[{x_,y_}]={Re[f[x+I y]],Im[f[x+I y]]};n=40;
{ParametricPlot[Table[{{t(1-j/n)+j/n,j/n},{j/n,t j/n}},{j,0,n}],{t,0,1}],
ParametricPlot[Table[{F[{t(1-j/n)+j/n,j/n}],F[{j/n,t j/n}]},{j,0,n}],{t,0,1}]}
```

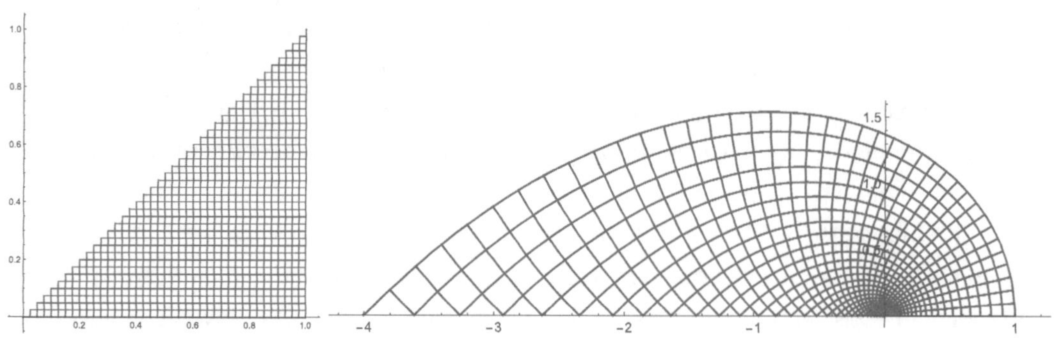

Here is the heuristic. (References to rigorous arguments along these lines are given in [155].) Suppose one does a simple random walk in a wedge of angle θ, with lattice size $1/\sqrt{n}$, and we want the probability of the walk starting at the origin and returning in n steps (without leaving the wedge). Using the fact that the random walk approximates Brownian motion, and Brownian motion is conformally invariant under the map shown above, it appears that the odds of escaping to a macroscopic distance within $n/3$ steps should be of order $(1/\sqrt{n})^{\pi/\theta} = n^{-\pi/(2\theta)}$. The same should hold if one selects the walk increments in reverse (starting at time n) for order $n/3$ steps; and then there is an order $1/n$ chance that the middle third lines up correctly, so overall one expects a probability of order $n^{-\pi/\theta-1}$.

The same should also hold if the simple random walk is replaced by another walk that has Brownian motion as a scaling limit. If $\rho \neq 0$ then it is necessary to *stretch* or *squash* the grid by some amount (in the $(1,-1)$ direction) in order to produce a walk with Brownian motion as a scaling limit, and one can easily work out the angle θ as a function of ρ.

The formula $n^{-\pi/\theta-1}$ corresponds to n^{-b+k} where $k = 1$ and $b = \pi/\theta + 2$ so that $\theta = \pi/(b-2)$. If we have $b = 4$ then $\theta = \pi/2$, which makes sense since the Mullin bijection corresponds to a simple random walk in a quadrant. More generally, if we have $X_t = \tilde{B}_t + aB_t$ and $Y_t = \tilde{B}_t - aB_t$ (where B_t and $\tilde{B}_t$ are independent standard one-dimensional Brownian motions), we have a correlation coefficient $\rho = (1-a^2)/(1+a^2)$ where a is the factor by which the space is scaled in the $(1,-1)$ direction. Then θ is the angle obtained when we squash the standard positive quadrant by a factor of a in that direction. Precisely, $1/a = \tan(\theta/2)$ so $a = \cot(\theta/2)$ and $\rho = (1-\cot(\theta/2)^2)/(1+\cot(\theta/2))^2 = -\cos(\theta) = -\cos(\frac{\pi}{b-2})$. Plotting this relationship, we see that the b value increases from 3 to ∞ as ρ decreases from 1 to -1.

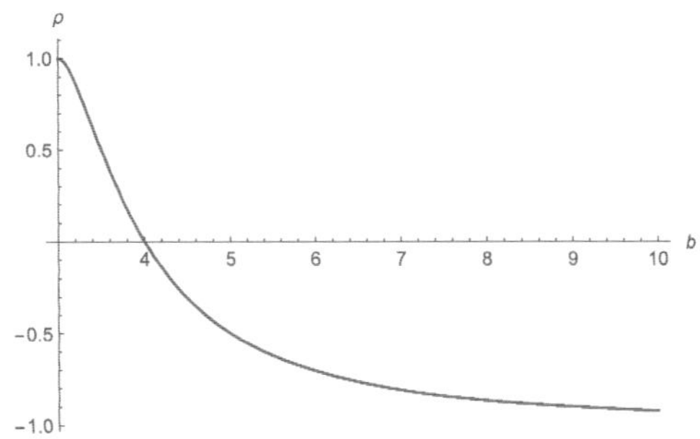

2.5. Schramm–Loewner evolution

Schramm–Loewner evolution is by itself a large subject. It has been the topic of several other sets of ICM lecture notes, see, e.g., the ICM notes by Duplantier, Lawler, Schramm, Smirnov, and Werner [101,162,229,240,242,258]. Longer expository introductions to the subject include, e.g., [36,152,161,257].

Here we will briefly explain what SLE curves are. First, let us present the axioms that motivated the definition. In 1999 [228], Schramm set out to construct—for any simply connected domain D with boundary points a and b—a random non-self-crossing chordal curve η connecting a and b. Schramm insisted that the definition of an SLE curve have two properties. First, the definition had to be *conformally invariant* meaning that if ψ is a conformal (i.e., analytic and one-to-one) map taking D to a domain $\psi(D)$ then the image of η under ψ should have the law of an SLE in $\psi(D)$ from $\psi(a)$ to $\psi(b)$ (up to a time change). Second, the path should be *Markovian* in the sense that *given* η up to a stopping time τ, the conditional law of *the rest* of η is (up to a time change) that of an SLE in $D \setminus \eta([0, \tau])$ from $\eta(\tau)$ to b. Schramm showed that there was only a one-parameter family of ways to define SLE if one insists on these properties. Schramm indexed this family by a parameter $\kappa \in [0, \infty)$.

For completeness, let us now give Schramm's more explicit definition of SLE (though we will not say too much more about it here). By conformal invariance, it is enough to define the law of η for one domain and one pair of boundary points. It turns out to be convenient to work with the upper half plane $\mathbb{H} \subset \mathbb{C}$ with $a = 0$ and $b = \infty$. For any time t, we define the function g_t to be the unique conformal map from the unbounded component of $\mathbb{H} \setminus \eta([0, t])$ to $\mathbb{H}$ that satisfies

$$\lim_{|z| \to \infty} g_t(z) - z = 0.$$

Schramm defined SLE in a rather indirect way: namely, he constructed the analytic functions g_t, and then used these functions to deduce what η must be. The g_t are defined by setting $g_0(z) = z$ and then requiring that for any fixed $z \in \mathbb{H}$, the ODE

$$\partial_t g_t(z) = \frac{2}{g_t - W_t}$$

is satisfied up until the smallest time t at which z is hit (or cut off from infinity by) the curve $\eta([0, t])$, where $W_t := B_{\kappa t}$ is a standard Brownian motion sped up by a factor of κ. This requirement determines the functions g_t which in turn determine η.

In some sense, the larger the κ (and hence the faster the W_t is moving up and down), the "windier" the curve becomes. In fact, Rohde and Schramm showed in [224] that η is a.s. a simple curve when $\kappa \in [0, 4]$, that η a.s. hits (but does not cross) itself when $\eta \in (4, 8)$ and that η is a.s. a *space-filling* curve when $\kappa \geq 8$. Beffara showed that the Hausdorff dimension of the range of η is linear in κ, a.s. given by $\min\{1 + \kappa/8, 2\}$ [29]. (Beffara builds on a related dimension calculation in [224].) In particular, since the scaling limit of the percolation picture above corresponds to $\kappa = 6$, the Hausdorff dimension of the curve is almost surely $1 + 6/8 = 7/4$.

At least on the square lattice, scaling limits of critical Ising model interfaces and the FK–Ising interfaces (which correspond to the FK cluster model for a particular choice of parameters) have scaling limits given by SLE_3 and $\mathrm{SLE}_{16/3}$, respectively, as shown in a remarkable series of papers by Smirnov and coauthors including Chelkak, Duminil-Copin, Hongler, Kemppainen [75, 76, 241]. Loop-erased random walks and uniform spanning tree boundaries have scaling limits given by forms of SLE_2 and SLE_8, respectively, as shown in a remarkable paper by Lawler, Schramm, and Werner [163]. Level lines of the so-called Gaussian free field are given by SLE_4 as shown by Schramm and the author [230, 231].

2.6. Conformal loop ensembles

What happens if we consider a percolation model with all blue boundary conditions, and then consider the collection of *all* of the loops that form boundaries between blue and yellow regions?

Camia and Newman showed in [66] that in the fine-mesh scaling limit, these loops converge to a random collection of continuum loops, called a *conformal loop ensemble* (CLE) with parameter $\kappa = 6$. Conformal loop ensembles are defined for any $\kappa \in (8/3, 8]$, see the CLE construction by the author in [234] and/or the axiomatic characterization of simple CLE loops by the author and Werner in [238].

We will not give a formal definition of CLE here. But we wish to stress one point. As explained in [238], it turns out that there is a natural way to use an instance of CLE_κ for $\kappa \in (4, 8)$ to generate a *space-filling* version of SLE_κ (even though ordinary SLE_κ is not space-filling for κ in this range) which is somehow the continuum version of the procedure used in Section 2.3 to combine multiple red loops into one loop. One can also reverse the procedure and recover the loops from the space-filling curve. Very roughly speaking, one tries to follow the CLE interfaces, but any time the curve separates a region from the target point, one violates the rules and fills up that region before continuing. The space-filling curve divides space into a continuum tree-and-dual-tree pair. Although we are not giving details here, this implies that, at least in the continuum, if one wants to understand *loop-decorated* random surfaces one can equivalently try to understand *tree-decorated* random surfaces. We remark that these continuum trees also have an interpretation as coalescing rays within a so-called *imaginary geometry* [98,190–193] where the dual tree corresponds to the coalescing tree of rays drawn in the opposite direction.

3. LIOUVILLE QUANTUM GRAVITY SPHERE: A RANDOM RIEMANNIAN GEOMETRY

The *Liouville quantum gravity sphere* is a random sphere-homeomorphic space whose law depends on the single parameter d (or Q or γ) as mentioned earlier. When $d = 0$ the LQG sphere is called the *pure LQG sphere*. The pure LQG sphere differs from the Brownian sphere in that the former *a priori* comes endowed with a conformal structure but no metric space structure, while the latter is *a priori* endowed with a metric space structure but no conformal structure.

A recent summary of LQG surfaces and the Gaussian free field can be found in the lecture notes by Berestycki and Powell [33, 37]. See also the Astérisque summary of LQG surfaces and the KPZ formula by Garban [113], the *Notices of the AMS* overview article by Gwynne [125], and the ICM proceedings article by Miller. A more recent set of ICM lecture notes by Ding, Dubédat and Gwynne gives an overview of recent works that have established a metric space structure for general values of $d < 25$ [91].

To summarize the latter point, recently various researchers (such as Ang, Basu, Bhatia, Ding, Dubédat, Dunlap, Falconet, Ganguly, Gwynne, Holden, Miller, Pfeffer, Remy, Sepúlveda, and Sun) have contributed to a spectacular international program to show that LQG spheres with $d \neq 0$ can also be given a canonical metric measure space structure, and to prove some basic properties about the resulting random metric spaces [13, 28, 90, 92, 93, 99, 100, 124, 127, 132, 140, 213]. The reader might start by looking at the metric constructions by Gwynne and Miller or by Ding and Gwynne [92, 133]. Viewed as random metric spaces, general LQG spheres can thus also be viewed as generalizations of the Brownian map. On the other hand, we stress that when $d \neq 0$ there is no reason to believe that the geodesic-tree/dual-tree pair is described by anything as simple as the Brownian snake. That is, as far as we know, the simple Brownian map construction given in Section 1 has no simple analog corresponding to $d \neq 0$.

3.1. The Gaussian free field

Brownian motion is a natural random function from $\mathbb{R}$ to $\mathbb{R}$. A generalization of Brownian motion called the *Gaussian free field* (GFF) is a random (generalized) function from $\mathbb{R}^d$ to $\mathbb{R}$ for any d. Our use of the term GFF in this paper will be limited to $d = 2$.

There are many ways to define the Gaussian free field, see the author's survey [233]. It is a scaling limit of discrete random functions from $\mathbb{Z}^2$ to $\mathbb{R}$ much as Brownian motion is the scaling limit of random functions from $\mathbb{Z}$ to $\mathbb{R}$. One particularly concise definition is as follows.

Fix a bounded planar domain D. If f and g are functions on D whose gradients lie in L^2 then we can write $(f, g)_\nabla = (2\pi)^{-1} \int_D \nabla f(z) \cdot \nabla g(z) dz$ for the *Dirichlet inner product* of f and g. Let $H(D)$ be the Hilbert space closure of the space of compactly supported smooth functions with respect to this inner product. Then the *Gaussian free field* on D with zero boundary conditions is the sum $\sum \alpha_i f_i$ where the f_i are an orthonormal basis for $H(D)$ and the α_i are independent standard normal random variables (mean zero, variance one). The sum a.s. does not converge pointwise or in $H(D)$ but it a.s. does converge in the space of generalized functions (a.k.a. *distributions*) [233].

The GFF can also be defined as the Gaussian random distribution with covariance given by Green's function $G(x, y)$. Here $G(x, \cdot)$ is given by $-\log|x - \cdot|$ minus the harmonic extension of $-\log|x - \cdot|$ from ∂D to D. The symbols h and ϕ are both commonly used to describe an instance of the Gaussian free field, depending on the context. We will use ϕ in this paper.

The GFF can also be defined on the whole plane, where one simply has $G(x, y) = -\log|x - y|$. In this setting the field ϕ is only defined up to additive constant. But one may nonetheless write $\mathrm{Cov}((\phi, f), (\phi, g)) = \iint -\log|x - y| f(x)g(y)dxdy$ as long as both f and g have mean zero. The integration-by-parts identity

$$(f, g)_\nabla = \frac{1}{2\pi}\int \nabla f(z) \cdot \nabla g(z)dz = \frac{-1}{2\pi} f(z)\Delta g(z)dz = \frac{-1}{2\pi}(f, -\Delta g)$$

is frequently used.

3.2. Conformal parameterizations

The LQG-sphere has a long history. On the physics side, LQG surfaces come up in certain formulations of string theory and 2D quantum field theories based on the Einstein equations (which in two dimensions reduce to the very simple Liouville equation). This literature is rich and complex, with foundational contributions by Belavin, Brézin, David, Di Francesco, Distler, Dorn, Duplantier, Eynard, Fateev, Itzykson, Kawai, Kazakov, Knizhnik, Kostov, Migdal, Otto, Parisi, Polchinski, Polyakov, Segal, Seiberg, Teschner, Witten, the Zamolodchikov brothers, Zinn-Justin, Zuber, and many others. (This list is far from exhaustive.) We will not attempt to properly survey the physics literature in this paper, but we point the reader to the long list of references in [185] (or the articles cited in Section 4) as a place to start.

On the mathematics side, one might begin with Gauss [115] who explained in 1827 how curvature could be understood as an intrinsic property of a two-dimensional surface, independently of how the surface was "embedded" in a higher-dimensional space. The Riemann mapping theorem (formulated by Riemann in 1851, proved by Osgood in 1900) and the more general Riemann uniformization theorem (conjectured by Klein in 1893, Poincaré in 1892, proved by Poincaré in 1907, Koebe in 1907) also play a central role [119, 255].

It is standard in differentiable geometry to define a surface (or two-dimensional manifold) by covering the surface with a "chart" of open sets that can each be diffeomorphically mapped to a planar domain. Within one of these open sets, parameterized by pairs (x, y), the "metric" can be written $A(x, y)dx^2 + B(x, y)dxdy + C(x, y)dy^2$. The parameterization is said to be *conformal* if $A = C$ and $B = 0$, and the Riemann uniformization theorem implies that one can always find a conformal parameterization. If we treat the parameterizing domain U as a subset of $\mathbb{C}$ and write $z = x + iy$ then we can write the metric as $e^{\rho(z)}(dx^2 + dy^2)$ and the associated area measure as $e^{\rho(z)}dz$ where dz is Lebesgue measure on U. (In some conventions the definition of ρ may differ by a factor of two; the issue is whether $e^{\rho(z)}$ is interpreted as the length multiplier or the area multiplier.)

The Gaussian curvature is $-e^{-\rho(z)}\Delta\rho(z)$ so the integral of the Gaussian curvature over the set parameterized by a region R is equal to the integral of $\Delta\rho$ over R. In particular, the Gaussian curvature is zero if and only if $\Delta\rho = 0$ so that ρ is harmonic. A function ρ on a bounded domain D is harmonic if and only if it minimizes $\int(\nabla\rho(z)\cdot\nabla\rho(z))dz = \int(-\rho(z)\Delta\rho(z))dz$ given its boundary conditions. More generally, the Gaussian curvature is equal to the constant K if and only if $\Delta\rho = -Ke^{\rho}$. The latter equation is called *Liouville's equation* and was formulated by Liouville in 1838 **[179]**. A function with constant curvature minimizes $\int(\nabla\rho(z)\cdot\nabla\rho(z)) + Ke^{\rho}$, which is a linear combination of the Dirichlet energy $(\rho,\rho)_\nabla$ and the overall surface area. (Polyakov used the latter quantity to define the so-called *Liouville action*, see Section 3.5.2.)

If we want to choose a random perturbation of a flat metric, we need to find a random function ρ that is in some sense a random perturbation of a harmonic function. Roughly speaking, one "randomly perturbs" a harmonic function by replacing the assertion that "the Dirichlet energy of ρ is minimal given its boundary values" with "the probability of a given ρ is proportional to the exponential of minus the Dirichlet energy of ρ." Formally, this means taking ρ to be a constant $\gamma \in (0,2)$ times the Gaussian free field. The induced area measure then takes the form $\nu_\phi := e^{\gamma\phi(z)}dz$. This cannot make literal sense (since ϕ is a random distribution) but there are various ways to make sense of this through regularization.

The first rigorous construction of a random measure with the law of ν_ϕ is due to Høegh-Krohn **[158]** who constructed the object for $\gamma \in [0,\sqrt{2}]$. Høegh-Krohn was motivated by earlier works in the quantum field theory literature that made sense of $V(\phi)$ where ϕ was the Gaussian free field and V was a polynomial. The exponential of the free field (viewed as a quantum field theory) was studied in several papers over the subsequent decade, and was cited by prominent quantum field theorists such as Glimm, Jaffe, and Simon.

In the 1980s Kahane derived a similar construction for all $\gamma \in [0,2)$ and used the term *Gaussian multiplicative chaos* to describe the associated random measures **[153]**. Kahane was motivated by the multiplicative cascades popularized by Mandelbrot. (See, e.g., **[103]** for $\gamma = 2$.) Neither Høegh-Krohn nor Kahane *interpreted* these random measures as the pullback to a planar domain of the area measure on a random surface parameterized by that domain. The author's work with Duplantier **[105]** constructed the measure ν_ϕ as a weak limit as $\varepsilon \to 0$ of the random measures $\varepsilon^{\gamma^2/2}e^{\gamma\phi_\varepsilon(z)}dz$, where $\phi_\varepsilon(z)$ is the mean value of ϕ on the ball boundary $\partial B_\varepsilon(z)$, see also **[237]**. This construction showed that the measure ν_ϕ is a function of the Gaussian free field ϕ.

Computer visualization can help us generate some intuition about what the measures ν_ϕ look like. The following Mathematica code generates the two figures shown below.

```
K = 8; fieldmultiplier = 1.5; squarefraction = .001;
phi=Re[Fourier[Table[(InverseErf[2 Random[]-1]+I InverseErf[2 Random[]-1])*If[j+k == 2,0,
     1/Sqrt[(Sin[(j-1)*Pi/2^K]^2+Sin[(k-1)*Pi/2^K]^2)]],{j,2^K},{k,2^K}]]];
CO = squarefraction Sum[M[[i,j]], {i,1,2^K}, {j,1,2^K}]; M=Exp[fieldmultiplier phi];
{ListPlot3D[phi],Graphics[Table[Table[If[Sum[M[[2^k m+i,2^k n+j]],{i,1,2^k},{j,1,2^k}]<CO,
    {Hue[k/8],EdgeForm[Thin],Rectangle[{2^k m, 2^k n},{2^k m+2^k,2^k n+2^k}]]}],
  {m,0,2^(K-k)-1},{n,0,2^(K-k)-1}], {k,0,K-1}]]}
```

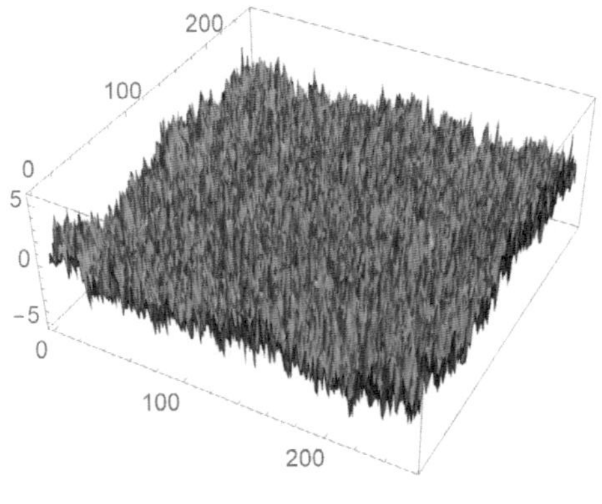

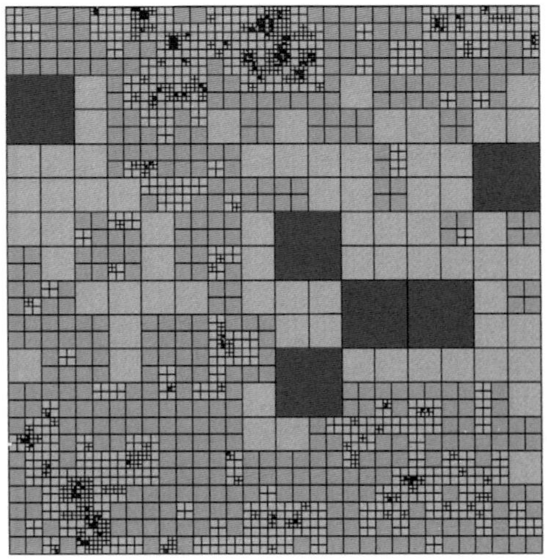

The left figure is an instance ϕ of a discrete version of the Gaussian free field on the 256×256 torus (lines 2 and 3 encode the discrete GFF; see the explanation in [233]). The exponential of *fieldmultiplier* times ϕ (called M in the code) describes a random measure on the torus. To obtain the picture on the right one starts with the whole square, then divides it into four equal squares, then divides each of those into four equal squares and so on, except that one stops dividing whenever one reaches a square where the area in the square is less than some constant cutoff (taken in the code to be *squarefraction* times the total area). The squares are colored according to their Euclidean size. Each square S shown in the picture above has area less than the cutoff—but its dyadic parent S' must have area greater than the cutoff (as otherwise S' would have been drawn after S, and would have covered S). In this sense one expects that all of the squares shown are "about the same size" in the random geometry. The blue squares correspond to regions where ϕ is smaller on average (and the measure is less dense) while the orange and red squares correspond to high density areas. One can easily paste the code above into Mathematica and experiment with different variants: for example, shown below are the figures obtained by taking a fieldmultiplier of 0.5 (left) and 1 (right) instead of 1.5 above. The measure represented by the left figure is much closer to Euclidean measure.

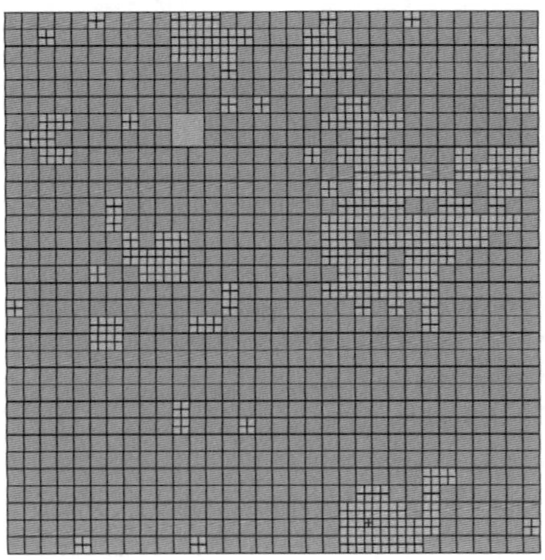

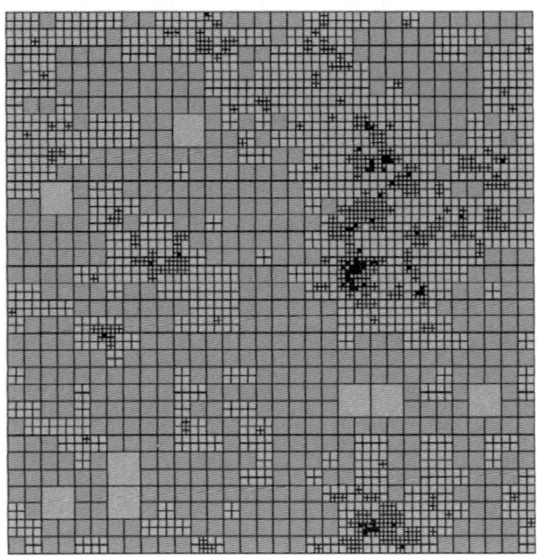

3.3. LQG surfaces

We use the term *LQG surface* broadly to describe any surface whose area measure is $e^{\gamma\phi(z)}dz$ where ϕ has a law that is locally absolutely continuous with respect to the Gaussian free field. Formally, an LQG surface is an equivalence class of pairs (D, ϕ), where (D, ϕ) and $(\tilde{D}, \tilde{\phi})$ are equivalent if they are related as in the following diagram with $Q = \gamma/2 + 2/\gamma$.

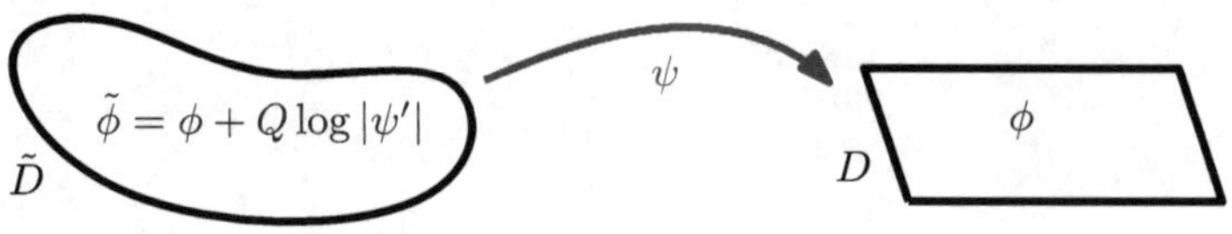

Using the definition $\nu_\phi := \lim_{\varepsilon\to 0} \varepsilon^{\gamma^2/2} e^{\gamma\phi_\varepsilon(z)}dz$, one can show that the $\nu_{\tilde{\phi}}$ area measure on $\tilde{D}$ above must a.s. agree with the pullback of the ν_ϕ area measure on D. Roughly speaking, this is because $\exp(\gamma Q \log|\psi'|) = |\psi'|^2|\psi'|^{\gamma^2/2}$. The $|\psi'|^2$ is part of the usual change of variables formula, and the $|\psi'|^{\gamma^2/2}$ accounts for the fact that ψ maps circles of diameter ε to (approximate) circles of size diameter $|\psi'|\varepsilon$, which affects the $\varepsilon^{\gamma^2/2}$ factor in the regularization.

In addition to the area measure ν_ϕ, it is possible to define a boundary length measure (in the case that ϕ is a GFF with free boundary [105]) or a fractal length associated to a line or an SLE curve or other fractal set, see [38]. All of these measures, as well as the LQG *distance function* discussed above, are preserved by the coordinate change—which makes sense since (D, ϕ) and $(\tilde{D}, \tilde{\phi})$ literally represent the same surface. The image under ψ of the so-called *Liouville Brownian motion* in D as defined in [34,114] is a Liouville Brownian motion in $\tilde{D}$.

Furthermore, if $x_1, \dots, x_k$ are points in D, and we write $\tilde{x}_i = \psi(x_i)$, then we say that $(D, \phi, x_1, \dots, x_k)$ and $(\tilde{D}, \tilde{\phi}, \tilde{x}_1, \dots, \tilde{x}_k)$ represent the same "LQG surface with k marked points." Finally, let us stress that the above definition of LQG surface makes sense for any $Q \geq 0$, not only for $Q \geq 2$. (The values $Q \geq 2$ are those that have the form $Q = \gamma/2 + 2/\gamma$ for some $\gamma > 0$ or, equivalently, those for which $d = 25 - 6Q^2 < 1$.) One just has to accept that the *area measure* is not well defined when $Q < 2$. However, as we mentioned earlier, the metric space structure (along with the fractal measure of some other sets) continues to be well defined when $Q < 2$ (although the diameter of the surface becomes infinite when $Q < 2$, which corresponds to $d > 1$).

3.4. Constructing the LQG sphere

One simple way to define a unit-area LQG sphere is to consider a GFF ϕ on a simply connected bounded domain D with boundary conditions given by a constant C, and to condition on $\nu_\phi(D) = 1$. As $C \to -\infty$ this object converges in law to the unit area LQG sphere (with the boundary somehow shrinking—in the metric sense—to a single marked point in the limit) as explained in [235]. Another approach involves starting with an infinite cylinder and two marked points (the cylinder's endpoints), as described in [102]. Yet another approach involves starting with the infinite-volume Polyakov measure (see the next subsection) and "pinning it down" at three points, in a manner described in [87]. The paper by Aru, Huang, and Sun [23] established the equivalence of the approaches in [102] and [87]. (See also the alternate proof in [16] and the disk analog in [70].) A chart, presenting several equivalent definitions and the relationships between them, is included in [194].

All of these approaches have analogous constructions that produce *unrestricted area* LQG spheres (instead of unit area LQG spheres). In the $C \to -\infty$ construction above, for example, instead of conditioning on $\nu_\phi(D) = 1$, one can (for any C value) multiply the measure by a constant factor to ensure that the measure assigned to ϕ distributions with $\nu_\phi(D) \in [1, 2]$ is constant, and then take the vague $C \to -\infty$ limit.

3.5. Polyakov's infinite measure on embedded LQG surfaces

The Polyakov measure is an *infinite* measure on the space of unmarked, unrestricted-area LQG spheres *embedded* in $\mathbb{C}$. As mentioned in the introduction, it corresponds to an unrestricted-area LQG sphere embedded in all possible ways (with the embedding chosen from Haar measure on the Möbius group). The measure is infinite for two reasons: first, we recall from the introduction that the law of the area for an unrestricted-area LQG sphere has the form $A^{-b} dA$ (or $A^{-b} e^{-\mu A} dA$ for some constant μ in the "off-critical" case) which is an infinite measure for the b values that we will encounter (namely $b > 3$, recall Section 3.7). Second, the embedding is chosen from the Haar measure on the Möbius group, which itself has infinite volume. Also, just to avoid confusion, let us clarify that it is the Polyakov measure on the space of surfaces—not the area measure on any individual surface—that is infinite. In the Polyakov measure, almost all surfaces have finite area (assuming $d \leq 1$; the total area is not defined if $d > 1$). Each embedded surface is described by a generalized function ϕ, so the Polyakov measure can be viewed as an infinite measure on the set of generalized functions.

The Polyakov measure is usually defined in a slightly different way. It is presented as a way to make sense of the expression "$e^{-S(\phi)} d\phi$" where S is the so-called *Liouville action*, which we will discuss below, see the presentation by David, Kupiainen, Rhodes, and Vargas at [87] or the lecture notes by Kupianen at [159]. The simplest way to describe it (in the case $\mu = 0$) is to say that ϕ is an instance of the zero-mean GFF on the sphere plus an independent constant C chosen from the infinite measure $e^{-2Qc} dc$. To obtain the general-μ measure, one simply weights the zero-μ measure by $e^{-\mu A}$, where A is the area of the LQG surface.

The fact that the Liouville action produces a measure on embedded LQG surfaces of the form described above (in particular, a measure that is Möbius invariant) is counterintuitive at first glance, but it is carefully explained, e.g., by Ang, Holden, and Sun in [16], see also the conformal invariance discussion in [159]. There is a certain miracle (related to what we will call "semi-Gaussian" measures) that makes everything work out. Before we discuss the specifics, let us present a couple of simpler semi-Gaussian measures as a warmup.

3.5.1. Semi-Gaussian measures

A *semi-Gaussian* measure is a constant times $e^{-F(v)} dv$ where F is a quadratic that is strictly convex in all directions except one. For example, $\frac{1}{\sqrt{2\pi}} e^{y-x^2} dx dy$ is semi-

Gaussian. It is a product of a normal measure $\frac{1}{\sqrt{2\pi}}e^{-x^2/2}dx$ and an infinite measure $e^y dy$. If we restrict this measure to any nonvertical line, we obtain a finite measure, which is a constant times a normal probability measure. If we restrict to any vertical line, we obtain an infinite measure.

In the figures below, the orange lines are level sets of the function $y - x^2$ (and hence also the function $\frac{1}{\sqrt{2\pi}}e^{y-x^2}dxdy$). In each figure, one can imagine "sampling" (X, Y) from the infinite measure in two steps. First one decides which blue line (X, Y) belongs to. (The blue lines in the three figures are the lines of slope 0, -1, and 1, respectively. By integrating the density function along the blue lines, we find that the law of the y-intercept of the blue line is given by a constant multiple of $e^y dy$.) Then *given that* one chooses the location *on* the blue line. The *conditional law* of the location on the blue line is that of a Gaussian random variable centered at a point on the vertical green line (which is the point on the blue line where $y - x^2$ is largest).

```
n = 4;{ParametricPlot[{Table[{t,j/n},{j,-2n,2n}],Table[{t,t^2+ j/n},{j,-2n,2n}],{0,t}},{t,-2,2},
PlotRange->{{-1,1},{-1,1}}],ParametricPlot[{Table[{t,-t+j/n},{j,-2n,2n}],Table[{t,t^2+j/n},
{j,-2n,2n}],{-.5,t}},{t,-2,2},PlotRange->{{-1,1},{-1,1}}],ParametricPlot[{Table[{t,t+j/n},
{j,-2n,2n}],Table[{t,t^2+j/n},{j,-2n,2n}],{.5,t}},{t,-2,2},PlotRange->{{-1,1},{-1,1}}]}
```

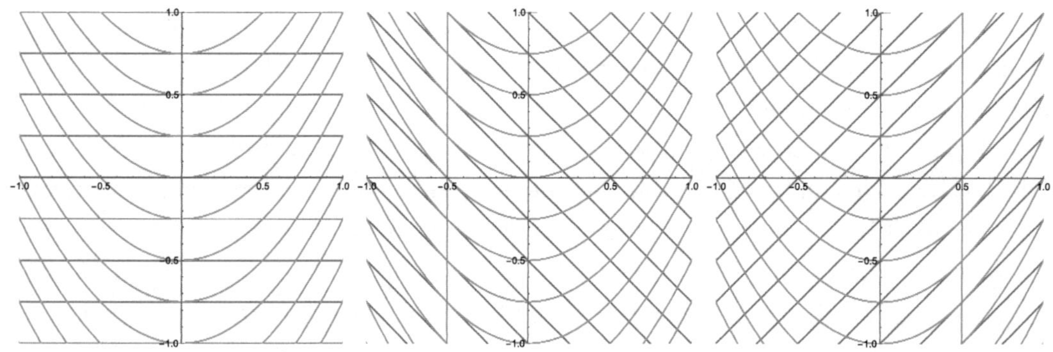

In the left figure, *no matter what* blue line we choose, the conditional expectation of X is 0. In the second and third figures, *no matter what* blue line we choose, the conditional expectation of X is (respectively) $-1/2$ or $1/2$. Changing the blue-line slope somehow has the effect of "shifting" the conditional law of X.

This is a bit counterintuitive. For a closely related example (essentially a rotation of the one above by 45 degrees), suppose

$$e^{-(a-b)^2/2t}e^{(a+b)/2}dadb$$

is the (semi-Gaussian) density function for a pair (A, B). Then we can formally compute $\mathbb{E}[B|A] = A + t/2$ and $E[A|B] = B + t/2$. This is because if we restrict to a fixed value of a, we obtain a multiple of a Gaussian measure on b values centered at $a + t/2$ (and similarly with a and b reversed). Checking this fact is an elementary complete-the-square calculation: if b is fixed then $-(a^2 - 2ab + b^2)/2t + (a + b)/2$ is $-(a^2 - 2a(b + t/2) + (b + t/2)^2)/2t$ plus a term that does not depend on a.

At a glance, the above seems to suggest that A is $t/2$ units bigger than B on average, and B is $t/2$ units bigger than A on average, which in turn seems contradictory. This is somehow reminiscent of "envelope switching" paradoxes, where after one observes the amount of money in either *one* of two envelopes, one always expects the *other* to contain more, see, e.g., **[62]**. In fact, this is just the sort of paradox that one encounters when dealing with infinite measures and/or infinite expectations. We find that the following are equivalent:

- First "sample" A from the infinite measure $e^a da$. Then choose B as a normal with variance t and mean $A + t/2$.
- First "sample" B from the infinite measure $e^b db$. Then choose A as a normal with variance t and mean $B + t/2$.

One can imagine similar constructions in higher dimension. For example, we could replace the X in the figures above by a vector $(X_1, X_2, \dots, X_n)$ and replace $y - x^2$ with $y - \sum_{i=1}^n x_i^2$, and replace the blue lines by blue hyperplanes. One could then argue in a similar way that changing the slope of the blue hyperplanes has the effect of shifting the location of the vertical green line.

By taking a limit of an increasing sequence of finite-dimensional subpaces, we can also make sense of an infinite dimensional semi-Gaussian—precisely the same way we make sense of the GFF as an infinite dimensional Gaussian, using the quadratic function $-\frac{1}{2}(\phi, \phi)_\nabla$, or the way we define Brownian motion, using the quadratic function $\frac{1}{2}\int_{-\infty}^{\infty}(\frac{\partial}{\partial t}B(t))^2 dt$.

Let us give an example. Define a process (whose law is an infinite measure) by $M(t) = B(t) + |t|/2 + Y$ where Y has density $e^y dy$ and $B(t)$ is an independent standard Brownian motion (defined for all $t \in \mathbb{R}$, with $B(0) = 0$). This is an infinite measure on paths that at first glance seems to be "mostly supported" on paths that have their minima near 0. If we tried to define an "action" for $M(t)$ it would have the form

$$S(M) = e^{M(0)} + \frac{1}{2}\int_{-\infty}^{\infty}\left(\frac{\partial}{\partial t}\big(M(t) - |t|/2\big)^2\right)dt.$$

On the other hand, some thought reveals that, when t is fixed, the (infinite-measure) law of $B(0)$ and $B(t)$ is equivalent to the law of A and B in the example above, and in fact the law is symmetric with respect to swapping the roles 0 and t. One can also show that *given* $M(0)$ and $M(t)$ the conditional law of the rest of the process is given by a Brownian bridge on $(0, t)$ and Brownian motions with drift on (t, ∞) and (running time backward) on $(-\infty, 0)$, with the given values of $M(0)$ and $M(t)$ as endpoints. One can make a similar argument using $-t$ and 0 in place of 0 and t and use this to show that the the law of M is invariant w.r.t. to translation by t units (for any t).

This is a rather remarkable fact. The definition of $M(t)$ does not look translation invariant at all—it is clearly centered at 0. But somehow translating the definition by r units to the right does two things: it effectively *weights* the law of M by $e^{M(r)-M(0)} = e^{B(r)-B(0)}$ (this puts a bias in favor of functions that increase on the interval $[0, r]$) and it deterministically adds the function $f(t) = \frac{|t-r|}{2} - \frac{|t|}{2}$ (which decreases on the interval $[0, r]$ and is flat elsewhere). And these two changes magically cancel each other out.

We remark that another way to construct this measure is by considering a Brownian bridge measure on processes $B(t)$ indexed by $[-T, T]$, with boundary values $B(T) = B(-T) = T/2$, then multiplying the measure by a constant to ensure that the measure of paths with $B(0) < 0$ is some fixed constant, then taking the vague $T \to \infty$ limit. This construction might make the translation invariance a bit more intuitive.

To extend the story to two dimensions, note that the above construction is still translation invariant if we take the law of Y to be $e^{-2Qy}dy$ and write $M(t) = B(t) - Q|t| + Y$. This just corresponds to taking $\mathcal{B}(T) = \mathcal{B}(-T) = -QT$ in the limiting procedure mentioned above. On the cylinder $\mathcal{C} = \{t + \theta i : t \in \mathbb{R}, i \in [0, 2\pi)\}$ one can then define $\phi(t + \theta i)$ to be $M(t)$ plus the projection of the GFF (on the cylinder) onto the space of functions with mean zero on every fixed-t slice of the cylinder. If restrict to $M(t)$ that have their maximum at $t = 0$ (to eliminate the translation symmetry) we obtain the so -called "$\alpha = 0$ quantum sphere" **[102]**. If we *don't* restrict in this way, we obtain the Polyakov measure (with $\mu = 0$), an infinite measure on surfaces that turns out to be invariant w.r.t. *all* Möbius coordinate changes, not only translations.

To see why, note that there are three "mostly flat" surfaces often used to parameterize an LQG sphere: the Riemann sphere $\mathbb{C} \cup \{0\}$, the cylinder $\mathcal{C} \cup \{\pm\infty\}$ and the gluing $\mathcal{D}$ of two unit disks together their boundaries. There are obvious unit-circle- preserving conformal maps between these three spaces. If ψ is the obvious conformal map from $\mathcal{C}$ to $\mathcal{D}$ then $Q \log |\psi'(t + i\theta)| = -Q|t|$. This is the term that comes up in the change of coordinates from $\mathcal{D}$ to $\mathcal{C}$. If we instead change coordinates from $\mathcal{D}$ to $\mathbb{C}$, then this term is replaced by $-2Q \max\{\log |z|, 0\}$. Working in $\mathbb{C}$, the proof that the law of the construction is invariant under translations and rotations of $\mathbb{C}$ (as well as inversions and dilations) is very similar to the one dimensional argument above, and since these operations generate the Möbius group, full Möbius invariance follows, see **[16, 159]**.

3.5.2. Embedded Polyakov sphere

When building physical field theories it is often natural to define a measure on fields ϕ by writing formally $e^{-S(\phi)}d\phi$ where $d\phi$ represents some sort of uniform measure on the space of all functions. In some cases this is hard to make mathematically precise because it is not clear what the measure $d\phi$ corresponds to. One might attempt to approximate $d\phi$ by something well defined (like Lebesgue measure on a finite-dimensional space of piecewise-linear functions) but it may be unclear how to normalize the construction to obtain a non-trivial limit, or how to prove such a limit exists. In the special case of the Liouville action, however, these concerns can be overcome. Let us tell the story with a playful dialog.

INSTRUCTOR: Consider the measure $e^{-A(\phi)-B(\phi)}d\phi$ where $d\phi$ is the uniform measure on the space of all functions.

MATH POLICE: Sorry, there is no such thing as the uniform measure on all functions. Your object is not defined.

INSTRUCTOR: But my A is a quadratic function on the space of ϕ for which it is finite. If I restrict A to a codimension-one subspace then A is the norm for a Hilbert space. So $e^{-A(\phi)}d\phi$ is just a Gaussian Hilbert space cross an infinite measure that looks something like $e^y dy$. Check out Janson's book on Gaussian Hilbert spaces **[150]**. These things are certainly well defined.

MATH POLICE: Okay fine, but you still need to weight by the $e^{-B(\phi)}$ factor.

INSTRUCTOR: My friends proved that (with appropriate normalizing) $B(\phi)$ is well-defined and finite for almost all ϕ taken from the measure $e^{-A(\phi)}d\phi$.

MATH POLICE: So $e^{-B(\phi)}$ is the Radon–Nikodym derivative with respect to the semi-Gaussian measure? And B is obviously measurable?

INSTRUCTOR: That's right.

MATH POLICE: Okay, you're free to go.

We can take the S in the above dialog to be the *Liouville action* defined as follows:

$$S(\phi) = \int \left(\frac{1}{2} \cdot \frac{1}{2\pi} |\nabla \phi(z)|^2 + \frac{1}{4\pi} QR(z)\phi(z) + \mu e^{\gamma\phi(z)} \right) dz,$$

where R is the Ricci curvature associated to the "reference metric," see e.g. **[87]**. For example, we may assume that the reference metric is the ordinary sphere so that R is constant (with total integral 8π by Gauss-Bonnet—recall that the Ricci curvature is twice the Gaussian curvature) so that the second term is $2Q$ times the mean value of ϕ. Alternatively, if the reference metric is the glued pair of disks $\mathcal{D}$ from the previous subsection, then the second term becomes $2Q$ times the mean value of ϕ on the unit circle.

The integral of the first two terms is the quadratic part $A(\phi)$ while the integral of the last term is the $B(\phi)$. The fact that the latter is defined (provided $Q > 2$) follows essentially from the ideas of Høegh -Krohn and Kahane—see also the author's later work with Duplantier **[105]** which constructs $e^{\gamma\phi(z)}dz$ as a measure-valued function of the field ϕ and explains the LQG context. In light of the above dialog, the Polyakov measure $e^{-S(\phi)}d\phi$ is in fact rigorously defined; this is explained in more detail in **[87]** (see also the higher genus version in **[88]**).

Stories like the above, where S has a quadratic term (relatively simple to handle) and a nonquadratic term (requiring more thought) are relatively common in quantum field theory, and it is certainly not *always* the case that the nonquadratic part simply modifies the Gaussian part in an absolutely continuous way.

The $\mu = 0$ measure is already rather interesting; for example, it can be easily shown to be Möbius invariant (following the LQG coordinate change rules), despite appearing at first glance to be "centered" at a specific location within the Möbius group, a phenomenon analogous to the example at the end of Section 3.5.1. The miracle is that when one applies a Möbius transformation ψ, the $Q \log |\psi'|$ factor one has to add exactly compensates for the effect of the recentering (swapping the curvature measure $R(z)dz$ for its image under ψ), see

[16]. This implies that the Polyakov measure factors as the product of a measure on (unembedded) LQG surfaces and a measure on embeddings (given by the infinite-volume Haar measure on the Möbius group) see [16]. The same is true if one considers a nonzero μ. However, the Polyakov measure with a nonzero μ only makes sense as described above if $d \leq 1$ (since otherwise the natural volume measure is infinite), while the construction of the zero-μ measure makes sense for any $Q \geq 0$. Changing the choice of reference metric is analogous to changing the slope of the blue lines/hyperplanes in the previous section and (up to a constant multiplicative factor) it leaves the measure construction unchanged [87].

On a flat reference metric, the quadratic action is simply the Dirichlet energy of ϕ while the $\mu e^{\gamma\phi(z)}$ term corresponds to the Ke^{ρ} term associated to the Liouville equation with a nonzero K. Polyakov was not the first to work with variants of the free field action, and in [214] and [216] he attributes a closely related "quadratic action" to Douglas, who used it in his work on Plateau's problem for minimal surfaces. (Douglas was awarded one of the first two Fields medals for this work in 1936 [97].) See also [60,89] which introduce the action now called the *Polyakov action*, which Polyakov used later in [215].

3.6. More history

The Brownian map and peanosphere constructions were directly motivated by discrete objects—namely random planar map models. The construction of LQG surfaces, on the other hand, was motivated within physics as a "quantization" based on the Dirichlet energy or on the Liouville equation. Why would one expect these objects to be equivalent?

The discrete planar map models were well studied in physics due to their relationship with random matrices and the random particle systems corresponding to random matrix eigenvalues, see the seminal papers from the 1970s by 't Hooft and by Breézin, Itzykson, Parisi, and Zuber [1,59], along with more recent overviews by Eynard, Guionnet, and Maurel-Segala [107,122]. However, it took some time for people to be persuaded that the Liouville theory corresponded to the scaling limit of these models. For example, Polyakov wrote in a memoir [216] (see also his previous memoir [214]) that he did not become convinced of the connection between the discrete models and Liouville quantum gravity until the late 1980s after jointly deriving, with Knizhnik and Zamolodchikov, the so-called *KPZ formula* for certain Liouville quantum gravity scaling dimensions [156] and comparing them with known combinatorial results for the discrete models (for a rigorous approach, see also [32, 103,105,113,220]).

At this point the relationship between the planar map models and the continuum objects is on more solid mathematical ground. We have convergence results of various kinds and formal equivalence proofs between the different continuum objects, see Section 5. But these rely on a lot of machinery that was not yet available in the 1980s. The quantum gravity zipper perhaps gives the cleanest way to rigorously relate the peanosphere construction and the Liouville quantum gravity construction, as described in [102,235], see also [26].

3.7. Computing the scaling exponent

On the other hand, just as in the peanosphere case, one can compute the scaling exponent quite easily. This is already enough to show that *if* for some γ the unrestricted-area LQG sphere is the scaling limit of the unrestricted-area discrete models, then we must have $\gamma = \sqrt{8/3}$ in the undecorated case and $\gamma = \sqrt{2}$ in the tree-decorated case.

Say we fix the Gaussian free field up to the mean value X on the unit circle. Then this mean value is sampled from the infinite measure $e^{-2Qx}dx$. The total area of $e^{\gamma\phi(z)}dz$ goes like $e^{\gamma x}$ times a constant. So we have to consider the image of the measure $e^{-2Qx}dx$ under the map $\psi(x) = e^{\gamma x}$. By standard change of variables the new measure is (up to a constant factor)

$$\frac{1}{(\psi^{-1})'(A)} e^{-2Q\phi^{-1}(A)} dA = \frac{1}{A} e^{-2Q\log(A)/\gamma} = A^{-1-2Q/\gamma} dA.$$

Here $1 + 2Q/\gamma = (4/\gamma^2 + 2)$ which comes to $7/2$ if $\gamma = \sqrt{8/3}$ and 4 if $\gamma = \sqrt{2}$.

3.8. Random surfaces embedded in d-dimensional space

Suppose we accept, based on the previous discussion, that the $\gamma = \sqrt{8/3}$ theory (which corresponds to $25 - 6Q^2 = 0$) describes the scaling limit of the undecorated $d = 0$ model. We can call this *pure Liouville quantum gravity* model. We would then like to argue that if we *weight* the law of the pure model by the dth power of the GFF partition function (or the corresponding and equivalent "loop soup" partition function) we obtain a new LQG model with a Q parameter that satisfies $d = 25 - 6Q^2$.

This has been explained heuristically in various ways over the years. The author with Ang, Park, and Pfeffer gave a rigorous version of this statement that applied to a certain way of regularizing the random surfaces [17]. We won't gvie details here but we mention below a few of the ingredients used to make the connection between loop soup weightings and changes to $d = 25 - 6Q^2$.

On a compact surface with boundary, the heat kernel trace can be written $Z = Z(t) = \mathrm{sp}\, e^{t\Delta} = \sum e^{t\lambda_n}$ where λ_n are the eigenvalues of the Laplace–Beltrami operator Δ. By standard Tauberian theory, the asymptotics of Z (as $t \to 0$) are closely related to the asymptotics of λ_n (as $n \to \infty$). Weyl addressed the latter for bounded planar domains D in 1911 [260] (see discussion in [182]) by showing $-\lambda_n \sim \frac{2\pi n}{\mathrm{area}(D)}$ as $n \to \infty$ which is equivalent to

$$Z \sim \frac{\mathrm{area}(D)}{4\pi t}$$

as $t \to 0$. In 1966 Kac gave higher-order correction terms for Z on domains with piecewise linear boundaries (accounting for boundary length and corners) in his famously titled "Can you hear the shape of a drum?" which asks what features of the geometry of D can be deduced from the λ_n or, equivalently, from Z [151]. (Short answer is some but not all.)

McKean and Singer (among others) extended these asymptotics from planar domains to smooth manifolds with nonzero curvature [182] where the constant order correction term is a certain curvature integral. For two-dimensional surfaces with boundary, the integral $\int_\delta^\infty Z(t)/t\,dt$ turns out to describe the *Brownian loop soup measure* of the set of loops longer than δ (as developed and explored by Lawler, Werner, Dubédat, and others; see [17] for further explanation and references). When the metric takes the form e^{ρ} times a flat metric, the small δ asymptotics have a constant order correction term that corresponds to the Dirichlet energy of ρ. This is the so-called Polyakov–Alvarez formula, also known as the Polyakov–Ray–Singer or Weyl anomaly formula, and it has played a major role in conformal field theory. See e.g. the early discussion of Laplacian determinants and this formula by Osgood, Phillips and Sarnak [212, 225].

4. CONFORMAL FIELD THEORY AND MULTIPOINT CORRELATIONS

Conformal field theory is a huge subject. For a broader overview of conformal field theory, beyond just Liouville theory, the reader might begin with the well-known (and very long) textbook by Di Francesco, Mathieu, and Sénéchal [95]. Alternatively, the overview at [221] begins with a list of several conformal field theory textbooks.

Liouville theory is one of many conformal field theories, but it is by itself a large and highly studied subject. Here one might begin with the 1990 survey by Seiberg [232], the 1993 lecture notes by Ginsparg and Moore [118], the 1995 survey by Di Francesco, Ginsparg, and Zinn-Justin [94], the 2004 retrospective by Nakayama [210], or the 2014 textbook by Ribault [221] (which among other things explains why Liouville conformal field theory is uniquely characterized by certain axioms).

What is a conformal field theory? An internet search for "A conformal field theory is" reveals several definitions, the first two from Wikipedia:

- a quantum field theory that is invariant under conformal transformations,
- a set of correlation functions that obey a number of axioms,
- a functor [204] between categories satisfying certain "sewing axioms,"
- a Virasoro module [227]

  $$V = \bigoplus_{i=B_1} W(c_i, \phi_i) \otimes W(\bar{c}_i, \bar{\phi}_i)$$

  with unitary highest weight modules $W(c_i, \phi_i)$, $W(\bar{c}_i, \bar{\phi}_i)$ subject to [certain] axioms.

The first definition is standard but its precise meaning depends on how one defines a quantum field theory. The third and fourth definitions represent formalization efforts that would go somewhat beyond the scope of this note. So let us focus on the second definition, sometimes called the *conformal bootstrap* approach to CFT. Here a CFT is no more or less than a set of *correlation functions*, and the interpretation of these functions is that they represent (in some sense) expected products of random generalized functions called *fields* evaluated at different points.

The famous 1984 paper by Belavin, Polyakov, and Zamolodchikov (BPZ) [30] argued that conformal invariance symmetries should imply certain properties for these correlation functions, and that this should be sufficient to allow one to explicitly compute the correlation functions for *some* conformal field theories (the so called *minimal models*) including a theory that one would expect to describe the scaling limit of the Ising model. About a decade later Dorn, Otto, Zamolodchikov, and Zamolodchikov were able to compute certain three-point correlations for the Liouville theory [96, 262]. Other correlation functions could then be deduced from these using the BPZ theory [30] and further input proposed by Teschner [244–246]. These arguments involved mathematically nonrigorous steps, such as assuming without proof that formulas defined in one setting could be analytically continued and applied in other settings.

Building on [87], Guillarmou, Kupianen, Rhodes, and Vargas have produced a series of papers that define and derive the correlation functions for Liouville theory mathematically, building on earlier derivations from the physics literature that we mentioned above. The expressions describing the correlation functions are complicated (integrals, special functions, recursive definitions, etc.) but nonetheless explicit. The first work in this series is a proof of the DOZZ formula [160]. The second paper derives an analog of Plancharel's theorem (which states that the Fourier transform preserves L^2-norm) along with a certain "spectrum" relevant to this context [120]. The final paper completes the bootstrap program [120] with an extension to n-point functions and higher genus surfaces [121]. We recommend that the reader take a look at the introduction to [120], which summarizes this viewpoint and situates it within the larger enterprise of quantum field theory. Here we will give a much shorter overview of this viewpoint, aiming only to give a simple account of the relationship to the other perspectives in this paper.

The physics literature on conformal field theory can be challenging for mathematicians to follow. It comes with a large and very specialized jargon, and it does not always proceed in the order a mathematician would expect (where one first produces the measure space and the σ-algebra, then constructs the measure, then begins doing calculations). Partly this is because (when quantum wave functions are involved) not everything in quantum field theory *can* be described in a simple probabilistic way—sometimes "observables" are noncommuting operators that can only be defined in indirect ways. Fortunately, Liouville theory does have a simple probabilistic interpretation.

Let us make one more comment on nomenclature. In quantum field theory, a (generalized) function on $\mathbb{R}^{n+1}$ can be interpreted as a *path* on the space of functions on $\mathbb{R}^n$, and the term *integral* is often a shorthand for *a measure with respect to which one integrates*. In this context, the object we call the Polyakov measure (on the space of surfaces) in Section 3.5 is also called the *Polyakov path integral* (with n taken to be 1). This language evokes the Feynman path integral, an integral over particle trajectories that appears in quantum mechanics. Similar interpretations arise in string theory, where an integral over the space of string trajectories is seen as an integral over the space of embedded surfaces (a.k.a. "worldsheets"). A very informal overview of this viewpoint (featuring Vargas) appears in an online video produced by Quantum Magazine (starting at time 7:20) https://youtu.be/9uASADiYe_8.

4.1. Gaussian case

It is generally instructive to do something easy before doing something hard. So let us start in the simple setting where ϕ is a whole plane Gaussian field with additive constant chosen so that the mean value on the unit circle is 0. (This is also the starting point in [120], for example.) In this case the Green's function is given by

$$G(x,y) = \ln \frac{1}{|x-y|} + \ln |x|_+ + \ln |y|_+,$$

where $|x|_+ = \max\{1, |x|\}$ and the construction of "random fields" of the form $V_\alpha(x) := e^{\alpha\phi(x)}$ is relatively straightforward. In some sense it can already be seen in the work of quantum field theorists of the 1970s, beginning with the work of Høegh-Krohn. As we have already discussed, one way to make sense of this is by writing $V_\alpha^\varepsilon(x) := \varepsilon^{\alpha^2/2} e^{\alpha\phi_\varepsilon(z)}$ and $V_\alpha(x) := \lim_{\varepsilon\to 0} V_\alpha^\varepsilon(x)$, where $\phi_\varepsilon(z)$ is the mean value of ϕ on $\partial B_\varepsilon(z)$ and the convergence holds locally a.s. in the space of random generalized functions (or in the space of random measures). Then for any sufficiently small ε (so that the balls $B_\varepsilon(x_i)$ do not overlap), we have

$$\left\langle \prod V_{\alpha_i}^\varepsilon(x_i) \right\rangle = e^{\sum \alpha_i \alpha_j \tilde{G}(x_i, x_j)}$$

where

$$\tilde{G}(x,y) = \begin{cases} G(x,y), & x \neq y, \\ \log |x|_+ + \log |y|_+, & x = y. \end{cases}$$

This comes up out to be

$$\prod_{i \neq j} |x_i - x_j|^{\alpha_i \alpha_j} \prod_{i,j} \left(\max\{|x_i|, 1\} \max\{|x_j|, 1\}\right)^{\alpha_i \alpha_j},$$

and, in the case that all the x_i lie inside the unit disc, the expression is simply

$$\prod_{i \neq j} |x_i - x_j|^{\alpha_i \alpha_j}.$$

By starting with $V_{\alpha_i}^\varepsilon$ and taking $\varepsilon \to 0$, we can give meaning to the multipoint correlation function (also known as *Schwinger function*) written as

$$\left\langle \prod V_{\alpha_i}(x_i) \right\rangle = \prod_{i \neq j} |x_i - x_j|^{\alpha_i \alpha_j}.$$

This can be interpreted as the density function for a two-dimensional "Coulomb gas" of particles with charges α_i. The overall integral over all $x = (x_1, \ldots, x_n)$ may be infinite (depending on the α_i values), but it is finite if one restricts to a subset of $(x_1, \ldots, x_n)$ values such that the $|x_i|$ are bounded above and the $|x_i - x_j|$ are bounded below. Just to clarify, for now we are making this Coulomb gas calculation only for the centered GFF ϕ, not for the full Polyakov measure from Section 3.5.

As a formal collection of correlation functions (obtained in the $\varepsilon \to 0$ limit) this expression makes perfect sense for any $\alpha_i \in \mathbb{C}$. Differentiation commutes with expectation, so one can use the same formula to compute expectations of products involving derivatives like $\frac{\partial}{\partial x_i} V_{\alpha_i}(x_i)$. Expectations of fields involving ϕ itself (or polynomials in ϕ) can also be defined mathematically.

4.2. Incorporating the Liouville term or area conditioning

The ϕ used in Section 4.1 (plus a deterministic function) can be obtained by restricting the Polyakov measure to the set of ϕ whose mean value on the unit circle is zero. But what if we instead restrict the Polyakov measure to the set of surfaces of total quantum area 1 (or weight by $e^{-\mu A}$ where A is the surface area)? The answer is that after such a conditioning or such a weighting, ϕ is no longer Gaussian, and the n-point correlation computation transforms from *easy* to *doable but only barely*. On the other hand, in order to understand some fundamental things like the law of conformal modulus of four points (or n points) sampled independently from the measure on an LQG sphere, one has to address the harder question, and this is precisely what is done in [120, 121, 160]. Although these results are rather recent, they have already inspired a tremendous amount of activity, establishing *exact solvability* for many problems that could previously only be addressed more qualitatively.

Papers by Ang and Sun, including some with coauthors Holden and Remy, give applications of LCFT results to various domains (SLE, CLE, and a variance formula for the peanosphere) [15,16,18,19]. These remarkable papers combine mating-of-trees and conformal-welding techniques with LCFT techniques, leading to rigorous proofs of physics results (such as the FZZ formula and the imaginary DOZZ formula), as well as entirely new results. They have to extend the welding/mating theory to finite-volume surfaces/trees, which is harder than the infinite-volume work, since the finite-volume surfaces lack scale invariance. Recent integrability achievements for the disk include works by Remy and Zhu [218,219] and a proof of the Fyodorov–Bouchaud formula by Remy [217]. See also the work of Ghosal, Remy, Sun, and Sun on the torus [117].

5. RELATIONSHIPS

Although all four viewpoints are equivalent in some sense, the relationships are somewhat involved. The following diagram summarizes a few of the associated keywords, which we will discuss briefly below.

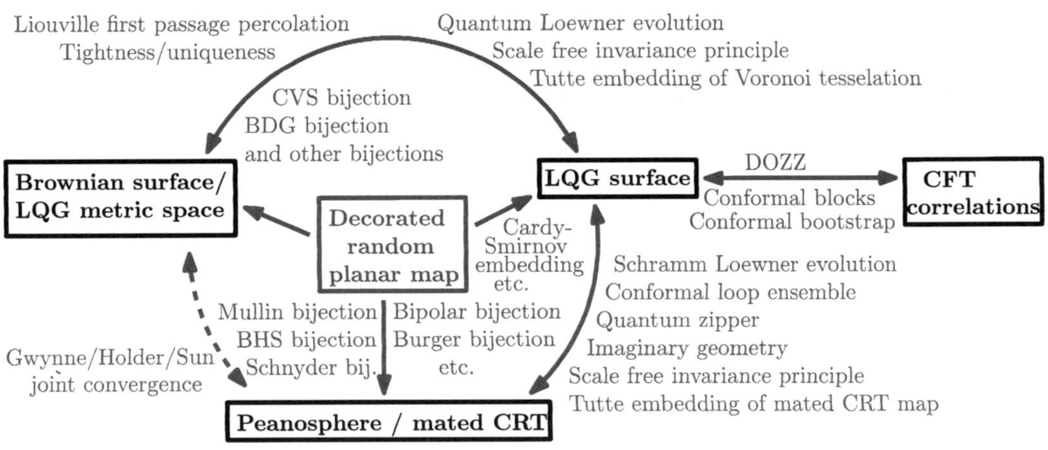

Miller and the current author recently showed that the LQG and Brownian spheres are *a posteriori* equivalent [195, 197, 198]. That is, we showed that there is a canonical way to endow each object with the other object's structure, and that once this is done the two objects agree in law. This is not all obvious. The papers are long and difficult, and build on hundreds of pages of prior work. The author's other joint works on the LQG construction include [38, 103–105, 154]. This establishes the upper arrow in the special case of the Brownian map. The construction of the metric space structure for general LQG surfaces (with $d \neq 0$) was discussed earlier.

In [235] the author showed that when two infinite half-plane-homeomorphic γ-LQG surfaces are "conformally welded" to one another along their boundaries—and the new surface is conformally mapped to the half-plane—the interface between these curves becomes an SLE_κ curve with $\kappa = \gamma^2$. The sphere version is established in a follow-up work [194], and a disk version appears in [14]. Establishing the correlation formula for $\kappa > 8$ was achieved in [126].

With Duplantier and Miller, the author showed the equivalence of the peanosphere approach and the SLE-decorated-LQG approach in [102], which draws from the imaginary geometry results in [190–193] and the quantum zipper construction in [235]. A remarkable series of papers by Holden and Sun has shown that if one embeds the uniformly random triangulation in the plane in a natural way (inspired by the conformal coordinates Smirnov developed for his proof of Cardy's formula) then the counting measure on the vertices converges (in an appropriate scaling limit) to the Liouville quantum gravity measure, see the overview at [147, 148].

Random planar maps converge to continuum objects in the limit, but there are also natural ways to generate random planar maps from the continuum constructions by "coarse graining" in some sense. The Poisson–Voronoi tesselation (constructed from the Brownian map) and mated-CRT map (constructed from the peanosphere) have been shown to converge to LQG when they are embedded according to the Tutte embedding. This was done in a series of four papers with Gwynne and Miller [136–139]. Effectively, this gives a way of putting a conformal structure on the Brownian map or the matings of trees that is more concrete than the one guaranteed by [195, 197, 198]. The heart of all of this is [138] which gives an invariance principle (i.e., Brownian motion convergence) for random walks in random environments that are "scale free" in the sense that there is no universally typical length scale. (All of the most natural discretizations of Liouville quantum gravity measures are scale free in this sense.)

Gwynne, Holden, and Sun established the *joint convergence* of random triangulations in the metric and peanosphere sense [129]. (Convergence in one topology coupled with convergence in another topology does not imply joint convergence in the product of the two topologies, but it was established in this particular case.) This result builds on earlier convergence work by Gwynne and Miller for percolation-interface-decorated surfaces [134, 135].

A series by the author plus Miller and Werner explores conformal loop ensembles on random surfaces, the surfaces obtained by cutting along the boundaries of these loops, and so forth [201, 202]. These papers extend the conformal welding stories described earlier and show that much of the intuition one derives from discrete planar maps is correct in the continuum as well (e.g., the surfaces inside and outside of a CLE loop are independent given the interface length) and this leads to a number of interesting computations. These papers build on another recent paper by the same authors [199] which concerns continuum versions of the classical Edwards–Sokal couplings involving FK-models and Ising/Potts models and their variants, see also [200].

6. GAUGE THEORY

Random surfaces are related to many areas of math and physics, including random matrix theory, two-dimensional statistical physics, string theory, and so on. But we should also note that Polyakov's influential 1981 paper began by mentioning an interest in gauge theory:

> *"In my opinion at the present time we have to develop an art of handling sums over random surfaces. These sums replace the old-fashioned (and extremely useful) sums over random paths. The replacement is necessary, because today gauge invariance plays the central role in physics. Elementary excitations in gauge theories are formed by the flux lines (closed in the absence of charges) and the time development of these lines forms the world surfaces. All transition amplitude[s] are given by the sums over all possible surfaces with fixed boundary."*
> *(A. M. Polyakov, Moscow, 1981)* [215]

Over 40 years later, many fundamental gauge theory problems remain unsolved, including the famous Clay Millenial Prize Problem, and it is unclear how much random surface theory will help—see, e.g., the "skeptic vs. enthusiast" dialog in Section 2 of [79], written in 1995. Nonetheless, there has been some progress on so-called *gauge string duality*. For more on these efforts, the reader may consult the literature on the AGT conjecture and the AdS/CFT correspondence, or see the recent works of Chatterjee and Jafarov on lattice string trajectories and Yang–Mills theory [71,72,74,149]. Good entry points to the subject for probabilists include Chatterjee's recent survey [73] and Thierry Lévy's recent books about continuum Yang–Mills theory in two dimensions [164–166]. The 1986 article by Bridges, Giffen, Durhuus, and Fröhlich may also be read as a first step toward realizing Polyakov's vision [64].

ACKNOWLEDGMENTS

Random surfaces and curves have featured prominently in most of my own work over the past decade, and I thank the many amazing coauthors and students who have collaborated with me on this subject, including Tom Alberts, Morris Ang, Nathanaël Berestycki, Manan Bhatia, Bertrand Duplantier, Ewain Gwynne, Nina Holden, Richard Kenyon, Sungwook Kim, Greg Lawler, Asad Lodhia, Oren Louidor, Jason Miller, Andrei Okounkov, Minjae Park, Yuval Peres, Joshua Pfeffer, Rémi Rhodes, Oded Schramm, Nike Sun, Xin Sun, Vincent Vargas, Sam Watson, Menglu Wang, Wendelin Werner, David Wilson, Catherine

Wolfram, Hao Wu and Pu Yu. We also thank Tom Alberts, Olivier Bernardi, Bertrand Duplantier, Morris Ang, Sky Cao, Grégory Miermont, Ewain Gwynne, Jason Miller, Minjae Park, Guillaume Remy, Rémi Rhodes, Steffen Rohde, Stanislav Smirnov, Yilin Wang, and Wendelin Werner for their help in reading and improving the manuscript. And finally we thank Thomas Budzinski for allowing us to use his wonderful illustration.

FUNDING

This work was partially supported by NSF Award: DMS 1712862.

REFERENCES

[1] G. 't Hooft, A planar diagram theory for strong interactions. *Nuclear Phys. B* **72** (1974), 461–473.

[2] C. Abraham, Rescaled bipartite planar maps converge to the Brownian map. *Ann. Inst. Henri Poincaré Probab. Stat.* **52** (2016), no. 2, 575–595.

[3] C. Abraham, J. Bettinelli, G. Collet, and I. Kortchemski, Random maps. *ESAIM Proc. Surv.* **51** (2015), 133–149.

[4] L. Addario Berry and M. Albenque, The scaling limit of random simple triangulations and random simple quadrangulations. *Ann. Probab.* **45** (2017), no. 5, 2767–2825.

[5] L. Addario Berry and M. Albenque, Convergence of odd-angulations via symmetrization of labeled trees. 2019, arXiv:1904.04786.

[6] L. Addario Berry and M. Albenque, Convergence of non-bipartite maps via symmetrization of labeled trees. *Ann. Henri Lebesgue* **4** (2021), 653–683.

[7] M. Albenque, N. Holden, and X. Sun, Scaling limit of triangulations of polygons. *Electron. J. Probab.* **25** (2020), 1–43.

[8] M. Albenque and D. Poulalhon, A generic method for bijections between blossoming trees and planar maps. *Electron. J. Combin.* **22** (2015), no. 2, P2–38.

[9] D. Aldous, The continuum random tree. I. *Ann. Probab.* (1991), 1–28.

[10] D. Aldous, The continuum random tree. II. An overview. *Stoch. Anal.* **167** (1991), 23–70.

[11] D. Aldous, The continuum random tree III. *Ann. Probab.* (1993), 248–289.

[12] J. Ambjørn and T. G. Budd, Trees and spatial topology change in causal dynamical triangulations. *J. Phys. A* **46** (2013), no. 31, 315201, 33.

[13] M. Ang, H. Falconet, and X. Sun, Volume of metric balls in Liouville quantum gravity. *Electron. J. Probab.* **25** (2020), 1–50.

[14] M. Ang and E. Gwynne, Liouville quantum gravity surfaces with boundary as matings of trees. *Ann. Inst. Henri Poincaré Probab. Stat.* **57** (2021), no. 1, 1–53.

[15] M. Ang, N. Holden, and X. Sun, Conformal welding of quantum disks. 2020, arXiv:2009.08389.

[16] M. Ang, N. Holden, and X. Sun, Integrability of SLE via conformal welding of random surfaces. 2021, arXiv:2104.09477.

[17] M. Ang, M. Park, J. Pfeffer, and S. Sheffield, Brownian loops and the central charge of a Liouville random surface. *Ann. Probab.* (2020, to appear).

[18] M. Ang, G. Remy, and X. Sun, FZZ formula of boundary Liouville CFT via conformal welding. 2021, arXiv:2104.09478.

[19] M. Ang and X. Sun, Integrability of the conformal loop ensemble. 2021, arXiv:2107.01788.

[20] O. Angel, B. Kolesnik, and G. Miermont, Stability of geodesics in the Brownian map. *Ann. Probab.* **45** (2017), no. 5, 3451–3479.

[21] O. Angel and O. Schramm, Uniform infinite planar triangulations. *Comm. Math. Phys.* **241** (2003), no. 2, 191–213.

[22] J. Aru, N. Holden, E. Powell, and X. Sun, Mating of trees for critical Liouville quantum gravity. 2021, arXiv:2109.00275.

[23] J. Aru, Y. Huang, and X. Sun, Two perspectives of the 2D unit area quantum sphere and their equivalence. *Comm. Math. Phys.* **356** (2017), no. 1, 261–283.

[24] J. Aru, E. Powell, and A. Sepúlveda, Critical Liouville measure as a limit of subcritical measures. *Electron. Commun. Probab.* **24** (2019), 1–16.

[25] M. Aspenberg and M. Yampolsky, Mating non-renormalizable quadratic polynomials. *Comm. Math. Phys.* **287** (2009), no. 1, 1–40.

[26] K. Astala, A. Kupiainen, E. Saksman, and P. Jones, Random conformal weldings. *Acta Math.* **207** (2011), no. 2, 203–254.

[27] J. C. Baez, The Brownian Map. *Not. Amer. Math. Soc.* **68** (2021), 801–803.

[28] R. Basu, M. Bhatia, and S. Ganguly, Environment seen from infinite geodesics in Liouville Quantum Gravity. 2021, arXiv:2107.12363.

[29] V. Beffara, The dimension of the SLE curves. *Ann. Probab.* **36** (2008), no. 4, 1421–1452.

[30] A. A. Belavin, A. M. Polyakov, and A. B. Zamolodchikov, Infinite conformal symmetry in two-dimensional quantum field theory. *Nuclear Phys. B* **241** (1984), no. 2, 333–380.

[31] J. Beltran and J.-F. Le Gall, Quadrangulations with no pendant vertices. *Bernoulli* **19** (2013), no. 4, 1150–1175.

[32] I. Benjamini and O. Schramm, KPZ in one dimensional random geometry of multiplicative cascades. *Comm. Math. Phys.* **289** (2009), no. 2, 653–662.

[33] N. Berestycki, Introduction to the Gaussian free field and Liouville quantum gravity. Lecture notes, 2015.

[34] N. Berestycki, Diffusion in planar Liouville quantum gravity. *Ann. Inst. Henri Poincaré Probab. Stat.* **51** (2015), no. 3, 947–964.

[35] N. Berestycki, B. Laslier, and G. Ray, Critical exponents on Fortuin–Kasteleyn weighted planar maps. *Comm. Math. Phys.* **355** (2017), no. 2, 427–462.

[36] N. Berestycki and J. Norris, Lectures on Schramm–Loewner Evolution. Lecture notes, available on the webpages of the authors, 2014.

[37] N. Berestycki and E. Powell, Gaussian free field, Liouville quantum gravity and Gaussian multiplicative chaos. *Lecture notes* (2021).

[38] N. Berestycki, S. Sheffield, and X. Sun, Equivalence of Liouville measure and Gaussian free field. 2014, arXiv:1410.5407.

[39] O. Bernardi, Bijective counting of Kreweras walks and loopless triangulations. *J. Combin. Theory Ser. A* **114** (2007), no. 5, 931–956.

[40] O. Bernardi, Bijective counting of tree-rooted maps and shuffles of parenthesis systems. *Electron. J. Combin.* **14** (2007), no. 1, Research Paper 9, 36 pp.

[41] O. Bernardi, Tutte polynomial, subgraphs, orientations and sandpile model: new connections via embeddings. *Electron. J. Combin.* **15** (2008), no. 1, Research Paper 109, 53 pp.

[42] O. Bernardi and N. Bonichon, Intervals in Catalan lattices and realizers of triangulations. *J. Combin. Theory Ser. A* **116** (2009), no. 1, 55–75.

[43] O. Bernardi and M. Bousquet-Mélou, Counting colored planar maps: algebraicity results. *J. Combin. Theory Ser. B* **101** (2011), no. 5, 315–377.

[44] O. Bernardi and E. Fusy, Unified bijections for maps with prescribed degrees and girth. *J. Combin. Theory Ser. A* **119** (2012), no. 6, 1351–1387.

[45] O. Bernardi, N. Holden, and X. Sun, Percolation on triangulations: a bijective path to Liouville quantum gravity. 2018, arXiv:1807.01684.

[46] J. Bertoin, T. Budd, N. Curien, and I. Kortchemski, Martingales in self-similar growth-fragmentations and their connections with random planar maps. *Probab. Theory Related Fields* **172** (2018), no. 3, 663–724.

[47] J. Bertoin, N. Curien, and I. Kortchemski, Random planar maps and growth-fragmentations. *Ann. Probab.* **46** (2018), no. 1, 207–260.

[48] J. Bettinelli, Geodesics in Brownian surfaces (Brownian maps). *Ann. Inst. Henri Poincaré Probab. Stat.* **52** (2016), no. 2, 612–646.

[49] J. Bettinelli, E. Jacob, and G. Miermont, The scaling limit of uniform random plane maps, via the Ambjørn–Budd bijection. *Electron. J. Probab.* **19** (2014), 1–16.

[50] J. Bettinelli and G. Miermont, Compact Brownian surfaces I: Brownian disks. *Probab. Theory Related Fields* **167** (2017), no. 3–4, 555–614.

[51] P. Biane, Mating of discrete trees and walks in the quarter-plane. *Electron. J. Combin.* **28** (2021), no. 3, Paper No. 3.56, 29 pp.

[52] G. Borot, J. Bouttier, and E. Guitter, A recursive approach to the $O(n)$ model on random maps via nested loops. *J. Phys. A* **45** (2011), no. 4, 045002.

[53] M. Bousquet-Mélou and G. Schaeffer, The degree distribution in bipartite planar maps: applications to the Ising model. In *Formal power series and algebraic combinatorics*, pp. 312–323, 2003.

[54] J. Bouttier, P. Di Francesco, and E. Guitter, Geodesic distance in planar graphs. *Nuclear Phys. B* **663** (2003), no. 3, 535–567.

[55] J. Bouttier, P. Di Francesco, and E. Guitter, Planar maps as labeled mobiles. *Electron. J. Combin.* **11** (2004), no. 1, R69.

[56] J. Bouttier, P. Di Francesco, and E. Guitter, Combinatorics of bicubic maps with hard particles. *J. Phys. A: Math. Gen.* **38** (2005), no. 21, 4529.

[57] J. Bouttier and E. Guitter, The three-point function of planar quadrangulations. *J. Stat. Mech. Theory Exp.* **2008** (2008), no. 07, P07020.

[58] J. Bouttier and E. Guitter, Confluence of geodesic paths and separating loops in large planar quadrangulations. *J. Stat. Mech. Theory Exp.* **2009** (2009), no. 03, P03001.

[59] E. Brézin, C. Itzykson, G. Parisi, and J. B. Zuber, Planar diagrams. *Comm. Math. Phys.* **59** (1978), no. 1, 35–51.

[60] L. Brink, P. Di Vecchia, and P. Howe, A locally supersymmetric and reparametrization invariant action for the spinning string. *Phys. Lett. B* **65** (1976), no. 5, 471–474.

[61] S. R. Broadbent and J. M. Hammersley, Percolation processes. I. Crystals and mazes. *Proc. Camb. Philos. Soc.* **53** (1957), 629–641.

[62] J. Broome, The two-envelope paradox. *Analysis* **55** (1995), no. 1, 6–11.

[63] S. G. Brush, History of the Lenz–Ising model. *Rev. Modern Phys.* **39** (1967), no. 4, 883.

[64] D. Brydges, C. Giffen, B. Durhuus, and J. Fröhlich, Surface representations of Wilson loop expectations in lattice gauge theory. *Nuclear Phys. B* **275** (1986), no. 3, 459–487.

[65] T. Budzinski and B. Louf, Local limits of uniform triangulations in high genus. *Invent. Math.* **223** (2021), no. 1, 1–47.

[66] F. Camia and C. M. Newman, Two-dimensional critical percolation: the full scaling limit. *Comm. Math. Phys.* **268** (2006), no. 1, 1–38.

[67] F. Camia and C. M. Newman, Critical percolation exploration path and SLE 6: a proof of convergence. *Probab. Theory Related Fields* **139** (2007), no. 3, 473–519.

[68] A. Caraceni and A. Stauffer, Polynomial mixing time of edge flips on quadrangulations. *Probab. Theory Related Fields* **176** (2020), no. 1, 35–76.

[69] A. Carrance, Convergence of Eulerian triangulations. *Electron. J. Probab.* **26** (2021), 1–48.

[70] B. Cerclé, Unit boundary length quantum disk: a study of two different perspectives and their equivalence. *ESAIM Probab. Stat.* **25** (2021), 433–459.

[71] S. Chatterjee, The leading term of the Yang–Mills free energy. *J. Funct. Anal.* **271** (2016), no. 10, 2944–3005.

[72] S. Chatterjee, Rigorous solution of strongly coupled $SO(N)$ lattice gauge theory in the large N limit. *Comm. Math. Phys.* **366** (2019), no. 1, 203–268.

[73] S. Chatterjee, Yang–Mills for probabilists. In *Probability and analysis in interacting physical systems*, pp. 1–16, Springer Proc. Math. Stat. 283, Springer, 2019.

[74] S. Chatterjee and J. Jafarov, The $1/N$ expansion for $SO(N)$ lattice gauge theory at strong coupling. 2016, arXiv:1604.04777.

[75] D. Chelkak, H. Duminil Copin, C. Hongler, A. Kemppainen, and S. Smirnov, Convergence of Ising interfaces to Schramm's SLE curves. *C. R. Math.* **352** (2014), no. 2, 157–161.

[76] D. Chelkak and S. Smirnov, Universality in the 2D Ising model and conformal invariance of fermionic observables. *Invent. Math.* **189** (2012), no. 3, 515–580.

[77] L. Chen, Basic properties of the infinite critical-FK random map. *Ann. Inst. Henri Poincaré D* **4** (2017), no. 3, 245–271.

[78] L. Chen, N. Curien, and P. Maillard, The perimeter cascade in critical Boltzmann quadrangulations decorated by an $O(n)$ loop model. *Ann. Inst. Henri Poincaré D* **7** (2020), no. 4, 535–584.

[79] S. Cordes, G. Moore, and S. Ramgoolam, Lectures on 2D Yang–Mills theory, equivariant cohomology and topological field theories. *Nuclear Phys. B Proc. Suppl.* **41** (1995), no. 1–3, 184–244.

[80] R. Cori and B. Vauquelin, Planar maps are well labeled trees. *Canad. J. Math.* **33** (1981), no. 5, 1023–1042.

[81] H. Coxeter, The four-color map problem, 1840-1890. *Math. Teach.* **52** (1959), no. 4, 283–289.

[82] H. H. Crapo, The Tutte polynomial. *Aequationes Math.* **3** (1969), 211–229.

[83] N. Curien, Peeling random planar maps. *Saint-Flour lecture notes*, 2019.

[84] N. Curien and J.-F. Le Gall, The Brownian plane. *J. Theor. Probab.* **27** (2014), no. 4, 1249–1291.

[85] N. Curien and J.-F. Le Gall, The hull process of the Brownian plane. *Probab. Theory Related Fields* **166** (2016), no. 1, 187–231.

[86] N. Curien, J.-F. Le Gall, and G. Miermont, The Brownian cactus I. Scaling limits of discrete cactuses. *Ann. Inst. Henri Poincaré Probab. Stat.* **49** (2013), no. 2, 340–373.

[87] F. David, A. Kupiainen, R. Rhodes, and V. Vargas, Liouville quantum gravity on the Riemann sphere. *Comm. Math. Phys.* **342** (2016), no. 3, 869–907.

[88] F. David, R. Rhodes, and V. Vargas, Liouville quantum gravity on complex tori. *J. Math. Phys.* **57** (2016), no. 2, 022302.

[89] S. Deser and B. Zumino, A complete action for the spinning string. *Phys. Lett. B* **65** (1976), no. 4, 369–373.

[90] J. Ding, J. Dubédat, A. Dunlap, and H. Falconet, Tightness of Liouville first passage percolation for $\gamma \in (0, 2)$. *Publ. Math. Inst. Hautes Études Sci.* **132** (2020), no. 1, 353–403.

[91] J. Ding, J. Dubédat, and E. Gwynne, Introduction to the Liouville quantum gravity metric. 2021, arXiv:2109.01252.

[92] J. Ding and E. Gwynne, Uniqueness of the critical and supercritical Liouville quantum gravity metrics. 2021, arXiv:2110.00177.

[93] J. Ding, E. Gwynne, and A. Sepúlveda, The distance exponent for Liouville first passage percolation is positive. *Probab. Theory Related Fields* **181** (2021), no. 4, 1035–1051.

[94] P. Di Francesco, P. Ginsparg, and J. Zinn Justin, 2D gravity and random matrices. *Phys. Rep.* **254** (1995), no. 1–2, 1–133.

[95] P. Di Francesco, P. Mathieu, and D. Sénéchal, *Conformal field theory*. Springer, 2012.

[96] H. Dorn and H.-J. Otto, Two-and three-point functions in Liouville theory. *Nuclear Phys. B* **429** (1994), no. 2, 375–388.

[97] J. Douglas, Solution of the problem of Plateau. *Trans. Amer. Math. Soc.* **33** (1931), no. 1, 263–321.

[98] J. Dubédat, SLE and the free field: partition functions and couplings. *J. Amer. Math. Soc.* **22** (2009), no. 4, 995–1054.

[99] J. Dubédat and H. Falconet, Metric growth dynamics in Liouville quantum gravity. 2021, arXiv:2112.13933.

[100] J. Dubédat, H. Falconet, E. Gwynne, J. Pfeffer, and X. Sun, Weak LQG metrics and Liouville first passage percolation. *Probab. Theory Related Fields* **178** (2020), no. 1, 369–436.

[101] B. Duplantier, Liouville Quantum Gravity, KPZ and Schramm–Loewner Evolution. In *Proceedings of the ICM*, 2014.

[102] B. Duplantier, J. Miller, and S. Sheffield, Liouville quantum gravity as a mating of trees. *Astérisque* (2021), no. 427, viii+257 pp.

[103] B. Duplantier, R. Rhodes, S. Sheffield, and V. Vargas, Critical Gaussian multiplicative chaos: convergence of the derivative martingale. *Ann. Probab.* **42** (2014), no. 5, 1769–1808.

[104] B. Duplantier, R. Rhodes, S. Sheffield, and V. Vargas, Renormalization of critical Gaussian multiplicative chaos and KPZ relation. *Comm. Math. Phys.* **330** (2014), no. 1, 283–330.

[105] B. Duplantier and S. Sheffield, Liouville quantum gravity and KPZ. *Invent. Math.* **185** (2011), no. 2, 333–393.

[106] E. Dynkin and S. Kuznetsov, Markov snakes and superprocesses. *Probab. Theory Related Fields* **103** (1995), no. 4, 433–473.

[107] B. Eynard, Counting surfaces. *Prog. Math. Phys.* **70** (2016).

[108] G. E. Farr, Tutte–Whitney polynomials: some history and generalizations. In *Combinatorics, complexity, and chance*, pp. 28–52, Oxford Lecture Ser. Math. Appl. 34, Oxford University Press, 2007.

[109] C. M. Fortuin, On the random-cluster model II. The percolation model. *Physica* **58** (1972), no. 3, 393–418.

[110] C. M. Fortuin, On the random-cluster model: III. The simple random-cluster model. *Physica* **59** (1972), no. 4, 545–570.

[111] C. M. Fortuin and P. W. Kasteleyn, On the random-cluster model: I. Introduction and relation to other models. *Physica* **57** (1972), no. 4, 536–564.

[112] R. Fritsch and G. Fritsch, *The four-color theorem*. Springer, New York, 1998.

[113] C. Garban, Quantum gravity and the KPZ formula [after Duplantier–Sheffield]. *Astérisque* **352** (2013), no. 1052, 315–354.

[114] C. Garban, R. Rhodes, and V. Vargas, Liouville Brownian motion. *Ann. Probab.* **44** (2016), no. 4, 3076–3110.

[115] K. F. Gauss and P. Pesic, *General investigations of curved surfaces*. Courier Corporation, 2005.

[116] I. M. Gessel and D. Zeilberger, Random walk in a Weyl chamber. *Proc. Amer. Math. Soc.* **115** (1992), no. 1, 27–31.

[117] P. Ghosal, G. Remy, X. Sun, and Y. Sun, Probabilistic conformal blocks for Liouville CFT on the torus. 2020, arXiv:2003.03802.

[118] P. Ginsparg and G. Moore, Lectures on 2D gravity and 2D string theory (TASI 1992). 1992, arXiv:hep-th/9304011.

[119] J. Gray, On the history of the Riemann mapping theorem. *Supp. Rend.* **19** (1994).

[120] C. Guillarmou, A. Kupiainen, R. Rhodes, and V. Vargas, Conformal bootstrap in Liouville Theory. 2020, arXiv:2005.11530.

[121] C. Guillarmou, A. Kupiainen, R. Rhodes, and V. Vargas, Segal's axioms and bootstrap for Liouville Theory. 2021, arXiv:2112.14859.

[122] A. Guionnet and E. Maurel-Segala, Combinatorial aspects of matrix models. 2005, arXiv:math/0503064.

[123] R. K. Guy, C. Krattenthaler, and B. E. Sagan, Lattice paths, reflections, & dimension-changing bijections. *Ars Combin.* **34** (1992), 3–15.

[124] E. Gwynne, The dimension of the boundary of a Liouville quantum gravity metric ball. *Comm. Math. Phys.* **378** (2020), no. 1, 625–689.

[125] E. Gwynne, Random surfaces and Liouville quantum gravity. *Not. Amer. Math. Soc.* **64** (2020), no. 4, 484–491.

[126] E. Gwynne, N. Holden, J. Miller, and X. Sun, Brownian motion correlation in the peanosphere for $\kappa > 8$. *Ann. Inst. Henri Poincaré Probab. Stat.* **53** (2017), no. 4, 1866–1889.

[127] E. Gwynne, N. Holden, J. Pfeffer, and G. Remy, Liouville quantum gravity with matter central charge in $(1, 25)$: a probabilistic approach. *Comm. Math. Phys.* (2020), 1–53.

[128] E. Gwynne, N. Holden, and X. Sun, Mating of trees for random planar maps and Liouville quantum gravity: a survey. 2019, arXiv:1910.04713.

[129] E. Gwynne, N. Holden, and X. Sun, Joint scaling limit of site percolation on random triangulations in the metric and peanosphere sense. *Electron. J. Probab.* **26** (2021), 1–58.

[130] E. Gwynne, A. Kassel, J. Miller, and D. B. Wilson, Active Spanning Trees with Bending Energy on Planar Maps and SLE-Decorated Liouville Quantum Gravity for $\kappa > 8$. *Comm. Math. Phys.* **358** (2018), no. 3, 1065–1115.

[131] E. Gwynne, C. Mao, and X. Sun, Scaling limits for the critical Fortuin–Kasteleyn model on a random planar map I: Cone times. *Ann. Inst. Henri Poincaré Probab. Stat.* **55** (2019), no. 1, 1–60.

[132] E. Gwynne and J. Miller, Conformal covariance of the Liouville quantum gravity metric for $\gamma \in (0, 2)$. *Ann. Inst. Henri Poincaré Probab. Stat.* **57** (2021), no. 2, 1016–1031.

[133] E. Gwynne and J. Miller, Existence and uniqueness of the Liouville quantum gravity metric for $\gamma \in (0, 2)$. *Invent. Math.* **223** (2021), no. 1, 213–333.

[134] E. Gwynne and J. Miller, Characterizations of SLE_κ for $\kappa \in (4, 8)$ on Liouville quantum gravity. 2021, arXiv:1701.05174.

[135] E. Gwynne and J. Miller, Convergence of percolation on uniform quadrangulations with boundary to SLE_6 on $\sqrt{8/3}$-Liouville quantum gravity. 2021, arXiv:1701.05175.

[136] E. Gwynne, J. Miller, and S. Sheffield, Harmonic functions on mated-CRT maps. *Electron. J. Probab.* **24** (2019), Paper No. 58, 55 pp.

[137] E. Gwynne, J. Miller, and S. Sheffield, The Tutte embedding of the Poisson-Voronoi tessellation of the Brownian disk converges to $\sqrt{8}/3$-Liouville quantum gravity. *Comm. Math. Phys.* **374** (2020), no. 2, 735–784.

[138] E. Gwynne, J. Miller, and S. Sheffield, An invariance principle for ergodic scale-free random environments. *Acta Math.* (2021, to appear).

[139] E. Gwynne, J. Miller, and S. Sheffield, The Tutte embedding of the mated-CRT map converges to Liouville quantum gravity. *Ann. Probab.* **49** (2021), no. 4, 1677–1717.

[140] E. Gwynne and J. Pfeffer, KPZ formulas for the Liouville quantum gravity metric. 2019, arXiv:1905.11790.

[141] E. Gwynne and X. Sun, Scaling limits for the critical Fortuin–Kastelyn model on a random planar map III: finite volume case. 2015, arXiv:1510.06346.

[142] E. Gwynne and X. Sun, Scaling limits for the critical Fortuin–Kasteleyn model on a random planar map II: local estimates and empty reduced word exponent. *Electron. J. Probab.* **22** (2017), 1–56.

[143] K. Hartnett, A unified theory of randomness, 2016, https://www.quantamagazine.org/20160802-unified_theory_of_randomness.

[144] K. Hartnett, Random surfaces hide an intricate order, 2019, https://www.quantamagazine.org/random-surfaces-hide-an-intricate-order-20190702.

[145] K. Hartnett, Mathematicians prove 2D version of quantum gravity really works, 2021, https://www.quantamagazine.org/mathematicians-prove-2d-version-of-quantum-gravity-really-works-20210617.

[146] P. J. Heawood, Map color theorems. *Q. J. Math.* **24** (1890), 332–338.

[147] N. Holden, X. Li, and X. Sun, Natural parametrization of percolation interface and pivotal points. *Ann. Inst. Henri Poincaré Probab. Stat.* **58** (2022), no. 1, 7–25.

[148] N. Holden and X. Sun, Convergence of uniform triangulations under the Cardy embedding. 2019, arXiv:1905.13207.

[149] J. Jafarov, Wilson loop expectations in $SU(N)$ lattice gauge theory. 2016, arXiv:1610.03821.

[150] S. Janson, *Gaussian Hilbert spaces. 129*. Cambridge University Press, Cambridge, 1997.

[151] M. Kac, Can one hear the shape of a drum? *Amer. Math. Monthly* **73** (1966), 1–23.

[152] W. Kager and B. Nienhuis, A guide to stochastic Löwner evolution and its applications. *J. Stat. Phys.* **115** (2004), no. 5, 1149–1229.

[153] J.-P. Kahane, Sur le chaos multiplicatif. *Ann. Sci. Math. Québec* **9** (1985), no. 2, 105–150.

[154] R. Kenyon, J. Miller, S. Sheffield, and D. B. Wilson, Six-vertex model and Schramm–Loewner evolution. *Phys. Rev. E* **95** (2017), no. 5, 052146.

[155] R. Kenyon, J. Miller, S. Sheffield, and D. B. Wilson, Bipolar orientations on planar maps and $\mathrm{SLE}_1 2$. *Ann. Probab.* **47** (2019), no. 3, 1240–1269.

[156] V. G. Knizhnik, A. M. Polyakov, and A. B. Zamolodchikov, Fractal structure of 2d-quantum gravity. *Modern Phys. Lett. A* **3** (1988), no. 08, 819–826.

[157] M. A. Krikun, Uniform infinite planar triangulation and related time-reversed critical branching process. *J. Math. Sci.* **131** (2005), no. 2, 5520–5537.

[158] R. H. Krohn, A general class of quantum fields without cut-offs in two space-time dimensions. *Comm. Math. Phys.* **21** (1971), no. 3, 244–255.

[159] A. Kupiainen, Constructive Liouville conformal field theory. 2016, arXiv:1611.05243.

[160] A. Kupiainen, R. Rhodes, and V. Vargas, Integrability of Liouville theory: proof of the DOZZ Formula. *Ann. of Math.* **191** (2020), no. 1, 81–166.

[161] G. F. Lawler, An introduction to the stochastic Loewner evolution. In *Random walks and geometry*, pp. 261–293, De Gruyter, 2004.

[162] G. F. Lawler, Conformally invariant loop measures. In *Proceedings of the International Congress of Mathematicians—Rio de Janeiro 2018. Vol. I. Plenary lectures*, pp. 669–703, 2018.

[163] G. F. Lawler, O. Schramm, and W. Werner, Conformal invariance of planar loop-erased random walks and uniform spanning trees. In *Selected works of Oded Schramm. Volume 1, 2*, pp. 931–987, Sel. Works Probab. Stat., Springer, 2011.

[164] T. Lévy, Yang–Mills measure on compact surfaces. *Mem. Amer. Math. Soc.* **166** (2003), no. 790, xiv+122 pp.

[165] T. Lévy, The master field on the plane. *Astérisque* **388** (2017).

[166] T. Lévy, Two-dimensional quantum Yang–Mills theory and the Makeenko–Migdal equations. In *Frontiers in analysis and probability—in the spirit of the Strasbourg–Zürich meetings*, pp. 275–325, 2020.

[167] J.-F. Le Gall, The Brownian snake and solutions of $\Delta u = u^2$ in a domain. *Probab. Theory Related Fields* **102** (1995), no. 3, 393–432.

[168] J.-F. Le Gall, The topological structure of scaling limits of large planar maps. *Invent. Math.* **169** (2007), no. 3, 621–670.

[169] J.-F. Le Gall, Geodesics in large planar maps and in the Brownian map. *Acta Math.* **205** (2010), no. 2, 287–360.

[170] J.-F. Le Gall, Uniqueness and universality of the Brownian map. *Ann. Probab.* **41** (2013), no. 4, 2880–2960.

[171] J.-F. Le Gall, Random geometry on the sphere. In *Proceedings of the ICM*, 2014.
[172] J.-F. Le Gall, The Brownian cactus II: Upcrossings and local times of super-Brownian motion. *Probab. Theory Related Fields* **162** (2015), no. 1, 199–231.
[173] J.-F. Le Gall, Subordination of trees and the Brownian map. *Probab. Theory Related Fields* **171** (2018), no. 3, 819–864.
[174] J.-F. Le Gall, Brownian disks and the Brownian snake. *Ann. Inst. Henri Poincaré Probab. Stat.* **55** (2019), no. 1, 237–313.
[175] J.-F. Le Gall, Brownian geometry. *Jpn. J. Math.* **14** (2019), no. 2, 135–174.
[176] J.-F. Le Gall and G. Miermont, Scaling limits of random trees and planar maps. In *Probability and statistical physics in two and more dimensions*, p. 155, 2010.
[177] J.-F. Le Gall and F. Paulin, Scaling limits of bipartite planar maps are homeomorphic to the 2-sphere. *Geom. Funct. Anal.* **18** (2008), no. 3, 893–918.
[178] Y. Li, X. Sun, and S. S. Watson, Schnyder woods, SLE(16), and Liouville quantum gravity. 2017, arXiv:1705.03573.
[179] J. Liouville, Note sur la Théorie de la Variation des constantes arbitraires. *J. Math. Pures Appl.* (1838), 342–349.
[180] J.-F. Marckert and A. Mokkadem, Limit of normalized quadrangulations: The Brownian map. *Ann. Probab.* **34** (2006), no. 6, 2144–2202.
[181] C. Marzouk, Scaling limits of random bipartite planar maps with a prescribed degree sequence. *Random Structures Algorithms* **53** (2018), no. 3, 448–503.
[182] H. P. Jr. McKean and I. M. Singer, Curvature and the eigenvalues of the Laplacian. *J. Differential Geom.* **1** (1967), 43–69.
[183] G. Miermont, On the sphericity of scaling limits of random planarquadrangulations. *Electron. Commun. Probab.* **13** (2008), 248–257.
[184] G. Miermont, Random maps and their scaling limits. In *Fractal geometry and stochastics IV*, pp. 197–224, Progr. Probab. 61, Springer, 2009.
[185] G. Miermont, The Brownian map is the scaling limit of uniform random plane quadrangulations. *Acta Math.* **210** (2013), no. 2, 319–401.
[186] G. Miermont, Aspects of random maps. *Saint-Flour lecture notes*, 2014.
[187] G. Miermont, Compact Brownian surfaces. In *Stochastic Analysis, Random Fields and Integrable Probability—Fukuoka 2019, vol. 87*, pp. 173–199, Mathematical Society of Japan, 2021.
[188] J. Miller, Liouville quantum gravity as a metric space and a scaling limit. In *Proceedings of the International Congress of Mathematicians—Rio de Janeiro 2018. Vol. IV. Invited lectures*, pp. 2945–2971, 2018.
[189] J. Miller and W. Qian, Geodesics in the Brownian map: Strong confluence and geometric structure. 2020, arXiv:2008.02242.
[190] J. Miller and S. Sheffield, Imaginary geometry I: Interacting SLEs. *Probab. Theory Related Fields* **164** (2016), no. 3–4, 553–705.
[191] J. Miller and S. Sheffield, Imaginary geometry II: Reversibility of $\mathrm{SLE}_\kappa(\rho_1;\rho_2)$ for $\kappa \in (0,4)$. *Ann. Probab.* **44** (2016), no. 3, 1647–1722.

[192] J. Miller and S. Sheffield, Imaginary geometry III: Reversibility of SLE_κ for $\kappa \in (4, 8)$. *Ann. of Math. (2)* **184** (2016), no. 2, 455–486.

[193] J. Miller and S. Sheffield, Imaginary geometry IV: Interior rays, whole-plane reversibility, and space-filling trees. *Probab. Theory Related Fields* **169** (2017), no. 3–4, 729–869.

[194] J. Miller and S. Sheffield, Liouville quantum gravity spheres as matings of finite-diameter trees. *Ann. Inst. Henri Poincaré Probab. Stat.* **55** (2019), no. 3, 1712–1750.

[195] J. Miller and S. Sheffield, Liouville quantum gravity and the Brownian map I: The $\mathrm{QLE}(8/3, 0)$ metric. *Invent. Math.* **219** (2020), no. 1, 75–152.

[196] J. Miller and S. Sheffield, An axiomatic characterization of the Brownian map. *J. Éc. Polytech. Math.* **8** (2021), 609–731.

[197] J. Miller and S. Sheffield, Liouville quantum gravity and the Brownian map II: Geodesics and continuity of the embedding. *Ann. Probab.* **49** (2021), no. 6, 2732–2829.

[198] J. Miller and S. Sheffield, Liouville quantum gravity and the Brownian map III: The conformal structure is determined. *Probab. Theory Related Fields* **179** (2021), no. 3–4, 1183–1211.

[199] J. Miller, S. Sheffield, and W. Werner, CLE percolations. *Forum Math. Pi* **5** (2017), e4, 102 pp.

[200] J. Miller, S. Sheffield, and W. Werner, Non-simple SLE curves are not determined by their range. *J. Eur. Math. Soc. (JEMS)* **22** (2020), no. 3, 669–716.

[201] J. Miller, S. Sheffield, and W. Werner, Simple conformal loop ensembles on Liouville quantum gravity. 2020, arXiv:2002.05698.

[202] J. Miller, S. Sheffield, and W. Werner, Non-simple conformal loop ensembles on Liouville quantum gravity and the law of CLE percolation interfaces. *Probab. Theory Related Fields* **181** (2021), no. 1–3, 669–710.

[203] J. Milnor, Pasting together Julia sets: a worked out example of mating. *Exp. Math.* **13** (2004), no. 1, 55–92.

[204] G. Moore and N. Seiberg, Classical and quantum conformal field theory. *Comm. Math. Phys.* **123** (1989), no. 2, 177–254.

[205] R. L. Moore, Concerning upper semi-continuous collections of continua. *Trans. Amer. Math. Soc.* **27** (1925), no. 4, 416–428.

[206] R. Mullin, Enumeration of rooted triangular maps. *Amer. Math. Monthly* **71** (1964), no. 9, 1007–1010.

[207] R. Mullin, The enumeration of Hamiltonian polygons in triangular maps. *Pacific J. Math.* **16** (1966), no. 1, 139–145.

[208] R. C. Mullin, On counting rooted triangular maps. *Canad. J. Math.* **17** (1965), 373–382.

[209] R. C. Mullin, On the enumeration of tree-rooted maps. *Canad. J. Math.* **19** (1967), 174–183.

[210] Y. Nakayama, Liouville field theory: A decade after the revolution. *Internat. J. Modern Phys. A* **19** (2004), no. 17–18, 2771–2930.
[211] M. Noy, C. Requilé, and J. Rué, On the expected number of perfect matchings in cubic planar graphs. *Publ. Mat.* **66** (2022), no. 1, 325–353.
[212] B. Osgood, R. Phillips, and P. Sarnak, Extremals of determinants of Laplacians. *J. Funct. Anal.* **80** (1988), no. 1, 148–211.
[213] J. Pfeffer, Weak Liouville quantum gravity metrics with matter central charge $c \in (-\infty, 25)$. 2021, arXiv:2104.04020.
[214] A. Polyakov, Confinement and liberation. In *50 Years of Yang–Mills Theory*, pp. 311–329, World Scientific, 2005.
[215] A. M. Polyakov, Quantum geometry of bosonic strings. *Phys. Lett. B* **103** (1981), no. 3, 207–210.
[216] A. M. Polyakov, From quarks to strings. 2008, arXiv:0812.0183.
[217] G. Remy, The Fyodorov–Bouchaud formula and Liouville conformal field theory. *Duke Math. J.* **169** (2020), no. 1, 177–211.
[218] G. Remy and T. Zhu, The distribution of Gaussian multiplicative chaos on the unit interval. *Ann. Probab.* **48** (2020), no. 2, 872–915.
[219] G. Remy and T. Zhu, Integrability of boundary Liouville conformal field theory. 2020, arXiv:2002.05625.
[220] R. Rhodes and V. Vargas, KPZ formula for log-infinitely divisible multifractal random measures. *ESAIM Probab. Stat.* **15** (2011), 358–371.
[221] S. Ribault, Conformal field theory on the plane. 2014, arXiv:1406.4290.
[222] A. Riera, *Brownian Geometry*. PhD thesis, Université Paris-Saclay, 2021.
[223] L. Rogers, The four colour theorem. *nrich, September*, 2008.
[224] S. Rohde and O. Schramm, Basic properties of SLE. *Ann. of Math.* (2005), 883–924.
[225] P. Sarnak, Determinants of Laplacians; heights and finiteness. In *Analysis, et cetera*, pp. 601–622, Elsevier, 1990.
[226] G. Schaeffer, *Conjugaison d'arbres et cartes combinatoires aléatoires*. PhD thesis, Bordeaux 1, 1998.
[227] M. Schottenloher, *A mathematical introduction to conformal field theory* 759, Springer, Berlin, 2008.
[228] O. Schramm, Scaling limits of loop-erased random walks and uniform spanning trees. *Israel J. Math.* **118** (2000), no. 1, 221–288.
[229] O. Schramm, Conformally invariant scaling limits: an overview and a collection of problems. In *Selected works of Oded Schramm*, pp. 1161–1191, Springer, 2011.
[230] O. Schramm and S. Sheffield, Contour lines of the two-dimensional discrete Gaussian free field. *Acta Math.* **202** (2009), no. 1, 21–137.
[231] O. Schramm and S. Sheffield, A contour line of the continuum Gaussian free field. *Probab. Theory Related Fields* **157** (2013), no. 1–2, 47–80.
[232] N. Seiberg, Notes on quantum Liouville theory and quantum gravity. *Progr. Theoret. Phys. Suppl.* **102** (1990), 319–349.

[233] S. Sheffield, Gaussian free fields for mathematicians. *Probab. Theory Related Fields* **139** (2007), no. 3–4, 521–541.

[234] S. Sheffield, Exploration trees and conformal loop ensembles. *Duke Math. J.* **147** (2009), no. 1, 79–129.

[235] S. Sheffield, Conformal weldings of random surfaces: SLE and the quantum gravity zipper. *Ann. Probab.* **44** (2016), no. 5, 3474–3545.

[236] S. Sheffield, Quantum gravity and inventory accumulation. *Ann. Probab.* **44** (2016), no. 6, 3804–3848.

[237] S. Sheffield and M. Wang, Field-measure correspondence in Liouville quantum gravity almost surely commutes with all conformal maps simultaneously. 2016, arXiv:1605.06171.

[238] S. Sheffield and W. Werner, Conformal loop ensembles: the Markovian characterization and the loop-soup construction. *Ann. of Math. (2)* **176** (2012), no. 3, 1827–1917.

[239] S. Smirnov, Critical percolation in the plane: conformal invariance, Cardy's formula, scaling limits. *C. R. Acad. Sci., Sér. 1 Math.* **333** (2001), no. 3, 239–244.

[240] S. Smirnov, Towards conformal invariance of 2D lattice models. In *International Congress of Mathematicians. Vol. II*, pp. 1421–1451, 2006.

[241] S. Smirnov, Conformal invariance in random cluster models. I. Holmorphic fermions in the Ising model. *Ann. of Math.* (2010), 1435–1467.

[242] S. Smirnov, Discrete complex analysis and probability. In *Proceedings of the International Congress of Mathematicians. Volume I*, pp. 595–621, 2010.

[243] N. Sun, Conformally invariant scaling limits in planar critical percolation. *Probab. Surv.* **8** (2011), 155–209.

[244] J. Teschner, On the Liouville three-point function. *Phys. Lett. B* **363** (1995), no. 1–2, 65–70.

[245] J. Teschner, Liouville theory revisited. *Classical Quantum Gravity* **18** (2001), no. 23, R153.

[246] J. Teschner, A lecture on the Liouville vertex operators. *Internat. J. Modern Phys. A* **19** (2004), 436–458.

[247] V. Timorin, Topological regluing of rational functions. *Invent. Math.* **179** (2010), no. 3, 461–506.

[248] W. T. Tutte, A contribution to the theory of chromatic polynomials. *Canad. J. Math.* **6** (1954), 80–91.

[249] W. T. Tutte, A census of Hamiltonian polygons. *Canad. J. Math.* **14** (1962), 402–417.

[250] W. T. Tutte, A census of planar triangulations. *Canad. J. Math.* **14** (1962), 21–38.

[251] W. T. Tutte, A census of slicings. *Canad. J. Math.* **14** (1962), 708–722.

[252] W. T. Tutte, A census of planar maps. *Canad. J. Math.* **15** (1963), 249–271.

[253] W. T. Tutte, *Graph theory as I have known it. Vol. 11*. Oxford University Press, Oxford, 2012.

[254] V. Vargas, Lecture notes on Liouville theory and the DOZZ formula. 2017, arXiv:1712.00829.

[255] J. Walsh, History of the Riemann mapping theorem. *Amer. Math. Monthly* **80** (1973), no. 3, 270–276.

[256] D. J. Welsh and C. Merino, The Potts model and the Tutte polynomial. *J. Math. Phys.* **41** (2000), no. 3, 1127–1152.

[257] W. Werner, Random planar curves and Schramm–Loewner evolutions. In *Lectures on probability theory and statistics*, pp. 107–195, Lecture Notes in Math. 1840, Springer, 2004.

[258] W. Werner, Conformal restriction properties. In *International Congress of Mathematicians. Vol. III*, pp. 741–762, 2006.

[259] W. Werner and E. Powell, Lecture notes on the Gaussian free field. *Cours Spéc.* **28** (2022).

[260] H. Weyl, Über die asymptotische Verteilung der Eigenwerte. *Nachr. Ges. Wiss. Gött., Math.-Phys. Kl.* **1911** (1911), 110–117.

[261] M. Yampolsky and S. Zakeri, Mating Siegel quadratic polynomials. *J. Amer. Math. Soc.* **14** (2001), no. 1, 25–78.

[262] A. Zamolodchikov and A. Zamolodchikov, Conformal bootstrap in Liouville field theory. *Nuclear Phys. B* **477** (1996), no. 2, 577–605.

SCOTT SHEFFIELD

77 Mass. Ave., Cambridge, MA, USA, sheffield@math.mit.edu

This Page Intentionally Left Blank

THE DISTRIBUTION OF VALUES OF ZETA AND L-FUNCTIONS

KANNAN SOUNDARARAJAN

ABSTRACT

We survey recent progress on understanding the distribution of values of zeta and L-functions. In particular, we discuss the problem of moments of $|\zeta(\frac{1}{2}+it)|$ and moments of central L-values in families, where the last 25 years have seen a conjectural understanding of the asymptotics of these moments, together with progress in obtaining good upper and lower bounds in many situations.

MATHEMATICS SUBJECT CLASSIFICATION 2020

Primary 11M06; Secondary 11M26, 11M50, 60F05, 60F10

KEYWORDS

Zeta and L-functions, value distribution, moments, extreme values

ICM INTERNATIONAL CONGRESS OF MATHEMATICIANS 2022 JULY 6–14

Proc. Int. Cong. Math. 2022, Vol. 2, pp. 1260–1310

DOI 10.4171/ICM2022/148

This article concerns the distribution of values of the Riemann zeta-function and related L-functions. We begin with a brief discussion of L-values at the edge of the critical strip, which give information on arithmetic invariants such as class numbers. The remainder of the article is concerned with the value distribution of $\zeta(\frac{1}{2}+it)$ and the distribution of central values in families of L-functions. The typical behavior of $\zeta(\frac{1}{2}+it)$ is described by a fundamental theorem of Selberg (discussed in Section 2) which asserts that $\log\zeta(\frac{1}{2}+it)$ is distributed like a complex Gaussian with prescribed mean and variance. Analogues of Selberg's theorem for central values in families of L-functions were conjectured by Keating and Snaith, and we motivate these conjectures and the progress towards them in Section 3. Section 4 begins our treatment of the problem of understanding the moments of $|\zeta(\frac{1}{2}+it)|$ and analogous questions for central L-values. While this is a classical topic, going back to work of Hardy and Littlewood, it is only in the last 25 years that even a good conjectural understanding of the problem has emerged. The Keating–Snaith conjectures for the asymptotics of moments were first developed by pursuing an analogy between values of the zeta function and the values of the characteristic polynomial of large random matrices. These conjectures are described in Section 5, which also shows how the problem of understanding moments is tied up with understanding the *large deviations* range in Selberg's theorem. Progress towards the moment conjectures (see Section 6) has been of three types: (i) understanding asymptotics for small moments in a number of examples, (ii) obtaining lower bounds of the correct order of magnitude (which are known in many cases), and (iii) obtaining in great generality upper bounds of the correct order of magnitude assuming the Generalized Riemann Hypothesis. In Section 7 we discuss what is known about the maximal size of $|\zeta(\frac{1}{2}+it)|$ and central L-values, and speculate on what the truth might be. Finally, in Section 8 we consider briefly an intriguing problem of Fyodorov–Hiary–Keating on understanding the "local maximum" of $|\zeta(\frac{1}{2}+it)|$ for t in intervals of length 1, which is closely connected to problems in *branching Brownian motion* and *Gaussian multiplicative chaos*.

1. VALUES AT THE EDGE OF THE CRITICAL STRIP

It was already observed by Gauss and Dirichlet that certain special values of L-functions encode interesting arithmetic information. Recall that a *discriminant* is an integer $d \equiv 0, 1 \pmod 4$, and d is called a *fundamental discriminant* if d/m^2 is not a discriminant for any divisor m^2 of d larger than 1. Fundamental discriminants are in one-to-one correspondence with discriminants of quadratic fields $\mathbb{Q}(\sqrt{d})$. Associated to a fundamental discriminant d is the Kronecker–Legendre symbol $\chi_d(n) = (\frac{d}{n})$, which is a primitive Dirichlet character (mod $|d|$). For example, if p is an odd prime then either p or $-p$ is a fundamental discriminant (depending on whether p is 1 or 3 (mod 4)), and in either case the associated quadratic character is the familiar Legendre symbol (mod p). Associated to the primitive character χ_d is the Dirichlet L-function

$$L(s,\chi_d) = \sum_{n=1}^{\infty} \frac{\chi_d(n)}{n^s} = \prod_p \left(1 - \frac{\chi_d(p)}{p^s}\right)^{-1}.$$

Although $d = 1$ is permitted in our definition of fundamental discriminants (and corresponds to the Riemann zeta-function), it is an anomalous case and we shall mainly be interested in fundamental discriminants $d \neq 1$. Like the Riemann zeta-function, the Dirichlet L-function $L(s, \chi_d)$ converges absolutely for $\mathrm{Re}(s) > 1$, extends analytically to the entire complex plane (unlike $\zeta(s)$, there is no pole at $s = 1$ here), and satisfies a functional equation connecting values at s to values at $1 - s$. The nontrivial zeros of $L(s, \chi_d)$ lie in the *critical strip* $0 < \mathrm{Re}(s) < 1$, with the Generalized Riemann Hypothesis (GRH) predicting that they lie on the critical line $\mathrm{Re}(s) = \frac{1}{2}$. For background on Dirichlet L-functions see Davenport **[58]**, and for a general comprehensive treatment of analytic number theory (including information on many other families of L-functions that will be considered here) see Iwaniec and Kowalski **[92]**.

In this family of quadratic Dirichlet L-functions, the values $L(1, \chi_d)$ (lying at the edge of the critical strip) are of great arithmetical interest. A key step in Dirichlet's proof that there are infinitely many primes in arithmetic progressions involves showing that $L(1, \chi_d) \neq 0$. Dirichlet established this by finding a beautiful connection between $L(1, \chi_d)$ and the group of equivalence classes of binary quadratic forms of discriminant d which had earlier been studied by Gauss. For example, if d is a negative fundamental discriminant, then Dirichlet's class number formula states that

$$L(1, \chi_d) = \frac{2\pi}{w} \frac{h(d)}{\sqrt{|d|}},$$

where $h(d)$ is a positive integer, namely the class number of the imaginary quadratic field $\mathbb{Q}(\sqrt{d})$, and w counts the number of roots of unity in $\mathbb{Q}(\sqrt{d})$ (so that $w = 2$ for $d < -4$, and $w = 4$ for $d = -4$, and $w = 6$ for $d = -3$). The special case $d = -4$ of the Dirichlet class number formula is widely familiar as the Madhava–Leibniz–Gregory series $1 - 1/3 + 1/5 - 1/7 + \cdots = \pi/4$. Another classical connection to these special L-values arises in the Gauss–Legendre three squares theorem. If n is a square-free integer with $n \equiv 3 \pmod 8$, then the number of ways of writing n as a sum of three squares, $r(n)$, equals $24h(-n)$; a result known to Gauss, together with variants when $n \equiv 1, 2 \pmod 4$.

These connections motivate the study of the distribution of the values $L(1, \chi_d)$. Here are some natural questions that arise. If fundamental discriminants d are chosen uniformly with $|d| \leq X$, (i) what is the statistical distribution of the values $L(1, \chi_d)$, and (ii) what are the largest and smallest possible values of $L(1, \chi_d)$? As we shall see, the problem of the statistical distribution of $L(1, \chi_d)$ can be understood quite precisely, but there are still large gaps in our understanding of the extreme values.

Let us begin with the simpler situation of $L(2, \chi_d)$ where both the Dirichlet series and Euler product in the definition of $L(s, \chi_d)$ converge absolutely. If the values $\chi_d(p)$ are known for all primes $p \leq z$ then

$$\left| L(2, \chi_d) - \prod_{p \leq z} \left(1 - \frac{\chi_d(p)}{p^2}\right)^{-1} \right| \leq \sum_{n > z} \frac{1}{n^2} = O\left(\frac{1}{z}\right).$$

The value $\chi_d(p) = (\frac{d}{p})$ is determined by $d \pmod p$ for odd primes p, and for $p = 2$ the value of $\chi_d(p)$ is determined by $d \pmod 8$. Thus, by the Chinese Remainder Theorem, the values of $\chi_d(p)$ for $p \le z$ are determined by d modulo $4\prod_{p\le z} p$. One way to view this is as a kind of *almost periodicity*: if two fundamental discriminants d_1 and d_2 are congruent modulo $4\prod_{p\le z} p$ then $L(2, \chi_{d_1}) = L(2, \chi_{d_2}) + O(1/z)$.

If p is an odd prime and X is large, then a little calculation shows that a proportion $\frac{1}{p+1}$ of the fundamental discriminants d with $|d| \le X$ are multiples of p (this is essentially the proportion of square-free integers that are multiples of p) and $\chi_d(p) = 0$ here. The remaining proportion $\frac{p}{p+1}$ of fundamental discriminants are evenly split among the possible values $\chi_d(p) = 1$ or -1. Pleasantly, it turns out that for $p = 2$ also a proportion $\frac{1}{3}$ of the fundamental discriminants $|d| \le X$ satisfy each of the three cases $\chi_d(2) = 0, 1$ or -1. Moreover, the Chinese Remainder Theorem tells us that for different primes p, the values $\chi_d(p)$ are distributed "independently" of each other, at least if we restrict to primes $p \le z$ with $\prod_{p\le z} p$ being small in comparison with X. This motivates us to define for prime numbers p, independent random variables $\mathbb{X}(p)$ taking the values 0 with probability $1/(p+1)$ and the values ± 1 with probability $p/(2(p+1))$. Then the distribution of $\prod_{p\le z}(1 - \chi_d(p)/p^2)^{-1}$ is the same as the distribution of the random Euler product $\prod_{p\le z}(1 - \mathbb{X}(p)/p^2)^{-1}$. Letting $z \to \infty$, we have described the distribution of $L(2, \chi_d)$ as being precisely the distribution of $\prod_p (1 - \mathbb{X}(p)/p^2)^{-1}$.

The story for extreme values is also clear:

$$\frac{\zeta(4)}{\zeta(2)} = \prod_p \left(1 + \frac{1}{p^2}\right)^{-1} \le \prod_p \left(1 - \frac{\chi_d(p)}{p^2}\right)^{-1} = L(2, \chi_d) \le \prod_p \left(1 - \frac{1}{p^2}\right)^{-1} = \zeta(2).$$

Moreover, we may find values $L(2, \chi_d)$ arbitrarily close to $\zeta(4)/\zeta(2)$ by choosing d with $\chi_d(p) = -1$ for all primes $p \le z$, and we may find values arbitrarily close to $\zeta(2)$ by choosing d with $\chi_d(p) = 1$ for all primes $p \le z$.

Let us now turn to the distribution of $L(1, \chi_d)$ where there is a similar story but with some added complications since the series and product defining $L(s, \chi_d)$ are no longer absolutely convergent. For example, one can show that if $z \le (\log X)^{10}$ then $L(1, \chi_d) = \prod_{p\le z}(1 - \chi_d(p)/p)^{-1} + O(1/z^{\frac{1}{4}})$ for all but $O(X/z^{\frac{1}{4}})$ of the fundamental discriminants $|d| \le X$. This again may be viewed as a kind of almost periodicity: allowing z to tend slowly to infinity with X, for almost all pairs of discriminants d_1 and d_2 with $d_1 \equiv d_2 \pmod{4\prod_{p\le z} p}$ one has $L(1, \chi_{d_1}) \approx L(1, \chi_{d_2})$.

For primes p, let $\mathbb{X}(p)$ denote the random variables described earlier, and extend $\mathbb{X}$ to all integers using (complete) multiplicativity; thus, if $n = p_1^{e_1} \cdots p_k^{e_k}$ then $\mathbb{X}(n) = \mathbb{X}(p_1)^{e_1} \cdots \mathbb{X}(p_k)^{e_k}$. This is an example of a *random multiplicative function*, and we may correspondingly consider the random L-function

$$L(s, \mathbb{X}) = \sum_{n=1}^{\infty} \frac{\mathbb{X}(n)}{n^s} = \prod_p \left(1 - \frac{\mathbb{X}(p)}{p^s}\right)^{-1}. \tag{1.1}$$

Both the series and product above converge almost surely provided $\mathrm{Re}(s) > \frac{1}{2}$; this follows essentially from the fact that the variance of $\sum_p \mathbb{X}(p)/p^s$ is $\sum_p \frac{p}{p+1} \frac{1}{p^{2\,\mathrm{Re}(s)}}$, which is a convergent sum when $\mathrm{Re}(s) > \frac{1}{2}$. In particular, the random Euler product $L(1, \mathbb{X})$ converges almost surely, and the values $L(1, \chi_d)$ are distributed like $L(1, \mathbb{X})$. We may see this by first approximating most $L(1, \chi_d)$ by $\prod_{p \le z}(1 - \chi_d(p)/p)^{-1}$, noting that this truncated Euler product is distributed exactly like $\prod_{p \le z}(1 - \mathbb{X}(p)/p)^{-1}$, and finally letting $z \to \infty$.

Let us state the result discussed above more precisely. Given any $\tau > 0$, the proportion of fundamental discriminants $|d| \le X$ with $L(1, \chi_d) \ge e^\gamma \tau$ tends as $X \to \infty$ to $\mathrm{Prob}(L(1, \mathbb{X}) > e^\gamma \tau)$. Here γ is Euler's constant, and we have normalized in this fashion in view of Mertens's theorem $\prod_{p \le z}(1 - 1/p)^{-1} \sim e^\gamma \log z$. If τ is large, and we seek values of $L(1, \chi_d)$ larger than $e^\gamma \tau$, the most likely way in which such large values arise is when $\chi_d(p) = 1$ for all primes up to about e^τ. Similarly, the proportion of fundamental discriminants $|d| \le X$ with $L(1, \chi_d) < \zeta(2)/(e^\gamma \tau)$ tends as $X \to \infty$ to $\mathrm{Prob}(L(1, \mathbb{X}) < \zeta(2)/(e^\gamma \tau))$. The normalization here is made in view of $\prod_{p \le z}(1 + 1/p)^{-1} \sim \zeta(2)/(e^\gamma \log z)$. The distribution of $L(1, \mathbb{X})$ is continuous—it is more natural to think of the distribution of $\log L(1, \mathbb{X})$ which is smooth—and its tails $\mathrm{Prob}(L(1, \mathbb{X}) > e^\gamma \tau)$ and $\mathrm{Prob}(L(1, \mathbb{X}) < \zeta(2)/(e^\gamma \tau))$ decay double exponentially, behaving like $\exp(-(1 + o(1))e^{\tau - C_1}/\tau)$ for a suitable constant C_1 (see **[71]**). With high likelihood one has $1/10 \le L(1, \mathbb{X}) \le 10$, although there is a small positive probability of finding arbitrarily large or arbitrarily small values.

The qualitative results mentioned above were obtained by Chowla and Erdős **[39]**, and with some uniformity in τ by Elliott **[58]**. The question of uniformity in τ is studied in more detail by Montgomery and Vaughan **[126]**, and Granville and Soundararajan **[71]**, with the aim of understanding the extreme values of $L(1, \chi_d)$. By "uniformity in τ," we mean the problem of allowing τ to depend on X while still guaranteeing that the proportion of $|d| \le X$ with $L(1, \chi_d) > e^\gamma \tau$ is comparable to the tail probability that $L(1, \mathbb{X}) > e^\gamma \tau$ (and similarly for small values of $L(1, \chi_d)$). In view of the double exponential decay of the tails of the distribution of $L(1, \mathbb{X})$ mentioned above, the largest viable range for uniformity in τ is $\tau \le \tau_{\max} + \varepsilon$, with $\tau_{\max} = \log\log X + \log\log\log X + C_1$ and any fixed $\varepsilon > 0$—at this point one has $\mathrm{Prob}(L(1, \mathbb{X}) > e^\gamma \tau_{\max}) < 1/X$. The results in **[71]** show excellent agreement between the distribution of $L(1, \chi_d)$ and the probabilistic model $L(1, \mathbb{X})$ in almost the entire viable range. These results suggest the following conjectures on the extreme values of $L(1, \chi_d)$:

$$\max_{|d| \le X} L(1, \chi_d) = e^\gamma \big(\tau_{\max} + o(1)\big), \quad \text{and} \quad \min_{|d| \le X} L(1, \chi_d) = \zeta(2)/\big(e^\gamma \big(\tau_{\max} + o(1)\big)\big). \tag{1.2}$$

In **[71]** it is shown that there are values of $L(1, \chi_d)$ nearly as large as the conjecture in (1.2) (for example, assuming the truth of GRH one can find values as large as $e^\gamma(\tau_{\max} - C)$ for some constant C) and values almost as small as in (1.2). However, as we shall discuss next, there are large gaps in our understanding of why the extreme values cannot be much larger or smaller.

How large can z be such that for some fundamental discriminant $|d| \le X$ one has $\chi_d(p) = 1$ for all primes $p \le z$? This problem is intimately related to finding large values of $L(1, \chi_d)$. Correspondingly, the problem of finding small values of $L(1, \chi_d)$ may be thought of as wanting $\chi_d(p) = -1$ for all primes $p \le z$. We noted already that the values of $\chi_d(p)$ for $p \le z$ may be determined by knowing $d \pmod{4\prod_{p \le z} p}$. The prime number theorem gives $\prod_{p \le z} p = e^{z(1+o(1))}$, so that with $z = \frac{1}{2}\log X$ (say) we can find $|d| \le X$ with any given signs $\chi_d(p)$ for $p \le z$—for example we may make them all 1, or all -1. If we think of the probabilistic model $\mathbb{X}$ which treats $\chi_d(p)$ as essentially being a "coin toss" we may expect that the primes up to about $z = \log X \log\log X$ (there are about $\log X$ primes below this z) may take any prescribed signs. This dovetails nicely with the conjectured size of extreme values in (1.2), since (in the case of large values) $\prod_{p \le z}(1 - 1/p)^{-1} \sim e^{\gamma} \log z \approx e^{\gamma}(\log\log X + \log\log\log X)$. For primes p larger than about $\log X \log\log X$, we expect randomness to kick in, and to find an equal number of positive and negative values of $\chi_d(p)$.

Our current knowledge is very far from these probabilistic considerations. Given a prime ℓ, Vinogradov conjectured that the least quadratic nonresidue (mod ℓ) lies below $C(\varepsilon)\ell^{\varepsilon}$ for some constant $C(\varepsilon)$. That is, there must be a prime $p \le C(\varepsilon)\ell^{\varepsilon}$ with $(\frac{p}{\ell}) = -1$, which is a weak version of the prediction from the random model that there exists such p with $p \le C \log \ell \log\log \ell$ for some constant C. Toward Vinogradov's conjecture, we know, as a consequence of the Burgess bounds for character sums, that the least quadratic nonresidue lies below $\ell^{1/(4\sqrt{e})+o(1)}$ (see **[29]**), and no improvement over this exponent has been made in more than 50 years. In terms of $L(1, \chi_d)$, the work towards Vinogradov's conjecture may be used to show that (see **[70, 168]**)

$$L(1, \chi_d) < \left(\frac{1}{4}\left(2 - \frac{2}{\sqrt{e}}\right) + o(1)\right) \log |d|.$$

This is far from the conjecture in (1.2), and even an improvement in the constant above would be significant and lead to an improvement on the bound for the least quadratic nonresidue (see also **[20, 72, 169]** for related work).

Even less is known about the problem of bounding the least prime p such that p is a quadratic residue (mod ℓ). To give a sense of the interest of this problem, we note that if $\ell \equiv 3 \pmod 4$ is a prime, then the imaginary quadratic field $\mathbb{Q}(\sqrt{-\ell})$ has class number 1 if and only if $(\frac{p}{\ell}) = -1$ for all $p < (1 + \ell)/4$. For such a prime ℓ, the polynomial $n^2 + n + (1 + \ell)/4$ takes prime values for $0 \le n < (\ell - 3)/4$. Euler's famous polynomial $n^2 + n + 41$ is the largest example of this phenomenon, corresponding to the prime $\ell = 163$ for which the first 12 primes (the primes below 41) are all quadratic nonresidues. Toward this problem, we know that the least prime quadratic residue (mod ℓ) lies below $C(\varepsilon)\ell^{\frac{1}{4}+\varepsilon}$ for any $\varepsilon > 0$ (see **[82]**), but with a constant $C(\varepsilon)$ that is *ineffective* (meaning the proof only shows the existence of $C(\varepsilon)$, but without any way to compute it, even in principle). This is related to Siegel's ineffective lower bound (see **[50]**): for any $\varepsilon > 0$ there exists $C(\varepsilon) > 0$ with

$$L(1, \chi_d) > C(\varepsilon)|d|^{-\varepsilon}.$$

Thus our knowledge of small values of $L(1, \chi_d)$ is even further from the conjecture in (1.2).

If we assume the truth of GRH, then much better results are known. On GRH, the least quadratic nonresidue (mod ℓ) can be shown to be $< (\log \ell)^2$, and the least prime quadratic residue also lies below $(1 + o(1))(\log \ell)^2$ (see **[114]**). Moreover, for any fundamental discriminant d, one has

$$L(1, \chi_d) \sim \prod_{p \le (\log |d|)^2} \left(1 - \frac{\chi_d(p)}{p}\right)^{-1}, \tag{1.3}$$

so that the extreme values of $L(1, \chi_d)$ over all $|d| \leq X$ are bounded above by $(2 + o(1))e^\gamma \tau_{\max}$ and below by $(\frac{1}{2} + o(1))\zeta(2)/(e^\gamma \tau_{\max})$. There is still a gap between these GRH bounds and the probabilistic conjecture in (1.2), but now one is off only by a factor of 2, corresponding to the expectation based on the random model that in (1.3) we only need to take the product over primes $p \leq (\log |d|)$ in order to approximate $L(1, \chi_d)$.

To summarize our discussion, the values of $L(1, \chi_d)$ have an almost periodic structure in d, and these values may be accurately modeled by random Euler products. The random model gives a satisfactory description of the statistical distribution of $L(1, \chi_d)$. It also makes predictions on the largest and smallest possible values of $L(1, \chi_d)$, but there is a large gap between these predictions and our current unconditional knowledge, and even assuming GRH there is still a factor of 2 at issue.

Similar results may be established for the distribution at the edge of the critical strip for values in other families of L-functions. For example, consider the distribution of $\zeta(1 + it)$, where t is chosen uniformly from $[T, 2T]$ with $T \to \infty$. These values may be modeled by the random Euler product

$$\zeta(s, \mathbb{X}) = \prod_p \left(1 - \frac{\mathbb{X}(p)}{p^s}\right)^{-1} = \sum_{n=1}^{\infty} \frac{\mathbb{X}(n)}{n^s}, \tag{1.4}$$

where the random variables $\mathbb{X}(p)$ are independent for different primes p, and are all chosen uniformly from the unit circle $\{|z| = 1\}$, and extended to random variables $\mathbb{X}(n)$ over all natural numbers n by multiplicativity. As before, the product and series both converge almost surely when $\mathrm{Re}(s) > \frac{1}{2}$. Then the statistical distribution of $\zeta(1 + it)$ is identical to that of $\zeta(1, \mathbb{X})$ (equivalently, of $\zeta(1 + iy, \mathbb{X})$ for any real y). We can also formulate an *almost periodicity* result: For any $\varepsilon > 0$, we can find a sequence of *almost periods* τ_n, with $\tau_n \to \infty$ and $|\tau_{n+1} - \tau_n|$ bounded, such that for T sufficiently large (in terms of any fixed almost period τ) one has $|\zeta(1 + it + i\tau) - \zeta(1 + it)| < \varepsilon$ for almost all $t \in [T, 2T]$. The sequence of almost periods are found by requiring $p^{i\tau} \approx 1$ for all primes p up to some point. For a study of the distribution of $\zeta(1 + it)$, with a focus on uniformity, see Lamzouri **[109]**.

There is an extensive literature concerned with distribution at the edge of the critical strip, and we end this section with references to some further examples. We motivated our discussion of $L(1, \chi_d)$ with the class number formula, which (for negative fundamental discriminants) shows that $2\sqrt{|d|}L(1, \chi_d)/(2\pi)$ is quantized to be an integer. This raises questions on the *granularity* of the distribution of $L(1, \chi_d)$, and shows that in very short scales there must be arithmetic deviations from the random model. These questions are related to the problem of understanding how many imaginary quadratic fields there are with any given class number (see **[88, 111, 162]**). For positive fundamental discriminants, the class number formula relates $L(1, \chi_d)$ to the product of the class number and the regulator which cannot in general be separated from each other. One way to get around this problem is to order the real quadratic fields by the size of their regulator rather than by discriminant, and this ordering has a pleasing interpretation in terms of lengths of closed geodesics on the hyperbolic surface $\mathrm{PSL}(2, \mathbb{Z})\backslash\mathbb{H}$. The study of $L(1, \chi_d)$, or the class number $h(d)$, when d is ordered in this way was initiated by Sarnak **[152]**; it is closely related to specializing discriminants d in suitable quadratic sequences (for example, of the form $4n^2 + 1$, or $n^2 + 4$), and for recent investigations see [49, 110, 145]. For a small sample of investigations in other families of L-functions, see [40, 56, 117, 118, 123].

2. SELBERG'S CENTRAL LIMIT THEOREM

In the previous section we discussed the distribution of values of L-functions at the edge of the critical strip. In fact, similar results hold for the value distribution inside the critical strip, but keeping to the right of the critical line. As an illustration, consider the problem of the distribution of values of $\zeta(\sigma + it)$ where $\frac{1}{2} < \sigma \leq 1$ is fixed, and t is chosen uniformly from $[T, 2T]$ with $T \to \infty$. The random $\zeta(s, \mathbb{X})$ defined in (1.4) still converges when $\mathrm{Re}(s) = \sigma > \frac{1}{2}$, and one can show that $\zeta(\sigma + it)$ is distributed like $\zeta(\sigma, \mathbb{X})$. To give a very brief indication of the proof, one can show that for any parameter $1 \leq N \leq T$,

$$\frac{1}{T} \int_T^{2T} \left| \zeta(\sigma + it) - \sum_{n \leq N} \frac{1}{n^{\sigma + it}} \right|^2 dt = O\Big(\sum_{n > N} \frac{1}{n^{2\sigma}} \Big) = O\big(N^{1-2\sigma}\big), \tag{2.1}$$

which parallels

$$\mathbb{E}\left[\left| \zeta(\sigma, \mathbb{X}) - \sum_{n \leq N} \frac{\mathbb{X}(n)}{n^{\sigma}} \right|^2 \right] = \sum_{n > N} \frac{1}{n^{2\sigma}} = O\big(N^{1-2\sigma}\big).$$

Since $\sigma > \frac{1}{2}$, the term $N^{1-2\sigma}$ tends to 0 provided N tends to infinity with T, and for such N it follows that for most $t \in [T, 2T]$ one has $\zeta(\sigma + it) \approx \sum_{n \leq N} n^{-\sigma + it}$. If now N tends slowly to infinity with T, then we can show that $\sum_{n \leq N} n^{-\sigma + it}$ is distributed like $\sum_{n \leq N} \mathbb{X}(n)/n^{\sigma}$, by matching the moments of both quantities, for example. This is a classical result (see Chapter XI of [170]), and a recent quantitative study has been made in [113].

As with the distribution of $\zeta(1 + it)$, there is an almost periodic structure in the values of $\zeta(\sigma + it)$. The partial sums $\sum_{n \leq N} n^{-\sigma - it}$ clearly have an almost periodic structure—if $n^{i\tau} \approx 1$ for all $n \leq N$, then τ will be an almost period for these partial sums—and as we noted above $\zeta(\sigma + it)$ can often be approximated by such partial sums.

For $\frac{1}{2} < \sigma \leq 1$, the values $\zeta(\sigma, \mathbb{X})$ are distributed densely in the complex plane; indeed, for any given complex number z and any $\varepsilon > 0$, with positive probability (depending on z and ε) one has $|\zeta(\sigma, \mathbb{X}) - z| < \varepsilon$. This is not hard to show, starting with the fact that $\log \zeta(\sigma, \mathbb{X})$ is essentially $\sum_p \mathbb{X}(p)/p^{\sigma}$. It follows that the set $\{\zeta(\sigma + it) : t \in \mathbb{R}\}$ is dense in $\mathbb{C}$. A related striking *universality* result of Voronin [171] states that if f is any nonvanishing continuous function in $|z| \leq r$ with $0 < r < \frac{1}{4}$, then there exist arbitrarily large values $t \in \mathbb{R}$ such that $|\zeta(\frac{3}{4} + it + z) - f(z)| < \varepsilon$ for all $|z| \leq r$. In other words, the zeta function in a disc of radius r around $\frac{3}{4} + it$ can be made to mimic any given analytic function that does not take the value 0. The value 0 must be excluded in view of the Riemann Hypothesis! There are more precise versions of this result, but we do not pursue this direction further, pointing instead to [10, 106, 112] for recent related work.

We now turn to the distribution of values of $\zeta(\frac{1}{2} + it)$, which forms the main focus of this article. The random Euler product $\zeta(s, \mathbb{X})$ defined in (1.4) does not converge for $s = \frac{1}{2}$.

Indeed, there is no almost periodic structure to the values $\zeta(\frac{1}{2}+it)$, and on the critical line the zeta-function cannot typically be understood simply from a knowledge of p^{it} for small primes p. Instead, we have the following fundamental result of Selberg.

Theorem 2.1 (Selberg [155, 156]). *If T is large, and t is chosen uniformly from $[T, 2T]$, then $\log\zeta(\frac{1}{2}+it)$ is distributed like a complex Gaussian with mean 0 and variance $\log\log T$. In particular, $\operatorname{Re}(\log\zeta(\frac{1}{2}+it))$ and $\operatorname{Im}(\log\zeta(\frac{1}{2}+it))$ are distributed like real Gaussians with mean 0 and variance $\frac{1}{2}\log\log T$.*

To clarify normalizations, we recall that a standard complex Gaussian (of mean 0 and variance 1) has density $\frac{1}{\pi}e^{-|z|^2}$, and that its real and imaginary part are independent real Gaussians with mean 0 and variance $\frac{1}{2}$. Selberg's theorem gives that for any fixed box $\mathcal{B}$ in the complex plane, as $T\to\infty$ one has

$$\frac{1}{T}\operatorname{meas}\left\{T\le t\le 2T,\ \frac{\log\zeta(\frac{1}{2}+it)}{\sqrt{\log\log T}}\in\mathcal{B}\right\}\to\frac{1}{\pi}\int_{x+iy\in\mathcal{B}}e^{-x^2-y^2}\,dx\,dy.$$

In Selberg's theorem we may omit the countably many zeros of $\zeta(s)$ where the logarithm is not defined. For t not equalling the ordinate of a zero of $\zeta(s)$, the argument of $\zeta(\frac{1}{2}+it)$ (that is, $\operatorname{Im}(\log\zeta(\frac{1}{2}+it))$) is defined by continuous variation along the straight lines from 2 (where the argument is taken to be zero) to $2+it$ and thence to $1/2+it$.

Here is a striking illustration of the difference between the value distributions of $\zeta(\frac{1}{2}+it)$ and $\zeta(\sigma+it)$ for $1\ge\sigma>\frac{1}{2}$. Typically, $|\zeta(\sigma+it)|$ is of constant size, for example, taking values between $1/2$ and 2 with positive probability. On the other hand, Selberg's theorem implies that for any fixed V and large T,

$$\frac{1}{T}\operatorname{meas}\left\{T\le t\le 2T,\ \frac{\log|\zeta(\frac{1}{2}+it)|}{\sqrt{\frac{1}{2}\log\log T}}\ge V\right\}\sim\frac{1}{\sqrt{2\pi}}\int_V^\infty e^{-x^2/2}\,dx, \tag{2.2}$$

so that $|\zeta(\frac{1}{2}+it)|$ is large (say, $>\exp(\varepsilon\sqrt{\log\log T})$) nearly half the time, or $|\zeta(\frac{1}{2}+it)|$ is small (below $\exp(-\varepsilon\sqrt{\log\log T})$) nearly half the time. We noted earlier that the set $\{\zeta(\sigma+it): t\in\mathbb{R}\}$ is dense in the complex plane. It is rare to find values of $\zeta(\frac{1}{2}+it)$ of constant size, and whether the set $\{\zeta(\frac{1}{2}+it): t\in\mathbb{R}\}$ is dense in $\mathbb{C}$ remains an intriguing open problem. This question was raised first by Ramachandra; for partial progress, see [108].

The argument principle, together with the functional equation for $\zeta(s)$ and Stirling's formula, may be used to show that $N(t)$, the number of zeros of $\zeta(s)$ with real part between 0 and 1 and imaginary part between 0 and t, satisfies

$$N(t)=\frac{t}{2\pi}\log\frac{t}{2\pi}-\frac{t}{2\pi}+\frac{7}{8}+S(t)+O\left(\frac{1}{t}\right),\quad\text{where } S(t)=\frac{1}{\pi}\arg\zeta\left(\frac{1}{2}+it\right). \tag{2.3}$$

Thus Selberg's theorem for $\operatorname{Im}(\log\zeta(\frac{1}{2}+it))$ shows that the remainder term in the asymptotic formula for $N(t)$ has Gaussian fluctuations.

We now give a brief, oversimplified, description of the ideas behind Selberg's theorem; we caution the reader that some statements below should be taken as merely indicative, and not interpreted as being literally correct. Taking logarithms in the Euler product for $\zeta(s)$, we may write

$$\log \zeta(s) = \sum_{p,k} \frac{1}{kp^{ks}} = \sum_{n=2}^{\infty} \frac{\Lambda(n)}{\log n} \frac{1}{n^s},$$

where the sums above are over prime powers p^k, and $\Lambda(n)$ is the von Mangoldt function which equals $\log p$ if $n = p^k$ and 0 otherwise. The series above converges absolutely when $\text{Re}(s) > 1$, and it certainly does not converge on the critical line $\text{Re}(s) = \frac{1}{2}$. Nevertheless, we might hope that a truncated sum over prime powers might serve as an approximation to $\log \zeta(s)$ (thinking of $s = \frac{1}{2} + it$ with $T \le t \le 2T$). This forms the first step in Selberg's argument, who finds an expression of the form

$$\log \zeta(s) = \sum_{2 \le n \le x} \frac{\Lambda(n)}{n^s \log n} + Z_x(s), \tag{2.4}$$

where $Z_x(s)$ is a remainder term that may be thought of as the contribution from zeros ρ of $\zeta(s)$ with $|\rho - s| \le 1/\log x$. By a complicated argument, Selberg showed how the sum over zeros may in turn also be bounded in terms of sums over primes, and thus shown to be small on average. An alternative argument of Bombieri and Hejhal **[21]** avoids some of Selberg's difficulties by bounding the average values of $Z_x(s)$ instead of seeking pointwise bounds. Nevertheless, these arguments are technically involved; they are simpler if the Riemann hypothesis is assumed, but can be established unconditionally by relying on a subtle zero-density estimate for zeros of $\zeta(s)$ near the critical line (established by Selberg). Although we have not made the relation (2.4) precise, we give a couple of remarks that may be helpful in thinking about such relations. Firstly, one can think of such relations as variants of the explicit formula connecting zeros and primes. Secondly, in addition to the Euler product, the zeta function possesses a Hadamard product over its zeros

$$s(s-1)\pi^{-s/2}\Gamma(s/2)\zeta(s) = e^{Bs} \prod_{\rho} \left(1 - \frac{s}{\rho}\right) e^{s/\rho}, \tag{2.5}$$

where the product is over all nontrivial zeros of the zeta-function, and B is a constant. The relation (2.4) has the flavor of a hybrid Euler–Hadamard product (see **[69]** for work in this direction), taking some primes and some zeros, and it is natural to expect an inverse relationship (or uncertainty principle) between the number of primes that one must take versus the number of zeros that are needed.

Returning to the argument, in the range $x \le T$, the remainder term $Z_x(s)$ in (2.4) is typically of size $O(\log T/\log x)$; this corresponds to the expected number of zeros of $\zeta(s)$ within $1/\log x$ of $\frac{1}{2} + it$. If we choose $x = T^{1/(\log\log T)^{\frac{1}{4}}}$, for example, then $\log T/\log x = (\log\log T)^{\frac{1}{4}}$ is small in comparison to the typical expected size of $\log \zeta(s)$, which is $\sqrt{\log\log T}$, and therefore the remainder term is negligible. In other words, with this choice of x, the proof of Selberg's theorem reduces to establishing the Gaussian nature of

$$\sum_{2 \le n \le x} \frac{\Lambda(n)}{\log n} \frac{1}{n^s} = \sum_{p \le x} \frac{1}{p^s} + \frac{1}{2} \sum_{p \le \sqrt{x}} \frac{1}{p^{2s}} + \sum_{\substack{p^k \le x \\ k \ge 3}} \frac{1}{kp^{ks}}. \tag{2.6}$$

The contribution from prime powers p^k with $k \geq 3$ is $O(1)$ and may be omitted. The contribution from the squares of primes is also negligible; it is $\frac{1}{2}\sum_{p\leq\sqrt{x}} 1/p^{1+2it}$ which behaves roughly like $\frac{1}{2}\log\zeta(1+2it)$ and so is of constant size typically. We are left with the contribution of just the primes, which we may understand by computing moments. If k and ℓ are any natural numbers then, for large T,

$$\frac{1}{T}\int_T^{2T}\Big(\sum_{p\leq x}\frac{1}{p^{1/2+it}}\Big)^k\Big(\sum_{p\leq x}\frac{1}{p^{1/2-it}}\Big)^\ell dt = \begin{cases}(1+o(1))k!(\log\log T)^k & \text{if } k=\ell,\\ o(T) & \text{if } k\neq\ell.\end{cases} \tag{2.7}$$

These moments match asymptotically the moments of a complex Gaussian with mean 0 and variance $\log\log T$, from which Selberg's theorem would follow.

To give a justification for (2.7), we discuss an orthogonality relation for Dirichlet polynomials, which we shall find useful in the sequel. Roughly speaking, integrals over $[T, 2T]$ may be thought of as possessing T "harmonics" that can distinguish between the functions $f_n(t) = n^{it}$ for natural numbers n going up to about T. More precisely, suppose Φ is a smooth function approximating the indicator function of $[1, 2]$. Then, if $\max(M, N) \leq T/\log T$,

$$\begin{aligned}\int \sum_{m\leq M} a(m)m^{it}\overline{\sum_{n\leq N} b(n)n^{it}}\,\Phi\Big(\frac{t}{T}\Big)dt &= \sum_{m=n} a(m)\overline{b(n)}T\widehat{\Phi}(0)\\ &\quad+\sum_{m\neq n} a(m)\overline{b(n)}T\widehat{\Phi}\Big(T\log\frac{n}{m}\Big)\\ &\sim T\widehat{\Phi}(0)\sum_{m=n} a(m)\overline{b(n)},\end{aligned} \tag{2.8}$$

where the contribution of the "off-diagonal" terms $m \neq n$ is negligible because $T|\log(m/n)| \gg T|m-n|/|m+n| \geq T/(M+N)$ is large and the Fourier transform $\widehat{\Phi}$ decays rapidly.

Write $(\sum_{p\leq x} 1/p^{1/2+it})^k = \sum_{n\leq x^k} a_k(n)/n^{1/2+it}$, so that $a_k(n) = 0$ unless n has exactly k prime factors. If n has prime factorization $p_1^{e_1}\cdots p_r^{e_r}$ with $e_1+\cdots+e_r = k$ then $a_k(n) = k!/(e_1!\cdots e_r!)$. Then an application of (2.8) shows that the moment in (2.7) is

$$\sim T\sum_{m=n\leq x^k}\frac{a_k(n)a_\ell(n)}{n}.$$

If $k \neq \ell$ then either $a_k(n)$ or $a_\ell(n)$ must be zero, and this case of (2.7) follows. If $k = \ell$, then the diagonal terms are dominated by integers with k distinct prime factors, and so the above is

$$\sim Tk!\sum_{n\leq x^k}\frac{a_k(n)}{n} = Tk!\Big(\sum_{p\leq x}\frac{1}{p}\Big)^k \sim Tk!(\log\log x)^k,$$

and since $\log\log x$ and $\log\log T$ are close, the other case in (2.7) follows.

This concludes our sketch of the ideas behind Selberg's theorem. Two alternative approaches that work for $\log|\zeta(\frac{1}{2}+it)|$ are given in **[115, 142]**. These avoid the subtle zero-density estimates near the critical line, and it would be of interest to extend such approaches to $\mathrm{Im}(\log\zeta(\frac{1}{2}+it))$.

3. ANALOGUES OF SELBERG'S THEOREM IN FAMILIES OF L-FUNCTIONS

Selberg's theorem discussed above applies not only to the Riemann zeta-function, but more generally to a large class of L-functions. For example, in [156] Selberg introduced what is now known as the *Selberg class* of L-functions, which formalizes some of the observed properties of automorphic L-functions and is expected to coincide with this class. For a primitive L-function in the Selberg class (or, if one prefers, for a cuspidal automorphic L-function for $\mathrm{GL}_n(\mathbb{Q})$), one expects that $\log L(\frac{1}{2}+it)$ with $T \le t \le 2T$ is distributed like a complex Gaussian with mean 0 and variance $\log\log T$. The key ingredient needed to make this precise is an analogue of the zero-density estimate close to the critical line, and this is known for GL_1 and GL_2; in the general case, GRH must be assumed (see [21, 156] for more details).

Interesting differences arise when we consider analogues of Selberg's theorem for central values in families of L-functions. There are three categories into which families of L-functions fall, and we illustrate these with examples. Unlike Selberg's *Theorem*, the analogous central limit theorems that we formulate in these families are still *conjectural*, and these conjectures were first formulated by Keating and Snaith [181].

Unitary families. A typical example is the family of all Dirichlet characters $\chi \pmod q$, with q a large prime (for simplicity). The question is to understand the distribution of $\log L(\frac{1}{2},\chi)$ as χ ranges over all primitive characters $\chi \pmod q$ (if q is prime, this is equivalent to χ not being the principal character). We must discard potential characters with $L(\frac{1}{2},\chi)=0$, but in fact it is conjectured that $L(\frac{1}{2},\chi)\neq 0$ for all Dirichlet L-functions. This situation is expected to be exactly as in Selberg's theorem, and the Keating–Snaith conjecture for this family states that for large q the distribution of $\log L(\frac{1}{2},\chi)$ is approximately a complex Gaussian with mean 0 and variance $\log\log q$. In particular, $\log|L(\frac{1}{2},\chi)|$ is (conjecturally) distributed like a real Gaussian with mean 0 and variance $\frac{1}{2}\log\log q$, so that (like $|\zeta(\frac{1}{2}+it)|$) roughly half the time $|L(\frac{1}{2},\chi)|$ is as large as $\exp(\varepsilon\sqrt{\log\log q})$ and the other half of the time it is as small as $\exp(-\varepsilon\sqrt{\log\log q})$.

Another example of this type is the family of twists by Dirichlet characters of a fixed newform f. The family $\zeta(\frac{1}{2}+it)$ with $T \le t \le 2T$ may also be thought of as an example of a unitary family.

Symplectic families. Consider the family of quadratic Dirichlet L-functions $L(s,\chi_d)$, where d ranges over fundamental discriminants with $|d| \le X$. The values $L(\frac{1}{2},\chi_d)$ are real, and GRH predicts that they are all nonnegative (else there would be a real zero of $L(s,\chi_d)$ between $1/2$ and 1). Further, the values $L(\frac{1}{2},\chi_d)$ are all expected to be nonzero (a conjecture of Chowla, which is a special case of the belief that $L(\frac{1}{2},\chi)\neq 0$ for all Dirichlet characters χ). The Keating–Snaith conjecture for this family predicts that the values $\log L(\frac{1}{2},\chi_d)$ are distributed like a real Gaussian with mean $\frac{1}{2}\log\log X$ and variance $\log\log X$. Since the mean is positive, the values of $L(\frac{1}{2},\chi_d)$ are (conjecturally) of typical size $(\log X)^{\frac{1}{2}+o(1)}$.

Orthogonal families. These families arise naturally in the context of modular forms, and we give a couple of prototypical examples. Let k be an even integer, and consider the family $\mathcal{H}_k$ of all weight k modular forms for the full modular group $\mathrm{SL}_2(\mathbb{Z})$ that are also eigenfunctions of all Hecke operators. Associated to such a form f is its L-function, which we normalize so that the functional equation connects values at s to $1-s$:

$$\Lambda(s,f) = (2\pi)^{-s}\Gamma\left(s+\frac{k-1}{2}\right)L(s,f) = i^k\Lambda(1-s,f).$$

In the case $k \equiv 2 \pmod 4$, the sign of this functional equation is -1, and all the central values $L(\frac{1}{2},f)$ are zero. In the case $k \equiv 0 \pmod 4$, the sign of the functional equation is $+1$, and we ask for the distribution of $L(\frac{1}{2},f)$ (or, in keeping with Selberg's theorem, $\log L(\frac{1}{2},f)$). In this situation, a remarkable result of Waldspurger **[173]** (see also **[185]** for an explicit version) relates these central L-values to the squares of Fourier coefficients of a half-integer weight modular form associated to f (namely its Shimura correspondent). As a byproduct, we know that $L(\frac{1}{2},f)$ is nonnegative, and it is conjectured never to be zero. The Keating–Snaith conjectures predict that for large $k \equiv 0 \pmod 4$, the values $\log L(\frac{1}{2},f)$ are distributed like a real Gaussian with mean $-\frac{1}{2}\log\log k$ and variance $\log\log k$. Since the mean is negative, the values $L(\frac{1}{2},f)$ in this family are typically small, of size $(\log k)^{-\frac{1}{2}+o(1)}$.

A related example is to fix a newform f, and to consider the family of quadratic twists of f. Once again normalizing so that the functional equation connects s and $1-s$, our interest is in the central values $L(\frac{1}{2}, f\times\chi_d)$, where d runs over fundamental discriminants $|d| \le X$ with d coprime to the level of f for simplicity. As in the previous example, half of these twists will have a functional equation with $-$ sign (where the central L-value vanishes), and we restrict attention to the complementary case when the sign is $+$. Again Waldspurger's formula shows that the central L-values are nonnegative, but it is possible for these values to be 0. For example, if f corresponds to an elliptic curve, then the Birch–Swinnerton-Dyer conjectures predict that the central value is zero when the quadratic twist of this elliptic curve has positive rank (and the rank must also be even when the sign of the functional equation is $+$). However, one expects that typically $L(\frac{1}{2}, f\times\chi_d)\neq 0$, and the Keating–Snaith conjectures predict further that the distribution of $\log L(\frac{1}{2}, f\times\chi_d)$ (where $|d|\le X$ is coprime to the level of f and the twist has $+$ sign of the functional equation) is that of a real Gaussian with mean $-\frac{1}{2}\log\log X$ and variance $\log\log X$.

The classification of families into unitary, symplectic, and orthogonal is based on the philosophy of Katz and Sarnak **[98]** which connects (conjecturally) the distribution of low lying zeros in these families to the distribution of eigenvalues near 1 of large random matrices chosen from the corresponding classical groups—we shall discuss these links to random matrix theory later. We now give heuristic reasons to explain the three different Keating–Snaith conjectures, point out the obstructions to making these precise, and describe the partial progress that has been made.

Recall that in (2.4) we considered approximations to $\log\zeta(\frac{1}{2}+it)$ by Dirichlet series over prime powers of a flexible length x. In (2.6) we saw that for $\zeta(\frac{1}{2}+it)$ the contribution of prime powers p^k with $k\ge 3$ is bounded, and the contribution from prime squares is also typically small. Finally, the distribution of the sums over primes could be understood by computing moments. We now consider analogues of this calculation for the families discussed above, and the key difference in the orthogonal and symplectic cases will arise in the contribution of squares of primes.

Let us first look at the unitary family of Dirichlet characters (mod q) with q a large prime. Suppose that we have an approximation of the form

$$\log L\left(\frac{1}{2},\chi\right) \approx \sum_{n\le x} \frac{\Lambda(n)}{\sqrt{n}\log n}\chi(n) = \sum_{p\le x}\frac{\chi(p)}{\sqrt{p}} + \frac{1}{2}\sum_{p\le\sqrt{x}}\frac{\chi(p)^2}{p} + O(1). \tag{3.1}$$

A typical character χ (mod q) is not quadratic; χ^2 is then a nonprincipal character and the sum over prime squares above is typically of bounded size, behaving a lot like $\log L(1,\chi^2)$. We are left with the sum over primes, and if x is a small power of q, then we can understand the moments of this sum (much as in (2.7)) using the orthogonality relation for the characters (mod q) (in place of (2.8)). This gives a heuristic justification for the Keating–Snaith conjectures in this family, and the missing ingredient is the very first step which may fail badly, for example, if $L(\frac{1}{2},\chi) = 0$ for many characters χ (mod q).

Consider next the symplectic example of quadratic Dirichlet L-functions $L(s,\chi_d)$ with d ranging over fundamental discriminants $|d| \le X$. Suppose that an approximation as in (3.1) holds. Since χ_d is a quadratic character, note that the squares of primes in (3.1) have $\chi_d(p)^2 = 1$ (ignoring the primes p that divide d), and so these terms contribute

$$\frac{1}{2}\sum_{p\le\sqrt{x}}\frac{1}{p} \sim \frac{1}{2}\log\log x \sim \frac{1}{2}\log\log X,$$

if x is a small power of X. Thus the prime square terms account for the mean of $\log L(\frac{1}{2},\chi_d)$ being $\sim \frac{1}{2}\log\log X$ in the Keating–Snaith conjectures. If x is a small power of X, we may compute the moments of the sum over primes:

$$\sum_{|d|\le X}\left(\sum_{p\le x}\frac{\chi_d(p)}{\sqrt{p}}\right)^k = \sum_{p_1,\dots,p_k\le x}\frac{1}{\sqrt{p_1\cdots p_k}}\sum_{|d|\le X}\left(\frac{d}{p_1\cdots p_k}\right).$$

The inner sum over d may be viewed as a character sum (mod $p_1\cdots p_k$). This character is principal if $p_1\cdots p_k$ is a square, and we get a main term here, while if $p_1\cdots p_k$ is not a square we may expect the character sum to cancel out (and this can be justified if x^k is small in comparison to X). The product $p_1\cdots p_k$ can be a square only if k is even, and the primes $p_1,\dots,p_k$ can be paired off into $k/2$ equal pairs. With a little calculation, this shows that the moments of the sum over primes match the moments of a real Gaussian with mean 0 and variance $\sum_{p\le x} 1/p \sim \log\log X$. Taking into account the shift in mean arising from the prime square terms, this gives a heuristic justification for the Keating–Snaith conjecture.

Finally, let us look at the orthogonal family of quadratic twists of a newform in the case where the sign of the functional equation is $+$. The L-function $L(s, f\times\chi_d)$ is given by an Euler product, the pth factor of which (for a prime p not dividing the level of the form) takes the shape

$$\left(1-\frac{\alpha_p\chi_d(p)}{p^s}\right)^{-1}\left(1-\frac{\beta_p\chi_d(p)}{p^s}\right)^{-1},$$

where $\alpha_p \beta_p = 1$ and $\alpha_p + \beta_p = \lambda(p)$ is the normalized Hecke eigenvalue of f (normalized so that the Deligne bound gives $|\lambda(p)| \le 2$). The logarithm of this Euler factor is

$$\sum_{k=1}^{\infty} (\alpha_p^k + \beta_p^k) \frac{\chi_d(p^k)}{kp^{ks}},$$

and in analogy with (2.4), (2.6), (3.1), we may hope to approximate $\log L(\frac{1}{2}, f \times \chi_d)$ by

$$\begin{aligned} &\sum_{p \le x} \frac{(\alpha_p + \beta_p)\chi_d(p)}{\sqrt{p}} + \frac{1}{2} \sum_{p \le \sqrt{x}} \frac{(\alpha_p^2 + \beta_p^2)\chi_d(p)^2}{p} + O(1) \\ &= \sum_{p \le x} \frac{\lambda(p)\chi_d(p)}{\sqrt{p}} + \frac{1}{2} \sum_{p \le \sqrt{x}} \frac{\lambda(p)^2 - 2}{p} + O(1). \end{aligned} \tag{3.2}$$

If the discriminants d go up to size X, and x is a small power of X, then the distribution of $\sum_{p \le x} \lambda(p)\chi_d(p)/\sqrt{p}$ may be determined by computing moments (similarly to the discussion for $L(\frac{1}{2}, \chi_d)$). The prime terms in (3.2) are distributed like a real Gaussian with mean 0 and variance

$$\sum_{p \le x} \frac{\lambda(p)^2}{p} \sim \log\log x \sim \log\log X, \tag{3.3}$$

by Rankin–Selberg theory. In view of (3.3), the prime square terms in (3.2) contribute

$$\frac{1}{2} \sum_{p \le \sqrt{x}} \frac{\lambda(p)^2 - 2}{p} \sim -\frac{1}{2} \log\log \sqrt{x} \sim -\frac{1}{2} \log\log X.$$

This justifies the Keating–Snaith conjecture for this family.

In all these heuristics, it is the first step of connecting $\log L(\frac{1}{2})$ to sums over prime powers that is a serious stumbling-block. Indeed, if $L(\frac{1}{2})$ is zero (or if there is a zero very close to $\frac{1}{2}$) for many elements in the family, then the Keating–Snaith conjectures would not hold. This problem does not arise in the continuous Selberg theorem, since the points t with $\frac{1}{2} + it$ very close to a zero of $\zeta(s)$ have small measure and thus do not affect the distribution.

The problem of nonvanishing of L-functions has been investigated extensively, but in general it remains a challenge to show that almost all elements in a family have nonzero central value. More often, progress towards this problem focusses on showing that a positive proportion of L-functions in a family have nonzero central value. To give a few examples: in the family of Dirichlet characters $\chi \pmod q$, Khan and Ngo **[103]** have shown that at least $\frac{3}{8}$ of these characters have $L(\frac{1}{2}, \chi) \ne 0$; in the family of quadratic Dirichlet L-functions, Soundararajan **[161]** shows that a proportion at least $\frac{7}{8}$ of such central values are nonzero; in the family $\mathcal{H}_k$ of all Hecke eigenforms of weight $k \equiv 0 \pmod 4$ for the full modular group, with $k \le K$, Iwaniec and Sarnak **[93]** show that at least $\frac{1}{2}$ of the central values are non-zero, and improving this proportion (in a certain sense) would have consequences for the existence of Landau–Siegel zeros of Dirichlet L-functions.

There are some situations where, for deep algebraic reasons, one can show that most central values in a family are nonzero, but these arguments do not appear to control the size of the central value, or to deal with the possibility that there might be a zero very near $\frac{1}{2}$. For example, Chinta [38] (following work of Rohrlich [147]) has shown that if E is an elliptic curve over $\mathbb{Q}$ then for all but $O(q^{\frac{7}{8}})$ of the Dirichlet characters mod q (with q a large prime) one has $L(\frac{1}{2}, E \times \chi) \neq 0$. This exploits the fact (established by Shimura) that if χ^σ is a Galois conjugate of the character χ, then the vanishing of $L(\frac{1}{2}, E \times \chi)$ is equivalent to the vanishing of $L(\frac{1}{2}, E \times \chi^\sigma)$ (the algebraic parts of these L-values are Galois conjugate). Another example where algebraic techniques are very successful concerns the family of quadratic twists of an elliptic curve. In special cases, Smith [160] has shown that the (algebraic) rank of quadratic twists of elliptic curves is typically 0 (when the sign of the functional equation is $+$) or 1 (when the sign is $-$). The Birch–Swinnerton-Dyer conjecture (on which there has been a lot of progress in the cases of rank 0 and 1) would then yield Goldfeld's conjecture that the central L-values are typically nonzero when the sign of the functional equation is $+$.

If there is a zero at or very near $\frac{1}{2}$, we might expect that its effect is to make $|L(\frac{1}{2})|$ unusually small. This observation was made in Soundararajan [164], where it was shown (assuming GRH) that $\log |L(\frac{1}{2})|$ can be bounded from above using Dirichlet series over prime powers of flexible length; we shall discuss this in more detail in Section 6. It was also observed in [164] that one could (assuming a suitable GRH) establish a one sided version of the Keating–Snaith conjecture, showing that the frequency with which $\log |L(\frac{1}{2})| \geq \text{Mean} + \lambda \sqrt{\text{Var}}$ is bounded above by the expected Gaussian $\frac{1}{\sqrt{2\pi}} \int_\lambda^\infty e^{-x^2/2} dx$; here λ is a fixed real number, the size of the family is assumed to grow. Further, if one knew that most elements in the family did not have a zero near $\frac{1}{2}$ (which, for example, would follow from the "one level density" conjectures in Katz and Sarnak [98]) then the Keating–Snaith conjecture for $\log |L(\frac{1}{2})|$ would follow.

Such one sided central limit theorems were first made precise (and unconditional) by Hough [89] in certain families of L-functions. Hough's approach relies on knowledge of a zero density estimate putting most low lying zeros of L-functions in the family close to the critical line—an analogue of Selberg's zero density estimate for the zeta function, mentioned in Section 2. For example, Hough's approach would work for $\log |L(\frac{1}{2}, \chi)|$ in the unitary family of Dirichlet characters $\chi \pmod q$, or $\log |L(\frac{1}{2}, \chi_d)|$ in the symplectic family of quadratic Dirichlet L-functions, or in the orthogonal family $\log L(\frac{1}{2}, f)$ where f ranges over Hecke eigenforms of weight $k \equiv 0 \pmod 4$ for the full modular group.

An alternative approach to this half of the Keating–Snaith conjectures is developed in Radziwiłł and Soundararajan [141]. This method is arguably simpler and also more widely applicable, relying only on knowledge of the first moment "+ epsilon" in the family, and avoiding zero density estimates (which require knowledge of the second moment "+ epsilon"). In [141] the method is illustrated for the family of quadratic twists of an elliptic curve (with positive sign of the functional equation), where the zero density estimates required in Hough's approach are not known. Conjecturally, the central values in this family (when nonzero) measure (after accounting for quantities such as Tamagawa factors that are relatively easy to understand) the size of the Tate–Shafarevich group for the twisted elliptic curve. The Keating–Snaith conjecture thus predicts that the sizes of Tate–Shafarevich groups in the family of quadratic twists have a log normal distribution, with prescribed means and variance (see Conjecture 1 in [141]). The method applies to quadratic twists of any newform (holomorphic or Maass form), and thus (by Waldspurger's formula) also gives information on the size of Fourier coefficients of half-integer weight modular forms, establishing that these are typically a little bit smaller than the conjectured Ramanujan bounds.

Another application where this method works is to the problem of the fluctuations of a quantum observable for the modular surface. Let ψ denote a fixed even Hecke–Maass form for the full modular group, and let ϕ_j denote an even Hecke–Maass form with eigenvalue λ_j. The problem is to understand $\mu_j(\psi) = \int_{\mathrm{PSL}_2(\mathbb{Z})\backslash\mathbb{H}} \psi(z)|\phi_j(z)|^2 \frac{dxdy}{y^2}$ for large eigenvalue λ_j. For generic hyperbolic surfaces, it has been suggested in the physics literature [57] that similar quantum fluctuations have a Gaussian distribution. In the case of the modular group, $|\mu_j(\psi)|^2$ is related to the central value $L(\frac{1}{2}, \psi \times \phi_j \times \phi_j)$, so that the Keating–Snaith conjectures predict that it is in fact $\log|\mu_j(\psi)|$ (rather than $\mu_j(\psi)$ itself) that has a normal distribution. A one sided central limit theorem for $\log|\mu_j(\psi)|$ is obtained in Siu [159], and in particular it follows that $\lambda_j^{\frac{1}{4}}|\mu_j(\psi)| = o(1)$ for almost all eigenfunctions ϕ_j.

We have already discussed that the problem of nonvanishing of central L-values is a barrier to obtaining lower bounds towards the Keating–Snaith conjectures. There are two analytic techniques that produce a positive proportion of nonzero central values of L-functions in families: (i) the mollifier method, which is unconditional and relies on knowledge of two moments ("+ epsilon") and (ii) understanding 1-level densities of low lying zeros, which is conditional on GRH and is not always guaranteed to yield a nonzero proportion. Both of these methods may be refined to permit an understanding of the typical size of nonzero L-values that are produced [165]. Here are two such sample results. In the family of quadratic Dirichlet L-functions, where we know [161] that $\frac{7}{8}$ of the fundamental discriminants $|d| \le X$ satisfy $L(\frac{1}{2}, \chi_d) \ne 0$, we may establish that for any interval (α, β) of $\mathbb{R}$ and large X,

$$\#\left\{|d| \le X : \frac{\log|L(\frac{1}{2}, \chi_d)| - \frac{1}{2}\log\log X}{\sqrt{\log\log X}} \in (\alpha, \beta)\right\}$$
$$\ge \left(\frac{7}{8}\frac{1}{\sqrt{2\pi}}\int_\alpha^\beta e^{-x^2/2}dx + o(1)\right)\#\{|d| \le X\}.$$

In the family of quadratic twists of a fixed newform f with positive sign of the functional equation, on GRH it is known that a proportion $\ge \frac{1}{4}$ of such L-values are nonzero (see [85]), and we may refine this to yield (with $\mathcal{E}(X)$ denoting the set of fundamental discriminants $|d| \le X$ with the quadratic twist of f has positive sign)

$$\#\left\{d \in \mathcal{E}(X) : \frac{\log L(\frac{1}{2}, f \times \chi_d) + \frac{1}{2}\log\log X}{\sqrt{\log\log X}} \in (\alpha, \beta)\right\}$$
$$\ge \left(\frac{1}{4}\frac{1}{\sqrt{2\pi}}\int_\alpha^\beta e^{-x^2/2}dx + o(1)\right)|\mathcal{E}(X)|.$$

Finally, we mention recent work of Bui et al. [26] which considers a variant of the Keating–Snaith conjectures when L-values are counted with suitable weights (which depend on "mollified L-values").

4. MOMENTS OF ZETA AND L-FUNCTIONS

A classical problem, going back to Hardy and Littlewood, asks for an understanding of the moments of $\zeta(\frac{1}{2}+it)$,

$$M_k(T) = \int_0^T \left|\zeta\left(\frac{1}{2}+it\right)\right|^{2k} dt, \tag{4.1}$$

where k is a natural number. Hardy and Littlewood established that $M_1(T) \sim T\log T$ (see [170]), and this was later refined by Ingham who showed that

$$M_1(T) = \int_0^T \left|\zeta\left(\frac{1}{2}+it\right)\right|^{2} dt = T\log\frac{T}{2\pi} + (2\gamma-1)T + E(T), \tag{4.2}$$

with $E(T) = O(T^{\frac{1}{2}}\log T)$, with a further refinement in Balasubramanian [13] yielding $E(T) = O(T^{\frac{1}{3}+\varepsilon})$. Ingham also established an asymptotic for the fourth moment: $M_2(T) \sim \frac{1}{2\pi^2}T(\log T)^4$, which was refined by Heath-Brown [81] to

$$M_2(T) = \int_0^T \left|\zeta\left(\frac{1}{2}+it\right)\right|^{4} dt = TP_4(\log T) + O\big(T^{\frac{7}{8}+\varepsilon}\big), \tag{4.3}$$

for a polynomial P_4 of degree 4 with leading coefficient $1/(2\pi^2)$.

Despite much effort, these remain the only two cases in which an asymptotic formula for $M_k(T)$ is known. To explain why, we recall that Hardy and Littlewood gave an "approximate functional equation" (in fact, Riemann's unpublished notes had a more precise version, known now as the Riemann–Siegel formula)

$$\zeta\left(\frac{1}{2}+it\right) \approx \sum_{n\le\sqrt{|t|/2\pi}} \frac{1}{n^{\frac{1}{2}+it}} + e^{i\vartheta(t)} \sum_{n\le\sqrt{|t|/2\pi}} \frac{1}{n^{\frac{1}{2}-it}}, \tag{4.4}$$

where $e^{i\vartheta(t)} = \pi^{it/2}\Gamma((\frac{1}{2}-it)/2)/(\pi^{-it/2}\Gamma((\frac{1}{2}+it)/2))$ is the ratio of Γ-factors in the functional equation for $\zeta(s)$. Thus $\zeta(\frac{1}{2}+it)$ can be approximated by two Dirichlet polynomials of length about $\sqrt{|t|}$. We saw in (2.8) that the mean square of Dirichlet polynomials of length up to T could be evaluated, with the diagonal terms making the dominant contribution. This permits the evaluation of the second moment (4.2) with Ingham's bound on the remainder term $E(T)$ (we have not discussed the cross terms that arise in squaring (4.4) but these turn out to be negligible). Similarly, we can approximate $\zeta(\frac{1}{2}+it)^2$ by two Dirichlet polynomials of length about $|t|/2\pi$, and this leads to Ingham's asymptotic for $M_2(T)$, although the more precise form in (4.3) requires further ideas. When $k\ge 3$, the complexity of $\zeta(\frac{1}{2}+it)^k$ becomes too great; to approximate it, we require Dirichlet polynomials of length about $|t|^{k/2}$ (which is now larger than $|t|$), and (2.8) is no longer sufficient to estimate the mean-square of such long Dirichlet polynomials. Let $d_k(n)$ denote the k-divisor function, which arises as the Dirichlet series coefficients of $\zeta(s)^k = \sum_{n=1}^{\infty} d_k(n)/n^s$ (valid for $\mathrm{Re}(s)>1$). One new problem that arises when considering higher moments involves the correlations

$$\sum_{n\le x} d_k(n)d_k(n+h). \tag{4.5}$$

One would like asymptotics for such quantities, uniformly in a range for h, and while this problem has been solved for $k = 2$ (and underlies the precise asymptotics given in (4.3)), when $k = 3$ or larger, asymptotics for the quantity in (4.5) remain unknown (even in the case $h = 1$).

Indeed, until the late 1990s it was not even clear what the conjectural asymptotics for $M_k(T)$ should be. However, in the last 25 years, much progress has been made in understanding conjecturally the nature of these moments, obtaining lower bounds of the correct conjectured value (for all positive real k), and obtaining complementary upper bounds of the correct order conditional on the Riemann Hypothesis. Similar progress has been made for moments in a number of different families of L-functions. We shall discuss these conjectures and the progress towards them in the following sections, but first give some motivation for considering such moments.

One motivation for considering the moments of $\zeta(s)$ is that they capture information about the large values of $|\zeta(\frac{1}{2} + it)|$. The Lindelöf hypothesis that $|\zeta(\frac{1}{2} + it)| \ll_\varepsilon (1 + |t|)^\varepsilon$ (which is a consequence of RH) is equivalent to the bound $M_k(T) \ll_{k,\varepsilon} T^{1+\varepsilon}$ for all $k \in \mathbb{N}$. From the approximate functional equation (4.4) it follows that $|\zeta(\frac{1}{2} + it)| \ll (1 + |t|)^{\frac{1}{4}}$, a bound known as the *convexity bound*. Going beyond the convexity bound involves showing cancelation in the exponential sums in (4.4), and has remained an active problem from its initiation by Weyl, and Hardy and Littlewood who showed early on that $|\zeta(\frac{1}{2} + it)| \ll (1 + |t|)^{\frac{1}{6}+\varepsilon}$ (see **[170]** and the best current exponent may be found in **[25]**). Sharp moment estimates encode Lindelöf bounds on average, and in some cases can also yield pointwise subconvexity estimates. For example, we note that Ingham's bound $E(T) \ll T^{\frac{1}{2}} \log T$ (for the error term in the second moment (4.2)) implies that $\int_T^{T+1} |\zeta(\frac{1}{2} + it)|^2 dt \ll T^{\frac{1}{2}} \log T$ from which the convexity bound $|\zeta(\frac{1}{2} + it)| \ll |t|^{\frac{1}{4}+\varepsilon}$ may be deduced. Similarly Balasubramanian's improved estimate for $E(T)$ implies the Hardy–Littlewood–Weyl subconvexity bound $|\zeta(\frac{1}{2} + it)| \ll (1 + |t|)^{\frac{1}{6}+\varepsilon}$. Similarly, Ingham's asymptotic for the fourth moment yields the convexity bound, while the more precise result (4.3) of Heath-Brown gives a subconvexity bound for $\zeta(s)$. As a third example of bounds for moments that encode good pointwise bounds, we mention Heath-Brown's **[80]** estimate for the 12th moment

$$\int_0^T \left|\zeta\left(\frac{1}{2} + it\right)\right|^{12} dt \ll T^{2+\varepsilon},$$

which again contains the bound $|\zeta(\frac{1}{2} + it)| \ll |t|^{\frac{1}{6}+\varepsilon}$.

Ingham's work on the fourth moment of $\zeta(\frac{1}{2} + it)$ is also crucial in establishing "zero density estimates" which are bounds for the number of potential exceptions to the Riemann hypothesis. These have arithmetic applications, for example, playing a key role in showing that the prime number theorem holds in short intervals: $\pi(x + h) - \pi(x) \sim h/\log x$ provided $x^{\frac{7}{12}+\varepsilon} < h \le x$. A sharp bound for the sixth moment (for instance) would lead to improvements in zero-density results and in the application to the prime number theorem. We refer to Chapter 10 of **[92]** for a discussion of these themes.

There is a large body of work studying analogous problems for moments of central values in families of L-functions, and in many cases asymptotics for small moments are known. We give a few examples here, and discuss some more in Section 6. Two motivations for studying such questions are (i) the problem of showing that many central values are nonzero, which can be attacked analytically if we know two moments with a little room to spare (we gave a few examples of such results in the previous section), and (ii) obtaining subconvexity bounds for L-functions (there is a vast literature here, and we content ourselves to pointing to earlier surveys on this topic [63,94,119,128] and to Nelson [130,132] for very recent progress).

The unitary family of Dirichlet characters (mod q) (for a large prime q) is closest in spirit to $\zeta(\frac{1}{2}+it)$, but there are still some differences. It is easy to evaluate the second moment $\sum^*_{\chi \pmod q} |L(\frac{1}{2},\chi)|^2$ (where the $*$ indicates that the sum is restricted to primitive characters) and, in analogy with (4.1), this is, $\sim q\log q$. The fourth moment can also be evaluated, and in analogy with Ingham's result, Heath-Brown [83] established that $\sum^*_{\chi \pmod q} |L(\frac{1}{2},\chi)|^4 \sim \frac{1}{2\pi^2} q(\log q)^4$. However, an analogue of (4.3), obtaining lower order terms in the asymptotic formula with a "power saving" in the error term, proved substantially more difficult, and was first achieved in the work of Young [174]. Higher moments remain unknown, although one can make progress by averaging over q (see Section 6). Another natural unitary family is the twists of a fixed Hecke eigenform by Dirichlet characters (mod q). The complexity of the second moment in this family is naively comparable to the fourth moment of Dirichlet L-functions, but there are further formidable difficulties. An extensive discussion of this problem, with variants and applications, may be found in the memoir of Blomer et al. [18].

In the symplectic case of quadratic Dirichlet L-functions, the first three moments $\sum_{|d|\le x} L(\frac{1}{2},\chi_d)^k$ are known (see [96,161], and for interesting work on the error term in the cubic moment see [55,175]), and the asymptotics in these cases ($k=1,2,3$) take the shape of $xP_k(\log x)$ for a polynomial P_k of degree $k(k+1)/2$. We shall explain in the next section how this ties in with the Keating–Snaith conjecture for the distribution of $\log L(\frac{1}{2},\chi_d)$. The techniques behind evaluating these moments also establish that a proportion at least $\frac{7}{8}$ of these values are nonzero (see [161]).

As an example of an orthogonal family, consider the set $\mathcal{H}_k$ of Hecke eigenforms for the full modular group with large weight $k \equiv 0 \pmod 4$. Here the moments $\sum_{f\in\mathcal{H}_k} L(\frac{1}{2},f)^r$ may be evaluated for $r=1$, 2, and if an extra averaging over $K \le k \le 2K$ is taken, then in the cases $r=3$ and 4 also (this follows from the techniques in [93]). The asymptotic answers here are of the shape $|\mathcal{H}_k|P_r(\log k)$ for a polynomial P_r of degree $r(r-1)/2$. A sharp bound for the third moment (without an average in k) is established in Peng [136]; this permits a subconvexity bound $L(\frac{1}{2},f) \ll k^{\frac{1}{3}+\varepsilon}$, which is comparable in strength to the Hardy–Littlewood–Weyl subconvexity bound for $\zeta(\frac{1}{2}+it)$. An analogous cubic moment (with such a subconvexity bound) has been studied in the case of Maass forms by Ivic [91]; interestingly, these cubic moments are also connected by a beautiful formula of Motohashi [127] to the fourth moment of $\zeta(\frac{1}{2}+it)$. Substantial progress has been made towards obtaining estimates for the fifth moment for modular forms (in the weight and level aspects) and in finding "reciprocity relations" among the fourth moments in different families; see [19,102,104].

We mention one more striking example: the work of Conrey–Iwaniec [47] gives sharp estimates for the cubic moment of $L(\frac{1}{2}, f \times \chi)$ where f runs over modular forms of level dividing q (an odd square-free integer) and χ denotes the quadratic character (mod q). This gives a good Weyl-type subconvexity bound for such L-values, and an analogous calculation for Maass forms gives Weyl-type subconvexity bounds for quadratic Dirichlet L-functions (improving upon classical results of Burgess). Further spectacular work in this direction may be found in Petrow and Young [137], and Nelson [131].

5. CONJECTURES FOR THE ASYMPTOTICS OF MOMENTS

Before discussing in detail the moments on the critical line, let us consider the moments on the line $\mathrm{Re}(s) = \sigma > \frac{1}{2}$. We mentioned in Section 2 that $\zeta(\sigma + it)$ is distributed like the random object $\zeta(\sigma, \mathbb{X})$ defined in (1.4). We may therefore expect that for any $k \in \mathbb{N}$ and as $T \to \infty$,

$$\frac{1}{T} \int_0^T \left|\zeta(\sigma + it)\right|^{2k} dt \sim \mathbb{E}\left[\left|\zeta(\sigma, \mathbb{X})\right|^{2k}\right] = \sum_{n=1}^{\infty} \frac{d_k(n)^2}{n^{2\sigma}}, \tag{5.1}$$

since $\zeta(\sigma, \mathbb{X})^k = \sum_{n=1}^{\infty} d_k(n)\mathbb{X}(n)/n^{\sigma}$ with $d_k(n)$ being the k-divisor function (the series converges almost surely for $\sigma > \frac{1}{2}$). When $\sigma > 1$, it is clear that (5.1) holds (indeed, for any real number k), since the values $|\zeta(\sigma + it)|$ lie in a compact subset of $(0, \infty)$ and the distributions match. The case $\sigma = 1$ is more delicate, but with a little more effort one can justify (5.1) here as well. Moving now into the critical strip, there is no known value of $\frac{1}{2} < \sigma < 1$ where the asymptotic (5.1) is known to hold for all $k \in \mathbb{N}$. Indeed, such a result would imply that $|\zeta(\sigma + it)| \ll |t|^{\varepsilon}$, which remains unknown for any $\frac{1}{2} < \sigma < 1$. However, if one is willing to assume RH, then it is possible to approximate $\zeta(\sigma + it)^k$ by short Dirichlet polynomials, and then (5.1) follows for all real numbers k.

Returning to moments on the critical line, as mentioned previously, asymptotic formulae for $M_k(T)$ are known only in the cases $k = 1$ and 2. But, using (5.1) as a guide, we may guess the order of magnitude of $M_k(T)$. The series on the right-hand side of (5.1) diverges when $\sigma = \frac{1}{2}$, but we might consider truncating that sum around size T. It is easy to show that for any real number k,

$$\sum_{n \le T} \frac{d_k(n)^2}{n} \sim \frac{a_k}{\Gamma(k^2 + 1)} (\log T)^{k^2}, \tag{5.2}$$

where

$$a_k = \prod_p \left(1 - \frac{1}{p}\right)^{k^2} \left(\sum_{a=0}^{\infty} \frac{d_k(p^a)^2}{p^a}\right). \tag{5.3}$$

Thus one might guess that for all positive real numbers k, $M_k(T) \sim C_k T (\log T)^{k^2}$ for some constant C_k. Conrey and Ghosh suggested that it might be instructive to write C_k as $g_k a_k / \Gamma(k^2 + 1)$, and expected that the unknown factor g_k might have nice properties (for example, that g_k would be a natural number when k is a natural number). The Hardy–Littlewood asymptotics for the second moment (see (4.2)) is in keeping with this conjecture, and gives $g_1 = 1$. Similarly, Ingham's result on the fourth moment (see (4.3)) yields $g_2 = 2$.

Another way to guess at the order of magnitude for $M_k(T)$ arises from extrapolations of Selberg's central limit theorem. If X is a random variable that is normally distributed with mean μ and variance σ^2, then for any real number t we have

$$\begin{aligned}\mathbb{E}[e^{tX}] &= \frac{1}{\sqrt{2\pi}\sigma}\int_{-\infty}^{\infty}\exp\biggl(tu - \frac{(u-\mu)^2}{2\sigma^2}\biggr)du \\ &= e^{t\mu+t^2\sigma^2/2}\frac{1}{\sqrt{2\pi}\sigma}\int_{-\infty}^{\infty}\exp\biggl(-\frac{(u-\mu-t\sigma^2)^2}{2\sigma^2}\biggr)du = e^{t\mu+t^2\sigma^2/2}. \end{aligned} \tag{5.4}$$

Further, the dominant contribution above comes from values of X that are about $\mu + t\sigma^2 + O(\sigma)$. Selberg's theorem tells us that $\log|\zeta(\frac{1}{2}+it)|$ is distributed like a Gaussian with mean 0 and variance $\sim \frac{1}{2}\log\log T$. The calculation in (5.4) therefore suggests that

$$M_k(T) = \int_0^T \exp\biggl(2k\log\biggl|\zeta\biggl(\frac{1}{2}+it\biggr)\biggr|\biggr)dt = T\exp\biggl((2k)^2\frac{\frac{1}{2}\log\log T}{2}\biggr) = T(\log T)^{k^2}.$$

Moreover, the dominant contribution to the $(2k)$th moment should arise from values of $\zeta(\frac{1}{2}+it)$ of size $(\log T)^k$ and the set on which such values are attained has measure about $T/(\log T)^{k^2}$. We should clarify that Selberg's theorem is concerned with typical values of $\log|\zeta(\frac{1}{2}+it)|$, which are on the scale of $\sqrt{\log\log T}$, whereas the moments $M_k(T)$ are concerned with the large deviations regime where $\log|\zeta(\frac{1}{2}+it)|$ is of size $k\log\log T$. In this regime Selberg's result does not immediately apply, and indeed we should expect some deviations from the Gaussian, which are reflected in the constant C_k appearing in the conjecture for $M_k(T)$ (see **[60, 139]**). Later we shall discuss a coarse version of Selberg's theorem in this large deviations regime **[164]**, conditional on RH, which leads to good (conditional) upper bounds for $M_k(T)$. To give an analogy, both $\omega(n)$ (the number of distinct prime factors of n) and $\log d(n)/\log 2$ (with $d(n)$ being the divisor function) are additive functions that are distributed (if n is chosen uniformly in $[1, N]$) like a Poisson random variable with parameter $\log\log N$; this is the Erdős–Kac theorem (noting that Poisson with large parameter approximates a Gaussian). This suggests that both $\sum_{n\le N} 2^{\omega(n)}$ and $\sum_{n\le N} d(n)$ are on the scale of $N\log N$, but the constants involved in the asymptotics are not immediate (and are different in the two cases).

Neither of the two heuristics given above makes a prediction for the constant $C_k = a_k g_k/\Gamma(k^2+1)$. Indeed, until the 1990s there was no clear conjecture as to the value of g_k for any $k \neq 1$, or 2. Then Conrey and Ghosh **[44, 45]**, based on an earlier conjecture of Balasubramanian, Conrey, and Heath-Brown **[15]**, advanced the conjecture that $g_3 = 42$. A little later Conrey and Gonek **[46]**, based on conjectures on the asymptotics of divisor correlation sums (as in (4.5)), arrived again at the conjecture that $g_3 = 42$ (see Ng **[133]** for further work on making this precise), while also advancing the conjecture that $g_4 = 24024$. These methods did not extend to produce conjectures for larger k, and the problem once again seemed stuck. A great advance was made when Keating and Snaith **[100]**, using ideas from random matrix theory, arrived at the following remarkable conjecture for $M_k(T)$ for all positive real numbers k.

Conjecture 5.1 (Keating and Snaith). *For any positive real number k, as $T \to \infty$, we have $M_k(T) \sim g_k \frac{a_k}{\Gamma(k^2+1)} T(\log T)^{k^2}$, with*

$$g_k = \Gamma(k^2+1)\frac{G(1+k)^2}{G(1+2k)},$$

where G is the Barnes G-function. In particular, if $k \in \mathbb{N}$ then

$$g_k = (k^2)! \prod_{j=0}^{k-1} \frac{j!}{(k+j)!},$$

so that $g_1 = 1$, $g_2 = 2$, $g_3 = 42$, and $g_4 = 24024$.

We recall that the Barnes G-function is an entire function of order 2 which satisfies the functional equation $G(z+1) = \Gamma(z)G(z)$ with the normalization $G(1) = 1$. Thus for a natural number n, one has $G(n) = \prod_{j=0}^{n-2} j!$.

The key insight of Keating and Snaith was to quantify and develop in the context of value distribution problems a conjectural connection between the distribution of zeros of the Riemann zeta function and the distribution of eigenvalues of large random matrices. Nearly 50 years back, Montgomery [124] initiated a study of the spacings between the ordinates of zeros of the Riemann zeta function, and a chance conversation with Dyson revealed that his partial results on this question matched corresponding statistics in the study of spacings between eigenvalues of large random matrices. Assuming RH for clarity, let $\gamma_1 \le \gamma_2 \le \cdots$ denote the sequence of nonnegative ordinates of zeros of $\zeta(s)$ (written with multiplicity), so that from (2.3) it follows that $\gamma_n \sim 2\pi n/\log n$. The question then is to determine the distribution (as $n \to \infty$) of $(\gamma_{n+1} - \gamma_n)(\log \gamma_n)/(2\pi)$, which has been normalized to have mean spacing 1. For example, with what frequency does this normalized spacing lie in a given interval $(\alpha, \beta) \subset (0, \infty)$? One way to express the (amazing!) conjectured answer is as follows. Consider a random element g drawn from the unitary group $\mathrm{U}(N)$ with respect to the Haar measure dg (normalized so that $U(N)$ has volume 1). Each such g has eigenvalues $e^{i\theta_1}, \dots, e^{i\theta_N}$ with the angles ordered $0 \le \theta_1 \le \theta_2 \le \cdots \le \theta_N < 2\pi$, and consider the spacings $(\theta_{n+1} - \theta_n)N/(2\pi)$ (normalized to have average approximately 1). Average this spacing distribution over the whole group $\mathrm{U}(N)$, and finally let $N \to \infty$. For example, we could count the frequency with which $(\theta_{n+1} - \theta_n)N/(2\pi)$ lies in (α, β), average that frequency over $\mathrm{U}(N)$, and take the limiting frequency as $N \to \infty$. The model that we have described is known as the *Circular Unitary Ensemble* (CUE), and the same distribution for nearest neighbor spacings arises in other models of random matrices such as the Gaussian Unitary Ensemble (GUE).

Theoretical support for this link between zeros of $\zeta(s)$ and random matrix theory arose first with Montgomery's calculation of the *pair correlation* of zeros (in certain ranges) mentioned earlier, and this was generalized to general n-level correlations in the work of Rudnick and Sarnak [148]. Experimental support for this link comes from extensive computations of Odlyzko [134] who considered the spacing distribution of about 175 million zeros around the 10^{20}th zero (which occurs at height around 1.5×10^{19}), and found an astonishingly close match between the empirical data and the predicted answer. Yet, Odlyzko's data found that the numerical data did not match closely some other statistics for $\zeta(s)$ such as Selberg's theorem on $\log \zeta(\frac{1}{2}+it)$. One might attribute such deviations to the slow growth of the variance $\log\log T$, which even at height 10^{19} is only about 3.7, but Keating and Snaith [100] suggested a much more insightful explanation. They posited that properties of $\zeta(\frac{1}{2}+it)$ for t around a specific height T may be modeled by analogous objects for random matrices of a specific size N, refining the expectation that the large T and large N limits coincide. The relation between N and T is suggested by the average spacing between the zeros at height T, which is about $(2\pi)/\log(T/2\pi)$ by (2.3), and the average spacing between eigenvalues, which is about $(2\pi)/N$. Setting these equal, we arrive at the correspondence $N \approx \log(T/2\pi)$. The analogue of the zeta function, which is determined by its zeros, is the characteristic polynomial of a random matrix, which is determined in a similar fashion by its eigenvalues. Keating and Snaith determined the distribution of $\log\det(I - ge^{-i\theta})$ for a random matrix $g \in \mathrm{U}(N)$, and found that in the large N limit this tends to a complex Gaussian with mean 0 and variance $\log N$ (analogously to Selberg's theorem), but there are lower order terms that are significant for finite N. The range of Odlyzko's computations, $T \approx 1.5 \times 10^{19}$, corresponds to matrices of size $N = 42$, and Keating and Snaith found an excellent fit between Odlyzko's numerical data for $\log \zeta(\frac{1}{2}+it)$ and the distribution of $\log\det(I - ge^{-i\theta})$ for random $g \in \mathrm{U}(42)$ (see Figure 1 in [100]).

Returning to the moments, one might now hope to understand the asymptotic behavior of $M_k(T)$ by computing the analogous moments in the context of $\mathrm{U}(N)$: namely

$$\int_{g\in\mathrm{U}(N)} \frac{1}{2\pi}\int_0^{2\pi} \left|\det\left(I - ge^{-i\theta}\right)\right|^{2k} d\theta\, dg = \int_{g\in\mathrm{U}(N)} \left|\det(I-g)\right|^{2k} dg. \tag{5.5}$$

By the Weyl integration formula expressing the measure dg in terms of the eigenvalues of g, this equals the multiple integral

$$\frac{1}{(2\pi)^N N!}\int_0^{2\pi}\cdots\int_0^{2\pi} \left|\prod_{j=1}^{N}\left(1-e^{i\theta_j}\right)\right|^{2k} \prod_{1\le j<m\le N} \left|e^{i\theta_j}-e^{i\theta_m}\right|^2 d\theta_1\cdots d\theta_N. \tag{5.6}$$

It turns out that the integral in (5.6) may be evaluated exactly using a remarkable formula of Selberg [154] (see [62] for many developments arising from the Selberg integral) and it equals

$$\prod_{j=1}^{N} \frac{\Gamma(j)\Gamma(2k+j)}{(\Gamma(j+k))^2} \sim g_k \frac{N^{k^2}}{\Gamma(k^2+1)}, \tag{5.7}$$

where g_k is as in Conjecture 5.1, and the asymptotic holds for large N. The constant g_k has an intriguing combinatorial interpretation as the number of standard Young tableaux of shape $k \times k$ (that is, the number of ways of filling a $k \times k$ array with the numbers $1, \dots, k^2$ such that the entries along each row and column are in increasing order). See [28, 53, 99] for related combinatorial discussions, and [42, 67] for discussions on the divisibility properties of g_k and related constants.

This calculation motivates Conjecture 5.1, but note that no primes appear in the random matrix model, and so the constant a_k must be "put in by hand." Here we note that the Euler product for a_k in (5.3) arises naturally upon considering $\mathbb{E}[|(1 - X(p)/\sqrt{p})|^{-2k}] = \sum_{a=0}^{\infty} d_k(p^a)^2/p^a$ with $X(p)$ chosen uniformly from the unit circle. Thus the constant a_k may be thought of as arising from a version of the random Euler product, while the g_k term arises from the local behavior of zeros of the zeta function. For an exploration of Conjecture 5.1 along these lines, developing a hybrid Euler–Hadamard product, see the work of Gonek, Hughes, and Keating **[69]**. We mentioned earlier the analogy with determining asymptotics for multiplicative functions such as $k^{\omega(n)}$ or $d_k(n)$, and here the known asymptotic formulae (going back to Landau, Selberg, and Delange) factor as a "local" product over primes together with a "global" term determined by the Poisson behavior of $\omega(n)$; for an interesting discussion of this analogy, see **[95]**.

Random matrix theory also informs our understanding of moments of central values of L-functions in families. While the distribution of spacings between zeros at large height for any given L-function is expected to follow the same law that we described for $\zeta(s)$ (see **[148]**), the distribution of the zeros close to the central point $\frac{1}{2}$ can vary depending on the particular family. Based on analogies with the function field case, Katz and Sarnak **[98]** found (conjecturally) that the distribution of zeros near $\frac{1}{2}$ in families of L-functions fell into the three categories *unitary, symplectic, and orthogonal* (which we discussed in Section 3), and that these distributions matched the distribution of the eigenvalues close to 1 of large random matrices chosen from $\mathrm{U}(N)$, $\mathrm{USp}(2N)$, or $\mathrm{SO}(2N)$ (or $\mathrm{SO}(2N+1)$ depending on the sign of the functional equation). To give an illustration of the Katz–Sarnak conjectures, consider the family of quadratic Dirichlet L-functions $L(s, \chi_d)$ as d ranges over fundamental discriminants, which is expected to have symplectic symmetry. The density of zeros of $L(s, \chi_d)$ near $\frac{1}{2}$ is about $(\log |d|)/(2\pi)$, and a sample question is to understand the distribution of $\gamma_1 \frac{\log |d|}{2\pi}$ where γ_1 is the least nonnegative ordinate of a zero of $L(s, \chi_d)$. To describe the conjectured answer, consider a random matrix $g \in \mathrm{USp}(2N)$ (chosen with respect to Haar measure normalized to have total volume 1) and write its eigenvalues as $e^{\pm i\theta_1}, e^{\pm i\theta_2}, \dots, e^{\pm i\theta_N}$ with $0 \le \theta_1 \le \cdots \le \theta_N \le \pi$. Then as d varies over fundamental discriminants $|d| \le X$ with $X \to \infty$, the distribution of $\gamma_1 \frac{\log |d|}{2\pi}$ is identical to the limiting distribution of $\theta_1 \frac{2N}{2\pi}$ for randomly chosen $g \in \mathrm{USp}(2N)$ as $N \to \infty$.

Conrey and Farmer **[42]** proposed that the moments of central values of L-functions in families are also dictated by the symmetry type in the Katz–Sarnak conjectures. In particular, the analogue of the factor g_k should depend only on the symmetry type and not on the particular family, whereas the analogue of the factor a_k will be sensitive to the particular family (in a straightforward way). This was developed further by Keating and Snaith **[101]**, who modeled properties of the central L-values by the characteristic polynomial $\det(I - ge^{-i\theta})$ evaluated at $\theta = 0$, with the size parameter N of the random matrix ensemble chosen to match with the density of zeros in the family. Indeed, it is a consideration of the behavior of $\log \det(I - g)$ in $\mathrm{USp}(2N)$ or $\mathrm{SO}(2N)$ that informed their conjectures for the analogues of Selberg's theorem in symplectic and orthogonal families (discussed in Section 3).

Just as extrapolating Selberg's theorem allows us to guess the order of magnitude of moments of $\zeta(s)$, the Keating–Snaith lognormality conjectures, together with the calculation in (5.4), give an understanding of the order of magnitude of moments in families. For example, in the symplectic example of moments of $L(\frac{1}{2}, \chi_d)$ with $|d| \le X$, since $\log L(\frac{1}{2}, \chi_d)$ is conjectured to be normal with mean $\sim \frac{1}{2} \log\log X$ and variance $\sim \log\log X$, the calculation in (5.4) suggests that $\sum_{|d| \le X} L(\frac{1}{2}, \chi_d)^k$ is of size $X(\log X)^{\frac{k(k+1)}{2}}$. Similarly, in the orthogonal case of Hecke eigenforms $f \in \mathcal{H}_k$, since $\log L(\frac{1}{2}, f)$ is expected to be normal with mean $\sim -\frac{1}{2} \log\log k$ and variance $\sim \log\log k$, the moments $\sum_{f \in \mathcal{H}_k} L(\frac{1}{2}, f)^r$ may be expected to be of order $k(\log k)^{\frac{r(r-1)}{2}}$.

Further, by considering moments of $\det(I - g)$ in the appropriate matrix group, Keating and Snaith [101] formulated analogues of Conjecture 5.1 in families of L-functions. For example, in the family of quadratic Dirichlet L-functions $L(s, \chi_d)$, the analogue of the constant g_k is predicted by considering

$$\int_{g \in \mathrm{USp}(2N)} \det(I - g)^k dg = 2^{2Nk} \prod_{j=1}^{N} \frac{\Gamma(1 + N + j)\Gamma(1/2 + k + j)}{\Gamma(1/2 + j)\Gamma(1 + k + N + j)}$$
$$\sim f_k \frac{N^{k(k+1)/2}}{\Gamma(k(k+1)/2 + 1)}.$$

This calculation again reduces to the Selberg integral, and the constant f_k may be expressed in terms of the Barnes G-function. If k is a natural number then f_k takes the pleasant form $(k(k+1)/2)!/\prod_{j=1}^{k}(2j-1)!!$. After incorporating an analogue of the constant a_k in (5.3), which here is (with $\mathbb{X}(p)$ denoting the random variables modeling quadratic characters discussed in Section 1)

$$\prod_p \Big(1 - \frac{1}{p}\Big)^{\frac{k(k+1)}{2}} \mathbb{E}\Big[\Big(1 - \frac{\mathbb{X}(p)}{\sqrt{p}}\Big)^{-k}\Big]$$
$$= \prod_p \Big(1 - \frac{1}{p}\Big)^{\frac{k(k+1)}{2}} \Big(\frac{p}{2(p+1)}\Big(\Big(1 + \frac{1}{\sqrt{p}}\Big)^{-k} + \Big(1 - \frac{1}{\sqrt{p}}\Big)^{-k}\Big) + \frac{1}{p+1}\Big),$$

we arrive at a conjecture for the moments of $L(\frac{1}{2}, \chi_d)$, which matches the known asymptotics for the first three moments.

The Keating–Snaith conjectures identify the leading order term in the asymptotics for moments, but there will be lower order terms (just a logarithm smaller) which are not identified. We may see this already in the asymptotics for the second and fourth moments of $\zeta(\frac{1}{2} + it)$ (see (4.2) and (4.3)), and other examples in families given in Section 4. Identifying such lower order terms is of interest because the leading order constant in Conjecture 5.1, $a_k g_k / \Gamma(k^2 + 1)$ tends rapidly to zero as k grows, and so for the ranges of T in which numerical investigations may be carried out, the lower order terms may dominate the eventual main term. When k is a positive integer, Conrey et al. [43] conjectured that $M_k(T) = \int_0^T P_k(\log t/2\pi) dt + O(T^{1-\delta})$ (for some $\delta > 0$, and perhaps even any $\delta < \frac{1}{2}$ is permissible) for a polynomial P_k of degree k^2 with leading coefficient $a_k g_k/(k^2!)$, and they gave a "recipe" for determining all the coefficients of P_k. Their recipe predicts the full main term for integral moments in many families of L-functions, but it remains open to give an asymptotic expansion when k is not an integer. The paper [43] also gives numerical evidence towards the full moment conjecture, and further data may be found in [86]. A related approach via *multiple Dirichlet series* is described in the work of Diaconu, Goldfeld, and Hoffstein [54] who develop conjectures for the integral moments of quadratic Dirichlet L-functions (which are in agreement with [43]).

We give a brief illustration of the recipe from [43] in the unitary family of Dirichlet L-functions $\chi \pmod q$ with q a large prime. For simplicity, we consider only even characters (thus $\chi(-1) = 1$), where the functional equation reads $\Lambda(s,\chi) = (q/\pi)^{s/2}\Gamma(s/2)L(s,\chi) = \varepsilon_\chi \Lambda(1-s,\overline{\chi})$ with ε_χ satisfying $|\varepsilon_\chi| = 1$ and $\varepsilon_\chi\varepsilon_{\overline{\chi}} = 1$. Let $\underline{\alpha} = (\alpha_1,\dots,\alpha_k)$, and $\underline{\beta} = (\beta_1,\dots,\beta_k)$ denote two k-tuples of complex numbers (thought of as small), and we also find it convenient to write $\alpha_{k+j} = \beta_j$ and think of $(\underline{\alpha},\underline{\beta})$ as the $2k$-tuple $(\alpha_1,\dots,\alpha_{2k})$. Instead of considering $|L(\frac{1}{2},\chi)|^{2k}$ directly, we work with

$$\Lambda(\chi;\underline{\alpha},\underline{\beta}) := \prod_{j=1}^{k} \Lambda\Big(\frac{1}{2}+\alpha_j,\chi\Big)\Lambda\Big(\frac{1}{2}-\beta_j,\overline{\chi}\Big)$$

and finally let all the parameters α_j and β_j tend to zero (which would then equal $|L(\frac{1}{2},\chi)|^{2k}$ multiplied by the constant $(q/\pi)^{k/2}\Gamma(1/4)^{2k}$). Permuting the k entries in $\underline{\alpha}$ or the k entries in $\underline{\beta}$ does not change $\Lambda(\chi;\underline{\alpha},\underline{\beta})$. Less obviously, it turns out that $\Lambda(\chi;\underline{\alpha},\underline{\beta})$ is invariant under any permutation of the $2k$-entries in $(\underline{\alpha},\underline{\beta})$; this is because any such permutation must change some ℓ of the α's to β's and a corresponding number of β's to α's and 2ℓ applications of the functional equation (ℓ of them with a factor ε_χ and ℓ with a factor $\varepsilon_{\overline{\chi}}$) justify the claim. Thus any conjecture that we propose for $\sum_\chi \Lambda(\chi;\underline{\alpha},\underline{\beta})$ must satisfy this S_{2k} symmetry.

Now if $\mathrm{Re}(s)$ is large, expanding the L-functions into their Dirichlet series, we may write

$$\prod_{j=1}^{k} \Lambda(s+\alpha_j,\chi)\Lambda(s-\beta_j,\chi) = \prod_{j=1}^{k}\Big(\frac{q}{\pi}\Big)^{s+\frac{\alpha_j-\beta_j}{2}} \Gamma\Big(\frac{s+\alpha_j}{2}\Big)\Gamma\Big(\frac{s-\beta_j}{2}\Big)$$
$$\times \sum_{m,n=1}^{\infty} \frac{\sigma(m;\underline{\alpha})}{m^s}\chi(m)\frac{\sigma(n;-\underline{\beta})}{n^s}\overline{\chi}(n), \tag{5.8}$$

where $\sigma(m;\underline{\alpha}) = \sum_{m=m_1\cdots m_k} m_1^{-\alpha_1}\cdots m_k^{-\alpha_k}$ and similarly $\sigma(n;-\underline{\beta}) = \sum_{n=n_1\cdots n_k} n_1^{\beta_1}\cdots n_k^{\beta_k}$, so that if $\alpha_i = \beta_i = 0$ these would simply be the k-divisor function. We average this over all the even characters mod q (omitting the trivial character), and hypothesize that only the diagonal terms $m = n$ survive this averaging. This is of course not justified, but is similar to the first heuristic we gave in this section for the order of magnitude of moments. After a computation with Euler products, these terms give (for the sum over m,n in (5.8))

$$\sum_{n=1}^{\infty} \frac{\sigma(n;\underline{\alpha})\sigma(n;-\underline{\beta})}{n^{2s}} = \mathcal{A}(s;\underline{\alpha},\underline{\beta}) \prod_{j,\ell=1}^{k} \zeta(2s+\alpha_j-\beta_\ell), \tag{5.9}$$

where $\mathcal{A}$ is given by an Euler product that converges absolutely in $\mathrm{Re}(s) > \frac{1}{2}-\delta$ if α_j,β_j are small enough. This factor $\mathcal{A}$ is similar to the a_k appearing in (5.3). Evaluating this at

$s = \frac{1}{2}$, we arrive at a candidate for the average value of $\Lambda(\chi; \underline{\alpha}, \underline{\beta})$, namely

$$\mathcal{C}(\underline{\alpha}, \underline{\beta}) = \prod_{j=1}^{k} \left(\frac{q}{\pi}\right)^{\frac{1+\alpha_j-\beta_j}{2}} \Gamma\left(\frac{\frac{1}{2}+\alpha_j}{2}\right)\Gamma\left(\frac{\frac{1}{2}-\beta_j}{2}\right)\mathcal{A}\left(\frac{1}{2}; \underline{\alpha}, \underline{\beta}\right) \prod_{j,\ell=1}^{k} \zeta(1+\alpha_j-\beta_\ell). \tag{5.10}$$

The candidate answer $\mathcal{C}(\underline{\alpha}, \underline{\beta})$ is invariant when the entries of $\underline{\alpha}$ are permuted, or when the entries of $\underline{\beta}$ are permuted, but does not have the S_{2k} symmetry we require of being allowed to permute the $2k$-entries of $(\underline{\alpha}, \underline{\beta})$. The beautifully simple answer proposed in **[43]** is to symmetrize $\mathcal{C}(\underline{\alpha}, \underline{\beta})$ by summing over all $\binom{2k}{k}$ cosets of $S_{2k}/(S_k \times S_k)$,

$$\sum_{\pi \in S_{2k}/S_k \times S_k} \mathcal{C}(\pi(\underline{\alpha}, \underline{\beta})). \tag{5.11}$$

While the expression in (5.10) has singularities whenever $\alpha_j = \beta_\ell$, the symmetrized expression in (5.11) turns out to be regular when $|\alpha_j|, |\beta_j|$ are small. Now setting $\alpha_1 = \cdots = \alpha_k = \beta_1 = \cdots = \beta_k = 0$, we arrive at the conjectured answer for the average of $|L(\frac{1}{2}, \chi)|^{2k}$. The leading term matches the Keating–Snaith conjecture, but now we also have the full polynomial of degree k^2.

To end our discussion of the moment conjectures, we mention recent work of Conrey and Keating **[41]** which aims to give a heuristic derivation of the moment conjectures of $\zeta(s)$ from correlations of divisor functions (as in **[46]** for the sixth and eighth moments). It would be of interest to develop their work in other families of L-functions. Sawin **[153]** develops a heuristic approach based on representation theory which (conditional on some hypotheses) recovers the recipe in Conrey et al. **[43]** in the function field setting (with a fixed field of constants).

6. PROGRESS TOWARDS UNDERSTANDING THE MOMENTS

In Section 4 we gave a number of examples where asymptotics for low moments are known, and all of these are in agreement with the conjectures described in the previous section. A rule of thumb suggests that an asymptotic for a moment may be computed if there are more elements in the family compared to the complexity of approximating the required power of the L-value (what we have informally called the complexity can be thought of as the square-root of the *analytic conductor*, see **[94]**). For example, as we saw in (4.4) $\zeta(\frac{1}{2}+it)$ may be approximated by (two) Dirichlet polynomials of length about $\sqrt{t}$, allowing for the calculation of the second and fourth moments. This rule of thumb is only a rough guide, and can be difficult to attain. For example, the fourth moment of Dirichlet L-functions mod q (evaluated in **[174]**), or the mean square of twists of a modular form by Dirichlet characters mod q (see **[18, 107]**) may seem of comparable difficulty to the fourth moment of the zeta function, but the first two problems turn out to be substantially harder. The largest moment that may be computed by this rule of thumb recovers the convexity bound for the L-value, and so there is great interest in going beyond this range, either by shrinking suitably the family over which we average, or by adding an extra short Dirichlet polynomial to the moment.

From the viewpoint of verifying the moment conjectures (for example to check the constants 42 and 24024 appearing in the sixth and eighth moments) one might look for large families where the complexity is still small. The family of primitive Dirichlet characters $\chi \pmod q$ ranging over all moduli $q \le Q$ is a good example, where the size of the family is about Q^2 whereas the complexity of such $L(\frac{1}{2}, \chi)$ is about $\sqrt{Q}$. This suggests the possibility of evaluating the sixth and eighth moments in this family, and indeed the large sieve gives a quick upper bound of the correct order of magnitude for these moments (see [90]). By developing an asymptotic version of the large sieve, Conrey, Iwaniec, and Soundararajan [48] obtained an asymptotic formula for

$$\sum_{q \le Q} \sideset{}{^\flat}\sum_{\chi \pmod q} \int_{-\infty}^{\infty} \left| \Lambda\left(\frac{1}{2} + iy, \chi\right)\right|^6 dy, \tag{6.1}$$

where $\Lambda(s, \chi) = (q/\pi)^{s/2}\Gamma(s/2)L(s, \chi)$ denotes the completed L-function, and the $\flat$ indicates a sum over even primitive characters χ. Here the averaging over y is a technical defect, needed for the proof, which (owing to the rapid decay of the Γ-function) may be thought of as an integral over essentially a bounded range of y. This asymptotic formula verified the predicted constant $g_3 = 42$ in this instance, and, moreover, [48] obtained a similar asymptotic formula with shifts $(\alpha_1, \alpha_2, \alpha_3)$ and $(\beta_1, \beta_2, \beta_3)$ which verified the recipe of [43] in this situation and yielded the full polynomial of degree 9 in $\log Q$ for (6.1). Chandee and Li [35] tackle the analogue of (6.1) for the eighth moment, and obtain an asymptotic formula conditional on the Generalized Riemann Hypothesis. Their work confirmed that $g_4 = 24024$ in this instance, but they could only verify the leading order term in the asymptotic and not the full polynomial of degree 16. Forthcoming work of Chandee, Li, Matomäki, and Radziwiłł (see [138] for an announcement) removes the imperfection of the average over y in (6.1) for the sixth moment while still obtaining the full asymptotic formula with power saving. They also obtain the leading order behavior of the eighth moment without invoking GRH, and without the integral in y.

The family of newforms of a fixed weight k for the group $\Gamma_1(q)$ with q a large prime offers another instance of a large family where the complexity (or analytic conductor) remains small. These correspond to newforms for $\Gamma_0(q)$ with character $\chi \pmod q$. This is a family of about q^2 elements, and is unitary since almost all of the characters $\chi \pmod q$ are not real. The complexity of the L-values is about size $\sqrt{q}$, and we may hope to address the sixth and eighth moments. Chandee and Li [34] give an asymptotic for the sixth moment analogous to (6.1) in this family (confirming again $g_3 = 42$), and obtain in [33] a good upper bound for the eighth moment. It would be of interest to find further examples of families where one can compute higher moments, and in particular to obtain such examples of symplectic and orthogonal families. The recent work of Nelson [130] on subconvexity for automorphic L-functions raises the hope that one might be able to compute high moments in $\mathrm{GL}(n)$ families for suitably large n.

In addition to examples where asymptotics for moments are known, substantial progress has been made in obtaining upper and lower bounds of the conjectured order of magnitude in a good deal of generality. Summarizing the work of many researchers, here is our knowledge of such bounds for the moments of $\zeta(\frac{1}{2}+it)$.

Theorem 6.1. *Let $k > 0$ and $T \geq e$ be real numbers. Then there are positive constants c_k and C_k such that*

$$c_k T(\log T)^{k^2} \leq \int_0^T \left|\zeta\left(\frac{1}{2}+it\right)\right|^{2k} dt \leq C_k T(\log T)^{k^2}.$$

Here the lower bound holds unconditionally for all k, while the upper bound holds unconditionally in the range $0 < k \leq 2$, and the upper bound holds assuming the truth of the Riemann Hypothesis for all $k > 2$.

We shall now discuss this result and its extensions in families of L-functions. The discussion splits naturally into three parts (i) lower bounds for moments, (ii) unconditional upper bounds for moments, and (iii) upper bounds assuming RH or GRH.

The lower bound stated in Theorem 6.1 was first established by Ramachandra [143, 144] in the case when $2k$ is a natural number. This was then extended by Heath-Brown [84] to the case when k is any positive rational number, but the constants c_k in his result depended upon the height of the rational number k, so that the method did not extend to irrational k. Further, the techniques in these works were specific to the "t-aspect" and did not extend to moments in families of L-functions. Rudnick and Soundararajan [149, 150] developed an alternative approach, which worked in general families. For example, their method would show that $\sum_{|d| \leq X} |L(\frac{1}{2}, \chi_d)|^k \geq c_k X(\log X)^{k(k+1)/2}$ for all rational $k \geq 1$ and a suitable positive constant c_k, which again did not vary continuously with k but depended on the height of the rational number k. This was further refined by Radziwiłł and Soundararajan [140], who obtained the lower bounds in Theorem 6.1 for all real $k \geq 1$ with e^{-30k^4} being a permissible value for c_k if T is large. A further round of simplification is carried out in Heap and Soundararajan [79], which also gives the lower bound in Theorem 6.1 for real $0 < k \leq 1$.

The story for lower bounds may be encapsulated in the following broad principle. Whenever we can compute the mean value of $L(\frac{1}{2})$ multiplied by short Dirichlet polynomials in a family, we can obtain lower bounds of the right order of magnitude for the moments $|L(\frac{1}{2})|^k$ for any real $k \geq 1$. Of course, in general Hölder's inequality will give lower bounds for higher moments in terms of smaller moments, but those would not be of the conjectured order of magnitude since the exponent of the logarithm in the moment conjectures is quadratic in k. If we can also compute the mean value of $|L(\frac{1}{2})|^2$ multiplied by short Dirichlet polynomials, then we can obtain lower bounds of the right order of magnitude for the moments $|L(\frac{1}{2})|^k$ in the range $0 < k \leq 1$ as well. It may seem puzzling why the problem for small k should require more information than for large k, but in fact this is natural. Consider letting $k \to 0^+$. Then the moments $|L(\frac{1}{2})|^k$ essentially pick up whether $L(\frac{1}{2})$ is zero or not, so that lower bounds for the small moments encode lower bounds for non-vanishing. The analytic methods for producing nonzero values of $L(\frac{1}{2})$ (the *mollifier method*) rely on knowledge of the first two moments in the family (with a little room to spare). Thus we may establish (using the methods of either [148] or [79]) that for all real $k > 0$,

$$\sum_{\chi \pmod q} \left| L\left(\frac{1}{2}, \chi\right) \right|^{2k} \gg_k q(\log q)^{k^2}, \tag{6.2}$$

where q is a large prime, and that

$$\sum_{|d| \leq X} \left| L\left(\frac{1}{2}, \chi\right) \right|^{k} \gg_k X(\log X)^{\frac{k(k+1)}{2}}. \tag{6.3}$$

In the family of quadratic twists of a fixed Hecke eigenform f, we only have access to the first moment and not the second, and therefore we only know in the range $k \geq 1$ that

$$\sum_{|d| \leq X} L\left(\frac{1}{2}, f \times \chi_d\right)^{k} \gg_k X(\log X)^{\frac{k(k-1)}{2}}. \tag{6.4}$$

We now turn to the unconditional upper bounds in Theorem 6.1, which were established in the special cases $k = 1/n$ or $k = 1 + 1/n$ (for natural numbers n) by Heath-Brown [84] and Bettin, Chandee, and Radziwiłł [17], respectively. Then in Heap, Radziwiłł, and Soundararajan [78] the bound was established for all $0 < k \leq 2$, as an illustration of an upper bound principle (complementing the one for lower bounds above) enunciated by Radziwiłł and Soundararajan [141]. Whenever we can compute a moment $|L(\frac{1}{2})|^k$ (usually with k being a positive integer) together with flexibility to introduce a short Dirichlet polynomial, we can obtain upper bounds of the conjectured order of magnitude for the moments $|L(\frac{1}{2})|^r$ for all $0 < r \leq k$. Thus one can obtain complementary upper bounds in (6.2) for $k \leq 1$ (with more effort, using Young's work [174], this could perhaps be extended to the range $k \leq 2$), matching upper bounds in (6.3) in the range $k \leq 2$ (if one knew the positivity of $L(\frac{1}{2}, \chi_d)$ this would also follow in the range $k \leq 3$ and it would be interesting to attain that range unconditionally), and for the family in (6.3) for $k \leq 1$ (this is the example carried out in [141]).

The conditional bounds in Theorem 6.1 originated from work of Soundararajan [164] who established (assuming RH) the nearly sharp bound $M_k(T) \ll_{k,\varepsilon} T(\log T)^{k^2+\varepsilon}$. This was then refined in the beautiful work of Harper [77] to its present sharp form. The method is very general and applies in any family where we are able to compute the mean values of short Dirichlet polynomials. Thus (assuming GRH in the relevant families) one can obtain upper bounds of the correct order of magnitude for all nonnegative k in the examples (6.2), (6.3), and (6.4).

The main idea behind the conditional upper bounds in Theorem 6.1 is that on RH (or GRH) one can obtain an *upper bound* for $\log |\zeta(\frac{1}{2} + it)|$ (or more generally the logarithm of central L-values) just in terms of sums over primes. This is related to the ideas behind Selberg's central limit theorem and the one sided versions for L-values that we discussed in Sections 2 and 3. A barrier to approximating $\log |\zeta(\frac{1}{2} + it)|$ by a suitable Dirichlet polynomial is the presence of zeros near $\frac{1}{2} + it$; the crucial point is that these zeros should only make $|\zeta(\frac{1}{2} + it)|$ smaller, so that such Dirichlet polynomials could serve as an upper bound.

One way to see this is to note that RH is equivalent to the property that, with $s = \sigma + it$,

$$|\xi(s)| = |s(s-1)\pi^{-s/2}\Gamma(s/2)\zeta(s)| = \prod_{\rho} \left|1 - \frac{s}{\rho}\right|$$

is an increasing function of σ in $\sigma \geq \frac{1}{2}$ for any fixed t. This permits bounding $|\zeta(\frac{1}{2}+it)|$ in terms of $|\zeta(\sigma_0 + it)|$ for any $\sigma_0 > \frac{1}{2}$, and one can adapt Selberg's ideas to approximate $\log|\zeta(\sigma_0+it)|$. In this manner, it was shown in **[164]** that for $T \leq t \leq 2T$ and any $2 \leq x \leq T^2$ one has, assuming RH and with $\sigma_0 = \frac{1}{2} + \frac{1}{\log x}$,

$$\log\left|\zeta\left(\frac{1}{2}+it\right)\right| \leq \operatorname{Re} \sum_{2 \leq n \leq x} \frac{\Lambda(n)}{n^{\sigma_0+it}\log n} \frac{\log(x/n)}{\log x} + \frac{\log T}{\log x} + O\left(\frac{1}{\log x}\right). \tag{6.5}$$

Analogous bounds hold for $\log|L(\frac{1}{2})|$ if a corresponding GRH is assumed.

The usefulness of (6.5) lies in its flexibility with choosing the parameter x. If x is suitably small, then the distribution of the sum (which is essentially $\operatorname{Re}\sum_{p\leq x} 1/p^{\frac{1}{2}+it}$) in (6.5) can be understood accurately by studying its moments (as we discussed in Sections 2 and 3), but we lose some information in the $\log T/\log x$ term. Here it is also useful to split the sum over p into different ranges (say $p \leq z$ and $z < p \leq x$); for small ranges of p, more moments may be computed so that a finer understanding of the sum is possible, while for the larger ranges the slow growth of the variance (which is roughly $\sum_{z\leq p\leq x} 1/p \sim \log\frac{\log x}{\log z}$) permits a good understanding with fewer moments. In this way **[164]** established a coarse version of Selberg's central limit theorem in the large deviations regime, showing that in the range $\sqrt{\log\log T} \leq V = o(\log\log T \log\log\log T)$ one has

$$\operatorname{meas}\left\{T \leq t \leq 2T : \log\left|\zeta\left(\frac{1}{2}+it\right)\right| \geq V\right\} \ll T\exp\left(-\frac{V^2}{\log\log T}(1+o(1))\right). \tag{6.6}$$

As we mentioned in Section 5, the $2k$-th moment of zeta should be dominated by values of $|\zeta(\frac{1}{2}+it)|$ of size $(\log T)^k$, and the (6.6) shows that this set has measure $\ll T(\log T)^{-k^2+o(1)}$, which yields $M_k(T) \ll T(\log T)^{k^2+\varepsilon}$.

Harper's sharp upper bound for $M_k(T)$ builds on some of these ideas, but deals directly with the moments rather than going through the intermediary of the large deviations in Selberg's theorem (6.6). Instead there is an elaborate decomposition of the sum over primes in (6.5) into many ranges, and then the exponentials of such sums are handled by approximating these by suitable truncations of their Taylor expansion. Similar ideas were developed independently around the same time in **[141]** for bounding small moments unconditionally, and the recent paper **[79]** develops these ideas in the context of lower bounds. Thus the proofs of all three aspects of Theorem 6.1 have a unified feel, and the spirit of the proofs may be described as thinking in terms of Euler products but performing computations by replacing Euler products by short Dirichlet series obtained from their Taylor expansions. These proofs were also influenced by ideas from sieve theory. For example, in analogy with (6.5) we may note that $\omega(n)$ (the number of prime factors of n) may be bounded above by $\sum_{p|n, p\leq y} 1 + (\log n)/\log y$ for any y, and this could be used to give upper bounds for the mean-value of $d_k(n)$ (which is roughly $k^{\omega(n)}$) in short intervals.

The ideas behind obtaining conditional bounds for moments have found diverse applications. Soundararajan and Young [166] used such bounds for "shifted moments" (see also [32]) to obtain an asymptotic formula (on GRH) for the second moment of quadratic twists of an eigenform $\sum_{|d|\le X} L(\frac{1}{2}, f \times \chi_d)^2$. This is a tantalizing problem, which falls within the purview of the rule of thumb described at the beginning of this section, but an unconditional asymptotic has so far been elusive. A similar problem is to compute the asymptotic for the fourth moment of quadratic Dirichlet L-functions $\sum_{|d|\le X} L(\frac{1}{2}, \chi_d)^4$, and recently Shen [157] has extended the method in [166] to obtain (on GRH) such an asymptotic. Analogues of these two problems over function fields have been established in [27, 61], and since GRH is known in this setting, the corresponding results hold unconditionally.

In a very different direction, Lester and Radziwiłł [116] showed on GRH that the Fourier coefficients of half-integer weight Hecke cusp forms exhibit a positive proportion of sign changes as we range over fundamental discriminants. Among the many innovations in their beautiful proof, is an application of the ideas discussed above to obtain sharp upper bounds for the second mollified moment of quadratic twists of the Shimura correspondent of the given half-integer weight form. This realization that sharp upper bounds for the second mollified moment suffice has led to another striking result in the work of David, Florea, and Lalin [51], who show that a positive proportion of L-functions attached to cubic characters (in the function field setting) have nonzero central value. Two other recent applications include Zenz [176] to bounding the L^4 norm of Hecke eigenforms of large weight k for the full modular group, and Shubin [158] to bounding the variance of lattice points on the sphere in random small spherical caps. See [66, 121, 122] for further examples.

7. EXTREME VALUES

In Sections 2 and 3 we discussed the typical size of $|\zeta(\frac{1}{2}+it)|$ and central values of L-functions, which are governed by Selberg's central limit theorem, and the analogous Keating–Snaith conjectures. In Sections 5 and 6 we discussed how the moment problem aims for an understanding of the *large deviations* range of values of $|\zeta(\frac{1}{2}+it)|$ (or $|L(\frac{1}{2})|$). We now discuss the maximal size of $|\zeta(\frac{1}{2}+it)|$ (for $0 \le t \le T$) and analogous problems in families of L-functions.

As we mentioned in Section 4, our unconditional knowledge is far from the Lindelöf hypothesis that $|\zeta(\frac{1}{2}+it)| \ll (1+|t|)^{\varepsilon}$, and for general L-functions already the subconvexity problem poses formidable difficulties. In 1924 Littlewood established that the Riemann Hypothesis implies the Lindelöf hypothesis in the quantitative form

$$\left|\zeta\left(\frac{1}{2}+it\right)\right| \ll \exp\left(\frac{C \log|t|}{\log\log|t|}\right) \tag{7.1}$$

for some constant C. The estimate (6.5) yields such a result, upon taking $x = (\log t)^2$ there, and bounding the sum over prime powers trivially. This strategy was optimized in [36] which showed that one may take any $C > \frac{\log 2}{2}$ in (7.1). Apart from this refinement of the constant C, no improvement has been made over Littlewood's estimate. Corresponding results hold for general L-functions, and explicit versions of such bounds (which are useful in computational applications) may be found in [31].

Complementing (7.1), one may ask for lower bounds on $\max_{T\le t\le 2T} |\zeta(\frac{1}{2}+it)|$. Recall that in Section 1 we discussed the extreme values of zeta and L-functions at the edge of the critical strip, and already there was a gap in our knowledge between the extreme values that may be exhibited and the bounds that follow from GRH (see the discussion surrounding (1.2) and (1.3)). This gap becomes much more pronounced on the critical line. By using lower bounds for integer moments of $\zeta(\frac{1}{2}+it)$, with attention to the uniformity in k, Balasubramanian and Ramachandra [16] (optimized in [14]) established that

$$\max_{T\le t\le 2T}\left|\zeta\left(\frac{1}{2}+it\right)\right| \ge \max_{k}\left(\frac{1}{T}\int_T^{2T}\left|\zeta\left(\frac{1}{2}+it\right)\right|^{2k}dt\right)^{\frac{1}{2k}}$$

$$\gg \max_{k\in\mathbb{N}}\left(\sum_{n\le T}\frac{d_k(n)^2}{n}\right)^{\frac{1}{2k}} = \exp\left((B+o(1))\frac{\sqrt{\log T}}{\sqrt{\log\log T}}\right), \quad (7.2)$$

with $B \approx 0.53$. With the development of lower bounds for moments in families of L-functions (discussed in Section 6), such bounds also became available for central L-values. However, a different *resonance method* developed in [163] has proved to be still more efficient. The main idea in [163] is to find a Dirichlet polynomial $R(t) = \sum_n r(n)n^{-it}$ which "resonates" with $\zeta(\frac{1}{2}+it)$ and picks out its large values. This is based on computing

$$I_1 = \int_T^{2T} |R(t)|^2 dt \quad \text{and} \quad I_2 = \int_T^{2T} \zeta\left(\frac{1}{2}+it\right)|R(t)|^2 dt, \quad (7.3)$$

and noting that

$$\max_{T\le t\le 2T}\left|\zeta\left(\frac{1}{2}+it\right)\right| \ge \frac{|I_2|}{I_1}. \quad (7.4)$$

If the resonator Dirichlet polynomial $R(t)$ is short, in the sense that $r(n) = 0$ unless $n \le T^{1-\varepsilon}$, then I_1 and I_2 in (7.3) may be evaluated asymptotically, and these quantities give two quadratic forms in the unknown coefficients $r(n)$. The ratio of these two quadratic forms is maximized in [163], yielding

$$\max_{T\le t\le 2T}\left|\zeta\left(\frac{1}{2}+it\right)\right| \ge \exp\left((1+o(1))\frac{\sqrt{\log T}}{\sqrt{\log\log T}}\right). \quad (7.5)$$

While this is only a little bit better than (7.2), the method also yields lower bounds on the measure of the set on which large values are attained:

$$\operatorname{meas}\left\{t\in[T,2T] : \left|\zeta\left(\frac{1}{2}+it\right)\right| \ge e^V\right\} \gg \frac{T}{(\log T)^4}\exp\left(-10\frac{V^2}{\log\frac{\log T}{8V^2\log V}}\right), \quad (7.6)$$

uniformly for $3 \le V \le \frac{1}{5}\sqrt{\log T/\log\log T}$. There is some scope to improve such bounds, especially when V is of size $C\log\log T$, where one would like to match the upper bound in (6.6) which would be in keeping with Selberg's theorem (see [79] for more precise results when $V \le (2-\varepsilon)\log\log T$). The estimate (7.6) shows that large values on the scale of (7.5) occur fairly often (on a set of measure $\ge T^{1-C/\log\log T}$) suggesting that still larger values might exist. Furthermore, the resonance method extends readily to families of L-functions, and thus we may show (for example) that

$$\max_{X\le|d|\le 2X} L\Big(\frac{1}{2},\chi_d\Big) \ge \exp\bigg(\Big(\frac{1}{\sqrt{5}}+o(1)\Big)\frac{\sqrt{\log X}}{\sqrt{\log\log X}}\bigg), \tag{7.7}$$

and that, for any Hecke eigenform f,

$$\max_{X\le|d|\le 2X} L\Big(\frac{1}{2},f\times\chi_d\Big) \ge \exp\bigg(c\frac{\sqrt{\log X}}{\sqrt{\log\log X}}\bigg), \tag{7.8}$$

for a suitable positive constant c. Indeed, the large values in (7.7) and (7.8) are attained for more than $X^{1-\varepsilon}$ discriminants d with $X\le|d|\le 2X$. By Waldspurger's formula, the large values produced in (7.8) show that fundamental Fourier coefficients of half-integer weight eigencuspforms must get large, and the resonance method has been adapted in **[73]** to show that this holds more generally for half-integer weight cusp forms (not necessarily an eigenform). Another application of this resonance method may be found in the work of Milicevic **[120]** who obtains large values of Hecke–Maass cusps forms on arithmetic hyperbolic surfaces.

Bondarenko and Seip **[23]** recently made a breakthrough on this problem, by exhibiting still larger values of $|\zeta(\frac{1}{2}+it)|$. The key ingredient is a beautiful result on *GCD sums* or *Gál sums*: The problem is to find

$$\max_{|\mathcal{N}|=N}\sum_{m,n\in\mathcal{N}}\frac{(m,n)}{\sqrt{mn}}, \tag{7.9}$$

where the maximum is over all N element subsets of the natural numbers. This elegant combinatorial problem turns out to be closely related to maximizing the ratio of quadratic forms (see **[2]**)

$$\max_{|\mathcal{N}|=N}\sup_{\underline{x}\in\mathbb{C}^N\neq 0}\bigg(\sum_{m,n\in\mathcal{N}}x_m\overline{x_n}\frac{(m,n)}{\sqrt{mn}}\bigg)\Big/\bigg(\sum_n|x_n|^2\bigg). \tag{7.10}$$

Bondarenko and Seip **[22, 23]** established that the maximum in (7.9) (and also (7.10)) lies between

$$N\exp\bigg((1-\varepsilon)\frac{\sqrt{\log N\log\log\log N}}{\sqrt{\log\log N}}\bigg)\quad\text{and}\quad N\exp\bigg((7+\varepsilon)\frac{\sqrt{\log N\log\log\log N}}{\sqrt{\log\log N}}\bigg),$$

De la Bretèche and Tenenbaum **[52]** refined this to show that the maximums in (7.9) and (7.10) equal

$$N\exp\bigg(\big(2\sqrt{2}+o(1)\big)\frac{\sqrt{\log N\log\log\log N}}{\sqrt{\log\log N}}\bigg). \tag{7.11}$$

The relevance of the bounds for (related) GCD sums to large values of $|\zeta(\sigma+it)|$ was first appreciated by Aistleitner **[1]** who showed that for fixed $\sigma\in(\frac{1}{2},1)$ and T large one has (for some $c_\sigma>0$)

$$\max_{0<t\le T}|\zeta(\sigma+it)|\ge\exp\bigg(\frac{c_\sigma(\log T)^{1-\sigma}}{(\log\log T)^\sigma}\bigg),$$

which improved upon earlier applications of the resonance method (see **[87, 172]**) but only matched the results obtained by Montgomery **[125]** using very different ideas (see also **[3]** for large values on the 1-line, and **[4]** for analogous results for Dirichlet L-functions). On the critical line, Bondarenko and Seip **[23]** obtained a substantial improvement over the previously known large values of $|\zeta(\frac{1}{2}+it)|$ (see (7.5)) by establishing that

$$\max_{0<t\le T}\left|\zeta\Big(\frac{1}{2}+it\Big)\right| \ge \exp\Big(\big(c+o(1)\big)\frac{\sqrt{\log T\log\log\log T}}{\sqrt{\log\log T}}\Big), \tag{7.12}$$

for a positive constant c (in **[23]** $c=1/\sqrt{2}$ is permissible, while **[52]** allows for the improved $c=\sqrt{2}$). The key insight is that in the resonance method one can choose "long resonators" where $R(t)$ is no longer constrained to be a short Dirichlet polynomial ($r(n)=0$ unless $n\le T^{1-\varepsilon}$) but instead $R(t)$ is allowed to have $T^{1-\varepsilon}$ nonzero coefficients $r(n)$ so long as these are *positive*. This leads to an optimization problem closely related to the GCD/Gál sums discussed above, and permits the stronger bound in (7.12). Why is it possible to take such long resonators? Consider a smooth nonnegative function Φ whose Fourier transform $\widehat{\Phi}$ is also nonnegative; for example, we could take $\Phi(t)=e^{-t^2/2}$. In place of I_1 and I_2 in (7.3) consider the smoothed integrals

$$\int_{-\infty}^{\infty}|R(t)|^2\Phi(t/T)dt \quad\text{and}\quad \int_{-\infty}^{\infty}\zeta\Big(\frac{1}{2}+it\Big)|R(t)|^2\Phi(t/T)dt. \tag{7.13}$$

Replacing $\zeta(\frac{1}{2}+it)$ with its approximation $\sum_{k\le T}k^{-\frac{1}{2}-it}$, the second quantity above is approximately

$$\begin{aligned}&\sum_{k\le T}\frac{1}{\sqrt{k}}\sum_{m,n}r(m)r(n)\int_{-\infty}^{\infty}\Big(\frac{n}{mk}\Big)^{it}\Phi(t/T)dt\\ &\quad=T\sum_{k\le T}\frac{1}{\sqrt{k}}\sum_{m,n}r(m)r(n)\widehat{\Phi}\big(T\log(n/mk)\big).\end{aligned}$$

Since m and n may be much larger than T, we are unable to restrict just to the "diagonal terms" $n=mk$, but the crucial point is that the positivity of $\widehat{\Phi}$, the resonator coefficients $r(m)$, $r(n)$, and the "coefficients of ζ" (namely, the function taking 1 on all positive integers) all allow us to keep any terms that we please on the right side above, and ignore other contributions. In this way, one can get a satisfactory lower bound for the ratio of the quantities in (7.13), without needing to evaluate each of these integrals. The restriction on the number of terms allowed in the resonator arises from the fact that $\sum_{k\le T}k^{-\frac{1}{2}-it}$ is a poor approximation to $\zeta(\frac{1}{2}+it)$ if t is small. These small values of t are unavoidable because the condition that $\widehat{\Phi}$ is nonnegative forces $\Phi(0)$ to be strictly positive.

Unlike the resonance method which applies in great generality, there are (at present) limitations on when the Bondarenko–Seip method of using long resonators applies. In the first place, as we noted above small t must be included, and therefore the maximum in (7.12) is over $t\in[0,T]$ (this can be refined to the interval $[T^\beta,T]$ for any $\beta<1$ at the cost of weakening the constant c in (7.12)), rather than the dyadic intervals $[T,2T]$ seen in (7.5). More significantly, the method requires the positivity of the Dirichlet series coefficients of the L-functions in question (analogously to ζ having coefficients 1), and also the positivity of the right-hand side of any orthogonality relation or trace formula (analogously to $\widehat{\Phi}$ being nonnegative). Apart from $\zeta(s)$, there is one other example in which the Bondarenko–Seip method has been successfully implemented, and this is the work of de la Breteche and Tenenbaum **[52]** which produces large values of $|L(\frac{1}{2}, \chi)|$ as χ varies over Dirichlet characters (mod q) with q a large prime. To illustrate the subtleties involved, we note that **[52]** exhibits large values of $|L(\frac{1}{2}, \chi)|$ for *even* characters χ, but the method does not work for *odd* character. This is because in the even case the orthogonality relation

$$\sum_{\substack{\chi \ (\mathrm{mod}\ q) \\ \chi \mathrm{even}}} \chi(a) = \begin{cases} \frac{\phi(q)}{2} & \text{if } a \equiv \pm 1 \ (\mathrm{mod}\ q), \\ 0 & \text{otherwise} \end{cases}$$

involves only nonnegative terms on the right-hand side, whereas this is not the situation for odd characters

$$\sum_{\substack{\chi \ (\mathrm{mod}\ q) \\ \chi \mathrm{odd}}} \chi(a) = \begin{cases} \pm\frac{\phi(q)}{2} & \text{if } a \equiv \pm 1 \ (\mathrm{mod}\ q), \\ 0 & \text{otherwise.} \end{cases}$$

In particular, the results in (7.7) and (7.8) remain the best currently known, and it would be of great interest to see if the Bondarenko–Seip method could be extended to more general situations.

There is a vast gulf between the conditional upper bounds for $|\zeta(\frac{1}{2} + it)|$ in (7.1) and the large values exhibited in (7.12), and it is natural to ask which of these is closer to the truth. Already in Section 1 we saw a gap (of a factor of 2) between the extreme values of $L(1, \chi_d)$ that may be exhibited (see (1.2)) and the conditional bounds on these extreme values (see (1.3)). There the probabilistic models suggested that the extreme values exhibited in (1.2) represented the truth, and on the critical line too we expect the large values exhibited in (7.12) to be closer to the truth than the bounds in (7.1). For example, if we use Selberg's central limit theorem as a guide and extrapolate, then the measure of $t \in [0, T]$ with $|\zeta(\frac{1}{2} + it)| \geq e^V$ may be expected to be $\ll T \exp(-(1 + o(1))V^2/\log\log T)$ (confer (6.6)). If $V = (1 + \varepsilon)\sqrt{\log T \log\log T}$, this measure becomes $\leq T^{-\varepsilon}$, but one can show that if $|\zeta(\frac{1}{2} + it)|$ attains its maximum for $t \in [0, T]$ at $t = t_0$ then in an interval $|t - t_0| \leq c/\log T$ its values are at least of size $\frac{1}{2}|\zeta(\frac{1}{2} + it_0)|$ (see Lemma 2.2 of **[59]**). This suggests that

$$\max_{t \in [0,T]} \left|\zeta\left(\frac{1}{2} + it\right)\right| \leq \exp\left((1 + o(1))\sqrt{\log T \log\log T}\right).$$

Farmer, Gonek, and Hughes **[59]** have conjectured that even this overestimates the true size of the maximum, and that possibly

$$\max_{0 \leq t \leq T} \left|\zeta\left(\frac{1}{2} + it\right)\right| = \exp\left(\left(\frac{1}{\sqrt{2}} + o(1)\right)\sqrt{\log T \log\log T}\right). \tag{7.14}$$

To give one indication of why this might hold, consider (6.5) which gives (on RH) an upper bound for $\log|\zeta(\frac{1}{2} + it)|$ in terms of essentially a sum over primes going up to x, accepting an error term of size $\log T/\log x$. If we choose $x = \exp(\sqrt{\log T})$ then this error term is negligible, and now $\operatorname{Re}\sum_{p\le x} 1/p^{\frac{1}{2}+it}$ behaves like a Gaussian with mean 0 and variance $\frac{1}{2}\sum_{p\le x} 1/p \sim \frac{1}{2}\log\log x \sim \frac{1}{4}\log\log T$. Extrapolating this Gaussian behavior, we arrive at the conjectured behavior in (7.14). The conjecture in **[59]** is based upon a more careful analysis of the hybrid Euler–Hadamard formula developed in **[69]**, which decomposes $\log|\zeta(\frac{1}{2}+it)|$ into terms arising from both primes and zeros in suitable ranges. Probabilistic models for both these terms are analyzed (with the contribution of zeros being modeled using random matrix theory), and the conjecture (7.14) is consistent with many different ways of splitting into primes and zeros. Similar conjectures may be formulated in other families of L-functions, and for example **[59]** conjectures that

$$\max_{|d|\le X} L\left(\frac{1}{2}, \chi_d\right) = \exp\big((1+o(1))\sqrt{\log X \log\log X}\big), \tag{7.15}$$

which again is a little smaller (by a factor $\sqrt{2}$ in the exponent) than what might be guessed from extrapolating the Keating–Snaith conjectures for $\log L(\frac{1}{2}, \chi_d)$.

As we discussed in Section 4, one motivation for studying the moments of $|\zeta(\frac{1}{2}+it)|$ is to gain an understanding of its extreme values. In order to do so, one would need an understanding of how $M_k(T)$ behaves with uniformity in k, and a first step might be to examine the asymptotic behavior of the constants a_k and g_k appearing in Conjecture 5.1. One can show that $\log a_k \sim -k^2\log(2e^{\gamma}\log k)$, and that $\log g_k \sim k^2\log(k/4\sqrt{e})$ (see **[46]**), so that it may seem tempting to speculate that for $T \ge 10$ (say) and uniformly for all $k \ge 2$ one has (for some positive constant c)

$$T\left(\frac{c\log T}{k\log k}\right)^{k^2} \le \int_0^T \left|\zeta\left(\frac{1}{2}+it\right)\right|^{2k} dt \le T(\log T)^{k^2}.$$

But there is a curious paradox, and the upper and lower bounds above are inconsistent! If the upper bound above holds uniformly, then it follows that

$$\max_{0\le t\le T}\left|\zeta\left(\frac{1}{2}+it\right)\right| \le \exp\big((1+o(1))\sqrt{\log T\log\log T}\big).$$

Whereas if the lower bound above holds uniformly, then one must have

$$\max_{0\le t\le T}\left|\zeta\left(\frac{1}{2}+it\right)\right| \ge \exp(C\log T/\log\log T)$$

for some positive constant C. This is an instance where the leading order asymptotic in the moment conjecture does not capture the full story, and one should look instead at the recipe in **[43]** which (for natural numbers k) gives the entire (conjectural) polynomial P_k of degree k^2. An analysis of this full moment conjecture suggests that the uniform upper bound stated above might hold: thus, for $T \ge 10$ and natural numbers $k \ge 1$ we conjecture that

$$\int_0^T \left|\zeta\left(\frac{1}{2}+it\right)\right|^{2k} dt \le T(\log T)^{k^2}. \tag{7.16}$$

In other words, we guess that $\log|\zeta(\frac{1}{2}+it)|$ is *sub-Gaussian* (when thinking of the frequency of its large values), and this gives a weaker version of the Farmer, Gonek, Hughes conjecture (7.14).

While we have confined our discussion above to large values of $|\zeta(\frac{1}{2}+it)|$, or equivalently $\mathrm{Re}(\log\zeta(\frac{1}{2}+it))$, similar considerations apply also to $\mathrm{Im}(\log\zeta(\frac{1}{2}+it))$; see for example [24, 30, 68].

8. FYODOROV–HIARY–KEATING CONJECTURE

A fascinating set of problems has emerged recently with the work of Fyodorov and Keating [65], and Fyodorov, Hiary, and Keating [64], who initiated a study of the distribution of "local maxima" of the Riemann zeta function. More precisely, if t is chosen uniformly from $[T, 2T]$, what is the distribution of

$$\max_{0\le h\le 1}\left|\zeta\left(\frac{1}{2}+it+ih\right)\right|?$$

Although it does not make much of a difference, [64] considers the maximum over intervals of length 2π instead of 1 since this has a natural analogue in random matrix theory. If a matrix g is chosen randomly from $\mathrm{U}(N)$ (with respect to Haar measure), what is the distribution of

$$\max_{\theta\in[0,2\pi)}\left|\det\left(I-ge^{-i\theta}\right)\right|?$$

In the context of $\zeta(\frac{1}{2}+it)$, one initial motivation for considering this problem was that it might shed new light on the global maximum over the long interval $[0, T]$ (discussed in the previous section). While the distribution of the local maxima leads to striking new and subtle phenomena involving the local correlations of the zeta function, it does not seem to inform the behavior of the global maximum.

Conjecture 8.1 (Fyodorov–Hiary–Keating [64]). *For any real number y, as $T\to\infty$ one has*

$$\frac{1}{T}\,\mathrm{meas}\left\{T\le t\le 2T : \max_{0\le h\le 1}\left|\zeta\left(\frac{1}{2}+it+ih\right)\right|\le e^{y}\frac{\log T}{(\log\log T)^{\frac{3}{4}}}\right\}\to F(y), \quad (8.1)$$

where the cumulative distribution function F satisfies $F(y)\to 0$ as $y\to-\infty$, and satisfies $1-F(y)\sim Cye^{-2y}$ as $y\to\infty$ for some constant $C>0$. In particular, for any function $g(T)$ tending to infinity with T, one has

$$\mathrm{meas}\left\{T\le t\le 2T : \left|\max_{0\le h\le 1}\log\left|\zeta\left(\frac{1}{2}+it+ih\right)\right|-\log\log T+\frac{3}{4}\log\log\log T\right|\le g(T)\right\} \sim T. \quad (8.2)$$

Let us first explain what is striking and unexpected about this conjecture. Roughly speaking, in an interval of length 1 we may think of the zeta function as being determined by about $\log T$ values—this is about the number of zeros we expect to find in such an interval, and we may guess that if $|t_1-t_2|\le 1/\log T$ then $\log|\zeta(\frac{1}{2}+it_1)|$ and $\log|\zeta(\frac{1}{2}+it_2)|$ are about the same. Selberg's theorem tells us that the values $\log|\zeta(\frac{1}{2}+it)|$ are distributed like a normal variable with mean 0 and variance $\frac{1}{2}\log\log T$. Thus a first guess for the distribution of $\max_{0\le h\le 1}\log|\zeta(\frac{1}{2}+it+ih)|$ might be that it behaves like the maximum of about $\log T$ independently drawn normal random variables with mean 0 and variance $\frac{1}{2}\log\log T$. The maximum of N independent normal variables with mean 0 and variance 1 is very sharply concentrated around $\sqrt{2\log N} - (\log\sqrt{4\pi\log N})/\sqrt{2\log N}$ (the precise distribution is known as the Gumbel distribution, and has been extensively studied in view of its enormous significance in practical assessments of the risk of rare events). After scaling by the standard deviation $\sqrt{\frac{1}{2}\log\log T}$ in Selberg's theorem, this naive model would indicate that $\max_{0\le h\le 1}\log|\zeta(\frac{1}{2}+it+ih)|$ should typically be around

$$\log\log T - \frac{1}{4}\log\log\log T + O(1).$$

In contrast, Conjecture 8.1 predicts that $\max_{0\le h\le 1}|\zeta(\frac{1}{2}+it)|$ is usually a bit smaller, of size $(\log T)/(\log\log T)^{\frac{3}{4}}$. There is also a subtle difference in the decay of $1-F(y)$ in (8.1), which is predicted to decay like ye^{-2y}, whereas the Gumbel distribution would have predicted a decay rate of e^{-2y}.

The flaw in the naive heuristic presented above is that nearby values of the zeta function are not independent, but are correlated. Suppose t is randomly chosen from $[T, 2T]$ and $0\le h\le 1$, and consider the covariance of $\log|\zeta(\frac{1}{2}+it)|$ and $\log|\zeta(\frac{1}{2}+it+ih)|$. As in our discussion of Selberg's theorem in Section 2, we may often approximate these values by corresponding sums over primes $\operatorname{Re}\sum_{p\le x} 1/p^{\frac{1}{2}+it}$ and $\operatorname{Re}\sum_{p\le x} 1/p^{\frac{1}{2}+it+ih}$ with x a suitable small power of T. If p is small in comparison to $e^{1/h}$ then $p^{ih}\approx 1$, and the corresponding terms in our prime sums are strongly correlated. The terms with p much larger than $e^{1/h}$ are largely uncorrelated, since as p varies in such large ranges p^{ih} will become equidistributed on the unit circle. Thus one may see that

$$\frac{1}{T}\int_T^{2T}\log\left|\zeta\Big(\frac{1}{2}+it\Big)\right|\log\left|\zeta\Big(\frac{1}{2}+it+ih\Big)\right|dt \sim \frac{1}{2}\sum_{p\le x}\frac{\cos(h\log p)}{p}$$
$$\sim \frac{1}{2}\log\min\big(h^{-1},\log T\big). \qquad (8.3)$$

This correlation structure of nearby values must be taken into account when trying to predict the behavior of local maxima.

To gain a rough idea of how to model the local behavior of $\log|\zeta(\frac{1}{2}+it)|$, put for each $1\le k\le\log\log T-1$,

$$\mathcal{P}_k(t) = \operatorname{Re}\sum_{e^{e^{k-1}}\le p\le e^{e^k}}\frac{1}{p^{1/2+it}}, \qquad (8.4)$$

so that we may think of $\log|\zeta(\frac{1}{2}+it)|$ as something like $\sum_k \mathcal{P}_k(t)$. Each $\mathcal{P}_k(t)$ is distributed like a normal random variable with mean 0 and variance $\sim\frac{1}{2}\sum_{e^{e^{k-1}}\le p\le e^{e^k}} 1/p\sim\frac{1}{2}$. Moreover, for different values of k, the sums $\mathcal{P}_k(t)$ involve primes in disjoint ranges, and therefore behave independently of each other. Notice further that if $|t_1-t_2|\le e^{-k}$ then $\mathcal{P}_k(t_1)$ and $\mathcal{P}_k(t_2)$ are more or less the same. Thus instead of modeling $\log|\zeta(\frac{1}{2}+it)|$ in intervals of length 1 by about $\log T$ independent samples of a normal random variable, we are led to the following more nuanced model. For each k, let P_k denote any one of about e^k independent drawings of a normal random variable with mean 0 and variance $\frac{1}{2}$. Then $\log|\zeta(\frac{1}{2}+it)|$ in an interval of length 1 is modeled by all the possibilities for $\sum_k P_k$.

The model described above has been analyzed in the probability literature surrounding branching random walks and branching Brownian motion. Consider a particle starting at time 0 and moving as a standard Brownian motion. At time t, with probability e^{-t} the particle might split into two particles, that move according to independent standard Brownian motions starting from that position. These particles may again split (independently of each other) at a future time, giving rise to more daughter particles, and so on. After time T, how is the maximum value of all these particles distributed? This problem was resolved by Bramson who established that the maximum is almost surely $\sqrt{2}(T - \frac{3}{4}\log T) + O(1)$. Notice the $\frac{3}{4}$ term here, which exactly parallels the $\frac{3}{4}$ terms appearing in Conjecture 8.1!

In recent years there has been a lot of progress towards understanding Conjecture 8.1. In [7] Arguin, Belius, and Harper considered $\max_{0\le h\le 1} \operatorname{Re}\sum_{p\le T} X(p)/p^{\frac{1}{2}+ih}$ where the $X(p)$'s are independent random variables chosen uniformly on the unit circle (a randomized model for $\log|\zeta(\frac{1}{2}+it+ih)|$), and established that almost surely this is $\log\log T - (\frac{3}{4}+o(1))\log\log\log T$. Najnudel [129] established that on RH the set of $t\in[T,2T]$ with $\max_{0\le h\le 1}|\zeta(\frac{1}{2}+it+ih)| = (\log T)^{1+o(1)}$ has measure $\sim T$. Independently this result was also established unconditionally by Arguin, Belius, Bourgade, Radziwiłł, and Soundararajan [6]. A lovely exposition of Conjecture 8.1 and the results mentioned so far may be found in Harper's Bourbaki seminar [75]. Still more recently, Harper [76] established that if t is not in an exceptional subset of $[T,2T]$ with measure $o(T)$, then

$$\max_{0\le h\le 1}\log\left|\zeta\left(\frac{1}{2}+it+ih\right)\right| \le \log\log T - \frac{3}{4}\log\log\log T + O(\log\log\log\log T),$$

so that at least in one direction, the difference between the naive constant $\frac{1}{4}$ and the refined prediction $\frac{3}{4}$ could be established. Independently Arguin, Bourgade, and Radziwiłł [8] established the shaper result that for any $y\ge 1$,

$$\frac{1}{T}\operatorname{meas}\left\{t\in[T,2T]:\max_{0\le h\le 1}\left|\zeta\left(\frac{1}{2}+it+ih\right)\right| > \frac{e^y\log T}{(\log\log T)^{\frac{3}{4}}}\right\} \le Cye^{-2y},$$

for some constant C. This beautiful result establishes part of Conjecture 8.1, and the decay in y above matches (up to constants) the conjectured behavior of $1-F(y)$. There has also been substantial progress toward the analogue of Conjecture 8.1 in random matrix theory; see [5, 37, 135].

Instead of considering the maximum of the zeta function in intervals of length 1, one may also examine other "local moments" $\int_0^1|\zeta(\frac{1}{2}+it+ih)|^\beta\,dh$. This was already suggested in [64], who conjectured that a transition in the behavior of these local moments occurs at the critical exponent $\beta=2$—for $\beta<2$ these local moments are typically of size $(\log T)^{\beta^2/4}$ (the size of the global moment $\frac{1}{T}\int_T^{2T}|\zeta(\frac{1}{2}+it)|^\beta\,dt$), whereas for $\beta>2$ they are typically of size $(\log T)^{\beta-1}$ corresponding to the largest value of zeta in that interval (about size $\log T$) which might be expected to occur on an interval of length about $1/\log T$. For work in this direction see [9, 11, 76]. We mention a lovely result of Harper [76] for the critical exponent $\beta=2$:

$$\frac{1}{T}\int_T^{2T}\left(\frac{1}{\log T}\int_0^1\left|\zeta\left(\frac{1}{2}+it+ih\right)\right|^2 dh\right)^{\frac{1}{2}} dt \ll \frac{1}{(\log\log T)^{\frac{1}{4}}}.$$

A simple application of Cauchy's inequality together with the second moment of $\zeta(\frac{1}{2}+it)$ shows that the above quantity is $\ll 1$, and the fact that it is a little bit smaller is a reflection of the correlation structure of nearby values of $\zeta(s)$ that also underlies Conjecture 8.1.

The ideas discussed here are closely connected to what is termed *Gaussian multiplicative chaos*, which was initiated by Kahane [97], and which has been extensively studied in the probability literature [146]. In number theory, these ideas are closely related to the study of mean values of random multiplicative functions. We content ourselves with giving a few pointers to surveys and related work [12, 74, 151, 167].

ACKNOWLEDGMENTS

I am grateful to Brian Conrey, Jon Keating, Emmanuel Kowalski, Vivian Kuperberg, Maksym Radziwiłł, Matt Tyler, and Max Xu for their careful reading and many valuable suggestions.

FUNDING

This work was partially supported by grants from the National Science Foundation, and a Simons Investigator Award from the Simons Foundation.

REFERENCES

[1] C. Aistleitner, Lower bounds for the maximum of the Riemann zeta function along vertical lines. *Math. Ann.* **365** (2016), no. 1–2, 473–496.

[2] C. Aistleitner, I. Berkes, and K. Seip, GCD sums from Poisson integrals and systems of dilated functions. *J. Eur. Math. Soc. (JEMS)* **17** (2015), no. 6, 1517–1546.

[3] C. Aistleitner, K. Mahatab, and M. Munsch, Extreme values of the Riemann zeta function on the 1-line. *Int. Math. Res. Not. IMRN* **22** (2019), 6924–6932.

[4] C. Aistleitner, K. Mahatab, M. Munsch, and A. Peyrot, On large values of $L(\sigma, \chi)$. *Q. J. Math.* **70** (2019), no. 3, 831–848.

[5] L.-P. Arguin, D. Belius, and P. Bourgade, Maximum of the characteristic polynomial of random unitary matrices. *Comm. Math. Phys.* **349** (2017), no. 2, 703–751.

[6] L.-P. Arguin, D. Belius, P. Bourgade, M. Radziwiłł, and K. Soundararajan, Maximum of the Riemann zeta function on a short interval of the critical line. *Comm. Pure Appl. Math.* **72** (2019), no. 3, 500–535.

[7] L.-P. Arguin, D. Belius, and A. J. Harper, Maxima of a randomized Riemann zeta function, and branching random walks. *Ann. Appl. Probab.* **27** (2017), no. 1, 178–215.

[8] L.-P. Arguin, P. Bourgade, and M. Radziwiłł, The Fyodorov–Hiary–Keating conjecture. I. 2020, arXiv:2007.00988.

[9] L.-P. Arguin, F. Ouimet, and M. Radziwiłł, Moments of the Riemann zeta function on short intervals of the critical line. 2021, arXiv:1901.04061.

[10] B. Bagchi, *Statistical behaviour and universality properties of the Riemann zeta-function and other allied Dirichlet series*. Ph.D. thesis, Indian Statistical Institute, Kolkata, 1981.

[11] E. C. Bailey and J. P. Keating, On the moments of the moments of $\zeta(1/2 + it)$. *J. Number Theory* **223** (2021), 79–100.

[12] E. C. Bailey and J. P. Keating, Maxima of log-correlated fields: some recent developments. 2021, arXiv:2106.15141.

[13] R. Balasubramanian, An improvement on a theorem of Titchmarsh on the mean square of $|\zeta(\frac{1}{2} + it)|$. *Proc. Lond. Math. Soc. (3)* **36** (1978), no. 3, 540–576.

[14] R. Balasubramanian, On the frequency of Titchmarsh's phenomenon for $\zeta(s)$. IV. *Hardy-Ramanujan J.* **9** (1986), 1–10.

[15] R. Balasubramanian, J. B. Conrey, and D. R. Heath-Brown, Asymptotic mean square of the product of the Riemann zeta-function and a Dirichlet polynomial. *J. Reine Angew. Math.* **357** (1985), 161–181.

[16] R. Balasubramanian and K. Ramachandra, On the frequency of Titchmarsh's phenomenon for $\zeta(s)$. III. *Proc. Indian Acad. Sci., Sect. A, Phys. Sci.* **86** (1977), no. 4, 341–351.

[17] S. Bettin, V. Chandee, and M. Radziwiłł, The mean square of the product of the Riemann zeta-function with Dirichlet polynomials. *J. Reine Angew. Math.* **729** (2017), 51–79.

[18] V. Blomer, É. Fouvry, E. Kowalski, P. Michel, D. Milićević, and W. Sawin, The second moment theory of families of L-functions. 2019, arXiv:1804.01450.

[19] V. Blomer and R. Khan, Twisted moments of L-functions and spectral reciprocity. *Duke Math. J.* **168** (2019), no. 6, 1109–1177.

[20] J. W. Bober and L. Goldmakher, Pólya–Vinogradov and the least quadratic non-residue. *Math. Ann.* **366** (2016), no. 1–2, 853–863.

[21] E. Bombieri and D. A. Hejhal, On the distribution of zeros of linear combinations of Euler products. *Duke Math. J.* **80** (1995), no. 3, 821–862.

[22] A. Bondarenko and K. Seip, GCD sums and complete sets of square-free numbers. *Bull. Lond. Math. Soc.* **47** (2015), no. 1, 29–41.

[23] A. Bondarenko and K. Seip, Large greatest common divisor sums and extreme values of the Riemann zeta function. *Duke Math. J.* **166** (2017), no. 9, 1685–1701.

[24] A. Bondarenko and K. Seip, Extreme values of the Riemann zeta function and its argument. *Math. Ann.* **372** (2018), no. 3–4, 999–1015.

[25] J. Bourgain, Decoupling, exponential sums and the Riemann zeta function. *J. Amer. Math. Soc.* **30** (2017), no. 1, 205–224.

[26] H. M. Bui, N. Evans, S. Lester, and K. Pratt, Weighted central limit theorems for central values of L-functions. 2021, arXiv:2109.06829.

[27] H. M. Bui, A. Florea, J. P. Keating, and E. Roditty-Gershon, Moments of quadratic twists of elliptic curve L-functions over function fields. *Algebra Number Theory* **14** (2020), no. 7, 1853–1893.

[28] D. Bump and A. Gamburd, On the averages of characteristic polynomials from classical groups. *Comm. Math. Phys.* **265** (2006), no. 1, 227–274.

[29] D. A. Burgess, The distribution of quadratic residues and non-residues. *Mathematika* **4** (1957), 106–112.

[30] E. Carneiro, V. Chandee, and M. B. Milinovich, Bounding $S(t)$ and $S_1(t)$ on the Riemann hypothesis. *Math. Ann.* **356** (2013), no. 3, 939–968.

[31] V. Chandee, Explicit upper bounds for L-functions on the critical line. *Proc. Amer. Math. Soc.* **137** (2009), no. 12, 4049–4063.

[32] V. Chandee, On the correlation of shifted values of the Riemann zeta function. *Q. J. Math.* **62** (2011), no. 3, 545–572.

[33] V. Chandee and X. Li, The eighth moment of Dirichlet L-functions. *Adv. Math.* **259** (2014), 339–375.

[34] V. Chandee and X. Li, The sixth moment of automorphic L-functions. *Algebra Number Theory* **11** (2017), no. 3, 583–633.

[35] V. Chandee and X. Li, The 8th moment of the family of $\Gamma_1(q)$-automorphic L-functions. *Int. Math. Res. Not. IMRN* **22** (2020), 8443–8485.

[36] V. Chandee and K. Soundararajan, Bounding $|\zeta(\frac{1}{2}+it)|$ on the Riemann hypothesis. *Bull. Lond. Math. Soc.* **43** (2011), no. 2, 243–250.

[37] R. Chhaibi, T. Madaule, and J. Najnudel, On the maximum of the CβE field. *Duke Math. J.* **167** (2018), no. 12, 2243–2345.

[38] G. Chinta, Analytic ranks of elliptic curves over cyclotomic fields. *J. Reine Angew. Math.* **544** (2002), 13–24.

[39] S. Chowla and P. Erdős, A theorem on the distribution of the values of L-functions. *J. Indian Math. Soc. (N.S.)* **15** (1951), 11–18.

[40] J. Cogdell and P. Michel, On the complex moments of symmetric power L-functions at $s=1$. *Int. Math. Res. Not.* **31** (2004), 1561–1617.

[41] B. Conrey and J. P. Keating, Moments of zeta and correlations of divisor-sums: V. *Proc. Lond. Math. Soc. (3)* **118** (2019), no. 4, 729–752.

[42] J. B. Conrey and D. W. Farmer, Mean values of L-functions and symmetry. *Int. Math. Res. Not.* **17** (2000), 883–908.

[43] J. B. Conrey, D. W. Farmer, J. P. Keating, M. O. Rubinstein, and N. C. Snaith, Integral moments of L-functions. *Proc. Lond. Math. Soc. (3)* **91** (2005), no. 1, 33–104.

[44] J. B. Conrey and A. Ghosh, Mean values of the Riemann zeta-function. III. In *Proceedings of the Amalfi conference on analytic number theory (Maiori, 1989)*, pp. 35–59, Univ. Salerno, Salerno, 1992.

[45] J. B. Conrey and A. Ghosh, A conjecture for the sixth power moment of the Riemann zeta-function. *Int. Math. Res. Not.* **15** (1998), 775–780.

[46] J. B. Conrey and S. M. Gonek, High moments of the Riemann zeta-function. *Duke Math. J.* **107** (2001), no. 3, 577–604.

[47] J. B. Conrey and H. Iwaniec, The cubic moment of central values of automorphic L-functions. *Ann. of Math. (2)* **151** (2000), no. 3, 1175–1216.

[48] J. B. Conrey, H. Iwaniec, and K. Soundararajan, The sixth power moment of Dirichlet L-functions. *Geom. Funct. Anal.* **22** (2012), no. 5, 1257–1288.

[49] A. Dahl and Y. Lamzouri, The distribution of class numbers in a special family of real quadratic fields. *Trans. Amer. Math. Soc.* **370** (2018), no. 9, 6331–6356.

[50] H. Davenport, *Multiplicative number theory. Second edn*. Grad. Texts in Math. 74, Springer, New York–Berlin, 1980.

[51] C. David, A. Florea, and M. Lalin, Non-vanishing for cubic L-functions. 2020, arXiv:2006.15661.

[52] R. de la Bretèche and G. Tenenbaum, Sommes de Gál et applications. *Proc. Lond. Math. Soc. (3)* **119** (2019), no. 1, 104–134.

[53] P.-O. Dehaye, Combinatorics of lower order terms in the moment conjectures for the Riemann zeta function. 2012, arXiv:1201.4478.

[54] A. Diaconu, D. Goldfeld, and J. Hoffstein, Multiple Dirichlet series and moments of zeta and L-functions. *Compos. Math.* **139** (2003), no. 3, 297–360.

[55] A. Diaconu and I. Whitehead, On the third moment of $L(\frac{1}{2}, \chi_d)$ II: the number field case. *J. Eur. Math. Soc. (JEMS)* **23** (2021), no. 6, 2051–2070.

[56] W. Duke, Extreme values of Artin L-functions and class numbers. *Compos. Math.* **136** (2003), no. 1, 103–115.

[57] B. Eckhardt, S. Fishman, J. Keating, O. Agam, J. Main, and K. Müller, Approach to ergodicity in quantum wave functions. *Phys. Rev. E* **52** (1995), 5893–5903.

[58] P. D. T. A. Elliott, On the distribution of the values of quadratic L-series in the half-plane $\sigma > \frac{1}{2}$. *Invent. Math.* **21** (1973), 319–338.

[59] D. W. Farmer, S. M. Gonek, and C. P. Hughes, The maximum size of L-functions. *J. Reine Angew. Math.* **609** (2007), 215–236.

[60] A. Fazzari, A weighted central limit theorem for $\log|\zeta(1/2 + it)|$. *Mathematika* **67** (2021), no. 2, 324–341.

[61] A. Florea, The fourth moment of quadratic Dirichlet L-functions over function fields. *Geom. Funct. Anal.* **27** (2017), no. 3, 541–595.

[62] P. J. Forrester and S. O. Warnaar, The importance of the Selberg integral. *Bull. Amer. Math. Soc. (N.S.)* **45** (2008), no. 4, 489–534.

[63] J. B. Friedlander, Bounds for L-functions. In *Proceedings of the international congress of mathematicians, Vol. 1, 2 (Zürich, 1994)*, pp. 363–373, Birkhäuser, Basel, 1995.

[64] Y. V. Fyodorov, G. Hiary, and J. Keating, Freezing transition, characteristic polynomials of random matrices, and the Riemann zeta function. *Phys. Rev. Lett.* **108** (2012), 170601.

[65] Y. V. Fyodorov and J. Keating, Freezing transitions and extreme values: random matrix theory, and disordered landscapes. *Philos. Trans. R. Soc. Lond. Ser. A Math. Phys. Eng. Sci.* **372** (2014), no. 2007, 20120503.

[66] P. Gao and L. Zhao, Bounds for moments of cubic and quartic Dirichlet L-functions. 2021, arXiv:2104.09909.

[67] J. Germain, Factorization of constants involved in conjectural moments of zeta-functions. *Integers* **8** (2008), A47, 18.

[68] D. A. Goldston and S. M. Gonek, A note on $S(t)$ and the zeros of the Riemann zeta-function. *Bull. Lond. Math. Soc.* **39** (2007), no. 3, 482–486.

[69] S. M. Gonek, C. P. Hughes, and J. P. Keating, A hybrid Euler–Hadamard product for the Riemann zeta function. *Duke Math. J.* **136** (2007), no. 3, 507–549.

[70] A. Granville and K. Soundararajan, Upper bounds for $|L(1,\chi)|$. *Q. J. Math.* **53** (2002), no. 3, 265–284.

[71] A. Granville and K. Soundararajan, The distribution of values of $L(1,\chi_d)$. *Geom. Funct. Anal.* **13** (2003), no. 5, 992–1028.

[72] A. Granville and K. Soundararajan, Large character sums: Burgess's theorem and zeros of L-functions. *J. Eur. Math. Soc. (JEMS)* **20** (2018), no. 1, 1–14.

[73] S. Gun, W. Kohnen, and K. Soundararajan, Large Fourier coefficients of half-integer weight modular forms. 2020, arXiv:2004.14450.

[74] A. J. Harper, Moments of random multiplicative functions, I: Low moments, better than square-root cancellation, and critical multiplicative chaos. *Forum Math. Pi* **8** (2020), e1, 95.

[75] A. J. Harper, The Riemann zeta function in short intervals [after Najnudel, and Arguin, Belius, Bourgade, Radziwiłł and Soundararajan]. *Astérisque* **422** (2020). Séminaire Bourbaki. Vol. 2018/2019. Exposés 1151–1165, 391–414.

[76] A. J. Harper, On the partition function of the Riemann zeta function, and the Fyodorov–Hiary–Keating conjecture. 2019, arXiv:1906.05783.

[77] A. J. Harper, Sharp conditional bounds for moments of the Riemann zeta function. 2013, arXiv:1305.4618.

[78] W. Heap, M. Radziwiłł, and K. Soundararajan, Sharp upper bounds for fractional moments of the Riemann zeta function. *Q. J. Math.* **70** (2019), no. 4, 1387–1396.

[79] W. Heap and K. Soundararajan, Lower bounds for moments of zeta and L-functions revisited. 2020, arXiv:2007.13154.

[80] D. R. Heath-Brown, The twelfth power moment of the Riemann zeta-function. *Quart. J. Math. Oxf. Ser. (2)* **29** (1978), no. 116, 443–462.

[81] D. R. Heath-Brown, The fourth power moment of the Riemann zeta function. *Proc. Lond. Math. Soc. (3)* **38** (1979), no. 3, 385–422.

[82] D. R. Heath-Brown, On a paper of Baker and Schinzel. *Acta Arith.* **35** (1979), no. 2, 203–207.

[83] D. R. Heath-Brown, The fourth power mean of Dirichlet's L-functions. *Analysis* **1** (1981), no. 1, 25–32.

[84] D. R. Heath-Brown, Fractional moments of the Riemann zeta function. *J. Lond. Math. Soc. (2)* **24** (1981), no. 1, 65–78.

[85] D. R. Heath-Brown, The average analytic rank of elliptic curves. *Duke Math. J.* **122** (2004), no. 3, 591–623.

[86] G. A. Hiary and A. M. Odlyzko, The zeta function on the critical line: numerical evidence for moments and random matrix theory models. *Math. Comp.* **81** (2012), no. 279, 1723–1752.

[87] T. Hilberdink, An arithmetical mapping and applications to Ω-results for the Riemann zeta function. *Acta Arith.* **139** (2009), no. 4, 341–367.

[88] S. Holmin, N. Jones, P. Kurlberg, C. McLeman, and K. Petersen, Missing class groups and class number statistics for imaginary quadratic fields. *Exp. Math.* **28** (2019), no. 2, 233–254.

[89] B. Hough, The distribution of the logarithm in an orthogonal and a symplectic family of L-functions. *Forum Math.* **26** (2014), no. 2, 523–546.

[90] M. N. Huxley, The large sieve inequality for algebraic number fields. II. Means of moments of Hecke zeta-functions. *Proc. Lond. Math. Soc. (3)* **21** (1970), 108–128.

[91] A. Ivić, On sums of Hecke series in short intervals. *J. Théor. Nombres Bordeaux* **13** (2001), no. 2, 453–468.

[92] H. Iwaniec and E. Kowalski, *Analytic number theory*. Amer. Math. Soc. Colloq. Publ. 53, American Mathematical Society, Providence, RI, 2004.

[93] H. Iwaniec and P. Sarnak, The non-vanishing of central values of automorphic L-functions and Landau–Siegel zeros. *Israel J. Math. (A)* **120** (2000), 155–177.

[94] H. Iwaniec and P. Sarnak, Perspectives on the analytic theory of L-functions. *Geom. Funct. Anal.* **Special Volume, Part II** (2000), 705–741.

[95] J. Jacod, E. Kowalski, and A. Nikeghbali, Mod-Gaussian convergence: new limit theorems in probability and number theory. *Forum Math.* **23** (2011), no. 4, 835–873.

[96] M. Jutila, On the mean value of $L(\frac{1}{2}, \chi)$ for real characters. *Analysis* **1** (1981), no. 2, 149–161.

[97] J.-P. Kahane, Sur le chaos multiplicatif. *Ann. Sci. Math. Québec* **9** (1985), no. 2, 105–150.

[98] N. M. Katz and P. Sarnak, *Random matrices, Frobenius eigenvalues, and monodromy*. Amer. Math. Soc. Colloq. Publ. 45, American Mathematical Society, Providence, RI, 1999.

[99] J. P. Keating, B. Rodgers, E. Roditty-Gershon, and Z. Rudnick, Sums of divisor functions in $\mathbb{F}_q[t]$ and matrix integrals. *Math. Z.* **288** (2018), no. 1–2, 167–198.

[100] J. P. Keating and N. C. Snaith, Random matrix theory and $\zeta(1/2 + it)$. *Comm. Math. Phys.* **214** (2000), no. 1, 57–89.

[101] J. P. Keating and N. C. Snaith, Random matrix theory and L-functions at $s = 1/2$. *Comm. Math. Phys.* **214** (2000), no. 1, 91–110.

[102] R. Khan, The fifth moment of Hecke L-functions in the weight aspect. *Math. Proc. Cambridge Philos. Soc.* **168** (2020), no. 3, 543–566.

[103] R. Khan and H. T. Ngo, Nonvanishing of Dirichlet L-functions. *Algebra Number Theory* **10** (2016), no. 10, 2081–2091.

[104] E. M. Kıral and M. Young, The fifth moment of modular L-functions. *J. Eur. Math. Soc. (JEMS)* **23** (2021), no. 1, 237–314.

[105] W. Kohnen and D. Zagier, Values of L-series of modular forms at the center of the critical strip. *Invent. Math.* **64** (1981), no. 2, 175–198.

[106] E. Kowalski, Bagchi's theorem for families of automorphic forms. In *Exploring the Riemann zeta function*, pp. 181–199, Springer, Cham, 2017.

[107] E. Kowalski, P. Michel, and W. Sawin, Bilinear forms with Kloosterman sums and applications. *Ann. of Math. (2)* **186** (2017), no. 2, 413–500.

[108] E. Kowalski and A. Nikeghbali, Mod-Gaussian convergence and the value distribution of $\zeta(\frac{1}{2}+it)$ and related quantities. *J. Lond. Math. Soc. (2)* **86** (2012), no. 1, 291–319.

[109] Y. Lamzouri, The two-dimensional distribution of values of $\zeta(1+it)$. *Int. Math. Res. Not. IMRN* **2008** (2008), rnn 106, 48 pp.

[110] Y. Lamzouri, Large moments and extreme values of class numbers of indefinite binary quadratic forms. *Mathematika* **63** (2017), no. 2, 564–586.

[111] Y. Lamzouri, On the average of the number of imaginary quadratic fields with a given class number. *Ramanujan J.* **44** (2017), no. 2, 411–416.

[112] Y. Lamzouri, S. Lester, and M. Radziwiłł, An effective universality theorem for the Riemann zeta function. *Comment. Math. Helv.* **93** (2018), no. 4, 709–736.

[113] Y. Lamzouri, S. Lester, and M. Radziwiłł, Discrepancy bounds for the distribution of the Riemann zeta-function and applications. *J. Anal. Math.* **139** (2019), no. 2, 453–494.

[114] Y. Lamzouri, X. Li, and K. Soundararajan, Conditional bounds for the least quadratic non-residue and related problems. *Math. Comp.* **84** (2015), no. 295, 2391–2412.

[115] A. Laurinchikas, A limit theorem for the Riemann zeta-function on the critical line. II. *Litov. Mat. Sb.* **27** (1987), no. 3, 489–500.

[116] S. Lester and M. Radziwiłł, Signs of Fourier coefficients of half-integral weight modular forms. *Math. Ann.* **379** (2021), no. 3–4, 1553–1604.

[117] X. Li, Upper bounds on L-functions at the edge of the critical strip. *Int. Math. Res. Not. IMRN* **4** (2010), 727–755.

[118] W. Luo, Values of symmetric square L-functions at 1. *J. Reine Angew. Math.* **506** (1999), 215–235.

[119] P. Michel and A. Venkatesh, Equidistribution, L-functions and ergodic theory: on some problems of Yu. Linnik. In *International congress of mathematicians. Vol. II*, pp. 421–457, Eur. Math. Soc., Zürich, 2006.

[120] D. Milićević, Large values of eigenfunctions on arithmetic hyperbolic surfaces. *Duke Math. J.* **155** (2010), no. 2, 365–401.

[121] M. B. Milinovich, Upper bounds for moments of $\zeta'(\rho)$. *Bull. Lond. Math. Soc.* **42** (2010), no. 1, 28–44.

[122] M. B. Milinovich and N. Ng, Lower bounds for moments of $\zeta'(\rho)$. *Int. Math. Res. Not. IMRN* **12** (2014), 3190–3216.

[123] G. Molteni, Upper and lower bounds at $s = 1$ for certain Dirichlet series with Euler product. *Duke Math. J.* **111** (2002), no. 1, 133–158.
[124] H. L. Montgomery, The pair correlation of zeros of the zeta function. In *Analytic number theory (Proc. Sympos. Pure Math., Vol. XXIV, St. Louis Univ., St. Louis, Mo., 1972)*, pp. 181–193, American Math Society, 1973.
[125] H. L. Montgomery, Extreme values of the Riemann zeta function. *Comment. Math. Helv.* **52** (1977), no. 4, 511–518.
[126] H. L. Montgomery and R. C. Vaughan, Extreme values of Dirichlet L-functions at 1. In *Number theory in progress, Vol. 2 (Zakopane–Kościelisko, 1997)*, pp. 1039–1052, de Gruyter, Berlin, 1999.
[127] Y. Motohashi, *Spectral theory of the Riemann zeta-function*. Cambridge Tracts in Math. 127, Cambridge University Press, Cambridge, 1997.
[128] R. Munshi, The subconvexity problem for L-functions. In *Proceedings of the International Congress of Mathematicians—Rio de Janeiro 2018. Vol. II. Invited lectures*, pp. 363–376, World Sci. Publ., Hackensack, NJ, 2018.
[129] J. Najnudel, On the extreme values of the Riemann zeta function on random intervals of the critical line. *Probab. Theory Related Fields* **172** (2018), no. 1–2, 387–452.
[130] P. D. Nelson, Bounds for standard L-functions. 2021, arXiv:2109.15230.
[131] P. D. Nelson, Eisenstein series and the cubic moment for PGL(2). 2020, arXiv:1911.06310.
[132] P. D. Nelson, Spectral aspect subconvex bounds for $\mathrm{U}_{n+1} \times \mathrm{U}_n$. 2021, arXiv:2012.02187.
[133] N. Ng, The sixth moment of the Riemann zeta function and ternary additive divisor sums. *Discrete Anal.* **6** (2021), 60.
[134] A. M. Odlyzko, The 10^{20}-th zero of the Riemann zeta function and 175 million of its neighbors. 1992, http://www.dtc.umn.edu/~odlyzko/unpublished/zeta.10to20.1992.pdf.
[135] E. Paquette and O. Zeitouni, The maximum of the CUE field. *Int. Math. Res. Not. IMRN* **16** (2018), 5028–5119.
[136] Z. Peng, *Zeros and central values of automorphic L-functions*. Ph.D. thesis, Princeton University, 2001.
[137] I. Petrow and M. P. Young, The Weyl bound for Dirichlet L-functions of cube-free conductor. *Ann. of Math. (2)* **192** (2020), no. 2, 437–486.
[138] M. Radziwiłł, High moments of Dirichlet L-functions. In *Automorphic forms and arithmetic*, pp. 2520–2521, Oberwolfach Rep. 40, EMS Press, 2017.
[139] M. Radziwiłł, Large deviations in Selberg's central limit theorem. 2011, arXiv:1108.5092.
[140] M. Radziwiłł and K. Soundararajan, Continuous lower bounds for moments of zeta and L-functions. *Mathematika* **59** (2013), no. 1, 119–128.
[141] M. Radziwiłł and K. Soundararajan, Moments and distribution of central L-values of quadratic twists of elliptic curves. *Invent. Math.* **202** (2015), no. 3, 1029–1068.

[142] M. Radziwiłł and K. Soundararajan, Selberg's central limit theorem for $\log|\zeta(1/2+it)|$. *Enseign. Math.* **63** (2017), no. 1–2, 1–19.

[143] K. Ramachandra, Some remarks on the mean value of the Riemann zeta function and other Dirichlet series. I. *Hardy-Ramanujan J.* **1** (1978), 15.

[144] K. Ramachandra, Some remarks on the mean value of the Riemann zeta function and other Dirichlet series. II. *Hardy-Ramanujan J.* **3** (1980), 1–24.

[145] N. Raulf, Limit distribution of class numbers for discriminants in progressions and fundamental discriminants. *Int. J. Number Theory* **12** (2016), no. 5, 1237–1258.

[146] R. Rhodes and V. Vargas, Gaussian multiplicative chaos and applications: a review. *Probab. Surv.* **11** (2014), 315–392.

[147] D. E. Rohrlich, On L-functions of elliptic curves and cyclotomic towers. *Invent. Math.* **75** (1984), no. 3, 409–423.

[148] Z. Rudnick and P. Sarnak, Zeros of principal L-functions and random matrix theory. *Duke Math. J.* **81** (1996), no. 2, 269–322.

[149] Z. Rudnick and K. Soundararajan, Lower bounds for moments of L-functions. *Proc. Natl. Acad. Sci. USA* **102** (2005), no. 19, 6837–6838.

[150] Z. Rudnick and K. Soundararajan, Lower bounds for moments of L-functions: symplectic and orthogonal examples. In *Multiple Dirichlet series, automorphic forms, and analytic number theory*, pp. 293–303, Proc. Sympos. Pure Math. 75, Amer. Math. Soc., Providence, RI, 2006.

[151] E. Saksman and C. Webb, The Riemann zeta function and Gaussian multiplicative chaos: statistics on the critical line. *Ann. Probab.* **48** (2020), no. 6, 2680–2754.

[152] P. Sarnak, Class numbers of indefinite binary quadratic forms. *J. Number Theory* **15** (1982), no. 2, 229–247.

[153] W. Sawin, A representation theory approach to integral moments of L-functions over function fields. *Algebra Number Theory* **14** (2020), no. 4, 867–906.

[154] A. Selberg, Remarks on a multiple integral. *Norsk Mat. Tidsskr.* **26** (1944), 71–78.

[155] A. Selberg, Contributions to the theory of the Riemann zeta-function. *Arch. Math. Naturvid.* **48** (1946), no. 5, 89–155.

[156] A. Selberg, Old and new conjectures and results about a class of Dirichlet series. In *Proceedings of the Amalfi conference on analytic number theory (Maiori, 1989)*, pp. 367–385, Univ. Salerno, Salerno, 1992.

[157] Q. Shen, The fourth moment of quadratic Dirichlet L-functions. *Math. Z.* **298** (2021), no. 1–2, 713–745.

[158] A. Shubin, Variance estimates in Linnik's problem. 2021, arXiv:2108.00726.

[159] H. C. Siu, *Value distribution of automorphic forms in a family*. Ph.D. thesis, Stanford University, 2016.

[160] A. Smith, 2^∞-Selmer groups, 2^∞-class groups, and Goldfeld's conjecture. 2017, arXiv:1702.02325.

[161] K. Soundararajan, Nonvanishing of quadratic Dirichlet L-functions at $s = \frac{1}{2}$. *Ann. of Math. (2)* **152** (2000), no. 2, 447–488.

[162] K. Soundararajan, The number of imaginary quadratic fields with a given class number. *Hardy-Ramanujan J.* **30** (2007), 13–18.

[163] K. Soundararajan, Extreme values of zeta and L-functions. *Math. Ann.* **342** (2008), no. 2, 467–486.

[164] K. Soundararajan, Moments of the Riemann zeta function. *Ann. of Math. (2)* **170** (2009), no. 2, 981–993.

[165] K. Soundararajan, Value distribution of L-functions. In *Automorphic forms and arithmetic*, pp. 2528–2529, Oberwolfach Rep. 40, EMS Press, 2017.

[166] K. Soundararajan and M. P. Young, The second moment of quadratic twists of modular L-functions. *J. Eur. Math. Soc. (JEMS)* **12** (2010), no. 5, 1097–1116.

[167] K. Soundararajan and A. Zaman, A model problem for multiplicative chaos in number theory. 2021, arXiv:2108.07264.

[168] P. J. Stephens, Optimizing the size of $L(1, \chi)$. *Proc. Lond. Math. Soc. (3)* **24** (1972), 1–14.

[169] T. Tao, The Elliott–Halberstam conjecture implies the Vinogradov least quadratic nonresidue conjecture. *Algebra Number Theory* **9** (2015), no. 4, 1005–1034.

[170] E. C. Titchmarsh, *The theory of the Riemann zeta-function. Second edn.* The Clarendon Press Oxford University Press, New York, 1986.

[171] S. M. Voronin, A theorem on the distribution of values of the Riemann zeta-function. *Dokl. Akad. Nauk SSSR* **221** (1975), no. 4, 771.

[172] S. M. Voronin, Lower bounds in Riemann zeta-function theory. *Izv. Akad. Nauk SSSR Ser. Mat.* **52** (1988), no. 4, 882–892, 896.

[173] J.-L. Waldspurger, Sur les coefficients de Fourier des formes modulaires de poids demi-entier. *J. Math. Pures Appl. (9)* **60** (1981), no. 4, 375–484.

[174] M. P. Young, The fourth moment of Dirichlet L-functions. *Ann. of Math. (2)* **173** (2011), no. 1, 1–50.

[175] M. P. Young, The third moment of quadratic Dirichlet L-functions. *Selecta Math. (N.S.)* **19** (2013), no. 2, 509–543.

[176] P. Zenz, Sharp bound for the fourth moment of holomorphic Hecke cusp forms. 2021, arXiv:2108.13868.

KANNAN SOUNDARARAJAN

Department of Mathematics, Stanford University, Stanford, CA, USA,
ksound@stanford.edu

This Page Intentionally Left Blank

CATEGORIFICATION: TANGLE INVARIANTS AND TQFTS

CATHARINA STROPPEL

Dedicated to Igor Frenkel who introduced me to the world of categorification.

ABSTRACT

Based on different views on the Jones polynomial, we review representation theoretic categorified link and tangle invariants. We unify them in a common combinatorial framework and connect them via the theory of Soergel bimodules. The influence of these categorifications on the development of 2-representation theory and the interaction between topological invariants and 2-categorical structures is discussed. Finally, we indicate how categorified representations of quantum groups, on the one hand, and monoidal 2-categories of Soergel bimodules, on the other hand, might lead to new interesting 4-dimensional TQFTs.

MATHEMATICS SUBJECT CLASSIFICATION 2020

Primary 18N25; Secondary 17B20, 17B10, 05E10, 57K16

KEYWORDS

Jones polynomial, knot invariants, categorification, TQFT, Khovanov homology, category O, skew Howe duality, quantum groups

ICM INTERNATIONAL CONGRESS OF MATHEMATICIANS 2022 JULY 6–14

Proc. Int. Cong. Math. 2022, Vol. 2, pp. 1312–1353
DOI 10.4171/ICM2022/109

1. INTRODUCTION

The study of Topological Quantum Field Theories (TQFTs) is a fruitful interaction between physics and mathematics. The search for interesting TQFTs leads to many developments in mathematical theories which are interesting on their own and also motivates constructions presented in this report. A first mathematical formulation of TQFTs goes back to Atiyah [6], influenced by Segal [129] and Witten [143]. A d-(dimensional) TQFT is a symmetric monoidal functor F from a bordism category with objects being closed $(d-1)$-dimensional manifolds and morphisms d-dimensional bordisms to some symmetric monoidal category, e.g., categories of vector spaces or chain complexes, or more complicated categories.

Representation theory is a good source for TQFTs and monoidal categories. For example, categories of group representations give Dijkgraaf–Witten TQFTs [38] for any d. For $d = 3$, representations of quantum groups, i.e., quantized representations of Lie algebras, provide rich and interesting TQFTs due to Reshetikhin–Turaev [119] and Turaev–Viro [141]. They are often viewed as mathematical formulations of Chern–Simons theory [144] or of a form of Ponzano–Regge state sum model from Quantum Gravity [112]. These theories are closely related to (Laurent-)polynomial invariants of knots and links. In Chern–Simons theory [144], for instance, the partition function is a 3-manifold invariant, but the expectation values of nonlocal observables supported on one-dimensional defects, the Wilson lines, give such an invariant of links. The *Jones polynomial* $\mathtt{J}(L)$ arises in this way for the gauge group $\mathrm{SU}(2)$. In our setting $\mathtt{J}(L)$ appears as the special $\mathfrak{sl}_2$ example of the Reshetikhin–Turaev-*link invariants*. While 3-manifolds are rather well-understood, new 4-TQFTs might help to solve open 4-dimensional (smoothness) problems.

TQFTs provide not only numerical invariants for closed manifolds, but also enjoy good locality properties. In order to compute their values on a complicated closed manifold, one usually *cuts* along lower dimensional submanifolds, assigns data to them and *recombines* this simpler data in a clever way. A cutting principle is common in representation theory: representations are described by decomposing into smaller pieces, by finding simple constituents and their multiplicities in a direct sum decomposition or a Jordan–Hölder filtration and by studying functors on these pieces. Encoding such information combinatorially as character formulas, Poincaré polynomials or Kazhdan–Lusztig polynomials, etc., has a long successful history. We call this process *decategorification*.

(Re)Combining or gluing conceptually is a rather new focus motivated partially by TQFTs. In *algebraic categorification*, combinatorial data gets interpreted categorically by gluing simple constituents in a predicted way, by realizing (Laurent-)polynomials as Poincaré polynomials or Euler characteristics, groups as Grothendieck groups of categories and group homomorphisms as the image of exact functors on a Grothendieck group, etc. Moreover, functors are considered in *families* with relations between them described algebraically in terms of (quantised) *Lie algebra* or *Hecke algebra actions*. Classical representation theoretic categories are now viewed as *higher categories* equipped with *categorical actions*. As a byproduct, new invariants of links, surfaces or higher-dimensional manifolds emerge.

We will summarize and bring together known categorifications of (Laurent-)polynomial link invariants based on the *Jones polynomial* $\mathrm{J}(L)$ and its colored version $\mathrm{J}_{\mathrm{col}}(L)$ focusing on algebraic-representation theoretic constructions around *Soergel (bi)modules* [131]. The precise meaning of *categorification* will depend on the specific construction:

- Section 3.1: L turns into a complex of graded vector spaces with Euler characteristic $\mathrm{J}(L)$, link cobordisms turn into linear maps;
- Section 3.2: L turns into a complex of bigraded vector spaces whose Euler characteristic is a 2-parameter polynomial which specializes to $\mathrm{J}(L)$;
- Section 3.3: L is viewed as a tangle. Boundaries of tangles turn into graded linear categories, tangles into functors, tangle cobordisms into natural transformations, and $\mathrm{J}(L)$ is the value at a specific element of a map between Grothendieck groups;
- Section 3.4: boundaries of *colored* tangles turn into graded linear categories of possibly infinite global dimension, tangles into functors, and $\mathrm{J}_{\mathrm{col}}(L)$ is the value at a specific element of a map between *completed* Grothendieck groups.

In Section 2 we set up a framework on the decategorified level with different approaches to the Jones polynomial, all coming from quantum groups and Hecke algebras. It unifies and also stresses the differences of the later categorifications. The material is known, but combined from several sources and carefully adapted. An unusual parameter η is introduced in order to fit all the normalizations and categorified theories into one common setup.

In Section 3 the pioneering Khovanov invariant Kh [75] is described first. Recent advances in categorified link invariants indicate that this theory has interesting topological applications and the chance to provide a 4-TQFT [101]. The second categorification we deal with is the triply graded Khovanov–Rozansky link invariant KR [78, 81], presented in representation-theoretic terms. Its values are Laurent series in v over a polynomial ring in two variables. A three parameter *superpolynomial* invariant of links was predicted on the physics side in [39] and constructed for torus links via refined Chern–Simons theory [3]. Connections to double affine Hecke algebras indicated by the appearance of generalized Verlinde algebras were explored mathematically, e.g., in [30, 58]. For torus links, superpolynomials can be matched with the KR invariants by explicit calculations which substantially use categorified Young projectors. Such projectors were introduced in [33, 52, 125] in the context of categorifications of colored Reshetikhin–Turaev link invariants [33, 52] and are now important tools in the categorified representation theory of Hecke algebras. The third categorification we describe are the Lie theoretic Mazorchuk–Stroppel–Sussan tangle invariants $\mathrm{MSS}^{\pm}$ [105, 136, 139] which implicitly include the $\mathfrak{sl}_k$ Khovanov–Rozansky link invariant [81] via MSS^{-}. Up to some sign issues which appear when passing from webs to matrix factorizations, the two constructions are even connected by a functor. This follows from the Uniqueness Theorem 3.45, Theorem 3.44, and [97]. The two Lie-theoretic constructions $\mathrm{MSS}^{\pm}$ (connected by Koszul duality) go one step further: tensor products of representations of quantum $\mathfrak{gl}_k$ are lifted to categories, the action of quantum $\mathfrak{gl}_k$ to functors, and the resulting invariant of tangles has values in the homotopy category of some exact functors. Via a categorified *q-skew Howe duality*, the action of tangles can again be expressed in terms of a quantum group action. This allows putting the construction into the setup from **[31,123]**. This is an *axiomatic definition* of categorifications of representations of Lie algebras and provides a conceptual 2-categorical framework, where also uniqueness results are established. Quantum group representations and their tangle invariants are then, finally, turned into *2-representations of categorified quantum groups* introduced by Khovanov–Lauda **[80]** and Rouquier **[123]**.

The MSS^+-invariant is equivalent to the quantum $\mathfrak{sl}_k$ version from **[142]** which deals more generally with quantum invariants for *any* reductive Lie algebra. In comparison, **[142]** *is defined to fit into* and *substantially further developed* the framework of 2-representations, whereas MSS^+ *leads naturally to* and *motivated* the framework of 2-representations. In the MSS-theory one should not expect generalizations to other Lie algebra as in **[142]**, but rather to braid groups of general type and to categorified representation theory of certain quantum symmetric pairs via **[41]** and probably to Khovanov–Rozansky invariants for orthogonal Lie algebras. Important for us is that $\mathrm{MSS}^\pm$ *directly* connects to Soergel bimodules, a possible source for an intriguing connection between KR and categorified colored tangle invariants described as the fourth example. This connection and the complicated combinatorics of categorified colored link and tangle invariants **[33,53,138,142]** needs still to be explored.

In Section 4 we return to our motivation: we indicate two (partially conjectural) new approaches towards potentially rich 4-TQFTs, one via categorified representations of quantum groups, the other via semistrict monoidal 2-categories of Soergel bimodules.

Conventions. We denote $\mathbb{N} = \mathbb{Z}_{>0}$, $\mathbb{N}_0 = \mathbb{Z}_{\geq 0}$. We fix $\mathbb{C}$ as ground field. Let S_n be the symmetric group on n letters with standard generators $s_i = (i, i+1)$, $1 \leq i \leq n-1$, and length function ℓ. For a variable v and $a \in \mathbb{Z}$, let

$$[a] := \frac{v^a - v^{-a}}{v - v^{-1}} = v^{a-1} + v^{a-3} + \cdots + v^{1-a} \in \mathbb{Z}[v^{\pm 1}]$$

be the *v-quantum number*, a Laurent polynomial in v. By *graded* we mean $\mathbb{Z}$-graded, and $\langle i \rangle$ denotes the shift up in the grading, i.e., $(M\langle i \rangle)_n = (\langle i \rangle M)_n = M_{n-i}$. Similarly, we write $[i]$ for the shift of complexes by i in the direction of the differential. When displaying complexes, we indicate the homological degree zero by putting a box around the component. For an additive category $\mathcal{A}$, we denote by $K^b(\mathcal{A})$ the homotopy category of bounded complexes in $\mathcal{A}$. When describing morphisms or functors diagrammatically, we read from bottom to top, and composition is vertical stacking, whereas a monoidal product $\otimes$ is denoted by horizontal juxtaposition, and identities are usually displayed by a vertical strand.

2. FOUR APPROACHES TO THE JONES POLYNOMIAL

We summarize four similar, but different, algebraic approaches to knot or link invariants giving rise to the $\mathbb{Z}[v^{\pm 1}]$-valued Jones polynomial. These approaches will later be connected with four theories in the context of categorification. The third and fourth, RT and wRT, are more involved and cover also tangles (a common generalization of links and braids).

The first is best for computations, but the passage to tangles requires extra adjustments like the use of skein algebras. The second does not cover tangles at all, but is probably the most intuitive approach for categorifications. It works with link closures instead of planar projections of links. In the following, v denotes a (generic) variable.

I. Kauffman bracket of links. We fix an orientation of $\mathbb{R}^3$ and consider oriented knots or links L in $\mathbb{R}^3$. Following Kauffman [72], we first ignore the orientation and assign to any generic, i.e., with no triple intersections, no tangencies and no cusps, planar projection D of L, the *Kauffman bracket* $[\![D]\!] \in \mathbb{Z}[v^{\pm 1}]$. It is characterized by the *multiplicativity property* $[\![D_1 \sqcup D_2]\!] := [\![D_1]\!][\![D_2]\!]$, i.e., the bracket of a disjoint union is the product of the brackets, and the following *normalization* and *local smoothing* relation (which removes crossings):

$[\![\bigcirc]\!] = v + v^{-1} = [2]$ and 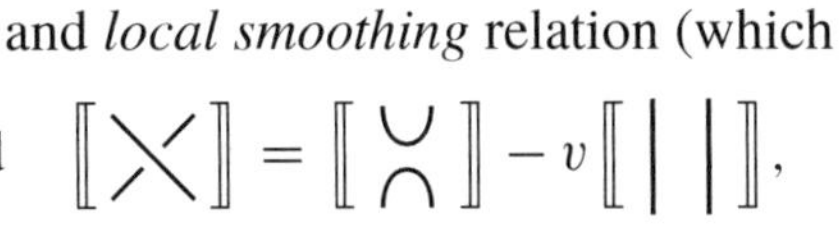 respectively. (2.1)

The assignment $D \mapsto \mathrm{J}(D) := (-1)^{n_-(D)} v^{n_+(D)-2n_-(D)} [\![D]\!] \in \mathbb{Z}[v^{\pm 1}]$, where $n_\pm(D)$ denotes the number of positive respectively negative crossings in D, defines then an invariant of oriented links, the *Jones polynomial* $\mathrm{J}(D)$. It fulfils the following skein relation, with $\mathbf{a} = v^2$ and $\mathrm{P}(D) = \mathrm{J}(D)$,

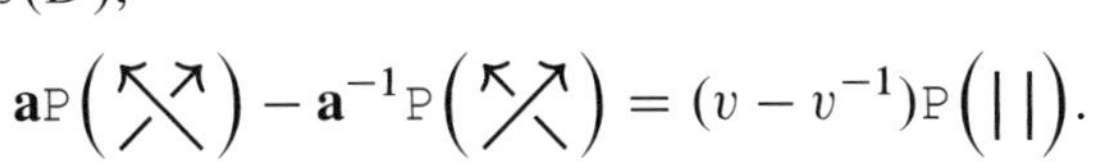 (2.2)

Example 2.1. For the Hopf link diagram $D =$  the Kauffman bracket has the value

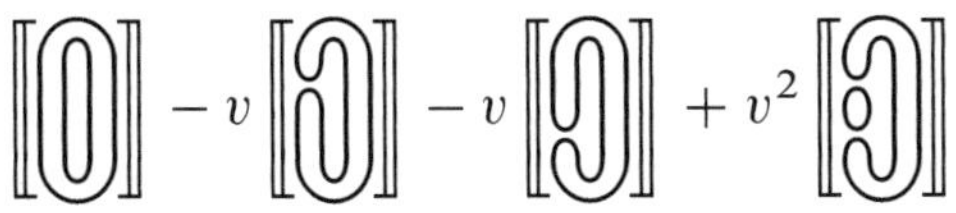 $= \underbrace{[2]^2 - v[2] - v[2] + v^2[2]^2 = v[4]}_{\Rightarrow \text{ Jones polynomial } \mathrm{J}(D) = v^3[4]}$. (2.3)

II. Closures of braids. Due to Alexander's theorem [4], every oriented link can be realized as the closure of some upwards oriented braid, i.e., of an element in the usual braid group $\mathrm{Br}_n = \langle \beta_1, \ldots, \beta_{n-1} \rangle$ for some n (see the Hopf link above with $n = 2$). A *Markov trace* Tr with values in some target Inv is a function $\mathrm{Tr} : \coprod_{n \geq 1} \mathrm{Br}_n \to \mathrm{Inv}$ satisfying the *trace condition* $\mathrm{Tr}(\alpha\alpha') = \mathrm{Tr}(\alpha'\alpha)$ and $\mathrm{Tr}(\alpha) = \mathrm{Tr}(\alpha\beta_n^{\pm 1})$ for every $\alpha, \alpha' \in \mathrm{Br}_n$, $n \geq 1$. By Markov's theorem (announced in [102], proved in [18]), Tr induces a well-defined map on isomorphism classes of closures of braids, hence defines an invariant of oriented links. This is a conceptual method to pass from braid invariants to (families) of link invariants. There is an important Markov trace, the *Ocneanu trace* (3.6). Its link invariant is the HOMFLY-PT polynomial $P(L)(v, \mathbf{a}) \in \mathbb{C}(v)[\mathbf{a}^{\pm 1}]$ introduced in [54, 113]. It satisfies (2.2) and $P(L)(v, v^2) = \mathrm{J}(v)$, see Remark 3.11.

III. Quantum invariants. The Jones polynomial of oriented links can also arise as the *(Witten–)Reshetikhin–Turaev (RT) invariant* [118] associated with quantum $\mathfrak{gl}_k$ in the special case $k = 2$. An oriented link is a special case of an oriented tangle, i.e., a disjoint embedding of finitely many arcs and circles into $\mathbb{R}^2 \times [0, 1]$ (sending endpoints of arcs to boundary points) modulo ambient isotopy fixing the boundary points. The RT invariant assigns to each generic horizontal cut of a tangle a tensor product of modules for the quantum group $U_v(\mathfrak{gl}_k)$, and to each tangle a homomorphism in a consistent way, see Overview 1.

OVERVIEW 1

Hopf link: The RT invariant (see Section 2.III) and its web version (see Section 2.IV) with categorifications (see Section 3.3)

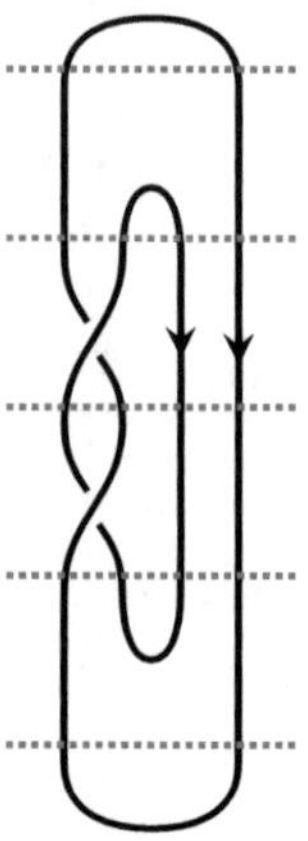

RT	wRT	MSS$^-$	MSS$^+$
	$(\beta = -v^k H_1^{-1})$	for $W = V_-$:	for $W = V_+$:
$\mathbb{C}(v)$	$\bigwedge^k W \otimes \bigwedge^k W$	$\hat{\mathcal{O}}_{\mathrm{c}}^{(k,k)}(\mathfrak{g}_{2k})$	$\hat{\mathcal{O}}^{\mathrm{c}}_{(k,k)}(\mathfrak{g}_{2k})$
$\uparrow$ ev	$\uparrow \curlywedge^k_{1,k-1}$	$\uparrow \hat{\mathrm{Z}}$	$\uparrow$
$W \otimes W^*$	$W \otimes \bigwedge^k W \otimes \bigwedge^{k-1} W$	$\hat{\mathcal{O}}_{\mathrm{c}}^{1,k,k-1}(\mathfrak{g}_{2k})$	$\hat{\mathcal{O}}^{\mathrm{c}}_{k-1,k,1}(\mathfrak{g}_{2k})$
$\uparrow$ id $\otimes$ ev $\otimes$ id	$\uparrow$ id $\otimes \curlywedge^k_{1,k-1} \otimes$ id	$\uparrow \hat{\mathrm{Z}}$	$\uparrow$
$W \otimes W \otimes W^* \otimes W^*$	$W \otimes W \otimes \bigwedge^{k-1} W \otimes \bigwedge^{k-1} W$	$\hat{\mathcal{O}}_{\mathrm{c}}^{1,1,k-1,k-1}(\mathfrak{g}_{2k})$	$\hat{\mathcal{O}}^{\mathrm{c}}_{k-1,k-1,1,1}(\mathfrak{g}_{2k})$
$\uparrow \beta$	$\uparrow \beta$	$\uparrow$ (3.16)	$\uparrow$
$W \otimes W \otimes W^* \otimes W^*$	$W \otimes W \otimes \wedge^{k-1} W \otimes \bigwedge^{k-1} W$	$\hat{\mathcal{O}}_{\mathrm{c}}^{1,1,k-1,k-1}(\mathfrak{g}_{2k})$	$\hat{\mathcal{O}}^{\mathrm{c}}_{k-1,k-1,1,1}(\mathfrak{g}_{2k})$
$\uparrow \beta$	$\uparrow \beta$	$\uparrow$ (3.16)	$\uparrow$
$W \otimes W \otimes W^* \otimes W^*$	$W \otimes W \otimes \bigwedge^{k-1} W \otimes \bigwedge^{k-1} W$	$\hat{\mathcal{O}}_{\mathrm{c}}^{1,1,k-1,k-1}(\mathfrak{g}_{2k})$	$\hat{\mathcal{O}}^{\mathrm{c}}_{k-1,k-1,1,1}(\mathfrak{g}_{2k})$
$\uparrow$ id $\otimes$ coev $\otimes$ id	$\uparrow$ id $\otimes \curlyvee^{1,k-1}_k \otimes$ id	$\uparrow \widehat{\mathrm{incl}}$	$\uparrow$
$W \otimes W^*$	$W \otimes \bigwedge^k W \otimes \bigwedge^{k-1} W$	$\hat{\mathcal{O}}_{\mathrm{c}}^{1,k,k-1}(\mathfrak{g}_{2k})$	$\hat{\mathcal{O}}^{\mathrm{c}}_{k-1,k,1}(\mathfrak{g}_{2k})$
$\uparrow$ coev	$\uparrow \curlyvee^{1,k-1}_k$	$\uparrow \widehat{\mathrm{incl}}$	$\uparrow$
$\mathbb{C}(v)$	$\bigwedge^k W \otimes \bigwedge^k W$	$\hat{\mathcal{O}}_{\mathrm{c}}^{(k,k)}(\mathfrak{g}_{2k})$	$\hat{\mathcal{O}}^{\mathrm{c}}_{(k,k)}(\mathfrak{g}_{2k})$

To make this more precise, we consider a tangle as a morphism in the monoidal category $\mathcal{T}an$ of oriented tangles with stacking as composition and juxtaposition as tensor product. The *quantum group* $U_v(\mathfrak{gl}_k)$ is a deformation of the universal enveloping Hopf algebra of $\mathfrak{gl}_k$ and is often described as the $\mathbb{C}(v)$-algebra with generators E_i, F_i, $D_j^{\pm 1}$, $1 \leq i \leq k-1$, $1 \leq j \leq k$ "quantizing" the usual matrix units $\mathrm{E}_{i,i+1}$, $\mathrm{E}_{i+1,i}$, $\pm\mathrm{E}_{i,i}$, modulo quantized Serre relations, see, e.g., **[24]** for a definition. It is still a Hopf algebra, but now with an interesting noncocommutative comultiplication (due to the appearance of some D_j's):

$$\begin{gathered} \Delta(E_i) = E_i \otimes 1 + D_i D_{i+1}^{-1} \otimes E_i, \quad \Delta(F_i) = 1 \otimes F_i + F_i \otimes D_i^{-1} D_{i+1}, \\ \Delta(D_j^{\pm 1}) = D_j^{\pm 1} \otimes D_j^{\pm 1}. \end{gathered} \tag{2.4}$$

Every finite-dimensional representation of $\mathfrak{gl}_k$ quantizes to a $U_v(\mathfrak{gl}_k)$-module. As often in quantum algebra, there are different choices for such a quantization, but for irreducible representations they only differ by a one-dimensional twist. We encode the choice by a function $\eta : \{1, \ldots, k\} \to \{\pm 1\}$ such that the spectrum of D_j is contained in $\eta(j) v^{\mathbb{Z}}$. To capture different normalizations of link invariants, we at least need to consider the additive monoidal subcategory generated by the irreducibles corresponding to constant $\eta = \pm 1$. These signs, although annoying in practice, often have a deeper meaning in categorifications.

Example 2.2. The quantization $V_\pm = V_{\pm,\mathfrak{gl}_k}$ of the natural representation of $\mathfrak{gl}_k$ for the constant functions $\eta = \pm 1$ can be realized as the k-dimensional $\mathbb{C}(v)$-vector space with basis e_r, $1 \leq r \leq k$, and the following $U_v(\mathfrak{gl}_k)$-actions

$$\begin{array}{llll} V_+: & E_i e_r = \delta_{i,r} e_{r+1}, & F_i e_{r+1} = \delta_{i,r} e_r, & D_j e_r = v^{\delta_{j,r}} e_r, \\ V_-: & E_i e_{r+1} = \delta_{i,r} e_r, & F_i e_r = -\delta_{i,r} e_{r+1}, & D_j e_r = -v^{\delta_{j,r}} e_r. \end{array} \tag{2.5}$$

A crucial observation behind the invention of quantum groups was that the permutation action of the symmetric group on tensor products of representations quantizes (i.e., lifts) to an action of the braid group. A modern formulation is that $\mathcal{R}\mathrm{ep}_k$ is (non symmetric!) *braided monoidal*. In particular, Br_n acts on ${V_\eta}^{\otimes n}$ by $U_v(\mathfrak{gl}_k)$-homomorphisms. Explicitly, β_i acts on the ith and $(i+1)$th tensor factor of ${V_\eta}^{\otimes n}$ for constant $\eta = \pm 1$ as

$$H_i: \quad e_a \otimes e_b \mapsto \begin{cases} e_b \otimes e_a & \text{if } a > b, \\ e_b \otimes e_a + (v^{-1} - v) e_a \otimes e_b & \text{if } a < b, \\ \gamma e_a \otimes e_a & \text{if } a = b, \quad \text{with } \gamma := \eta v^{-\eta}. \end{cases} \tag{2.6}$$

These actions factor through $\mathbb{C}(v) \otimes_{\mathbb{Z}[v^{\pm 1}]} \mathbb{H}_n$, where $\mathbb{H}_n$ is the *Hecke algebra*. We define $\mathbb{H}_n$ as the $\mathbb{Z}[v^{\pm 1}]$-algebra quotient of the group algebra $\mathbb{Z}[v^{\pm 1}][\mathrm{Br}_n]$ by the following *quadratic relation*, and denote the image of β_i in $\mathbb{H}_n$ or $\mathbb{C}(v) \otimes_{\mathbb{Z}[v^{\pm 1}]} \mathbb{H}_n$ by abuse of notation also H_i:

$$-\beta_i + \beta_i^{-1} = v - v^{-1}, \quad \text{or equivalently,} \quad (\beta_i + v)(\beta_i - v^{-1}) = 0. \tag{2.7}$$

Set $W = V_\eta$. Following **[118]**, the *(Witten)–Reshetikhin–Turaev functor* associated with W is now a monoidal functor $\mathrm{RT} = \mathrm{RT}_W : \mathcal{T}an \to \mathcal{R}\mathrm{ep}_k$. It sends an oriented tangle t with, say, m endpoints at the bottom and n endpoints at the top to a $U_v(\mathfrak{gl}_k)$-homomorphism

$$\mathrm{RT}(t): \quad W^{\varepsilon_1} \otimes \cdots \otimes W^{\varepsilon_m} \to W^{\varepsilon'_1} \otimes \cdots \otimes W^{\varepsilon'_n}, \tag{2.8}$$

where $W^{\varepsilon_i} = W$ or $W^{\varepsilon_i} = W^*$, respectively, depending whether the ith strand on the bottom of t is oriented up- or downwards, similarly for the top using $W^{\varepsilon'_i}$. Now RT_W is determined by the values on the *elementary* tangles . These are sent to the corresponding evaluations, coevaluations, and to the morphism $-v^{-k} H_i$ from (2.6) and its inverse $-v^k H_i^{-1}$, respectively. To compute $\mathrm{RT}(t)$, one first reads a chosen generic tangle projection from bottom to top as a vertical composition of *basic tangle diagrams*, i.e., of those which differ from an elementary one just by adding some strands to the left or right, see Overview 1. Each basic tangle diagram is sent to the value of the elementary diagram with identities tensored on the left or right. Finally, $\mathrm{RT}(t)$ is the composition of the values of the basic tangle diagrams. If t is a link, the result is an endomorphism f of $\mathbb{C}(v)$. Evaluating at 1 gives $f(1) \in \mathbb{C}(v)$ which equals $\mathrm{J}(L)$ in case $k = 2$, $W = V_-$. The Hecke relation (2.7) implies the $\mathfrak{gl}_k$-skein relation (2.2) with $\mathbf{a} = v^k$. The RT constructions work for arbitrary reductive Lie algebras, not only $\mathfrak{gl}_k$, and thus provide several families of tangle invariants.

Remark 2.3. The (unusual) choice of V_- over V_+ has the advantage that the unknot has the value $[k] \in \mathbb{N}[v^{\pm 1}]$ (with nonnegative coefficients!) instead of $(-1)^{k-1}[k]$.

Remark 2.4. The construction also works if we pick an irreducible representation for each component of t and gives the *colored* RT tangle invariant of framed tangles from **[118]** and the *colored Jones polynomial* for links if $k = 2$. Coloring only with V_η makes life easier, e.g., one can avoid framings and all constructions are defined over $\mathbb{Z}[v^{\pm 1}]$, see Example 3.53.

IV. Webs and spin networks. A fourth way to get the Jones polynomial is via webs or spin networks and their evaluations. Following Penrose [111], a web is a certain labeled graph built from trivalent vertices, where a vertex may be interpreted as an event in which either a single unit splits into two or two units collide and join into a single one. More precisely, let $\eta \in \{\pm 1\}$. The *universal* $\mathfrak{gl}$*-web category* is the monoidal $\mathbb{C}(v)$-linear category $\mathcal{W}^\eta$ which is the linear additive closure of the strict monoidal category generated (as monoidal category) by the set of objects $\mathbb{N}$, and on the level of morphisms by diagrams

 (from $a + b$ to $a \otimes b$),  (from $a \otimes b$ to $a + b$), (2.9)

modulo the following *associativity* and *coassociativity* relations and *thin square switches*

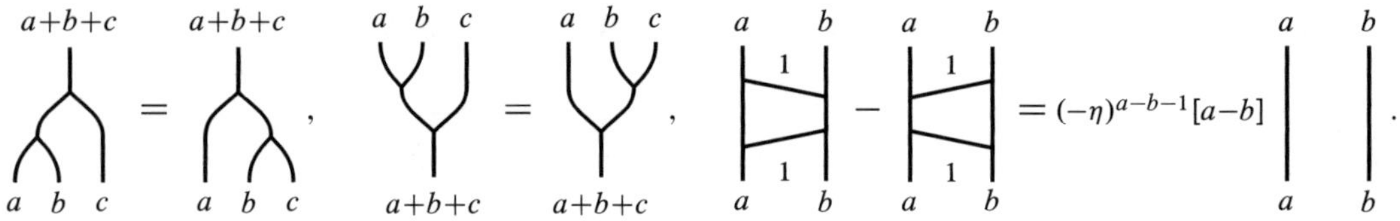

By convention, thin square switches include the *digon removals* (2.10) as degenerate $a = 0$ (or $b = 0$) cases. Together with (co)associativity, one obtains *thick square switches* expressing $\boxminus(a, r, b) - \boxminus(a, r, b)$ from (2.11) as a sum over thinner squares, see, e.g., [27].

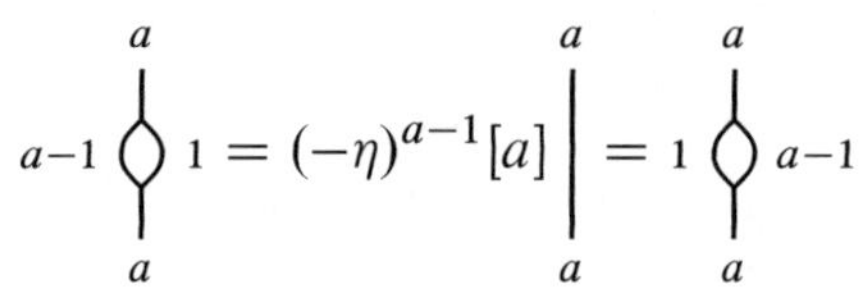 (2.10)

$\boxminus(a, r, b) :=$ 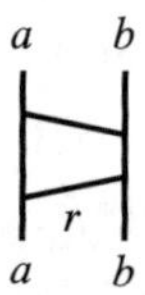 $\boxminus(a, r, b) :=$ 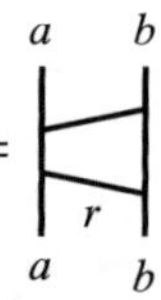 . (2.11)

An object in $\mathcal{W}^\eta$ is just a finite sequence of nonnegative integers including the empty sequence as tensor unit; a morphism is a linear combination of *webs* obtained by gluing horizontally and vertically the generating pieces (2.9) with identities drawn by vertical lines. For fixed $k \in \mathbb{N}$, let $\mathcal{W}_k^\eta$ be the quotient of $\mathcal{W}^\eta$ by all morphisms factoring through an object involving a number $> k$. We will see that this category provides a concrete graphical presentation of the monoidal category of $U_v(\mathfrak{gl}_k)$-modules generated by quantizations of the fundamental representations of $\mathfrak{gl}_k$. Thus it continues pioneering works on graphical presentations, e.g., via spiders [85], spin networks [73], or plane graphs [107].

Remark 2.5. Digon removal is used to *evaluate closed webs* in $\mathcal{W}_k^\eta$, i.e., diagrams with boundary labels equal to k only can be simplified to a $\mathbb{Z}[v^{\pm 1}]$-multiple of identity diagrams.

Remark 2.6. The category $\mathcal{W}_2^\eta$ (allowing labels 1 and 2) is a $\mathfrak{gl}_2$-analogue of the usual Temperley–Lieb category attached to $\mathfrak{sl}_2$ (where 2 equals the (empty) unit object).

We connect now $\mathcal{W}^\eta$ with *(quantized) fundamental representations* or *(quantized) exterior powers* $\bigwedge^d W$, $d \geq 1$, of the $U_v(\mathfrak{gl}_k)$-modules $W = V_\pm$. The latter is zero if $d > k$ and otherwise defined as the simultaneous $(-\gamma^{-1})$-eigenspace inside $W^{\otimes d}$ for the action of the braid group generators via (2.6). It has the expected explicit basis, namely

$$e_{\mathbf{i}} := e_{i_1} \wedge \cdots \wedge e_{i_d} := \sum_{w \in S_d} (-\gamma)^{-\ell(w)} e_{w(i_1)} \otimes \cdots \otimes e_{w(i_d)} \quad (k \geq i_1 > \cdots > i_d \geq 1), \tag{2.12}$$

indexed by d-tuples $\mathbf{i}$. For $\eta = \pm 1$, let $\mathcal{F}und_k^\eta$ be the monoidal category generated by all nontrivial exterior powers of W, i.e., objects are tensor products of $\bigwedge^d W$, $1 \leq d \leq k$ inclusively the empty product as monoidal unit. Important morphisms are *q-wedging* and *q-shuffling*:

$$\text{“}q\text{-wedging”} \quad \curlywedge_{a,b}^{a+b} : \bigwedge^a W \otimes \bigwedge^b W \leftrightarrows \bigwedge^{a+b} W : \curlyvee_{a+b}^{a,b} \quad \text{“}q\text{-shuffling.”} \tag{2.13}$$

For example, with $\nu := \delta_{\eta,-1} v^{k-1}$, $e := e_{(1,\ldots,k)}$, and $e(s)$ the same tuple but with s omitted,

$$\begin{aligned} \curlywedge_{1,k-1}^k\big(e_s \otimes e(j)\big) &= \delta_{s,j}(-\gamma)^{s-k}\nu e, \quad \curlyvee_k^{1,k-1}(e) = \sum_{s=1}^{k} (-\gamma)^{s-1}\nu^{-1} e_s \otimes e(s), \\ \curlywedge_{k-1,1}^k\big(e(j) \otimes e_s\big) &= \delta_{s,j}(-\gamma)^{1-s}\nu e, \quad \curlyvee_k^{k-1,1}(e) = \sum_{s=1}^{k} (-\gamma)^{k-s}\nu^{-1} e(s) \otimes e_s. \end{aligned} \tag{2.14}$$

We have (for any k) the *smoothing relation* $\curlyvee_2^{1,1} \circ \curlywedge_{1,1}^2 + \gamma\,\mathrm{id} = H_1$, see (2.6). This directly implies with quantized Schur–Weyl duality the first part of the following (where $\eta \in \{\pm 1\}$):

Proposition 2.7. *There is a dense full monoidal functor $\Phi_\eta : \mathcal{W}^\eta \to \mathcal{F}und_k^\eta$ which sends a generating object d to $\bigwedge^d W$ and a generating web from (2.9) to the corresponding q-wedging respectively q-shuffling. It induces a monoidal equivalence $\mathcal{W}_k^\eta \simeq \mathcal{F}und_k^\eta$.*

This result provides a purely diagrammatic description of $\mathcal{F}und_k^\eta$. It implies in particular that the asymmetric braiding morphisms can be expressed in terms of webs, e.g.,

$$\beta_{a,b} : \bigwedge^a W \otimes \bigwedge^b W \to \bigwedge^b W \otimes \bigwedge^a W, \quad \begin{cases} \sum_{r=0}^{a} \gamma^{a-r}\, \mathsf{H}(a,r,b) & \text{if } a \leq b, \\ \sum_{r=0}^{b} \gamma^{b-r}\, \mathsf{H}(a,r,b) & \text{if } a \geq b. \end{cases} \tag{2.15}$$

Proposition 2.7 is a reformulation of results from [27]. The authors work, in fact, with a larger pivotal version, where in the target category one also includes the duals and in the source additionally incorporates flow-lines on webs. From the perspective of tangle invariants, it suffices to work with $\mathcal{F}und_k^\eta$ by a clever trick. Namely, we can copy the RT construction above, but replace W^* with $\bigwedge^{k-1} W$, the trivial representation with $\bigwedge^k W$ and the cup and cap by the morphisms (2.14). This then provides a monoidal functor $\mathtt{wRT}^\eta$ from oriented tangles to $\mathcal{F}und_k^\eta$. An advantage of this construction is that it stays completely inside $\mathcal{F}und_k^\eta$ and avoids taking duals. This simplifies the situation from a categorification point of view, see Remark 3.51. The invariant of an oriented link L is an endomorphism f of a tensor product of the kth exterior powers. Evaluation at the tensor product of the top degree wedges e gives an element in $\mathbb{C}(v)$ which agrees with $\mathrm{J}(L)$ in case of $\mathcal{F}und_k^-$ for $k = 2$.

Example 2.8. To compute $f = f_{\mathrm{Hopf}}$ for the Hopf link, we first translate nested cups (and similarly caps) into webs. Each cup gives rise to copies of $\bigwedge^k W$ depending on the size (indicated by dotted lines). Reading through the webs, see Overview 1, defines the morphism

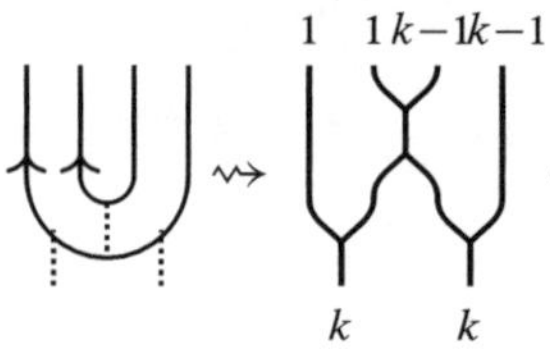

$$\rightsquigarrow f_{\mathrm{Hopf}}\colon \ \textstyle\bigwedge^2 W \otimes \bigwedge^2 W \xrightarrow{\Cup} W^{\otimes 4} \xrightarrow{H_1^{-2}} W^{\otimes 4} \xrightarrow{\Cap} \bigwedge^2 W \otimes \bigwedge^2 W$$
$$e \otimes e \longmapsto v^3[4]e \otimes e. \tag{2.16}$$

The construction of this invariant and the more general HOMFLY-PT polynomial via webs and exterior powers is due to **[107, 113]**. As common in the literature, they work with quantum $\mathfrak{sl}_k$ which produces the same invariant as WRT$^-$. We prefer to use $\mathfrak{gl}_k$, mainly to make categorifications functorial, see, e.g., **[42]**. In addition, the weight combinatorics and branching rules are much easier, but most importantly, skew Howe duality holds.

Skew Howe duality. The crucial observation behind Proposition 2.7 is that a quantum version, established in **[27, 89]**, of a classical tool from invariant theory, namely *skew Howe duality*, can be used to describe all morphisms in $\mathcal{F}und_k^{\eta}$:

Proposition 2.9 (q-skew Howe duality). *There is an isomorphism of $U_v(\mathfrak{gl}_k)$-modules*

$$\textstyle\bigwedge^{\bullet}(W \otimes \mathbb{C}(v)^m) \cong \bigoplus_{\mathbf{d}\in\mathbb{N}_0^m} \bigwedge^{\mathbf{d}} W \quad \text{with } \bigwedge^{\mathbf{d}} W := \bigwedge^{d_1} W \otimes \cdots \otimes \bigwedge^{d_m} W. \tag{2.17}$$

By turning $\mathbb{C}(v)^m$ into a $U_v(\mathfrak{gl}_m)$-module, one gets commuting mutually centralizing actions

$$U_v(\mathfrak{gl}_k) \ \curvearrowright \ X_{\pm} := \textstyle\bigwedge^{\bullet}(V_{\pm,\mathfrak{gl}_k} \otimes V_{\pm,\mathfrak{gl}_m}) \ \curvearrowleft \ U_v(\mathfrak{gl}_m)^{\mathrm{op}}. \tag{2.18}$$

Example 2.10. Let $k = 3$ and $m = 2$. Then $\bigwedge^3(V_{\pm,\mathfrak{gl}_3} \otimes V_{\pm,\mathfrak{gl}_2})$ is, as $U_v(\mathfrak{gl}_k)$-module, isomorphic to the following direct sum of modules. Each summand becomes a weight space for $U_v(\mathfrak{gl}_m)$ with the action of the generators $E = E_1$, $F = F_1$ indicated via webs:

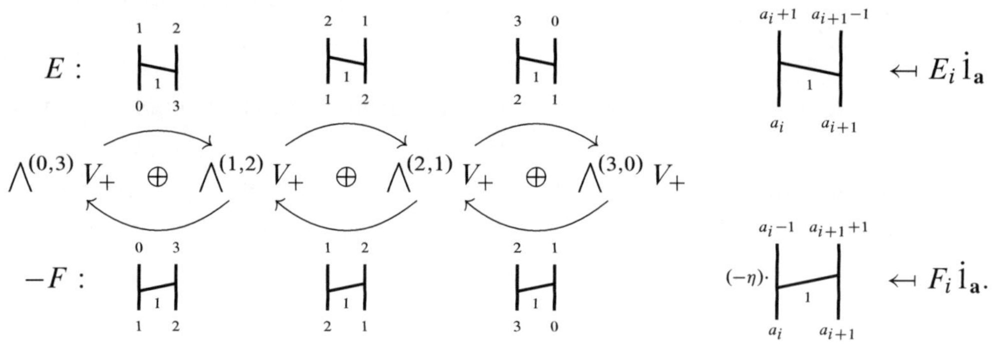

(2.19)

The labels on the webs encode the $\mathfrak{gl}_m$-weights.

The bimodules $X := X_{\pm}$ inherit some nice symmetry. Namely, the weight spaces for the $U_v(\mathfrak{gl}_m)$-action are direct summands for the $U_v(\mathfrak{gl}_k)$-action, and vice versa. As labeling sets we can use their classical weight, i.e., m-tuples (respectively k-tuples) $\mathbf{c}$ of integers. Such tuples, in fact, also index the summands of X from (2.17) and, indeed, there is an isomorphism of vectors spaces ${}^{\mathbf{d}^{\pm}}X_{\mathbf{c}} \cong {}_{\mathbf{d}}X^{\mathbf{c}^{\pm}}$. Here, the indices at the top encode the summands and at the bottom the weight space. The left and right position refers to $U_v(\mathfrak{gl}_k)$ and $U_v(\mathfrak{gl}_m)$, respectively, and $\mathbf{d}^{\pm}$ just means we reverse the tuple $\mathbf{d}$ if the module is V_{-}.

Remark 2.11. The q-skew Howe duality describes naturally the action of E_i and F_i after projection $1_{\mathbf{d}}$ onto a weight space. These projections can be encoded conceptually by passing from $U_v(\mathfrak{gl}_m)$ to Lusztig's idempotent version $\dot{U}_v(\mathfrak{gl}_m)$ of $U_v(\mathfrak{gl}_m)$ [94], where idempotent generators $\dot{1}_{\mathbf{d}}$ are added such that weight modules of $U_v(\mathfrak{gl}_m)$ correspond to modules for $\dot{U}_v(\mathfrak{gl}_m)$. The *fundamental problem* in invariant theory of determining the kernels of the actions is easy in terms of $\dot{U}_v(\mathfrak{gl}_n)$, $n \in \{k, m\}$. By [27], the kernels are the ideals I_k and I_m respectively, generated by all $\dot{1}_{\mathbf{d}}$, where $\mathbf{d}$ falls outside the respective weight support of X.

Altogether, $\mathcal{F}und_k^{\eta}$ including its action, braiding, and corresponding wRT*-tangle invariants is completely controlled by actions of (Lusztig's idempotent version) of quantum groups.*

Remark 2.12. There exist variants of q-skew Howe dualities, e.g., versions for (i) symmetric powers [121], (ii) general linear Lie *super*algebras [115, 140], (iii) orthogonal and symplectic Lie algebras [127] (replacing $\mathbb{H}_n$ by some Brauer algebra), or (iv) quantum symmetric pairs (replacing $\mathbb{H}_n$ with a Hecke algebras of Coxeter types BCD) [41]. In (iii) the dual partner is only a quantum symmetric pair for a fixed point Lie algebra of Langlands dual type inside $\mathfrak{gl}_{2m}$, see [127]; and (iv) involves two quantum symmetric pairs for the fixed point Lie algebras $\mathfrak{gl}_k \oplus \mathfrak{gl}_k \subset \mathfrak{gl}_{2k}$ and $\mathfrak{gl}_m \oplus \mathfrak{gl}_m \subset \mathfrak{gl}_{2m}$, [41]. This version sits nicely between the ones from Proposition 2.9 for $(\mathfrak{gl}_k, \mathfrak{gl}_{2m})$ and $(\mathfrak{gl}_{2k}, \mathfrak{gl}_m)$ via restriction/inclusion. It is connected via Hecke algebras of types BC with invariants of knots in an annulus or a disc with a puncture [56, 57]. A disc with an order-two orbifold point can be treated using type D following [5].

Q1: *Can Hecke algebras of complex reflection groups treat orbifold points of any order?*

3. FOUR APPROACHES TO CATEGORIFICATIONS

We now sketch representation theoretic categorifications of link and tangle invariants related to the four different views on the Jones polynomial.

3.1. Ad I: Khovanov homology

The first categorification of link invariants is given in the work of Khovanov [75] and assigns to an oriented link L a complex $\mathrm{Kh}(L)$ of finite-dimensional graded $\mathbb{C}$-vector spaces. It realizes the Jones polynomial $\mathrm{J}(L)$ as the graded Euler characteristic $\chi(\mathrm{Kh}(L))$ of $\mathrm{Kh}(L)$. Thus, $\mathrm{Kh}(L)$ relates to the Jones polynomial of L as a topological space relates to its Betti numbers. Stipulated by the Kauffman bracket, Kh assigns to the unknot a graded vector space A, viewed as complex concentrated in homological degree zero, with Poincaré polynomial $v + v^{-1} = [2]$. Each additional crossing produces a complex one step longer. To make the assignment well defined, one has to work in the homotopy category $K^b(\mathbb{C}\text{-mod}^{\mathbb{Z}(v)})$ of the category $\mathbb{C}\text{-mod}^{\mathbb{Z}(v)}$ of finite-dimensional $\mathbb{Z}$-graded vector spaces (with grading shift v).

Then the *Khovanov invariant* is an assignment

$$\mathrm{Kh} : \{\text{oriented links in } \mathbb{R}^3 \text{ up to isotopy}\} \to K^b(\mathbb{C}\text{-mod}^{\mathbb{Z}(v)}) \text{ with } \chi(\mathrm{Kh}(L)) = \mathrm{J}(L).$$

Its cohomology, the *Khovanov (co)homology*, and the *Khovanov polynomial*

$$\mathrm{P}_{\mathrm{Kh}}(L) := \sum_{d,j\in\mathbb{Z}} \dim \mathrm{H}^j(\mathrm{Kh}(L))_d t^j v^d \in \mathbb{Z}[v^{\pm 1}, t^{\pm 1}],$$

are invariants as well. The definition of Kh relies on a *categorified Kauffman bracket* $D \mapsto [\![D]\!]_{\mathrm{cat}}$ with values in $K^b(\mathbb{C}\text{-mod}^{\mathbb{Z}(v)})$ whose Euler characteristic is the Kauffman bracket from Section 2.I. This bracket is characterized by

(i) the *multiplicativity property* $[\![D_1 \sqcup D_2]\!]_{\mathrm{cat}} = [\![D_1]\!]_{\mathrm{cat}} \otimes_{\mathbb{C}} [\![D_2]\!]_{\mathrm{cat}}$,

(ii) the *normalization* $[\![\bigcirc]\!]_{\mathrm{cat}} = A$, and

(iii) the *local smoothing complex* $[\![\times]\!]_{\mathrm{cat}} = [\![\asymp]\!]_{\mathrm{cat}} \xleftarrow{\delta} [\![\,|\,|\,]\!]_{\mathrm{cat}}[1]\langle 1\rangle$.

Local smoothing means that the bracket of a diagram involving a crossing can be expressed as the total complex of a 2-term complex with entries in $K^b(\mathbb{C}\text{-mod}^{\mathbb{Z}(v)})$ given respectively by the bracket of the first and second smoothing (with the shift suggested by $-v = (-1)v^1$ in (2.1)). Since this decreases the number of crossings, by induction one may reduce to the case of no crossing (i.e., circles only), where the functor is specified by (i) and (ii). For the construction of the differential δ, Khovanov identifies $A\langle 1\rangle$ with $\mathbb{C}[x]/(x^2) = H(\mathbb{CP}^1)$ (with x in degree 2), which has additionally a Frobenius algebra structure. The (co)multiplication provide maps $A \otimes A \overset{m^*}{\leftarrow} A\langle 1\rangle$, $A \overset{m}{\leftarrow} A \otimes A\langle 1\rangle$. Applied locally (with appropriate sign rules) they define a map δ which is then a differential due to the Frobenius algebra properties and sign choices. As for the Jones polynomial one obtains from the bracket a link invariant after incorporating appropriate shifts, i.e., $\mathrm{Kh}(D) = [\![D]\!][n_-]\langle n_+ - 2n_-\rangle$.

Khovanov homology is, as expected, a stronger invariant than the Jones polynomial. Even more striking, Khovanov [77] and Jacobsson [67] could prove that a surface bounded by two links induces a chain map between the Khovanov complexes defining an invariant of the surface, up to signs. The sign issue is fixed in various ways, in [19] via foams, in [32] via surfaces with disorientation lines, and in [42] via a sign adaption of Khovanov's construction. The latter, see also [13], provides an *explicit* sign adaption of the involved differential.

Theorem 3.1. *The sign-adjusted construction of the Khovanov invariant defines a* functor

$$\mathrm{Kh}_{\mathrm{sgn}} : \{\textit{oriented links in } \mathbb{R}^3\} \to K^b(\mathbb{C}\text{-mod}^{\mathbb{Z}(v)}) \quad \textit{with homology } \mathrm{P}_{\mathrm{Kh}_{\mathrm{sgn}}} = \mathrm{P}_{\mathrm{Kh}}. \tag{3.1}$$

This *functoriality* is crucial for topological applications, e.g., to prove Milnor's conjecture on slice genus of torus knots [116] or unknot detection by Khovanov homology [84].

Remark 3.2. The (categorified) Kauffman bracket works well for links. For tangles, an additional direction of composition has to be reflected in the target category of a possible invariant. Instead of working with the category of vector spaces, one has to pass to, e.g., categories of bimodules over (generalized) Khovanov arc algebras [29, 76, 137], operads and canopolis [11], or various topological incarnations related to foam categories. An analogue, although not very practical, of the Kauffman bracket for $\mathfrak{gl}_k$, $k > 2$, can be given via (2.15).

Remark 3.3. In practice one often considers all complete smoothings at once and arranges their values as vertices in the famous cube of resolutions [10, 75], with the differential on the edges. However, the interpretation of (2.1) in terms of a 2-term complex was chosen to highlight the important role played by such complexes in algebraic(-geometric) categorifications. They do not only appear in crucial definitions (like coherent sheaves or other Serre quotient categories), but also provide technical toolkits for categorical actions, for instance, in form of spherical twists, spherical functors, or Rickard complexes.

Remark 3.4. Although *odd Khovanov homology* and *Lee homology* are often called variants of Kh, they are rather different theories from our point of view. Instead of $\mathfrak{gl}_2$, they are connected with the Lie (super) algebras $\mathfrak{osp}(1|2)$ [49] and $\mathfrak{gl}_1 \times \mathfrak{gl}_1$ [122], respectively.

Q2: *Which surfaces are distinguished by Khovanov homology? Does it provide a* 4-*TQFT?*

3.2. Ad II: Triply graded link homology

A categorification of link invariants using braid closures and traces is the *triply graded Khovanov–Rozansky homology*. It was originally constructed using matrix factorizations [81] and then reinterpreted [78] in representation theoretic terms via Soergel bimodules. To sketch the construction, we view $\beta \in \mathrm{Br}_n$ as a special tangle with n bottom and n top points or as a "map" from n inputs to n outputs. We associate variables $x_1, \dots, x_n$ to the inputs and consider the category R_n-mod of modules over the polynomial ring $R_n = \mathbb{C}[x_1, \dots, x_n]$. To β we assign a certain (complex of) R_n-bimodule(s) $X(\beta)$ which defines a "map" $X(\beta) \otimes_{R_n} \_ : R_n\text{-mod} \to R_n\text{-mod}$. Taking the closure $\hat{\beta}$ of β connects or identifies the points at the bottom with those at the top, see Example 2.1. Categorically this corresponds to identifying the left with the right action of R_n on $X(\beta)$. Algebraically one takes (derived) coinvariants, i.e., Hochschild homology of $X(\beta)$. This Hochschild homology is a bigraded vector space with gradings coming from the Hochschild and homological grading. It is even triply graded if one works with graded R_n-modules.

To be more rigorous, consider $R := R_n$ as a graded ring with $\deg(x_i) = 2$ and let S_n act on R by permuting the variables. Given any subset $I \subset S_n$ of simple transpositions, let $R^I = R^{\mathrm{W}_I} \subset R$ be the ring of invariants under the action of I or, equivalently, under the parabolic subgroup W_I generated by I inside $\mathrm{W} = S_n$.

Example 3.5. Obviously, R^W is the ring of symmetric polynomials and $R^{\emptyset} = R$. In case $I = \{s_i\}$, we obtain $R^{s_i} := R^{\{s_i\}} = \mathbb{C}[x_1, \dots, x_{i-1}, x_i + x_{i+1}, x_i x_{i+1}, x_{i+2}, \dots, x_n]$.

To any word $\ddot{w} = s_{i_1} s_{i_2} \dots s_{i_r}$ in simple transpositions from S_n, there is an associated *Soergel bimodule* $\mathrm{BS}(\ddot{w}) = R \otimes_{R^{s_{i_1}}} R \otimes_{R^{s_{i_2}}} \cdots \otimes_{R^{s_{i_r}}} R\langle -\ell(\ddot{w})\rangle$ which is the *Bott–Samelson bimodule for* $\ddot{w}$, see Remark 3.18. We have in particular, $\mathrm{BS}(s_i) = R \otimes_{R^{s_i}} R\langle -1\rangle$ and $\mathrm{BS}(\emptyset) = R$.

The *category of Soergel bimodules* $\mathcal{SB}im_n$ **[131, 132]** is defined as the Karoubian closure of the additive category generated by Bott–Samelson bimodules and its grading shifts (inside the category of all graded bimodules with degree zero maps). It is an additive category and closed under $\_ \otimes_R \_$, i.e., it is monoidal with unit R. Next, the generating *Rouquier complexes* associated with β_i and $\beta_i^{-1} \in \mathrm{Br}_n$ are

$$X(\beta_i): \quad \big(R\langle 1\rangle \to \boxed{\mathrm{BS}(s_i)}\big), \quad \text{and} \quad X(\beta_i^{-1}): \quad \big(\boxed{\mathrm{BS}(s_i)} \to R\langle -1\rangle\big)$$

with differentials given by $1 \mapsto (x_i - x_{i+1}) \otimes 1 + 1 \otimes (x_i - x_{i+1})$ and $1 \otimes 1 \mapsto 1$, respectively.

To an element $\beta \in Br_n$ written as a word $\ddot{\beta} = \beta_{i_1}^{\varepsilon_1} \cdots \beta_{i_r}^{\varepsilon_r}$ in the $\beta_i^{\pm 1}$, Rouquier **[124]** attaches the corresponding tensor product $X(\ddot{\beta}) := X(\beta_{i_1}^{\varepsilon_1}) \otimes_R \cdots \otimes_R X(\beta_{i_r}^{\varepsilon_r})$ (with the convention that the identity braid in Br_n is sent to R) in $K^b(\mathcal{SB}im_n)$. He then proves the following important result which allows one to use the notation $X(\beta)$.

Theorem 3.6. *If two words $\ddot{\beta}$ and $\ddot{\beta}'$ represent the same element in Br_n, then the Rouquier complexes $X(\ddot{\beta})$ and $X(\ddot{\beta}')$ are* canonically *isomorphic in $K^b(\mathcal{SB}im_n)$.*

Remark 3.7. Rouquier **[124]** in fact constructed a *genuine braid group action* on $K^b(\mathcal{SB}im_n)$ by these Rouquier complexes. Explicit rigidity maps (even over $\mathbb{Z}$) were determined in **[46]**.

To categorify braid closures and traces, consider the Hochschild homology functor

$$\mathrm{HH}(\_) := \bigoplus_{i \in \mathbb{N}_0} \mathrm{HH}_i(R, \_) := \bigoplus_{i \in \mathbb{N}_0} \mathrm{Tor}_i(R, \_)$$

from the category of $\mathbb{Z}$-graded R-bimodules to the category of $(\mathbb{N}_0 \times \mathbb{Z})$-graded (viewed as $\mathbb{Z} \times \mathbb{Z}$-graded) vector spaces. For a complex C of finitely generated graded R-bimodules, let $\mathrm{HH}(C)$ denote the complex of bigraded abelian groups obtained by applying the functor HH to the components and differentials of C. Set $\mathbb{HHH}(C) := \mathrm{H}^\bullet(\mathrm{HH}(C))$. This is an object in the category $\mathbb{C}\text{-mod}^{\mathbb{Z}(t,v,h)}$ of triply graded vector spaces with t, v and h referring to the homological, internal and Hochschild degree, respectively (with shift functors $[\_]$, $\langle \_\rangle$, $\{\_\}$). Its *(3-parameter) Poincaré series* is a Laurent series in v with coefficients in $\mathbb{Z}[h^{\pm 1}, t^{\pm 1}]$:

$$\mathrm{P}\big(\mathbb{HHH}(C)\big) := \sum_{d,i,j \in \mathbb{Z}} \dim\big(\mathrm{H}^j\big(\mathrm{HH}_i(C)\big)_d\big) t^j h^i v^d \in \mathbb{Z}\big[h^{\pm 1}, t^{\pm 1}\big]((v)). \qquad (3.2)$$

We have to work here with *Laurent series* in v, indicated by $((v))$, since R is infinite-dimensional, but the expression makes sense since the components are finite in each fixed triple degree. Evaluating $t = -1$ gives the *graded Euler characteristic* $\chi(\mathbb{HHH}(C)) \in \mathbb{Z}[h^{\pm}]((v))$.

Khovanov showed in **[78]** that (3.2) gives, up to some rescaling, an invariant of oriented links (here $\varepsilon(\ddot{\beta})$ denotes the sum of the exponents of the β_i appearing in $\ddot{\beta}$):

Theorem 3.8. *For a braid word $\ddot{\beta}$ in Br_n, the normalized Poincaré series*

$$\mathrm{KR}(\hat{\beta}) := (th)^{\frac{1}{2}(\varepsilon(\ddot{\beta})-n)} v^{\varepsilon(\ddot{\beta})}\, \mathrm{P}\big(\mathrm{HHH}\big(X(\ddot{\beta})\big)\big) \tag{3.3}$$

only depends on the braid closure $\hat{\beta}$. Thus there is a well-defined assignment

$$\mathrm{KR} : \{\textit{oriented links in } \mathbb{R}^3 \textit{ up to isotopy}\} \to \mathbb{Z}\big[h^{\pm\frac{1}{2}}, t^{\pm\frac{1}{2}}\big]((v)), \quad \hat{\beta} \mapsto \mathrm{KR}(\hat{\beta}). \tag{3.4}$$

The invariant $\mathrm{KR}(\hat{\beta})$ is called the triply graded Khovanov–Rozansky homology *of $\hat{\beta}$.*

Example 3.9. We calculate $\mathrm{KR}(\bigcirc)$, i.e., β is for $n = 1$ the identity braid in Br_n with $X(\beta) = R_n = R = \mathbb{C}[x_1]$. The Hochschild homology can easily be computed as $\mathrm{HH}_0(R) = R$, $\mathrm{HH}_1(R) = R\langle 2\rangle$, $\mathrm{HH}_{\geq 2}(R) = \{0\}$ from the Koszul resolution $\mathbb{C}[y] \otimes \mathbb{C}[y'] \overset{(y-y')\cdot}{\rightarrow} \mathbb{C}[y] \otimes \mathbb{C}[y']$ of $R = \mathbb{C}[x_1]$ (with $y, y' \mapsto x_1$). Since the Poincaré series of R equals $\mathrm{P}(R) = \frac{1}{1-v^2}$, we obtain

$$\mathrm{KR}(\bigcirc) = t^{-\frac{1}{2}} h^{-\frac{1}{2}}\, \mathrm{P}\big(\mathrm{HHH}(R)\big) = t^{-\frac{1}{2}} h^{-\frac{1}{2}} \frac{1+hv^2}{1-v^2} \in \mathbb{Z}[h^{\pm}, t^{\pm}]((v)). \tag{3.5}$$

For general n, $\mathrm{HH}(R_n) \cong R_n \otimes \bigwedge^{\bullet}(\xi_1, \ldots, \xi_n)$, where each ξ_i is of v-degree 2 and h-degree 1, and $\mathrm{HHH}(R_n) = \mathrm{HH}(R_n)$. As expected, the identity braid gives with (3.3), $\mathrm{KR}(\bigcirc)^n$.

Crucial for the proof of Theorem 3.8 are isomorphisms $\mathrm{HHH}(X(\alpha) \otimes_R X(\beta_n)) \cong \mathrm{HHH}(X(\alpha))\langle -1\rangle$, $\mathrm{HHH}(X(\alpha) \otimes X(\beta_n^{-1})) \cong \mathrm{HHH}(X(\alpha))\langle 1\rangle[1]\{1\}$ and the trace property $\mathrm{HHH}(X(\alpha) \otimes_R X(\alpha')) \cong \mathrm{HHH}(X(\alpha') \otimes_R X(\alpha))$ for $\alpha, \alpha' \in \mathrm{Br}_n \subset \mathrm{Br}_{n+1}$. The latter follows directly from the canonical isomorphism $\mathrm{HH}(M \otimes_R N) = \mathrm{HH}(N \otimes_R M)$ noting that for Soergel bimodules one does not need to derive the tensor product, since they are by standard invariant theory free as one-sided modules. The formulas imply that $\hat{\beta} \to \chi(\mathrm{HHH}(X(\beta)))$ factors through a $\mathbb{Z}[v^{\pm 1}]$-linear trace function $\tau : \coprod_{n\geq 1} \mathbb{H}_n \to \mathbb{Z}[h^{\pm 1}]((v))$ such that

$$\tau(1) = \frac{1+hv^2}{1-v^2}, \quad \tau(xH_n^{\pm 1}) = z_{\pm}\tau(x)\ \forall x \in \mathbb{H}_n \text{ with } z_+ = v^{-1} \text{ and } z_- = -hv, \tag{3.6}$$

where H_i is the image of β_i in $\mathbb{H}_n$. With the normalization (3.3), τ becomes a Markov trace. By introducing a homological $\frac{1}{2}\mathbb{Z}$-grading, even HHH can be turned into an invariant, see **[124]**.

Remark 3.10. Since trace functions τ on $\coprod_{n\geq 1} \mathbb{H}_n$ are classified by the pair $(\tau(1), z_+)$ as in (3.6), one can identify τ from (3.6), up to normalization of $\tau(1)$ with the (Jones–)-Ocneanu trace **[69]**. In **[86]** it is proved that for any finitely generated Coxeter group, with the more general definition of Soergel bimodules and Hecke algebra from Theorem 3.19, the Euler characteristic of the KR-homology provides a Markov trace on the Hecke algebra.

Remark 3.11. To get nicer formulas, we make a change of variables by setting $\mathbf{a} = v(ht)^{\frac{1}{2}}$. Then $\mathrm{KR}(\hat{\beta}) \in \mathbb{Z}[\mathbf{a}^{\pm 1}, t^{\pm 1}]((v))$. For example, $\mathrm{KR}(\bigcirc) = v\frac{\mathbf{a}^{-1}+\mathbf{a}t^{-1}}{1-v^2} = \frac{\mathbf{a}^{-1}+\mathbf{a}t^{-1}}{v^{-1}-v}$. Setting $t = -1$ gives $\frac{\mathbf{a}-\mathbf{a}^{-1}}{v-v^{-1}}$ and the characterizing skein relation (2.2) of the HOMFLY-PT polynomial holds. With $\mathbf{a} = v^2$, we obtain the Jones polynomial, e.g., $v + v^{-1}$ here and $v^3[4]$ in Example 3.13.

Theorem 3.12. $\mathrm{KR}(\hat{\beta}) \in \mathbb{Z}[\mathbf{a}^{\pm 1}, t^{\pm 1}]((v))$ *specializes for $t = -1$ to the HOMFLY-PT polynomial associated with $\hat{\beta}$.*

Example 3.13. For L the Hopf link from (2.1), we get $\chi(\mathrm{KR}(L)) = v^2\tau(H_1^2)$. By (2.7), $\tau(H_1^2) = \tau(1) + (v^{-1} - v)\tau(H_1) = \sigma^2 + (v^{-1} - v)v^{-1}\sigma$ with $\sigma := \tau(1)$.

Remark 3.14. The appearance of $\mathbb{H}_n$ here is not surprising because Soergel originally invented his bimodules to understand the Kazhdan–Lusztig basis in the Hecke algebra $\mathbb{H}_n$. To formulate this more precisely, let $\mathrm{K}_0^{\oplus}(\mathcal{SB}im_n)$ be the split Grothendieck ring of the additive category of Soergel bimodules. That is the free $\mathbb{Z}[v^{\pm 1}]$-module generated by isomorphism classes $[M]$ of objects M in $\mathcal{SB}im_n$ modulo relations $[M \oplus N] = [M] + [N]$, $v[M] = [M\langle 1\rangle]$, and multiplication $[M][N] = [M \otimes_R N]$. In **[131]**, Soergel proved an influential categorification theorem which is crucial for all representation theoretic constructions of categorified link invariants: there is an isomorphism of $\mathbb{Z}[v^{\pm 1}]$-algebras

$$\Upsilon_n : \mathbb{H}_n \to \mathrm{K}_0^{\oplus}(\mathcal{SB}im_n), \quad H_i + v \longmapsto \big[B(i)\big]. \tag{3.7}$$

which moreover identifies the Kazhdan–Lusztig basis with classes of indecomposables bimodules. We observe that H_i corresponds hereby to a virtual object only. This can be fixed by identifying $\mathrm{K}_0^{\oplus}(\mathcal{SB}im_n)$ with the Grothendieck group of the triangulated category $K^b(\mathcal{SB}im_n)$, since then $[X(\beta_i)] = \Upsilon(H_i)$. This shows that Rouquier's braid group action, despite its faithfulness **[82]**, is honestly a categorical Hecke algebra action which also descends to a Hecke algebra action on the Grothendieck group. The relations (2.1), (2.2), (2.7) indicate that the presented invariants should be rather connected with the Hecke algebra instead of the braid group.

In contrast to Kh, computing KR is usually hard, although the resulting values might be more conceptual and expressible using generating series. Important progress was however made recently for the *torus links* $t_{(p,q)}$ which are the closure of $(\beta_{p-1} \cdots \beta_2\beta_1)^q$ (with the Hopf link as special case $(p, q) = (2, 2)$). An important first step is done in **[65]** with the observation that $\mathrm{KR}(t_{(n,q)})$ stabilizes for $q \to \infty$ to a limit isomorphic to $\mathbb{C}[u_1, \ldots, u_n] \otimes \bigwedge^{\bullet}[\xi_1, \ldots \xi_n]$ with u_i in h-degree zero and ξ_i in h-degree 1 (cf. $\mathrm{HH}(R_n)$ in Example 3.9). This limit is identified in **[65]** with the derived endomorphism ring of a certain categorified Young idempotent in the Hecke algebra $\mathbb{H}_n$. This idempotent provides a bridge to categorified colored RT-invariants, Remark 3.54 and Conjecture 3.58, since it acts on $(V_{-,\mathfrak{gl}_2})^{\otimes n}$ as a projector, the $V_{-,\mathfrak{gl}_2}$-version of the Jones-Wenzl projector (3.21) below.

In **[44]**, $\mathrm{KR}(t_{(p,q)})$ is determined via a beautiful recursive formula in case $p = q$, and extended to general (p, q) in **[66]**. They both use categorifications of idempotents in $\mathbb{H}_n$ which are interesting tools on their own, e.g., for developing a categorified representation theory of $\mathbb{H}_n$. For general links, computing KR seems still to be out of reach. Instead of studying the invariant via its original definitions **[78, 81]**, alternative constructions were proposed, e.g., the following involving Hilbert schemes and Cherednik algebras with their underlying combinatorics of symmetric functions and Macdonald polynomials.

Remark 3.15. The approach of **[108, 109]** starts by viewing a torus link $L = t_{(p,q)}$ as an algebraic link, i.e., as the intersection of a planar curve $C := C_{p,q} \subset \mathbb{C}^2$ (defined by the polynomial $f = x^p - y^q \in \mathbb{C}[x, y]$) with a sufficiently large sphere around the origin in $\mathbb{C}^2$.

Attached to C is the *Hilbert scheme* $C^{[r]}$ of r points on C which, as a set, is given by all ideals $I \subset \mathbb{C}[x, y]$ of codimension r containing f. In [109] it is proved for coprime p, q that the Euler characteristic of $\mathrm{KR}(t_{(p,q)})$, i.e., the HOMFLY-PT polynomial, equals up to a normalization the Euler characteristic of the disjoint union of all *nested Hilbert schemes*

$$C^{[d,d+i]} := \{(I, J) \mid I \cdot (x, y) \subset J \subset I\} \subset C^{[d]} \times C^{[d+i]}$$

with d and i encoding the v- respectively $\mathbf{a}$-degree. For a generalization to algebraic links, see [103]. In [108] it is conjectured that replacing the Euler characteristic with the virtual Poincaré polynomial (see [108] for the definition) provides the triply graded Khovanov homology. For torus links, this is proved in [110]. In general, this conjecture is still open.

Remark 3.16. As indicated in the introduction, KR is related to double affine Hecke algebras (DAHAs) and their rational degenerations from [50]. The rational DAHA $\mathrm{H}_c = \mathrm{H}_c(S_n)$ with parameter $c \in \mathbb{C}$ is the quotient of $\mathbb{C}\langle x_i, y_i \mid 1 \leq i \leq n\rangle \rtimes S_n$ modulo

$$[x_i, x_j] = 0 = [y_i, y_j], \quad [x_i, y_j] = c \cdot (i, j), \quad [x_i, y_i] = 1 - c \sum_{j \neq i} (i, j)$$

for any $i \neq j$ with $(i, j) \in S_n$. It is a flat deformation of $\mathrm{H}_0 = \mathcal{D} \rtimes S_n$, where $\mathcal{D}$ is the algebra of differential operators on $\mathbb{C}^n$. If $c = \frac{p}{q}$ with $(p, q) \in \mathbb{Z}^2$ coprime, there is a unique irreducible finite-dimensional $\mathrm{H}_{\frac{p}{q}}$-module $\mathrm{L}_{p,q}$ [15]. When restricting the H_c-action to S_n, this module decomposes into direct summands. Let $1_i \mathrm{L}_{p,q}$ be the isotypic component of $\bigwedge^i \mathbb{C}^{n-1}$ using the reflection representation $\mathbb{C}^{n-1}$. The internal grading on H_c realised by the eigenvalues of the Euler operator $\mathrm{eu} = \sum_{i=1}^n x_i y_i$ (and encoding the difference of the polynomial degree in the x's and the y's) induces a grading on $\mathrm{L}_{p,q}$ and $1_i \mathrm{L}_{p,q}$. In [60], the Poincaré polynomial $P(M)$ of $M := \bigoplus_i 1_i \mathrm{L}_{p,q}$ is identified with the HOMFLY-PT polynomial of $t_{(p,q)}$ up to renormalization. Here, i contributes to the $\mathbf{a}$-degree and eu to the v-degree. The identification is achieved by matching known formulas for the HOMFLY-PT polynomial with the character formula for $\mathrm{L}_{p,q}$ from [15]. In [60], a filtration on M is predicted such that $\mathrm{KR}(t_{(p,q)})$ arises as $P(\mathrm{gr}M)$ for the associated graded $\mathrm{gr}M$. This is verified in [110] in terms of a geometric perverse filtration, after realizing $1_i \mathrm{L}_{p,q}$ (with the action of the spherical Hecke algebra $1_i \mathrm{H}_{\frac{p}{q}} 1_i$) as $\bigoplus_d H(C^{[d,d+i]})$ (with the action of certain Hecke–Nakajima operators). The comparison and identification of $P(\mathrm{gr}M)$ with $\mathrm{KR}(t_{(p,q)})$ is again done by matching explicit formulas from [44, 66].

Q3: *Is there a combinatorial model to compute* KR*? For which cobordisms is* KR *functorial?*

3.2.1. Interlude: Hecke categories

The quantum $\mathfrak{gl}_k$-invariants and the construction of the fundamental representations (2.12) use heavily the monoidal structure of $\mathcal{R}\mathrm{ep}_k$. By (2.7), the action of the braid group on $(V_\pm)^{\otimes n}$ factors through an $\mathbb{H}_n$-action preserving the weight spaces of $(V_\pm)^{\otimes n}$. To get categorified tangle invariants, one might therefore categorify these Hecke algebra actions in terms of a monoidal category acting via functors on a category, ideally with an extension to categorified quantum group actions and q-skew Howe duality (2.18). To motivate the

OVERVIEW 2

The geometric, algebraic, and Lie-theoretic Hecke categories

Hecke algebra	⇝	Hecke category	$\overset{\text{Theorem 3.19}}{\simeq}$	Soergel bimodules	$\overset{\text{Remark 3.24}}{\simeq}$	Projective functors
$\mathbb{H}_n, \mathrm{H}_n(q)$		$\mathcal{H}_n^{\mathrm{geo}}$		$\mathcal{SB}im_n$		$\mathcal{P}_n$

origin of such actions we go back to the original definition of Hecke algebras arising from split reductive groups G defined over a finite field $\mathbb{F}_q$ with a choice $\mathrm{T} \subset \mathrm{B} \subset \mathrm{G}$ of a maximal torus and Borel subgroup and the finite group $\mathrm{G}(\mathbb{F}_q)$ of $\mathbb{F}_q$-points. Most finite simple groups, in particular of Lie type, arise in this way. For us the case of $\mathrm{G} = \mathrm{GL}_n$ suffices with the choice of diagonal matrices inside the upper triangular matrices and their corresponding finite groups $G_q := \mathrm{GL}_n(\mathbb{F}_q) \supset B_q \supset T_q$. The *Weyl group* $\mathrm{W} = N_{G_q}(T_q)/T_q$ can be identified with the group $S_n \subset G_q$ of permutation matrices.

The associated *Iwahori–Hecke algebra* $\mathrm{H}_n(q)$ is the vector space $\mathrm{Func}_{B_q \times B_q}(G_q, \mathbb{C})$ of complex valued functions f on G_q invariant under both the left and the right action of B_q, i.e., $f(bg) = f(g) = f(gb)$ for all $g \in G_q, b \in B_q$, equipped with the convolution product

$$(f \star g)(x) = \frac{1}{|B_q|} \sum_{y \in G_q} f(xy^{-1})g(y). \tag{3.8}$$

The indicator functions h_w, $w \in W$, for the double cosets $B_q w B_q$ form a basis of $\mathrm{H}_n(q)$ by the *Bruhat decomposition* (or just Gauss elimination) $G_q = \bigsqcup_{w \in W} B_q w B_q$. In this basis, the structure constants of the multiplication are polynomial in $q = |\mathbb{F}_q|$ and thus one can replace q by a generic variable and "treat all q at once." Then the resulting algebra $\mathrm{H}_n(q)$ becomes isomorphic to $\mathbb{H}_n$ via $q \mapsto v^{-2}$, $h_{s_i} \mapsto v^{-1}H_i$ after adjoining a square root of q.

Remark 3.17. The construction allows vast generalizations, e.g., by replacing $\mathbb{F}_q$ by a local field with finite residue field (to get Iwahori–Hecke algebras arising in number theory), by working with topological groups, or with convolution products in homology theories.

The usual Grothendieck function–sheaf correspondence, see, e.g., **[88]**, indicates that a categorification is given by a certain category of $(\mathrm{B} \times \mathrm{B})$-equivariant sheaves on G. Since $(\mathrm{B} \times \mathrm{B})$-equivariant functions on G can be identified with B-equivariant functions on G/B, a categorification might therefore work with B-equivariant sheaves on G/B.

For a first categorification, see Overview 2, we use the related geometry over $\mathbb{C}$ with $G := \mathrm{GL}_n(\mathbb{C})$, $B := \mathrm{B}(\mathbb{C})$, $T := \mathrm{T}(\mathbb{C})$, and the algebraic variety $\mathcal{F} = G/B$ of all full flags $\{F_1 \subset \cdots \subset F_n = \mathbb{C}^n \mid \dim(F_i) = i\}$ of vector subspaces in $\mathbb{C}^n$. The bounded equivariant derived category $\mathcal{D}_B^b(\mathcal{F}, \mathbb{C})$ of sheaves of $\mathbb{C}$-vector spaces **[17]**, is a monoidal category with a convolution product $\star$, **[133]**.

The *geometric Hecke category* $\mathcal{H}_n^{\mathrm{geo}}$ is defined as the full subcategory of $\mathcal{D}_B^b(\mathcal{F}, \mathbb{C})$ generated by the constant sheaves $\underline{\mathbb{C}}_{P_i}$ on $P_i = \overline{Bs_iB} = Bs_iB \cup B \subset G$ under convolution $\star$, homological shifts [1], finite direct sums and direct summands. Concretely, the objects in $\mathcal{H}_n^{\mathrm{geo}}$ are shifts of objects $\mathrm{BS}^{\mathrm{geo}}(\ddot{w}) = \underline{\mathbb{C}}_{P_{i_1}} \star \cdots \star \underline{\mathbb{C}}_{P_{i_r}}[-r]$ for any word $\ddot{w} = s_{i_1} \ldots s_{i_r}$ in simple transpositions, and finite direct sums and summands of those. The shift functors $[i]$ turn $\mathcal{H}_n^{\mathrm{geo}}$ into a graded category. Note the similarity to $\mathcal{SBim}_n$ with shift functors $\langle i\rangle$.

Remark 3.18. The objects $\mathrm{BS}^{\mathrm{geo}}(\ddot{w})$ have a nice alternative description in terms of the *Bott–Samelson varieties* $Z(\ddot{w}) = P_{i_1} \times \cdots \times P_{i_{r+1}}/B^r$, where $y = (y_1, \ldots, y_r) \in B^r$ acts as $y.(p_1, \ldots, p_r) = (p_1 y_1^{-1}, y_1 p_2 y_2^{-1}, \ldots, y_r p_{r+1})$. If $\ddot{w}$ is a reduced expression for $w \in W$, then the multiplication map $\pi : Z(\ddot{w}) \to G/B$, $(p_1, \ldots, p_r) \mapsto p_1 \cdots p_r$ is known to be a resolution of singularities for the Schubert variety $\overline{BwB}/B$, first studied in the context of compact Lie groups by Bott and Samelson. It is not hard to see that $\mathrm{BS}^{\mathrm{geo}}(\ddot{w}) \cong \pi_* \underline{\mathbb{C}}_{Z(\ddot{w})}$ and that the Bott–Samelson bimodules arise as T-equivariant cohomology $H_T(Z(\ddot{w})) \cong \mathrm{BS}(\ddot{w})$.

Soergel's categorification result, Remark 3.14, arises now naturally:

Theorem 3.19. *There is an equivalence* $\mathcal{H}_n^{\mathrm{geo}} \simeq \mathcal{SBim}_n$ *of graded monoidal categories sending* $\mathrm{BS}^{\mathrm{geo}}(\ddot{w})$ *to* $\mathrm{BS}(\ddot{w})$. *In particular,* $\mathrm{K}_0^{\oplus}(\mathcal{H}_n^{\mathrm{geo}}) \cong \mathbb{H}_n$ *as* $\mathbb{Z}[v^{\pm 1}]$*-algebras.*

Remark 3.20. Theorem 3.19 can be proved by identifying both, $\mathcal{H}_n^{\mathrm{geo}}$ [120] and $\mathcal{SBim}_n$ [48], with the Karoubian closure $\mathcal{DBim}_n$ of a *diagrammatic monoidal category* $\mathcal{DBim}_n'$ invented in [45, 48] and proved to be equivalent to the full subcategory of $\mathcal{SBim}_n$ given by Bott–Samelson bimodules. Strikingly, this category $\mathcal{DBim}_n'$ has a *presentation* with generators and relations. Prominently applied is this in the proof of the long outstanding positivity conjecture for Kazhdan–Lusztig polynomials of an arbitrary Coxeter system and an algebraic proof of the Kazhdan–Lusztig conjectures for reductive complex Lie algebras in [47].

3.3. Ad IV: Categorification of the web calculus and its tangle invariant

We reverse the order from Section 2 and pass to categorifications for $\mathtt{wRT}^{\pm}$ which are further developed than for $\mathtt{RT}$. A categorification of the quantum $\mathfrak{gl}_n$ tangle invariant $\mathtt{wRT}^{\pm}$ is constructed by Mazorchuk and the author [105, 136] and Sussan [139], using highest weight categories of infinite-dimensional representations of the (again, but now in a different role!) general linear Lie algebras $\mathfrak{gl}_N(\mathbb{C})$. It categorifies $\mathcal{Fund}^{\pm}$ and even skew Howe duality: objects $\bigwedge^{\mathbf{d}} V_{\pm}$ as in (2.17) are realized as Grothendieck groups of categories, and actions and morphisms are lifted to functors with relations realized by specific natural transformations. This construction is part of a major change of perspective in representation theory in recent years. The starting point goes back to Crane and Frenkel [34], who proposed the idea of *Hopf categories* to construct 4-TQFTs based on categorified quantum groups and canonical bases. Categorified quantum groups were then defined in [80, 123] as certain 2-categories. We will indicate later how they arise naturally in the context of categorified tangle invariants.

Let $\mathfrak{h} \subset \mathfrak{b} \subset \mathfrak{g} = \mathfrak{g}_N := \mathfrak{gl}_N(\mathbb{C})$ be the Cartan and Borel subalgebra given by all diagonal respectively upper triangular matrices. Equip $\mathfrak{h}^*$ with the standard basis $\delta_1, \ldots, \delta_N$, such that δ_i picks out the ith diagonal matrix entry, and with the symmetric bilinear form $(\delta_i, \delta_j) = \delta_{i,j}$. We identify the lattice $\mathfrak{h}_{\mathrm{int}} := \mathbb{Z}\delta_1 \oplus \cdots \oplus \mathbb{Z}\delta_n$ of integral weights with $\mathbb{Z}^N$ via $\lambda \leftrightarrow (\lambda_1, \ldots, \lambda_N)$, where $\lambda_i = (\lambda + \rho, \delta_i)$ with $\rho = \sum_{j=1}^{N}(N - j + 1)\delta_j$. The group S_N acts on $\mathbb{Z}^N = \mathfrak{h}_{\mathrm{int}}$ by permuting components and defines the *Bruhat ordering* generated by $\mu < \lambda$ if λ differs from μ by swapping a pair μ_i, μ_j such that $\mu_i < \mu_j$ and $i < j$.

We set up now a dictionary between standard basis vectors $\vec{e} \in \bigwedge^{\mathbf{d}} V_{\pm}$ of $\bigwedge^{\mathbf{d}} V_{\pm}$ (for fixed $\pm$) and a subset $\Lambda^{\mathbf{d}} \subset \mathbb{Z}^N$ of $\mathfrak{g}_N$-weights (with $N = \sum_{i=1}^m d_i$). Each tensor product $\vec{e}$ of basis vectors (2.12) is identified with an element $\mathrm{wt}(\vec{e}) \in \{1, 2, \ldots, k\}^N \subset \mathbb{Z}^N$ via

$$\vec{e} = e_{\mathbf{i}}^{(1)} \otimes \cdots \otimes e_{\mathbf{i}}^{(m)} \mapsto \mathrm{wt}(\vec{e}) := (i_1^{(1)}, \ldots, i_{d_1}^{(1)}, i_1^{(2)}, \ldots, i_{d_m}^{(m)}) \in \Lambda^{\mathbf{d}}. \tag{3.9}$$

Let $\Lambda^{\mathbf{d}}$ be the image. Note that a weight space in $\bigwedge^{\mathbf{d}} V_{\pm}$ corresponds to an S_N-orbit $\mathbf{c}$ in $\Lambda^{\mathbf{d}}$.

Now we construct a category $\mathcal{O}^{\mathbf{d}}$ whose Grothendieck group has a basis naturally labeled by $\Lambda^{\mathbf{d}}$. For this consider the BGG category $\mathcal{O}$ of all finitely generated $\mathfrak{g}$-modules M which are locally finite over $\mathfrak{b}$ and have a weight space decomposition with only integral weights $\lambda \in \mathfrak{h}_{\mathrm{int}}$. This is an abelian finite length category, where the irreducible objects are exactly the irreducible highest weight modules $L(\lambda)$ of highest weight $\lambda \in \mathfrak{h}_{\mathrm{int}}$, i.e., the irreducible quotients of the *Verma modules* $\Delta(\lambda)$ for $\lambda \in \mathfrak{h}_{\mathrm{int}}$. Objects in $\mathcal{O}$ which have a Δ-flag, i.e., a finite filtration with subquotients isomorphic to Verma modules, form an exact additive subcategory $\mathcal{O}^{\Delta}$ which is closed under direct summands and contains all projective objects. Even more, category $\mathcal{O}$ is a *highest weight category*, see, e.g., **[26]**, for the set $\mathfrak{h}_{\mathrm{int}} = \mathbb{Z}^N$ viewed as poset with *standard objects* the $\Delta(\lambda)$. Technically this means that the projective cover of $L(\lambda)$ surjects onto $\Delta(\lambda)$, and $\Delta(\lambda)$ surjects onto $L(\lambda)$, and the kernel has a Δ-flag with subquotients some $\Delta(\mu)$ where $\mu > \lambda$, respectively a Jordan–Hölder filtration with subquotients $L(\mu)$'s with $\mu < \lambda$. As a consequence, the canonical maps induce isomorphisms between Grothendieck groups for (i) the additive category of projectives, (ii) the exact category $\mathcal{O}^{\Delta}$, (iii) the abelian category $\mathcal{O}$, and (iv) the triangulated bounded derived category $D^b(\mathcal{O})$:

$$\mathrm{K}_0^{\oplus}\big(\mathrm{Proj}(\mathcal{O})\big) = \mathrm{K}_0(\mathcal{O}^{\Delta}) = \mathrm{K}_0(\mathcal{O}) = \mathrm{K}_0\big(D^b(\mathcal{O})\big). \tag{3.10}$$

To $\Lambda^{\mathbf{d}}$ we associate simply the Serre subcategory $\mathcal{O}^{\mathbf{d}}$ of $\mathcal{O}$ generated by all $L(\lambda)$ with $\lambda \in \Lambda^{\mathbf{d}}$. More concretely, this is just a direct summand, specified by k, of the full subcategory $\mathcal{O}^{\mathfrak{p}_{\mathbf{d}}}$ of $\mathcal{O}$ of all modules which are locally finite over the standard parabolic subalgebra $\mathfrak{p}_{\mathbf{d}}$ with Levi factor $\mathfrak{gl}_{d_1} \oplus \cdots \oplus \mathfrak{gl}_{d_m}$. Sending $\vec{e} \in \bigwedge^{\mathbf{d}} V_{\pm}$ from (3.9) to the class of the $\mathfrak{p}_{\mathbf{d}}$-parabolic Verma module $\Delta^{\mathfrak{p}_{\mathbf{d}}}(\mathrm{wt}(\vec{e}))$ (a standard object for the induced highest weight structure on $\mathcal{O}^{\mathbf{d}}$) with highest weight $\mathrm{wt}(\vec{e})$ defines an isomorphism of abelian groups

$$\Big(\bigwedge^{\mathbf{d}} V_{\pm}^{\mathbb{Z}}\Big) \otimes_{\mathbb{Z}[v^{\pm 1}]} \mathbb{Z} \cong \mathrm{K}_0(\mathcal{O}^{\mathbf{d}}) = \mathrm{K}_0\big(D^b(\mathcal{O}^{\mathbf{d}})\big), \quad \vec{e} \mapsto \big[\Delta^{\mathfrak{p}_{\mathbf{d}}}\big(\mathrm{wt}(\vec{e})\big)\big]. \tag{3.11}$$

Here $\mathbb{Z}$ is a $\mathbb{Z}[v^{\pm 1}]$-module via $v \mapsto 1 \in \mathbb{Z}$ and $V_{\pm}^{\mathbb{Z}}$ denotes the $\mathbb{Z}[v^{\pm 1}]$-module in $V_{\pm}$ spanned by the vectors e_i. We like to find functors realizing the $U_v(\mathfrak{gl}_k)$-action and also incorporate v.

Tensoring with finite-dimensional representations of $\mathfrak{g}$ provides exact endofunctors of $\mathcal{O}$. These functors and their direct summands form the monoidal category $\mathcal{P}_N$ of *projective functors*. Describing their effect on Verma modules is easy (but hard on other objects):

Example 3.21. If U is a finite-dimensional representation of $\mathfrak{g}$, then $\Delta(\lambda) \otimes U \in \mathcal{O}^{\Delta}$. The subquotients in a Δ-flag are the $\Delta(\lambda + \nu)$, where ν runs through the multiset $P(U)$ of weights ν of U with multiplicity $\dim U_{\nu}$. Thus, $[\Delta(\lambda) \otimes U] = \sum_{\nu \in P(U)} [\Delta(\lambda + \nu)]$

in $K_0(\mathcal{O}^{\Delta})$. Important examples are $U = \mathbb{C}^N$ with $[\Delta(\lambda) \otimes U] = \sum_{i=1}^{N} [\Delta(\lambda + \delta_i)]$ or $U = (\mathbb{C}^N)^*$ with $[\Delta(\lambda) \otimes U] = \sum_{i=1}^{N} [\Delta(\lambda - \delta_i)]$.

Lemma 3.22. *The functors* $\mathrm{E}, \mathrm{F} : \mathcal{O} \mapsto \mathcal{O}, M \mapsto M \otimes U$ *with* $U = \mathbb{C}^N$ *and* $U = (\mathbb{C}^N)^*$, *respectively, decompose into direct summands* $\mathrm{E} = \bigoplus_{i \in \mathbb{Z}} \mathrm{E}_i$, $\mathrm{F} = \bigoplus_{i \in \mathbb{Z}} \mathrm{F}_i$ *with*

$$[\mathrm{E}_i\, \Delta(\lambda)] = \sum_{\{j \mid \lambda_j = i\}} [\Delta(\lambda + \delta_j)], \quad [\mathrm{F}_i\, \Delta(\lambda)] = \sum_{\{j \mid \lambda_j = i+1\}} [\Delta(\lambda - \delta_j)]. \tag{3.12}$$

This is an easy consequence of Example 3.21 and the fact that $\mathcal{O}$ decomposes into summands $\mathcal{O}_{\mathbf{c}}$ labeled by S_N-orbits $\mathbf{c}$ in $\mathbb{Z}^N$. Here $\mathcal{O}_{\mathbf{c}}$ denotes the Serre subcategory of $\mathcal{O}$ generated by $L(\lambda)$ with $\lambda \in \mathbf{c}$. Under (3.11), $\mathcal{O}^{\mathbf{d}} \cap \mathcal{O}_{\mathbf{c}}$ corresponds to a weight space. By definition, the functors E_i and F_i preserve $\mathcal{O}^{\mathfrak{p}_{\mathbf{d}}}$, even $\mathcal{O}^{\mathbf{d}}$ if $1 \leq i \leq k-1$. Formulas (3.12) resemble Lie algebra actions. Generalized formulas from Example 3.21 for $\mathcal{O}^{\mathbf{d}}$ imply that the induced action on $K_0(\mathcal{O}^{\mathbf{d}})$ agrees via (3.11) with the $(v \mapsto 1)$ specialized $U_v(\mathfrak{gl}_k)$-action on $\bigwedge^{\mathbf{d}} V_+^{\mathbb{Z}}$. (Note the positive sign $+$ here!)

Remark 3.23. Inside $\mathcal{P}_N$, the Hecke category appears naturally: It is known that $\mathcal{P}_N$ is the Karoubian closure of the additive monoidal category generated by E_i, F_i. The proof relies on a monoidal equivalence $\mathcal{P}_N \simeq \mathcal{HC}_N$ with a certain category $\mathcal{HC}_N$ of Harish-Chandra bimodules. Via Soergel's functor $\mathbb{V}$ from **[131]** and its extension in **[135]**, $\mathcal{P}_N$ is equivalent to a category of *singular Soergel bimodules*. By restriction to endofunctors of $\mathcal{O}_0 := \mathcal{O}_{\mathbf{c}}$ with $0 \in \mathbf{c}$, one gets $\mathrm{f}(\mathcal{SB}im_N) \simeq \mathrm{f}(\mathcal{H}_N{}^{\mathrm{geo}})$ as a full monoidal subcategory, where f means that we forget the grading. Remarkably, a classification of indecomposable projective functors for the categories $\mathcal{O}^{\mathbf{d}}$ was only recently obtained **[83]**, based on advances in the 2-representation theory of Hecke algebras, i.e., the representation theory of categorified Hecke algebras.

To incorporate v, we work with a *graded version* $\hat{\mathcal{O}}$ of $\mathcal{O}$ and its Serre subcategories as defined in **[12]**, i.e., with graded modules over the endomorphism ring A of a minimal projective generator of $\mathcal{O}$ equipped with the Koszul grading from **[12]**.

Remark 3.24. The origin of the grading is an equivalence of additive categories between $\mathrm{Proj}(\mathcal{O}_0)$ and the full subcategory of R_n-mod of *Soergel modules* $\mathbb{C} \otimes_{R^W} M$ for $M \in \mathrm{f}(\mathcal{SB}im_N)$ which has an obvious graded lift. We get a graded version of $\mathrm{Proj}(\mathcal{O}_0)$ and then also of $\mathcal{O}_0$. Note that $\mathcal{SB}im_N$ obviously acts on this category by tensoring over R from the right. With some extra work, all Lie theoretic categories and functors used here can be lifted to a graded version. A general approach to lift modules (e.g., (parabolic) Verma modules) and the above functors to the graded setting is developed in **[134]**.

Lemma 3.25. *Any choice of graded lift* $\hat{\Delta}^{\mathfrak{p}_{\mathbf{d}}}(\mathrm{wt}(\vec{e}))$ *of* $\Delta^{\mathfrak{p}_{\mathbf{d}}}(\mathrm{wt}(\vec{e}))$ *lifts* (3.11) *to an isomorphism of* $\mathbb{Z}[v^{\pm 1}]$*-modules* ($V_{\pm}^{\mathbb{Z}}$ *denotes the* $\mathbb{Z}[v^{\pm}]$*-submodule of* $V_{\pm}$ *spanned by the* e_i):

$$\Psi: \quad \bigwedge^{\mathbf{d}} V_{\pm}^{\mathbb{Z}} \cong K_0(\hat{\mathcal{O}}^{\mathbf{d}}) = K_0\big(D^b(\hat{\mathcal{O}}^{\mathbf{d}})\big), \quad \vec{e} \mapsto \big[\hat{\Delta}^{\mathfrak{p}_{\mathbf{d}}}\big(\mathrm{wt}(\vec{e})\big)\big]. \tag{3.13}$$

We realized now $\bigwedge^{\mathbf{d}} V_{\pm}^{\mathbb{Z}}$ as the Grothendieck group of a category and want to lift morphisms and the $U_v(\mathfrak{gl}_m)$-action from (2.19) to functors. We first consider $\bigwedge^{\mathbf{d}} V_{+}^{\mathbb{Z}}$. If $\mathfrak{p}_{\mathbf{d}'} \subset \mathfrak{p}_{\mathbf{d}}$ are two standard parabolic subalgebras in $\mathfrak{gl}_N$, then there is the exact inclusion functor incl and its left adjoint *Zuckerman functor* Z of taking the largest quotient in the target category, incl : $\mathcal{O}^{\mathbf{d}} \rightleftarrows \mathcal{O}^{\mathbf{d}'}$: Z. Now incl and the derived functor $\mathcal{L}Z$ induce morphisms on K_0 which we connect to (2.14). Recall Proposition 2.7 and observe that $\mathcal{W}^+$ is generated as category by *basic webs* which look like a generator from (2.9) with identities to the left and right. To each basic web t we associate a functor $\mathrm{MSS}^+(t)$, which is, up to an overall shift, the obvious graded lift $\widehat{\mathrm{incl}}$ or $\mathcal{L}\hat{Z}$ of the inclusion respectively the derived Zuckerman functor (with hopefully self-explanatory notation)

$$\curlyvee_{\mathbf{d}}^{\mathbf{d}'} := \widehat{\mathrm{incl}}[-ab]\langle -ab\rangle : \quad D^b(\hat{\mathcal{O}}^{\mathbf{d}}) \rightleftarrows D^b(\hat{\mathcal{O}}^{\mathbf{d}'}) \quad : \mathcal{L}\hat{Z} =: \curlywedge_{\mathbf{d}'}^{\mathbf{d}} . \tag{3.14}$$

To each composition t of basic web diagrams assign the composition $\mathrm{MSS}^+(t)$ of functors.

Example 3.26. Let $k = 2$ and consider the webs (2.9) for $a = b = 1$ with induced morphisms $\curlywedge_{(2,0)}^{(1,1)} : \bigwedge^{(2,0)} V_+^{\mathbb{Z}} \rightleftarrows \bigwedge^{(1,1)} V_+^{\mathbb{Z}} : \curlywedge_{(1,1)}^{(2,0)}$. To $e_2 \wedge e_1$ we associate $\Delta^{\mathfrak{p}(2,0)}((2,1))$ which is just the trivial $\mathfrak{gl}_2$-module $\mathbb{C}$. The BGG resolution $\Delta^{\mathfrak{p}(1,1)}((1,2)) \to \boxed{\Delta^{\mathfrak{p}(1,1)}((2,1))}$ of $\mathbb{C}$ implies that incl$[-1]$ induces the linear map $e_2 \wedge e_1 \mapsto -v^{-1}(e_2 \otimes e_1 - v e_1 \otimes e_2)$ with $v = 1$ on the Grothendieck group. On the other hand, $\mathcal{L}Z\,\Delta^{\mathfrak{p}(1,1)}((2,1)) = Z\,\Delta^{\mathfrak{p}(1,1)}((2,1)) = \Delta^{\mathfrak{p}(2,0)}((2,1))$ and $\mathcal{L}Z\,\Delta^{(1,1)}((1,2)) = \Delta^{\mathfrak{p}(1,1)}((2,1))[-1]$ induce $e_2 \otimes e_1 \mapsto e_2 \wedge e_1$, $e_1 \otimes e_2 \mapsto -v$ with $v = 1$. We can obtain now formulas (2.14) by picking appropriate graded lifts $\hat{\Delta}^{\mathfrak{p}(2,0)}((2,1))$, $\hat{\Delta}^{\mathfrak{p}(1,1)}((1,2))$, $\hat{\Delta}^{\mathfrak{p}(1,1)}((2,1))$ with a morphism $\hat{\Delta}^{\mathfrak{p}(1,1)}((1,2))\langle 1\rangle \to \boxed{\hat{\Delta}^{\mathfrak{p}(1,1)}((2,1))}$ lifting the BGG resolution.

The following summarizes results from **[41, 105, 139]** and categorifies q-skew Howe duality from Proposition 2.9:

Theorem 3.27. *Let t be a basic web from $\mathbf{d}'$ to $\mathbf{d}$ with corresponding homomorphism $\Phi_+(t)$ from Proposition 2.7. Then there are choices of graded lifts in (3.13) and of (3.12) such that the following diagram commutes for $1 \le i \le k-1$ (also for E_i replaced by F_i):*

$$\begin{array}{ccccccccccc}
 & & & & & \overset{\Psi}{\frown\!\!\!\longrightarrow} & & & & & \\
\bigwedge^{\mathbf{d}'} V_+^{\mathbb{Z}} & \xrightarrow{\Phi_+(t)} & \bigwedge^{\mathbf{d}} V_+^{\mathbb{Z}} & \xrightarrow{\Psi} & K_0(\hat{\mathcal{O}}^{\mathbf{d}}) & = & K_0(D^b(\hat{\mathcal{O}}^{\mathbf{d}})) & \xleftarrow{[\mathrm{MSS}^+(t)]} & K_0(D^b(\hat{\mathcal{O}}^{\mathbf{d}'})) & = & K_0(\hat{\mathcal{O}}^{\mathbf{d}'}) \\
\downarrow{\scriptstyle E_i} & & \downarrow{\scriptstyle E_i} & & \downarrow{\scriptstyle [\hat{E}_i]} & & \downarrow{\scriptstyle [\hat{E}_i]} & & \downarrow{\scriptstyle [\hat{E}_i]} & & \downarrow{\scriptstyle [\hat{E}_i]} \\
\bigwedge^{\mathbf{d}'} V_+^{\mathbb{Z}} & \xrightarrow{\Phi_+(t)} & \bigwedge^{\mathbf{d}} V_+^{\mathbb{Z}} & \xrightarrow{\Psi} & K_0(\hat{\mathcal{O}}^{\mathbf{d}}) & = & K_0(D^b(\hat{\mathcal{O}}^{\mathbf{d}})) & \xleftarrow{[\mathrm{MSS}^+(t)]} & K_0(D^b(\hat{\mathcal{O}}^{\mathbf{d}'})) & = & K_0(\hat{\mathcal{O}}^{\mathbf{d}'}). \\
 & & & & & \underset{\Psi}{\smile\!\!\!\longrightarrow} & & & & &
\end{array}$$

Moreover, the family of functors $\hat{E}_i$, $\hat{F}_i$ naturally commutes with the functors $\mathrm{MSS}^+(t)$ associated with webs. On the Grothendieck group they induce skew Howe duality (2.18)

$$U_v^{\mathbb{Z}}(\mathfrak{gl}_k) \quad \curvearrowright \quad X := \bigwedge^{\bullet}(V_+^{\mathbb{Z}}(k) \otimes V_-^{\mathbb{Z}}(m)) \quad \curvearrowleft \quad (U_v^{\mathbb{Z}}(\mathfrak{gl}_m))^{\mathrm{op}} \tag{3.15}$$

(with $\mathbb{Z}$ referring to Lusztig's integral version of the quantum group). The action of D_j is hereby categorified by an appropriate grading shift on each categorified weight space.

Example 3.28. We turned each summand from (2.19) into a category $D^{\mathbf{d}} := D^b(\hat{\mathcal{O}}^{\mathbf{d}})$, the $U_v(\mathfrak{gl}_k)$-action into functors $\hat{\mathrm{E}}_i, \hat{\mathrm{F}}_i$ and the action by E and $-F$ into functors from (3.14):

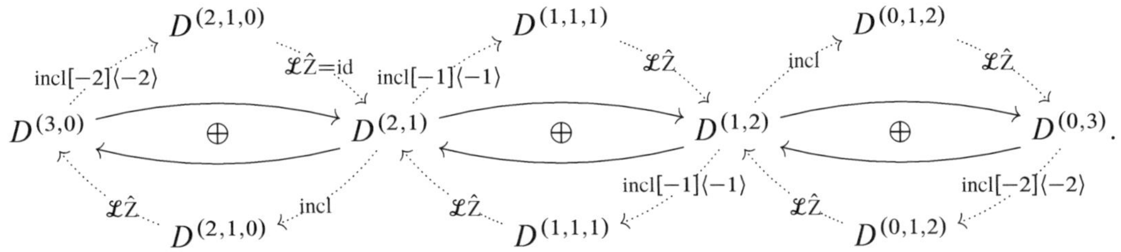

In this categorified q-skew Howe duality, the two sides seem to be asymmetric. The action of $U_v^{\mathbb{Z}}(\mathfrak{gl}_k)$ is given by exact functors on the abelian categories, whereas $U_v^{\mathbb{Z}}(\mathfrak{gl}_m)$ acts by derived functors (note V_+ versus V_-). This asymmetry is explained nicely via Koszul (self-)duality **[12, 104]**. This directly gives an analogue of the theorem for V_- instead of V_+.

Remark 3.29. *Koszul duality* means an equivalence $D^b(\hat{\mathcal{O}}_{\mathbf{c}}^{\mathfrak{p}_{\mathbf{d}}}) \simeq D^b(\hat{\mathcal{O}}_{\mathbf{d}}^{\mathfrak{p}_{\mathbf{c}}})$ which swaps the two types of functors **[104]**. Passing to the Grothendieck groups, it induces an isomorphism of groups $\bigwedge^{\bullet}(V_+^{\mathbb{Z}}(k) \otimes V_-^{\mathbb{Z}}(m)) \cong \bigwedge^{\bullet}(V_-^{\mathbb{Z}}(k) \otimes V_+^{\mathbb{Z}}(m))$. The parameters v, γ, η from Section 2.III reflect important properties of this duality: it does not commute with grading shifts ($v \mapsto -v$ encoded by η) nor preserves the standard t-structures **[104]** (encoded by γ).

Under Koszul duality, the derived functors $\mathrm{MSS}^+(t)$ from (3.14) turn into exact projective functors $\mathrm{MSS}^-(t)$ between the corresponding abelian categories. We use now these easier functors to construct tangle invariants with values in the homotopy categories $K^b(\hat{\mathcal{O}}^{\mathbf{d}}, \hat{\mathcal{O}}^{\mathbf{d}'})$ of exact functors from $\hat{\mathcal{O}}^{\mathbf{d}}$ to $\hat{\mathcal{O}}^{\mathbf{d}'}$. From **[105]** it follows that the relations in $\mathcal{W}^{\eta}$ can be interpreted in terms of isomorphisms of functors $\mathrm{MSS}^-(t)$. Thus we have (cf. (2.8)) an exact functor assigned to any basic tangle diagram except the crossings to which we assign the following complexes (possibly with identity strands added) given by canonical adjunction morphisms:

$$\mathrm{MSS}^-\big(\text{positive crossing}\big) : \big(\boxed{\mathrm{id}\langle 1\rangle} \to \mathrm{MSS}^-(\asymp)\big)\langle -k\rangle, \quad \mathrm{MSS}^-\big(\text{negative crossing}\big) : \big(\mathrm{MSS}^-(\asymp) \to \boxed{\mathrm{id}\langle -1\rangle}\big)\langle k\rangle. \quad (3.16)$$

The following is proved in **[105]** and the Koszul dual version in **[139]**:

Theorem 3.30. *Let t be an oriented tangle with a planar projection $t_1 \cdots t_r$ written in terms of basic tangle diagrams. Let $\mathrm{wRT}^-(t_1 \cdots t_r) : \bigwedge^{\mathbf{d}} V_-^{\mathbb{Z}} \to \bigwedge^{\mathbf{d}'} V_-^{\mathbb{Z}}$. Then the composition*

$$\mathrm{MSS}^-(t_1) \cdots \mathrm{MSS}^-(t_r) \in K^b(\hat{\mathcal{O}}^{\mathbf{d}}, \hat{\mathcal{O}}^{\mathbf{d}'}) \qquad (3.17)$$

is independent of the chosen projection. Thus, $t \mapsto \mathrm{MSS}^-(t)$ provides an invariant of oriented tangles. The induced morphism $\mathrm{K}_0(D^b(\mathcal{O}^{\mathbf{d}})) \to \mathrm{K}_0(D^b(\mathcal{O}^{\mathbf{d}'}))$ agrees via (3.13) with appropriate graded lifts with $\mathrm{wRT}^-(t)$. Analogously for $\mathrm{wRT}^+(t)$ using the Koszul dual functors.

Corollary 3.31. *In case t is a link, the categories $\hat{\mathcal{O}}^{\mathbf{d}}$, $\hat{\mathcal{O}}^{\mathbf{d}'}$ in (3.17) can be identified canonically with the category of graded vector spaces. Thus we obtain a bigraded link homology.*

Remark 3.32. Let $k = 2$. Then the invariant MSS^- was first defined in **[136]** based on **[16]**, where it was observed that for nonquantized $\mathfrak{gl}_2$, the action of the Temperley–Lieb algebra on $(\mathbb{C}^2)^{\otimes n}$ can be categorified using category $\mathcal{O}$. In **[137]** it is shown that the Khovanov complex for an oriented link agrees with the value of MSS^- by an explicit description of the involved categories as modules over (an extension of) *Khovanov's arc algebra*. Using [24], one can also match MSS^- with Khovanov's tangle invariant [76] via an equivalence of categories. Via [1,98] which realizes the extended arc algebra from [137] in terms of Fukaya–Seidel categories, a rigorous categorical equivalence from MSS^- to the symplectic Khovanov invariant from [138] holds. A weaker combinatorial identification, the equality of the bigraded homology groups, is established (in fact, for all known algebraic-geometric link homologies) in [96].

Remark 3.33. For $k = 2$, functoriality (as in (3.1)) of $\mathrm{MSS}^{\pm}$ is reduced to that of $\mathrm{Kh}_{\mathrm{sgn}}$. For general k, we expect functoriality to follow from the functoriality results in [43].

Remark 3.34. We focused here on defining the involved functors and describing their action on the Grothendieck group, although all defining relations in the quantum group or web category can, in fact, be turned into actual relations, i.e., isomorphisms, between functors.

Remark 3.35. Formula (2.15) is implicitly categorified via $\mathrm{MSS}^-(\beta_{a,b})$: For $a = b = 1$, this holds by (3.16). A composition of those, cf. illustration in (4.1), gives the braiding morphism for $W^{\otimes a} \otimes W^{\otimes b}$ and restriction to $\hat{\mathcal{O}}^{(a,b)}$ then $\mathrm{MSS}^-(\beta_{a,b})$. One can verify purely combinatorially based on [105] that the complex $\mathrm{MSS}^-(\beta_{a,b})$ of exact functors can be written with entries encoded by (2.15). Lie-theoretically $\mathrm{MSS}^-(\beta_{a,b})$ is easy to describe as the derived functor of a classical shuffling functor [105,136] which gets *reinterpreted* in terms of the explicit complexes. This is opposite to most categorifications, where the braiding is *defined* by explicit complexes indicated by (2.15), e.g., [114,145]. The construction of categorified braid group actions from categorified Lie algebra actions using Rickard complexes goes back to [31].

Remark 3.36. Categorifications of (parts of) q-skew Howe duality were obtained and used in many ways in recent years. The significance of the above construction is the fact that *both* quantum group actions are visible. This is in particular not the case in diagrammatic or foam based categorifications, since a (diagrammatic) replacement of the derived functors is missing. It would be nice to find a general theory towards categorifications of dualities, in particular for those in Remark 2.12 where a categorification so far only exists for (iv) [41].

Q4: *Are there other interesting Koszul self-dual categories? What do they categorify?*

3.3.1. Towards 2-representation theory: categorical actions

Two basic questions arise from the above construction: is there a conceptual source for isomorphisms specifying the desired relations between the functors (topologically speaking the values for tangle cobordisms)? To which extent are such categorifications unique? Both questions are addressed with the concept of *categorical Lie algebra actions* [31,123], which we try to motivate based on our example. The categorified quantum groups due to Khovanov–Lauda [80] and Rouquier [123] occur in this context naturally. Adjunction morphisms between functors are used to specify commutation relations (e.g., for E_i and F_i) and most tangle cobordisms. More involved are the Serre relations between the E_i which arise from endomorphisms (= natural transformations) of (powers of) $\mathrm{E} = \_ \otimes \mathbb{C}^N$ which we construct below. In the general construction **[80, 123]**, such morphisms constitute 2-morphisms of a 2-category.

An obvious choice for $\mathbb{s} \in \mathrm{End}(\mathrm{E}^2)$ is the *flip* morphism given on $\mathfrak{g}$-modules M by $\mathbb{s}_M : M \otimes \mathbb{C}^N \otimes \mathbb{C}^N \to M \otimes \mathbb{C}^N \otimes \mathbb{C}^N, m \otimes u_1 \otimes u_2 \mapsto m \otimes u_2 \otimes u_1$. The action maps $\mathbb{x}_M : M \otimes \mathbb{C}^N \to M \otimes \mathbb{C}^N$ of the *Casimir element* $\Omega = \sum_{i,j=1}^N \mathrm{E}_{i,j} \otimes \mathrm{E}_{j,i} \in \mathfrak{g} \otimes \mathfrak{g}$ define an endomorphism $\mathbb{x} \in \mathrm{End}(\mathrm{E})$. More generally, define endomorphisms $\mathbb{x}_j, \mathbb{s}_j$ of E^n, $n \geq 0$, via

$$(\mathbb{x}_j)_M := \mathrm{E}^{n-j}(\mathbb{x}_{\mathrm{E}^{j-1}(M)}) \quad \text{and} \quad (\mathbb{s}_j)_M := \mathrm{E}^{n-j-1}(\mathbb{s}_{\mathrm{E}^{j-1}(M)}). \tag{3.18}$$

One easily verifies that these endomorphisms satisfy the defining relations of a *degenerate affine Hecke algebra* $\mathrm{H}_n^{\mathrm{daff}}$. This means that the $\mathbb{x}_i$ commute (defining a subalgebra $\mathbb{C}[\mathbb{x}_1, \ldots, \mathbb{x}_n]$), the $\mathbb{s}_i$ satisfy the Coxeter relations of the symmetric groups (defining a subalgebra $\mathbb{C}[S_n]$), and the two sets of generators interact via the degenerate semidirect product relations $\mathbb{s}_j \mathbb{x}_{j+1} = \mathbb{x}_j \mathbb{s}_j + 1$ and $\mathbb{s}_j \mathbb{x}_l = \mathbb{x}_l \mathbb{s}_j$ for $l \neq j, j+1$.

Amazingly (although easy to verify with Example 3.21), E_i from (3.12) equals the *(generalized) i-eigenspace subfunctor* for $\mathbb{x}$ of E, i.e., $\mathrm{E}_i(M) = \sum_{l \geq 0} \ker((\mathbb{x}_M - i)^l)$. To (re)define F_i consistently, we use that F is *right* adjoint to E. With a *fixed* counit $c : \mathrm{E}\,\mathrm{F} \to \mathrm{id}$ and unit $c^* : \mathrm{id} \to \mathrm{F}\,\mathrm{E}$, we define elements $\mathbb{x}' \in \mathrm{End}(\mathrm{F})$ and $\mathbb{s}' \in \mathrm{End}(\mathrm{F}^2)$ following **[123]**:

$$\mathbb{x}' := \mathrm{F}(c) \circ \mathrm{F}(\mathbb{x})_{\mathrm{F}} \circ c^*{}_{\mathrm{F}}, \quad \mathbb{s}' := \mathrm{F}^2(c) \circ \mathrm{F}^2\,\mathrm{E}(c)_{\mathrm{F}} \circ \mathrm{F}^2(\mathbb{s})_{\mathrm{F}^2} \circ \mathrm{E}(c^*)_{\mathrm{E}\,\mathrm{F}^2} f \circ c_{\mathrm{F}^2}. \tag{3.19}$$

Then F inherits a decomposition into i-eigenspace functors F_i, $1 \leq i \leq k$, for $\mathbb{x}'$. By **[123]**, the *bi*adjointness of E_i and F_i follows. Thus, these functors are exact, send projectives to projectives, and provide a *based categorification*, i.e., induce on the Grothendieck group $\mathrm{K}_0^{\oplus}(\mathrm{Proj}(\mathcal{O}^{\mathbf{d}})) \otimes_{\mathbb{Z}} \mathbb{C}$ the structure of an integrable $\mathfrak{sl}_k$-module $\bigwedge^{\mathbf{d}} \overline{V}_+$, where the classes of the indecomposable projectives are (a basis of) weight vectors. We work here with $\mathfrak{sl}_k$ to agree with the existing literature.

These ingredients and properties listed here were axiomatized in **[123]**:

Definition 3.37. Let $\mathfrak{C}$ be a $\mathbb{C}$-linear abelian finite length category with enough projectives. Then, a *categorical* sl_k*-action* on $\mathfrak{C}$ (categorifying C) consists of

- an endofunctor E with a right adjoint F specified by a counit c and a unit c^*, and
- an element $\mathbb{s} \in \mathrm{End}(\mathrm{E}^2)$ and an endomorphism $\mathbb{x} \in \mathrm{End}(\mathrm{E})$

which satisfy $\mathrm{E} = \bigoplus_{i=1}^k \mathrm{E}_i$, where E_i is the i-eigenspace subfunctor for $\mathbb{x}$, the endomorphisms $\mathbb{s}_j, \mathbb{x}_j$ defined via (3.18) satisfy the relations of $\mathrm{H}_n^{\mathrm{daff}}$, the functor F is right adjoint to E, and finally with the definition of F_i as i-eigenspace functor for $\mathbb{x}'$ as in (3.19), the functors E_i and F_i define a based categorification of an integrable $\mathfrak{sl}_k$-module C.

We get categorifications of the tensor products $\bigwedge^{\mathbf{d}} \overline{V}_+$ of $\mathfrak{sl}_k$ exterior powers:

Theorem 3.38. *The constructions* (3.18) *and* (3.11) *define a categorical* sl_k*-action on* $\mathcal{O}^{\mathbf{d}}$.

Remark 3.39. Definition 3.37 is the easiest example from the theory of categorical actions of Kac–Moody Lie algebras $\mathfrak{g}_{\mathrm{KM}}$ **[80, 123]**. The degenerate affine Hecke algebra gets replaced by a more general *quiver Hecke algebra* (or KLR algebra) which is used to define a certain graded 2-category ${}^2\dot{U}_v(\mathfrak{g}_{\mathrm{KM}})$ categorifying $\dot{U}_v(\mathfrak{g}_{\mathrm{KM}})$, cf. Remark 2.11. The definition is via generators and relations, algebraically [123] or diagrammatically [80], and matched in [21].

Application 3.40. A nice situation occurs when the morphisms (3.18) generate the endomorphism ring of an object $\mathrm{E}^n(M)$ and the kernel is controlled by a cyclotomic quotient of $\mathrm{H}_n^{\mathrm{daff}}$. Then this ring can be determined explicitly. If, moreover, every indecomposable projective object in $\mathfrak{C}$ arises as a summand of $\mathrm{E}^n(M)$ for $n \gg 0$, one might construct equivalences by determining and matching endomorphism rings of projective generators instead of providing a functor. This idea is applied, e.g., in [25] to the category $\mathcal{F}(a|b)$ of finite-dimensional representations of the linear supergroup $\mathrm{GL}(a|b)$: $\mathcal{F}(a|b)$ is equivalent to the category of modules for an infinite-dimensional analogue of Khovanov's arc algebra, Remarks 3.2, 3.32. The notion *higher Schur–Weyl duality* [22] formalizes such nice Lie-theoretic situations.

Definition 3.37 significantly rigidifies the involved category $\mathfrak{C}$. If C is finite-dimensional irreducible of highest weight ξ, its weight space decomposition implies a decomposition of $\mathfrak{C}$ into direct summands $\mathfrak{C}_\lambda$ (cf. with $\mathcal{O}^{\mathbf{d}}$) and the *Uniqueness Theorem*, a very special case of Rouquier's *Universality Theorem*, holds [31,123]: a *minimal* (i.e., $\mathfrak{C}_\xi \simeq \mathrm{Vect}$) categorification of such C is unique up to *strong equivalence,* meaning an equivalence of categories $\Gamma : \mathfrak{C} \to \mathfrak{C}'$ with an isomorphism $\phi : \Gamma\,\mathrm{E} \cong \mathrm{E}\,\Gamma$ satisfying the expected compatibilities with $\mathtt{x}$, $\mathtt{s}$. Uniqueness allows establishing abstractly equivalences of categories.

Application 3.41. The Uniqueness Theorem is powerful even for $\mathfrak{sl}_2$-modules. It is used in [31] to prove Broué's abelian defect group conjecture for symmetric groups, one of the most famous conjectures in modular representation theory of finite groups.

Application 3.42. By the Universality Theorem, the $k \geq 3$ generalizations [95,97] of Khovanov's arc algebras are Morita equivalent to certain cyclotomic quotients of $\mathrm{H}_n^{\mathrm{daff}}$. These algebras should provide an algebraic construction of the MSS^- invariants as in Remark 3.32.

3.3.2. Tensor product categorifications

The Uniqueness Theorem heavily relies on the fact that finite-dimensional irreducible modules are generated by their highest weight vectors and thus does not directly apply to tensor products as in Theorem 3.38. A general theory for *the process of taking* tensor products of categorifications is still missing. The *naive* outer tensor product of categorical $\mathfrak{sl}_k$-actions has the desired K_0, but only a categorical action of $\mathfrak{sl}_k \oplus \mathfrak{sl}_k$ instead of $\mathfrak{sl}_k$. For a *given* tensor product, an axiomatic definition of a categorification was first formulated by Losev and Webster in [91].

Their definition uses the *reverse dominance ordering* on weights in a tensor product. Concretely, consider again the $\mathfrak{sl}_k$-module $\bigwedge^{\mathbf{d}} \overline{V}_+$ with $\mathbf{d} = (d_1, \dots, d_m)$. View its weights as tuples $\lambda = (\lambda_1, \dots, \lambda_m)$ of $\mathfrak{sl}_k$-weights, and $\lambda \geq \mu$ if $\lambda_1 + \cdots + \lambda_m = \mu_1 + \cdots + \mu_m$ and $\lambda_1 + \cdots + \lambda_i \leq \mu_1 + \cdots + \mu_i$ (in the usual ordering on $\mathfrak{sl}_k$-weights) for each $i < m$.

Via (3.9), this ordering translates into the Bruhat ordering on $\Lambda^{\mathbf{d}}$. The following is a reformulation of the original definition [91] following closely [23].

Definition 3.43. A *tensor product categorification* of the $\mathfrak{sl}_k$-module $\bigwedge^{\mathbf{d}} \overline{V}_+$ is the same data as in Definition 3.37, but with the last property on based categorification replaced by

- $\mathfrak{C}$ is a highest weight category with respect to the poset $\Lambda^{\mathbf{d}}$,
- the exact functors E_i and F_i send objects with Δ-flags to objects with Δ-flags,
- under the isomorphism $\bigwedge^{\mathbf{d}} \overline{V}_+ \cong \mathrm{K}_0(\mathfrak{C}) \otimes_{\mathbb{Z}} \mathbb{C}$, $\vec{e} \mapsto [\Delta^{\mathfrak{C}}(\mathrm{wt}(\vec{e}))]$ as in (3.11), the actions of E_i and F_i correspond to the actions of $[\mathrm{E}_i]$ and $[\mathrm{F}_i]$, respectively.

Theorem 3.44. *The highest weight category $\mathcal{O}^{\mathbf{d}}$ defines with the data from Theorem* 3.38, *the poset $\Lambda^{\mathbf{d}}$,* (3.12) *and* (3.11) *a tensor product categorification of the $\mathfrak{sl}_k$-module $\bigwedge^{\mathbf{d}} \overline{V}_+$.*

By the following result from [91] this is the only one up to strong equivalence:

Theorem 3.45. *A tensor product categorification of the $\mathfrak{sl}_k$-module $\bigwedge^{\mathbf{d}} \overline{V}_+$ is unique.*

Remark 3.46. Definition 3.43 is again a special case of a more general definition [91] which works for any Kac–Moody Lie algebra $\mathfrak{g}_{\mathrm{KM}}$ instead of $\mathfrak{sl}_k$ and any integrable highest weight module of $\mathfrak{g}_{\mathrm{KM}}$ for each tensor factor. It requires the following adjustments. On the one hand, the action of the degenerate affine Hecke algebra gets replaced by a quiver Hecke algebra from Remark 3.39 or an even more general *Webster algebra* [142]. On the other hand, the highest weight category gets replaced by a *fully stratified* category. For the general theory of such generalisations of highest weight categories see [26].

Remark 3.47. Let us return to the graded setting to obtain categorifications of the $U_v(\mathfrak{gl}_k)$-modules $\bigwedge^{\mathbf{d}} V_+$. One can turn $\hat{\mathcal{O}}^{\mathbf{d}}$ into a graded additive $\mathbb{C}$-linear 2-category. For this, note that $\mathrm{Hom}_{\mathcal{O}^{\mathbf{d}}}(M, N) = \bigoplus_{j \in \mathbb{Z}} \mathrm{Hom}_{\hat{\mathcal{O}}^{\mathbf{d}}}(\hat{M}, \hat{N}\langle j \rangle)$ for $M, N \in \mathcal{O}^{\mathbf{d}}$ which have graded lifts $\hat{M}, \hat{N} \in \hat{\mathcal{O}}^{\mathbf{d}}$; similarly for functors. Objects in this 2-category are weights $\mathbf{c}$ of $\bigwedge^{\mathbf{d}} V_+$, but thought of as the corresponding summands in $\hat{\mathcal{O}}^{\mathbf{d}}$ via (3.13). Morphisms are generated by (i) the functors $\mathbb{1}_{\mathbf{c}}$ which are the identity on $\mathbf{c}$ and zero otherwise, (ii) the functors $\hat{\mathrm{E}}_i \mathbb{1}_{\mathbf{c}}$ viewed as morphisms from $\mathbf{c}$ to $\mathbf{c} + \alpha_i$ where α_i is the corresponding simple root for $\mathfrak{gl}_k$, and (iii) fixed right adjoints of (ii) which are the $\mathbb{1}_{\mathbf{c}}\hat{\mathrm{F}}_i$ up to shifts. The 2-morphisms are generated by the homogeneous components of the natural transformations (3.18). From [123] it follows that this data defines a *(strong) 2-representation* of $\mathfrak{sl}_k$. By [28], it extends to an *action of* ${}^2\dot{U}_v(\mathfrak{q})$, called a *2-representation of* $\dot{U}_v(\mathfrak{q})$, for $\mathfrak{q} = \mathfrak{sl}_k$ and also for $\mathfrak{gl}_k$ by adding a grading shifting operator.

Remark 3.48. The original definition in [91] connects Theorem 3.44 with naive outer tensor products: For $\lambda \in \Lambda^d$, there are Serre subcategories $\mathcal{O}^{\mathbf{d}}[< \lambda] \subset \mathcal{O}^{\mathbf{d}}[\leq \lambda]$ in $\mathcal{O}^{\mathbf{d}}$ generated by all $L(\mu)$ with $\mu < \lambda$ respectively $\mu \leq \lambda$. The *associated graded* $\bigoplus_{\lambda} \mathcal{O}^{\mathbf{d}}[\leq \lambda]/\mathcal{O}^{\mathbf{d}}[< \lambda]$ of $\mathcal{O}^{\mathbf{d}}$ formed from the subquotients can be identified with the naive tensor product of the categorifications of the factors $\bigwedge^{d_i} \overline{V}_+$, see [126] for an explicit identification. The highest weight structure, explicitly the poset $\Lambda^{\mathbf{d}}$, creates a desired asymmetry, namely the asymmetry in the tensor factors (2.4) when passing to the graded/quantized setting as in Remark 3.47.

Application 3.49. Categorical actions are often used in (specifically modular and super) representation theory to create interesting gradings or to determine decomposition numbers. We sketch an example directly connected to our framework. In [23], tensor product categorifications were defined for the limit Lie algebra $\mathfrak{sl}_{\mathbb{Z}}$ and constructed for $M := \overline{V}_{\infty}^{\otimes a} \otimes (\overline{V}_{\infty}^{*})^{\otimes b}$, very similar to above, using category $\mathcal{O}$ for the Lie superalgebra $\mathfrak{gl}_{a|b}$. Here, $\overline{V}_{\infty} = \mathbb{C}^{\mathbb{Z}}$ is the natural representation of $\mathfrak{sl}_{\mathbb{Z}}$ and $\overline{V}_{\infty}^{*}$ its restricted dual. The basis vectors in $\mathbb{C}^{\mathbb{Z}}$ labeled by a length k interval in $\mathbb{Z}$ span an $\mathfrak{sl}_k$-module $\overline{V}_{+}^{\otimes a} \otimes (\bigwedge^{k-1} \overline{V}_{+})^{\otimes b}$. Theorems 3.44 and 3.45 allow translating properties from $\mathcal{O}(\mathfrak{gl}_{a+(k-1)b})$ to the super side [23]. This finally implies that the (integral blocks of) category $\mathcal{O}$ for $\mathfrak{gl}(a|b)$ and the category $\mathcal{F}(a|b)$ of finite-dimensional representation of $\mathrm{GL}(a|b)$ can be equipped with a Koszul grading. Moreover, the graded decomposition numbers are given by parabolic Kazhdan–Lusztig polynomials. In case of $\mathcal{F}(a|b)$, this grading agrees with the explicit construction in [25] from Application 3.40. For a generalization to the more involved orthosymplectic supergroups, see [40].

3.4. Ad III: Categorified colored tangle invariants and projectors

The colored framed tangle invariant RT from Remark 2.4 involves ultimately tensor products of arbitrary finite-dimensional irreducible $U_v(\mathfrak{gl}_k)$-modules, not only exterior powers. A categorification of all such tensor products exists for $\mathfrak{sl}_k$ and $U_v(\mathfrak{sl}_k)$ [52,126,142], and by [142] even for any simple complex Lie algebra $\mathfrak{g}_{\mathrm{s}}$. Webster [142] also ensures the existence of tensor product categorifications for $\mathfrak{g}_{\mathrm{s}}$. He uses categories of graded modules over graded algebras which generalize quiver Hecke and quiver Schur algebras. These algebras are defined diagrammatically, so that all calculations are elementary, but usually not easy. The grading allows to get categorifications of $U_v(\mathfrak{g}_{\mathrm{s}})$-modules as in Remark 3.47 [142] with a direct generalization of Theorem 3.30 to arbitrary $\mathfrak{g}_{\mathrm{s}}$. Instead of formulating this in detail, we indicate phenomena which occur, even for $\mathfrak{g}_{\mathrm{s}} = \mathfrak{sl}_k$, when passing from tensor products of fundamental representations to arbitrary irreducible ones. On the way we construct tensor product categorifications for $\mathfrak{sl}_k$ using the results from the previous section. The $\mathfrak{sl}_k$-action extends by construction to a $\mathfrak{gl}_k$-action, even to a 2-representation of $\dot{U}_v(\mathfrak{gl}_k)$ when invoking gradings. However, not all irreducible $\mathfrak{gl}_k$-modules occur in this way as tensor factors.

Any irreducible finite-dimensional $\mathfrak{sl}_k$-module is a quotient of some $\bigwedge^{\mathbf{d}} \overline{V}_{+}$ as in Section 3.3.2 such that its highest weight is the sum of the highest weights of the tensor factors. Taking tensor products $\bigwedge^{\mathbf{d}(1)} \overline{V}_{+} \otimes \cdots \otimes \bigwedge^{\mathbf{d}(r)} \overline{V}_{+} \twoheadrightarrow \overline{V}(\xi_1) \otimes \cdots \otimes \overline{V}(\xi_r) =: \overline{V}(\xi)$ realizes finite tensor products $\overline{V}(\xi)$ also as quotients of some $\bigwedge^{\mathbf{d}} \overline{V}_{+}$ which we consider now. Via (3.11), the irreducible objects in $\mathcal{O}^{\mathbf{d}}$ give rise to a *special basis* (in fact, the $v = 1$-specialized Lusztig dual canonical basis) of $\bigwedge^{\mathbf{d}} \overline{V}_{+}$. It turns out that the kernel of the quotient to $\overline{V}(\xi)$ is spanned by a subset of these special basis vectors. Fix $1 \leq j \leq r$. Combinatorially, one can label standard basis vectors (2.12) in $\bigwedge^{\mathbf{d}(j)} \overline{V}_{+}$ canonically by column strict tableaux and then basis vectors in $\overline{V}(\xi_j)$ by the set I_j of semistandard tableaux, i.e., those which are additionally weakly row strict. The shape is determined by $\mathbf{d}_{(j)}$ or, equivalently,

ξ_j and fillings are from $\{1, \ldots, k\}$. Consider now the standard basis vectors $\vec{e}$ as in (3.9) which correspond to m-tuples not in $I_1 \times \cdots \times I_r$. They define (by taking the irreducible quotient of the corresponding parabolic Verma module in (3.13)) a set of irreducible objects $L(\mathrm{wt}(\vec{e})) \in \mathcal{O}^{\mathbf{d}}$, thus a Serre subcategory $\mathcal{S}_\xi$ in $\mathcal{O}^{\mathbf{d}}$.

We obtain a categorification of $\overline{V}(\xi)$ as constructed in [126] and implicitly in [142]:

Theorem 3.50. *The Serre quotient $\mathcal{O}^{\mathbf{d}}/\mathcal{S}_\xi$ inherits a categorical $\mathfrak{sl}_k$-action from $\mathcal{O}^{\mathbf{d}}$. This is a tensor product categorification in the sense of [91] categorifying $\overline{V}(\xi)$ with the ordering on the labeling set of irreducible objects induced from $\Lambda^{\mathbf{d}}$. From $\hat{\mathcal{O}}^{\mathbf{d}}$ as in Remark 3.47, the graded version $\hat{\mathcal{O}}^{\mathbf{d}}/\hat{\mathcal{S}}_\xi$ inherits an action of ${}^2\dot{U}_v(\mathfrak{gl}_k)$.*

The quotient functors $\pi_\xi : \hat{\mathcal{O}}^{\mathbf{d}} \to \hat{\mathcal{O}}^{\mathbf{d}}/\hat{\mathcal{S}}_\xi$ are exact and induce $\mathbb{Z}[v^{\pm 1}]$-linear morphism on the Grothendieck groups (which, however, usually do not split over $\mathbb{Z}[v^{\pm 1}]$):

$$\begin{array}{ccc} \hat{\mathcal{O}}^{\mathbf{d}} & \xrightarrow{\pi_\xi} & \hat{\mathcal{O}}^{\mathbf{d}}/\hat{\mathcal{S}}_\xi \\ \downarrow{\scriptstyle \mathrm{K}_0} & & \downarrow{\scriptstyle \mathrm{K}_0} \\ \bigwedge^{\mathbf{d}} V_+^{\mathbb{Z}} & \xrightarrow{[\pi_\xi]} & V(\xi)^{\mathbb{Z}} \end{array} \tag{3.20}$$

This categorifies in particular for $k = 2$ any $\overline{V}(\xi)$ with $\xi = m\omega_1$ where ω_1 is the fundamental weight and $m \in \mathbb{Z}_{\geq 0}$, by realising it as a categorification of the *Jones-Wenzl quotient*

$$(V_{+,\mathfrak{gl}_2}^{\mathbb{Z}})^{\otimes m} \underset{\text{split ?}}{\xrightarrow{[\pi_m]:=[\pi_\xi]}} V_+^{\mathbb{Z}}(\xi). \tag{3.21}$$

with categorificaton of the quotient map. Note that this quotient map splits over $\mathbb{C}(v)$.

Remark 3.51. Theorem 3.50 requires a more general version of Definition 3.43 from [91], see Remark 3.46, as the quotient category $\mathcal{O}^{\mathbf{d}}/\mathcal{S}_\xi$ might not be highest weight, but only fully stratified. Combinatorially, this is reflected in higher-dimensional weight spaces of the tensor factors of $\overline{V}(\xi)$. Using only fundamental representations avoids this problem as they are minuscule, and also avoids taking duals or inverses of determinant representations.

In contrast to $\hat{\mathcal{O}}^{\mathbf{d}}$, the quotients $\mathcal{O}^{\mathbf{d}}/\mathcal{S}_\xi$ usually have *infinite global dimension*. Thus, the computation of a derived left adjoint $\mathcal{L}\iota_\xi$ to the quotient functor π_ξ requires *infinite* resolutions and *unbounded* derived categories. This becomes relevant in categorifications of colored tangle invariants following the knot-theoretic coloring via *cabling* and *projectors*. The idempotent functor $\mathrm{pr}_\xi := \mathcal{L}\iota_\xi\, \pi_\xi$ is a categorified projector.

Remark 3.52. Working with infinite complexes is delicate, in particular when Grothendieck groups or Euler characteristics are involved. To avoid an Eilenberg swindle and the collapse of the Grothendieck groups, we work in the graded setting with certain subcategories $D^\nabla(\hat{\mathcal{O}})$ of the unbounded derived category such that $K_0(D^\nabla(\hat{\mathcal{O}})) \cong K_0(\hat{\mathcal{O}}) \otimes_{\mathbb{Z}[v^{\pm 1}]} \mathbb{Z}((v))$, see [2] for a precise definition. The functors π_ξ, $\mathcal{L}\iota_\xi$ induce then $\mathbb{Z}((v))$-linear maps

$$[\pi_\xi] : (V_+^{\mathbb{Z}})^{\otimes m} \otimes_{\mathbb{Z}[v^{\pm 1}]} \mathbb{Z}((v)) \rightleftarrows V(\xi)^{\mathbb{Z}} \otimes_{\mathbb{Z}[v^{\pm 1}]} \mathbb{Z}((v)) : [\mathcal{L}\iota_\xi]. \tag{3.22}$$

Example 3.53. In case of $U_v(\mathfrak{gl}_2)$, there is the quotient map $[\pi_2]$ from (3.21) to the biggest irreducible quotient. Explicitly for $m = 2$ we have a basis of the quotient:

$$b_1 := [\pi_2](e_1 \otimes e_1), \quad b := [\pi_2](e_1 \otimes e_2) = v^{-1}[\pi_2](e_2 \otimes e_1), \quad b_2 := [\pi_2](e_2 \otimes e_2).$$

A split of $[\pi_2]$ over $\mathbb{C}(v)$ is given by $b_i \mapsto e_i \otimes e_i$ and $b \mapsto \frac{1}{[2]}(e_2 \otimes e_1 + v^{-1} e_1 \otimes e_2)$. Interpreting the latter as $(ve_2 \otimes e_1 + e_1 \otimes e_2)(1 - v^2 + v^4 - \cdots) \in (V_+^{\mathbb{Z}})^{\otimes 2} \otimes_{\mathbb{Z}[v^{\pm 1}]} \mathbb{Z}((v))$, we obtain the morphism $[\mathcal{L}\iota_2]$ induced via (3.22) from $\mathcal{L}\iota_2$ (in fact without explicitly constructing $\mathcal{L}\iota_2$). Over $\mathbb{Z}((v))$, a split (3.21) exists with a categorification. The functor pr_2 then categorifies $[\mathrm{pr}_2]$, which is the easiest example of a Jones-Wenzl projector.

Remark 3.54. In case of $U_v(\mathfrak{gl}_2)$, the projector (3.21) from $V_{+,\mathfrak{gl}_2}^{\otimes m}$ onto the biggest irreducible summand is the famous *Jones–Wenzl projector* JW_m. This was categorified the first time in [52] using a Serre quotient functor, and independently in its Koszul dual $V_{-,\mathfrak{gl}_2}$-version in [33] using Bar-Natan's approach to Khovanov homology, and in [125] using iterated categorified full twists.

As a special case, the RT value of the unknot colored by the $U_v(\mathfrak{gl}_k)$-module $V^{\mathbb{Z}}(\xi)$ as in (3.20) can be categorified by taking the MSS^+-value of $\sum_i d_i$ nested cups (viewed as a derived functor) followed by pr_ξ and followed by the value of $\sum_i d_i$ nested caps (the projector pr_ξ is displayed on the left and the categorified value of the colored unknot on the right):

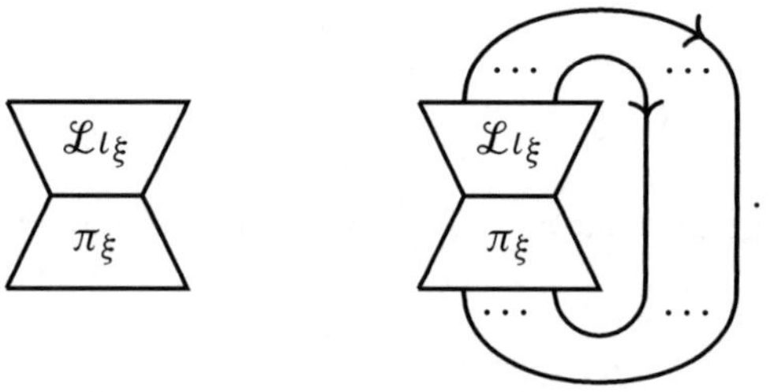

(3.23)

Example 3.55. In case $k = 2$ and $V(\xi)$ is 3-dimensional, the value of (3.23) can be realized as a complex in the unbounded homotopy category $K^-(\mathbb{C}\text{-mod}^{\mathbb{Z}(v)})$. A lengthy calculation gives the graded Poincaré polynomial $v^2t^2 + 1 + v^{-2} + \frac{v^{-6}t^{-2}(1+t^{-1})}{1-t^{-2}v^{-4}} \in \mathbb{Z}[t^{\pm 1}]((v))$ [138]. Its Euler characteristic equals $v^2 + 1 + v^{-2} = [3]$, which is indeed the $\mathrm{RT}_{V(\xi)}$-value of the unknot. By a uniqueness result of categorified Jones–Wenzl projectors from [33], the value of the unknot from (3.23) or from Theorem 3.56 agrees with the Cooper–Krushkal categorified value [33] of the colored unknot up to Koszul duality (i.e., a transformation $v \mapsto t^{-1}v^{-1}$).

Let L be an oriented link with planar projection D and coloring col assigning some $\overline{V}(\xi_c)$ to each components c of D. Assume $(V_+^{\mathbb{Z}})^{\otimes m_c} \twoheadrightarrow V^{\mathbb{Z}}(\xi_c)$ as in (3.20). To D we attach its *color-cabled version* D_{cc}: we first replace each strand in a component c by its cabling, i.e., by m_c parallel strands oriented as before. Then we write the result as a composition $t_1 \cdots t_r$ of basic tangle diagrams, and finally place for one upwards pointing original strand in D a projector (3.23) on its cabling. Let $\mathrm{MSS}^+(D_{\mathrm{cc}})$ be the associated composition of derived functors given by MSS^+ with additionally pr_ξ included when the projector occurs. Using the identification from Corollary 3.31, we can apply this functor to the vector space $\mathbb{C}$ concentrated in bidegree zero to get an object $\mathrm{MSS}^+(D_{\mathrm{cc}})(\mathbb{C})$ in $D^\nabla(\mathbb{C}\text{-mod}^{\mathbb{Z}(v)}) \subset K^-(\mathbb{C}\text{-mod}^{\mathbb{Z}(v)})$.

The categorification [138] of the colored framed oriented tangle invariant with colors irreducible $U_v(\mathfrak{sl}_k)$-modules (or their $U_v(\mathfrak{gl}_k)$-versions (3.20)) implies for links:

Theorem 3.56. *The assignment $D \mapsto \mathrm{MSS}^+(D_{\mathrm{cc}})(\mathbb{C})$ defines an invariant $\mathrm{MSS}^+_{\mathrm{col}}$ of colored framed oriented links. It induces on K_0 the colored* RT*-invariant from Remark 2.4 for $\mathfrak{sl}_k$.*

Remark 3.57. The colored knot invariants from Theorem 3.56 are usually infinite complexes, even for the unknot. The Poincaré series of $\mathrm{MSS}^+(D_{\mathrm{cc}})(\mathbb{C})$ has values in $\mathbb{Z}[t^{\pm 1}]((v))$. This is similar to the HHH-invariant, but we believe it is even harder to compute. For $k = 2$, these invariants should be directly connected to the invariants constructed in [38], where impressive explicit examples are computed. The occurring infinite series are secretly rewriting quotients $\frac{[a]}{[b]}$ of quantum numbers, see Example 3.53. A realization of such quotients as Euler characteristic of an infinite complex is called *fractional Euler characteristic* in [53].

Recall from Section 2 that the HOMFLY-PT polynomial recovers the quantum $\mathfrak{gl}_k$ link invariants RT_{V_-} by specialization of $\mathbf{a}$ to v^k. One might expect a similar connection for the categorifications, i.e., between the triply graded KR link homology which, by Theorem 3.12, is a categorification of the HOMFLY-PT-polynomial and MSS^-. Naive specialization does not work, but there is a spectral sequence connecting the two theories, predicted in [39] and established in [117]. Also, recall from Section 3.2 that the approach to compute $\mathrm{KR}(t_{(n,q)})$ and its limit $\mathrm{KR}(t_{(n,\infty)})$ for torus links uses categorified projectors. On the other hand, $\mathrm{MSS}^+(\bigcirc_{\mathrm{cc}})(\mathbb{C})$, or its Koszul dual version $\mathrm{MSS}^-(\bigcirc_{\mathrm{cc}})(\mathbb{C})$, can be seen as a categorification of the closure of a projector. One again might expect a connection between $\mathrm{KR}(\bigcirc)^n$ from Example 3.9, $\mathrm{KR}(t_{(n,\infty)})$, and $\mathrm{MSS}^\pm(\bigcirc_{\mathrm{cc}})(\mathbb{C})$. The following reformulates conjectures from [59]:

Conjecture 3.58. *The algebra $B = \mathbb{C}[u_1, \ldots, u_n] \otimes \bigwedge^\bullet[\xi_1, \ldots \xi_n]$ can be turned into a differential bigraded algebra $(B, d_{k,\pm})$ with homology isomorphic to $\mathrm{MSS}^\pm(\bigcirc_{\mathrm{cc}})(\mathbb{C})$ where* col $= V_\pm(n\omega_1)$. *The grading on B and the differential $d_{k,\pm}$ depends on k and the sign $\pm$.*

Remark 3.59. A conjectural grading and differential is formulated in [59] for $-$. In case $k = n = 2$, Conjecture 3.58 follows up to an overall grading shift by a comparison of [33] with the formulas in [59], see [138] for a precise statement. In general, the conjecture is still open.

Q5: *Is there a conceptual method to compute the categorified colored invariants?*
Q6: *To which extent is $\mathrm{MSS}^\pm_{\mathrm{col}}$ and its extension to framed tangles functorial?*

Motivated by and based on constructions of link homologies in physics, invariants of 3-manifolds are developed in, e.g., [61, 62]. On the mathematical side, first steps in this direction are done in [53] by constructing categorified $3j$- and $6j$-symbols via fractional Euler characteristics.
Q7: *Do these colored $\mathfrak{sl}_k$-invariants give rise to some invariant of 3-manifolds?*

4. TWO PROPOSALS TOWARD 4-TQFTS

We sketch two promising routes towards 4-TQFT based on Soergel bimodules. The first one is based on tensor product categorifications, the second one on the categorification of Hecke algebras and braid groups using Soergel bimodules and Rouquier complexes.

Braided monoidal structure on 2-representations. Recall the starting point of algebraic categorification: the proposal [34] for constructing a 4-dimensional TQFT via Hopf categories. We like to interpret this as the wish of constructing, via categorified representation theory of quantum groups, a 0–1–2–3–4-*theory* [51], i.e., a theory for $d = 4$ which not only evaluates at d- and $(d-1)$-, but also at $(d-2)$-, ..., 1- and 0-dimensional manifolds. To express the gluing laws between these levels, one has to work [8], [92] in general with an n-category of bordisms (viewed as (∞, n)-category) and define a *fully extended n-TQFT* as a functor from this symmetric monoidal category into some symmetric monoidal n- (respectively (∞, n))-category. According to the *cobordism hypothesis* [8,92] such a fully extended TQFT F is determined by the value $F(\mathrm{pt})$ at a point, see [7,92,128] for partial proofs.

Already the question *what Chern–Simons theory attaches to a point* is subtle and depends on the perspective. Following [51,144], Chern–Simons theory or the related Witten–Reshetikhin–Turaev theory can be viewed as an *anomalous* 0–1–2–3 *theory* of oriented 4-manifolds, i.e., a morphism from the trivial theory to an invertible fully extended 4-TQFT F defined on oriented manifolds. A similar interpretation was proposed by Walker, and the related invariant of a 4-manifold was combinatorially described in [35]. These interpretations propose attaching a certain *braided monoidal category* $F(\mathrm{pt})$ to a point [51]. Coming back to our setting, this suggests that a *categorification of the braided monoidal category of representations of a quantum group* might arise as the value $F_{\mathrm{cat}}(\mathrm{pt})$ of a point of an anomaly F_{cat}, some fully extended (possibly partial) 5-TQFT with an anomalous 0–1–2–3–4-theory.

Remark 4.1. Some relevance [101] for 4-dimensional topology is already visible in Kh and MSS^{-}, i.e., in categorified intertwiners of $\mathcal{F}und_2^{-}$ as in Remark 3.32, in particular via tangle cobordisms for surfaces in dimension 4 [67,77] and for invariants of 4-manifolds [106].

Concretely, one seeks a monoidal structure on the 2-category of 2-representations of $U_v(\mathfrak{q})$ as in Remark 3.47 for $\mathfrak{q} = \mathfrak{gl}_k$ or $\mathfrak{sl}_k$ and say $\eta = 1$. Sections 3.3 and 3.4 presented tensor product categorifications and indicated categorifications of the duals and the braiding morphisms. The *process of taking tensor products*, i.e., the construction of a *tensor product* or an inner hom for 2-representations is, however, more involved. Inspired by (bordered) Heegard Floer theory, Manion and Rouquier [99,100] give such a construction in case $\mathfrak{q}$ is the positive part $\mathfrak{gl}(1|1)_+$ of the Lie superalgebra $\mathfrak{gl}(1|1)$ (for the analogue of Remark 3.47 see [79]). The passage to $\mathfrak{gl}(1|1)_+$ surprisingly simplifies the situation. In contrast to $\mathfrak{q} = \mathfrak{gl}_2$ or $\mathfrak{sl}_2$, homotopical complications disappear. The result of [100] is supposed to connect (as the value at an interval) to a slightly different type of TQFT and the theory predicted in [62].

Remark 4.2. This seemingly very different $\mathfrak{gl}(1|1)_+$-theory is still related to Section 3.3 via an interpretation in terms of subquotients of category $\mathcal{O}$ [87]. Only $\mathfrak{gl}(1|1)_+$ appears, since categorical actions of $\mathfrak{gl}(1|1)$ have not yet been defined. This might be connected with the nonsemisimplicity of the finite-dimensional representation theory of $\mathfrak{gl}(1|1)$, see, e.g., [25].

Rouquier, however, announced (in an appropriate A_∞-setting) the existence of a monoidal structure on the 2-category of 2-representations of $U_v(\mathfrak{q})$ for an arbitrary Kac–Moody Lie algebra $\mathfrak{q}$ and a candidate for a braiding.

This result should provide the desired value $F_{\text{cat}}(\text{pt})$. In the spirit of [34], we *propose to call* the resulting 2-category with its braided monoidal structure *the Hopf category of* $\mathfrak{q}$ and reformulate ideas from [34] as:

Prediction 4.3. The Hopf category of $\mathfrak{q}$ is the value $F_{\text{cat}}(\text{pt})$ for an anomaly fully extended (partially defined at the top) 5-TQFT with an anomalous 0–1–2–3–4-theory.

Soergel bimodules, braided monoidal 2-categories, and TQFT. We finish by proposing another approach towards 4-TQFTs using more directly categories of Soergel bimodules. This is again motivated by the idea that a braided monoidal category might occur as the value $F(\text{pt})$ at a point [51] in a 0–1–2–3-theory. We seek to increase the dimensions to a 0–1–2–3–4-theory with a *braided monoidal bicategory* as the value $F(\text{pt})$ of some fully extended 5-TQFT F. We sketch some first steps. This is current work with Paul Wedrich.

Remark 4.4. The first definition of a *semistrict monoidal* and a *semistrict braided monoidal 2-category* is due to [70, 71]. It was then improved and put into a more concise definition in [9] with a technical adjustment in [36]. The concepts also appear as (braided) Gray monoids in [37]. By a *braided monoidal bicategory* we mean the less strict version from [63].

In the following let $m, n \in \mathbb{N}_0$. Recall the category $\mathcal{SB}im_n$ of Soergel bimodules from Section 3.2 with $R_0 := \mathbb{C}$ and $\mathcal{SB}im_0$ finite-dimensional graded vector spaces. We view $\mathcal{SB}im_n$ as a graded monoidal category with tensor product $\circ_1 : (M, N) \mapsto M \otimes_{R_n} N$. If now $M \in \mathcal{SB}im_m$, $N \in \mathcal{SB}im_n$, then $M \boxtimes N := M \otimes_{\mathbb{C}} N$ is an $R_m \otimes_{\mathbb{C}} R_n = R_{m+n}$-bimodule and by construction an object in $\mathcal{SB}im_{m+n}$. For morphisms f and g in $\mathcal{SB}im_m$ and $\mathcal{SB}im_n$, respectively, we define then $f \boxtimes g$ in the obvious way and set $m \boxtimes n = m + n$.

To get the desired semistrictness we use the monoidal category $\mathcal{DB}im_n \simeq \mathcal{SB}im_n$ [45, 48] from Remark 3.20. We omit giving the definition of $\mathcal{DB}im_n$ (it would not even fit on a page) and just recall that $\mathcal{DB}im_n$ is the Karoubian closure of a graded monoidal category $\mathcal{DB}im'_n$ [48]. The definition of $\mathcal{DB}im'_n$ is via generators and relations in terms of diagrams (similar to the usual string diagrams for higher categories). The morphism spaces come with distinguished bases, often called light leaves bases. A picked basis allows one to mimic the concept of coordinatized vector spaces from [70] and (semi)strictify the setup. Implicitly we assume this now, not altering the notation. We obtain categories *enriched in* $\mathbb{C}$*-linear categories* (we use *bicategories* as in [14] and *monoidal bicategories* as, e.g., in [128]):

Theorem 4.5. *There is a bicategory* ${}^{(2)}\mathcal{SB}im$ *with objects* $\mathbb{N}_0$ *and nontrivial* $\mathbb{C}$*-linear morphism categories* ${}^{(2)}\mathcal{SB}im(m,n)$ *only in case* $m = n$*, in which case* ${}^{(2)}\mathcal{SB}im(n,n) = \mathcal{SB}im_n$ *with composition* $\circ_1$*. Similarly for* ${}^{(2)}\mathcal{DB}im$*, but with* $\mathcal{SB}im_n$ *replaced by* $\mathcal{DB}im_n$*.*

Moreover, ${}^{(2)}\mathcal{SB}im$ *and* ${}^{(2)}\mathcal{DB}im$ *can be turned into monoidal bicategories with tensor functor* $\boxtimes$, *even into a semistrict monoidal* 2*-category in the sense of* **[9,36]** *in case of* ${}^{(2)}\mathcal{DB}im$.

Remark 4.6. An analogue of ${}^{(2)}\mathcal{SB}im$ for singular Soergel bimodules as in Remark 3.23 exists as well (with the expected definition). For simplicity, we do not discuss this here.

The proof is done by *explicitly* constructing the required data and checking the coherence relations. Replacing $\mathcal{SB}im_n$ with the graded dg-category $C^b(\mathcal{SB}im_n)$ of bounded chain complexes of Soergel bimodules we get a category *enriched in graded* $\mathbb{C}$*-linear dg-categories*, similarly with $C^b(\mathcal{DB}im_n)$ instead of $\mathcal{DB}im_n$. Theorem 4.5 directly extends and provides bicategories, denoted ${}^2\mathcal{SB}im$ and ${}^2\mathcal{DB}im$, now realized as categories *enriched* **[55]** *in the monoidal category of* $\mathbb{C}$*-linear dg-categories* **[74]**.

We consider from now on only the stricter version ${}^2\mathcal{DB}im$. To define a braiding, we need in particular an adjoint equivalence $\mathbb{B} : \boxtimes \Rightarrow \boxtimes^{\mathrm{op}}$ **[9,63]**. This data includes a *braiding* 1*-morphism* $\mathbb{B}((a,b))$ in ${}^2\mathcal{DB}im(a \boxtimes b = a+b, b \boxtimes a = b+a)$ for any $a, b \in \mathbb{N}_0$. Thinking intuitively about this braiding 1-morphism gives us a candidate:

$\beta =$ 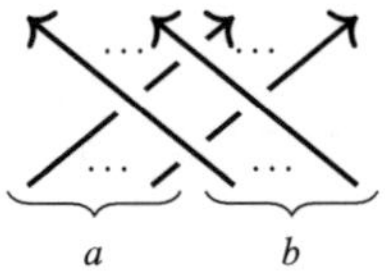

$\rightsquigarrow$ Rouquier complex $X(\beta)$ with

$$\ddot{\beta} = (\beta_b \cdots \beta_1) \cdots (\beta_{i+b-1} \cdots \beta_i) \cdots (\beta_{a+b-1} \cdots \beta_a), \quad (4.1)$$

namely the Rouquier complex $X(\ddot{\beta}) \in C^b(\mathcal{SB}im_{a+b})$ from Section 3.2 with $\ddot{\beta}$ as in (4.1) translated via the above equivalence to an object $\mathbb{B}((a,b))$ in $C^b(\mathcal{DB}im_{a+b})$.

Theorem 4.7. *The proposed adjoint equivalence* $\mathbb{B}$ *satisfies the required naturality conditions* **[90]** *for the generating* 1*- and* 2*-morphisms of* $\mathcal{DB}im$ up to canonical homotopy.

To obtain an honest braiding, however, one has to pass to the homotopy categories which loses quite a lot of information or to a category ${}^2_\infty\mathcal{DB}im$ *enriched in* ∞*-categories* **[55]**. We construct such a category ${}^2_\infty\mathcal{DB}im$ by applying a (rather technical and not standard) dg-nerve construction to the morphism categories. We expect this construction to satisfy:

Conjecture 4.8. ${}^2_\infty\mathcal{DB}im$ *is a braided monoidal bicategory.*

Remark 4.9. Braided monoidal 2-categories with linear hom-categories and finiteness conditions should be objects in some symmetric monoidal 5-category which arises as next step in the ladder of symmetric monoidal n-categories (made explicit in **[20]**: objects are certain monoidal categories for $n = 3$ and certain braided monoidal categories for $n = 4$).

Remark 4.10. We can view ${}^2_\infty\mathcal{DB}im$ as a category object in ∞-categories $\mathbf{Cat}_\infty$. We expect this to be an E_2-algebra in the ∞-category of $(\infty, 2)$-categories **[64,93]**. Higher Morita theory of E_n-algebras **[68]** provides a possible ambient $(\infty, 5)$-category for our hoped for TQFT.

Because of lacking finiteness conditions one should not expect n-dualisability **[92]** of ${}^2_\infty\mathcal{DB}im$ for $n > 3$, but we hope it holds for $n = 3, 4$ for quotients arising from actions on the $\mathfrak{gl}_k$-theories MSS_ for fixed $k \in \mathbb{N}$: An analogue of ${}^2_\infty\mathcal{DB}im$ defined using singular Soergel bimodules, Remark 4.6, acts by Remark 3.24 on the 2-categories $\hat{\mathcal{O}}^{\mathbf{d}}$ from Remark 3.47 for any fixed k. We conjecture that the largest quotient ${}_{\infty}^{2}\mathcal{DB}im(k)$ which still acts (for fixed k) has the desired finiteness properties to provide a fully extended (partial) 5-TQFT:

Conjecture 4.11. *Soergel bimodules give rise to a braided monoidal bicategory* ${}_{\infty}^{2}\mathcal{DB}im(k)$, $k \in \mathbb{N}$, *which is the value at a point of an anomaly with an anomalous* 0–1–2–3–4-*theory.*

ACKNOWLEDGMENTS

It is a pleasure to thank J. Brundan, A. Mathas, R. Rouquier, M. Stroppel, and J. Sussan for many mathematical and (non)mathematical discussions and for openly sharing ideas over the years. I am grateful to the Bonn representation theory group, in particular J. Eberhardt, G. Jasso, T. Heidersdorf, J. Matherne, J. Meinel, D. Tubbenhauer, P. Wedrich, and T. Wehrhan for feedback and constructive criticism on draft versions of this article.

FUNDING

This work was supported by the Hausdorff Center of Mathematics (HCM) in Bonn.

REFERENCES

[1] M. Abouzaid and I. Smith, The symplectic arc algebra is formal. *Duke Math. J.* **165** (2016), no. 6, 985–1060.

[2] P. Achar and C. Stroppel, Completions of Grothendieck groups. *Bull. Lond. Math. Soc.* **45** (2013), no. 1, 200–212.

[3] M. Aganagic and S. Shakirov, Knot homology and refined Chern–Simons index. *Comm. Math. Phys.* **333** (2015), 187–228.

[4] J. Alexander, A lemma on a system of knotted curves. *Proc. Natl. Acad. Sci. USA* **9** (1923), no. 3, 93–95.

[5] D. Allcock, Braid pictures for Artin groups. *Trans. Amer. Math. Soc.* **354** (2002), no. 9, 3455–3474.

[6] M. Atiyah, Topological quantum field theories. *Publ. Math. Inst. Hautes Études Sci.* **68** (1989), 175–189.

[7] D. Ayala and J. Francis, The cobordism hypothesis. 2017, arXiv:1705.02240.

[8] J. Baez and J. Dolan, Higher-dimensional algebra and topological quantum field theory. *J. Math. Phys.* **36** (1995), no. 11, 6073–6105.

[9] J. Baez and M. Neuchl, Higher-dimensional algebra. I. Braided monoidal 2-categories. *Adv. Math.* **121** (1996), no. 2, 196–244.

[10] D. Bar-Natan, On Khovanov's categorification of the Jones polynomial. *Algebr. Geom. Topol.* **2** (2002), 337–370.

[11] D. Bar-Natan, Khovanov's homology for tangles and cobordisms. *Geom. Topol.* **9** (2005), 1443–1499.

[12] A. Beilinson, V. Ginzburg, and W. Soergel, Koszul duality patterns in representation theory. *J. Amer. Math. Soc.* **9** (1996), no. 2, 473–527.

[13] A. Beliakova, M. Hogancamp, K. Putyra, and S. Wehrli, On the functoriality of $\mathfrak{sl}(2)$ tangle homology. 2019, arXiv:1903.12194.

[14] J. Bénabou, Introduction to bicategories. In *Reports of the midwest category seminar*, pp. 1–77, Springer, Berlin, 1967.

[15] Y. Berest, P. Etingof, and V. Ginzburg, Finite-dimensional representations of rational Cherednik algebras. *Int. Math. Res. Not.* **19** (2003), 1053–1088.

[16] J. Bernstein, I. Frenkel, and M. Khovanov, A categorification of the Temperley–Lieb algebra and Schur quotients of $U(\mathfrak{sl}_2)$ via projective and Zuckerman functors. *Selecta Math. (N.S.)* **5** (1999), no. 2, 199–241.

[17] J. Bernstein and V. Lunts, *Equivariant sheaves and functors*. Lecture Notes in Math. 1578, Springer, Berlin, 1994.

[18] J. Birman, *Braids, links and mapping class groups*. Ann. of Math. Stud. 82, Princeton University Press, Princeton, NJ, 1974.

[19] C. Blanchet, An oriented model for Khovanov homology. *J. Knot Theory Ramifications* **19** (2010), no. 2, 291–312.

[20] A. Brochier, D. Jordan, and N. Snyder, On dualizability of braided tensor categories. *Compos. Math.* **157** (2021), no. 3, 435–483.

[21] J. Brundan, On the definition of Kac–Moody 2-category. *Math. Ann.* **364** (2016), no. 1–2, 353–372.

[22] J. Brundan and A. Kleshchev, Schur–Weyl duality for higher levels. *Selecta Math. (N.S.)* **14** (2008), no. 1, 1–57.

[23] J. Brundan, I. Losev, and B. Webster, Tensor product categorifications and the super Kazhdan–Lusztig conjecture. *Int. Math. Res. Not.* **20** (2017), 6329–6410.

[24] J. Brundan and C. Stroppel, Highest weight categories arising from Khovanov's diagram algebra III: category $\mathcal{O}$. *Represent. Theory* **15** (2011), 170–243.

[25] J. Brundan and C. Stroppel, Highest weight categories arising from Khovanov's diagram algebra IV: the general linear supergroup. *J. Eur. Math. Soc. (JEMS)* **14** (2012), no. 2, 373–419.

[26] J. Brundan and C. Stroppel, Semi-infinite highest weight categories. 2021, arXiv:1808.08022v4. To appear in *Mem. Amer. Math. Soc.*

[27] S. Cautis, J. Kamnitzer, and S. Morrison, Webs and quantum skew Howe duality. *Math. Ann.* **360** (2014), no. 1–2, 351–390.

[28] S. Cautis and A. Lauda, Implicit structure in 2-representations of quantum groups. *Selecta Math. (N.S.)* **21** (2015), no. 1, 201–244.

[29] Y. Chen and M. Khovanov, An invariant of tangle cobordisms via subquotients of arc rings. *Fund. Math.* **225** (2014), no. 1, 23–44.

[30] I. Cherednik, Jones polynomials of torus knots via DAHA. *Int. Math. Res. Not.* **23** (2013), 5366–5425.

[31] J. Chuang and R. Rouquier, Derived equivalences for symmetric groups and $\mathfrak{sl}_2$-categorification. *Ann. of Math. (2)* **167** (2008), no. 1, 245–298.

[32] D. Clark, S. Morrison, and K. Walker, Fixing the functoriality of Khovanov homology. *Geom. Topol.* **13** (2009), no. 3, 1499–1582.

[33] B. Cooper and V. Krushkal, Categorification of the Jones–Wenzl projectors. *Quantum Topol.* **3** (2012), no. 2, 139–180.

[34] L. Crane and I. Frenkel, Four-dimensional topological quantum field theory, Hopf categories, and the canonical bases. *J. Math. Phys.* **35** (1994), no. 10, 5136–5154.

[35] L. Crane and D. Yetter, On algebraic structures implicit in topological quantum field theories. *J. Knot Theory Ramifications* **8** (1999), no. 2, 125–163.

[36] S. Crans, Generalized centers of braided and sylleptic monoidal 2-categories. *Adv. Math.* **136** (1998), 183–223.

[37] B. Day and R. Street, Monoidal bicategories and Hopf algebroids. *Adv. Math.* **129** (1997), 99–157.

[38] R. Dijkgraaf and E. Witten, Topological gauge theories and group cohomology. *Comm. Math. Phys.* **129** (1990), no. 2, 393–429.

[39] N. Dunfield, S. Gukov, and J. Rasmussen, The superpolynomial for knot homologies. *Exp. Math.* **15** (2006), 129–159.

[40] M. Ehrig and C. Stroppel, On the category of finite-dimensional representations of $\mathrm{OSp}(r|2n)$: Part I. Representation theory—current trends and perspectives. In *EMS Ser. Congr. Rep., Eur. Math., Zürich*, pp. 109–170, Amer. Math. Soc. Providence, RI, 2017.

[41] M. Ehrig and C. Stroppel, Nazarov–Wenzl algebras, coideal subalgebras and categorified skew Howe duality. *Adv. Math.* **331** (2018), 58–142.

[42] M. Ehrig, C. Stroppel, and D. Tubbenhauer, The Blanchet–Khovanov algebras. In *Categorification and higher representation theory*, pp. 183–226 Contemp. Math. 683, 2017.

[43] M. Ehrig, D. Tubbenhauer, and P. Wedrich, Functoriality of colored link homologies. *Proc. Lond. Math. Soc.* **117** (2018), no. 5, 996–1040.

[44] B. Elias and M. Hogancamp, On the computation of torus link homology. *Compos. Math.* **155** (2019), no. 1, 164–205.

[45] B. Elias and M. Khovanov, Diagrammatics for Soergel categories. *Int. J. Math. Math. Sci.* **978635** (2010).

[46] B. Elias and D. Krasner, Rouquier complexes are functorial over braid cobordisms. *Homology, Homotopy Appl.* **12** (2010), no. 2, 109–146.

[47] B. Elias and G. Williamson, The Hodge theory of Soergel bimodules. *Ann. of Math. (2)* **180** (2014), no. 3, 1089–1136.

[48] B. Elias and G. Williamson, Soergel calculus. *Represent. Theory* **20** (2016), 295–374.

[49] A. Ellis and A. Lauda, An odd categorification of $U_q(\mathfrak{sl}_2)$. *Quantum Topol.* **7** (2016), no. 2, 329–433.

[50] P. Etingof and V. Ginzburg, Symplectic reflection algebras, Calogero–Moser space, and deformed Harish-Chandra homomorphism. *Invent. Math.* **147** (2002), no. 2, 243–348.

[51] D. Freed, M. Hopkins, J. Lurie, and C. Teleman, A celebration of the mathematical legacy of Raoul Bott. In *Topological quantum field theories from compact Lie groups*, pp. 367–403, CRM Proc. Lecture Notes 50, AMS, 2010.

[52] I. Frenkel, M. Khovanov, and C. Stroppel, A categorification of finite-dimensional irreducible representations of quantum $\mathfrak{sl}_2$ and their tensor products. *Selecta Math. (N.S.)* **12** (2006), no. 3–4, 379–431.

[53] I. Frenkel, C. Stroppel, and J. Sussan, Categorifying fractional Euler characteristics, Jones–Wenzl projectors and 3j-symbols. *Quantum Topol.* **3** (2012), no. 2, 181–253.

[54] P. Freyd, D. Yetter, J. Hoste, W. Lickorish, K. Millett, and A. Ocneanu, A new polynomial invariant of knots and links. *Bull. Am. Meteorol. Soc.* **12** (1985), no. 2, 239–246.

[55] R. Garner and M. Shulman, Enriched categories as a free cocompletion. *Adv. Math.* **289** (2016), 1–94.

[56] M. Geck, Trace functions on Iwahori–Hecke algebras. In *Knot theory (Warsaw, 1995)*, pp. 87–109, Banach Center Publ. 42, Polish Acad. Sci. Inst. Math., 1998.

[57] M. Geck and S. Lambropoulou, Markov traces and knot invariants related to Iwahori–Hecke algebras of type B. *J. Reine Angew. Math.* **482** (1997), 191–213.

[58] E. Gorsky and A. Negut, Refined knot invariants and Hilbert schemes. *J. Math. Pures Appl. (9)* **104** (2015), no. 3, 403–435.

[59] E. Gorsky, A. Oblomkov, and J. Rasmussen, On stable Khovanov homology of torus knots. *Exp. Math.* **22** (2013), no. 3, 265–281.

[60] E. Gorsky, A. Oblomkov, J. Rasmussen, and V. Shende, Torus knots and the rational DAHA. *Duke Math. J.* **163** (2014), no. 14, 2709–2794.

[61] S. Gukov, D. Pei, and C. Vafa, BPS spectra and 3-manifold invariants. *J. Knot Theory Ramifications* **29** (2020), no. 2, 2040003, 85 pp.

[62] S. Gukov, P. Putrov, and C. Vafa, Fivebranes and 3-manifold homology. *J. High Energy Phys.* **71** (2017).

[63] N. Gurski, Loop spaces, and coherence for monoidal and braided monoidal bicategories. *Adv. Math.* **226** (2011), no. 5, 4225–4265.

[64] R. Haugseng, *Weakly enriched higher categories*. Ph.D. thesis, MIT, 2013.

[65] M. Hogancamp, Categorified Young symmetrizers and stable homology of torus links. *Geom. Topol.* **22** (2018), no. 5, 2943–3002.

[66] M. Hogancamp and A. Mellit, Torus link homology. 2019, arXiv:1909.00418.

[67] M. Jacobsson, An invariant of link cobordisms from Khovanov homology. *Algebr. Geom. Topol.* **4** (2004), 1211–1251.

[68] T. Johnson-Freyd and C. Scheimbauer, (Op)lax natural transformations, twisted quantum field theories, and "even higher" Morita categories. *Adv. Math.* **307** (2017), 147–223.

[69] V. Jones, Hecke algebra representations of braid groups and link polynomials. *Ann. of Math.* **126** (1987), 335–388.

[70] M. Kapranov and V. Voevodsky, 2-categories and Zamolodchikov tetrahedra equations. In *Algebraic groups and their generalizations: quantum and infinite-dimensional methods (University Park, PA, 1991)*, pp. 177–259, Proc. Sympos. Pure Math. 56, Amer. Math. Soc., Providence, RI, 1994.

[71] M. Kapranov and V. Voevodsky, Braided monoidal 2-categories and Manin-Schechtman higher braid groups. *J. Pure Appl. Algebra* **92** (1994), no. 3, 241–267.

[72] L. Kauffman, State models and the Jones polynomial. *Topology* **26** (1987), no. 3, 395–407.

[73] L. Kauffman and S. Lins, *Temperley–Lieb recoupling theory and invariants of 3-manifolds*. Ann. of Math. Stud. 134, Princeton University Press, 1974.

[74] B. Keller, Deriving DG categories. *Ann. Sci. Éc. Norm. Supér.* **27** (1994), no. 1, 63–102.

[75] M. Khovanov, A categorification of the Jones polynomial. *Duke Math. J.* **101** (2000), no. 3, 359–426.

[76] M. Khovanov, A functor-valued invariant of tangles. *Algebr. Geom. Topol.* **2** (2002), 665–741.

[77] M. Khovanov, An invariant of tangle cobordisms. *Trans. Amer. Math. Soc.* **358** (2006), 315–327.

[78] M. Khovanov, Triply-graded link homology and Hochschild homology of Soergel bimodules. *Internat. J. Math.* **18** (2007), no. 8, 869–885.

[79] M. Khovanov, How to categorify one-half of quantum $\mathfrak{gl}(1|2)$. In *Knots in Poland III*, pp. 211–232, Banach Center Publ. 103, Polish Acad. Sci. Inst. Math., 2014.

[80] M. Khovanov and A. Lauda, A diagrammatic approach to categorification of quantum groups. I. *Represent. Theory* **13** (2009), 309–347.

[81] M. Khovanov and L. Rozansky, Matrix factorizations and link homology. *Fund. Math.* **199** (2008), no. 1, 1–91.

[82] M. Khovanov and R. Thomas, Braid cobordisms, triangulated categories, and flag varieties. *Homology, Homotopy Appl.* **9** (2007), no. 2, 19–94.

[83] T. Kildetoft and V. Mazorchuk, Parabolic projective functors in type A. *Adv. Math.* **301** (2016), 785–803.

[84] P. Kronheimer and T. Mrowka, Khovanov homology is an unknot-detector. *Publ. Math. Inst. Hautes Études Sci.* **113** (2011), 97–208.

[85] G. Kuperberg, Spiders for rank 2 Lie algebras. *Comm. Math. Phys.* **180** (1996), no. 1, 109–151.

[86] T. Lasy, Markov property and Khovanov–Rozansky homology: Coxeter case. 2012, arXiv:1202.5547.

[87] A. Lauda and A. Manion, Ozsváth–Szabó bordered algebras and subquotients of category $\mathcal{O}$. *Adv. Math.* **376** (2021), 107455, 59 pp.

[88] G. Laumon, Transformation de Fourier, constantes d'équations fonctionnelles et conjecture de Weil. *Publ. Math. Inst. Hautes Études Sci.* **65** (1987), 131–210.

[89] G. Lehrer, H. Zhang, and R. Zhang, A quantum analogue of the first fundamental theorem of classical invariant theory. *Comm. Math. Phys.* **301** (2011), no. 1, 131–174.

[90] T. Leinster, Basic bicategories. 1998, arXiv:math/9810017.

[91] I. Losev and B. Webster, On uniqueness of tensor products of irreducible categorifications. *Selecta Math. (N.S.)* **21** (2015), no. 2, 345–377.

[92] J. Lurie, In *On the classification of topological field theories*, pp. 129–280, Curr. Dev. Math. 2008, Int. Press, Somerville, MA, 2009.

[93] J. Lurie, Higher algebra. 2017, https://www.math.ias.edu/~lurie/papers/HA.pdf.

[94] G. Lusztig, *Introduction to quantum groups*. Modern Birkhäuser Classics, 2010.

[95] M. Mackaay, W. Pan, and D. Tubbenhauer, The $\mathfrak{sl}_3$-web algebra. *Math. Z.* **277** (2014), no. 1–2, 401–479.

[96] M. Mackaay and B. Webster, Categorified skew Howe duality and comparison of knot homologies. *Adv. Math.* **330** (2018), 876–945.

[97] M. Mackaay and Y. Yonezawa, $\mathfrak{sl}_N$-web categories and categorified skew Howe duality. *J. Pure Appl. Algebra* **223** (2019), no. 5, 2173–2229.

[98] C. Mak and I. Smith, Fukaya–Seidel categories of Hilbert schemes and parabolic category $\mathcal{O}$. 2019, arXiv:1907.07624.

[99] A. Manion, Trivalent vertices and bordered knot Floer homology in the standard basis. 2020, arXiv:2012.07184.

[100] A. Manion and R. Rouquier, Higher representations and cornered Heegaard Floer homology. 2020, arXiv:2009.09627.

[101] C. Manolescu, Four dimensional topology. 2020, https://web.stanford.edu/~cm5/4D.pdf.

[102] A. Markov, Über die freie Äquivalenz der geschlossenen Zöpfe. *Rec. Math. Mosc.* **43** (1936), 73–78.

[103] D. Maulik, Stable pairs and the HOMFLY polynomial. *Invent. Math.* **204** (2016), no. 3, 787–831.

[104] V. Mazorchuk, S. Ovsienko, and C. Stroppel, Quadratic duals, Koszul dual functors, and applications. *Trans. Amer. Math. Soc.* **361** (2009), no. 3, 1129–1172.

[105] V. Mazorchuk and C. Stroppel, A combinatorial approach to functorial quantum $\mathfrak{sl}_k$ knot invariants. *Amer. J. Math.* **131** (2009), no. 6, 1679–1713.

[106] S. Morrison, K. Walker, and P. Wedrich, Invariants of 4-manifolds from Khovanov–Rozansky link homology. 2019, arXiv:1907.12194.

[107] H. Murakami, T. Ohtsuki, and S. Yamada, Homfly polynomial via an invariant of colored plane graphs. *Enseign. Math. (2)* **44** (1998), no. 3–4, 325–360.

[108] A. Oblomkov, J. Rasmussen, and V. Shende, The Hilbert scheme of a plane curve singularity and the HOMFLY homology of its link. With an appendix by E. Gorsky. *Geom. Topol.* **22** (2018), no. 2, 645–691.

[109] A. Oblomkov and V. Shende, The Hilbert scheme of a plane curve singularity and the HOMFLY polynomial of its link. *Duke Math. J.* **161** (2012), no. 7, 1277–1303.

[110] A. Oblomkov and Z. Yun, The cohomology ring of certain compactified Jacobians. 2017, arXiv:1710.05391.

[111] R. Penrose, Angular momentum: an approach to combinatorial spacetime. In *Quantum theory and beyond*, pp. 151–180, Cambridge University Press, 1971.

[112] G. Ponzano and T. Regge, Semiclassical limit of Racah coefficients. In *Spectroscopic and group theoretical methods in physics*, edited by F. Block et al., pp. 1–58, North Holland, Amsterdam, 1968.

[113] J. Przytycki and P. Traczyk, Conway algebras and Skein equivalence of links. *Proc. Amer. Math. Soc.* **100** (1987), 744–748.

[114] H. Queffelec and D. Rose, The $\mathfrak{sl}_n$ foam 2-category: a combinatorial formulation of Khovanov–Rozansky homology via categorical skew Howe duality. *Adv. Math.* **302** (2016), 1251–1339.

[115] H. Queffelec and A. Sartori, Mixed quantum skew Howe duality and link invariants of type A. *J. Pure Appl. Algebra* **223** (2019), no. 7, 2733–2779.

[116] J. Rasmussen, Khovanov homology and the slice genus. *Invent. Math.* **182** (2010), no. 2, 419–447.

[117] J. Rasmussen, Some differentials on Khovanov–Rozansky homology. *Geom. Topol.* **19** (2015), 3031–3104.

[118] N. Reshetikhin and V. Turaev, Ribbon graphs and their invariants derived from quantum groups. *Comm. Math. Phys.* **127** (1990), 1–26.

[119] N. Reshetikhin and V. Turaev, Invariants of 3-manifolds via link polynomials and quantum groups. *Invent. Math.* **103** (1991), no. 3, 547–597.

[120] S. Riche and G. Williamson, Tilting modules and the p-canonical basis. *Astérisque* **397** (2018).

[121] D. Rose and D. Tubbenhauer, Symmetric webs, Jones–Wenzl recursions, and q-Howe duality. *Int. Math. Res.* **17** (2016), 5249–5290.

[122] D. Rose and P. Wedrich, Deformations of colored $\mathfrak{sl}_N$ link homologies via foams. *Geom. Topol.* **20** (2016), no. 6, 3431–3517.

[123] R. Rouquier, 2-Kac–Moody algebras. 2008, arXiv:0812.5023.

[124] R. Rouquier, Khovanov–Rozansky homology and 2-braid groups. In *Categorification in geometry, topology, and physics*, pp. 147–157, Contemp. Math. 68, AMS, 2017.

[125] L. Rozansky, An infinite torus braid yields a categorified Jones–Wenzl projector. *Fund. Math.* **225** (2014), no. 1, 305–326.

[126] A. Sartori and C. Stroppel, Categorification of tensor product representations of $\mathfrak{sl}_k$ and category $\mathcal{O}$. *J. Algebra* **428** (2015), 256–291.

[127] A. Sartori and D. Tubbenhauer, Webs and q-Howe dualities in types BCD. *Trans. Amer. Math. Soc.* **371** (2019), no. 10, 7387–7431.

[128] C. Schommer-Pries, *The classification of two-dimensional extended topological field theories*. Ph.D. thesis, University of California, Berkeley, 2009.

[129] G. Segal, Topological structures in string theory. *Philos. Trans. R. Soc. Lond. Ser. A Math. Phys. Eng. Sci.* **359** (2001), no. 1784, 1389–1398.

[130] P. Seidel and I. Smith, A link invariant from the symplectic geometry of nilpotent slices. *Duke Math. J.* **134** (2006), no. 3, 453–514.

[131] W. Soergel, The combinatorics of Harish-Chandra bimodules. *J. Reine Angew. Math.* **429** (1992), 49–74.

[132] W. Soergel, Kazhdan–Lusztig-Polynome und unzerlegbare Bimoduln über Polynomringen. *J. Inst. Math. Jussieu* **6** (2007), no. 3, 501–525.

[133] T. Springer, Quelques applications de la cohomologie d'intersection. *Bourbaki Seminar*, Vol. 1981/1982. *Astérisque* **92** (Astérisque), 249–273.

[134] C. Stroppel, Category $\mathcal{O}$: gradings and translation functors. *J. Algebra* **268** (2003), no. 1, 301–326.

[135] C. Stroppel, A structure theorem for Harish-Chandra bimodules via coinvariants and Golod rings. *J. Algebra* **282** (2004), no. 1, 349–367.

[136] C. Stroppel, Categorification of the Temperley–Lieb category, tangles, and cobordisms via projective functors. *Duke Math. J.* **126** (2005), no. 3, 547–596.

[137] C. Stroppel, Parabolic category $\mathcal{O}$, perverse sheaves on Grassmannians, Springer fibres and Khovanov homology. *Compos. Math.* **145** (2009), no. 4, 954–992.

[138] C. Stroppel and J. Sussan, A Lie theoretic categorification of the coloured Jones polynomial. 2021, arXiv:2109.12889.

[139] J. Sussan, *Category $\mathcal{O}$ and $\mathfrak{sl}_k$ link invariants*. Ph.D. thesis, Yale University, 2007.

[140] D. Tubbenhauer, P. Vaz, and P. Wedrich, Super q-Howe duality and web categories. *Algebr. Geom. Topol.* **17** (2017), no. 6, 3703–3749.

[141] V. Turaev and O. Viro, State sum invariants of 3-manifolds and quantum $6j$-symbols. *Topology* **31** (1992), no. 4, 865–902.

[142] B. Webster, Knot invariants and higher representation theory. *Mem. Amer. Math. Soc.* **250** (2017), no. 1191.

[143] E. Witten, Super-symmetry and theory. *J. Differential Geom.* **17** (1982), no. 4, 661–692.

[144] E. Witten, Quantum field theory and the Jones polynomial. *Comm. Math. Phys.* **121** (1989), 351–399.

[145] H. Wu, A colored $\mathfrak{sl}(N)$ homology for links in S^3. *Dissertationes Math.* **499** (2014), 1–217.

CATHARINA STROPPEL

Mathematical Institute, Endenicher Allee 60, 53115 Bonn, Germany,
stroppel@math.uni-bonn.de

NONCOMMUTATIVE CREPANT RESOLUTIONS, AN OVERVIEW

MICHEL VAN DEN BERGH

ABSTRACT

Noncommutative crepant resolutions (NCCRs) are noncommutative analogues of the usual crepant resolutions that appear in algebraic geometry. In this paper we survey some results around NCCRs.

MATHEMATICS SUBJECT CLASSIFICATION 2020

Primary 14A22; Secondary 14E30, 18E30

KEYWORDS

Noncommutative crepant resolutions, tilting objects, quotient singularities

ICM INTERNATIONAL CONGRESS OF MATHEMATICIANS 2022 JULY 6–14

Proc. Int. Cong. Math. 2022, Vol. 2, pp. 1354–1391
DOI 10.4171/ICM2022/177

Published by EMS Press

1. INTRODUCTION

In this paper we will give an introduction to noncommutative crepant resolutions with some emphasis on our joint work with Špela Špenko about quotient singularities of reductive groups. Other surveys are [96, 115, 131].

1.1. Notation and conventions

We fix a few notations and definitions which are mostly self explanatory. For simplicity, we assume throughout that k is an algebraically closed field of characteristic zero, although this is often not necessary. In Section 5 we put $k = \mathbb{C}$ when invoking Hodge theory. For us an *algebraic variety* is a possibly singular integral separated scheme of finite type over k.

Modules over rings or sheaves of rings are left modules. Right modules are indicated by $(-)^\circ$. If Λ is a ring then we denote by $D(\Lambda)$ the unbounded derived category of complexes of Λ-modules and by $\mathrm{Perf}(\Lambda)$ its full subcategory of perfect Λ-complexes. If Λ is noetherian then we write $\mathrm{mod}(\Lambda)$ for the category of finitely generated Λ-modules. We also put $\mathcal{D}(\Lambda) = D^b(\mathrm{mod}(\Lambda))$. We use similar notations in the geometric context. If $\mathcal{X}$ is an Artin stack and Λ is a quasicoherent sheaf of rings on $\mathcal{X}$ then $D_{\mathrm{Qch}}(\Lambda)$ is the unbounded derived category of complexes of left Λ-modules with quasicoherent cohomology. The category of perfect Λ-complexes is denoted by $\mathrm{Perf}(\Lambda)$, and we also put $\mathcal{D}(\Lambda) = D^b(\mathrm{coh}(\Lambda))$ when Λ is noetherian. If $\Lambda = \mathcal{O}_{\mathcal{X}}$ then we replace Λ in the notations by $\mathcal{X}$.

A finitely generated R-module M over a normal noetherian domain is said to be *reflexive* if the canonical map $M \mapsto M^{\vee\vee}$ is an isomorphism where $M^\vee = \mathrm{Hom}_R(M, R)$. This implies in particular that M is torsion free. If R a commutative noetherian domain then a *maximal Cohen–Macaulay* R-module is an R-module M such that M_m is maximal Cohen–Macaulay as R_m-module for every maximal ideal m. If R is has finite injective dimension then we say that R is *Gorenstein*. This implies that R is maximal Cohen–Macaulay.

1.2. Crepant resolutions and derived equivalences

Let X be a normal algebraic variety with Gorenstein singularities. A resolution of singularities $\pi : Y \to X$ is said to be *crepant* if $\pi^*\omega_X = \omega_Y$. In some sense, a crepant resolution is the tightest possible smooth approximation of an algebraic variety. Such crepant resolutions need not exist, however. For starters, their existence implies that X has rational singularities [89, COROLLARY 5.24] and this already strong restriction is far from sufficient. For example, the three-dimensional hypersurface singularities

$$x^2 + y^2 + z^2 + w^n = 0 \quad (n \geq 2) \tag{1.1}$$

have crepant resolutions if and only if n is even [111, COROLLARY 1.16]. When crepant resolutions do exist they are generally not unique. For example,

$$xy - zw = 0, \tag{1.2}$$

which corresponds to $n = 2$ in (1.1), has two distinct crepant resolutions given by blowing up (x, z) and (x, w). This is the so-called "Atiyah flop."

Nonetheless, experience has shown that such different crepant resolutions are strongly related. In particular, we have the following result:

Theorem 1.1 ([6, 90], see also Section 5 below). *Assume that X has canonical Gorenstein singularities. Then the Hodge numbers of Y for a crepant resolution $Y \to X$ are independent of Y.*

Kawamata and independently Bondal and Orlov in their lecture at ICM2002 conjectured an analogous categorical result, a variant[1] of which we state below.

Conjecture 1.2 ([22, CONJECTURE 4.4], [76, CONJECTURE 1.2]). Assume X is a normal algebraic variety with Gorenstein singularities and $\pi_i : Y_i \to X$ for $i = 1, 2$ are two crepant resolutions (by schemes or DM-stacks). Then there is an equivalence of triangulated categories $F : \mathcal{D}(Y_1) \cong \mathcal{D}(Y_2)$, *linear over* X (cf. Remark 1.5 below).

The conjecture is known (under some probably unnecessary projectivity hypotheses) in dimension ≤ 3, by the work of Bridgeland [27] (see Section 1.4 below), and for toric varieties, by the work of Kawamata [79]. For symplectic singularities [10], it is true, up to an étale covering of X, by [72, THEOREM 1.6]. Furthermore, it is known for many specific crepant resolutions, e.g., those related by variation of GIT [5, 59, 60] (see also Section 4.2 below).

Remark 1.3. Conjecture 1.2 makes no statement about the nature of the equivalence $\mathcal{D}(Y_1) \cong \mathcal{D}(Y_2)$. In the case of the Atiyah flop, one possible equivalence is given by the Fourier–Mukai functor for the "fiber product kernel" $\mathcal{O}_{Y_1 \times_X Y_2}$ [21, THEOREM 3.6] (see also [18]) but this is far from the only possibility. Furthermore, $\mathcal{O}_{Y_1 \times_X Y_2}$ does not always work as Example 1.4 below shows.

It is now understood, thanks to intuition from mirror symmetry, that the equivalences in Conjecture 1.2 should be canonically associated to paths connecting two points in a topological space called the "stringy Kähler moduli space" (SKMS). In the case of the Atiyah flop, the SKMS is given by $\mathbb{P}^1 - \{0, 1, \infty\}$ [48, 59]. See also [63] and Section 4.3 below. The fact that the asserted equivalence in Conjecture 1.2 is expected to be noncanonical by itself might be the reason that the conjecture seems difficult to prove.

Below $\mathrm{Gr}(d, n)$ is the Grassmannian of d-dimensional subspaces of the n-dimensional vector space k^n.

Example 1.4. The cotangent bundles $T^* \mathrm{Gr}(d, n)$ and $T^* \mathrm{Gr}(n - d, n)$, for complementary Grassmannians with $d \leq n/2$ are crepant resolutions of $\overline{B(d)} := \{X \in M_n(k) \mid X^2 = 0, \mathrm{rk}\, X \leq d\}$ (e.g., [37, §6.1]). According to [37, §6], there is an equivalence $F : \mathcal{D}(T^* \mathrm{Gr}(d, n)) \to \mathcal{D}(T^* \mathrm{Gr}(n - d, n))$, but it is not given by the fiber product kernel (see [77, 103] for the case $(k, n) = (2, 4)$).

1 We have omitted the projectivity hypotheses which appear in the original context and extended the conjecture to DM-stacks which is the natural context as will become clear below.

Remark 1.5. As said, one requires the derived equivalence F in Conjecture 1.2 to be linear over X. On the most basic level, this means the following: let $\mathrm{Perf}(X)$ be the category of perfect complexes on X. Then $\mathcal{D}(Y_1)$, $\mathcal{D}(Y_2)$ are $\mathrm{Perf}(X)$-modules, where $A \in \mathrm{Perf}(X)$ acts as $L\pi_i^* A \overset{L}{\otimes}_{Y_i} -$, for $i = 1, 2$, and we want natural isomorphisms $F(L\pi_1^* A \overset{L}{\otimes}_{Y_1} -) \cong L\pi_2^* A \overset{L}{\otimes}_{Y_2} F(-)$ satisfying the appropriate compatibilities. *To simplify the exposition, we will implicitly assume in the rest of this paper that all constructions satisfy the appropriate linearity hypotheses.*

1.3. Noncommutative rings

Most of the results below will be based on the interplay between algebraic geometry and noncommutative rings. The relation between those subjects was first observed by Beilinson [11]. The connection is via tilting complexes.

Definition 1.6. Let Y be a noetherian scheme. A *partial tilting complex* $\mathcal{T}$ on Y is a perfect complex such that $\mathrm{Ext}^i_Y(\mathcal{T}, \mathcal{T}) = 0$ for $i \neq 0$. A *tilting complex* is a partial tilting complex that *generates* $D_{\mathrm{Qch}}(Y)$ in the sense that its right orthogonal is zero, i.e., $\mathrm{RHom}_Y(\mathcal{T}, \mathcal{F}) = 0$ implies $\mathcal{F} = 0$. A *(partial) tilting bundle* is a (partial) tilting complex which is a vector bundle.

Below we will also use tilting complexes in slightly more general contexts (e.g., DM-stacks). Very general results concerning tilting complexes are [82, THEOREMS 1,2]. For simplicity, we state a slightly dumbed down version of them, although below we will sometimes silently rely on the stronger results in [82]. See also [19,114].

Theorem 1.7 ([82, THEOREMS 1,2]). *If $\mathcal{T}$ is a tilting complex on a noetherian scheme Y then* $\mathrm{RHom}_Y(\mathcal{T}, -)$ *defines an equivalence of categories between $D_{\mathrm{Qch}}(Y)$ and $D(\Lambda^\circ)$ for* $\Lambda = \mathrm{End}_Y(\mathcal{T})$. *Moreover, if Y is regular then Λ has finite global dimension. If, furthermore, Λ is right noetherian then* $\mathrm{RHom}_Y(\mathcal{T}, -)$ *restricts to an equivalence of categories* $\mathcal{D}(Y) \cong \mathcal{D}(\Lambda^\circ)$.

So a tilting complex reduces the homological algebra of Y to the usually noncommutative ring $\Lambda = \mathrm{End}_Y(\mathcal{T})$. In the case of projective space $\mathbb{P}^n$, one can take $\mathcal{T} = \mathcal{O} \oplus \mathcal{O}(1) \oplus \cdots \oplus \mathcal{O}(n)$ [11].

1.4. Bridgeland's result

1.4.1. Flops

Let us return to Conjecture 1.2. In the absence of any specific conjectural construction of the asserted derived equivalence (see Remark 1.3), one may try to use the fact that if π_1, π_2 are projective then Y_1, Y_2 are connected by a sequence of "flops" [78, THEOREM 1], so that it is then sufficient to prove the conjecture for flops. Recall that crepant resolutions $\pi_1 : Y_1 \to X$, $\pi_2 : Y_2 \to X$ form a flop if X has terminal singularities [89, DEFINITION 2.12] and there is a line bundle $\mathcal{L}$ on Y_1, relatively ample for π_1, such that the corresponding[2] line bundle $\mathcal{L}'$ on Y_2 is antiample.

In [27] Bridgeland proves that Conjecture 1.2 is true for three-dimensional flops (see also [38]). The key point is that the fibers of π_1, π_2 have dimension ≤ 1. In the next section we explain a reinterpretation of Bridgeland's proof, following [129].

1.4.2. Maps with fibers of dimension ≤ 1

Assume that $\pi : Y \to X$ is a projective map between noetherian schemes. We impose the following conditions:

(1) $R\pi_*\mathcal{O}_Y = \mathcal{O}_X$.

(2) The fibers of π have dimension ≤ 1.

To simplify the discussion, we will restrict ourselves, furthermore, to the case that $X = \operatorname{Spec} R$ is affine.[3] It turns out that in this case $\operatorname{coh}(Y)$ contains a tilting bundle which is of the form $\mathcal{T} := \mathcal{O}_Y \oplus \mathcal{T}_0$ where $\mathcal{T}_0$ is obtained as an extension

$$0 \to \mathcal{O}_Y^r \to \mathcal{T}_0 \to \mathcal{L} \to 0, \tag{1.3}$$

where $\mathcal{L}$ is an ample line bundle on $\mathcal{O}_Y$ generated by global sections and (1.3) is associated to an arbitrary finite set of generators of $H^1(Y, \mathcal{L}^{-1})$ as R-module (see [129, (3.1)]).

Remark 1.8. Note that by hypothesis (1), $\mathcal{O}_X$, $\mathcal{L}$ are partial tilting bundles on Y such that $\mathcal{O}_X \oplus \mathcal{L}$ generates $D_{\mathrm{Qch}}(Y)$ [129, LEMMA 3.2.2]. Moreover, (2) and the fact that $\mathcal{L}$ is generated by global sections imply $\operatorname{Ext}_Y^{>0}(\mathcal{O}_Y, \mathcal{L}) = 0$. Likewise, (2) implies $\operatorname{Ext}_Y^{>1}(\mathcal{L}, \mathcal{O}_Y) = 0$. The construction of the tilting bundle $\mathcal{T}$ is based on the principle of "killing the remaining backward Ext^1" in the sequence $(\mathcal{O}_X, \mathcal{L})$ by a so-called "semiuniversal extension." This principle extends to longer sequences. See, e.g., [61, LEMMA 2.4], [62, LEMMA 3.1]. See also Section 3.4 below for another application.

So if we put $\Lambda = \operatorname{End}_Y(\mathcal{T})$, then we have $\operatorname{End}_Y(\mathcal{T}^\vee) = \Lambda^\circ$, and from Theorem 1.7 we obtain equivalences[4]

$$\operatorname{RHom}_Y(\mathcal{T}, -) : \mathcal{D}(Y) \cong \mathcal{D}(\Lambda^\circ), \quad \operatorname{RHom}_Y(\mathcal{T}^\vee, -) : \mathcal{D}(Y) \cong \mathcal{D}(\Lambda). \tag{1.4}$$

To understand (1.4), we can ask what Λ looks like.

Example 1.9. Consider again the Atiyah flop (1.2). In this case $R = k[x, y, z, w]/(xy - zw)$. This is a toric singularity, and one can check that its class group is $\mathbb{Z}$ with

2 This makes sense since Y_1 and Y_2 are isomorphic in codimension one.

3 In [129] X is assumed to be quasiprojective.

4 It is a fact that $\mathcal{T}$ is tilting if and only if $\mathcal{T}^\vee$ is tilting. The only nontrivial part is the generation property. To this end one may use that $\mathcal{T}$ generates $D_{\mathrm{Qch}}(Y)$ if and only if $\operatorname{Perf}(Y)$ is the smallest épaisse subcategory of $D_{\mathrm{Qch}}(Y)$ containing $\mathcal{T}$ [105, LEMMA 2.2], together with the fact that $(-)^\vee$ is an autoequivalence of $\operatorname{Perf}(Y)$.

generator $I = (x, z)$. The inverse of I is the fractional ideal $I^{-1} = x^{-1}(x, w)$. The ring Λ turns out to be the same (up to isomorphism) for both crepant resolutions of Spec R,

$$\Lambda = \begin{pmatrix} R & I \\ I^{-1} & R \end{pmatrix}. \tag{1.5}$$

Interestingly, Λ is built up from the three indecomposable graded maximal Cohen–Macaulay R-modules: R, I, and I^{-1}. In particular, Λ is itself Cohen–Macaulay as R-module. This last fact turns out to be true more generally.

Theorem 1.10. *Assume that $X =$ Spec R is a normal Gorenstein variety. Assume that there exists a projective crepant resolution of singularities $\pi : Y \to X$ such that the dimensions of the fibers of π are ≤ 1. Let $\mathcal{T}$ be the tilting bundle defined above*[5] *and put $T = \Gamma(Y, \mathcal{T})$. Then $\Lambda = \mathrm{End}_Y(\mathcal{T}) = \mathrm{End}_R(T)$. Furthermore, Λ and T are maximal Cohen–Macaulay R-modules.*

Proof. The fact that $\Lambda = \mathrm{End}_Y(\mathcal{T})$ is maximal Cohen–Macaulay follows from **[129, LEMMA 3.2.9]** (see also **[70, THEOREM 1.5]**, stated as Theorem 2.6 below). Now T is maximal Cohen–Macaulay because it is a direct summand of Λ as R-modules. Functoriality yields a map $i : \Lambda \to \mathrm{End}_R(T)$ which is an isomorphism in codimension one (since the singular locus of X has codimension ≥ 2, as X is normal). Since Λ is maximal Cohen–Macaulay, it is reflexive and hence i must be an isomorphism. ∎

This result applies in particular if X has dimension 2 or if it is of dimension 3 with terminal singularities since then the condition on the dimension of the fibers is automatic.

Let us now assume that X in Conjecture 1.2 is 3-dimensional and π_1, π_2 form a flop (see Section 1.4.1). We will still be assuming that $X =$ Spec R is affine for simplicity. For $i = 1, 2$, we then have tilting bundles $\mathcal{T}_i$ on Y_i defined via (1.3), using $\mathcal{L}$ on Y_1 and $(\mathcal{L}')^{-1}$ on Y_2 (see Section 1.4.1 for $\mathcal{L}, \mathcal{L}'$). Let $(\Lambda_i)_{i=1,2}$ be the corresponding endomorphism rings. In this case Conjecture 1.2 follows from

$$\mathcal{D}(Y_1) \overset{(1.4)}{\cong} \mathcal{D}(\Lambda_1^\circ), \quad \mathcal{D}(Y_2) \overset{(1.4)}{\cong} \mathcal{D}(\Lambda_2), \quad \Lambda_1^\circ \overset{\text{Morita}}{\cong} \Lambda_2.$$

The asserted Morita equivalence is obtained in **[129, §4.4]** using the local structure of 3-dimensional terminal singularities (see **[88, EXAMPLE 2.3]**). Nowadays we may use **[67, COROLLARY 8.8]** (see also **[68, THEOREM 1.5]**) combined with **[70, THEOREM 1.5]** (stated as Theorem 2.6 below) to obtain that in any case $\Lambda_1, \Lambda_2^\circ$ are derived equivalent.

At the end of the day, we find that the two crepant resolutions Y_1, Y_2 of X are derived equivalent to the same noncommutative ring (either Λ_1° or Λ_2). It turns out to be fruitful to think of this intermediate noncommutative ring as a *third crepant resolution* of X, or of R, namely a *noncommutative crepant resolution*.

5 As we have stated in Section 1.2, the fact that X has a crepant resolution implies that it has rational singularities by **[89, COROLLARY 5.24]**. Thus in particular $R\pi_*\mathcal{O}_Y = \mathcal{O}_X$.

2. NONCOMMUTATIVE (CREPANT) RESOLUTIONS

2.1. Generalities

Below R is a normal noetherian domain with quotient field K. We denote by $\mathrm{ref}(R)$ the category of reflexive R-modules and if Λ is a reflexive R-algebra then $\mathrm{ref}(\Lambda)$ is the category of Λ-modules which are reflexive as R-modules. A *reflexive Azumaya algebra* [98] Λ is a reflexive R-algebra which is Azumaya in codimension one. A reflexive Azumaya algebra Λ is said to be trivial if it is of the form $\mathrm{End}_R(M)$ for M a reflexive R-module. In that case $\mathrm{ref}(R)$ and $\mathrm{ref}(\Lambda)$ are equivalent. This is a particular case of "reflexive Morita equivalence" which is defined in the obvious way.

Definition 2.1. A *twisted noncommutative resolution* of R is a reflexive Azumaya algebra Λ over R such that $\operatorname{gl\,dim} \Lambda < \infty$. If Λ is trivial then Λ is said to be a *noncommutative resolution (NCR)* of R.

Definition 2.2. Assume that R is Gorenstein. A *twisted noncommutative crepant resolution* Λ of R is a twisted NCR of R which is in addition a Cohen–Macaulay R-module. If Λ is an NCR then such a Λ is said to be a *noncommutative crepant resolution (NCCR)* of R.

The point of these definitions is that they provide reasonable noncommutative substitutes for "regularity," "birationality," and "crepancy." This is explained in more detail in [129, §4].

Remark 2.3. We will sometimes use the concepts introduced in Definitions 2.1, 2.2 for schemes, possibly nonaffine. It is then understood that they reduce to the affine concepts, when restricting to open affine subschemes.

Remark 2.4. In the sequel we will be mostly concerned with NCCRs and thus the other definitions are mainly provided for context. Twisted NCCRs are natural generalizations of NCCRs, but the good properties of NCCRs (sometimes conjectural) are usually not shared by twisted NCCRs. See, e.g., Example 5.11 below. The definition of a (twisted) NCR is more tentative. In particular, the normality and reflexivity hypotheses do not seem very relevant. For example, there is a nice theory of noncommutative resolutions of nonnormal singularities in dimension one [96].

Example 2.5. It follows from Theorem 1.10 and Theorem 1.7 that if there exists a projective crepant resolution of singularities $\pi : Y \to X$ such that the dimensions of the fibers of π are ≤ 1 then R has an NCCR.

We mention the following theorem which gives another indication that the definition of an NCCR is the "correct one."

Theorem 2.6 ([70, THEOREM 1.5]). *Let $f : Y \to \operatorname{Spec} R$ be a projective birational morphism between Gorenstein varieties. Suppose that Y is derived equivalent to some ring Λ, then f is a crepant resolution if and only if Λ is an NCCR of R.*

The following conjecture is a natural extension of Conjecture 1.2.

Conjecture 2.7 ([128, CONJECTURE 4.6]). All crepant resolutions of X (commutative as well as noncommutative) are derived equivalent.

We have the following result which is proved in the same way as the 3-dimensional McKay correspondence [28].

Proposition 2.8 ([128, THEOREM 6.3.1, PROPOSITION 6.2.1]). *If X has three-dimensional Gorenstein singularities and it has an NCCR Λ, then it has a projective crepant resolution $Y \to X$ such that Λ and Y are derived equivalent.*

Proposition 2.9. *Conjecture 2.7 is true if X has dimension three, if we restrict to projective crepant resolutions.*

Proof. If X has an NCCR Λ then by Proposition 2.8 Λ is derived equivalent to a crepant resolution. Hence we are reduced to Bridgeland's result (see Section 1.4.1). Alternatively, to have a very nice direct argument that any two NCCRs are derived equivalent in dimension three, we may use [67, COROLLARY 8.8] (see also [68, THEOREM 1.5]). ■

Proposition 2.8 is false for arbitrary three-dimensional Gorenstein singularities as was shown by Dao [41].

Proposition 2.10 ([41, THEOREM 3.1, REMARK 3.2]). *Assume S is a regular local ring which is equicharacteristic or unramified, $0 \neq f \in S$ and $R = S/(f)$ is normal. If $\dim R = 3$ and R is factorial then R has no NCCR.*

Example 2.11. It turns out that there are 3-dimensional factorial hypersurface singularities that admit a crepant resolution. A concrete example is given by $R = k[[x_0, x_1, x_2, x_3]]/(x_0^4 + x_1^3 + x_2^3 + x_3^3)$ [100, THEOREM A,B]. In particular, a crepant resolution of such R does not admit a tilting complex by Theorem 2.6.

If X is a normal Gorenstein algebraic variety with a crepant resolution then it has rational singularities [89, COROLLARY 5.24]. A similar result is true for NCCRs.

Theorem 2.12 ([122, THEOREM 1.1]). *Let R be a normal finitely generated Gorenstein k-algebra. If R has a twisted NCCR then it has rational singularities.*

The actual result proved in [122] applies in a more general context and this has been further exploited in [64, 65] (see also [42, COROLLARY 1.7]).

Remark 2.13. In order to deal with singularities with a singular minimal model, Iyama and Wemyss generalize the definition of an NCCR [69, 70, 132] to certain rings, of possibly infinite global dimension, called *maximal modification algebras* (MMAs). Remarkably, many of the results about NCCRs extend to MMAs. However, in this overview we will restrict ourselves for simplicity to NCCRs.

2.2. Relation with crepant categorical resolutions

We conjecture that noncommutative crepant resolutions are examples of "strongly crepant categorical resolutions" as introduced by Kuznetsov in [93]. However, we can only prove this in special cases.

Let X be an algebraic variety. A *categorical* resolution [93] of $\mathcal{D}(X)$ is a "smooth" triangulated category $\tilde{\mathcal{D}}$ together with functors

$$\pi_* : \tilde{\mathcal{D}} \to \mathcal{D}(X), \quad \pi^* : \mathrm{Perf}(X) \to \tilde{\mathcal{D}}$$

which are adjoint (i.e., $\mathrm{Hom}_{\tilde{\mathcal{D}}}(\pi^* A, B) \cong \mathrm{Hom}_{\mathcal{D}(X)}(A, \pi_* B)$ for $A \in \mathrm{Perf}(X)$, $B \in \tilde{\mathcal{D}}$) such that the natural transformation $\mathrm{id}_{\mathrm{Perf}(X)} \to \pi_*\pi^*$, obtained by putting $B = \pi^* A$, is an isomorphism. This implies in particular that π^* is fully faithful. There is some variation possible in the definition of smoothness. For us it means that $\tilde{\mathcal{D}}$ is equivalent to the derived category of perfect modules over a smooth DG-algebra [75, DEFINITION 2.23].

Remark 2.14. If $\pi : Y \to X$ is a resolution of singularities of X then $(\mathcal{D}(Y), R\pi_*, L\pi^*)$ is a categorical resolution of $\mathcal{D}(X)$ if and only if X has rational singularities. Remarkably, however, it has been shown in [94] that $\mathcal{D}(Y)$ can be suitably enlarged to yield a categorical resolution. On the other hand, this result cannot be extended to more general dg-categories [54].

Following [93], we say that a categorical resolution $(\tilde{\mathcal{D}}, \pi_*, \pi^*)$ of $\mathcal{D}(X)$ is *weakly crepant* if π^* is both a left and a right adjoint to π_*.

There is also a notion of a *strongly crepant categorical* resolution for which we need the notion of a *relative Serre functor*. To define this, assume that X is Gorenstein and that $\tilde{\mathcal{E}}$ is a smooth triangulated category which is a $\mathrm{Perf}(X)$-module. We will denote the action of $A \in \mathrm{Perf}(X)$ on $B \in \tilde{\mathcal{E}}$ as $A \otimes_X B$ and we assume that $- \otimes -$ is exact in both arguments. We also assume that the functor $\mathrm{Perf}(X) \to \tilde{\mathcal{E}} : A \mapsto A \otimes_X B$ has a right adjoint $\tilde{\mathcal{E}} \to \mathcal{D}(X)$ which we denote by $\mathrm{R}\mathcal{H}om_{\tilde{\mathcal{E}}/X}(B, -)$. That is, for $C \in \tilde{\mathcal{E}}$ we have functorial isomorphisms

$$\mathrm{Hom}_{\tilde{\mathcal{E}}}(A \otimes_X B, C) \cong \mathrm{Hom}_X\big(A, \mathrm{R}\mathcal{H}om_{\tilde{\mathcal{E}}/X}(B, C)\big).$$

An autoequivalence $S_{\tilde{\mathcal{E}}/X} : \tilde{\mathcal{E}} \to \tilde{\mathcal{E}}$ is said to be a *relative Serre functor* for $\tilde{\mathcal{E}}/X$ if there are functorial isomorphisms

$$\mathrm{R}\mathcal{H}om_X\big(\mathrm{R}\mathcal{H}om_{\tilde{\mathcal{E}}/X}(B, C), \mathcal{O}_X\big) \cong \mathrm{R}\mathcal{H}om_{\tilde{\mathcal{E}}/X}(C, S_{\tilde{\mathcal{E}}/X} B)$$

for $B, C \in \tilde{\mathcal{E}}$. We say that $\tilde{\mathcal{E}}/X$ is *strongly crepant* if the identity functor $\tilde{\mathcal{E}} \to \tilde{\mathcal{E}}$ is a relative Serre functor.

A *strongly crepant categorical resolution* of X is a quadruple $(\tilde{\mathcal{D}}, \pi_*, \pi^*, \otimes_X)$ such that $(\tilde{\mathcal{D}}, \pi_*, \pi^*)$ is a categorical resolution of X, $- \otimes_X -$ is a $\mathrm{Perf}(X)$-module structure on $\tilde{\mathcal{D}}$ such that $\tilde{\mathcal{D}}/X$ is strongly crepant and π^* is $\otimes_X$-linear. The last condition means that for $A, B \in \mathrm{Perf}(X)$ we have functorial isomorphisms $A \otimes_X \pi^* B \cong \pi^*(A \otimes_X B)$ satisfying the appropriate compatibilities.

It is shown in [93, §3] that a strongly crepant categorical resolution is weakly crepant, and, moreover, that if $\pi : Y \to X$ is a crepant resolution in the usual sense then $(\mathcal{D}(Y), R\pi_*, L\pi^*, L\pi^*(-) \otimes_Y -)$ is a strongly crepant categorical resolution of X.

The following easy lemma, which is an extension of [102, EXAMPLE 5.3], shows that, under suitable conditions, rings of the form $\operatorname{End}_R(M)$ form crepant categorical resolutions. If M is an R-module then $\operatorname{add}(M)$ is the category spanned by modules which are direct summands of some $M^{\oplus n}$.

Lemma 2.2.1. *Assume that $X = \operatorname{Spec} R$ is an algebraic variety and let M be a finitely generated R-module such that $\Lambda = \operatorname{End}_R(M)$ has finite global dimension. Then $\mathcal{D}(\Lambda)$ is smooth. Assume in addition that $R \in \operatorname{add}(M)$. Then*

$$\operatorname{Perf}(R) \to \mathcal{D}(\Lambda) : N \mapsto M \overset{L}{\otimes}_R N \tag{2.1}$$

yields a categorical resolution of singularities of X (since $\operatorname{Perf}(R) \cong \operatorname{Perf}(X)$). Moreover, assuming furthermore that R is normal Gorenstein:

(1) *if M is maximal Cohen–Macaulay then this categorical resolution is weakly crepant;*

(2) *if Λ is an NCCR then this categorical resolution is strongly crepant.*

Note that if (2) holds then M is maximal Cohen–Macaulay since we have assumed that $R \in \operatorname{add}(M)$.

The hypotheses of Lemma 2.2.1 are actually too strong. For example, an NCCR is always a strongly crepant categorical resolution in dimension ≤ 3. This follows from Proposition 2.15 below which can be proved using the methods of [67, 68].

Proposition 2.15. *Assume that $\Lambda = \operatorname{End}_R(M)$ is an NCCR and* $\dim R \leq 3$ *then*

$$\operatorname{Ext}^i_\Lambda(M, M) = 0 \quad \text{for } i > 0. \tag{2.2}$$

Proof. For the benefit of the reader, we give a proof. We may assume that R is local of dimension 3 (the case dim ≤ 2 is easy). By the Auslander–Buchsbaum formula [67, PROPOSITION 2.3] Λ has global dimension 3. Since M is reflexive, it has depth ≥ 2, and hence, again by the Auslander–Buchsbaum formula, it has projective dimension ≤ 1 over Λ and, moreover, it is projective over Λ in codimension 2.

Hence we have a projective resolution of M as Λ-module

$$0 \to P_1 \to P_0 \to M \to 0.$$

Applying $\operatorname{Hom}_\Lambda(-, M)$, we get a long exact sequence of R-modules

$$0 \to R \to \operatorname{Hom}_\Lambda(P_0, M) \to \operatorname{Hom}_\Lambda(P_1, M) \to \operatorname{Ext}^1_\Lambda(M, M) \to 0. \tag{2.3}$$

Assume $\operatorname{Ext}^1_\Lambda(M, M) \neq 0$. Since M is projective over Λ in codimension two, $\operatorname{Ext}^1_\Lambda(M, M)$ is finite dimensional and hence it has depth 0 as R-module. On the other hand, since $\operatorname{Hom}_\Lambda(P_i, M)$ is reflexive as R-module, it has depth ≥ 2. Finally, R being maximal Cohen–Macaulay has depth 3. One may verify that these depth restrictions are incompatible with (2.3). ■

It seems too much to hope for that (2.2) would always be true, but the lack of time has prevented us from seriously looking for a counterexample. On the other hand, we are sufficiently optimistic to make the following conjecture.

Conjecture 2.16. If $X = \operatorname{Spec} R$ is a normal algebraic variety with Gorenstein singularities then an NCCR of R always yields a strongly crepant categorical resolution of X.

To prove this conjecture, one would have to construct for an NCCR Λ of R a partial tiling complex $P^\bullet$ of Λ-modules such that $\operatorname{RHom}_\Lambda(P^\bullet, P^\bullet) = R$.

Remark 2.17. The strongly crepantness of $\mathcal{E}/X$ as defined above is independent of the resolution property. One may check that if Λ/R is a twisted NCCR then $\mathcal{D}(\Lambda)$ is strongly crepant over Spec R. But one may also check that it is not a categorical resolution.

3. CONSTRUCTIONS OF NONCOMMUTATIVE CREPANT RESOLUTIONS

3.1. Quotient singularities

Here we will restrict ourselves to quotient singularities for finite groups. Quotient singularities for (infinite) reductive groups will be covered in Section 4.

If G is a finite group and W is a faithful finite-dimensional unimodular (i.e., $\det W = k$) representation of G then the skew group ring

$$\operatorname{Sym}(W)\#G = \operatorname{End}_{\operatorname{Sym}(W)^G}\big(\operatorname{Sym}(W)\big)$$

is an NCCR for $R = \operatorname{Sym}(W)^G$ (which is Gorenstein because of the unimodularity hypothesis).

In dimension ≤ 3 such quotient singularities always have a crepant resolution by the celebrated BKR-theorem [28]. In higher dimension this is not so. The simplest counterexample is given by $\mathbb{Z}_2$ acting with weights $(-1, -1, -1, -1)$ on $W = k^4$ because in that case R is $\mathbb{Q}$-factorial and terminal. See, e.g., [1].

3.2. Crepant resolutions with tilting complexes

In case R is a normal Gorenstein domain and $Y \to \operatorname{Spec} R$ is a crepant resolution and $\mathcal{T}$ is a tilting complex on Y then $\mathcal{E}nd_Y(\mathcal{T})$ is an NCCR of R by Theorem 2.6. Conversely, assuming a crepant resolution exists, any NCCR has to be of this form if we accept Conjecture 2.7 ($\mathcal{T}$ is the dual of the image of Λ under the asserted derived equivalence $\mathcal{D}(\Lambda) \cong \mathcal{D}(Y)$).

This is a very general method for constructing NCCRs. Note, however, that even in dimension three there may be crepant resolutions without tilting complex. See Example 2.11. Furthermore, as indicated in Section 3.1, there are normal Gorenstein singularities that admit an NCCR but not a crepant resolution.

Example 3.1. A textbook example where this method works very well is the case of determinantal varieties [34, 36]. Let $n \geq 1$ and $0 \leq l < n$. Let $X_{l,n} = \operatorname{Spec} R$ be the varieties of matrices in $\mathrm{Hom}_k(k^n, k^n) = M_{n\times n}(k)$ which have rank $\leq l$. It is a classical result that X is Gorenstein. It is also well known that X has a crepant Springer type resolution given by

$$Y = \{(\phi, V) \mid V \in \mathrm{Gr}(l,n), \phi \in \mathrm{Hom}_k(k^n, V)\},$$

where $\pi : Y \to X$ sends (ϕ, V) to the composition of ϕ with the inclusion $V \hookrightarrow k^n$. If $\mathcal{R}$ denotes the universal subbundle on $\mathrm{Gr}(l,n)$ then Y is the vector bundle $\mathrm{Hom}(k^n, \mathcal{R})$ (i.e., $Y = \underline{\mathrm{Spec}}\,\mathrm{Sym}((\mathcal{R}^{\oplus n})^\vee)$). Using Bott's theorem, one computes that the Kapranov tilting bundle on $\mathrm{Gr}(l,n)$ [73] (see also [35][53] for the case of finite characteristic) pulls back to a tilting bundle on Y, which then gives an NCCR of R. For other approaches to this example, see [49] and Theorem 4.10 below.

Alas, things are often more complicated. For determinantal varieties associated to symmetric of skew-symmetric matrices, the Springer type resolutions are not crepant so a tilting bundle on them only gives an NCR (see [136]). NCCRs of such generalized determinantal varieties will be obtained in Section 4 using a different approach.

Example 3.2. Another beautiful and much deeper example [13, 16] is given by cotangent bundles of (partial) flag varieties $T^*(G/P)$. If P is a Borel subgroup of G then this is a crepant resolution of the nilpotent cone in $\mathrm{Lie}(G)$. In general, they are crepant resolutions of closures of Richardson orbits [104]. It is shown in [13, 16] that $T^*(G/P)$ has a tilting bundle but it is not obtained as the pullback of a tilting bundle on G/P. In fact, the construction of the tilting bundle is highly nontrivial. To explain the construction, it is useful to exhibit a slightly different point of view on tilting bundles.

Let Y be a noetherian scheme. If $\mathcal{A}$ is a quasicoherent sheaf of algebras on Y and $A = \Gamma(Y, \mathcal{A})$ then we say that $\mathcal{A}$ is *derived affine* if $A = R\Gamma(Y, \mathcal{A})$ and the right orthogonal to $\mathcal{A}$ in $D_{\mathrm{Qch}}(\mathcal{A})$ is zero. In that case $R\Gamma(Y, -)$ defines an equivalence of categories between $D_{\mathrm{Qch}}(\mathcal{A})$ and $D(A)$. It is not difficult to see that a vector bundle $\mathcal{T}$ on Y which has everywhere nonzero rank is a tilting bundle, provided $\mathcal{E}nd_Y(\mathcal{T})$ is derived affine.

We say that Y is derived $\mathcal{D}$-affine if $\mathcal{D}_Y$ is derived affine where $\mathcal{D}_Y$ is the sheaf of differential operators on Y. In characteristic > 0 we mean by $\mathcal{D}_Y$ the sheaf of *crystalline* differential operators, i.e., differential operators which may be expressed in terms of derivations, without using divided powers.

Now let $Z = G/P$. The Bernstein–Beilinson theorem [12], valid in characteristic zero, states that Z is even "$\mathcal{D}$-affine" meaning that the equivalence $R\Gamma(Z, -)$ is also compatible with the natural t-structures. This is false in characteristic > 0. However, Z is still derived $\mathcal{D}$-affine [16, THEOREM 3.2] whenever p is strictly bigger than the Coxeter number, which we will assume now.

We will give a rough sketch how this is used in [13, 16] to construct a tilting bundle on $Y = T^*Z$. Let us first assume that the characteristic of k is $p > 0$. To indicate this, we will adorn our notations with $(-)_p$. In that case $\mathcal{D}_{Z_p}$ is coherent as a module over its center which is equal to $(\mathrm{Sym}_{Z_p}\, \mathcal{E}_p)^{(1)}$ where $(-)^{(1)}$ denotes the Frobenius twist, and $\mathcal{E}_p$ is the tangent bundle on Z_p. Hence we may view $\mathcal{D}_{Z_p}$ as a sheaf of coherent algebras $\tilde{\mathcal{D}}$ on $\underline{\mathrm{Spec}}(\mathrm{Sym}_{Z_p}\, \mathcal{E}_p)^{(1)} = Y_p^{(1)}$ where $Y_p = T^*Z_p$. The sheaf $\tilde{\mathcal{D}}$ is still derived affine.

Now $\tilde{\mathcal{D}}$ is not of the form $\mathcal{E}nd_{Y_p^{(1)}}(\mathcal{T}_p^{(1)})$. However, if we let $\hat{Y}_p$ be the formal completion of Y_p at the zero section then it turns out that the restriction $\hat{\mathcal{D}}$ of $\tilde{\mathcal{D}}_p$ to $\hat{Y}_p^{(1)}$ is of the form $\mathcal{E}nd_{\hat{Y}_p}(\hat{\mathcal{T}}_p)^{(1)}$ for a vector bundle $\hat{\mathcal{T}}_p$ on $\hat{Y}_p$. Moreover, $\hat{\mathcal{D}}$ is still derived affine and so $\hat{\mathcal{T}}_p$ is a tilting bundle on $\hat{Y}_p$. Then one uses deformation theory[6] to lift $\hat{\mathcal{T}}_p$ to a tilting bundle $\hat{\mathcal{T}}$ in characteristic zero. Finally, one may use the fact that $Y = T^*Z$ (as a vector bundle) admits a nice G_m action to conclude by [72, THEOREM 1.8] that $\hat{\mathcal{T}}$ is actually the completion of a tilting bundle $\mathcal{T}$ on Y.

Hidden behind this construction is the fact that $\mathcal{D}_Z$ is, in some sense, a canonical noncommutative deformation of the symplectic variety T^*Z. If Y is a general symplectic variety then one may try to construct a noncommutative deformation using Fedosov quantization. This general idea has been used by Bezrukavnikov and Kaledin to prove an analogue of the BKR theorem [28] for crepant resolutions of symplectic quotient singularities [14] and by Kaledin to prove a suitable version of Conjecture 1.2 [72] for general symplectic singularities. To apply the method, one needs to be able to do Fedosov quantization in finite characteristic, a problem which has been solved to some extent in [15].

3.3. Resolutions with partial tilting complexes

Assume R is a normal Gorenstein domain with rational singularities and $Y \to \operatorname{Spec} R$ is a resolution which is not crepant. A strengthening of Conjecture 2.7 inspired by [93] is that NCCRs are minimal in a categorical sense, i.e., their derived category embeds inside $\mathcal{D}(Y)$. This means that they are obtained as $\mathcal{E}nd_Y(\mathcal{T})$ for a partial tilting complex $\mathcal{T}$ on Y. For a very general result in this direction, see [93, THEOREM 2]. We will restrict ourselves to a special case which will be useful in Section 5 and which can be easily proved directly.

Proposition 3.3 ([93]). *Let Z be a smooth projective variety with ample line bundle $\mathcal{O}_Z(1)$ and let $X = \operatorname{Spec} R$ be the corresponding cone. Assume $\omega_Z = \mathcal{O}_Z(-n)$ for $n \geq 1$. Then R is Gorenstein. Moreover, a resolution of singularities $\pi : Y \to X$ of X is given by the line bundle over Z associated to $\mathcal{O}_Z(1)$. Assume $\mathcal{E} \in \mathcal{D}(Z)$ is such that:*

(1) $\operatorname{Ext}^i_Z(\mathcal{E}, \mathcal{E}(m)) = 0$ *for* $i > 0$ *and* $m \geq 0$;

(2) $\operatorname{Ext}^i_Z(\mathcal{E}, \mathcal{E}(m)) = 0$ *for* $i \geq 0$ *and* $m \in \{-1, \dots, -n+1\}$;

(3) $\mathcal{E} \oplus \mathcal{E}(1) \oplus \cdots \oplus \mathcal{E}(n-1)$ *is a generator for* $D_{\mathrm{Qch}}(Z)$.

Let $\gamma : Y \to Z$ be the projection map and put $\mathcal{T} = \gamma^*\mathcal{E}$. Then $\operatorname{End}_Y(\mathcal{T})$ is an NCCR of R.

Proof. We write $\mathcal{T}(m) = \gamma^*(\mathcal{E}(m))$. Then we have

$$\operatorname{RHom}_Y(\mathcal{T}, \mathcal{T}(m)) = \bigoplus_{l \geq 0} \operatorname{RHom}_Z(\mathcal{E}, \mathcal{E}(m+l)). \tag{3.1}$$

6 Tilting bundles have in particular vanishing $\operatorname{Ext}^{1,2}$. Hence by classical deformation theory they are unobstructed and rigid.

Using (1) and (2), we deduce in particular that $\bar{\mathcal{T}} := \mathcal{T} \oplus \cdots \oplus \mathcal{T}(-n+1)$ is partial tilting (and hence this is also the case for $\mathcal{T}$). Furthermore, from (3) we obtain $\bar{\mathcal{T}}^{\perp} = 0$. So $\bar{\mathcal{T}}$ is in fact tilting. Put $\Lambda = \operatorname{End}_Y(\mathcal{T})$, $\bar{\Lambda} = \operatorname{End}_Y(\bar{\mathcal{T}})$. By Theorem 1.7, $\bar{\Lambda}$ has finite global dimension.

Via the decomposition (3.1) Λ is an $\mathbb{N}$-graded ring. Put $\Lambda_{\geq u} = \bigoplus_{m \geq u} \Lambda_m$. Then (as ungraded rings) we have

$$\bar{\Lambda} = \begin{pmatrix} \Lambda & \Lambda_{\geq 1} & \dots & \Lambda_{\geq n-1} \\ \Lambda & \Lambda & \dots & \Lambda_{\geq n-2} \\ \vdots & \vdots & \ddots & \vdots \\ \Lambda & \Lambda & \dots & \Lambda_{\geq 1} \\ \Lambda & \Lambda & \dots & \Lambda \end{pmatrix}.$$

If we put $\Gamma = M_n(\Lambda)$ then $\bar{\Lambda} \subset \Gamma$ and, moreover, Γ is (left and right) projective over $\bar{\Lambda}$ and in addition the multiplication map $\Gamma \otimes_{\bar{\Lambda}} \Gamma \to \Gamma$ is an isomorphism (it is a surjective map between projective Γ-modules of the same rank). We claim that Γ (and hence Λ) has finite global dimension. Indeed, if M is a right Γ-module and $P^{\bullet} \to M$ is a finite projective resolution of M as $\bar{\Lambda}$-module (which exists since $\operatorname{gl\,dim} \bar{\Lambda} < \infty$) then $\Gamma \otimes_{\bar{\Lambda}} P^{\bullet}$ is a finite Γ-projective resolution of $\Gamma \otimes_{\bar{\Lambda}} M = \Gamma \otimes_{\bar{\Lambda}} \Gamma \otimes_{\Gamma} M \cong M$.

Moreover, for $i > 0$,

$$\begin{aligned} \operatorname{Ext}^i_R\big(\operatorname{End}_Y(\mathcal{T}), \omega_R\big) &= \operatorname{Ext}^i_X\big(\pi_* \mathcal{E}nd_Y(\mathcal{T}), \omega_X\big) \\ &= \operatorname{Ext}^i_Y\big(\mathcal{E}nd_Y(\mathcal{T}), \omega_Y\big) \\ &= \operatorname{Ext}^i_Y\big(\mathcal{T}, \mathcal{T}(-n+1)\big) \\ &= 0, \end{aligned}$$

where in the second line we have used Grothendieck duality, in the third line the easily verified fact that $\omega_Y = \gamma^*(\omega_Z(1))$, and in the fourth line (3.1) and (1)–(2). It follows that $\operatorname{End}_Y(\mathcal{T})$ is maximal Cohen–Macaulay over R. ■

3.4. Three-dimensional affine toric varieties

For simplicity, we define an affine toric variety as $X = \operatorname{Spec} R$ where $R = k[\sigma^{\vee} \cap M]$ where M is a lattice and $\sigma^{\vee}$ is a strongly convex full dimensional lattice cone in $M_{\mathbb{R}}$. Such an R is Gorenstein if there exists $m \in M$ such that σ (the dual cone of $\sigma^{\vee}$) is spanned by lattice vectors $x \in M^{\vee}$ satisfying $\langle x, m \rangle = 1$. The lattice polytope associated to R is defined as $P = \sigma \cap \langle m, - \rangle$.

In this case there is the following beautiful result by Broomhead [32, THEOREM 8.6].

Theorem 3.4. *The coordinate ring of a 3-dimensional Gorenstein affine toric variety admits a toric NCCR.*

By a toric NCCR we mean that the reflexive module defining the NCCR is isomorphic to a sum of ideals. Broomhead's proof uses the theory of "dimer models" which is possible thanks to the combinatorics [57, 66]. A proof not using dimer models but using this combinatorics directly was given in [120].

A different method for constructing NCCRs for affine Gorenstein toric varieties was given in [119] and is based on a standard fact from toric geometry:

Lemma 3.5. *A subdivision of σ obtained by a regular triangulation of P with no extra vertices yields a projective crepant resolution of* Spec R *by a toric Deligne–Mumford stack* [24]. *If* $\dim X \leq 3$ *then such a crepant resolution has fibers of dimension* ≤ 1.

In dimension ≤ 3 one may then, starting from a sequence of generating line bundles, construct a tilting bundle using the principle of "killing backward Ext^1's" (see Remark 1.8).

While this method yields an NCCR, it generally does not yield a toric one. On the other hand, it is also applicable to some higher dimensional toric singularities which do not have a toric NCCR.

Example 3.6 ([116, §9.1], [119, EXAMPLE 6.4]). Let $T = G_m^2$ be the two-dimensional torus and (after the identifying the character group $X(T)$ of T with $\mathbb{Z}^2$) consider the vector space W with weights $(3,0)$, $(1,1)$, $(0,3)$, $(-1,0)$, $(-3,-3)$, $(0,-1)$. Put $R := \mathrm{Sym}(W)^T = k[x_1, x_2, x_3, x_4, x_5, x_6]^T = k[x_2x_4x_6, x_1x_3x_5, x_1x_4^3, x_3x_6^3, x_2^3x_5] \cong k[a,b,c,d,e]/(a^3b - cde)$. Clearly, R is the coordinate ring of a 4-dimensional affine toric variety, but it was shown in [116, §9.1] that R does not have a toric NCCR.

On the other hand, by [119, PROPOSITION 6.1], R does have a nontoric NCCR. In [119, EXAMPLE 6.4] an explicit NCCR is constructed which is given by a reflexive module which is the direct sum of 12 modules of rank 1 and 1 module of rank 2.

We conjecture:

Conjecture 3.7. An affine Gorenstein toric variety always has an NCCR.

Besides Theorem 3.4 this conjecture is also true for "quasisymmetric GIT quotients" for tori. See Corollary 4.7 below.

By [119, THEOREM A.1], the Grothendieck group of the DM-stack exhibited in Lemma 3.5 has rank $\mathrm{Vol}(P)$. This suggests the following conjecture:

Conjecture 3.8. The number of indecomposable summands in the reflexive module defining an NCCR of R is equal to $\mathrm{Vol}(P)$.

3.5. Mutations

It follows from the minimal model program that the number of crepant resolutions of an algebraic variety is finite.[7] On the other hand, NCCRs can be modified by a process called "mutation" which is closely related to flopping of crepant resolutions. The difference is that the mutation process generally leads to an infinite number of different NCCRs (see, however, Example 3.12 below).

The following definitions and results are taken from [69]. Let R be a normal Gorenstein ring. Let M be a reflexive R-module such that $\Lambda = \mathrm{End}_R(M)$ is an NCCR and let

7 I thank Shinnosuke Okawa for explaining to me how this follows from [17].

$0 \neq N \in \mathrm{add}(M)$. Let K_0 be defined by the short exact sequence

$$0 \to K_0 \to N_0 \to M,$$

where $N_0 \in \mathrm{add}(N)$ is a *right approximation* of M, i.e., any other map $N_0' \to M$ with $N_0' \in \mathrm{add}(N)$ factors through N_0. One defines the *right mutation* of M at N to be $\mu_N^+(M) := N \oplus K_0$. The *left mutation* of M at N is defined via duality as $\mu_N^-(M) = (\mu_{N^\vee}^+(M^\vee))^\vee$. We also put $\mu_N^\pm(\Lambda) = \mathrm{End}_R(\mu_N^\pm(M))$. Note, however, that the passage from $\mu_N^\pm(M)$ to $\mu_N^\pm(\Lambda)$ loses some information.

Remark 3.9. Needless to say that $\mu_N^\pm(M)$ is only determined up to additive closure (i.e., up to taking $\mathrm{add}(-)$). However, if R is complete local then we can make a minimal choice for $\mu_N^+(M)$ which we will do silently.

Theorem 3.10 ([69, THEOREMS 1.22, 1.23]). *Let M, N, Λ be as above.*

(1) *$\mu_N^\pm(M)$ define NCCRs.*

(2) *Λ, $\mu_N^+(\Lambda)$ and $\mu_N^-(\Lambda)$ are all derived equivalent.*

(3) *μ_N^+ and μ_N^- are mutually inverse operations (this statement makes sense since $N \in \mathrm{add}(\mu_N^\pm(M))$).*

If R is complete local of dimension 3, things simplify. Let us call a reflexive R-module *basic* if every indecomposable summand appears occurs only once.

Theorem 3.11 ([69, THEOREMS 1.25]). *Assume that R is complete local of dimension 3. Let M be a basic reflexive R-module defining an NCCR, having at least two nonisomorphic indecomposable summands and let M_i be such an indecomposable summand. Then $\mu_{M/M_i}^+(M) \cong \mu_{M/M_i}^-(M)$.*

Example 3.12 ([63,71,133]). If R is complete local ring with a 3-dimensional terminal Gorenstein singularity then the basic reflexive modules yielding an NCCR correspond to the maximal cells in an affine hyperplane arrangement of dimension $\mathrm{rk}\,\mathrm{Cl}(R)$ with mutations at indecomposable summands corresponding to wall crossings [133, THEOREM 4.4]. The group Cl(R) acts by translation on this hyperplane arrangement and the quotient consists of a finite number of cells which correspond to the NCCRs of R. The number of such NCCRs is generally higher than the number of crepant resolutions.

It is an interesting problem to understand this for other types of 3-dimensional singularities.

Remark 3.13. If $\Lambda = \mathrm{End}_R(M)$ is an NCCR then because of the reflexive Morita equivalence $\mathrm{ref}(\Lambda) = \mathrm{ref}(R)$ the mutation procedure may also be defined on the level of reflexive Λ-modules (see [67, §5]). The resulting procedure also works for twisted NCCRs, where there is no reflexive Morita equivalence.

We now describe a different point of view on mutations, taken from [47]. For Q a quiver with n vertices let $\widehat{kQ}$ be the completion of the path algebra of Q at path length.

A *potential* $w \in \widehat{kQ}$ is a convergent sum of cycles considered up to rotation (or, equivalently, $w \in \widehat{kQ}/[\widehat{kQ}, \widehat{kQ}]\hat{}$). If w is a potential then (∂w) denotes the (completed) two sided ideal generated by the cyclic derivatives $\partial_x w$ of w with respect to the arrows in Q, where for a cyclic path m we have $\partial_x m := \sum_{m=uxv} vu$ (note that this is invariant under path rotation). The *completed Jacobi algebra* associated to (Q, w) is defined as $\hat{J}(Q, w) := \widehat{kQ}/(\partial w)$. We say that w is *reduced* if it only contains cycles of length ≥ 3. We can also consider the uncompleted version $J(Q, w) := kQ/(\partial w)$, in case w is a finite sum. We have the following result.

Theorem 3.14 ([130, THEOREMS A&B]). *If Λ is a basic (i.e., $\Lambda / \operatorname{rad} \Lambda \cong k^{\oplus n}$) twisted NCCR of a 3-dimensional normal Gorenstein complete local ring then Λ is a completed Jacobi algebra $\hat{J}(Q, w)$ with w reduced.*

If Q does not have loops or 2-cycles then the mutations of $\Lambda := \hat{J}(Q, w)$ can be obtained by an alternative procedure described in [47]. The procedure to mutate at a vertex i of Q yields a new Jacobi algebra $\hat{J}(Q', w')$ defined as follows (see [83, §2.4]).

(1) For each arrow β with target i and each arrow α with source i, add a new arrow $[\alpha\beta]$ from the source of β to the target of α.

(2) Replace each arrow α with source or target i with an arrow α^* in the opposite direction.

The new potential w' is the sum of two potentials w'_1 and w'_2. The potential w'_1 is obtained from w by replacing each composition $\alpha\beta$ (up to cyclic rotation) by $[\alpha\beta]$, where β is an arrow with target i. The potential w'_2 is given by

$$w'_2 = \sum_{\alpha,\beta} [\alpha\beta]\beta^*\alpha^*,$$

where the sum ranges over all pairs of arrows α and β such that β ends at i and α starts at i. It follows from [83, THEOREM 3.2] that this mutation coincides with the mutation defined in [69] and described above.

It may be that w' is not reduced, i.e., it contains 2-cycles. In that case the corresponding relations allow one to eliminate some arrows in Q'. By doing this, we find that the Jacobi algebra $\hat{J}(Q', w')$ can be more economically written as $J((Q')^{\mathrm{red}}, (w')^{\mathrm{red}})$ where $(w')^{\mathrm{red}}$ is reduced.

If we are lucky that $(Q')^{\mathrm{red}}$ does not contain any 2-cycles (it cannot contain loops) then we can repeat the mutation procedure at arbitrary vertices. If we can keep doing this forever then we call the original potential w *nondegenerate*.

Note that if $(Q')^{\mathrm{red}}$ does not contain 2-cycles, it can be obtained from Q' by simply deleting all 2-cycles, so that the mutation procedure becomes to some extent combinatorial [81]. For a nondegenerate potential, this nice property persists under iterated mutations. The catch, however, is that in general it is not clear how to check that a potential is nondegenerate. A useful criterion, based on the theory of graded mutations [2], is given in [44].

Theorem 3.15 ([44, COROLLARY 1.3]). *Assume:*

(1) *Q is a $\mathbb{Z}$-graded quiver such that $(kQ)_{\leq 0}$ is finite dimensional.*

(2) *Q has at least three vertices.*

(3) *w is a homogeneous reduced potential of degree r (in particular, it is a finite sum).*

(4) *$\Lambda = J(Q, w)$ is a twisted NCCR whose center is 3-dimensional with an isolated singularity.*

(5) *$\Lambda/[\Lambda, \Lambda]$ does not contain any elements whose degree is in the interval $[1, r/2]$.*

Then w is nondegenerate.

Note that (5) is automatic if $r = 1$. This gives an alternative proof why the potentials associated to "rolled up helix algebras" of Del Pezzo surfaces are nondegenerate (see [30, THEOREM 1.7], [44, THEOREM 4.2.1]). Theorem 3.15 also applies to many skew group rings $\Lambda = k[x, y, z]\#(\mathbb{Z}/n\mathbb{Z})$. For example, $n = 5$ and $\bar{1}$ acting with weights $(1/5, 2/5, 2/5)$ (see [67, §7]).

4. QUOTIENT SINGULARITIES FOR REDUCTIVE GROUPS

4.1. NCCRs via modules of covariants

In this section we discuss some results from [116]. Also G will always be a reductive group. Let S be the coordinate ring of a smooth affine G-variety X. Then S^G is the coordinate ring of the categorical quotient $X/\!\!/G$. We will be interested in constructing (twisted) NCCRs for S^G. In the particular case when G is finite and X is a faithful unimodular G-representation, this was discussed in Section 3.1. An NCCR for S^G is given by the skew group ring $\Lambda = S\#G$. However, Λ can be described in a different way. For U a finite dimensional G-representation, put $M(U) := (U \otimes S)^G$. Then $M(U)$ is a reflexive S^G module (in fact, it is maximal Cohen–Macaulay). If every irreducible representation of G occurs at least once in U then Λ is Morita equivalent to $\operatorname{End}_{S^G}(M(U))$. Hence $M(U)$ defines an NCCR of S^G.

The modules $M(U)$ we introduced are the so-called *modules of covariants* [31] and they make perfect sense for general reductive groups. A mild obstacle is that modules of covariants do not have to be reflexive in general [31]. This is not a serious problem, but if we want to avoid it anyway, we can restrict the pairs (G, X) we consider. We will say that G acts *generically* on a smooth affine variety if the locus of points with closed orbit and trivial stabilizer is nonempty and its complement has codimension ≥ 2. If W is a G-representation then we will say that (G, W) is *generic* if G acts generically on $\operatorname{Spec} \operatorname{Sym} W \cong W^*$. We then have in particular

$$\operatorname{End}_{S^G}\big(M(U)\big) = M\big(\operatorname{End}(U)\big). \tag{4.1}$$

It is reasonable to search for NC(C)Rs of the form $\operatorname{End}_{S^G}(M(U))$. However, if G is not finite there are nontrivial obstacles:

(1) There are an infinite number irreducible representations so we cannot just take the sum of all of them. We need to make a careful selection.

(2) Modules of covariants are usually not Cohen–Macaulay and so demanding that $\operatorname{End}_{S^G}(M(U)) = M(\operatorname{End}(U))$ (cf. (4.1)) is Cohen–Macaualay is a severe restriction on U.

The first issue is handled in [116, §10] where we construct certain nice complexes relating different modules of covariants (see also [135, CHAPTER 5]). The second issue is handled using results from [124] (see also [123,125–127]).

Before we discuss NCCRs let us give a result on NCRs.

Proposition 4.1 ([116, COROLLARY 1.3.5]). *Assume that (G, W) is generic. Then there exists a finite dimensional G-representation U containing the trivial representation such that $\Lambda = \operatorname{End}_{S^G}(M(U))$ is an NCR for S^G.*

Remark 4.2. The fact that U contains the trivial representation implies that Λ defines a categorical resolution by Lemma 2.2.1. It turns out that NCRs are easier to construct than NCCRs since it is sufficient to take U big enough, in a suitable sense.

To state our results about (twisted) NCCRs, we need to introduce some notation. Let G be a connected[8] reductive group. Let $T \subset B \subset G$ be respectively a maximal torus and a Borel subgroup of G, with $\mathcal{W} = N(T)/T$ being the corresponding Weyl group. Put $X(T) = \operatorname{Hom}(T, G_m)$ and let $\Phi \subset X(T)$ be the roots of G. By convention the roots of B are the negative roots Φ^- and $\Phi^+ = \Phi - \Phi^-$ is the set of positive roots. We write $\bar{\rho} \in X(T)_{\mathbb{R}}$ for half the sum of the positive roots. Let $X(T)^+_{\mathbb{R}}$ be the dominant cone in $X(T)_{\mathbb{R}}$ and let $X(T)^+ = X(T)^+_{\mathbb{R}} \cap X(T)$ be the set of dominant weights. For $\chi \in X(T)^+$, we denote the simple G-representation with highest weight χ by $V(\chi)$.

Let W be a finite-dimensional G-representation of dimension d and put $S = \operatorname{Sym}(W)$, $X = \operatorname{Spec}\operatorname{Sym}(W) = W^*$. Let $(\beta_i)_{i=1}^d \in X(T)$ be the T-weights of W.

Put

$$\Sigma = \left\{ \sum_i a_i \beta_i \mid a_i \in]-1, 0] \right\} \subset X(T)_{\mathbb{R}}.$$

The elements of the intersection $X(T)^+ \cap (-2\bar{\rho} + \Sigma)$ are called *strongly critical (dominant) weights* for G.

Theorem 4.3 ([116, THEOREM 3.4.3][124]). *Assume that X contains a point with closed orbit and finite stabilizer. Let $\chi \in X(T)^+$ be a strongly critical weight and $U = V(\chi)$. Then $M(U^*)$ is a Cohen–Macaulay S^G-module.*

8 In [116] we also consider the nonconnected case.

If we look at (4.1) and observe that the weights of $\mathrm{End}(U)$ are very roughly speaking about twice those of U, then Theorem 4.3 suggests that to construct an NCCR we should restrict ourselves to representations whose highest weights are approximately contained in $-\bar{\rho} + (1/2)\Sigma$. This idea works for the class of "quasisymmetric" representations, which includes the class of self dual representations.

We say that W is *quasisymmetric* if for every line $\ell \subset X(T)_{\mathbb{R}}$ through the origin we have $\sum_{\beta_i \in \ell} \beta_i = 0$. This implies in particular that W is unimodular and hence S^G is Gorenstein if W is generic by a result of Knop [87].

From now on we assume that (G, W) is generic and W is quasisymmetric. Deviating slightly from [116], following [59], we introduce a certain affine hyperplane arrangement on $X(T)_{\mathbb{R}}^{\mathcal{W}}$. Let $\bar{\mathcal{H}}$ be the collection of affine hyperplanes spanned by the facets of $-\bar{\rho} + (1/2)\bar{\Sigma}$. We consider the hyperplane arrangement in[9] $X(T)_{\mathbb{R}}^{\mathcal{W}}$ given by

$$\mathcal{H} = \bigcup_{H \in \bar{\mathcal{H}}} (-H + X(T)) \cap X(T)_{\mathbb{R}}^{\mathcal{W}}. \tag{4.2}$$

Remark 4.4. The hyperplane arrangement (4.2) may be degenerate in the sense that $X(T)_{\mathbb{R}}^{\mathcal{W}} \subset -H + \chi$ for some $\chi \in X(T)$.

Example 4.5. We give a simple example where degeneration occurs. Let $G = \mathrm{SL}(2)$. If V is the standard representation and $W = V^n$ with n even, then $(1/2)\Sigma$ is the interval $]-n/2, n/2[$ (identifying $X(T) \cong \mathbb{Z}$). Moreover, $\bar{\rho} = 1$. Hence the "hyperplanes" $-H + X(T)$ are given by the integers. Furthermore, $X(T)_{\mathbb{R}}^{\mathcal{W}} = 0$. Thus the induced hyperplane arrangement in $X(T)_{\mathbb{R}}^{\mathcal{W}}$ is indeed degenerate. If n is odd, on the other hand, then it is nondegenerate. See [116, THEOREM 1.4.5] for a complete treatment of the case $G = \mathrm{SL}(2)$.

This hyperplane arrangement is such that if δ is the complement of $\mathcal{H}$ then

$$(-\bar{\rho} + \delta + 1/2\partial\bar{\Sigma}) \cap X(T) = \emptyset.$$

The following result is a slight variation on [116, THEOREM 1.6.4].

Theorem 4.6. *Let (G, W) be generic and assume that W is quasisymmetric. Let δ be an element of the complement of $\mathcal{H}$. Put*

$$\mathcal{L}_\delta = X(T)^+ \cap (-\bar{\rho} + \delta + (1/2)\bar{\Sigma}), \tag{4.3}$$

$$U_\delta = \bigoplus_{\chi \in \mathcal{L}_\delta} V(\chi), \tag{4.4}$$

$$\Lambda_\delta = \mathrm{End}_{S^G}(M(U_\delta)). \tag{4.5}$$

If $\mathcal{L}_\delta \neq \emptyset$ then Λ is an NCCR for $\mathrm{Sym}(W)^G$.

It is easy to see that $\mathcal{L}_\delta$ and hence U_δ depend only on the connected component of the complement of $\mathcal{H}$ to which δ belongs.

---

9 Note that $X(T)^{\mathcal{W}}$ is just the character group $X(G)$ of G.

We obtain some evidence for Conjecture 3.7.[10]

Corollary 4.7 ([116, THEOREM 1.6.2]). *If $G = T$ is a torus and W is quasisymmetric then* $\mathrm{Sym}(W)^T$ *has a (toric) NCCR.*

Remark 4.8. For reference we note that there is extension of Theorem 4.6 that may allow one to construct twisted NCCRs [116, THEOREM 1.6.4].

We now state some consequences of these results for determinantal varieties.

Theorem 4.9 ([116, THEOREM 1.4.1]). *For $l < n$, let $X_{l,n}$ be the variety of $n \times n$-matrices of rank $\leq l$. Then $k[X_{l,n}]$ has an NCCR.*

The variety $X_{l,n}$ was already discussed in Example 3.1 and the NCCR obtained in [116] is the same as that we obtain in Theorem 4.9. To prove Theorem 4.9, we use the classical description of $k[X_{l,n}]$ as an invariant ring [134]. Put $G = \mathrm{GL}(l)$ and let V be the standard representation of G. Put $W = V^n \oplus (V^*)^n$. Then $k[X_{l,n}] = \mathrm{Sym}(W)^G$, and we show in [116] that Theorem 4.9 follows from Theorem 4.6. For the benefit of the reader, we describe the actual module of covariants that gives the NCCR. Let $B_{l,n-l}$ be the set of partitions that fit in a rectangle of size $l \times (n-l)$. In [116] it is shown that the following module of covariants defines an NCCR for R:

$$M = \bigoplus_{\lambda \in B_{l,n-l}} M(S^\lambda V), \tag{4.6}$$

where $S^\lambda V$ denotes the Schur functor indexed by λ applied to V.

Theorem 4.10. *For $2l < n$, let $X^-_{2l,n}$ be the variety of skew-symmetric $n \times n$ matrices of rank $\leq 2l$. If n is odd then $k[X^-_{2l,n}]$ has an NCCR.*

This time we put $G = \mathrm{Sp}(2l)$ and $W = V^n$ where V is the standard representation of G.

Theorem 4.11 ([116, THEOREM 1.4.1]). *For $l < n$, let $X^+_{l,n}$ be the variety of symmetric $n \times n$ matrices of rank $\leq l$. If l and n have opposite parity then $k[X^+_{l,n}]$ has an NCCR. If l and n have the same parity then $k[X^+_{l,n}]$ has a twisted NCCR.*

Here we put $G = \mathrm{O}(l)$ and again $W = V^n$ where V is the standard representation of G. A complication arises since $\mathrm{O}(l)$ is not connected, so we cannot directly apply Theorem 4.6. So we have to perform a more refined analysis which is carried out in [116, §6]. Twisted NCCRs appear because $\mathrm{SO}(l)$, the connected component of $\mathrm{O}(l)$, is not simply connected.

The NCCRs given in Theorems 4.10, 4.11 have been crucial for establishing homological projective duality [91] for determinantal varieties of skew-symmetric matrices by Rennemo and Segal [113]. The corresponding results for symmetric matrices are work in progress by the same authors [112].

---

10 This is stated in [116] for W generic. However, one easily reduces to this case.

Even if W is not quasisymmetric then it is still possible that $\mathrm{Sym}(W)^G$ has an NCCR given by a module of covariants but we are unaware of a general rule like Theorem 4.6 for constructing them. Three-dimensional affine toric varieties (see Section 3.4) are an example of this, since they can be written as $\mathrm{Sym}(W)^G$ where W is generally not quasisymmetric. Another example is given by the recent work of Doyle:

Example 4.12 ([51, THEOREM 3.11]). Let $0 < l < n$ be integers such that $\gcd(l, n) = 1$. Let V be the standard representation of $G = \mathrm{SL}(l)$ and put $W = V^n$.

Then $R := \mathrm{Sym}(W)^G$ is the homogeneous coordinate ring of the Grassmannian $\mathrm{Gr}(l, n)$ for the Plücker embedding. Let $P_{l,n-l}$ be the set of partitions whose young tableaux are above the diagonal in a rectangle of size $l \times (n - l)$. In [51] it is shown that the following module of covariants defines an NCCR for R:

$$M = \bigoplus_{\lambda \in P_{l,n-l}} M(S^\lambda V)$$

(compare with (4.6)).

We reiterate that even if an NCCR exists, there does not have to be one given by a module of covariants. See Example 3.6.

4.2. NCCRs via crepant resolutions obtained by GIT

Here we discuss some results from [59] that shows that in certain cases the NCCRs for the categorical quotients $X /\!/ G$ that we constructed in Section 4 can be obtained as the endomorphisms of a tilting bundle on a crepant resolution, i.e., the method of Section 3.2. This crepant resolution is constructed using a geometric invariant theory. It turns out that we have to allow crepant resolutions by Deligne–Mumford stacks. This occurred already before in Lemma 3.5 and Example 3.6.

Remark 4.13. To construct a resolution of $X /\!/ G$ using geometric invariant theory, one needs a linearized line bundle on X. Since here X is a representation, the only G-equivariant line bundles on X are those obtained from characters of G. If G is semisimple then there are no (nontrivial) characters so we cannot proceed. Thus we can, for example, not deal with determinantal varieties of symmetric and skew-symmetric matrices (see Theorems 4.10 and 4.11). In those cases the relevant groups were respectively $\mathrm{Sp}(2l)$ and the connected component $\mathrm{SO}(l)$ of $\mathrm{O}(l)$, both of which are semisimple. On the other hand, ordinary determinantal varieties are fine since in that case $G = \mathrm{GL}(l)$ which has a nontrivial character given by the determinant, which may be used to construct a crepant resolution.

Remark 4.14. Geometric invariant theory is still helpful for constructing a resolution of $X /\!/ G$ via a procedure invented by Kirwan [52, 85, 110]. However, these resolutions are usually not crepant. Consistent with expected minimality of NCCRs (see Section 3.3), we are able to show that some NCCRs embed inside them [118].

We retain the notations and assumptions of the previous section. We assume that W is a quasisymmetric representation of G and $X = \mathrm{Spec}\,\mathrm{Sym}(W) = W^*$. Recall that for a character $\mu \in X(G) = X(T)^{\mathcal{W}}$ we may define a G-invariant open subset of X as

$$X^{ss,\mu} = \{x \in X \mid \exists k > 0 \text{ and } s \in \Gamma(\mathcal{O}_X \otimes \mu^k)^G \text{ such that } s(x) \neq 0\}.$$

The variety $X^{ss,\mu}$ admits a good quotient[11] $X^{ss,\mu}/\!/G$ which is proper over $X/\!/G$. For U a representation of G, we write $\mathcal{M}(U)$ for the vector bundle on $X^{ss,\mu}/G$ given by $U \otimes \mathcal{O}_{X^{ss,\mu}/G}$. The global sections of $\mathcal{M}(U)$ are equal to $M(U)$.

Below we let $\mathcal{H}_0$ be the *central* hyperplane arrangement on $X(G)_{\mathbb{R}} = X(T)^{\mathcal{W}}_{\mathbb{R}}$ corresponding to the affine hyperplane arrangement $\mathcal{H}$ introduced in (4.2). Thus the hyperplanes in $\mathcal{H}_0$ are the hyperplanes which are induced from central hyperplanes in $X(T)_{\mathbb{R}}$ which are parallel to the facets of $\bar{\Sigma}$.

Proposition 4.15 ([59, PROPOSITION 2.1]). *Assume that the action of T on X has generically finite stabilizers and let $\mu \in X(G)$ be in the complement of $\mathcal{H}_0$. Then $X^{ss,\mu}/G$ is a Deligne–Mumford stack.*

Lemma 4.16. *Assume that (G, W) is generic. Then the canonical map $X^{ss,\mu}/G \to X/\!/G$ is crepant.*

Proof. This is proved in [119, LEMMA 4.5] in the case that G is a torus, but this assumption is not relevant for the proof. ■

The following is one of the main results of [59]. It is proved using similar combinatorics as in [116].

Theorem 4.17 ([59]). *Assume that the action of T on X has generically finite stabilizers and let $\mu \in X(G)$ be in the complement of $\mathcal{H}_0$. Let U_δ be as in the statement of Theorem 4.6. Then $\mathcal{M}(U_\delta)$ is a tilting bundle on $X^{ss,\mu}/G$ such that* $\operatorname{End}_{X^{ss,\mu}/G}(\mathcal{M}(U_\delta)) = M(\operatorname{End}(U_\delta))$.

Proof. This follows from combining [59, THEOREM 1.2] with [59, LEMMA 2.9]. ■

In this way we obtain more evidence for Conjecture 1.2.

Corollary 4.18 ([59, COROLLARY 1.3]). *Under the hypotheses of Theorem 4.17, if $\mu, \mu' \in X(G)$ are in the complement of $\mathcal{H}_0$ and the complement of $\mathcal{H}$ is nonempty (i.e., $\mathcal{H}$ is nondegenerate) then $\mathcal{D}(X^{ss,\mu}/G) \cong \mathcal{D}(X^{ss,\mu'}/G)$.*

We also obtained the promised description of NCCRs via resolutions.

Corollary 4.19. *Assume that (G, W) is generic and let Λ be an NCCR constructed via Theorem 4.6. Let μ be in the complement of $\mathcal{H}_0$. Then Λ is the endomorphism ring of a tilting bundle on the DM stack $X^{ss,\mu}/G$.*

Remark 4.20. Hidden behind what is discussed in Sections 4.1 and 4.2 is the idea of *windows*, pioneered in [49]. This is based on the fact that we have a restriction map

$$\operatorname{Res} : \mathcal{D}(X/G) \to \mathcal{D}(X^{ss,\mu}/G).$$

---

11 A G-equivariant map $Z \to Y$ is a good quotient if locally on Y it is of the form $U \to U/\!/G$ for U affine.

It is then natural to try to find a full subcategory $\mathcal{D} \subset \mathcal{D}(X/G)$ such that the restriction of Res to $\mathcal{D}$ yields an equivalence $\mathcal{D} \cong \mathcal{D}(X^{ss,\mu}/G)$. A very general result in this direction is [58, THEOREM 1.1], see also [5].

In concrete cases one may hope to define $\mathcal{D}$ as the full subcategory of $\mathcal{D}(X/G)$ which is split generated by $U \otimes \mathcal{O}_{X/G}$ for a suitable G-representation U whose highest weights are restricted to a certain subset $\mathcal{L}$ of $X(T)^+$ (a "window"). This is precisely what happens in Theorem 4.17, where we take $\mathcal{L} = \mathcal{L}_\delta$. The resulting category $\mathcal{D}$ is a concrete realization of [58, THEOREM 1.1], see [59, LEMMA 3.5].

One does not actually need to have nontrivial $X^{ss,\mu}/G$ to apply the window principle. The proof of Theorem 4.6, is based on the fact that $\operatorname{mod}(\Lambda_\delta)$ embeds in $\operatorname{coh}(X/G)$ as the abelian category with a projective generator $U_\delta \otimes \mathcal{O}_{X/G}$.

4.3. Local systems, the SKMS, and schobers

In this slightly informal section we assume that the hypotheses of Theorem 4.17 hold. While Corollary 4.18 implies that two different $\mathcal{D}(X^{ss,\mu}/G)$, $\mathcal{D}(X^{ss,\mu'}/G)$ are derived equivalent, the actual derived equivalence depends on the choice of δ in the complement of $\mathcal{H}$. Moreover, by considering compositions $\mathcal{D}(X^{ss,\mu}/G) \xrightarrow{\delta} \mathcal{D}(X^{ss,\mu''}/G) \xrightarrow{\delta'} \mathcal{D}(X^{ss,\mu'}/G)$, we may produce more derived equivalences. This is consistent with the assertion in Remark 1.3 that there is no "god-given" derived equivalence between different crepant resolutions. A different way of saying this is that a crepant resolution may have a large group of derived autoequivalences.

If M is a Calabi–Yau variety then homological mirror symmetry predicts the existence of a space S (the "stringy Kähler moduli space," or SKMS) such that $\pi_1(S)$ acts on $\mathcal{D}(M)$. More precisely, S is the moduli space of complex structures on the mirror dual $M^\vee$ of M. In many cases there are good heuristic descriptions of $M^\vee$ and S.

Even without access to the full mirror symmetric context, which may be technically challenging or even only heuristic, it turns out to be very illuminating to represent the derived autoequivalences of an algebraic variety (or stack) as elements of $\pi_1(S)$ for a suitable topological space S. Alternatively, we may think of such a representation as a *local system of triangulated categories on* S. Understanding this for $\mathcal{D}(X^{ss,\mu}/G)$ was, according to the authors, one of the main motivations for writing [59]. Indeed, when X is a quasisymmetric representation, under hypotheses of Theorem 4.17, one may take

$$S = \big(X(G)_\mathbb{C} - \mathcal{H}_\mathbb{C}\big)/X(G),$$

where $\mathcal{H}_\mathbb{C}$ denotes the complexification of the real hyperplane arrangement $\mathcal{H}$ [59, PROPOSITION 6.6].

In this case there is a nice way to understand that action of $\pi_1(S)$ on $\mathcal{D}(X^{ss}/G)$ [59, §6], [117]. Using Theorem 4.17 again, we may just as well describe the action of $\pi_1(S)$ on $\mathcal{D}(\Lambda_\delta)$ for δ contained in the complement of $\mathcal{H}$ and $\Lambda_\delta = M(\operatorname{End}(U_\delta))$.

For $\delta \in X(G)_\mathbb{R}$, define U_δ as in (4.4) and put $\mathcal{D}_\delta = \mathcal{D}(\Lambda_\delta)$. Now $\mathcal{H}$ defines a cell decomposition of $X(G)_\mathbb{R}$ and it is easy to see that U_δ only depends on the cell to which δ belongs. Hence for a cell C let us write $\Lambda_C := \Lambda_\delta$, $\mathcal{D}_C := \mathcal{D}_\delta$ for $\delta \in C$. We will refer to the cells of maximal dimension as chambers. These are also the connected components of the complement of $\mathcal{H}$.

If C' is a face of C then there is an idempotent $e_{C,C'} \in \Lambda_{C'}$ such that $\Lambda_C = e_{C,C'}\Lambda_{C'}e_{C,C'}$. If $C \neq C''$ are distinct adjacent chambers, sharing a codimension one face C' then the functor $e_{C'',C'}\Lambda_{C'}e_{C,C'} \overset{L}{\otimes}_{\Lambda_C} -$ defines an equivalence of categories $\phi_{C,C''}$: $\mathcal{D}_C \to \mathcal{D}_{C''}$.

Put $\tilde{S} = X(G)_{\mathbb{C}} - \mathcal{H}_{\mathbb{C}}$ and let $\Pi_1(\tilde{S})$ be the groupoid whose objects are the chambers and whose morphisms are given by the homotopy classes of paths in $X(G)_{\mathbb{C}} - \mathcal{H}_{\mathbb{C}}$ connecting the chambers. Then $\Pi_1(\tilde{S})$ is equivalent to the fundamental groupoid of $X(G)_{\mathbb{C}} - \mathcal{H}_{\mathbb{C}}$. If C, C'' are adjacent chambers separated by a hyperplane $H \in \mathcal{H}$ such that $H(C'') > 0$ then there is a canonical (up to homotopy) minimal path $\nu_{C,C''}$ in $X(G)_{\mathbb{C}} - \mathcal{H}_{\mathbb{C}}$ going from C to C'' and passing through $\{\mathrm{Im}\, H_{\mathbb{C}} > 0\}$. Sending C to $\mathcal{D}_C$ and $\nu_{C,C''}$ to $\phi_{C,C''}$ defines a representation of the groupoid $\Pi_1(\tilde{S})$ in triangulated categories.

If $\chi \in X(G)$ then tensoring by χ defines an equivalence $\mathcal{D}_C \to \mathcal{D}_{C+\chi}$ and in this way the representation of $\Pi_1(\tilde{S})$ may be extended to a representation of $\Pi_1(\tilde{S}) \rtimes X(G)$ and the latter is equivalent to the fundamental groupoid $\Pi_1(S)$ of $S = \tilde{S}/X(G)$ [33, CHAPTER 11], [59, §6]. Hence, fixing a "base chamber" C, we get an action of $\pi_{1,C}(S)$ on $\mathcal{D}_C$.

Remark 4.21. It is shown in [117] that the family of triangulated categories $(\mathcal{D}_C)_C$ for *all* cells C is a so-called $X(G)$-equivariant *perverse schober*. This is a categorification of a perverse sheaf on $X(G)_{\mathbb{C}}/X(G)$ [74] (see also [20, 48]). Note that $X(G)_{\mathbb{C}}/X(G)$ is a torus and S is the complement of a "toric hyperplane arrangement." If G is itself a torus T then $X(T)_{\mathbb{C}}/X(T)$ may be identified with the dual torus $T^{\vee}$.

Remark 4.22. The $X(G)$-equivariant hyperplane arrangement constructed by Halpern–Leistner and Sam in [59] is very similar to the $\mathrm{Cl}(R)$-equivariant hyperplane arrangement associated to a 3-dimensional terminal complete Gorenstein ring R constructed by Iyama and Wemyss (see Example 4.12). One would expect there to be an associated equivariant schober also in this case. In the case of a single curve flop this is essentially contained in [50, §3].

Remark 4.23. As explained we have an action of $\pi_1(S)$ on $\mathcal{D}_C$ for a chamber C and hence also an action of $\pi_1(S)$ on $K_0(\mathcal{D}_C)_{\mathbb{C}}$. In other words, we have a local system L on S. It is then a natural question if this local system occurs as the solutions of a natural system of differential equations. In the case that G is a torus we show in [121] that a generic "equivariant" deformation of L is obtained as the solution of a well-known system of differential equations introduced by Gel'fand, Kapranov, and Zelevinsky [55]. This starts from a computation by Kite [86] which shows that the hyperplane arrangement constructed in [59] is up to translation defined by the so-called "principal A-determinant," an important ingredient in the theory developed Gel'fand, Kapranov, and Zelevinsky. For more information, see [115].

Remark 4.24. The themes touched upon in this section occur in many different contexts. See, e.g., [3, 23, 29].

5. NCCRS AND STRINGY E-FUNCTIONS

In this section we discuss some ongoing work of Timothy De Deyn (see [43]). Let X be an algebraic variety over $\mathbb{C}$. The cohomology groups $H_c^i(X,\mathbb{C})$ carry a natural mixed Hodge structure. We denote by $h^{p,q}(H_c^i(X,\mathbb{C}))$ the dimension of the (p,q)-type component of $H_c^i(X,\mathbb{C})$. The *Hodge polynomial* of X is defined by

$$E(X,u,v)=\sum_{p,q,i}(-1)^i h^{p,q}\big(H_c^i(X,\mathbb{C})\big)u^p v^p.$$

The Hodge polynomial defines a ring homomorphism from the Grothendieck ring of algebraic varieties $K_0(\mathrm{Var}/\mathbb{C})$ to $\mathbb{Z}[u,v]$.

We put $e(X)=E(X,1,1)$, i.e.,

$$e(X)=\sum_i(-1)^i\sum_{p,q}h^{p,q}\big(H_c^i(X,\mathbb{C})\big)=\sum_i(-1)^i\dim H_c^i(X,\mathbb{C}).$$

In other words, $e(X)$ is the *Euler characteristic* (with compact support[12]) of X. It defines a ring homomorphism from $K_0(\mathrm{Var}/\mathbb{C})$ to $\mathbb{Z}$.

Definition 5.1 ([6, DEFINITION 3.1]). Assume that X is a normal $\mathbb{Q}$-Gorenstein algebraic variety/$\mathbb{C}$ with at most log-terminal singularities and let $\pi: Y\to X$ be a resolution of singularities whose exceptional locus is a normal crossing divisor. Let $D_1,\dots,D_r$ be the irreducible components of the exceptional locus and put $I=\{1,\dots,r\}$. For any subset $J\subset I$ we set $D_J=\bigcap_{j\in J}D_j$, $D_J^\circ:=D_J\setminus\bigcup_{j\in I\setminus J}D_j$. The *stringy E-function* of X is defined as

$$E_{st}(X,u,v):=\sum_{J\subset I}E(D_J^\circ,u,v)\prod_{j\in J}\frac{uv-1}{(uv)^{a_j+1}-1},\tag{5.1}$$

where the numbers $a_j\in\mathbb{Q}\cap\,]-1,\infty[$ are defined by

$$K_Y=\pi^*K_X+\sum_{j=1}^r a_jD_j.$$

Putting $e_{st}(X)=\lim_{u,v\to 1}E_{st}(X,u,v)$ defines the *stringy Euler characteristic* of X, with the formula

$$e_{st}(X)=\sum_{J\subset I}e(D_J^\circ)\prod_{j\in J}\frac{1}{a_j+1}.$$

It follows from the theory of motivic integration (see, e.g., [8, 40, 45, 90, 95]) that $E_{st}(X,u,v)$ is independent of the chosen resolution Y [6, THEOREM 3.4]. Indeed, $E_{st}(X,u,v)$ may be obtained by integrating over the arc space associated to X [45]. In a similar vein, $E_{st}(X,u,v)=E_{st}(Y,u,v)$ holds for birational maps $\pi: Y\to X$ satisfying $\pi^*K_X=K_Y$ [6, THEOREM 3.12].

If X is smooth then the stringy E-function coincides with the Hodge polynomial. Hence one has

---

12 If X is smooth then, by Poincaré duality, the Euler characteristic with compact support coincides with the usual Euler characteristic $\sum_i(-1)^i\dim H^i(X,\mathbb{C})$.

Theorem 5.2 ([6, THEOREM 3.12]). *If X has a crepant resolution Y then the stringy E-function of X coincides with the Hodge polynomial of Y. In particular, it is a polynomial. Similarly, the stringy Euler characteristic of X coincides with the usual Euler characteristic of Y and hence it is an integer.*

The following conjecture seems natural:

Conjecture 5.3 ([43]). If X is a normal Gorenstein variety$/\mathbb{C}$ with an NCCR then its stringy E-function is a polynomial.

We give some evidence for this conjecture below, but at this point it is probably safer to regard it as a question. We illustrate below in Remark 5.8 and Example 5.11 that reasonable extensions of this conjecture are false.

Example 5.4. Quotient varieties of the form $\mathbb{C}^n /\!/ G$ for $G \subset \mathrm{SL}(n)$ finite always have a stringy E-function which is a polynomial by [9], [46, THEOREM 3.6]. They also have an NCCR by Section 3.1. So in this case Conjecture 5.3 is true.

Example 5.5. Batyrev proves in [8, PROPOSITION 4.4] that the stringy E-function of any toric variety with Gorenstein singularities is a polynomial. Hence Conjecture 5.3 is compatible with Conjecture 3.7.

A good test for Conjecture 5.3 is given by cones over Fano varieties.

Proposition 5.6 ([43]). *Let Z be a smooth projective variety$/\mathbb{C}$ with ample line bundle $\mathcal{O}_Z(1)$ and let $X = \operatorname{Spec} R$ be the corresponding cone. Assume $\omega_Z = \mathcal{O}_Z(-n)$ for $n \geq 1$. Then R is Gorenstein and*

$$E_{st}(X,u,v) = E(Z,u,v)\frac{(q-1)q^n}{q^n-1} \tag{5.2}$$

with $q = uv$. In particular,

$$e_{st}(X) = \frac{e(Z)}{n}. \tag{5.3}$$

Example 5.7. Consider the Grassmannian $Z := \mathrm{Gr}(d,n)$. Then (e.g., [25, PROPOSITION A.4])

$$E_{st}(Z,u,v) = \binom{n}{d}_q, \tag{5.4}$$

where

$$\binom{n}{d}_q = \frac{(q^n-1)(q^{n-1}-1)\cdots(q^{n-d+1}-1)}{(q-1)(q^2-1)\cdots(q^d-1)}.$$

Hence

$$e_{st}(Z) = \binom{n}{d}. \tag{5.5}$$

Let X be the cone over Z with respect to the Plucker embedding and let R be the coordinate ring of X. Using (5.2) and (5.4), it is shown in [43] that in this case $E_{st}(X,u,v)$ is a polynomial precisely when $\gcd(d,n) = 1$. On the other hand, by [51, THEOREM 3.11] (see

Example 4.12 above), R has an NCCR when $\gcd(d,n) = 1$. So Conjecture 5.3 is true in this case.

Remark 5.8. It is shown in [43, §4.2] that the cone over an arbitrary Grassmannian $\mathrm{Gr}(d,n)$ always has a weakly crepant categorical resolution (see Section 2.2). Hence it follows that Conjecture 5.3 is false for weakly crepant categorical resolutions. On the other hand, it seems reasonable to extend Conjecture 5.3 to strongly crepant categorical resolutions.

Remark 5.9. In view of Theorem 5.2, one may naively ask if it is true that $E_{st}(X,u,v)$ being a polynomial implies that X has a crepant resolution. Not unexpectedly, this fails drastically. Finite group quotients and affine toric varieties have a polynomial stringy E-function, as we have seen above, but they need not have a crepant resolution. In fact, the example $\mathbb{C}^4/\!/\mathbb{Z}_2$ given in Section 3.1 of a Gorenstein singularity with an NCCR but without a crepant resolution, lives in both classes. On the other hand, these classes of counterexamples are not very convincing since they admit crepant resolutions by smooth Deligne–Mumford stacks, which is just as good. In the case of finite quotient singularities this is clear, and for toric varieties it follows from Lemma 3.5.

In contrast, one may show that for X as in Example 5.7 there is no crepant resolution by a smooth DM-stack (this is mainly because R is factorial). So in some sense it is a "better" counterexample (when $\gcd(d,n) \neq 1$).

GIT quotients form an important class of toric varieties (see [39, COROLLARY 14.2.16]). So in view of Example 5.5, as well as Example 5.4, the following question by Batyrev suggests itself:

Question 5.10 ([6, QUESTION 5.5]). Does a GIT quotient of $\mathbb{C}^n$ for a linear action of $G \subset \mathrm{SL}(n)$ always have a stringy E-function that is a polynomial?

Alas, the answer is negative. Indeed, Example 5.7 for $\gcd(d,n) \neq 1$ gives a simple counterexample since the cone over a Grassmannian is a GIT quotient for $\mathrm{SL}(d)$ acting on n copies of its standard representation.

The first counterexample, however, was constructed much earlier in [84].

Example 5.11. Let W be given by three copies of the adjoint representation of $G = \mathrm{SL}(2)$. Then by a quite involved computation it is shown in [84, COROLLARY 1.2] that the stringy E-function of $\mathrm{Spec}\, R$ for $R = \mathrm{Sym}(W)^G$ is not a polynomial. This example is interesting since, by [116, THEOREM 1.4.5], R has a twisted NCCR. In other words, Conjecture 5.3 is also false for twisted NCCRs.

As a side remark, we note that this twisted NCCR is a rather classical object. It is the trace ring generated by 3 generic traceless 2×2 matrices [4, 97, 99, 107–109].

In the setting of Theorem 5.2, the Euler characteristic of Y can be computed using periodic cyclic homology, thanks to the Hochschild–Kostant–Rosenberg theorem. Below we define the Euler characteristic $e(\Lambda)$ of an algebra Λ or a sheaf of algebras as dim $\mathrm{HP}^{\mathrm{even}}(\Lambda) - \dim \mathrm{HP}^{\mathrm{odd}}(\Lambda)$. If Λ is a quasicoherent sheaf of algebras then we use $\mathrm{HP}^*(\Lambda) := \mathrm{HP}^*(\mathrm{Perf}_{dg}\,\Lambda)$ (where $\mathrm{Perf}_{dg}(\Lambda)$ is a standard dg-enhancement of $\mathrm{Perf}(\Lambda)$). The following conjecture appears plausible.

Conjecture 5.12. The stringy Euler characteristic of a normal Gorenstein variety can be computed as the Euler characteristic of an NCCR, computed via periodic cyclic homology.

Remark 5.13. One can again not expect this conjecture to hold for twisted NCCRs. An interesting example is given in [26]. It was shown by [22, 92] that for a generic complete intersection Y of n quadrics in $\mathbb{P}^{2n-1}$ one has a derived equivalence between Y and $(\mathbb{P}^{n-1}, \mathcal{B}_0)$ where $\mathcal{B}_0$ is the even part of the universal Clifford algebra corresponding to the quadrics defining Y. Because of the derived equivalence, we then have $e(Y) = e(\mathcal{B}_0)$ [80]. One may show that $\mathcal{B}_0$ is a twisted NCCR of its center which is a double cover Z of $\mathbb{P}^{n-1}$ [26, §1]. It is shown in [26] that in general $e(Y) \neq e_{st}(Z)$ and hence $e(\mathcal{B}_0) \neq e_{st}(Z)$. So Conjecture 5.12 does not extend to twisted NCCRs.

In suitable "local" contexts (e.g., [130, THEOREM 9.1]) Conjecture 5.12 leads to a more concrete conjecture:[13]

Conjecture 5.14 ([43]). Let R a normal Gorenstein ring which is either a complete local ring, or else connected $\mathbb{N}$-graded (i.e., $R_0 = \mathbb{C}$). Assume that R has an NCCR $\mathrm{End}_R(M)$. Then the number of nonisomorphic indecomposable summands of M is equal to $e_{st}(X)$ for $X = \mathrm{Spec}\,R$.

Example 5.15. If $X = \mathrm{Spec}\,R$ is an affine toric variety as in Section 3.4, with Gorenstein singularities, then Batyrev [6, PROPOSITION 4.10] proves that $e_{st}(X)$ is equal to the volume of the associated polytope P (see Section 3.4). So Conjecture 5.14 is compatible with Conjecture 3.8.

Example 5.16. For varieties of the form $X = W/\!\!/G = \mathrm{Spec}\,\mathbb{C}[W]^G$ for $G \subset \mathrm{SL}(n)$ finite and W a finite dimensional representation of G, it follows from [7, THEOREM 8.4] that $e_{st}(X)$ is equal to the number of conjugacy classes in G. This number is in turn equal to the number of irreducible representations of G and hence equal to the number of nonisomorphic indecomposable summands of the reflective $\mathbb{C}[W]^G$-module $\mathbb{C}[W]$ which defines an NCCR for $\mathbb{C}[W]^G$ by Section 3.1. So Conjecture 5.14 is true in this specific example.

Example 5.17. Let X be the cone over $\mathrm{Gr}(d, n)$ as in Example 5.7. Then by (5.3) and (5.5) we have

$$e_{st}(X) = \frac{1}{n}\binom{n}{d}.$$

13 To handle the complete case, one has to use a "completed" version of periodic cyclic homology. See [130].

We check that Conjecture 5.14 is compatible with the NCCR constructed by Doyle in [51, THEOREM 3.11] (see Example 4.12 above). Conjecture 5.14 amounts to

$$|P_{d,n-d}| = \frac{1}{n}\binom{n}{d},$$

which is indeed true by [101, §12.1].

Example 5.17 can be put in a more general context. Let us first state a lemma.

Lemma 5.18. *Let Z be a smooth projective variety with a tilting complex. Then one has $e(Z) = \operatorname{rk} K_0(Z)$.*

Proof. Let $\mathcal{T}$ be the tilting complex and put $A = \operatorname{End}_Z(\mathcal{T})$. We have:

(1) Euler characteristics may be computed with periodic cyclic homology.

(2) Periodic cyclic homology is invariant under derived equivalence [80], and so are Grothendieck groups;

(3) $\mathrm{HP}^*(A) = \mathrm{HP}^*(A/\operatorname{rad} A)$ by Goodwillie's theorem [56, THEOREM II.5.1], and the standard fact that $K_0(A) = K_0(A/\operatorname{rad} A)$.

So we conclude

$$e(Z) \overset{(1,2)}{=} e(A) \overset{(3)}{=} \operatorname{rk} K_0(A) \overset{(2)}{=} \operatorname{rk} K_0(Z). \qquad \blacksquare$$

Let us go back to the setting of Proposition 5.6 but assume now in addition that Z has a tilting complex. Then by (5.3) combined with Lemma 5.18, we get

$$e_{st}(X) = \frac{\operatorname{rk} K_0(Z)}{n}.$$

Assuming that R has a graded NCCR Λ, Conjecture 5.14 implies

$$\operatorname{rk} K_0(\Lambda) = \frac{\operatorname{rk} K_0(Z)}{n}. \tag{5.6}$$

Example 5.19. This formula holds for the NCCRs constructed via Proposition 3.3. Indeed, $\operatorname{rk} K_0(\Lambda)$ is given by the number u of nonisomorphic indecomposable summands of $\mathcal{E}$. On the other hand, $\mathcal{D}(Z)$ has a semiorthogonal decomposition consisting of n parts whose K_0 also has rank u. So (5.6) does indeed hold, and we obtain again some evidence for Conjecture 5.14.

Remark 5.20. One way to think of this example as the realization of the (conjectured) "motivic" identity (5.6) via semiorthogonal decompositions of derived categories. See [106] for another (deeper) instance of this principle.

ACKNOWLEDGMENTS

First and foremost, I am grateful to my coauthor and friend Špela Špenko for contributing much of the mathematics of our joint work. Without her input this survey would have been a lot shorter.

Furthermore, I thank Shinnosuke Okawa for readily answering all my questions about the minimal model program. Likewise I thank Michael Wemyss for input on the noncommutative geometry of cDV singularities.

FUNDING

This work was partially supported by the Advanced ERC grant 885203 "Schobers, Mutations and Stability".

REFERENCES

[1] Quotient singularities with no crepant resolution?, 2011, https://mathoverflow.net/questions/66657/quotient-singularities-with-no-crepant-resolution/66702.

[2] C. Amiot and S. Oppermann, Cluster equivalence and graded derived equivalence. *Doc. Math.* **19** (2014), 1155–1206.

[3] R. Anno, R. Bezrukavnikov, and I. Mirković, Stability conditions for Slodowy slices and real variations of stability. *Mosc. Math. J.* **15** (2015), no. 2, 187–203, 403.

[4] M. Artin, On Azumaya algebras and finite dimensional representations of rings. *J. Algebra* **11** (1969), 532–563.

[5] M. Ballard, D. Favero, and L. Katzarkov, Variation of geometric invariant theory quotients and derived categories. *J. Reine Angew. Math.* **746** (2019), 235–303.

[6] V. V. Batyrev, Stringy Hodge numbers of varieties with Gorenstein canonical singularities. In *Integrable systems and algebraic geometry (Kobe/Kyoto, 1997)*, pp. 1–32, World Sci. Publ., River Edge, NJ, 1998.

[7] V. V. Batyrev, Non-Archimedean integrals and stringy Euler numbers of log-terminal pairs. *J. Eur. Math. Soc. (JEMS)* **1** (1999), no. 1, 5–33.

[8] V. V. Batyrev, Stringy Hodge numbers and Virasoro algebra. *Math. Res. Lett.* **7** (2000), no. 2–3, 155–164.

[9] V. V. Batyrev and D. I. Dais, Strong McKay correspondence, string-theoretic Hodge numbers and mirror symmetry. *Topology* **35** (1996), no. 4, 901–929.

[10] A. Beauville, Symplectic singularities. *Invent. Math.* **139** (2000), no. 3, 541–549.

[11] A. Beilinson, Coherent sheaves on $\mathbb{P}^n$ and problems of linear algebra. *Funct. Anal. Appl.* **12** (1978), 214–216.

[12] A. Beilinson and J. Bernstein, Localisation de $\mathfrak{g}$-modules. *C. R. Acad. Sci. Paris Sér. I Math.* **292** (1981), 15–18.

[13] R. Bezrukavnikov, Noncommutative counterparts of the Springer resolution. In *International Congress of Mathematicians. Vol. II*, pp. 1119–1144, Eur. Math. Soc., Zürich, 2006.

[14] R. Bezrukavnikov and D. Kaledin, McKay equivalence for symplectic resolutions of quotient singularities. *Tr. Mat. Inst. Steklova* **246** (2004), no. Algebr. Geom. Metody, Svyazi i Prilozh., 20–42.

[15] R. Bezrukavnikov and D. Kaledin, Fedosov quantization in positive characteristic. *J. Amer. Math. Soc.* **21** (2008), no. 2, 409–438.

[16] R. Bezrukavnikov, I. Mirković, and D. Rumynin, Localization of modules for a semisimple Lie algebra in prime characteristic. *Ann. of Math. (2)* **167** (2008), no. 3, 945–991, With an appendix by Bezrukavnikov and Simon Riche.

[17] C. Birkar, P. Cascini, C. D. Hacon, and J. McKernan, Existence of minimal models for varieties of log general type. *J. Amer. Math. Soc.* **23** (2010), no. 2, 405–468.

[18] A. Bodzenta and A. Bondal, Flops and spherical functors. 2015, arXiv:1511.00665v2.

[19] A. Bondal, Representations of associative algebras and coherent sheaves. *Math. USSR, Izv.* **34** (1990), no. 1, 23–42.

[20] A. Bondal, M. Kapranov, and V. Schechtman, Perverse schobers and birational geometry. *Selecta Math. (N.S.)* **24** (2018), no. 1, 85–143.

[21] A. Bondal and D. Orlov, Semiorthogonal decomposition for algebraic varieties. 1995, arXiv:alg-geom/9506012.

[22] A. Bondal and D. Orlov, Derived categories of coherent sheaves. In *Proceedings of the International Congress of Mathematicians, Vol. II (Beijing, 2002)*, pp. 47–56, Higher Ed. Press, Beijing, 2002.

[23] L. A. Borisov and R. P. Horja, Mellin–Barnes integrals as Fourier–Mukai transforms. *Adv. Math.* **207** (2006), no. 2, 876–927.

[24] L. Borisov, L. Chen, and G. Smith, The orbifold Chow ring of toric Deligne–Mumford stacks. *J. Amer. Math. Soc.* **18** (2005), no. 1, 193–215.

[25] L. Borisov and A. Libgober, Stringy E-functions of Pfaffian-Grassmannian double mirrors. *Algebr. Geom.* **6** (2019), no. 4, 486–515.

[26] L. Borisov and C. Wang, On stringy Euler characteristics of Clifford noncommutative varieties. *Adv. Math.* **349** (2019), 1117–1150.

[27] T. Bridgeland, Flops and derived categories. *Invent. Math.* **147** (2002), no. 3, 613–632.

[28] T. Bridgeland, A. King, and M. Reid, The McKay correspondence as an equivalence of derived categories. *J. Amer. Math. Soc.* **14** (2001), no. 3, 535–554.

[29] T. Bridgeland, Y. Qiu, and T. Sutherland, Stability conditions and the A2 quiver. *Adv. Math.* **365** (2020), 107049.

[30] T. Bridgeland and D. Stern, Helices on del Pezzo surfaces and tilting Calabi–Yau algebras. *Adv. Math.* **224** (2010), no. 4, 1672–1716.

[31] M. Brion, Sur les modules de covariants. *Ann. Sci. Éc. Norm. Supér. (4)* **26** (1993), 1–21.

[32] N. Broomhead, Dimer models and Calabi–Yau algebras. *Mem. Amer. Math. Soc.* **215** (2012), no. 1011, viii+86 pp.

[33] R. Brown, *Topology and groupoids*. BookSurge, LLC, Charleston, SC, 2006.

[34] R.-O. Buchweitz, G. J. Leuschke, and M. Van den Bergh, Non-commutative desingularization of determinantal varieties I. *Invent. Math.* **182** (2010), no. 1, 47–115.

[35] R.-O. Buchweitz, G. J. Leuschke, and M. Van den Bergh, On the derived category of Grassmannians in arbitrary characteristic. *Compos. Math.* **151** (2015), no. 7, 1242–1264.

[36] R.-O. Buchweitz, G. J. Leuschke, and M. Van den Bergh, Non-commutative desingularization of determinantal varieties, II: arbitrary minors. *Int. Math. Res. Not. IMRN* **9** (2016), 2748–2812.

[37] S. Cautis, J. Kamnitzer, and A. Licata, Derived equivalences for cotangent bundles of Grassmannians via categorical sl_2 actions. *J. Reine Angew. Math.* **675** (2013), 53–99.

[38] J.-C. Chen, Flops and equivalences of derived categories for threefolds with only terminal Gorenstein singularities. *J. Differential Geom.* **61** (2002), no. 2, 227–261.

[39] D. A. Cox, J. B. Little, and H. K. Schenck, *Toric varieties*. Grad. Stud. Math. 124, American Mathematical Society, Providence, RI, 2011.

[40] A. Craw, An introduction to motivic integration. In *Strings and geometry*, pp. 203–225, Clay Math. Proc. 3, Amer. Math. Soc., Providence, RI, 2004.

[41] H. Dao, Remarks on non-commutative crepant resolutions of complete intersections. *Adv. Math.* **224** (2010), no. 3, 1021–1030.

[42] H. Dao, O. Iyama, R. Takahashi, and M. Wemyss, Gorenstein modifications and $\mathbb{Q}$-Gorenstein rings. *J. Algebraic Geom.* **29** (2020), no. 4, 729–751.

[43] T. De Deyn, A note on affine cones over Grassmannians and their stringy E-functions. 2022, arXiv:2203.06040.

[44] L. de Thanhoffer de Völcsey and M. Van den Bergh, Some new examples of nondegenerate quiver potentials. *Int. Math. Res. Not. IMRN* **20** (2013), 4672–4686.

[45] J. Denef and F. Loeser, Germs of arcs on singular algebraic varieties and motivic integration. *Invent. Math.* **135** (1999), no. 1, 201–232.

[46] J. Denef and F. Loeser, Motivic integration, quotient singularities and the McKay correspondence. *Compos. Math.* **131** (2002), no. 3, 267–290.

[47] H. Derksen, J. Weyman, and A. Zelevinsky, Quivers with potentials and their representations. I. Mutations. *Selecta Math. (N.S.)* **14** (2008), no. 1, 59–119.

[48] W. Donovan, Perverse schobers on Riemann surfaces: constructions and examples. *Eur. J. Math.* **5** (2019), no. 3, 771–797.

[49] W. Donovan and E. Segal, Window shifts, flop equivalences and Grassmannian twists. *Compos. Math.* **150** (2014), no. 6, 942–978.

[50] W. Donovan and M. Wemyss, Stringy Kähler moduli, mutation and monodromy. 2019, arXiv:1907.10891.

[51] B. Doyle, Homological Projective Duality for the Plücker embedding of the Grassmannian. 2021, arXiv:2110.10589.

[52] D. Edidin and D. Rydh, Canonical reduction of stabilizers for Artin stacks with good moduli spaces. *Duke Math. J.* **170** (2021), no. 5, 827–880.

[53] A. I. Efimov, Derived categories of Grassmannians over integers and modular representation theory. *Adv. Math.* **304** (2017), 179–226.

[54] A. I. Efimov, Categorical smooth compactifications and generalized Hodge-to-de Rham degeneration. *Invent. Math.* **222** (2020), no. 2, 667–694.

[55] I. M. Gel'fand, M. M. Kapranov, and A. V. Zelevinsky, Hypergeometric functions and toric varieties. *Funktsional. Anal. i Prilozhen.* **23** (1989), no. 2, 12–26.

[56] T. G. Goodwillie, Cyclic homology, derivations, and the free loopspace. *Topology* **24** (1985), no. 2, 187–215.

[57] D. R. Gulotta, Properly ordered dimers, R-charges, and an efficient inverse algorithm. *J. High Energy Phys.* **10** (2008), 14.

[58] D. Halpern-Leistner, The derived category of a GIT quotient. *J. Amer. Math. Soc.* **28** (2015), no. 3, 871–912.

[59] D. Halpern-Leistner and S. Sam, Combinatorial constructions of derived equivalences. *J. Amer. Math. Soc.* **33** (2020), 735–773.

[60] D. Halpern-Leistner and I. Shipman, Autoequivalences of derived categories via geometric invariant theory. *Adv. Math.* **303** (2016), 1264–1299.

[61] W. Hara, On the Abuaf–Ueda flop via non-commutative crepant resolutions, SIGMA Symmetry Integrability Geom. *Methods Appl.* **17** (2021), Paper No. 044, 22 pp.

[62] L. Hille and M. Perling, Tilting bundles on rational surfaces and quasi-hereditary algebras. *Ann. Inst. Fourier (Grenoble)* **64** (2014), no. 2, 625–644.

[63] Y. Hirano and M. Wemyss, Stability conditions for 3-fold flops. 2019, arXiv:1907.09742.

[64] C. Ingalls and T. Yasuda, Log centres of twisted noncommutative crepant resolutions are Kawamata log terminal: remarks on a paper of Stafford and Van den Bergh. Preprint, 2013.

[65] C. Ingalls and T. Yasuda, Logarithmic centres of twisted noncommutative crepant resolutions are Kawamata log terminal, 2013, https://people.math.carleton.ca/~cingalls/research/presentations/cmsWinter2013.pdf.

[66] A. Ishii and K. Ueda, Dimer models and the special McKay correspondence. *Geom. Topol.* **19** (2015), no. 6, 3405–3466.

[67] O. Iyama and I. Reiten, Fomin–Zelevinsky mutation and tilting modules over Calabi–Yau algebras. *Amer. J. Math.* **130** (2008), no. 4, 1087–1149.

[68] O. Iyama and M. Wemyss, On the noncommutative Bondal–Orlov conjecture. *J. Reine Angew. Math.* **683** (2013), 119–128.

[69] O. Iyama and M. Wemyss, Maximal modifications and Auslander–Reiten duality for non-isolated singularities. *Invent. Math.* **197** (2014), no. 3, 521–586.

[70] O. Iyama and M. Wemyss, Singular derived categories of $\mathbb{Q}$-factorial terminalizations and maximal modification algebras. *Adv. Math.* **261** (2014), 85–121.

[71] O. Iyama and M. Wemyss, Tits cones intersections and applications, 2021, https://www.maths.gla.ac.uk/~mwemyss/MainFile_for_web.pdf.

[72] D. Kaledin, Derived equivalences by quantization. *Geom. Funct. Anal.* **17** (2008), no. 6, 1968–2004.

[73] M. M. Kapranov, On the derived categories of coherent sheaves on some homogeneous spaces. *Invent. Math.* **92** (1988), no. 3, 479–508.

[74] M. Kapranov and V. Schechtman. Perverse schobers 2015, arXiv:1411.2772.

[75] L. Katzarkov, M. Kontsevich, and T. Pantev, Hodge theoretic aspects of mirror symmetry. In *From Hodge theory to integrability and TQFT* tt*-*geometry*, pp. 87–174, Proc. Sympos. Pure Math. 78, Amer. Math. Soc., Providence, RI, 2008.

[76] Y. Kawamata, D-equivalence and K-equivalence. *J. Differential Geom.* **61** (2002), no. 1, 147–171.

[77] Y. Kawamata, In *Derived equivalence for stratified Mukai flop on* $G(2,4)$, *Mirror symmetry. V*, pp. 285–294, AMS/IP Stud. Adv. Math. 38, Amer. Math. Soc., Providence, RI, 2006.

[78] Y. Kawamata, Flops connect minimal models. *Publ. Res. Inst. Math. Sci.* **44** (2008), no. 2, 419–423.

[79] Y. Kawamata, Derived categories of toric varieties III. *Eur. J. Math.* **2** (2016), no. 1, 196–207.

[80] B. Keller, On the cyclic homology of exact categories. *J. Pure Appl. Algebra* **136** (1999), no. 1, 1–56.

[81] B. Keller, Quiver mutation in JavaScript and Java, https://webusers.imj-prg.fr/~bernhard.keller/quivermutation/.

[82] B. Keller and H. Krause, Tilting preserves finite global dimension. *C. R. Math. Acad. Sci. Paris* **358** (2020), no. 5, 563–570.

[83] B. Keller and D. Yang, Derived equivalences from mutations of quivers with potential. *Adv. Math.* **226** (2011), no. 3, 2118–2168.

[84] Y.-H. Kiem, The stringy E-function of the moduli space of rank 2 bundles over a Riemann surface of genus 3. *Trans. Amer. Math. Soc.* **355** (2003), no. 5, 1843–1856.

[85] F. C. Kirwan, Partial desingularisations of quotients of nonsingular varieties and their Betti numbers. *Ann. of Math. (2)* **122** (1985), no. 1, 41–85.

[86] A. Kite, Discriminants and quasi-symmetry. 2017, arXiv:1711.08940.

[87] F. Knop, Über die Glattheit von Quotientenabbildungen. *Manuscripta Math.* **56** (1986), no. 4, 419–427.

[88] J. Kollár, Flops. *Nagoya Math. J.* **113** (1989), 15–36.

[89] J. Kollár and S. Mori, *Birational geometry of algebraic varieties*. Cambridge Tracts in Math. 134, Cambridge University Press, Cambridge, 1998. With the collaboration of C. H. Clemens and A. Corti, Translated from the 1998 Japanese original

[90] M. Kontsevich, *Motivic integration*. Lecture at Orsay, December 1995.

[91] A. Kuznetsov, Homological projective duality. *Publ. Math. Inst. Hautes Études Sci.* **105** (2007), 157–220.

[92] A. Kuznetsov, Derived categories of quadric fibrations and intersections of quadrics. *Adv. Math.* **218** (2008), no. 5, 1340–1369.

[93] A. Kuznetsov, Lefschetz decompositions and categorical resolutions of singularities. *Selecta Math. (N.S.)* **13** (2008), no. 4, 661–696.

[94] A. Kuznetsov and V. A. Lunts, Categorical resolutions of irrational singularities. *Int. Math. Res. Not.* **13** (2015), 4536–4625.

[95] E. León-Cardenal, J. Martín-Morales, W. Veys, and J. Viu-Sos, Motivic zeta functions on Q-Gorenstein varieties. *Adv. Math.* **370** (2020), 107192.

[96] G. J. Leuschke, Non-commutative crepant resolutions: scenes from categorical geometry. In *Progress in commutative algebra 1*, pp. 293–361, de Gruyter, Berlin, 2012.

[97] L. Le Bruyn, Trace rings of generic 2 by 2 matrices. *Mem. Amer. Math. Soc.* **66** (1987), no. 363, vi+100 pp.

[98] L. Le Bruyn, Quiver concomitants are often reflexive Azumaya. *Proc. Amer. Math. Soc.* **105** (1989), no. 1, 10–16.

[99] L. Le Bruyn and M. Van den Bergh, Regularity of trace rings of generic matrices. *J. Algebra* **117** (1988), no. 1, 19–29.

[100] H.-W. Lin, On crepant resolution of some hypersurface singularities and a criterion for UFD. *Trans. Amer. Math. Soc.* **354** (2002), no. 5, 1861–1868.

[101] N. A. Loehr, In *Bijective combinatorics*, Discrete Math. Appl. (Boca Raton), CRC Press, Boca Raton, FL, 2011.

[102] V. A. Lunts, Categorical resolution of singularities. *J. Algebra* **323** (2010), no. 10, 2977–3003.

[103] Y. Namikawa, Mukai flops and derived categories. II. In *Algebraic structures and moduli spaces*, pp. 149–175, CRM Proc. Lecture Notes 38, Amer. Math. Soc., Providence, RI, 2004.

[104] Y. Namikawa, Induced nilpotent orbits and birational geometry. *Adv. Math.* **222** (2009), no. 2, 547–564.

[105] A. Neeman, The connection between the K-theory localization theorem of Thomason, Trobaugh and Yao and the smashing subcategories of Bousfield and Ravenel. *Ann. Sci. Éc. Norm. Supér. (4)* **25** (1992), no. 5, 547–566.

[106] A. Polishchuk and M. Van den Bergh, Semiorthogonal decompositions of the categories of equivariant coherent sheaves for some reflection groups. *J. Eur. Math. Soc. (JEMS)* **21** (2019), no. 9, 2653–2749.

[107] C. Procesi, The invariant theory of $n \times n$ matrices. *Adv. Math.* **19** (1976), no. 3, 306–381.

[108] C. Procesi, A formal inverse to the Cayley–Hamilton theorem. *J. Algebra* **107** (1987), no. 1, 63–74.

[109] Ju. P. Razmyslov, Identities with trace in full matrix algebras over a field of characteristic zero. *Izv. Akad. Nauk SSSR Ser. Mat.* **38** (1974), 723–756.

[110] Z. Reichstein, Stability and equivariant maps. *Invent. Math.* **96** (1989), no. 2, 349–383.

[111] M. Reid, Minimal models of canonical 3-folds. In *Algebraic varieties and analytic varieties (Tokyo, 1981)*, pp. 131–180, Adv. Stud. Pure Math. 1, North-Holland, Amsterdam, 1983.

[112] J. V. Rennemo, *Homological projective duality for symmetric rank loci*, lecture at the conference "Workshop on Derived Categories, Moduli Spaces, and Deformation Theory", June 2019.

[113] J. V. Rennemo and E. Segal, Hori-mological projective duality. *Duke Math. J.* **168** (2019), no. 11, 2127–2205.

[114] J. Rickard, Morita theory for derived categories. *J. Lond. Math. Soc. (2)* **39** (1989), 436–456.

[115] Š. Špenko, HMS symmetries and hypergeometric systems. In *Proceeding of the 8th European Congres of Mathematics*, 2021, to appear.

[116] Š. Špenko and M. Van den Bergh, Non-commutative resolutions of quotient singularities for reductive groups. *Invent. Math.* **210** (2017), no. 1, 3–67.

[117] Š. Špenko and M. Van den Bergh, A class of perverse schobers in geometric invariant theory. 2019, arXiv:1908.04213.

[118] Š. Špenko and M. Van den Bergh, Comparing the Kirwan and noncommutative resolutions of quotient varieties. 2019, arXiv:1912.01689.

[119] Š. Špenko and M. Van den Bergh, Non-commutative crepant resolutions for some toric singularities I. *Int. Math. Res. Not. IMRN* **21** (2020), 8120–8138.

[120] Š. Špenko and M. Van den Bergh, Non-commutative crepant resolutions for some toric singularities. II. *J. Noncommut. Geom.* **14** (2020), no. 1, 73–103.

[121] Š. Špenko and M. Van den Bergh, Perverse schobers and GKZ systems, arXiv:2007.04924. To appear in *Adv. Math.*, 2020.

[122] J. T. Stafford and M. Van den Bergh, Noncommutative resolutions and rational singularities. *Michigan Math. J.* **57** (2008), 659–674.

[123] M. Van den Bergh, Modules of covariants. In *Proceedings of the International Congress of Mathematicians, Vol. 1, 2 (Zürich, 1994)*, pp. 352–362, Birkhäuser, 1995.

[124] M. Van den Bergh, Cohen–Macaulayness of modules of covariants. *Invent. Math.* **106** (1991), 389–409.

[125] M. Van den Bergh, Cohen–Macaulayness of semi-invariants for tori. *Trans. Amer. Math. Soc.* **336** (1993), no. 2, 557–580.

[126] M. Van den Bergh, A converse to Stanley's conjecture for Sl_2. *Proc. Amer. Math. Soc.* **121** (1994), no. 1, 47–51.

[127] M. Van den Bergh, Local cohomology of modules of covariants. *Adv. Math.* **144** (1999), no. 2, 161–220.

[128] M. Van den Bergh, Non-commutative crepant resolutions. In *The legacy of Niels Henrik Abel*, pp. 749–770, Springer, Berlin, 2004.

[129] M. Van den Bergh, Three-dimensional flops and noncommutative rings. *Duke Math. J.* **122** (2004), no. 3, 423–455.

[130] M. Van den Bergh, Calabi—Yau algebras and superpotentials. *Selecta Math. (N.S.)* **21** (2015), no. 2, 555–603.

[131] M. Wemyss, Noncommutative resolutions. In *Noncommutative algebraic geometry*, pp. 239–306, Math. Sci. Res. Inst. Publ. 64, Cambridge Univ. Press, New York, 2016.

[132] M. Wemyss, Flops and clusters in the homological minimal model programme. *Invent. Math.* **211** (2018), no. 2, 435–521.

[133] M. Wemyss, A lockdown survey on cDV singularities. 2021, arXiv:2103.16990.

[134] H. Weyl, *The classical groups*. Princeton University Press, 1946.

[135] J. Weyman, *Cohomology of vector bundles and syzygies* 149, Cambridge University Press, 2003.

[136] J. Weyman and G. Zhao, Noncommutative desingularization of orbit closures for some representations of GL_n. 2012, arXiv:1204.0488.

MICHEL VAN DEN BERGH

Free University of Brussels, Pleinlaan 2, 1050 Brussel, Belgium, and
Hasselt University, Martelarenlaan 42, 3500 Hasselt, Belgium, and
Research Foundation – Flanders, Egmontstraat 5, 1000 Brussel, Belgium,
michel.vandenbergh@uhasselt.be, michel.van.den.bergh@vub.be

INTERACTIONS OF COMPUTATIONAL COMPLEXITY THEORY AND MATHEMATICS

AVI WIGDERSON

ABSTRACT

[This paper is a (modified, self contained) chapter in my recent book on computational complexity theory [176], called *Mathematics and Computation*, available online at https://www.math.ias.edu/avi/book].

We survey some concrete interaction areas between computational complexity theory and different fields of mathematics. We hope to demonstrate here that hardly any area of modern mathematics is untouched by the computational connection (which in some cases is completely natural and in others may seem quite surprising). In my view, the breadth, depth, beauty, and novelty of these connections is inspiring, and speaks to a great potential of future interactions (which indeed, are quickly expanding). We aim for *variety*. We give short, simple descriptions (without proofs or much technical detail) of ideas, motivations, results, and connections; this will hopefully entice the reader to dig deeper. Each vignette focuses only on a single topic within a large mathematical field, and is meant to be illustrative rather that comprehensive. We cover the following:

- Number Theory: *Primality testing*
- Combinatorial Geometry: *Point-line incidences*
- Operator Theory: *The Kadison–Singer problem*
- Metric Geometry: *Distortion of embeddings*
- Group Theory: *Generation and random generation*
- Statistical Physics: *Monte Carlo Markov chains*
- Analysis and Probability: *Noise stability*
- Lattice Theory: *Short vectors*
- Invariant Theory: *Group actions on matrix tuples (and beyond)*

MATHEMATICS SUBJECT CLASSIFICATION 2020

68Q01

KEYWORDS

Computational complexity

ICM INTERNATIONAL CONGRESS OF MATHEMATICIANS 2022 JULY 6–14

Proc. Int. Cong. Math. 2022, Vol. 2, pp. 1392–1432
DOI 10.4171/ICM2022/190

1. INTRODUCTION

The Theory of Computation (ToC) lays out the mathematical foundations of computer science. I am often asked if ToC is a branch of Mathematics, or of Computer Science. The answer is easy: it is clearly both (and in fact, much more). Ever since Turing's 1936 definition of the *Turing machine*, we have had a formal mathematical model of computation that enables the rigorous mathematical study of computational tasks, algorithms to solve them, and the resources these require. At the same time, the simple description of the Turing machine allowed its simple logical structure to be implemented in hardware, and its universal applicability fueled the rapid development of computer technology, which now dominates our life.

Computation was part mathematics from its origins, and motivated many of its developments. Algorithmic questions have occupied mathematicians throughout history (as elaborated in the introduction to the book [176]), and this naturally grew considerably when computers arrived. However, the advent of *computational complexity theory* over the past few decades has greatly expanded and deepened these connections. The study of new diverse models generated and studied in complexity theory broadened the nature of mathematical problems it encountered and formulated, and the mathematical areas and tools which bear upon these problems. This expansion has led to numerous new interactions that enrich both disciplines. This survey tells the stories of some of these interactions with different mathematical fields, illustrating their diversity.

We note in passing that a similar explosion of connections and interactions is underway between ToC and practically *all* sciences. These stem from computational aspects of diverse natural processes, which beg for algorithmic modeling and analysis. As with mathematics, these interactions of ToC with the sciences enrich both sides, expose *computation* as a central notion of intellectual thought, and highlight its study as an independent discipline, whose mission and goals expand way beyond those emanating from its parent fields of Math and CS. But this is the subject of a different survey (which I partly provide in the last chapter of [176]).

Back to the interactions of computational complexity theory and different areas of math. I have chosen to focus on essentially one problem or development within each mathematical field. Typically, this touches only a small subarea, and does not do justice to a wealth of other connections. Thus each vignette should be viewed as a demonstration of a larger body of work and even bigger potential. Indeed, while in some areas the collaborations are quite well established, in others they are just budding, with lots of exciting problems waiting to be solved and theories to be developed. Furthermore, the connections to algorithms and complexity (which I explain in each) are quite natural in some areas, but quite surprising in others. While the descriptions of each topic are relatively short, they include background and intuition, as well as further reading material. Indeed, I hope these vignettes will tempt the reader to explore further.

We note that new connections are discovered at a rapid pace. A strong case in point is the recent complexity-theoretic breakthrough of $\mathcal{MIP}^* = \mathcal{RE}$ [100], establishing the surprising power of quantum, multiprover interactive proof systems. This paper had already discussed several surprising applications resolving key conjectures in different mathematical areas, including operator algebras, quantum information theory, and group theory, and now more implications of the techniques and results are being pursued.

The sections below can be read in any order. The selection of fields and foci was affected by my personal taste and limited knowledge. More connections to other fields like Combinatorics, Optimization, Logic, Topology, Coding Theory, and Information Theory appear in parts of the book [176].

2. NUMBER THEORY

As mentioned, the need to efficiently compute mathematical objects has been central to mathematicians and scientists throughout history, and, of course, the earliest subject is arithmetic. Perhaps the most radical demonstration is the place value system we use to represent integers, which is in place for millenia precisely due to the fact that it supports extremely efficient manipulation of arithmetic operations. The next computational challenge in arithmetic, since antiquity, was accessing the multiplicative structure of integers represented this way.

Here is an except from C. F. Gauss' appeal[1] to the mathematics community of his time (in article 329 of *Disquisitiones Arithmeticae* (1801)), regarding the computational complexity of *testing primality* and *integer factorization*. The importance Gauss assigns to this computational challenge, his frustration of the state-of-the-art, and his imploring the mathematical community to resolve it shine through!

The problem of distinguishing prime numbers from composite numbers, and of resolving the latter into their prime factors is known to be one of the most important and useful in arithmetic. It has engaged the industry and wisdom of ancient and modern geometers to such an extent that it would be superfluous to discuss the problem at length. Nevertheless, we must confess that all methods that have been proposed thus far are either restricted to very special cases or are so laborious and difficult that even for numbers that do not exceed the limits of tables constructed by estimable men, they try the patience of even the practiced calculator. And these methods do not apply at all to larger numbers ... the dignity of the science itself seems to require that every possible means be explored for the solution of a problem so elegant and so celebrated.

We briefly recount the state-of-the-art of these two basic algorithmic problems in number theory. A remarkable response to Gauss' first question, *efficiently deciding primality*, was found in 2002 by Agrawal, Kayal, and Saxena [8]. The use of symbolic polynomials for this problem is completely novel. Here is their elegant characterization of prime numbers.

Theorem 2.1 ([8]). *An integer $N \geq 2$ is prime if and only if*

1 Which is, of course, in Latin. I copied this English translation from a wonderful survey of Granville [82] on the subject matter of this section.

- *N is not a perfect power,*
- *N does not have any prime factor $\leq (\log N)^4$,*
- *For every $r, a < (\log N)^4$, we have the following equivalence of polynomials over $\mathbb{Z}_N[X]$:*

$$(X + a)^N \equiv X^N + a \bmod (X^r - 1).$$

It is not hard to see that this characterization gives rise to a simple algorithm for testing primality that is deterministic, and runs in time that is *polynomial* in the binary description length of N. Previous deterministic algorithms either assumed the generalize Riemann hypothesis **[133]** or required slightly superpolynomial time **[5]**. The AKS deterministic algorithm came after a sequence of efficient *probabilistic* algorithms **[4,80,150,163]**, some elementary and some requiring sophisticated use and development of number-theoretic techniques. These probabilistic and deterministic algorithms were partly motivated by and are important to the field of cryptography.

What is not so well-known, even for those who did read the beautiful, ingenious proof in **[8]**, is that AKS developed their deterministic algorithm by carefully "derandomizing" a previous probabilistic algorithm for primality of **[7]** (which uses polynomials). We note that *derandomization*, the conversion of probabilistic algorithms into deterministic ones, is by now a major area in computational complexity with a rich theory, and many other similar successes as well as challenges. The stunning possibility that *every* efficient probabilistic algorithm has a deterministic counterpart is one of the major problems of computational complexity, and there is strong evidence supporting it (see **[94]**). Much more on this can be found in the randomness chapters of **[176]**.

Gauss' second challenge, of whether efficiently factoring integers is possible, remains open. But this very challenge has enriched computer science, both practical and theoretical, in several major ways. Indeed, the assumed hardness of factoring is the main guarantee of security in almost all cryptographic and e-commerce systems around the world (showing that difficult problems can be useful!). More generally, cryptography is an avid consumer of number theoretic notions, including elliptic curves, Weil pairings, and more, which are critical to a variety of cryptographic primitives and applications. These developments shatter Hardy's view of number theory as a completely useless intellectual endeavor.

There are several problems on integers whose natural definitions depend on factorization, but can, nevertheless, be solved efficiently, bypassing the seeming need to factor. Perhaps the earliest algorithm ever formally described is Euclid's algorithm for computing the GCD (greatest common divisor) of two given integers[2] m and n. Another famous such algorithm is for computing the Legendre–Jacobi symbol $(\frac{m}{n})$ via Gauss' law of quadratic reciprocity.

A fast algorithm for factoring may come out of left-field with the new development of quantum computing, the study of computers based on quantum-mechanical principles,

2 It extends to polynomials, and allows for an efficient way of computing multiplicative inverses in quotient rings of $\mathbb{Z}$ and $\mathbb{F}[x]$.

which we discussed in the quantum chapter of the book [176]. Shor has shown in [158] that such computers are capable of factoring integers in polynomial time. This result led governments, companies, and academia to invest billions in developing technologies which will enable building large-scale quantum computers, and the jury is still out on the feasibility of this project. There is no known theoretical impediment for doing so, but one possible reason for failure of this project is the existence of yet-undiscovered principles of quantum mechanics.

Other central computational problems include solving polynomial equations in finite fields, for which one of the earliest efficient (probabilistic) algorithms was developed by Berlekamp [32] (it remains a great challenge to derandomize this algorithm!). Many other examples can be found in the Algorithmic Number Theory book [26].

3. COMBINATORIAL GEOMETRY

What is the smallest area of a planar region which contains a unit length segment in *every* direction? This is the Kakeya needle problem (and such sets are called *Kakeya sets*), which was solved surprisingly by Besicovich [33] who showed that this area can be arbitrarily close to zero! Slight variation on his method produces a Kakeya set of Lebesque measure zero. It makes sense to replace "area" (namely, Lesbegue measure) by the more robust measures, such as the Hausdorff and Minkowski dimensions. This changes the picture: Davies [48] proved that a Kakeya set in the plane must have full dimension ($= 2$) in both measures, despite being so sparse in Lebesgue measure.

It is natural to extend this problem to higher dimensions. However, obtaining analogous results (namely, that the Hausdorff and Minkowski dimensions are full) turns out to be extremely difficult. Despite the seemingly recreational flavor, this problem has significant importance in a number of mathematical areas (Fourier analysis, wave equations, analytic number theory, and randomness extraction), and has been attacked through a considerable diversity of mathematical ideas (see [169]).

The following finite field analogue of the above Euclidean problem was suggested by Wolff [177]. Let $\mathbb{F}$ denote a finite field of size q. A set $K \subseteq \mathbb{F}^n$ is called Kakeya if it contains a line in every direction. More precisely, for every direction $b \in \mathbb{F}^n$ there is a point $a \in \mathbb{F}^n$ such that the line $\{a + bt : t \in \mathbb{F}\}$ is contained in K. As above, we would like to show that any such K must be large (think of the dimension n as a large constant, and the field size q as going to infinity).

Conjecture 3.1. Let $K \subseteq \mathbb{F}^n$ be a Kakeya set. Then $|K| \geq C_n q^n$, where C_n is a constant depending only on the dimension n.

The best exponent of q in such a lower bound intuitively corresponds to the Hausdorff and Minkowski dimensions in the Euclidean setting. Using sophisticated techniques from arithmetic combinatorics, Bourgain, Tao, and others improved the trivial bound of $n/2$ to about $4n/7$.

Curiously, the exact same conjecture arose, completely independently, within ToC, from the work [122] on *randomness extractors*, an area which studies the "purification" of "weak random sources" (see, e.g., the survey [171] on this important notion). In [122] Wolff's conjecture takes a probabilistic form, asking about the (min)-entropy of a random point on a random line in a Kakeya set. With this motivation, Dvir [61] brilliantly proved the Wolff conjecture (sometimes called the Finite Field Kakeya conjecture), using the (algebraic-geometric) "polynomial method" (which is inspired by techniques in decoding algebraic error-correcting codes). Many other applications of this technique to other geometric problems quickly followed, including the Guth–Katz [87] resolution of the famous Erdős distance problem, as well as for optimal randomness extraction and more (some are listed in Dvir's survey [62]).

Subsequent work determined the exact value of the constant C_n above (up to a factor of 2) [63].

Theorem 3.2 ([63]). *Let $K \subseteq \mathbb{F}^n$ be a Kakeya set. Then $|K| \geq (q/2)^n$. On the other hand, there exist Kakeya sets of size $\leq 2 \cdot (q/2)^n$.*

Many other problems regarding incidences of points and lines (and higher-dimensional geometric objects) have been the source of much activity and collaboration between geometers, algebraists, combinatorialists, and computer scientists. The motivation for these questions in the computer science side come from various sources, e.g., problems on local correction of errors [27] and derandomization [64, 105]. Other incidence theorems, e.g., Szemerédi–Trotter [168] and its finite field version of Bourgain–Katz–Tao [37] have been used, e.g., in randomness extraction [28] and compressed sensing [85].

4. OPERATOR THEORY

The following basic mathematical problem of Kadison and Singer from 1959 [102] was intended to formalize a basic question of Dirac concerning the "universality" of measurements in quantum mechanics. We need a few definitions. Consider $B(\mathcal{H})$, the algebra of continuous linear operators on a Hilbert space $\mathcal{H}$. Define a *state* to be a linear functional f on $B(\mathcal{H})$, normalized to $f(I) = 1$, which takes nonnegative values on positive semidefinite operators. The states form a convex set, and a state is called *pure* if it is not a convex combination of other states. Finally, let D be the subalgebra of $B(\mathcal{H})$ consisting of all *diagonal* operators (after fixing some basis).

Kadison and Singer asked if every pure state on D has a *unique* extension to $B(\mathcal{H})$. This problem on infinite-dimensional operators found a host of equivalent formulations in finite dimensions, with motivations and intuitions from operator theory, discrepancy theory, Banach space theory, signal processing, and probability. All of them were solved affirmatively in recent work of Marcus, Spielman, and Srivastava [127] (which also surveys the many related conjectures). Here is one statement they prove, which implies the others.

Theorem 4.1 ([127]). *For every $\varepsilon > 0$, there is an integer $k = k(\varepsilon)$ so that the following holds. Fix any n and any $n \times n$ matrix A with zeros on the diagonal and of spectral norm 1. Then there is a partition of $\{1, 2, \ldots, n\}$ into k subsets, $S_1, S_2, \ldots, S_k$, so that each of the principal minors A_i (namely A restricted to rows and columns in S_i) has spectral norm at most ε.*

This statement clearly implies that one of the minors has linear size, at least n/k. This consequence is known as the *Restricted Invertibility* Theorem of Bourgain and Tzafriri [38], itself an important result in operator theory.

How did computer scientists get interested in this problem? Without getting into too many details, here is a sketchy description of the meandering path which led to this spectacular result.

A central computational problem, at the heart of numerous applications, is solving a linear system of equations. While Gaussian elimination does the job quite efficiently (the number of arithmetic operations is about n^3 for $n \times n$ matrices), for large n this is still inefficient. Thus faster methods are sought, hopefully nearly linear in the number of nonzero entries of the given matrix. For *Laplacian*[3] linear systems (arising in many graph theory applications, such as computing electrical flows and random walks), Spielman and Teng [165] achieved precisely that! A major notion they introduced was *spectral sparsifiers* of matrices (or equivalently, weighted graphs).

A sparsifier of a given matrix is another matrix, with far fewer (indeed, linear) nonzero entries, which, nevertheless, has essentially the same (normalized) spectrum as the original (it is not even obvious that such a sparse matrix exists). We note that a very special case of sparsifiers of complete graphs are by definition *expander graphs*[4] (see much more about this central concept of expanders in [93, 176]). The algorithmic applications led to a quest for optimal constructions of sparsifiers for arbitrary Laplacian matrices (in terms of trade-off between sparsity and approximation), and these were beautifully achieved in [29] (who also provided a deterministic polynomial time algorithm to construct such sparsifiers). This in turn has led [164] to a new proof, with better analysis, of the Restricted Invertibility theorem mentioned above, making the connection to the Kadison–Singer problem.

However, the solution to Kadison–Singer seemed to require another detour. The same team [126] first resolved a bold conjecture of Bilu and Linial [34] on the spectrum of "signings" of matrices.[5] This conjecture was part of a plan for a *simple*, iterative construction of Ramanujan graphs, the best[6] possible expander graphs. Ramanujan graphs were introduced and constructed in [124, 128], but rely on deep results in number theory and alge-

---

3 Simply, symmetric PSD matrices with zero row sum.

4 All nontrivial eigenvalues of the complete graph (or constant matrix) are 0, and an expander is a sparse graph in which all nontrivial eigenvalues are tiny.

5 Simply, this beautiful conjecture states that for *every* d-regular graph, there exist $\{-1, 1\}$ signs of the edges which make all eigenvalues of the resulting signed adjacency matrix lie in the "Ramanujan interval" $[-2\sqrt{d-1}, 2\sqrt{d-1}]$.

6 With respect to the spectral gap. This is one of a few important expansion parameters to optimize.

braic geometry (believed by some to be essential for *any* such construction). Bilu and Linial sought instead an elementary construction, and made progress on their conjecture, showing how their iterative approach gives yet another way to construct "close to" Ramanujan expanders.

To prove the Bilu–Linial conjecture (and indeed produce bipartite Ramanujan graphs of every possible degree—something the algebraic constructions could not provide), [126] developed a theory of *interlacing polynomials* that turned out to be the key technical tool for resolving Kadison–Singer in [127]. In both cases, the novel view is to think of these conjectures probabilistically, and analyze the norm of a random operator by analyzing the average characteristic polynomial. That this method makes sense and actually works is deep and mysterious. Moreover, it provides a new kind of existence proofs for which no efficient algorithm (even probabilistic) of finding the desired objects is known. The analysis makes heavy use of the theory of *real stable* polynomials, and the inductive process underlying it is reminiscent (and inspired by) Gurvits' [86] remarkable proof of the van der Waerden conjecture and its generalizations.[7]

5. METRIC GEOMETRY

How close one metric space is to another is captured by the notion of *distortion*, measuring how distorted distances of one become when embedded into the other. More precisely,

Definition 5.1. Let (X, d) and (X', d') be two metric spaces. An embedding $f : X \to X'$ has distortion $\leq c$ if for every pair of points $x, y \in X$ we have

$$d(x, y) \leq d'\big(f(x), f(y)\big) \leq c \cdot d(x, y).$$

When X is finite and of size n, we allow $c = c(n)$ to depend on n.

Understanding the best embeddings between various metric and normed spaces has been a long endeavor in Banach space theory and metric geometry. An example of one major result in this area is Bourgain's embedding theorem [36].

Theorem 5.2 ([36]). *Every metric space of size n can be embedded into Euclidean space L_2 with distortion $O(\log n)$.*

The first connection between these structural questions and computational complexity was made in the important paper of Linial, London, and Rabinovich [120]. They asked for efficient algorithms for actually finding embeddings of low distortion, and noticed that for some such problems it is natural to use semidefinite programming. They applied

7 This is yet another example of a structural result (on doubly stochastic matrices) whose proof was partly motivated by algorithmic ideas. The connection is the use of hyperbolic polynomials in optimization (more specifically, as barrier functions in interior point methods.

this geometric connection to get old and new results for algorithmic problems on graphs (in particular, the sparsest cut problem we will soon discuss. Another motivation they discuss (which quickly developed into a major direction in approximation algorithms) is that some computations (e.g., finding nearest neighbors) are more efficient in some spaces than others, and so *efficient*, low-distortion embedding may provide useful reductions from harder to easier space. They describe such an efficient algorithm implementing Bourgain's Theorem 5.2 above, and also prove that his bound is best possible (the metric proving it is simply the distance between points in any constant-degree *expander* graph[8]).

The next shift in the evolution of this field, and in the level of interactions between geometers and ToC researchers, came from trying to prove "hardness of approximation" results. One example is the Goemans–Linial conjecture [78, 119], studying the sparsest cut problem, about the relation between L_1 and the "negative type" metric space L_2^2 (a general class of metrics which arise naturally in several contexts). Roughly, these are metrics on $\mathbb{R}^n$ in which Euclidean distances are squared. More precisely, a metric (X, d) is of negative type (namely, in L_2^2), if $(X, \sqrt{d})$, is isometric (has no distortion) to a subset of L_2.

Conjecture 5.3. Every L_2^2 metric can be embedded into L_1 with constant distortion.

This conjecture was proved false by Khot and Vishnoi [110]:

Theorem 5.4 ([110]). *For every n, there are n-point subsets of L_2^2 for which every embedding to L_1 requires distortion $\Omega(\log\log n)^{1/6}$.*

Far more interesting than the result itself is its origin. Khot and Vishnoi were trying to prove that the (weighted) "sparsest cut" problem is hard to approximate. They managed to do so under a computational assumption, known as the *Unique Games* conjecture of Khot [107] via a so-called *PCP*-reduction (see also [108, 176]). The elimination of this computational assumption is the magical part that demonstrates the power and versatility of reductions between computational problems. They apply their PCP reduction to a *particular*, carefully chosen unique games instance, which cannot be well approximated by a certain semidefinite program. The outcome was an instance of the sparsest cut problem which the same reduction ensures is hard to approximate by a semidefinite program. As discussed above, that outcome instance could be understood as a metric space, and the hardness of approximation translates to the required distortion bound!

The exact distortion of embedding L_2^2 into L_1 has been determined precisely to be $\sqrt{\log n}$ (up to lower order factors) in two beautiful sequences of works developing new algorithmic and geometric tools; we mention only the final word for each, as these papers contain a detailed history. On the upper bound side, the efficient algorithm approximating nonuniform sparsest cut to a factor $\sqrt{\log n}\log\log n$, which yields the same distortion bound, was obtained by Arora, Lee, and Naor [20] via a combination of the so-called "chaining argument" of [21] and the "measured descent" embedding method of [112]. A lower bound

8 The presence of such graphs in different sections illustrate how fundamental they are in diverse mathematical areas, and the same holds for algorithms and complexity theory.

of $\sqrt{\log n}$ on the distortion was very recently proved by Naor and Young [144] using a new isoperimetric inequality on the Heisenberg group.

Another powerful connection between such questions and ToC is through (again) expander graphs. A basic example is that the graph metric of any constant-degree expander proves that Bourgain's embedding theorem above is optimal! Much more sophisticated examples arise from trying to understand (and perhaps disprove) the Novikov and the Baum–Connes conjectures (see [104]). This program relies on another, much weaker notion of *coarse* embedding.

Definition 5.5. (X, d) has a coarse embedding into (X', d') if there is a map $f : X \to X'$ and two increasing, unbounded real functions α, β such that for every two points $x, y \in X$,

$$\alpha\big(d(x, y)\big) \leq d'\big(f(x), f(y)\big) \leq \beta\big(d(x, y)\big).$$

Gromov [83] was the first to construct a metric (the word metric of a group) which cannot be coarsely embedded into a Hilbert space. His construction uses an infinite family of *Cayley* expanders (graphs defined by groups). This result was greatly generalized by Lafforgue [114] and Mendel–Naor [130], who constructed graph metrics that cannot be coarsely embedded into any *uniformly convex* space. It is interesting that while Lafforgue's method is algebraic, the Mendel–Naor construction follows the combinatorial *zigzag* construction of expanders [155] from computational complexity.

Many other interaction projects regarding metric embeddings and distortion we did not touch on include their use in numerous algorithmic and data structure problems like clustering, distance oracles the k-server problem, as well as the fundamental interplay between distortion and *dimension reduction* relevant to both geometry and CS, where so many basic problems are open.

6. GROUP THEORY

Group theorists, much like number theorists, have been intrinsically interested in computational problems since the origin of the field. For example, the *word problem* (given a word in the generators of some group, does it evaluate to the trivial element?) is so fundamental to understanding any group one studies, that as soon as language was created to formally discuss the computational complexity of this problem, hosts of results followed trying to pinpoint that complexity. These include decidability and undecidability results once Turing set up the theory of computation and provided the first undecidable problems, and these were followed with $\mathcal{NP}$-completeness results and efficient algorithms once $\mathcal{P}$ and $\mathcal{NP}$ were introduced around 1970. Needless to say, these *algorithmic results* inform of *structural* complexity of the groups at hand. And the word problem is but the first example. Another demonstration is the beautiful interplay between algorithmic and structural advances over decades, on the *graph isomorphism problem*, recently leading to breakthrough of Babai [24]! A huge body of work is devoted to finding efficient algorithms for computing commutator subgroups, Sylow subgroups, centralizers, bases, representations, characters, and a host of other important substructures of a group from some natural description of it. Excellent textbooks include [92, 157].

Here we focus on two related problems, the *generation* and *random generation* problems, and new conceptual notions borrowed from computational complexity which are essential for studying them. Before defining them formally (below), let us consider an example. Assume I hand you 10 invertible matrices, say 100×100 in size, over the field of size 3. Can you tell me if they generate another such given matrix? Can you even produce convincing evidence of this before we both perish? How about generating a random matrix in the subgroup spanned by these generators? The problem, of course, is that this subgroup will have size far larger than the number of atoms in the known universe, so its elements cannot be listed, and typical words generating elements in the group may need to be prohibitively long. Indeed, even the extremely special cases, for elements in $\mathbb{Z}_p^*$ (namely one, 1×1 matrix), the first question is related to the *discrete logarithm* problem, and for $\mathbb{Z}_{p \cdot q}^*$ it is related to the *integer factoring* problem, both currently requiring exponential time to solve (as a function of the description length).

Let us consider any finite group G and let $n \approx \log |G|$ be roughly the length of a description of an element of G. Assume we are given k elements in G, $S = \{s_1, s_2, \ldots, s_k\}$. It would be ideal if the procedures we describe would work in time polynomial in n and k (which prohibits enumerating the elements of G, whose size is exponential in n).

The *generation problem* asks if a given element $g \in G$ is generated by S. How does one prove such a fact? A standard certificate for a positive answer is a *word* in the elements of S (and their inverses) which evaluates to g. However, even if G is cyclic, the shortest such word may be exponential in n. An alternative, computationally motivated description, is to give a *program* for g. Its definition shows that the term "program" suits it perfectly, as it has the same structure as usual computer programs, only that instead of applying some standard Boolean or arithmetic operations, we use the group operations of multiplication and inverse.

Definition 6.1. A *program* (over S) is a finite sequence of elements $g_1, g_2, \ldots, g_m$, where every element g_i is either in S, or is the inverse of a previous g_j, or is the product of previous g_j, g_ℓ. We say that it computes g simply if $g = g_m$.

In the cyclic case, programs afford exponential savings over words in description length, as a program allows us to write large powers by repeatedly squaring elements. What is remarkable is that such savings are possible for *every* group. This discovery of Babai and Szemerédi [25] says that every element of every group has an extremely succinct description in terms of any set of elements generating it.

Theorem 6.2 ([25]). *For every group G, if a subset of elements S generates another element g, then there is a program of length at most $n^2 \approx (\log |G|)^2$ which computes g from S.*

It is interesting to note that the proof uses a structure which is very combinatorial and counterintuitive for group theorists, namely that of a *cube*, which we will see again later. For a sequence $(h_1, h_2, \ldots, h_t)$ of elements from G, the cube $C(h_1, h_2, \ldots, h_t)$ is the (multi)set of 2^t elements $\{h_1^{\varepsilon_1}, h_2^{\varepsilon_2}, \ldots, h_t^{\varepsilon_t}\}$, with $\varepsilon_i \in \{0, 1\}$. Another important feature of the proof is that it works in a very general setting of "black-box" groups—it never needs an explicit description of the host group, only the ability to multiply elements and take their inverses. This is a very important paradigm for arguing about groups, and will be used again below.

How does one prove that an element g is *not* generated by S? It is possible that there is no short "classical" proof! This question motivated Babai to define Arthur–Merlin games—a new notion of probabilistic, interactive proofs (simultaneously with Goldwasser, Micali, and Rackoff [81], who proposed a similar notion for cryptographic reasons), and showed how nonmembership can be certified in this new framework. The impact of the definition of interactive proofs on the theory of computation has been immense, and is discussed in, e.g., in the books [19,79,176].

Returning to the generation problem, let us now consider the problem of *random generation*. Here we are given S, and would like a randomized procedure which will quickly output an (almost) uniform distribution on the subgroup H of G generated by S. This problem, besides its natural appeal, is often faced by computational group theorists, being a subroutine in many group-theoretic algorithms. In practice often heuristics are used, like the famous "product replacement algorithm" and its variants, which often work well in practice (see, e.g., the recent [22] and references). We will discuss here provable bounds.

It is clear that sufficiently long random words in the elements of S and its inverses will do the job, but just as with certificates, sufficiently long is often prohibitively long. In a beautiful paper, Babai [23] describes a certain process generating a random program which computes a nearly-uniform element of H, and runs in time $n^5 \approx (\log |G|)^5$ steps. It again uses cubes, and works in the full generality of black-box groups. This paper was followed by even faster algorithms with simpler analysis by Cooperman and by Dixon [45,58], and the state-of-the-art is an algorithm whose number of steps is remarkably the same as the length of proofs of generation above—in other words, randomness roughly achieves the efficiency of nondeterminism for this problem. Summarizing:

Theorem 6.3 ([23,45,58]). *For every group G, there is a* probabilistic *program of length* $\mathrm{poly}(n) \approx \mathrm{poly}(\log |G|)$ *that, given* any *generating set S for G, produces with high probability a (nearly) uniformly random element of G.*

7. STATISTICAL PHYSICS

The field of statistical physics is huge, and we focus here mainly on connections of statistical mechanics with the theory of computation. Numerous mathematical models exist of various physical and chemical systems, designed to understand basic properties of different materials and the dynamics of basic processes. These include such familiar models as Ising, Potts, monomer–dimer, spin-glass, percolation, etc. A typical example explaining the connection of such mathematical models to physics and chemistry, and the basic problems studied is the seminal paper of Heilmann and Lieb [90].

Many of the problems studied can be viewed in the following general setting. We have a huge (exponential) space of objects called Ω (these objects may be viewed as the different configurations of a system). Each object is assigned a nonnegative weight (which may be viewed as the "energy" of that state). Scaling these weights gives rise to a probability distribution (often called the Gibbs distribution) on Ω, and to study its properties (phase transitions, critical temperatures, free energy, etc.) one attempts to generate samples from this distribution. Note that if the description of a state takes n bits, then brute-force listing of all probabilities in question is exponentially prohibitive. Thus efficiency of the sampling procedure is essential to this study.

As Ω may be highly unstructured, the most common approach to this sampling problem is known as "Monte Carlo Markov Chain" (or "MCMC") method. The idea is to build a graph on the objects of Ω, with a pair of objects connected by an edge if they are similar in some sense (e.g., sequences which differ only in a few coordinates). Next, one starts from any object, and performs a biased random walk on this graph for some time, and the object reached is the sample produced. In many settings it is not hard to set up the random walk (often called Glauber dynamics or the Metropolis algorithm) so that the *limiting* distribution of the Markov chain is indeed the desired distribution. The main question in this approach is *when* to stop the walk and output a sample; *when* are we close enough to the limit? In other words, how long does it take the chain to converge to the limit? In most cases, these decisions were taken on intuitive, heuristic grounds, without rigorous analysis of convergence time. The exceptions where rigorous bounds were known were typically structured, e.g., where the chain was a Cayley graph of a group (e.g., [11, 56]).

This state of affairs has changed considerably since the interaction in the past couple of decades with the theory of computation. Before describing it, let us see where computational problems even arise in this field. The two major sources are *optimization* and *counting*. That the setting above suits many instances of optimization problems is easy to see. Think of Ω as the set of solutions to a given optimization problem (e.g., the values of certain parameters designed to satisfy a set of constraints), and the weights representing the quality of a solution (e.g., the number of constraints satisfied). So, picking at random from the associated distribution favors high-quality solutions. The counting connection is more subtle. Here Ω represents a set of combinatorial objects one wants to count or approximate (e.g., the set of perfect matchings in a graph, or satisfying assignments to a set of constraints). It turns out that for very general situations of this type, sampling an object (approximately) at random is tightly connected to counting their number; it often allows a recursive procedure to approximate the size of the set [99]. An additional observation is that viewing a finite set as a fine discretization of a continuous object (e.g., fine lattice points in a convex set) allows one to compute volumes and more generally integrate functions over such domains.

Around 1990, rigorous techniques were introduced [12, 39, 65, 161] to analyze the convergence rates of such general Markov chains arising from different approximation algorithms. They establish *conductance* bounds on the Markov chains, mainly via *canonical paths* or *coupling* arguments (a survey of this early work is [96]). Collaborative work was soon able to formally justify the physical intuition behind some of the suggested heuristics for many models, and, moreover, drew physicists to suggest such ingenious chains for optimization problems. The field drew in probabilists and geometers as well, and by now is highly active and diverse. We mention two results to illustrate rigorous convergence bounds for important problems of this type.

Theorem 7.1 ([97]). *The permanent of any nonnegative $n \times n$ matrix can be approximated, to any multiplicative factor $(1+\varepsilon)$, in polynomial time in n/ε.*

The importance of this approximation algorithm stems from the seminal result of Valiant [173] about the permanent polynomial (that notorious sibling of the determinant polynomial, that looks identical except that the permanent has no signs; for more see [159, 176]). Valiant proved that the permanent is *universal*, capturing (via efficient reductions) essentially all natural counting problems, including those arising in the statistical physics models and optimization and counting problems above. So, unlike determinant, computing the permanent *exactly* is extremely difficult (harder than $\mathcal{NP}$-complete).

Theorem 7.2 ([65]). *The volume of any convex set in n dimensions can be approximated, to any multiplicative factor $(1+\varepsilon)$, in polynomial time in n/ε.*

The volume, besides its intrinsic interest, captures as well natural counting problems, e.g., the number of linear extensions of a given partially ordered set. The analysis of this algorithm, as well as its many subsequent improvements, has used and developed purely structural results of independent interest in differential and convex geometry. It also led to generalizations, like efficiently sampling from any log-concave distribution (see the survey [174]).

Another consequence of this collaboration was a deeper understanding of the relation between *spacial* properties (such as phase transitions, and long-range correlations between distant sites in the Gibbs distribution) and *temporal* properties (such as speed of convergence of the sampling or approximately counting algorithms, like Glauber dynamics). This connection (surveyed, e.g., in [66]) was established by physicists for spin systems since the 1970s. The breakthrough work of Weitz [175] on the *hard core* model gave an *deterministic* algorithm which is efficient up to the phase transition, and this was complemented by a hardness result of Sly [162] beyond the phase transition. These phase transition of computational complexity, at the same point as the phase transition of the Gibbs distribution are striking, and the generality of this phenomenon is still investigated.

More generally, the close similarity between statistical physics models and optimization problems, especially on random instances, is benefitting both sides. Let us mention a few exciting developments. It has unraveled the fine geometric structure of the space of solutions at the phase transition, pinpointing it, e.g., for k-SAT in [1]. At the same time, physics intuition based on such ideas as renormalization, annealing, and replica symmetry breaking has led to new algorithms for optimization problems, some of them now rigorously analyzed, e.g., as in [98]. Others, like one of the fastest (yet unproven) heuristics for such problems as Boolean Satisfiability (which is $\mathcal{NP}$-complete in general) are based on the physics method of "survey propagation" of [131]. Finally, new algorithmic techniques for similar physics and optimization problems, originate from an unexpected source, the *Lovasz Local Lemma* (LLL). The LLL is a probabilistic proof technique for the existence rare events in a probability space. Its efficient versions, formulating it algorithmically as a *directed, nonreversible* Markov chains, starting with the works of Moser [136,137], have led to approximate counting and sampling versions for such events (see, e.g., [84]). A completely different, *deterministic* algorithm of Moitra [135] for the LLL regime (of rare events) promises many more applications: it works even when the solution space (and hence the natural Markov chain) is not connected!

We conclude this story with the recent breakthrough connection between high-dimensional expanders and the analysis of MCMC of Anari, Liu, Gharan, and Vinzant [18]. The theory of high-dimensional expanders (generalizing that of expander graphs to higher dimensional complexes – see the survey [123]), which easily merits a separate vignette, has been rapidly developing in the past decade within combinatorics and complexity theory, following deep roots in the theory of Bruhat–Tits buildings, connections with several areas of math, and new applications. Anari et al. realized that the local-to-global principle underlying high-dimensional expansion can be used for an inductive analysis of the convergence rate of many families of Markov chain-based algorithms. Their first application resolves a 30-year old conjecture, proving

Theorem 7.3 ([18]). *The number of bases of any matroid on n elements can be approximated, to any multiplicative factor $(1+\varepsilon)$, in polynomial time in n/ε.*

The revolutionary impact and future potential of this connection (and further ideas), in just a couple of years, to problems of approximate counting and random sampling in optimization and statistical physics, can be appreciated, e.g., from these papers and the references therein [16,17,121].

8. ANALYSIS AND PROBABILITY

This section gives a taste of a growing number of families of inequalities—large deviation inequalities, isoperimetric inequalities, etc.—that have been generalized beyond their classical origins due to a variety of motivations in the theory of computing and discrete mathematics. Further, the applications sometimes call for *stability* versions of these inequalities, namely an understanding of the structures which make an inequality nearly sharp. Here too these motivations pushed for generalizations of classical results and many new ones. Most of the material below, and much more on the motivations, applications, and developments in this exciting area of the analysis of Boolean functions, can be found in the book [145] by O'Donnell.

The following story can be told from several angles. One is the *noise sensitivity* of functions. We restrict ourselves to the Boolean cube endowed with the uniform probability measure, but many of the questions and results extend to arbitrary product probability spaces. Let $f : \{-1,1\}^n \to \mathbb{R}$, which we assume is balanced, namely $E[f] = 0$. When the image of f is $\{-1,1\}$, we can think of f as a voting scheme, translating the binary votes of n individuals into a binary outcome. One natural desire from such a voting scheme may be *noise stability*—that typically very similar inputs (vote vectors) will yield the same outcome.

While natural in this social science setting, such questions also arise in statistical physics settings, where natural functions such as bond percolation turn out to be extremely sensitive to noise [31]. Let us formally define noise stability.

Definition 8.1. Let $\rho \in [0, 1]$ be a correlation parameter. We say two vectors $x, y \in \{-1, 1\}^n$ are ρ-correlated if they are distributed as follows. The vector x is drawn uniformly at random, and y is obtained from x by flipping each bit x_i independently with probability $(1-\rho)/2$. Note that for every i the correlation $E[x_i y_i] = \rho$. The *noise sensitivity* of f at ρ, $S_\rho(f)$, is simply defined as the correlation of the outputs, $E[f(x)f(y)]$.

It is not hard to see that the function maximizing noise stability is any *dictatorship* function, e.g., $f(x) = x_1$, for which $S_\rho(f) = \rho$. But another natural social scientific concern is the *influence* of players in voting schemes [30], which prohibits such solutions (in democratic environments). The influence of a single voter[9] is the probability with which it can change the outcome given that all other votes are uniformly random (so, in a dictatorship it is 1 for the dictator and 0 for all others). A fair voting scheme should have no voter with high influence. As we define influence for real-valued functions, we will use the (conditional) *variance* to measure a player's potential effect given all other (random) votes.

Definition 8.2. A function $f : \{-1, 1\}^n \to \mathbb{R}$ has influence τ if for every i, $\mathrm{Var}[x_i | x_{-i}] \le \tau$ for all i (where x_{-i} denotes the vector x without the ith coordinate).

For example, the majority function has influence $O(1/\sqrt{n})$. The question of how small the influence of a balanced function can be is extremely interesting, and leads to a highly relevant inequality for our story (both in content and techniques). As it turns out, ultimate fairness (influence $1/n$ per player) is impossible— the authors of [103] show that every function has a player with nonproportional influence, at least $\Omega(\log n/n)$. At any rate, one can ask which of the functions with *small* influence is most stable, and it is natural to guess that majority should be the best.[10]

The conjecture that this is the case, called the *Majority is Stablest* conjecture, arose from a completely different and surprising angle—the field of optimization, specifically "hardness of approximation." A remarkable paper [109] has shown that this conjecture implies[11] the optimality of a certain natural algorithm for approximating the *maximum cut* of a graph (i.e., the partition of vertices that maximizes the number of edges between them).[12] This connection is highly nontrivial, but by now we have many examples showing how the analysis of certain (semidefinite programming-based) approximation algorithms for a vari-

---

9 This seminal paper [30] also studies the influences of coalitions of players, extremely natural in game theory, which arises in and contributes to other areas of computational complexity (including circuit complexity, learning and pseudorandomness), and raises other analytic questions which we will not discuss here.

10 This noise sensitivity tends, as n grows, to $S_\rho(\mathrm{Majority}_n) = \frac{2}{\pi} \arcsin \rho$.

11 Assuming another, complexity-theoretic, conjecture called the "Unique Games" conjecture of [107] (discussed already in the metric geometry section above; see also [108, 176]).

12 Maximum Cut is a basic optimization problem whose exact complexity is $\mathcal{NP}$-complete.

ety of optimization problems raise many new isoperimetric questions,[13] greatly enriching this field.

The Majority is Stablest conjecture was proved in a strong form by **[138]** shortly after it was posed. Here is a formal statement (which actually works for bounded functions).

Theorem 8.3 (**[138]**). *For every (positive correlation parameter) $\rho \geq 0$ and $\varepsilon > 0$, there exists (an influence bound) $\tau = \tau(\rho, \varepsilon)$ such that for every n and every $f : \{-1, 1\}^n \to [-1, 1]$ of influence at most τ, $S_\rho(f) \leq S_\rho(\text{Majority}_n) + \varepsilon$.*

The proof reveals another angle on the story—large deviation inequalities and invariance principles. To see the connection, recall the Berry–Esseen theorem **[70]**, generalizing the standard central limit theorem to *weighted* sums of independent random signs. In this theorem, influences arise very naturally. Consider $\sum_{i=1}^n c_i x_i$. If we normalize the weights c_i to satisfy $\sum_i c_i^2 = 1$, then c_i is the influence of the ith voter, and $\tau = \max_i |c_i|$. The quality of this central limit theorem deteriorates linearly with the influence τ. Lindeberg's proof of Berry–Esseen uses an invariance principle, showing that for linear functions, the cumulative probability distribution $\Pr[\sum_{i=1}^n c_i x_i \leq t]$ (for every t) is unchanged (up to τ), *regardless* of the distribution of the variables x_i, as long as they are independent and have expectation 0 and variance 1. Thus, in particular, they can be taken to be standard Gaussian, which trivializes the problem, as the weighted sum is a Gaussian as well!

To prove their theorem, **[138]** first observed that also in the noise stability problem, the Gaussian case is simple. If the x_i, y_i are standard Gaussians with correlation ρ, the stability problem reduces to a classical result of Borell **[35]**: that noise stability is maximized by any hyperplane through the origin. Note that here the rotational symmetry of multidimensional Gaussians, which also aids the proof, does not distinguish "dictator" functions from majority—both are such hyperplanes. Given this theorem, an invariance principle whose quality depends on τ would do the job. They next show that it is sufficient to prove the principle only for *low degree* multilinear polynomials (as the effect of noise decays with the degree). Finally, they prove this nonlinear extension of Berry–Esseen for such polynomials, a form of which we state below. They also use their invariance principle to prove other conjectures, and since the publication of their paper, quite a number of further generalizations and applications were found.

Theorem 8.4 (**[138]**). *Let x_i be any n independent random variables with mean 0, variance 1, and bounded 3rd moments. Let g_i be n independent standard Gaussians. Let Q be any degree d multilinear n-variate polynomial of influence τ. Then for any t,*

$$\left|\Pr\left[Q(x) \leq t\right] - \Pr\left[Q(g) \leq t\right]\right| \leq O(d\tau^{1/d}).$$

We now only seem to be switching gears... To conclude this section, let me give one more, very different demonstration of the surprising questions (and answers) regarding

13 Many over continuous domains, like the unit cube or Gaussian space, where the connection between noise stability and isoperimetry may be even clearer.

noise stability and isoperimetry, arising from the very same computational considerations of optimization of hardness of approximation. Here is the question: *What is the smallest surface area of a (volume 1) body which tiles $\mathbb{R}^d$ periodically along the integer lattice $\mathbb{Z}^d$?* Namely, we seek a d-dimensional volume 1 subset $B \subseteq \mathbb{R}^d$ such that $B + \mathbb{Z}^d = \mathbb{R}^d$, such that its boundary has minimal $(d-1)$-dimensional volume.[14] Let us denote this infimum by $s(d)$. The curious reader can stop here a bit and test your intuition, what do you expect the answer to be, asymptotically in d?

Such questions originate from the late 19th century study by Thomson (later Lord Kelvin) of *foams* in 3 dimensions [170], further studied, generalized, and applied in mathematics, physics, chemistry, material science, and even architecture. However, for this very basic question, where periodicity is defined by the simplest integer lattice, it seems that, for large d, the trivial upper and lower bounds on $s(d)$ were not improved on for over a century. The trivial upper bound on $s(d)$ is provided by the unit cube, which has surface area $2d$. The trivial lower bound on $s(d)$ comes from ignoring the tiling restriction, and considering only the volume – here the unit volume ball has the smallest surface area, $\sqrt{2\pi e d}$. Where in this quadratic range does $s(d)$ lie? In particular, can there be *"spherical cubes,"* with $s(d) = O(\sqrt{d})$?

The last question became a central issue for complexity theorists when [69] related it directly to the important Unique Games conjecture, and optimal inapproximability proofs of combinatorial problems (in particular, the maximum cut problem) discussed above. The nontrivial connection, which the paper elaborates and motivates, goes through attempts to find the tightest version of Raz' [151] celebrated parallel repetition theorem.[15] A limit on how "strong" a parallel repetition theorem can get was again provided by Raz [152]. Extending his techniques [111] to the geometric setting, resolved the question above, proving that "spherical cubes" do exist!

Theorem 8.5 ([111]). *For all d, $s(d) \leq \sqrt{4\pi d}$.*

A simple proof, and various extensions of this result were given subsequently in [14]. We note that all known proofs are probabilistic. Giving an explicit construction, that might better illustrate how a "spherical cube" (even with much worse parameters) looks like, seems a challenging problem.

9. LATTICE THEORY

Lattices in Euclidean space are among the most "universal" objects in mathematics, in that besides being natural (e.g., arising in crystalline structures) and worthy of study in their own right, they capture a variety of problems in different fields such as number theory,

---

14 Note that the volume of B ensures that the interiors of $B+v$ and $B+u$ are disjoint for any two distinct integer vectors $u, v \in \mathbb{Z}^d$, so this gives a tiling.

15 A fundamental information theoretic inequality of central importance to "amplification" of Probabilistically Checkable Proofs (PCPs).

analysis, approximation theory, Lie algebras, convex geometry, and more. Many of the basic results in lattice theory, as we shall see, are *existential* (namely supply no efficient means for obtaining the objects whose existence is proved), which in some cases has limited progress on these applications.

This section tells the story of one algorithm, of Lenstra, Lenstra, and Lovász [117], often called the LLL algorithm, and some of its implications on these classical applications as well as modern ones in cryptography, optimization, number theory, symbolic algebra, and more. But we had better define a lattice[16] first.

Let $B = \{b_1, b_2, \dots, b_n\}$ be a basis of $\mathbb{R}^n$. Then the *lattice* $L(B)$ denotes the set (indeed, Abelian group) of all *integer* linear combinations of these vectors, i.e., $L(B) = \{\sum_i z_i b_i : z_i \in \mathbb{Z}\}$; B is also called a basis of the lattice. Naturally, a given lattice can have many different bases, e.g., the standard integer lattice in the plane, generated by $\{(0,1),(1,0)\}$, is equally well generated by $\{(999,1),(1000,1)\}$. A basic invariant associated with a lattice L is its determinant $d(L)$, which is the absolute value of $\det(B)$ for any basis B of L (this is also the volume of the fundamental parallelpiped of the lattice). For simplicity and without loss of generality, we will assume that B is normalized so that we only consider lattices L of $d(L) = 1$.

The most basic result about lattices, namely that they must contain *short* vectors (in any norm) was proved by Minkowski (who initiated Lattice Theory, and with it, the Geometry of Numbers) [134].

Theorem 9.1 ([134]). *Consider an arbitrary convex set K in $\mathbb{R}^n$ which is centrally symmetric*[17] *and has volume $> 2^n$. Then, every lattice L (of determinant 1) has a nonzero point in K.*

This innocent theorem, which has a simple, but *existential* (pigeonhole) proof, turns out to have numerous fundamental applications in geometry, algebra, and number theory. Among famous examples this theorem yields with appropriate choice of norms and lattices, results like Dirichlet's Diophantine approximation theorem and Lagrange's four-squares theorem, and (with much more work) the finiteness of class numbers of number fields (see, e.g., [148]).

From now on we will focus on short vectors in the (most natural) Euclidean norm. A direct corollary of Minkowski's theorem when applying it to the cube $K = [-1,1]^n$ yields:

Corollary 9.2. *Every lattice L of determinant 1 has a nonzero point of Euclidean norm at most $\sqrt{n}$.*

Digressing a bit, we note that very recently, a century after Minkowski, a strong converse of the above corollary[18] conjectured by Dadush (see [47]) for *computational* motivation has been proved in [154]. This converse has many structural consequences, on the covering

16 We only define full-rank lattices here, which suffice for this exposition.

17 Namely, $x \in K$ implies that also $-x \in K$. Such sets are precisely balls of arbitrary norms.

18 Which has to be precisely formulated.

radius of lattices, arithmetic combinatorics, Brownian motion, and others. We will not elaborate here on this new interaction of computational complexity and optimization with lattice theory and convex geometry. The papers above beautifully motivate these connections and applications, and the history of ideas and technical work needed for this complex proof.

Returning to Minkowski's corollary for the Euclidean norm, the proof is still existential, and the obvious algorithm for finding such a short vector requires exponential time in n. The breakthrough paper [117] describe the LLL algorithm, an efficient, polynomial-time algorithm, which approximates the length of the shortest vector in any n-dimensional lattice by a 2^n factor.

Theorem 9.3 ([117]). *There is a polynomial time algorithm, which, given any lattice L, produces a vector in L of Euclidean length at most 2^n factor longer than the shortest vector in L.*

This exponential bound may seem excessive at first, but the number and diversity of applications is staggering. First, in many problems, the dimension n is a small constant (so the actual input length arises from the bit-size of the given basis). This leads, for instance, to Lenstra's algorithm for (exactly solving) Integer Programming [118] in constant dimensions. It also leads to Odlyzko and Riele's refutation [146] of Mertens' conjecture about cancellations in the Möbius function, and to the long list of number-theoretic examples in [160]. But it turns out that even when n is arbitrarily large, many problems can be solved in poly(n)-time as well. Here is a list of examples of old and new problems representing this variety, some going back to the original paper [117]. In all, it suffices that real number inputs are approximated to poly(n) digits in dimension n.

- *Diophantine approximation.* While the best possible approximation of one real number by rationals with bounded denominator is readily solved by its (efficiently computable) continued fraction expansion, no such procedure is known for *simultaneous* approximation. Formally, given a *set* of real numbers, say $\{r_1, r_2, \ldots, r_n\}$, a bound Q and $\varepsilon > 0$, find integers $q \leq Q$ and $p_1, \ldots, p_n$ such that all $|r_i - p_i/q| \leq \varepsilon$. Existentially (using Minkowski), the Dirichlet "box-principle" shows that $\varepsilon < Q^{1/n}$ is possible. Using LLL, one efficiently obtains $\varepsilon < 2^{n^2} Q^{1/n}$ which is meaningful for Q described by poly(n) many bits.

- *Minimal polynomials of algebraic numbers.* Here we are given a single real number r and a degree bound n, and are asked if there is a polynomial $g(x)$ with integer coefficients, of degree at most n of which r is a root (and also to produce such a polynomial g if it exists). Indeed, this is a special case of the problem above with $r_i = r^i$. While the algorithm only outputs g for which $g(r) \approx 0$, it is often easy to check that it actually vanishes. Note that by varying n we can find the minimal such polynomial.

- *Polynomial factorization over rationals.* Here the input is an integer polynomial h of degree n, and we want to factor it over $\mathbb{Q}$. The high level idea is to first find an (approximate) root r of h (e.g., using Newton's method), feed it to the problem above, which will return a minimal g having r as a root, and thus divides h. We stress that this algorithm produces the exact factorization, not an approximate one!

- *Small integer relations between reals.* Given reals $r_1, r_2, \ldots, r_n$, and a bound Q, determine if there exist integers $|z_i| < Q$ such that $\sum_i z_i r_i = 0$ (and if so, find these integers). As a famous example, LLL can find an integer relation among $\arctan(1) \approx 0.785398$, $\arctan(1/5) \approx 0.197395$, and $\arctan(1/239) \approx 0.004184$, yielding Machin's formula

$$\arctan(1) - 4\arctan(1/5) + \arctan(1/239) = 0.$$

- *Cryptanalysis.* Note that a very special case of the problem above (in which the coefficients z_i must be Boolean) is the "Knapsack problem," a famous $\mathcal{NP}$-complete problem. The point here is that in the early days of cryptography, some systems were based on the assumed "average case" hardness of Knapsack. Many such systems were broken by using LLL, e.g., [115]. LLL was also used to break some versions of the RSA cryptosystem (with "small public exponents").

It is perhaps a fitting epilogue to the last item that lattices cannot only destroy cryptosystems, but also create them. The problem of efficiently approximating short vectors up to polynomial (as opposed to exponential, as LLL produces) factors is believed to be computationally hard. Here are some major consequences of this assumption. First, Ajtai showed in a remarkable paper [9] that such hardness is preserved "on average", over a cleverly-chosen distribution of random lattices. This led to a new public-key encryption scheme by Ajtai and Dwork [10] based on this hardness, which is arguably the only one known that can potentially sustain quantum attacks (Shor's efficient quantum algorithms can factor integers and compute discrete logarithms [158]). In another breakthrough work of Gentry [77], this hardness assumption is used to devise *fully homomorphic* encryption, a scheme which allows not only to encrypt data, but to perform arbitrary computations directly with encrypted data. See more in this excellent survey [147].

10. INVARIANT THEORY (AND MORE)

This section is somewhat longer than the rest. One reason is that much of it has been a primary research interest of mine in recent years, and indeed is the subject of my ICM lecture. Another reason is that the connections revealed here are considerably richer. On the one hand, several different areas within the theory of computation play a role, including algebraic complexity theory, derandomization, and optimization. On the other, while invariant theory is central in these developments, connections and implications are revealed to and between other mathematical areas, including noncommutative algebra, analysis, representation theory, quantum information theory, statistics, and operator theory. We will explore some of these here.

Invariant theory, born in an 1845 paper of Cayley [43], is major branch of algebra, with important natural connections to algebraic geometry and representation theory, but also to many other areas of mathematics. We will see some here, as well as some new connections with computational complexity, leading to new questions and results in this field. We note that computational efficiency was always important in invariant theory, which is rife with ingenious algorithms (starting with Cayley's *Omega process*), as is evident from the books [46, 49, 167].

Invariants are familiar enough, from examples like the following:

- In high school physics we learn that energy and momentum are conserved (namely, are *invariants*) in the dynamics of general physical systems.
- In chemical reactions the number of atoms of each element is preserved as one mixture of molecules is transformed to yield another (e.g., as combining sodium hydroxide (NaOH) and hydrochloric acid (HCl) yields the common salt sodium chloride (NaCl) and water (H_2O)).
- In geometry, a classical puzzle asks when can a plane polygon be "cut and pasted" along straight lines to another polygon. Here the obvious invariant, *area*, is the only one![19] However, in generalizing this puzzle to 3-dimensional polyhedra, it turns out that besides the obvious invariant, *volume*, there is another invariant, discovered by Dehn.[20]

More generally, questions about the "topological equivalence" of two topological objects (e.g., knots), whether two groups are isomorphic, or whether two points are in the same orbit of a dynamical system, etc., all give rise to similar questions and treatment. A canonical way to give negative answers to such questions is through *invariants*, namely quantities preserved under some action on an underlying space.

We will focus on invariants of *linear groups* acting linearly on *vector spaces*. Let us present some notation. Fix a field $\mathbb{F}$ (while problems are interesting in every field, results mostly work for infinite fields only, and sometimes just for characteristic zero or algebraically closed ones). We will also suppress some technicalities. Let G be a group, and V a representation of G, namely an $\mathbb{F}$-vector space on which G acts: for every $g, h \in G$ and $v \in V$, we have $gv \in V$ and $g(hv) = (gh)v$.

The *orbit* under G of a vector (or point) $v \in V$, denoted Gv, is the set of all other points that v can be moved to by this action, namely $\{gv : g \in G\}$. Understanding the orbits of a group objects is a central task of this field. A basic question capturing many of the examples above is, given two points $u, v \in V$, do they lie in the same G-orbit, namely if $u \in Gv$. A related basic question, which is even more natural in algebraic geometry (when

19 And so, every two polygons of the same area can be cut to produce an *identical* (multi)sets of triangles.

20 So there are pairs of 3-dimensional polyhedra of the same volume, which cannot be cut to identical (multi)sets of tetrahedra.

the field $\mathbb{F}$ is algebraically closed of characteristic zero) is whether the *closures*[21] of the two orbits intersect, namely if some point in V can be approximated arbitrarily well by points in both Gu and Gv. We will return to specific incarnations of these questions.

When G acts on V, it also acts on $\mathbb{F}[V]$, the polynomial functions on V, also called the *coordinate ring* of V. In our setting V will have finite dimension (say m), and so $\mathbb{F}[V]$ is simply $\mathbb{F}[x_1, x_2, \ldots, x_m] = \mathbb{F}[X]$, the polynomial ring over $\mathbb{F}$ in m variables. We will denote by gp the action of a group element $g \in G$ on a polynomial $p \in \mathbb{F}[V]$.

A polynomial $p(X) \in \mathbb{F}[X]$ is *invariant* if it is unchanged by this action, namely for every $g \in G$ we have $gp = p$. All invariant polynomials clearly form a subring of $\mathbb{F}[X]$, denoted $\mathbb{F}[X]^G$, called the *ring of invariants* of this action. Understanding the invariants of group actions is the main subject of Invariant Theory. A fundamental result of Hilbert **[91]** shows that in our linear setting,[22] *all* invariant rings will be *finitely generated* as an algebra.[23] Finding the "simplest" such generating set of invariants is our main concern here.

Two familiar examples of perfect solutions to this problem follow:

- In the first, $G = S_m$, the symmetric group on m letters, is acting on the set of m formal variables X (and hence the vector space they generate) by simply permuting them. Then the ring of invariants is simply all symmetric polynomials, and a (minimal) set of generating invariants is the first m *elementary* symmetric polynomials in X.

- In the second, $G = \mathrm{SL}_n(\mathbb{F})$, the special linear group of matrices with determinant 1, is acting on the vector space $M_n(\mathbb{F})$ of $n \times n$ matrices (so $m = n^2$), simply by left matrix multiplication. In this case all polynomial invariants are generated by a single polynomial, the determinant of this m-variable matrix X.

In these two cases, which really supply a complete understanding of the invariant ring $\mathbb{F}[X]^G$, the generating sets are *good* in several senses. There are *few* generating invariants, they all have *low* degree, and they are *easy* to compute[24]—all these quantities are bounded by a polynomial in m, the dimension of the vector space.[25] In such good cases, one has efficient algorithms for the basic problems regarding orbits of group actions. For example, a fundamental duality theorem of Geometric Invariant Theory **[143]** (see Theorem A.1.1), shows how generating sets of the invariant ring can be used for the orbit closure intersection problem.

---

21 One can take closure in either the Euclidean or the Zariski topology (the equivalence in this setting was proved by Mumford **[142]**).

22 The full generality under which this result holds is actions of *reductive* groups, which we will not define here, but includes all examples we discuss.

23 This means that there is a finite set of polynomials $\{q_1, q_2, \ldots, q_t\}$ in $\mathbb{F}[X]^G$ so that for every polynomial $p \in \mathbb{F}[X]^G$ there is a t-variate polynomial r over $\mathbb{F}$ so that $p = r(q_1, q_2, \ldots, q_t)$.

24 For example, have *small* arithmetic circuits or formulae.

25 There are additional desirable structural qualities of generating sets that we will not discuss, e.g., completely understanding algebraic relations between these polynomials (called *syzygies*).

Theorem 10.1 ([143]). *For an algebraically closed field $\mathbb{F}$ of characteristic 0, the following are equivalent for any two $u, v \in V$ and generating set P of the invariant ring $\mathbb{F}[X]^G$:*

- *The orbit closures of u and v intersect.*
- *For every polynomial $p \in P$, $p(v) = p(u)$.*

10.1. Geometric complexity theory

We now briefly explain one direction from which computational complexity became interested in these algebraic problems, in work that has generated many new questions and collaboration between the fields. First, some quick background on the main problem of arithmetic complexity theory (see that chapter in [176] for definitions and more discussion). In [173], Valiant defined arithmetic analogs $\mathcal{VP}$ and $\mathcal{VNP}$ of the complexity classes $\mathcal{P}$ and $\mathcal{NP}$, respectively, and conjectured that these two arithmetic classes are different. He further proved (via surprising completeness results) that to separate these classes it is sufficient to prove that the *permanent* polynomial on $n \times n$ matrices does not project to the *determinant* polynomial on $m \times m$ matrices for any $m = \text{poly}(n)$. Note that this is a pure and concrete algebraic formulation of a central computational conjecture.

In a series of papers, Mulmuley and Sohoni introduced *Geometric Complexity Theory* (GCT) to tackle this major open problem.[26] This program is surveyed by Mulmuley here [139,140], as well as in Landsberg's book [116]. Very concisely, the GCT program starts off as follows. First, a simple "padding" of the $n \times n$ permanent polynomial makes it have degree m and act on the entries of an $m \times m$ matrix. Consider the linear group SL_{m^2} action on all entries of such $m \times m$ matrices. This action extends to polynomials in those variables, and so in particular the two we care about: determinant and modified permanent. *The main connection is that the permanent projects to the determinant (in Valiant's sense) if and only if the orbit closures of these two polynomials intersect.* Establishing that they do not intersect (for $m = \text{poly}(n)$) naturally leads to questions about finding representation-theoretic obstructions to such intersection (and hence, to the required computational lower bound). This is where things get very complicated, and describing them is beyond the scope of this survey. We note that to date, the tools of algebraic geometry and representation theory were not sufficient even to improve the quadratic bound on m of Mignon and Ressayre [132]. Indeed, some recent developments show severe limitations to the original GCT approach (and perhaps guiding it in more fruitful directions); see [42] and its historical account. Nevertheless, this line of attack (among others in computational complexity) has lead to many new questions in computational commutative algebra and to growing collaborations between algebraists and complexity theorists—we will describe some of these now.

To do so, we will focus on two natural actions of linear groups on *tuples* of matrices, simultaneous conjugation and the left–right action. Both are special cases of *quiver*

26 Origins of using invariant theory to argue computational difficulty via similar techniques go back to Strassen [166].

representations (see [55, 74]).[27] For these two group actions, we will discuss, respectively in Sections 10.2, 10.3, the classical questions and results on the rings of invariants, and recent advances motivated by computational considerations. Section 10.4 will be devoted to significant extensions of the algorithmic technique developed for the left–right action. And in Section 10.5 we will close full circle and discuss yet another central problem on matrix tuples, namely the Symbolic Determinant Identity Testing (SDIT) problem, which ties together many aspects we have seen and suggest further interesting challenges in the interface of computational complexity with invariant theory and algebraic geometry.

10.2. Simultaneous conjugation

Consider the following action of $\mathrm{SL}_n(\mathbb{F})$ on d-tuples of $n \times n$ matrices. We have $m = dn^2$ variables arranged as d $n \times n$ matrices $X = (X_1, X_2, \ldots, X_d)$. The action of a matrix $Z \in \mathrm{SL}_n(\mathbb{F})$ on this tuple is by simultaneous conjugation, by transforming it to the tuple $(Z^{-1}X_1Z, Z^{-1}X_2Z, \ldots, Z^{-1}X_dZ)$. Now, the general question above, for this action, is which polynomials in the variables X are invariant under this action?

The work of Procesi, Formanek, Razmyslov, and Donkin [60, 72, 149, 153] provides a good set (in most aspects discussed above) of generating invariants (over algebraically closed fields of characteristic zero). The generators are simply the traces of products of length at most n^2 of the given matrices,[28] namely the set

$$\left\{\mathrm{Tr}(X_{i_1}X_{i_2}\cdots X_{i_t}) : t \le n^2, i_j \in [d]\right\}.$$

These polynomials are explicit, have small degree, and are easily computable. The one shortcoming is the *exponential* size of this generating set. For example, using it to decide the intersection of orbit closures will only lead to an exponential time algorithm.

By Hilbert's so-called "Noether's normalization lemma" [91],[29] we know that the size of this set of generating invariants can, in principle (as the proof is existential), be reduced to $dn^2 + 1$. Indeed, when the group action is on a vector space of dimension m, taking $m + 1$ "random" linear combinations of any finite generating set will result (with probability 1) in a small generating set. However, as we start with an exponential number of generators above, this procedure is both inefficient and also not explicit (it is not clear how to make it deterministic). One can get an explicit generating set of minimal size deterministically using the Gröbner basis algorithm (see [129] for the best known complexity bounds) but this will take doubly exponential time in n.

---

27 We will not elaborate on the theory of quiver representations here, but only remark that reductions and completeness occur in this study as well! The left–right quiver is *complete* in a well defined sense (see [50, SECTION 5]). Informally, this means understanding its (semi)invariants implies the same understanding of the (semi)invariants of *all* acyclic quivers.

28 Convince yourself that such polynomials are indeed invariant.

29 We remark that this is the same foundational paper which proved the *finite basis* and *Nullstellensatz* theorems. It is interesting that Hilbert's initial motivation to formulate and prove these cornerstones of commutative algebra was the search for invariants of linear actions.

The works above [71,141] reduce this complexity to polynomial time! This happened in two stages. First, Mulmuley [141] gave a probabilistic polynomial time algorithm, by cleverly using the structure of the exponentially many invariants above (using which one can obtain sufficiently random linear combinations using only polynomially many random bits and in polynomial time). He then argues that using conditional derandomization results (discussed in the chapter on randomness in [176]), one can derive a deterministic polynomial time algorithm under natural computational hardness assumptions. Shortly afterwards, Forbes and Shpilka [71] derandomized a variant of Mulmuley's algorithm *without* any unproven assumption, yielding an unconditional deterministic polynomial time algorithm for the problem! Their algorithm uses the derandomization methodology: very roughly speaking, they first notice that Mulmuley's probabilistic algorithm can be implemented by a very restricted computational model (a certain read-once branching program), and then use an efficient pseudorandom generator for this computational model. Here is one important algorithmic corollary (which can be extended to other quivers).

Theorem 10.2 ([71, 141]). *There is a deterministic polynomial time algorithm to solve the following problem: given two tuples of* rational *matrices* $(A_1, A_2, \dots, A_d), (B_1, B_2, \dots, B_d)$, *determine if the closure of their orbits under simultaneous conjugation intersect.*

It is interesting to remark that if we only consider the orbits themselves (as opposed to their closure), namely ask if there is $Z \in \mathrm{SL}_n(\mathbb{F})$ such that for all $i \in [d]$ we have $Z^{-1} A_i Z = B_i$, this becomes the *module isomorphism* problem over $\mathbb{F}$. For this important problem, there is a deterministic algorithm (of a very different nature than above, using other algebraic tools) that can solve the problem over any field $\mathbb{F}$ using only a polynomial number of arithmetic operations over $\mathbb{F}$ [40].

10.3. Left–right action

Consider now the following action of two copies, $\mathrm{SL}_n(\mathbb{F}) \times \mathrm{SL}_n(\mathbb{F})$ on d-tuples of $n \times n$ matrices. We still have $m = dn^2$ variables arranged as d $n \times n$ matrices $X = (X_1, X_2, \dots, X_d)$. The action of a pair of matrices $(Z, W) \in \mathrm{SL}_n(\mathbb{F}) \times \mathrm{SL}_n(\mathbb{F})$ on this tuple is by left–right action, transforming it to the tuple $(Z^{-1}X_1W, Z^{-1}X_2W, \dots, Z^{-1}X_dW)$. Again, for this action, which polynomials in the variables X are invariant under this action? Despite the superficial similarity to the to simultaneous conjugation, the invariants here have entirely different structure, and bounding their size required different arguments.

The works of [6, 54, 59, 156] provides an infinite set of generating invariants. The generators (again, over algebraically closed fields) are determinants of linear forms of the d matrices, with *matrix* coefficients C_i of arbitrary dimension. Namely the following set generates all invariants:

$$\left\{\det(C_1 \otimes X_1 + C_2 \otimes X_2 + \cdots + C_d \otimes X_d) : C_i \in M_k(\mathbb{F}), k \in \mathbb{N}\right\}.$$

These generators, while concisely described, fall short on most goodness aspects above, and we now discuss improvements. First, by Hilbert's finite generation, we know in particular that some finite bound k on the dimension of the matrix coefficients C_i exist. A quest to find explicit bounds on k ensued. A quadratic upper bound $k \leq n^2$ was obtained by Derksen and Makam [50] after a long sequence of improvements described there. Still, there is an exponential number[30] of possible matrix coefficients of this size exist. However, it is easy to see that picking the C_i at random leads to a *probabilistic* polynomial time algorithm for the orbit closure intersection for this left–right action. A sequence of developments which we describe below and in the next subsection, eventually led to a *deterministic* polytime algorithm for this problem over the complex numbers by Allen-Zhu, Garg, Li, Oliveira, and Wigderson [13]. A different, simpler algorithm which works for all fields was later found by Derksen and Makam [51]).

Theorem 10.3 ([13,51]). *There is a deterministic polynomial time algorithm to solve the following problem: given two tuples of matrices $(A_1, A_2, \ldots, A_d)$, $(B_1, B_2, \ldots, B_d)$, determine if the closure of their orbits under the left–right action intersect.*

In the remainder we discuss an important special case of this problem, namely when all $B_i = 0$, for which *deterministic* polynomial time algorithms were found first, which were key to the general result above. While this problem is in commutative algebra, this algorithm surprisingly has implications in analysis, noncommutative algebra, computational complexity, quantum information theory, and other areas. We will mention some of these, but let us start by defining the problem.

For an action of a linear group G on a vector space V, define the *nullcone* of the action to be the set of all points $v \in V$ such that the closure of the orbit Gv contains 0. The points in the nullcone are sometimes called *unstable*. The nullcone is of fundamental importance in invariant theory! Some examples of nullcones for actions we have discussed are the following. For the action of $\mathrm{SL}_n(\mathbb{C})$ on $M_n(\mathbb{C})$ by left multiplication, it is the set of *singular* matrices. For the action of $\mathrm{SL}_n(\mathbb{C})$ on $M_n(\mathbb{C})$ by conjugation, it is the set of *nilpotent* matrices. As you would guess (and follows from Theorem 10.1), the nullcone is precisely the set of points in V which vanish under all invariant polynomials. Thus if we have a good generating set, one can use them to efficiently test membership in the nullcone. However, we are not in this situation for the left–right action. Despite that, deterministic polynomial-time algorithms were obtained, independently, by Garg, Gurvits, Oliveira, and Wigderson [75] (which is analytic in nature) over the complex numbers, and by Ivanyos, Qiao, and Subrahmanyam [95] (which is algebraic in nature) and works for all fields. These two algorithms have different properties, and use in different ways the upper bounds on the dimension of matrix coefficients in the invariants.[31]

Theorem 10.4 ([75,95]). *There is a deterministic polynomial time algorithm that, on a given tuple of matrices $(A_1, A_2, \ldots, A_d)$ in $M_n(\mathbb{F})$, determines if it is in the nullcone of the left–right action.*

30 Well, a possibly infinite number, but it can be reduced to exponential.

31 Yet a third algorithm, quite different than the two above, was very recently developed by Hamada and Hirai [88].

We will focus in what follows on the first algorithm. We discuss its broad extensions in the next subsection. Here we discuss some of its diverse consequences to basic problems in different fields (reflecting the many different mathematical objects that can be represented by matrix tuples). All the precise definitions of the notions below, as well as the proofs, interconnections and the meandering story leading to it can be found in [75, 76].

Theorem 10.5 ([75, 76]). *There are deterministic polynomial-time algorithms to solve the following problems:*

- (Analysis) *The feasibility problem for Brascamp–Lieb inequalities, and more generally, computing the optimal constant for each one.*
- (Noncommutative algebra) *The word problem over the free skew field (of rational functions in noncommuting variables).*
- (Quantum information theory) *Testing if a completely positive quantum operator is rank-decreasing.*
- (Arithmetic complexity) *Approximating the commutative rank of a symbolic matrix to within a factor of two.*[32]

We note that this algorithm also inspired purely structural results, both in the areas mentioned above, but also in others. In frame theory, it led to the complete resolution of the central Paulsen problem [89, 113]. In statistics, it has led to complete understanding of when Maximum Likelihood Estimates (MLE) exist, and when they are unique, first for matrix random models [15, 52] and then for tensor random models [53].

10.4. Nullcones, moment polytopes, geodesic convexity, and noncommutative optimization

Reflecting on the algorithm of [75] from the previous section marked several features which merited further investigation. For one, it is an analytic/numerical algorithm, very different that the typical algebraic/symbolic algorithms so common for problems of invariant theory and algebraic geometry, and in the applications above. This algorithm is a special case of a general heuristic called *alternate minimization*, common in optimization, statistics, and machine learning, where the input evolves via a sequence of local, greedy steps. In general, convergence of such algorithms, let alone fast convergence, may not be guaranteed or is hard to establish, whereas here it always converges, and in polynomial time! The analysis uses the fact that the evolution above happens along the orbit of the input by the left–right group action, and tracks a measure called *capacity* which this evolution minimizes. And fast convergence to a unique optimum occurs despite the fact that both the domain (a pair of continuous linear groups) and the optimized function are patently nonconvex.

Understanding the power of such continuous optimization algorithms for a larger and larger classes of nullcone problems (capturing other problems, in discrete optimization,

32 Computing this rank exactly is the central PIT problem, discussed at the last subsection.

quantum information, representation theory, and other areas) progressed in a series of papers, culminating in a general theory, that applies in principle to *any* linear (reductive) group action [41]. The paper contains a detailed account of the history, background, theory, and applications, and we relate below just the highlights, partly explaining the mysteries above.

First, nullcone problems may be viewed as optimization problems for general group actions, where the *capacity* being minimized is simply the minimum norm of any element in the orbit of the input. This viewpoint then benefits from a beautiful noncommutative duality theory (the Kempf–Ness theorem [106], which greatly expends linear programming duality in the commutative case). Underlying this theorem are notions of *geodesic convexity* (extending the Euclidean one) and *moment maps* (extending the Euclidean gradient). Thus, the seeming nonconvexity of these problems mentioned above only stems from the wrong representation: viewed with the correct metric on the Riemannian manifold which is the acting group makes both the domain and optimization goal (geodesically) convex, explaining convergence to a unique optimal point, which determines membership in the nullcone.

Using these tools, it turns out that the most basic tools of convex optimization in Euclidean space extend to the far more general setting of Riemannian manifolds that arise from the symmetries of noncommutative groups. The paper develops "geodesic" first- and second-order algorithms in this setting, and analyzes their performance in general. Proving convergence bounds requires making quantitative the duality theory above, which uses significant algebraic and analytic machinery. However, the bounds themselves depend in an elegant way on few natural geometric "smoothness" parameters (arising from the given group action), in analogy with the Euclidean (commutative) case.

These algorithms can actually be modified to solve a significant generalization of the nullcone membership problem, namely computing membership in so-called *moment polytopes*, implicitly defined polyhedral bodies associated with any linear group action. These capture a variety of "scaling problems," such as marginal problems in classical and quantum information theory, as well as basic combinatorial optimization problem such as the matroid intersection problem.

10.5. Symbolic determinants, varieties, and circuit lower bounds

We now return to a another basic computational question on matrix tuples, the *Symbolic Determinant Identity Testing* (SDIT) problem (of interest over any field $\mathbb{F}$): Given a tuple of $n \times n$ matrices $(A_1, A_2, \dots, A_d)$, determine if the symbolic determinant $\det(\sum_i x_i A_i)$ vanishes as a polynomial in the variables x_i. This problem has a several different formulations, and has arisen independently in different fields. We mention a few.

One equivalent formulation comes from considering the linear space $\{\sum_i c_i A_i : c_i \in \mathbb{F}\}$ arising from all possible evaluations of the variables x_i. Then SDIT asks if this matrix space contains only singular matrices. In algebraic geometry, it arises in close connection with certain sheaves on projective space [68]. In topology, it arises naturally in connection to linearly independent vector fields on spheres, which led to the development of the Adams operations on topological K-theory [2, 3]. In invariant theory, they were used by Dieudonné [57] to classify the symmetries of the determinant, recovering a result of Frobenius [73].

Another equivalent formulation brings up beautiful connections between the cases of commuting and noncommuting variables x_i. Consider the symbolic matrix $A(x) = \sum_i x_i A_i$. In this terminology SDIT becomes the question of the invertibility of this symbolic matrix $A(x)$ over the field $\mathbb{F}(x)$ of rational functions in the (commuting) variables x_i. In his seminal work on the (noncommutative) free skew field, Cohn [44] proved that the elements of this field can be described as inverses of such symbolic matrices in *noncommuting* variables x_i. Thus the noncommutative analog of SDIT is the word problem for this field.[33] Another connection mentioned in Section 10.3 above is that the noncommutative SDIT problem is equivalent to the nullcone problem for the left–right action! Recalling the generating invariants for this group action from Section 10.3, one observes that (commutative) SDIT is the question of vanishing of the lowest possible invariant, $k = 1$.

SDIT played a crucial role in algorithms and computational complexity. It was initially raised by Edmonds [67] in the context of combinatorial optimization. Another interpretation of Valiant's completeness result is that SDIT captures the general Polynomial Identity Testing (PIT) problem (see the survey [159]). Noting that SDIT has a simple fast probabilistic algorithm over large fields (namely, assign random values to the variables and evaluate the resulting numeric determinant), finding an efficient deterministic algorithm became one of the most basic *derandomization* challenges, which has been under attack now for half a century. The difficulty (and importance) of finding such a deterministic algorithm was clarified (bigtime) by the following remarkable result of Kabanets and Impagliazzo [101].

Theorem 10.6 ([101]). *If there is a deterministic polynomial time algorithm for SDIT, then either $\mathcal{VP} \neq \mathcal{VNP}$, or $\mathcal{NEXP}$ has no polynomial size Boolean circuits.*

In simpler words, such a derandomization will result in a major breakthrough in computational complexity, providing explicit lower bounds either in arithmetic or Boolean complexity, each in the ballpark of proving $\mathcal{P} \neq \mathcal{NP}$. Even the logical nature of this theorem statement demands attention: it states that an efficient algorithm for one problem (SDIT) will mean that host of other natural problems have no efficient algorithm!

On the other hand, this theorem suggests a concrete algorithmic attack on these lower bound questions (and in particular, $\mathcal{VP}$ vs. $\mathcal{VNP}$) discussed in Section 10.1—simply design a deterministic algorithm for SDIT. The past decades have seen much progress on designing such algorithms for a variety of special cases of SDIT (and the more general PIT), which is far too large to survey here. We conclude here with the possibility of finding such an algorithm via the new algorithmic techniques described in Section 10.4 above. This is explored in [125], and we only summarize what is currently known.

First, let us note that the set of singular matrices is an algebraic variety. Thus SDIT is a special case of a very large class of natural problems. Fix an algebraic variety in $U \subset \mathbb{F}^m$

33 Valiant's completeness result [172], mentioned in Section 10.1, analogously makes SDIT the word problem for the commutative field $\mathbb{F}(x)$.

(e.g., for SDIT, $m = dn^2$). Given a point $u \in \mathbb{F}^m$ (e.g., for SDIT u is a matrix tuple), determine if $u \in U$. Of course, it is natural to work here with algebraically closed fields, e.g., $\mathbb{F} = \mathbb{C}$.

Such membership problems in algebraic varieties obviously arise naturally in numerous settings. One way to view the developments of the previous section is that if the variety U is actually the nullcone of a (nice) group action, then continuous, convex optimization methods (extended to the geodesic setting), such a gradient descent, may be far more efficient than symbolic, algebraic algorithms, and indeed in some cases may have polynomial-time convergence. Thus, a first question to ask is whether SDIT itself is the nullcone of some group action. Unfortunately, it is not (unless $d \leq 2$ or $n \leq 2$), again possibly helping to understand its difficulty.

Theorem 10.7 ([125]). *For $d, n \geq 3$, SDIT for a d-tuple of $n \times n$ matrices is not the nullcone of any linear group action.*

A central part of the proof of this theorem is the characterization the symmetries of the SDIT variety, extending to d-tuples for any d the aforementioned theorem of Frobenuis [73] for the case $d = 1$. Among some of the natural directions suggested by this work we name three basic ones: (1) Find methods of determining the symmetries of naturally given algebraic varieties; (2) Find methods to determine if a given algebraic variety is the nullcone of a linear group action; (3) Extend the convex optimization methods of Section 10.4 to prove membership in other algebraic varieties, beside nullcones.

FUNDING

This work was partially supported by NSF grants CCF-1412958 and CCF-1900460.

REFERENCES

[1] D. Achlioptas, A. Coja-Oghlan, and F. Ricci-Tersenghi, On the solution-space geometry of random constraint satisfaction problems. *Random Structures Algorithms* **38** (2011), no. 3, 251–268.

[2] J. F. Adams, Vector fields on spheres. *Ann. of Math. (2)* **75** (1962), 603–632.

[3] J. F. Adams, P. D. Lax, and R. S. Phillips, On matrices whose real linear combinations are nonsingular. *Proc. Amer. Math. Soc.* **16** (1965), no. 2, 318–322.

[4] L. Adleman and M. Huang, *Primality testing and abelian varieties over finite fields*. Lecture Notes in Math. 1512, Springer, 1992.

[5] L. Adleman, C. Pomerance, and R. Rumely, On distinguishing prime numbers from composite numbers. *Ann. of Math.* (1983), 173–206.

[6] B. Adsul, S. Nayak, and K. Subrahmanyam, *A geometric approach to the Kronecker problem II: rectangular shapes, invariants of $n \times n$ matrices, and a generalization of the Artin-Procesi theorem*. Citeseer, 2010.

[7] M. Agrawal and S. Biswas, Primality and identity testing via Chinese remaindering. *J. ACM* **50** (2003), no. 4, 429–443.

[8] M. Agrawal, N. Kayal, and N. Saxena, Primes is in P. *Ann. of Math.* **160** (2004), no. 2, 781–793.

[9] M. Ajtai, Generating hard instances of lattice problems. In *Proceedings of the 28th annual ACM symposium on theory of computing*, pp. 99–108, ACM, 1996.

[10] M. Ajtai and C. Dwork, A public-key cryptosystem with worst-case/average-case equivalence. In *Proceedings of the 29th annual ACM symposium on theory of computing*, pp. 284–293, ACM, 1997.

[11] D. J. Aldous, Random walks on finite groups and rapidly mixing Markov chains. In *Séminaire de probabilités xvii 1981/82*, pp. 243–297, Springer, 1983.

[12] D. J. Aldous, The random walk construction of uniform spanning trees and uniform labelled trees. *SIAM J. Discrete Math.* **3** (1990), no. 4, 450–465.

[13] Z. Allen-Zhu, A. Garg, Y. Li, R. Oliveira, and A. Wigderson, Operator scaling via geodesically convex optimization, invariant theory and polynomial identity testing. In *Proceedings of the 50th annual ACM SIGACT symposium on theory of computing*, pp. 172–181, ACM, 2018.

[14] N. Alon and B. Klartag, Economical toric spines via Cheeger's inequality. *J. Topol. Anal.* **1** (2009), no. 02, 101–111.

[15] C. Améndola, K. Kohn, P. Reichenbach, and A. Seigal, Invariant theory and scaling algorithms for maximum likelihood estimation. *SIAM J. Appl. Algebra Geom.* **5** (2021), no. 2, 304–337.

[16] N. Anari, V. Jain, F. Koehler, H. T. Pham, and T.-D. Vuong, Entropic independence in high-dimensional expanders: Modified log-Sobolev inequalities for fractionally log-concave polynomials and the Ising model. 2021, arXiv:2106.04105.

[17] N. Anari, K. Liu, and S. O. Gharan, Spectral independence in high-dimensional expanders and applications to the hardcore model. *SIAM J. Comput.* (2021), FOCS20-1–FOCS20-37. DOI 10.1137/20M1367696.

[18] N. Anari, K. Liu, S. O. Gharan, and C. Vinzant, Log-concave polynomials II: High-dimensional walks and an FPRAS for counting bases of a matroid. In *Proceedings of the 51st annual ACM SIGACT symposium on theory of computing*, pp. 1–12, ACM, 2019.

[19] S. Arora and B. Barak, *Computational complexity: a modern approach.* Cambridge University Press, 2009.

[20] S. Arora, J. Lee, and A. Naor, Euclidean distortion and the sparsest cut. *J. Amer. Math. Soc.* **21** (2008), no. 1, 1–21.

[21] S. Arora, S. Rao, and U. Vazirani, Expander flows, geometric embeddings and graph partitioning. *J. ACM* **56** (2009), no. 2, 1–37.

[22] H. Bäärnhielm and C. Leedham-Green, The product replacement prospector. *J. Symbolic Comput.* **47** (2012), no. 1, 64–75.

[23] L. Babai, Local expansion of vertex-transitive graphs and random generation in finite groups. In *Proceedings of the 23rd annual ACM symposium on theory of computing, STOC 1991*, pp. 164–174, Citeseer, 1991.

[24] L. Babai, Graph isomorphism in quasipolynomial time. 2015, arXiv:1512.03547.

[25] L. Babai and E. Szemerédi, On the complexity of matrix group problems I. In *Proceedings of 25th annual IEEE symposium on foundations of computer science*, pp. 229–240, IEEE, 1984.

[26] E. Bach and J. Shallit, *Algorithmic number theory: Efficient algorithms. 1.* MIT Press, Cambridge, 1997.

[27] B. Barak, Z. Dvir, A. Wigderson, and A. Yehudayoff, Fractional Sylvester–Gallai theorems. *Proc. Natl. Acad. Sci.* **110** (2013), no. 48, 19213–19219.

[28] B. Barak, R. Impagliazzo, and A. Wigderson, Extracting randomness using few independent sources. *SIAM J. Comput.* **36** (2006), no. 4, 1095–1118.

[29] J. Batson, D. A. Spielman, and N. Srivastava, Twice-Ramanujan sparsifiers. *SIAM Rev.* **56** (2014), no. 2, 315–334.

[30] M. Ben-Or and N. Linial, Collective coin flipping, robust voting schemes and minima of Banzhaf values. In *Proceedings of 26th annual IEEE symposium on foundations of computer science*, pp. 408–416, IEEE, 1985.

[31] I. Benjamini, G. Kalai, and O. Schramm, Noise sensitivity of Boolean functions and applications to percolation. *Publ. Math. IHÉS* **90** (1999), no. 1, 5–43.

[32] E. Berlekamp, Factoring polynomials over finite fields. *Bell Syst. Tech. J.* **46** (1967), no. 8, 1853–1859.

[33] A. Besicovitch, Sur deux questions d'integrabilité des fonctions. *J. Soc. Phys. Math.* **2** (1919), 105–123.

[34] Y. Bilu and N. Linial, Lifts, discrepancy and nearly optimal spectral gap. *Combinatorica* **26** (2006), no. 5, 495–519.

[35] C. Borell, Geometric bounds on the Ornstein–Uhlenbeck velocity process. *Z. Wahrsch. Verw. Gebiete* **70** (1985), no. 1, 1–13.

[36] J. Bourgain, On Lipschitz embedding of finite metric spaces in Hilbert space. *Israel J. Math.* **52** (1985), no. 1–2, 46–52.

[37] J. Bourgain, N. Katz, and T. Tao, A sum-product estimate in finite fields, and applications. *Geom. Funct. Anal.* **14** (2004), no. 1, 27–57.

[38] J. Bourgain and L. Tzafriri, On a problem of Kadison and Singer. *J. Reine Angew. Math.* **420** (1991), 1–43.

[39] A. Broder, Generating random spanning trees. In *Proceedings of 30th annual IEEE symposium on foundations of computer science*, pp. 442–447, IEEE, 1989.

[40] P. A. Brooksbank and E. M. Luks, Testing isomorphism of modules. *J. Algebra* **320** (2008), no. 11, 4020–4029.

[41] P. Bürgisser, C. Franks, A. Garg, R. Oliveira, M. Walter, and A. Wigderson, Towards a theory of non-commutative optimization: geodesic first and second order methods for moment maps and polytopes. 2019, arXiv:1910.12375.

[42] P. Bürgisser, C. Ikenmeyer, and G. Panova, No occurrence obstructions in geometric complexity theory. 2016, arXiv:1604.06431.

[43] A. Cayley, *On the theory of linear transformations*. E. Johnson, 1845.

[44] P. M. Cohn, *Skew field constructions*. 27, CUP Archive, 1977.

[45] G. Cooperman, Towards a practical, theoretically sound algorithm for random generation in finite groups. 2002, arXiv:math/0205203.

[46] D. Cox, J. Little, and D. O'Shea, *Ideals, varieties, and algorithms: an introduction to computational algebraic geometry and commutative algebra*. Undergrad. Texts Math., Springer, New York, 1992.

[47] D. Dadush and O. Regev, Towards strong reverse Minkowski-type inequalities for lattices. In *Proceedings of 57th annual IEEE symposium on foundations of computer science*, pp. 447–456, IEEE, 2016.

[48] R. Davies, Some remarks on the Kakeya problem. *Math. Proc. Cambridge Philos. Soc.* **69** (1971), 417–421.

[49] H. Derksen and G. Kemper, *Computational invariant theory*. Springer, 2015.

[50] H. Derksen and V. Makam, Polynomial degree bounds for matrix semi-invariants. 2015, arXiv:1512.03393.

[51] H. Derksen and V. Makam, Algorithms for orbit closure separation for invariants and semi-invariants of matrices. *Algebra Number Theory* **14** (2020), no. 10, 2791–2813.

[52] H. Derksen and V. Makam, Maximum likelihood estimation for matrix normal models via quiver representations. *SIAM J. Appl. Algebra Geom.* **5** (2021), no. 2, 338–365.

[53] H. Derksen, V. Makam, and M. Walter, Maximum likelihood estimation for tensor normal models via castling transforms. 2020, arXiv:2011.03849.

[54] H. Derksen and J. Weyman, Semi-invariants of quivers and saturation for Littlewood–Richardson coefficients. *J. Amer. Math. Soc.* **13** (2000), no. 3, 467–479.

[55] H. Derksen and J. Weyman, The combinatorics of quiver representations. 2006, arXiv:math/0608288.

[56] P. Diaconis, Group representations in probability and statistics. *Lecture Notes Monogr. Ser.* **11** (1988), i–192.

[57] J. Dieudonné, Sur une généralisation du groupe orthogonal à quatre variables. *Arch. Math.* **1** (1948), no. 4, 282–287.

[58] J. Dixon, Generating random elements in finite groups. *Electron. J. Combin.* **13** (2008), no. R94, 1.

[59] M. Domokos and A. Zubkov, Semi-invariants of quivers as determinants. *Transform. Groups* **6** (2001), no. 1, 9–24.

[60] S. Donkin, Invariants of several matrices. *Invent. Math.* **110** (1992), no. 1, 389–401.

[61] Z. Dvir, On the size of Kakeya sets in finite fields. *J. Amer. Math. Soc.* **22** (2009), no. 4, 1093–1097.

[62] Z. Dvir, Guest column: From randomness extraction to rotating needles. *ACM SIGACT News* **40** (2010), no. 4, 46–61.

[63] Z. Dvir, S. Kopparty, S. Saraf, and M. Sudan, Extensions to the method of multiplicities, with applications to Kakeya sets and mergers. *SIAM J. Comput.* **42** (2013), no. 6, 2305–2328.

[64] Z. Dvir and A. Shpilka, Locally decodable codes with two queries and polynomial identity testing for depth 3 circuits. *SIAM J. Comput.* **36** (2007), no. 5, 1404–1434.

[65] M. Dyer, A. Frieze, and R. Kannan, A random polynomial-time algorithm for approximating the volume of convex bodies. *J. ACM* **38** (1991), no. 1, 1–17.

[66] M. Dyer, A. Sinclair, E. Vigoda, and D. Weitz, Mixing in time and space for lattice spin systems: a combinatorial view. *Random Structures Algorithms* **24** (2004), no. 4, 461–479.

[67] J. Edmonds, Systems of distinct representatives and linear algebra. *J. Res. Natl. Bur. Stand., B* **71** (1967), 241–245.

[68] D. Eisenbud and J. Harris, Vector spaces of matrices of low rank. *Adv. Math.* **70** (1988), no. 2, 135–155.

[69] U. Feige, G. Kindler, and R. O'Donnell, Understanding parallel repetition requires understanding foams. In *Proceedings of the 22nd annual IEEE conference on computational complexity*, pp. 179–192, IEEE, 2007.

[70] W. Feller, *An introduction to probability theory and its applications*. Wiley Ser. Prob. Math. Stat., Wiley, 1971.

[71] M. Forbes and A. Shpilka, Explicit Noether normalization for simultaneous conjugation via polynomial identity testing. In *Approximation, randomization, and combinatorial optimization. Algorithms and techniques*, pp. 527–542, Springer, 2013.

[72] E. Formanek, Invariants and the ring of generic matrices. *J. Algebra* **89** (1984), no. 1, 178–223.

[73] F. G. Frobenius, Über die Darstellung der endlichen Gruppen durch lineare Substitutionen. *Sitz. Kön. Preuss. Akad. Wiss. Berlin* (1897), 944–1015.

[74] P. Gabriel, Unzerlegbare darstellungen I. *Manuscripta Math.* **6** (1972), no. 1, 71–103.

[75] A. Garg, L. Gurvits, R. Oliveira, and A. Wigderson, A deterministic polynomial time algorithm for non-commutative rational identity testing. 2015, arXiv:1511.03730.

[76] A. Garg, L. Gurvits, R. Oliveira, and A. Wigderson, Algorithmic aspects of Brascamp–Lieb inequalities. 2016, arXiv:1607.06711.

[77] C. Gentry, Fully homomorphic encryption using ideal lattices. In *Proceedings of the 41st annual ACM symposium on theory of computing, STOC '09*, pp. 169–178, ACM, 2009.

[78] M. Goemans, Semidefinite programming in combinatorial optimization. *Math. Program.* **79** (1997), no. 1–3, 143–161.

[79] O. Goldreich, *Computational complexity: a conceptual perspective*. Cambridge University Press, Cambridge, 2008.

[80] S. Goldwasser and J. Kilian, Almost all primes can be quickly certified. In *Proceedings of the 18th annual ACM symposium on theory of computing*, pp. 316–329, ACM, 1986.

[81] S. Goldwasser, S. Micali, and C. Rackoff, The knowledge complexity of interactive proof systems. *SIAM J. Comput.* **18** (1989), no. 1, 186–208.

[82] A. Granville, It is easy to determine whether a given integer is prime. *Bull. Amer. Math. Soc.* **42** (2005), 3–38.

[83] M. Gromov, Essays in group theory. *Math. Sci. Res. Inst. Publ.* **8** (1987), 75–263.

[84] H. Guo, M. Jerrum, and J. Liu, Uniform sampling through the Lovász local lemma. 2016, arXiv:1611.01647.

[85] V. Guruswami, J. Lee, and A. Razborov, Almost Euclidean subspaces of ℓ_1^n via expander codes. *Combinatorica* **30** (2010), no. 1, 47–68.

[86] L. Gurvits, Van der Waerden/Schrijver–Valiant like conjectures and stable (aka hyperbolic) homogeneous polynomials: One theorem for all. *Electron. J. Combin.* **15** (2008), no. 1, R66.

[87] L. Guth and N. Katz, On the Erdős distinct distance problem in the plane. 2010, arXiv:1011.4105.

[88] M. Hamada and H. Hirai, Computing the NC-rank via discrete convex optimization on CAT(0) spaces. *SIAM J. Appl. Algebra Geom.* **5** (2021), no. 3, 455–478.

[89] L. Hamilton and A. Moitra, The Paulsen problem made simple. *Israel J. Math.* **246** (2021), no. 1, 299–313.

[90] O. Heilmann and E. Lieb, Theory of monomer-dimer systems. *Comm. Math. Phys.* **25** (1972), no. 3, 190–232.

[91] D. Hilbert, Über die vollen Invariantensysteme. *Math. Ann.* **42** (1893), no. 3, 313–373.

[92] D. Holt, B. Eick, and E. O'Brien, *Handbook of computational group theory*. CRC Press, 2005.

[93] S. Hoory, N. Linial, and A. Wigderson, Expander graphs and their applications. *Bull. Amer. Math. Soc.* **43** (2006), no. 4, 439–561.

[94] R. Impagliazzo and A. Wigderson, $P = \mathrm{BPP}$ unless E has subexponential circuits: Derandomizing the XOR lemma. In *Proceedings of the 29th annual ACM symposium on theory of computing*, pp. 220–229, ACM, 1997.

[95] G. Ivanyos, Y. Qiao, and K. Subrahmanyam, Non-commutative Edmonds' problem and matrix semi-invariants. 2015, arXiv:1508.00690.

[96] M. Jerrum, The Markov chain Monte Carlo method: An approach to approximate counting and integration. In *Approximation algorithm for NP-hard problems*, pp. 482–520, PWS Publishing, 1996.

[97] M. Jerrum, A. Sinclair, and E. Vigoda, A polynomial-time approximation algorithm for the permanent of a matrix with nonnegative entries. *J. ACM* **51** (2004), no. 4, 671–697.

[98] M. Jerrum and G. Sorkin, Simulated annealing for graph bisection. In *Proceedings of 34th annual IEEE symposium on foundations of computer science*, pp. 94–103, IEEE, 1993.

[99] M. Jerrum, L. Valiant, and V. Vazirani, Random generation of combinatorial structures from a uniform distribution. *Theoret. Comput. Sci.* **43** (1986), 169–188.

[100] Z. Ji, A. Natarajan, T. Vidick, J. Wright, and H. Yuen, MIP* = RE. 2020, arXiv:2001.04383.

[101] V. Kabanets and R. Impagliazzo, Identity tests means proving circuit lower bounds. *Comput. Complexity* **13** (2004), no. 1–2, 1–46.

[102] R. Kadison and I. Singer, Extensions of pure states. *Amer. J. Math.* **81** (1959), no. 2, 383–400.

[103] J. Kahn, G. Kalai, and N. Linial, The influence of variables on Boolean functions. In *Proceedings of 29th annual IEEE symposium on foundations of computer science*, pp. 68–80, IEEE, 1988.

[104] G. Kasparov and G. Yu, The coarse geometric Novikov conjecture and uniform convexity. *Adv. Math.* **206** (2006), no. 1, 1–56.

[105] N. Kayal and S. Saraf, Blackbox polynomial identity testing for depth 3 circuits. In *Proceedings of 50th annual IEEE symposium on foundations of computer science*, pp. 198–207, IEEE, 2009.

[106] G. Kempf and L. Ness, The length of vectors in representation spaces. In *Algebraic geometry*, pp. 233–243, Springer, 1979.

[107] S. Khot, On the power of unique 2-prover 1-round games. In *Proceedings of the 34th annual ACM symposium on theory of computing*, pp. 767–775, ACM, 2002.

[108] S. Khot, Inapproximability of NP-complete problems, discrete Fourier analysis, and geometry. In *Proceedings of the international congress of mathematicians, vol. 5*, pp. 2676–2697, Hindustan Book Agency, New Delhi, 2010.

[109] S. Khot, G. Kindler, E. Mossel, and R. O'Donnell, Optimal inapproximability results for MAX-CUT and other 2-variable CSPs? *SIAM J. Comput.* **37** (2007), no. 1, 319–357.

[110] S. Khot and N. Vishnoi, The unique games conjecture, integrality gap for cut problems and embeddability of negative-type metrics into ℓ_1. In *Proceedings of 46th annual IEEE symposium on foundations of computer science*, pp. 53–62, IEEE, 2005.

[111] G. Kindler, R. O'Donnell, A. Rao, and A. Wigderson, Spherical cubes and rounding in high dimensions. In *Proceedings of 49th annual IEEE symposium on foundations of computer science*, pp. 189–198, IEEE, 2008.

[112] R. Krauthgamer, J. Lee, M. Mendel, and A. Naor, Measured descent: A new embedding method for finite metrics. *Geom. Funct. Anal.* **15** (2005), no. 4, 839–858.

[113] T. C. Kwok, L. C. Lau, Y. T. Lee, and A. Ramachandran, The Paulsen problem, continuous operator scaling, and smoothed analysis. In *Proceedings of the 50th annual ACM SIGACT symposium on theory of computing*, pp. 182–189, ACM, 2018.

[114] V. Lafforgue, Un renforcement de la propriété (T). *Duke Math. J.* **143** (2008), no. 3, 559–602.

[115] J. Lagarias, Knapsack public key cryptosystems and Diophantine approximation. In *Advances in cryptology*, pp. 3–23, Springer, 1984.

[116] J. Landsberg, *Geometry and complexity theory*. Cambridge University Press, 2017.

[117] A. Lenstra, H. Lenstra, and L. Lovász, Factoring polynomials with rational coefficients. *Math. Ann.* **261** (1982), no. 4, 515–534.

[118] H. Lenstra, Integer programming with a fixed number of variables. *Math. Oper. Res.* **8** (1983), no. 4, 538–548.

[119] N. Linial, Finite metric spaces–combinatorics, geometry and algorithms. In *Proceedings of the international congress of mathematicians III*, pp. 573–586, 2002.

[120] N. Linial, E. London, and Y. Rabinovich, The geometry of graphs and some of its algorithmic applications. *Combinatorica* **15** (1995), no. 2, 215–245.

[121] K. Liu, From coupling to spectral independence and blackbox comparison with the down-up walk. 2021, arXiv:2103.11609.

[122] C. Lu, O. Reingold, S. Vadhan, and A. Wigderson, Extractors: Optimal up to constant factors. In *Proceedings of the 35th annual ACM symposium on theory of computing*, pp. 602–611, ACM, 2003.

[123] A. Lubotzky, High dimensional expanders. In *Proceedings of the international congress of mathematicians: Rio de Janeiro 2018*, pp. 705–730, World Scientific, 2018.

[124] A. Lubotzky, R. Phillips, and P. Sarnak, Ramanujan graphs. *Combinatorica* **8** (1988), no. 3, 261–277.

[125] V. Makam and A. Wigderson, Singular tuples of matrices is not a null cone (and the symmetries of algebraic varieties). *J. Reine Angew. Math.* **2021** (2021), no. 780, 79–131.

[126] A. Marcus, D. Spielman, and N. Srivastava, Interlacing families I: Bipartite Ramanujan graphs of all degrees. In *Proceedings of 54th annual IEEE symposium on foundations of computer science*, pp. 529–537, IEEE, 2013.

[127] A. Marcus, D. Spielman, and N. Srivastava, Interlacing families II: Mixed characteristic polynomials and the Kadison–Singer problem. 2013, arXiv:1306.3969.

[128] G. Margulis, Explicit group-theoretic constructions of combinatorial schemes and their applications in the construction of expanders and concentrators. *Probl. Inf. Transm.* **24** (1988), 39–46.

[129] E. Mayr and S. Ritscher, Space-efficient Gröbner basis computation without degree bounds. In *Proceedings of the 36th international symposium on symbolic and algebraic computation*, pp. 257–264, ACM, 2011.

[130] M. Mendel and A. Naor, Nonlinear spectral calculus and super-expanders. *Publ. Math. IHÉS* **119** (2014), no. 1, 1–95.

[131] M. Mézard, G. Parisi, and R. Zecchina, Analytic and algorithmic solution of random satisfiability problems. *Science* **297** (2002), no. 5582, 812–815.

[132] T. Mignon and N. Ressayre, A quadratic bound for the determinant and permanent problem. *Int. Math. Res. Not.* **2004** (2004), no. 79, 4241–4253.

[133] G. Miller, Riemann's hypothesis and tests for primality. *J. Comput. System Sci.* **13** (1976), no. 3, 300–317.

[134] H. Minkowski, *Geometrie der Zahlen*. Teubner, 1910.

[135] A. Moitra, Approximate counting, the Lovász local lemma and inference in graphical models. 2016, arXiv:1610.04317.

[136] R. Moser, A constructive proof of the Lovász local lemma. In *Proceedings of the 41st annual ACM symposium on theory of computing*, pp. 343–350, ACM, 2009.

[137] R. Moser and G. Tardos, A constructive proof of the general Lovász local lemma. *J. ACM* **57** (2010), no. 2, 11.

[138] E. Mossel, R. O'Donnell, and K. Oleszkiewicz, Noise stability of functions with low influences: Invariance and optimality. *Ann. of Math.* **171** (2010), no. 1, 295–341.

[139] K. Mulmuley, On P vs. NP and geometric complexity theory. *J. ACM* **58** (2011), no. 2, 5.

[140] K. Mulmuley, The GCT program toward the P vs. NP problem. *Commun. ACM* **55** (2012), no. 6, 98–107.

[141] K. Mulmuley, Geometric complexity theory V: Equivalence between blackbox derandomization of polynomial identity testing and derandomization of Noether's normalization lemma. In *Proceedings of 53rd annual IEEE symposium on foundations of computer science*, pp. 629–638, IEEE, 2012.

[142] D. Mumford, *Algebraic geometry: Complex projective varieties. 1*. Springer, 1995.

[143] D. Mumford, J. Fogarty, and F. Kirwan, *Geometric invariant theory*, 2nd edn. Ergeb. Math. Grenzgeb. 34, Springer, 1982.

[144] A. Naor and R. Young, Vertical perimeter versus horizontal perimeter. 2017, arXiv:1701.00620.

[145] R. O'Donnell, *Analysis of Boolean functions*. Cambridge University Press, 2014.

[146] A. M. Odlyzko and H. J. J. te Riele, Disproof of the Mertens conjecture. *J. Reine Angew. Math.* **357** (1985), 138–160.

[147] C. Peikert, A decade of lattice cryptography. *Found. Trends Theor. Comput. Sci.* **10** (2016), no. 4, 283–424.

[148] M. Pohst and H. Zassenhaus, *Algorithmic algebraic number theory*. Cambridge University Press, 1989.

[149] C. Procesi, The invariant theory of $n \times n$ matrices. *Adv. Math.* **19** (1976), no. 3, 306–381.

[150] M. Rabin, Probabilistic algorithm for testing primality. *J. Number Theory* **12** (1980), no. 1, 128–138.

[151] R. Raz, A parallel repetition theorem. *SIAM J. Comput.* **27** (1998), no. 3, 763–803.

[152] R. Raz, A counterexample to strong parallel repetition. *SIAM J. Comput.* **40** (2011), no. 3, 771–777.

[153] J. Razmyslov, Trace identities of full matrix algebras over a field of characteristic zero. *Math. USSR, Izv.* **8** (1974), no. 4, 727.

[154] O. Regev and N. Stephens-Davidowitz, A reverse Minkowski theorem. 2016, arXiv:1611.05979.

[155] O. Reingold, S. Vadhan, and A. Wigderson, Entropy waves, the zig-zag graph product, and new constant-degree expanders. *Ann. of Math.* (2002), 157–187.

[156] A. Schofield and M. Van den Bergh, Semi-invariants of quivers for arbitrary dimension vectors. *Indag. Math.* **12** (2001), no. 1, 125–138.

[157] Á. Seress, *Permutation group algorithms*. Cambridge Tracts in Math. 152, Cambridge University Press, 2003.

[158] P. Shor, Algorithms for quantum computation: discrete logarithms and factoring. In *Proceedings of 35th annual IEEE symposium on foundations of computer science*, pp. 124–134, IEEE, 1994.

[159] A. Shpilka and A. Yehudayoff, Arithmetic circuits: a survey of recent results and open questions. *Found. Trends Theor. Comput. Sci.* **5** (2010), no. 3–4, 207–388.

[160] D. Simon, Selected applications of LLL in number theory. In *The LLL algorithm*, pp. 265–282, Springer, 2010.

[161] A. Sinclair and M. Jerrum, Approximate counting, uniform generation and rapidly mixing Markov chains. *Inform. and Comput.* **82** (1989), no. 1, 93–133.

[162] A. Sly, Computational transition at the uniqueness threshold. In *Proceedings of 51st annual IEEE symposium on foundations of computer science*, pp. 287–296, IEEE, 2010.

[163] R. Solovay and V. Strassen, A fast Monte-Carlo test for primality. *SIAM J. Comput.* **6** (1977), no. 1, 84–85.

[164] D. Spielman and N. Srivastava, An elementary proof of the restricted invertibility theorem. *Israel J. Math.* **190** (2012), no. 1, 83–91.

[165] D. Spielman and S. Teng, Spectral sparsification of graphs. *SIAM J. Comput.* **40** (2011), no. 4, 981–1025.

[166] V. Strassen, Relative bilinear complexity and matrix multiplication. *J. Reine Angew. Math.* **375** (1987), 406–443.

[167] B. Sturmfels, *Algorithms in invariant theory*. Springer, 2008.

[168] E. Szemerédi and W. Jr. Trotter, Extremal problems in discrete geometry. *Combinatorica* **3** (1983), no. 3–4, 381–392.

[169] T. Tao, Recent progress on the Kakeya conjecture. 2009, http://terrytao.wordpress.com/2009/05/11/.

[170] S. W. Thomson, On the division of space with minimum partitional area. *Acta Math.* **11** (1887), no. 1–4, 121–134.

[171] S. Vadhan. Pseudorandomness. *Found. Trends Theor. Comput. Sci.* **7** (2011), no. 1–3, 1–336.

[172] L. Valiant, Completeness classes in algebra. In *Proceedings of the 11th annual ACM symposium on theory of computing*, pp. 249–261, ACM, 1979.

[173] L. Valiant, The complexity of computing the permanent. *Theoret. Comput. Sci.* **8** (1979), no. 2, 189–201.

[174] S. Vempala, Geometric random walks: a survey. *Comb. Comput. Geom.* **52** (2005), no. 2, 573–612.

[175] D. Weitz, Counting independent sets up to the tree threshold. In *Proceedings of the 38th annual ACM symposium on theory of computing*, pp. 140–149, ACM, 2006.

[176] A. Wigderson, *Mathematics and computation*. Princeton University Press, 2019.

[177] T. Wolff, Recent work connected with the Kakeya problem. *Prospects Math.* **2** (1999), 129–162.

AVI WIGDERSON

School of Mathematics, Institute for Advanced Study, Princeton, NJ 08540, USA, avi@ias.edu

LIST OF CONTRIBUTORS